Essentials of GEOLOGY

13e

Essentials of GEOLOGY

13e

Frederick K. Lutgens

Edward J. Tarbuck

Illustrated by

Dennis Tasa

330 Hudson Street, NY NY 10013

Executive Editor, Geosciences Courseware: Christian Botting
Director, Courseware Portfolio Management: Beth Wilbur
Content Producer: Lizette Faraji
Managing Producer: Mike Early
Courseware Director, Content Development: Ginnie Simione Jutson
Courseware Sr. Analyst: Margot Otway
Geosciences Courseware Editorial Assistant: Emily Bornhop
Rich Media Content Producer: Mia Sullivan
Full Service Vendor: SPi Global
Full Service Project Manager: Patty Donovan
Copyeditor: Kitty Wilson
Design Manager: Mark Ong
Cover and Interior Designer: Jeff Puda
Photo and Illustration Support: Kevin Lear, International Mapping
Rights and Permissions Project Manager: Kathleen Zander
Rights and Permissions Management: Ben Ferrini
Manufacturing Buyer: Maura Zaldivar-Garcia
Marketing Managers: Neena Bali/Mary Salzman
Cover Image Credit: © Tim Kemple

Credits and acknowledgments borrowed from other sources and reproduced, with permission, in this textbook appear on the appropriate page within text or are listed below.

Page 6: Quote by Aristotle from The Birth and Development of the Geological Sciences by Frank Dawson Adams. Published by Dover Publications, © 1954; page 7: Excerpt from Essentials of Earth History, 3e by William Lee Stokes. Published by Pearson Education Inc., © 1973; page 7: Quote from Transactions of the Royal Society of Edinburgh by James Hutton. Published by The Royal Society of Edinburgh, © 1788; page 10: Quote from The Common Sense of Science by Jacob Bronowski. Published by Harvard University Press, © 1953; page 10: Quote from Science for All Americans by F. James Rutherford and Andrew Ahlgren. Published by Oxford University Press, © 1990; page 11: Quote by Louis Pasteur from Pasteur Vallery-Radot. Published by Masson et cie, © 1939; page 37: Quote from The Origin of Continents and Oceans by Alfred Wegener. Published by Methuen Publishing, Ltd., © 1966; page 14: Quote by R. T. Chamberlain from A Revolution in the Earth Sciences by Anthony Hallam. Published by Oxford University Press, © 1973; page 356: Quote from Exploration of the Colorado River of the West and Its Tributaries. Published by U.S. Government Printing Office, © 1875; page 417: Excerpt from Variations in the Earth's Orbit: Pacemaker of the Ice Ages by J.D. Hays, John Imbrie and N.J. Shackleton in Science, Vol 194, Issue 4270, pp.1121–1132. Published by American Association for the Advancement of Science, © 1976; page 438: Quote from The Physics of Blown Sand and Desert Dunes by R.A. Bagnold. Published by Courier Corporation, © 2005; page 474: Quote from James Hutton, Transactions of the Royal Society of Edinburgh, 1805.

Library of Congress Cataloging-in-Publication Data

Names: Lutgens, Frederick K. | Tarbuck, Edward J. | Tasa, Dennis.
Title: Essentials of geology / Frederick K. Lutgens, Edward J. Tarbuck; illustrated by Dennis Tasa.
Description: 13e. [13th edition]. | Hoboken, New Jersey : Pearson Education,
2016.
Identifiers: LCCN 2016042061| ISBN 9780134446622 | ISBN 0134446623
Subjects: LCSH: Geology—Textbooks.
Classification: LCC QE26.3 .L87 2016 | DDC 551—dc23 LC record available at https://lccn.loc.gov/2016042061

3 17

ISBN-10: 0-13-444662-3
ISBN-13: 978-0-13-444662-2

www.pearsonhighered.com

BRIEF CONTENTS

CONTENTS

15 Glaciers & Glaciation 394

16 Deserts & Wind 422

17 Shorelines 440

SMARTFIGURES

Use your mobile device to scan a SmartFigure identified by a Quick Response (QR) code, and a video or animation illustrating the SmartFigure's concept launches immediately. No slow websites or hard-to-remember logins required. These mobile media transform textbooks into convenient digital platforms, breathe life into your learning experience, and help you grasp difficult geology concepts.

CONDOR VIDEO
https://goo.gl/dPwXf4

Chapter 9

Chapter 10

Chapter 11

Chapter 12

Chapter 13

Chapter 14

Chapter 15

Chapter 16

Chapter 17

Chapter 18

Chapter 19

Chapter 20

PREFACE

The thirteenth edition of *Essentials of Geology*, like its predecessors, is a college-level text that is intended to be a meaningful, nontechnical survey for students taking their first course in geology. In addition to being informative and up-to-date, a major goal of this book is to meet the need of students for a readable and user-friendly text that is a valuable tool for learning the basic principles and concepts of geology.

Although many topical issues are treated in the 13th edition of *Essentials*, it should be emphasized that the main focus of this new edition remains the same as the focus of each of its predecessors: to promote student understanding of basic principles. As much as possible, we have attempted to provide the reader with a sense of the observational techniques and reasoning processes that constitute the science of geology.

New & Important Features

This 13th edition is an extensive and thorough revision of *Essentials of Geology* that integrates improved textbook resources with new online features to enhance the learning experience:

- **Significant updating and revision of content.** A basic function of a college science textbook is to provide clear, understandable presentations that are accurate, engaging, and up-to-date. In the long history of this textbook, our number-one goal has always been to keep *Essentials of Geology* current, relevant, and highly readable for beginning students. With this goal as a priority, every part of this text has been examined carefully. The following are a few examples. In Chapter 9, the text and figures for Section 9.3, "Locating the Source of an Earthquake," are substantially revised, and a discussion of the USGS Community Internet Intensity Map project is added. In Chapter 11, the treatment of stress, strain, and rock deformation are substantially revised, as is the final section on isostatic balance. In Chapter 12, the mechanism responsible for long-runout landslides is updated, with reference to the occurrence of such landslides on Mars, and the 2015 Nepal earthquake is used as a landslide-triggering event. In Chapter 13, a section on the loss of wetlands in coastal Louisiana is added, and the treatment of flood control is updated and tightened. Many discussions, case studies, examples, and illustrations have been updated and revised.
- **SmartFigures make this 13th edition much more than a traditional textbook.** Through its many editions, an important strength of *Essentials* has always been clear, logically organized, and well-illustrated explanations. Now complementing and reinforcing this strength are a series of SmartFigures. Simply by scanning the Quick Response (QR) code next to a SmartFigure with a mobile device, students can link to hundreds of unique and innovative digital learning opportunities that will increase their understanding of important ideas. Each SmartFigure also displays a short URL for students who may lack a smartphone. SmartFigures are truly media that teach! The more than 200 SmartFigures in the 13th edition of *Essentials of Geology* are of five types:

  1. **SmartFigure Tutorials.** Each of these 2- to 4-minute tutorials, prepared and narrated by Professor Callan Bentley, is a mini-lesson that examines and explains the concepts illustrated by the figure.
  2. **SmartFigure Mobile Field Trips.** Scattered throughout this new edition are 24 video field trips that explore classic geologic sites from Iceland to Hawaii. On each trip you will accompany geologist/pilot/photographer Michael Collier in the air and on the ground to see and learn about landscapes that relate to discussions in the chapter.
  3. **SmartFigures Condor.** The 10 *Project Condor* videos take you to sites in the American Mountain West. By coupling videos acquired by a quadcopter aircraft with ground-level views, effective narrative, and helpful animations, these videos will engage you in real-life case studies.
  4. **SmartFigure Animations.** These animations bring the art to life, illustrating and explaining difficult-to-visualize topics more effectively than static art alone.
  5. **SmartFigure Videos.** Rather than providing a single image to illustrate an idea, these figures include short video clips that help illustrate such diverse subjects as mineral properties and the structure of ice sheets.

- **Objective-driven active learning path.** Each chapter in this 13th edition begins with *Focus on Concepts:* a set of learning objectives that correspond to the chapter's major sections. By identifying key knowledge and skills, these objectives help students prioritize the material. Each major section concludes with *Concept Checks* so that students can check their learning. Two end-of-chapter features complete the learning path. *Concepts in Review* is coordinated with the *Focus on Concepts* at the beginning of the chapter and with the numbered sections within the chapter. It is a readable and concise overview of key ideas, with photos, diagrams, and questions. Finally, the questions and problems in *Give It Some Thought* challenge learners by requiring higher-order thinking skills to analyze, synthesize, and apply the material.
- **An unparalleled visual program.** In addition to more than 100 new high-quality photos and satellite images, dozens of figures are new or have been redrawn by the gifted and highly respected geoscience illustrator Dennis Tasa. Maps and diagrams are frequently paired with photographs for greater effectiveness. Further, many new and revised figures have additional labels that narrate the process being illustrated and guide students as they examine the figures, resulting in a visual program that is clear and easy to understand.

Digital & Print Resources

MasteringGeology™ with Pearson eText

Used by over 1 million science students, the Mastering platform is the most effective and widely used online tutorial, homework, and assessment system for the sciences. Now available with *Essentials of Geology*, 13th edition, **MasteringGeology**™ offers tools for use before, during, and after class:

- **Before class:** Assign adaptive Dynamic Study Modules and reading assignments from the eText with Reading Quizzes to ensure that students come prepared to class, having done the reading.
- **During class:** Learning Catalytics, a "bring your own device" student engagement, assessment, and classroom intelligence system, allows students to use smartphones, tablets, or laptops to respond to questions in class. With Learning Catalytics, you can assess students in real-time, using open-ended question formats to uncover student misconceptions and adjust lectures accordingly.
- **After class:** Assign an array of assignments such as Mobile Field Trips, Project Condor videos, GigaPan activities, Google Earth Encounter Activities, Geoscience Animations, and much more. Students receive wrong-answer feedback personalized to their answers, which will help them get back on track.

MasteringGeology Student Study Area also provides students with self-study material including videos, geoscience animations, *In the News* articles, Self Study Quizzes, Web Links, Glossary, and Flashcards.

Pearson eText 2.0 gives students access to the text whenever and wherever they can access the Internet. Features of Pearson eText include:

- Now available on smartphones and tablets using the Pearson eText 2.0 app
- Seamlessly integrated videos and other rich media
- Fully accessible (screen-reader ready)
- Configurable reading settings, including resizable type and night reading mode
- Instructor and student note-taking, highlighting, bookmarking, and search

For more information or access to MasteringGeology, please visit www.masteringgeology.com.

For Instructors

Instructor Resource Center (Download Only)

The IRC puts all of your lecture resources in one easy-to-reach place:

- The IRC provides all the line art, tables, and photos from the text in JPEG files.
- PowerPoint™ Presentations: Found in the IRC are three PowerPoint files for each chapter. Cut down on your preparation time, no matter what your lecture needs, by taking advantage of these components of the PowerPoint files:
  - **Exclusive art.** All the photos, art, and tables from the text, in order, have been loaded into PowerPoint slides.
  - **Lecture outline.** This set averages 50 slides per chapter and includes customizable lecture outlines with supporting art.
  - **Classroom Response System (CRS) questions.** Authored for use in conjunction with classroom response systems, these PowerPoint files allow you to electronically poll your class for responses to questions, pop quizzes, attendance, and more.
- The IRC provides Word and PDF versions of the *Instructor Resource Manual.*

Instructor Resource Manual (Download Only)

The *Instructor Resource Manual* has been designed to help seasoned and new instructors alike, offering the following sections in each chapter: an introduction to the chapter, outline, learning objectives/focus on concepts; teaching strategies; teacher resources; and answers to *Concept Checks* and *Give It Some Thought* questions from the textbook. www.pearsonhighered.com/irc

TestGen Computerized Test Bank (Download Only)

TestGen is a computerized test generator that lets instructors view and edit Test Bank questions, transfer questions to tests, and print the test in a variety of customized formats. This Test Bank includes more than 2,000 multiple-choice, matching, and essay questions. Questions are correlated to Bloom's Taxonomy, each chapter's learning objectives, the Earth Science Learning Objectives, and the Pearson Science Global Outcomes to help instructors better map the assessments against both broad and specific teaching and learning objectives. The Test Bank is also available in Microsoft Word and can be imported into Blackboard, www.pearsonhighered.com/irc

For Students

***Laboratory Manual in Physical Geology*, 11th Edition** by the American Geological Institute and the National Association of Geoscience Teachers, edited by Vincent Cronin, illustrated by Dennis G. Tasa (0134446607)

This user-friendly, best-selling lab manual examines the basic processes of geology and their applications to everyday life. Featuring contributions from more than 170 highly regarded geologists and geoscience educators, along with an exceptional illustration program by Dennis Tasa, *Laboratory Manual in Physical Geology*, 11th edition, offers an inquiry- and activities-based approach that builds skills and gives students a more complete learning experience in the lab. Pre-lab videos linked from the print labs introduce students to the content, materials, and techniques they will use each lab. These teaching videos help TAs prepare for lab setup and learn new teaching skills. The lab manual is available in MasteringGeology with Pearson eText, allowing teachers to use activity-based exercises to build students' lab skills.

***Dire Predictions: Understanding Global Climate Change*, 2nd Edition** by Michael Mann, Lee R. Kump (0133909778)

Periodic reports from the Intergovernmental Panel on Climate Change (IPCC) evaluate the risk of climate change brought on by humans. But the sheer volume of scientific data remains inscrutable to the general public, particularly to those who may still question the validity of climate change. In just over 200 pages, this practical text presents and expands upon the latest climate change data and scientific consensus of the IPCC's *Fifth Assessment Report* in a visually stunning and undeniably powerful way to the lay reader. Scientific findings that provide validity to the implications of climate change are presented in clear-cut graphic elements, striking images, and understandable analogies. The second edition integrates mobile media links to online media. The text is also available in various eText formats, including an eText upgrade option from MasteringGeology courses.

Acknowledgments

Writing a college textbook requires the talents and cooperation of many people. It is truly a team effort, and the authors are fortunate to be part of an extraordinary team at Pearson Education. In addition to being great people to work with, all of them are committed to producing the best textbooks possible. Special thanks to our geology editor, Christian Botting. We appreciate his enthusiasm, hard work, and quest for excellence. We also appreciate our conscientious project manager, Lizette Faraji, whose job it was to keep track of all that was going on—and a lot was going on. As always, our marketing managers, Neena Bali and Mary Salzman, who talk with faculty daily, provide us with helpful advice and many good ideas. The 13th edition of *Essentials of Geology* was certainly improved by the talents of our developmental editor, Margot Otway. Our sincere thanks to Margot for her fine work. It was the job of the production team, led by Patty Donovan at SPi Global, to turn our manuscript into a finished product. The team also included copyeditor Kitty Wilson, proofreader Erika Jordan, and photo researcher Kristin Piljay. We think these talented people did great work. All are true professionals, with whom we are very fortunate to be associated.

The authors owe special thanks to four people who were very important contributors to this project:

- **Dennis Tasa.** Working with Dennis Tasa, who is responsible for all of the text's outstanding illustrations and several of its animations, is always special for us. He has been part of our team for more than 30 years. We value not only his artistic talents, hard work, patience, and imagination but his friendship as well.
- **Michael Collier.** As you read this text, you will see dozens of extraordinary photographs by Michael Collier. Most are aerial shots taken from his nearly 60-year-old Cessna 180. Michael was also responsible for preparing the remarkable Mobile Field Trips that are scattered through the text. Among his many awards is the American Geological Institute Award for Outstanding Contribution to the Public Understanding of Geosciences. We think that Michael's photographs and field trips are the next best thing to being there. We were very fortunate to have had Michael's assistance on *Essentials of Geology*, 13th edition. Thanks, Michael.
- **Callan Bentley.** Callan Bentley made many contributions to the new edition of *Essentials*. Callan is a professor of geology at Northern Virginia Community College in Annandale, where he has been honored many times as an outstanding teacher. He is a frequent contributor to *EARTH* magazine and is author of the popular geology blog *Mountain Beltway*. Callan assisted with the revision of Chapter 11, "Crustal Deformation & Mountain Building," and was responsible for preparing the SmartFigure Tutorials that appear throughout the text. As you take advantage of these outstanding learning aids, you will hear his voice explaining the ideas.
- **Scott Linneman.** We were fortunate to have Scott Linneman join the *Essentials of Geology* team as we prepared the 13th edition. Scott provided many thoughtful suggestions and ideas and was responsible for revising Chapter 12, "Mass Movement on Slopes: The Work of Gravity." Scott is an award-winning professor of geology and science education and director of the Honors Program at Western Washington University in Bellingham.

Great thanks also go to our colleagues who prepared in-depth reviews. Their critical comments and thoughtful input helped guide our work and clearly strengthened the text. Special thanks to:

Jessica Barone, Monroe Community College
Paul Belasky, Ohlone College
Larry Braile, Purdue University
Alan Coulson, Clemson University
Nels Forsman, University of North Dakota
Edward Garnero, Arizona State University
Maria Mercedes Gonzales, Central Michigan University
Callum Hetherington, Texas Tech University
Uwe Richard Kackstaetter, Metropolitan State University of Denver
Haraldur Karlsson, Texas Tech University
Johnny MacLean, Southern Utah University
Jennifer Nelson, Indiana University–Purdue University Indianapolis
Cassiopeia Paslick, Rock Valley College
Jeff Richardson, Columbus State Community College
Jennifer Stempien, University of Colorado–Boulder
Donald Thieme, Valdosta State University

Last but certainly not least, we gratefully acknowledge the support and encouragement of our wives, Nancy Lutgens and Joanne Bannon. Preparation of this edition of *Essentials* would have been far more difficult without their patience and understanding.

Fred Lutgens
Ed Tarbuck

Use Dynamic Media to Bring Geology to Life

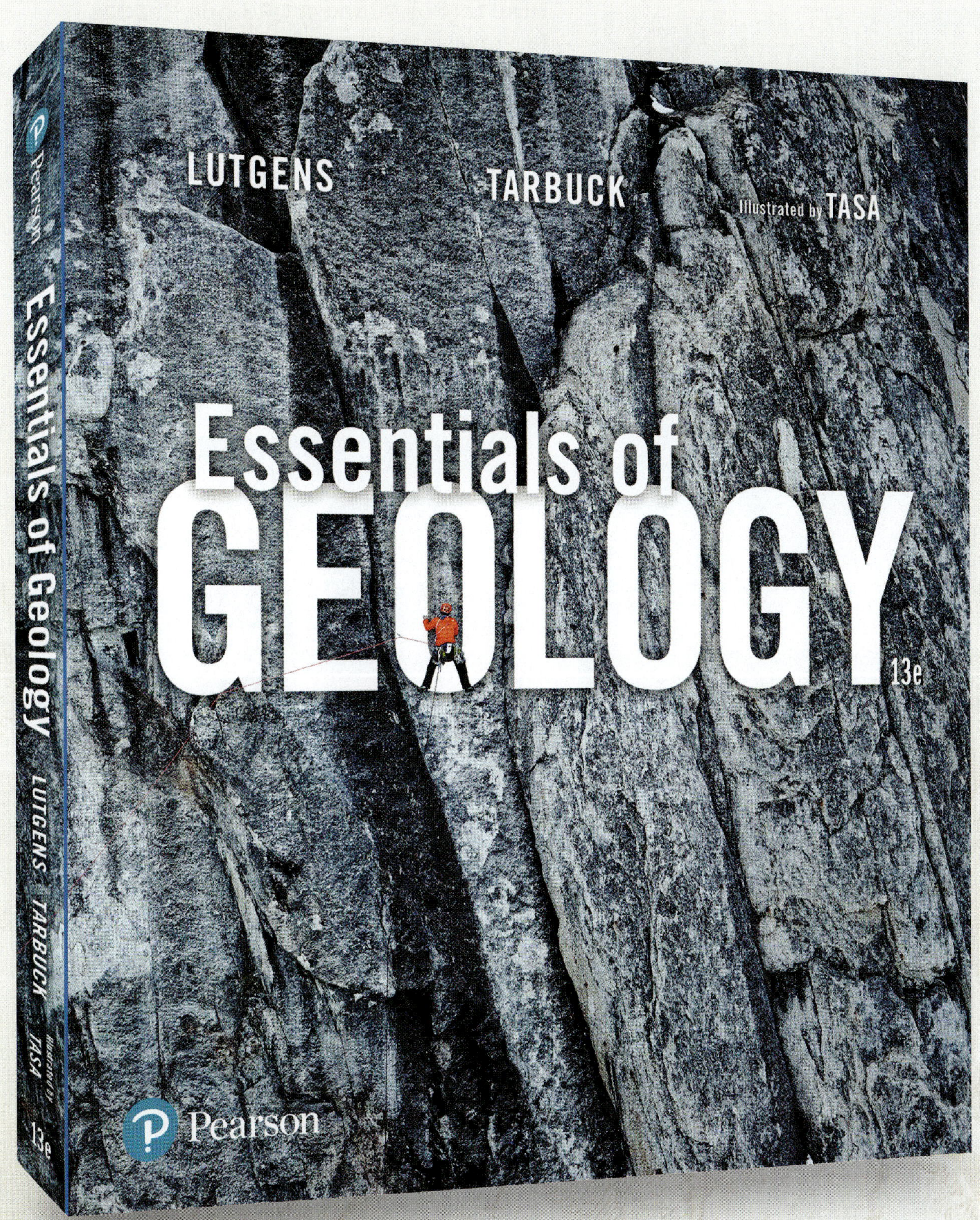

Bring Field Experience to Students' Fingertips...

▶ SmartFigure 2.14
East African Rift valley
The East African Rift valley represents the early stage in the breakup of a continent. Areas shown in red consist of lithosphere that has been stretched and thinned, allowing magma to well up from the mantle.

CONDOR VIDEO
https://goo.gl/RXv8qH

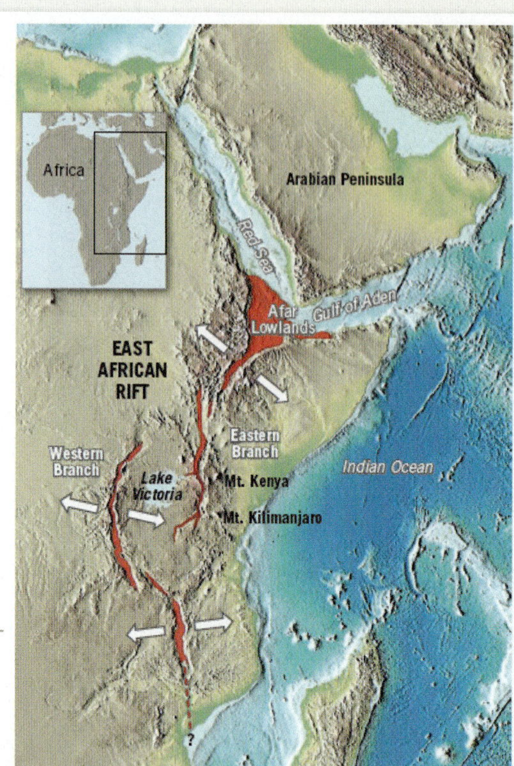

◀ SmartFigure 2.13
Continental rifting: Formation of new ocean basins

TUTORIAL
https://goo.gl/9CokZD

NEW! QR Codes link out to SmartFigures

Quick Response (QR) codes link out to over 200 videos and animations, giving readers immediate access to five types of dynamic media: Project Condor Quadcopter Videos, Mobile Field Trips, Tutorials, Animations, and Videos to help visualize physical processes and concepts. SmartFigures extend the print book to bring geology to life.

How to download a QR Code Reader

Using a smartphone, students are encouraged to download a QR Code reader app from Google Play or the Apple App Store. Many are available for free. Once downloaded, students open the app and point the camera to a QR Code. Once scanned, they're prompted to open the url to immediately be connected to the digital world and deepen their learning experience with the printed text.

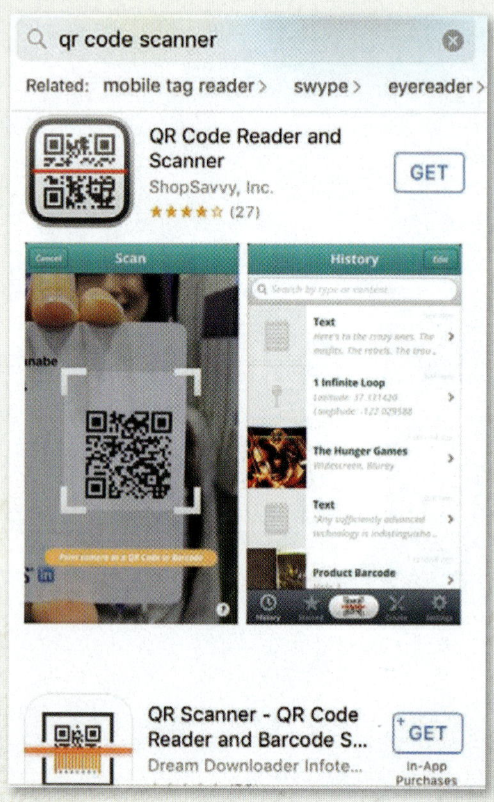

NEW! SmartFigure: Project Condor Quadcopter Videos

Bringing Physical Geology to life for geology students, three geologists, using a quadcopter-mounted GoPro camera, have ventured into the field to film 10 key geologic locations and processes. These process-oriented videos, accessed through QR codes, are designed to bring the field to the classroom and improve the learning experience within the text.

...with SmartFigures

NEW! SmartFigure: Mobile Field Trips
On each trip, students will accompany geologist-pilot-photographer Michael Collier in the air and on the ground to see and learn about iconic landscapes that relate to discussions in the chapter. These extraordinary field trips are accessed by using QR codes throughout the text. New Mobile Field Trips for the 13th edition include *Formation of a Water Gap, Ice Sculpts Yosemite, Fire and Ice Land, Dendrochronology*, and *Desert Geomorphology*.

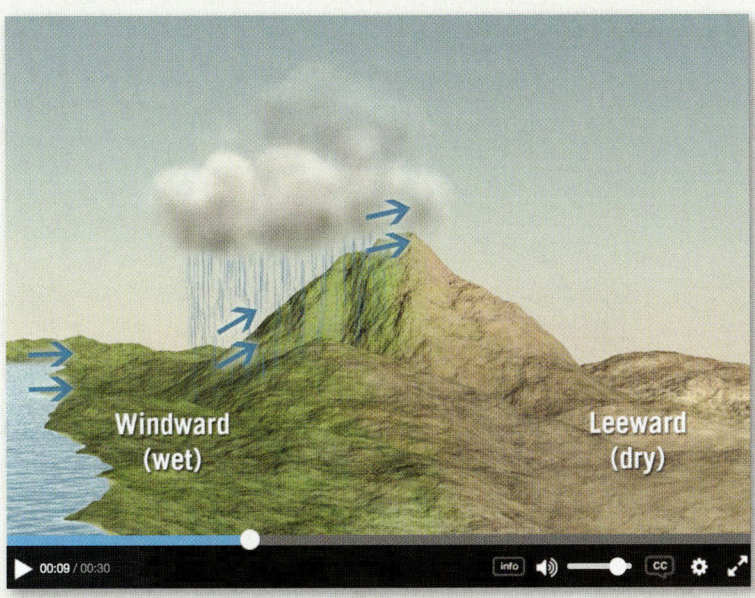

NEW! SmartFigure: Animations
Brief animations created by text illustrator Dennis Tasa animate a process or concept depicted in the textbook's figures. With QR codes, students are given a view of moving figures rather than static art to depict how geologic processes move throughout time.

HALLMARK! SmartFigure: Tutorials
These brief tutorial videos present the student with a 3- to 4-minute feature (mini-lesson) narrated and annotated by Professor Callan Bentley. Each lesson examines and explains the concepts illustrated by the figure. With over 100 SmartFigure Tutorials inside the text, students have a multitude of ways to enjoy art that teaches.

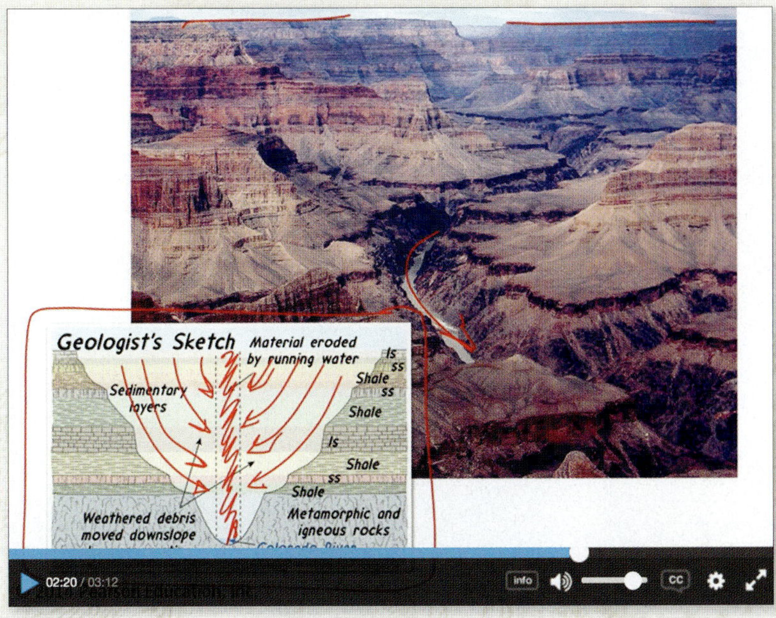

Award-Winning Contributing Authors

The language of this text is straightforward and written to be understood. Clear, readable discussions with a minimum of technical language is the rule. In the 13th edition, we have continued to improve readability with the addition of two new contributing authors, Scott Linneman and Callan Bentley.

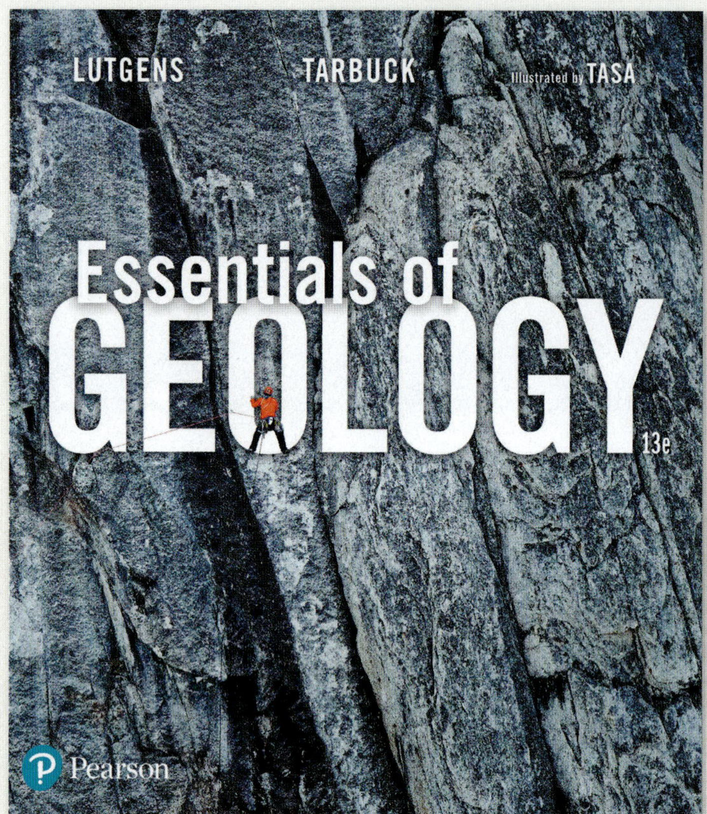

Scott Linneman provided many thoughtful suggestions and ideas throughout the text and was responsible **for revising Chapter 12: Mass Movement on Slopes: The Work of Gravity.** Linneman is an award-winning Professor of Geology and Science Education and director of the Honors Program at Western Washington University in Bellingham.

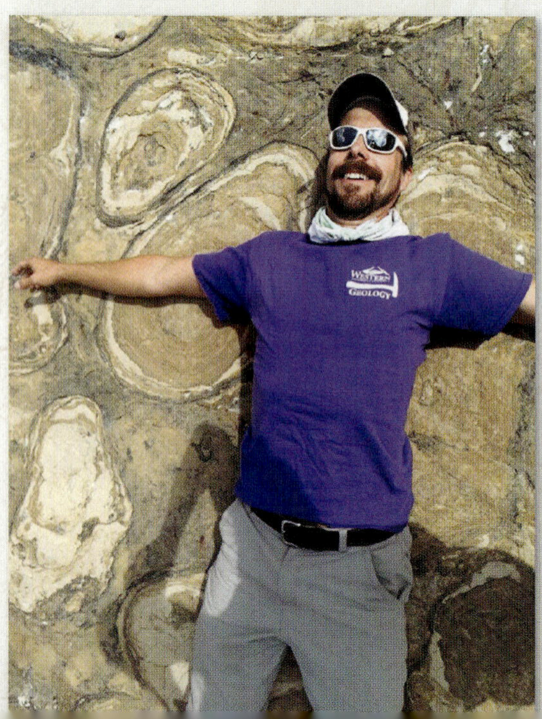

Callan Bentley is Professor of Geology at Northern Virginia Community College in Annandale, where he has been honored many times as an outstanding teacher. He is a frequent contributor to EARTH magazine and is author of the popular geology blog Mountain Beltway. Bentley assisted with the **revision of Chapter 11: Crustal Deformation and Mountain Building** and created the SmartFigure Tutorials that appear throughout the text. As students take advantage of these outstanding learning aids, they will hear his voice explaining the ideas.

Objective-Driven Active Learning

Most chapters have been designed to be self-contained so that materials may be taught in a different sequence, according to the preference of the instructor or the needs of the laboratory. Thus, an instructor who wishes to discuss erosional processes prior to earthquakes, plate tectonics, and mountain building may do so without difficulty.

The chapter-opening **Focus on Concepts** lists the learning objectives for each chapter. Each section of the chapter is tied to a specific learning objective, providing students with a clear learning path to the chapter content.

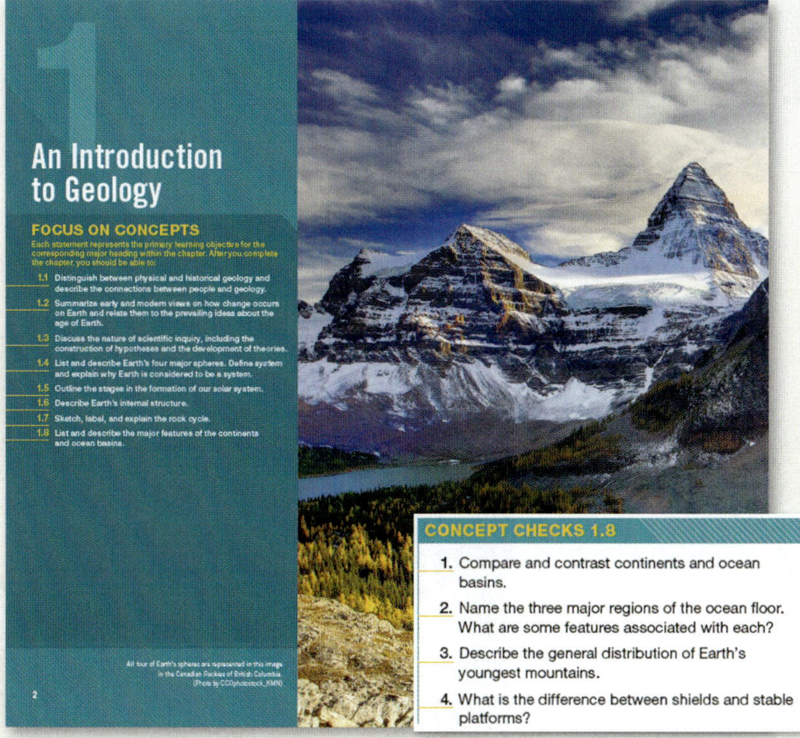

1
An Introduction to Geology
FOCUS ON CONCEPTS
Each statement represents the primary learning objective for the corresponding major heading within the chapter. After you complete the chapter, you should be able to:
1.1 Distinguish between physical and historical geology and describe the connections between people and geology.
1.2 Summarize early and modern views on how change occurs on Earth and relate them to the prevailing ideas about the age of Earth.
1.3 Discuss the nature of scientific inquiry, including the construction of hypotheses and the development of theories.
1.4 List and describe Earth's four major spheres. Define system and explain why Earth is considered to be a system.
1.5 Outline the stages in the formation of our solar system.
1.6 Describe Earth's internal structure.
1.7 Sketch, label, and explain the rock cycle.
1.8 List and describe the major features of the continents and ocean basins.
All four of Earth's spheres are represented in this image in the Canadian Rockies of British Columbia. (Photo by CCOphotostock_KMN)
2

CONCEPT CHECKS 1.8

1. Compare and contrast continents and ocean basins.
2. Name the three major regions of the ocean floor. What are some features associated with each?
3. Describe the general distribution of Earth's youngest mountains.
4. What is the difference between shields and stable platforms?

Each chapter section concludes with **Concept Checks**, a set of questions that is tied to the section's learning objective and allows students to monitor their grasp of significant facts and ideas.

Give It Some Thought activities challenge learners by requiring higher-order thinking skills to analyze, synthesize, and apply the material.

GIVE IT SOME THOUGHT

1 The length of recorded history for humankind is about 5000 years. Clearly, most people view this span as being very long. How does it compare to the length of geologic time? Calculate the percentage or fraction of geologic time that is represented by recorded history. To make calculations easier, round the age of Earth to the nearest billion.

2 After entering a dark room, you turn on a wall switch, but the light does not come on. Suggest at least three hypotheses that might explain this observation. Once you have formulated your hypotheses, what is the next logical step?

b. If you are flying in a commercial jet at an altitude of 12 kilometers (about 39,000 feet), about what percentage of the atmosphere's mass is below you?

CONCEPTS IN REVIEW

Plate Tectonics: A Scientific Revolution Unfolds

2.1 From Continental Drift to Plate Tectonics

Summarize the view that most geologists held prior to the 1960s regarding the geographic positions of the ocean basins and continents.

- Fifty years ago, most geologists thought that ocean basins were very old and that continents were fixed in place. Those ideas were discarded with a scientific revolution that revitalized geology: the theory of plate tectonics. Supported by multiple kinds of evidence, plate tectonics is the foundation of modern Earth science.

Concepts in Review provides students with a structured review of the chapter. Consistent with the Focus on Concepts and Concept Checks, the **Concepts in Review** is structured around the learning objective for each section.

Continuous Learning
Before, During, and After Class

BEFORE CLASS

Mobile Media and Reading Assignments Ensure Students Come to Class Prepared

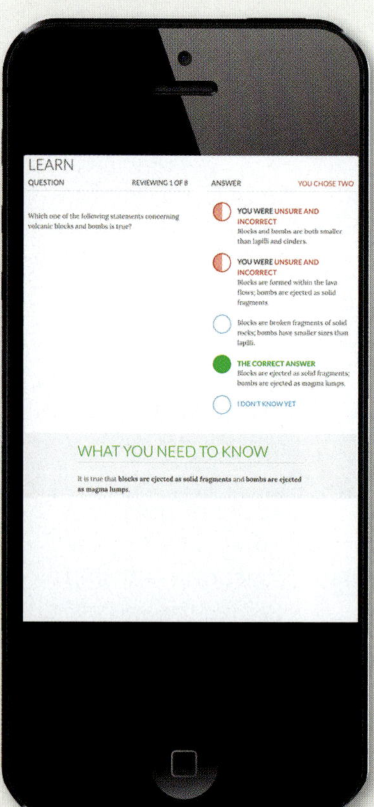

Updated! Dynamic Study Modules help students study effectively by continuously assessing student performance and providing practice in areas where students struggle the most. Each Dynamic Study Module, accessed by computer, smartphone, or tablet, promotes fast learning and long-term retention.

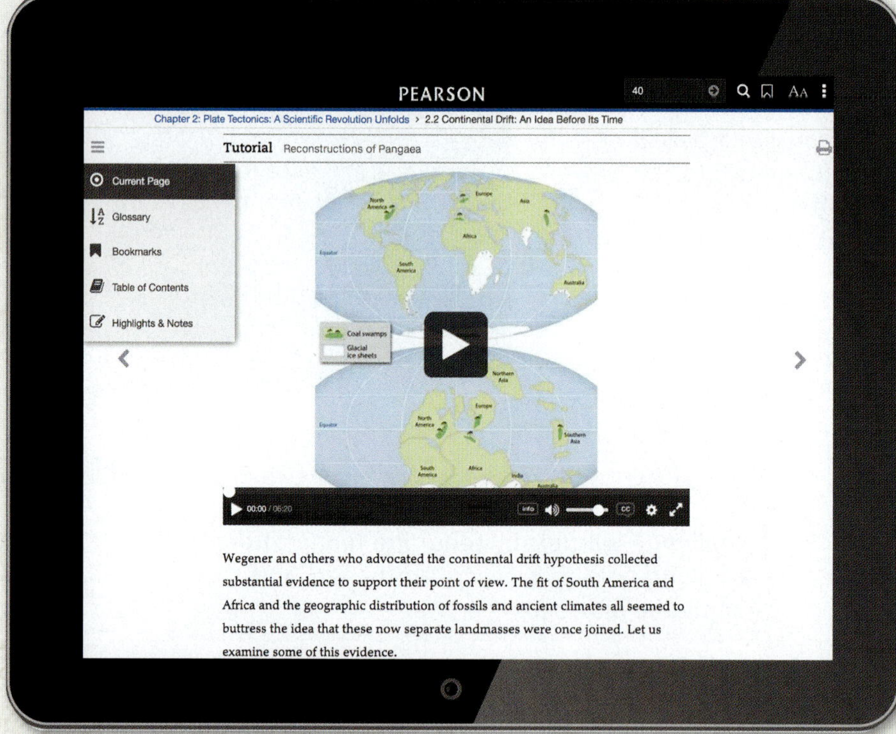

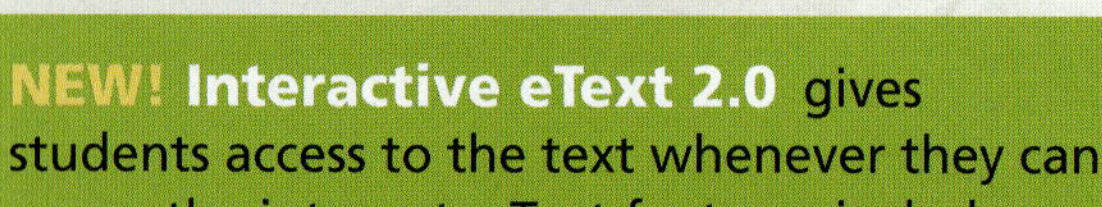

NEW! Interactive eText 2.0 gives students access to the text whenever they can access the internet. eText features include:

- Now available on smartphones and tablets.
- Seamlessly integrated videos and other rich media.
- Accessible (screen-reader ready).
- Configurable reading settings, including resizable type and night reading mode.
- Instructor and student note-taking, highlighting, bookmarking, and search.

Pre-Lecture Reading Quizzes are easy to customize and assign

Reading Questions ensure that students complete the assigned reading before class and stay on track with reading assignments. Reading Questions are 100% mobile ready and can be completed by students on mobile devices.

with MasteringGeology™

DURING CLASS

Engage students with Learning Catalytics

What has teachers and students excited? Learning Catalytics, a 'bring your own device' student engagement, assessment, and classroom intelligence system, allows students to use their smartphone, tablet, or laptop to respond to questions in class. With Learning Catalytics, you can:

- Assess students in real time using open-ended question formats to uncover student misconceptions and adjust lecture accordingly.
- Automatically create groups for peer instruction based on student response patterns, to optimize discussion productivity.

"My students are so busy and engaged answering Learning Catalytics questions during lecture that they don't have time for Facebook."

Declan De Paor, Old Dominion University

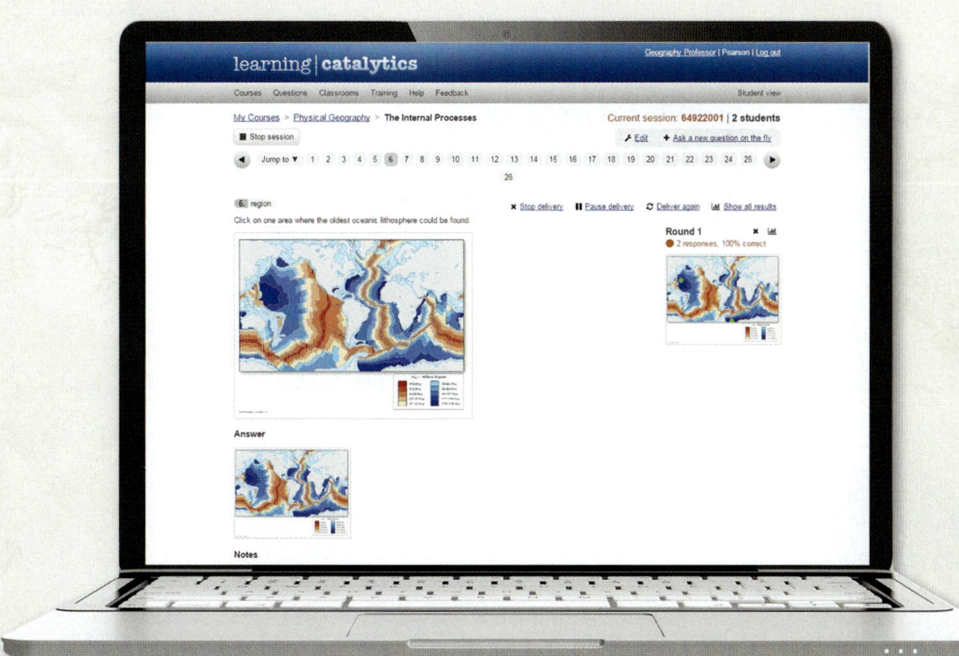

AFTER CLASS

Easy to Assign, Customizable, Media-Rich, and Automatically Graded Assignments

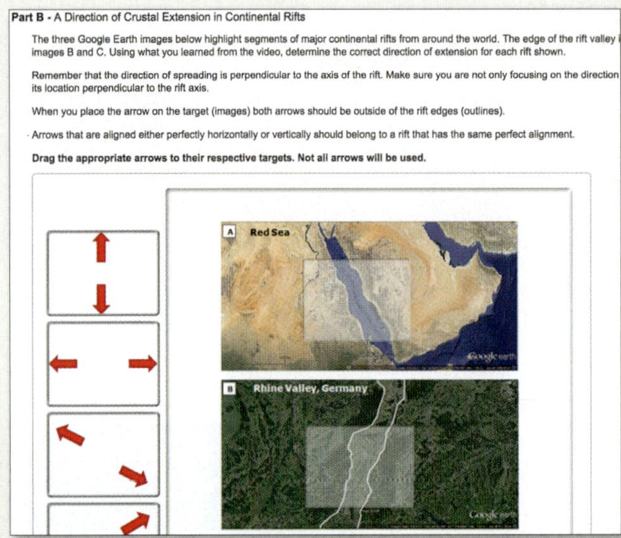

NEW! Project Condor Quadcopter Videos
A series of quadcopter videos with annotations, sketching, and narration help improve the way students learn about monoclines, streams and terraces, and so much more. In MasteringGeology™, these videos are accompanied by assessments to test student understanding.

NEW! 24 Mobile Field Trips take students to iconic geological locations with Michael Collier in the air and on the ground to see and learn about geologic locations that relate to concepts in the chapter. In Mastering, these videos are accompanied by auto-gradable assessments that will track what students have learned.

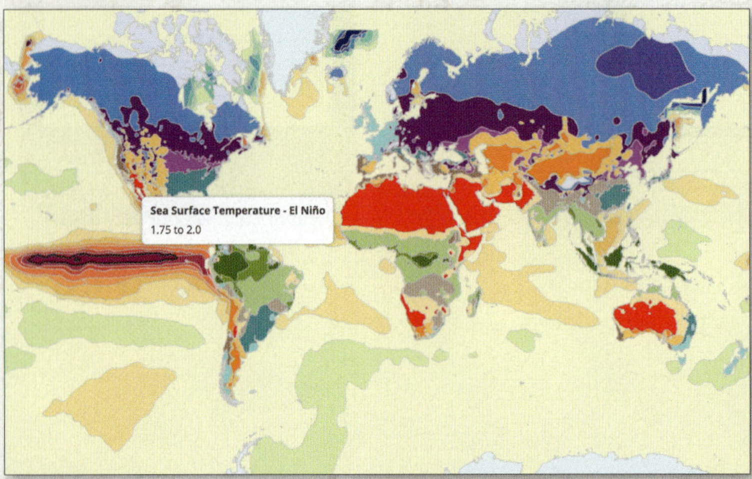

NEW! MapMaster 2.0
GIS-inspired interactive map activities help to enhance students' data analysis and spatial reasoning skills, and overall geologic literacy.

www.masteringgeology.com

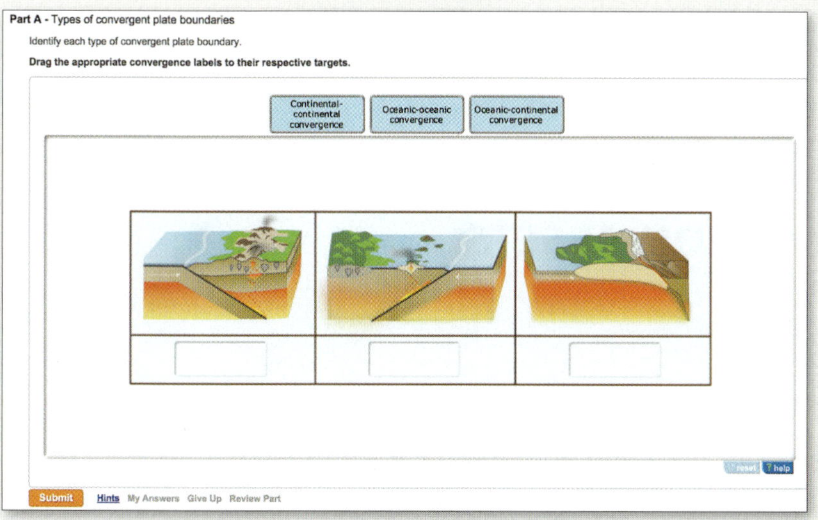

GeoTutors

These coaching activities help students master important physical geoscience concepts with highly visual, kinesthetic activities focused on critical thinking and application of core geoscience concepts.

GigaPan Activities allow students to take advantage of a virtual field experience with high-resolution imaging technology developed by Carnegie Mellon University in conjunction with NASA.

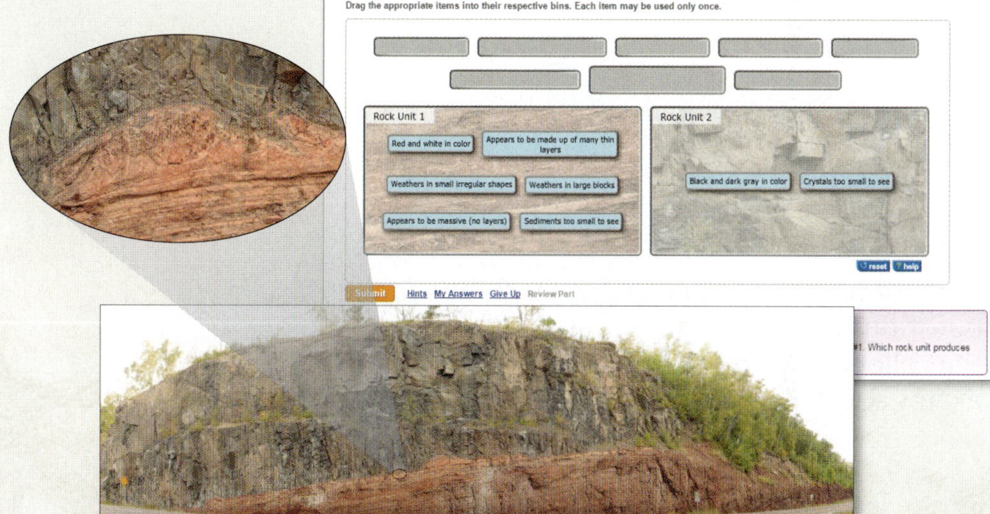

Encounter Activities

Using Google Earth™ to visualize and explore Earth's physical landscape, Encounter activities provide rich, interactive explorations of geology and earth science concepts. Dynamic assessments include questions related to core geology concepts. All explorations include corresponding Google Earth KMZ media files, and questions include hints and specific wrong-answer feedback to help coach students toward mastery of the concepts.

Resources for YOU, the Instructor

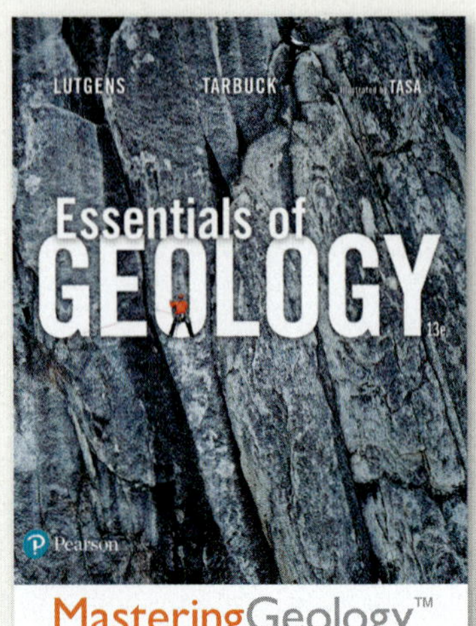

MasteringGeology™ provides you with everything you need to prep for your course and deliver a dynamic lecture, all in one convenient place. Resources include:

LECTURE PRESENTATION ASSETS FOR EACH CHAPTER

- PowerPoint Lecture Outlines
- PowerPoint clicker questions and Jeopardy-style quiz show questions
- All book images and tables in JPEG and PowerPoint formats

TEST BANK

- The Test Bank in Microsoft Word formats
- Computerized Test Bank, which includes all the questions from the printed test bank in a format that allows you to easily and intuitively build exams and quizzes.

TEACHING RESOURCES

- Instructor Resource Manual in Microsoft Word and PDF formats
- Pearson Community Website (https://communities.pearson.com/northamerica/s/)

Measuring Student Learning Outcomes?

All MasteringGeology assignable content is tagged to learning outcomes from the book and Bloom's Taxonomy. You also have the ability to add your own learning outcomes, helping you track student performance against your learning outcomes. You can view class performance against the specified learning outcomes and share those results quickly and easily by exporting to a spreadsheet.

SELECT MAJOR CHANGES IN *ESSENTIALS OF GEOLOGY* 13E

Global

- QR codes for additional SmartFigures added, including Mobile Field Trips; SmartFigure types indicated in figure captions

Chapter 1

- Subsection "Origin of Planet Earth" substantially revised
- New *Did You Know* feature added (Section 1.5)
- Two *Give It Some Thought* questions modified
- Substantively revised figures: 1.13, 1.17, 1.18, 1.20, 1.23, 1.24
- Eleven new photographs

Chapter 2

- *Concept Check* questions for Section 2.6 revised
- Treatment of whole-mantle convection and plumes substantially rewritten for clarity and currency (Section 2.10)
- One *Give It Some Thought* question added and one modified
- Substantively revised figures: 2.9, 2.11, 2.17–2.19, 2.29, 2.30, 2.31, 2.35
- Two new photographs

Chapter 3

- Introduction to mineral properties revised (Section 3.4)
- One new *Give It Some Thought* question added; one modified
- Figure 3.33 now combines illustration and tabular data
- New figures: 3.26, 3.28, 3.33. Figures that have been revised substantively: 3.5 (atomic weight changed to atomic mass), 3.8, 3.9. 3.11, 3.12
- Three new photographs

Chapter 4

- Treatment of magmatic volatiles revised for clarity (Section 4.1)
- Subsection "Compositional Categories" rewritten for clarity; replaces former subsection "Granitic (Felsic) versus Basaltic (Mafic) Compositions" (Section 4.2)
- Terminology "felsic/intermediate/mafic" given priority over "granitic/andesitic/basaltic" (Section 4.4)
- Subsection "Temperature Increase: Melting Crustal Rocks" substantially rewritten for clarity (Section 4.5)
- Improved description of how mineral grains interact with a melt of changing composition
- Footnote added noting complex formation history of Palisades Sill (under "Magmatic Differentiation and Crystal Settling" in Section 4.6)
- Stocks now treated in the section on batholiths (paragraph 4 under "Batholiths" in Section 4.8)
- One *Give It Some Thought* question modified
- Substantively revised figures: 4.5, 4.12, 4.16, 4.17, 4.33
- Eight new photographs

Chapter 5

- Section 5.2, "The Nature of Volcanic Eruptions," largely rewritten
- Paragraph added to cover silica-rich pyroclastic intraplate volcanism
- In Section 5.10, volcanism at divergent boundaries now treated before volcanism at divergent boundaries
- Two new *Give It Some Thought* questions added; one modified
- New figures: 5.3 (replaces 12e Table 5.1), 5.8 (replaces 12e Figure 5.7). Figures that have been revised substantively: 5.5, 5.12, 5.16, 519, 5.21, 5.32
- Twelve new photographs

Chapter 6

- New discussion of oxidation as an agent of weathering ("Oxidation" in Section 6.3)
- In the subsection "Controls of Soil Formation," order of topics changed to put "Time" later (Section 6.5)
- Two new *Give It Some Thought* questions added
- Substantively revised figures: 6.11, 6.24
- Five new photographs

Chapter 7

- Updated treatment of energy resources, including expanded discussion of emissions from coal combustion and changes in oil and gas production due to fracking (Section 7.8)
- Revised treatment of the slowest limb of the carbon cycle (Section 7.9, including Figure 7.34)
- One new *Give It Some Thought* question added
- New figure, 7.33. Figure 7.30 substantively expanded
- Five new photographs

Chapter 8

- New contextual paragraph added at start of Section 8.1
- Improved introduction of temperature and pressure as agents of metamorphism at the end of Section 8.1
- Description and figure of a stretched pebble conglomerate added to help students understand the concept of differential stress (subection "Differential Stress" in Section 8.2)
- In subsection"Other Metamorphic Textures," improved treatment of nonfoliated metamorphic rocks, including coverage of hornfels (Section 8.3)
- One new *Give It Some Thought* question
- Four figures added: 8.5, 8.23, 8.27, 8.29. Figures that have been modified substantively: 8.4, 8.6, 8.10, 8.11, 8.24, 8.26
- Eight new photographs

Chapter 9

- Subsection "Faults & Large Earthquakes" substantially rewritten for clarity and conciseness (Section 9.1)
- Section 9.3, "Locating the Source of an Earthquake," substantially revised, including three figures
- Discussion added for the U.S.G.S. Community Internet Intensity Map project, including a figure (within "Intensity Scales" in Section 9.4)
- Section 9.8 reorganized to put the subsection "Probing Earth's Interior: "Seeing" Seismic Waves" first; treatment of Earth's layered structure substantially revised
- Two new *Give It Some Thought* questions added;
- Two figures added: 9.16, 9.23. Figures that have been modified substantively: 9.10, 9.13–9.15, 9.27 (completely redrawn)
- Two new photographs

Chapter 10

- One *Give It Some Thought* question replaced with a new one
- One new figure added: 10.4 (two-page global sea-floor map). Figures that have been modified substantively: 10.12, 10.16, 10.21
- Two new photographs

Chapter 11

- Treatment of deformation, stress, and strain in Section 11.1 significantly clarified
- Discussion of the factors that affect how rocks deform significantly clarified (Section 11.1)
- Distinction between faults and joints now covered at the start of Section 11.3
- Description of thrust faulting in the formation of the Himalayas improved (paragraph 4 under "The Himalayas" in Section 11.6)
- Description of isostatic balance and its effects rewritten (Section 11.7)
- One new *Give It Some Thought* question added

- Three figures added: 11.4, 11.5, 11.21. Figures that have been modified substantively: 11.3, 11.6–11.8, 11.10, 11.12, 11.14–11.16, 11.18, 11.19, 11.23, 11.27, 11.29, 11.30
- Six new photographs

Chapter 12

- "Mass movement" introduced in place of older term "mass wasting."
- Landslides introduced more thoroughly at the start of Section 12.1
- Treatment of mass movements that lack an obvious trigger clarified and moved to the start of section 12.2
- Treatment of mechanism for long-runout landslides updated (subsection "Rate of Movement" in Section 12.3)
- Definition and description of normal faults made clearer (first paragraph of section "Normal Faults" in Section 11.3)
- 2015 Nepal earthquake added as example of a landslide-triggering event (subsection "Examples from Plate Boundaries: California and Nepal" in Section 12.2)
- New *Did You Know* about 2013 Bingham Canyon Copper Mine landslide added (Section 12.2)
- One new *Give It Some Thought* question added
- Figure 12.11 modified substantively
- Six new photographs

Chapter 13

- Section 13.1 largely rewritten
- Selected paragraphs of Section 13.2 tightened; headward erosion added as final paragraph in section "Drainage Basins; formation of a water gap added at the end of "Drainage Patterns."
- Section on the loss of wetlands from the Mississipi delta and coastal Louisiana added (subsection "Vanishing Wetlands" in Section 13.7)
- Treatment of flood control updated and tightened (Section 13.8)
- One new *Give It Some Thought* question added
- Figure 13.29 added; "Floods & Flood Control" now supported by four new figures 13.31–13.33; Figure 13.24 substantively changed
- Three new photographs

Chapter 14

- Section added on Geothermal Energy (p. 385 in Section 14.5)
- Section added on the impact of prolonged drought on groundwater resources (p. 387 of Section 14.5)
- Three figures added: 14.21, 14.23, 14.29. Figures that have been modified substantively: 14.1, 14,3, 14,22
- Three new photographs

Chapter 15

- Information on Larsen B ice shelf updated (p. 402 in Section 15.1)
- New *Give It Some Thought* question
- Figures that have been replaced or modified substantively: 15.3, 15.4, 15.6, 15.9, 15.11, 15.22
- Five new photographs

Chapter 16

- New *Give It Some Thought* question
- Figures that have been modified substantively: 16.2, 16,3, 16.9
- Three new photographs

Chapter 17

- Section 17.1 ("The Shoreline & Ocean Waves") has been revised to cover both the basic features of shorelines and the behavior of ocean waves. Beaches are now covered along with shoreline processes in Section 17.2 ("Beaches & Shoreline Processes"). Both sections have been tightened to focus more on processes and less on terminology
- Explanation of wave refraction reworded for greater clarity
- Section 17.6 ("Stabilizing the Shore") moved to later in the chapter than in the preceding edition; it now follows Sections 17.4 ("Contrasting America's Coasts") and 17.5 ("Hurricanes: The Ultimate Hazard")
- Section 17.4 ("Contrasting America's Coasts") reorganized to start with the the basic classification of coasts as emergent or submergent. This section also now uses cliff retreat at Pacifica, CA as a topical example of erosion on an emergent coast
- Section 17.5 ("Hurricanes: The Ultimate Hazard") now uses Superstorm Sandy as an example and covers the effect of sea-level rise on vulnerability
- The response of Staten Island to Superstorm Sandy added as an example of a decision to change how coastal land is used ("Changing Land Use" in Section 17.6)
- Four new photographs

Chapter 18

- Section "Correlation within Limited Areas" tightened (in Section 18.3)
- Text and figures for Section 18.4, "Numerical Dating with Nuclear Decay," substantially revised for better clarity and effectiveness
- Section 18.5, "Determining Numerical Dates for Sedimentary Strata," moved so that it now immediately follows Section 18.4
- Two *Give It Some Thought* questions added
- Figures that have been modified substantively: 18.19–18.22, 18.24
- Two new photographs

Chapter 19

- In the section "Oxygen in the Atmosphere," updated treatment of the effects on land organisms of the apparent high levels of oxygen in the Pennsylvanian (in Section 19.3)
- Acasta Gneiss added to discussion of Earth's oldest dated rocks (in Section 19.4)
- Section "Supercontinents and Climate" substantially revised (in Section 19.4)
- Figure 19.17 added, illustrating the major provinces of the Appalachian Mountains (in Section 19.5)
- Paragraphs on the origin of prokaryotes, eukaryotes, and photosynthesis substantively revised ("Earth's First Life: Prokaryotes" in Section 19.6)
- Updated discussion of the origin of tetrapods ("Vertebrates Move to Land" in Section 19.7)
- Updated treatment of the extinction of nonavian dinosaurs ("Demise of the Dinosaurs" in Section 19.7)
- Updated treatment of hominin evolution ("Humans: Mammals with Large Brains & Bipedal Locomotion" in Section 19.9)
- New *Give It Some Thought* question
- Five new photographs

Chapter 20

- Within Section 20.2 ("Detecting Climate Change,") section "Climates Change" added, including Figures 20.2 and 20.3
- In Section 20.5, context-setting second paragraph added
- Section "Rising CO_2 Levels" updated to include current data, including updated discussion of tropical deforestation
- Section "The Atmosphere's Response" updated to reflect the 2013–2014 IPCC 5th Assessment Report
- Section "The Role of Trace Gases" updated to reflect current science, and section "How Aerosols Influence Climate" moved into this section
- Section 20.7, "Climate Feedback Mechanisms," updated to reflect current science
- Table 20.1, "IPCC Projections for the Late Twenty-First Century," added to Section 20.8, and section updated to reflect current science
- Section "The Changing Arctic" largely revised
- Section "The Potential for Surprises" updated
- Three new *Give it Some Thought* questions added
- New figures added: 20.2, 20.3, 20.8, 20.34. Figures modified or updated substantively: 20.21, 20.23, 20.25, 20.26, 20.31, 20.25. Several new photographs.

Essentials of GEOLOGY

13e

1

An Introduction to Geology

FOCUS ON CONCEPTS

Each statement represents the primary learning objective for the corresponding major heading within the chapter. After you complete the chapter, you should be able to:

1.1 Distinguish between physical and historical geology and describe the connections between people and geology.

1.2 Summarize early and modern views on how change occurs on Earth and relate them to the prevailing ideas about the age of Earth.

1.3 Discuss the nature of scientific inquiry, including the construction of hypotheses and the development of theories.

1.4 List and describe Earth's four major spheres. Define *system* and explain why Earth is considered to be a system.

1.5 Outline the stages in the formation of our solar system.

1.6 Describe Earth's internal structure.

1.7 Sketch, label, and explain the rock cycle.

1.8 List and describe the major features of the continents and ocean basins.

All four of Earth's spheres are represented in this image in the Canadian Rockies of British Columbia. (Photo by CCOphotostock_KMN)

THE SPECTACULAR ERUPTION OF A VOLCANO, the terror brought by an earthquake, the magnificent scenery of a mountain range, and the destruction created by a landslide or flood are all subjects for a geologist. The study of geology deals with many fascinating and practical questions about our physical environment. What forces produce mountains? When will the next major earthquake occur in California? What are ice ages like, and will there be another? How were ore deposits formed? Where should we search for water? Will plentiful oil be found if a well is drilled in a particular location? Geologists seek to answer these and many other questions about Earth, its history, and its resources.

1.1 Geology: The Science of Earth

Distinguish between physical and historical geology and describe the connections between people and geology.

The subject of this text is **geology**, from the Greek *geo* (Earth) and *logos* (discourse). Geology is the science that pursues an understanding of planet Earth. Understanding Earth is challenging because our planet is a dynamic body with many interacting parts and a complex history. Throughout its long existence, Earth has been changing. In fact, it is changing as you read this page and will continue to do so. Sometimes the changes are rapid and violent, as when landslides or volcanic eruptions occur. Just as often, change takes place so slowly that it goes unnoticed during a lifetime. Scales of size and space also vary greatly among the phenomena that geologists study. Sometimes geologists must focus on phenomena that are microscopic, such as the crystalline structure of minerals, and at other times they must deal with features that are continental or global in scale, such as the formation of major mountain ranges.

Physical and Historical Geology

Geology is traditionally divided into two broad areas—physical and historical. **Physical geology**, which is the primary focus of this book, examines the materials composing Earth and seeks to understand the many processes that operate beneath and upon its surface (**Figure 1.1**). The aim of **historical geology**, on the other hand, is to understand the origin of Earth and its development through time. Thus, it strives to establish an orderly chronological arrangement of the multitude of physical and biological changes that have occurred in the geologic past. The study of physical geology logically precedes the study of Earth history because we must first understand how Earth works before we attempt to unravel its past. It should also be pointed out that physical and historical geology are divided into many areas of specialization. Every chapter of this book represents one or more areas of specialization in geology.

Geology is perceived as a science that is done outdoors—and rightly so. A great deal of geology is based on observations, measurements, and experiments conducted in the field. But geology is also done in the laboratory, where, for example, analysis of minerals and rocks provides insights into many basic processes and the microscopic study of fossils unlocks clues to past environments (**Figure 1.2**). Geologists must also understand and apply knowledge and principles from physics,

▼ **Figure 1.1 Internal and external processes** The processes that operate beneath and upon Earth's surface are an important focus of physical geology. (River photo by Michael Collier; volcano photo by AM Design/Alamy Live News/Alamy Images)

A.

External processes, such as landslides, rivers, and glaciers, erode and sculpt surface features. The Colorado River played a major role in creating the Grand Canyon.

B.

Internal processes are those that occur beneath Earth's surface. Sometimes they lead to the formation of major features at the surface, such as Italy's Mt. Etna.

A.

B.

▲ **Figure 1.2 In the field and in the lab** Geology involves not only outdoor fieldwork but work in the laboratory as well. **A.** This research team is gathering data at Mount Nyiragongo, an active volcano in the Democratic Republic of the Congo. (Photo by Carsten Peter/National Geographic Image Collection/Alamy) **B.** This researcher is using a petrographic microscope to study the mineral compositions of rock samples. (Photo by Jon Wilson/Science Source)

Did You Know?
Each year an average American requires huge quantities of Earth materials. Imagine receiving your annual share in a single delivery. A large truck would pull up to your home and unload 12,965 lb of stone, 8945 lb of sand and gravel, 895 lb of cement, 395 lb of salt, 361 lb of phosphate, and 974 lb of other nonmetals. In addition, there would be 709 lb of metals, including iron, aluminum, and copper.

chemistry, and biology. Geology is a science that seeks to expand our knowledge of the natural world and our place in it.

Geology, People, and the Environment

The primary focus of this book is to develop your understanding of basic geologic principles, but along the way we will explore numerous important relationships between people and the natural environment. Many of the problems and issues addressed by geology are of practical value to people.

Natural hazards are a part of living on Earth. Every day they adversely affect millions of people worldwide and are responsible for staggering damages. Among the hazardous Earth processes that geologists study are volcanoes, floods, tsunamis, earthquakes, and landslides. Of course, geologic hazards are *natural* processes. They become hazards only when people try to live where these processes occur (**Figure 1.3**).

According to the United Nations, more people now live in cities than in rural areas. This global trend toward urbanization concentrates millions of people into megacities, many of which are vulnerable to natural hazards. Coastal sites are becoming more vulnerable because development often destroys natural defenses such as wetlands and sand dunes. In addition, there is a growing threat associated with human influences on the Earth system; one example is sea-level rise that is linked to global climate change. Some megacities are exposed to seismic (earthquake) and volcanic hazards, the threat of which may be compounded by inappropriate land use, poor construction practices, and rapid population growth.

Resources are another important focus of geology that is of great practical value to people. Resources include water and soil, a great variety of metallic and nonmetallic minerals, and energy (**Figure 1.4**). Together they form the very foundation of modern civilization. Geology deals not only with the formation and occurrence of these vital resources but also with maintaining

▼ **Figure 1.3 Earthquake destruction** During a three-week span in spring 2015, the small Himalayan country of Nepal experienced two major earthquakes. There were more than 8000 fatalities and nearly a half million homes destroyed. Geologic hazards are natural processes. They become hazards only when people try to live where these processes occur. The debris flow shown in Figure 1.15 and the volcanic eruption related to Figure 1.17 are also examples of geologic hazards that had deadly consequences. (Photo by Roberto Schmidt/AFP/Getty Images)

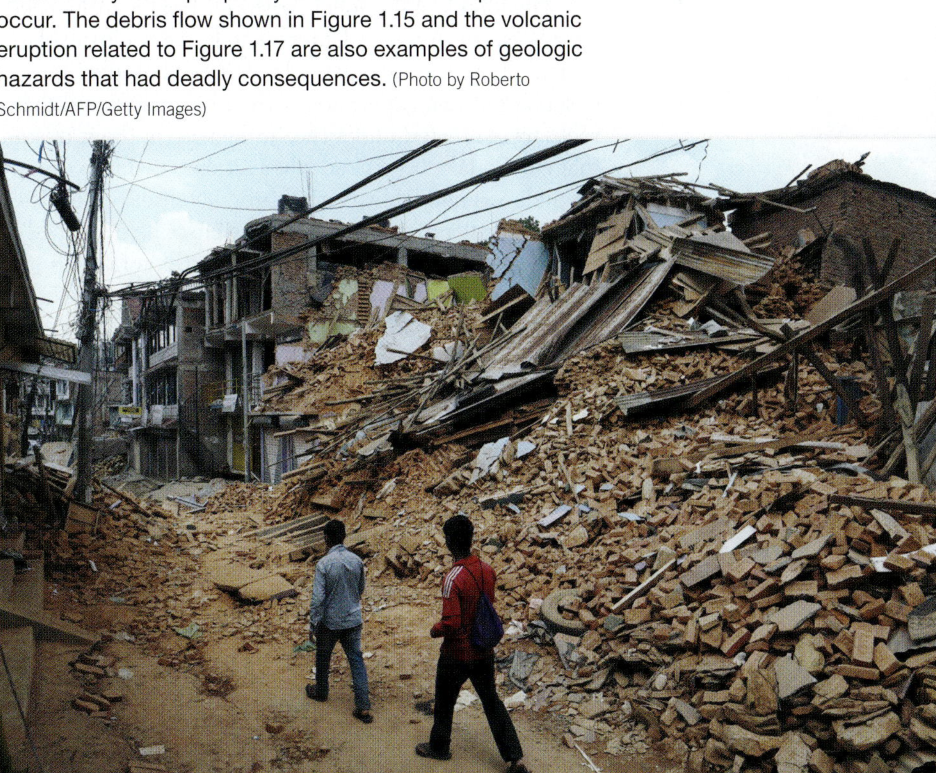

Did You Know?
It took until about the year 1800 for the world population to reach 1 billion. By 1927, the number had doubled to 2 billion. According to United Nations estimates, world population reached 7 billion in late October 2011. We are currently adding about 80 million people per year to the planet.

supplies and with the environmental impact of their extraction and use.

Geologic processes clearly have an impact on people. In addition, we humans can dramatically influence geologic processes. For example, landslides and river flooding occur naturally, but the magnitude and frequency of these processes can be affected significantly by human activities such as clearing forests, building cities, and constructing dams. Unfortunately, natural systems do not always adjust to artificial changes in ways that we can anticipate. Thus, an alteration to the environment that was intended to benefit society sometimes has the opposite effect.

At appropriate places throughout this textbook, you will have opportunities to examine different aspects of our relationship with the physical environment. Nearly every chapter addresses some aspect of natural hazards, resources, and the environmental issues associated with each. Significant parts of some chapters provide the basic geologic knowledge and principles needed to understand environmental problems.

▲ **Figure 1.4 Copper mining** Mineral and energy resources represent an important link between people and geology. This large open pit mine is in Arizona. (Photo by Ball Miwako/Alamy)

CONCEPT CHECKS 1.1

1. Name and distinguish between the two broad subdivisions of geology.
2. List at least three different geologic hazards.
3. Aside from geologic hazards, describe another important connection between people and geology.

1.2 The Development of Geology

Summarize early and modern views on how change occurs on Earth and relate them to the prevailing ideas about the age of Earth.

The nature of our Earth—its materials and processes—has been a focus of study for centuries. Writings about such topics as fossils, gems, earthquakes, and volcanoes date back to the early Greeks, more than 2300 years ago.

The Greek philosopher Aristotle strongly influenced later Western thinking. Unfortunately, Aristotle's explanations about the natural world were not based on keen observations and experiments. He arbitrarily stated that rocks were created under the "influence" of the stars and that earthquakes occurred when air crowded into the ground, was heated by central fires, and escaped explosively. When confronted with a fossil fish, he explained that "a great many fishes live in the earth motionless and are found when excavations are made." Although Aristotle's explanations may have been adequate for his day, they unfortunately continued to be viewed as authoritative for many centuries, thus inhibiting the acceptance of more up-to-date ideas. After the Renaissance of the 1500s, however, more people became interested in finding answers to questions about Earth.

Catastrophism

In the mid-1600s, James Ussher, Anglican Archbishop of Armagh, Primate of all Ireland, published a major work that had immediate and profound influences. A respected scholar of the Bible, Ussher constructed a chronology of human and Earth history in which he calculated that Earth was only a few thousand years old, having been created in 4004 B.C.E. Ussher's treatise earned widespread acceptance among Europe's scientific and religious leaders, and his chronology was soon printed in the margins of the Bible itself.

During the seventeenth and eighteenth centuries, Western thought about Earth's features and processes was strongly influenced by Ussher's calculation. The result was a guiding doctrine called **catastrophism**. Catastrophists believed that Earth's landscapes were shaped primarily by great catastrophes. Features such as mountains and canyons, which today we know take great spans of time to form, were explained as resulting from sudden and often worldwide disasters produced by unknowable causes that no longer operate. This philosophy was an attempt to fit the rates of Earth processes to the then-current ideas about the age of Earth.

The Birth of Modern Geology

Against the backdrop of Aristotle's views and the idea of an Earth created in 4004 B.C.E. a Scottish physician and gentleman farmer named James Hutton published *Theory of the Earth* in 1795. In this work, Hutton put forth a fundamental principle that is a pillar of geology today: **uniformitarianism**. It states that the *physical, chemical, and biological laws that operate today have also operated in the geologic past.* This means that the forces and processes that we observe presently shaping our planet have been at work for a very long time. Thus, to understand ancient rocks, we must first understand present-day processes and their results. This idea is commonly stated as *the present is the key to the past.*

Prior to Hutton's *Theory of the Earth*, no one had effectively demonstrated that geologic processes occur over extremely long periods of time. However, Hutton persuasively argued that forces that appear small can, over long spans of time, produce effects that are just as great as those resulting from sudden catastrophic events. Unlike his predecessors, Hutton carefully cited verifiable observations to support his ideas. For example, when Hutton argued that mountains are sculpted and ultimately destroyed by weathering and the work of running water and that the products are carried to the oceans by observable processes, he said, "We have a chain of facts which clearly demonstrate . . . that the materials of the wasted mountains have traveled through the rivers"; and further, "There is not one step in all this progress . . . that is not to be actually perceived." He then went on to summarize this thought by asking a question and immediately providing the answer: "What more can we require? Nothing but time."

Geology Today

Today the basic tenets of uniformitarianism are just as viable as in Hutton's day. Indeed, today we realize more strongly than ever before that the present gives us insight into the past and that the physical, chemical, and biological laws that govern geologic processes remain unchanging through time. However, we also understand that the doctrine should not be taken too literally. To say that geologic processes in the past were the same as those occurring today is not to suggest that they have always had the same relative importance or that they have operated at precisely the same rate. Moreover, some important geologic processes are not currently observable, but evidence that they occur is well established. For example, we know that impacts from large meteorites have altered Earth's climate and influenced the history of life, even though we have no historical accounts of such impacts.

The acceptance of uniformitarianism meant the acceptance of a very long history for Earth. Although Earth processes vary in intensity, they almost always take a very long time to create or destroy major landscape features. The Grand Canyon provides a good example (**Figure 1.5**).

The rock record contains evidence which shows that Earth has experienced many cycles of mountain building and erosion. Concerning the ever-changing nature of Earth through great expanses of geologic time, Hutton famously stated in 1788: "The results, therefore, of our present enquiry is, that we find no vestige of a beginning—no prospect of an end."

In the chapters that follow, we will be examining the materials that compose our planet and the processes that modify it. It is important to remember that, although many features of our physical landscape may seem to be unchanging over the decades we observe them, they are nevertheless changing—but on time scales of hundreds, thousands, or even many millions of years.

Did You Know?
Shortly after Archbishop Ussher determined an age for Earth, another biblical scholar, Dr. John Lightfoot of Cambridge, felt he could be even more specific. He wrote that Earth was created "on the 26th of October 4004 BC at 9 o'clock in the morning." (As quoted in William L. Stokes, *Essentials of Earth History*, Prentice Hall, Inc. 1973, p. 20.)

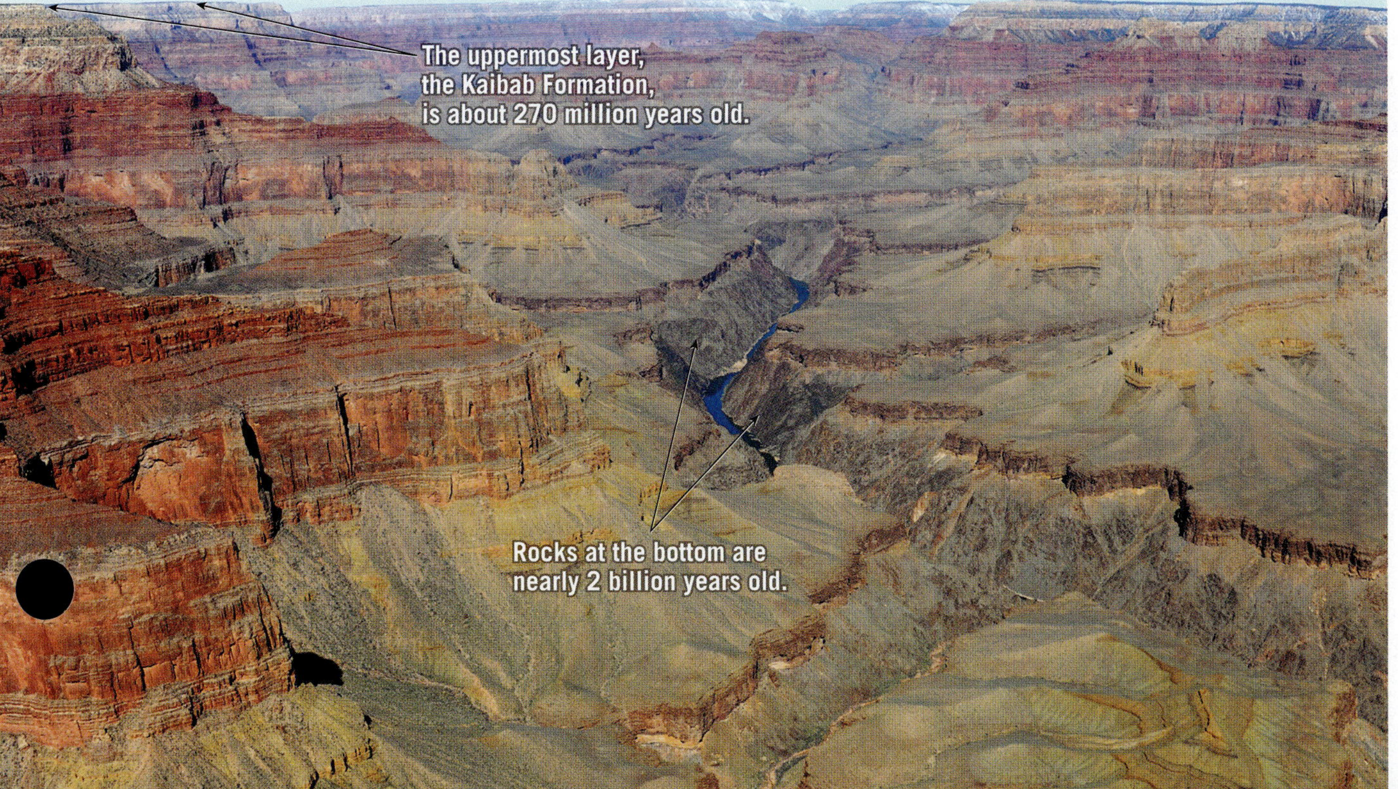
Grand Canyon rocks span more than 1.5 billion years of Earth history.
The uppermost layer, the Kaibab Formation, is about 270 million years old.
Rocks at the bottom are nearly 2 billion years old.

SmartFigure 1.5
Earth history—Written in the rocks The Grand Canyon of the Colorado River in northern Arizona. (Photo by Dennis Tasa)

MOBILE FIELD TRIP
https://goo.gl/kECNV1

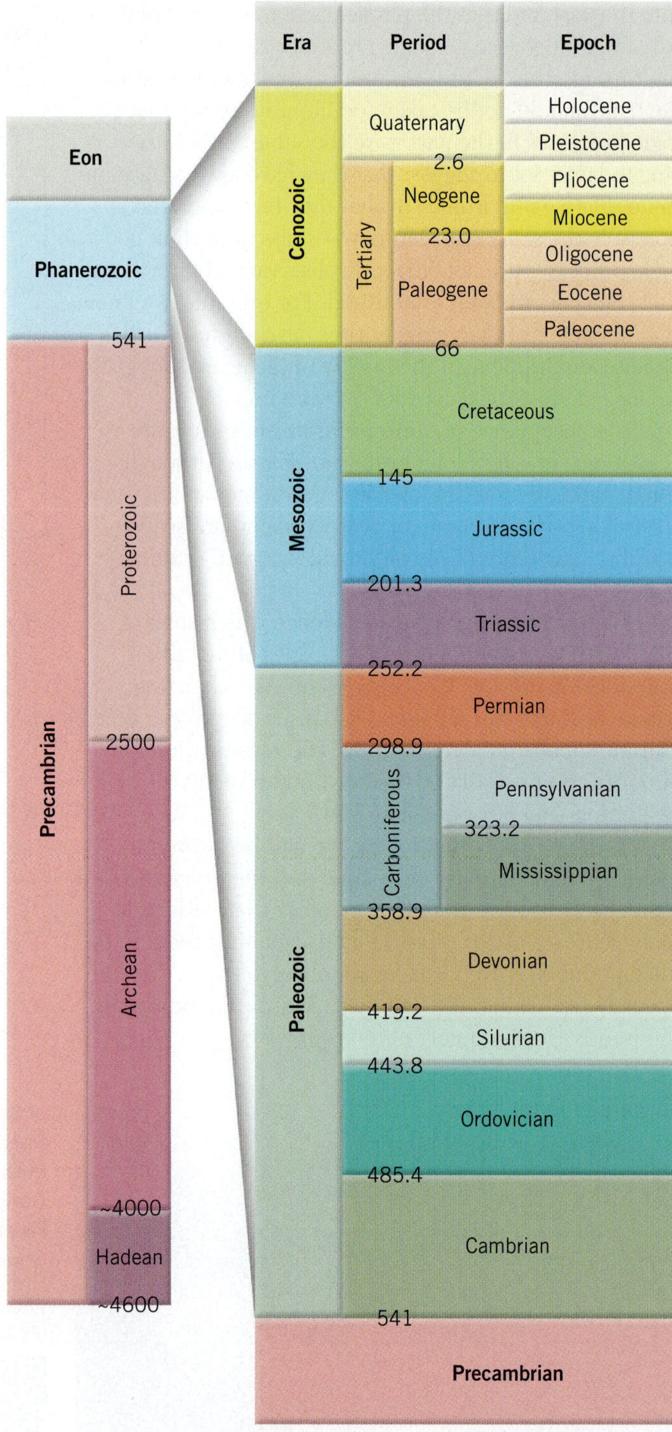

▲ **Figure 1.6 Geologic time scale: A basic reference** The time scale divides the vast 4.6-billion-year history of Earth into eons, eras, periods, and epochs. Numbers on the time scale represent time in millions of years before the present. The Precambrian accounts for more than 88 percent of geologic time. The geologic time scale is a dynamic tool that is periodically updated. Numerical ages appearing on this time scale are those that were currently accepted by the International Commission on Stratigraphy (ICS) in 2015. The color scheme used on this chart was selected because it is similar to that used by the ICS. The ICS is responsible for establishing global standards for the time scale.

The Magnitude of Geologic Time

Among geology's important contributions to human knowledge is the discovery that Earth has a very long and complex history. Although James Hutton and others recognized that geologic time is exceedingly long, they had no methods to accurately determine the age of Earth. Early time scales simply placed the events of Earth history in the proper sequence or order, without knowledge of how long ago in years they occurred.

Today our understanding of radioactivity allows us to accurately determine numerical dates for rocks that represent important events in Earth's distant past (**Figure 1.6**). For example, we know that the dinosaurs died out about 66 million years ago. Today the age of Earth is put at about 4.6 billion years. Chapter 18 is devoted to a much more complete discussion of geologic time and the geologic time scale.

The concept of geologic time is new to many non-geologists. People are accustomed to dealing with increments of time that are measured in hours, days, weeks, and years. Our history books often examine events over spans of centuries, but even a century is difficult to appreciate fully. For most of us, someone or something that is 90 years old is *very old*, and a 1000-year-old artifact is *ancient*.

By contrast, those who study geology must routinely deal with vast time periods—millions or billions (thousands of millions) of years. When viewed in the context of Earth's 4.6-billion-year history, a geologic event that occurred 100 million years ago may be characterized as "recent" by a geologist, and a rock sample that has been dated at 10 million years may be called "young." An appreciation for the magnitude of geologic time is important in the study of geology because many processes are so gradual that vast spans of time are needed before significant changes occur. How long is 4.6 billion years? If you were to begin counting at the rate of one number per second and continued 24 hours a day, 7 days a week and never stopped, it would take about two lifetimes (150 years) to reach 4.6 billion! **Figure 1.7** provides another interesting way of viewing the expanse of geologic time. This is just one of many analogies that have been conceived in an attempt to convey the magnitude of geologic time. Although helpful, all of them, no matter how clever, only begin to help us comprehend the vast expanse of Earth history.

CONCEPT CHECKS 1.2

1. Describe Aristotle's influence on geology.
2. Contrast catastrophism and uniformitarianism. How did each view the age of Earth?
3. How old is Earth?
4. Refer to Figure 1.6 and list the eon, era, period, and epoch in which we live.

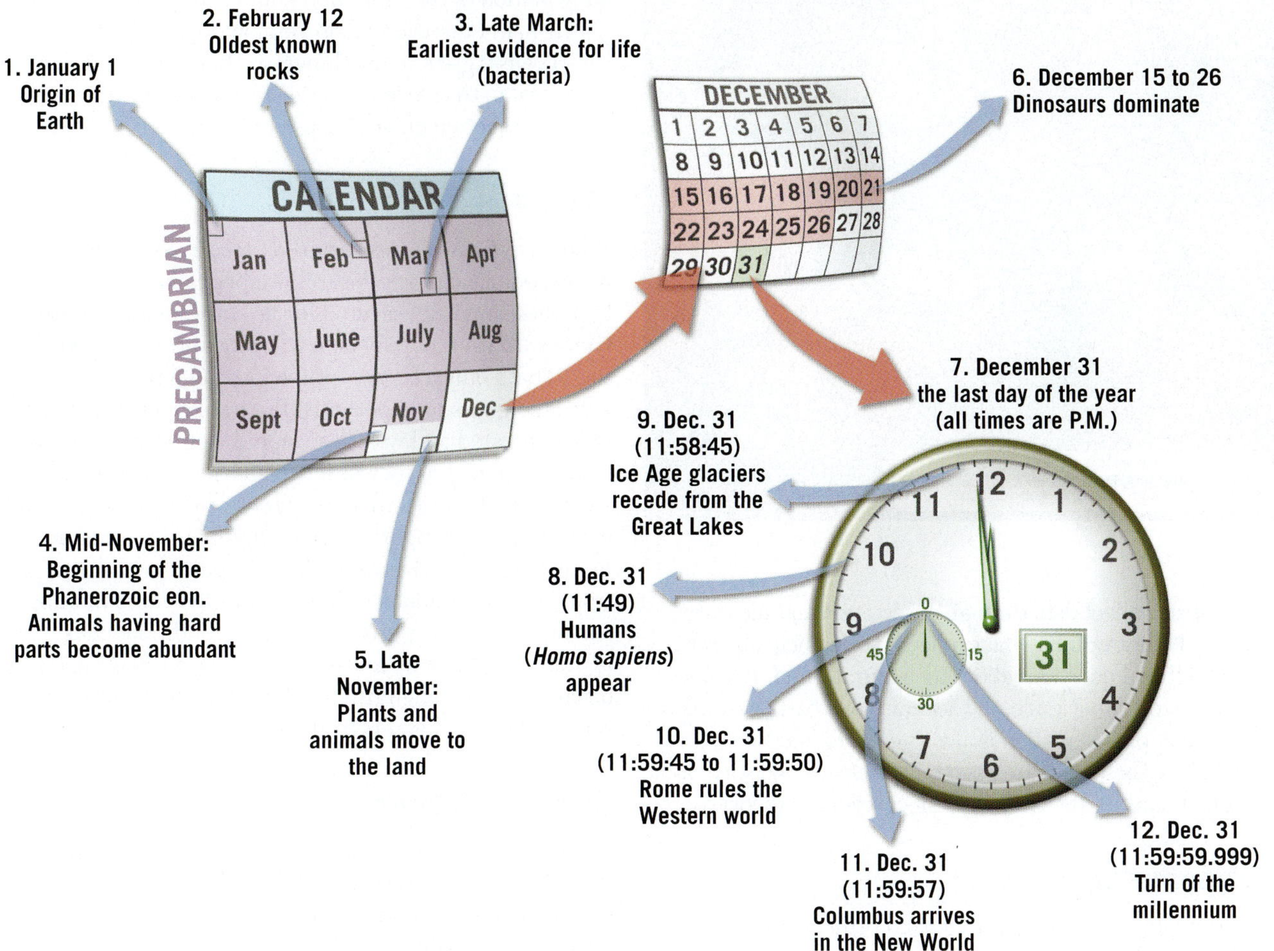
What if we compress the 4.6 billion years of Earth history into a single year?
1. January 1 Origin of Earth
2. February 12 Oldest known rocks
3. Late March: Earliest evidence for life (bacteria)
CALENDAR
PRECAMBRIAN
Jan
Feb
Mar
Apr
May
June
July
Aug
Sept
Oct
Nov
Dec
DECEMBER
6. December 15 to 26 Dinosaurs dominate
7. December 31 the last day of the year (all times are P.M.)
4. Mid-November: Beginning of the Phanerozoic eon. Animals having hard parts become abundant
5. Late November: Plants and animals move to the land
9. Dec. 31 (11:58:45) Ice Age glaciers recede from the Great Lakes
8. Dec. 31 (11:49) Humans (*Homo sapiens*) appear
10. Dec. 31 (11:59:45 to 11:59:50) Rome rules the Western world
11. Dec. 31 (11:59:57) Columbus arrives in the New World
12. Dec. 31 (11:59:59.999) Turn of the millennium

◀ SmartFigure 1.7
Magnitude of geologic time

TUTORIAL
https://goo.gl/V1WFRd

1.3 The Nature of Scientific Inquiry

Discuss the nature of scientific inquiry, including the construction of hypotheses and the development of theories.

In our modern society, we are constantly reminded of the benefits derived from science. But what exactly is the nature of scientific inquiry? Science is a process of producing knowledge, based on making careful observations and on creating explanations that make sense of the observations. Developing an understanding of how science is done and how scientists work is an important theme that appears throughout this textbook. You will explore the difficulties in gathering data and some of the ingenious methods that have been developed to overcome these difficulties. You will also see many examples of how hypotheses are formulated and tested, and you will learn about the evolution and development of some major scientific theories.

All science is based on the assumption that the natural world behaves in a consistent and predictable manner that is comprehensible through careful, systematic study. The overall goal of science is to discover the underlying patterns in nature and then to use that knowledge to make predictions about what should or should not be expected, given certain facts or circumstances. For example, by knowing how oil deposits form, geologists are able to predict the most favorable sites for exploration and, perhaps as importantly, how to avoid regions that have little or no potential.

The development of new scientific knowledge involves some basic logical processes that are universally accepted. To determine what is occurring in the natural world,

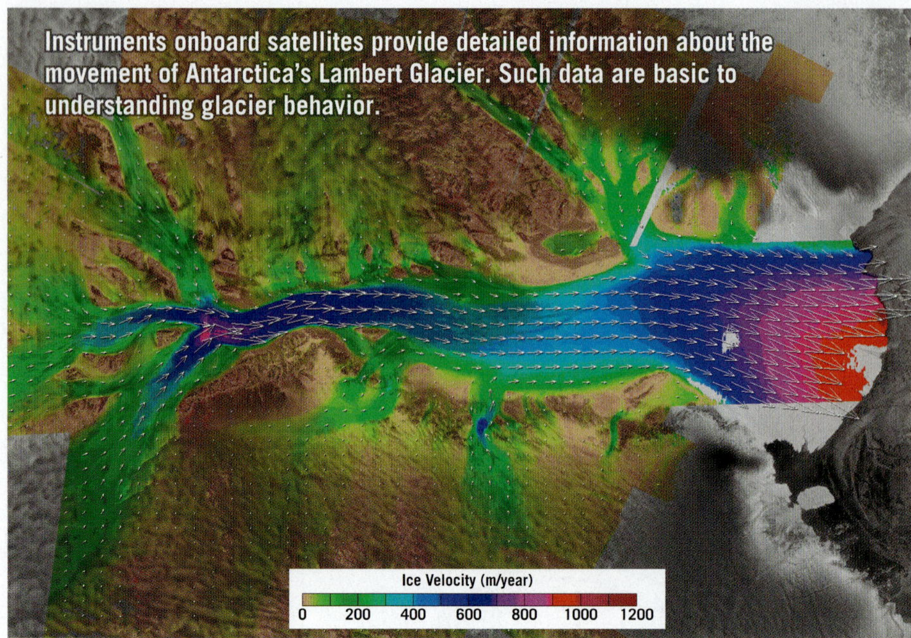

Instruments onboard satellites provide detailed information about the movement of Antarctica's Lambert Glacier. Such data are basic to understanding glacier behavior.
Ice Velocity (m/year)
0
200
400
600
800
1000
1200

▲ **Figure 1.8 Observation and measurement** Scientific facts are gathered in many ways. (Satellite image by NASA)

scientists collect data through observation and measurement (**Figure 1.8**). The data collected often help answer well-defined questions about the natural world. Because some error is inevitable, the accuracy of a particular measurement or observation is always open to question. Nevertheless, these data are essential to science and serve as a springboard for the development of scientific theories.

Hypothesis

Once data have been gathered and principles have been formulated to describe a natural phenomenon, investigators try to explain how or why things happen in the manner observed. They often do this by constructing a tentative (untested) explanation, which is called a scientific **hypothesis**. It is best if an investigator can formulate more than one hypothesis to explain a given set of observations. If an individual scientist is unable to devise multiple hypotheses, others in the scientific community will almost always develop alternative explanations. A spirited debate frequently ensues. As a result, extensive research is conducted by proponents of opposing hypotheses, and the results are made available to the wider scientific community in scientific journals.

Before a hypothesis can become an accepted part of scientific knowledge, it must pass objective testing and analysis. If a hypothesis cannot be tested, it is not scientifically useful, no matter how interesting it might seem. The verification process requires that *predictions* be made, based on the hypothesis being considered, and that the predictions be tested through comparison against objective observations of nature. Put another way, hypotheses must fit observations other than those used to formulate them in the first place. Hypotheses that fail rigorous testing are ultimately discarded. The history of science is littered with discarded hypotheses. One of the best known is the Earth-centered model of the universe—a proposal that was supported by the apparent daily motion of the Sun, Moon, and stars around Earth. As the mathematician Jacob Bronowski so ably stated, "Science is a great many things, . . . but in the end they all return to this: Science is the acceptance of what works and the rejection of what does not."

Theory

When a hypothesis has survived extensive scrutiny and when competing hypotheses have been eliminated, a hypothesis may be elevated to the status of a scientific **theory**. In everyday speech, we frequently hear people say, "That's only a theory," implying that a theory is an educated guess or hypothesis. But to a scientist, a theory is a well-tested and widely accepted view that the scientific community agrees best explains certain observable facts. Some theories that are extensively documented and extremely well supported are comprehensive in scope. For example, the theory of plate tectonics provides a framework for understanding the origins of mountains, earthquakes, and volcanic activity. In addition, plate tectonics explains the evolution of the continents and the ocean basins through time—ideas that are explored in some detail in Chapters 2, 10, and 11.

Scientific Methods

The process just described, in which researchers gather facts through observations and formulate scientific hypotheses, is called the **scientific method**. Contrary to popular belief, the scientific method is not a standard recipe that scientists apply in a routine manner to unravel the secrets of our natural world; rather, it is an endeavor that involves creativity and insight. Rutherford and Ahlgren put it this way: "Inventing hypotheses or theories to imagine how the world works and then figuring out how they can be put to the test of reality is as creative as writing poetry, composing music, or designing skyscrapers."*

There is no fixed path that scientists always follow that leads unerringly to scientific knowledge. However, many scientific investigations involve the steps outlined in **Figure 1.9**. In addition, some scientific discoveries result from purely theoretical ideas that stand up to extensive examination. Some researchers use high-speed computers to create models that simulate what is happening in the "real" world. These models are useful when dealing with natural processes that occur on very long time scales or take place in extreme or inaccessible locations. Still other scientific advancements are made when a totally unexpected happening occurs during an experiment. These serendipitous

*F. James Rutherford and Andrew Ahlgren, *Science for All Americans* (New York: Oxford University Press, 1990), p. 7.

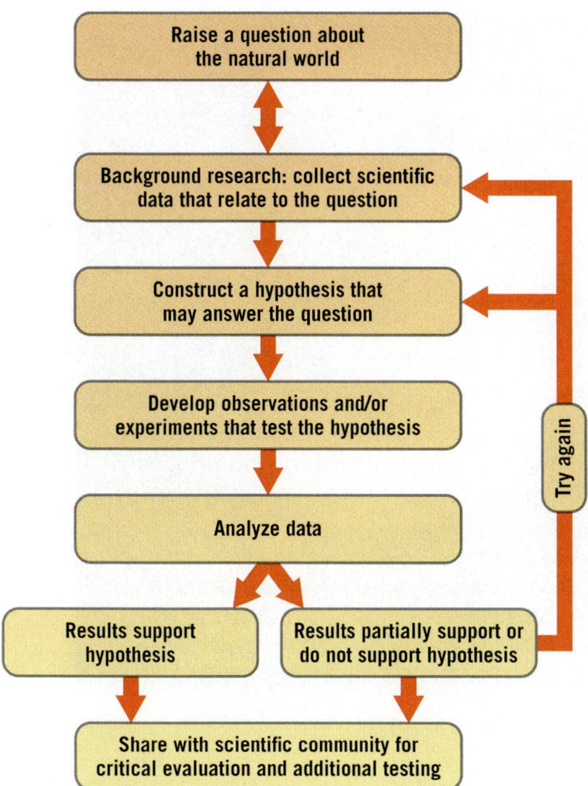

▲ **Figure 1.9 Steps frequently followed in scientific investigations** The diagram depicts the steps involved in the process many refer to as the *scientific method*.

discoveries are more than pure luck, for as the nineteenth-century French scientist Louis Pasteur said, "In the field of observation, chance favors only the prepared mind."

Scientific knowledge is acquired through several avenues, so it might be best to describe the nature of scientific inquiry as the methods of science rather than as the scientific method. In addition, it should always be remembered that even the most compelling scientific theories are still simplified explanations of the natural world.

Plate Tectonics and Scientific Inquiry

This textbook offers many opportunities to develop and reinforce your understanding of how science works and, in particular, how the science of geology works. You will learn about data-gathering methods and the observational techniques and reasoning processes used by geologists.

Chapter 2 provides an excellent example. Over the past 50 years, we have learned a great deal about the workings of our dynamic planet. This period has seen an unequaled revolution in our understanding of Earth. The revolution began in the early part of the twentieth century, with the radical proposal of *continental drift*—the idea that the continents move about the face of the planet. This hypothesis contradicted the established view that the continents and ocean basins are permanent and stationary features on the face of Earth. For that reason, the notion of drifting continents was received with great skepticism and even ridicule. More than 50 years passed before enough data were gathered to transform this controversial hypothesis into a sound theory that wove together the basic processes known to operate on Earth. The theory that finally emerged, called the *theory of plate tectonics*, provided geologists with the first comprehensive model of Earth's internal workings.

In Chapter 2, you will not only gain insights into the workings of our planet but also see an excellent example of the way geologic "truths" are uncovered and reworked.

Did You Know?
A scientific *law* is a basic principle that describes a particular behavior of nature that is generally narrow in scope and can be stated briefly—often as a simple mathematical equation.

CONCEPT CHECKS 1.3

1. How is a scientific hypothesis different from a scientific theory?
2. Summarize the basic steps followed in many scientific investigations.

1.4 Earth as a System

List and describe Earth's four major spheres. Define *system* and explain why Earth is considered to be a system.

Anyone who studies Earth soon learns that our planet is a dynamic body with many separate but interacting parts, or *spheres*. The hydrosphere, atmosphere, biosphere, and geosphere and all of their components can be studied separately. However, the parts are *not* isolated. Each is related in some way to the others, producing a complex and continuously interacting whole that we call the *Earth system*.

Earth's Spheres

The images in **Figure 1.10** are considered to be classics because they let humanity see Earth differently than ever before. These early views profoundly altered our conceptualizations of Earth and remain powerful images decades after they were first viewed. Seen from space, Earth is breathtaking in its beauty and startling in its solitude. The photos remind us that our home is, after all, a planet—small, self-contained, and in some ways even fragile.

As we look closely at our planet from space, it becomes apparent that Earth is much more than rock and soil. In fact, the most conspicuous features of Earth in Figure 1.10A are swirling clouds suspended above the surface of the vast global ocean. These features emphasize the importance of water on our planet.

▶ **SmartFigure 1.10** **Two classic views of Earth from space** The accompanying video commemorates the forty-fifth anniversary of *Apollo 8*'s historic flight by re-creating the moment when the crew first saw and photographed the Earth rising from behind the Moon. (NASA)

VIDEO
https://goo.gl/AQKqaa

View called "Earthrise" that greeted *Apollo 8* astronauts as their spacecraft emerged from behind the Moon in December 1968. This classic image let people see Earth differently than ever before.

A.

This image taken from *Apollo 17* in December 1972 is perhaps the first to be called "The Blue Marble." The dark blue ocean and swirling cloud patterns remind us of the importance of the oceans and atmosphere.

B.

Did You Know?
The volume of ocean water is so large that if Earth's solid mass were perfectly smooth (level) and spherical, the oceans would cover Earth's entire surface to a uniform depth of more than 2000 m (1.2 mi).

The closer view of Earth from space shown in Figure 1.10B helps us appreciate why the physical environment is traditionally divided into three major parts: the water portion of our planet, the *hydrosphere*; Earth's gaseous envelope, the *atmosphere*; and, of course, the solid Earth, or *geosphere*. It needs to be emphasized that our environment is highly integrated and not dominated by rock, water, or air alone. Rather, it is characterized by continuous interactions as air comes in contact with rock, rock with water, and water with air. Moreover, the *biosphere*, which is the totality of all life on our planet, interacts with each of the three physical realms and is an equally integral part of the planet. Thus, Earth can be thought of as consisting of four major spheres: the hydrosphere, atmosphere, geosphere, and biosphere. All four spheres are represented in the chapter-opening photo.

The interactions among Earth's spheres are incalculable. **Figure 1.11** provides us with one easy-to-visualize example. The shoreline is an obvious meeting place for rock, water, and air. In this scene, ocean waves created by the drag of air moving across the water are breaking against the rocky shore.

▼ **Figure 1.11** **Interactions among Earth's spheres** The shoreline is one obvious interface—a common boundary where different parts of a system interact. In this scene, ocean waves (hydrosphere) that were created by the force of moving air (atmosphere) break against a rocky shore (geosphere). The force of the water can be powerful, and the erosional work that is accomplished can be great. (Photo by Michael Collier)

Hydrosphere

Earth is sometimes called the *blue* planet. Water, more than anything else, makes Earth unique. The **hydrosphere** is a dynamic mass of water that is continually on the move, evaporating from the oceans to the atmosphere, precipitating to the land, and running back to the ocean again. The global ocean is certainly the most prominent feature of the hydrosphere, blanketing nearly 71 percent of Earth's surface to an average depth of about 3800 meters (12,500 feet). It accounts for about 97 percent of Earth's water (**Figure 1.12**). However, the hydrosphere also includes the freshwater found underground and in streams, lakes, and glaciers. Moreover, water is an important component of all living things.

Even though freshwater constitutes only a small fraction of Earth's hydrosphere, it plays an outsized role in Earth's external processes. Streams, glaciers, and groundwater sculpt many of our planet's varied landforms, and freshwater is vital for life on land.

Atmosphere

Earth is surrounded by a life-giving gaseous envelope called the **atmosphere** (Figure 1.13). When we watch a high-flying jet plane cross the sky, it seems that the atmosphere extends upward for a great distance. However, when compared to the thickness (radius) of the solid Earth (about 6400 kilometers [4000 miles]), the atmosphere is a very shallow layer. Despite its modest dimensions, this thin blanket of air is an integral part of the planet. It not only provides the air we breathe but also protects us from the Sun's intense heat and dangerous ultraviolet radiation. The energy exchanges that continually occur between the atmosphere and Earth's surface and between the atmosphere and space produce the effects we call *weather* and *climate*. Climate has a strong influence on the nature and intensity of Earth's external processes. When climate changes, these processes respond.

If, like the Moon, Earth had no atmosphere, our planet would be lifeless, and many of the processes and interactions that make the surface such a dynamic place could not operate. Without weathering and erosion, the face of our planet might more closely resemble the lunar surface, which has not changed appreciably in nearly 3 billion years.

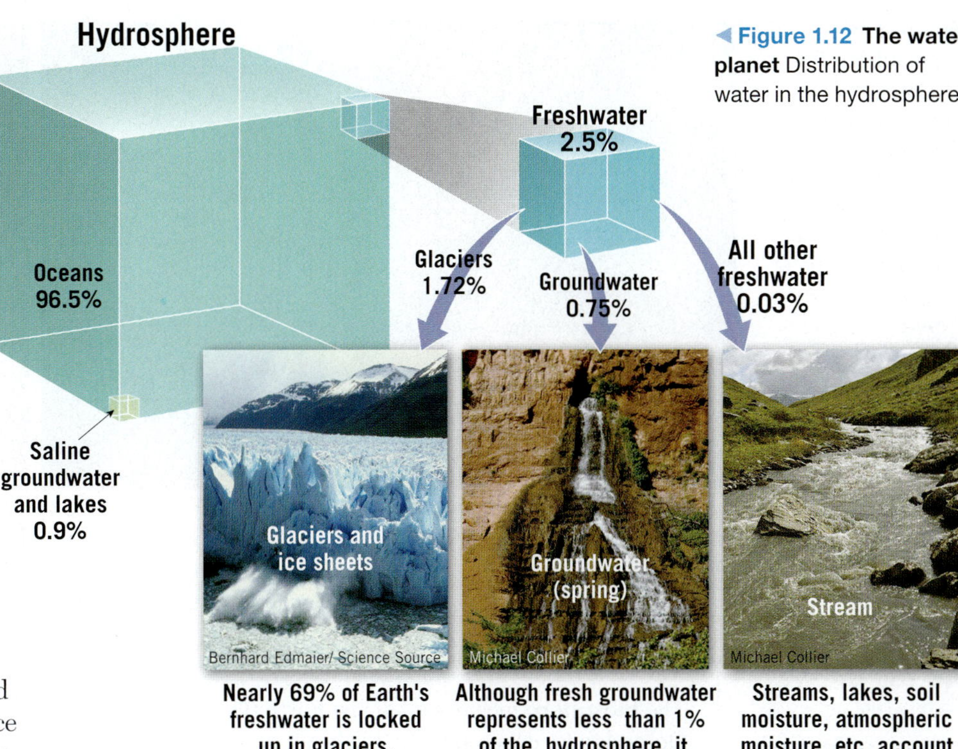
Hydrosphere
Oceans 96.5%
Saline groundwater and lakes 0.9%
Freshwater 2.5%
Glaciers 1.72%
Groundwater 0.75%
All other freshwater 0.03%
Glaciers and ice sheets
Groundwater (spring)
Stream
Bernhard Edmaier/Science Source
Michael Collier
Michael Collier
Nearly 69% of Earth's freshwater is locked up in glaciers.
Although fresh groundwater represents less than 1% of the hydrosphere, it accounts for 30% of all freshwater and about 96% of all liquid freshwater.
Streams, lakes, soil moisture, atmospheric moisture, etc. account for 0.03% (3/100 of 1%)

Figure 1.12 **The water planet** Distribution of water in the hydrosphere.

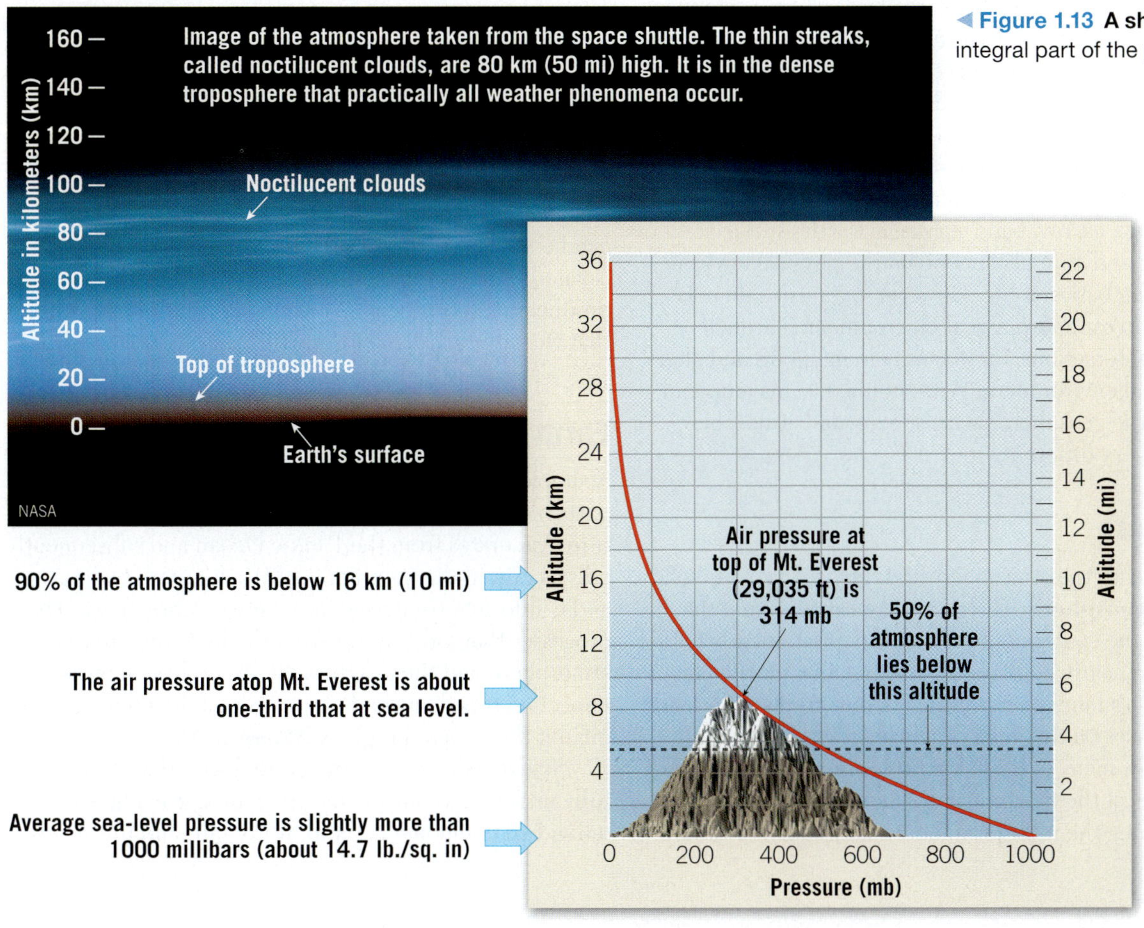
Image of the atmosphere taken from the space shuttle. The thin streaks, called noctilucent clouds, are 80 km (50 mi) high. It is in the dense troposphere that practically all weather phenomena occur.
Altitude in kilometers (km)
160
140
120
100
80
60
40
20
0
Noctilucent clouds
Top of troposphere
Earth's surface
NASA
90% of the atmosphere is below 16 km (10 mi)
The air pressure atop Mt. Everest is about one-third that at sea level.
Average sea-level pressure is slightly more than 1000 millibars (about 14.7 lb./sq. in)
Altitude (km)
36
32
28
24
20
16
12
8
4
Altitude (mi)
22
20
18
16
14
12
10
8
6
4
2
Air pressure at top of Mt. Everest (29,035 ft) is 314 mb
50% of atmosphere lies below this altitude
0
200
400
600
800
1000
Pressure (mb)

Figure 1.13 **A shallow layer** The atmosphere is an integral part of the planet. (NASA)

A. Tropical rain forests are characterized by hundreds of different species per square kilometer.

B. Microorganisms are nourished by hot, mineral-rich fluids spewing from vents on the deep-ocean floor. The microbes support larger organisms such as tube worms.

▶ Figure 1.14 **The biosphere** The biosphere, one of Earth's four spheres, includes all life. **A.** Tropical rain forests are teeming with life and occur in the vicinity of the equator. (Photo by AGE Fotostock/ Superstock) **B.** Some life is found in extreme environments such as the absolute darkness of the deep ocean. (Photo by Fisheries and Oceans Canada/Verena Tunnicliffe/Newscom)

Did You Know?
Primitive life first appeared in the oceans about 4 billion years ago and has been spreading and diversifying ever since.

Biosphere

The **biosphere** includes all life on Earth (Figure 1.14). Ocean life is concentrated in the sunlit upper waters. Most life on land is also concentrated near the surface, with tree roots and burrowing animals reaching a few meters underground and flying insects and birds reaching a kilometer or so into the atmosphere. A surprising variety of life-forms are also adapted to extreme environments. For example, on the ocean floor, where pressures are extreme and no light penetrates, there are places where vents spew hot, mineral-rich fluids that support communities of exotic life-forms. On land, some bacteria thrive in rocks as deep as 4 kilometers (2.5 miles) and in boiling hot springs. Moreover, air currents can carry microorganisms many kilometers into the atmosphere. But even when we consider these extremes, life still must be thought of as being confined to a narrow band very near Earth's surface.

Plants and animals depend on the physical environment for the basics of life. However, organisms do not just respond to their physical environment. Through countless interactions, life-forms help maintain and alter the physical environment. Without life, the makeup and nature of the geosphere, hydrosphere, and atmosphere would be very different.

Did You Know?
Since 1970, Earth's average surface temperature has increased by about 0.6°C (1°F). By the end of the twenty-first century, the average global temperature may increase by an additional 2° to 4.5°C (3.5° to 8.1°F).

Geosphere

Lying beneath the atmosphere and the oceans is the solid Earth, or **geosphere**. The geosphere extends from the surface to the center of the planet, a depth of nearly 6400 kilometers (nearly 4000 miles), making it by far the largest of Earth's four spheres. Much of our study of the solid Earth focuses on the more accessible surface features. Fortunately, many of these features represent the outward expressions of the dynamic behavior of Earth's interior. By examining the most prominent surface features and their global extent, we can obtain clues to the dynamic processes that have shaped our planet. A first look at the structure of Earth's interior and at the major surface features of the geosphere will come later in the chapter.

Soil, the thin veneer of material at Earth's surface that supports the growth of plants, may be thought of as part of all four spheres. The solid portion is a mixture of weathered rock debris (geosphere) and organic matter from decayed plant and animal life (biosphere). The decomposed and disintegrated rock debris is the product of weathering processes that require air (atmosphere) and water (hydrosphere). Air and water also occupy the open spaces between the solid particles.

Anyone who studies Earth soon learns that our planet is a dynamic body with many separate but interacting parts, or *spheres*. The hydrosphere, atmosphere, biosphere, and geosphere and all of their components can be studied separately. However, the parts are *not* isolated. Each is related in some way to the others, producing a complex and continuously interacting whole that we call the *Earth system*.

Earth System Science

A simple example of the interactions among different parts of the Earth system occurs every winter, as moisture evaporates from the Pacific Ocean and subsequently falls as rain in the mountains of Washington, Oregon, and California, triggering destructive debris flows. The processes that move water from the hydrosphere to the atmosphere and then to the solid Earth have a profound impact on the plants and animals (including humans) that inhabit the affected regions (Figure 1.15).

Scientists have recognized that in order to more fully understand our planet, they must learn how its individual components (land, water, air, and life-forms) are interconnected. This endeavor, called **Earth system science**, aims to study Earth as a *system* composed of numerous interacting parts, or *subsystems*. Rather than look through the limited lens of only one of the traditional sciences—geology, atmospheric science, chemistry, biology, and so on—Earth system science attempts to integrate the knowledge of several academic fields. Using an interdisciplinary approach, those engaged in Earth system science attempt to achieve the level of understanding necessary to comprehend and solve many of our global environmental problems.

A **system** is a group of interacting, or interdependent, parts that form a complex whole. Most of us hear and use the term *system* frequently. We may service our car's cooling *system*, make use of the city's transportation *system*, and be a participant in the political *system*. A news report might inform us of an approaching weather *system*. Further, we know that Earth is just a small part of a larger system known as the *solar system*, which in turn is a subsystem of an even larger system called the Milky Way Galaxy.

The Earth System

The Earth system has a nearly endless array of subsystems in which matter is recycled over and over. One familiar loop or subsystem is the *hydrologic cycle*. It represents the unending circulation of Earth's water among the hydrosphere, atmosphere, biosphere, and geosphere (**Figure 1.16**). Water enters the atmosphere during volcanic eruptions and through evaporation from Earth's surface and transpiration from plants. Water vapor condenses in the atmosphere to form clouds, which in turn produce precipitation that falls back to Earth's surface. Some of the rain that falls onto the land infiltrates (soaks in) and is taken up by plants or becomes groundwater, and some flows across the surface toward the ocean.

Viewed over long time spans, the rocks of the geosphere are constantly forming, changing, and re-forming. The loop that involves the processes by which one rock changes to another is called the *rock cycle* and will be discussed at some length later in the chapter. The cycles of the Earth system are not independent; to the contrary, these cycles come in contact and interact in many places.

The parts of the Earth system are linked so that a change in one part can produce changes in any or all of the other parts. For example, when a volcano erupts, lava from Earth's interior may flow out at the surface and block a nearby valley. This new obstruction influences the region's drainage system by creating a lake or causing streams to change course. The large quantities of volcanic ash and gases that can be emitted during an eruption alter the composition of the atmosphere and influence the amount of solar energy that reaches Earth's surface. The result could be a drop in air temperatures over the entire hemisphere.

▲ **Figure 1.15 Deadly debris flow** This image provides an example of interactions among different parts of the Earth system. Extraordinary rains triggered this debris flow (popularly called a mudslide) on March 22, 2014, near Oso, Washington. The mass of mud and debris blocked the North Fork of the Stillaguamish River and engulfed an area of about 2.6 square kilometers (1 square mile). Forty-three people perished. (Photo by Michael Collier)

▼ **Figure 1.16 The hydrologic cycle** Water readily changes state from liquid, to gas (vapor), to solid at the temperatures and pressures occurring on Earth. This cycle traces the movements of water among Earth's four spheres. It is one of many subsystems that collectively make up the Earth system.

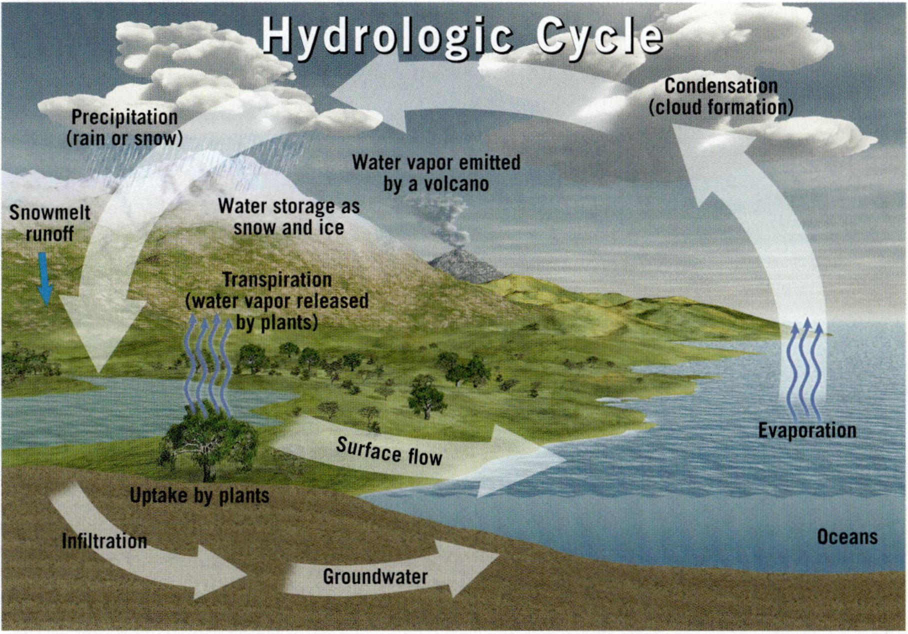

▲ **Figure 1.17 Change is a geologic constant** When Mount St. Helens, Washington, erupted in May 1980 (inset photo), the area shown here was buried by a volcanic mudflow. Now plants are reestablished, and new soil is forming. (Photo by Terry Donnelly/Alamy Images; inset photo by U.S. Geological Survey)

Where the surface is covered by lava flows or a thick layer of volcanic ash, existing soils are buried. This causes the soil-forming processes to begin anew to transform the new surface material into soil (**Figure 1.17**). The soil that eventually forms will reflect the interactions among many parts of the Earth system—the volcanic parent material, the climate, and the impact of biological activity. Of course, there would also be significant changes in the biosphere. Some organisms and their habitats would be eliminated by the lava and ash, whereas new settings for life, such as a lake formed by a lava dam, would be created. The potential climate change could also impact sensitive life-forms.

The Earth system is characterized by processes that vary on spatial scales from fractions of millimeters to thousands of kilometers. Time scales for Earth's processes range from milliseconds to billions of years. As we learn about Earth, it becomes increasingly clear that despite significant separations in distance or time, many processes are connected, and a change in one component can influence the entire system.

The Earth system is powered by energy from two sources. The Sun drives external processes that occur in the atmosphere, in the hydrosphere, and at Earth's surface. Weather and climate, ocean circulation, and erosional processes are driven by energy from the Sun. Earth's interior is the second source of energy. Heat remaining from when our planet formed and heat that is continuously generated by radioactive decay power the internal processes that produce volcanoes, earthquakes, and mountains.

Humans are *part of* the Earth system, a system in which the living and nonliving components are entwined and interconnected. Therefore, our actions produce changes in all the other parts. When we burn gasoline and coal, dispose of our wastes, and clear the land, we cause other parts of the system to respond, often in unforeseen ways. Throughout this textbook you will learn about many of Earth's subsystems, including the hydrologic system, the tectonic (mountain-building) system, the rock cycle, and the climate system. Remember that these components *and we humans* are all part of the complex interacting whole we call the Earth system.

Did You Know?
Estimates indicate that erosional processes are lowering the North American continent at a rate of about 3 cm (1.2 in) per 1000 years. At this rate, it would take 100 million years to level a 3000 m (10,000 ft) high peak.

CONCEPT CHECKS 1.4

1. List and briefly describe the four spheres that constitute the Earth system.
2. Compare the height of the atmosphere to the thickness of the geosphere.
3. How much of Earth's surface do oceans cover? What percentage of Earth's water supply do oceans represent?
4. What is a system? List three examples.
5. What are the two sources of energy for the Earth system?

1.5 Origin and Early Evolution of Earth

Outline the stages in the formation of our solar system.

Recent earthquakes caused by displacements of Earth's crust and lavas spewed from active volcanoes represent only the latest in a long line of events by which our planet has attained its present form and structure. The geologic processes operating in Earth's interior can be best understood when viewed in the context of much earlier events in Earth history.

Origin of Planet Earth

This section describes the most widely accepted views on the origin of our solar system. The theory described here represents the most consistent set of ideas we have to explain what we know about our solar system today.

The Universe Begins Our scenario begins about 13.7 billion years ago, with the *Big Bang*, an almost incomprehensible event in which space itself, along with all the matter and energy of the universe, exploded in an instant from tiny to huge dimensions. As the universe continued to expand, subatomic particles condensed to form hydrogen and helium gas, which later cooled and clumped to form the first stars and galaxies. It was in one of these galaxies, the Milky Way, that our solar system, including planet Earth, took form.

The Solar System Forms Earth is one of eight planets that, along with dozens of moons and numerous smaller bodies, revolve around the Sun. The orderly nature of our solar system helped scientists determine that Earth and the other planets formed at essentially the same time and from the same primordial material as the Sun. The **nebular theory** proposes that the bodies of our solar system evolved from an enormous rotating cloud called the **solar nebula** (Figure 1.18). Besides the hydrogen and helium atoms generated during the Big Bang, the solar

Did You Know?
The circumference of Earth is slightly more than 40,000 km (nearly 25,000 mi). It would take a jet plane traveling at 1000 km/hr (620 mi/hr) 40 hours (1.7 days) to circle the planet.

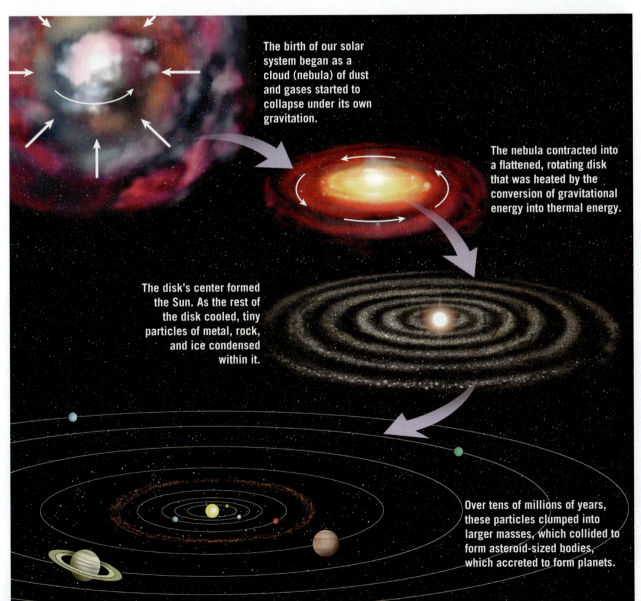

The birth of our solar system began as a cloud (nebula) of dust and gases started to collapse under its own gravitation.

The nebula contracted into a flattened, rotating disk that was heated by the conversion of gravitational energy into thermal energy.

The disk's center formed the Sun. As the rest of the disk cooled, tiny particles of metal, rock, and ice condensed within it.

Over tens of millions of years, these particles clumped into larger masses, which collided to form asteroid-sized bodies, which accreted to form planets.

SmartFigure 1.18
Nebular theory
The nebular theory explains the formation of the solar system.

TUTORIAL
https://goo.gl/HbtC0S

Did You Know?
The Sun contains 99.86 percent of the mass of the solar system. The circumference of the Sun is 109 times that of Earth. A jet plane traveling at 1000 km/hr would require nearly 182 days to circle the Sun.

nebula consisted of microscopic dust grains and other matter ejected ultimately from long-dead stars. (Nuclear fusion in stars converts hydrogen and helium into the other elements found in the universe.)

Nearly 5 billion years ago, something—perhaps a shock wave from an exploding star (*supernova*)—caused this nebula to start collapsing in response to its own gravitation. As it collapsed, it evolved from a huge, vaguely rotating cloud to a much smaller, fast-spinning disk. The cloud flattened into a disk for the same reason that it is easier to move along with a crowd of circling ice skaters than to cross their path. The orbital plane within the cloud that started out with the largest amount of matter gradually, through collisions and other interactions, incorporated gas and particles that originally had other orbits until all the matter orbited in one plane. The disk spun faster as it shrank for the same reason ice skaters spin faster when they draw their arms toward their bodies. Most of the cloud's matter ended up in the center of the disk, where it formed the *protosun* (pre-Sun). Astronomers have observed many such disks around newborn stars in neighboring regions of our Galaxy.

The protosun and inner disk were heated by the gravitational energy of infalling matter. In the inner disk, temperatures became high enough to cause the dust grains to evaporate. However, at distances beyond the orbit of Mars, the temperatures probably remained quite low. At −200°C (−328°F), the tiny particles in the outer portion of the nebula were likely covered with a thick layer of frozen water, carbon dioxide, ammonia, and methane. The disk also contained appreciable amounts of the lighter gases hydrogen and helium.

The Inner Planets Form The formation of the Sun marked the end of the period of contraction and thus the end of gravitational heating. Temperatures in the region where the inner planets now reside began to decline. The decrease in temperature caused those substances with high melting points to condense into tiny particles that began to coalesce (join together). Materials such as iron and nickel and the elements of which the rock-forming minerals are composed—silicon, calcium, sodium, and so forth—formed metallic and rocky clumps that orbited the Sun (see Figure 1.18). Repeated collisions caused these masses to coalesce into larger asteroid-size bodies, called *planetesimals*, which in a few tens of millions of years accreted into the four inner planets we call Mercury, Venus, Earth, and Mars (**Figure 1.19**). Not all of these clumps of matter were incorporated into the planetesimals. Those rocky and metallic pieces that remained in orbit are called *meteorites* when they survive an impact with Earth.

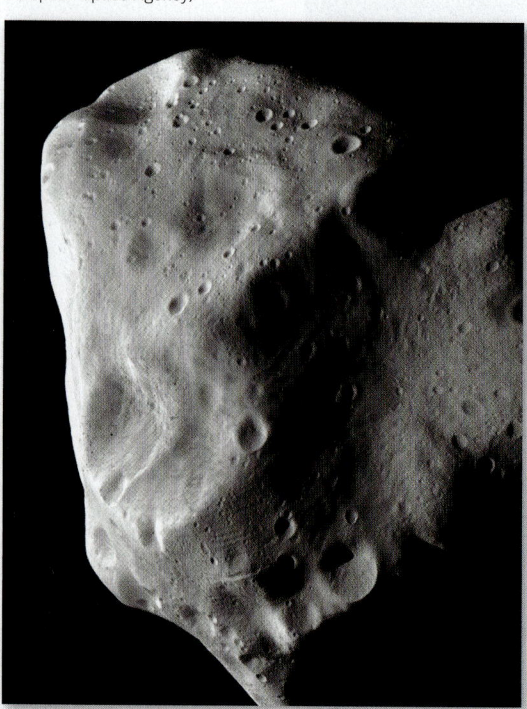
▼ **Figure 1.19 A remnant planetesimal** This image of Asteroid 21 Lutetia was obtained by special cameras aboard the *Rosetta* spacecraft on July 10, 2010. Spacecraft instruments showed that Lutetia is a primitive body (planetesimal) left over from when the solar system formed. (Image courtesy of European Space Agency)

As more and more material was swept up by the planets, the high-velocity impact of nebular debris caused the temperatures of these bodies to rise. Because of their relatively high temperatures and weak gravitational fields, the inner planets were unable to accumulate much of the lighter components of the nebular cloud. The lightest of these, hydrogen and helium, were eventually whisked from the inner solar system by the solar wind.

The Outer Planets Develop At the same time that the inner planets were forming, the larger, outer planets (Jupiter, Saturn, Uranus, and Neptune), along with their extensive satellite systems, were also developing. Because of low temperatures far from the Sun, the material from which these planets formed contained a high percentage of ices—water, carbon dioxide, ammonia, and methane—as well as rocky and metallic debris. The accumulation of ices accounts, in part, for the large size and low density of the outer planets. The two most massive planets, Jupiter and Saturn, had a surface gravity sufficient to attract and hold large quantities of even the lightest elements—hydrogen and helium.

Formation of Earth's Layered Structure

As material accumulated to form Earth (and for a short period afterward), the high-velocity impact of nebular debris and the decay of radioactive elements caused the temperature of our planet to increase steadily. During this time of intense heating, Earth became hot enough that iron and nickel began to melt. Melting produced liquid blobs of dense metal that sank toward the center of the planet. This process occurred rapidly on the scale of geologic time and produced Earth's dense iron-rich core.

Chemical Differentiation and Earth's Layers The early period of heating resulted in another process of chemical differentiation, whereby melting formed buoyant masses of molten rock that rose toward the surface, where they solidified to produce a primitive crust. These rocky materials were enriched in oxygen and "oxygen-seeking" elements, particularly silicon and aluminum, along with lesser amounts of calcium, sodium, potassium, iron, and magnesium. In addition, some heavy metals such as gold, lead, and uranium, which have low melting points or were highly soluble in the ascending molten masses, were scavenged from Earth's interior and concentrated in the developing crust. This early period of chemical differentiation established the three basic divisions of Earth's interior: the iron-rich *core*; the thin *primitive crust*; and Earth's largest layer, called the *mantle*, which is located between the core and crust.

An Atmosphere Develops An important consequence of the early period of chemical differentiation is that large quantities of gaseous materials were allowed to escape from Earth's interior, as happens today during volcanic eruptions. By this process, a primitive atmosphere gradually evolved. It is on this planet, with this atmosphere, that life as we know it came into existence.

Continents and Ocean Basins Evolve Following the events that established Earth's basic structure, the primitive crust was lost to erosion and other geologic processes, so we have no direct record of its makeup. When and exactly how the continental crust—and thus Earth's first landmasses—came into existence is a matter of ongoing research. Nevertheless, there is general agreement that the continental crust formed gradually over the past 4 billion years. (The oldest rocks yet discovered are isolated fragments found in the Northwest Territories of Canada that have radiometric dates of about 4 billion years.) In addition, as you will see in subsequent chapters, Earth is an evolving planet whose continents and ocean basins have continually changed shape and even location.

CONCEPT CHECKS 1.5

1. Name and briefly outline the theory that describes the formation of our solar system.
2. List the inner planets and outer planets. Describe basic differences in size and composition.
3. Explain why density and buoyancy were important in the development of Earth's layered structure.

Did You Know?

The *light-year* is a unit for measuring distances to stars. Such distances are so large that familiar units such as kilometers or miles are cumbersome to use. One light-year is the distance light travels in 1 Earth year—about 9.5 trillion km (5.8 trillion mi)!

1.6 Earth's Internal Structure

Describe Earth's internal structure.

In the preceding section, you learned that the differentiation of material that began early in Earth's history resulted in the formation of three major layers defined by their chemical composition: the crust, mantle, and core. In addition to these compositionally distinct layers, Earth is divided into layers based on physical properties. The physical properties used to define such zones include whether the layer is solid or liquid and how weak or strong it is. Important examples include the lithosphere, asthenosphere, outer core, and inner core. Knowledge of both chemical and physical layers is important to our understanding of many geologic processes, including volcanism, earthquakes, and mountain building. Figure 1.20 shows different views of Earth's layered structure.

Earth's Crust

The **crust**, Earth's relatively thin, rocky outer skin, is of two different types—continental crust and oceanic crust. Both share the word *crust*, but the similarity ends there. The oceanic crust is roughly 7 kilometers (4.5 miles) thick and composed of the dark igneous rock *basalt*. By contrast, the continental crust averages about 35 kilometers (22 miles) thick but may exceed 70 kilometers (40 miles) in some mountainous regions such as the Rockies and Himalayas. Unlike the oceanic crust, which has a relatively homogeneous chemical composition, the continental crust consists of many rock types. Although the upper crust has an average composition of a *granitic rock* called *granodiorite*, it varies considerably from place to place.

Continental rocks have an average density of about 2.7 g/cm^3, and some have been discovered that are more than 4 billion years old. The rocks of the oceanic crust are younger (180 million years or less) and denser (about 3.0 g/cm^3) than continental rocks. For comparison, liquid water has a density of 1 g/cm^3; therefore, the density of basalt, the primary rock composing oceanic crust, is three times that of water.

Earth's Mantle

More than 82 percent of Earth's volume is contained in the **mantle**, a solid, rocky shell that extends to a depth of about 2900 kilometers (1800 miles). The boundary between the crust and mantle represents a marked change in chemical composition. The dominant rock type in the uppermost mantle is *peridotite*, which is richer in the metals magnesium and iron than the minerals found in either the continental or oceanic crust.

The Upper Mantle The upper mantle extends from the crust–mantle boundary down to a depth of about 660 kilometers (410 miles). The upper mantle can be divided into three different parts. The top portion of the upper mantle is part of the stronger *lithosphere*, and beneath that is the weaker *asthenosphere*. The bottom part of the upper mantle is called the *transition zone*.

The **lithosphere** ("sphere of rock") consists of the entire crust plus the uppermost mantle and forms Earth's relatively cool, rigid outer shell (see Figure 1.20). Averaging about 100 kilometers (60 miles) thick, the lithosphere is more than 250 kilometers (155 miles) thick below the oldest portions of the continents. Beneath this rigid layer to a depth of about 410 kilometers (255 miles) lies a comparatively weak layer known as the **asthenosphere** ("weak sphere"). The top portion of the asthenosphere has a temperature/pressure regime that results in a small amount of melting. Within this very weak zone, the lithosphere is mechanically detached from the layer below. The result is that the lithosphere is able to move independently of the asthenosphere, a fact we will consider in the next chapter.

It is important to emphasize that the strength of various Earth materials is a function of both their composition and the temperature and pressure of their environment. You should not get the idea that the entire lithosphere behaves like a rigid or brittle solid similar to rocks found on the surface. Rather, the rocks of the

Did You Know?

Geologists have never sampled the mantle or core directly, so how did we learn about the composition and structure of Earth's interior? The structure of Earth's interior is determined by analyzing seismic waves from earthquakes. As these waves of energy penetrate Earth's interior, they change speed and are bent and reflected as they move through zones having different properties. Monitoring stations around the world detect and record this energy. There is more about this in Chapter 9.

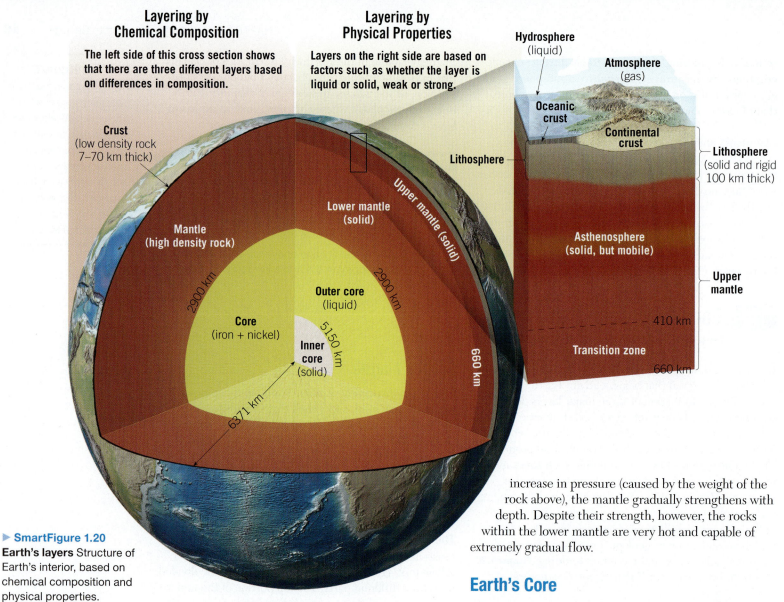

Layering by Chemical Composition

The left side of this cross section shows that there are three different layers based on differences in composition.

Layering by Physical Properties

Layers on the right side are based on factors such as whether the layer is liquid or solid, weak or strong.

Crust (low density rock 7–70 km thick)
Mantle (high density rock)
Core (iron + nickel)
2900 km
6371 km
Lower mantle (solid)
Upper mantle (solid)
Outer core (liquid)
Inner core (solid)
5150 km
2900 km
660 km
Hydrosphere (liquid)
Atmosphere (gas)
Oceanic crust
Continental crust
Lithosphere
Lithosphere (solid and rigid 100 km thick)
Asthenosphere (solid, but mobile)
Upper mantle
410 km
Transition zone
660 km

▶ **SmartFigure 1.20**

Earth's layers Structure of Earth's interior, based on chemical composition and physical properties.

TUTORIAL
https://goo.gl/8IwyPV

lithosphere get progressively hotter and weaker (more easily deformed) with increasing depth. At the depth of the uppermost asthenosphere, the rocks are close enough to their melting temperature (some melting may actually occur) that they are very easily deformed. Thus, the uppermost asthenosphere is weak because it is near its melting point, just as hot wax is weaker than cold wax.

From about 410 kilometers (255 miles) to about 660 kilometers (410 miles) in depth is the part of the upper mantle called the **transition zone**. The top of the transition zone is identified by a sudden increase in density from about 3.5 to 3.7 g/cm^3. This change occurs because minerals in the rock peridotite respond to the increase in pressure by forming new minerals with closely packed atomic structures.

The Lower Mantle From a depth of 660 kilometers (410 miles) to the top of the core, at a depth of 2900 kilometers (1800 miles), is the **lower mantle**. Because of an increase in pressure (caused by the weight of the rock above), the mantle gradually strengthens with depth. Despite their strength, however, the rocks within the lower mantle are very hot and capable of extremely gradual flow.

Earth's Core

The **core** is composed of an iron–nickel alloy with minor amounts of oxygen, silicon, and sulfur—elements that readily form compounds with iron. At the extreme pressure found in the core, this iron-rich material has an average density of nearly 11 g/cm^3 and approaches 14 times the density of water at Earth's center.

The core is divided into two regions that exhibit very different mechanical strengths. The **outer core** is a *liquid layer* 2270 kilometers (1410 miles) thick. The movement of metallic iron within this zone generates Earth's magnetic field. The **inner core** is a sphere that has a radius of 1216 kilometers (754 miles). Despite its higher temperature, the iron in the inner core is *solid* due to the immense pressures that exist in the center of the planet.

CONCEPT CHECKS 1.6

1. List and describe the three major layers defined by their chemical composition.
2. Contrast the lithosphere and asthenosphere.
3. Distinguish between the outer core and the inner core.

1.7 Rocks and the Rock Cycle

Sketch, label, and explain the rock cycle.

Rock is the most common and abundant material on Earth. To a curious traveler, the variety seems nearly endless. When a rock is examined closely, we find that it usually consists of smaller crystals called minerals. *Minerals* are chemical compounds (or sometimes single elements), each with its own composition and physical properties. The grains or crystals may be microscopically small or easily seen with the unaided eye.

The minerals that compose a rock strongly influence its nature and appearance. In addition, a rock's *texture*—the size, shape, and/or arrangement of its constituent minerals—also has a significant effect on its appearance. A rock's mineral composition and texture, in turn, reflect the geologic processes that created it (**Figure 1.21**). Such analyses are critical to an understanding of our planet. This understanding has many practical applications, as in the search for energy and mineral resources and the solution of environmental problems.

Geologists divide rocks into three major groups: igneous, sedimentary, and metamorphic. **Figure 1.22** provides some examples. As you will learn, each group is linked to the others by the processes that act upon and within the planet.

Earlier in this chapter, you learned that Earth is a system. This means that our planet consists of many interacting parts that form a complex whole. Nowhere is this idea better illustrated than in the rock cycle (**Figure 1.23**). The **rock cycle** allows us to view many of the interrelationships among different parts of the Earth system. It helps us understand the origin of igneous, sedimentary, and metamorphic rocks and to see that each type is linked to the others by external and internal processes that act upon and within the planet. Consider the rock cycle to be a simplified but useful overview of physical geology. Learn the rock cycle well; you will be examining its interrelationships in greater detail throughout this textbook.

The Basic Cycle

Magma is molten rock that forms deep beneath Earth's surface. Over time, magma cools and solidifies. This process, called *crystallization*, may occur either beneath the surface or, following a volcanic eruption, at the surface. In either situation, the resulting rocks are called **igneous rocks**.

If igneous rocks are exposed at the surface, they undergo *weathering*, in which the day-in and day-out influences of the atmosphere slowly disintegrate and decompose rocks. The materials that result are often moved downslope by gravity before being picked up and transported by any of a number of erosional agents, such as rivers, glaciers, wind, or waves. Eventually these particles and dissolved substances, called **sediment**, are deposited. Although most sediment ultimately comes to rest in the ocean, other sites of deposition include river floodplains, desert basins, swamps, and sand dunes.

Next, the sediments undergo *lithification*, a term meaning "conversion into rock." Sediment is usually lithified into **sedimentary rock** when compacted by the weight of overlying layers or when cemented as percolating groundwater fills the pores with mineral matter.

If the resulting sedimentary rock is buried deep within Earth and involved in the dynamics of mountain building or intruded by a mass of magma, it is subjected to great pressures and/or intense heat. The sedimentary rock reacts to the changing environment and turns into the third rock type, **metamorphic rock**. When metamorphic rock is subjected to additional pressure changes or to still higher temperatures, it melts, creating magma, which eventually crystallizes into igneous rock, starting the cycle all over again.

Where does the energy that drives Earth's rock cycle come from? Processes driven by heat from Earth's interior are responsible for creating igneous and metamorphic rocks. Weathering and erosion, external processes powered by energy from the Sun, produce the sediment from which sedimentary rocks form.

Alternative Paths

The paths shown in the basic cycle are not the only ones that are possible. To the contrary, other paths are just as likely to be followed as those described in the preceding section. These alternatives are indicated by the light blue arrows in Figure 1.23.

Rather than being exposed to weathering and erosion at Earth's surface, igneous rocks may remain deeply buried. Eventually these masses may be subjected to the strong compressional forces and high temperatures

A. The large crystals of light-colored minerals in granite result from the slow cooling of molten rock deep beneath the surface. Granite is abundant in the continental crust.
B. Basalt is rich in dark minerals. Rapid cooling of molten rock at Earth's surface is responsible for the rock's microscopically small crystals.
The oceanic crust is a basalt-rich layer.

▲ **Figure 1.21 Two basic rock characteristics** Texture and mineral composition are basic rock features. These two samples are the common igneous rocks granite **A.** and basalt **B.** (Photo A by geoz/Alamy Images; photo B by Tyler Boyes/Shutterstock)

Igneous rocks form when molten rock solidifies at the surface (extrusive) or beneath the surface (intrusive). The lava flow in the foreground is the fine-grained rock basalt and came from SP Crater in northern Arizona.

Sedimentary rocks consist of particles derived from the weathering of other rocks. This layer consists of durable sand-size grains of the mineral quartz that are cemented into a solid rock. The grains were once a part of extensive dunes. This rock layer, called the Navajo Sandstone, is prominent in southern Utah.

The metamorphic rock pictured here, known as the Vishnu Schist, is exposed in the inner gorge of the Grand Canyon. It formed deep below Earth's surface where temperatures and pressures are high and in association with mountain-building episodes in Precambrian time.

▶ **Figure 1.22 Three rock groups** Geologists divide rocks into three groups—igneous, sedimentary, and metamorphic.

associated with mountain building. When this occurs, they are transformed directly into metamorphic rocks.

Metamorphic and sedimentary rocks, as well as sediment, do not always remain buried. Rather, overlying layers may be stripped away, exposing the once-buried rock. When this happens, the material is attacked by weathering processes and turned into new raw materials for sedimentary rocks.

Although rocks may seem to be unchanging masses, the rock cycle shows that they are not. The changes, however, take time—great amounts of time. We can observe different parts of the cycle operating all over the world. Today new magma is forming beneath the island of Hawaii. When it erupts at the surface, the lava flows add to the size of the island. Meanwhile, the Colorado Rockies are gradually being worn down by weathering and erosion. Some of this weathered debris will eventually be carried to the Gulf of Mexico, where it will add to the already substantial mass of sediment that has accumulated there.

CONCEPT CHECKS 1.7

1. List two rock characteristics that are used to determine the processes that created a rock.
2. Sketch and label a basic rock cycle. Make sure to include alternate paths.

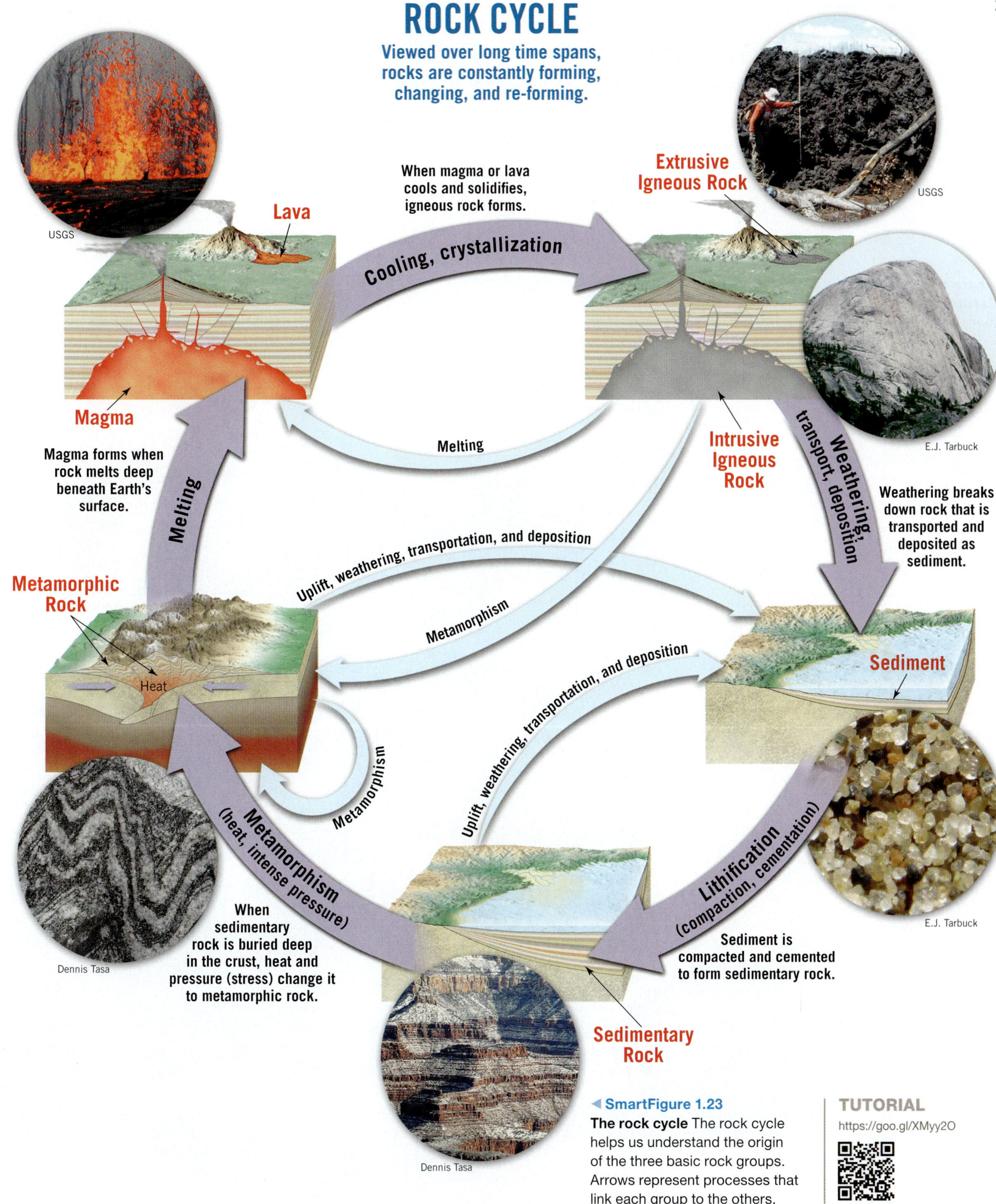
ROCK CYCLE
Viewed over long time spans, rocks are constantly forming, changing, and re-forming.
Lava
USGS
When magma or lava cools and solidifies, igneous rock forms.
Cooling, crystallization
Extrusive Igneous Rock
USGS
Magma
Melting
Magma forms when rock melts deep beneath Earth's surface.
Melting
Intrusive Igneous Rock
Weathering, transport, deposition
E.J. Tarbuck
Weathering breaks down rock that is transported and deposited as sediment.
Uplift, weathering, transportation, and deposition
Metamorphic Rock
Metamorphism
Heat
Sediment
Uplift, weathering, transportation, and deposition
Metamorphism
Metamorphism (heat, intense pressure)
Lithification (compaction, cementation)
E.J. Tarbuck
When sedimentary rock is buried deep in the crust, heat and pressure (stress) change it to metamorphic rock.
Dennis Tasa
Sediment is compacted and cemented to form sedimentary rock.
Sedimentary Rock
Dennis Tasa
TUTORIAL
https://goo.gl/XMyy2O

◄ **SmartFigure 1.23**
The rock cycle The rock cycle helps us understand the origin of the three basic rock groups. Arrows represent processes that link each group to the others.

1.8 The Face of Earth

List and describe the major features of the continents and ocean basins.

The two principal divisions of Earth's surface are the **ocean basins** and the **continents** (Figure 1.24). A significant difference between these two areas is their relative elevation. This difference results primarily from differences in their respective densities and thicknesses:

- **Ocean basins.** The average depth of the ocean floor is about 3.8 kilometers (2.4 miles) below sea level, or about 4.5 kilometers (2.8 miles) lower than the average elevation of the continents. The basaltic rocks that comprise the oceanic crust average only 7 kilometers

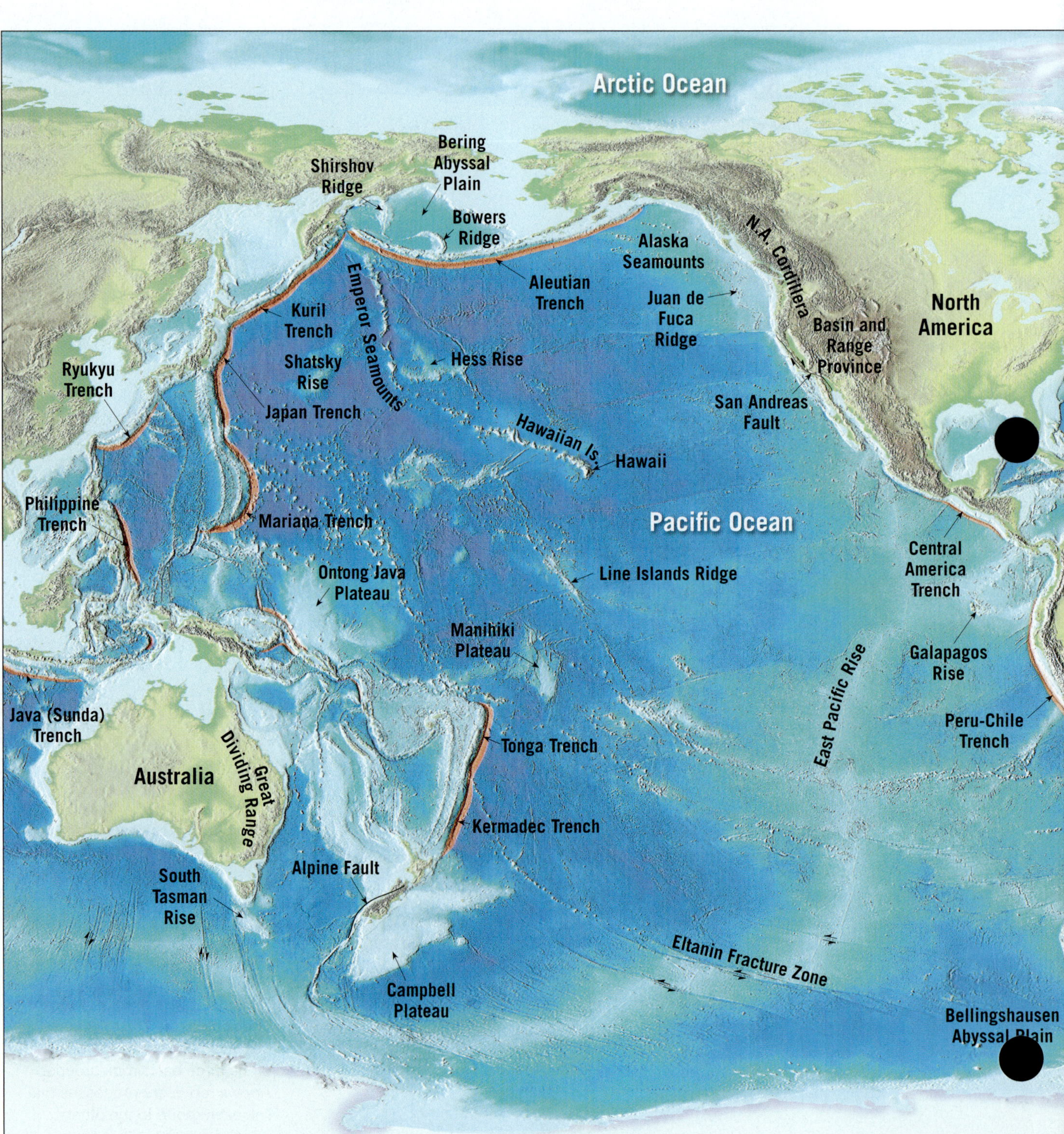

▶ **Figure 1.24 The face of Earth** Major surface features of the geosphere.

(5 miles) thick and have an average density of about 3.0 g/cm^3.

- **Continents.** The continents are remarkably flat features that have the appearance of plateaus protruding above sea level. With an average elevation of about 0.8 kilometer (0.5 mile), continental blocks lie close to sea level, except for limited areas of mountainous terrain. Recall that the continents average about 35 kilometers (22 miles) thick and are composed of granitic rocks that have a density of about 2.7 g/cm^3.

The thicker and less dense continental crust is more buoyant than the oceanic crust. As a result, continental crust floats on top of the deformable rocks of the mantle at a higher level than oceanic crust for the same reason that a large, empty (less dense) cargo ship rides higher than a small, loaded (denser) one.

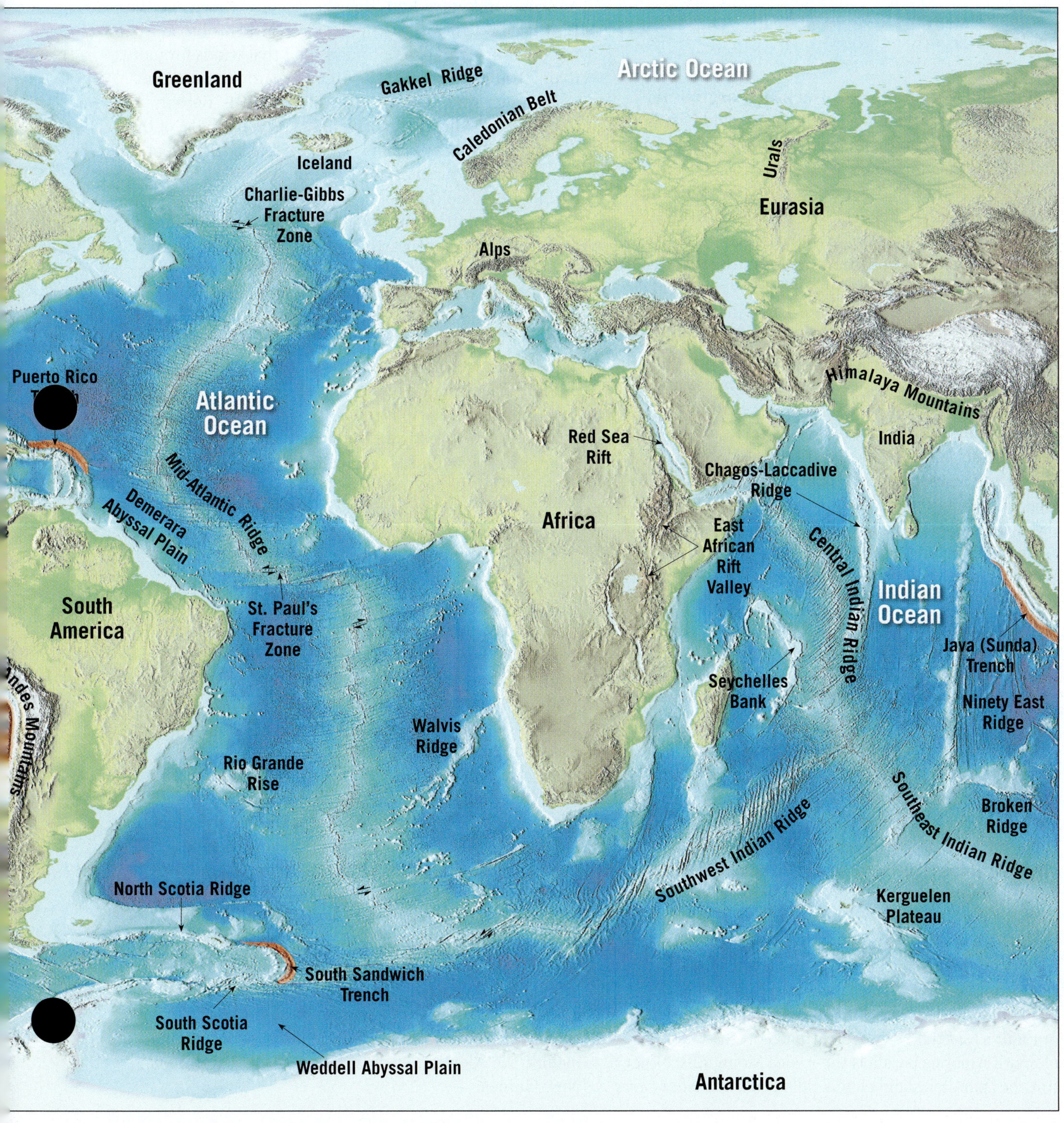

Did You Know?
Ocean depths are often expressed in *fathoms*. One fathom equals 1.8 m or 6 ft, which is about the distance of a person's outstretched arms. The term is derived from how depth-sounding lines were brought back on board a vessel by hand. As the line was hauled in, a worker counted the number of arm lengths collected. By knowing the length of the person's outstretched arms, the amount of line taken in could be calculated. The length of 1 fathom was later standardized to 6 ft.

Major Features of the Ocean Floor

If all water were drained from the ocean basins, a great variety of features would be visible, including chains of volcanoes, deep canyons, plateaus, and large expanses of monotonously flat plains. In fact, the scenery would be nearly as diverse as that on the continents (see Figure 1.24).

During the past 70 years, oceanographers have used modern depth-sounding equipment and satellite technology to map significant portions of the ocean floor. These studies have led them to identify three major regions: *continental margins*, *deep-ocean basins*, and *oceanic (mid-ocean) ridges*.

Continental Margin

The **continental margin** is the portion of the seafloor adjacent to major landmasses. It may include the *continental shelf*, the *continental slope*, and the *continental rise*.

Although land and sea meet at the shoreline, this is *not* the boundary between the continents and the ocean basins. Rather, along most coasts, a gently sloping platform, called the **continental shelf**, extends seaward from the shore. Because it is underlain by continental crust, it is clearly a flooded extension of the continents. A glance at Figure 1.24 shows that the width of the continental shelf is variable. For example, it is broad along the east and Gulf coasts of the United States but relatively narrow along the Pacific margin of the continent.

The boundary between the continents and the deep-ocean basins lies along the **continental slope**, which is a relatively steep dropoff that extends from the outer edge of the continental shelf, called the *shelf break*, to the floor of the deep ocean (see Figure 1.24). Using this as the dividing line, we find that about 60 percent of Earth's surface is represented by ocean basins and the remaining 40 percent by continents.

In regions where trenches do not exist, the steep continental slope merges into a more gradual incline known as the **continental rise**. The continental rise consists of a thick wedge of sediment that moved downslope from the continental shelf and accumulated on the deep-ocean floor.

Deep-Ocean Basins

Between the continental margins and oceanic ridges are **deep-ocean basins**. Parts of these regions consist of incredibly flat features called **abyssal plains**. The ocean floor also contains extremely deep depressions that are occasionally more than 11,000 meters (36,000 feet) deep. Although these **deep-ocean trenches** are relatively narrow and represent only a small fraction of the ocean floor, they are nevertheless very significant features. Some trenches are located adjacent to young mountains that flank the continents. For example, in Figure 1.24 the Peru–Chile trench off the west coast of South America parallels the Andes Mountains. Other trenches parallel island chains called *volcanic island arcs*.

Dotting the ocean floor are submerged volcanic structures called **seamounts**, which sometimes form long, narrow chains. Volcanic activity has also produced several large *lava plateaus*, such as the Ontong Java Plateau located northeast of New Guinea. In addition, some submerged plateaus are composed of continental-type crust. Examples include the Campbell Plateau southeast of New Zealand and the Seychelles Bank northeast of Madagascar.

Oceanic Ridges

The most prominent feature on the ocean floor is the **oceanic ridge**, or **mid-ocean ridge**. As shown in Figure 1.24, the Mid-Atlantic Ridge and the East Pacific Rise are parts of this system. This broad elevated feature forms a continuous belt that winds for more than 70,000 kilometers (43,000 miles) around the globe, in a manner similar to the seam of a baseball. Rather than consist of highly deformed rock, such as most of the mountains on the continents, the oceanic ridge system consists of layer upon layer of igneous rock that has been fractured and uplifted.

Being familiar with the topographic features that comprise the face of Earth is essential to understanding the mechanisms that have shaped our planet. What is the significance of the enormous ridge system that extends through all the world's oceans? What is the connection, if any, between young, active mountain belts and oceanic trenches? What forces crumple rocks to produce majestic mountain ranges? These are a few of the questions that will be addressed in the next chapter, as we begin to investigate the dynamic processes that shaped our planet in the geologic past and will continue to shape it in the future.

Major Features of the Continents

The major features of the continents can be grouped into two distinct categories: uplifted regions of deformed rocks that make up present-day mountain belts and extensive flat, stable areas that have eroded nearly to sea level. Notice in **Figure 1.25** that the young mountain belts tend to be long, narrow features at the margins of continents and that the flat, stable areas are typically located in the interior of the continents.

Mountain Belts

The most prominent features of the continents are mountains. Although the distribution of mountains appears to be random, this is not the case. The youngest mountains (those less than 100 million years old) are located principally in two major zones. The circum-Pacific belt (the region surrounding the Pacific Ocean) includes the mountains of the western Americas and continues into the western Pacific, in the form of volcanic island arcs (see Figure 1.24). Island arcs are active mountainous regions composed largely of volcanic rocks and deformed sedimentary rocks. Examples include the Aleutian Islands, Japan, the Philippines, and New Guinea.

The other major **mountain belt** extends eastward from the Alps through Iran and the Himalayas and then dips southward into Indonesia. Careful examination of mountainous terrains reveals that most are places where thick sequences of rocks have been squeezed and highly deformed, as if placed in a gigantic vise. Older mountains are also found on the continents. Examples include the Appalachians in the eastern United States and the Urals in Russia. Their once lofty peaks are now worn low, as a result of millions of years of weathering and erosion.

The Stable Interior Unlike the young mountain belts, which have formed within the past 100 million years, the interiors of the continents, called **cratons**, have been relatively stable (undisturbed) for the past 600 million years or even longer. Typically these regions were involved in mountain-building episodes much earlier in Earth's history.

Within the stable interiors are areas known as **shields**, which are expansive, flat regions composed largely of deformed igneous and metamorphic rocks. Notice in Figure 1.25 that the Canadian Shield is exposed in much of the northeastern part of North America. Radiometric dating of various shields has revealed that they are truly ancient regions. All contain Precambrian-age rocks that are more than 1 billion years old, with some samples approaching 4 billion years in age. Even these oldest-known rocks exhibit evidence of enormous forces that have folded, faulted, and metamorphosed them. Thus, we conclude that these rocks were once part of an ancient mountain system that has

▼ SmartFigure 1.25
The continents Distribution of mountain belts, stable platforms, and shields.

TUTORIAL
https://goo.gl/tFkUPM

The Canadian Shield is an expansive region of ancient Precambrian rocks, some more than 4 billion years old. It was recently scoured by Ice Age glaciers.

Superstock

The Appalachians are old mountains. Mountain building began about 480 million years ago and continued for more than 200 million years. Erosion has lowered these once lofty peaks.

Michael Collier

The rugged Himalayas are the highest mountains on Earth and are geologically young. They began forming about 50 million years ago and uplift continues today.

Alamy Images

since been eroded away to produce these expansive, flat regions.

In other flat areas of the craton, highly deformed rocks, like those found in the shields, are covered by a relatively thin veneer of sedimentary rocks. These areas are called **stable platforms**. The sedimentary rocks in stable platforms are nearly horizontal, except where they have been warped to form large basins or domes. In North America a major portion of the stable platform is located between the Canadian Shield and the Rocky Mountains.

CONCEPT CHECKS 1.8

1. Compare and contrast continents and ocean basins.
2. Name the three major regions of the ocean floor. What are some features associated with each?
3. Describe the general distribution of Earth's youngest mountains.
4. What is the difference between shields and stable platforms?

CONCEPTS IN REVIEW

An Introduction to Geology

1.1 Geology: The Science of Earth

Distinguish between physical and historical geology and describe the connections between people and geology.

KEY TERMS: geology, physical geology, historical geology

- Geologists study Earth. Physical geologists focus on the processes by which Earth operates and the materials that result from those processes. Historical geologists apply an understanding of Earth materials and processes to reconstruct the history of our planet.
- People have a relationship with planet Earth that can be positive and negative. Earth processes and products sustain us every day, but they can also harm us. Similarly, people have the ability to alter or harm natural systems, including those that sustain civilization.

1.2 The Development of Geology

Summarize early and modern views on how change occurs on Earth and relate them to the prevailing ideas about the age of Earth.

KEY TERMS: catastrophism, uniformitarianism

- Early ideas about the nature of Earth were based on religious traditions and notions of great catastrophes. In 1795, James Hutton emphasized that the same slow processes have acted over great spans of time and are responsible for Earth's rocks, mountains, and landforms. This similarity of process over vast spans of time led to this principle being dubbed "uniformitarianism."
- Based on the rate of radioactive decay of certain elements, the age of Earth has been calculated to be about 4,600,000,000 (4.6 billion) years. That is an incredibly vast amount of time.

? In what eon, era, period, and epoch do we live?

1.3 The Nature of Scientific Inquiry

Discuss the nature of scientific inquiry, including the construction of hypotheses and the development of theories.

KEY TERMS: hypothesis, theory, scientific method

- Geologists make observations, construct tentative explanations for those observations (hypotheses), and then test those hypotheses with field investigations and laboratory work. In science, a theory is a well-tested and widely accepted view that the scientific community agrees best explains certain observable facts.

(1.3 continued)

- As flawed hypotheses are discarded, scientific knowledge moves closer to a correct understanding, but we can never be fully confident that we know all the answers. Scientists must always be open to new information that forces changes in our model of the world.

1.4 Earth as a System

List and describe Earth's four major spheres. Define *system* and explain why Earth is considered to be a system.

KEY TERMS: hydrosphere, atmosphere, biosphere, geosphere, Earth system science, system

- Earth's physical environment is traditionally divided into three major parts: the solid Earth, called the geosphere; the water portion of our planet, called the hydrosphere; and Earth's gaseous envelope, called the atmosphere.
- A fourth Earth sphere is the biosphere, the totality of life on Earth. It is concentrated in a relatively thin zone that extends a few kilometers into the hydrosphere and geosphere and a few kilometers up into the atmosphere.
- Of all the water on Earth, more than 96 percent is in the oceans, which cover nearly 71 percent of the planet's surface.
- Although each of Earth's four spheres can be studied separately, they are all related in a complex and continuously interacting whole that is called the Earth system.
- Earth system science uses an interdisciplinary approach to integrate the knowledge of several academic fields in the study of our planet and its global environmental problems.
- The two sources of energy that power the Earth system are (1) the Sun, which drives the external processes that occur in the atmosphere, hydrosphere, and at Earth's surface, and (2) heat from Earth's interior that powers the internal processes that produce volcanoes, earthquakes, and mountains.

(1.4 continued)

? Is glacial ice part of the geosphere, or does it belong to the hydrosphere? Explain your answer.

Michael Collier

1.5 Origin and Early Evolution of Earth

Outline the stages in the formation of our solar system.

KEY TERMS: nebular theory, solar nebula

- The nebular theory describes the formation of the solar system. The planets and Sun began forming about 5 billion years ago from a large cloud of dust and gases.
- As the cloud contracted, it began to rotate and assume a disk shape. Material that was gravitationally pulled toward the center became the protosun. Within the rotating disk, small centers, called planetesimals, swept up more and more of the cloud's debris.
- Because of their high temperatures and weak gravitational fields, the inner planets were unable to accumulate and retain many of the lighter components. Because of the very cold temperatures far from the Sun, the large outer planets consist of huge amounts of lighter materials. These gaseous substances account for the comparatively large sizes and low densities of the outer planets.

? Earth is about 4.6 billion years old. If all of the planets in our solar system formed at about the same time, How old would you expect Mars to be? Jupiter? The Sun?

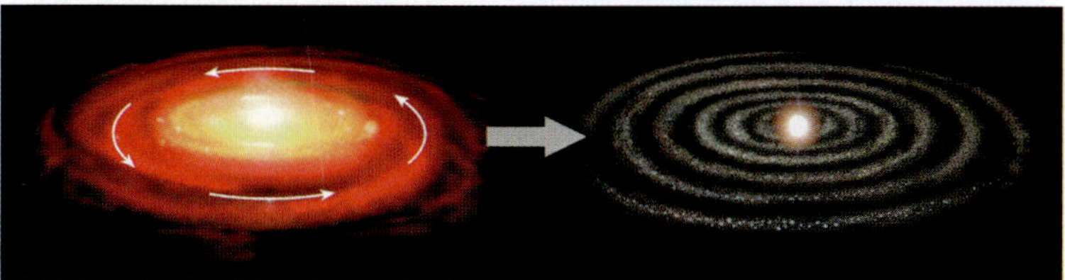

1.6 Earth's Internal Structure

Describe Earth's internal structure.

KEY TERMS: crust, mantle, lithosphere, asthenosphere, transition zone, lower mantle, core, outer core, inner core

- Compositionally, the solid Earth has three layers: core, mantle, and crust. The core is most dense, and the crust is least dense.
- Earth's interior can also be divided into layers based on physical properties. The crust and upper mantle make a two-part layer called the lithosphere, which is broken into the plates of plate tectonics. Beneath that is the "weak" asthenosphere. The lower mantle is stronger than the asthenosphere and overlies the molten outer core. This liquid is made of the same iron–nickel alloy as the inner core, but the extremely high pressure of Earth's center compacts the inner core into a solid form.

A.
B.
C.
2900 km
D.
5150 km
E.
6371 km

? The diagram represents Earth's layered structure. Does it show layering based on physical properties or layering based on composition? Identify the lettered layers.

1.7 Rocks and the Rock Cycle

Sketch, label, and explain the rock cycle.

KEY TERMS: rock cycle, igneous rock, sediment, sedimentary rock, metamorphic rock

- The rock cycle is a good model for thinking about the transformation of one rock to another due to Earth processes. All igneous rocks are made from molten rock. All sedimentary rocks are made from weathered products of other rocks. All metamorphic rocks are the products of preexisting rocks that are transformed at high temperatures or pressures. Given the right conditions, any kind of rock can be transformed into any other kind of rock.

? Name the processes that are represented by each of the letters in this simplified rock cycle diagram.

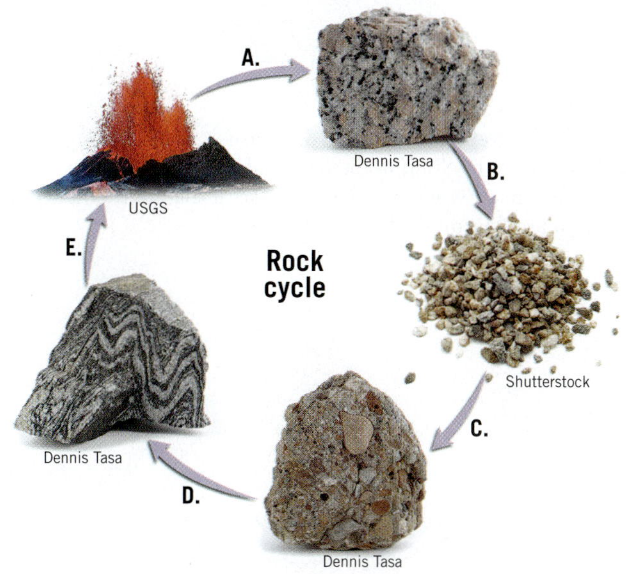

USGS

Dennis Tasa

Shutterstock

Dennis Tasa

Dennis Tasa

1.8 The Face of Earth

List and describe the major features of the continents and ocean basins.

KEY TERMS: ocean basin, continent, continental margin, continental shelf, continental slope, continental rise, deep-ocean basin, abyssal plain, deep-ocean trench, seamount, oceanic ridge (mid-ocean ridge), mountain belt, craton, shield, stable platform

- Two principal divisions of Earth's surface are the continents and ocean basins. A significant difference is their relative elevations, which results primarily from differences in their respective densities and thicknesses.
- Continents consist of relatively flat, stable areas called cratons. Where a craton is blanketed by a relatively thin layer of sediment or sedimentary rock, it is called a stable platform. Where a craton is exposed at the surface, it is known as a shield. Wrapping around the edges of some cratons are mountain belts, linear zones of intense deformation and metamorphism.
- Shallow portions of the oceans are essentially flooded margins of the continents, and deeper portions include vast abyssal plains and deep ocean trenches. Seamounts and lava plateaus interrupt the abyssal plain in some places.

? Put these features of the ocean floor in order from shallowest to deepest: continental slope, deep-ocean trench, continental shelf, abyssal plain, and continental rise.

GIVE IT SOME THOUGHT

1 The length of recorded history for humankind is about 5000 years. Clearly, most people view this span as being very long. How does it compare to the length of geologic time? Calculate the percentage or fraction of geologic time that is represented by recorded history. To make calculations easier, round the age of Earth to the nearest billion.

2 After entering a dark room, you turn on a wall switch, but the light does not come on. Suggest at least three hypotheses that might explain this observation. Once you have formulated your hypotheses, what is the next logical step?

3 Refer to the graph in Figure 1.13 to answer the following questions.

a. If you were to climb to the top of Mount Everest, how many breaths of air would you have to take at that altitude to equal the amount of air in one breath at sea level?

b. If you are flying in a commercial jet at an altitude of 12 kilometers (about 39,000 feet), about what percentage of the atmosphere's mass is below you?

Interpixels/Shutterstock

4 Making accurate measurements and observations is a basic part of scientific inquiry. Identify two images in this chapter that illustrate a way in which scientific data are gathered. Suggest an advantage that might be associated with the examples you select.

5 The accompanying photo provides an example of interactions among different parts of the Earth system. It is a view of a debris flow that was triggered by extraordinary rains in January 2005. Describe how each of Earth's four spheres was influenced and/or involved in this natural disaster that buried a portion of La Conchita, California.

Kevork Djansezian/Associated Press

6 Refer to Figure 1.23. How does the rock cycle diagram, particularly the process arrows, support the fact that sedimentary rock is the most abundant rock type on the surface of Earth?

7 This photo shows the picturesque coastal bluffs and rocky shoreline along a portion of the California coast south of San Simeon State Park. This area, like other shorelines, is described as an *interface*. What does this mean? Does the shoreline represent the boundary between the continent and ocean basin? Explain.

Michael Collier

MasteringGeology™

Looking for additional review and test prep materials? Visit the Study Area in MasteringGeology to enhance your understanding of this chapter's content by accessing a variety of resources, including Self-Study Quizzes, Geoscience Animations, SmartFigures, Mobile Field Trips, *Project Condor* Quadcopter videos, *In the News* RSS feeds, flashcards, web links, and an optional Pearson eText.

www.masteringgeology.com

2 Plate Tectonics: A Scientific Revolution Unfolds

FOCUS ON CONCEPTS

Each statement represents the primary learning objective for the corresponding major heading within the chapter. After you complete the chapter, you should be able to:

2.1 Summarize the view that most geologists held prior to the 1960s regarding the geographic positions of the ocean basins and continents.

2.2 List and explain the evidence Wegener presented to support his continental drift hypothesis.

2.3 List the major differences between Earth's lithosphere and asthenosphere and explain the importance of each in the plate tectonics theory.

2.4 Sketch and describe the movement along a divergent plate boundary that results in the formation of new oceanic lithosphere.

2.5 Compare and contrast the three types of convergent plate boundaries and name a location where each type can be found.

2.6 Describe the relative motion along a transform fault boundary and locate several examples of transform faults on a plate boundary map.

2.7 Explain why plates such as the African and Antarctic plates are increasing in size, while the Pacific plate is decreasing in size.

2.8 List and explain the evidence used to support the plate tectonics theory.

2.9 Describe two methods researchers use to measure relative plate motion.

2.10 Describe plate-mantle convection and explain two of the primary driving forces of plate motion.

Hikers crossing a crevasse in Khumbu Glacier, Mount Everest, Nepal. (Photo by Christian Kober/Robert Harding)

PLATE TECTONICS IS THE FIRST THEORY to provide a comprehensive view of the processes that produced Earth's major surface features, including the continents and ocean basins. Within the framework of this model, geologists have found explanations for the basic causes and distribution of earthquakes, volcanoes, and mountain belts. Further, the plate tectonics theory helps explain the formation and distribution of igneous and metamorphic rocks and their relationship with the rock cycle.

2.1 From Continental Drift to Plate Tectonics

Summarize the view that most geologists held prior to the 1960s regarding the geographic positions of the ocean basins and continents.

Until the late 1960s most geologists held the view that the ocean basins and continents had fixed geographic positions and were of great antiquity. Over the following decade, scientists came to realize that Earth's continents are not static; instead, they gradually migrate across the globe. These movements cause blocks of continental material to collide, deforming the intervening crust and thereby creating Earth's great mountain chains (**Figure 2.1**). Furthermore, landmasses occasionally split apart. As continental blocks separate, a new ocean basin emerges between them. Meanwhile, other portions of the seafloor plunge into the mantle. In short, a dramatically different model of Earth's tectonic processes emerged. *Tectonic processes* (*tekto* = to build) are processes that deform Earth's crust to create major structural features, such as mountains, continents, and ocean basins.

This profound reversal in scientific thought has been appropriately called a *scientific revolution*.

▼ **Figure 2.1 Himalayan mountain range as seen from northern India** The tallest mountains on Earth, the Himalayas, were created when the subcontinent of India collided with southeastern Asia. (Photo by Hartmut Postges/ Robert Harding)

The revolution began early in the twentieth century as a relatively straightforward proposal termed *continental drift*. For more than 50 years, the scientific community categorically rejected the idea that continents are capable of movement. North American geologists in particular had difficulty accepting continental drift, perhaps because much of the supporting evidence had been gathered from Africa, South America, and Australia, continents with which most North American geologists were unfamiliar.

After World War II, modern instruments replaced rock hammers as the tools of choice for many Earth scientists. Armed with more advanced tools, geologists and a new breed of researchers, including *geophysicists* and *geochemists*, made several surprising discoveries that rekindled interest in the drift hypothesis. By 1968 these developments had led to the unfolding of a far more encompassing explanation known as the *theory of plate tectonics*.

In this chapter, we will examine the events that led to this dramatic reversal of scientific opinion. We will also briefly trace the development of the *continental drift hypothesis*, examine why it was initially rejected, and consider the evidence that finally led to the acceptance of its direct descendant—the theory of plate tectonics.

CONCEPT CHECKS 2.1

1. Briefly describe the view held by most geologists prior to the 1960s regarding the ocean basins and continents.
2. What group of geologists were the least receptive to the continental drift hypothesis, and why?

Did You Know?
Alfred Wegener, best known for his continental drift hypothesis, also wrote numerous scientific papers on weather and climate. Following his interest in meteorology, Wegener made four extended trips to the Greenland Ice Sheet to study its harsh winter weather. In November 1930, while making a month-long trek across the ice sheet, Wegener and a companion perished.

2.2 Continental Drift: An Idea Before Its Time

List and explain the evidence Wegener presented to support his continental drift hypothesis.

During the 1600s, as better world maps became available, people noticed that continents, particularly South America and Africa, could be fit together like pieces of a jigsaw puzzle. However, little significance was given to this observation until 1915, when Alfred Wegener (1880–1930), a German meteorologist and geophysicist, wrote *The Origin of Continents and Oceans*. This book outlined Wegener's hypothesis, called **continental drift**, which dared to challenge the long-held assumption that the continents and ocean basins had fixed geographic positions.

Wegener suggested that a single **supercontinent** consisting of all Earth's landmasses once existed.* He named this giant landmass **Pangaea** (pronounced "Pan-jee-ah," meaning "all lands") (**Figure 2.2**). Wegener further hypothesized that about 200 million years ago, during a time period called the *Mesozoic era* (see Figure 1.6, page 8), this supercontinent began to fragment into smaller landmasses. These continental blocks then "drifted" to their present positions over a span of millions of years.

Wegener and others who advocated the continental drift hypothesis collected substantial evidence to support their point of view. The fit of South America and Africa and the geographic distribution of fossils and ancient climates all seemed to buttress the idea that these now separate landmasses had once been joined. Let us examine some of this evidence.

Evidence: The Continental Jigsaw Puzzle

Like a few others before him, Wegener suspected that the continents might once have been joined when he noticed the remarkable similarity between the coastlines

▼ **SmartFigure 2.2 Reconstructions of Pangaea** The supercontinent of Pangaea, as it is thought to have formed in the late Paleozoic and early Mesozoic eras more than 200 million years ago.

TUTORIAL
https://goo.gl/qWfzr2

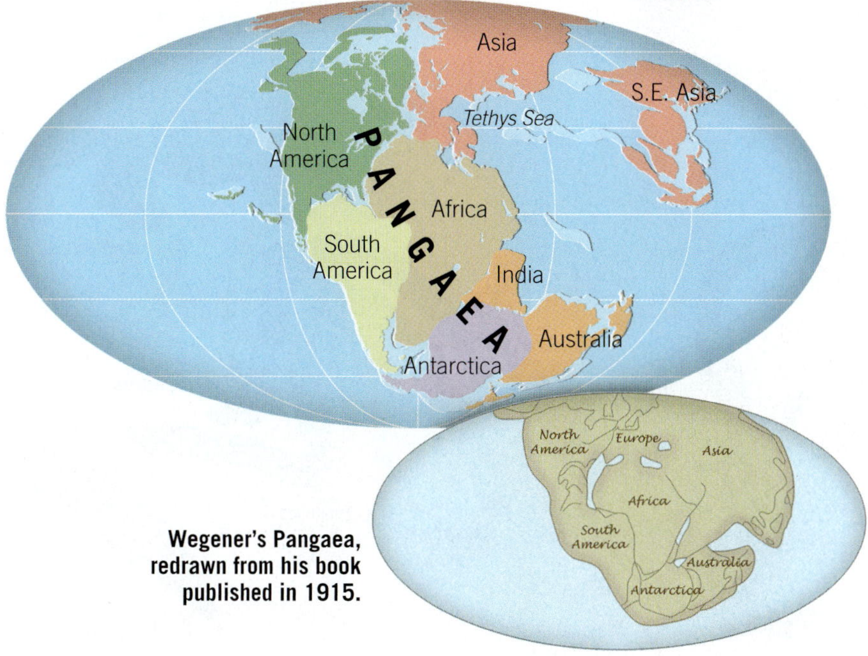

*Wegener was not the first to conceive of a long-vanished supercontinent. Eduard Suess (1831–1914), a distinguished nineteenth-century Austrian geologist, pieced together evidence for a giant landmass comprising South America, Africa, India, and Australia.

on opposite sides of the Atlantic Ocean. However, other Earth scientists challenged Wegener's use of present-day shorelines to "fit" these continents together. These opponents correctly argued that wave erosion and depositional processes continually modify shorelines. Even if continental displacement had taken place, a good fit today would be unlikely. Because Wegener's original jigsaw fit of the continents was crude, it is assumed that he was aware of this problem (see Figure 2.2).

Scientists later determined that a much better approximation of the outer boundary of a continent is the seaward edge of its continental shelf, which lies submerged a few hundred meters below sea level. In the early 1960s, Sir Edward Bullard and two associates constructed a map that pieced together the edges of the continental shelves of South America and Africa at a depth of about 900 meters (3000 feet) (**Figure 2.3**). The remarkable fit obtained was more precise than even these researchers had expected.

▲ **Figure 2.3 Two of the puzzle pieces** The best fit of South America and Africa occurs along the continental slope at a depth of 500 fathoms (about 900 meters [3000 feet]).

Evidence: Fossils Matching Across the Seas

Although the seed for Wegener's hypothesis came from the remarkable similarities of the continental margins on opposite sides of the Atlantic, it was when he learned that identical fossil organisms had been discovered in rocks from both South America and Africa that his pursuit of continental drift became more focused. Wegener learned that most paleontologists (scientists who study the fossilized remains of ancient organisms) agreed that some type of land connection was needed to explain the existence of similar Mesozoic-age life-forms on widely separated landmasses. Just as modern life-forms native to North America are not the same as those of Africa and Australia, Mesozoic-era organisms on widely separated continents should have been distinctly different.

Mesosaurus To add credibility to his argument, Wegener documented several cases in which the same fossil organism is found only on landmasses that are now widely separated, even though it is unlikely that the living organism could have crossed the barrier of a broad ocean (**Figure 2.4**). A classic example is *Mesosaurus*, a small aquatic freshwater reptile whose fossil remains are limited to rocks of Permian age (about 260 million years ago) in eastern South America and southwestern Africa. If *Mesosaurus* had been able to make the long journey across the South Atlantic, its remains should be more widely distributed. As this is not the case, Wegener asserted that South America and Africa must have been joined during that period of Earth history.

How did opponents of continental drift explain the existence of identical fossil organisms in places separated by thousands of kilometers of open ocean? Rafting, transoceanic land bridges (isthmian links), and island stepping stones were the most widely invoked explanations for these migrations (**Figure 2.5**). We know, for example, that during the Ice Age that ended about 8000 years ago, the lowering of sea level allowed mammals (including humans) to cross the narrow Bering Strait that separates Russia and Alaska. Was it possible that land bridges once connected Africa and South America but later subsided below sea level? Modern maps of the seafloor substantiate Wegner's views and show no such sunken land bridges.

Glossopteris Wegener also cited the distribution of the fossil "seed fern" *Glossopteris* as evidence for Pangaea's existence (see Figure 2.4). With tongue-shaped leaves and seeds too large to be carried by the wind, this plant was known to be widely dispersed thoughtout Africa, Australia, India, and South America. Later, fossil remains of *Glossopteris* were also discovered in Antarctica.*

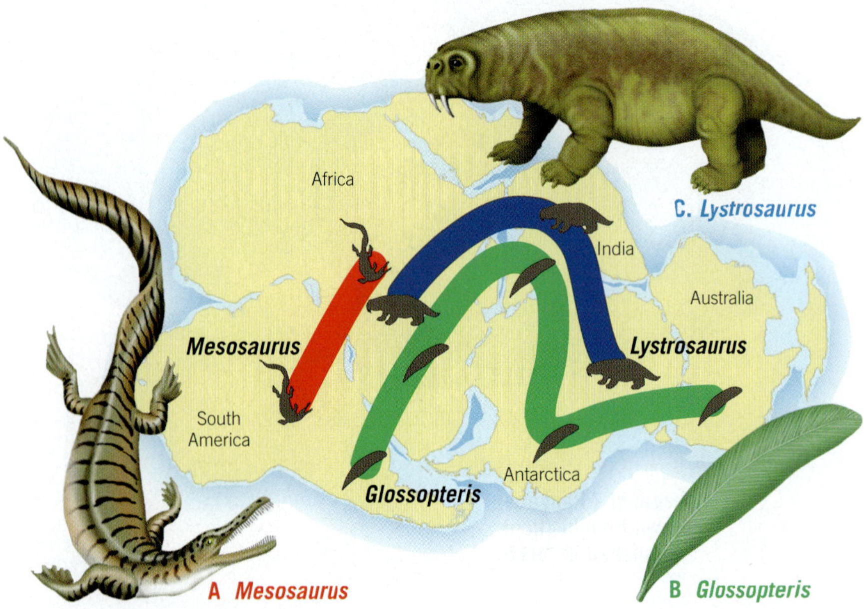

*In 1912 Captain Robert Scott and two companions froze to death lying beside 16 kilograms (35 pounds) of rock on their return from a failed attempt to be the first to reach the South Pole. These samples, collected on Beardmore Glacier, contained fossil remains of *Glossopteris*.

◄ **Figure 2.4 Fossil evidence supporting continental drift** Fossils of identical organisms have been discovered in rocks of similar age in Australia, Africa, South America, Antarctica, and India—continents that are currently widely separated by ocean barriers. Wegener accounted for these occurrences by placing these continents in their pre-drift locations.

Wegener also learned that these seed ferns and associated flora grew only in cool climates—similar to central Canada. Therefore, he concluded that when these landmasses were joined, they were located much closer to the South Pole.

Evidence: Rock Types and Geologic Features

You know that successfully completing a jigsaw puzzle requires maintaining the continuity of the picture while fitting the pieces together. In the case of continental drift, this means that the rocks on either side of the Atlantic that predate the proposed Mesozoic split should match up to form a continuous "picture" when the continents are fitted together as Wegener proposed.

Indeed, Wegener found such "matches" across the Atlantic. For instance, highly deformed igneous rocks in Brazil closely resemble similar rocks of the same age in Africa. Also, the mountain belt that includes the Appalachians trends northeastward through the eastern United States and disappears off the coast of Newfoundland (**Figure 2.6A**). Mountains of comparable age and structure are found in the British Isles and Scandinavia. When these landmasses are positioned as Wegener proposed (**Figure 2.6B**), the mountain chains form a nearly continuous belt. As Wegener wrote, "It is just as if we were to refit the torn pieces of a newspaper by matching their edges and then check whether the lines of print run smoothly across. If they do, there is nothing left but to conclude that the pieces were in fact joined in this way."*

Figure 2.5 How do land animals cross vast oceans? These sketches illustrate various early proposals to explain the occurrence of similar species on landmasses now separated by vast oceans. (Used by permission of John C. Holden)

Evidence: Ancient Climates

Because Alfred Wegener was a student of world climates, he suspected that paleoclimatic (*paleo* = ancient, *climatic* = climate) data might also support the idea of mobile continents. His assertion was bolstered by the discovery of evidence for a glacial period dating to the late *Paleozoic era* (see Figure 1.6, page 8) in southern Africa, South America, Australia, and India. This meant that about 300 million years ago, vast ice sheets covered

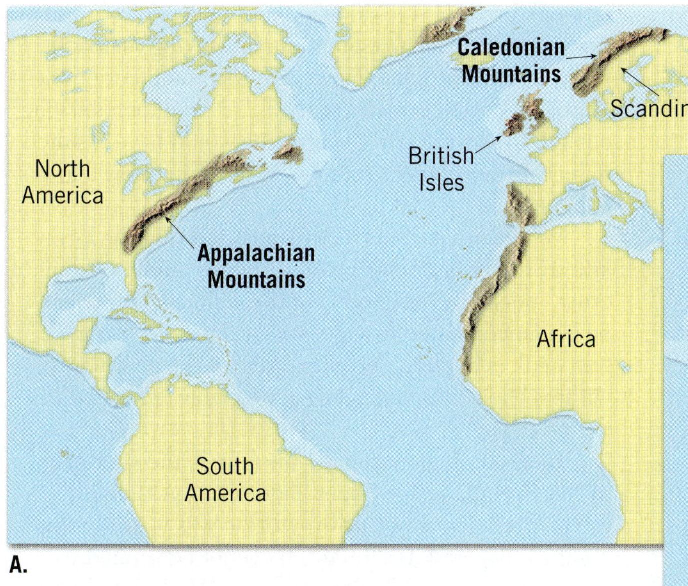

Figure 2.6 Matching mountain ranges across the North Atlantic A. The current locations of the continents surrounding the Atlantic. **B.** The configuration of the continents about 200 million years ago.

*Alfred Wegener, *The Origin of Continents and Oceans*, translated from the fourth revised German ed. of 1929 by J. Birman (London: Methuen, 1966).

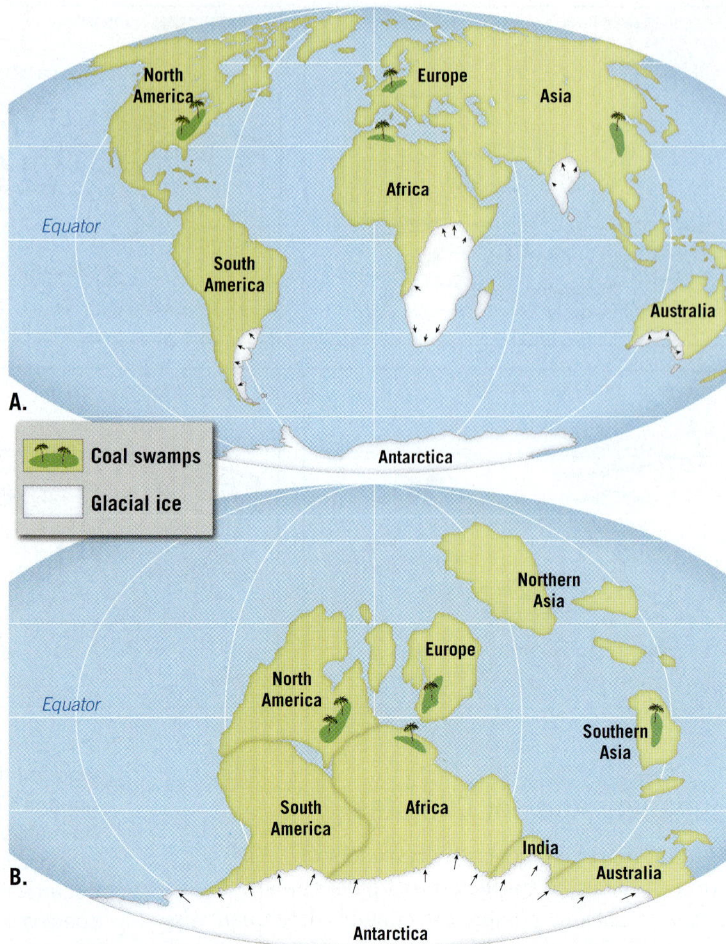

▲ **Figure 2.7 Paleoclimatic evidence for continental drift A.** About 300 million years ago, ice sheets covered extensive areas of the Southern Hemisphere and India. Arrows show the direction of ice movement that can be inferred from the pattern of glacial scratches and grooves found in the bedrock. Tropical coal swamps also existed in areas that are now temperate. **B.** Restoring the continents to their pre-drift positions creates a single glaciation centered on the South Pole and puts the coal swamps near the equator.

extensive portions of the Southern Hemisphere as well as India (**Figure 2.7A**). Much of the land area that contains evidence of this Paleozoic glaciation presently lies within 30 degrees of the equator, in subtropical or tropical climates.

How could extensive ice sheets form near the equator? One proposal suggested that our planet experienced a period of extreme global cooling. Wegener rejected this explanation because during the same span of geologic time, large tropical swamps existed in several locations in the Northern Hemisphere. The lush vegetation in those swamps was eventually buried and converted to coal (**Figure 2.7B**). Today these deposits comprise major coal fields in the eastern United States and Northern Europe. Many of the fossils found in these coal-bearing rocks were produced by tree ferns with large fronds—ferns that would have grown in warm, moist climates.* The existence of these large tropical swamps, Wegener argued, was inconsistent with the proposal that extreme global cooling caused glaciers to form in areas that are currently tropical.

Wegener suggested a more plausible explanation for the late Paleozoic glaciation: The southern continents were joined together in the supercontinent of Pangaea and located near the South Pole (see Figure 2.7B). This would account for the polar conditions required to generate extensive expanses of glacial ice over much of these landmasses. At the same time, this geography places today's northern continents nearer the equator and accounts for the tropical swamps that generated the vast coal deposits.

As compelling as this evidence may have been, 50 years passed before most of the scientific community accepted the concept of continental drift.

The Great Debate

From 1924, when Wegener's book was translated into English, French, Spanish, and Russian, until his death in 1930, his proposed drift hypothesis encountered a great deal of hostile criticism. The respected American geologist R. T. Chamberlain stated, "Wegener's hypothesis in general is of the foot-loose type, in that it takes considerable liberty with our globe, and is less bound by restrictions or tied down by awkward, ugly facts than most of its rival theories."

One of the main objections to Wegener's hypothesis stemmed from his inability to identify a credible mechanism for continental drift. Wegener proposed that gravitational forces of the Moon and Sun that produce Earth's tides were also capable of gradually moving the continents across the globe. However, the prominent physicist Harold Jeffreys correctly argued that tidal forces strong enough to move Earth's continents would have resulted in halting our planet's rotation, which, of course, has not happened.

Wegener also incorrectly suggested that the larger and sturdier continents broke through thinner oceanic crust, much as icebreakers cut through ice. However, no evidence existed to suggest that the ocean floor was weak enough to permit passage of the continents without the continents being appreciably deformed in the process.

In 1930, Wegener made his fourth and final trip to the Greenland Ice Sheet (**Figure 2.8**). Although the primary focus of this expedition was to study this great ice cap and its climate, Wegener continued to test his continental drift hypothesis. While returning from Eismitte, an experimental station located in the center of Greenland, Wegener perished along with his

*It is important to note that coal can form in a variety of climates, provided that large quantities of plant life are buried.

Alfred Wegener shown waiting out the 1912–1913 Arctic winter during an expedition to Greenland, where he made a 1200-kilometer traverse across the widest part of the island's ice sheet.

▲ **Figure 2.8 Alfred Wegener during an expedition to Greenland** (Photo courtesy of Archive of Alfred Wegener Institute)

Greenland companion. His intriguing idea, however, did not die.

Why was Wegener unable to overturn the established scientific views of his day? Foremost was the fact that, although the central theme of Wegener's drift hypothesis was correct, some details were incorrect. For example, continents do not break through the ocean floor, and tidal energy is much too weak to move continents. Moreover, for any comprehensive scientific theory to gain wide acceptance, it must withstand critical testing from all areas of science. Despite Wegener's great contribution to our understanding of Earth, not *all* of the evidence supported the continental drift hypothesis as he had proposed it.

As a result, most of the scientific community (particularly in North America) rejected continental drift or at least treated it with considerable skepticism. However, some scientists recognized the strength of the evidence Wegner had accumulated and continued to pursue the idea.

CONCEPT CHECKS 2.2

1. What was the first line of evidence that led early investigators to suspect that the continents were once connected?
2. Explain why the discovery of the fossil remains of *Mesosaurus* in both South America and Africa, but nowhere else, supports the continental drift hypothesis.
3. Early in the twentieth century, what was the prevailing view of how land animals migrated across vast expanses of open ocean?
4. How did Wegener account for evidence of glaciers in portions of South America, Africa, and India, when areas in North America, Europe, and Asia supported lush tropical swamps?
5. Describe two aspects of Wegener's continental drift hypothesis that were objectionable to most Earth scientists.

2.3 The Theory of Plate Tectonics

List the major differences between Earth's lithosphere and asthenosphere and explain the importance of each in the plate tectonics theory.

Following World War II, oceanographers equipped with new marine tools and ample funding from the U.S. Office of Naval Research embarked on an unprecedented period of oceanographic exploration. Over the next two decades, a much better picture of large expanses of the seafloor slowly and painstakingly began to emerge. From this work came the discovery of a global oceanic ridge system that winds through all the major oceans.

In other parts of the ocean, more discoveries were being made. Studies conducted in the western Pacific demonstrated that earthquakes were occurring at great depths beneath deep-ocean trenches. Of equal importance was the fact that dredging of the seafloor did not bring up any oceanic crust that was older than 180 million years. Further, sediment accumulations in the deep-ocean basins were found to be thin, not the thousands of meters that had been predicted. By 1968 these developments, among others, had led to the unfolding of a far more encompassing theory than continental drift, known as the **theory of plate tectonics**.

Rigid Lithosphere Overlies Weak Asthenosphere

According to the plate tectonics model, the crust and the uppermost, and therefore coolest, part of the mantle constitute Earth's strong outer layer, the **lithosphere** (*lithos* = stone). The lithosphere varies in both thickness and density, depending on whether it is oceanic or

Did You Know?

A group of scientists proposed an interesting but incorrect explanation for continental drift. They suggested that early in Earth's history, our planet was only about half its current diameter and completely covered by continental crust. Through time, Earth expanded, causing the continents to split into their current configurations, while new seafloor "filled in" the spaces as they drifted apart.

▶ SmartFigure 2.9 **The rigid lithosphere overlies the weak asthenosphere**

TUTORIAL

https://goo.gl/ujkFfZ

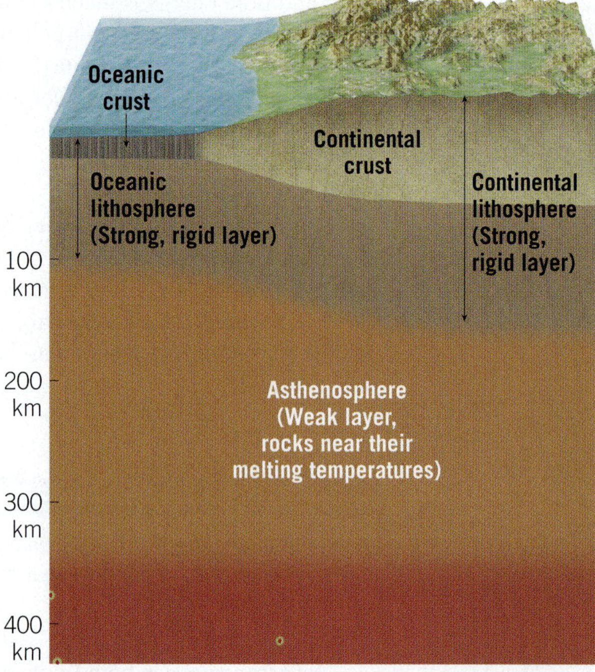

continental (**Figure 2.9**). Oceanic lithosphere is about 100 kilometers (60 miles) thick in the deep-ocean basins but is considerably thinner along the crest of the oceanic ridge system—a topic we will consider later. In contrast, continental lithosphere averages about 150 kilometers (90 miles) thick but may extend to depths of 200 kilometers (125 miles) or more beneath the stable interiors of the continents. Further, oceanic and continental crust differ in density. Oceanic crust is composed of basalt, a rock rich in dense iron and magnesium, whereas continental crust is composed largely of less dense granitic rocks. Because of these differences, the overall density of oceanic lithosphere (crust and upper mantle) is greater than the overall density of continental lithosphere. This important difference will be considered in greater detail later in this chapter.

The **asthenosphere** (*asthenos* = weak) is a hotter, weaker region in the mantle that lies below the lithosphere (see Figure 2.9). In the upper asthenosphere (located between 100 and 200 kilometers [60 to 125 miles] depth), the pressure and temperature bring rock very near to melting. Consequently, although the rock remains largely solid, it responds to forces by *flowing*, similarly to the way clay may deform if you compress it slowly. By contrast, the relatively cool and rigid lithosphere tends to respond to forces acting on it by *bending or breaking but not flowing*. Because of these differences, Earth's rigid outer shell is effectively detached from the asthenosphere, which allows these layers to move independently.

Earth's Major Plates

The lithosphere is broken into about two dozen segments of irregular size and shape called **lithospheric plates**, or simply **plates**, that are in constant motion with respect to one another (**Figure 2.10**). Seven major lithospheric plates are recognized and account for 94 percent of Earth's surface area: the *North American, South American, Pacific, African, Eurasian, Australian-Indian*, and *Antarctic plates*. The largest is the Pacific plate, which encompasses a significant portion of the Pacific basin. Each of the six other large plates consists of an entire continent, as well as a significant amount of oceanic crust. Notice in Figure 2.10 that the South American plate encompasses almost all of South America and about one-half of the floor of the South Atlantic. Note also that none of the plates are defined entirely by the margins of a single continent. This is a major departure from Wegener's continental drift hypothesis, which proposed that the continents move through the ocean floor, not with it.

Intermediate-sized plates include the *Caribbean, Nazca, Philippine, Arabian, Cocos, Scotia*, and *Juan de Fuca plates*. These plates, with the exception of the Arabian plate, are composed mostly of oceanic lithosphere. In addition, several smaller plates (*microplates*) have been identified but are not shown in Figure 2.10.

Plate Movement

One of the main tenets of the plate tectonics theory is that plates move as somewhat rigid units relative to all other plates. As plates move, the distance between two locations on different plates, such as New York and London, gradually changes, whereas the distance between sites on the same plate—New York and Denver, for example—remains relatively constant. However, parts of some plates are comparatively "weak," such as southern China, which is literally being squeezed as the Indian subcontinent rams into Asia proper.

Because plates are in constant motion relative to each other, most major interactions among them (and, therefore, most deformation) occur along their *boundaries*. In fact, plate boundaries were first established by plotting the locations of earthquakes and volcanoes. Plates are delimited by three distinct types of boundaries, which are differentiated by the type of movement they exhibit. These boundaries are depicted at the bottom of Figure 2.10 and are briefly described here:

- Divergent plate boundaries—where two plates move apart, resulting in upwelling and partial melting of hot material from the mantle to create new seafloor (Figure 2.10A).
- Convergent plate boundaries—where two plates move towards each other, resulting either in oceanic lithosphere descending beneath an overriding plate, eventually to be reabsorbed into the mantle, or possibly in the collision of two continental blocks to create a mountain belt (Figure 2.10B).

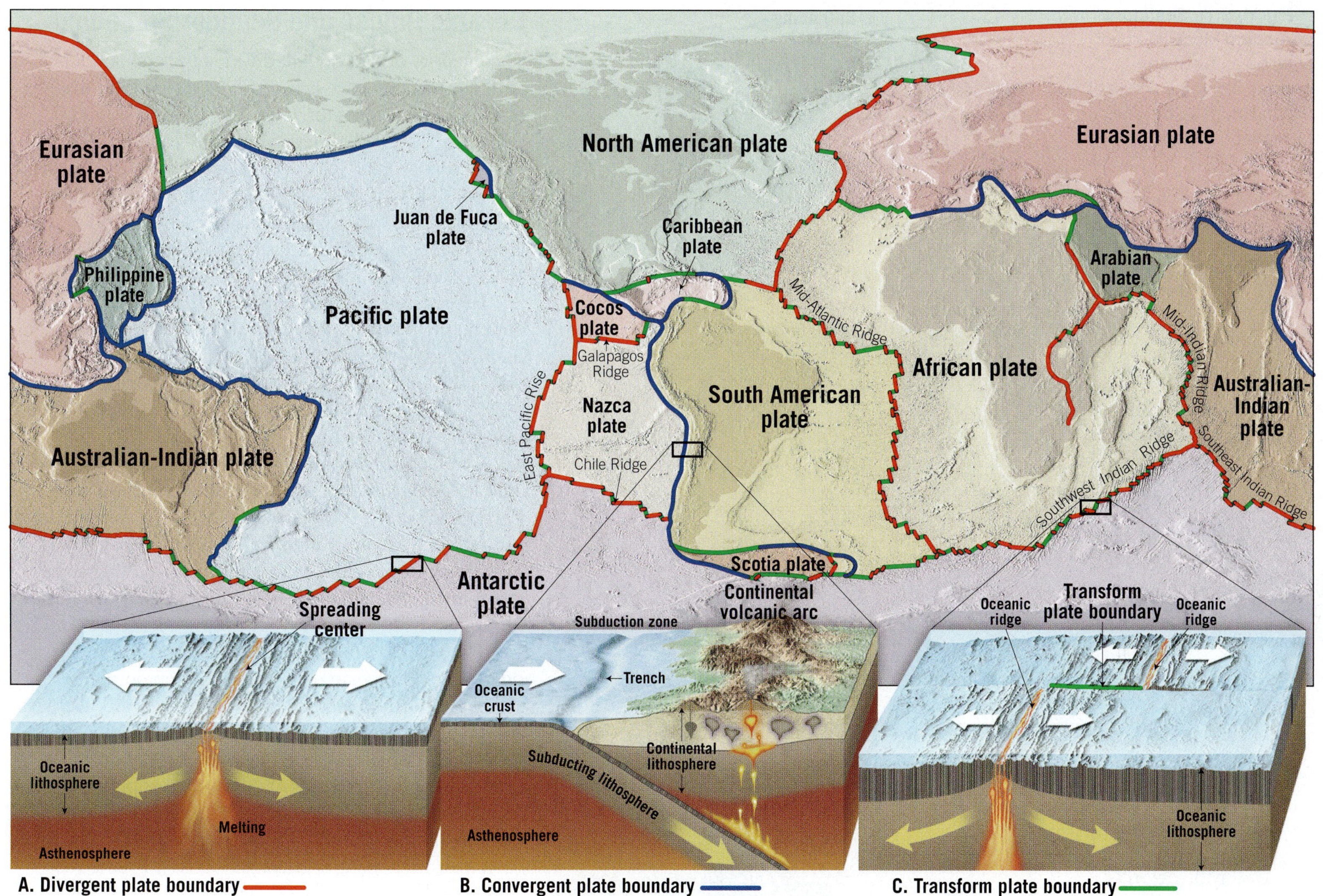

▲ **Figure 2.10 Earth's major lithospheric plates** The block diagrams below the map illustrate divergent, convergent, and transform plate boundaries.

- Transform plate boundaries—where two plates grind past each other without the production or destruction of lithosphere (Figure 2.10C).

Divergent and convergent plate boundaries each account for about 40 percent of all plate boundaries. Transform boundaries account for the remaining 20 percent. In the following sections we will discuss the three types of plate boundaries.

CONCEPT CHECKS 2.3

1. What new findings about the ocean floor did oceanographers discover after World War II?
2. Compare and contrast Earth's lithosphere and asthenosphere.
3. List the seven largest lithospheric plates.
4. List the three types of plate boundaries and describe the relative motion along each.

Did You Know?

An observer on another planet would notice, after only a few million years, that all of Earth's continents and ocean basins are indeed moving. The Moon, on the other hand, is tectonically "dead" (inactive), so it would look virtually unchanged millions of years in the future.

2.4 Divergent Plate Boundaries and Seafloor Spreading

Sketch and describe the movement along a divergent plate boundary that results in the formation of new oceanic lithosphere.

Most **divergent plate boundaries** (*di* = apart, *vergere* = to move) are located along the crests of oceanic ridges and can be thought of as *constructive plate margins* because this is where new ocean floor is generated (**Figure 2.11**). Here, two adjacent plates move away from each other, producing long, narrow fractures in the oceanic crust. As a result, hot molten rock from the mantle below migrates upward to fill the voids left as the crust is being ripped apart. This molten material gradually cools to produce new slivers of seafloor. In a slow yet unending manner, adjacent plates spread apart, and new oceanic lithosphere forms between them. For this reason, divergent plate boundaries are also called **spreading centers**.

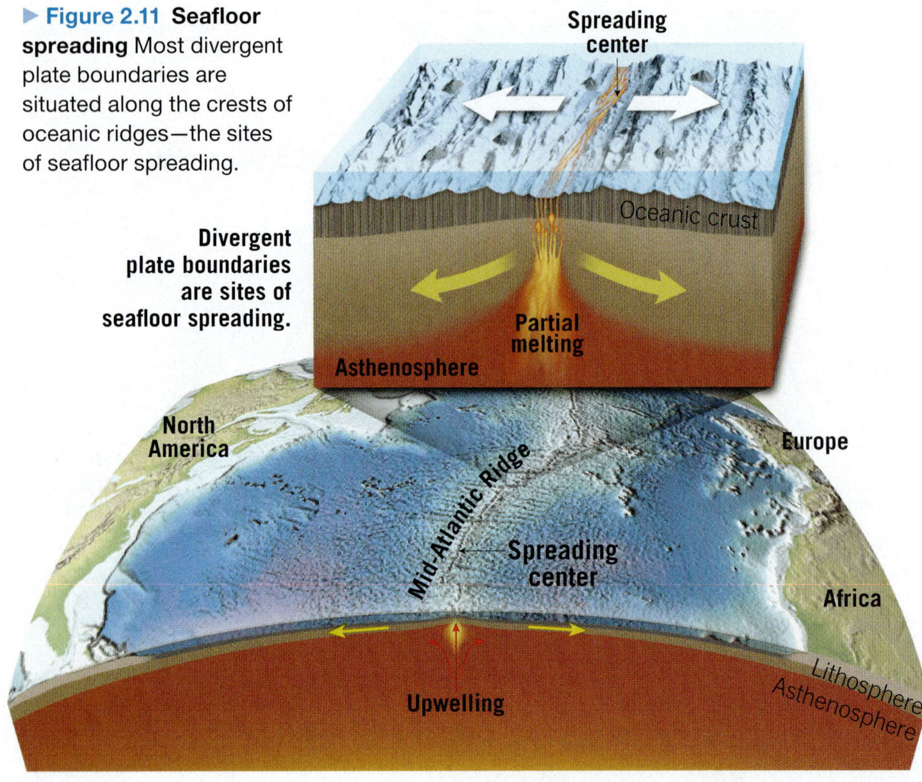

▶ **Figure 2.11 Seafloor spreading** Most divergent plate boundaries are situated along the crests of oceanic ridges—the sites of seafloor spreading.

Oceanic Ridges and Seafloor Spreading

The majority of, but not all, divergent plate boundaries are associated with *oceanic ridges*: elevated areas of the seafloor characterized by high heat flow and volcanism. The global **oceanic ridge system** is the longest topographic feature on Earth's surface, exceeding 70,000 kilometers (43,000 miles) in length. As shown in Figure 2.10, various segments of the global ridge system have been named, including the Southwest Atlantic Ridge, East Pacific Rise, and Southwest Indian Ridge.

Representing 20 percent of Earth's surface, the oceanic ridge system winds through all major ocean basins, like the seams on a baseball. Although the crest of the oceanic ridge is commonly 2 to 3 kilometers (1 to 2 miles) higher than the adjacent ocean basins, the term *ridge* may be misleading because it implies "narrow" when, in fact, ridges vary in width from 1000 kilometers (600 miles) to more than 4000 kilometers (2500 miles). Further, along the crest of some ridge segments is a deep canyonlike structure called a **rift valley** (Figure 2.12). This structure is evidence that tensional (pulling apart) forces are actively pulling the oceanic crust apart at the ridge crest.

The mechanism that operates along the oceanic ridge system to create new seafloor is appropriately called **seafloor spreading**. Spreading typically averages around 5 centimeters (2 inches) per year, roughly the same rate at which human fingernails grow. Comparatively slow spreading rates of 2 centimeters per year are found along the Mid-Atlantic Ridge, whereas spreading rates exceeding 15 centimeters (6 inches) per year have been measured along sections of the East Pacific Rise. Although these rates of seafloor production are slow on a human time scale, they are rapid enough to have generated all of Earth's current oceanic lithosphere within the past 200 million years.

The primary reason for the elevated position of the oceanic ridge is that newly created oceanic lithosphere is hot and, therefore, less dense than cooler rocks located away from the ridge axis. (Geologists use the term *axis* to refer to a line that follows the general trend of the ridge crest.) As soon as new lithosphere forms, it is slowly yet continually displaced away from the zone of mantle upwelling.

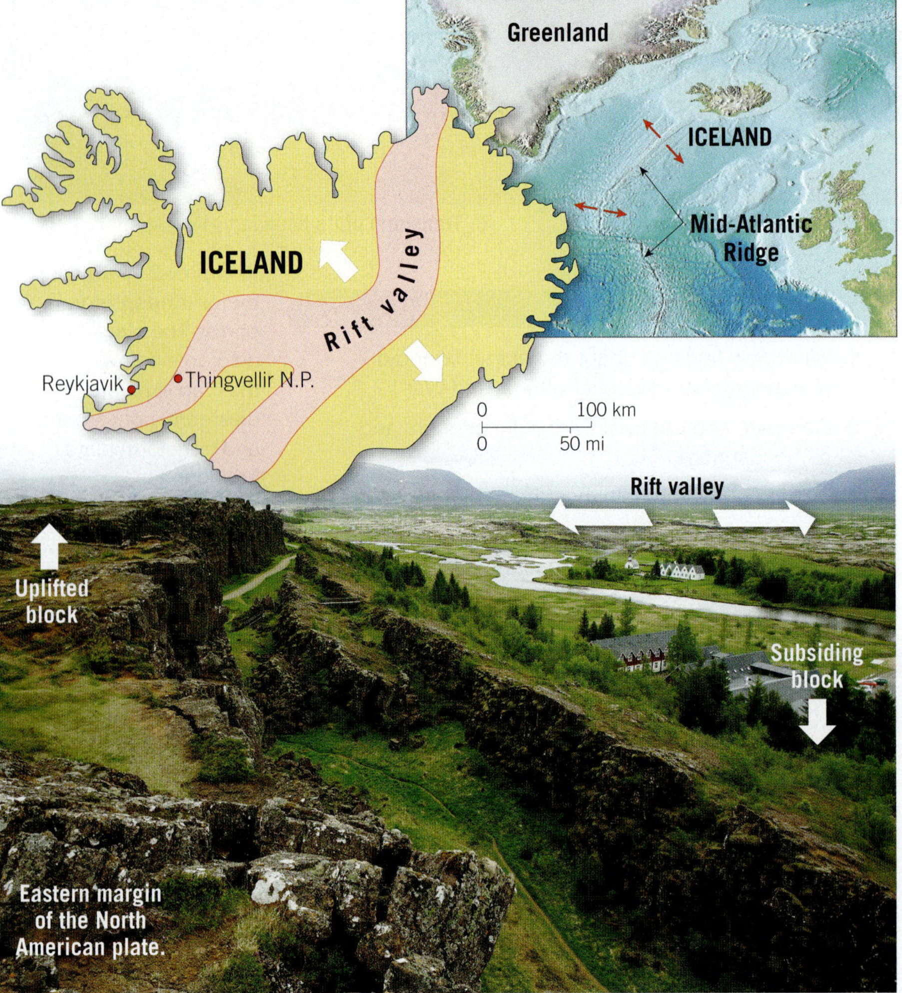

◀ **SmartFigure 2.12 Rift valley in Iceland** Thingvellir National Park, Iceland, is located on the western margin of a rift valley roughly 30 kilometers (20 mile) wide. This rift valley is connected to a similar feature that extends along the crest of the Mid-Atlantic Ridge. The cliff in the left half of the image approximates the eastern edge of the North American plate. (Photo by Ragnar Sigurdsson/Arctic/Alamy)

MOBILE FIELD TRIP
https://goo.gl/D7KNKF

Thus, it begins to cool and contract, thereby increasing in density. This thermal contraction accounts for the increase in ocean depth away from the ridge crest. It takes about 80 million years for the temperature of oceanic lithosphere to stabilize and contraction to cease. By this time, rock that was once part of the elevated oceanic ridge system is located in the deep-ocean basin, where it may be buried by substantial accumulations of sediment.

In addition, as the plate moves away from the ridge, cooling of the underlying asthenosphere causes its upper layers to become increasingly rigid. Thus, oceanic lithosphere is generated by cooling of the asthenosphere from the top down. Stated another way, the thickness of oceanic lithosphere is age dependent. The older (cooler) it is, the greater its thickness. Oceanic lithosphere that exceeds 80 million years in age is about 100 kilometers (60 miles) thick—approximately its maximum thickness.

Continental Rifting

Divergent boundaries can develop within a continent and may cause the landmass to split into two or more smaller segments separated by an ocean basin. Continental rifting begins when plate motions produce tensional forces that pull and stretch the lithosphere. This stretching, in turn, promotes mantle upwelling and broad upwarping of the overlying lithosphere (**Figure 2.13A**). This process thins the lithosphere and breaks the brittle crustal rocks into large blocks. As the tectonic forces continue to pull apart the crust, the broken crustal fragments sink, generating an elongated depression called a **continental rift**, which can widen to form a narrow sea (**Figure 2.13B,C**) and eventually a new ocean basin (**Figure 2.13D**). The formation of new oceans is discussed further in Chapter 10.

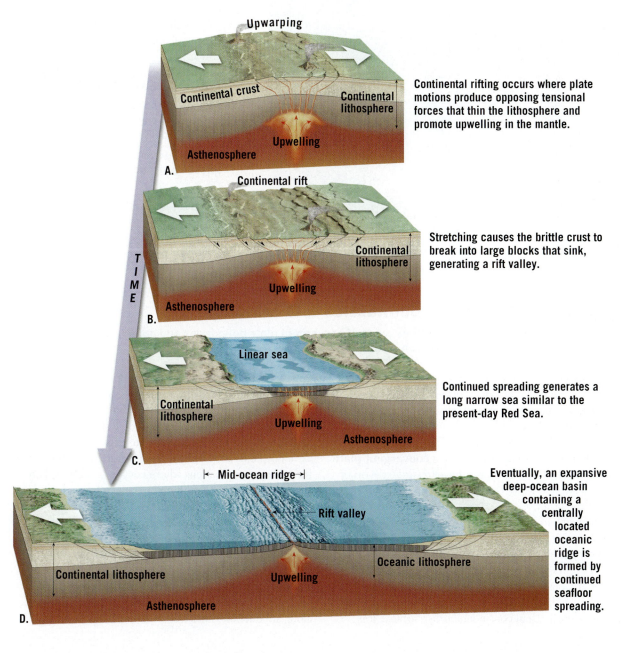

◀ SmartFigure 2.13 Continental rifting: Formation of new ocean basins

TUTORIAL

https://goo.gl/9CokZD

Continued spreading generates a long narrow sea similar to the present-day Red Sea.

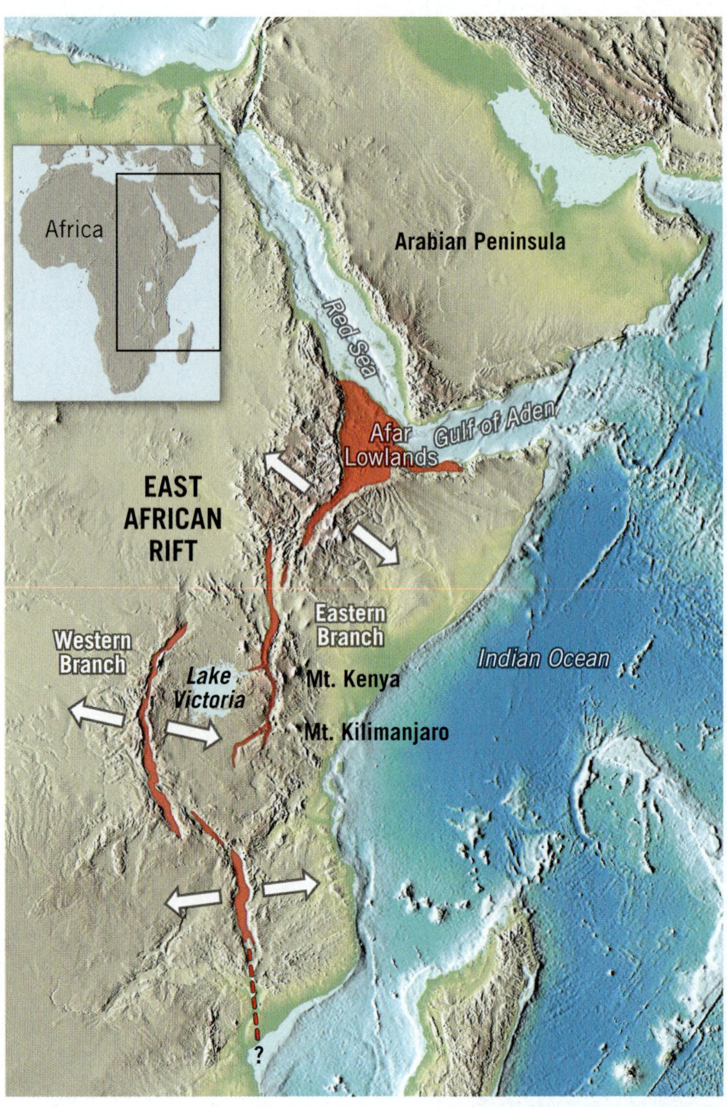

▶ **SmartFigure 2.14**
East African Rift valley The East African Rift valley represents the early stage in the breakup of a continent. Areas shown in red consist of lithosphere that has been stretched and thinned, allowing magma to well up from the mantle.

CONDOR VIDEO
https://goo.gl/RXv8qH

An example of an active continental rift is the East African Rift (**Figure 2.14**). Whether this rift will eventually result in the breakup of Africa is a topic of ongoing research. Nevertheless, the East African Rift is an excellent model of the initial stage in the breakup of a continent. Here, tensional forces have stretched and thinned the lithosphere, allowing molten rock to ascend from the mantle. Evidence for this upwelling includes several large volcanic mountains, including Mount Kilimanjaro and Mount Kenya, the tallest peaks in Africa. Research suggests that if rifting continues, the rift valley will lengthen and deepen (see Figure 2.13C). At some point, the rift valley will become a narrow sea with an outlet to the ocean. The Red Sea, formed when the Arabian Peninsula split from Africa, is a modern example of such a feature and provides us with a view of how the Atlantic Ocean may have looked in its infancy (see Figure 2.13D).

CONCEPT CHECKS 2.4

1. Sketch or describe how two plates move in relation to each other along divergent plate boundaries.
2. What is the average rate of seafloor spreading in modern oceans?
3. List four features that characterize the oceanic ridge system.
4. Briefly describe the process of continental rifting. Name a location where is it occurring today.

2.5 Convergent Plate Boundaries and Subduction

Compare and contrast the three types of convergent plate boundaries and name a location where each type can be found.

New lithosphere is constantly being produced at the oceanic ridges. However, our planet is not growing larger; its total surface area remains constant. A balance is maintained because older, denser portions of oceanic lithosphere descend into the mantle at a rate equal to seafloor production. This activity occurs along **convergent plate boundaries**, where two plates move toward each other and the leading edge of one is bent downward as it slides beneath the other.

Convergent boundaries are also called **subduction zones** because they are sites where lithosphere is descending (being subducted) into the mantle. Subduction occurs because the density of the descending lithospheric plate is greater than the density of the underlying asthenosphere. Recall that oceanic crust has a greater density than continental crust because it is largely composed of dense ferromagnesian-rich mineral. In general, old oceanic lithosphere is about 2 percent more dense than the underlying asthenosphere, causing it to sink much like an anchor on a ship. Continental lithosphere, in contrast, is less dense than the underlying asthenosphere and tends to resist subduction. However, there are a few locations where continental lithosphere is thought to have been forced below an overriding plate, albeit to relatively shallow depths.

Deep-ocean trenches are long, linear depressions in the seafloor that are generally located only a few hundred kilometers offshore of either a continent or a chain of volcanic islands such as the Aleutian chain. These underwater surface features are produced where oceanic lithosphere bends as it descends into the mantle along subduction zones (see Figure 1.24, page 24). An

Did You Know?
The remains of some of the earliest humans, *Homo habilis* and *Homo erectus*, were discovered by anthropologists Louis and Mary Leakey in the East African Rift. Scientists consider this region to be the "birthplace" of the human race.

example is the Peru–Chile trench, located along the west coast of South America. It is more than 4500 kilometers (3000 miles) long, and its floor is as much as 8 kilometers (5 miles) below sea level. Western Pacific trenches, including the Mariana and Tonga trenches, are even deeper than those of the eastern Pacific.

Slabs of oceanic lithosphere descend into the mantle at angles that vary from a few degrees to nearly vertical (90 degrees). The angle at which oceanic lithosphere subducts depends largely on its age and, therefore, its density. For example, when seafloor spreading occurs relatively near a subduction zone, as is the case along the coast of Chile (see Figure 2.10), the subducting lithosphere is young and buoyant, which results in a low angle of descent. As the two plates converge, the overriding plate scrapes over the top of the subducting plate below—a type of forced subduction. Consequently, the region around the Peru–Chile trench experiences great earthquakes, including the 2010 Chilean earthquake—one of the 10 largest on record.

As oceanic lithosphere ages (moves farther from the spreading center), it gradually cools, which causes it to thicken and increase in density. In parts of the western Pacific, some oceanic lithosphere is 180 million years old—the thickest and densest in today's oceans. The very dense slabs in this region typically plunge into the mantle at angles approaching 90 degrees. This largely explains why most trenches in the western Pacific are deeper than trenches in the eastern Pacific.

Although all convergent zones have the same basic characteristics, they may vary considerably depending on the type of crustal material involved and the tectonic setting. Convergent boundaries can form *between one oceanic plate and one continental plate*, *between two oceanic plates*, or *between two continental plates* (**Figure 2.15**).

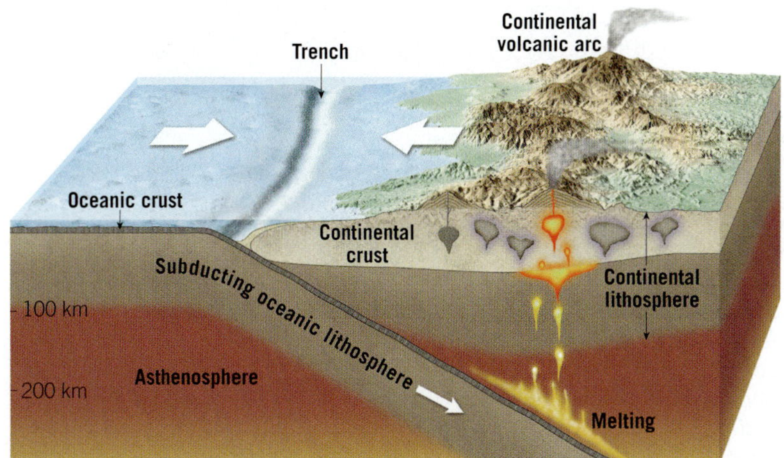

A. Convergent plate boundary where oceanic lithosphere is subducting beneath continental lithosphere.

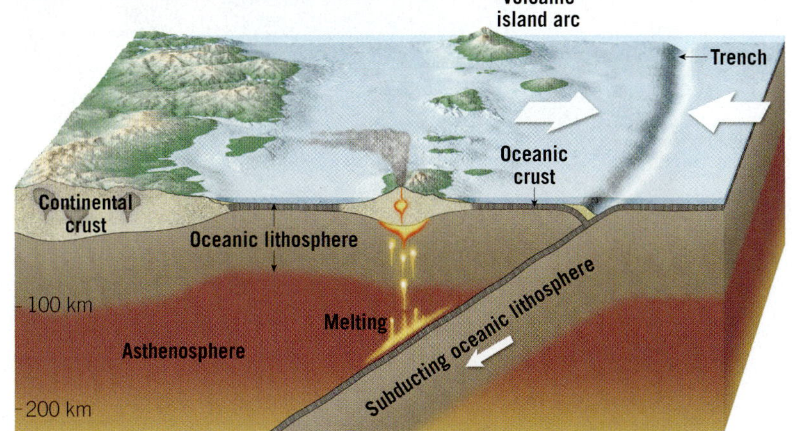

B. Convergent plate boundary involving two slabs of oceanic lithosphere.

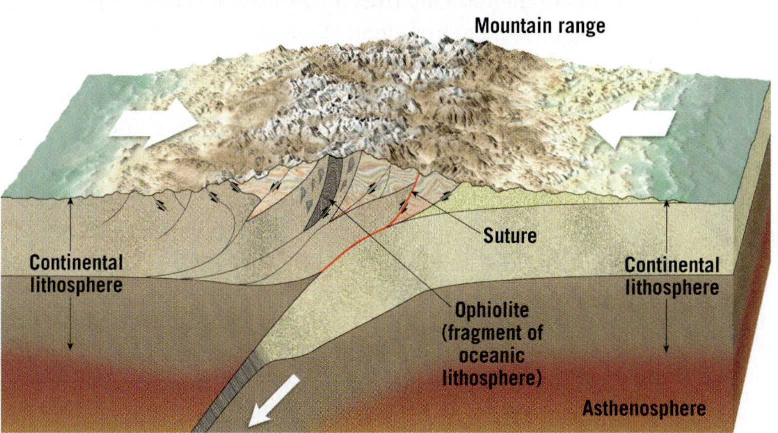

C. Continental collisions occur along convergent plate boundaries when both plates are capped with continental crust.

▲ **SmartFigure 2.15**
Three types of convergent plate boundaries

TUTORIAL
https://goo.gl/TDOFNu

Oceanic–Continental Convergence

When the leading edge of a plate capped with continental crust converges with a slab of oceanic lithosphere, the buoyant continental block remains "floating," while the denser oceanic slab sinks into the mantle (see Figure 2.15A). When a descending oceanic slab reaches a depth of about 100 kilometers (60 miles), melting is triggered within the wedge of hot asthenosphere that lies above it. But how does the subduction of a cool slab of oceanic lithosphere cause mantle rock to melt? The answer lies in the fact that water contained in the descending plates acts the way salt does to melt ice. That is, "wet" rock in a high-pressure environment melts at substantially lower temperatures than does "dry" rock of the same composition.

Sediments and oceanic crust contain large amounts of water, which is carried to great depths by a subducting plate. As the plate plunges downward, heat and pressure drive out water from the hydrated (water-rich) minerals in the subducting slab. At a depth of roughly 100 kilometers (60 miles), the wedge of mantle rock is sufficiently hot that the introduction of water from the slab below leads to some melting. This process, called **partial melting**, is thought to generate some molten material, which is mixed with unmelted mantle rock. Being less dense than the surrounding mantle, this hot mobile

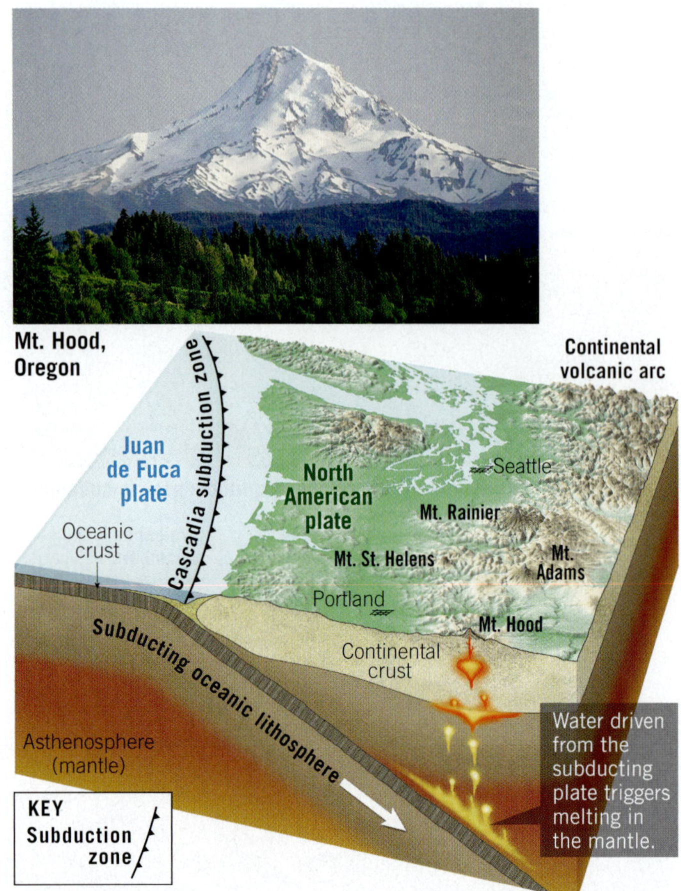

▶ **Figure 2.16** **Example of an oceanic–continental convergent plate boundary** The Cascade Range is a continental volcanic arc formed by the subduction of the Juan de Fuca plate below the North American plate. Mount Hood, Oregon, is one of more than a dozen large composite volcanoes in the Cascade Range. (Photo by Wallace Garrison/Getty Images)

material gradually rises toward the surface. Depending on the environment, these mantle-derived masses of molten rock may ascend through the crust and give rise to a volcanic eruption. However, much of this material never reaches the surface but solidifies at depth—a process that thickens the crust.

The volcanoes of the towering Andes were produced by molten rock generated by the subduction of the Nazca plate beneath the South American continent (see Figure 2.10). Mountain systems like the Andes, which are produced in part by volcanic activity associated with the subduction of oceanic lithosphere, are called **continental volcanic arcs**. The Cascade Range in Washington, Oregon, and California is another mountain system consisting of several well-known volcanoes, including Mount Rainier, Mount Shasta, Mount St. Helens, and Mount Hood (**Figure 2.16**). This active volcanic arc also extends into Canada, where it includes Mount Garibaldi and Mount Silverthrone.

Oceanic–Oceanic Convergence

An *oceanic–oceanic convergent boundary* has many features in common with oceanic–continental plate margins (see Figure 2.15A,B). Where two oceanic slabs converge, one descends beneath the other, initiating volcanic activity by the same mechanism that operates at all subduction zones (see Figure 2.10). Water released from the subducting slab of oceanic lithosphere triggers melting in the hot wedge of mantle rock above. In this setting, volcanoes grow up from the ocean floor rather than upon a continental platform. Sustained subduction eventually results in a chain of volcanic structures large enough to emerge as islands. The newly formed land, consisting of an arc-shaped chain of volcanic islands, is called a **volcanic island arc** or simply an **island arc** (**Figure 2.17**).

The Aleutian, Mariana, and Tonga Islands are examples of relatively young volcanic island arcs. Island arcs are generally located 120 to 360 kilometers (75 to 225 miles) from a deep-ocean trench. Located adjacent to the island arcs just mentioned are the Aleutian trench, the Mariana trench, and the Tonga trench.

Most volcanic island arcs are located in the western Pacific. Only two are located in the Atlantic—the Lesser Antilles arc, on the eastern margin of the Caribbean Sea, and the Sandwich Islands, located off the tip of South America. The Lesser Antilles are a product of the subduction of the Atlantic seafloor beneath the Caribbean plate. Located within this volcanic arc are the Virgin Islands of the United States and Britain as well as Martinique, where Mount Pelée erupted in 1902, destroying the town of St. Pierre and killing an estimated 28,000 people. This chain of islands also includes Montserrat, where volcanic activity has occurred as recently as 2010.

Island arcs are typically simple structures made of numerous volcanic cones underlain by oceanic crust that is generally less than 20 kilometers (12 miles) thick. Some island arcs, however, are more complex and are underlain by highly deformed crust that may reach 35 kilometers (22 miles) in thickness. Examples include Japan, Indonesia, and the Alaskan Peninsula. These island arcs are built on material generated by earlier episodes of subduction or on small slivers of continental crust that have rafted away from the mainland.

Continental–Continental Convergence

The third type of convergent boundary results when one landmass moves toward the margin of another because of subduction of the intervening seafloor (**Figure 2.18A**). Whereas oceanic lithosphere tends to be dense and readily sinks into the mantle, the buoyancy of continental material generally inhibits it from being subducted, at least to any great depth. Consequently, a collision between two converging continental fragments ensues (**Figure 2.18B**). This process folds and deforms the accumulation of sediments and sedimentary rocks along the continental margins as if they had been placed in a gigantic vise. The result is the formation of a new

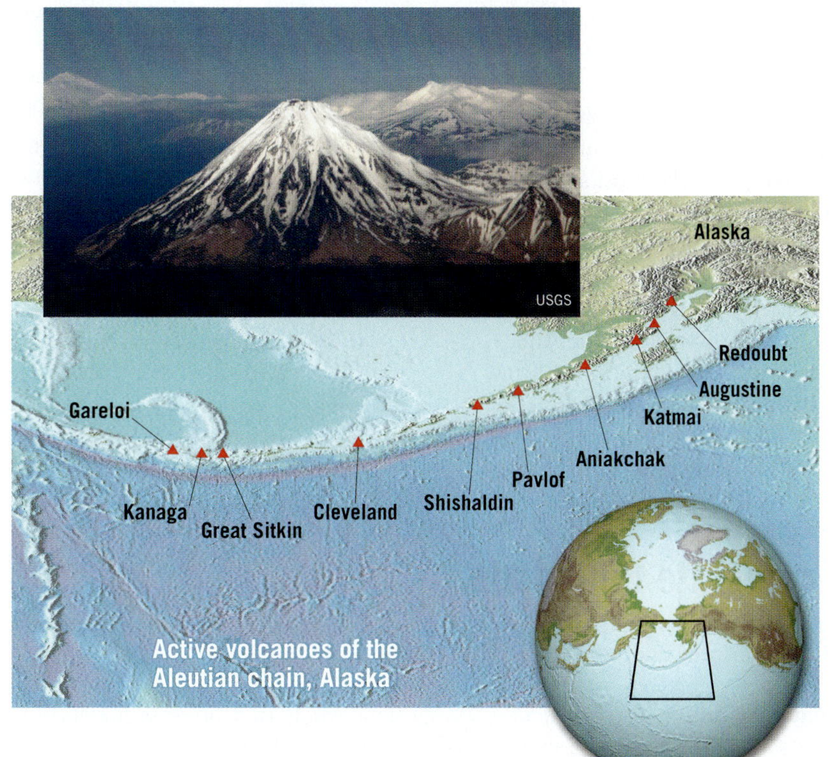

Figure 2.17 Volcanoes in the Aleutian chain The Aleutian Islands are a volcanic island arc produced by the subduction of the Pacific plate beneath the North American plate. Notice that the volcanoes of the Aleutian chain extend into Alaska proper.

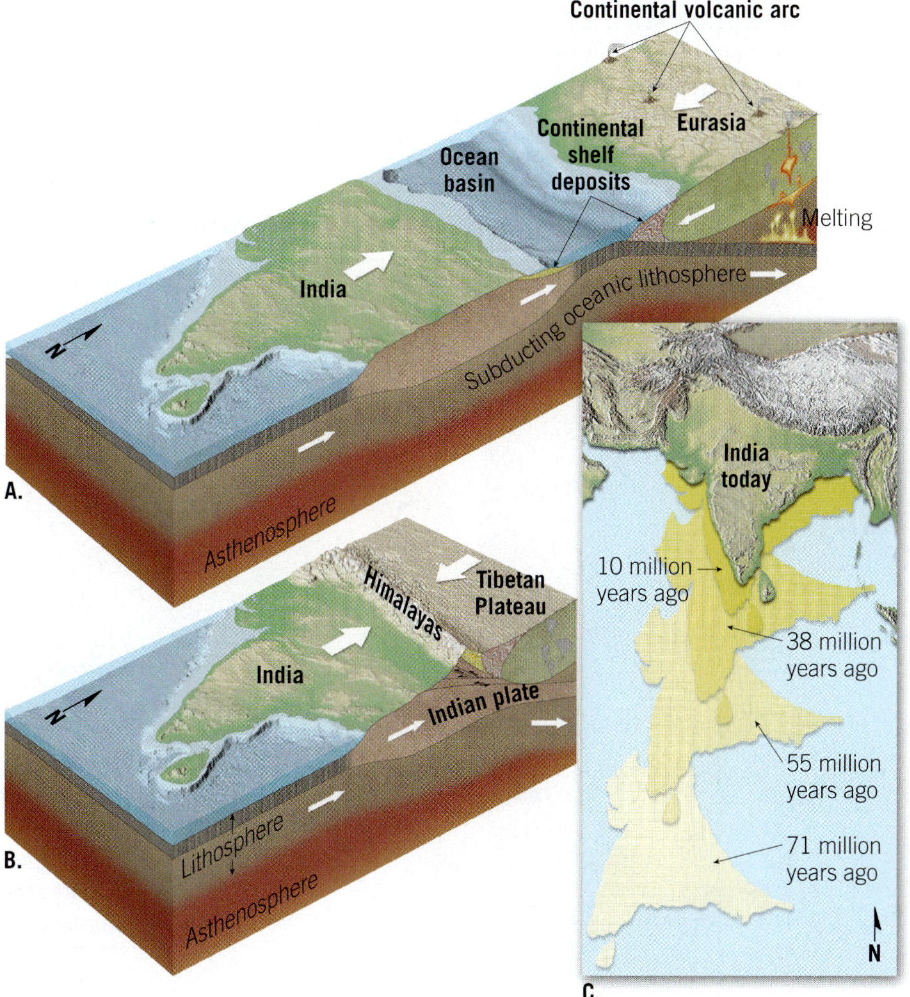

SmartFigure 2.18 The collision of India and Eurasia formed the Himalayas The ongoing collision of the subcontinent of India with Eurasia began about 50 million years ago and produced the majestic Himalayas. It should be noted that both India and Eurasia were moving as these landmasses collided. The map in part C illustrates only the movement of India.

ANIMATION
https://goo.gl/SU79OW

mountain belt composed of deformed sedimentary and metamorphic rocks that often contain slivers of oceanic lithosphere.

Such a collision began about 50 million years ago, when the subcontinent of India "rammed" into Asia, producing the Himalayas—the most spectacular mountain range on Earth (Figure 2.18C). During this collision, the continental crust buckled and fractured and was generally shortened horizontally and thickened vertically. In addition to the Himalayas, several other major mountain systems, including the Alps, Appalachians, and Urals, formed as continental fragments collided. This topic will be considered further in Chapter 11.

CONCEPT CHECKS 2.5

1. Explain why the rate of lithosphere production is roughly equal to the rate of lithosphere destruction.
2. Why does oceanic lithosphere subduct, while continental lithosphere does not?
3. What characteristic of a slab of oceanic lithosphere leads to the formation of a deep oceanic trench as opposed to one that is less deep?
4. What distinguishes a continental volcanic arc from a volcanic island arc?
5. Briefly describe how mountain belts such as the Himalayas form.

2.6 Transform Plate Boundaries

Describe the relative motion along a transform fault boundary and locate several examples of transform faults on a plate boundary map.

Along a **transform plate boundary**, also called a **transform fault**, plates slide horizontally past one another without the production or destruction of lithosphere. The nature of transform faults was discovered in 1965 by Canadian geologist J. Tuzo Wilson, who proposed that these large faults connected two spreading centers (divergent boundaries) or, less commonly, two trenches (convergent boundaries). Most transform faults are found on the ocean floor, where they offset segments of the oceanic ridge system, producing a steplike plate margin (Figure 2.19A). Notice that the zigzag shape of the Mid-Atlantic Ridge in Figure 2.10 (see page 41) roughly reflects the shape of the original rifting that caused the breakup of the supercontinent of Pangaea. (Compare the shapes of the continental margins of the landmasses on both sides of the Atlantic with the shape of the Mid-Atlantic Ridge.)

▼ **SmartFigure 2.19**
Transform plate boundaries Most transform faults offset segments of a spreading center, producing a plate margin that exhibits a zigzag pattern.

TUTORIAL
https://goo.gl/ZT7a9i

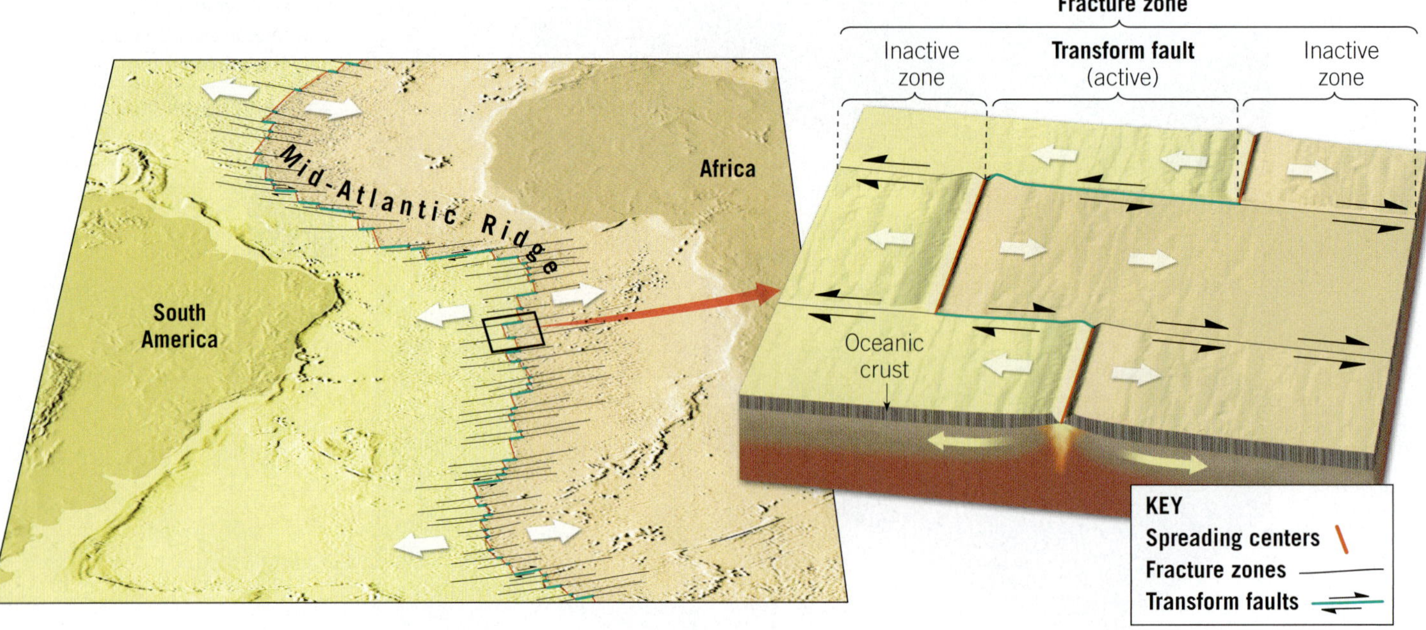

A. The Mid-Atlantic Ridge, with its zigzag pattern, roughly reflects the shape of the rifting zone that resulted in the breakup of Pangaea.

B. Fracture zones are long, narrow scar-like features in the seafloor that are roughly perpendicular to the offset ridge segments. They include both the active transform fault and its "fossilized" trace.

Typically, transform faults are part of prominent linear breaks in the seafloor known as **fracture zones**, which include both active transform faults and their inactive extensions into the plate interior (Figure 2.19B). In a fracture zone, the active transform fault lies *only between* the two offset ridge segments; it is generally defined by weak, shallow earthquakes. On each side of the fault, the seafloor moves away from the corresponding ridge segment. Thus, between the ridge segments, these adjacent slabs of oceanic crust are grinding past each other along a transform fault. Beyond the ridge crests, these faults are inactive because the rock on either side moves in the same direction. However, these inactive faults are preserved as linear topographic depressions. The trend (orientation) of these fracture zones roughly parallels the direction of plate motion at the time of their formation. Thus, these structures help geologists map the direction of plate motion in the geologic past.

Transform faults also provide the means by which the oceanic crust created at ridge crests can be transported to a site of destruction—the deep-ocean trenches. Figure 2.20 illustrates this situation. Notice that the Juan de Fuca plate moves in a southeasterly direction, eventually being subducted under the west coast of the United States and Canada. The southern end of this plate is bounded by a transform fault called the Mendocino Fault. This transform boundary connects the Juan de Fuca Ridge to the Cascadia subduction zone. Therefore, it facilitates the movement of the crustal material created at the Juan de Fuca Ridge to its destination beneath the North American continent.

Like the Mendocino Fault, most other transform fault boundaries are located within the ocean basins; however, a few cut through continental crust. Two examples are the earthquake-prone San Andreas Fault of California and New Zealand's Alpine Fault. Notice in Figure 2.20 that the San Andreas Fault connects a spreading center located in the Gulf of California to the Cascadia subduction zone and the Mendocino Fault. Along the San Andreas Fault, the Pacific plate is moving toward the northwest, past the North American plate (Figure 2.21). If this movement continues, the part of California west of the fault zone, including Mexico's Baja Peninsula, will become an island off the West Coast of the United States and Canada. However, a more immediate concern is the earthquake activity triggered by movements along this fault system.

CONCEPT CHECKS 2.6

1. Sketch or describe how two plates move in relation to each other along a transform plate boundary.
2. List two characteristics that differentiate transform faults from the two other types of plate boundaries.

Did You Know?

The Great Alpine Fault is a transform fault that runs through New Zealand's South Island. The northwestern part of the South Island sits on the Australian plate, whereas the rest of the island lies on the Pacific plate. As with its sister fault, California's San Andreas, the rocks on one side of this fault have moved several hundred miles relative to the rocks on the other side.

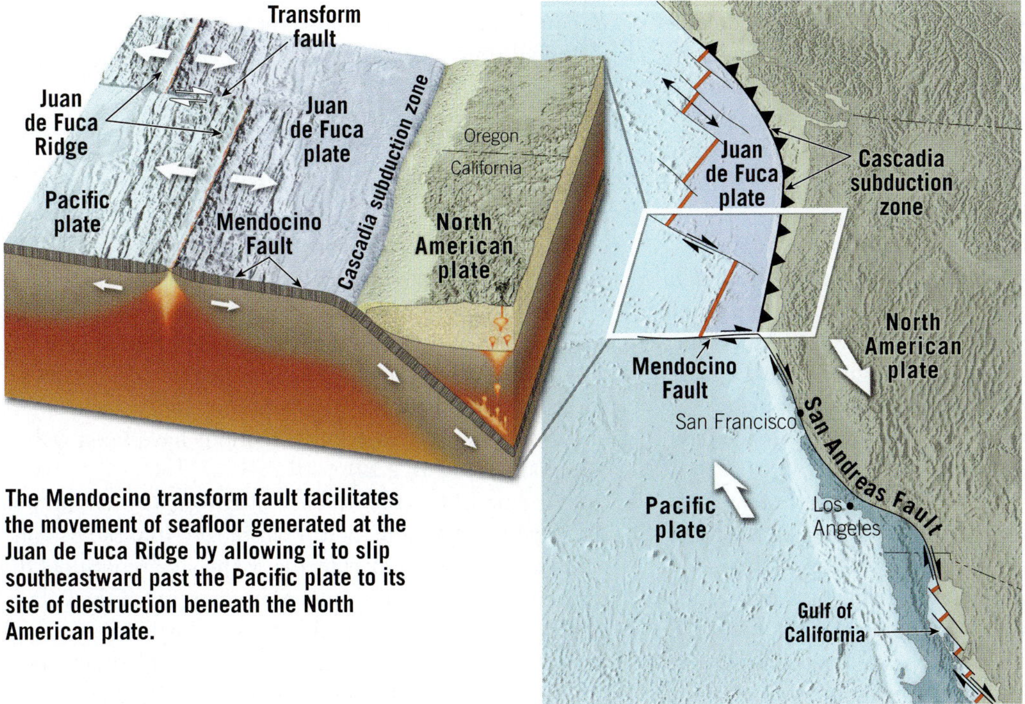

Figure 2.20 Transform faults facilitate plate motion Seafloor generated along the Juan de Fuca Ridge moves southeastward, past the Pacific plate. Eventually it subducts beneath the North American plate. Thus, this transform fault connects a spreading center (divergent boundary) to a subduction zone (convergent boundary). Also shown is the San Andreas Fault, a transform fault connecting a spreading center located in the Gulf of California with the Mendocino Fault.

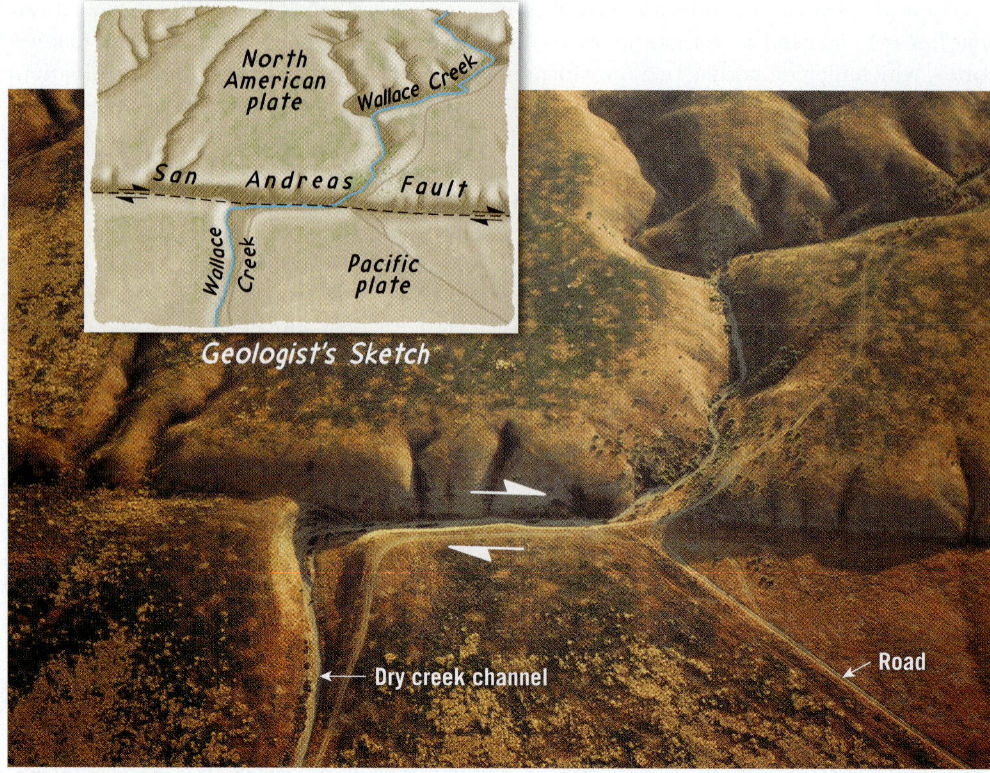

▶ **SmartFigure 2.21**
Movement along the San Andreas Fault This aerial view shows the offset in the dry channel of Wallace Creek near Taft, California. (Photo by Michael Collier)

MOBILE FIELD TRIP
https://goo.gl/fW4cFE

2.7 How Do Plates and Plate Boundaries Change?

Explain why plates such as the African and Antarctic plates are increasing in size, while the Pacific plate is decreasing in size.

Although Earth's total surface area does not change, the size and shape of individual plates are constantly changing. For example, the African and Antarctic plates, which are mainly bounded by divergent boundaries—sites of seafloor production—are continually growing in size as new lithosphere is added to their margins. By contrast, the Pacific plate is being consumed into the mantle along much of its flanks faster that it is being generated along the East Pacific Rise and thus is diminishing in size.

Another result of plate motion is that boundaries also migrate. For example, the position of the Peru–Chile trench, which is the result of the Nazca plate being bent downward as it descends beneath the South American plate, has changed over time (see Figure 2.10). Because of the westward drift of the South American plate relative to the Nazca plate, the Peru–Chile trench has migrated in a westerly direction as well.

Plate boundaries can also be created or destroyed in response to changes in the forces acting on the lithosphere. For example, some plates carrying continental crust are presently moving toward one another. In the South Pacific, Australia is moving northward, toward southern Asia. If Australia continues its northward migration, the boundary separating it from Asia will eventually disappear as these plates become one. Other plates are moving apart. Recall that the Red Sea is the site of a relatively new spreading center that came into existence less than 20 million years ago, when the Arabian Peninsula began to break apart from Africa. The breakup of Pangaea is a classic example of how plate boundaries change through geologic time.

The Breakup of Pangaea

Wegener used evidence from fossils, rock types, and ancient climates to create a jigsaw-puzzle fit of the continents, thereby creating his supercontinent of Pangaea. By employing modern tools not available to Wegener, geologists have re-created the steps in the breakup of this supercontinent, an event that began about 180 million years ago. From this work, the dates when individual crustal fragments separated from one another and their relative motions have been well established (**Figure 2.22**).

An important consequence of Pangaea's breakup was the creation of a "new" ocean basin: the Atlantic. As you can see in Figure 2.22, splitting of the supercontinent did not occur simultaneously along the margins of the Atlantic. The first split developed between North America and Africa. Here, the continental crust was highly fractured, providing pathways for huge quantities of fluid lavas to reach the surface. Today, these lavas are represented by weathered igneous rocks found along the eastern

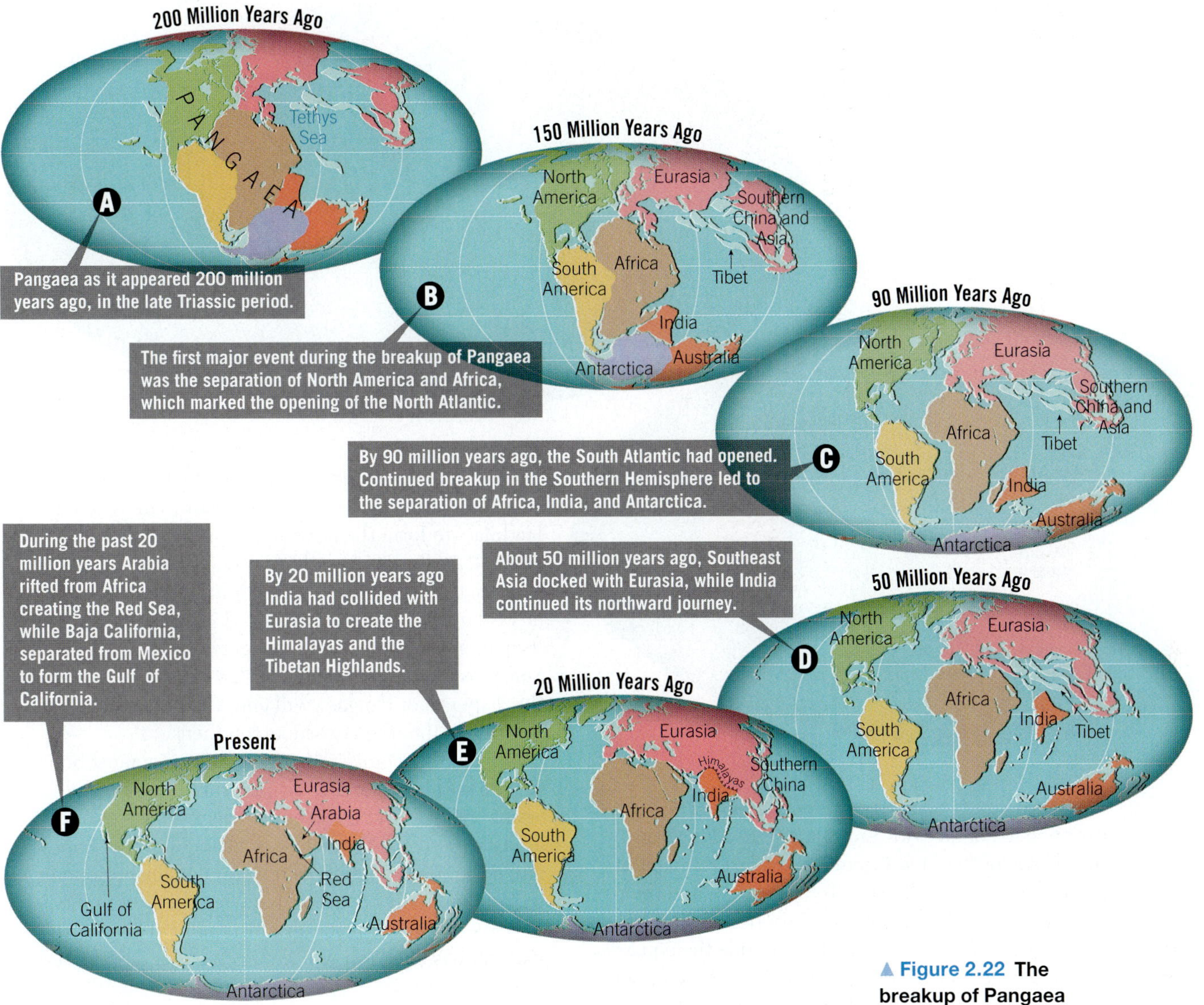
200 Million Years Ago
PANGAEA
Tethys Sea
A
Pangaea as it appeared 200 million years ago, in the late Triassic period.
150 Million Years Ago
North America
Eurasia
Southern China and Asia
South America
Africa
Tibet
India
Antarctica
Australia
B
The first major event during the breakup of Pangaea was the separation of North America and Africa, which marked the opening of the North Atlantic.
90 Million Years Ago
North America
Eurasia
Southern China and Asia
Africa
Tibet
South America
India
Australia
Antarctica
C
By 90 million years ago, the South Atlantic had opened. Continued breakup in the Southern Hemisphere led to the separation of Africa, India, and Antarctica.
During the past 20 million years Arabia rifted from Africa creating the Red Sea, while Baja California, separated from Mexico to form the Gulf of California.
By 20 million years ago India had collided with Eurasia to create the Himalayas and the Tibetan Highlands.
About 50 million years ago, Southeast Asia docked with Eurasia, while India continued its northward journey.
50 Million Years Ago
North America
Eurasia
D
Africa
India
Tibet
South America
Australia
Antarctica
20 Million Years Ago
E
North America
Eurasia
Himalayas
Southern China
India
Africa
South America
Australia
Antarctica
Present
F
North America
Eurasia
Arabia
India
Africa
Red Sea
South America
Gulf of California
Australia
Antarctica

▲ **Figure 2.22 The breakup of Pangaea**

seaboard of the United States—primarily buried beneath the sedimentary rocks that form the continental shelf. Radiometric dating of these solidified lavas indicates that rifting began between 200 million and 190 million years ago. This time span represents the "birth date" for this section of the North Atlantic.

By 130 million years ago, the South Atlantic began to open near the tip of what is now South Africa. As this zone of rifting migrated northward, it gradually opened the South Atlantic (**Figures 2.22B,C**). Continued breakup of the southern landmass led to the separation of Africa and Antarctica and sent India on a northward journey. By the early Cenozoic era, about 50 million years ago, Australia had separated from Antarctica, and the South Atlantic had become a full-fledged ocean (**Figure 2.22D**).

India eventually collided with Asia (**Figure 2.22E**), an event that began about 50 million years ago and created the Himalayas and the Tibetan Highlands. About the same time, the separation of Greenland from Eurasia completed the breakup of the northern landmass. During the past 20 million years or so of Earth's history, Arabia has rifted from Africa to form the Red Sea, and Baja California has separated from Mexico to form the Gulf of California (**Figure 2.22F**). Meanwhile, the Panama Arc joined North America and South America to produce our globe's familiar modern appearance.

Plate Tectonics in the Future

Geologists have extrapolated present-day plate movements into the future. **Figure 2.23** illustrates where Earth's landmasses may be 50 million years from now if present plate movements persist during this time span.

In North America we see that the Baja Peninsula and the portion of southern California that lies west of the San Andreas Fault will have slid past the North American plate. If this northward migration continues, Los Angeles and San Francisco will pass each other in about 10 million years, and in about 60 million years the Baja Peninsula will begin to collide with the Aleutian Islands.

Did You Know?
Researchers estimate that continents join to form supercontinents roughly every 500 million years. Since it has been about 200 million years since Pangaea broke up, we have only 300 million years to wait before the next supercontinent is completed.

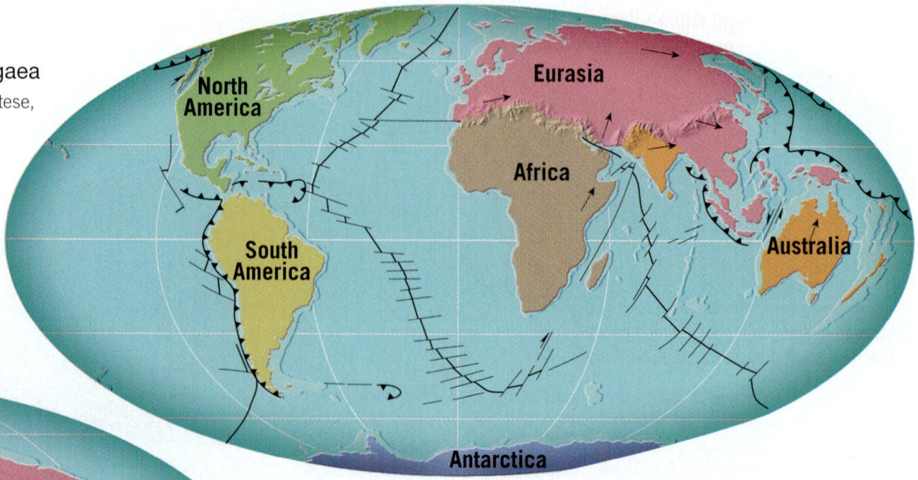

▶ Figure 2.23 **The world as it may look 50 million years from now** This reconstruction is highly idealized and based on the assumption that the processes that caused the breakup of Pangaea will continue to operate. (Based on Robert S. Dietz, John C. Holden, C. Scotese, and others)

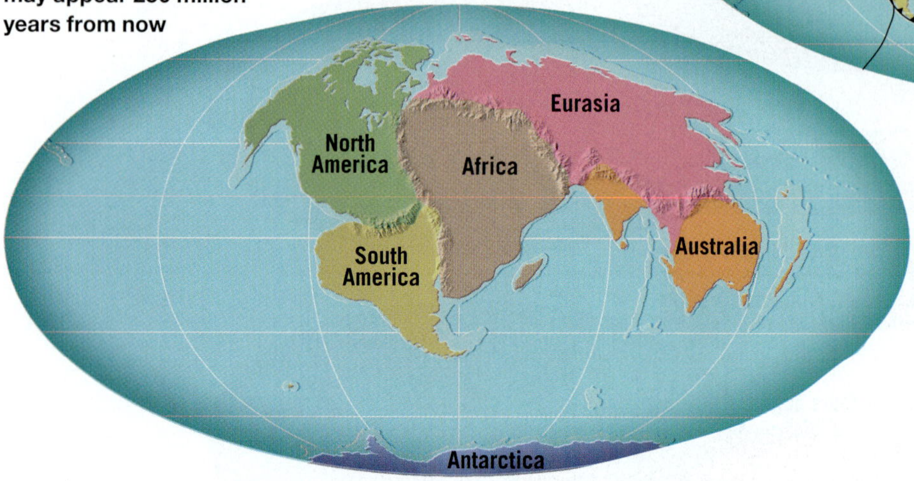

▼ Figure 2.24 **Earth as it may appear 250 million years from now**

Did You Know?
When all the continents were joined to form Pangaea, the rest of Earth's surface was covered with a huge ocean called *Panthalassa* (*pan* = all, *thalassa* = sea). Its modern descendant is the Pacific Ocean, which has been decreasing in size since the breakup of Pangaea.

If Africa maintains its northward path, it will continue to collide with Eurasia. The result will be the closing of the Mediterranean, the last remnant of a once-vast ocean called the Tethys Ocean, and the initiation of another major mountain-building episode (see Figure 2.23). Australia will be astride the equator and, along with New Guinea, will be on a collision course with Asia. Meanwhile, North and South America will begin to separate, while the Atlantic and Indian Oceans will continue to grow, at the expense of the Pacific Ocean.

A few geologists have even speculated on the nature of the globe 250 million years in the future. In this scenario the Atlantic seafloor will eventually become old and dense enough to form subduction zones around much of its margins, not unlike the present-day Pacific basin. Continued subduction of the Atlantic Ocean floor will result in the closing of the Atlantic basin and the collision of the Americas with the Eurasian–African landmass to form the next supercontinent, shown in Figure 2.24. Support for the possible closing of the Atlantic comes from evidence for a similar event, when an ocean predating the Atlantic closed during Pangaea's formation. Australia is also projected to collide with Southeast Asia by that time. If this scenario is accurate, the dispersal of Pangaea will end when the continents reorganize into the next supercontinent.

Such projections, although interesting, must be viewed with considerable skepticism because many assumptions must be correct for these events to unfold as just described. Nevertheless, changes in the shapes and positions of continents that are equally profound will undoubtedly occur for many hundreds of millions of years to come. Only after much more of Earth's internal heat has been lost will the engine that drives plate motions cease.

CONCEPT CHECKS 2.7

1. Name two plates that are growing in size. Name a plate that is shrinking in size.
2. What new ocean basin was created by the breakup of Pangaea?
3. Briefly describe changes in the positions of the continents if we assume that the plate motions we see today continue 50 million years into the future.

2.8 Testing the Plate Tectonics Model

List and explain the evidence used to support the plate tectonics theory.

Some of the evidence supporting continental drift was presented earlier in this chapter. With the development of plate tectonics theory, researchers began testing this new model of how Earth works. In addition to new supporting data, new interpretations of already existing data often swayed the tide of opinion.

Evidence: Ocean Drilling

Some of the most convincing evidence for seafloor spreading came from the Deep Sea Drilling Project, which operated from 1966 until 1983. One of the early goals of the project was to gather samples of the ocean floor in order to establish its age. To accomplish this,

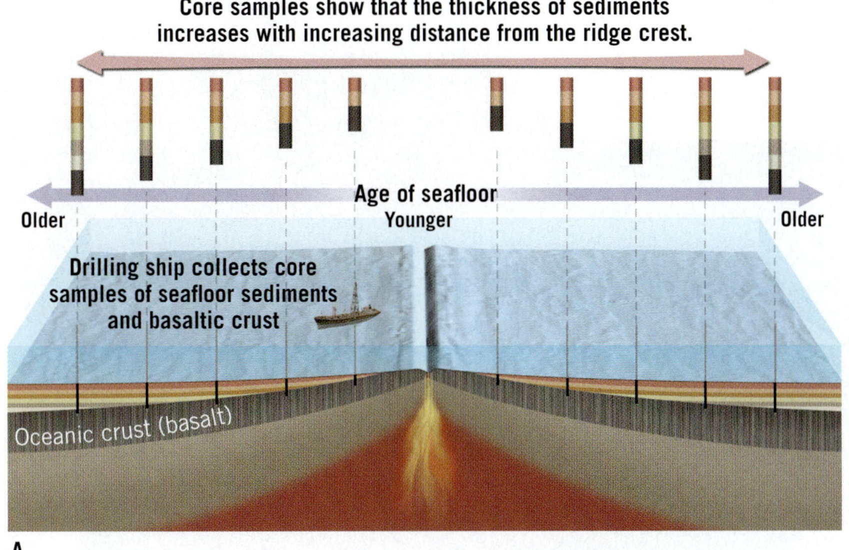

Chikyu is a state-of-the-art drilling ship designed to drill up to 7000 meters (more than 4 miles) below the seafloor.
B.

Figure 2.25 Deep-sea drilling A. Data collected through deep-sea drilling have shown that the ocean floor is indeed youngest at the ridge axis. **B.** The Japanese deep-sea drilling ship *Chikyu* became operational in 2007. (Photo by AP Photo/Itsuo Inouye)

the *Glomar Challenger*, a drilling ship capable of working in water thousands of meters deep, was built. Hundreds of holes were drilled through the layers of sediments that blanket the oceanic crust, as well as into the basaltic rocks below. Rather than use radiometric dating, which can be unreliable on oceanic rocks because of the alteration of basalt by seawater, researchers dated the seafloor by examining the fossil remains of microorganisms found in the sediments resting directly on the crust at each site.

When researchers recorded the age of the sediment from each drill site and its distance from the ridge crest, they found that the sediments increased in age with increasing distance from the ridge. This finding supported the seafloor-spreading hypothesis, which predicted that the youngest oceanic crust would be found at the ridge crest—the site of seafloor production—and the oldest oceanic crust would be located adjacent to the continents.

The distribution and thickness of ocean-floor sediments provided additional verification of seafloor spreading. Drill cores from the *Glomar Challenger* revealed that sediments are almost entirely absent on the ridge crest and that sediment thickness increases with increasing distance from the ridge (**Figure 2.25A**). This pattern of sediment distribution should be expected if the seafloor-spreading hypothesis is correct.

The data collected by the Deep Sea Drilling Project also reinforced the idea that the ocean basins are geologically young because no seafloor older than 180 million years was found. By comparison, most continental crust exceeds several hundred million years in age, and some samples are more than 4 billion years old.

In 1983, a new ocean-drilling program was launched by the Joint Oceanographic Institutions for Deep Earth Sampling (JOIDES). Now the International Ocean Discovery Program (IODP), this ongoing international effort uses multiple vessels for exploration, including the massive 210-meter-long (nearly 690-foot-long) *Chikyu* ("planet Earth" in Japanese), which began operations in 2007 (**Figure 2.25B**). One of the goals of the IODP is to recover a complete section of the oceanic crust, from top to bottom.

Evidence: Mantle Plumes and Hot Spots

Mapping volcanic islands and *seamounts* (submarine volcanoes) in the Pacific Ocean revealed several linear chains of volcanic structures. One of the most-studied chains consists of at least 129 volcanoes that extend from the Hawaiian Islands to Midway Island and continue northwestward toward the Aleutian trench (**Figure 2.26**). Radiometric dating of this linear structure, called the Hawaiian Island–Emperor Seamount chain, showed that the volcanoes increase in age with increasing distance from the Big Island of Hawaii. The youngest volcanic island in the chain (Hawaii) rose from the ocean floor less than 1 million years ago, whereas Midway Island is 27 million years old, and Detroit Seamount, near the Aleutian trench, is about 80 million years old (see Figure 2.26).

One widely accepted hypothesis* proposes that a roughly cylindrically shaped upwelling of hot rock, called a **mantle plume**, is located beneath the island of Hawaii. As the hot, rocky plume ascends through the mantle, the confining pressure drops, which triggers partial melting. (This process, called *decompression melting*, is discussed in Chapter 4.) The surface manifestation of this activity is a **hot spot**, an area of volcanism, high heat flow, and crustal uplifting that is a few hundred kilometers across. As the Pacific plate moved over a hot spot, a chain of volcanic structures known as a **hot-spot track** was built. As shown in Figure 2.26, the age of each volcano indicates how much time has elapsed since it was situated over

*Recall from Section 1.3 that a *hypothesis* is a tentative scientific explanation for a given set of observations. Although widely accepted, the validity of the plume hypothesis, unlike the theory of plate tectonics, remains unresolved.

Did You Know?
Olympus Mons is a huge volcano on Mars that strongly resembles the Hawaiian volcanoes. Rising 25 km (16 mi) above the surrounding plains, Olympus Mons owes its massive size to the *lack* of plate tectonics on Mars. Instead of being carried away from the hot spot by plate motion, like the Hawaiian volcanoes, Olympus Mons remained fixed and grew to a gigantic size.

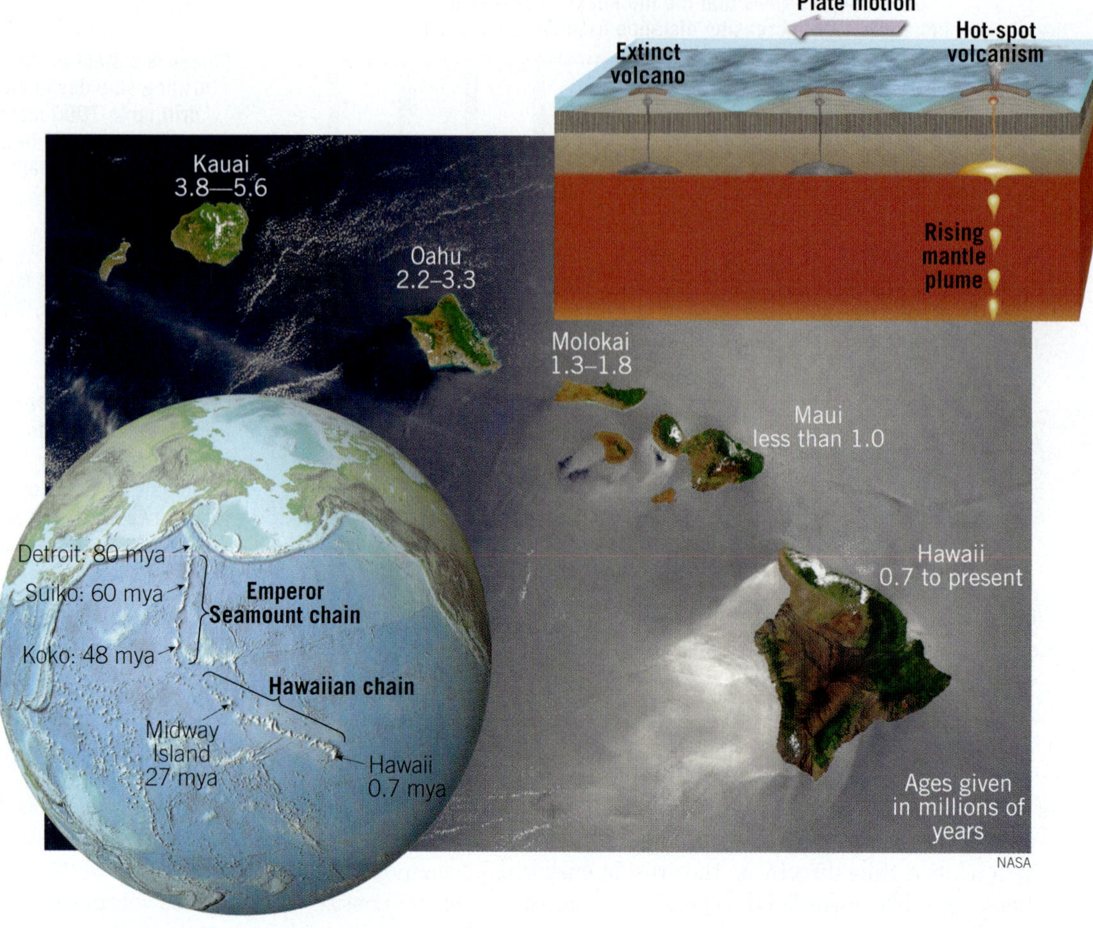

Figure 2.26 Hot-spot volcanism and the formation of the Hawaiian chain Radiometric dating of the Hawaiian Islands shows that volcanic activity increases in age moving away from the Big Island of Hawaii.

the mantle plume. Of approximately 40 hot spots that are thought to have formed because of upwelling of hot mantle plumes, most, but not all, have hot-spot tracks.

A closer look at the five largest Hawaiian Islands reveals a similar pattern of ages, from the volcanically active island of Hawaii to the inactive volcanoes that make up the oldest island, Kauai (see Figure 2.26). Five million years ago, when Kauai was positioned over the hot spot, it was the *only* modern Hawaiian island in existence. Kauai's age is evident in the island's extinct volcanoes, which have been eroded into jagged peaks and vast canyons. By contrast, the relatively young island of Hawaii exhibits many fresh lava flows, and one of its five major volcanoes, Kilauea, remains active today.

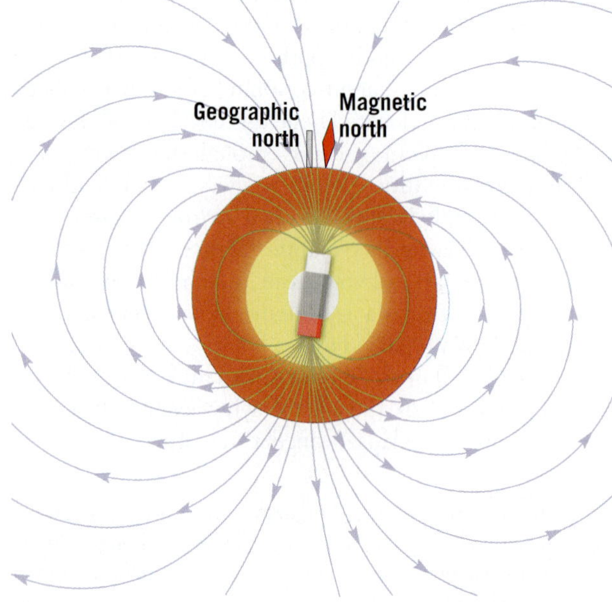

Figure 2.27 Earth's magnetic field Earth's magnetic field consists of lines of force much like those a giant bar magnet would produce if placed at the center of Earth.

Evidence: Paleomagnetism

You are probably familiar with how a compass operates and know that Earth's magnetic field has north and south magnetic poles. Today these magnetic poles roughly align with the geographic poles that are located where Earth's rotational axis intersects the surface. Earth's magnetic field is somewhat similar to that produced by a simple bar magnet. Invisible lines of force pass through the planet and extend from one magnetic pole to the other (**Figure 2.27**). A compass needle, itself a small magnet free to rotate on an axis, becomes aligned with the magnetic lines of force and points to the magnetic poles.

Earth's magnetic field is less obvious to us than the pull of gravity because we cannot feel it. Movement of a compass needle, however, confirms its presence. In addition, some naturally occurring minerals are magnetic and are influenced by Earth's magnetic field. One of the most common is the iron-rich mineral *magnetite*, which is abundant in lava flows of basaltic composition.* Basaltic lavas erupt at the surface at temperatures greater than 1000°C (1800°F), exceeding a threshold temperature for magnetism known as the **Curie point** (about 585°C [1085°F]). The magnetite grains in molten lava are nonmagnetic, but as the lava cools, these iron-rich grains become magnetized and align themselves in the direction of the existing magnetic lines of force. Once the minerals solidify, the magnetism they possess usually remains "frozen" in this position. Thus, they act like a compass needle because they "point" toward the position of the magnetic poles at the time of their formation. Rocks that formed thousands or millions of years ago and contain a "record" of the direction of the magnetic poles at the time of their formation are said to possess **paleomagnetism**, or **fossil magnetism**.

Apparent Polar Wandering A study of paleomagnetism in ancient lava flows throughout Europe led to an interesting discovery. Taken at face value, the magnetic alignment of iron-rich minerals in lava flows of different ages would indicate that the position of the paleomagnetic poles had changed through time. A plot of the location of the magnetic north pole, as measured from Europe, seemed to indicate that during the past 500 million years, the pole had gradually "wandered" from a location near Hawaii northeastward to its present location over the Arctic Ocean (**Figure 2.28**). This was strong evidence that either the magnetic north pole had migrated, an idea known as *polar wandering*, or that the poles had remained in place and the continents had drifted beneath them—in other words, Europe had drifted relative to the magnetic north pole.

Although the magnetic poles are known to move in a somewhat erratic path, studies of paleomagnetism from numerous locations show that the positions of the magnetic poles, averaged over thousands of years, correspond closely to the positions of the geographic poles. Therefore, a more acceptable explanation for the apparent polar wandering was provided by Wegener's hypothesis: If the magnetic poles remain stationary, their *apparent movement* is produced by the drift of the seemingly fixed continents.

Further evidence for continental drift came a few years later, when a polar-wandering path was constructed for North America (see **Figure 2.28A**). For the first 300 million years or so, the paths for North America and Europe were found to be similar in direction—but separated by about 5000 kilometers (3000 miles). Then, during the middle of the Mesozoic era (180 million years ago), they began to converge on the present North Pole. The explanation for these curves is that North America and Europe were joined until the Mesozoic, when the Atlantic began to open. From this time forward, these continents continuously moved apart. When North America and Europe are moved back to their pre-drift positions, as shown in **Figure 2.28B**, these paths of apparent polar wandering coincide. This is evidence that North America and Europe were once joined and moved relative to the poles as part of the same continent.

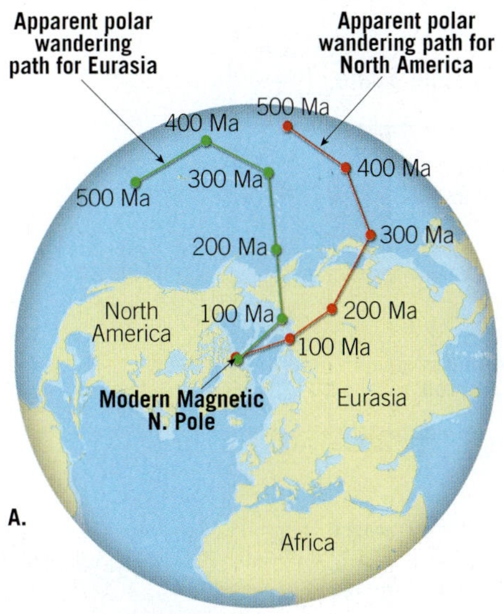

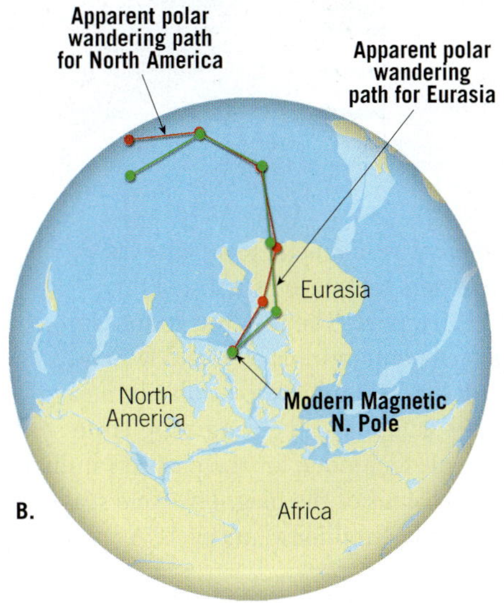

Figure 2.28 Apparent polar-wandering path **A.** Scientists believe that the more westerly path determined from North American data was caused by the westward drift of North America by about 24 degrees from Eurasia. **B.** The positions of the wandering paths when the landmasses are reassembled in their pre-drift locations.

Magnetic Reversals and Seafloor Spreading More evidence emerged when geophysicists learned that over periods of hundreds of thousands of years, Earth's magnetic field periodically reverses polarity. During a **magnetic reversal**, the magnetic north pole becomes the magnetic south pole and vice versa. Lava that solidified during a period of reverse polarity is magnetized with the polarity opposite that of volcanic rocks being formed today. When rocks exhibit the same magnetism as the present magnetic field, they are said to possess **normal polarity**, whereas rocks exhibiting the opposite magnetism are said to have **reverse polarity**.

*Some sediments and sedimentary rocks also contain enough iron-bearing mineral grains to acquire a measurable amount of magnetization.

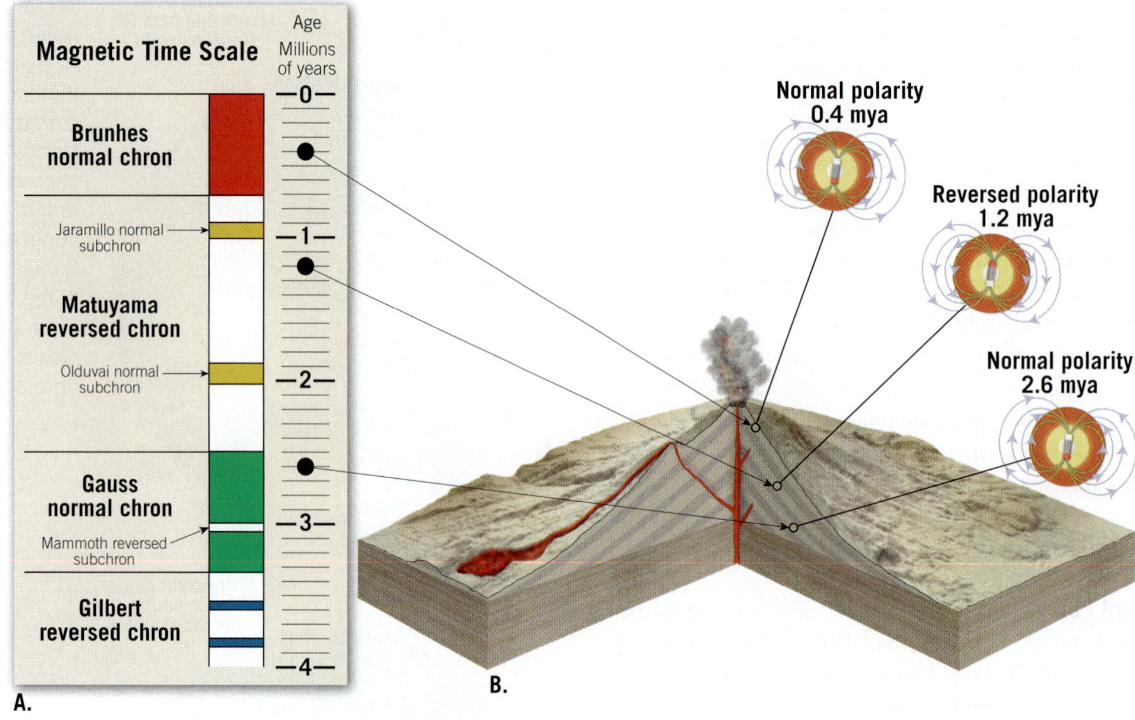

▶ SmartFigure 2.29 **Time scale of magnetic reversals A.** Time scale of Earth's magnetic reversals for the past 4 million years. **B.** This time scale was developed by establishing the magnetic polarity for lava flows of known age. (Data from Allen Cox and G. B. Dalrymple)

TUTORIAL

https://goo.gl/KNenyq

Once the concept of magnetic reversals was confirmed, researchers set out to establish a time scale for these occurrences. The task was to measure the magnetic polarity of hundreds of lava flows and use radiometric dating techniques to establish the age of each flow. Figure 2.29 shows the **magnetic time scale** established using this technique for the past few million years. The major divisions of the magnetic time scale, *chrons*, last for roughly 1 million years each. As more measurements became available, researchers realized that several short-lived reversals (less than 200,000 years long) often occurred during a single chron.

Meanwhile, oceanographers had begun magnetic surveys of the ocean floor in conjunction with their efforts to construct detailed maps of seafloor topography. These magnetic surveys were accomplished by towing very sensitive instruments, called **magnetometers**, behind research vessels (Figure 2.30A). The goal

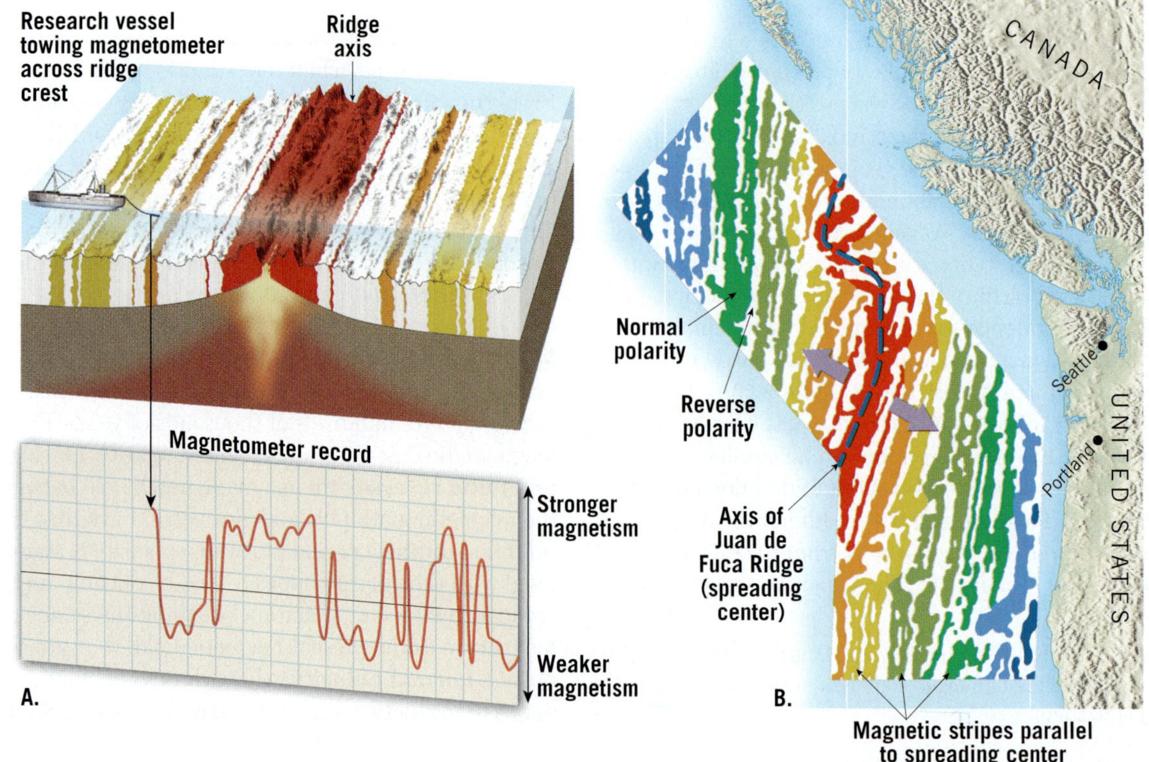

▶ Figure 2.30 **Ocean floor as a magnetic recorder A.** Magnetic intensities are recorded when a magnetometer is towed across a segment of the oceanic floor. **B.** Notice the symmetrical stripes of low- and high-intensity magnetism that parallel the axis of the Juan de Fuca Ridge. The colored stripes of high-intensity magnetism occur where normally magnetized oceanic rocks enhance the existing magnetic field. Conversely, the white low-intensity stripes are regions where the crust is polarized in the reverse direction, which weakens the existing magnetic field.

of these geophysical surveys was to map variations in the strength of Earth's magnetic field that arise from differences in the magnetic properties of the underlying crustal rocks.

The first comprehensive study of this type was performed off the Pacific coast of North America and had an unexpected outcome. Researchers discovered alternating stripes of high- and low-intensity magnetism, as shown in **Figure 2.30B**. This relatively simple pattern of magnetic variation defied explanation until 1963, when Fred Vine and D. H. Matthews demonstrated that the high- and low-intensity stripes supported the concept of seafloor spreading. Vine and Matthews suggested that the stripes of high-intensity magnetism are regions where the paleomagnetism of the oceanic crust exhibits normal polarity (see Figure 2.29A). Consequently, these rocks *enhance* (reinforce) Earth's magnetic field. Conversely, the low-intensity stripes are regions where the oceanic crust is polarized in the reverse direction and therefore *weaken* the existing magnetic field. But how do parallel stripes of normally and reversely magnetized rock become distributed across the ocean floor?

Vine and Matthews reasoned that as magma solidifies at the crest of an oceanic ridge, it is magnetized with the polarity of Earth's magnetic field at that time (**Figure 2.31**). Because of seafloor spreading, this strip of magnetized crust would gradually increase in width. When Earth's magnetic field reverses polarity, any newly formed seafloor having the opposite polarity would form in the middle of the old strip. Gradually, the two halves of the old strip would be carried in opposite directions, away from the ridge crest. Subsequent reversals would build a pattern of normal and reverse magnetic stripes, as shown in Figure 2.31. Because new rock is added in equal amounts to both trailing edges of the spreading ocean floor, we should expect the pattern of stripes (width and polarity) found on one side of an oceanic ridge to be a mirror image of those on the other side. In fact, a survey across the Mid-Atlantic Ridge just south of Iceland reveals a pattern of magnetic stripes exhibiting a remarkable degree of symmetry in relation to the ridge axis.

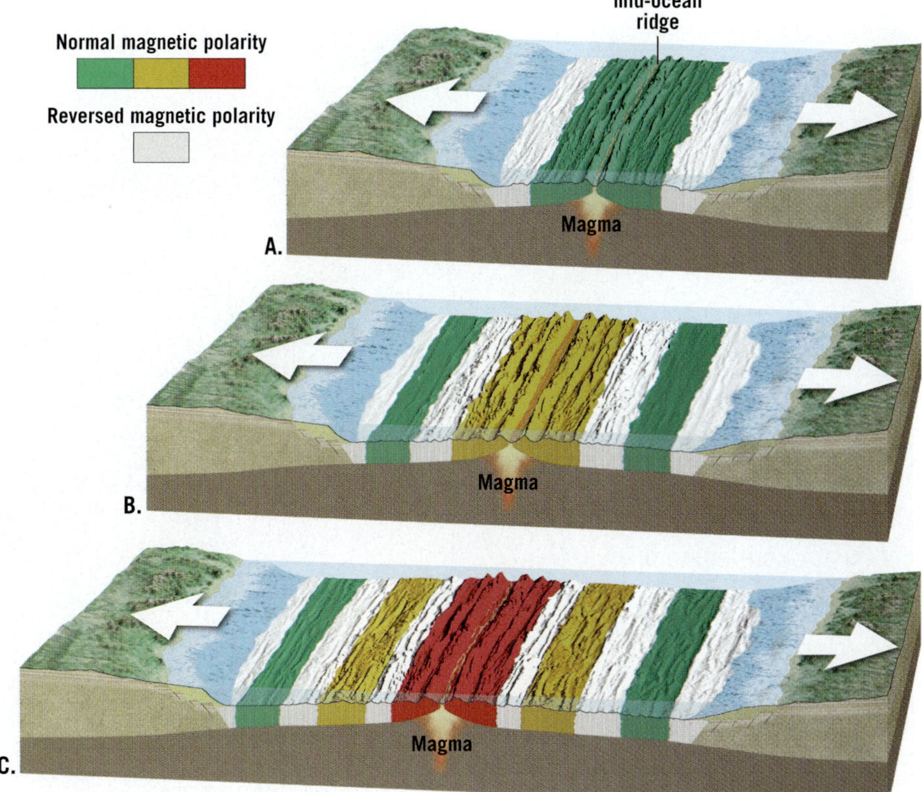

▲ **SmartFigure 2.31**
Magnetic reversals and seafloor spreading
When new basaltic rocks form at mid-ocean ridges, they magnetize according to Earth's existing magnetic field. Hence, oceanic crust provides a permanent record of each reversal of our planet's magnetic field over the past 200 million years.

ANIMATION
https://goo.gl/35qNFg

CONCEPT CHECKS 2.8

1. What is the age of the oldest sediments recovered using deep-ocean drilling? How do the ages of these sediments compare to the ages of the oldest continental rocks?
2. How do sedimentary cores from the ocean floor support the concept of seafloor spreading?
3. Assuming that hot spots remain fixed, in what direction was the Pacific plate moving while the Hawaiian Islands were forming?
4. Describe how Fred Vine and D. H. Matthews related the seafloor-spreading hypothesis to magnetic reversals.

2.9 How Is Plate Motion Measured?

Describe two methods researchers use to measure relative plate motion.

A number of methods are used to establish the direction and rate of plate motion. Some of these techniques not only confirm that lithospheric plates move but allow us to trace those movements back in geologic time.

Geologic Measurement of Plate Motion

Using ocean-drilling ships, researchers have obtained dates for hundreds of locations on the ocean floor. By knowing the age of a rock sample and its distance from the ridge axis where it was generated, an average rate of plate motion can be calculated.

Scientists used these data, combined with their knowledge of paleomagnetism stored in hardened lavas on the ocean floor and seafloor topography, to create maps that show the age of the ocean floor. The reddish-orange bands shown in **Figure 2.32** range in age from the present to about 30 million years ago. The width of the bands indicates how much crust formed during that time period. For example, the reddish-orange band

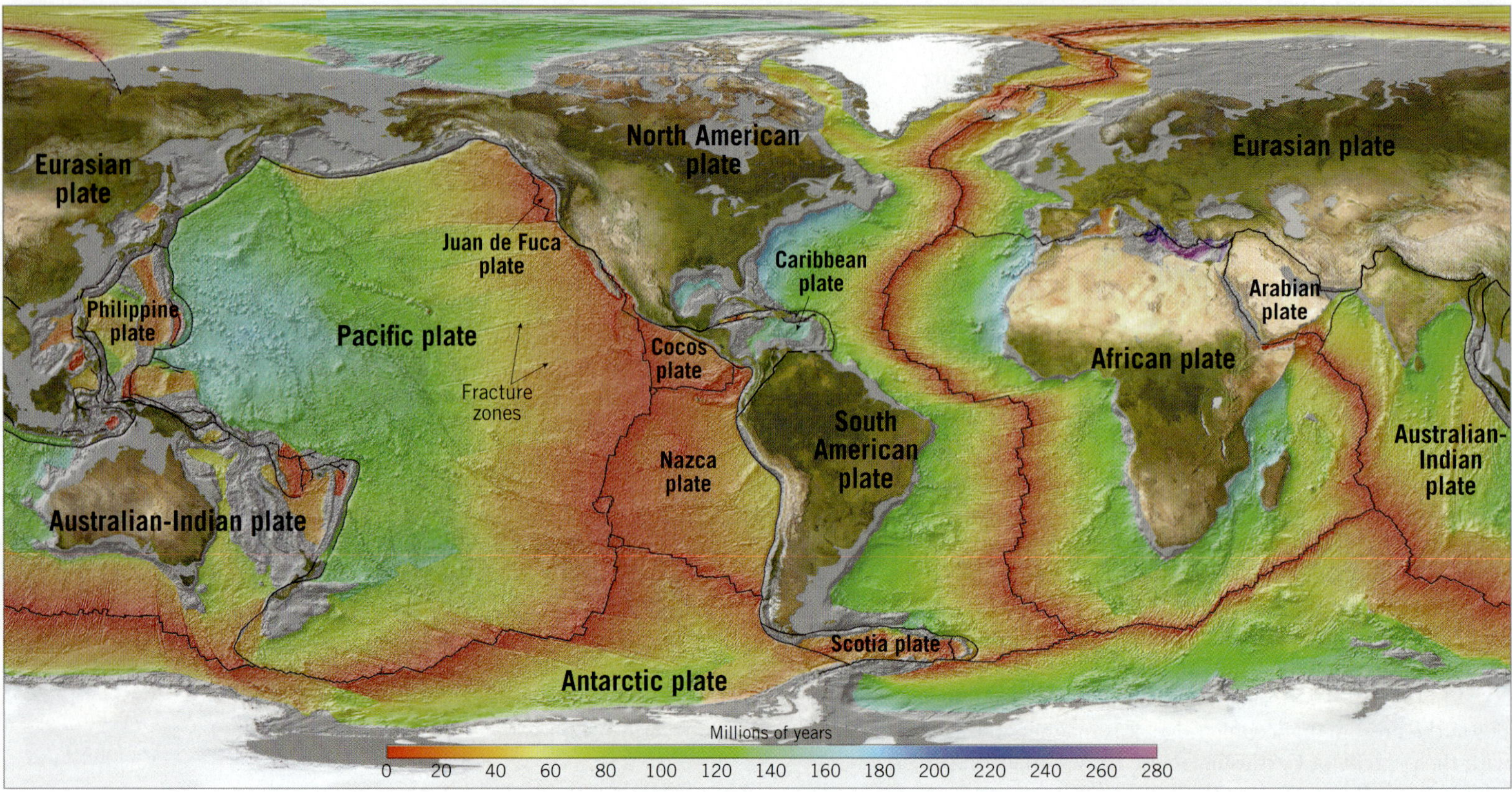

▲ **Figure 2.32 Age of the ocean floor**

along the East Pacific Rise is more than three times wider than the same-color band along the Mid-Atlantic Ridge. Therefore, the rate of seafloor spreading has been approximately three times faster in the Pacific basin than in the Atlantic.

Maps of this type also provide clues to the current direction of plate movement. Notice the offsets in the ridges; these are transform faults that connect the spreading centers. Recall that transform faults are aligned parallel to the direction of spreading. Careful measurement of transform faults reveals the direction of plate movement.

To establish the direction of plate motion in the past, geologists can examine the long *fracture zones* that extend for hundreds or even thousands of kilometers from ridge crests. Fracture zones are inactive extensions of transform faults and are therefore a record of past directions of plate motion. Unfortunately, most of the ocean floor is less than 180 million years old, so to look deeper into the past, researchers must rely on paleomagnetic evidence provided by continental rocks.

Measuring Plate Motion from Space

You are likely familiar with the Global Positioning System (GPS) used to locate one's position in order to provide directions to some other location. The GPS employs satellites that send radio signals that are intercepted by GPS receivers located at Earth's surface. The exact position of a site is determined by simultaneously establishing the distance from the receiver to four or more satellites. Researchers use specially designed equipment to locate a point on Earth to within a few millimeters (about the diameter of a small pea). To establish plate motions, GPS data are collected at numerous sites repeatedly over a number of years.

Data obtained from GPS and other techniques are shown in **Figure 2.33**. Calculations show that Hawaii is moving in a northwesterly direction toward Japan at 8.3 centimeters per year. A location in Maryland is retreating from a location in England at a speed of 1.7 centimeters per year—a value close to the 2.0-centimeters-per-year spreading rate established from paleomagnetic evidence obtained for the North Atlantic. Techniques involving GPS devices have also been useful in confirming small-scale crustal movements, like those occurring along faults in regions known to be tectonically active (for example, the San Andreas Fault).

CONCEPT CHECKS 2.9

1. What do transform faults that connect spreading centers indicate about plate motion?
2. Refer to Figure 2.33 to determine which three plates appear to exhibit the highest rates of motion.

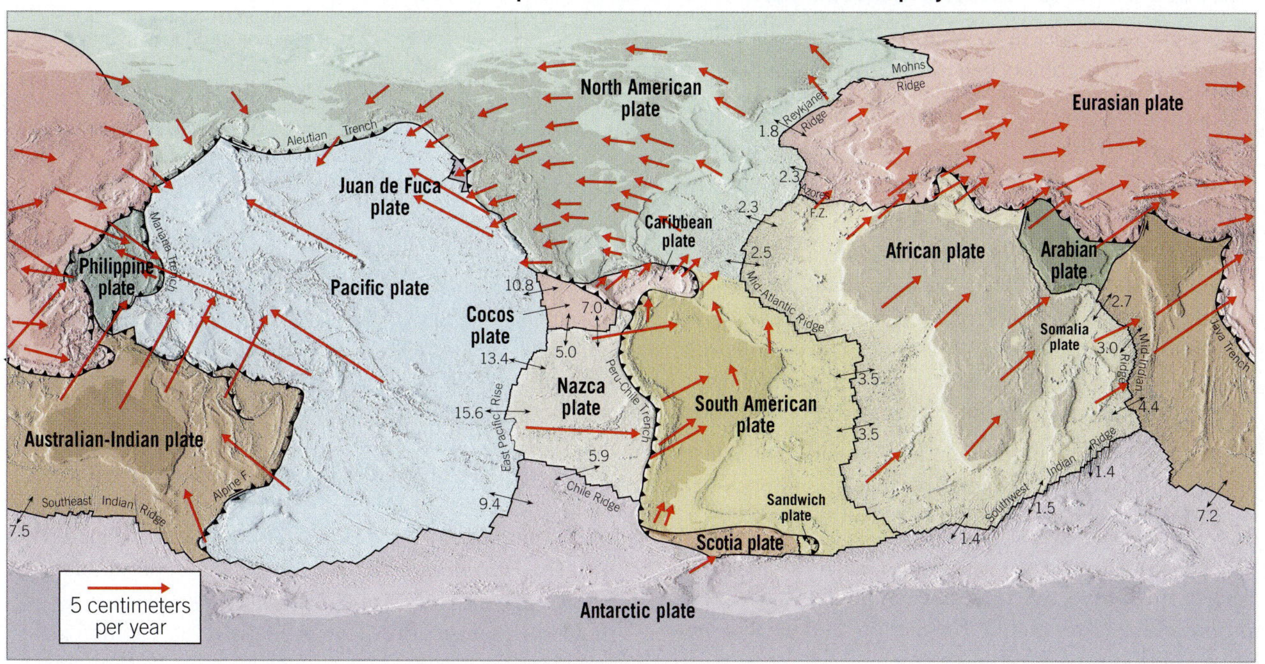

◀ **Figure 2.33 Rates of plate motion** The red arrows show plate motion at selected locations, based on GPS data. Longer arrows represent faster spreading rates. The small black arrows and labels show seafloor spreading velocities based mainly on paleomagnetic data. (Seafloor data from DeMets and others; GPS data from Jet Propulsion Laboratory)

2.10 What Drives Plate Motions?

Describe plate–mantle convection and explain two of the primary driving forces of plate motion.

Researchers are in general agreement that some type of *convection*—with hot mantle rocks rising and cold, dense oceanic lithosphere sinking—is the ultimate driver of plate tectonics. Many of the details of this convective flow, however, remain topics of debate in the scientific community.

Forces That Drive Plate Motion

Geophysical evidence confirms that although the mantle consists almost entirely of solid rock, it is hot and weak enough to exhibit a slow, fluid-like convective flow. The simplest type of **convection** is analogous to heating a pot of water on a stove (**Figure 2.34**). Heating the base of a pot warms the water, making it less dense (more buoyant) and causing it to rise in relatively thin sheets or blobs that spread out at the surface. As the surface layer cools, its density increases, and the cooler water sinks back to the bottom of the pot, where it is reheated until it achieves enough buoyancy to rise again. Mantle convection is similar to, but considerably more complex than, the model just described.

Geologists generally agree that subduction of cold, dense slabs of oceanic lithosphere is a major driving force of plate motion (**Figure 2.35**). This phenomenon, called **slab pull**, occurs because cold slabs of oceanic lithosphere are more dense than the underlying warm asthenosphere and hence "sink like a rock"—meaning that they are pulled down into the mantle by gravity.

Another important driving force is **ridge push** (see Figure 2.35). This gravity-driven mechanism results from the elevated position of the oceanic ridge, which causes slabs of lithosphere to "slide" down the flanks of the ridge. Despite its importance, ridge push contributes far less to plate motions than slab pull. The primary evidence for this is that the fastest-moving plates—the Pacific, Nazca, and Cocos plates—have extensive subduction zones along their margins. By contrast, the spreading rate in the North Atlantic basin, which is nearly devoid of subduction zones, is one of the lowest, at about 2.5 centimeters (1 inch) per year.

◀ **Figure 2.34 Convection in a cooking pot** As a stove warms the water in the bottom of a cooking pot, the heated water expands, becomes less dense (more buoyant), and rises. Simultaneously, the cooler, denser water near the top sinks.

Did You Know?
Because tectonic processes are powered by heat from Earth's interior, the forces that drive plate motion will cease sometime in the distant future. The work of external forces (such as wind, water, and ice), however, will continue to erode Earth's surface. Eventually, landmasses will be nearly flat. What a different world it will be—an Earth with no earthquakes, no volcanoes, and no mountains.

Models of Plate–Mantle Convection

Although *convection* in the mantle has yet to be fully understood, researchers generally agree on the following:

- Convective flow—in which warm, buoyant mantle rocks rise while cool, dense lithospheric plates sink—is the underlying driving force for plate movement.
- Mantle convection and plate tectonics are part of the same system. Subducting oceanic plates drive the cold downward-moving portion of convective flow, while shallow upwelling of hot rock along the oceanic ridge and buoyant mantle plumes are the upward-flowing arms of the convective mechanism.
- Convective flow in the mantle is a major mechanism for transporting heat away from Earth's interior to the surface, where it is eventually radiated into space.

What is not known with certainty is the exact structure of this convective flow. Several models have been proposed for plate–mantle convection, and we will look at two of them.

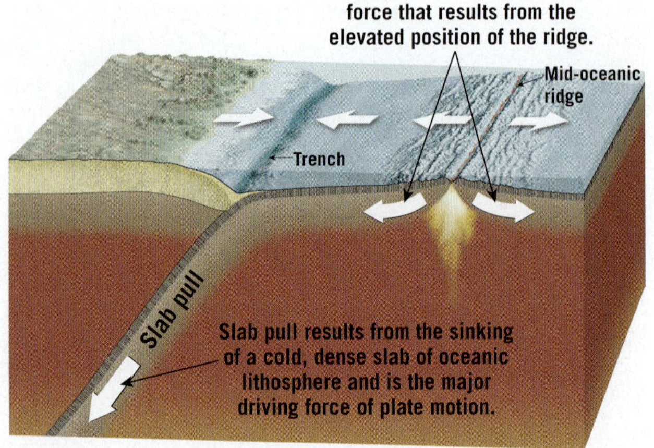

▲ **Figure 2.35 Forces that act on lithospheric plates**

Whole-Mantle Convection One group of researchers favor some type of *whole-mantle convection* model, also called the *plume model*, in which cold oceanic lithosphere sinks to great depths and stirs the entire mantle (**Figure 2.36A**). The whole-mantle model suggests that the ultimate burial ground for these subducting lithospheric slabs is the core–mantle boundary. The downward flow of these subducting slabs is balanced by buoyantly rising mantle plumes that transport hot mantle rock toward the surface.

Two kinds of plumes have been proposed—narrow tube-like plumes and giant upwellings, often referred to as *mega-plumes*. The long, narrow plumes are thought to originate from the core–mantle boundary and produce hot-spot volcanism of the type associated with the Hawaiian Islands, Iceland, and Yellowstone. Scientists believe that areas of large mega-plumes, as shown in Figure 2.36A, occur beneath the Pacific basin and southern Africa. These mega-plumes are thought to explain why southern Africa has an elevation much higher than would be predicted for a stable continental landmass. In the whole-mantle convection model, heat for both the narrow plumes and the mega-plumes is thought to arise mainly from Earth's core, while the deep mantle provides a source for chemically distinct magmas. However, some researchers have questioned that idea and instead propose that the source of magma for most hot-spot volcanism is found in the upper mantle (asthenosphere).

Layer Cake Model Some researchers argue that the mantle resembles a "layer cake" divided at a depth of perhaps 660 kilometers (410 miles) but no deeper than 1000 kilometers (620 miles). As shown in **Figure 2.36B**,

▼ **Figure 2.36 Models of mantle convection**

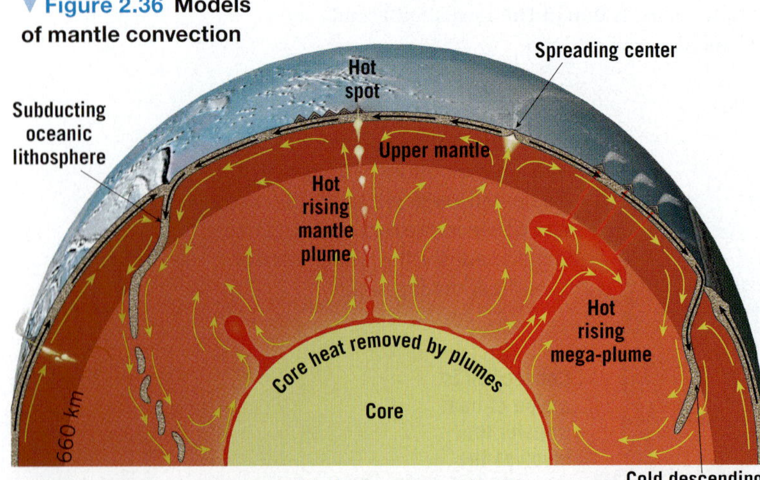

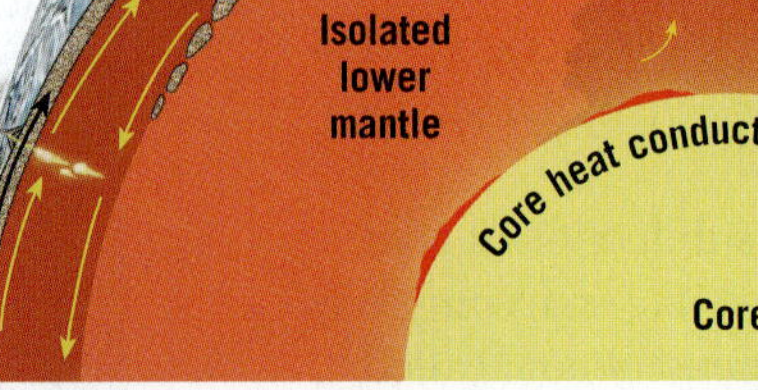

A. In the "whole-mantle model," sinking slabs of cold oceanic lithosphere are the downward limbs of convection cells, while rising mantle plumes carry hot material from the core–mantle boundary toward the surface.

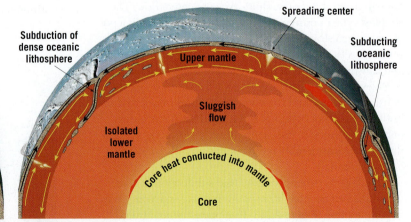

B. The "layer cake model" has two largely disconnected convective layers; a dynamic upper layer driven by descending slabs of cold oceanic lithosphere and a sluggish lower layer that carries heat upward without appreciably mixing with the layer above.

this layered model has two zones of convection—a thin, dynamic layer in the upper mantle and a thick, larger, sluggish one located below. As with the whole-mantle model, the downward convective flow is driven by the subduction of cold, dense oceanic lithosphere. However, rather than reach the lower mantle, these subducting slabs penetrate to depths of no more than 1000 kilometers (620 miles). Notice in Figure 2.36B that the upper layer in the layer cake model is littered with recycled oceanic lithosphere of various ages. Melting of these fragments is thought to be the source of magma for some of the volcanism that occurs away from plate boundaries, such as the hot-spot volcanism of Hawaii.

In contrast to the active upper mantle, the lower mantle is sluggish and does not provide material to support volcanism at the surface. Very slow convection within this layer likely carries heat upward, but very little mixing occurs between these two layers.

Geologists continue to debate the nature of the convective flow in the mantle. As they investigate the possibilities, perhaps a widely accepted hypothesis that combines features from the layer cake model and the whole-mantle convection model will emerge.

CONCEPT CHECKS 2.10

1. Define *slab pull* and *ridge push*. Which of these forces contributes more to plate motion?
2. Briefly describe the two models of plate–mantle convection.
3. What geologic processes are associated with the upward and downward circulation in the mantle?

CONCEPTS IN REVIEW

Plate Tectonics: A Scientific Revolution Unfolds

2.1 From Continental Drift to Plate Tectonics

Summarize the view that most geologists held prior to the 1960s regarding the geographic positions of the ocean basins and continents.

- Fifty years ago, most geologists thought that ocean basins were very old and that continents were fixed in place. Those ideas were discarded with a scientific revolution that revitalized geology: the theory of plate tectonics. Supported by multiple kinds of evidence, plate tectonics is the foundation of modern Earth science.

2.2 Continental Drift: An Idea Before Its Time

List and explain the evidence Wegener presented to support his continental drift hypothesis.

KEY TERMS: continental drift, supercontinent, Pangaea

- German meteorologist Alfred Wegener formulated the continental drift hypothesis in 1912. He suggested that Earth's continents are not fixed in place but move slowly over geologic time.
- Wegener proposed a supercontinent called Pangaea that existed about 200 million years ago, during the late Paleozoic and early Mesozoic eras.
- Wegener's evidence that Pangaea existed and later broke into pieces that drifted apart included (1) the shapes of the continents, (2) continental fossil organisms that matched across oceans, (3) matching rock types and modern mountain belts, and (4) sedimentary rocks that recorded ancient climates, including glaciers on the southern portion of Pangaea.
- Wegener's hypothesis suffered from two flaws: It proposed tidal forces as the mechanism for the motion of continents, and it implied that the continents would have plowed their way through weaker oceanic crust, like boats cutting through a thin layer of sea ice. Most geologists rejected the idea of continental drift when Wegener proposed it, and it wasn't resurrected for another 50 years.

(2.2 continued)

? Why did Wegener choose organisms such as *Glossopteris* and *Mesosaurus* as evidence for continental drift, as opposed to other fossil organisms such as sharks or jellyfish?

/AGE Fotostock

2.3 The Theory of Plate Tectonics

List the major differences between Earth's lithosphere and asthenosphere and explain the importance of each in the plate tectonics theory.

KEY TERMS: theory of plate tectonics, lithosphere, asthenosphere, lithospheric plate (plate)

- Research conducted after World War II led to new insights that helped revive Wegener's hypothesis of continental drift. Exploration of the seafloor uncovered previously unknown features, including an extremely long mid-ocean ridge system. Sampling of the oceanic crust revealed that it was quite young relative to the continents.
- The lithosphere, Earth's outermost rocky layer, is relatively stiff and deforms by bending or breaking. The lithosphere consists both of crust (either oceanic or continental) and underlying upper mantle. Beneath the lithosphere is the asthenosphere, a relatively weak layer that deforms by flowing.
- The lithosphere consists of about two dozen segments of irregular size and shape. There are seven large lithospheric plates, another seven intermediate-size plates, and numerous relatively small microplates. Plates meet along boundaries that may be divergent (moving apart from each other), convergent (moving toward each other), or transform (moving laterally past each other).

2.4 Divergent Plate Boundaries and Seafloor Spreading

Sketch and describe the movement along a divergent plate boundary that results in the formation of new oceanic lithosphere.

KEY TERMS: divergent plate boundary (spreading center), oceanic ridge system, rift valley, seafloor spreading, continental rift

- Seafloor spreading leads to the formation of new oceanic lithosphere at mid-ocean ridge systems. As two plates move apart from one another, tensional forces open cracks in the plates, allowing magma to well up and generate new slivers of seafloor. This process generates new oceanic lithosphere at a rate of 2 to 15 centimeters (1 to 6 inches) each year.
- As it ages, oceanic lithosphere cools and becomes denser. It therefore subsides as it is transported away from the mid-ocean ridge. At the same time, the underlying asthenosphere cools, adding new material to the underside of the plate, which consequently thickens.
- Divergent boundaries are not limited to the seafloor. Continents can break apart, too, starting with a continental rift (as in modern-day east Africa) and potentially producing a new ocean basin between the two sides of the rift.

2.5 Convergent Plate Boundaries and Subduction

Compare and contrast the three types of convergent plate boundaries and name a location where each type can be found.

KEY TERMS: convergent plate boundary (subduction zone), deep-ocean trench, partial melting, continental volcanic arc, volcanic island arc (island arc)

- When plates move toward one another, oceanic lithosphere is subducted into the mantle, where it is recycled. Subduction manifests itself on the ocean floor as a deep linear trench. The subducting slab of oceanic lithosphere can descend at a variety of angles, from nearly horizontal to nearly vertical.
- Aided by the presence of water, the subducted oceanic lithosphere triggers melting in the mantle, which produces magma. The magma is less dense than the surrounding rock and will rise. It may cool at depth, thickening the crust, or it may make it all the way to Earth's surface, where it erupts as a volcano.

(2.5 continued)

- A line of volcanoes that emerge through continental crust is termed a continental volcanic arc, while a line of volcanoes that emerge through an overriding plate of oceanic lithosphere is a volcanic island arc.
- Continental crust resists subduction due to its relatively low density, and so when an intervening ocean basin is completely destroyed through subduction, the continents on either side collide, generating a new mountain range.

? Sketch a typical continental volcanic arc and label the key parts. Then repeat the drawing with an overriding plate made of oceanic lithosphere.

2.6 Transform Plate Boundaries

Describe the relative motion along a transform fault boundary and locate several examples of transform faults on a plate boundary map.

KEY TERMS: transform plate boundary (transform fault), fracture zone

- At a transform boundary, lithospheric plates slide horizontally past one another. No new lithosphere is generated, and no old lithosphere is consumed. Shallow earthquakes signal the movement of these slabs of rock as they grind past their neighbors.
- The San Andreas Fault in California is an example of a transform boundary in continental crust, while the fracture zones between segments of the Mid-Atlantic Ridge are transform faults in oceanic crust.

? On the accompanying tectonic map of the Caribbean, find the Enriquillo Fault. (The location of the 2010 Haiti earthquake is shown as a yellow star.) What kind of plate boundary is shown here? Are there any other faults in the area that show the same type of motion?

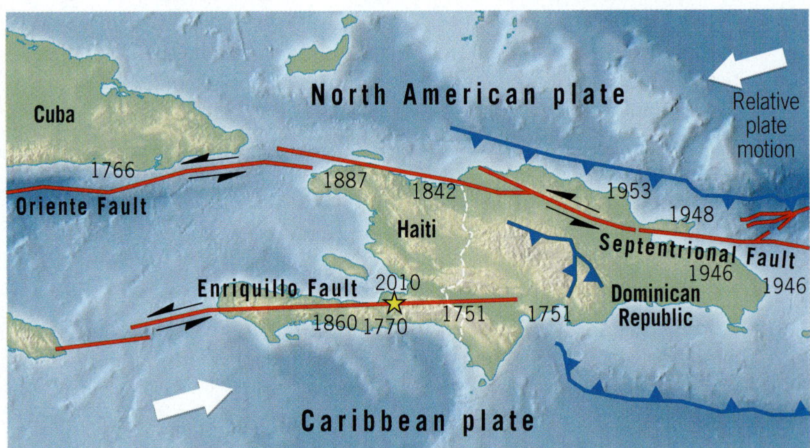

2.7 How Do Plates and Plate Boundaries Change?

Explain why plates such as the African and Antarctic plates are increasing in size, while the Pacific plate is decreasing in size.

- Although the total surface area of Earth does not change, the shapes and sizes of individual plates are constantly changing as a result of subduction and seafloor spreading. Plate boundaries can also be created or destroyed in response to changes in the forces acting on the lithosphere.
- The breakup of Pangaea and the collision of India with Eurasia are two examples of how plates change through geologic time.

2.8 Testing the Plate Tectonics Model

List and explain the evidence used to support the plate tectonics theory.

KEY TERMS: mantle plume, hot spot, hot-spot track, Curie point, paleomagnetism (fossil magnetism), magnetic reversal, normal polarity, reverse polarity, magnetic time scale, magnetometer

- Multiple lines of evidence have verified the plate tectonics model. For instance, the Deep Sea Drilling Project found that the age of the seafloor increases with distance from a mid-ocean ridge. The thickness of sediment atop this seafloor is also proportional to distance from the ridge: Older lithosphere has had more time to accumulate sediment.
- A hot spot is an area of volcanic activity where a mantle plume reaches Earth's surface. Volcanic rocks generated by hot-spot volcanism provide evidence of both the direction and rate of plate movement over time.
- Magnetic minerals such as magnetite align themselves with Earth's magnetic field as rock forms. These fossil magnets are records of the ancient orientation of Earth's magnetic field. This is useful to geologists in two ways: (1) It allows a given stack of rock layers to be interpreted in terms of their orientation relative to the magnetic poles through time, and (2) reversals in the orientation of the magnetic field are preserved as "stripes" of normal and reversed polarity in the oceanic crust. Magnetometers reveal this signature of seafloor spreading as a symmetrical pattern of magnetic stripes parallel to the axis of the mid-ocean ridge.

2.9 How Is Plate Motion Measured?

Describe two methods researchers use to measure relative plate motion.

- Data collected from the ocean floor has established the direction and rate of motion of lithospheric plates. Transform faults point in the direction the plate is moving. Establishing dates for seafloor rocks helps to calibrate the rate of motion.

(2.9 continued)

- GPS satellites can be used to accurately measure the motion of special receivers to within a few millimeters. These "real-time" data support the inferences made from seafloor observations. On average, plates move at about the same rate human fingernails grow: about 5 centimeters (2 inches) per year.

2.10 What Drives Plate Motions?

Describe plate–mantle convection and explain two of the primary driving forces of plate motion.

KEY TERMS: convection, slab pull, ridge push

- Some kind of convection (upward movement of less dense material and downward movement of more dense material) appears to drive the motion of plates.
- Slabs of oceanic lithosphere sink at subduction zones because the subducted slab is denser than the underlying asthenosphere. In this process, called slab pull, Earth's gravity tugs at the slab, drawing the rest of the plate toward the subduction zone. As oceanic lithosphere slides down the mid-ocean ridge, it exerts a small additional force, called ridge push.
- Convection may occur throughout the entire mantle, as suggested by the whole-mantle model. Or it may occur in two layers within the mantle—the active upper mantle and the sluggish lower mantle—as proposed in the layer cake model.

? Compare and contrast mantle convection with the operation of a lava lamp.

GIVE IT SOME THOUGHT

1 Refer to Section 1.3, titled "The Nature of Scientific Inquiry," to answer the following:
- **a.** What observations led Alfred Wegener to develop his continental drift hypothesis?
- **b.** Why did most of the scientific community reject the continental drift hypothesis?
- **c.** Do you think Wegener followed the basic principles of scientific inquiry? Support your answer.

2 Refer to the accompanying diagrams illustrating the three types of convergent plate boundaries and complete the following:
- **a.** Identify each type of convergent boundary.
- **b.** On what type of crust do volcanic island arcs develop?
- **c.** Why are volcanoes largely absent where two continental blocks collide?
- **d.** Describe two ways that oceanic–oceanic convergent boundaries are different from oceanic–continental boundaries. How are they similar?

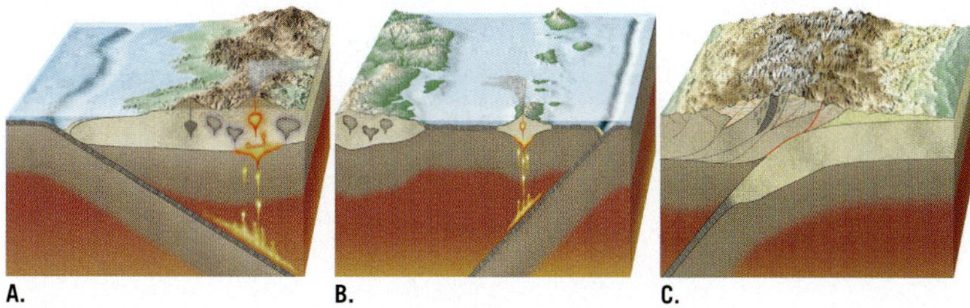

3 Some people predict that California will sink into the ocean. Is this idea consistent with the theory of plate tectonics? Explain.

4 Volcanic islands that form over mantle plumes, such as the Hawaiian chain, are home to some of Earth's largest volcanoes. However, several volcanoes on Mars are gigantic compared to any on Earth. What does this difference tell us about the role of plate motion in shaping the Martian surface?

5 Refer to the accompanying hypothetical plate map to answer the following questions:

a. How many portions of plates are shown?

b. Explain why active volcanoes are more likely to be found on continents A and B than on continent C.

c. Provide one scenario in which volcanic activity might be triggered on continent C.

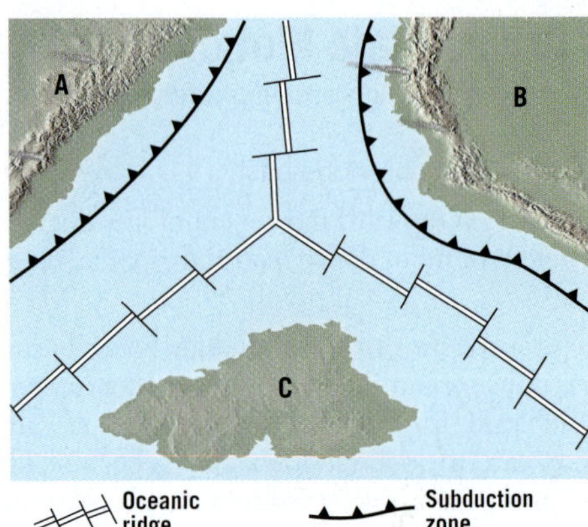

6 Australian marsupials (kangaroos, koalas, etc.) have direct fossil links to marsupial opossums found in the Americas. Yet the modern marsupials in Australia are markedly different from their American relatives. How does the breakup of Pangaea help to explain these differences? (*Hint:* See Figure 2.22.)

7 Density is a key component in the behavior of Earth materials and is essential to understanding important aspects of the plate tectonics model. Describe three different ways that density and/or density differences play a role in plate tectonics.

8 Explain how the processes that create hot-spot volcanic chains differ from the processes that generate volcanic island arcs.

9 Imagine that you are studying seafloor spreading along two different oceanic ridges. Using data from a magnetometer, you produced the two accompanying maps. From these maps, what can you determine about the relative rates of seafloor spreading along these two ridges? Explain.

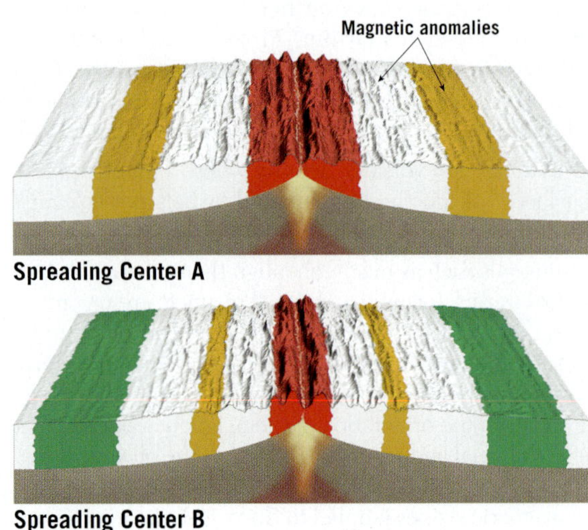

10 Refer to the accompanying plate motion map and these pairs of cities to complete the following:

(Boston, Denver), (London, Boston), (Honolulu, Beijing)

a. Which pair of cities is moving apart as a result of plate motion?

b. Which pair of cities is moving closer as a result of plate motion?

c. Which pair of cities is not presently moving relative to each other?

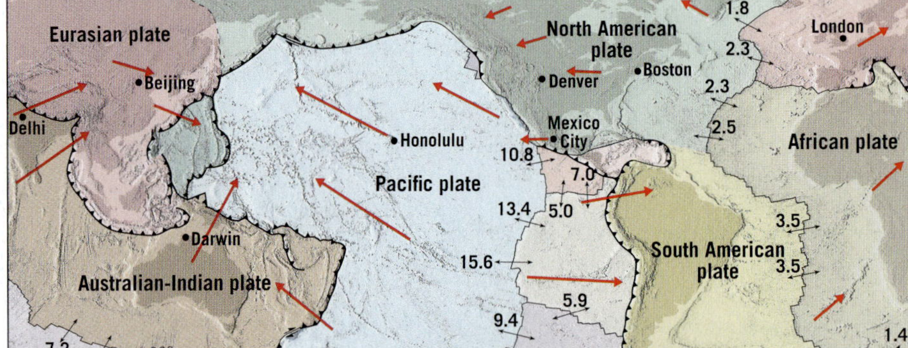

MasteringGeology™

Looking for additional review and test prep materials? Visit the Study Area in MasteringGeology to enhance your understanding of this chapter's content by accessing a variety of resources, including Self-Study Quizzes, Geoscience Animations, SmartFigures, Mobile Field Trips, *Project Condor* Quadcopter videos, *In the News* RSS feeds, flashcards, web links, and an optional Pearson eText.

www.masteringgeology.com

3

Matter & Minerals

FOCUS ON CONCEPTS

Each statement represents the primary learning objective for the corresponding major heading within the chapter. After you complete the chapter, you should be able to:

3.1 List the main characteristics that an Earth material must possess to be considered a mineral and describe each characteristic.

3.2 Compare and contrast the three primary particles contained in atoms.

3.3 Distinguish among ionic bonds, covalent bonds, and metallic bonds.

3.4 List and describe the properties used in mineral identification.

3.5 Explain how minerals are classified and name the most abundant mineral group in Earth's crust.

3.6 Sketch the silicon–oxygen tetrahedron and explain how this fundamental building block joins together to form various silicate structures.

3.7 Compare and contrast the light (nonferromagnesian) silicates with the dark (ferromagnesian) silicates and list three common minerals in each group.

3.8 List the common nonsilicate minerals and explain why each is important.

3.9 Discuss Earth's mineral resources in terms of renewability. Differentiate between mineral resources and ore deposits.

The Cave of Crystals, Chihuahua, Mexico, contains giant gypsum crystals, some of the largest natural crystals ever found. (Photo by Geographic Stock/Getty Images)

EARTH'S CRUST AND OCEANS are home to a wide variety of useful and essential minerals. Most people are familiar with the common uses of many basic metals, including aluminum in beverage cans, copper in electrical wiring, and gold and silver in jewelry. But some people are not aware that pencil "lead" contains the greasy-feeling mineral graphite and that bath powders and many cosmetics contain the mineral talc. Moreover, many do not know that dentists use drill bits impregnated with diamonds to drill through tooth enamel. In fact, practically every manufactured product contains materials obtained from minerals.

In addition to the economic uses of rocks and minerals, every geologic process in some way depends on the properties of these basic Earth materials. Events such as volcanic eruptions, mountain building, weathering and erosion, and even earthquakes involve rocks and minerals. Consequently, a basic knowledge of Earth materials is essential to understanding all geologic phenomena.

3.1 Minerals: Building Blocks of Rocks

List the main characteristics that an Earth material must possess to be considered a mineral and describe each characteristic.

We begin our discussion of Earth materials with an overview of **mineralogy** (*mineral* = mineral, *ology* = study of) because minerals are the building blocks of rocks. Humans have used minerals for both practical and decorative purposes for thousands of years. For example, the common mineral quartz is the source of silicon for computer chips. The first Earth materials mined were flint and chert, which humans fashioned into weapons and cutting tools. As early as 3700 B.C.E., Egyptians began mining gold, silver, and copper. By 2200 B.C.E. humans had discovered how to combine copper with tin to make bronze, a strong, hard alloy. Later, a process was developed to extract iron from minerals such as hematite—a discovery that marked the decline of the Bronze Age. During the Middle Ages, mining of a variety of minerals became common, and the impetus for the formal study of minerals was in place.

The term *mineral* is used in several different ways. For example, those concerned with health and fitness extol the benefits of vitamins and minerals. The mining industry typically uses the word *mineral* to refer to anything extracted from Earth, such as coal, iron ore, or sand and gravel. The guessing game *Twenty Questions* usually begins with the question *Is it animal, vegetable, or mineral?* What criteria do geologists use to determine whether something is a mineral (**Figure 3.1**)?

Defining a Mineral

Geologists define **mineral** as *any naturally occurring inorganic solid that possesses an orderly crystalline structure and a definite chemical composition that allows for some variation*. Thus, Earth materials that are classified as minerals exhibit the following characteristics:

1. **Naturally occurring.** Minerals form by natural geologic processes. Synthetic materials, meaning those produced in a laboratory or by human intervention, are not considered minerals.
2. **Generally inorganic.** Inorganic crystalline solids, such as ordinary table salt (halite), that are found naturally in the ground are considered minerals. (Organic compounds, which are chemical compounds of living things that contain carbon are generally not considered minerals. Sugar, a crystalline solid like

◀ **Figure 3.1 Quartz crystals** A collection of well-developed quartz crystals found near Hot Springs, Arkansas. (Photo by Jeff Scovil)

◀ **Figure 3.2 Arrangement of sodium and chloride ions in the mineral halite** The arrangement of atoms (ions) into basic building blocks that have a cubic shape results in regularly shaped cubic crystals. (Photo by Dennis Tasa)

salt but extracted from sugarcane or sugar beets, is a common example of such an organic compound.) Many marine animals secrete inorganic compounds, such as calcium carbonate (calcite), in the form of shells and coral reefs. If these materials are buried and become part of the rock record, geologists consider them minerals.

3. **Solid substance.** Only solid crystalline substances are considered minerals. Ice (frozen water) fits this criterion and is considered a mineral, whereas liquid water and water vapor do not.
4. **Orderly crystalline structure.** Minerals are crystalline substances, made up of atoms (or ions) that are arranged in an orderly, repetitive manner (Figure 3.2). This orderly packing of atoms is reflected in regularly shaped objects called *crystals*. Some naturally occurring solids, such as volcanic glass (obsidian), lack a repetitive atomic structure and are not considered minerals.
5. **Definite chemical composition that allows for some variation.** Most minerals are chemical compounds having compositions that can be expressed by a chemical formula. For example, the common mineral quartz has the formula SiO_2, which indicates that quartz consists of silicon (Si) and oxygen (O) atoms, in a 1:2 ratio. This proportion of silicon to oxygen is true for any sample of pure quartz, regardless of its origin. However, the compositions of some minerals can vary *within specific, well-defined limits*. This occurs because certain elements can substitute for others of similar size without changing the mineral's internal structure.

What Is a Rock?

In contrast to minerals, rocks are more loosely defined. Simply, a **rock** is any solid mass of mineral, or mineral-like, matter that occurs naturally as part of our planet. Most rocks, like the sample of granite shown in Figure 3.3, occur as aggregates of several different minerals. The term *aggregate* implies that the minerals are joined in such a way that their individual properties are retained. Note that the different minerals that make up granite can be easily identified. However, some rocks

▼ **SmartFigure 3.3 Most rocks are aggregates of minerals** Shown here is a hand sample of the igneous rock granite and three of its major constituent minerals. (Photos by E. J. Tarbuck)

TUTORIAL
https://goo.gl/k5qL1l

Did You Know?
Archaeologists discovered that the Romans transported water in lead pipes more than 2000 years ago. In fact, Roman smelting of lead and copper ores between 500 B.C.E. and C.E. 300 caused a small rise in atmospheric pollution that was recorded in Greenland ice cores.

are composed almost entirely of one mineral. A common example is the sedimentary rock *limestone*, which is an impure mass of the mineral calcite.

In addition, some rocks are composed of nonmineral matter. These include the volcanic rocks *obsidian* and *pumice*, which are noncrystalline glassy substances, and *coal*, which consists of solid organic debris.

Although this chapter deals primarily with the nature of minerals, keep in mind that most rocks are simply aggregates of minerals. Because the properties of rocks are determined largely by the chemical composition and crystalline structure of the minerals contained within them, we will first consider these Earth materials.

CONCEPT CHECKS 3.1

1. List five characteristics of a mineral.
2. Based on the definition of mineral, which of the following—gold, liquid water, synthetic diamonds, ice, and wood—are not classified as minerals?
3. Define the term *rock*. How do rocks differ from minerals?

3.2 Atoms: Building Blocks of Minerals

Compare and contrast the three primary particles contained in atoms.

When minerals are carefully examined, even under optical microscopes, the innumerable tiny particles of their internal structures are not visible. Nevertheless, scientists have discovered that all matter, including minerals, is composed of minute building blocks called **atoms**—the smallest particles that cannot be chemically split. Atoms, in turn, contain even smaller particles—*protons* and *neutrons* located in a central **nucleus** that is surrounded by *electrons* (**Figure 3.4**).

Properties of Protons, Neutrons, & Electrons

Protons and **neutrons** are very dense particles with almost identical masses. By contrast, **electrons** have a negligible mass, about 1/2000 that of a proton. To visualize this difference, imagine a scale on which a proton or neutron has the mass of a baseball, whereas an electron has the mass of a single grain of rice.

Both protons and electrons share a fundamental property called *electrical charge*. Protons have an electrical charge of +1, and electrons have a charge of −1. Neutrons, as the name suggests, have no charge. The charges of protons and electrons are equal in magnitude but opposite in polarity, so when these two particles are paired, the charges cancel each other out. Since matter typically contains equal numbers of positively charged protons and negatively charged electrons, most substances are electrically neutral.

Illustrations sometimes show electrons orbiting the nucleus in a manner that resembles the planets of our solar system orbiting the Sun (see Figure 3.4A). However, electrons do not actually behave this way. A more realistic depiction would show electrons as a cloud of negative charges surrounding the nucleus (see Figure 3.4B). Studies of the arrangements of electrons show that they move about the nucleus in regions called *principal shells*, each with an associated energy level. In addition, each shell can hold a specific number of electrons, with the outermost shell generally containing **valence electrons**. These electrons can be transferred to or shared with other atoms to form chemical bonds.

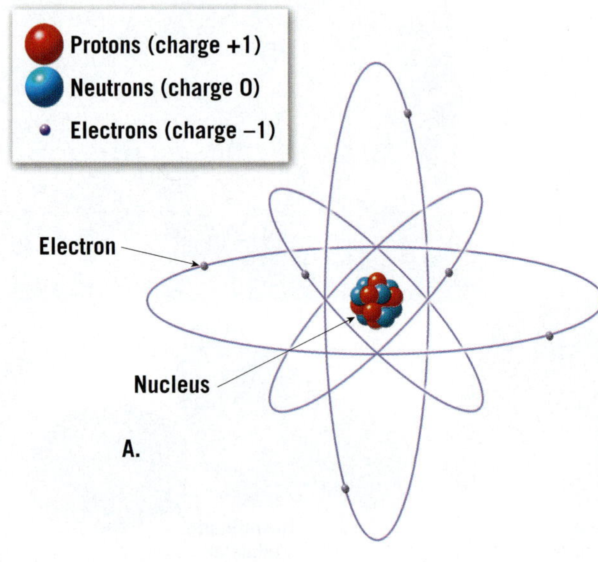

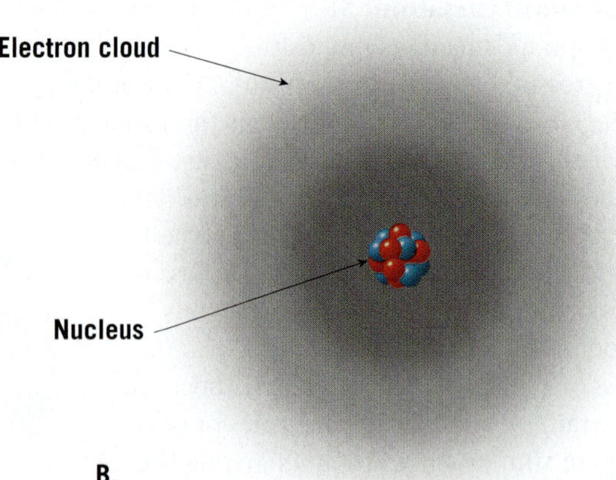

Figure 3.4 Two models of an atom A. Simplified view of an atom having a central nucleus composed of protons and neutrons, encircled by high-speed electrons. **B.** This model of an atom shows spherically shaped electron clouds (shells) surrounding a central nucleus. The nucleus contains virtually all of the mass of the atom. The remainder of the atom is the space occupied by negatively charged electrons. (The relative sizes of the nuclei shown are greatly exaggerated.)

Most of the atoms in the universe (except hydrogen and helium) were created inside massive stars by nuclear fusion and then released into interstellar space during hot, fiery supernova explosions. As this ejected material cooled, the newly formed nuclei attracted electrons to complete their atomic structure. At the temperatures found at Earth's surface, free atoms (those not bonded to other atoms) generally have a full complement of electrons—one for each proton in the nucleus.

Elements: Defined by Their Number of Protons

The simplest atoms have only 1 proton in their nuclei, whereas others have more than 100. The number of protons in the nucleus of an atom, called the **atomic number**, determines its chemical nature. All atoms with the same number of protons have the same chemical and physical properties; collectively they constitute an **element**. There are about 90 naturally occurring elements, and several more have been synthesized in the laboratory. You are probably familiar with the names of many elements, including carbon, nitrogen, and oxygen. All carbon atoms have six protons, whereas all nitrogen atoms have seven protons, and all oxygen atoms have eight.

The **periodic table**, shown in **Figure 3.5**, is a tool scientists use to organize the known elements. In it, elements with similar properties line up in columns, referred to as groups. Each element is assigned a one- or two-letter symbol. The atomic number and atomic mass for each element are also included in the periodic table.

Atoms of the naturally occurring elements are the basic building blocks of Earth's minerals. Most elements join with other elements to form **chemical compounds**. Therefore, most minerals are chemical compounds composed of atoms of two or more elements. These include the minerals quartz (SiO_2), halite ($NaCl$), and calcite ($CaCO_3$). However, a few minerals,

Did You Know?
The purity of gold is expressed by the number of *karats*. Twenty-four karats is pure gold. Gold less than 24 karats is an alloy (mixture) of gold and another metal, usually copper or silver. For example, 14-karat gold contains 14 parts gold (by weight) mixed with 10 parts of other metals.

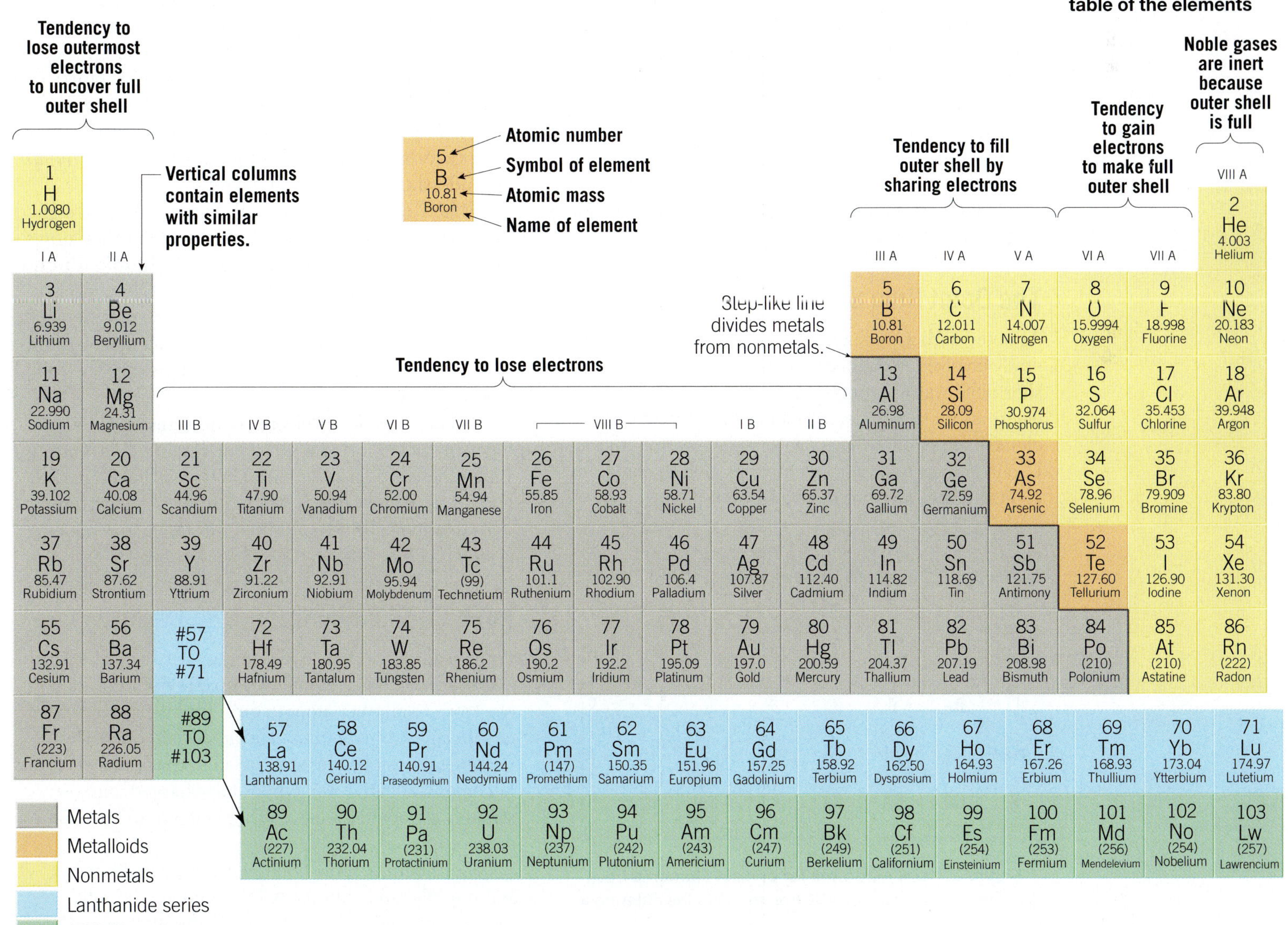

▼ **Figure 3.5** **Periodic table of the elements**

A. Gold on quartz

B. Sulfur

C. Copper

▲ **Figure 3.6 Examples of minerals composed of a single element** (Photos by Dennis Tasa)

such as diamonds, sulfur, and native gold and copper, are made entirely of atoms of only one element (**Figure 3.6**). (A metal is called "native" when it is found in its pure form in nature.)

CONCEPT CHECKS 3.2

1. Make a simple sketch of an atom and label its three main particles. Explain how these particles differ from one another.
2. What is the significance of valence electrons?

3.3 Why Atoms Bond

Distinguish among ionic bonds, covalent bonds, and metallic bonds.

Under the temperature and pressure conditions found on Earth, most elements do not occur in the form of individual atoms; instead, their atoms bond with other atoms. (A group of elements known as the noble gases are an exception.) Some atoms bond to form *ionic compounds*, some form *molecules*, and still others form *metallic substances*. Why does this happen? Experiments show that electrical forces hold atoms together and bond them to each other. These electrical attractions lower the total energy of the bonded atoms, and this, in turn, generally makes them more stable. Consequently, atoms that are bonded in compounds tend to be more stable than atoms that are free (not bonded).

The Octet Rule & Chemical Bonds

As noted earlier, valence (outer shell) electrons are generally involved in chemical bonding. **Figure 3.7** shows a shorthand way of representing the number of valence electrons for some selected elements. Notice that the elements in Group I have one valence electron, those in Group II have two valence electrons, and so on, up to eight valence electrons in Group VIII.

Electron Dot Diagrams for Some Representative Elements

I	II	III	IV	V	VI	VII	VIII
H							He
Li	Be	B	C	N	O	F	Ne
Na	Mg	Al	Si	P	S	Cl	Ar
K	Ca	Ga	Ge	As	Se	Br	Kr

▲ **Figure 3.7 Dot diagrams for certain elements** Each dot represents a valence electron found in the outermost principal shell.

The noble gases have very stable electron arrangements with eight valence electrons (except helium, which has two) and, therefore, tend to lack chemical reactivity. Many other atoms gain, lose, or share electrons during chemical reactions, ending up with electron arrangements of the noble gases. This observation led to a chemical guideline known as the **octet rule**: *Atoms tend to gain, lose, or share electrons until they are surrounded by eight valence electrons.* Although there are exceptions to the octet rule, it is a useful *rule of thumb* for understanding chemical bonding.

When an atom's outer shell does not contain eight electrons, it is likely to chemically bond to other atoms to achieve an octet in its outer shell. A **chemical bond** is a transfer or sharing of electrons that allows each atom to attain a full valence shell of electrons. Some atoms do this by transferring all their valence electrons to other atoms so that an inner shell becomes the full valence shell.

When the valence electrons are transferred between the elements to form ions, the bond is an *ionic bond*. When the electrons are shared between the atoms, the bond is a *covalent bond*. When the valence electrons are shared among all the atoms in a substance, the bonding is *metallic*.

Ionic Bonds: Electrons Transferred

Perhaps the easiest type of bond to visualize is the **ionic bond**, in which one atom gives up one or more valence electrons to another atom to form **ions**—*positively and negatively charged atoms*. The atom that loses electrons becomes a positive ion, and the atom that gains electrons becomes a negative ion. Oppositely charged ions are strongly attracted to one another and join to form *ionic compounds*.

Consider the ionic bonding that occurs between sodium (Na) and chlorine (Cl) to produce the solid ionic compound sodium chloride—the mineral halite

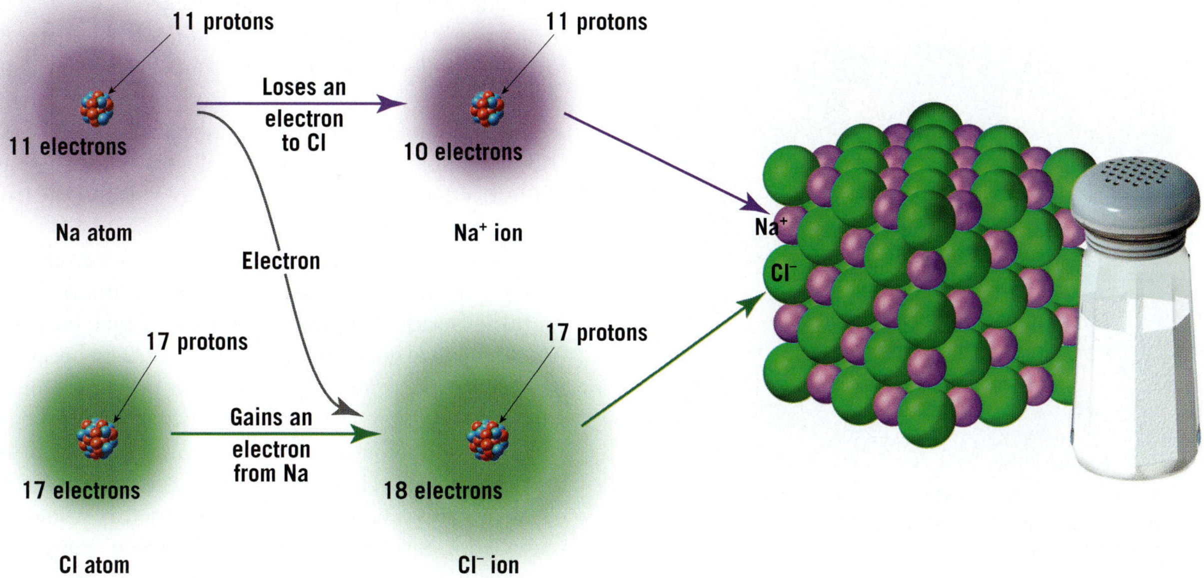

◀ **Figure 3.8 Formation of the ionic compound sodium chloride**

(common table salt). Notice in **Figure 3.8A** that a sodium atom gives up its single valence electron to chlorine and, as a result, becomes a positively charged sodium ion (Na^+). Chlorine, on the other hand, gains one electron and becomes a negatively charged chloride ion (Cl^-). We know that ions having unlike charges attract. Thus, an ionic bond is an attraction of oppositely charged ions to one another that produces an electrically neutral ionic compound.

Figure 3.8B illustrates the arrangement of sodium and chlorine ions in ordinary table salt. Notice that salt consists of alternating sodium and chlorine ions, positioned so that each positive ion is attracted to and surrounded on all sides by negative ions and vice versa. This arrangement maximizes the attraction between ions with opposite charges while minimizing the repulsion between ions with identical charges. Thus, ionic compounds consist of an orderly arrangement of oppositely charged ions assembled in a definite ratio that provides overall electrical neutrality.

The properties of a chemical compound are dramatically different from the properties of the various elements comprising it. For example, sodium is a soft silvery metal that is extremely reactive and poisonous. If you were to consume even a small amount of elemental sodium, you would need immediate medical attention. Chlorine, a green poisonous gas, is so toxic that it was used as a chemical weapon during World War I. Together, however, these elements produce sodium chloride, the harmless flavor enhancer that we call table salt. Thus, when elements combine to form compounds, their properties change significantly.

Covalent Bonds: Electron Sharing

Sometimes the forces that hold atoms together cannot be understood on the basis of the attraction of oppositely charged ions. One example is the hydrogen molecule (H_2), in which the two hydrogen atoms are held together tightly and no ions are present. The strong attractive force that holds two hydrogen atoms together results from a **covalent bond**, a chemical bond formed by the *sharing* of one or more valance electrons between a pair of atoms. (Hydrogen is one of the exceptions to the octet rule: Its single shell is full with just two electrons.)

Imagine two hydrogen atoms (each with one proton and one electron) approaching one another, as shown in **Figure 3.9**. Once they meet, the electron configuration changes so that both electrons primarily occupy the space between the atoms. In other words, the two electrons are

▼ **Figure 3.9 Formation of a covalent bond** When hydrogen atoms bond, the negatively charged electrons are shared by both hydrogen atoms and attracted simultaneously by the positive charge of the proton in the nucleus of each atom.

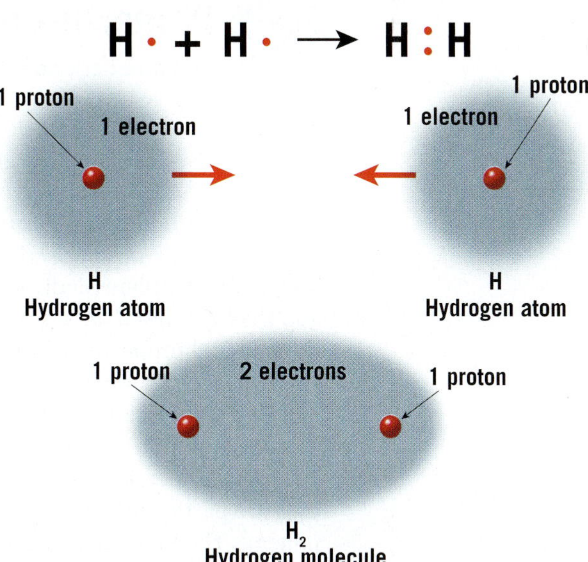

Two hydrogen atoms combine to form a hydrogen molecule, held together by the attraction of oppositely charged particles—positively charged protons in each nucleus and negatively charged electrons that surround these nuclei.

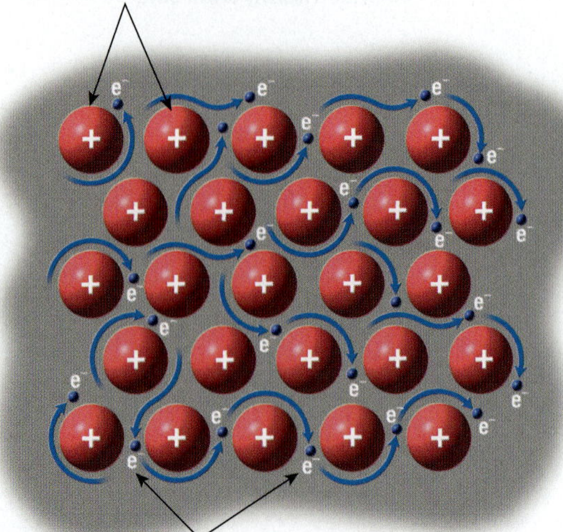

The central core of each metallic atom, which has an overall positive charge, consists of the nucleus and inner electrons.
A "sea" of negatively charged outer electrons, that are free to move throughout the structure, surrounds the metallic atoms.

▶ Figure 3.10 **Metallic bonding** Metallic bonding is the result of each atom contributing its valence electrons to a common pool of electrons that are free to move throughout the entire metallic structure. The attraction between the "sea" of negatively charged electrons and the positive ions produces the metallic bonds that give metals their unique properties.

shared by both hydrogen atoms and are attracted simultaneously by the positive charge of the proton in the nucleus of each atom. In this situation, the hydrogen atoms do not form ions; instead, the force that holds these atoms together arises from the attraction of oppositely charged particles—positively charged protons in the nuclei and negatively charged electrons that surround these nuclei.

Metallic Bonds: Electrons Free to Move

A few minerals, such as native gold, silver, and copper, are made entirely of metal atoms packed tightly together in an orderly way. The bonding that holds these atoms together results from each atom contributing its valence electrons to a common pool of electrons, which freely move throughout the entire metallic structure. The contribution of one or more valence electrons leaves an array of positive ions immersed in a "sea" of valence electrons, as shown in Figure 3.10.

The attraction between this "sea" of negatively charged electrons and the positive ions produces the **metallic bonds** that give metals their unique properties. Metals are good conductors of electricity because the valence electrons are free to move from one atom to another. Metals are also *malleable*, which means they can be hammered into thin sheets, and *ductile*, which means they can be drawn into thin wires. By contrast, ionic and covalent solids tend to be *brittle* and fracture when stress is applied. Consider the difference between dropping a metal frying pan and a ceramic plate onto a concrete floor.

CONCEPT CHECKS 3.3

1. What is the difference between an atom and an ion?
2. How does an atom become a positive ion? A negative ion?
3. Briefly distinguish between ionic, covalent, and metallic bonding and discuss the role that electrons play in each.

3.4 Properties of Minerals

List and describe the properties used in mineral identification.

Minerals have definite crystalline structures and chemical compositions that give them unique sets of physical and chemical properties shared by all specimens of that mineral, regardless of when or where they formed. For example, two samples of the mineral quartz will be equally hard and equally dense, and they will break in a similar manner. However, the physical properties of individual samples may vary within specific limits due to ionic substitutions, inclusions of foreign elements (impurities), and defects in the crystalline structure.

Some mineral properties, called **diagnostic properties**, are particularly useful in identifying an unknown mineral. The mineral halite, for example, has a salty taste. Because so few minerals share this property, a salty taste is considered a diagnostic property of halite. Other properties of certain minerals, particularly color, vary among different specimens of the same mineral. These properties are referred to as **ambiguous properties**.

Optical Properties

Of the many diagnostic properties of minerals, their optical characteristics such as luster, color, streak, and ability to transmit light are most frequently used for mineral identification.

Luster The appearance or quality of light reflected from the surface of a mineral is known as **luster**. Minerals that are shiny like a metal, regardless of color, are said to have a *metallic luster* (Figure 3.11A). Some metallic minerals, such as native copper and galena, develop a dull coating or tarnish when exposed to the atmosphere. Because they are not as shiny as samples with freshly broken surfaces, these samples are often said to exhibit a *submetallic luster* (Figure 3.11B).

Most minerals have a *nonmetallic luster* and are described using various adjectives. For example, some minerals are described as being *vitreous*, or *glassy*. Other

▲ **Figure 3.11 Metallic versus submetallic luster**
(Photo courtesy of E. J. Tarbuck)

Mineral (Pyrite)
Streak (Black)
Color (Brass yellow)
Although the color of a mineral is not always helpful in identification, the streak, which is the color of the powdered mineral, can be very useful.

◄ **SmartFigure 3.13**
Streak (Photo by Dennis Tasa)

VIDEO
https://goo.gl/MdH5j9

nonmetallic minerals are described as having a *dull*, or *earthy*, *luster* (a dull appearance like soil) or a *pearly luster* (such as a pearl or the inside of a clamshell). Still others exhibit a *silky luster* (like satin cloth) or a *greasy luster* (as though coated in oil).

Color Although **color** is generally the most conspicuous characteristic of any mineral, it is considered a diagnostic property of only a few minerals. Slight impurities in the common minerals fluorite and quartz, for example, give them a variety of tints, including pink, purple, yellow, white, gray, and even black (Figure 3.12). Other minerals, such as tourmaline, also exhibit a variety of hues, with multiple colors sometimes occurring in the same sample. Thus, the use of color as a means of identification is often ambiguous or even misleading.

Streak The color of a mineral in powdered form, called **streak**, is often useful in identification. A mineral's streak is obtained by rubbing it across a *streak plate* (a piece of unglazed porcelain) and observing the color of the mark it leaves (Figure 3.13). Although a mineral's color may vary from sample to sample, its streak is usually consistent in color. (Note that not all minerals produce a streak when rubbed across a streak plate. Quartz, for example, is harder than a porcelain streak plate and therefore leaves no streak.)

Streak can also help distinguish between minerals with metallic luster and those with nonmetallic luster. Metallic minerals generally have a dense, dark streak, whereas minerals with nonmetallic luster typically have a light-colored streak.

▲ **SmartFigure 3.12 Color variations in minerals** Some minerals, such as fluorite shown here, exhibit a variety of colors. (Photo by E. J. Tarbuck)

TUTORIAL
https://goo.gl/LWOfho

Ability to Transmit Light Another optical property used to identify minerals is the ability to transmit light. When no light is transmitted through a mineral sample, that mineral is described as *opaque*; when light, but not an image, is transmitted, the mineral is said to be *translucent*. When both light and an image are visible through the sample, the mineral is described as *transparent*.

Crystal Shape, or Habit

Mineralogists use the term **crystal shape**, or **habit**, to refer to the common or characteristic shape of individual crystals or aggregates of crystals. Some minerals tend to grow equally in all three dimensions, whereas others tend to be elongated in one direction or flattened if growth in one dimension is suppressed. The crystals of a few minerals can have a regular polygonal shape that is helpful in identification. For example, magnetite crystals sometimes occur as octahedrons, garnets often form dodecahedrons, and halite and fluorite crystals tend to grow as cubes or near-cubes. Most minerals have just one common crystal shape, but a few, such as the pyrite samples shown in Figure 3.14, have two or more characteristic crystal shapes.

In addition, some mineral samples consist of numerous intergrown crystals exhibiting characteristic shapes

Did You Know?
The name *crystal* is derived from the Greek (*krystallos* = ice) and was originally applied to quartz crystals. The ancient Greeks thought quartz was water that had crystallized at high pressures deep inside Earth.

▶ **Figure 3.14 Common crystal shapes of pyrite**

that are useful for identification. Terms commonly used to describe these and other crystal habits include *equant* (equidimensional), *bladed*, *fibrous*, *tabular*, *cubic*, *prismatic*, *platy*, *blocky*, and *banded*. Some of these habits are pictured in **Figure 3.15**.

Although most minerals exhibit only one common crystal shape, some, such as pyrite, have two or more characteristic habits.
Dennis Tasa

Mineral Strength

How easily minerals break or deform under stress is determined by the type and strength of the chemical bonds that hold the crystals together. Mineralogists use terms including *hardness*, *cleavage*, *fracture*, and *tenacity* to describe mineral strength and how minerals break when stress is applied.

Hardness One of the most useful diagnostic properties is **hardness**, a measure of the resistance of a mineral to abrasion or scratching. This property is determined by rubbing a mineral of unknown hardness against one of known hardness or vice versa. A numerical value of hardness can be obtained by using the **Mohs scale** of hardness, which consists of 10 minerals arranged in order from 1 (softest) to 10 (hardest), as shown in **Figure 3.16A**.

▼ **SmartFigure 3.15 Some common crystal habits A.** Thin, rounded crystals that break into fibers. **B.** Elongated crystals that are flattened in one direction. **C.** Minerals that have stripes or bands of different color or texture. **D.** Groups of crystals that are cube shaped.

TUTORIAL
https://goo.gl/vaVDiS

A. Fibrous

B. Bladed

C. Banded

D. Cubic crystals

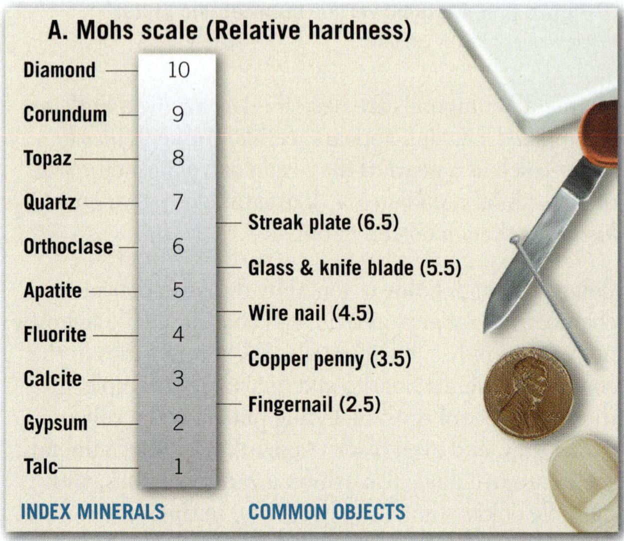

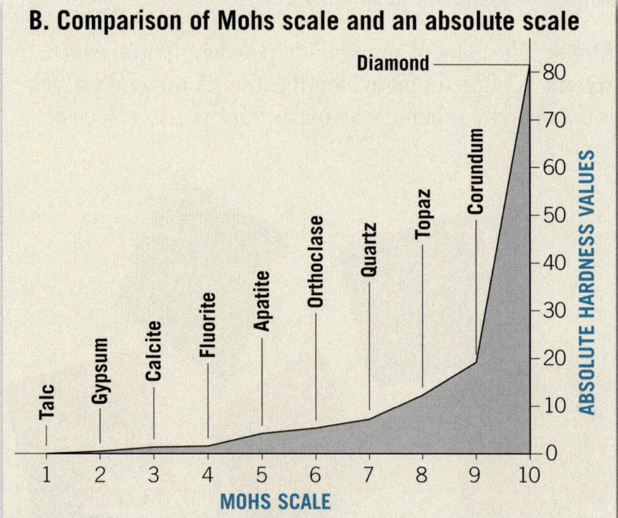

▲ **SmartFigure 3.16 Hardness scales A.** The Mohs scale of hardness, with the hardness of some common objects. **B.** Relationship between the Mohs relative hardness scale and an absolute hardness scale.

TUTORIAL
https://goo.gl/dvaqIF

It should be noted that the Mohs scale is a relative ranking and does not imply that a mineral with a hardness of 2, such as gypsum, is twice as hard as mineral with a hardness of 1, like talc. In fact, gypsum is only slightly harder than talc, as Figure 3.16B indicates.

In the laboratory, common objects used to determine the hardness of a mineral can include a human fingernail, which has a hardness of about 2.5, a copper penny (3.5), and a piece of glass (5.5). The mineral gypsum, which has a hardness of 2, can be easily scratched with a fingernail. On the other hand, the mineral calcite, which has a hardness of 3, will scratch a fingernail but will not scratch glass. Quartz, one of the hardest common minerals, will easily scratch glass. Diamonds, hardest of all, scratch anything, including other diamonds.

Cleavage In the crystal structure of many minerals, some atomic bonds are weaker than others. It is along these weak bonds that minerals tend to break when they are stressed. **Cleavage** (*kleiben* = carve) is the tendency of a mineral to break (cleave) along planes of weak bonding. Not all minerals have cleavage, but those that do can be identified by the relatively smooth, flat surfaces that are produced when the mineral is broken.

The simplest type of cleavage is exhibited by the micas (Figure 3.17). Because these minerals have very weak bonds in one direction, they cleave to form thin, flat sheets. Some minerals have excellent cleavage in one, two, three, or more directions, whereas others exhibit fair or poor cleavage, and still others have no cleavage at all. When minerals break evenly in more than one direction, cleavage is described by *the number of cleavage directions and the angle(s) at which they meet* (Figure 3.18).

SmartFigure 3.17 **Micas exhibit perfect cleavage** The thin sheets shown here exhibit one plane of cleavage. (Photo by Chip Clark/Fundamental Photographs)

ANIMATION
https://goo.gl/LvL2OD

Each cleavage surface that has a different orientation is counted as a different direction of cleavage. For example, some minerals, such as halite, cleave to form six-sided cubes. Because cubes are defined by three different sets of parallel planes that intersect at 90-degree angles, cleavage for the mineral halite is described as *three directions of cleavage that meet at 90 degrees*.

Do not confuse cleavage with crystal shape. When a mineral exhibits cleavage, it breaks into pieces that all have the same geometry. By contrast, the smooth-sided

A. Cleavage in one direction. Example: Muscovite

B. Cleavage in two directions at 90° angles. Example: Feldspar

C. Cleavage in two directions not at 90° angles. Example: Hornblende

D. Cleavage in three directions at 90° angles. Example: Halite

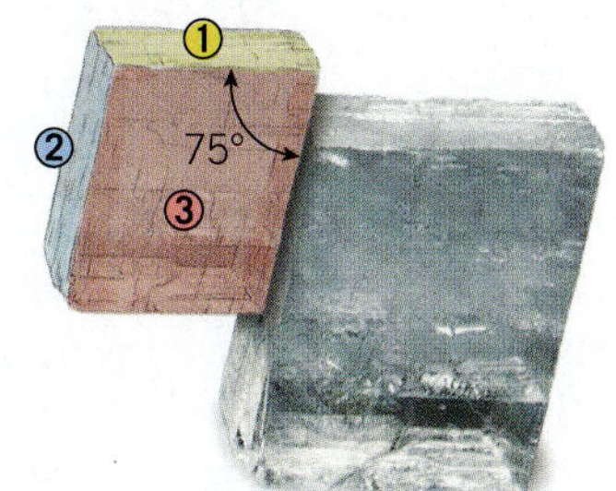

E. Cleavage in three directions not at 90° angles. Example: Calcite

F. Cleavage in four directions. Example: Fluorite

SmartFigure 3.18 **Cleavage directions exhibited by minerals** (Photos by E. J. Tarbuck and Dennis Tasa)

TUTORIAL
https://goo.gl/MN1wy7

◀ **Figure 3.19 Irregular versus conchoidal fracture** (Photos by E. J. Tarbuck)

quartz crystals shown in Figure 3.1 do not have cleavage. If broken, they fracture into shapes that do not resemble one another or the original crystals.

Fracture Minerals having chemical bonds that are equally, or nearly equally, strong in all directions exhibit a property called **fracture** (Figure 3.19A). When minerals fracture, most produce uneven surfaces and are described as exhibiting *irregular fracture*. However, some minerals, including quartz, sometimes break into smooth, curved surfaces resembling broken glass. Such breaks are called *conchoidal fractures* (Figure 3.19B). Still other minerals exhibit fractures that produce splinters or fibers referred to as *splintery fracture* and *fibrous fracture*, respectively.

Tenacity The term **tenacity** describes a mineral's resistance to breaking, bending, cutting, or other forms of deformation. As mentioned earlier, nonmetallic minerals such as quartz and minerals that are ionically bonded, such as fluorite and halite, tend to be *brittle* and fracture or exhibit cleavage when struck. By contrast, native metals, such as copper and gold, are *malleable*, which means they can be hammered without breaking. In addition, minerals that can be cut into thin shavings, including gypsum and talc, are described as *sectile*. Still others, notably the micas, are *elastic* and bend and snap back to their original shape after stress is released.

Did You Know?
The mineral pyrite is commonly called "fool's gold" because its golden-yellow color closely resembles gold. The name *pyrite* is derived from the Greek *pyros* ("fire") because it gives off sparks when struck sharply.

Density & Specific Gravity

Density, an important property of matter, is defined as mass per unit volume. Mineralogists often use a related measure called **specific gravity** to describe the density of minerals. Specific gravity is a number representing the ratio of a mineral's weight to the weight of an equal volume of water.

Most common minerals have a specific gravity between 2 and 3. For example, quartz has a specific gravity of 2.65. By contrast, some metallic minerals, such as pyrite, native copper, and magnetite, are more than twice as dense and thus have more than twice the specific gravity of quartz. Galena, an ore from which lead is extracted, has a specific gravity of roughly 7.5, whereas 24-karat gold has a specific gravity of approximately 20.

With a little practice, you can estimate the specific gravity of a mineral by hefting it in your hand. Does this mineral feel about as "heavy" as similarly sized rocks you have handled? If the answer is "yes," the specific gravity of the sample will likely be between 2.5 and 3.

Other Properties of Minerals

In addition to the properties discussed thus far, some minerals can be recognized by other distinctive properties. For example, halite is ordinary salt, so it can be quickly identified through taste. Talc and graphite both have distinctive feels: Talc feels soapy, and graphite feels greasy. Further, the streaks of many sulfur-bearing minerals smell like rotten eggs. A few minerals, such as magnetite, have high iron content and can be picked up with a magnet, while some varieties (such as lodestone) are themselves natural magnets and will pick up small iron-based objects such as pins and paper clips (see Figure 3.33F, page XXX).

Moreover, some minerals exhibit special optical properties. For example, when a transparent piece of calcite is placed over printed text, the letters appear twice. This optical property is known as *double refraction* (Figure 3.20).

One very simple chemical test to detect carbonate mineral involves placing a drop of dilute hydrochloric acid from a dropper bottle onto a freshly broken mineral surface. Samples containing carbonate minerals will effervesce (fizz) as carbon dioxide gas is released (Figure 3.21). This test is especially useful in identifying calcite, a common carbonate mineral.

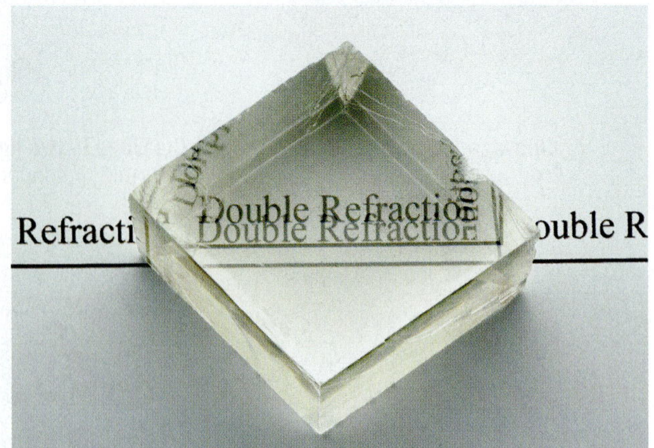

▲ **Figure 3.20 Double refraction** This sample of calcite exhibits double refraction. (Photo by Chip Clark/Fundamental Photographs)

▶ SmartFigure 3.21 **Calcite reacting with a weak acid** (Photo by Chip Clark/Fundamental Photographs)

VIDEO
https://goo.gl/5L3gns

CONCEPT CHECKS 3.4

1. Define *luster*.
2. Why is color not always a useful property in mineral identification? Give an example of a mineral that supports your answer.
3. What differentiates cleavage from fracture?
4. What is meant by a mineral's *tenacity*? List three terms that describe tenacity.
5. Describe a simple chemical test that is useful in identifying the mineral calcite.

3.5 Mineral Groups

Explain how minerals are classified and name the most abundant mineral group in Earth's crust.

More than 4000 minerals have been named, and several new ones are identified each year. Fortunately for students who are beginning to study minerals, no more than a few dozen are abundant. Collectively, these few make up most of the rocks of Earth's crust and, as such, they are often referred to as the **rock-forming minerals**.

Although less abundant, many other minerals are used extensively in the manufacture of products and are called **economic minerals**. However, rock-forming minerals and economic minerals are not mutually exclusive groups. When found in large deposits, some rock-forming minerals are economically significant. One example is calcite, the primary component of the sedimentary rock limestone. Calcite has many uses, including the production of concrete.

Classifying Minerals

Minerals are placed into categories in much the same way that plants and animals are classified. Mineralogists use the term *mineral species* for a collection of specimens that exhibit similar internal structures and chemical compositions. Some common mineral species are quartz, calcite, galena, and pyrite. However, just as individual plants and animals within a species differ somewhat from one another, so do most specimens of the same mineral.

Some mineral species are further subdivided into *mineral varieties*. For example, pure quartz (SiO_2) is colorless and transparent. However, when small amounts of aluminum are incorporated into its atomic structure, quartz appears quite dark, in a variety called *smoky quartz*. *Amethyst*, another variety of quartz, owes its violet color to the presence of trace amounts of iron.

Mineral species are assigned to *mineral groups*. Some important mineral groups are the silicates, carbonates, halides, and sulfates. Minerals within each class tend to have similar internal structures and, hence, similar properties. For example, minerals belonging to the carbonate group react chemically with acid—albeit to varying degrees—and many exhibit similar cleavage. Furthermore, minerals within the same group are often found together in the same rock. For example, halite (NaCl) and silvite (KCl) belong to the halide class and commonly occur together in evaporite deposits.

Silicate Versus Nonsilicate Minerals

It is worth noting that only *eight elements* make up the vast majority of the rock-forming minerals and represent more than 98 percent (by weight) of Earth's continental crust (Figure 3.22). These elements, in order of most to least abundant, are oxygen (O), silicon (Si), aluminum (Al), iron (Fe), calcium (Ca), sodium (Na), potassium (K),

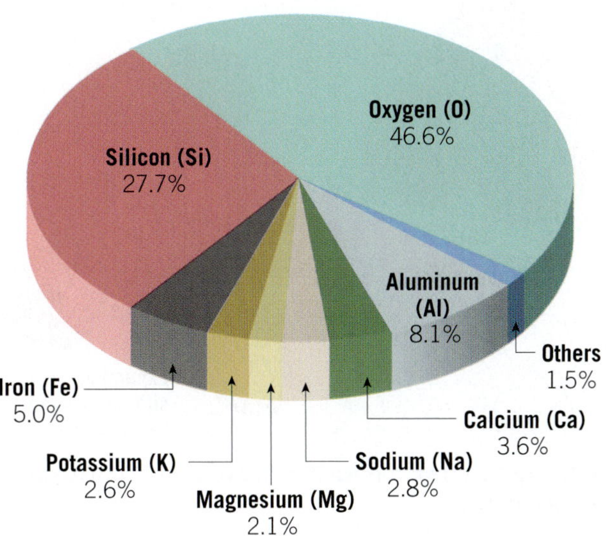

◀ Figure 3.22 **The eight most abundant elements in the continental crust** The numbers represent percentages by weight.

and magnesium (Mg). As shown in Figure 3.22, silicon and oxygen are by far the most common elements in Earth's crust. Furthermore, these two elements readily combine to form the basic "building block" for the most common mineral group, the **silicates**. More than 800 silicate minerals are known, and they account for about 92 percent of Earth's crust (see Figure 3.26, page 84).

Because other mineral groups are far less abundant in Earth's crust than the silicates, they are often grouped together under the heading **nonsilicates**. Although these minerals are not as common as silicates, some nonsilicates are very important economically. They provide us with iron and aluminum to build automobiles, gypsum for plaster and drywall for home construction, and copper wire that carries electricity and connects us to the Internet. Common nonsilicate mineral groups include the carbonates, sulfates, and halides. In addition to their economic importance, these groups include minerals that are major constituents in sediments and sedimentary rocks.

CONCEPT CHECKS 3.5

1. Distinguish between *rock-forming minerals* and *economic minerals*.
2. List the eight most common elements in Earth's crust.
3. Distinguish between a mineral *species* and a *variety*.

3.6 The Silicates

Sketch the silicon–oxygen tetrahedron and explain how this fundamental building block joins together to form various silicate structures.

Every silicate mineral contains the two most abundant elements in Earth's crust: oxygen and silicon. Further, most contain one or more of the other common elements. Together, these elements give rise to hundreds of silicate minerals with a wide variety of properties, including hard quartz, soft talc, sheetlike mica, fibrous asbestos, green olivine, and blood-red garnet.

Silicate Structures

All silicate minerals have the same fundamental building block, the **silicon–oxygen tetrahedron** (SiO_4^{4-}). This structure consists of four oxygen ions that are covalently bonded to one comparatively small silicon ion, forming a *tetrahedron*—a pyramid shape with four identical planar surfaces, or faces (**Figure 3.23**). These tetrahedrons are not chemical compounds but rather complex ions (SiO_4^{4-}) having a net charge of −4. To become electrically balanced, these complex ions bond to positively charged metal ions. Specifically, each O^{2-} has one of its valence electrons bonding with the Si^{4+} located at the center of the tetrahedron. The remaining negative charge on each oxygen is available to bond with another positive ion or with the silicon ion in an adjacent tetrahedron.

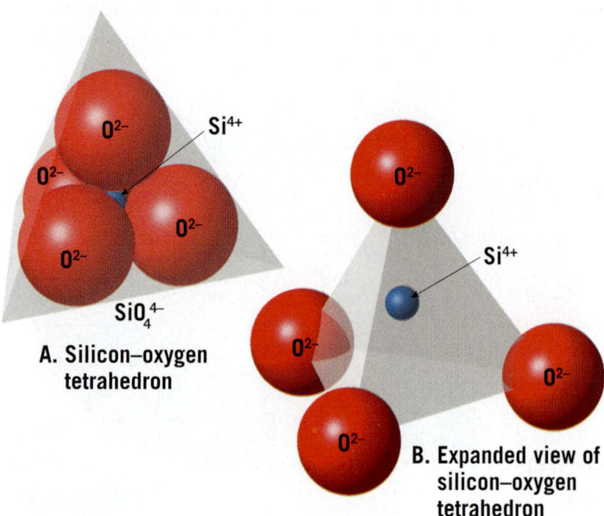

▲ **Figure 3.23 Two representations of the silicon–oxygen tetrahedron**

Minerals with Independent Tetrahedrons One of the simplest silicate structures consists of independent tetrahedrons that have their four oxygen ions bonded to positive ions, such as Mg^{2+}, Fe^{2+}, and Ca^{2+}. The mineral olivine, with the formula $(Mg,Fe)_2SiO_4$, is a good example. In olivine, magnesium (Mg^{2+}) and/or iron (Fe^{2+}) ions pack between comparatively large independent SiO_4 tetrahedrons, forming a dense, three-dimensional structure. Garnet, another common silicate, is also composed of independent tetrahedrons ionically bonded to positive ions. Both olivine and garnet form dense, hard, equidimensional crystals that lack cleavage.

Minerals with Chain or Sheet Structures One reason for the great variety of silicate minerals is the ability of SiO_4 tetrahedrons to link to one another in a variety of configurations. This important phenomenon, called **polymerization**, is achieved by the sharing of one, two, three, or all four of the oxygen atoms with adjacent tetrahedrons. Vast numbers of tetrahedrons join together to form single chains, double chains, sheet structures, or three-dimensional frameworks, as shown in **Figure 3.24**.

To see how oxygen atoms are shared between adjacent tetrahedrons, select one of the silicon ions (small

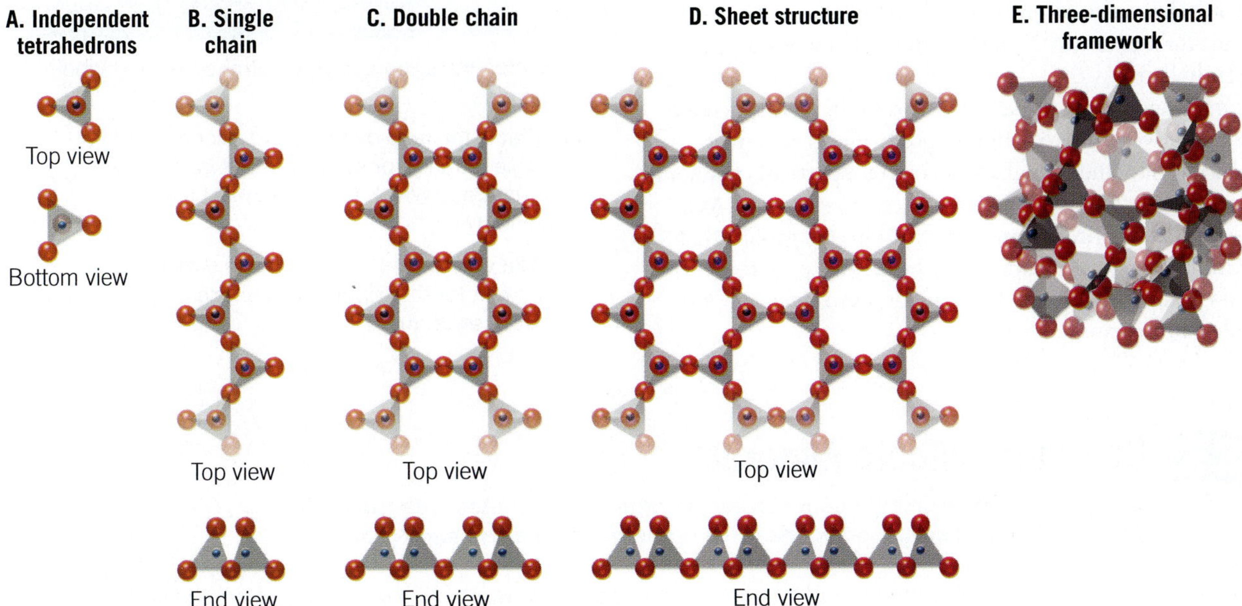

◀ SmartFigure 3.24
Five basic silicate structures

TUTORIAL
https://goo.gl/ZPuhGA

blue spheres) near the middle of the single chain shown in Figure 3.24B. Notice that this silicon ion is completely surrounded by four larger oxygen ions (red spheres). Also notice that two of the four oxygen atoms are bonded to two silicon atoms, whereas the other two are not shared in this manner. It is the linkage across the shared oxygen ions that joins the tetrahedrons into a chain structure. Now examine a silicon ion near the middle of the sheet structure (see Figure 3.24D) and count the number of shared and unshared oxygen ions surrounding it. As you likely observed, the sheet structure is the result of three of the four oxygen atoms being shared by adjacent tetrahedrons.

Minerals with Three-Dimensional Frameworks In the most common silicate structure, all four oxygen ions are shared, producing a complex three-dimensional framework (see Figure 3.24E). Quartz and the most common mineral group, the feldspars, exhibit this type of structure.

The ratio of oxygen ions to silicon ions differs in each type of silicate structure. In independent tetrahedrons (SiO_4) there are four oxygen ions for every silicon ion. In single chains, the oxygen-to-silicon ratio is 3:1 (SiO_3), and in three-dimensional frameworks, as found in quartz, the ratio is 2:1 (SiO_2). As more oxygen ions are shared, the percentage of silicon in the structure increases. Silicate minerals are therefore described as having a low or high silicon content, based on their ratio of oxygen to silicon. Silicate minerals with three-dimensional structures have the highest silicon content, while those composed of independent tetrahedrons have the lowest.

Joining Silicate Structures

Except for quartz (SiO_2), the basic structure (chains, sheets, or three-dimensional frameworks) of most silicate minerals has a net negative charge. Therefore, metal ions are required to bring the overall charge into balance and to serve as the "mortar" that holds these structures together. The positive ions that most often link silicate structures are iron (Fe^{2+}), magnesium (Mg^{2+}), potassium (K^{+}), sodium (Na^{+}), aluminum (Al^{3+}), and calcium (Ca^{2+}). These positively charged ions bond with the unshared oxygen ions that occupy the corners of the silicate tetrahedrons.

As a general rule, the covalent bonds between silicon and oxygen are stronger than the ionic bonds that hold one silicate structure to the next. Consequently, properties such as cleavage and, to some extent, hardness are controlled by the nature of the silicate framework. Quartz (SiO_2), which has only silicon–oxygen bonds, has great hardness and lacks cleavage, mainly because it has equally strong bonds in all directions. By contrast, the mineral talc (the source of talcum powder), has a sheet structure. Magnesium ions occur between the sheets and weakly join them together. The slippery feel of talcum powder is due to the silicate sheets sliding relative to one another, in much the same way that sheets of carbon atoms in graphite slide, giving graphite its lubricating properties.

Recall that atoms of similar size can substitute freely for one another without altering a mineral's structure. For example, in olivine, iron (Fe^{2+}) and magnesium (Mg^{2+}) substitute for each other. This also holds true for the third-most-common element in Earth's crust, aluminum (Al^{3+}), which often substitutes for silicon (Si^{4+}) in the center of silicon–oxygen tetrahedrons.

Because most silicate structures will readily accommodate two or more different positive ions at a given bonding site, individual specimens of a particular mineral may contain varying amounts of certain elements. As a result, many silicate minerals form mineral groups that exhibit a range of compositions between two end members. Examples include the olivines, pyroxenes, amphiboles, micas, and feldspars.

CONCEPT CHECKS 3.6

1. Sketch the silicon–oxygen tetrahedron and label its parts.
2. What is the ratio of oxygen to silicon found in single tetrahedrons? How about framework structures? Which has the highest silicon content?
3. What differences in their silicate structures account for the slipperiness of talc and the hardness of quartz?

3.7 Common Silicate Minerals

Compare and contrast the light (nonferromagnesian) silicates with the dark (ferromagnesian) silicates and list three common minerals in each group.

The major groups of silicate minerals and common examples are given in Figure 3.25. Most silicate minerals, including those shown in Figure 3.25, form when molten rock cools and crystallizes. Cooling can occur at or near Earth's surface (low temperature and pressure) or at great depths (high temperature and pressure). The environment during crystallization and the chemical composition of the molten rock determine, to a large degree, which minerals are produced. For example, the silicate mineral olivine crystallizes early, whereas quartz forms much later in the crystallization process.

In addition to silicate minerals that crystallize from molten rock, some form at Earth's surface from other silicate minerals through the process of weathering. Still others are formed under the extreme pressures associated with mountain building. Each silicate mineral, therefore, has a structure and a chemical composition that *indicate the conditions under which it formed.* By carefully examining the mineral constituents of rocks, geologists can usually determine the circumstances under which the rocks formed.

We will now examine some of the most common silicate minerals, which we divide into two major groups on the basis of their chemical makeup: the light silicates and the dark silicates.

The Light Silicates

The **light** (or **nonferromagnesian**) **silicates** are generally light in color and have a specific gravity of about 2.7, less than that of the dark (ferromagnesian) silicates. These differences are mainly attributable to the presence or absence of iron and magnesium, which are "heavy" elements. The light silicates contain varying amounts of aluminum, potassium, calcium, and sodium rather than iron and magnesium.

Feldspar Group *Feldspar minerals* are by far the most plentiful silicate group in Earth's crust, comprising about 51 percent of the crust (Figure 3.26). Their abundance can be partially explained by the fact that they can form under a wide range of temperatures and pressures. Two different feldspar structures exist (Figure 3.27). One group of feldspar minerals contains potassium ions in its structure and is therefore termed **potassium feldspar** (see Figure 3.27A,B). (*Orthoclase* and *microcline* are common members of the potassium feldspar group.) The other group, called **plagioclase feldspar**, contains both sodium and calcium ions that freely substitute for one another, depending on the environment during crystallization (see Figure 3.27C,D). Despite these differences, all feldspar minerals have similar physical properties. They have two planes of cleavage meeting at or near 90-degree angles, are relatively hard (6 on the Mohs scale), and have a luster that ranges from glassy to pearly. As a component in igneous rocks, feldspar crystals can be identified by their rectangular shape and rather smooth, shiny faces.

Potassium feldspar is usually light cream, salmon pink, or occasionally blue-green in color. The plagioclase feldspars, on the other hand, range in color from gray to blue-gray or sometimes black. However, color should not be used to distinguish these groups, as the only way to distinguish the feldspars by looking at them is through the presence of a multitude of fine parallel lines, called *striations*. Striations are found on some cleavage planes of plagioclase feldspar but are not present on potassium feldspar (see Figure 3.27B,D).

Quartz **Quartz** (SiO_2) is the second-most-abundant mineral in the continental crust and the only common silicate mineral that consists entirely of silicon and oxygen. Because quartz contains a ratio of two oxygen ions (O^{2-}) for every silicon ion (Si^{4+}), no other positive ions

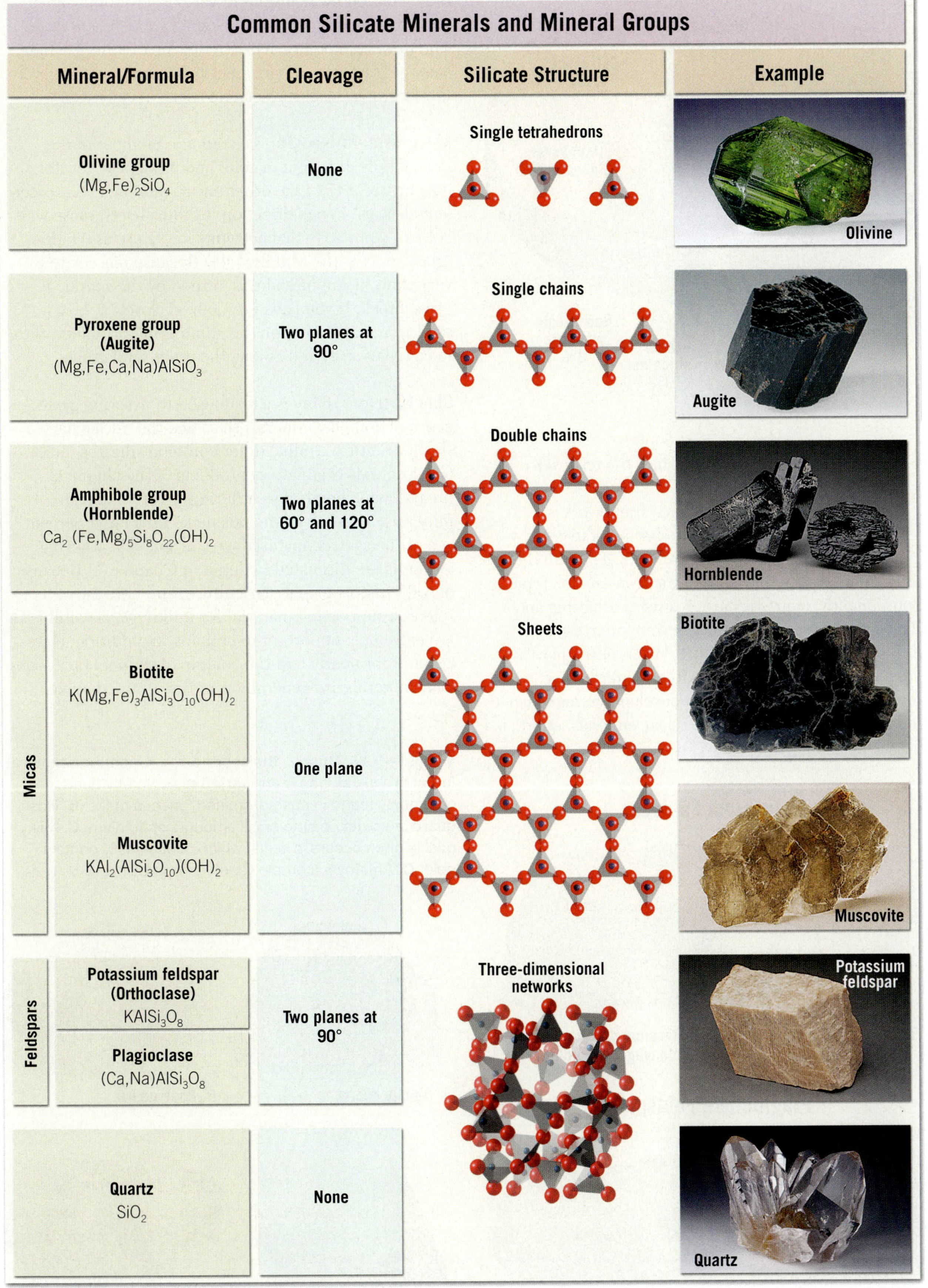

Common Silicate Minerals and Mineral Groups

Mineral/Formula		Cleavage	Silicate Structure	Example
Olivine group $(Mg,Fe)_2SiO_4$		None	Single tetrahedrons	Olivine
Pyroxene group (Augite) $(Mg,Fe,Ca,Na)AlSiO_3$		Two planes at 90°	Single chains	Augite
Amphibole group (Hornblende) $Ca_2(Fe,Mg)_5Si_8O_{22}(OH)_2$		Two planes at 60° and 120°	Double chains	Hornblende
Micas	Biotite $K(Mg,Fe)_3AlSi_3O_{10}(OH)_2$	One plane	Sheets	Biotite
Micas	Muscovite $KAl_2(AlSi_3O_{10})(OH)_2$	One plane	Sheets	Muscovite
Feldspars	Potassium feldspar (Orthoclase) $KAlSi_3O_8$	Two planes at 90°	Three-dimensional networks	Potassium feldspar
Feldspars	Plagioclase $(Ca,Na)AlSi_3O_8$	Two planes at 90°	Three-dimensional networks	
Quartz SiO_2		None	Three-dimensional networks	Quartz

◀ **Figure 3.25 Common silicate minerals** Note that the complexity of the silicate structure increases from the top of the chart to the bottom. (Photos by Dennis Tasa and E. J. Tarbuck)

▶ **Figure 3.26 Feldspar minerals make up about 51 percent of Earth's crust** Also, note from this graph that the silicate minerals make up about 92 percent of Earth's crust.

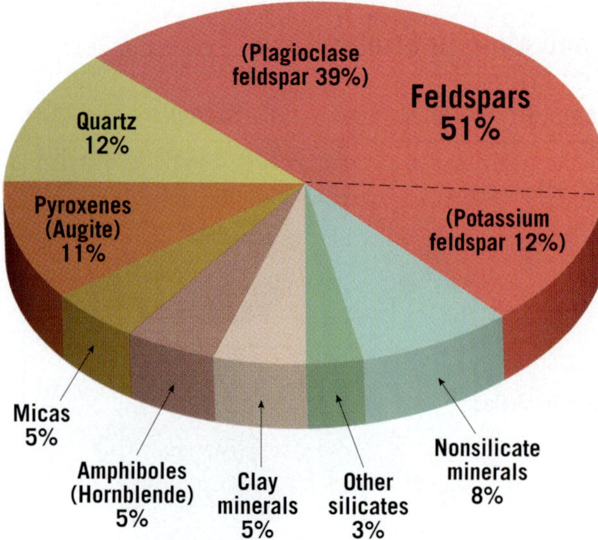

are needed to attain neutrality. Thus, the term *silica* is commonly applied to quartz.

In quartz, a three-dimensional framework is developed through the complete sharing of oxygen by adjacent silicon atoms (see Figure 3.25). Thus, all the bonds in quartz are of the strong silicon–oxygen type. Consequently, quartz is hard, resists weathering, and does not have cleavage. When broken, quartz generally exhibits conchoidal fracture. When pure, quartz is clear and, if allowed to grow without interference, will develop hexagonal crystals that develop pyramid-shaped ends. However, like most other clear minerals, quartz is often colored by inclusions of various ions (impurities) and often forms without developing good crystal faces. The most common varieties of quartz are milky (white), smoky (gray), rose (pink), amethyst (purple), citrine (yellow to brown), and rock crystal (clear) (Figure 3.28).

Muscovite **Muscovite** is a common member of the mica family. It is light in color and has a pearly luster (see Figure 3.17). Like other micas, muscovite has excellent cleavage in one direction. In thin sheets, muscovite is clear, a property that accounts for its use as window "glass" during the Middle Ages. Because muscovite is very shiny, it can often be identified by the sparkle it gives a rock. If you have ever looked closely at beach sand, you may have seen the glimmering brilliance of the mica flakes scattered among the other sand grains.

Clay Minerals **Clay** is a term used to describe a category of complex minerals that, like the micas, have a sheet structure. Unlike other common silicates, most clay minerals originate as products of the chemical breakdown (chemical weathering) of other silicate minerals. Thus, clay minerals make up a large percentage of the surface material we call soil. (Weathering and soils are discussed in detail in Chapter 6.) Because of soil's importance to agriculture, and because of its role as a supporting material for buildings, clay minerals are extremely important to humans. In addition, clays account for nearly half the volume of sedimentary rocks. Clay minerals are generally very fine grained, which

Potassium Feldspar

A. Potassium feldspar crystal (orthoclase)

B. Potassium feldspar showing cleavage (orthoclase)

Plagioclase Feldspar

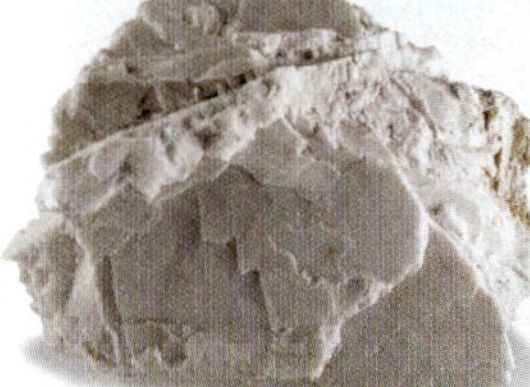

C. Sodium-rich plagioclase feldspar (albite)

D. Plagioclase feldspar showing striations (labradorite)

▶ **Figure 3.27 Some common feldspar minerals A.** Characteristic crystal form of potassium feldspar. **B.** Most salmon-colored feldspar belongs to the potassium feldspar subgroup. **C.** Sodium-rich plagioclase feldspar tends to be light in color with a pearly luster. **D.** Calcium-rich plagioclase feldspar tends to be gray, blue-gray, or black in color. Labradorite, the sample shown here, exhibits striations on one of its crystal faces. (Photos by Dennis Tasa and E. J. Tarbuck)

▼ **Figure 3.28 Quartz, the second-most-common mineral in Earth's crust, has many varieties A.** Smoky quartz is commonly found in coarse-grained igneous rocks. **B.** Rose quartz owes its color to small amounts of titanium. **C.** Milky quartz often occurs in veins, which occasionally contain gold. **D.** Amethyst a purple variety of quartz. (Photos by Dennis Tasa and E. J. Tarbuck)

A. Smoky quartz

B. Rose quartz

C. Milky quartz

D. Amethyst

▲ **Figure 3.29 Kaolinite** Kaolinite is a common clay mineral formed by weathering of feldspar minerals.

▲ **Figure 3.30 Olivine** Commonly black to olive green in color, olivine has a glassy luster and is often granular in appearance. Olivine is commonly found in the igneous rock basalt.

makes them difficult to identify unless they are studied microscopically. Their layered structure and the weak bonding between layers give them a characteristic feel when wet. Clays are common in shales, mudstones, and other sedimentary rocks.

One of the most common clay minerals is *kaolinite* (Figure 3.29), which is used in the manufacture of fine china and as a coating for high-gloss paper, such as that used in this textbook. Further, some clay minerals absorb large amounts of water, which allows them to swell to several times their normal size. These clays have been used commercially in a variety of ingenious ways, including as an additive to thicken milkshakes in fast-food restaurants.

The Dark Silicates

The **dark** (or **ferromagnesian**) **silicates** are minerals containing ions of iron (*ferro* = iron) and/or magnesium in their structure. Because of their iron content, ferromagnesian silicates are dark in color and have a greater specific gravity, between 3.2 and 3.6, than nonferromagnesian silicates. The most common dark silicate minerals are olivine, the pyroxenes, the amphiboles, dark mica (biotite), and garnet.

Olivine Group **Olivine**, a family of high-temperature silicate minerals, are black to olive green in color and have a glassy luster and a conchoidal fracture (see Figure 3.25). Transparent olivine is occasionally used as a gemstone called peridot. Rather than developing large crystals, olivine commonly forms small, rounded crystals that give olivine-rich rocks a granular appearance (Figure 3.30). Olivine and related forms are typically found in basalt, a common igneous rock of the oceanic crust and volcanic areas on the continents, and are thought to constitute up to 50 percent of Earth's upper mantle.

Pyroxene Group The *pyroxenes* are a group of diverse minerals that are important components of dark-colored igneous rocks. The most common member, **augite**, is a black, opaque mineral with two directions of cleavage that meet at nearly a 90-degree angle (Figure 3.31A). Augite is one of the dominant minerals in basalt.

Amphibole Group **Hornblende** is the most common member of a chemically complex group of minerals called *amphiboles*. Hornblende is usually dark green to black in color, and except for its cleavage angles, which are about 60 degrees and 120 degrees, it is very similar in appearance to augite (Figure 3.31B). In a rock, hornblende often forms elongated crystals. This helps distinguish it from pyroxene, which forms rather blocky crystals. Hornblende is found in igneous rocks, where it often makes up the dark portion of an otherwise light-colored rock (see Figure 3.3).

Biotite **Biotite** is a dark, iron-rich member of the mica family (see Figure 3.25). Like other micas, biotite possesses a sheet structure that gives it excellent cleavage in one direction. Biotite also has a shiny black appearance that helps distinguish it from the other dark ferromagnesian minerals. Like hornblende, biotite is a common constituent of igneous rocks, including the rock granite.

◀ **Figure 3.31 Augite and hornblende** These dark-colored silicate minerals are common constituents of a variety of igneous rocks. (Photos by E. J. Tarbuck)

▲ **Figure 3.32 Well-formed garnet crystal** Garnets come in a variety of colors and are commonly found in mica-rich metamorphic rocks. (Photo by E. J. Tarbuck)

Garnet **Garnet** is similar to olivine in that its structure is composed of individual tetrahedrons linked by metallic ions. Also like olivine, garnet has a glassy luster, lacks cleavage, and exhibits conchoidal fracture. Although the colors of garnet are varied, this mineral is most often brown to deep red. Well-developed garnet crystals have 12 diamond-shaped faces and are most commonly found in metamorphic rocks (Figure 3.32). When transparent, garnets are prized as semiprecious gemstones.

CONCEPT CHECKS 3.7

1. Apart from their difference in color, what is one main distinction between light and dark silicates? What accounts for this difference?
2. Based on the chart in Figure 3.25, what do muscovite and biotite have in common? How do they differ?
3. Is color a good way to distinguish between orthoclase and plagioclase feldspar? If not, what is a more effective means of distinguishing them?

3.8 Important Nonsilicate Minerals

List the common nonsilicate minerals and explain why each is important.

Although the nonsilicates make up only about 8 percent of Earth's crust, some nonsilicate minerals, such as gypsum, calcite, and halite, occur as constituents in sedimentary rocks in significant amounts. Many nonsilicates are also economically important.

Nonsilicate minerals are typically divided into groups based on the negatively charged ion or complex ion that the members have in common. For example, the *oxides* contain negative oxygen ions (O^{2-}), which bond to one or more kinds of positive ions. Thus, within each mineral group, the basic structure and type of bonding is similar. As a result, the minerals in each group have similar physical properties that are useful in mineral identification. Figure 3.33 lists some of the major nonsilicate mineral groups and includes a few examples of each.

Some of the most common nonsilicate minerals belong to one of three groups of minerals: the carbonates (CO_3^{2-}), the sulfates (SO_4^{2-}), and the halides (Cl^{1-}, F^{1-}, Br^{1-}). The carbonate minerals are much simpler structurally than the silicates. This mineral group is composed of the carbonate ion (CO_3^{2-}) and one or more kinds of positive ions. The two most common carbonate minerals are **calcite**, $CaCO_3$ (calcium carbonate), and **dolomite**, $CaMg(CO_3)_2$ (calcium/magnesium carbonate) (Figure 3.33A,B). Calcite and dolomite are usually found together as the primary constituents in the sedimentary rocks limestone and dolostone. When calcite is the dominant mineral, the rock is called *limestone*, whereas *dolostone* results from a predominance of dolomite. Limestone is used in road aggregate and as a building stone, and it is the main ingredient in Portland cement.

Two other nonsilicate minerals frequently found in sedimentary rocks are **halite** and **gypsum** (Figure 3.33C,I). Both of these minerals are commonly found in thick layers that are the last vestiges of ancient seas that have long since evaporated (Figure 3.34). Like limestone, both halite and gypsum are important nonmetallic resources. Halite is the mineral name for common table salt (NaCl). Gypsum ($CaSO_4 \cdot 2H_2O$), which is calcium sulfate with water bound into the structure, is the mineral from which plaster and other similar building materials are composed.

Most nonsilicate mineral classes contain members that are prized for their economic value. This includes the oxides, whose members *hematite* and *magnetite* are important ores of iron (Figure 3.33E,F). Also significant are the sulfides, which are basically compounds of sulfur (S) and one or more metals. Important sulfide

Did You Know?
Gypsum, a white to transparent mineral, was first used as a building material in Anatolia (present-day Turkey) around 6000 B.C.E. It is also found on the interiors of the great pyramids in Egypt, which were erected in about 3700 B.C.E. Today, the average new American home contains more than 7 metric tons of gypsum in the form of 6000 sq ft of wallboard.

Common Nonsilicate Mineral Groups

Mineral Group (key ion(s) or element(s))	Mineral Name	Chemical Formula	Economic Use	Examples
Carbonates (CO_3^{2-})	Calcite Dolomite	$CaCO_3$ $CaMg(CO_3)_2$	Portland cement, lime Portland cement, lime	A. Calcite B. Dolomite
Halides (Cl^{1-}, F^{1-}, Br^{1-})	Halite Fluorite Sylvite	$NaCl$ CaF_2 KCl	Common salt Used in steelmaking Used as fertilizer	C. Halite D. Fluorite
Oxides (O^{2-})	Hematite Magnetite Corundum Ice	Fe_2O_3 Fe_3O_4 Al_2O_3 H_2O	Ore of iron, pigment Ore of iron Gemstone, abrasive Solid form of water	E. Hematite F. Magnetite
Sulfides (S^{2-})	Galena Sphalerite Pyrite Chalcopyrite Cinnabar	PbS ZnS FeS_2 $CuFeS_2$ HgS	Ore of lead Ore of zinc Sulfuric acid production Ore of copper Ore of mercury	G. Galena H. Chalcopyrite
Sulfates (SO_4^{2-})	Gypsum Anhydrite Barite	$CaSO_4 \cdot 2H_2O$ $CaSO_4$ $BaSO_4$	Plaster Plaster Drilling mud	I. Gypsum J. Anhydrite
Native elements (single elements)	Gold Copper Diamond Graphite Sulfur Silver	Au Cu C C S Ag	Trade, jewelry Electrical conductor Gemstone, abrasive Pencil lead Sulfadrugs, chemicals Jewelry, photography	K. Copper L. Sulfur

▲ **Figure 3.33 Important nonsilicate mineral groups** (Photos by Dennis Tasa and E. J. Tarbuck)

▶ **Figure 3.34 Thick bed of halite exposed in an underground mine** Halite (salt) mine in Grand Saline, Texas. Note the person for scale. (Photo by Tom Bochsler)

minerals include galena (lead sulfide), sphalerite (zinc sulfide), and chalcopyrite (copper sulfide). In addition, native elements—including gold, silver, and carbon (diamonds)— are economically important, as are a host of other nonsilicate minerals—fluorite (used as a flux in making steel), corundum (gemstone, abrasive), and uraninite (a uranium source).

CONCEPT CHECKS 3.8

1. List six common nonsilicate mineral groups. What key ions or elements define each group?
2. What is the most common carbonate mineral?
3. List eight common nonsilicate minerals and their economic uses.

3.9 Minerals: A Nonrenewable Resource

Discuss Earth's mineral resources in terms of renewability. Differentiate between mineral resources and ore deposits.

Earth's crust and oceans are the source of a wide variety of useful and valuable materials. From the first use of clay to make pottery nearly 10,000 years ago, the use of Earth materials has expanded, resulting in more complex societies and our modern civilization. The mineral and energy resources we extract from Earth's crust are the raw materials from which we make all the products we use.

Natural resources are typically grouped into broad categories according to (1) their ability to be regenerated (renewable or nonrenewable) or (2) their origin or type. Here we will consider mineral resources. However, other natural resources are indispensable to humans, including air, water, and solar energy.

Did You Know?
One of the world's heaviest cut and polished gemstones is a 22,892.5-carat golden-yellow topaz. Currently housed in the Smithsonian Institution, this roughly 10-lb gem is about the size of an automobile headlight and could hardly be used as a piece of jewelry, except perhaps by an elephant.

Renewable Versus Nonrenewable Resources

Resources classified as **renewable** can be replenished over relatively short time spans. Common examples are corn used for food and for making ethanol, natural fibers such as cotton for clothing, and forest products for lumber and paper. Energy from flowing water, wind, and the Sun are also considered renewable (**Figure 3.35**).

By contrast, many other basic resources are classified as **nonrenewable**. Important metals such as iron, aluminum, and copper fall into this category, as do our many widely used fuels: oil, natural gas, and coal. Although these and other resources form continuously, the processes that create them are so slow that significant deposits take millions of years to accumulate. Thus, for all practical purposes, Earth contains fixed quantities of these substances. The present supplies will be depleted as they are mined or pumped from the ground. Although some nonrenewable resources, such as the aluminum we use for containers, can be recycled, others, such as the oil burned for fuel, cannot.

Mineral Resources & Ore Deposits

Today, practically every manufactured product contains materials obtained from minerals. Figure 3.33 lists some of the most economically important mineral groups. **Mineral resources** are occurrences of useful minerals that are formed in such quantities that eventual extraction is reasonably certain. Mineral resources include deposits of metallic minerals that can be presently extracted profitably, as well as known deposits that are not yet economically or technologically recoverable. Materials used for such purposes as building stone, road aggregate, abrasives, ceramics, and fertilizers are not usually called mineral resources; rather, they are classified as *industrial rocks and minerals*.

◀ **Figure 3.35 Solar energy is renewable** The Ivanpah Solar Electric Generating System is a solar thermal plant located in California's Mojave Desert, southwest of Las Vegas. It consists of 173,500 heliostats (mirrors that move so they reflect sunlight at a target), each with two mirrors that focus solar energy on boilers located on one of three centralized towers. The boilers, in turn, generate steam which turns turbines that generate electricity. (Photo by Steve Proehl/Getty Images/Corbis Documentary)

An **ore deposit** is a naturally occurring concentration of one or more metallic minerals that can be extracted economically. In common usage, the term *ore* is also applied to some nonmetallic minerals such as fluorite and sulfur. Recall that more than 98 percent of Earth's crust is composed of only eight elements, and except for oxygen and silicon, all other elements make up a relatively small fraction of common crustal rocks (see Figure 3.22). Indeed, the natural concentrations of many elements are exceedingly small. A deposit containing the average concentration of an element such as gold has no economic value because the cost of extracting it greatly exceeds the value of the gold that could be recovered.

In order to have economic value, an ore deposit must be highly concentrated. For example, copper makes up about 0.0068 percent of the crust. For a deposit to be considered a copper ore, it must contain a concentration of copper that is about 100 times this amount, or about 0.68 percent. Aluminum, on the other hand, represents about 8.1 percent of the crust and can be extracted profitably when it is found in concentrations 3 or 4 times that amount.

It is important to understand that due to economic or technological changes, a deposit may either become profitable to extract or lose its profitability. If the demand for a metal increases and its value rises sufficiently, the status of a previously unprofitable deposit can be upgraded from a mineral to an ore. Technological advances that allow a resource to be extracted more efficiently and, thus, more profitably than before may also trigger a change of status.

Conversely, changing economic factors can turn what was once a profitable ore deposit into an unprofitable mineral deposit. This situation was illustrated at the copper mining operation located at Bingham Canyon, Utah, one of the largest open-pit mines on Earth (**Figure 3.36**). Mining was halted there in 1985 because outmoded equipment had driven the cost of extracting the copper beyond the current selling price. In 1989 new owners responded by replacing an antiquated 1,000-car railroad with more modern conveyor belts and dump trucks for efficiently transporting the ore and waste. The advanced equipment reduced extraction costs by nearly 30 percent, ultimately returning the copper mine operation to profitability. Today the Bingham Canyon mine produces nearly 25 percent of the refined copper in the United States. In addition to producing 300,000 metric tons of copper, the Bingham Canyon mine produces about 400,000 ounces of gold, 4 million ounces of silver, and 25 million pounds of molybdenum.

Over the years, geologists have been keenly interested in learning how natural processes produce localized concentrations of essential minerals. One well-established fact is that occurrences of valuable mineral resources are closely related to the rock cycle. That is, the mechanisms that generate igneous, sedimentary, and metamorphic rocks, including the processes of

Did You Know?

The names of precious gems often differ from the names of parent minerals. For example, *sapphire* is one of two gems that are varieties of the same mineral, *corundum*. Tiny amounts of the elements titanium and iron in corundum produce the most prized blue sapphires. When the mineral corundum contains chromium, it exhibits a brilliant red color, and the gem is called *ruby*.

▲ **Figure 3.36** **Aerial view of Bingham Canyon copper mine near Salt Lake City, Utah** Although the amount of copper in the rock is less than 0.5 percent, the huge volume of material removed and processed each day (over 400,000 tons) yields enough metal to be profitable. In addition to copper, this mine produces gold, silver, and molybdenum. (Photo by Michael Collier)

weathering and erosion, play a major role in producing concentrated accumulations of useful elements.

Moreover, with the development of the theory of plate tectonics, geologists have added another tool for understanding the processes by which one rock is transformed into another. As these rock-forming processes are examined in the following chapters, we consider their role in producing some of our important mineral resources.

CONCEPT CHECKS 3.9

1. List three examples of renewable resources and three examples of nonrenewable resources.
2. Compare and contrast a *mineral resource* and an *ore deposit*.
3. Explain how a mineral deposit that previously could not be mined profitably might be upgraded to an ore deposit.

CONCEPTS IN REVIEW

Matter & Minerals

3.1 Minerals: Building Blocks of Rocks

List the main characteristics that an Earth material must possess to be considered a mineral and describe each characteristic.

KEY TERMS: mineralogy, mineral, rock

- In Earth science, the word *mineral* refers to naturally occurring inorganic solids that possess an orderly crystalline structure and a characteristic chemical composition. The study of minerals is called *mineralogy*.
- Minerals are the building blocks of rocks. Rocks are naturally occurring masses of minerals or mineral-like matter such as natural glass or organic material.

3.2 Atoms: Building Blocks of Minerals

Compare and contrast the three primary particles contained in atoms.

KEY TERMS: atom, nucleus, proton, neutron, electron, valence electron, atomic number, element, periodic table, chemical compound

- Minerals are composed of atoms of one or more elements. All atoms consist of the same three basic components: protons, neutrons, and electrons.
- The atomic number represents the number of protons found in the nucleus of an atom of a particular element. For example, an oxygen atom has eight protons, so its atomic number is eight. Protons and neutrons have approximately the same size and mass, but protons are positively charged, whereas neutrons have no charge.
- Electrons weigh only about 1/2000 as much as protons or neutrons. They occupy the space around the nucleus, where they form what can be thought of as a cloud that is structured into several distinct energy levels called principal shells. The electrons in the outermost principal shell, called valence electrons, are responsible for the bonds that hold atoms together to form chemical compounds.
- Elements that have the same number of valence electrons tend to behave similarly. The periodic table is organized so that elements with the same number of valence electrons form a column, called a group.

? Use the periodic table (see Figure 3.5) to identify the geologically important elements that have the following numbers of protons: (A) 14, (B) 6, (C) 13, (D) 17, and (E) 26.

3.3 Why Atoms Bond

Distinguish among ionic bonds, covalent bonds, and metallic bonds.

KEY TERMS: octet rule, chemical bond, ionic bond, ion, covalent bond, metallic bond

- When atoms are attracted to other atoms, they can form chemical bonds, which generally involve the transfer or sharing of valence electrons. The most stable arrangement for most atoms is to have eight electrons in the outermost principal shell. This concept is called the octet rule.
- To form ionic bonds, atoms of one element give up one or more valence electrons to atoms of another element, forming positively and negatively charged atoms called ions. The ionic bond results from the attraction between oppositely charged ions.
- Covalent bonds form when adjacent atoms share valence electrons.
- In metallic bonds, the sharing is more extensive: the shared valence electrons can move freely through the substance.

(3.3 continued)

? Which of the accompanying diagrams (A, B, or C) best illustrates ionic bonding? What are the distinguishing characteristics of ionic versus covalent bonding?

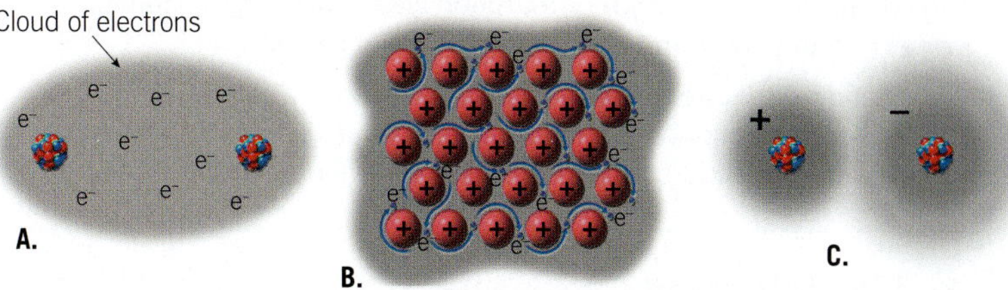

3.4 Properties of Minerals

List and describe the properties used in mineral identification.

KEY TERMS: diagnostic property, ambiguous property, luster, color, streak, crystal shape, (habit), hardness, Mohs scale, cleavage, fracture, tenacity, density, specific gravity

- The composition and internal crystalline structure of a mineral give it specific physical properties. Mineral properties useful in identifying minerals are termed *diagnostic properties*.
- Luster is a mineral's ability to reflect light. The terms *transparent*, *translucent*, and *opaque* describe the degree to which a mineral can transmit light. Color can be unreliable for mineral identification, as impurities can "stain" minerals with diverse colors. A more reliable identifier is streak, the color of the powder generated by scraping a mineral against a porcelain streak plate.
- Crystal shape, also called crystal habit, is often useful for mineral identification.
- Variations in the strength of chemical bonds give minerals properties such as hardness (resistance to being scratched) and tenacity (tendency to break in a brittle fashion or bend when stressed). Cleavage, the preferential breakage of a mineral along planes of weakly bonded atoms, is very useful in identifying minerals.
- The amount of matter packed into a given volume determines a mineral's density. To compare the densities of minerals, mineralogists use a related quantity, known as specific gravity, which is the ratio between a mineral's density and the density of water.
- Other properties are diagnostic for certain minerals but rare in most others—examples include smell, taste, feel, reaction to hydrochloric acid, magnetism, and double refraction.

? Research the minerals *quartz* and *calcite*. List five physical characteristics that may be used to distinguish one from the other.

Quartz Dennis Tasa **Calcite** Dennis Tasa

3.5 Mineral Groups

Explain how minerals are classified and name the most abundant mineral group in Earth's crust.

KEY TERMS: rock-forming mineral, economic mineral, silicate, nonsilicate

- More than 4000 different minerals have been identified, but only a few dozen are common in Earth's crust: These are the rock-forming minerals. Many minerals have economic value.
- Minerals are placed into classes on the basis of similar crystal structures and compositions. Minerals of the same class tend to have similar properties and are found in similar geologic settings.
- Silicon and oxygen are the most common elements in Earth's crust, and so the most common minerals in the crust are silicate minerals. In comparison, nonsilicate minerals make up only about 8 percent of the crust.

3.6 The Silicates

Sketch the silicon–oxygen tetrahedron and explain how this fundamental building block joins together to form various silicate structures.

KEY TERMS: silicon–oxygen tetrahedron, polymerization

- Silicate minerals have a basic building block in common: a small pyramid-shaped structure consisting of one silicon atom surrounded by four oxygen atoms. Because this structure has four sides, it is called the *silicon–oxygen tetrahedron*. Individual tetrahedrons can be bonded to other elements, such as aluminum, iron, or potassium. Neighboring tetrahedrons can share some of their oxygen atoms, causing them to develop long chains. This is the process of polymerization.
- Polymerization can produce silicate mineral structures with high or low degrees of oxygen sharing. The more sharing there is, the higher the ratio of silicon to oxygen. Polymerization can produce unit cells that are single "chains" of tetrahedrons, double chains, sheets of shared tetrahedrons, or even complicated three-dimensional networks of tetrahedrons that share all the oxygen atoms in the mineral.

3.7 Common Silicate Minerals

Compare and contrast the light (nonferromagnesian) silicates with the dark (ferromagnesian) silicates and list three common minerals in each group.

KEY TERMS: light, (nonferromagnesian), silicate, potassium feldspar, plagioclase feldspar, quartz, muscovite, clay, dark, (ferromagnesian), silicate, olivine, augite, hornblende, biotite, garnet

- Silicate minerals are the most common mineral class on Earth. They are subdivided into minerals that contain iron and/or magnesium (dark, or ferromagnesian, silicates) and those that do not (light, or nonferromagnesian, silicates).
- Nonferromagnesian silicates are generally light in color and generally of relatively low specific gravity. Feldspar, quartz, muscovite, and clays are all examples.

(3.7 continued)

- Ferromagnesian silicates are generally dark in color and relatively dense. Olivine, pyroxene, amphibole, biotite, and garnet are all examples.

? In general, nonferromagnesian silicates are light in color: shades of peach, tan, clear, or white. What could account for the fact that some nonferromagnesian silicates are dark colored, like the smoky quartz in this photo?

Dennis Tasa

Smoky quartz

3.8 Important Nonsilicate Minerals

List the common nonsilicate minerals and explain why each is important.

KEY TERMS: calcite, dolomite, halite, gypsum

- Nonsilicate mineral groups don't include the silicon–oxygen tetrahedron. Instead, these minerals use other negatively charged ions or complex ions.
- Nonsilicate minerals are grouped on the basis of their negatively charged ion or ion complex. Common groups are the oxides (O^{2-}), carbonates (CO_3^{2-}), sulfates (SO_4^{2-}), and halides (Cl^-, Br^-, F^-).
- Nonsilicate minerals are often economic minerals. Hematite is an important source of industrial iron, while calcite is a critical component of cement. Halite (table salt) makes popcorn taste good.

3.9 Minerals: A Nonrenewable Resource

Discuss Earth's mineral resources in terms of renewability. Differentiate between mineral resources and ore deposits.

KEY TERMS: renewable, nonrenewable, mineral resource, ore deposit

- Resources are classified as renewable when they can be replenished over short time spans and nonrenewable when they can't.
- Ore deposits are naturally occurring concentrations of one or more metallic minerals that can be extracted economically using current technology. A mineral resource can be upgraded to an ore deposit if the price of the commodity increases sufficiently or if the cost of extraction decreases. The reverse can also happen.

GIVE IT SOME THOUGHT

1 Using the geologic definition of *mineral* as your guide, determine which of the items in this list are minerals and which are not. If something in this list is not a mineral, explain.

a. Gold nugget
b. Seawater
c. Quartz
d. Cubic zirconia
e. Obsidian
f. Ruby
g. Glacial ice
h. Amber

2 Assume that the number of protons in a neutral atom is 92 and the atomic mass is 238.03. (*Hint:* Refer to the periodic table in Figure 3.5 to answer this question.)

a. What is the name of the element?
b. How many electrons does it have?

3 Referring to the accompanying photos of five minerals, determine which of these specimens exhibit a metallic luster and which have a nonmetallic luster.

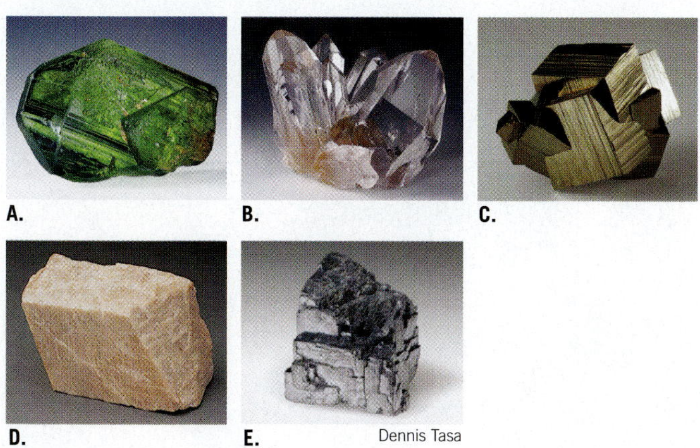

Dennis Tasa

4 Gold has a specific gravity of almost 20. A 5-gallon bucket of water weighs 40 pounds. How much would a 5-gallon bucket of gold weigh?

5 Examine the accompanying photo of a mineral that has several smooth, flat surfaces that resulted when the specimen was broken.

a. How many flat surfaces are present on this specimen?

b. How many different directions of cleavage does this specimen have?

c. Do the cleavage directions meet at 90-degree angles?

E. J. Tarbuck

Cleaved sample

6 Each of the following statements describes a silicate mineral or mineral group. In each case, provide the appropriate name:

a. The most common member of the amphibole group

b. The most common light-colored member of the mica family

c. The only common silicate mineral made entirely of silicon and oxygen

d. A silicate mineral with a name based on its color

e. A silicate mineral characterized by striations

f. Silicate minerals that originate as a product of chemical weathering

7 What mineral property is illustrated in the accompanying photo?

Dennis Tasa

8 Do an Internet search to determine which minerals are used to manufacture the following products:

a. Stainless steel utensils

b. Cat litter

c. Tums brand antacid tablets

d. Lithium batteries

e. Aluminum beverage cans

9 Examine Figure 3.33. What would be the most abundant elements on a planet composed of mostly halide minerals instead of silicate minerals? What about a carbonate planet?

10 The accompanying diagram shows one of several possible ways that silicon–oxygen tetrahedrons can bond together. Describe the silica structure shown and name a mineral group that displays this type of silicate structure.

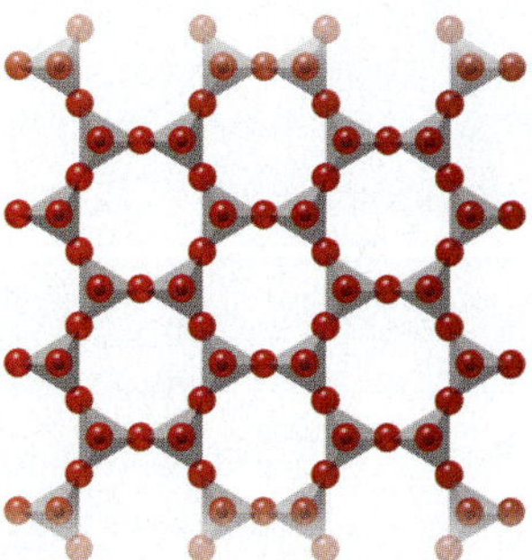

MasteringGeology™

Looking for additional review and test prep materials? Visit the Study Area in MasteringGeology to enhance your understanding of this chapter's content by accessing a variety of resources, including Self-Study Quizzes, Geoscience Animations, SmartFigures, Mobile Field Trips, *Project Condor* Quadcopter videos, *In the News* RSS feeds, flashcards, web links, and an optional Pearson eText.

www.masteringgeology.com

4

Igneous Rocks & Intrusive Activity

FOCUS ON CONCEPTS

Each statement represents the primary learning objective for the corresponding major heading within the chapter. After you complete the chapter, you should be able to:

4.1 List and describe the three major components of magma.

4.2 Compare and contrast the four basic igneous compositions: felsic, intermediate, mafic, and ultramafic.

4.3 Identify and describe the six major igneous textures.

4.4 Distinguish among the common igneous rocks based on texture and mineral composition.

4.5 Summarize the major processes that generate magma from solid rock.

4.6 Describe how magmatic differentiation can generate a magma body that has a mineralogy (chemical composition) that is different from its parent magma.

4.7 Describe how partial melting of the mantle rock peridotite can generate a basaltic (mafic) magma.

4.8 Compare and contrast these intrusive igneous structures: dikes, sills, batholiths, stocks, and laccoliths.

4.9 Explain how economic deposits of gold, silver, and many other metals form.

Granite outcrops reflected in Tenaya Lake, Yosemite National Park, California. (Photo by Adan Burton/Robert Harding)

UNDERSTANDING THE STRUCTURE, composition, and internal workings of our planet requires a basic knowledge of igneous rocks. Igneous rocks and metamorphic rocks derived from igneous "parents" make up most of Earth's crust and mantle. Thus, Earth can be described as a huge mass of igneous and metamorphic rocks that is covered with a thin veneer of sedimentary rock and has a relatively small iron-rich core.

Many prominent landforms are composed of igneous rocks, including volcanoes such as Mount Rainier and the large igneous bodies that make up the Sierra Nevada, the Black Hills, and the high peaks of the Adirondacks. Igneous rocks also make excellent building stones and are widely used as decorative materials, such as for monuments and household countertops.

4.1 Magma: Parent Material of Igneous Rock

List and describe the three major components of magma.

Our discussion of the rock cycle in Chapter 1 explained that **igneous rocks** (*ignis* = fire) form as molten rock cools and solidifies. Considerable evidence supports the idea that the parent material for igneous rocks, called **magma**, is formed by partial melting that occurs at various levels within Earth's crust and upper mantle to depths of about 250 kilometers (about 150 miles). Once formed, a magma body buoyantly rises toward the surface because it is less dense than the surrounding rocks. (When rock melts, it takes up more space and, hence, it becomes less dense than the surrounding solid rock.) Occasionally, molten rock reaches Earth's surface, where it is called **lava** (Figure 4.1). Sometimes lava is emitted as fountains that are produced when escaping gases propel it from a magma chamber. On other occasions, lava is explosively ejected, producing dramatic eruptions of steam and volcanic ash. However, not all eruptions are violent; many volcanoes emit quiet outpourings of fluid lava.

▼ **Figure 4.1 Eruption of Mount Etna, July 2014, Sicily, Italy.** (Photo courtesy of AM Design/Alamy)

The Nature of Magma

Magma is rock that is completely or partly molten, and when cooled it solidifies to form igneous rocks mainly composed of silicate minerals. Most magmas consist of three materials: a *liquid* component, a *solid* component, and a *gaseous* component.

The liquid portion, called **melt**, is composed mainly of mobile ions of the eight most common elements found in Earth's crust—silicon and oxygen, along with lesser amounts of aluminum, potassium, calcium, sodium, iron, and magnesium (see Figure 3.22, page 79).

The solid components (if any) in magma are crystals of silicate minerals. As a magma body cools, the size and number of crystals increase. During the last stage of cooling, a magma body is like a "crystalline mush" (resembling a bowl of very thick oatmeal) that contains only small amounts of melt.

The gaseous components of magma, called **volatiles**, are materials that vaporize (form a gas) at surface pressures. The most common volatiles found in magma are water vapor (H_2O), carbon dioxide (CO_2), and sulfur dioxide (SO_2). When magma is deep below the surface, the immense confining pressure keeps these volatiles dissolved in the melt, the way carbon dioxide is dissolved in soda before you open the can. As the melt rises toward the surface and the confining pressure is reduced, the volatiles begin to separate from the melt—again, similar to the way carbon dioxide forms bubbles when you reduce the pressure in a soda can by opening it. As the gases build up, they may eventually propel magma from the vent. When deeply buried magma bodies crystallize, the remaining volatiles collect as hot, water-rich fluids that migrate through openings in the surrounding rocks. These hot fluids play an important role in metamorphism and will be considered in Chapter 8.

From Magma to Crystalline Rock

To better understand how magma crystallizes, let us first consider how a simple crystalline solid melts. Recall that, in any crystalline solid, the ions are arranged in a closely packed regular pattern. However, they are not without some motion; they exhibit a restricted vibration about fixed points. As the temperature rises, ions vibrate more rapidly and consequently collide with ever-increasing vigor with their neighbors. Thus, heating causes the ions to occupy more space, which in turn causes the solid to expand. When the ions are vibrating rapidly enough to overcome the force of their chemical bonds, melting occurs. At this stage, the ions are able to slide past one another, and the orderly crystalline structure disintegrates. Thus, melting converts a solid consisting of tight, uniformly packed ions into a liquid composed of unordered ions moving randomly about.

In the process called **crystallization**, cooling reverses the events of melting. As the temperature of the liquid drops, ions pack more closely together as their rate of movement slows. When they are cooled sufficiently, the forces of the chemical bonds again confine the ions to an orderly crystalline arrangement.

When a magma body cools, the silicon and oxygen atoms link together first to form silicon–oxygen tetrahedra, the basic building blocks of the silicate minerals (see Figure 3.23, page 80). As magma continues to lose heat to its surroundings, the tetrahedra join with each other and with other ions to form embryonic crystal nuclei. Each nucleus slowly grows as ions lose their mobility and join the crystalline network.

The minerals that form the earliest have space to grow and tend to have better-developed crystal faces than do the ones that form later and occupy the remaining spaces. Eventually all of the melt is transformed into a solid mass of interlocking silicate minerals that we call an *igneous rock* (**Figure 4.2**).

◀ **Figure 4.2 Igneous rock composed of interlocking crystals** The largest crystals are about 1 centimeter in length.

Igneous Processes

Igneous rocks form in two basic settings. Molten rock may crystallize within Earth's crust over a range of depths, or it may solidify at Earth's surface (**Figure 4.3**).

▼ **SmartFigure 4.3 Intrusive versus extrusive igneous rocks**

TUTORIAL
https://goo.gl/ac0bu5

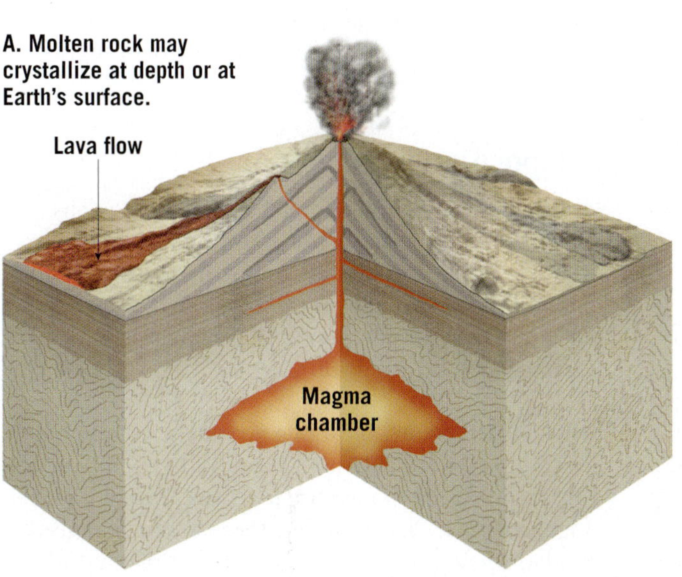

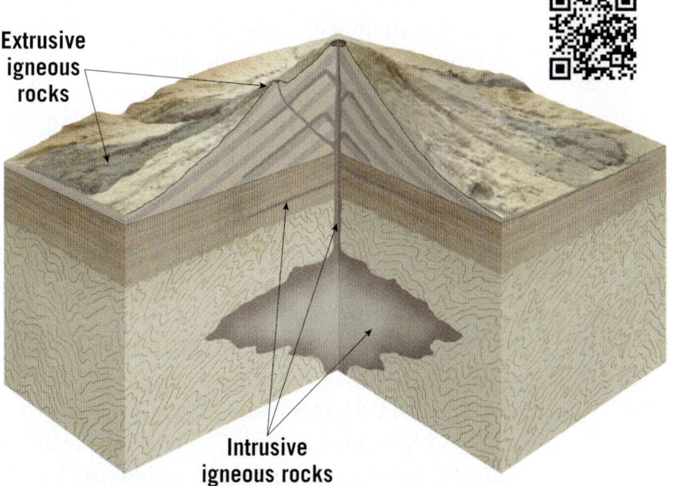

B. When magma crystallizes at depth, intrusive igneous rocks form. When magma solidifies on Earth's surface, extrusive igneous rocks form.

▲ **Figure 4.4 Mount Rushmore National Memorial** This memorial, located in the Black Hills of South Dakota, is carved from intrusive igneous rocks. (Photo by Barbara A. Harvey/Shutterstock)

When magma crystallizes *at depth*, it forms **intrusive igneous rocks**, also known as **plutonic rocks**—after Pluto, the god of the underworld in classical mythology. These rocks can be observed at the surface in locations where uplifting and erosion have stripped away the overlying rocks. Exposures of intrusive igneous rocks occur in many places, including the White Mountains, New Hampshire; Stone Mountain, Georgia; Mount Rushmore in the Black Hills of South Dakota; and Yosemite National Park, California (Figure 4.4).

Igneous rocks that form when molten rock solidifies *at the surface* are classified as **extrusive igneous rocks**. They are also called **volcanic rocks**—after Vulcan, the Roman fire god. Extrusive igneous rocks form when lava solidifies or when volcanic debris falls to Earth's surface. Extrusive igneous rocks are abundant in western portions of the Americas, where they make up the volcanic peaks of the Cascade Range and the Andes Mountains. In addition, many oceanic islands, including the Hawaiian chain and Alaska's Aleutian Islands, are composed almost entirely of extrusive igneous rocks. The nature of volcanic activity will be addressed in more detail in Chapter 5.

CONCEPT CHECKS 4.1

1. What is magma? How does magma differ from lava?
2. List and describe the three components of magma.
3. Describe the process of crystallization.
4. Compare and contrast extrusive and intrusive igneous rocks.

4.2 Igneous Compositions

Compare and contrast the four basic igneous compositions: felsic, intermediate, mafic, and ultramafic.

Igneous rocks are composed mainly of silicate minerals. Chemical analyses show that silicon (Si) and oxygen (O) are by far the most abundant constituents of igneous rocks. These two elements, plus ions of aluminum (Al), calcium (Ca), sodium (Na), potassium (K), magnesium (Mg), and iron (Fe), make up roughly 98 percent, by weight, of most magmas. In addition, magma contains small amounts of many other elements, including titanium and manganese, and trace amounts of rare elements, such as gold, silver, and uranium.

As magma cools and solidifies, these elements combine to form two major groups of silicate minerals. The *dark* (or *ferromagnesian*) *silicates* are rich in iron and/or magnesium and comparatively low in silica. *Olivine*, *pyroxene*, *amphibole*, and *biotite mica* are the common dark silicate minerals of Earth's crust. By contrast, the *light* (or *nonferromagnesian*) *silicates* contain greater amounts of potassium, sodium, and calcium. The light silicate minerals, including *quartz*, *muscovite mica*, and the most abundant mineral group, the *feldspars*, are richer in silica than the dark silicates.

Compositional Categories

Despite the great compositional diversity of igneous rocks, geologists classify these rocks (and the magmas from which they form) into four broad groups according to their proportions of light and dark minerals. As shown in Figure 4.5, these compositional groups are *felsic*, *intermediate*, *mafic*, and *ultramafic*.

Felsic Versus Mafic Near one end of the continuum are rocks composed almost entirely of light-colored silicates—quartz and potassium feldspar. The composition of igneous rocks dominated by these minerals is classified as **felsic**, a term derived from *fel*dspar and *si*lica (quartz). Because felsic magmas most commonly solidify to form *granite*, geologists also refer to this type of

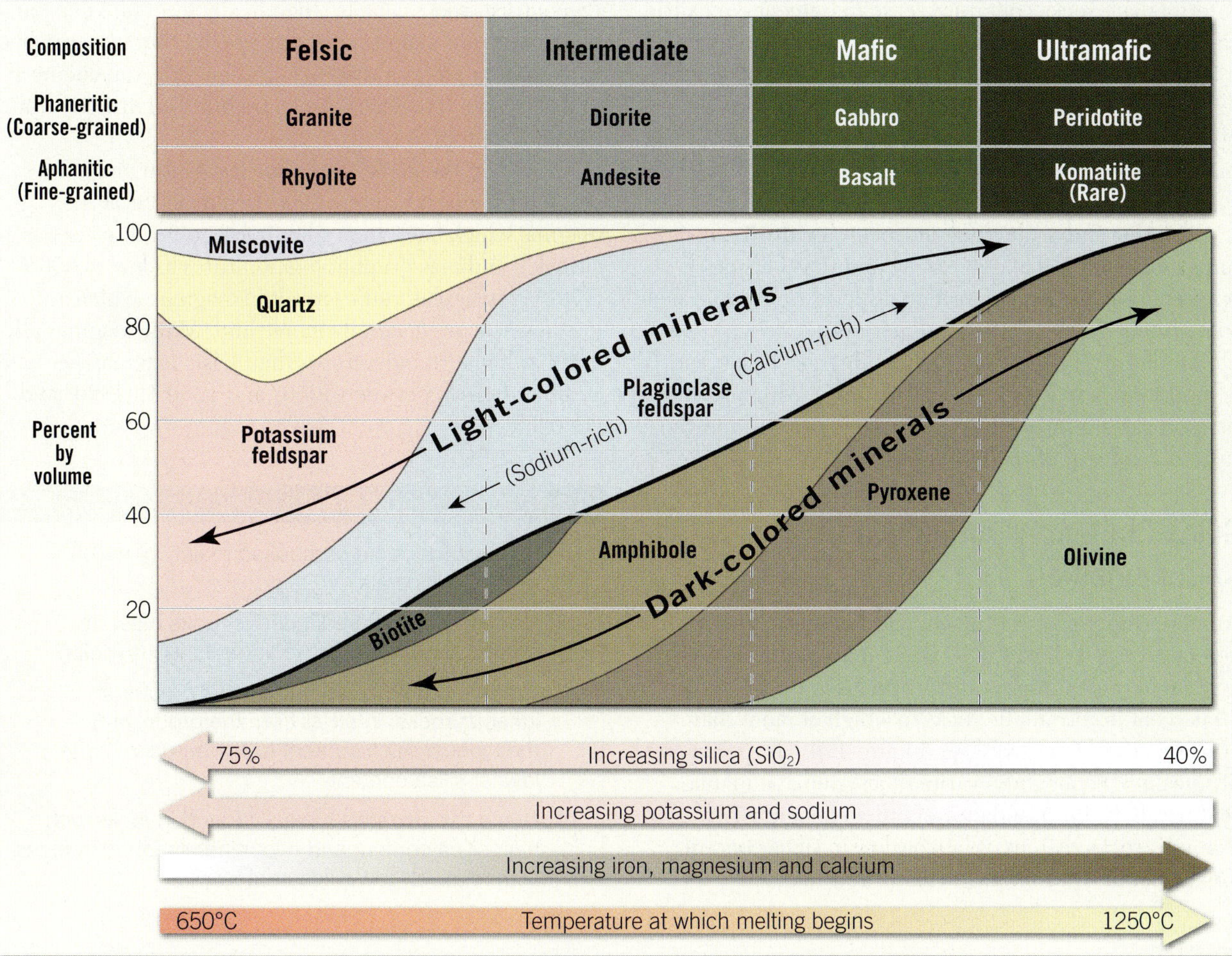

◀ SmartFigure 4.5 Mineral makeup of common igneous rocks

TUTORIAL
https://goo.gl/goy5yO

▼ **Figure 4.6 Granitic (felsic) versus basaltic (mafic) compositions** Inset images are photomicrographs that show the interlocking crystals that make up granite and basalt, respectively. (Photos provided by E. J. Tarbuck)

magma as having a **granitic composition**. In addition to quartz and feldspar, most granitic rocks contain about 10 percent dark silicate minerals, usually biotite mica and amphibole (Figure 4.6A). Other notable characteristics of felsic rocks and the magma from which they were derived is that they are rich in silica (about 70 percent, or more) and are major constituents of the continental crust.

Rocks that contain at least 45 percent dark silicates (ferromagnesian minerals) are classified as **mafic** (from *ma*gnesium and *f*errum, the Latin name for iron). As a result of their iron content, mafic rocks are typically darker and denser than felsic rocks. Because mafic magmas most often solidify to form the igneous rock *basalt*, geologists also refer to this type of magma as having a **basaltic composition** (Figure 4.6B). Basaltic rocks make up the ocean floor as well as many of the volcanic islands located within the ocean basins. Basalt also forms extensive lava flows on the continents.

Other Compositional Groups As illustrated in Figure 4.5, rocks with a composition between felsic and mafic rocks are said to have an **intermediate**, or **andesitic composition**, after the common volcanic rock *andesite*.

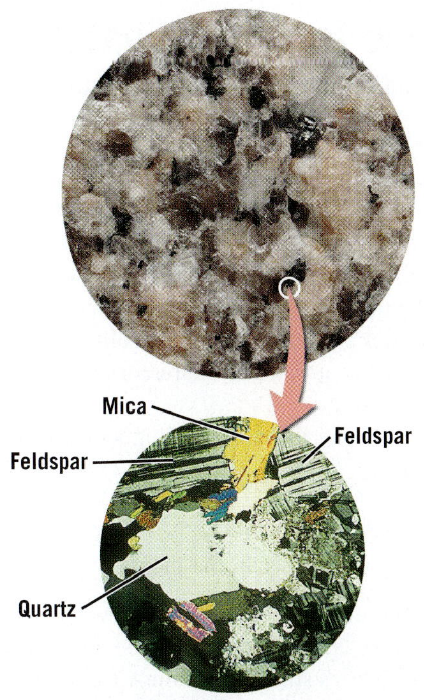

A. Granite is a felsic, coarse-grained igneous rock composed of light-colored silicates—quartz and potassium feldspar.

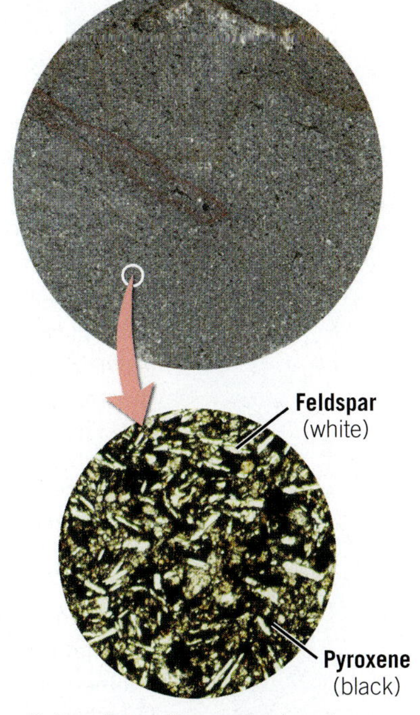

B. Basalt is a fine-grained mafic igneous rock containing substantial amounts of dark colored silicates and plagioclase feldspar.

Did You Know?
The most abundant element in Earth's crust is *oxygen*. It makes up 47 percent of Earth's crust by weight and 94 percent by volume. In fact, oxygen atoms are so much bigger than most of the other atoms in common minerals that Earth's crust is mainly oxygen atoms packed closely together, with smaller atoms such as silicon, aluminum, and potassium tucked in between.

Intermediate rocks contain at least 25 percent dark silicate minerals, mainly amphibole, pyroxene, and biotite mica, with the other dominant mineral being plagioclase feldspar. This important category of igneous rocks is often associated with volcanic activity on the seaward margins of continents and on volcanic island arcs such as the Aleutian chain.

Another important igneous rock, **peridotite**, contains mostly olivine and pyroxene and thus falls on the opposite side of the compositional spectrum from felsic rocks (see Figure 4.5). Because peridotite is composed almost entirely of ferromagnesian minerals, its chemical composition is referred to as **ultramafic**. Although ultramafic rocks are rare at Earth's surface, peridotite is the main constituent of the upper mantle.

Silica Content as an Indicator of Composition

An important aspect of the chemical composition of igneous rocks is silica (SiO_2) content. Typically, the silica content of crustal rocks ranges from as low as about 40 percent in ultramafic rocks to a high of more than 70 percent in felsic rocks (see Figure 4.5). The percentage of silica in igneous rocks varies in a systematic manner that parallels the abundance of other elements. For example, rocks that are relatively low in silica contain large amounts of iron, magnesium, and calcium. By contrast, rocks that are high in silica contain comparatively less iron, magnesium, and calcium but are enriched with sodium and potassium. Consequently, the chemical makeup of an igneous rock can be inferred directly from its silica content.

Further, the amount of silica present in magma strongly influences the magma's behavior. Felsic magma, which has a high silica content, is quite viscous ("thick") and may erupt at temperatures as low as 650°C (1200°F), whereas mafic (basaltic) magmas, which are low in silica, are generally more fluid. Mafic magmas also erupt at higher temperatures than felsic magmas—usually at temperatures between 1050° and 1250°C (1920° and 2280°F).

CONCEPT CHECKS 4.2

1. Igneous rocks are composed mainly of which group of minerals?
2. How do light-colored igneous rocks differ in composition from dark-colored igneous rocks?
3. List the four basic compositional groups of igneous rocks, in order from the group with the highest silica content to the group with the lowest silica content.
4. Name two minerals typically found in rocks with high silica content and two minerals found in rocks with relatively low silica content.

4.3 Igneous Textures: What Can They Tell Us?

Identify and describe the six major igneous textures.

The term **texture** is used to describe the overall appearance of a rock based on the size, shape, and arrangement of its mineral grains—not how it feels to touch. Texture is an important property because it reveals a great deal about the environment in which the rock formed (**Figure 4.7**). Geologists can make inferences about a rock's origin based on careful observations of grain size and other characteristics of the rock.

Three factors influence the textures of igneous rocks:

- The rate at which the molten rock cools
- The amount of silica in the magma
- The amount of dissolved gases in the magma

Among these, the rate of cooling tends to be the dominant factor. A very large magma body located many kilometers beneath Earth's surface remains insulated from lower surface temperatures by the surrounding rock and thus cools very slowly over a period of perhaps tens of thousands to millions of years. Initially, a relatively small number of crystal nuclei form. This slow cooling permits ions to migrate freely until they eventually join one of the existing crystals. Consequently, slow cooling promotes the growth of fewer but larger crystals.

On the other hand, when cooling occurs rapidly—for example, in a thin lava flow—the ions quickly lose their mobility and readily combine to form crystals. This results in the development of numerous embryonic crystal nuclei, all of which compete for the available ions. The result is a solid mass of many tiny intergrown crystals.

Types of Igneous Textures

In addition to cooling quickly or slowly, a magma body may migrate to a new location or erupt at the surface before it completely solidifies. As a result, several types of igneous textures exist, including aphanitic (fine-grained), phaneritic (coarse-grained), porphyritic, vesicular, glassy, and pyroclastic (fragmented).

A. Glassy texture
Composed of unordered atoms and resembles dark manufactured glass. (Obsidian is a natural glass that usually forms when highly silica-rich magmas solidify.)

B. Porphyritic texture
Composed of two distinctly different crystal sizes.

C. Phaneritic (coarse-grained) texture
Composed of mineral grains that are large enough to be identified without a microscope.

D. Vesicular texture
Extrusive rock containing voids left by gas bubbles that escape as lava solidifies. (Pumice is a frothy volcanic glass that displays a vesicular texture.)

E. Pyroclastic (fragmental) texture
Produced by the consolidation of fragments that may include ash, once molten blobs, or large angular blocks that were ejected during an explosive volcanic eruption.

F. Aphanitic (fine-grained) texture
Composed of crystals that are too small for the individual minerals to be identified without a microscope.

▲ SmartFigure 4.7
Igneous rock textures
(Photos by Dennis Tasa and E. J. Tarbuck)

TUTORIAL
https://goo.gl/pYPXWg

Aphanitic (Fine-Grained) Texture Igneous rocks that form at the surface or as small intrusive masses within the upper crust where cooling is relatively rapid exhibit a **fine-grained texture** termed **aphanitic** (*a* = not, *phaner* = visible). By definition, the crystals that make up aphanitic rocks are so small that individual minerals can be distinguished only with the aid of a polarizing microscope or using sophisticated techniques (see Figure 4.6B and Figure 4.7F). Therefore, we commonly characterize fine-grained rocks as being light, intermediate, or dark in color. Using this system, light-colored aphanitic rocks are those containing primarily light-colored nonferromagnesian silicate minerals.

Phaneritic (Coarse-Grained) Texture When large masses of magma slowly crystallize at great depth, they form igneous rocks that exhibit a **coarse-grained texture** described as **phaneritic** (*phaner* = visible). Coarse-grained rocks consist of a mass of intergrown crystals that are roughly equal in size and large enough for the individual minerals to be distinguished without the aid of a microscope (see Figure 4.6A and Figure 4.7C). Geologists often use a small magnifying lens to aid in identifying minerals in phaneritic rocks.

Porphyritic Texture A large mass of magma may require thousands or even millions of years to solidify. Because different minerals crystallize under different

▲ **Figure 4.8 Porphyritic texture** The large crystals in porphyritic rocks are called *phenocrysts*, and the matrix of smaller crystals is called *groundmass*. (Photo by Dennis Tasa)

environmental conditions (temperatures and pressure), it is possible for crystals of one mineral to become quite large before others even begin to form. If molten rock containing some large crystals moves to a different environment—for example, by erupting at the surface—the remaining liquid portion of the lava cools more quickly. The resulting rock, which has large crystals embedded in a matrix of smaller crystals, is said to have a **porphyritic texture** (see Figure 4.7B and Figure 4.8). The large crystals in porphyritic rocks are termed **phenocrysts** (*pheno* = show, *cryst* = crystal), whereas the matrix of smaller crystals is called **groundmass**. A rock with a *porphyritic* texture is termed a **porphyry**.

Vesicular Texture Common features of many extrusive rocks are the voids left by gas bubbles that escape as lava solidifies. These nearly spherical openings are called *vesicles*, and the rocks that contain them are said to have a **vesicular texture**. Rocks that exhibit a vesicular texture often form in the upper zone of a lava flow, where cooling occurs rapidly enough to preserve the openings produced by the expanding gas bubbles (Figure 4.9). Another common vesicular rock, called *pumice*, forms when silica-rich lava is ejected during an explosive eruption (see Figure 4.7D).

Glassy Texture During some volcanic eruptions, molten rock is ejected into the atmosphere, where it is quenched (very quickly cooled) to become a solid (see Figure 4.7A). Rapid cooling of this type may generate rocks having a **glassy texture**. Glass results when unordered ions are "frozen in place" before they are able to unite into an orderly crystalline structure.

Obsidian, a common type of natural glass, is similar in appearance to dark chunks of manufactured glass. Obsidian's excellent conchoidal fracture and ability to hold a sharp, hard edge made it a prized material from which Native Americans chipped arrowheads and cutting tools (Figure 4.10).

Obsidian flows, typically a few hundred feet thick, provide evidence that rapid cooling is not the only mechanism that produces a glassy texture. Magmas with high silica content tend to form long, chainlike structures (polymers) before crystallization is complete. These structures, in turn, slow the migration of ions, which impedes the formation of crystals. In addition. these long chainlike structures increase the magma's viscosity. (*Viscosity* is a measure of a fluid's resistance to flow.) So granitic magma, which is rich in silica, may be extruded as an extremely viscous mass that eventually solidifies to form obsidian.

By contrast, basaltic magma, which is low in silica, forms very fluid lavas that, upon cooling, usually generate fine-grained crystalline rocks. However, when a basaltic lava flow enters the ocean, its surface is quenched rapidly enough to form a thin, glassy skin.

▼ **Figure 4.9 Vesicular texture** The larger image shows a lava flow on Hawaii's Kilauea Volcano. The inset photo is a close-up showing the vesicular texture of hardened lava. Vesicles are small holes left by escaping gas bubbles. (Inset photo by E. J. Tarbuck)

▲ **Figure 4.10 Obsidian arrowhead** Native Americans made arrowheads and cutting tools from obsidian, a natural glass. (Photo by Jeffrey Scovil)

◀ **Figure 4.11 Pyroclastic rocks are the product of explosive eruptions** This eruptive column consists in part of volcanic fragments, which will fall out and may eventually consolidate to become rocks displaying a pyroclastic texture. (Photo by Richard Roscoe/Getty Images)

Pyroclastic (Fragmental) Texture Another group of igneous rocks is formed from the consolidation of individual rock fragments ejected during explosive volcanic eruptions. The ejected particles might be very fine volcanic ash, molten blobs, or large angular blocks torn from the walls of a vent during an eruption (Figure 4.11). Igneous rocks composed of these rock fragments are said to have a **pyroclastic texture**, or **fragmental texture** (see Figure 4.7E).

A common type of pyroclastic rock, called *welded tuff*, is composed of fine fragments of glass that remained hot enough to fuse together. Other pyroclastic rocks are composed of fragments that solidified before impact and became cemented together at some later time. Because pyroclastic rocks are made of individual particles or fragments rather than interlocking crystals, their textures often resemble those exhibited by sedimentary rocks rather than those associated with igneous rocks.

CONCEPT CHECKS 4.3

1. Define *texture*.
2. How does the rate of cooling influence crystal size? What other factors influence the texture of igneous rocks?
3. List the six major igneous rock textures.
4. What does a porphyritic texture indicate about the cooling history of an igneous rock?

4.4 Naming Igneous Rocks

Distinguish among the common igneous rocks based on texture and mineral composition.

Geologists classify igneous rocks on the basis of their texture and mineral composition (Figure 4.12). The various igneous textures described in the previous section result mainly from different cooling histories, whereas the mineral composition of an igneous rock depends on the chemical makeup of its parent magma. Because igneous rocks are classified on the basis of both mineral composition and texture, some rocks having similar mineral constituents but exhibiting different textures are given different names.

▶ SmartFigure 4.12
Classification of igneous rocks Igneous rocks are classified based on mineral composition and texture. (Photos by Dennis Tasa and E. J. Tarbuck)

TUTORIAL
https://goo.gl/VOzSR0

IGNEOUS ROCK CLASSIFICATION CHART

	MINERAL COMPOSITION			
	Felsic	**Intermediate**	**Mafic**	**Ultramafic**
Dominant Minerals	Quartz Potassium feldspar	Amphibole Plagioclase feldspar	Pyroxene Plagioclase feldspar	Olivine Pyroxene
Accessory Minerals	Plagioclase feldspar Amphibole Muscovite Biotite	Pyroxene Biotite	Amphibole Olivine	Plagioclase feldspar
TEXTURE: Phaneritic (coarse-grained)	Granite	Diorite	Gabbro	Peridotite
TEXTURE: Aphanitic (fine-grained)	Rhyolite	Andesite	Basalt	Komatiite (rare)
TEXTURE: Porphyritic (two distinct grain sizes)	Granite porphyry	Andesite porphyry	Basalt porphyry	Uncommon
TEXTURE: Glassy	Obsidian	Less common	Less common	Uncommon
TEXTURE: Vesicular (contains voids)	Pumice (also glassy)		Scoria	Uncommon
TEXTURE: Pyroclastic (fragmental)	Tuff or welded tuff (Most fragments < 4mm)	⇄	Volcanic breccia (Most fragments > 4mm)	Uncommon
Rock Color (based on % of dark minerals)	0% to 25%	25% to 45%	45% to 85%	85% to 100%

Cory Rich/Getty Images

Michael Collier

Granite

Dennis Tasa

▲ SmartFigure 4.13 **Rocks contain information about the processes that produced them** This massive granitic monolith (El Capitan) located in Yosemite National Park, California, was once a molten mass deep within Earth.

MOBILE FIELD TRIP
https://goo.gl/XvMfQ1

Felsic Igneous Rocks

Granite Of all the igneous rocks, **granite** is perhaps the best known. This is because of its natural beauty, which is enhanced when it is polished, and its abundance in the continental crust. Slabs of polished granite are commonly used for tombstones and monuments and as building stones. Well-known areas in the United States where granite is quarried include Barre, Vermont; Mount Airy, North Carolina; and St. Cloud, Minnesota.

Granite is a coarse-grained rock composed of about 10 to 20 percent quartz and roughly 50 percent feldspar. When examined close up, the quartz grains appear somewhat rounded in shape, glassy, and clear to gray in color. By contrast, feldspar crystals, which are generally white, gray, or salmon pink in color, are blocky or rectangular in shape. Other minor constituents of granite include small amounts of dark silicates, particularly biotite and amphibole, and sometimes muscovite. Although the dark components generally make up less than 10 percent of most granites, they stand out visually and give granite its speckled appearance.

From a distance, most granitic rocks appear gray in color (Figure 4.13). However, granite that is composed of dark pink feldspar grains exhibits a reddish color. In addition, granites commonly exhibit a porphyritic texture. These specimens contain elongated feldspar crystals a few centimeters in length that are scattered among smaller crystals of quartz and amphibole (see Figure 4.12).

Rhyolite **Rhyolite** is the fine-grained equivalent of granite and, like granite, is composed essentially of the light-colored silicates (see Figure 4.12). This fact accounts for its color, which is usually buff to pink or occasionally light gray. Rhyolite is fine grained and frequently contains glass fragments and voids, indicating that it cooled rapidly in a surface, or near-surface, environment. In contrast

Did You Know?
During the Stone Age, volcanic glass (obsidian) was used for making cutting tools. Today, scalpels made from obsidian are being employed for delicate plastic surgery because they leave less scarring than steel scalpels. "The steel scalpel has a rough edge, where the obsidian scalpel is smoother and sharper," explains Lee Green, MD, an associate professor at the University of Michigan Medical School.

Did You Know?
Although Earth's upper mantle is often depicted in diagrams as a reddish-colored layer to show its high temperature, its color is actually a dark green. But how do we know it is green? On occasion, unmelted fragments of mantle rocks are brought to Earth's surface by rapidly rising magma and are ejected from volcanoes. These mantle rocks, called *peridotite*, are composed mainly of olivine, which, as the name suggests, is olive green, and pyroxene, which is blackish green.

to granite, which is widely distributed as large intrusive masses, rhyolite deposits are less common and generally less voluminous. The thick rhyolite lava flows and extensive deposits of volcanic ash in and around Yellowstone National Park are well-known exceptions to this generalization.

Obsidian **Obsidian** is a dark-colored glassy rock that usually forms when highly silica-rich lava cools quickly at Earth's surface (see Figure 4.12). In contrast to the orderly arrangement of ions characteristic of minerals, the arrangement of ions in glass is unordered. Consequently, glassy rocks such as obsidian are not composed of minerals in the same sense as most other rocks.

Although generally black or reddish-brown in color, obsidian most often has a chemical composition that is roughly equivalent to that of the light-colored igneous rock granite. Obsidian's dark color results from small amounts of metallic ions in an otherwise relatively clear, glassy substance. If you examine a thin edge, obsidian will appear nearly transparent (see Figure 4.7).

Pumice **Pumice** is a glassy volcanic rock with a vesicular texture that forms when large amounts of gas escape through silica-rich lava to generate a gray, frothy mass. In some samples the voids are quite noticeable, whereas in others the pumice resembles fine shards of intertwined glass. Because of the large percentage of voids, many samples of pumice float when placed in water (**Figure 4.14**). Oftentimes, flow lines are visible in pumice, indicating that some movement occurred before solidification was complete. Moreover, pumice and obsidian can often be found in the same rock mass, existing in alternating layers.

Intermediate Igneous Rocks

Andesite **Andesite** is a medium-gray, fine-grained rock typically of volcanic origin. Its name comes from South America's Andes Mountains, where numerous volcanoes are composed of this rock type. The volcanoes of North America's Cascade Range and many of the volcanic structures occupying the continental margins that surround the Pacific Ocean are also of andesitic composition. Andesite commonly exhibits a porphyritic texture (see Figure 4.12). When this is the case, the phenocrysts are often light, rectangular crystals of plagioclase feldspar or black, elongated amphibole crystals. Andesite may also resemble rhyolite, so its identification usually requires microscopic examination to verify mineral makeup.

Diorite **Diorite** is the intrusive equivalent of andesite. It is a coarse-grained rock that looks somewhat like gray granite but can be distinguished from granite because it contains little or no visible quartz crystals and has a higher percentage of dark silicate minerals. The mineral makeup of diorite is primarily plagioclase feldspar and amphibole. Because the light-colored feldspar grains and dark amphibole crystals appear to be roughly equal in abundance, diorite has a salt-and-pepper appearance (see Figure 4.12).

Mafic Igneous Rocks

Basalt **Basalt** is a very dark green to black, fine-grained rock composed primarily of pyroxene and calcium-rich plagioclase feldspar, with lesser amounts of olivine and amphibole (see Figure 4.12). When it is porphyritic, basalt commonly contains small light-colored feldspar phenocrysts or green, glassy-appearing olivine grains embedded in a dark groundmass.

Basalt is the most common extrusive igneous rock. Many volcanic islands, such as the Hawaiian Islands and Iceland, are composed mainly of basalt (**Figure 4.15**). Further, the upper layers of the oceanic crust consist of basalt. In the United States, large portions of central Oregon and Washington were the sites of extensive basaltic outpourings (discussed in detail in Chapter 5). At some locations, these once-fluid basaltic flows have accumulated to combined thicknesses approaching 3 kilometers (2 miles).

Gabbro **Gabbro** is the intrusive equivalent of basalt (see Figure 4.12). Like basalt, it tends to be dark green to black in color and composed primarily of pyroxene and calcium-rich plagioclase feldspar. Although gabbro is uncommon in the continental crust, it makes up a significant percentage of oceanic crust.

Pyroclastic Rocks

Pyroclastic rocks are composed of fragments ejected during a volcanic eruption. One of the most common pyroclastic rocks, called *tuff*, is composed mainly of tiny, ash-size fragments that were later cemented together. In situations where the ash particles remained hot enough to fuse, the rock is called *welded tuff*. Although welded tuff consists mostly of tiny glass shards, it may contain walnut-size pieces of pumice and other rock fragments.

Welded tuff deposits cover vast portions of previously volcanically active areas of the western United

←2 cm→

▶ **Figure 4.14 Pumice, a vesicular (and also glassy) igneous rock** Most samples of pumice will float in water because they contain numerous vesicles. (Inset photo by Chip Clark/Fundamental Photos)

◀ **Figure 4.15 Basaltic lava flowing from Kilauea Volcano, Hawaii** (Photo by David Reggie/Getty Images)

▲ **Figure 4.16 Welded tuff, a pyroclastic igneous rock** Outcrop of welded tuff that erupted from Valles Caldera near Los Alamos, New Mexico. Tuff is composed mainly of ash-sized particles and may contain larger fragments of pumice or other volcanic rocks. (Photo by Marli Miller)

States (**Figure 4.16**). Some of these tuff deposits are hundreds of feet thick and extend for more than 100 kilometers (60 miles) from their source. Most formed millions of years ago as volcanic ash spewed from large volcanic structures (calderas), sometimes spreading laterally at speeds approaching 100 kilometers (60 miles) per hour. Early investigators of these deposits incorrectly classified them as rhyolite lava flows. Today, we know that silica-rich lava is too viscous (thick) to flow more than a few miles from a vent.

Pyroclastic rocks composed mainly of particles larger than ash are called *volcanic breccia*. The particles in volcanic breccia may consist of streamlined lava blobs that solidified in air, blocks broken from the walls of the vent, volcanic ash, and glass fragments.

Unlike most igneous rock names, such as granite and basalt, the terms *tuff* and *volcanic breccia* do not imply mineral composition. Instead, they are frequently identified with a modifier; for example, *rhyolite tuff* indicates a rock composed of ash-size particles having a felsic composition.

CONCEPT CHECKS 4.4

1. List the two criteria by which igneous rocks are classified.
2. How are granite and rhyolite different? In what way are they similar?
3. Classify each of the following rocks by their mineral composition (felsic, intermediate, or mafic): gabbro, obsidian, granite, and andesite.
4. Describe each of the following in terms of composition and texture: diorite, rhyolite, and basalt porphyry.
5. In what way do tuff and volcanic breccia differ from other igneous rocks such as granite and basalt?

4.5 Origin of Magma

Summarize the major processes that generate magma from solid rock.

Did You Know?
Since the 1930s carpenters, mechanics, and other people working with their hands have used *Lava Soap*, which contains powdered pumice. Because of the abrasiveness of pumice, Lava soap is particularly good at removing grease, dirt, paints, and adhesives.

Based on evidence from the study of earthquake waves, we know that *Earth's crust and mantle are composed primarily of solid, not molten, rock.* Although the outer core is fluid, this iron-rich material is very dense and remains deep within Earth. So where does magma come from?

Most magma originates in Earth's uppermost mantle. The greatest quantities are produced at divergent plate boundaries, in association with seafloor spreading, with lesser amounts forming at subduction zones, where oceanic lithosphere descends into the mantle. Magma also can be generated when crustal rocks are heated sufficiently to melt.

Generating Magma from Solid Rock

Workers in underground mines know that temperatures increase as they descend deeper below Earth's surface. Although the rate of temperature change varies considerably from place to place, it *averages* about 25°C (75°F) per kilometer in the *upper* crust. This increase in temperature with depth is known as the **geothermal gradient**. As shown in Figure 4.17, when a typical geothermal gradient is compared to the melting point curve for the mantle rock peridotite, the temperature at which peridotite melts is higher than the geothermal gradient. Thus, under normal conditions, the mantle is solid. However, tectonic processes trigger melting though various means, including reducing the mantle rock's melting point (the temperature at which a material changes from solid to liquid).

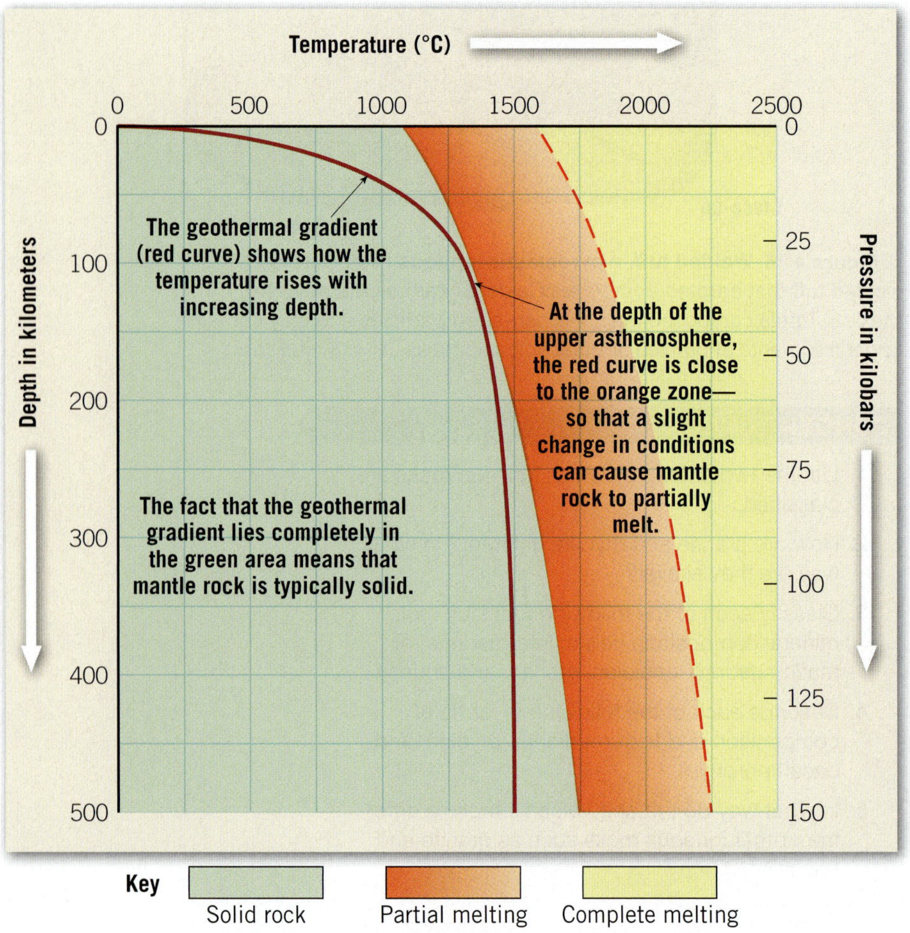

Decrease in Pressure: Decompression Melting If temperature were the only factor that determined whether rock melts, our planet would be a molten ball covered with a thin, solid outer shell. This is not the case because pressure, which also increases with depth, influences the melting temperatures of rocks.

Melting, which is accompanied by an increase in volume, occurs at progressively higher temperatures with increased depth. This is the result of the steady increase in confining pressure exerted by the weight of overlying rocks. Conversely, *reducing confining pressure lowers a rock's melting temperature.* When confining pressure drops sufficiently, **decompression melting** is triggered. Decompression melting occurs wherever hot, solid mantle rock ascends, thereby moving into regions of lower pressure.

Recall from Chapter 2 that tensional forces along spreading centers promote upwelling where plates diverge. This process is responsible for generating magma along oceanic ridges (divergent plate boundaries) where plates are rifting apart (Figure 4.18). Below the ridge crest, hot mantle rock rises and melts, generating a magma that replaces the material that shifted horizontally away from the ridge axis.

Decompression melting also occurs when ascending mantle plumes reach the uppermost mantle. If this rising magma reaches the surface, it triggers an episode of hot-spot volcanism.

Addition of Water Along with pressure, an important factor affecting the melting temperature of rock is its water content. Water and other volatiles, such as carbon dioxide, act in a similar way to salt melting ice. That is, water causes rock to melt at lower temperatures, just as putting rock salt on an icy sidewalk induces melting.

The introduction of water to generate magma occurs mainly at convergent plate boundaries, where cool slabs of oceanic lithosphere descend into the mantle (Figure 4.19). As an oceanic plate sinks, heat and pressure drive water from the subducting oceanic crust and overlying sediments. These fluids migrate into the wedge of hot mantle that lies directly above. At a depth of about 100 kilometers (60 miles), the wedge of mantle rock is sufficiently hot that the addition of water leads to some melting. Partial melting of the mantle rock peridotite generates hot basaltic magma whose temperatures may exceed 1250°C (nearly 2300°F).

▼ **Figure 4.17 Why the mantle is mainly solid** This diagram shows the geothermal gradient (the increase in temperature with depth) for the crust and upper mantle.

Temperature (°C)
0
500
1000
1500
2000
2500
Depth in kilometers
0
100
200
300
400
500
Pressure in kilobars
0
25
50
75
100
125
150
The geothermal gradient (red curve) shows how the temperature rises with increasing depth.
At the depth of the upper asthenosphere, the red curve is close to the orange zone—so that a slight change in conditions can cause mantle rock to partially melt.
The fact that the geothermal gradient lies completely in the green area means that mantle rock is typically solid.
Key
Solid rock
Partial melting
Complete melting

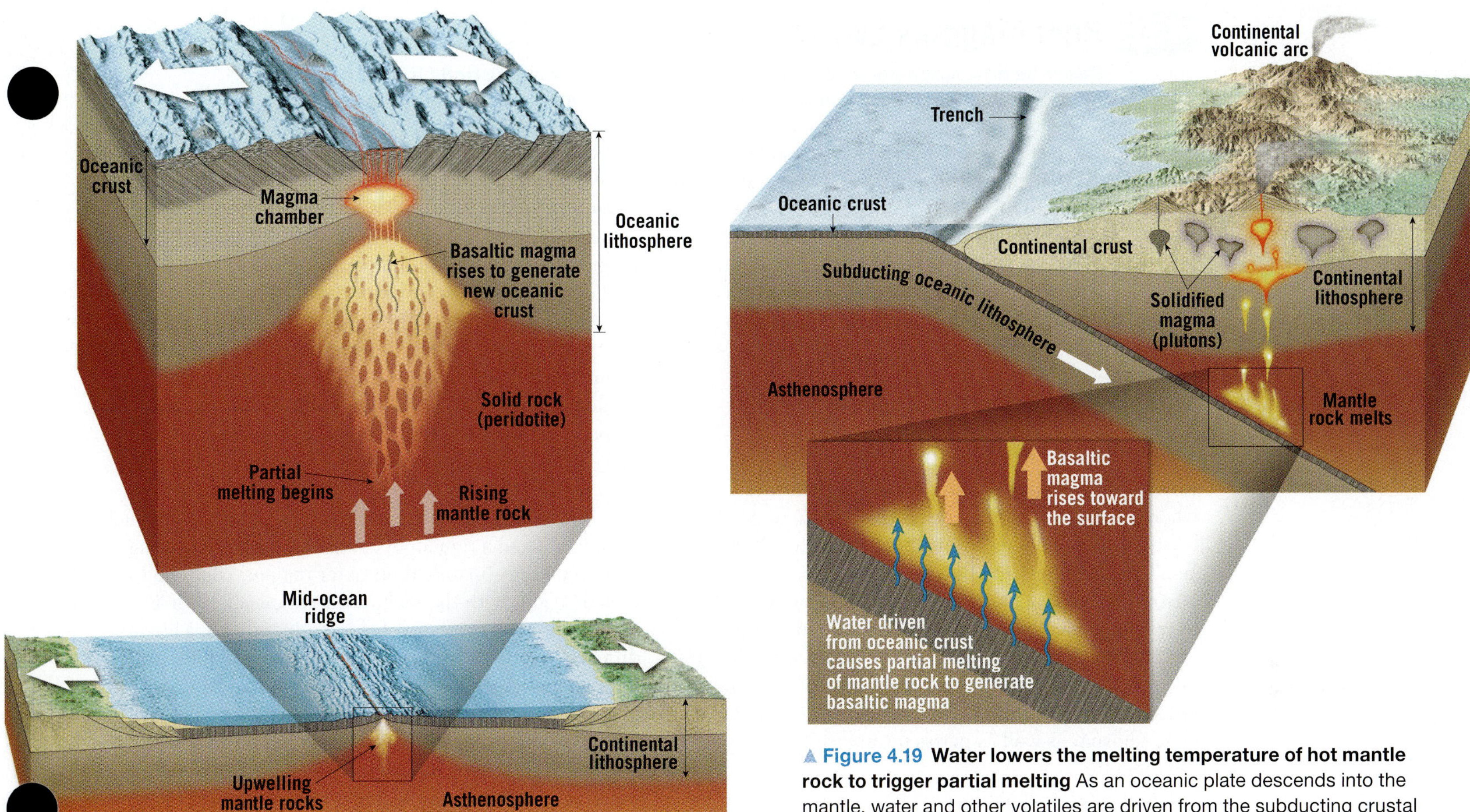

▲ **Figure 4.18 Decompression melting** As hot mantle rock ascends, it experiences continuously decreasing pressure. This drop in confining pressure usually initiates *decompression melting* in the upper mantle.

▲ **Figure 4.19 Water lowers the melting temperature of hot mantle rock to trigger partial melting** As an oceanic plate descends into the mantle, water and other volatiles are driven from the subducting crustal rocks into the mantle above.

Temperature Increase: Melting Crustal Rocks Mantle-derived basaltic (mafic) magma tends to be less dense than the surrounding rocks, which causes the magma to buoyantly rise toward the surface. In oceanic settings, these basaltic magmas often erupt on the ocean floor, generating seamounts, which may grow to form volcanic islands, as exemplified by the Hawaiian Islands. However, in continental settings, basaltic magma often "ponds" beneath low-density crustal rocks. Because the overlying crustal rocks have lower melting temperatures than basaltic magmas, the hot basaltic magma may heat them sufficiently to generate a secondary melt of silica-rich felsic magma. If these low-density, felsic magmas reach the surface, they tend to produce explosive eruptions; such eruptions occur most often at convergent plate boundaries.

Crustal rocks can also melt during continental collisions that result in the formation of a large mountain belt (discussed in detail in Chapter 11). During these events, the crust is greatly thickened, and some crustal rocks are carried to depths where the temperatures are high enough to cause partial melting. The felsic magmas produced in this manner usually solidify before reaching the surface, so volcanism is not typically associated with these collision-type mountain belts.

In summary, magma can be generated by (1) *decompression melting*, caused by a decrease in pressure as magma rises; (2) the *introduction of water*, which lowers the melting temperature of hot mantle rock; and (3) *heating of crustal rocks* above their melting temperature.

CONCEPT CHECKS 4.5

1. What is the geothermal gradient? Describe how the geothermal gradient compares with the melting temperatures of the mantle rock peridotite at various depths.
2. Explain the process of decompression melting.
3. What roles do water and other volatiles play in the formation of magma?
4. Name two plate tectonic settings in which you would expect magma to be generated.

4.6 How Magmas Evolve

Describe how magmatic differentiation can generate a magma body that has a mineralogy (chemical composition) that is different from its parent magma.

Geologists have observed that a single volcano may extrude lavas that change in composition over time. Such observations led to the idea that magma might change over time (evolve) and thus that one magma body could give rise to igneous rocks with a range of compositions. To explore this idea, N. L. Bowen carried out a pioneering investigation early in the twentieth century into the crystallization of magma.

Bowen's Reaction Series & the Composition of Igneous Rocks

Recall that ice freezes at a specific temperature, whereas basaltic magma crystallizes over a range of at least 200°C of cooling (from about 1200° to 1000°C). In a laboratory setting, Bowen and his coworkers demonstrated that as a basaltic magma cools, minerals tend to crystallize in a systematic fashion, based on their melting temperatures. As shown in **Figure 4.20**, the first mineral to crystallize is the ferromagnesian mineral olivine. Further cooling generates calcium-rich plagioclase feldspar as well as pyroxene, and so forth down the diagram.

During this crystallization process, the composition of the remaining liquid portion of the magma also continually changes. For example, at the stage when about one-third of the magma has solidified, the remaining molten material is nearly depleted of iron, magnesium, and calcium because these elements are major constituents of the minerals that form earliest in the process. The removal of these elements causes the melt to become enriched in sodium and potassium. Further, because the original basaltic magma contained about 50 percent silica (SiO_2), the crystallization of the earliest-formed mineral, olivine, which is only about 40 percent silica, leaves the remaining melt richer in SiO_2. Thus, the magma becomes progressively richer in silica as it evolves.

Bowen also demonstrated that when the crystals that form in a magma remain in contact with the remaining melt, then they (mainly their outer regions) continue to exchange ions with the melt (react chemically with it). As a result, the periphery of these mineral grains has a different, more evolved composition than the interiors. That is the significance of the arrows in Figure 4.20. Stated another way, minerals that remain in contact with a melt gradually change composition to become the next mineral in the series Bowen identified. This order of mineral formation became known as **Bowen's reaction series**. However, in nature, the earliest-formed minerals

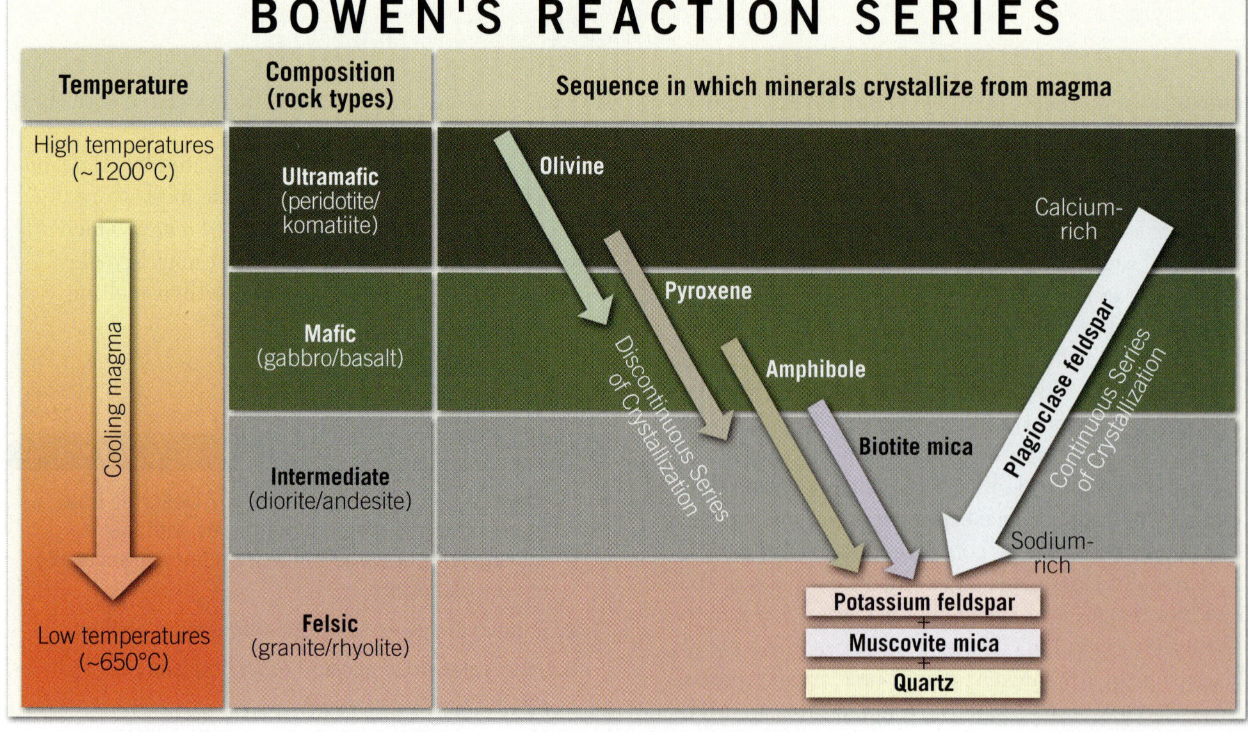

▶ **Figure 4.20 Bowen's reaction series** This diagram shows the sequence in which minerals crystallize from a basaltic magma. Compare this figure to the mineral composition of the rock groups in Figure 4.12. Note that each rock group consists of minerals that crystallize in the same temperature range.

can separate from the melt, thus halting further chemical reactions.

The diagram of Bowen's reaction series in Figure 4.20 depicts the sequence in which minerals crystallize from a magma of basaltic composition under laboratory conditions. Evidence that this highly idealized crystallization model approximates what can happen in nature comes from analysis of igneous rocks. In particular, scientists know that minerals that form in the same general temperature regime depicted in Bowen's reaction series are found together in the same igneous rocks. For example, notice in Figure 4.20 that the minerals quartz, potassium feldspar, and muscovite, which are located in the same region of Bowen's diagram, are typically found together as major constituents of the intrusive igneous rock granite.

Magmatic Differentiation & Crystal Settling

Bowen demonstrated that minerals crystallize from magma in a systematic fashion. But how do Bowen's findings account for the great diversity of igneous rocks? It has been shown that, at one or more stages during the crystallization of magma, a separation of various components can occur. One mechanism that causes this to happen is called **crystal settling**. This process occurs when the earlier-formed minerals are denser (heavier) than the melt and sink toward the bottom of the magma chamber, as shown in Figure 4.21. When the remaining melt solidifies—either in place or at another location, if it migrates into fractures in the surrounding rocks—it will form a rock with a mineral composition that is more felsic than the parent magma. The formation of a magma body having a mineralogy or chemical composition that is different than the parent magma is called **magmatic differentiation**.

A classic example of magmatic differentiation is found in the Palisades Sill, which is a 300-meter- (1000-foot-) thick slab of dark igneous rock exposed along the west bank of the lower Hudson River across from New York City. Because of its great thickness and consequent slow rate of solidification, crystals of olivine (the first mineral to form) sank and make up about 25 percent of the lower portion of the Palisades Sill. By contrast, near the top of this igneous body, where the last melt crystallized, olivine represents only 1 percent of the rock mass.*

Assimilation & Magma Mixing

Bowen successfully demonstrated that through magmatic differentiation, a single parent magma can generate several mineralogically different igneous rocks. However, more recent work indicates that magmatic differentiation involving crystal settling cannot, by itself, account for the entire compositional spectrum of igneous rocks.

Once a magma body forms, the incorporation of foreign material can also change its composition. For example, in near-surface environments where rocks

*Recent studies indicate that the Palisades Sill was produced by multiple injections of magma and does not represent a simple case of crystal settling. However, it is nonetheless an instructional example of that process.

Did You Know?

The formation of the most common chemical elements on Earth, such as oxygen, silicon, and iron, occurred billions of years ago inside distant massive stars. Through various processes of nuclear fusion, these stars converted the lightest elements, mostly hydrogen, into these heavier elements. In fact, most such elements found in the solar system, as well as in your body, are believed to have formed from debris scattered by stars that formed prior to the formation of the solar system.

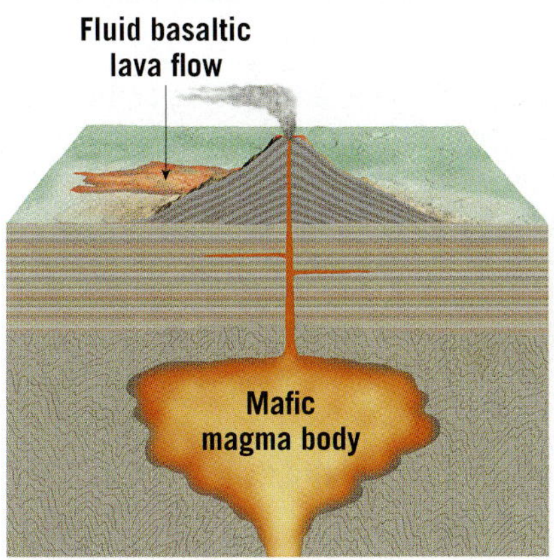

A. Magma having a mafic composition erupts fluid basaltic lavas.

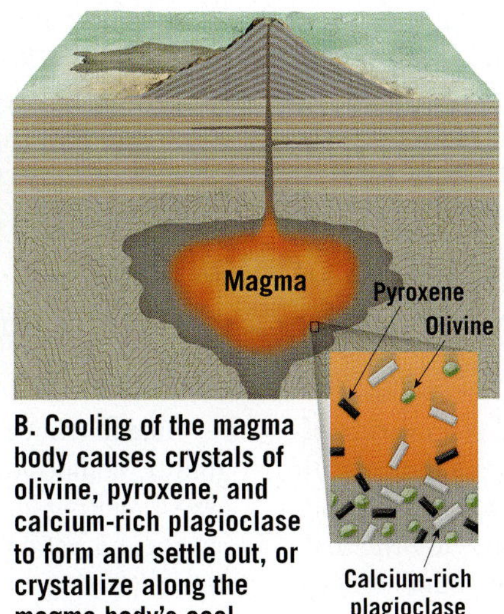

B. Cooling of the magma body causes crystals of olivine, pyroxene, and calcium-rich plagioclase to form and settle out, or crystallize along the magma body's cool margins.

Explosive eruption of silica-rich magma
Magma
Solid rock

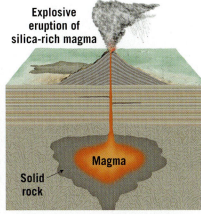

C. The remaining melt will be enriched with silica, and should a subsequent eruption occur, the rocks generated will be more silica-rich and closer to the felsic end of the compositional range than the initial magma.

◀ **Figure 4.21 Crystal settling results in a change in the composition of the remaining magma** A magma evolves as the earliest-formed minerals (those richer in iron, magnesium, and calcium) crystallize and settle to the bottom of the magma chamber, leaving the remaining melt richer in sodium, potassium, and silica (SiO_2).

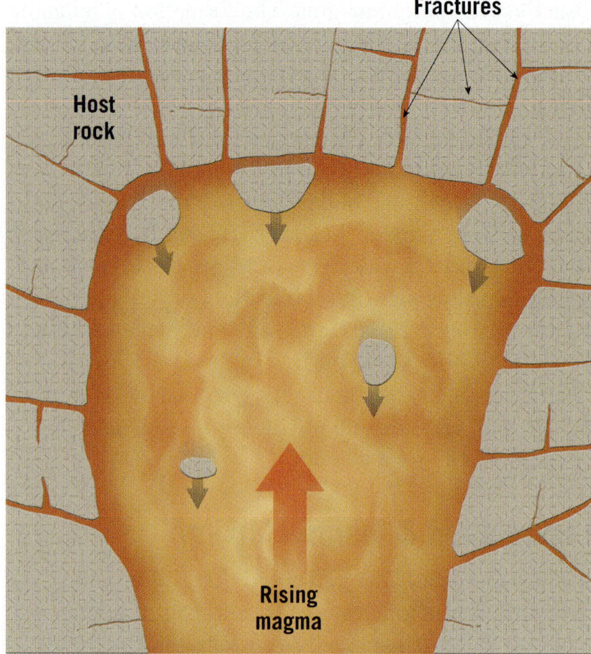

As magma rises through Earth's brittle upper crust, it may dislodge and incorporate the surrounding host rocks. Melting of these blocks, a process called *assimilation*, changes the overall composition of the rising magma body.

▶ **Figure 4.22 Assimilation of the host rock by a magma body**

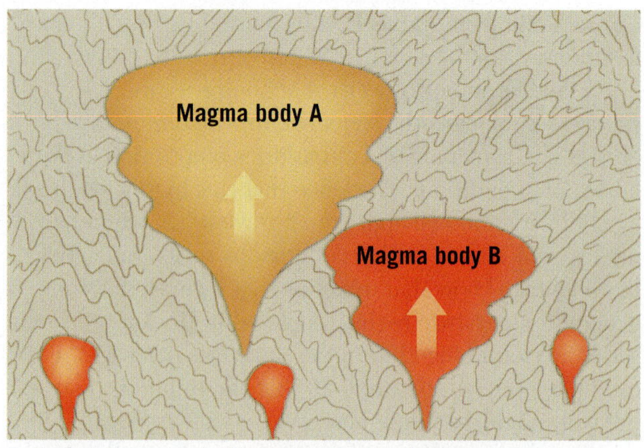

A. During the ascent of two chemically distinct magma bodies, the more buoyant mass may overtake the slower rising body.

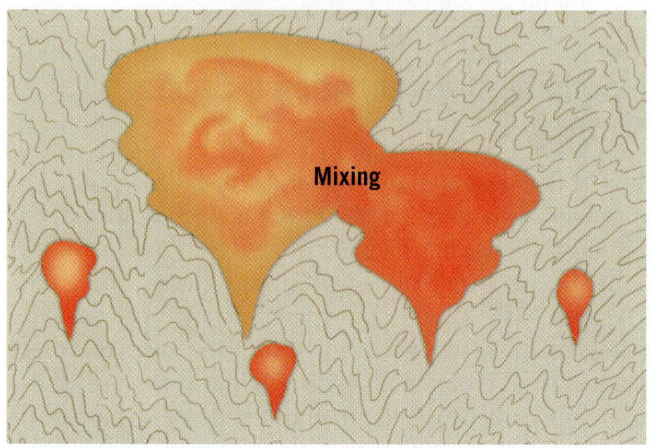

B. Once joined, convective flow mixes the two magmas, generating a mass that is a blend of the two magma bodies.

▲ **Figure 4.23 Magma mixing** This is one of the ways the composition of a magma body can change.

are brittle, the magma pushing upward can cause the overlying rock to fracture into numerous pieces. The force of the injected magma is often sufficient to dislodge and incorporate blocks of the surrounding host rock (Figure 4.22). Melting of these blocks, a process called **assimilation**, changes the overall chemical composition of the magma body.

Another means by which the composition of magma can be altered is called **magma mixing**. Magma mixing may occur during the ascent of two chemically distinct magma bodies as the more buoyant mass overtakes the more slowly rising body (Figure 4.23). Once they are joined, convective flow stirs the two magmas, generating a single mass that has an intermediate composition.

CONCEPT CHECKS 4.6

1. Define Bowen's reaction series.
2. How does the crystallization and settling of the earliest formed minerals affect the composition of the remaining magma?
3. Compare the processes of assimilation and magma mixing.

4.7 Partial Melting & Magma Composition

Describe how partial melting of the mantle rock peridotite can generate a basaltic (mafic) magma.

Recall that igneous rocks are composed of a mixture of minerals and, therefore, tend to melt over a temperature range of at least 200°C. As rock begins to melt, the minerals with the lowest melting temperatures are the first to melt. If melting continues, minerals with higher melting points begin to melt, and the composition of the melt steadily approaches the overall composition of the rock from which it was derived. Most often, however, melting is not complete, a process known as **partial melting**.

Recall from Bowen's reaction series that rocks with a granitic composition are composed of minerals with the lowest melting (crystallization) temperatures—namely, quartz and potassium feldspar (see Figure 4.20). Also note that as we move up Bowen's reaction series, the minerals have progressively higher melting temperatures, and that olivine, which is found at the top, has the highest melting point. When a rock undergoes partial melting, it forms a melt that is enriched in ions from minerals with the lowest melting temperatures, while the unmelted portion is composed of minerals with higher melting temperatures (**Figure 4.24**). Separation of these two fractions yields a melt with a chemical composition that is richer in silica and nearer the felsic (granitic) end of the spectrum than the rock from which it formed. In general, partial melting of *ultramafic* rocks tends to yield *mafic (basaltic) magmas*, partial melting of *mafic* rocks generally yields *intermediate (andesitic) magmas*, and partial melting of *intermediate* rocks can generate *felsic (granitic) magmas*.

Formation of Basaltic Magma

Most magma that erupts at Earth's surface is basaltic in composition and has a temperature range of 1000° to 1250°C. Experiments show that under the high-pressure conditions calculated for the upper mantle, partial melting of the ultramafic rock peridotite can generate a magma of basaltic composition. Further evidence that many basaltic magmas have a mantle source are the inclusions of peridotite, a rock that basaltic magmas often carry up to Earth's surface from the mantle.

Basaltic (mafic) magmas that originate from partial melting of mantle rocks are called *primary* or *primitive* magmas because they have not yet evolved. Recall that partial melting that produces mantle-derived magmas may be triggered by a reduction in confining pressure during the process of decompression melting. This can occur, for example, where hot mantle rock ascends as part of slow-moving convective flow at mid-ocean ridges (see Figure 4.18). Basaltic magmas are also generated at subduction zones, where water driven from the descending slab of oceanic crust promotes partial melting of the mantle rocks that lie above (see Figure 4.19).

Formation of Andesitic & Granitic Magmas

If partial melting of mantle rocks generates most basaltic magmas, what is the source of the magma that crystallizes to form andesitic (intermediate) and granitic (felsic) rocks? Recall that silica-rich magmas erupt mainly along the continental margins. This is strong evidence that continental crust, which is thicker and has a lower density than oceanic crust, must play a role in generating these more highly evolved magmas.

One way andesitic magma can form is when a rising mantle-derived basaltic magma undergoes magmatic differentiation as it slowly makes its way through the continental crust. Recall from our discussion of Bowen's reaction series that as basaltic magma solidifies, the silica-poor ferromagnesian minerals crystallize first. If these iron-rich components are separated from the liquid by crystal settling, the remaining melt has an andesitic composition (see Figure 4.20).

Andesitic magmas can also form when rising basaltic magmas assimilate crustal rocks that tend to be rich in silica. Partial melting of basaltic rocks is yet another way in which at least some andesitic magmas are thought to be produced.

Although granitic magmas can be formed through magmatic differentiation of andesitic magmas, most granitic magmas probably form when hot basaltic magma ponds (becomes trapped because of its greater density) below continental crust (**Figure 4.25**). When the heat

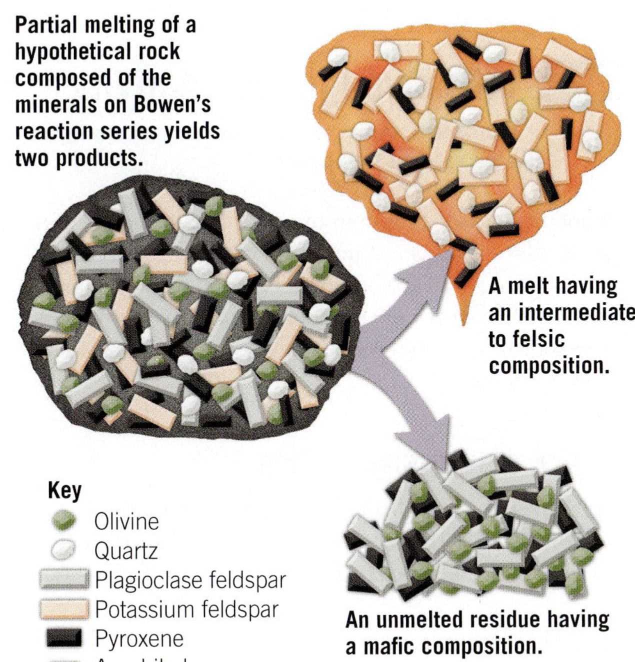

Partial melting of a hypothetical rock composed of the minerals on Bowen's reaction series yields two products.
A melt having an intermediate to felsic composition.
Key
Olivine
Quartz
Plagioclase feldspar
Potassium feldspar
Pyroxene
Amphibole
An unmelted residue having a mafic composition.

SmartFigure 4.24
Partial melting Partial melting generates a magma that is nearer the felsic (granitic) end of the compositional spectrum than the parent rock from which it was derived.

TUTORIAL
https://goo.gl/xGLCN6

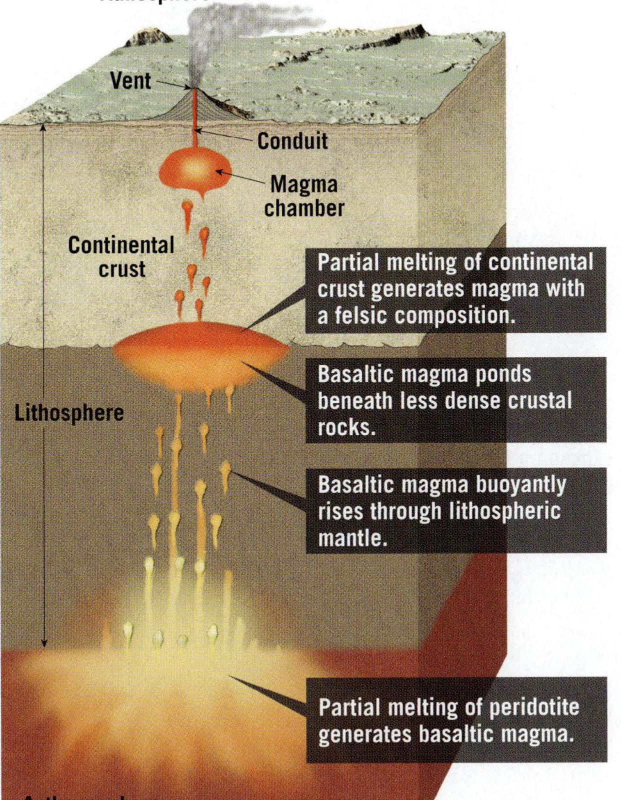

SmartFigure 4.25
Formation of granitic magma Granitic magmas are generated by the partial melting of continental crust.

ANIMATION
https://goo.gl/Wd57Cw

from the hot basaltic magma partially melts the overlying crustal rocks, which are silica rich and have a much lower melting temperature, the result can be the production of large quantities of granitic magmas. This process is thought to have been responsible for the volcanic activity in and around Yellowstone National Park in the distant past.

CONCEPT CHECKS 4.7

1. Briefly describe why partial melting results in a magma whose composition is different from that of the rock from which it was derived.
2. How are most basaltic magmas thought to have formed?
3. What is the process that is thought to generate most granitic magmas?

4.8 Intrusive Igneous Activity

Compare and contrast these intrusive igneous structures: dikes, sills, batholiths, stocks, and laccoliths.

Although volcanic eruptions are occasionally violent and spectacular events, most magma crystallizes at depth, without fanfare. Therefore, understanding the igneous processes that occur deep underground is as important to geologists as studying volcanic events, which are the focus of Chapter 5.

Nature of Intrusive Bodies

When magma rises through the crust, it forcefully displaces preexisting crustal rocks, termed **host rock** or **country rock**. The structures that result from the emplacement of magma into preexisting rocks are called **intrusions** or **plutons**. Because all intrusions form far below Earth's surface, they are studied primarily after uplifting and erosion (covered in later chapters) have exposed them. The challenge lies in reconstructing the events that generated these structures in vastly different conditions deep underground, millions of years ago.

Intrusions are known to occur in a great variety of sizes and shapes. Some of the most common types are illustrated in Figure 4.26. Notice that some plutons have a **tabular** (*tabula* = table) shape, whereas others are

▼ SmartFigure 4.26 **Intrusive igneous structures** (Photo: Belinda Images/SuperStock)

ANIMATION
https://goo.gl/2CGehV

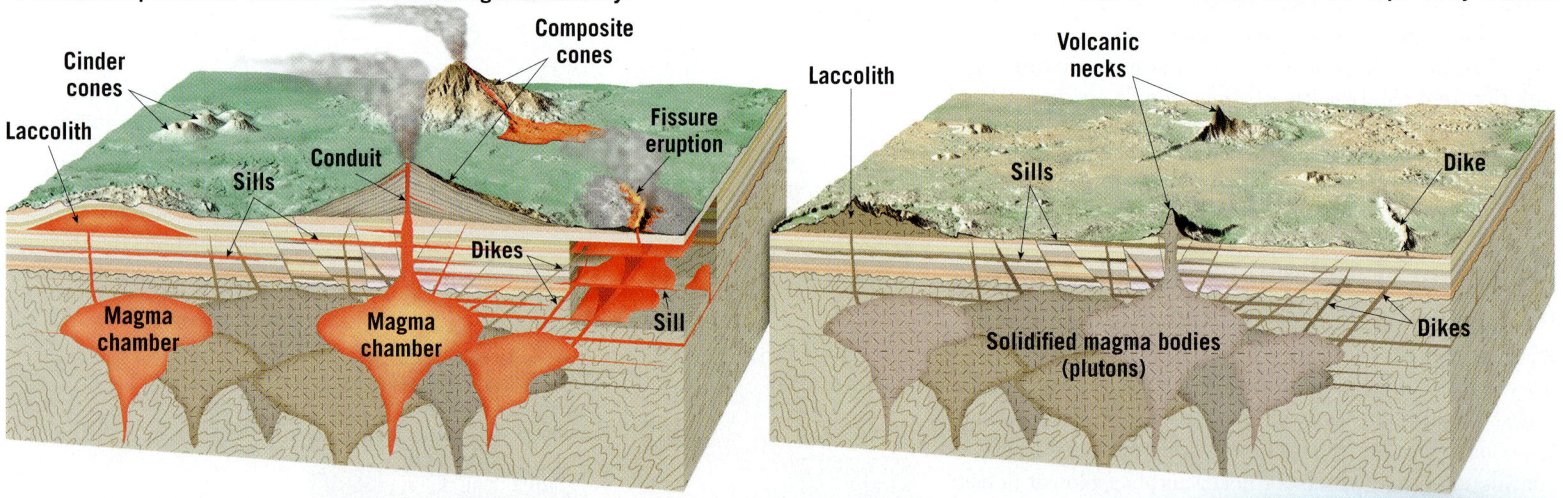

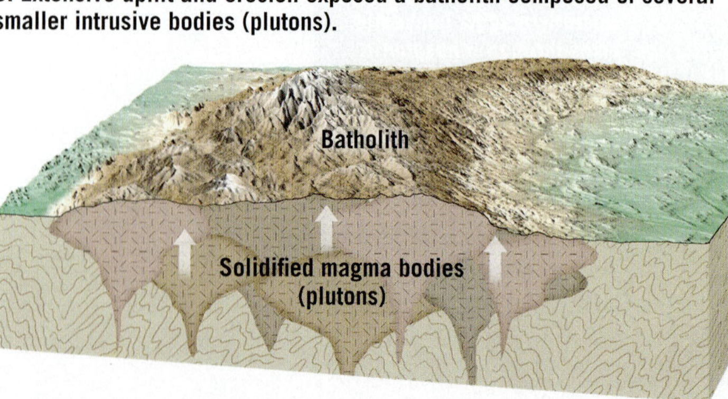

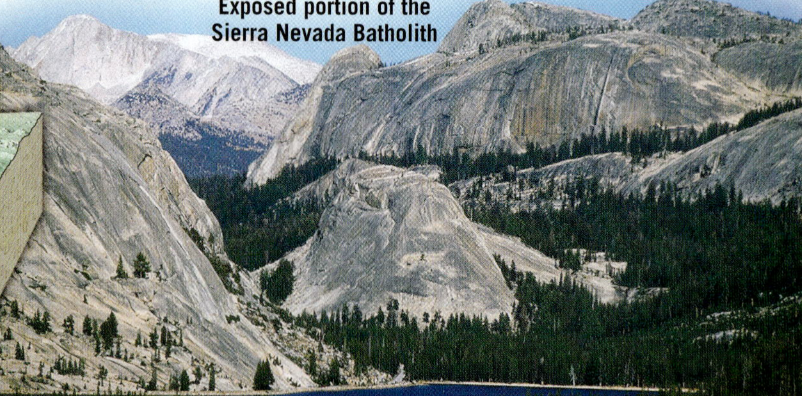

best described as **massive** (blob shaped). Also, observe that some of these bodies cut across existing structures, such as sedimentary strata, whereas others form when magma is injected between sedimentary layers. Because of these differences, intrusive igneous bodies are generally classified according to their shape as either tabular or massive, and by their orientation with respect to the host rock. Igneous bodies are said to be **discordant** (*discordare* = to disagree) if they cut across existing structures and **concordant** (*concordare* = to agree) if they inject parallel to features such as sedimentary strata.

Tabular Intrusive Bodies: Dikes & Sills

Dikes & Sills Tabular intrusive bodies are produced when magma is forcibly injected into a fracture or zone of weakness, such as a bedding surface (see Figure 4.26). **Dikes** are discordant bodies that form when magma is forcibly injected into fractures and cut across bedding surfaces and other structures in the host rock. By contrast, **sills** are nearly horizontal, concordant bodies that form when magma exploits weaknesses between sedimentary beds or other rock structures (**Figure 4.27**). In general, dikes serve as tabular conduits that transport magma upward, whereas sills tend to accumulate magma and increase in thickness.

Dikes and sills are typically shallow features, occurring where the country rocks are sufficiently brittle to fracture. They can range in thickness from less than 1 millimeter to more than 1 kilometer.

While dikes and sills can occur as solitary bodies, dikes tend to form in roughly parallel groups called *dike swarms*. These multiple structures reflect the tendency for fractures to form in sets when tensional forces pull apart brittle country rock. Dikes can also radiate from an eroded volcanic neck, like spokes on a wheel. In these situations, the active ascent of magma generated fissures in the volcanic cone, out of which lava flowed. Dikes frequently are more resistant and thus weather more slowly than the surrounding rock. Consequently, when exposed by erosion, dikes tend to have a wall-like appearance, as shown in **Figure 4.28**.

Because dikes and sills are relatively uniform in thickness and can extend for many kilometers, they are assumed to be the product of very fluid, and therefore mobile, magmas. One of the largest and most studied of all sills in the United States is the Palisades Sill. Exposed for 80 kilometers (50 miles) along the west bank of the Hudson River in southeastern New York and northeastern New Jersey, this sill is about 300 meters (1000 feet) thick. Because it is resistant to erosion, the Palisades Sill forms an imposing cliff that can be easily seen from the opposite side of the Hudson.

Columnar Jointing In many respects, sills closely resemble buried lava flows. Both are tabular and can extend over a wide area, and both may exhibit columnar jointing.

▲ **SmartFigure 4.27 Sill exposed in Sinbad County, Utah** The dark, essentially horizontal band is a sill of basaltic composition that intruded horizontal layers of sedimentary rock. (Photo by Michael Collier)

MOBILE FIELD TRIP
https://goo.gl/qC5DJE

◄ **SmartFigure 4.28 Dike exposed in the Spanish Peaks, Colorado** This wall-like dike is composed of igneous rock that is more resistant to weathering than the surrounding material. (Photo by Michael Collier)

CONDOR VIDEO
https://goo.gl/Qm3X6N

Columnar jointing occurs when igneous rocks cool and develop shrinkage fractures that produce elongated, pillar-like columns that most often have six sides (**Figure 4.29**). Further, because sills and dikes generally form in near-surface environments and may be only a few meters thick, the emplaced magma often cools quickly enough to generate a fine-grained texture. (Recall that most intrusive igneous bodies have a coarse-grained texture.)

Massive Intrusive Bodies: Batholiths, Stocks, & Laccoliths

Batholiths & Stocks By far the largest intrusive igneous bodies are **batholiths** (*bathos* = depth, *lithos* = stone). Batholiths occur as mammoth linear structures several hundred kilometers long and more than 100 kilometers wide (**Figure 4.30**). The Sierra Nevada batholith, for example, is a continuous granitic structure that forms much of the "backbone" of the Sierra Nevada in California. An even larger batholith extends for over 1800 kilometers (1100 miles) along the Coast Mountains of western Canada and into southern Alaska. Although batholiths can cover a large area, recent geophysical studies indicate that most are less than 10 kilometers (6 miles) thick. Some are even thinner; the coastal batholith of Peru, for example, is essentially a flat slab with an average thickness of only 2 to 3 kilometers (1 to 2 miles). Batholiths are typically composed of felsic (granitic) and intermediate rock types and are often called "granitic batholiths."

Early investigators thought the Sierra Nevada batholith was a huge single body of intrusive igneous rock.

▼ **Figure 4.29 Columnar jointing** Giant's Causeway in Northern Ireland is an excellent example of columnar jointing. (Photo by E. J. Tarbuck)

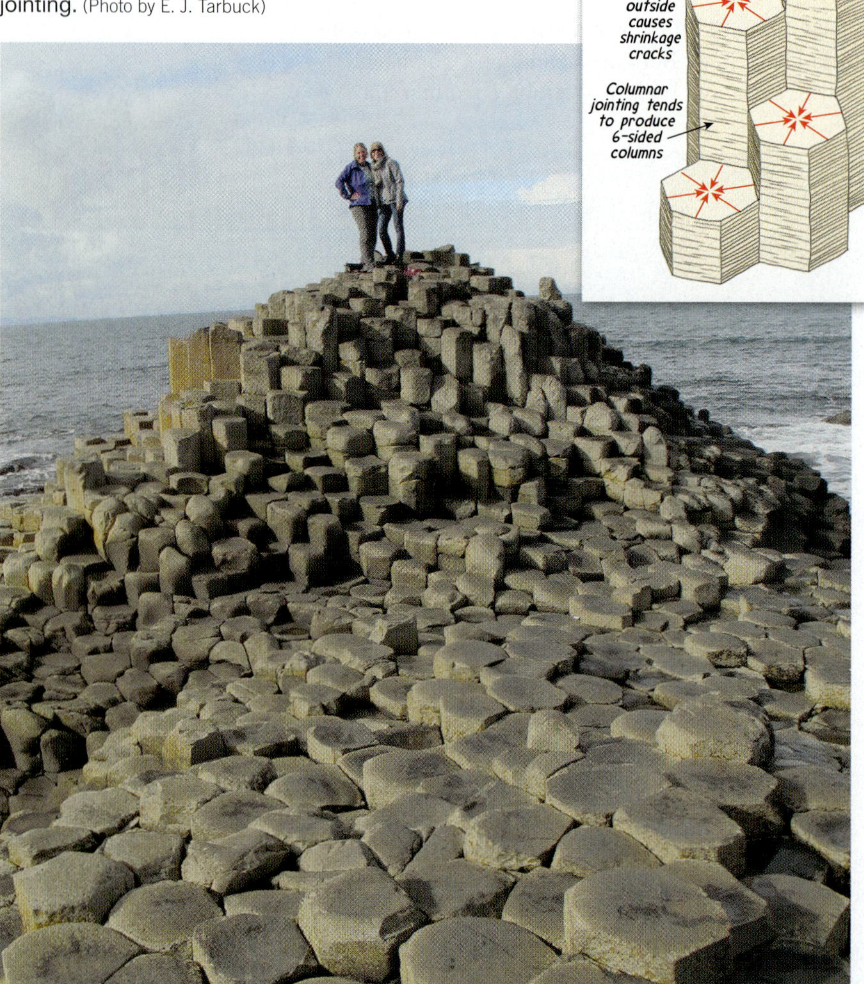

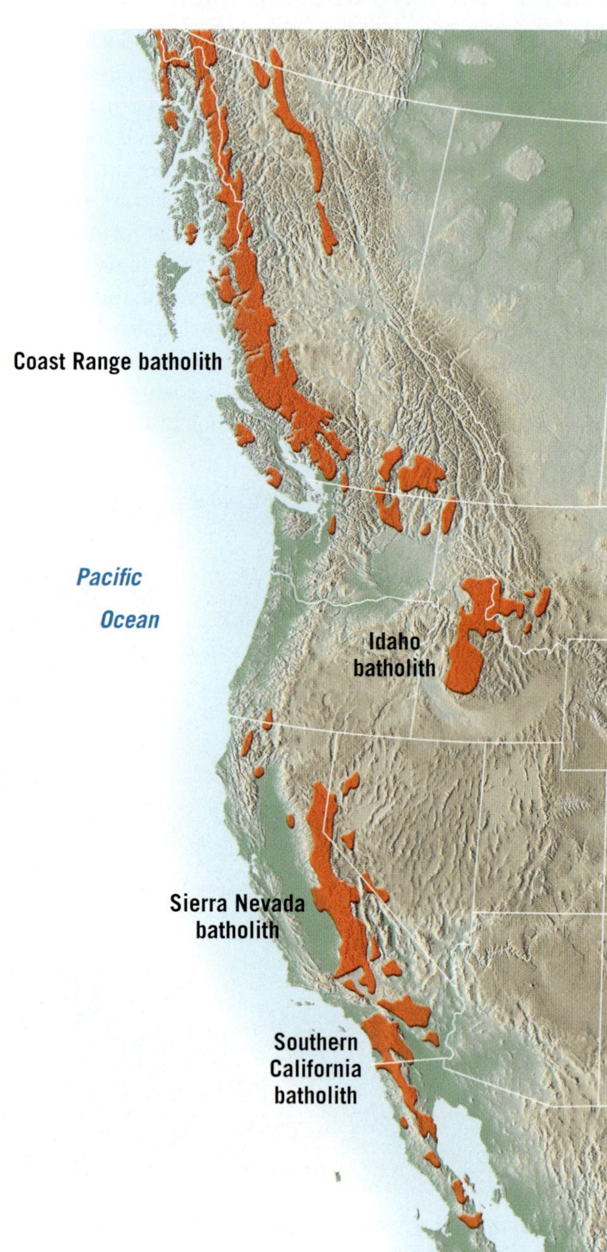

▲ **Figure 4.30 Granitic batholiths along the western margin of North America** These gigantic, elongated bodies consist of numerous plutons that were emplaced beginning about 150 million years ago.

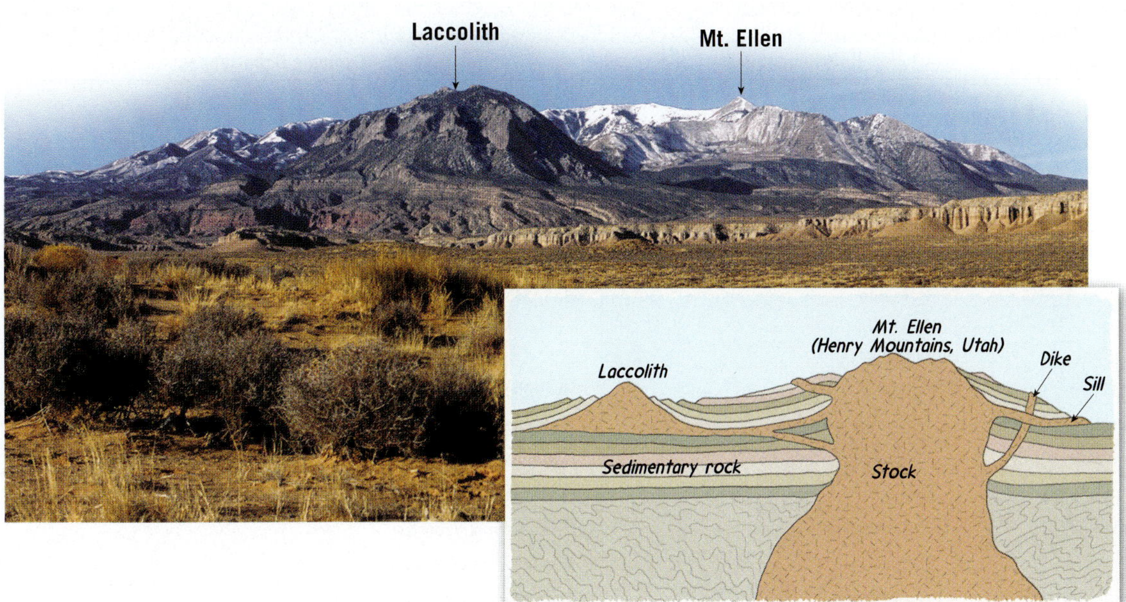

◀ **Figure 4.31**
Laccoliths Mount Ellen in Utah's Henry Mountains is one of five peaks that make up this small mountain range. Although the main intrusions in the Henry Mountains are stocks, numerous laccoliths formed as offshoots of these structures. (Photo by Michael DeFreitas North America/Alamy)

Today we know that large batholiths are produced by hundreds of discrete injections of magma that form smaller intrusive bodies (plutons) that intimately crowd against or penetrate one another. These bulbous masses are emplaced over spans of millions of years. The intrusive activity that created the Sierra Nevada batholith, for example, occurred nearly continuously over a 130-million-year period that ended about 80 million years ago (see Figure 4.30).

A batholith is generally defined as a plutonic body having a surface exposure greater than 100 square kilometers (40 square miles). Smaller plutons are termed **stocks**. However, many stocks appear to be portions of much larger intrusive bodies that would be classified as batholiths if they were fully exposed.

Laccoliths A nineteenth-century study by G. K. Gilbert of the U.S. Geological Survey in the Henry Mountains of Utah produced the first clear evidence that igneous intrusions can lift the sedimentary strata they penetrate. Gilbert named the igneous intrusions he observed **laccoliths**, which he envisioned as igneous rock forcibly injected between sedimentary strata, so as to arch the beds above while leaving those below relatively flat. It is now known that the five major peaks of the Henry Mountains are not laccoliths but stocks. However, these central magma bodies are the source material for branching offshoots that are true laccoliths, as Gilbert defined them (**Figure 4.31**).

Numerous other granitic laccoliths have since been identified in Utah. The largest is a part of the Pine Valley Mountains located north of St. George, Utah. Others are found in the La Sal Mountains near Arches National Park and in the Abajo Mountains directly to the south.

CONCEPT CHECKS 4.8

1. What is meant by the term *country rock*?
2. Describe *dikes* and *sills*, using the appropriate terms from the following list: massive, discordant, tabular, and concordant.
3. Distinguish among batholiths, stocks, and laccoliths in terms of size and shape.

Did You Know?
In the eastern United States, some exposed granitic intrusions have dome-shaped, nearly treeless summits—hence the name "summit balds." Examples include Cadillac Mountain in Maine, Mount Chocorua in New Hampshire, Black Mountain in Vermont, and Stone Mountain in Georgia.

4.9 Mineral Resources & Igneous Processes

Explain how economic deposits of gold, silver, and many other metals form.

Given the growth of the middle class in countries such as China, India, and Brazil, the demand for metallic natural resources has increased exponentially in recent years. Some of the most important accumulations of metals, such as gold, silver, copper, mercury, lead, platinum, and nickel, are produced by igneous processes (**Table 4.1**). These mineral resources result from processes that concentrate desirable materials to the point where they can be profitably extracted. Therefore, knowledge of how and where these important materials are likely to be concentrated is vital to our well-being.

Table 4.1 Occurrences of Metallic Minerals

Metal	Principal Ores	Geologic Occurrences
Aluminum	Bauxite	Residual product of weathering
Chromium	Chromite	Magmatic differentiation
Copper	Chalcopyrite Bornite Chalcocite	Hydrothermal deposits; contact metamorphism; enrichment by weathering processes
Gold	Native gold	Hydrothermal deposits; placers
Iron	Hematite Magnetite Limonite	Banded sedimentary formations; magmatic differentiation
Lead	Galena	Hydrothermal deposits
Magnesium	Magnesite	Hydrothermal deposits
Manganese	Pyrolusite	Residual product of weathering
Mercury	Cinnabar	Hydrothermal deposits
Molybdenum	Molybdenite	Hydrothermal deposits
Nickel	Pentlandite	Magmatic differentiation
Platinum	Native platinum	Magmatic differentiation; placers
Silver	Native silver Argentite	Hydrothermal deposits; enrichment by weathering processes
Tin	Cassiterite	Hydrothermal deposits; placers
Titanium	Ilmenite	Magmatic differentiation; placers
Tungsten	Wolframite	Pegmatites; contact metamorphic deposits; placers
Uranium	Uraninite (pitchblende)	Pegmatites; sedimentary deposits
Zinc	Sphalerite	Hydrothermal deposits

Magmatic Differentiation & Ore Deposits

The igneous processes that generate some important metal deposits are quite straightforward. For example, as a large basaltic magma body cools, the heavy minerals that crystallize early tend to settle to the lower portion of the magma chamber. This type of magmatic differentiation serves to concentrate some metals, producing major deposits of chromite (ore of chromium), magnetite, and platinum. Layers of chromite, interbedded with other heavy minerals, are mined at Montana's Stillwater Complex, whereas the Bushveld Complex in South Africa contains over 70 percent of the world's known reserves of platinum.

Pegmatite Deposits Magmatic differentiation during the late stages of the magmatic process is important in producing ores. This is particularly true of granitic magmas, in which the residual melt can become enriched in rare elements, including some heavy metals. Further, because water and other volatile substances do not crystallize along with the bulk of the magma body, these fluids make up a high percentage of the melt during the final phase of cooling and solidification. Crystallization in a fluid-rich environment, where ion migration is enhanced, results in the formation of crystals several centimeters, or even a few meters, in length. The resulting rocks, called **pegmatites**, are composed of these unusually large crystals (Figure 4.32).

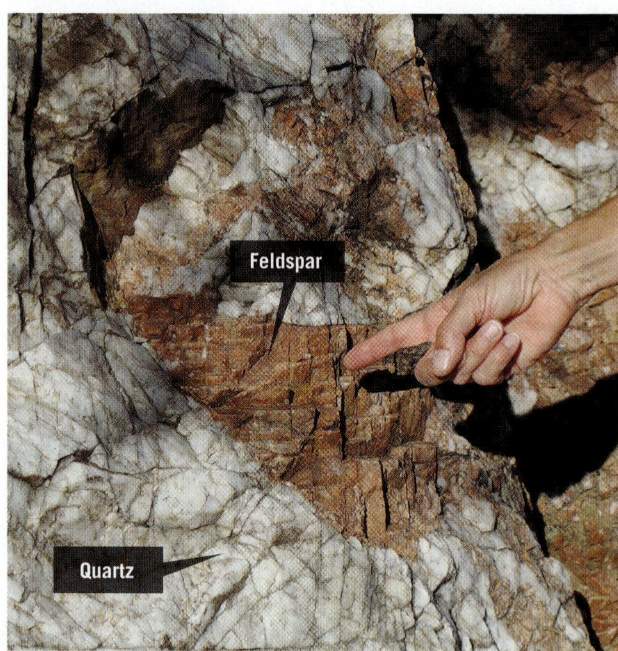

▲ **Figure 4.32 Pegmatite** This granite pegmatite, found in the inner gorge of the Grand Canyon, is composed mainly of quartz and feldspar. (Photo by Joanne Bannon/E. J. Tarbuck)

Feldspar masses the size of houses have been quarried from a pegmatite located in North Carolina. Gigantic hexagonal crystals of muscovite measuring a few

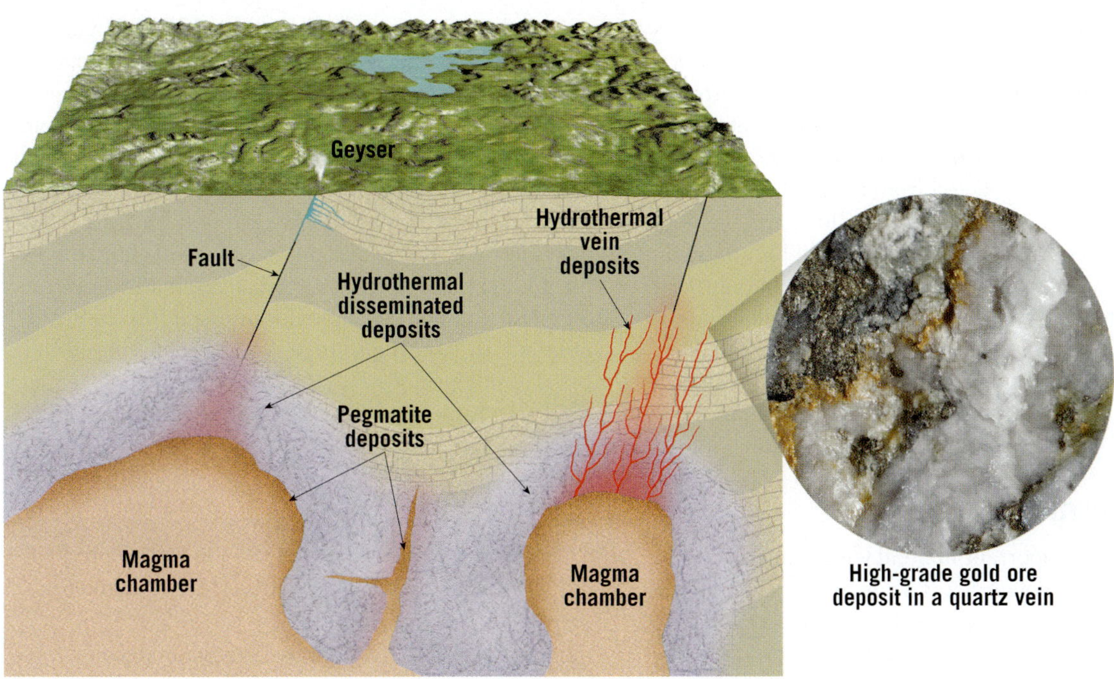

SmartFigure 4.33 Pegmatites and hydrothermal deposits Illustration of the relationship between an igneous body and associated pegmatites and hydrothermal mineral deposits. (Photo by Greenshoots Communications/Alamy)

TUTORIAL
http://goo.gl/Olczs

meters across have been found in Ontario, Canada. In the Black Hills, spodumene crystals as thick as telephone poles have been mined. The largest of these was more than 12 meters (40 feet) long. Not all pegmatites contain such large crystals, but these examples emphasize the special conditions that must exist during their formation.

Most pegmatites are granitic in composition and consist of unusually large crystals of quartz, feldspar, and muscovite. Feldspar is used in the production of ceramics, and muscovite is used for electrical insulation and glitter. Further, pegmatites often contain some of the least abundant elements. Minerals containing the elements lithium, cesium, uranium, and the rare earths are occasionally found. Moreover, some pegmatites contain semiprecious gems such as beryl, topaz, and tourmaline. Most pegmatites are located within large igneous masses, or as dikes or veins that cut into the host rock that surrounds the magma chamber (**Figure 4.33**).

Not all late-stage magmas produce pegmatites, nor do all have a granitic composition. Some magmas instead become enriched in iron or occasionally copper. For example, at Kiruna Sweden, magma composed of over 60 percent magnetite solidified to produce one of the largest iron deposits in the world.

Hydrothermal Deposits

Among the best-known and most important ore deposits are those generated from hot, ion-rich fluids called hydrothermal (hot-water) solutions.* Included in this group are the gold deposits of the Homestake mine in South Dakota; the lead, zinc, and silver ores near Coeur d'Alene, Idaho; the silver deposits of the Comstock Lode in Nevada; and the copper ores of Michigan's Keweenaw Peninsula.

*Because these hot, ion-rich fluids tend to chemically alter the host rock, this process is called hydrothermal metamorphism, and it is discussed in Chapter 8.

Hydrothermal Vein Deposits The majority of hydrothermal deposits originate from hot, metal-rich fluids that are remnants of late-stage magmatic processes. During the cooling process, liquids plus various metallic ions accumulate near the top of the magma chamber. Because of their mobility, these ion-rich solutions can migrate great distances through the surrounding rock before they are eventually deposited, usually as sulfides of various metals (see Figure 4.33). Some of these fluids move along openings such as fractures or bedding planes, where they cool and precipitate metallic ions to produce **vein deposits**. Many of the most productive deposits of gold, silver, copper, and mercury occur as hydrothermal vein deposits (**Figure 4.34**).

Figure 4.34 Native copper This nearly pure metal from northern Michigan's Keweenaw Peninsula is an excellent example of a hydrothermal deposit. This area was once an important source of copper that is now largely depleted. (Photo by E. J. Tarbuck)

Did You Know?
Although the United States is the largest consumer of industrial diamonds, it has no commercial source of this mineral. It does, however, manufacture large quantities of synthetic diamonds for industrial use.

Disseminated Deposits Another important type of accumulation generated by hydrothermal activity is called a **disseminated deposit**. Rather than being concentrated in narrow veins and dikes, these ores are distributed as minute masses throughout the entire rock mass. Much of the world's copper is extracted from disseminated deposits, including those at Chuquicamata, Chile, and the huge Bingham Canyon copper mine in Utah (see Figure 3.36, page 90). Because these accumulations contain only 0.4 to 0.8 percent copper, between 125 and 250 kilograms of ore must be mined for every 1 kilogram of metal recovered. The environmental impact of these large excavations is significant and includes problems related to waste disposal.

Origin of Diamonds

An economically important mineral with an igneous origin is diamond. Although best known as gems, diamonds are used extensively as abrasives. Diamonds are thought to originate at depths of nearly 200 kilometers (120 miles), where the confining pressure is great enough to generate this high-pressure form of carbon. Once crystallized, the diamonds are carried upward through pipe-shaped conduits that increase in diameter toward the surface. In diamond-bearing pipes, nearly the entire pipe contains diamond crystals that are disseminated throughout an ultramafic rock called *kimberlite*. The most productive kimberlite pipes are those in South Africa. The only equivalent source of diamonds in the United States is located near Murfreesboro, Arkansas, but several attempt at commercial diamond mining at this location failed. In 1972 the state of Arkansaa purchaced the property and developed it into Crater of Diamonds State Park. Park visitors are encourage to hunt for diamonds and can keep what they find.

CONCEPT CHECKS 4.9

1. Name and compare two types of hydrothermal deposits.
2. In what type of environment are pegmatites produced?

CONCEPTS IN REVIEW

Igneous Rocks & Intrusive Activity

4.1 Magma: Parent Material of Igneous Rock

List and describe the three major components of magma.

KEY TERMS: igneous rock magma, lava, melt, volatile, crystallization, intrusive igneous rock (plutonic rock), extrusive igneous rock (volcanic rock)

- Completely or partly molten rock is called magma if it is below Earth's surface and lava if it has erupted onto the surface. It consists of a liquid melt that may also contain solid mineral crystals and gases (volatiles), such as water vapor or carbon dioxide.
- As magma cools, silicate minerals begin to form from the "cocktail" of mobile ions in the melt. These tiny crystals grow through the addition of ions to their outer surface. As cooling proceeds, crystallization gradually transforms the magma into a solid mass of interlocking crystals—an igneous rock.
- Magmas that cool below the surface produce intrusive igneous rocks, whereas those that erupt onto Earth's surface produce extrusive igneous rocks.

4.2 Igneous Compositions

Compare and contrast the four basic igneous compositions: felsic, intermediate, mafic, and ultramafic.

KEY TERMS: felsic (granitic) composition, mafic (basaltic) composition, intermediate (andesitic) composition, peridotite, ultramafic

- Igneous rocks are composed mostly of silicate minerals. Igneous rocks that are composed mostly of light-colored silicate minerals are described as having a felsic composition. Rocks composed of greater than 45 percent ferromagnesian minerals are classified as mafic. Mafic rocks are generally darker in color and more dense than their felsic counterparts. Broadly, continental crust is felsic in composition, and oceanic crust is mafic.
- Intermediate rocks in which plagioclase feldspar predominates have a composition that is intermediate between felsic and mafic. They are typical of continental volcanic arcs. Ultramafic rocks, which are rich in the minerals olivine and pyroxene, dominate the upper mantle.
- The amount of silica (SiO_2) in an igneous rock is an indication of its overall composition. Rocks that are rich in silica (70 percent or more) are felsic, while rocks that are poor in silica (as low as 40 percent) are mafic or ultramafic. The amount of silica present in a magma determines the magma's viscosity and crystallization temperature.

? Describe igneous rocks having the compositions of samples A and D using terms such as mafic, felsic, etc. Would you ever expect to find quartz and olivine in the same rock? Why or why not?

(4.2 continued)

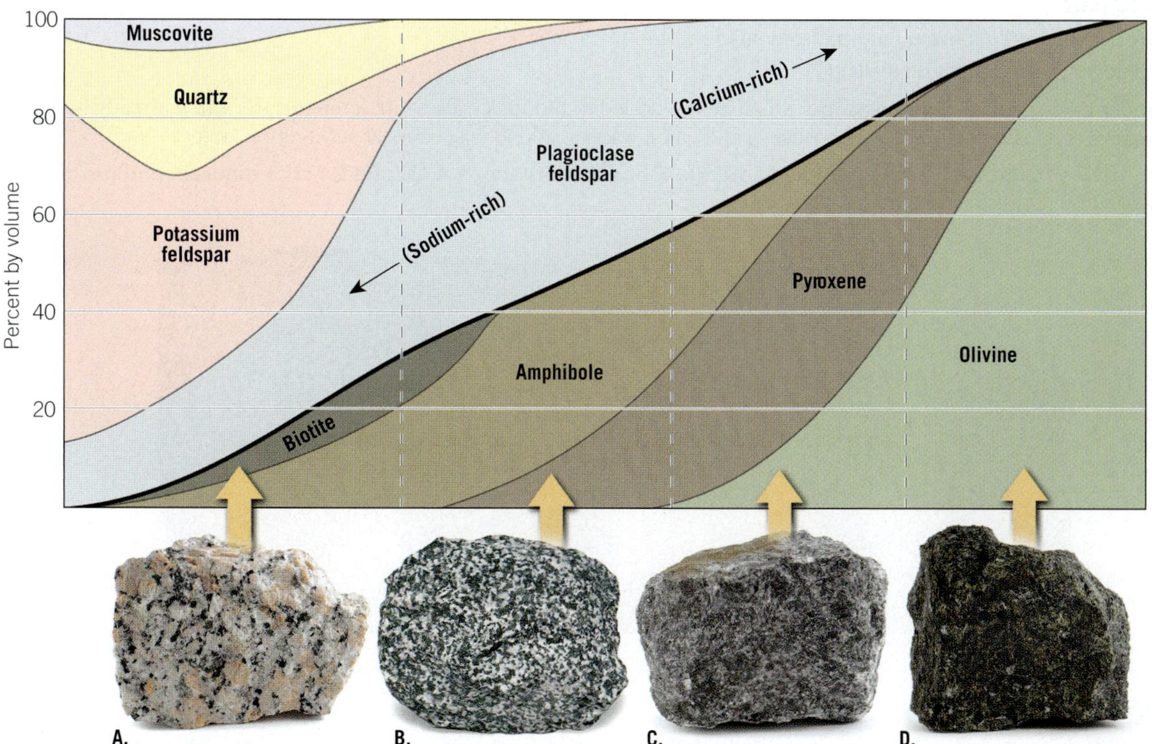

4.3 Igneous Textures: What Can They Tell Us?

Identify and describe the six major igneous textures.

KEY TERMS: texture, aphanitic (fine-grained texture), phaneritic (coarse-grained texture), porphyritic texture, phenocryst, groundmass, porphyry, vesicular texture, glassy texture, pyroclastic (fragmental) texture

- To geologists, "texture" is a description of the size, shape, and arrangement of mineral grains in a rock. Careful observation of the texture of igneous rocks can tell us about the conditions under which they formed. The rate at which magma or lava cools is an important factor in the rock's final texture.
- Lava cools quickly at or close to the surface, so crystallization is rapid and results in a large number of very small crystals. The result is a fine-grained texture. Magma cooling at depth loses heat more slowly. This allows sufficient time for the magma's ions to be organized into larger crystals, resulting in a rock with a coarse-grained texture. If crystals begin to form at depth and then the magma moves to a shallow depth or erupts at the surface, it will have a two-stage cooling history. The result is a rock with a porphyritic texture.
- Volcanic rocks may exhibit additional textures: vesicular if the lava had a high gas content, glassy if it was high in silica, or pyroclastic if the lava erupted explosively.

4.4 Naming Igneous Rocks

Distinguish among the common igneous rocks based on texture and mineral composition.

KEY TERMS: granite, rhyolite, obsidian, pumice, andesite, diorite, basalt, gabbro, pyroclastic rock

- Igneous rocks are classified on the basis of their textures and compositions. Figure 4.12 summarizes the naming system based on these two criteria. Two magmas with the same composition can cool at different rates, resulting in different final textures. On the other hand, two magmas that have different compositions may attain similar textures if they cool under similar circumstances.

? Is it possible for granite to be transformed into rhyolite? If so, what processes would have to be involved?

4.5 Origin of Magma

Summarize the major processes that generate magma from solid rock.

KEY TERMS: geothermal gradient, decompression melting

- Solid rock may melt under three geologic circumstances: when heat is added to the rock, raising its temperature; when already hot rock experiences lower pressures (decompression, as seen at mid-ocean ridges); and when water is added (as occurs at subduction zones).

? Different processes produce magma in various tectonic settings. Consider situations A, B, and C in the diagram and describe the processes that would be most likely to trigger melting in each.

4.6 How Magmas Evolve

Describe how magmatic differentiation can generate a magma body that has a mineralogy (chemical composition) that is different from its parent magma.

KEY TERMS: Bowen's reaction series, crystal settling, magmatic differentiation, assimilation, magma mixing

- Pioneering experimentation by N. L. Bowen revealed that in a cooling magma, minerals crystallize in a specific order. Ferromagnesian silicates such as olivine crystallize first, at the highest temperatures (1250°C), and nonferromagnesian silicates such as quartz crystallize last, at the lowest temperatures (650°C). Bowen found that in between these temperatures, chemical reactions take place between the crystallized silicates and the melt, resulting in compositional changes to each and the formation of new minerals.
- Various physical processes can cause changes in the composition of magma. For instance, if crystallized silicates are denser than the remaining magma, they will sink to the bottom of the magma chamber. Because these early-formed minerals are likely to be ferromagnesian, the magma has now differentiated toward a more felsic composition.
- As they migrate, magmas may assimilate fragments of their "host" rocks or mix with other magma bodies. These processes will alter the magma's composition.

? Consider the accompanying diagram, which shows a cross-sectional view of a hypothetical magma chamber. Using your understanding of Bowen's reaction series and magma evolution, interpret the layered structure by explaining how crystallization occurred.

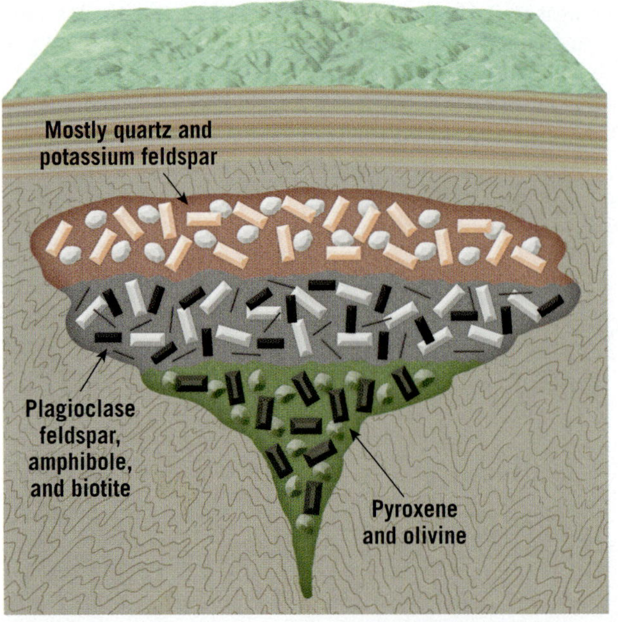

4.7 Partial Melting & Magma Composition

Describe how partial melting of the mantle rock peridotite can generate a basaltic (mafic) magma.

KEY TERMS: partial melting

- In most circumstances, when rocks melt, they do not melt completely. Different minerals have different temperatures at which they change state from solid to liquid (or liquid to solid). As rocks melt, minerals with the lowest melting temperatures melt first.
- Partial melting of the ultramafic mantle yields mafic oceanic crust. Partial melting of the lower continental crust at subduction zones produces magmas that have intermediate or felsic compositions.

? How is partial melting important for generating the different kinds of igneous rocks found on Earth?

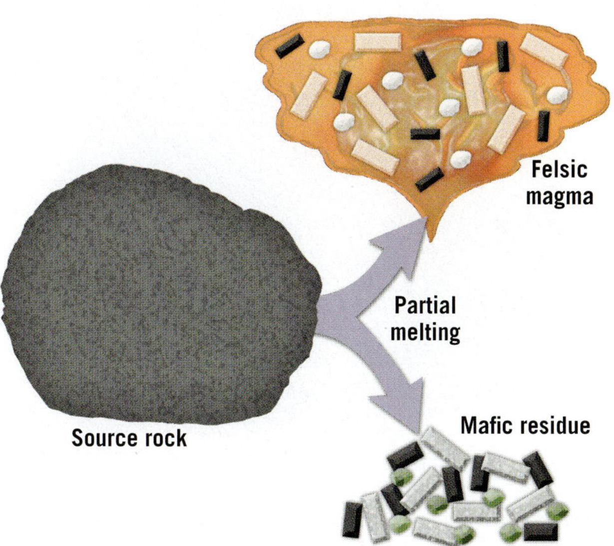

4.8 Intrusive Igneous Activity

Compare and contrast these intrusive igneous structures: dikes, sills, batholiths, stocks, and laccoliths.

KEY TERMS: host (country) rock, intrusion (pluton), tabular, massive, discordant, concordant, dike, sill, columnar jointing, batholith, stock, laccolith

- When magma intrudes other rocks, it may cool and crystallize before reaching the surface to produce intrusions called plutons. Plutons come in many shapes. They may cut across the host rocks without regard for preexisting structures, or the magma may flow along weak zones in the host rock, such as between the horizontal layers of sedimentary bedding.
- Tabular intrusions may be concordant (sills) or discordant (dikes). Massive plutons may be small (stocks) or very large (batholiths). A blister-like intrusion that lifts the overlying rock layers is a laccolith. As solid igneous rock cools, its volume decreases. Contraction can produce a distinctive fracture pattern called columnar jointing.

? Label the intrusive igneous structures in the accompanying diagram, using the following terms: volcanic neck, sill, batholith, laccolith.

(4.8 continued)

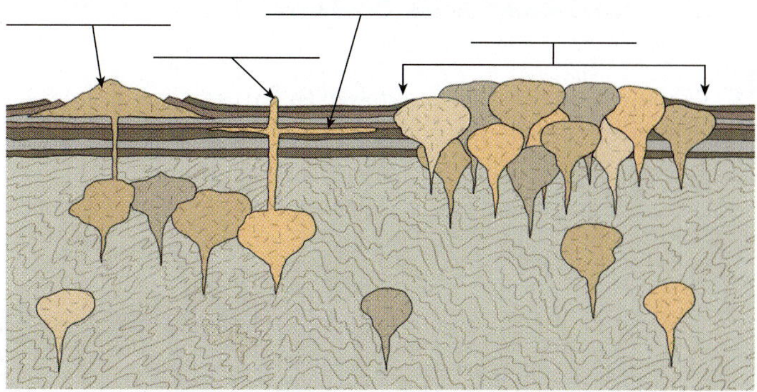

4.9 Mineral Resources & Igneous Processes

Explain how economic deposits of gold, silver, and many other metals form.

KEY TERMS: pegmatite, vein deposit, disseminated deposit

- Some of the most important accumulations of metallic resources, such as gold, silver, lead, and copper, are produced by igneous processes. Magmatic differentiation can concentrate some metals, producing major deposits. Crystallization in a fluid-rich environment, where ion migration is enhanced, results in the formation of unusually large crystals. The resulting rocks, which may become enriched in rare elements and metals, such as gold and silver, are called pegmatites.
- The best-known ore deposits are generated from hydrothermal (hot-water) solutions. Many hydrothermal deposits originate from hot, metal-rich fluids that are remnants of late-stage magmatic processes. Hydrothermal solutions move along fractures or bedding planes, cool, and precipitate the metallic ions to produce vein deposits. In a disseminated deposit (e.g., much of the world's copper deposits), the ores from hydrothermal solutions are distributed as minute masses throughout the entire rock mass.

GIVE IT SOME THOUGHT

1 Would you expect all the crystals in an intrusive igneous rock to be the same size? Explain why or why not.

2 Apply your understanding of igneous rock textures to describe the cooling history of each of the igneous rocks pictured here (Photos by E. J. Tarbuck).

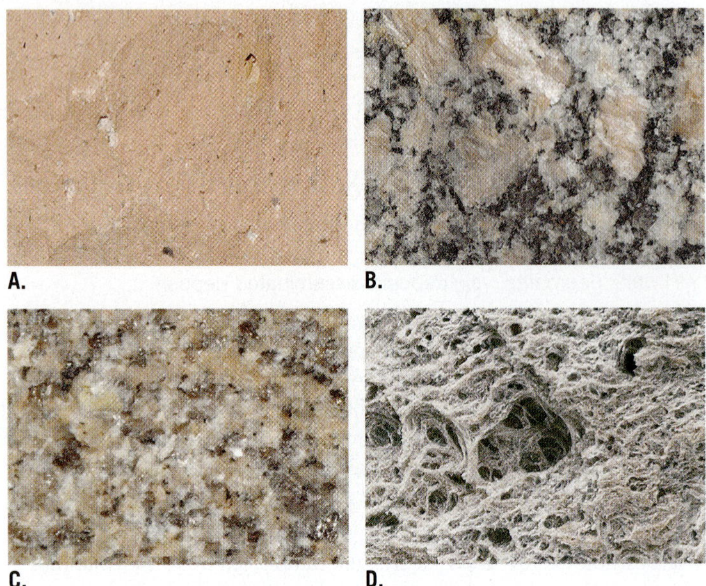

3 Use Figure 4.5 to classify the following igneous rocks:

a. An aphanitic rock containing about 30 percent calcium-rich plagioclase feldspar, 55 percent pyroxene, and 15 percent olivine

b. A phaneritic rock containing about 20 percent quartz, 40 percent potassium feldspar, 20 percent sodium-rich plagioclase feldspar, a few percent muscovite, and the remainder dark-colored silicate

c. An aphanitic rock containing about 50 percent plagioclase feldspar, 35 percent amphibole, 10 percent pyroxene, and minor amounts of other light-colored silicates

d. A phaneritic rock made mainly of olivine and pyroxene, with lesser amounts of calcium-rich plagioclase feldspar

4 Identify the igneous rock textures described by each of the following statements.

a. Openings produced by escaping gases

b. The texture of obsidian

c. A matrix of fine crystals surrounding phenocrysts

d. Consists of crystals that are too small to be seen without a microscope

e. A texture characterized by rock fragments welded together

f. Coarse grained, with crystals of roughly equal size

5 During a hike, you pick up the igneous rock shown in the accompanying photo.

a. What is the mineral name of the small, rounded, glassy green crystals?

b. Did the magma from which this rock formed likely originate in the mantle or in the crust? Explain.

c. Was the magma likely a high-temperature magma or a low-temperature magma? Explain.

d. Describe the texture of this rock.

Dennis Tasa

6 A common misconception about Earth's upper mantle is that it is a thick shell of molten rock. Explain why Earth's mantle is actually solid under most conditions.

7 Describe two mechanisms by which mantle rock can melt without an increase in temperature. How do these magma-generating mechanisms relate to plate tectonics?

8 Use your understanding of Bowen's reaction series (see Figure 4.20) to explain how partial melting can generate magmas that have different compositions.

9 During a field trip with your geology class, you visit an exposure of rock layers similar to the one sketched here. A fellow student suggests that the layer of basalt is a sill, but you disagree. Why do you think the other student is incorrect? What is a more likely explanation for the basalt layer?

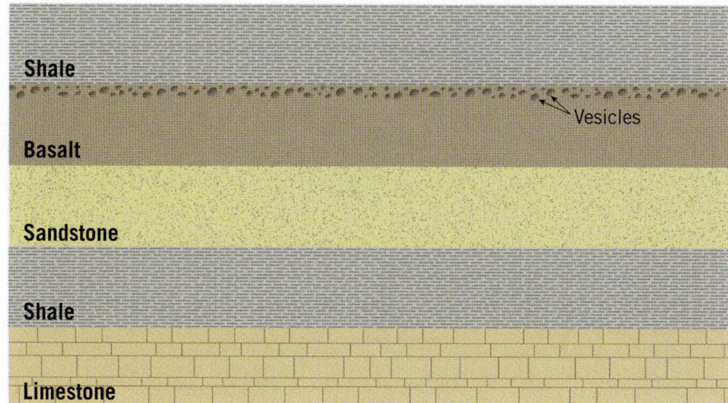

10 Each of the following statements describes how an intrusive feature appears when exposed at Earth's surface due to erosion. Name each feature.

a. A dome-shaped mountainous structure flanked by upturned layers of sedimentary rocks

b. A vertical wall-like feature a few meters wide and hundreds of meters long

c. A huge expanse of granitic rock forming a mountainous terrain tens of kilometers wide

d. A relatively thin layer of basalt sandwiched between horizontal layers of sedimentary rocks exposed along the walls of a river valley

11 Mount Whitney, the highest summit (4421 meters [14,505 feet]) in the contiguous United States, is located in the Sierra Nevada batholith. Based on its location, is Mount Whitney likely composed mainly of granitic, andesitic, or basaltic rocks?

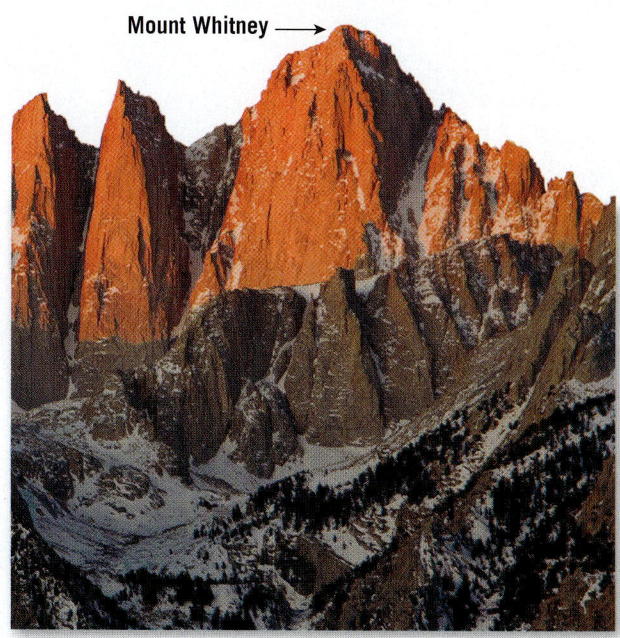

John Greim/Getty Images

MasteringGeology™

Looking for additional review and test prep materials? Visit the Study Area in MasteringGeology to enhance your understanding of this chapter's content by accessing a variety of resources, including Self-Study Quizzes, Geoscience Animations, SmartFigures, Mobile Field Trips, *Project Condor* Quadcopter videos, *In the News* RSS feeds, flashcards, web links, and an optional Pearson eText.

www.masteringgeology.com

5 Volcanoes & Volcanic Hazards

FOCUS ON CONCEPTS

Each statement represents the primary learning objective for the corresponding major heading within the chapter. After you complete the chapter, you should be able to:

- **5.1** Compare and contrast the 1980 eruption of Mount St. Helens with the most recent eruption of Kilauea, which began in 1983.
- **5.2** Explain why some volcanic eruptions are explosive and others are quiescent.
- **5.3** List and describe the three categories of materials extruded during volcanic eruptions.
- **5.4** Draw and label a diagram that illustrates the basic features of a typical volcanic cone.
- **5.5** Summarize the characteristics of shield volcanoes and provide one example of this type of volcano.
- **5.6** Describe the formation, size, and composition of cinder cones.
- **5.7** List the characteristics of composite volcanoes and describe how they form.
- **5.8** Describe the major geologic hazards associated with volcanoes.
- **5.9** List volcanic landforms other than shield, cinder cone, and composite volcanoes and describe their formation.
- **5.10** Explain how the global distribution of volcanic activity is related to plate tectonics.

Reventador ejecting volcanic bombs and incandescent ash at night, November 2015. This volcano, which is located in the Andes of Ecuador, has erupted more than 25 times since 1541. (Photo by Morely Read/Getty Images)

THE SIGNIFICANCE OF IGNEOUS ACTIVITY may not be obvious at first glance. However, because volcanoes extrude molten rock that formed at great depth, they provide our only means of directly observing processes that occur many kilometers below Earth's surface. Furthermore, Earth's atmosphere and oceans have evolved from gases emitted during volcanic eruptions. Either of these facts is reason enough for igneous activity to warrant our attention.

5.1 Mount St. Helens Versus Kilauea

Compare and contrast the 1980 eruption of Mount St. Helens with the most recent eruption of Kilauea, which began in 1983.

On May 18, 1980, the largest volcanic eruption to occur in North America in historic times transformed a picturesque volcano into a decapitated remnant (**Figure 5.1**). On that date in southwestern Washington State, Mount St. Helens erupted with tremendous force. The blast blew out the entire north flank of the volcano, leaving a gaping hole. In one brief moment, a prominent volcano whose summit had been more than 2900 meters (9500 feet) above sea level was lowered by more than 400 meters (1350 feet).

The event devastated a wide swath of timber-rich land on the north side of the mountain (**Figure 5.2**). Trees within a 400-square-kilometer (160-square-mile) area lay intertwined and flattened, stripped of their branches and appearing from the air like toothpicks strewn about. The accompanying mudflows carried ash, trees, and water-saturated rock debris 29 kilometers (18 miles) down the Toutle River. The eruption claimed 59 lives; some died from the intense heat and the suffocating cloud of ash and gases, others from the impact of the blast, and still others from being trapped in mudflows.

The eruption ejected nearly a cubic kilometer of ash and rock debris. Following the devastating explosion, Mount St. Helens continued to emit great quantities of hot gases and ash. The force of the blast was so strong that some ash was propelled more than 18 kilometers (over 11 miles) into the stratosphere. During the next few days, this very fine-grained material was carried around Earth by strong upper-air winds. Crops were damaged in central Montana, and measurable deposits were reported

▼ **Figure 5.1 Before-and-after photographs show the transformation of Mount St. Helens** The May 18, 1980, eruption of Mount St. Helens occurred in southwestern Washington.

The blast blew out the entire north flank of Mount St. Helens, leaving a gaping hole. In a brief moment, a prominent volcano was lowered by 1350 feet.

as far away as Oklahoma and Minnesota. Meanwhile, ash fallout in the immediate vicinity exceeded 2 meters (6 feet) in depth. The air over Yakima, Washington (130 kilometers [80 miles] to the east), was so filled with ash that residents experienced midnight-like darkness at noon.

Not all volcanic eruptions are as violent as the 1980 Mount St. Helens event. Some volcanoes, such as Hawaii's Kilauea Volcano, generate relatively quiet outpourings of fluid lavas. These quiescent (non-explosive) eruptions are not without some fiery displays; occasionally fountains of incandescent lava spray hundreds of meters into the air (see Figure 5.4), but most lava pours from the vent and flows downslope. During Kilauea's most recent active phase, which began in 1983, more than 180 homes and a national park visitor center have been destroyed by flowing lava igniting material in its path.

Testimony to the quiescent nature of Kilauea's eruptions is the fact that the Hawaiian Volcanoes Observatory has operated on its summit since 1912, despite the fact that Kilauea has had more than 50 eruptive phases since record keeping began in 1823.

▲ **Figure 5.2 Douglas fir trees snapped off or uprooted by the lateral blast of Mount St. Helens** (Large photo by Lyn Topinka/AP Photo/U.S. Geological Survey; inset photo by John M. Burnley/Photo Researchers, Inc.)

CONCEPT CHECKS 5.1

1. Briefly compare the 1980 eruption of Mount St. Helens to a typical eruption of Hawaii's Kilauea Volcano.

5.2 The Nature of Volcanic Eruptions

Explain why some volcanic eruptions are explosive and others are quiescent.

Volcanic activity is commonly perceived as a process that produces a picturesque, cone-shaped structure that periodically erupts in a violent manner. However, many eruptions are not explosive, as indicated by Kilauea's activity. What determines the manner in which volcanoes erupt?

Magma: Source Material for Volcanic Eruptions

Recall that **magma**, molten rock that may contain some solid crystalline material and also contains varying amounts of dissolved gas (mainly water vapor and carbon dioxide), is the parent material of igneous rocks. Erupted magma is called **lava**.

Composition of Magma As we discussed in Chapter 4, *mafic* (basaltic) igneous rocks contain a high percentage of dark silicate minerals and calcium-rich plagioclase feldspar, and as a result, they tend to be dark in color. By contrast, *felsic* rocks (granite and its extrusive equivalent, rhyolite) contain mainly light-colored silicate minerals—quartz and potassium feldspar. *Intermediate* (andesitic) rocks have a composition between mafic and felsic rocks. Correspondingly, mafic magmas contain a much *lower* percentage of silica (SiO_2) than do felsic magmas. The compositional differences between magmas also affect several other properties, as summarized in Figure 5.3. As shown in Figure 5.3 mafic (basaltic) magmas have the lowest silica content and the lowest gas content, and they erupt at the highest temperatures. By contrast, felsic (granitic or rhyolitic) magmas have the highest silica content and the highest gas content, and they erupt at the lowest temperatures. Intermediate magmas have characteristics between mafic and felsic magmas.

Where Is Magma Generated? Recall that most magma is generated in Earth's upper mantle (asthenosphere) by *partial melting of solid rock*. The magmas generated by melting mantle rocks tend to have basaltic (mafic) composition. Once formed, basaltic magma, which is less dense than the surrounding rocks, slowly rises toward Earth's surface. In some settings, this hot molten rock reaches the surface, where it usually produces fluid outflows of basaltic lavas. As a result of seafloor spreading, the largest quantity of basaltic magma erupts on the

Did You Know?

The eruption of Tambora, in Indonesia, in 1815 is the largest known volcanic event in modern history. About 20 times more ash and rock were explosively ejected during this eruption than were emitted during the 1980 Mount St. Helens event. The sound of the explosion was heard an incredible 4800 km (3000 mi) away, about the distance across the conterminous United States.

▶ Figure 5.3 Compositional differences of magma bodies cause their properties to vary

Properties of Magma Bodies with Differing Compositions

Composition	Silica Content (SiO_2)	Gas Content (% by weight)	Eruptive Temperature	Viscosity	Tendency to Form Pyroclastics	Volcanic Landform
Mafic (Basaltic) High in Fe, Mg, Ca, low in K, Na	**Least** (~50%)	**Least** (0.5–2%)	**Highest** 1000–1250°C	**Least**	**Least**	Shield volcanoes, basalt plateaus, cinder cones
Intermediate (Andesitic) Varying amounts of Fe, Mg, Ca, K, Na	**Intermediate** (~60%)	**Intermediate** (3–4%)	**Intermediate** 800–1050°C	**Intermediate**	**Intermediate**	Composite cones
Felsic (Granitic/Rhyolitic) High in K, Na, low in Fe, Mg, Ca	**Most** (~70%)	**Most** (5–8%)	**Lowest** 650–900°C	**Greatest**	**Greatest**	Pyroclastic flow deposits, lava domes

ocean floor along divergent plate boundaries. Extensive basaltic flows are also the product of hot-spot volcanism generated by rising hot mantle plumes (see Figure 2.26 on page 54).

In continental settings, however, overlying crustal rocks are usually less dense than the ascending basaltic magma, and as a result, the rising molten rock ponds at the crust–mantle boundary. Because the newly formed magma is much hotter than the melting temperature of crustal rocks, the rocks overlying the magma body begin to melt. This process generates a less dense, more silica-rich magma of intermediate or felsic composition, which then continues the journey toward Earth's surface.

Effusive Versus Explosive Eruptions

Volcanic eruptions exhibit a range of behavior from quiescent eruptions that produce outpourings of fluid lava to explosive eruptions. Geologists often refer to quiescent eruptions as **effusive** (meaning "pouring out") **eruptions**.

The two primary factors that determine how magma erupts are its *viscosity* and *gas content*. **Viscosity** (*viscos* = sticky) is a measure of a fluid's mobility. The more viscous a material, the greater its resistance to flow. For example, pancake syrup is more viscous, and thus more resistant to flow, than water.

Factors Affecting Viscosity Magma's viscosity depends primarily on its temperature and silica content: *The more silica in magma, the greater its viscosity.* Silicate structures begin to link together into long chains early in the crystallization process, which makes the magma more rigid and impedes its flow. Consequently, silica-rich felsic (rhyolitic) lavas are the most viscous and tend to travel at imperceptibly slow speeds to form comparatively short, thick flows. By contrast, mafic (basaltic) lavas, which contain much less silica, are relatively fluid and have been known to travel 150 kilometers (90 miles) or more before solidifying. Intermediate (andesitic) magmas have flow rates between these extremes.

Temperature affects the viscosity of magma in much the same way it affects the viscosity of pancake syrup: The hotter a magma, the more fluid (less viscous) it will be. As lava cools and begins to congeal, its viscosity increases, and the flow eventually halts.

Role of Gases The nature of volcanic eruptions also depends on the amount of dissolved gases held in the magma body by the pressure exerted by the overlying rock (the confining pressure). The most abundant gases in most magmas are water vapor and carbon dioxide. These dissolved gases tend to come out of solution when the confining pressure is reduced. This is analogous to how carbon dioxide is retained in cans and bottles of soft drinks. When the pressure is reduced on a soft drink by opening the cap, the dissolved carbon dioxide quickly separates from the solution to form bubbles that rise and escape.

The viscosity and gas content of magma are directly related to its composition, as shown in Figure 5.3. At one end of the spectrum are basaltic (mafic) magmas, which are very fluid and have a low gas content, sometimes as little as 0.5 percent by weight. At the other extreme are rhyolitic (felsic) magmas, which are highly viscous (sticky) and contain a lot of gas, as much as 8 percent by weight.

Effusive Hawaiian-Type Eruptions

All magmas contain some water vapor and other gases that are kept in solution by the immense pressure of the overlying rock. As magma rises (or the rocks confining the magma fail), the confining pressure drops, causing the dissolved gases to separate from the melt and form large numbers of tiny bubbles. When fluid basaltic magmas erupt, these pressurized gases readily escape. At temperatures that often exceed 1100°C (2000°F), these gases can quickly expand to occupy hundreds of times their original volumes. Occasionally, these expanding gases propel incandescent lava hundreds of meters into the air, producing lava fountains (**Figure 5.4**). Although spectacular, these fountains are usually harmless and generally not associated with major explosive events that cause great loss of life and property.

Eruptions that involve very fluid basaltic lavas, such as the recent eruptions of Kilauea on Hawaii's Big Island, are often triggered by the arrival of a new batch of molten rock, which accumulates in a near-surface magma chamber. Geologists can usually detect such an event because the summit of the volcano begins to inflate and rise months or even years before an eruption. The injection of a fresh supply of hot molten rock heats and remobilizes the semi-liquid magma in the chamber. In addition, swelling of the magma chamber fractures the rock above, allowing the fluid magma to move upward along the newly formed fissures, often generating effusions of fluid lava for weeks, months, or possibly years. The eruption of Kilauea that began in 1983 is ongoing.

How Explosive Eruptions Are Triggered

Recall that silica-rich rhyolitic magmas have a relatively high gas content and are quite viscous (sticky) compared to basaltic magmas. As rhyolitic magma rises, the gases remain dissolved until the confining pressure drops sufficiently, at which time tiny bubbles begin to form and increase in size. Because of the high viscosity of rhyolitic magma, gas bubbles tend to remain trapped in the magma, forming a sticky froth.

When the pressure exerted by the expanding magma exceeds the strength of the overlying rock, fracturing occurs. As the frothy magma moves up through the fractures, a further drop in confining pressure creates additional gas bubbles. This chain reaction often generates an explosive event in which magma is literally blown into fragments (ash and pumice) that are carried to great heights by the escaping hot gases. (The collapse of a volcano's flank can also greatly reduce the pressure on the magma below, causing an explosive eruption, as exemplified by the 1980 eruption of Mount St. Helens.)

Gases readily escape hot fluid basaltic flows, producing lava fountains. Although often spectacular, these features generally do not cause great loss of life or property.

▲ **Figure 5.4 Lava fountain produced by gases escaping fluid basaltic lava** Kilauea, on Hawaii's Big Island, is one of the most active volcanoes on Earth. (Photo by David Reggie/Getty Images)

Eruptions of highly viscous lavas may produce explosive clouds of hot ash and gases called eruption columns.

▲ **SmartFigure 5.5**
Eruption column generated by viscous, silica-rich magma Steam and ash eruption column from Mount Sinabung, Indonesia, 2014. A deadly cloud of fiery ash can be seen racing down the volcano's slope in the foreground. (Photo by REUTERS/Beawiharta)

VIDEO
https://goo.gl/RH97D7

When molten rock in the uppermost portion of the magma chamber is forcefully ejected by the escaping gases, the confining pressure on the magma directly below also drops suddenly. Thus, rather than being a single "bang," an explosive eruption is really a series of violent explosions that can last for a few days.

Because highly gaseous magmas expel fragmented lava at nearly supersonic speeds, they are associated with hot, buoyant **eruption columns** consisting mainly of volcanic ash and gases (Figure 5.5). Eruption columns can rise perhaps 40 kilometers (25 miles) into the atmosphere. It is not uncommon for a portion of an eruption column to collapse, sending hot ash rushing down the volcanic slope at speeds exceeding 100 kilometers (60 miles) per hour. As a result, volcanoes that erupt highly viscous magmas having a high gas content are the most destructive to property and human life.

Following explosive eruptions, partially degassed lava may slowly ooze out of the vent to form thick lava flows or dome-shaped lava bodies that grow over the vent.

CONCEPT CHECKS 5.2

1. List these magmas in order, from the most silica rich to the least silica rich: mafic (basaltic) magma, felsic (rhyolitic) magma, intermediate (andesitic) magma.
2. List the two primary factors that determine the manner in which magma erupts.
3. Define *viscosity*.
4. Are volcanoes fed by highly viscous magma *more* or *less likely* to be a greater threat to life and property than volcanoes supplied with very fluid magma?

5.3 Materials Extruded During an Eruption

List and describe the three categories of materials extruded during volcanic eruptions.

Volcanoes erupt lava, large volumes of gas, and pyroclastic materials (broken rock, lava "bombs," and ash). In this section we will examine each of these materials.

Lava Flows

The vast majority of Earth's lava, more than 90 percent of the total volume, is estimated to be mafic (basaltic) in composition. Most of mafic lavas erupt on the seafloor, via a process known as *submarine volcanism*. Lavas having an intermediate (andesitic) composition account for most of the rest, while felsic (rhyolitic) flows make up as little as 1 percent of the total. Rhyolitic magmas tend to extrude mostly volcanic ash rather than lava.

On land, hot basaltic lavas, which are usually very fluid, generally flow in thin, broad sheets or streamlike ribbons. Fluid basaltic lavas have been clocked at speeds exceeding 30 kilometers (19 miles) per hour down steep slopes. However, flow rates of 10 to 300 meters (30 to 1000 feet) per hour are more common. Silica-rich rhyolitic lava, by contrast, often moves too slowly to be perceived. Furthermore, rhyolitic lavas seldom travel more than a few kilometers from their vents. As you might expect, andesitic lavas, which are intermediate in composition, exhibit flow characteristics between these extremes.

Aa & Pahoehoe Flows Fluid basaltic magmas tend to generate two types of lava flows, which are known by their Hawaiian names. The first, called **aa** (pronounced "ah-ah") **flows**, have surfaces of rough jagged blocks with dangerously sharp edges and spiny projections (Figure 5.6A). Crossing a hardened aa flow can be a trying and miserable experience. The second type, **pahoehoe** (pronounced "pah-hoy-hoy") **flows**, exhibit smooth surfaces that sometimes resemble twisted braids of ropes (Figure 5.6B).

Although both lava types can erupt from the same volcano, pahoehoe lavas are hotter and more fluid than aa flows. In addition, pahoehoe lavas can change into aa lava flows, although the reverse (aa to pahoehoe) does not occur.

Cooling that occurs as the flow moves away from the vent is one factor that facilitates the change from pahoehoe to aa. The lower temperature increases viscosity and promotes bubble formation. Escaping gas bubbles produce numerous voids (vesicles) and sharp spines in the surface of the congealing lava. As the molten interior advances, the outer crust is broken, transforming the relatively smooth surface of a pahoehoe flow into an aa flow made up of an advancing mass of rough, sharp, broken lava blocks.

Pahoehoe flows often develop cave-like tunnels called **lava tubes** that start as conduits for carrying lava from an active vent to the flow's leading edge (Figure 5.7). Lava tubes form in the interior of a lava flow, where the temperature remains high long after the exposed surface cools and hardens. Because they serve as insulated pathways that allow lava to flow great distances from its source, lava tubes are important features of fluid lava flows.

Pillow Lavas Recall that most of Earth's volcanic output occurs along oceanic ridges (divergent plate boundaries), generating new oceanic crust. When outpourings of lava occur on the ocean floor, the flow's outer skin quickly

▼ **Figure 5.6 Lava flows A.** A slow-moving, basaltic aa flow advancing over hardened pahoehoe lava. **B.** A typical fluid pahoehoe (ropy) lava. Both of these lava flows erupted from a rift on the flank of Hawaii's Kilauea Volcano. (Photos courtesy of U.S. Geological Survey)

A. Active aa flow overriding an older pahoehoe flow.

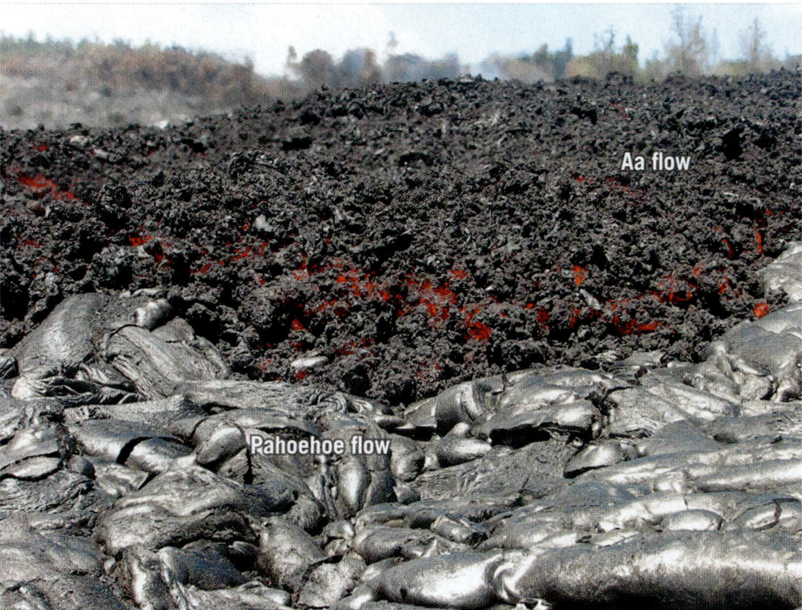

B. Pahoehoe flow displaying the characteristic ropy appearance.

A. Lava tubes are cave-like tunnels that once served as conduits carrying lava from an active vent to the flow's leading edge.

Valentine Cave, a lava tube at Lava Beds National Monument, California.

B. Skylights develop where the roofs of lava tubes collapse and reveal the hot lava flowing through the tube.

▲ **Figure 5.7 Lava tubes A.** A lava flow may develop a solid upper crust, while the molten lava below continues to advance in a conduit called a *lava tube*. Some lava tubes exhibit extraordinary dimensions. Kazumura Cave, located on the southeastern slope of Hawaii's Mauna Loa Volcano, is a lava tube extending more than 60 kilometers (40 miles). (Photo by Dave Bunell) **B.** The collapsed section of the roof of a lava tunnel results in a skylight. (Photo courtesy of U.S. Geological Survey)

freezes (solidifies) to form volcanic glass. However, the interior lava is able to move forward by breaking through the hardened surface. This process occurs over and over, as molten basalt is extruded like toothpaste from a tightly squeezed tube. The result is a lava flow composed of numerous tube-like structures called **pillow lavas**, stacked one atop the other (**Figure 5.8**). Pillow lavas are useful when reconstructing geologic history because their presence indicates that the lava flow formed below the surface of a water body.

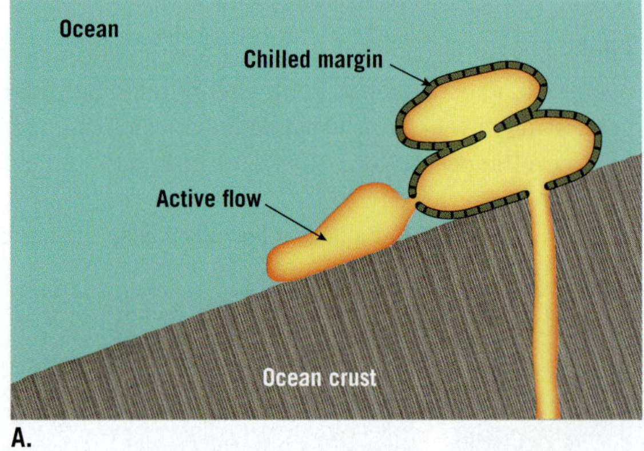

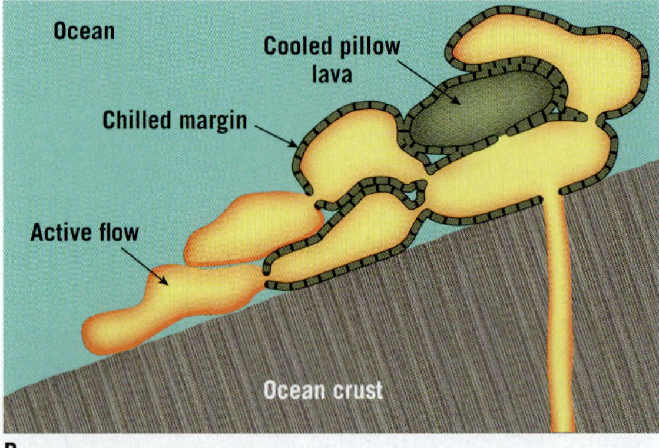

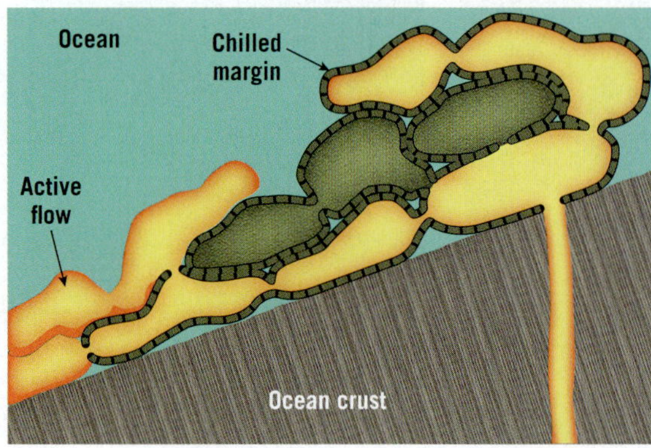

► **Figure 5.8 Pillow lava** These diagrams show the formation of pillow lava. The pillows tend to be elongated tube-like structures and are often variable in shape. The photo shows an undersea pillow lava flow off the coast of Hawaii. (Photo courtesy of U.S. Geological Survey)

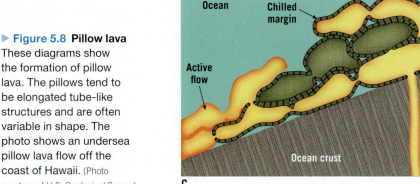

Block Lavas In contrast to fluid basaltic magmas that can travel many kilometers, viscous andesitic and rhyolitic magmas tend to generate relatively short prominent flows—a few hundred meters to a few kilometers long. Their upper surface consists largely of massive, detached blocks—hence the name **block lava**. Although similar to aa flows, these lavas consist of blocks with slightly curved, smooth surfaces rather than the rough, sharp, spiny surfaces typical of aa flows.

Gases

Recall that magmas contain varying amounts of dissolved gases, called **volatiles**. These gases are held in the molten rock by confining pressure, just as carbon dioxide is held in cans of soft drinks. As with soft drinks, as soon as the pressure is reduced, the gases begin to escape. Obtaining gas samples from an erupting volcano is difficult and dangerous, so geologists usually must estimate the amount of gas originally contained in the magma.

The gaseous portion of most magma bodies ranges from less than 1 percent to about 8 percent of the total weight, with most of this in the form of water vapor. Although the percentage may be small, the actual quantity of emitted gas can exceed thousands of tons per day. Occasionally, eruptions emit colossal amounts of volcanic gases that rise high into the atmosphere, where they may reside for several years.

The composition of volcanic gases is important because these gases contribute significantly to our planet's atmosphere. The most abundant gas typically released into the atmosphere from volcanoes is water vapor (H_2O), followed by carbon dioxide (CO_2) and sulfur dioxide (SO_2), with lesser amounts of hydrogen sulfide (H_2S), carbon monoxide (CO), and helium (H_2). (The relative proportion of each gas varies significantly from one volcanic region to another.) Sulfur compounds are easily recognized by their pungent odor. Volcanoes are also natural sources of air pollution; some emit large quantities of sulfur dioxide (SO_2), which readily combines with atmospheric gases to form toxic sulfuric acid and other sulfate compounds.

Pyroclastic Materials

When volcanoes erupt energetically, they eject pulverized rock and fragments of lava and glass from the vent. The particles produced, **pyroclastic materials** (*pyro* = fire, *clast* = fragment), are also called **tephra**. These fragments range in size from very fine dust and sand-sized volcanic ash (less than 2 millimeters) to pieces that weigh several tons (Figure 5.9).

Ash and *dust* particles are produced when gas-rich viscous magma erupts explosively. As magma moves up in the vent, the gases rapidly expand, generating a melt that resembles the froth that flows from a bottle of champagne. As the hot gases expand explosively, the froth is blown into very fine glassy fragments. When the hot ash falls, the glassy shards often fuse to form a rock called *welded tuff*. Sheets of this material, as well as ash deposits that later consolidate, cover vast portions of the western United States.

Somewhat larger pyroclasts that range in size from small beads to walnuts (2–64 millimeters [0.08–2.5 inches] in diameter) are known as *lapilli* ("little stones") or *cinders*. Particles larger than 64 millimeters (2.5 inches) in diameter are called *blocks* when they are made of hardened lava and *bombs* when they are ejected as incandescent lava (see Figure 5.7). Because bombs are semi-molten when ejected, they often take on a streamlined

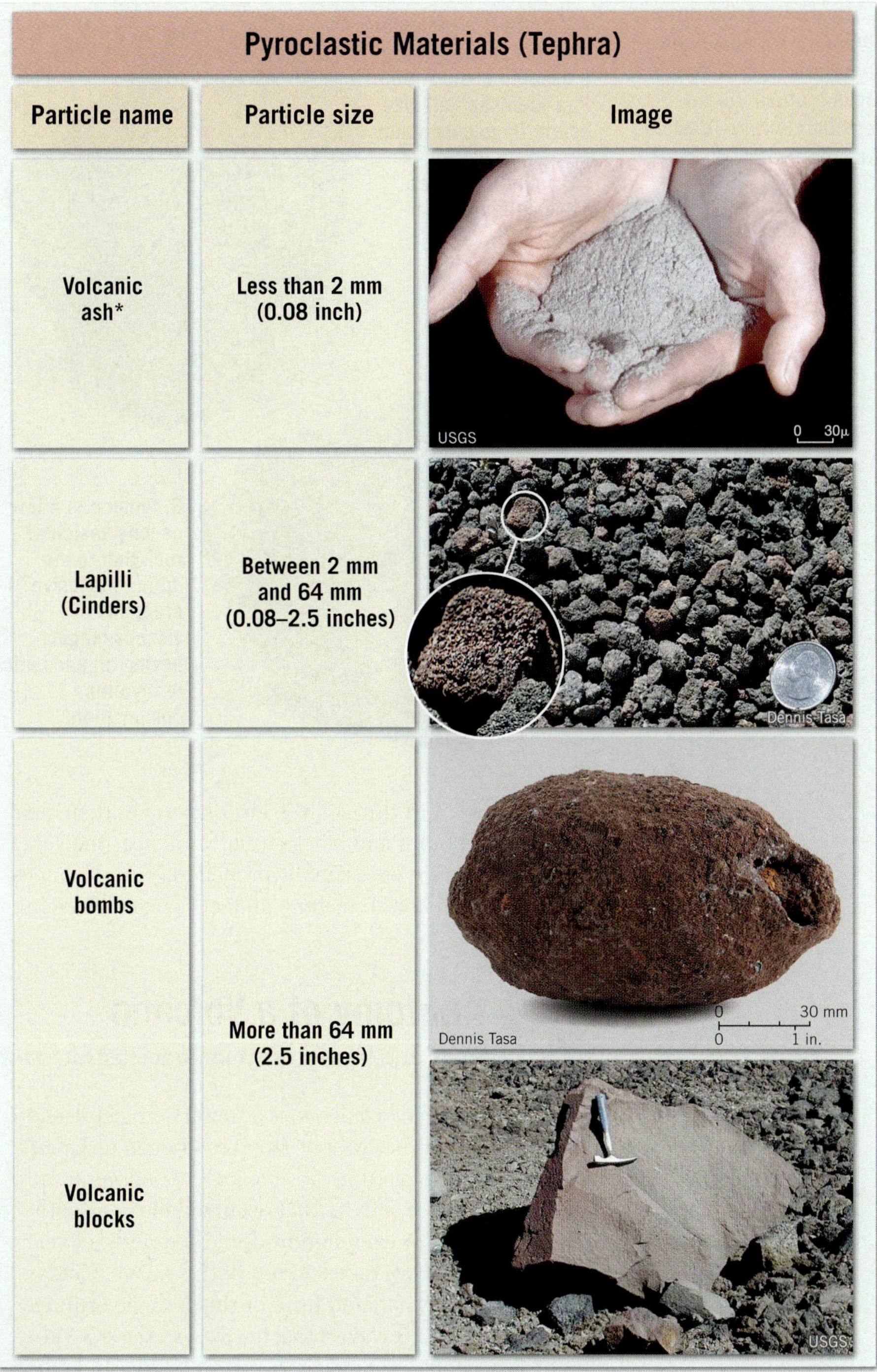
Pyroclastic Materials (Tephra)

Particle name	Particle size	Image
Volcanic ash*	Less than 2 mm (0.08 inch)	
Lapilli (Cinders)	Between 2 mm and 64 mm (0.08–2.5 inches)	
Volcanic bombs	More than 64 mm (2.5 inches)	
Volcanic blocks	More than 64 mm (2.5 inches)	

*The term volcanic dust is used for fine volcanic ash less than 0.063 mm (0.0025 inch).

▲ **Figure 5.9 Types of pyroclastic materials** Pyroclastic materials are also commonly referred to as tephra.

A. Scoria is a vesicular rock commonly having a basaltic composition. Pea-to-basketball size scoria fragments make up a large portion of most cinder cones (also called *scoria cones*).
B. Pumice is a low-density vesicular rock that forms during explosive eruptions of viscous magma having an andesitic to rhyolitic composition.

▶ **Figure 5.10 Common vesicular rocks** Scoria and pumice are volcanic rocks that exhibit a vesicular texture. Vesicles are small holes left by escaping gas bubbles. (Photos by E.J. Tarbuck)

shape as they hurl through the air. Because of their size and weight, bombs and blocks usually fall near the vent; however, they are occasionally propelled great distances. For instance, bombs 6 meters (20 feet) long and weighing about 200 tons were blown 600 meters (2000 feet) from the vent during an eruption of the Japanese volcano Asama.

Pyroclastic materials can be classified by texture and composition as well as by size. For instance, **scoria** is the term for vesicular ejecta produced most often during the eruption of basaltic magmas (**Figure 5.10**). These black to reddish-brown fragments are generally found in the size range of lapilli and resemble cinders and clinkers produced by furnaces used to smelt iron.

By contrast, when magmas with andesitic (intermediate) or rhyolitic (felsic) compositions erupt explosively, they emit ash and the vesicular rock **pumice** (**Figure 5.10B**). Pumice is usually lighter in color and less dense than scoria, and many pumice fragments have so many vesicles that they are light enough to float (see Figure 4.14, page 106).

CONCEPT CHECKS 5.3

1. Contrast pahoehoe and aa lava flows.
2. How do lava tubes form?
3. List the main gases released during a volcanic eruption.
4. How do volcanic bombs differ from blocks of pyroclastic debris?
5. What is scoria? How is scoria different from pumice?

5.4 Anatomy of a Volcano

Draw and label a diagram that illustrates the basic features of a typical volcanic cone.

A popular image of a volcano is a solitary, graceful, snow-capped cone, such as Mount Hood in Oregon or Japan's Fujiyama. These picturesque, conical mountains are produced by volcanic activity that occurred intermittently over thousands, or even hundreds of thousands, of years. However, many volcanoes do not fit this image. Cinder cones are quite small and form during a single eruptive phase that lasts a few days to a few years. Alaska's Valley of Ten Thousand Smokes is a flat-topped ash deposit that blanketed a river valley to a depth of 200 meters (600 feet). The eruption that produced it lasted less than 60 hours yet emitted more than 20 times more volcanic material than the 1980 Mount St. Helens eruption.

Volcanic landforms come in a wide variety of shapes and sizes, and each volcano has a unique eruptive history. Nevertheless, volcanologists have been able to classify volcanic landforms and determine their eruptive patterns. In this section we will consider the general anatomy of an idealized volcanic cone. We will follow this discussion by exploring the three major types of volcanic cones—shield volcanoes, cinder cones, and composite volcanoes—as well as their associated hazards.

Did You Know?
Chile, Peru, and Ecuador boast the highest volcanoes in the world. Dozens of cones exceed 6000 m (20,000 ft). Two volcanoes located in Ecuador, Chimborazo and Cotopaxi, were once considered the world's highest mountains. That distinction remained until the Himalayas were surveyed in the nineteenth century.

Volcanic activity frequently begins when a **fissure** (crack) develops in Earth's crust as magma moves forcefully toward the surface. As the gas-rich magma moves up through a fissure, its path is usually localized into a somewhat pipe-shaped **conduit** that terminates at a surface opening called a **vent** (**Figure 5.11**). The cone-shaped structure we call a **volcanic cone** is often created by successive eruptions of lava, pyroclastic material, or frequently a combination of both, often separated by long periods of inactivity.

Located at the summit of most volcanic cones is a somewhat funnel-shaped depression called a **crater** (*crater* = bowl). Volcanoes built primarily of pyroclastic materials typically have craters that form by gradual accumulation of volcanic debris on the surrounding rim. Other craters form during explosive eruptions, as the rapidly ejected particles erode the crater walls. Craters also form when the summit area of a volcano collapses following an eruption. Some volcanoes have very large circular depressions, called **calderas**, which have diameters that are greater than 1 kilometer (0.6 mile) and that in rare cases exceed 50 kilometers (30 miles). The formation of various

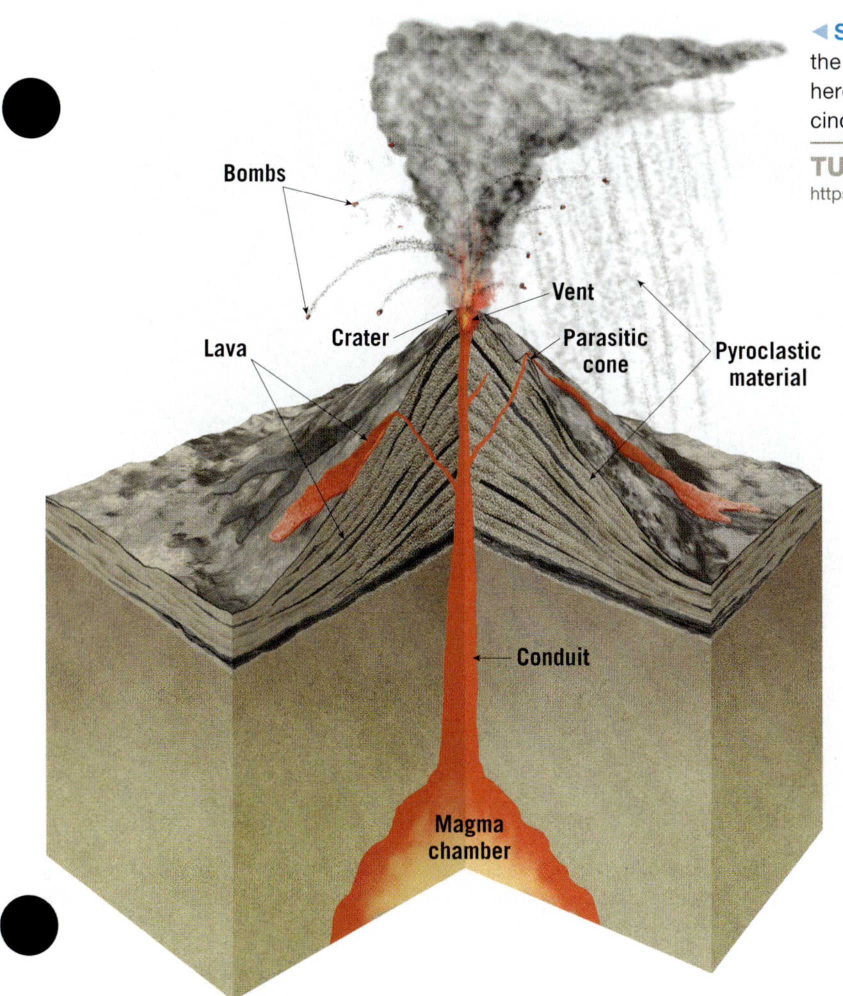

◄ **SmartFigure 5.11 Anatomy of a volcano** Compare the structure of the "typical" composite cone shown here to that of a shield volcano (Figure 5.12) and a cinder cone (Figure 5.15).

TUTORIAL
https://goo.gl/nbwG5k

types of calderas will be considered later in this chapter.

During early stages of growth, most volcanic discharges come from a central summit vent. As a volcano matures, material also tends to be emitted from fissures that develop along the flanks (sides) or at the base of the volcano. Continued activity from a flank eruption may produce one or more small **parasitic cones**. Italy's Mount Etna, for example, has more than 200 secondary vents, some of which have built parasitic cones. Many of these vents, however, emit only gases and are appropriately called **fumaroles** (*fumus* = smoke).

CONCEPT CHECKS 5.4

1. Distinguish among a conduit, a vent, and a crater.
2. How is a crater different from a caldera?
3. What is a parasitic cone, and where does it form?
4. What is emitted from a fumarole?

5.5 Shield Volcanoes

Summarize the characteristics of shield volcanoes and provide one example of this type of volcano.

Shield volcanoes are produced by the accumulation of fluid basaltic lavas and exhibit the shape of a broad, slightly domed structure that resembles a warrior's shield (**Figure 5.12**). Most shield volcanoes begin on the ocean floor as **seamounts** (submarine volcanoes), and a few of them grow large enough to form volcanic islands. In fact, many oceanic islands are either a single shield volcano or, more often, the coalescence of two or more shields built upon massive amounts of pillow lavas. Examples include the Hawaiian Islands, the Canary Islands, Iceland, the Galápagos Islands, and Easter Island. Although less common, some shield volcanoes form on continental crust. Included in this group are Nyamuragira, Africa's most active volcano, and Newberry Volcano, Oregon.

Mauna Loa: Earth's Largest Shield Volcano

Extensive study of the Hawaiian Islands has revealed that they are constructed of a myriad of thin basaltic lava flows, each averaging a few meters thick, intermixed with relatively minor amounts of ejected pyroclastic material. Mauna Loa is the largest of five overlapping shield volcanoes that comprise the Big Island of Hawaii (see Figure 5.12). From its base on the floor of the Pacific Ocean to its summit, Mauna Loa is over 9 kilometers (6 miles) high, exceeding the height of Mount Everest above sea level. The volume of material composing Mauna Loa is roughly 200 times greater than that of the large composite cone Mount Rainier, located in Washington (**Figure 5.13**).

Like Hawaii's other shield volcanoes, Mauna Loa has flanks with gentle slopes of only a few degrees. This low angle is due to the very hot, fluid lava that traveled "fast and far" from the vent. In addition, most of the lava (perhaps 80 percent) flowed through a well-developed system of lava tubes. Another feature common to active shield volcanoes is one or more large, steep-walled calderas that occupy the summit (see Figure 5.12). Calderas on shield volcanoes usually form when the roof above the

Did You Know?
According to legend, Pele, the Hawaiian goddess of volcanoes, makes her home at the summit of Kilauea Volcano. Evidence for her existence is "Pele's hair"—thin, delicate strands of glass, which are soft and flexible and have a golden-brown color. This threadlike volcanic glass forms when blobs of hot lava are spattered and shredded by escaping gases.

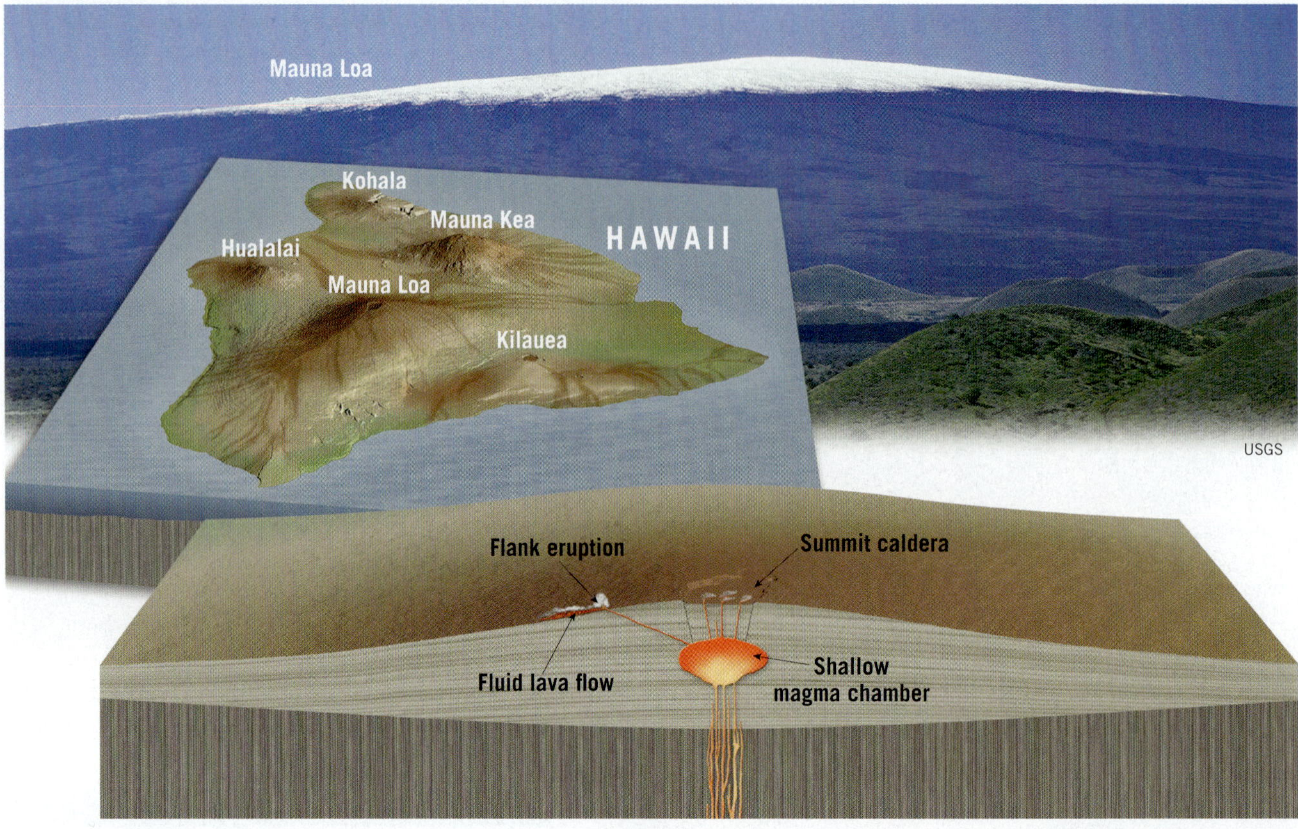

▲ **Figure 5.12 Volcanoes of Hawaii** Mauna Loa, Earth's largest volcano, is one of five shield volcanoes that collectively make up the Big Island of Hawaii. Shield volcanoes are built primarily of fluid basaltic lava flows and contain only a small percentage of pyroclastic materials.

magma chamber collapses. This occurs after the magma reservoir empties, either following a large eruption or as magma migrates to the flank of a volcano to feed a fissure eruption.

In their final stage of growth, shield volcanoes erupt more sporadically, and pyroclastic ejections are more common. The lava emitted later tends to be more viscous, resulting in thicker, shorter flows. These eruptions steepen the slope of the summit area, which often becomes capped with clusters of cinder cones. This explains why Mauna Kea, a more mature volcano that has not erupted in historic times, has a steeper summit than Mauna Loa, which erupted as recently as 1984. Astronomers are so certain that Mauna Kea is "over the hill" that they built an elaborate astronomical observatory on its summit to house some of the world's most advanced and expensive telescopes.

Kilauea: Hawaii's Most Active Volcano

Volcanic activity on the Big Island of Hawaii began on what is now the northwestern flank of the island and has gradually migrated southeastward. It is currently centered on Kilauea Volcano, one of the most active and intensely studied shield volcanoes in the world. Kilauea, located in the shadow of Mauna Loa, has experienced more than 50 eruptions since record keeping began in 1823.

Several months before each eruptive phase, Kilauea inflates as magma gradually migrates upward and accumulates in a central reservoir located a few kilometers

▼ **SmartFigure 5.13 Comparing scales of different volcanoes**
A. Profile of Mauna Loa, Hawaii, the largest shield volcano in the Hawaiian chain. Note the size comparison with Mount Rainier, Washington, a large composite cone. **B.** Profile of Mount Rainier, Washington. Note how it dwarfs a typical cinder cone. **C.** Profile of Sunset Crater, Arizona, a typical steep-sided cinder cone.

ANIMATION
https://goo.gl/awPZir

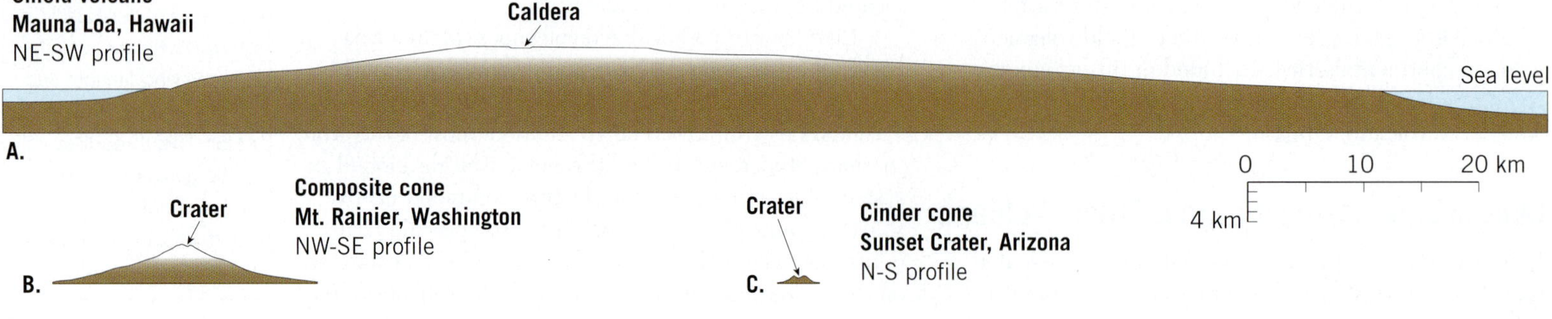

below the summit. For up to 24 hours before an eruption, swarms of small earthquakes warn of the impending activity. Most of the recent activity on Kilauea has occurred along the flanks of the volcano, in a region called the *East Rift Zone* (**Figure 5.14**). The longest and largest rift eruption ever recorded on Kilauea began in 1983 and continues to this day, with no signs of abating.

CONCEPT CHECKS 5.5

1. Describe the composition and viscosity of the lava associated with shield volcanoes.
2. Are pyroclastic materials a significant component of shield volcanoes?
3. Where do most shield volcanoes form—on the ocean floor or on the continents?
4. Where are the best-known shield volcanoes in the United States? Name some examples in other parts of the world.

▲ **SmartFigure 5.14 Lava "curtain" extruded along the East Rift Zone, Kilauea, Hawaii** (Photo by Greg Vaughn/Alamy)

MOBILE FIELD TRIP
https://goo.gl/TYC2Er

5.6 Cinder Cones

Describe the formation, size, and composition of cinder cones.

As the name suggests, **cinder cones** (also called **scoria cones**) are built from ejected lava fragments that begin to harden in flight to produce the vesicular rock *scoria* (**Figure 5.15**). These pyroclastic fragments range in size from fine ash to bombs that may exceed 1 meter (3 feet) in diameter. However, most of the volume of a cinder cone consists of pea- to walnut-sized fragments that are markedly vesicular and have a black to reddish-brown color (see Figure 5.10A). In addition, this pyroclastic material tends to have basaltic composition.

Although cinder cones are composed mostly of loose scoria fragments, some produce extensive lava fields. These lava flows generally form in the final stages of the volcano's life span, when the magma body has lost most of its gas content. Because cinder cones are composed of loose fragments rather than solid rock, the lava usually flows out from the unconsolidated base of the cone rather than from the crater.

Cinder cones have very simple, distinct shapes (see Figure 5.15). Because cinders have a high angle of repose (the steepest angle at which a pile of loose material

◄ **SmartFigure 5.15 Cinder cones** Cinder cones are built from ejected lava fragments (mostly cinders and bombs) and are relatively small—usually less than 300 meters (1000 feet) in height. (Photo by Michael Collier)

MOBILE FIELD TRIP
https://goo.gl/X9JvXE

remains stable), cinder cones are steep-sided, having slopes between 30 and 40 degrees. In addition, cinder cone have large, deep craters relative to the overall size of the structure. Although relatively symmetrical, some cinder cones are elongated and higher on the side that was downwind during the final eruptive phase.

Most cinder cones are produced by a single, short-lived eruptive event. One study found that half of all cinder cones examined were constructed in less than 1 month, and 95 percent of them formed in less than 1 year. Once the event ceases, the magma in the "plumbing" connecting the vent to the magma source solidifies, and the volcano usually does not erupt again. (One exception is Cerro Negro, a cinder cone in Nicaragua, which has erupted more than 20 times since it formed in 1850.) As a result of this typically short life span, cinder cones are small, usually between 30 and 300 meters (100 and 1000 feet) tall. A few rare examples exceed 700 meters (2300 feet) in height.

Cinder cones number in the thousands around the globe. Some occur in groups, such as the volcanic field near Flagstaff, Arizona, which consists of about 600 cones. Others are parasitic cones that are found on the flanks or within the calderas of larger volcanic structures.

Parícutin: Life of a Garden-Variety Cinder Cone

One of the very few volcanoes studied by geologists from its very beginning is the cinder cone called Parícutin, located about 320 kilometers (200 miles) west of Mexico City. In 1943, its eruptive phase began in a cornfield owned by Dionisio Pulido, who witnessed the event.

For 2 weeks prior to the first eruption, numerous tremors caused apprehension in the nearby village of Parícutin. Then, on February 20, sulfurous gases began billowing from a small depression that had been in the cornfield for as long as local residents could remember. During the night, hot, glowing rock fragments were ejected from the vent, producing a spectacular fireworks display. Explosive discharges continued, throwing hot fragments and ash occasionally as high as 6000 meters (20,000 feet) into the air. Larger fragments fell near the crater, some remaining incandescent as they rolled down the slope. These materials built an aesthetically pleasing cone, while finer ash fell over a much larger area, burning and eventually covering the village of Parícutin. In the first day, the cone grew to 40 meters (130 feet), and by the fifth day it was more than 100 meters (330 feet) high.

The first lava flow came from a fissure that opened just north of the cone, but after a few months flows began to emerge from the base of the cone. In June 1944, a clinkery aa flow 10 meters (30 feet) thick moved over much of the village of San Juan Parangaricutiro, leaving only remnants of the church exposed (**Figure 5.16**). After 9 years of intermittent pyroclastic explosions and nearly continuous discharge of lava from vents at its base, the activity ceased almost as quickly as it had begun. Today, Parícutin is just another one of the scores of cinder cones dotting the landscape in this region of Mexico. Like the others, it will not erupt again.

CONCEPT CHECKS 5.6

1. Describe the composition of a cinder cone.
2. How do cinder cones compare in size and steepness of their flanks with shield volcanoes?
3. Over what time span does a typical cinder cone form?

Parícutin, a cinder cone located in Mexico, erupted for 9 years.

An aa flow emanating from the base of the cone buried much of the village of San Juan Parangaricutiro, leaving only remnants of the village's church.

▶ **SmartFigure 5.16**
Parícutin, a well-known cinder cone The village of San Juan Parangaricutiro was engulfed by aa lava from Parícutin. Only portions of the church remain. (Photos by Michael Collier)

CONDOR VIDEO
https://goo.gl/Stb3bZ

5.7 Composite Volcanoes

List the characteristics of composite volcanoes and describe how they form.

Earth's most picturesque yet potentially dangerous volcanoes are **composite volcanoes**, also known as **stratovolcanoes**. Most are located in a relatively narrow zone that rims the Pacific Ocean, appropriately called the *Ring of Fire* (see Figure 5.29, page 151). This active zone includes a chain of continental volcanoes distributed along the west coast of the Americas, including the large cones of the Andes in South America and the Cascade Range of the western United States and Canada.

Classic composite cones are large, nearly symmetrical structures consisting of alternating layers of explosively erupted cinders and ash interbedded with lava flows. A few composite cones, notably Italy's Etna and Stromboli, display very persistent eruption activity, and molten lava has been observed in their summit craters for decades. Stromboli is so well known for eruptions that eject incandescent blobs of lava that it has been called the "Lighthouse of the Mediterranean." Mount Etna has erupted, on average, once every 2 years since 1979.

Just as shield volcanoes owe their shape to fluid basaltic lavas, composite cones reflect the viscous nature of the material from which they are made. In general, composite cones are the product of silica-rich magma having an andesitic composition. However, many composite cones also emit various amounts of fluid basaltic lava and, occasionally, pyroclastic material having a felsic (rhyolitic) composition. The silica-rich magmas typical of composite cones generate thick, viscous lavas that travel less than a few kilometers. Composite cones are also noted for generating explosive eruptions that eject huge quantities of pyroclastic material.

A conical shape, with a steep summit area and gradually sloping flanks, is typical of most large composite cones. This classic profile, which adorns calendars and postcards, is partially a result of the way viscous lavas and pyroclastic ejected materials contribute to the cone's growth. Coarse fragments ejected from the summit crater tend to accumulate near their source and contribute to the steep slopes around the summit. Finer ejected materials, on the other hand, are deposited as a thin layer over a large area and hence tend to flatten the flank of the cone. In addition, during the early stages of growth, lavas tend to be more abundant and flow greater distances from the vent than they do later in the volcano's history, which contributes to the cone's broad base. As a composite volcano matures, the shorter flows that come from the central vent serve to armor and strengthen the summit area. Consequently, steep slopes exceeding 40 degrees are possible. Two of the most perfect cones—Mount Mayon in the Philippines and Fujiyama in Japan—exhibit the classic form we expect of composite cones, with steep summits and gently sloping flanks (Figure 5.17).

Figure 5.17 Fujiyama, a classic composite volcano Japan's Fujiyama exhibits the classic form of a composite cone—a steep summit and gently sloping flanks. (Photo by Koji Nakano/Getty Images, Inc-Liaison)

Despite the symmetrical forms of many composite cones, most have complex histories. Many composite volcanoes have secondary vents on their flanks that have produced cinder cones or even much larger volcanic structures. Huge mounds of volcanic debris surrounding these structures provide evidence that large sections of these volcanoes slid downslope as massive landslides. Some develop amphitheater-shaped depressions at their summits as a result of explosive lateral eruptions—as occurred during the 1980 eruption of Mount St. Helens. Often, so much rebuilding has occurred since these eruptions that no trace of these amphitheater-shaped scars remain. Others, such as Crater Lake, have been truncated by the collapse of their summit (see Figure 5.23).

CONCEPT CHECKS 5.7

1. What name is given to the region having the greatest concentration of composite volcanoes?
2. Describe the materials that compose composite volcanoes.
3. How does the composition and viscosity of lava flows differ between composite volcanoes and shield volcanoes?

5.8 Volcanic Hazards

Describe the major geologic hazards associated with volcanoes.

Roughly 1500 of Earth's known volcanoes have erupted at least once, and some several times, in the past 10,000 years. Based on historical records and studies of active volcanoes, 70 volcanic eruptions can be expected each year. In addition, 1 large-volume eruption can be expected every decade; these large eruptions account for the vast majority of volcano-related human fatalities.

Today, an estimated 500 million people in places such as Japan, Indonesia, Italy, and Oregon live near active volcanoes. They face a number of volcanic hazards, such as destructive pyroclastic flows, molten lava flows, mudflows called lahars, and falling ash and volcanic bombs.

Pyroclastic Flow: A Deadly Force of Nature

Some of the most destructive forces of nature are **pyroclastic flows**, which consist of hot gases infused with incandescent ash and larger lava fragments. Also referred to as **nuée ardentes** ("glowing avalanches"), these fiery flows can race down steep volcanic slopes at speeds exceeding 100 kilometers (60 miles) per hour (**Figure 5.18**). Pyroclastic flows have two components—a low-density cloud of hot expanding gases containing fine ash particles and a ground-hugging portion composed of pumice and other vesicular pyroclastic material.

▼ **Figure 5.18 Pyroclastic flows, one of the most destructive volcanic forces A.** These pyroclastic flows occurred on Mount Mayon, Philippines, during the 1984 eruption. Pyroclastic flows are composed of hot ash and pumice and/or blocky lava fragments that race down the slope of volcanoes. (Photo courtesy of USGS) **B.** Residents running away from a pyroclastic flow that reached the base of Mound Sinabung, Indonesia, 2014. (Photo by Chaideer Mahyuddin/AFP/Getty Images)

Driven by Gravity Pyroclastic flows are propelled by the force of gravity and tend to move in a manner similar to snow avalanches. They are mobilized by expanding volcanic gases released from the lava fragments and by the expansion of heated air that is overtaken and trapped in the moving front. These gases reduce friction between ash and pumice fragments, which gravity propels downslope in a nearly frictionless environment. This is why some pyroclastic flow deposits are found many miles from their source.

Occasionally, powerful hot blasts that carry small amounts of ash separate from the main body of a pyroclastic flow. These low-density clouds, called *surges*, can be deadly but seldom have sufficient force to destroy buildings in their paths. Nevertheless, in 2014, a hot ash cloud from Japan's Mount Ontake killed 47 hikers and injured 69 more.

Pyroclastic flows may originate in a variety of volcanic settings. Some occur when a powerful eruption blasts pyroclastic material out of the side of a volcano. More frequently, however, pyroclastic flows are generated by the collapse of tall eruption columns during an explosive event. When gravity eventually overcomes the initial upward thrust provided by the escaping gases, the ejected materials begin to fall, sending massive amounts of incandescent blocks, ash, and pumice cascading downslope.

The Destruction of St. Pierre In 1902, an infamous pyroclastic flow and associated surge from Mount Pelée, a small volcano on the Caribbean island of Martinique, destroyed the port town of St. Pierre. Although the main pyroclastic flow was largely confined to the valley of Riviere Blanche, a low-density fiery surge spread south of the river and quickly engulfed the entire city. The destruction happened in moments and was so devastating that nearly all of St. Pierre's 28,000 inhabitants were killed. Only 1 person on the outskirts of town—a prisoner protected in a dungeon—and a few people on ships in the harbor were spared (**Figure 5.19**).

Scientists who arrived on the scene within days found that although St. Pierre was mantled by only a thin layer of volcanic debris, masonry walls nearly 1 meter (3 feet) thick had been knocked over like dominoes, large trees had been uprooted, and cannons had been torn from their mounts.

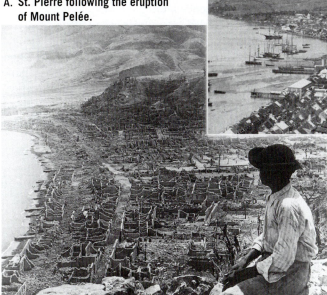

A. St. Pierre following the eruption of Mount Pelée.

B. St. Pierre before the 1902 eruption.

◀ **Figure 5.19 Destruction of St. Pierre A.** St. Pierre as it appeared shortly after the eruption of Mount Pelée in 1902. (Reproduced from the collection of the Library of Congress) **B.** St. Pierre before the eruption. Many vessels were anchored offshore when this photo was taken, as was the case on the day of the eruption. (Photo by UPPA/Photoshot)

The Destruction of Pompeii One well-documented event of historic proportions was the C.E. 79 eruption of the Italian volcano we now call Mount Vesuvius. For centuries prior to this eruption, Vesuvius had been dormant, with vineyards adorning its sunny slopes. Yet in less than 24 hours, the entire city of Pompeii (near Naples) and a few thousand of its residents were entombed beneath a layer of volcanic ash and pumice. The city and the victims of the eruption remained buried for nearly 17 centuries. The excavation of Pompeii gave archaeologists a superb picture of ancient Roman life (**Figure 5.20A**).

A.

B.

◀ **Figure 5.20 Pompeii was excavated nearly 17 centuries after the C.E. 79 eruption of Mount Vesuvius A.** The ruins of the Roman city of Pompeii as they appear today. In less than 24 hours, Pompeii and all of its residents were buried under a layer of volcanic ash and pumice that fell like rain. **B.** Plaster casts of some of the victims of the eruption of Mount Vesuvius.

Did You Know?
In 1902, when Mount Pelée erupted, the U.S. Senate was preparing to vote on the location of a canal connecting the Pacific and Atlantic Oceans. The choice was between Panama and Nicaragua, a country that featured a smoking volcano on its postage stamp. Supporters of the Panamanian route used the potential threat of volcanic eruptions in Nicaragua to bolster their argument in favor of Panama. The Senate subsequently approved the construction of the Panama Canal. In 2013, Nicaragua's National Assembly approved a bill to support the construction of a second canal connecting the Pacific Ocean and the Caribbean Sea (and therefore the Atlantic) to rival the Panama Canal. However, funding for this project has not been obtained.

By reconciling historic records with detailed scientific studies of the region, volcanologists reconstructed the sequence of events. During the first day of the eruption, a rain of ash and pumice accumulated at a rate of 12 to 15 centimeters (5 to 6 inches) per hour, causing most of the roofs in Pompeii to eventually give way. Then, suddenly, a surge of searing hot ash and gas swept rapidly down the flanks of Vesuvius. This deadly pyroclastic flow killed those who had somehow managed to survive the initial ash and pumice fall. Their remains were quickly buried by falling ash, and subsequent rainfall caused the ash to harden. Over the centuries, the remains decomposed, creating cavities that were discovered by nineteenth-century excavators. Casts were then produced by pouring plaster of Paris into the voids (Figure 5.20B). Mount Vesuvius has had more than two dozen explosive eruptions since C.E. 79, the most recent occurring in 1944. Today, Vesuvius towers over the Naples skyline, a region occupied by roughly 3 million people. Such an image should prompt us to consider how volcanic crises might be managed in the future.

Lahars: Mudflows on Active & Inactive Cones

In addition to violent eruptions, large composite cones may generate a type of fluid mudflow, known by its Indonesian name, **lahar**. These destructive flows occur when volcanic debris becomes saturated with water and rapidly moves down steep volcanic slopes, generally following stream valleys. Some lahars are triggered when magma nears the surface of a glacially clad volcano, causing large volumes of ice and snow to melt. Others are generated when heavy rains saturate weathered volcanic deposits. Thus, lahars may occur even when a volcano is *not* erupting.

When Mount St. Helens erupted in 1980, several lahars were generated. These flows and accompanying floodwaters raced down nearby river valleys at speeds exceeding 30 kilometers (20 miles) per hour. These raging rivers of mud destroyed or severely damaged nearly all the homes and bridges along their paths (Figure 5.21). Fortunately, the area was not densely populated.

In 1985, deadly lahars were produced during a small eruption of Nevado del Ruiz, a 5300-meter (17,400-foot) volcano in the Andes Mountains of Colombia. Hot pyroclastic material melted ice and snow that capped the mountain (*nevado* means "snowcap" in Spanish) and sent torrents of ash and debris down three major river valleys that flank the volcano. Reaching speeds of 100 kilometers (60 miles) per hour, these mudflows tragically claimed 25,000 lives.

Many consider Mount Rainier, Washington, to be America's most dangerous volcano because, like Nevado del Ruiz, it has a thick, year-round mantle of snow and glacial ice. Adding to the risk is the fact that more than 100,000 people live in the valleys around Rainier, and many homes are built on deposits left by lahars that flowed down the volcano hundreds or thousands of years ago. A future eruption, or perhaps just a period of heavier-than-average rainfall, may produce lahars that could be similarly destructive.

Other Volcanic Hazards

Volcanoes can be hazardous to human health and property in other ways. Ash and other pyroclastic material can collapse the roofs of buildings or may be drawn into the lungs of humans and other animals or into aircraft

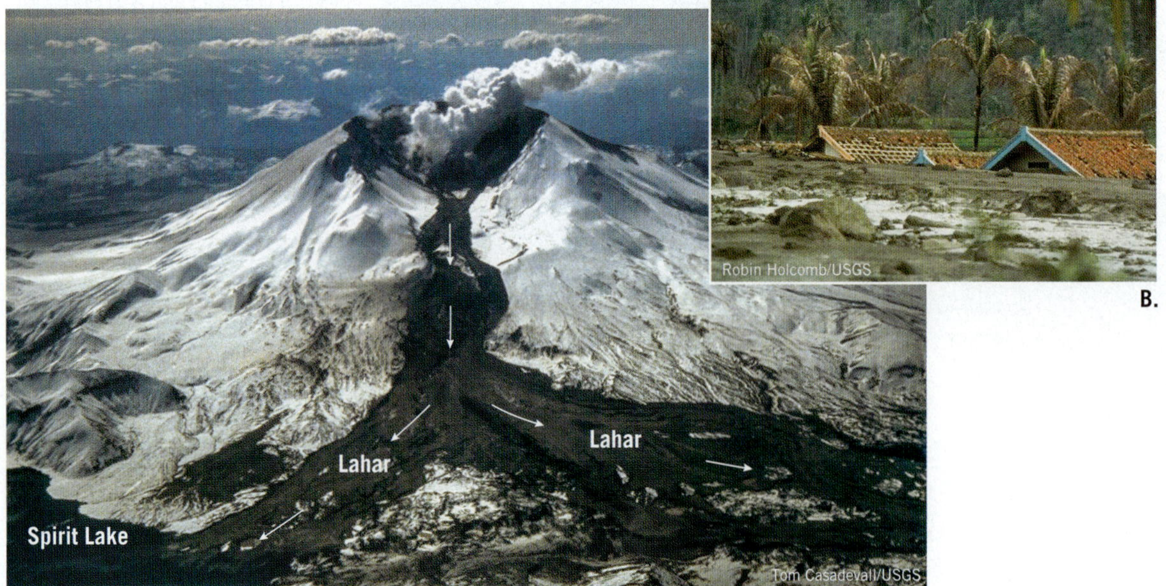

Figure 5.21 Lahars, mudflows that originate on volcanic slopes **A.** This lahar raced down the snow-covered slopes of Mount St. Helens following an explosive eruption on March 19, 1982. **B.** The aftermath of a lahar that formed following the 1982 eruption of Galunggung Volcano, Indonesia.

Ash and other pyroclastic materials can collapse roofs, or completely cover buildings.

Lava flows can destroy homes, roads, and other structures in their paths.

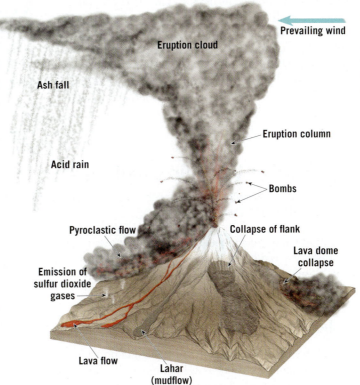

◀ **Figure 5.22 Volcanic hazards** In addition to generating destructive pyroclastic flows and lahars, volcanoes can be hazardous to human health and property in many other ways.

engines (**Figure 5.22**). Volcanic gases, most notably sulfur dioxide, pollute the air and, when mixed with rainwater, can destroy vegetation and reduce the quality of groundwater. Despite the known risks, millions of people live in close proximity to active volcanoes.

Volcano-Related Tsunamis Although **tsunamis** are most often associated with displacement along a fault located on the seafloor (see Chapter 9), some result from the collapse of a volcanic cone. This was dramatically demonstrated during the 1883 eruption on the Indonesian island of Krakatau, when the northern half of a volcano plunged into the Sunda Strait, creating a tsunami that exceeded 30 meters (100 ft) in height. Although Krakatau was uninhabited, an estimated 36,000 people were killed along the coastline of the islands of Java and Sumatra.

Volcanic Ash & Aviation During the past 20 years, at least 80 commercial jets have been damaged by inadvertently flying into clouds of volcanic ash. For example, in 1989, a Boeing 747 carrying more than 300 passengers encountered an ash cloud from Alaska's Redoubt Volcano; all four engines clogged with ash and stalled mid-air. Fortunately, the pilots were able to restart the engines and safely landed the aircraft in Anchorage.

More recently, the 2010 eruption of Iceland's Eyjafjallajökull Volcano sent ash high into the atmosphere. This thick plume of ash drifted over Europe, causing airlines to cancel thousands of flights and leaving hundreds of thousands of travelers stranded. Several weeks passed before air travel resumed its normal schedule.

Volcanic Gases & Respiratory Health One of the most destructive volcanic events, called the Laki eruptions, began along a large fissure in southern Iceland in 1783. An estimated 14 cubic kilometers of fluid basaltic lavas were released, along with 130 million tons of sulfur dioxide and other poisonous gases. When sulfur dioxide is inhaled, it reacts with moisture in the lungs to produce sulfuric acid, a deadly toxin. More than half of Iceland's livestock died, and the ensuing famine killed 25 percent of the island's human population.

This huge eruption also endangered people and property all across Europe. Crop failure occurred in parts of Western Europe, and thousands of residents perished from lung-related diseases. One report estimated that a similar eruption today would cause more than 140,000 cardiopulmonary fatalities in Europe alone.

Effects of Volcanic Ash & Gases on Weather & Climate Volcanic eruptions can eject dust-sized particles of volcanic ash and sulfur dioxide gas high into the atmosphere. The ash particles reflect sunlight back to space, producing temporary atmospheric cooling. The 1783 Laki eruptions in Iceland appear to have affected atmospheric circulation around the globe. Drought conditions prevailed in the Nile River valley and India, and the winter of 1784 saw the longest period of below-zero temperatures in New England's history.

Other eruptions that have produced significant effects on climate worldwide include the eruption of Indonesia's Mount Tambora in 1815, which produced the "year without a summer" (1816), and the eruption of

El Chichón in Mexico in 1982. El Chichón's eruption, although small, emitted an unusually large quantity of sulfur dioxide that reacted with water vapor in the atmosphere to produce a dense cloud of tiny sulfuric acid droplets. Such particles, called aerosols, take several years to settle out of the atmosphere. Like fine ash, these aerosols lower the mean temperature of the atmosphere by reflecting solar radiation back to space.

CONCEPT CHECKS 5.8

1. Describe pyroclastic flows and explain why they are capable of traveling great distances.
2. What is a lahar?
3. List at least three volcanic hazards besides pyroclastic flows and lahars.

5.9 Other Volcanic Landforms

List volcanic landforms other than shield, cinder cone, and composite volcanoes and describe their formation.

The most widely recognized volcanic structures are the cone-shaped edifices of composite volcanoes that dot Earth's surface. However, volcanic activity produces other distinctive and important landforms.

Calderas

Recall that *calderas* are large steep-sided depressions that have diameters exceeding 1 kilometer (0.6 miles) and have a somewhat circular form. Those that are less than 1 kilometer across are called *collapse pits*, or *craters*. Most calderas are formed by one of the following processes: (1) the collapse of the summit of a large composite volcano following an explosive eruption of silica-rich pumice and ash fragments (*Crater Lake–type calderas*); (2) the collapse of the top of a shield volcano caused by subterranean drainage from a central magma chamber (*Hawaiian-type calderas*); and (3) the collapse of a large area, caused by the discharge of colossal volumes of silica-rich pumice and ash along ring fractures (*Yellowstone-type calderas*).

Crater Lake–Type Calderas Crater Lake, Oregon, is situated in a caldera approximately 10 kilometers (6 miles) wide and 600 meters (1970 feet) deep. This caldera formed about 7000 years ago, when a composite cone named Mount Mazama violently extruded 50 to 70 cubic kilometers of pyroclastic material (Figure 5.23). With the loss of support, 1500 meters (nearly 1 mile) of the

SmartFigure 5.23
Formation of Crater Lake–type calderas
About 7000 years ago, a violent eruption partly emptied the magma chamber of former Mount Mazama, causing its summit to collapse. Precipitation and groundwater contributed to forming Crater Lake, the deepest lake in the United States—600 meters (1970 feet) deep—and the ninth-deepest in the world.

ANIMATION
https://goo.gl/kUCPNB

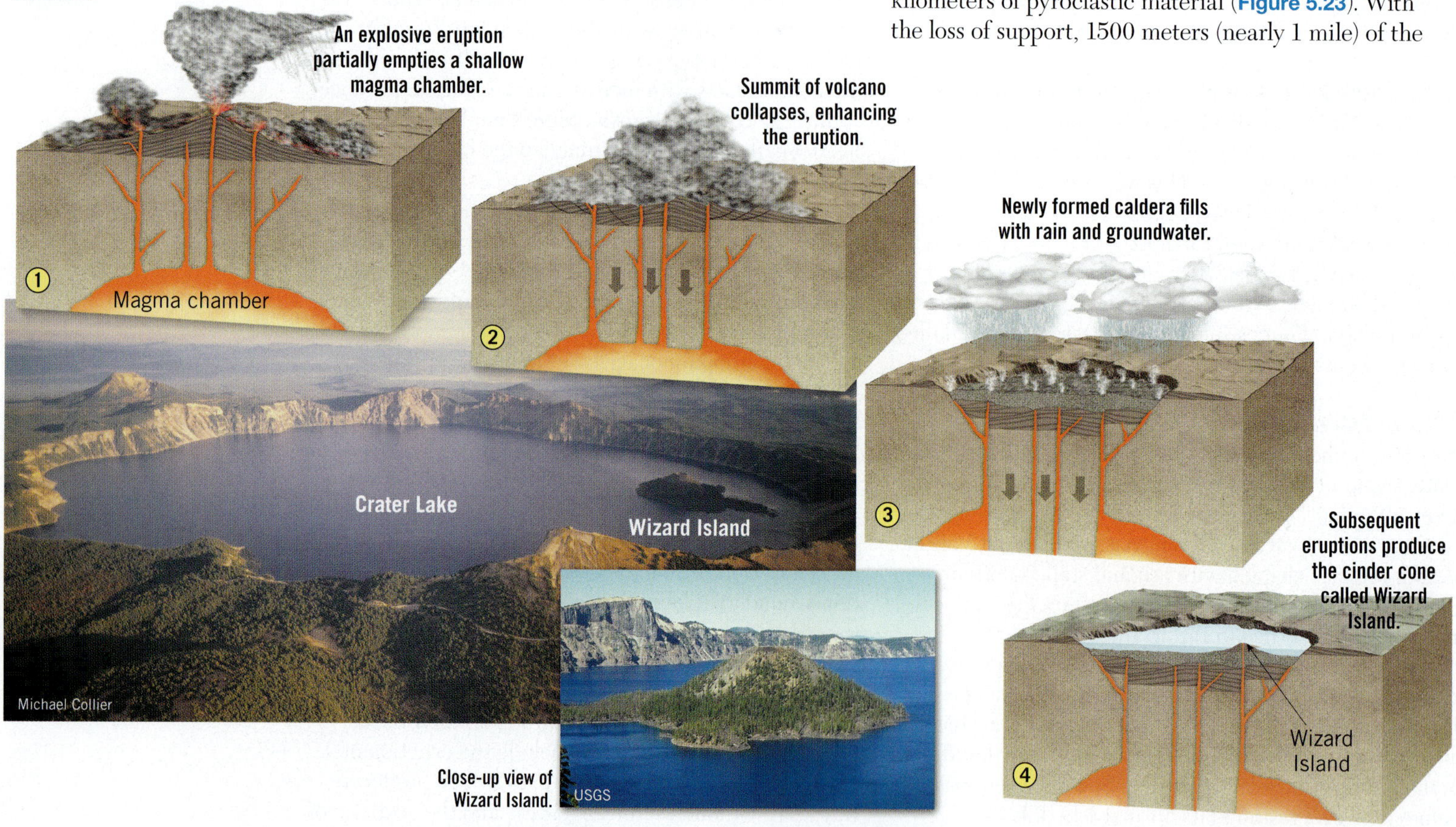

An explosive eruption partially empties a shallow magma chamber.
Magma chamber
Summit of volcano collapses, enhancing the eruption.
Newly formed caldera fills with rain and groundwater.
Subsequent eruptions produce the cinder cone called Wizard Island.
Wizard Island
Crater Lake
Wizard Island
Michael Collier
Close-up view of Wizard Island.
USGS

summit of this once-prominent cone collapsed, producing a caldera that eventually filled with water. Later, volcanic activity built a small cinder cone in the caldera. Today this cone, called Wizard Island, provides a mute reminder of past activity.

Hawaiian-Type Calderas Unlike Crater Lake–type calderas, many calderas form gradually because of the loss of lava from a shallow magma chamber underlying a volcano's summit. For example, Hawaii's active shield volcanoes, Mauna Loa and Kilauea, both have large calderas at their summits. Kilauea's measures 3.3 by 4.4 kilometers (about 2 by 3 miles) and is 150 meters (500 feet) deep. The walls are almost vertical, and as a result, the caldera looks like a vast, nearly flat-bottomed pit. Kilauea's caldera formed by gradual subsidence as magma slowly drained laterally from the underlying magma chamber, leaving the summit unsupported.

Yellowstone-Type Calderas Historic and destructive eruptions such as Mount St. Helens pale in comparison to what happened 630,000 years ago in the region now occupied by Yellowstone National Park, when approximately 1000 cubic kilometers of pyroclastic material erupted. This catastrophic eruption sent showers of ash as far as the Gulf of Mexico and formed a caldera 70 kilometers (43 miles) across (**Figure 5.24A**). Vestiges of this event are the many hot springs and geysers in the Yellowstone region.

Yellowstone-type eruptions eject huge volumes of pyroclastic materials, mainly in the form of ash and pumice fragments. Typically, these materials are ejected as *pyroclastic flows* that sweep across the landscape, destroying most living things in their paths. Upon coming to rest, the hot fragments of ash and pumice fuse together, forming a welded tuff that closely resembles a solidified lava flow. Despite the immense size of these calderas, the eruptions that produce them are brief, lasting hours to perhaps a few days.

Large calderas tend to exhibit a complex eruptive history. In the Yellowstone region, for example, three caldera-forming episodes are known to have occurred over the past 2.1 million years (**Figure 5.24B**). The most recent eruption (630,000 years ago) was followed by episodic outpourings of degassed rhyolitic and basaltic lavas. In the intervening years, a slow upheaval of the floor of the caldera has produced two elevated regions called *resurgent domes* (see Figure 5.24A). A recent study has determined that a huge magma reservoir still exists beneath Yellowstone; thus, another caldera-forming eruption is likely—but not necessarily imminent.

Unlike calderas associated with shield volcanoes or composite cones, Yellowstone-type calderas are so vast and poorly defined that many were undetected until high-quality aerial and satellite images became available. Other examples of Yellowstone-type calderas are California's Long Valley Caldera; LaGarita Caldera, located in the San Juan Mountains of southern Colorado; and the Valles Caldera, west of Los Alamos, New Mexico. These and similar calderas found around the globe are among the largest volcanic structures on Earth, hence the name *supervolcanoes*. Volcanologists compare their destructive force to that of the impact of a small asteroid. Fortunately, no Yellowstone-type eruption has occurred in historic times.

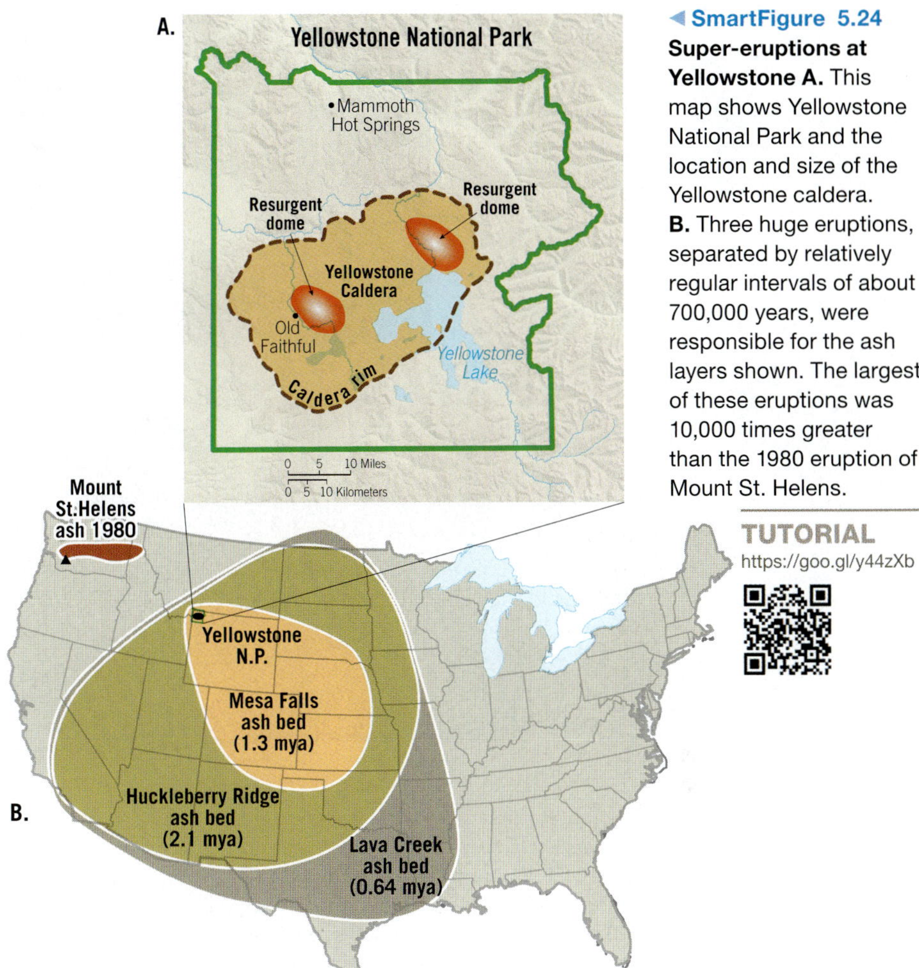

SmartFigure 5.24 Super-eruptions at Yellowstone A. This map shows Yellowstone National Park and the location and size of the Yellowstone caldera. **B.** Three huge eruptions, separated by relatively regular intervals of about 700,000 years, were responsible for the ash layers shown. The largest of these eruptions was 10,000 times greater than the 1980 eruption of Mount St. Helens.

TUTORIAL
https://goo.gl/y44zXb

Fissure Eruptions & Basalt Plateaus

The greatest volume of volcanic material is extruded from fractures in Earth's crust, called *fissures*. Rather than building cones, **fissure eruptions** usually emit fluid basaltic lavas that blanket wide areas (**Figure 5.25**). In some locations, extraordinary amounts of lava have been extruded along fissures in a relatively short time, geologically speaking. These voluminous accumulations are commonly referred to as **basalt plateaus** because most have a basaltic composition and tend to be rather flat and broad. The Columbia Plateau in the northwestern United States, which consists of the Columbia River basalts, is a product of this type of activity (**Figure 5.26**). Numerous fissure eruptions have buried

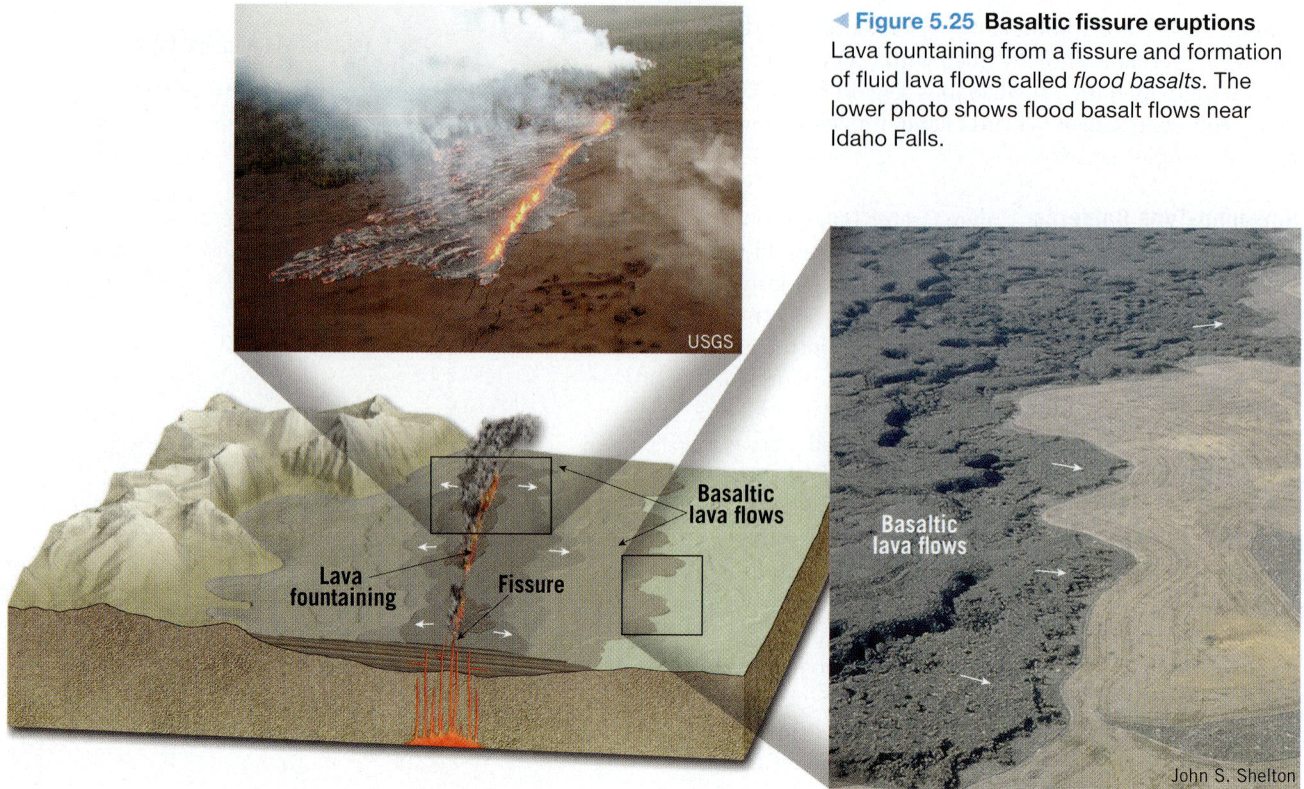

◄ **Figure 5.25 Basaltic fissure eruptions** Lava fountaining from a fissure and formation of fluid lava flows called *flood basalts*. The lower photo shows flood basalt flows near Idaho Falls.

▼ **Figure 5.26 Columbia River basalts A.** The Columbia River basalts cover an area of nearly 164,000 square kilometers (63,000 square miles) that is commonly called the *Columbia Plateau*. Activity here began about 17 million years ago, as lava began to pour out of large fissures, eventually producing a basalt plateau with an average thickness of more than 1 kilometer. **B.** Columbia River basalt flows exposed in the Palouse River Canyon in southwestern Washington State. (Photo by Williamborg)

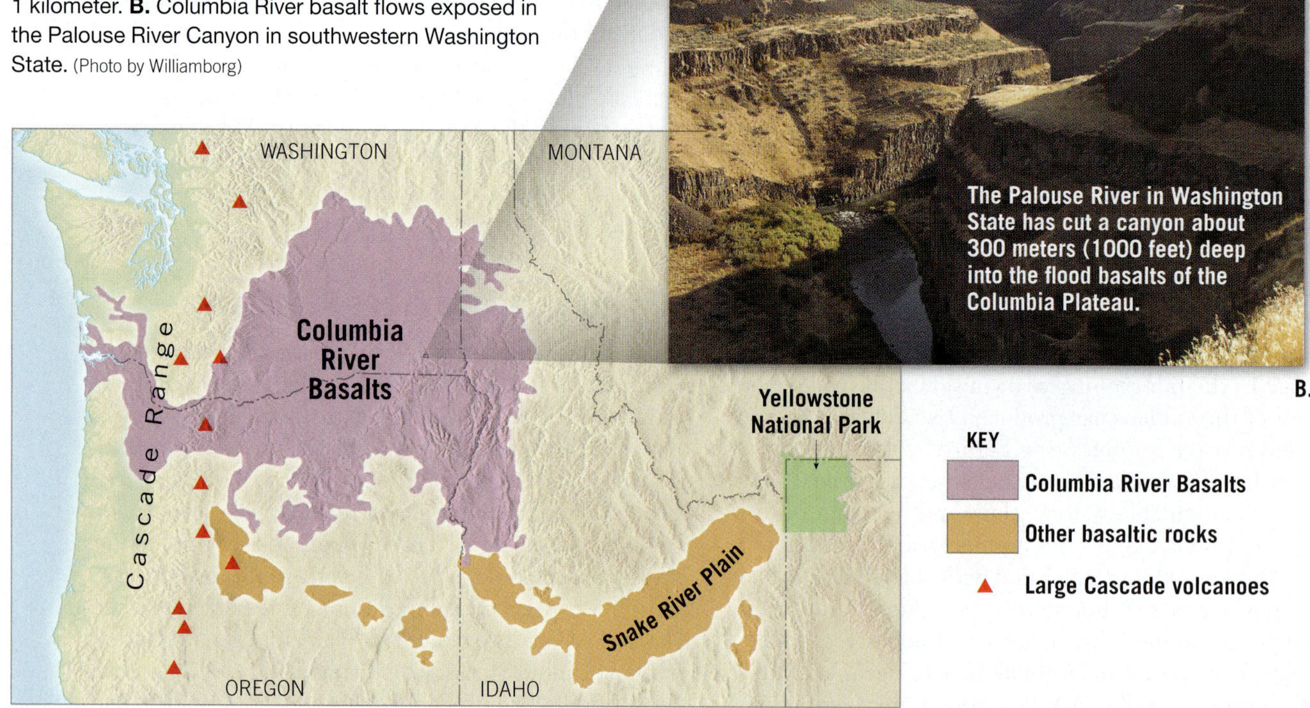

the landscape, creating a lava plateau nearly 1500 meters (1 mile) thick. Some of the lava remained molten long enough to flow 150 kilometers (90 miles) from its source. The term **flood basalts** appropriately describe these extrusions.

Massive accumulations of basaltic lava, similar to those of the Columbia Plateau, occur elsewhere in the world. One of the largest is known as the Deccan Plateau or Deccan Traps (*traps* = stairs), a thick sequence of flat-lying basalt flows covering nearly 500,000 square kilometers (195,000 square miles) of west-central India. When the Deccan Traps formed about 66 million years ago, nearly 2 million cubic kilometers of lava were extruded over a period of approximately 1 million years. Several other massive accumulations of flood basalts, including the Ontong Java Plateau, have been discovered in the deep ocean basins (see Figure 5.32, page 154).

Lava Domes

In contrast to hot basaltic lavas, cool silica-rich rhyolitic lavas are so viscous that they hardly flow at all. As the thick lava is "squeezed" out of a vent, it often produces a dome-shaped mass called a **lava dome**. Lava domes are usually only a few tens of meters high, and they come in a variety of shapes that range from pancake-like flows to steep-sided plugs that were pushed upward like pistons. Most lava domes grow over a period of several years, following an explosive eruption of silica-rich magma. A recent example is the dome that began to grow in the crater of Mount St. Helens immediately following the 1980 eruption (**Figure 5.27A**).

Collapsing lava domes, particularly those that form on the summit or along the steep flanks of composite cones, often produce powerful pyroclastic flows (**Figure 5.27B**). These flows result from highly viscous magma slowly entering the dome, causing it to expand and steepen its flanks. Over time, the cooler outer layer of the dome may start to crumble, producing relatively small pyroclastic flows consisting of dense blocks of lava. Occasionally, the rapid removal of the outer layer causes a significant decrease in pressure on the hot gaseous magma in the interior of the dome. Explosive degassing of the interior magma then triggers a fiery pyroclastic flow that races down the flanks of the volcano (see Figure 5.27B).

Since 1995, pyroclastic flows generated by the collapse of several lava domes on Soufrière Hills Volcano have rendered more than half of the Caribbean island of Montserrat uninhabitable. The capital city, Plymouth, was destroyed, and two-thirds of the population has evacuated. In 1991 a collapsed lava dome at the summit of Japan's Mount Unzen produced a pyroclastic flow that claimed 44 lives. Many of the victims were journalists and film makers who ventured too close to the volcano in order to obtain photographs and document the event.

Volcanic Necks

Most of the lava and materials erupted from a volcano travel through short conduits that connect shallow magma chambers to vents located at the surface. When a volcano becomes inactive, congealed magma is often preserved in the feeding conduit of the volcano as

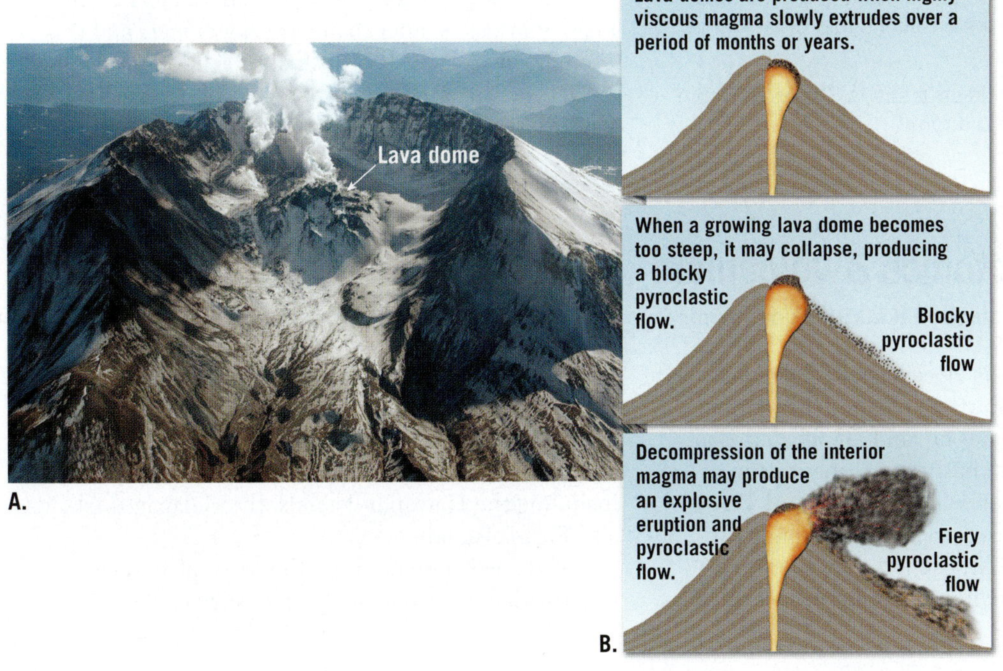
Lava dome
A.
Lava domes are produced when highly viscous magma slowly extrudes over a period of months or years.
When a growing lava dome becomes too steep, it may collapse, producing a blocky pyroclastic flow.
Blocky pyroclastic flow
Decompression of the interior magma may produce an explosive eruption and pyroclastic flow.
Fiery pyroclastic flow
B.

Figure 5.27 Lava domes can generate pyroclastic flows A. This lava dome began to develop in the vent of Mount St. Helens following the May 1980 eruption. **B.** The collapse of a lava dome often results in a powerful pyroclastic flow. (Photo by Lyn Topinka/U.S. Geological Survey)

▶ SmartFigure 5.28
Volcanic neck Shiprock, New Mexico, is a volcanic neck that stands about 520 meters (1700 feet) high. It consists of igneous rock that crystallized in the vent of a volcano that has long since been eroded.

TUTORIAL
https://goo.gl/TjW5uh

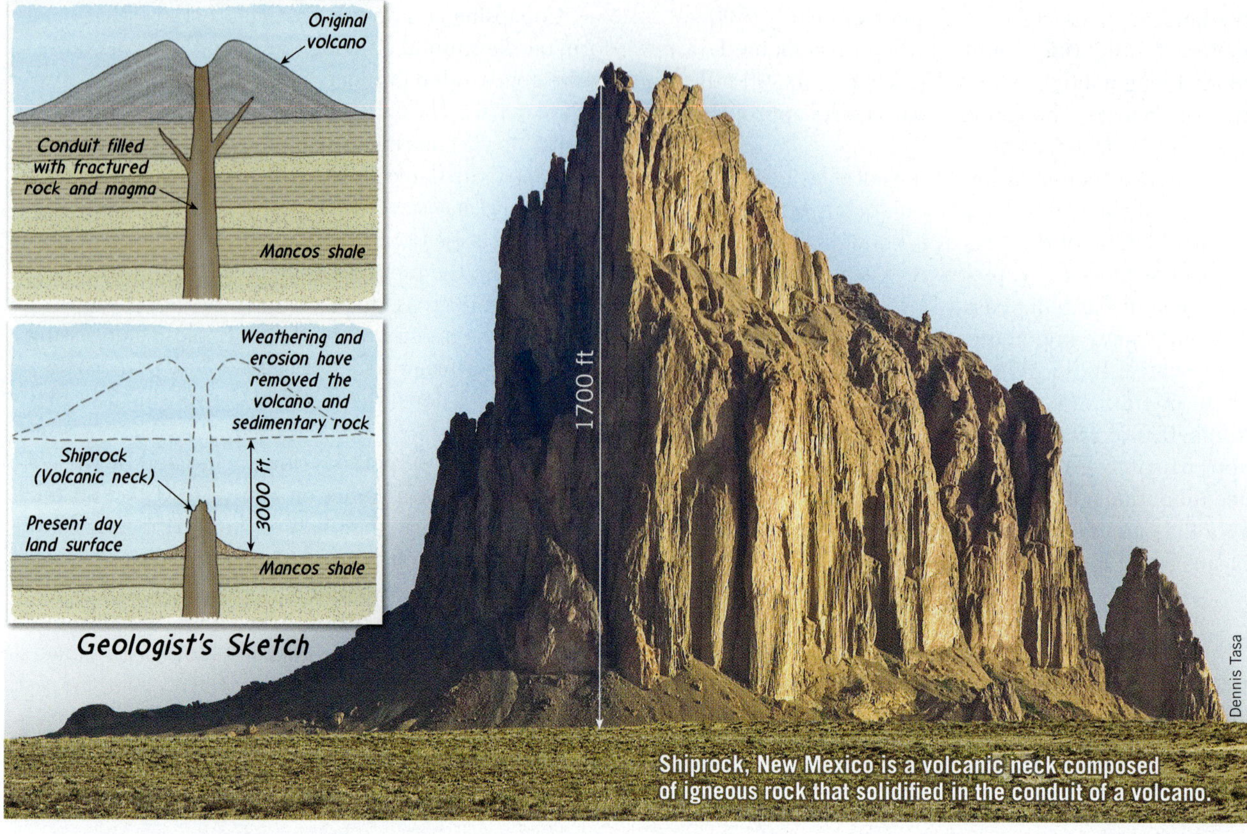

CONCEPT CHECKS 5.9

1. Describe the formation of Crater Lake. Compare it to the calderas found on shield volcanoes such as Kilauea.
2. Other than composite volcanoes, what volcanic landform can generate a pyroclastic flow?
3. How do the eruptions that created the Columbia Plateau differ from the eruptions that create large composite volcanoes?
4. What type of volcanic structure is Shiprock, New Mexico, and how did it form?

a crudely cylindrical mass. As the volcano succumbs to forces of weathering and erosion, the rock occupying the volcanic conduit, which is highly resistant to weathering, may remain standing above the surrounding terrain long after the cone has been worn away. Shiprock, New Mexico, is a widely recognized and spectacular example of these structures, which geologists call **volcanic necks** (or **plugs**) (Figure 5.28). More than 510 meters (1700 feet) high, Shiprock is taller than most skyscrapers and is one of many such landforms that protrude conspicuously from the red desert landscapes of the American Southwest.

5.10 Plate Tectonics & Volcanism

Explain how the global distribution of volcanic activity is related to plate tectonics.

Geologists have known for decades that the global distribution of most of Earth's volcanoes is not random. Most active volcanoes on land are located along the margins of the ocean basins—notably within the circum-Pacific belt known as the **Ring of Fire** (Figure 5.29), where denser oceanic lithosphere subducts under continental lithosphere. Another group of volcanoes includes the innumerable seamounts that form along the crest of the mid-ocean ridges. There are some volcanoes, however, that appear to be randomly distributed around the globe. These volcanic structures comprise most of the islands of the deep-ocean basins, including the Hawaiian Islands, the Galapagos Islands, and Easter Island.

The development of the theory of plate tectonics provided geologists with a plausible explanation

Figure 5.29 Ring of Fire Most of Earth's major volcanoes are located in a zone around the Pacific called *the Ring of Fire*. Another large group of active volcanoes lie unseen along the mid-ocean ridge system.

for the distribution of Earth's volcanoes and established the basic connection between plate tectonics and volcanism: *Plate motions provide the mechanisms by which mantle rocks undergo partial melting to generate magma.*

Volcanism at Divergent Plate Boundaries

The greatest volume of magma erupts along divergent plate boundaries associated with seafloor spreading—out of human sight (Figure 5.30B). Below the ridge axis where lithospheric plates are continually being pulled apart, the solid yet mobile mantle rises to fill the rift. Recall that as hot rock rises, it experiences a decrease in confining pressure and may undergo *decompression melting*. This activity continuously adds new basaltic rock to plate margins, temporarily welding them together, only to have them break again as spreading continues. Along some ridge segments, extrusions of pillow lavas build numerous volcanic structures, the largest of which is Iceland.

Although most spreading centers are located along the axis of an oceanic ridge, some are not. In particular, the East African Rift is a site where continental lithosphere is being pulled apart (see Figure 5.30F). Vast outpourings of fluid basaltic lavas as well as several active volcanoes are found in this region of the globe.

Volcanism at Convergent Plate Boundaries

Recall that along convergent plate boundaries, two plates move toward each other, and a slab of dense oceanic lithosphere descends into the mantle. In these settings, water driven from hydrated (water-rich) minerals found in the subducting oceanic crust and overlying sediments triggers partial melting in the hot mantle above (Figure 5.30A).

Volcanism at a convergent plate margin results in the development of a slightly curved chain of volcanoes called a *volcanic arc*. These volcanic chains develop roughly parallel to the associated trench—at distances of 200 to 300 kilometers (100 to 200 miles). Volcanic arcs that develop within the ocean and grow large enough for their tops to rise above the surface are labeled *archipelagos* in most atlases. Geologists prefer the more descriptive term **volcanic island arcs**, or simply **island arcs** (see Figure 5.30A). Several young volcanic island arcs border the western Pacific basin, including the Aleutians, the Tongas, and the Marianas.

Volcanism associated with convergent plate boundaries may also take place where slabs of oceanic lithosphere are subducted under continental lithosphere to produce a **continental volcanic arc** (Figure 5.30E). The mechanisms that generate these mantle-derived magmas are essentially the same as those that create volcanic island arcs. The most significant difference is that continental crust is much thicker and composed of rocks having higher silica content than oceanic crust. Hence, by melting the surrounding silica-rich crustal rocks, mantle-derived magma changes composition as it rises through the crust. The volcanoes of the Cascade Range in the northwestern United States, including Mount Hood, Mount Rainier, Mount Shasta, and Mount St. Helens, are examples of volcanoes generated

Did You Know?
At 4392 m (14,411 ft) in altitude, Washington's Mount Rainier is the tallest of the 15 great volcanoes that make up the backbone of the Cascade Range. Although Mount Rainier is considered an active volcano, its summit is covered by more than 25 alpine glaciers.

A. Convergent Plate Volcanism When an oceanic plate subducts, melting in the mantle produces magma that gives rise to a volcanic island arc on the overlying oceanic crust.

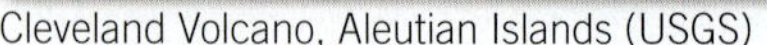
Cleveland Volcano, Aleutian Islands (USGS)

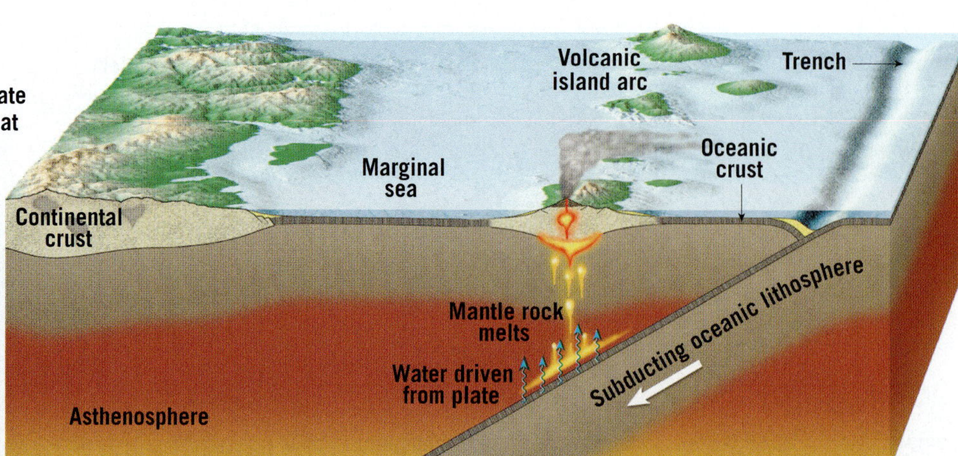

North America
Hawaii
South America

C. Intraplate Volcanism When an oceanic plate moves over a hot spot, a chain of volcanic structures such as the Hawaiian Islands is created.

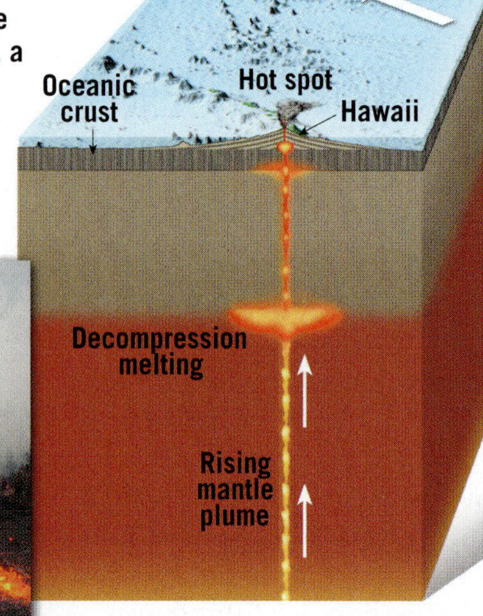

Kilauea, Hawaii (USGS)

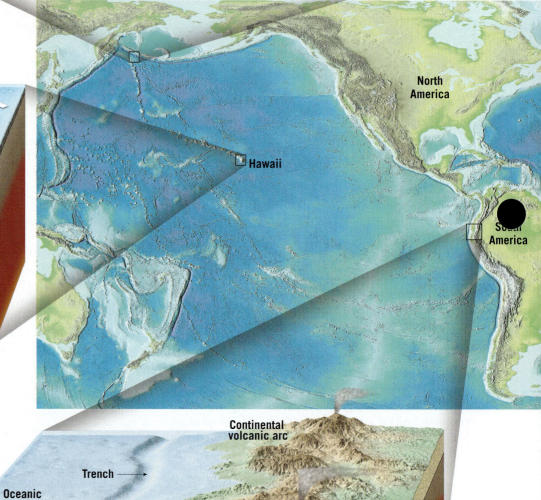

E. Convergent Plate Volcanism When oceanic lithosphere descends beneath a continent, magma generated in the mantle rises to form a continental volcanic arc.

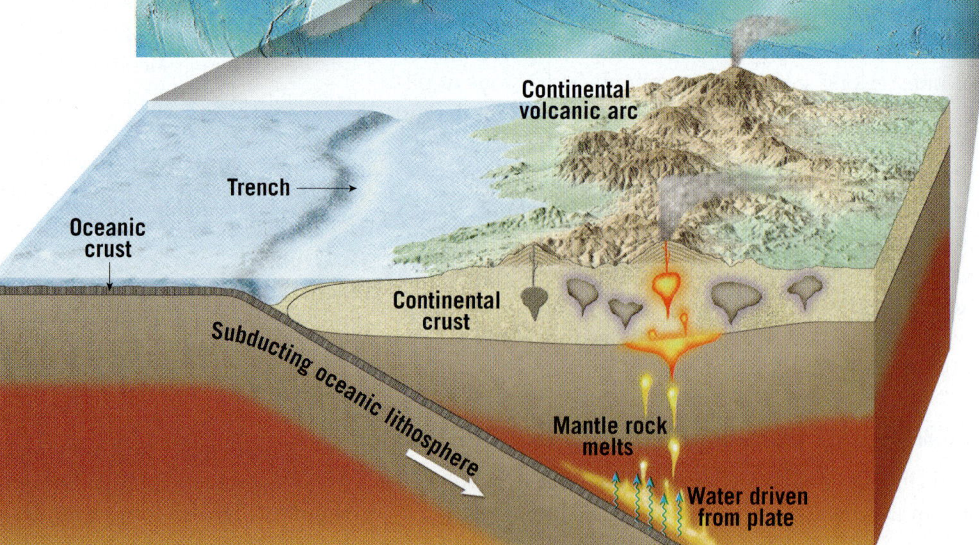

▲ **SmartFigure 5.30**
Earth's zones of volcanism

TUTORIAL
https://goo.gl/PSN9hc

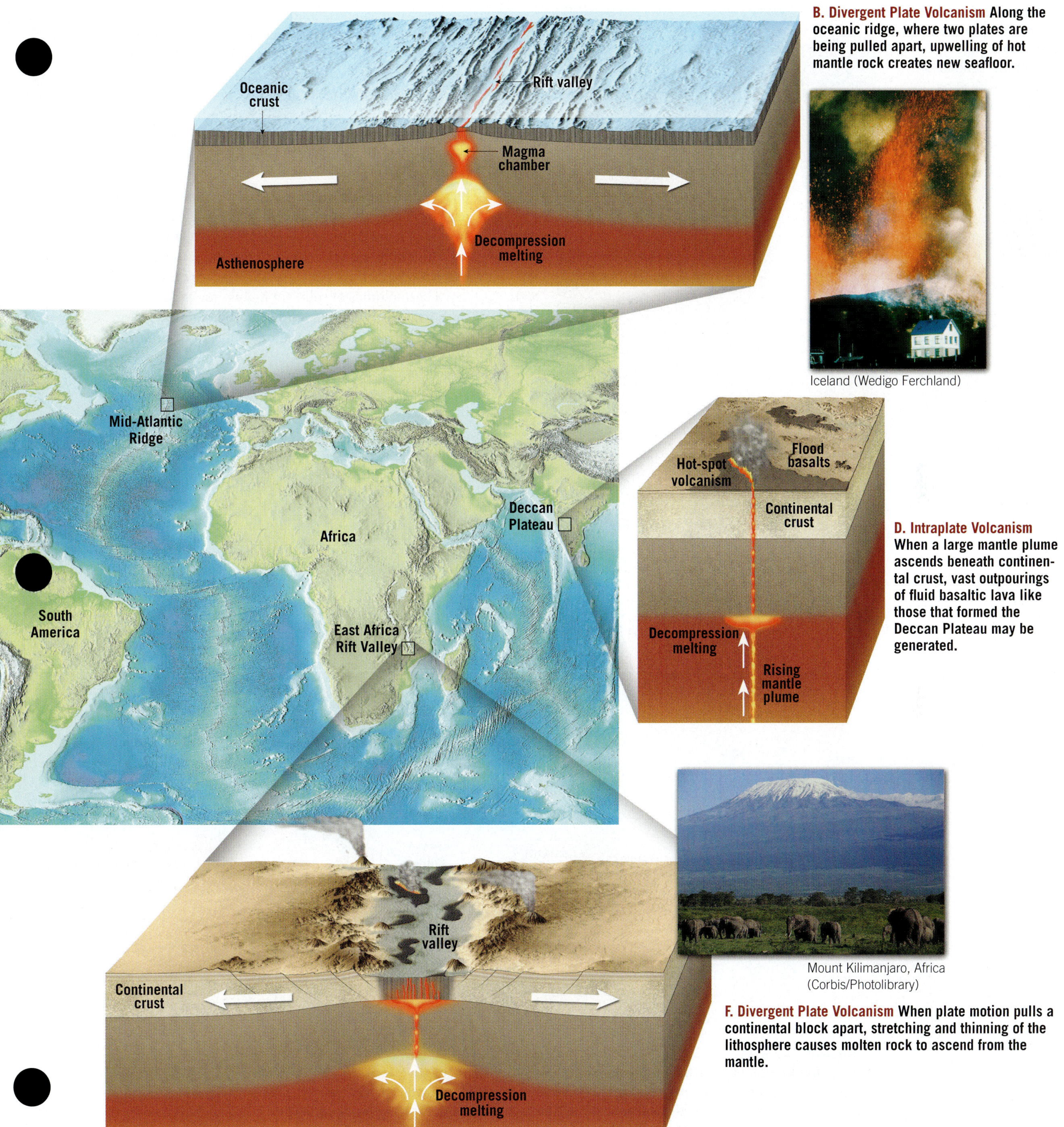

B. Divergent Plate Volcanism Along the oceanic ridge, where two plates are being pulled apart, upwelling of hot mantle rock creates new seafloor.

Iceland (Wedigo Ferchland)

D. Intraplate Volcanism When a large mantle plume ascends beneath continental crust, vast outpourings of fluid basaltic lava like those that formed the Deccan Plateau may be generated.

Mount Kilimanjaro, Africa (Corbis/Photolibrary)

F. Divergent Plate Volcanism When plate motion pulls a continental block apart, stretching and thinning of the lithosphere causes molten rock to ascend from the mantle.

▶ SmartFigure 5.31
Subduction-produced Cascade Range volcanoes Subduction of the Juan de Fuca plate along the Cascadia subduction zone produced the Cascade volcanoes.

TUTORIAL
https://goo.gl/gruZZD

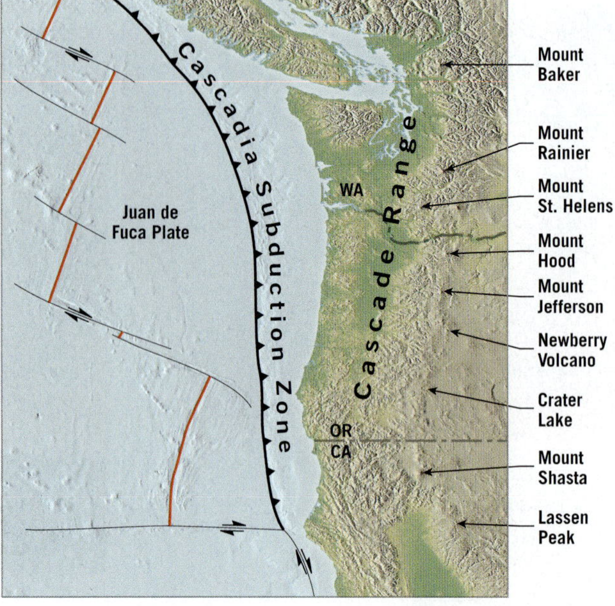

at a convergent plate boundary along a continental margin (Figure 5.31).

Intraplate Volcanism

We know why igneous activity is initiated along plate boundaries, but why do eruptions occur in the interiors of plates? Hawaii's Kilauea, considered one of the world's most active volcanoes, is situated thousands of kilometers from the nearest plate boundary, in the middle of the vast Pacific plate (Figure 5.30C). Sites of **intraplate** (meaning "within the plate") **volcanism** include the large outpourings of fluid basaltic lavas such as those that compose the Columbia Plateau, the Siberian Traps in Russia, India's Deccan Plateau, and several submerged oceanic plateaus, including the Ontong Java Plateau in the western Pacific (Figure 5.32).

Most intraplate volcanism occurs when a relatively narrow mass of hot **mantle plume** ascends toward the surface (Figure 5.33).* Although the depth at which mantle plumes originate is a topic of debate, some are thought to form deep within Earth, at the core–mantle boundary. These plumes of solid yet mobile rock rise toward the surface in a manner similar to the blobs that form within a lava lamp. Such a lamp contains two immiscible liquids in a glass container, and as the base of the lamp is heated, the denser liquid at the bottom becomes buoyant and forms blobs that rise to the top. Like the blobs in a lava lamp, a mantle plume has a bulbous head that draws out a narrow stalk beneath it as it rises. The surface manifestation of this activity is called a **hot spot**, an area of volcanism, high heat flow, and crustal uplifting a few hundred kilometers wide.

*Some geologists question the role of mantle plumes in the formation of Earth's volcanic landforms.

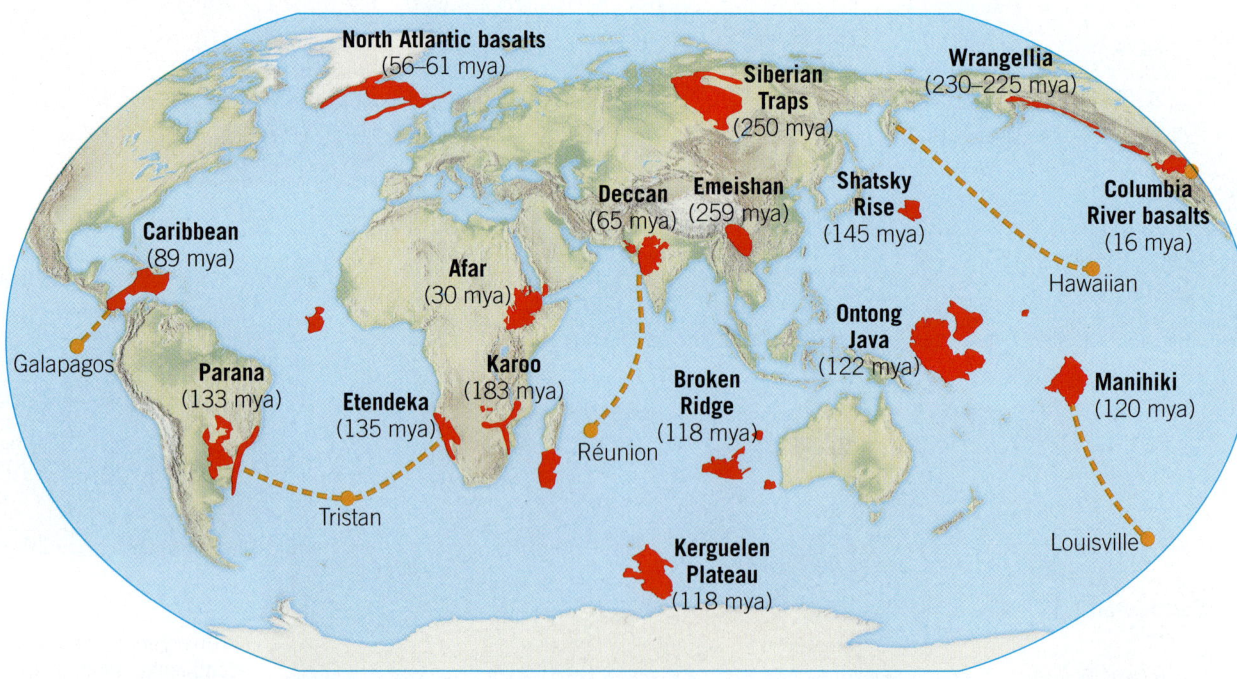

▲ SmartFigure 5.32
Global distribution of large basalt provinces. The basalt plateaus (shown in red) are thought to be the product of a burst of volcanism generated by partial melting of the bulbous head of a hot mantle plume. The orange dashed lines represent the chain of volcanic structures produced by partial melting of the plume tail. The orange dots are thought to be the current surface locations of the hot mantle plumes that generated the associated basalt plateaus.

TUTORIAL
https://goo.gl/O2Z6SI

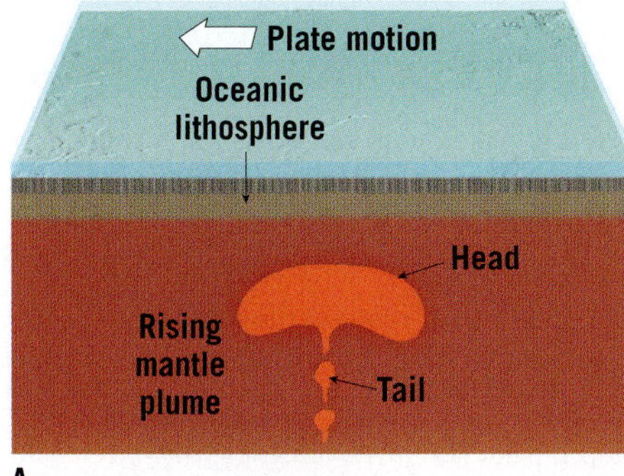
A rising mantle plume with a large bulbous head is thought to generate Earth's large basalt plateaus.
Plate motion
Oceanic lithosphere
Head
Rising mantle plume
Tail
A.

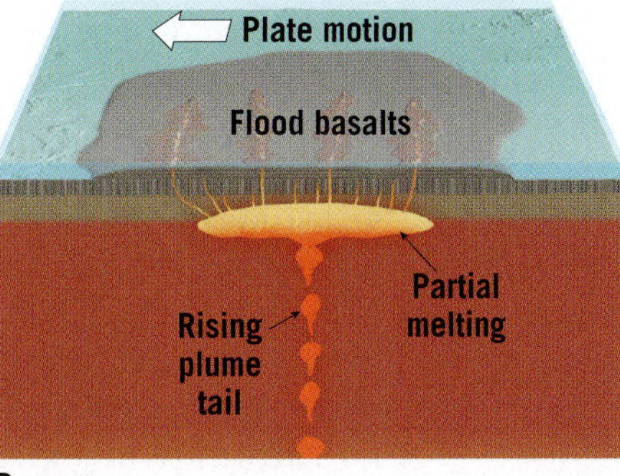
Rapid decompression melting of the plume head produces extensive outpourings of flood basalts over a relatively short time span.
Plate motion
Flood basalts
Partial melting
Rising plume tail
B.

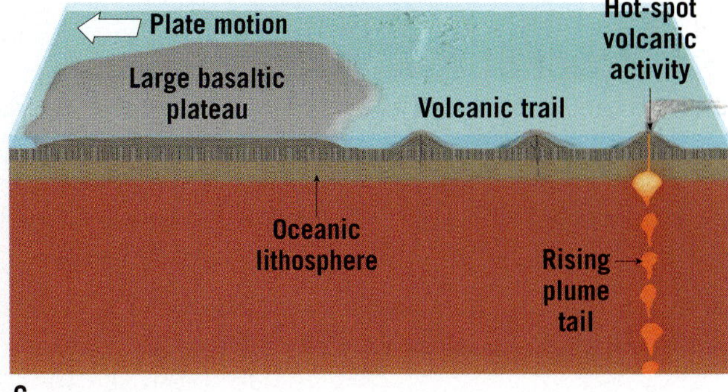
Because of plate movement, volcanic activity from the rising tail of the plume generates a linear chain of smaller volcanic structures.
Plate motion
Hot-spot volcanic activity
Large basaltic plateau
Volcanic trail
Oceanic lithosphere
Rising plume tail
C.

▲ **Figure 5.33 Mantle plumes and large basalt provinces** Model of hot-spot volcanism thought to explain the formation of large basalt plateaus and the chains of volcanic islands associated with these features.

Large mantle plumes, dubbed **superplumes**, are thought to be responsible for the vast outpourings of basaltic lava that created the large basalt plateaus. When the head of the plume reaches the base of the lithosphere, decompression melting progresses rapidly. This causes the burst of volcanism that emits voluminous flows of lava over a period of 1 million or so years (see **Figure 5.33B**). Extreme eruptions of this type would have affected Earth's climate, causing (or at least contributing to) the extinction events recorded in the fossil record.

The comparatively short initial eruptive phase is often followed by millions of years of less voluminous activity, as the plume tail slowly rises to the surface. Extending away from some large flood basalt plateaus is a chain of volcanic structures, similar to the Hawaiian chain (see **Figure 5.33C**).

Intraplate volcanism associated with mantle plumes is also thought to be responsible for the massive eruptions of silica-rich pyroclastic material that occurred in continental settings. Perhaps the best known of these hot-spot eruptions are the three caldera-forming eruptions that occurred in the Yellowstone region over the past 2.1 million years (see Figure 5.24).

CONCEPT CHECKS 5.10

1. Are volcanoes in the Ring of Fire generally described as effusive or explosive? Provide an example that supports your answer.
2. How is magma generated along convergent plate boundaries?
3. Volcanism at divergent plate boundaries is most often associated with which magma type? What causes rocks to melt in these settings?
4. What is thought to be the source of magma for most intraplate volcanism?
5. Which type of plate boundary generates the greatest quantity of magma?

CONCEPTS IN REVIEW

Volcanoes & Volcanic Hazards

5.1 Mount St. Helens Versus Kilauea

Compare and contrast the 1980 eruption of Mount St. Helens with the most recent eruption of Kilauea, which began in 1983.

- Volcanic eruptions cover a broad spectrum from explosive eruptions, like that of Mount St. Helens in 1980, to the quiescent eruptions of Kilauea.

5.2 The Nature of Volcanic Eruptions

Explain why some volcanic eruptions are explosive and others are quiescent.

KEY TERMS: magma, lava, effusive eruption, viscosity, eruption column

- The two primary factors determining the nature of a volcanic eruption are the viscosity (resistance to flow) of the magma and its gas content. In general, magmas that contain more silica are more viscous, while those with lower silica content are more fluid. Temperature also influences viscosity. Hot lavas are more fluid, while cool lavas are more viscous.
- Basaltic magmas, which are fluid and have low gas content, tend to generate effusive (non-explosive) eruptions. In contrast, silica-rich magmas (andesitic and rhyolitic), which are the most viscous and contain the greatest quantity of gases, are the most explosive.

? Although Kilauea mostly erupts in a gentle manner, what risks might you encounter if you chose to live nearby?

USGS

5.3 Materials Extruded During an Eruption

List and describe the three categories of materials extruded during volcanic eruptions.

KEY TERMS: aa flow, pahoehoe flow, lava tube, pillow lava, block lava, volatile, pyroclastic material, tephra, scoria, pumice

- Volcanoes erupt molten lava, gases, and solid pyroclastic materials.
- Low-viscosity basaltic lava flows can extend great distances from a volcano. On the surface, they travel as pahoehoe or aa flows. Sometimes the surface of the flow congeals, and lava continues to flow below in tunnels called lava tubes. When lava erupts underwater, the outer surface is chilled instantly to obsidian, while the inside continues to flow, producing pillow lavas.

(5.3 continued)

- The gases most commonly emitted by volcanoes are water vapor and carbon dioxide. Upon reaching the surface, these gases rapidly expand, leading to explosive eruptions that can generate a mass of lava fragments called pyroclastic materials.
- Pyroclastic materials come in several sizes. From smallest to largest, they are ash, lapilli, and blocks or bombs. Blocks exit the volcano as solid fragments, whereas bombs exit as liquid blobs.
- If bubbles of gas in lava don't pop before the lava solidifies, they are preserved as voids called vesicles. Especially frothy, silica-rich lava can cool to make lightweight pumice, while basaltic lava with lots of bubbles cools to make scoria.

? This photo shows layers of volcanic material ejected by a violent eruption and deposited roughly horizontally. What term is used to describe this type of volcanic material?

Erik Klemetti

5.4 Anatomy of a Volcano

Draw and label a diagram that illustrates the basic features of a typical volcanic cone.

KEY TERMS: fissure, conduit, vent, volcanic cone, crater, caldera, parasitic cone, fumarole

- Volcanoes vary in size and form but share a few common features. Most are roughly conical piles of extruded material that collect around a central vent. The vent is usually within a summit crater or caldera. On the flanks of the volcano, there may be smaller vents marked by small parasitic cones, or there may be fumaroles, spots where gas is expelled.

? Label the diagram using the following terms: conduit, vent, lava, parasitic cone, bombs, pyroclastic material.

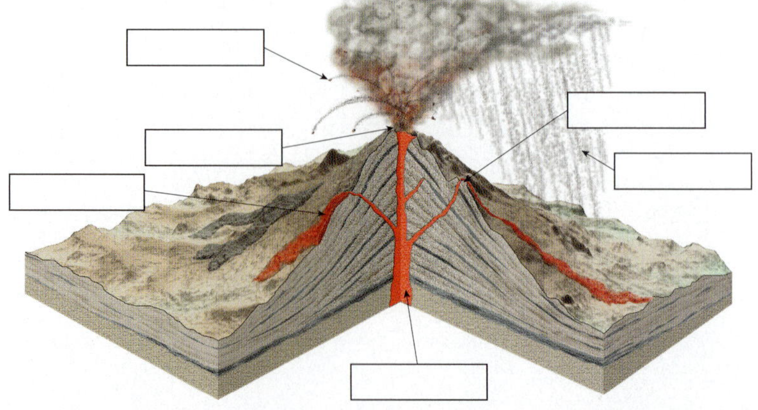

5.5 Shield Volcanoes

Summarize the characteristics of shield volcanoes and provide one example of this type of volcano.

KEY TERMS: shield volcano, seamount

- Shield volcanoes consist of many successive lava flows of low-viscosity basaltic lava but lack significant amounts of pyroclastic debris. Lava tubes help transport lava far from the main vent, resulting in very gentle, shield-like profiles.
- Most shield volcanoes begin as seamounts that grow from Earth's seafloor. Mauna Loa, Mauna Kea, and Kilauea in Hawaii are classic examples of the low, wide form characteristic of shield volcanoes.

5.6 Cinder Cones

Describe the formation, size, and composition of cinder cones.

KEY TERMS: cinder cone (scoria cone)

- Cinder cones are steep-sided structures composed mainly of pyroclastic debris, typically having a basaltic composition. Lava flows sometimes emerge from the base of a cinder cone but typically do not flow out of the crater.
- Cinder cones are small relative to the other major kinds of volcanoes, reflecting the fact that most form quickly, as single eruptive events. Because they are unconsolidated, cinder cones easily succumb to weathering and erosion.

5.7 Composite Volcanoes

List the characteristics of composite volcanoes and describe how they form.

KEY TERMS: composite volcano (stratovolcano)

- Composite volcanoes are called "composite" because they consist of both pyroclastic material and lava flows. They typically erupt silica-rich magmas of andesitic or rhyolitic composition. They are much larger than cinder cones and form from multiple eruptions over millions of years.
- Because andesitic and rhyolitic lavas are more viscous than basaltic lava, they accumulate at a steeper angle than does the lava from shield volcanoes. Over time, a composite volcano's combination of lava and cinders produces a towering volcano with a classic symmetrical shape.
- Mount Rainier and the other volcanoes of the Cascade Range in the northwest United States are good examples of composite volcanoes.

? If your family had to live next to a volcano, would you rather it be a shield volcano, cinder cone, or composite volcano? Explain.

5.8 Volcanic Hazards

Describe the major geologic hazards associated with volcanoes.

KEY TERMS: pyroclastic flow (nuée ardente), lahar, tsunami

- The greatest volcanic hazard to human life is the pyroclastic flow, or nuée ardente. This dense mix of hot gas and pyroclastic fragments races downhill at great speed and incinerates everything in its path. A pyroclastic flow can travel many kilometers from its source volcano. Because pyroclastic flows are hot, their deposits frequently "weld" together into a solid rock called welded tuff.
- Lahars are mudflows that form on volcanoes. These rapidly moving slurries of ash and debris suspended in water tend to follow stream valleys and can result in loss of life and/or significant damage to structures.

(5.8 continued)

- Volcanic ash in the atmosphere can be a risk to air travel when it is sucked into airplane engines. Volcanoes at sea level can generate tsunamis when they erupt or when their flanks collapse into the ocean. Those that spew large amounts of gas such as sulfur dioxide can cause respiratory problems. If volcanic gases reach the stratosphere, they screen out a portion of incoming solar radiation and can trigger short-term cooling at Earth's surface.

? What phenomenon is illustrated in the accompanying image?

(Ulet Infansasfi/Getty Images)

5.9 Other Volcanic Landforms

List volcanic landforms other than shield, cinder cone, and composite volcanoes and describe their formation.

KEY TERMS: fissure eruption, basalt plateau, flood basalt, lava dome, volcanic neck (plug)

- Calderas, which can be among the largest volcanic structures, form when the rigid, cold rock above a magma chamber cannot be supported and collapses, creating a broad, roughly circular depression. On shield volcanoes, calderas form slowly as lava drains from the magma chamber beneath the volcano. On a composite volcano, caldera collapse often follows an explosive eruption that can result in significant loss of life and destruction of property.
- Fissure eruptions occasionally produce massive floods of fluid basaltic lava from large cracks, called fissures, in the crust. Layer upon layer of these flood basalts may accumulate to significant thicknesses and blanket a wide area. The Columbia Plateau located in the northwestern United States is an example.
- Lava domes are thick masses of high-viscosity, silica-rich lava that accumulate in the summit crater or caldera of a composite volcano. When they collapse, lava domes can produce extensive pyroclastic flows.
- Shiprock, New Mexico, is an example of a volcanic neck where the lava in the "throat" of an ancient volcano crystallized to form a "plug" of solid rock that weathered more slowly than the surrounding volcanic rocks. The surrounding pyroclastic debris eroded, and the resistant neck remains as a distinctive landform.

5.10 Plate Tectonics & Volcanism

Explain how the global distribution of volcanic activity is related to plate tectonics.

KEY TERMS: Ring of Fire, volcanic island arc (island arc), continental volcanic arc, intraplate volcanism, mantle plume, hot spot, superplume

- Volcanoes occur at both convergent and divergent plate boundaries, as well as in intraplate settings.
- At divergent plate boundaries, where lithosphere is being rifted apart, decompression melting is the dominant generator of magma. As warm rock rises, it can begin to melt without the addition of heat.
- Convergent plate boundaries that involve the subduction of oceanic crust are the most common site for explosive volcanoes—most prominently in the Pacific Ring of Fire. The release of water from the subducting plate triggers melting in the overlying mantle. The ascending magma interacts with the lower crust of the overlying plate and can form a volcanic arc at the surface.
- In intraplate settings, the source of magma is a mantle plume—a column of mantle rock that is warmer and more buoyant than the surrounding mantle.

? The accompanying diagram shows one of the tectonic settings where volcanism is a dominant process. Name the tectonic setting and briefly explain how magma is generated in this setting.

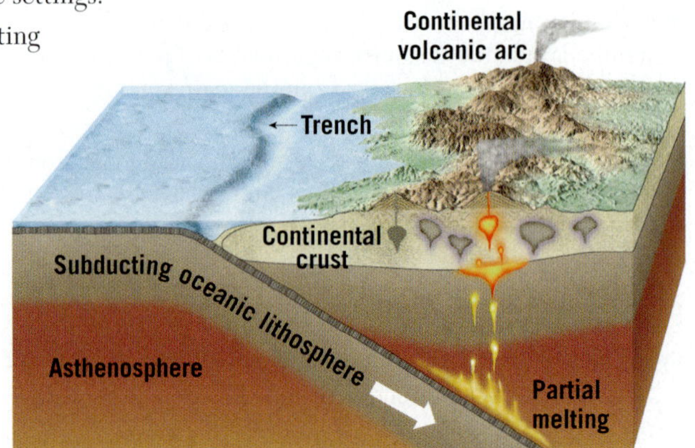

GIVE IT SOME THOUGHT

1 Examine the accompanying photo and complete the following:
- **a.** What type of volcano is it? What features helped you classify it as such?
- **b.** What is the eruptive style of such volcanoes? Describe the likely composition and viscosity of its magma.
- **c.** Which type of plate boundary is the likely setting for this volcano?
- **d.** Name a city that is vulnerable to the effects of a volcano of this type.

2 Answer the following questions about divergent boundaries, such as the Mid-Atlantic Ridge, and their associated lavas:
- **a.** Divergent boundaries are characterized by eruptions of what type of lava: andesitic, basaltic, or rhyolitic?
- **b.** What is the main source of the lavas that erupt at divergent plate boundaries?
- **c.** What process causes the source rocks to melt?

3 For each of the volcanoes or volcanic regions listed below, identify whether it is associated with a *convergent* or *divergent* plate boundary or with *intraplate volcanism*.
- **a.** Crater Lake
- **b.** Hawaii's Kilauea
- **c.** Mount St. Helens
- **d.** East African Rift
- **e.** Yellowstone
- **f.** Mount Pelée
- **g.** Deccan Traps
- **h.** Fujiyama

4 For each of the accompanying four sketches, identify the geologic setting (zone of volcanism). Which of these settings will most likely generate explosive eruptions? Which will produce outpouring of fluid basaltic lavas?

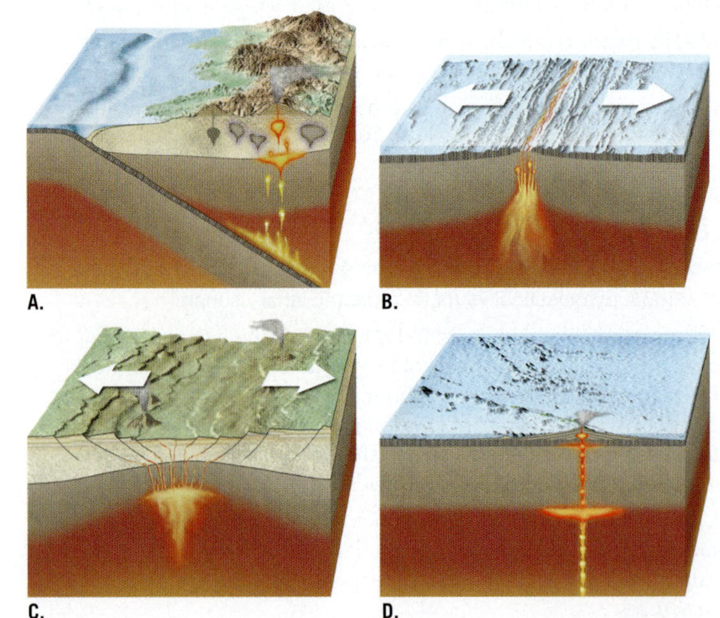

5 Explain why an eruption of Mount Rainier similar to the 1980 eruption of Mount St. Helens could be considerably more destructive.

6 This image shows the Buddhist monastery Taung Kalat, located in central Myanmar (Burma). The monastery sits high on a sheer-sided rock made mainly of magmas that solidified in the conduit of an ancient volcano. The volcano has since been worn away.

a. Based on this information, what igneous structure do you think is shown in this photo?

b. Would this volcanic structure most likely have been associated with a composite volcano or a cinder cone? Explain how you arrived at your answer.

Taolmor/Dreamstime

7 The formula for the volume of a cone is $V = 1/3\pi r^2 h$ (where V = volume, $\pi = 3.14$, r = radius, and h = height). If Mauna Loa is 9 kilometers high and has a radius of roughly 85 kilometers, what is its approximate total volume?

8 The accompanying image shows a geologist at the end of an unconsolidated flow consisting of lightweight lava blocks that rapidly descended the flank of Mount St. Helens.

a. What term best describes this type of flow: an aa flow, a pahoehoe flow, or a pyroclastic flow?

b. What lightweight (vesicular) igneous rock type is likely the main constituent of this flow?

Donald Swanson/USGS

9 Different processes produce magma in different tectonic settings. Consider magma bodies found at locations A, B, and C in the accompanying diagram and describe the process that most likely triggered the melting that produced each.

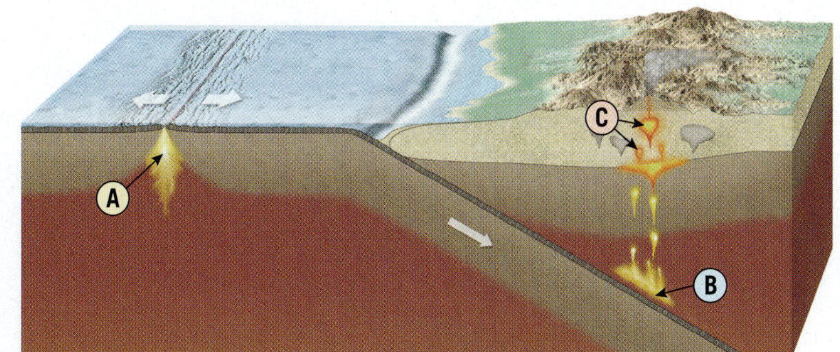

MasteringGeology™

Looking for additional review and test prep materials? Visit the Study Area in MasteringGeology to enhance your understanding of this chapter's content by accessing a variety of resources, including Self-Study Quizzes, Geoscience Animations, SmartFigures, Mobile Field Trips, *Project Condor* Quadcopter videos, *In the News* RSS feeds, flashcards, web links, and an optional Pearson eText.

www.masteringgeology.com

6

Weathering & Soils

FOCUS ON CONCEPTS

Each statement represents the primary learning objective for the corresponding major heading within the chapter. After you complete the chapter, you should be able to:

6.1 Define *weathering* and distinguish between the two main categories of weathering.

6.2 List and describe four examples of mechanical weathering.

6.3 Discuss the importance of water and carbonic acid in chemical weathering processes.

6.4 Summarize the factors that influence the type and rate of rock weathering.

6.5 Define *soil* and explain why soil is referred to as an *interface*. List and briefly discuss five controls of soil formation.

6.6 Sketch, label, and describe an idealized soil profile. Explain the need for classifying soils.

6.7 Explain the detrimental impact of human activities on soil.

6.8 Relate weathering to the formation of certain ore deposits.

Weathering processes helped shape the rock formations in California's Pinnacles National Park. (Photo by Spring Images/Alamy Stock Photo)

EARTH'S SURFACE IS CONSTANTLY CHANGING. Rock is disintegrated and decomposed, moved to lower elevations by gravity, and carried away by water, wind, or ice. In this manner, Earth's physical landscape is sculpted. This chapter focuses on the first step of this never-ending process—weathering. It looks at what causes solid rock to crumble and why the type and rate of weathering vary from place to place. Soil, an important product of the weathering process and a vital resource, is also examined.

6.1 Weathering

Define *weathering* and distinguish between the two main categories of weathering.

Weathering involves the physical breakdown (disintegration) and chemical alteration (decomposition) of rock at or near Earth's surface. Weathering goes on all around us, but it is such a slow and subtle process that its importance is easy to underestimate. Yet weathering is a basic part of the rock cycle and thus a key process in the Earth system. Weathering is also important to humans—even to those of us who are not studying geology. For example, many of the life-sustaining minerals and elements found in soil, and ultimately in the food we eat, were freed from solid rock by weathering processes. As the chapter-opening photo, Figure 6.1, and many other images in this book illustrate, weathering also contributes to the formation of some of Earth's most spectacular scenery. Of course, these same processes are also responsible for causing the deterioration of many of the structures we build.

There are two basic categories of weathering. **Mechanical weathering** is accomplished by physical forces that break rock into smaller and smaller pieces without changing the rock's mineral composition. **Chemical weathering** involves a chemical transformation of rock into one or more new compounds. These two concepts can be illustrated with a large log. The log disintegrates when it is split into smaller and smaller pieces, whereas decomposition occurs when the log is set afire and burned.

Why does rock weather? Simply, weathering is the response of Earth materials to a changing environment. For instance, after millions of years of uplift and erosion (the removal and transport of weathered rock material by water, wind, or ice), the rocks overlying a large, intrusive igneous body may be removed, exposing it at the surface. This mass of crystalline rock—formed deep below ground, where temperatures and pressures are high—is now subjected to a very different and comparatively hostile surface environment. In response, this rock mass will gradually change. This transformation of rock is what we call weathering.

In the following sections we will examine the various types of mechanical and chemical weathering. Although we will consider these two categories separately, keep in mind that mechanical and chemical weathering processes usually work simultaneously in nature and reinforce each other.

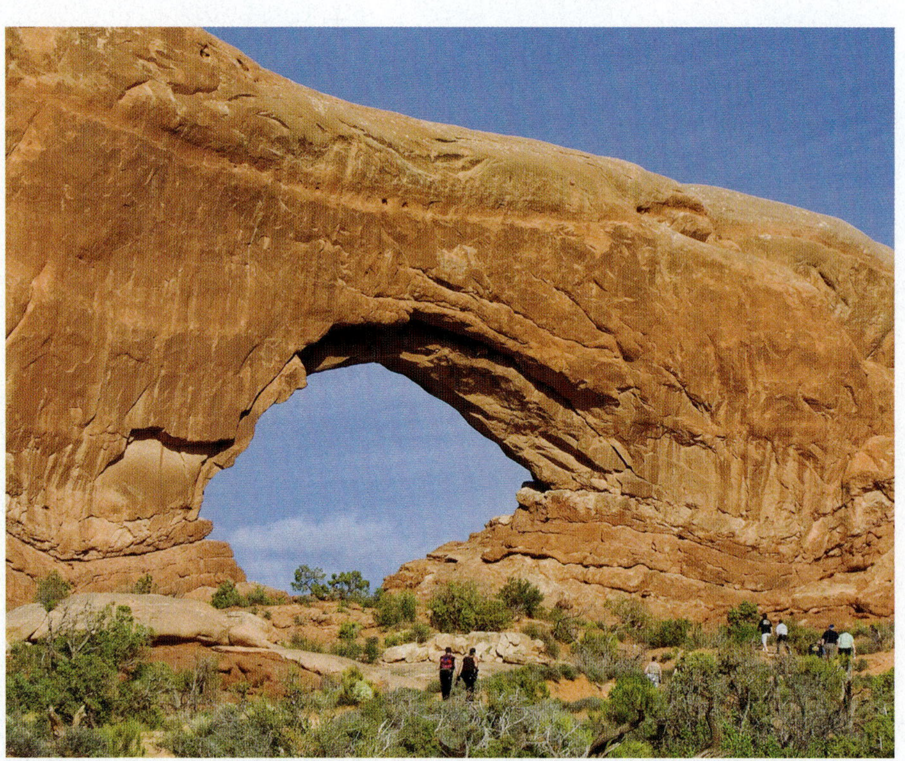

▲ SmartFigure 6.1 **Arches National Park** Mechanical and chemical weathering contributed greatly to the creation of North Window Arch and to all of the other arches and rock formations in Utah's Arches National Park. (Photo by Dennis Tasa)

ANIMATION
https://goo.gl/h9L7OO

CONCEPT CHECKS 6.1

1. What are the two basic categories of weathering?
2. How do the products of each category of weathering differ?

6.2 Mechanical Weathering

List and describe four examples of mechanical weathering.

When a rock undergoes *mechanical weathering*, it is broken into smaller and smaller pieces, each retaining the characteristics of the original material. The end result is many small pieces from a single large one. Figure 6.2 shows that breaking a rock into smaller pieces increases the surface area available for chemical attack. An analogous situation occurs when sugar is added to a liquid. A sugar cube dissolves much more slowly than an equal volume of sugar granules because the cube has much less surface area available for dissolution. Hence, by breaking rocks into smaller pieces, mechanical weathering increases the amount of surface area available for chemical weathering.

In nature, four physical processes are mainly responsible for fragmenting rock: frost wedging, salt crystal growth, sheeting, and biological activity. In addition, although the work of erosional agents such as wind, waves, glacial ice, and running water is usually considered separately from mechanical weathering, this work is nevertheless related. As these mobile agents (discussed in detail in later chapters) transport rock debris, particles continue to be broken and abraded.

Frost Wedging

If you leave a glass bottle of water in the freezer a bit too long, you will find the bottle fractured, as in Figure 6.3. The bottle breaks because liquid water has the unique property of expanding about 9 percent upon freezing. This is also the reason that poorly insulated or exposed water pipes rupture during frigid weather. You might also expect this same process to fracture rocks in nature. This is, in fact, the basis for the traditional explanation of **frost wedging**. After water works its way into the cracks in rock, the freezing water enlarges the cracks, and angular fragments break off (Figure 6.4).

For many years, the conventional wisdom was that most frost wedging occurred in this way. However, research has shown that frost wedging can also occur in a different way.* It has long been known that when moist soils freeze, they expand, or *frost heave*, due to the growth of ice lenses. These masses of ice grow larger because they are supplied with water migrating from unfrozen areas as thin liquid films. As more water accumulates and freezes, the soil is heaved upward. A similar process occurs within the cracks and pore spaces of rocks. Lenses of ice grow larger as they attract liquid water from surrounding pores. The growth of these ice masses gradually weakens the rock, causing it to fracture.

Salt Crystal Growth

Another expansive force that can split rocks is created by the growth of salt crystals. Rocky shorelines and arid regions are common settings for this process. It begins when sea spray from breaking waves or salty

*Bernard Hallet, "Why Do Freezing Rocks Break?" *Science*, 314(17): 1092–1093, November 2006.

Did You Know?
The intense heat from a brush or forest fire can cause flakes of rock to break from boulders or bedrock. As the rock surface becomes overheated, a thin layer expands and shatters.

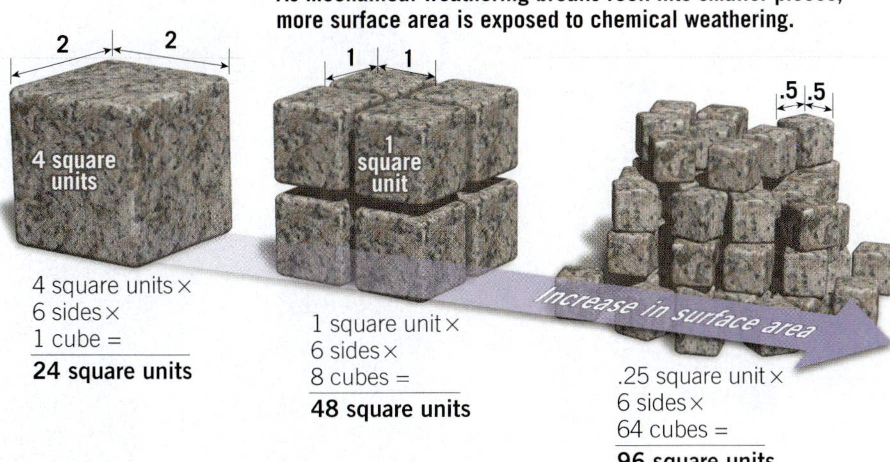

▲ SmartFigure 6.2 **Mechanical weathering increases surface area** Mechanical weathering adds to the effectiveness of chemical weathering because chemical weathering can occur only on exposed surfaces.

https://goo.gl/Gqd1iz

▲ Figure 6.3 **Ice breaks bottle** The bottle broke because water expands about 9 percent when it freezes. (Photo by Martyn F. Chillmaid/Science Source)

▲ **SmartFigure 6.4 Ice breaks rock** In mountainous areas, frost wedging creates angular rock fragments that accumulate to form talus slopes. (Photo by Marli Miller)

TUTORIAL
https://goo.gl/RZRD5W

groundwater penetrates crevices and pore spaces in rock. As this water evaporates, salt crystals form. As these crystals gradually grow larger, they weaken the rock by pushing apart the surrounding grains or enlarging tiny cracks.

This same process can also contribute to the crumbling of roadways where salt is spread to melt snow and ice in winter. The salt dissolves in water and seeps into cracks that quite likely originated from frost action. When the water evaporates, the growth of salt crystals further breaks the pavement.

Sheeting

When large masses of igneous rock, particularly granite, are exposed by erosion, concentric slabs begin to break loose. The process that generates these onion-like layers is called **sheeting**. It takes place, at least in part, due to the great reduction in pressure that occurs as the overlying rock is eroded away, a process called *unloading*. Figure 6.5 illustrates what happens: As the overburden is removed, the outer parts of the granitic mass expand more than the rock below and separate from the rock body. Continued weathering eventually causes the slabs to separate and peel off, creating an **exfoliation dome** (*ex* = off, *folium* = leaf). Excellent examples of exfoliation domes are Stone Mountain, Georgia, and Half Dome and Liberty Cap in Yosemite National Park.

A process analogous to sheeting can also occur when human activities reduce the confining pressure, similar

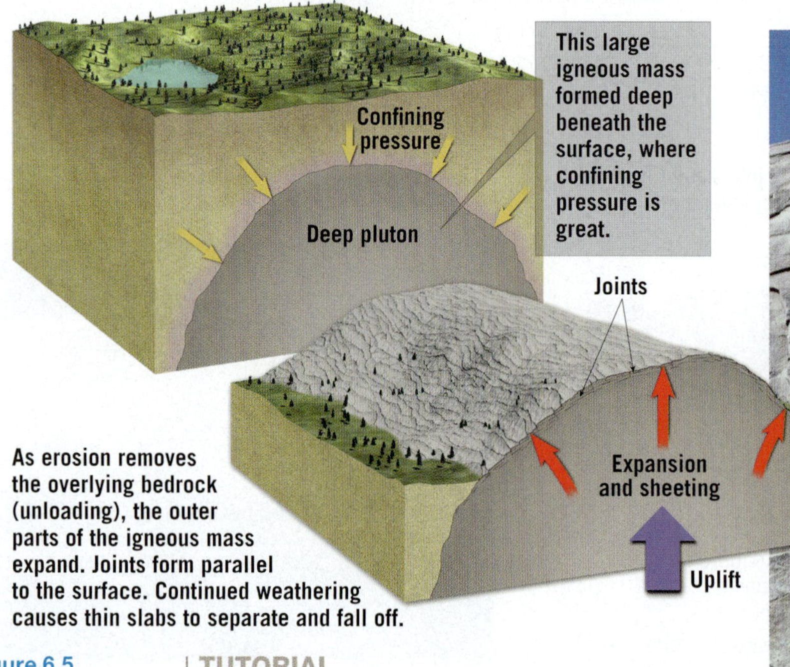

Confining pressure
Deep pluton
This large igneous mass formed deep beneath the surface, where confining pressure is great.
Joints
Expansion and sheeting
Uplift
As erosion removes the overlying bedrock (unloading), the outer parts of the igneous mass expand. Joints form parallel to the surface. Continued weathering causes thin slabs to separate and fall off.

The summit of Half Dome in California's Yosemite National Park is an exfoliation dome and illustrates the onion-like layers created by sheeting.

► **SmartFigure 6.5 Unloading leads to sheeting** Sheeting leads to the formation of an exfoliation dome. (Photo by Gary Moon/AGE Fotostock America, Inc.)

TUTORIAL
https://goo.gl/UFWTzX

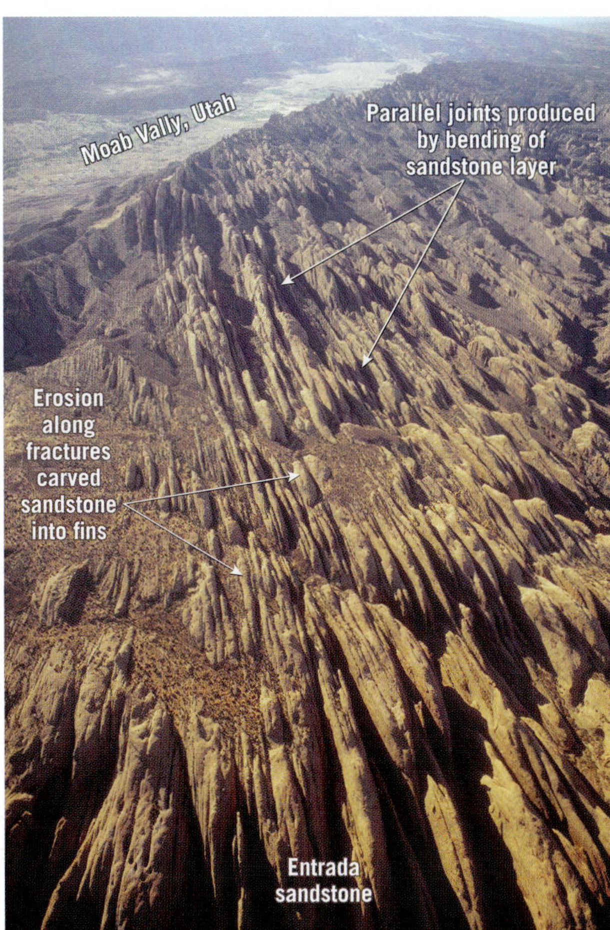

▲ **Figure 6.6 Joints aid weathering** Aerial view of nearly parallel joints near Moab, Utah. (Photo by Michael Collier)

to what occurs during unloading. For example, in deep mines, large rock slabs have been known to explode off the walls of newly cut tunnels. In quarries, fractures occur parallel to the floor when large blocks of rock are removed.

Although many fractures are created by expansion, others are produced by contraction during the crystallization of magma, and still others are produced by tectonic forces during mountain building. Fractures produced by these activities generally form a definite pattern and are called **joints** (Figure 6.6). Joints are important rock structures that allow water to penetrate to depth and start the process of weathering long before the rock is exposed.

Biological Activity

Weathering can be accomplished by the activities of organisms, including plants, burrowing animals, and humans. Plant roots in search of nutrients and water grow into fractures, and as the roots grow, they wedge apart the rock (Figure 6.7). Burrowing animals further break down rock by moving fresh material to the surface, where physical and chemical processes can more effectively attack it. Decaying organisms also produce acids that contribute to chemical weathering. Where rock has been blasted in search of minerals or for road construction, the impact of humans is particularly noticeable.

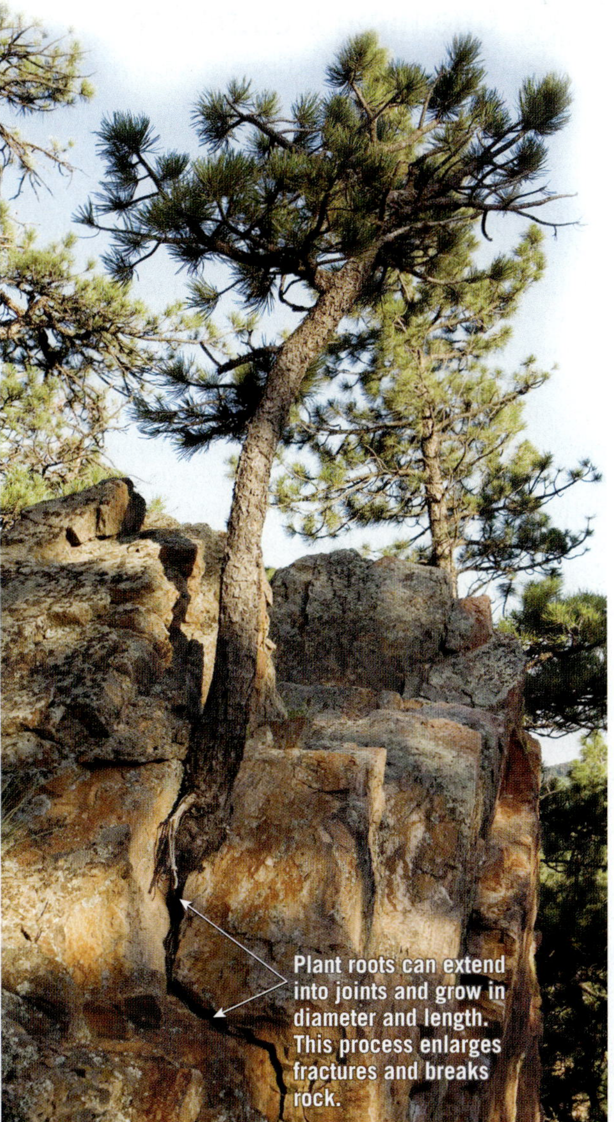

▲ **Figure 6.7 Plants can break rock** Root wedging near Boulder, Colorado. (Photo by Kristin Piljay)

CONCEPT CHECKS 6.2

1. When a rock is mechanically weathered, how does its surface area change? How does this influence chemical weathering?
2. Explain how water can cause mechanical weathering.
3. Describe how an exfoliation dome forms.
4. How do joints promote weathering?
5. How does biological activity contribute to weathering?

6.3 Chemical Weathering

Discuss the importance of water and carbonic acid in chemical weathering processes.

Did You Know? The only common mineral that is very resistant to both mechanical and chemical weathering is quartz.

In the preceding discussion of mechanical weathering, you learned that breaking rock into smaller pieces aids chemical weathering by increasing the surface area available for chemical attack. It should also be pointed out that chemical weathering contributes to mechanical weathering. It does so by weakening the outer portions of some rocks, which, in turn, makes them more susceptible to being broken by mechanical weathering processes.

Chemical weathering involves the complex processes that alter the internal structures of minerals by removing and/or adding elements. During this transformation, the original rock decomposes into substances that are stable in the surface environment. Consequently, the products of chemical weathering remain essentially unchanged as long as they remain in an environment similar to the one in which they formed.

The Importance of Water

Water is by far the most important agent of chemical weathering. Although pure water is nonreactive, a small amount of dissolved material is generally all that is needed to activate it.

▼ **Figure 6.8 Iron oxides add color** Many sedimentary rocks are very colorful. The most important "pigments" are small amounts of iron oxide. Just as iron oxide colors the rusty barrels in **A**, this product of chemical weathering is also responsible for the reds and oranges seen in the rocks composing the Supai Formation in the Grand Canyon in **B**. (Photo A by Vladimir Melnik/Shutterstock; photo B by Cedric Weber/Shutterstock)

Oxidation Everyone has seen iron and steel objects that have rusted when exposed to water. The same thing can happen to iron-rich minerals. The process of rusting occurs when oxygen combines with iron to form iron oxide, as follows:

$$\underset{\text{iron}}{4\,Fe} + \underset{\text{oxygen}}{3\,O_2} \longrightarrow \underset{\text{iron oxide (hematite)}}{2\,Fe_2O_3}$$

This type of chemical reaction, called **oxidation**, occurs when electrons are lost from one element during the reaction. In this case, we say that iron was oxidized because it lost electrons to oxygen. Although the oxidation of iron progresses very slowly in a dry environment, the addition of water greatly speeds the reaction.

Oxidation is important in decomposing such ferromagnesian minerals as olivine, pyroxene, hornblende, and biotite. Oxygen combines with the iron in these minerals to form the reddish-brown iron oxide called *hematite* (Fe_2O_3), or in other cases a yellowish-colored rust called *limonite* [$FeO(OH)$]. These products are responsible for the rusty color on the surfaces of dark igneous rocks, such as basalt, as they begin to weather. Hematite and limonite are also important cementing and coloring agents in many sedimentary rocks (**Figure 6.8**).

Carbonic Acid When carbon dioxide (CO_2) is dissolved in water (H_2O), it forms **carbonic acid** (H_2CO_3), the same weak acid produced when soft drinks are carbonated. Rain dissolves some carbon dioxide as it falls through the atmosphere, and additional amounts released by decaying organic matter are acquired as the water percolates through the soil. Carbonic acid ionizes to form the very reactive hydrogen ion (H^+) and the bicarbonate ion (HCO_3^-).

Acids such as carbonic acid readily decompose many rocks and produce certain products that are water soluble. For example, the mineral calcite ($CaCO_3$), which composes the common building stones marble and limestone, is easily attacked by even a weakly acidic solution.

The overall reaction by which calcite dissolves in water containing carbon dioxide is:

$$\underset{\text{calcite}}{CaCO_3} + \underset{\text{carbonic acid}}{(H^+ + HCO_3^-)} \longrightarrow \underset{\text{calcium ion}}{Ca^{2+}} + \underset{\text{bicarbonate ion}}{2HCO_3^-}$$

During this process, the insoluble calcium carbonate is transformed into soluble products. In nature, over periods of thousands of years, large quantities of limestone are dissolved and carried away by groundwater. This activity is clearly evidenced by the large number of caverns found in all of the contiguous 48 states

(Figure 6.9). Monuments and buildings made of limestone or marble are also subjected to the corrosive work of acids, particularly in urban and industrial areas that have smoggy, polluted air.

How Granite Weathers

To illustrate how rock chemically weathers when attacked by carbonic acid, we will consider the weathering of granite, the most abundant continental rock. Recall that granite consists mainly of quartz and potassium feldspar. The weathering of the potassium feldspar component of granite takes place as follows:

$$\underset{\text{potassium feldspar}}{2\,KAlSi_3O_8} + \underset{\text{carbonic acid}}{2(H^+ + HCO_3^-)} + \underset{\text{water}}{H_2O} \rightarrow$$

$$\underset{\text{clay mineral}}{Al_2Si_2O_5(OH)_4} + \underbrace{\underset{\text{potassium ion}}{2K^+} + \underset{\text{bicarbonate ion}}{2HCO_3^-} + \underset{\text{silica ion}}{4SiO_2}}_{\text{in solution}}$$

In this reaction, the hydrogen ions (H^+) attack and replace potassium ions (K^+) in the feldspar structure, thereby disrupting the crystalline network. Once the potassium is removed, it is available as a nutrient for plants or becomes the soluble salt potassium bicarbonate ($KHCO_3$), which may be incorporated into other minerals or carried to the ocean in dissolved form by streams.

The most abundant products of the chemical breakdown of feldspar are residual clay minerals. Clay minerals are the end products of weathering and are very stable under surface conditions. Consequently, clay minerals make up a high percentage of the inorganic material in soils. Moreover, the most abundant sedimentary rock, shale, contains a high proportion of clay minerals.

In addition to the formation of clay minerals during the weathering of feldspar, some silica is removed from the feldspar structure and is carried away by groundwater. This dissolved silica will eventually precipitate to produce nodules of chert or flint, or it will fill in the pore spaces between sediment grains, or it will be carried to the ocean, where microscopic animals remove it from the water to build hard silica shells.

▲ **Figure 6.9 Acidic waters create caves** The dissolving power of carbonic acid plays an important role in creating limestone caverns. This is an image of Baredine Cave in Croatia. (Photo by Gunter Lenz/imageBROKER)

To summarize, the weathering of potassium feldspar generates a residual clay mineral, a soluble salt (potassium bicarbonate), and some silica, which enters into solution.

Quartz, the other main component of granite, is *very resistant* to chemical weathering and remains substantially unaltered when attacked by weak acidic solutions. As a result, when granite weathers, the feldspar crystals dull and slowly turn to clay, releasing the once-interlocked quartz grains, which still retain their fresh, glassy appearance. Although some quartz remains in the soil, much is eventually transported to the sea or to other sites of deposition, where it becomes the main constituent of such features as sandy beaches and sand dunes. In time, these quartz grains may become lithified to form the sedimentary rock *sandstone*.

Weathering of Silicate Minerals

Table 6.1 lists the weathered products of some of the most common silicate minerals. Remember that silicate minerals make up most of Earth's crust and that these minerals are composed essentially of only eight

Table 6.1 Products of Chemical Weathering

Mineral	Residual Products	Material in Solution
Quartz	Quartz grains	Silica
Feldspars	Clay minerals	Silica, K^+, Na^+, Ca^{2+}
Amphibole (hornblende)	Clay minerals	Silica, Ca^{2+}, Mg^{2+}
	Limonite	
	Hematite	
Olivine	Limonite	Silica
	Hematite	Mg^{2+}

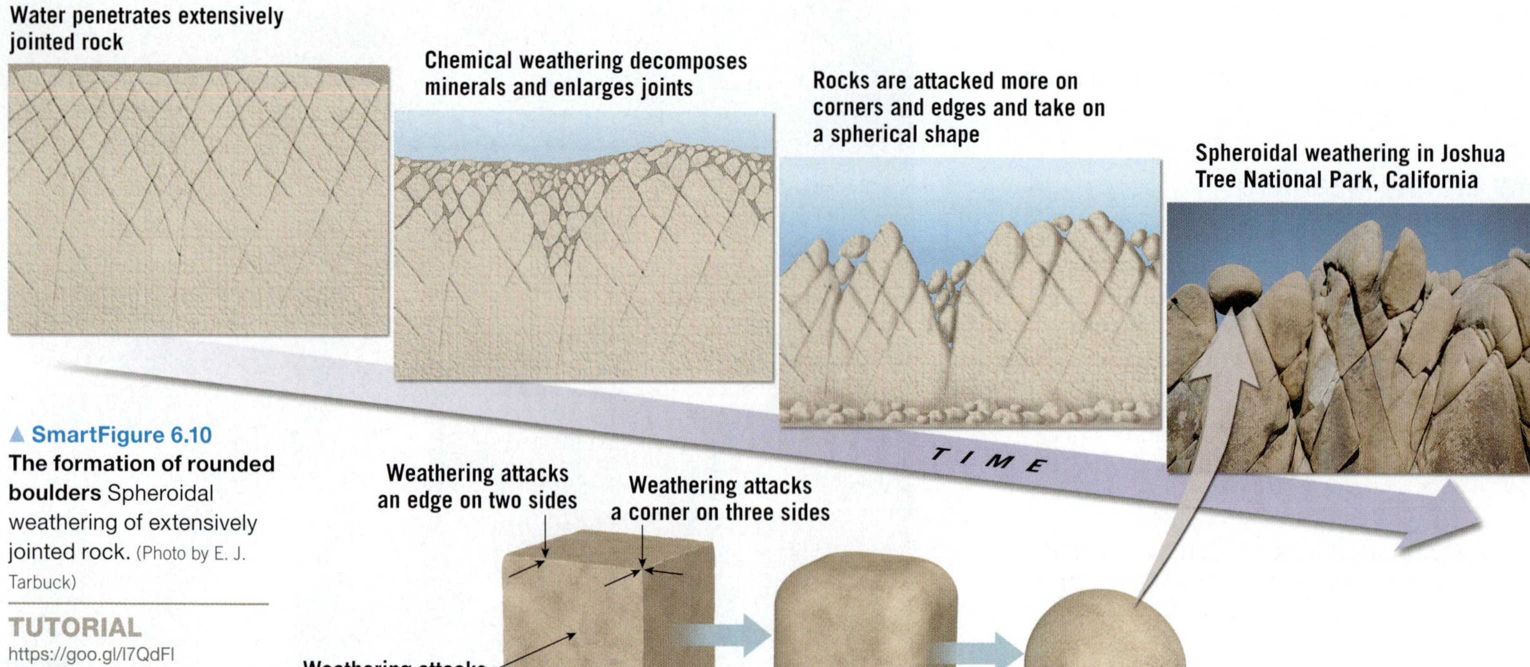
Water penetrates extensively jointed rock
Chemical weathering decomposes minerals and enlarges joints
Rocks are attacked more on corners and edges and take on a spherical shape
Spheroidal weathering in Joshua Tree National Park, California
TIME
Weathering attacks an edge on two sides
Weathering attacks a corner on three sides
Weathering attacks a face on one side

▲ SmartFigure 6.10 **The formation of rounded boulders** Spheroidal weathering of extensively jointed rock. (Photo by E. J. Tarbuck)

TUTORIAL
https://goo.gl/I7QdFI

elements. When chemically weathered, these silicate minerals yield sodium, calcium, potassium, and magnesium ions that form soluble products, which may be removed by groundwater. The element iron combines with oxygen, producing relatively insoluble iron oxides. Under most conditions the three remaining elements—aluminum, silicon, and oxygen—join with water to produce residual clay minerals. However, even the highly insoluble clay minerals are very slowly removed by subsurface water.

Spheroidal Weathering

Many rock outcrops have a rounded appearance. This occurs because chemical weathering works inward from exposed surfaces. Figure 6.10 illustrates how angular masses of jointed rock change through time. The process is aptly called **spheroidal weathering**. Because weathering attacks edges from two sides and corners from three sides, these areas wear down faster than a single flat surface. Gradually, sharp edges and corners become smooth and rounded. Eventually an angular block may evolve into a nearly spherical boulder. Once this occurs, the boulder's shape does not change, but the spherical mass continues to get smaller.

CONCEPT CHECKS 6.3

1. How is carbonic acid formed in nature?
2. What occurs when carbonic acid reacts with calcite-rich rocks such as limestone?
3. What products result when carbonic acid reacts with potassium feldspar?
4. Explain how angular masses of rock often become spherical boulders.

6.4 Rates of Weathering

Summarize the factors that influence the type and rate of rock weathering.

We have already seen how mechanical weathering affects the rate of weathering. When rock is broken into smaller pieces, the amount of surface area exposed to chemical weathering increases. Other important factors that influence the type and rate of rock weathering include rock characteristics and climate.

Rock Characteristics

Rock characteristics encompass all the chemical traits of rocks, including mineral composition and solubility. In addition, any physical features, such as joints, can be important because they influence the ability of water to penetrate rock.

This granite headstone was erected in 1868. The inscription still looks fresh.

This headstone of calcite-rich marble dates from 1874, six years after the granite stone. The inscription is barely legible.

◀ **SmartFigure 6.11** **Rock type influences weathering** An examination of headstones in the same cemetery shows that the rate of chemical weathering is influenced by rock type. (Photos by E. J. Tarbuck)

TUTORIAL
https://goo.gl/BjltRN

The variations in weathering rates due to the mineral constituents can be demonstrated by comparing old headstones made from different rock types. Headstones of granite, which is composed of silicate minerals, are relatively resistant to chemical weathering. In contrast, marble headstones show signs of extensive chemical alteration over a relatively short period. We can see this by examining the inscriptions on the headstones shown in **Figure 6.11**. Marble is composed of calcite (calcium carbonate), which readily dissolves even in a weakly acidic solution.

The silicates, the most abundant mineral group, chemically weather in essentially the same order in which they crystallize. By examining Bowen's reaction series (see Figure 4.20, page 110), you can see that olivine crystallizes first and is therefore least resistant to chemical weathering, whereas quartz, which crystallizes last, is the most resistant.

Climate

Climatic factors, particularly temperature and precipitation, are crucial to the rate of rock weathering. For example, the frequency of freeze–thaw cycles greatly affects the amount of frost wedging. Temperature and moisture also exert a strong influence on rates of chemical weathering and determine the kind and amount of vegetation present. Regions with lush vegetation often have a thick mantle of soil rich in decayed organic matter from which chemically active fluids such as carbonic acid and various organic acids are derived.

The optimum environment for chemical weathering is a combination of warm temperatures and abundant moisture. In polar regions, chemical weathering is ineffective because frigid temperatures keep the available moisture locked up as ice, whereas in arid regions there is insufficient moisture to promote rapid chemical weathering.

Human activities often produce pollutants that alter the composition of the atmosphere. Such changes can, in turn, influence the rate of chemical weathering. One well-known example is acid rain (**Figure 6.12**).

Differential Weathering

Masses of rock do not weather uniformly. Take a moment to look back at the photo of Shiprock, New Mexico, in Figure 5.28, page 150. This durable volcanic neck protrudes high above the surrounding terrain.

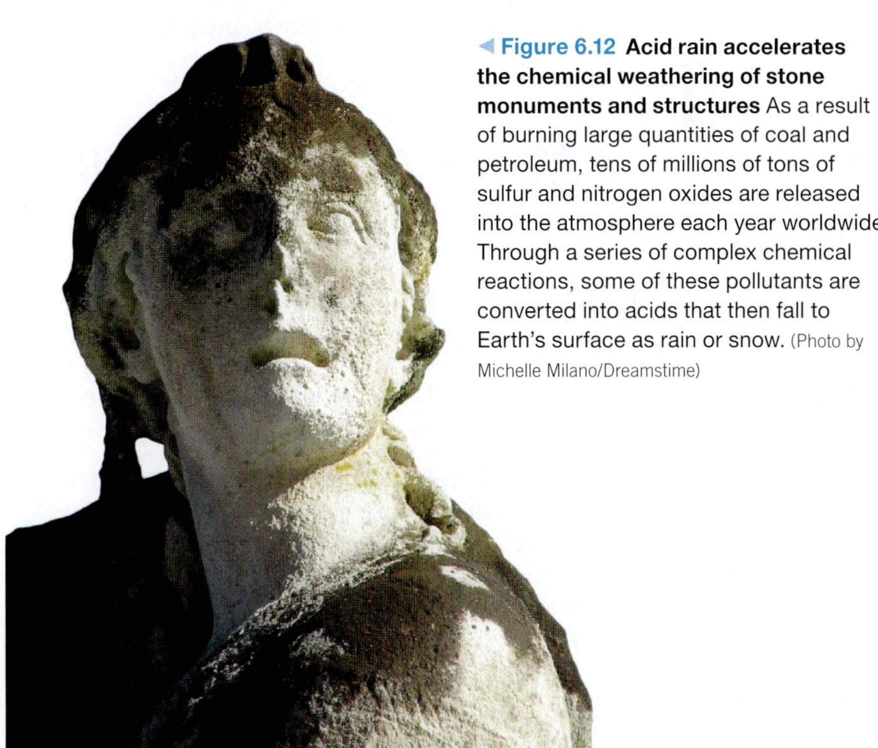

◀ **Figure 6.12** **Acid rain accelerates the chemical weathering of stone monuments and structures** As a result of burning large quantities of coal and petroleum, tens of millions of tons of sulfur and nitrogen oxides are released into the atmosphere each year worldwide. Through a series of complex chemical reactions, some of these pollutants are converted into acids that then fall to Earth's surface as rain or snow. (Photo by Michelle Milano/Dreamstime)

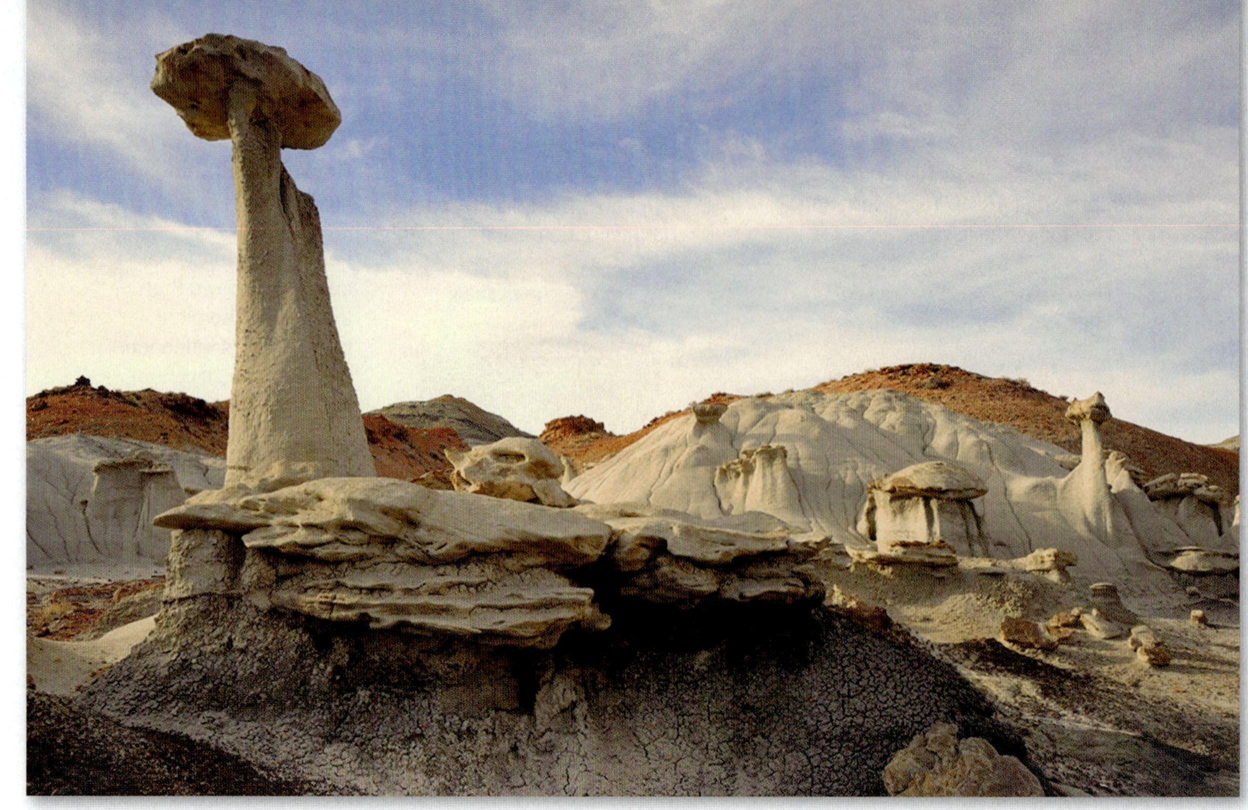

▶ SmartFigure 6.13
Monuments to weathering This example of differential weathering is in New Mexico's Bisti Badlands. When weathering accentuates differences in rocks, spectacular landforms are sometimes created. (Photo by Michael Collier)

MOBILE FIELD TRIP
https://goo.gl/MHpjx0

This phenomenon is called **differential weathering**. The results vary in scale from the rough, uneven surface of the marble headstone in Figure 6.11 to the boldly sculpted exposures of bedrock in New Mexico's Bisti Badlands (**Figure 6.13**).

Differential weathering and subsequent erosion are responsible for creating many unusual, often spectacular rock formations and landforms. Many factors influence the rate of rock weathering. Among the most important are variations in the composition of the rock. More resistant rock protrudes as ridges or pinnacles, or as steeper cliffs on an irregular hillside (see Figure 7.5, page 193). The number and spacing of joints can also be significant factors (see Figure 6.6 and 6.10).

CONCEPT CHECKS 6.4

1. Explain why the headstones in Figure 6.11 have weathered so differently.
2. How does climate influence weathering?

Did You Know?
The moon has no atmosphere, no water, and no biological activity. Therefore, the weathering processes we are familiar with on Earth are lacking on the Moon. However, all lunar terrains are covered with a layer of gray debris, called *lunar regolith*, derived from a few billion years of bombardment by meteorites. The rate of change at the lunar surface is so slow that the footprints left by *Apollo* astronauts will likely remain fresh looking for millions of years.

6.5 Soil: An Indispensable Resource

Define *soil* and explain why soil is referred to as an *interface*. List and briefly discuss five controls of soil formation.

Weathering is a key process in the formation of soil. Along with air and water, soil is one of our most indispensable resources. Also like air and water, soil is often taken for granted. The following quote helps put this vital layer in perspective:

> Science, in recent years, has focused more and more on the Earth as a planet, one that for all we know is unique—where a thin blanket of air, a thinner film of water, and the thinnest veneer of soil combine to support a web of life of wondrous diversity in continuous change.*

*Jack Eddy, "A Fragile Seam of Dark Blue Light," in *Proceedings of the Global Change Research Forum*. U.S. Geological Survey Circular 1086, 1993, p. 15.

Soil has accurately been called "the bridge between life and the inanimate world." All life—the entire biosphere—owes its existence to a dozen or so elements that must ultimately come from Earth's crust. Once weathering and other processes create soil, plants carry out the intermediary role of assimilating the necessary elements and making them available to animals, including humans.

When Earth is viewed as a system, as discussed in Chapter 1, soil is considered an *interface*—a common boundary where different parts of a system interact. This is an appropriate designation because soil forms where the geosphere, the atmosphere, the hydrosphere, and the biosphere meet. Soil is a material that develops in response to complex environmental interactions among different parts of the Earth system. Over time, soil gradually evolves to a state of equilibrium, or balance, with the environment. Soil is dynamic and sensitive to almost every aspect of its surroundings. Thus, when environmental changes occur, such as changes in climate, vegetative cover, and animal (including human) activity, the soil responds. Any such change gradually alters soil characteristics until a new balance is reached. Although thinly distributed over the land surface, soil functions as a fundamental interface, providing an excellent example of the integration among many parts of the Earth system.

What Is Soil?

With few exceptions, Earth's land surface is covered by **regolith**, a layer of rock and mineral fragments produced by weathering. Some would call this material soil, but soil is more than an accumulation of weathered debris. **Soil** is a combination of mineral and organic matter, water, and air—the portion of the regolith that supports the growth of plants. Although the proportions of the major components in soil vary, the same four components are always present to some extent (Figure 6.14). About one-half of the total volume of good-quality surface soil is a mixture of disintegrated and decomposed rock (mineral matter) and **humus**, the decayed remains of animal and plant life (organic matter). The remaining half consists of pore spaces among the solid particles where air and water circulate.

Although the mineral portion of the soil is usually much greater than the organic portion, humus is an essential component. In addition to being an important source of plant nutrients, humus enhances the soil's ability to retain water. Because plants require air and water to live and grow, the portion of the soil consisting of pore spaces that allow these fluids to circulate is as vital as the solid soil constituents.

▲ **Figure 6.14 What is soil?** The pie chart depicts the composition (by volume) of a soil in good condition for plant growth. Although percentages vary, each soil is composed of mineral and organic matter, water, and air. (Photo by i love images/gardening/Alamy Images)

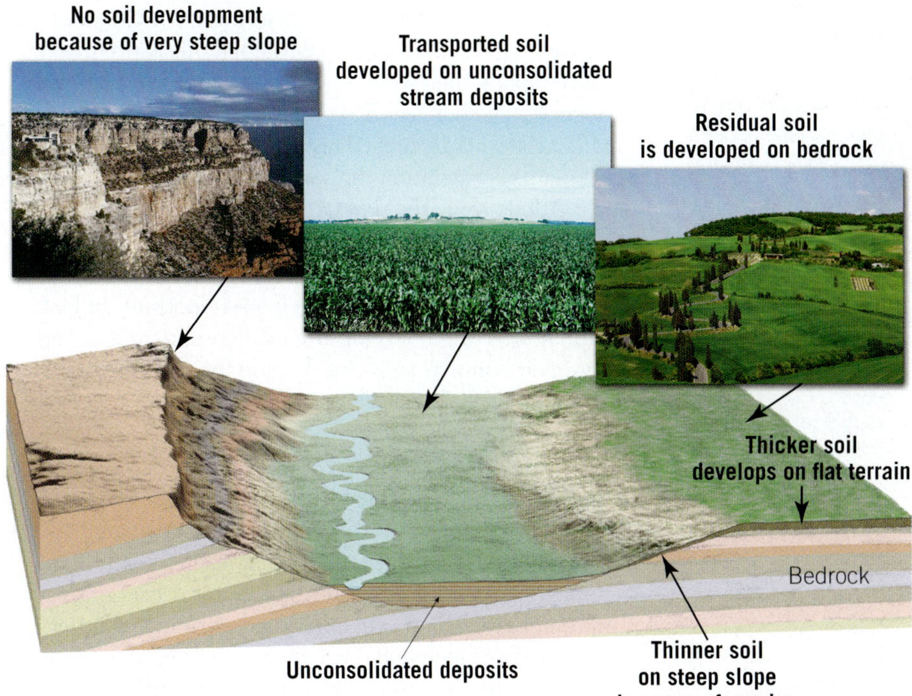

▲ **Figure 6.15 Slopes and soil development** The parent material for residual soils is the underlying bedrock. Transported soils form on unconsolidated deposits. Also note that as slopes become steeper, soil becomes thinner. (Left and center photos by E. J. Tarbuck; right photo by Lucarelli Temistocle/Shutterstock)

Soil water is far from "pure" water; instead, it is a complex solution that contains many soluble nutrients. Soil water not only provides the necessary moisture for the chemical reactions that sustain life, it also supplies plants with nutrients in a form they can use. The pore spaces that are not filled with water contain air. This air is the source of necessary oxygen and carbon dioxide for most microorganisms and plants that live in the soil.

Controls of Soil Formation

Soil is the product of the complex interplay of several factors, including parent material, climate, plants and animals, time, and topography. Although all these factors are interdependent, their roles will be examined separately.

Parent Material The source of the weathered mineral matter from which soils develop is called the **parent material** and is a major factor influencing newly forming soil. Gradually this weathered material undergoes physical and chemical changes as soil formation progresses. Parent material can either be the underlying bedrock or a layer of unconsolidated deposits. When the parent material is bedrock, the soils are termed *residual soils*. By contrast, those developed on unconsolidated sediment are called *transported soils* (Figure 6.15). It should be pointed out that transported soils form *in place* on parent materials that have been carried from elsewhere and deposited by gravity, water, wind, or ice.

Parent material influences soils in two ways. First, the type of parent material influences the rate of weathering and thus the rate of soil formation. Also, because

Did You Know?
Sometimes soils become buried and preserved. Later, if these ancient soils, called *paleosols*, are uncovered, they can provide useful clues to climates and the nature of landscapes thousands or millions of years ago.

unconsolidated deposits are already partly weathered, soil development on such material will likely progress more rapidly than when bedrock is the parent material. Second, the chemical makeup of the parent material will affect the soil's fertility. This influences the character of the natural vegetation the soil can support.

At one time, the parent material was thought to be the primary factor causing differences among soils. However, soil scientists have come to understand that other factors, especially climate, are more important. In fact, similar soils often develop from different parent materials, and dissimilar soils can develop from the same parent material. Such discoveries reinforce the importance of other soil-forming factors.

Climate Climate is considered to be the most influential control of soil formation. Temperature and precipitation are the elements that exert the strongest impact. As noted earlier in this chapter, variations in temperature and precipitation determine whether chemical or mechanical weathering will predominate and also greatly influence the rate and depth of weathering. For instance, a hot, wet climate may produce a thick layer of chemically weathered soil in the same amount of time that a cold, dry climate produces a thin mantle of mechanically weathered debris. Also, the amount of precipitation influences the degree to which various materials are removed from the soil by percolating water (a process called *leaching*), thereby affecting soil fertility. Finally, climatic conditions are an important control on the type of plant and animal life present.

Plants & Animals Plants and animals play a vital role in soil formation. The types and abundance of organisms strongly influence the physical and chemical properties of a soil (Figure 6.16). In fact, for well-developed soils in many regions, the significance of natural vegetation on soil type is frequently implied in the names used by soil scientists, such as *prairie soil*, *forest soil*, and *tundra soil*.

Plants and animals furnish organic matter to the soil. Certain bog soils are composed almost entirely of organic matter, whereas desert soils might contain as little as a small fraction of 1 percent. Although the quantity of organic matter varies substantially among soils, it is a rare soil that completely lacks it.

The primary source of organic matter in soil is plants, although animals and an infinite number of microorganisms also contribute. Decomposed organic matter supplies important nutrients to plants, as well as to animals and microorganisms living in the soil. Consequently, soil fertility is in part related to the amount of organic matter present. Furthermore, the decay of plant and animal remains causes the formation of various organic acids. These complex acids hasten the weathering process. Organic matter also has a high water-holding ability and thus aids water retention in a soil.

Microorganisms, including fungi, bacteria, and single-celled protozoa, play an active role in the decay of plant and animal remains. The end product is humus, a material that no longer resembles the plants and animals from which it is formed. In addition, certain microorganisms aid soil fertility by converting atmospheric nitrogen into soil nitrogen.

▼ Figure 6.16 **Plants influence soil** The nature of the vegetation in an area can have a significant influence on soil formation. (From left to right photos by Bill Brooks/Alamy Images, Nickolay Stanev/Shutterstock, and Elizabeth C. Doemer/Shutterstock)

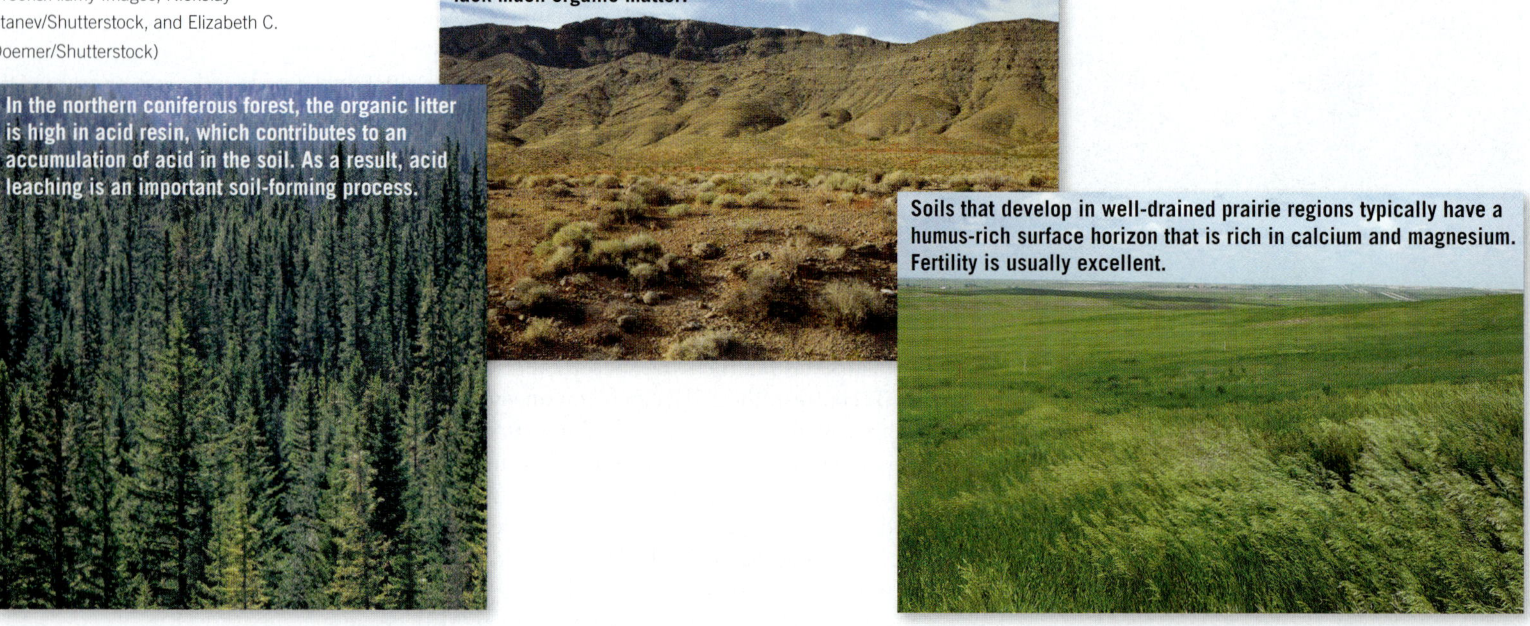

Meager desert rainfall means reduced rates of weathering and relatively meager vegetation. Desert soils are typically thin and lack much organic matter.
In the northern coniferous forest, the organic litter is high in acid resin, which contributes to an accumulation of acid in the soil. As a result, acid leaching is an important soil-forming process.
Soils that develop in well-drained prairie regions typically have a humus-rich surface horizon that is rich in calcium and magnesium. Fertility is usually excellent.

Earthworms and other burrowing animals act to mix the mineral and organic portions of a soil. Earthworms, for example, feed on organic matter and thoroughly mix soils in which they live, often moving and enriching many tons per acre each year. Burrows and holes also aid the passage of water and air through the soil.

Time Time is an important component of *every* geologic process, including soil formation. The nature of soil is strongly influenced by the length of time processes have been operating. If weathering has been going on for a comparatively short time, the character of the parent material strongly influences the characteristics of the soil. As weathering processes continue, the influence of parent material on soil is overshadowed by other soil-forming factors, especially climate. The amount of time required for various soils to evolve cannot be listed because the soil-forming processes act at varying rates under different circumstances. However, as a rule, the longer a soil has been forming, the thicker it becomes and the less it resembles the parent material.

Topography The lay of the land can vary greatly over short distances. Variations in topography can lead to the development of a variety of localized soil types. Many of the differences exist because the length and steepness of slopes significantly affect the amount of erosion and the water content of soil.

On steep slopes, soils are often poorly developed. Due to rapid runoff, the quantity of water soaking in is slight; as a result, the moisture content of the soil may not be sufficient for vigorous plant growth. Further, because of accelerated erosion on steep slopes, the soils are thin or in some cases nonexistent (see Figure 6.15).

In contrast, poorly drained and waterlogged soils found in bottomlands have a much different character. Such soils are usually thick and dark. The dark color results from the large quantity of organic matter that accumulates because saturated conditions retard the decay of vegetation. The optimum terrain for soil development is a flat-to-undulating upland surface. Here we find good drainage, minimum erosion, and sufficient infiltration of water into the soil.

Slope orientation, or the direction a slope is facing, is another consideration. In the midlatitudes of the Northern Hemisphere, a south-facing slope receives a great deal more sunlight than a north-facing slope. In fact, a steep north-facing slope may receive no direct sunlight at all. The difference in the amount of solar radiation received causes differences in soil temperature and moisture, which in turn influence the nature of the vegetation and the character of the soil.

Although this section deals separately with each of the soil-forming factors, remember that all of them work together to form soil. No single factor is responsible for a soil's character; rather, it is the combined influence of parent material, climate, plants and animals, time, and topography that determines a soil's character.

Did You Know?
It usually takes between 80 and 400 years for soil-forming processes to create 1 cm (less than 1/2 in) of topsoil.

CONCEPT CHECKS 6.5

1. Explain why soil is considered an interface in the Earth system.
2. How is regolith different from soil?
3. List the five basic controls of soil formation.
4. Which factor is most influential in soil formation?
5. How might the direction a slope is facing influence soil formation?

6.6 Describing & Classifying Soils

Sketch, label, and describe an idealized soil profile. Explain the need for classifying soils.

The factors controlling soil formation vary greatly from place to place and from time to time, leading to an amazing variety of soil types.

The Soil Profile

Because soil-forming processes operate from the surface downward, soil composition, texture, structure, and color gradually evolve differently at varying depths. These vertical differences, which usually become more pronounced as time passes, divide the soil into zones or layers known as **horizons**. If you were to dig a pit in soil, you would see that its walls are layered. Such a vertical section through all of the soil horizons constitutes the **soil profile**.

Figure 6.17 presents an idealized view of a well-developed soil profile in which five horizons are identified. From the surface downward, they are designated as *O*, *A*, *E*, *B*, and *C*. These five horizons are common to soils in temperate regions:

- *The O soil horizon* consists largely of organic material. In contrast, the layers beneath it consist mainly of mineral matter. The upper portion of the *O* horizon is primarily plant litter, such as loose leaves and other organic debris that is still recognizable. By contrast, the lower portion of the *O* horizon is made up of partly decomposed organic matter (humus) in which plant structures can no longer be identified. In addition to plants, the *O* horizon is teeming with microscopic life, including bacteria, fungi, algae, and insects. All these organisms contribute oxygen, carbon dioxide, and organic acids to the developing soil.

- *The A horizon* is largely mineral matter, yet biological activity is high, and humus is generally present—up to 30 percent in some instances. Together the *O* and *A* horizons make up what is commonly called the *topsoil*.
- *The E horizon* is a light-colored layer that contains little organic material. As water percolates downward through this zone, finer particles are carried away. This washing out of fine soil components is termed **eluviation**. Water percolating downward also dissolves soluble inorganic soil components and carries them to deeper zones. This depletion of soluble materials from the upper soil is termed **leaching**.
- *The B horizon*, or *subsoil*, is where much of the material removed from the *E* horizon by eluviation is deposited. Thus, the *B* horizon is often referred to as the *zone of accumulation*. The accumulation of the fine clay particles enhances water retention in this horizon. In extreme cases clay accumulation can form a very compact and impermeable layer called *hardpan*.
- The *O*, *A*, *E*, and *B* horizons together constitute the **solum**, or "true soil." It is in the solum that the soil-forming processes are active and that living roots and other plant and animal life are largely confined.
- *The C horizon* is characterized by partially altered parent material. Whereas the parent material is difficult to see in the *O*, *A*, *E*, and *B* horizons, it is easily identifiable in the *C* horizon. Although this material is undergoing changes that will eventually transform it into soil, it has not yet crossed the threshold that separates regolith from soil.

Not all soils have these five layers. The characteristics and extent of horizon development vary in different environments. Thus, different localities exhibit soil profiles that can contrast greatly with one another (**Figure 6.18**). The boundaries between soil horizons may be sharp, or the horizons may blend gradually from one to another. Consequently, a well-developed soil profile indicates that environmental conditions have been relatively stable over an extended time span and that the soil is *mature*. By contrast, some soils lack horizons altogether. Such soils are called *immature* because soil building has been going on for only a short time. Immature soils are also characteristic of steep slopes, where erosion continually strips away the soil, preventing full development.

Horizons are indistinct in this soil in Puerto Rico, giving it a relatively uniform appearance.

This profile shows a soil in southeastern South Dakota with well-developed horizons.

▲ **Figure 6.18 Contrasting soil profiles** Soil characteristics and development vary greatly in different environments. (Left photo by USDA; right photo courtesy of E. J. Tarbuck)

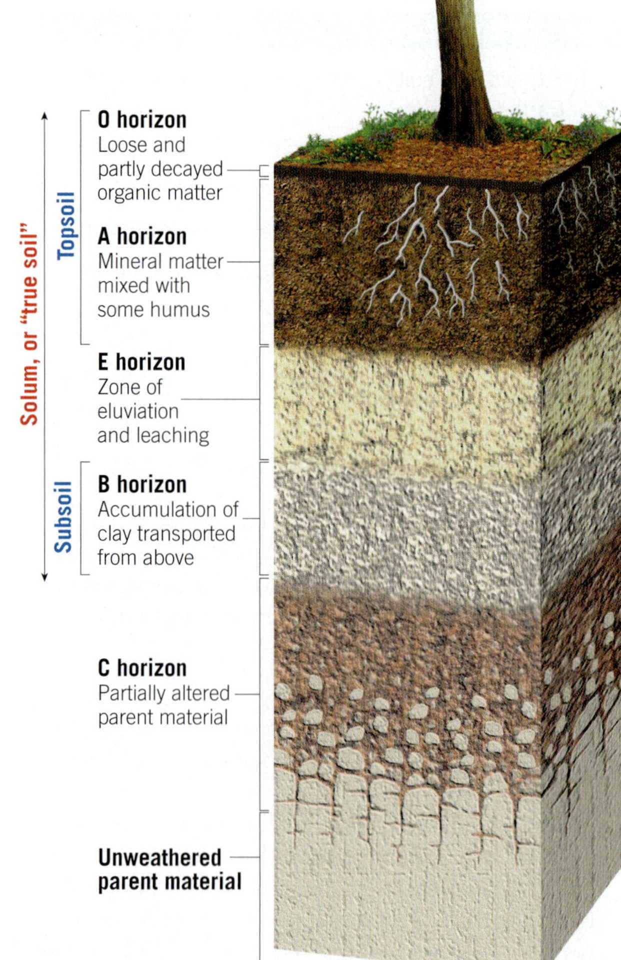

▶ **SmartFigure 6.17**
Soil horizons Idealized soil profile from a humid climate in the middle latitudes.

TUTORIAL
https://goo.gl/ASed3b

Table 6.2 Basic Soil Orders

Soil Order	Description	Percentage*
Alfisol	Moderately weathered soils that form under boreal forests or broadleaf deciduous forests, rich in iron and aluminum. Clay particles accumulate in a subsurface layer in response to leaching in moist environments. Fertile, productive soils because they are neither too wet nor too dry.	9.65
Andisol	Young soils in which the parent material is volcanic ash and cinders, deposited by recent volcanic activity.	0.7
Aridosol	Soils that develop in dry places where there is insufficient water to remove soluble minerals; may have an accumulation of calcium carbonate, gypsum, or salt in subsoil; low organic content.	12.02
Entisol	Young soils having limited development and exhibiting properties of the parent material. Productivity ranges from very high for those formed on recent river deposits to very low for those formed on shifting sand or rocky slopes.	16.16
Gelisol	Young soils with little profile development that occur in regions with permafrost. Low temperatures and frozen conditions for much of the year; slow soil-forming processes.	8.61
Histosol	Organic soils with little or no climatic implications. Can be found in any climate where organic debris can accumulate to form a bog soil. Dark, partially decomposed organic material commonly referred to as *peat*.	1.17
Inceptisol	Weakly developed young soils in which the beginning (inception) of profile development is evident. Most common in humid climates, they exist from the arctic to the tropics. Native vegetation is most often forest.	9.81
Mollisol	Dark, soft soils that have developed under grass vegetation, generally found in prairie areas. Humus-rich surface horizon that is rich in calcium and magnesium. Soil fertility is excellent. Also found in hardwood forests with significant earthworm activity. Climatic range is boreal or alpine to tropical. Dry seasons are normal.	6.89
Oxisol	Soils that occur on old land surfaces unless parent materials were strongly weathered before they were deposited. Generally found in the tropics and subtropical regions. Rich in iron and aluminum oxides, oxisols are heavily leached and hence are poor soils for agricultural activity.	7.5
Spodosol	Soils found only in humid regions on sandy material. Common in northern coniferous forests and cool humid forests. Beneath the dark upper horizon of weathered organic material lies a light-colored horizon of leached material, the distinctive property of this soil.	2.56
Ultisol	Soils that represent the products of long periods of weathering. Water percolating through the soil concentrates clay particles in the lower horizons (argillic horizons). Restricted to humid climates in the temperate regions and the tropics, where the growing season is long. Abundant water and a long frost-free period contribute to extensive leaching and, therefore, poorer soil quality.	8.45
Vertisol	Soils containing large amounts of clay, which shrink upon drying and swell with the addition of water. Found in subhumid to arid climates, provided that adequate supplies of water are available to saturate the soil after periods of drought. Soil expansion and contraction exert stresses on human structures.	2.24

*Percentages refer to the world's ice-free surface.

Classifying Soils

To understand the great variety of soils on Earth makes, scientists needed to devise some means of classifying the vast array of soil data. Establishing categories of items having certain important characteristics in common introduces order and simplicity, which not only aids comprehension and understanding but also facilitates analysis and explanation.

Soil scientists in the United States have devised a system for classifying soils known as the **Soil Taxonomy**. It emphasizes the physical and chemical properties of the soil profile and is organized on the basis of observable soil characteristics. There are 6 hierarchical categories of classification, ranging from *order*, the broadest category, to *series*, the most specific category. The system recognizes 12 soil orders and more than 19,000 soil series.

The names of the classification units are mostly combinations of Latin or Greek descriptive terms. For example, soils of the order aridosol (from the Latin *aridus* = dry and *solum* = soil) are characteristically dry soils in arid regions. Soils in the order inceptisol (from Latin *inceptum* = beginning and *solum* = soil) are soils with only the beginning, or inception, of profile development.

Table 6.2 provides brief descriptions of the 12 basic soil orders. Figure 6.19 shows the complex worldwide distribution pattern of the Soil Taxonomy's 12 soil orders. Like many other classification systems, the Soil Taxonomy is not suitable for every purpose. It is especially useful for agricultural and related land-use purposes, but it is not a useful system for engineers who are preparing evaluations of potential construction sites.

CONCEPT CHECKS 6.6

1. Sketch and label the main soil horizons in a well-developed soil profile.
2. Describe the following features or processes: eluviation, leaching, zone of accumulation, and hardpan.
3. Why are soils classified?
4. Refer to Figure 6.19 and identify three particularly extensive soil orders that occur in the contiguous 48 United States. Describe two soil orders in Alaska.

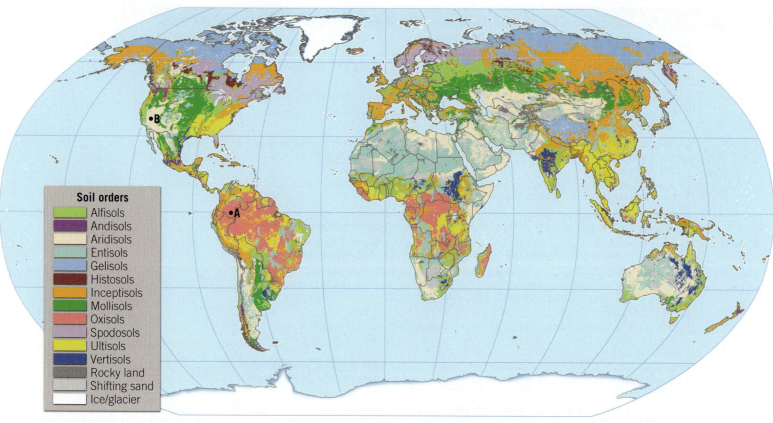

▲ **Figure 6.19 Global soil regions** Worldwide distribution of the Soil Taxonomy's 12 soil orders. Points A and B are references for a *Give It Some Thought* item at the end of the chapter. (Natural Resources Conservation Service/USDA)

6.7 The Impact of Human Activities on Soil

Explain the detrimental impact of human activities on soil.

Soils are just a tiny fraction of all Earth materials, yet they are vital. Soils are necessary for the growth of rooted plants and thus are a basic foundation of the human life-support system. Because soil forms very slowly, it must be considered a nonrenewable resource. Just as human ingenuity can increase the agricultural productivity of soils through fertilization and irrigation, soils can be damaged or destroyed by careless activities. Despite their role in providing food, fiber, and other basic materials, soils are among our most abused resources.

► **Figure 6.20 Tropical deforestation** The thick soils (oxisols) of the Amazon rain forest in Surinam are highly leached. Clearing the tropical rain forest is a serious environmental problem. (Photo by Wesley Bocxe/Science Source)

Clearing the Tropical Rain Forest: A Case Study of Human Impact on Soil

Over the past few decades, the destruction of tropical forests has become a serious environmental issue. Each year millions of acres are cleared for agriculture and logging (**Figure 6.20**). This clearing results in soil degradation, loss of biodiversity, and climate change.

Thick red-orange soils (oxisols) are common in the wet tropics and subtropics (see Figure 6.19). They are the end product of extreme chemical weathering. Because lush tropical rain forests are associated with these soils, many people assume that they are fertile and have great potential for agriculture. However, just the opposite is true: Oxisols are among the poorest soils for farming. How can this be?

Rain forest soils develop under conditions of high temperature and heavy rainfall, and they are therefore severely leached. Not only does leaching remove the soluble materials such as calcium carbonate, but the great quantities of percolating rainwater also remove much of the silica, and as a result, insoluble oxides of iron and aluminum become concentrated in the soil. Iron oxides give the soil its distinctive color. Because bacterial activity is high in the wet tropics, organic matter quickly breaks down, so rain forest soils contain very little humus. Moreover, leaching destroys fertility because most plant nutrients are removed by the large volume of downward-percolating water. Therefore, despite the dense and luxuriant vegetation, the soil itself contains few available nutrients.

Most nutrients that support the rain forest are locked up in the trees. As vegetation dies and decomposes, the roots of the rain forest trees quickly absorb the nutrients before they are leached from the soil. The nutrients are continuously recycled as trees die and decompose. Therefore, when forests are cleared to provide land for farming or to harvest timber, most of the nutrients are removed as well. What remains is a soil that contains little to nourish planted crops.

Rain forest clearing not only removes plant nutrients but also accelerates soil erosion. The roots of rain forest vegetation anchor the soil, and leaves and branches provide a canopy that protects the ground by deflecting the full force of the frequent heavy rains. When the protective vegetation is gone, soil erosion increases.

The removal of vegetation also exposes the ground to strong direct sunlight. When baked by the Sun, these tropical soils can harden to a bricklike consistency and become practically impenetrable to water and crop roots. In just a few years, a freshly cleared area may no longer be cultivable.

Soil Erosion: Losing a Vital Resource

Many people do not realize that soil erosion—the removal of topsoil—is a serious environmental problem. Perhaps this is the case because a substantial amount of soil seems to remain even where soil erosion is serious. Nevertheless, although the loss of fertile topsoil may not be obvious to the untrained eye, it is a significant and growing problem as human activities expand and disturb more and more of Earth's surface.

Raindrops may strike the surface at velocities approaching 35 km per hour. When a drop strikes an exposed surface, soil particles may splash as high as one meter and land more than a meter away from the point of raindrop impact.

Figure 6.21 Raindrop impact Soil dislodged by raindrop impact is more easily moved by sheet erosion. (Photo courtesy U.S. Department of the Navy/Soil Conservation Service/USDA)

Soil erosion is a natural process; it is part of the constant recycling of Earth materials that we call the *rock cycle*. Once soil forms, erosional forces, especially water and wind, move soil components from one place to another. Every time it rains, raindrops strike the land with surprising force (Figure 6.21). Each drop acts like a tiny bomb, blasting movable soil particles out of their positions in the soil mass. Then, water flowing across the surface carries away the dislodged soil particles. Because the soil is moved by thin sheets of water, this process is termed *sheet erosion*.

After the water flows as a thin, unconfined sheet for a relatively short distance, threads of current typically develop, and tiny channels called *rills* begin to form. Still deeper cuts in the soil, known as *gullies*, are created as rills enlarge (Figure 6.22). When normal farm cultivation cannot eliminate the channels, we know the rills have

Did You Know?
It has been estimated that between 3 and 5 million acres of prime U.S. farmland are lost each year through mismanagement (including soil erosion) and conversion to nonagricultural uses. According to the United Nations, since 1950 more than one-third of the world's farmable land has been lost to soil erosion.

Severe sheet and rill erosion on an Iowa farm following heavy rains. Just one millimeter of soil from a single acre amounts to about 5 tons.

A.

Gully erosion on unprotected soil on a Wisconsin farm.

B.

Figure 6.22 Soil erosion on unprotected soils A. Sheetflow and rills. **B.** Rills can grow into deep gullies. (Photo A by Lynn Betts/NRCS; photo B by D. P. Burnside/Science Source)

The man is pointing to where the ground surface was when the grasses began to grow.
Wind erosion lowered the land surface to the level of his feet.
Clumps of anchored soil
Unanchored soil
Sand dune
1.2 meters

▲ **Figure 6.23 Wind erosion** When the land is dry and largely unprotected by anchoring vegetation, soil erosion by wind can be significant. (Photo courtesy Natural Resources Conservation Service/USDA)

grown large enough to be called gullies. Although most dislodged soil particles move only a short distance during each rainfall, substantial quantities eventually leave the fields and make their way downslope to a stream. Once in the stream channel, these soil particles, which can now be called *sediment*, are transported downstream and eventually deposited.

Rates of Erosion We know that soil erosion is the ultimate fate of practically all soils. In the past, erosion occurred at slower rates than it does today because more of the land surface was covered and protected by trees, shrubs, grasses, and other plants. However, human activities such as farming, logging, and construction, which remove or disrupt the natural vegetation, have greatly accelerated the rate of soil erosion. Without the stabilizing effect of plants, the soil is more easily swept away by the wind or carried downslope by sheet wash.

Natural rates of soil erosion vary greatly from one place to another and depend on soil characteristics as well as factors such as climate, slope, and type of vegetation. Over a broad area, erosion caused by surface runoff may be estimated by determining how much sediment is carried by the streams that drain the region. Studies of this kind made on a global scale indicate that prior to the appearance of humans, sediment transport by rivers to the ocean amounted to just over 9 billion metric tons per year. In contrast, the amount of material currently transported to the sea by rivers is about 24 billion metric tons per year, or more than two and a half times the earlier rate.

It is estimated that flowing water is responsible for about two-thirds of the soil erosion in the United States. Much of the remainder is caused by wind. When dry conditions prevail, strong winds can remove large quantities of soil from unprotected fields (**Figure 6.23**). At present, it is estimated that topsoil is eroding faster than it forms on more than one-third of the world's croplands. The results are lower productivity, poorer crop quality, reduced agricultural income, and an ominous future.

The 1930s Dust Bowl During the 1930s, large dust storms plagued the Great Plains. Because of the size and severity of these storms, the region came to be called the *Dust Bowl*, and the time period was called the Dirty Thirties. The heart of the Dust Bowl was nearly 100 million acres in the panhandles of Texas and Oklahoma and adjacent parts of Colorado, New Mexico, and Kansas. At times, dust storms were so severe that they were called "black blizzards" and "black rollers" because visibility was sometimes reduced to only a few feet (**Figure 6.24**).

What caused the Dust Bowl? Clearly, the fact that portions of the Great Plains experience some of North America's strongest winds is important. However, it was the expansion of agriculture during an unusually wet period that set the stage for the disastrous period of soil erosion. Mechanization allowed the rapid transformation of the grass-covered prairies of this semiarid region into farms. As long as precipitation was adequate, the soil remained in place. However,

▲ **Figure 6.24 1930s Dust Bowl** Dust blackens the sky near Elkhart, Kansas, on May 21, 1937. Large dust storms like this one stripped topsoil from large parts of the Great Plains during the dry 1930s. In places, dust drifted like snow, covering farm buildings, fences, and fields. (Photo courtesy of the Library of Congress)

when a prolonged drought struck in the 1930s, the unprotected soils were vulnerable to the wind. Severe soil loss, crop failure, and economic hardship resulted.

Controlling Soil Erosion On every continent, unnecessary soil loss is occurring because appropriate conservation measures are not being taken. Although it is a recognized fact that soil erosion can never be completely eliminated, soil conservation programs can substantially reduce the loss of this basic resource.

Steepness of slope is an important factor in soil erosion. The steeper the slope, the faster the water runs off and the greater the erosion. It is best to leave steep slopes undisturbed. When such slopes are farmed, terraces can be constructed. These nearly flat, steplike surfaces slow runoff and thus decrease soil loss while allowing more water to soak into the ground.

Soil erosion by water also occurs on gentle slopes. Figure 6.25 illustrates one conservation method in which crops are planted parallel to the contours of the slope. This pattern reduces soil loss by slowing runoff. Strips of grass or cover crops such as hay slow runoff even more and act to promote water infiltration and trap sediment.

Creating grassed waterways is another common practice (Figure 6.26). Natural drainageways are shaped to form smooth, shallow channels and then planted with grass. The grass prevents the formation of gullies and traps soil washed from cropland. Frequently crop residues are also left on fields. This debris protects the surface from both water and wind erosion. To protect fields from excessive wind erosion, rows of trees and shrubs are planted as windbreaks that slow the wind and deflect it upward (Figure 6.27).

▲ **Figure 6.25 Soil conservation** Crops on a farm in northeastern Iowa are planted to decrease water erosion. (Photo courtesy of Erwin C. Cole/USDA/NRCS)

CONCEPT CHECKS 6.7

1. Why are soils in tropical rain forests not well suited for farming?
2. Place these phenomena related to soil erosion in the proper sequence: sheet erosion, gullies, rain drop impact, rills, stream.
3. Explain how human activities have affected the rate of soil erosion.
4. Briefly describe three ways to control soil erosion.

Did You Know?
In the United States it is estimated that the amount of soil annually eroded from farms exceeds the amount of soil that forms by more than 2 billion tons.

▲ **Figure 6.26 Reducing erosion by water** Grassed waterway on a Pennsylvania farm. (Photo courtesy Bob Nichols/NRCS/USDA)

These flat expanses are susceptible to wind erosion, especially when the fields are bare. The rows of trees slow and deflect the wind, which decreases the loss of top soil.

▲ **Figure 6.27 Reducing wind erosion** Windbreaks protect wheat fields in North Dakota. (Photo courtesy Natural Resources Conservation Service/USDA)

6.8 Weathering & Ore Deposits

Relate weathering to the formation of certain ore deposits.

Weathering creates many important mineral deposits by concentrating minor amounts of metals that are scattered through unweathered rock into economically valuable concentrations. Such a transformation, termed **secondary enrichment**, takes place in one of two ways. In one situation, chemical weathering coupled with downward-percolating water removes undesired materials from decomposing rock, leaving the desired elements enriched in the upper zones of the soil. The second way is basically the reverse of the first. That is, the desirable elements that are found in low concentrations near the surface are removed and carried to lower zones, where they are redeposited and become more concentrated.

Bauxite

Did You Know?
Bauxite is a useful indicator of past climates because it records periods of wet tropical climate in the geologic past.

The formation of *bauxite*, the principal ore of aluminum, is one important example of an ore created as a result of enrichment by weathering processes (**Figure 6.28**). Although aluminum is the third most abundant element in Earth's crust, economically valuable concentrations of this important metal are not common because most aluminum is tied up in silicate minerals, from which it is extremely difficult to extract.

▲ **Figure 6.28 Bauxite** This ore of aluminum forms as a result of weathering processes under tropical conditions. Its color varies from red or brown to nearly white.
(Photo by E. J. Tarbuck)

Bauxite forms in rainy tropical climates. When aluminum-rich source rocks are subjected to the intense and prolonged chemical weathering of the tropics, most of the common elements, including calcium, sodium, and potassium, are removed by leaching. Because aluminum is extremely insoluble, it becomes concentrated in the soil (as bauxite, a hydrated aluminum oxide). Thus, the formation of bauxite depends on climatic conditions in which chemical weathering and leaching are pronounced, plus, of course, the presence of aluminum-rich source rock. In a similar manner, important deposits of nickel and cobalt develop from igneous rocks rich in silicate minerals such as olivine.

There is significant concern regarding the mining of bauxite and other residual deposits because they tend to occur in environmentally sensitive areas of the tropics. Mining is preceded by the removal of tropical vegetation, thus destroying rain forest ecosystems. Moreover, the thin moisture-retaining layer of organic matter is also disturbed. When the soil dries out in the hot Sun, as has been mentioned, it becomes brick-like and loses its moisture-retaining qualities. Such soil cannot be productively farmed, nor can it support significant forest growth. The long-term consequences of bauxite mining are clearly of concern for developing countries in the tropics, where this important ore is mined.

Other Deposits

Many copper and silver deposits result when weathering processes concentrate metals that are dispersed through a low-grade primary ore. Usually such enrichment occurs in deposits containing pyrite (FeS_2), the most common and widespread sulfide mineral. Pyrite is important because when it chemically weathers, sulfuric acid forms, which enables percolating waters to dissolve the ore metals. Once dissolved, the metals gradually migrate downward through the primary ore body until they are precipitated. Deposition takes place because of changes that occur in the chemistry of the solution when it reaches the groundwater zone (the zone beneath the surface, where all pore spaces are filled with water). In this manner, the small percentage of dispersed metal can be removed from a large volume of rock and redeposited as a higher-grade ore in a smaller volume of rock.

CONCEPT CHECKS 6.8

1. How can weathering create an ore deposit?
2. Name an important ore that is associated with weathering processes.

CONCEPTS IN REVIEW
Weathering & Soils

6.1 Weathering

Define *weathering* and distinguish between the two main categories of weathering.

KEY TERMS: weathering, mechanical weathering, chemical weathering

- Weathering is the disintegration and decomposition of rocks on the surface of Earth. Rocks might break into many smaller pieces through physical processes called mechanical weathering. Rocks can also decompose as minerals react with environmental agents such as oxygen and water to produce new substances that are stable at Earth's surface. This is called chemical weathering.

? Would a shattered window be an example of mechanical weathering or chemical weathering? What about a rusty bicycle?

6.2 Mechanical Weathering

List and describe four examples of mechanical weathering.

KEY TERMS: frost wedging, sheeting, exfoliation dome, joint

- Mechanical weathering forces include the expansion of ice, the crystallization of salt, and the growth of plant roots. All work to pry apart grains and enlarge fractures.
- Rocks that form under lots of pressure deep in Earth expand when they are exposed at the surface, and sometimes this expansion is great enough to cause the rock to break into onion-like layers. This sheeting can generate broad dome-shaped exposures of rock called exfoliation domes.

? This is a close-up view of a massive granite feature in the Sierra Nevada of California. Notice the thin slabs of granite that are separating from this rock mass. Describe the process that caused this to occur. What term describes the dome-like feature that results?

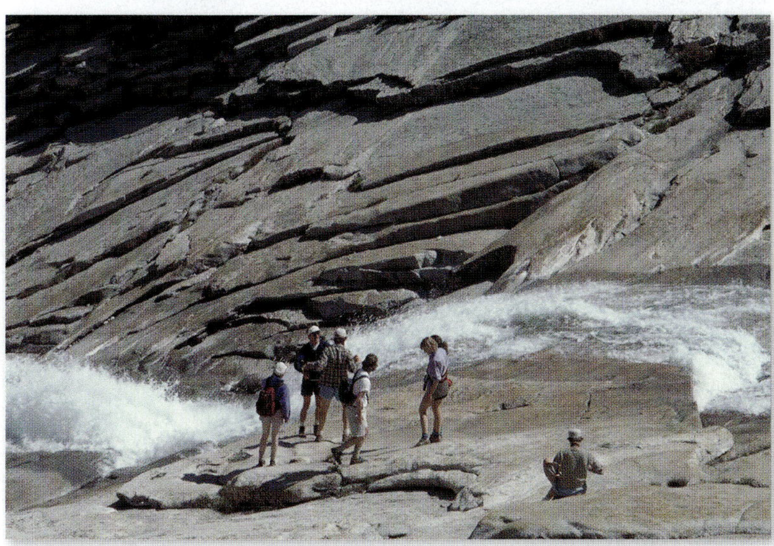

Marli Miller

6.3 Chemical Weathering

Discuss the importance of water and carbonic acid in chemical weathering processes.

KEY TERMS: oxidation, carbonic acid, spheroidal weathering

- Water is by far the most important agent of chemical weathering. Oxygen dissolved in water oxidizes iron-rich minerals. Carbon dioxide (CO_2) dissolved in water forms carbonic acid, which attacks and alters rock.
- The chemical weathering of silicate minerals produces soluble products containing sodium, calcium, potassium, and magnesium ions; silica in solution; insoluble iron oxides such as hematite and limonite; and clay minerals.
- Spheroidal weathering results when sharp edges and corners of rocks are chemically weathered more rapidly than flat rock faces. The higher proportion of surface area for a given volume of rock at the edges and corners means there is more mineral material exposed to chemical attack. Faster weathering at the corners produces weathered rocks that become increasingly sphere shaped over time.

? This sample of granite is rich in potassium feldspar and quartz. There are minor amounts of hornblende. If this rock were to undergo chemical weathering, how would its minerals change? List the products you would expect to result. Would all of the minerals decompose? If not, which mineral would likely remain relatively intact?

E.J. Tarbuck

6.4 Rates of Weathering

Summarize the factors that influence the type and rate of rock weathering.

KEY TERM: differential weathering

- Some types of rocks are more stable at Earth's surface than others, due to the minerals they contain. Different minerals break down at different rates under the same conditions. Quartz is the most stable silicate mineral, while minerals high in Bowen's reaction series such as olivine tend to decompose more rapidly.
- Rock weathers most rapidly in an environment with lots of heat to drive reactions and water to facilitate those reactions. Consequently, rocks decompose relatively quickly in hot, wet climates and slowly in cold, dry conditions.
- Frequently, rocks exposed at Earth's surface do not weather at the same rate. This differential weathering of rocks is influenced by factors such as mineral composition and degree of jointing. In addition, if a rock mass is protected from weathering by another, more resistant rock, then it will weather at a slower rate than a fully exposed equivalent rock. Differential weathering produces many of our most spectacular landforms.

6.5 Soil: An Indispensable Resource

Define *soil* and explain why soil is referred to as an *interface.* List and briefly discuss five controls of soil formation.

KEY TERMS: regolith, soil, humus, parent material

- Soils are vital combinations of organic and inorganic components found at the interface where the geosphere, atmosphere, hydrosphere, and biosphere meet. This dynamic zone is the overlap between different parts of the Earth system. It includes the regolith's rocky debris, mixed with humus, water, and air.
- Residual soils form in place due to the weathering of bedrock, whereas transported soils develop on unconsolidated sediment.
- Soils formed in different climates are different in part due to temperature and moisture differences but also due to the organisms that live in those different environments. These organisms can add organic matter or chemical compounds to the developing soil or can help mix the soil through their growth and movement.
- It takes time for soil to form. Soils that have developed for a longer period of time will have different characteristics than young soils. In addition, some minerals break down more readily than others. Soils produced from the weathering of different parent rocks are produced at different rates.
- The steepness of the slope on which a soil is forming is a key variable, with shallow slopes retaining their soils and steeper slopes shedding them to accumulate elsewhere.

? Label the four components of a soil on this pie chart.

6.6 Describing & Classifying Soils

Sketch, label, and describe an idealized soil profile. Explain the need for classifying soils.

KEY TERMS: horizon, soil profile, eluviation, leaching, solum, Soil Taxonomy

- Despite the great diversity of soils around the world, there are some broad patterns to the vertical anatomy of soil layers. Organic material, called humus, is added at the top (*O* horizon), mainly from plant sources. There, it mixes with mineral matter (*A* horizon). At the bottom, bedrock breaks down and contributes mineral matter (*C* horizon). In between, some materials are leached out or eluviated from higher levels (*E* horizon) and transported to lower levels (*B* horizon), where they may form an impermeable layer called hardpan.
- The need to bring order to huge quantities of data motivated the establishment of a classification scheme for the world's soils. This Soil Taxonomy features 12 broad orders.

? Which soil order would likely contain a higher proportion of humus: inceptisol or histosol? Which soil order would be more likely found in Brazil: gelisol or oxisol?

6.7 The Impact of Human Activities on Soil

Explain the detrimental impact of human activities on soil.

- The clearing of tropical rain forests is an issue of concern. Most of the nutrients in the tropical rain forest ecosystem are not in the soil but in the trees. When the trees are removed from the rain forest, most of the nutrients are removed as well. The loss of vegetation also makes the soils highly susceptible to erosion. Once cleared of vegetation, soils may also be baked by the Sun into a bricklike consistency.
- Soil erosion is a natural process, part of the constant recycling of Earth materials that we call the rock cycle. Human activities have increased soil erosion rates over the past several hundred years. Because natural soil production rates are constant, there is a net loss of soil at a time when a record-breaking number of people live on the planet.
- Using windbreaks, terracing, installing grassed waterways, and plowing the land along horizontal contour lines are all practices that have been shown to reduce soil erosion.

? Why was this row of evergreens planted on an Indiana farm?

Edwin C. Cole/NCRS

6.8 Weathering & Ore Deposits

Relate weathering to the formation of certain ore deposits.

KEY TERM: secondary enrichment

- Weathering creates ore deposits by concentrating minor amounts of metals into economically valuable deposits. The process, often called secondary enrichment, is accomplished by either removing undesirable materials and leaving the desired elements enriched in the upper zones of the soil or removing and carrying the desirable elements to lower soil zones, where they are redeposited and become more concentrated.
- Bauxite, the principal ore of aluminum, is one important ore created as a result of enrichment by weathering processes. In addition, many copper and silver deposits result when weathering processes concentrate metals that were formerly dispersed through low-grade primary ore.

GIVE IT SOME THOUGHT

1 How are the two main categories of weathering represented in this image that shows human-made objects?

Michael Collier

2 Describe how plants promote mechanical and chemical weathering but inhibit erosion.

3 Granite and basalt are exposed at Earth's surface in a hot, wet region. Will mechanical weathering or chemical weathering predominate? Which rock will weather more rapidly? Why?

4 The accompanying photo shows Shiprock, a well-known landmark in the northwestern corner of New Mexico. It is a mass of igneous rock that represents the "plumbing" of a now vanished volcanic feature. Extending toward the upper left is a related wall-like igneous structure known as a dike. The igneous features are surrounded by sedimentary rocks. Explain why these once deeply buried igneous features now stand high above the surrounding terrain. What term in Section 6.4, "Rates of Weathering," applies to this situation?

Michael Collier

5 Due to burning of fossil fuels such as coal and petroleum, the level of carbon dioxide (CO_2) in the atmosphere has been increasing for more than 150 years. Should this increase tend to accelerate or slow down the rate of chemical weathering of Earth's surface rocks? Explain how you arrived at your conclusion.

6 In Chapter 4, you learned that feldspars are very common minerals in igneous rocks. When you learn about the common minerals that compose sedimentary rocks in Chapter 7, you will find that feldspars are relatively rare. Applying what you have learned about chemical weathering, explain why this is true. Based on this explanation, what mineral might you expect to be common in sedimentary rocks that is not found in igneous rocks?

7 The accompanying photo shows a footprint on the Moon left by an *Apollo* astronaut in material popularly called *lunar soil*. Does this material satisfy the definition we use for soil on Earth? Explain why or why not. You may want to refer to Figure 6.14.

NASA

8 What might cause different soils to develop from the same kind of parent material or similar soils to form from different parent materials?

9 Using the map of global soil regions in Figure 6.19, identify the main soil order in the region adjacent to South America's Amazon River (point A on the map) and the predominant soil order in the American Southwest (point B). Briefly contrast these soils. Do they have anything in common? Referring to Table 6.2 might be helpful.

10 This soil sample is from a farm in the Midwest. From which horizon was the sample most likely taken—*A*, *E*, *B*, or *C*? Explain.

Lynn Betts/NRCS

MasteringGeology™

Looking for additional review and test prep materials? Visit the Study Area in MasteringGeology to enhance your understanding of this chapter's content by accessing a variety of resources, including Self-Study Quizzes, Geoscience Animations, SmartFigures, Mobile Field Trips, *Project Condor* Quadcopter videos, *In the News* RSS feeds, flashcards, web links, and an optional Pearson eText.

www.masteringgeology.com

7 Sedimentary Rocks

FOCUS ON CONCEPTS

Each statement represents the primary learning objective for the corresponding major heading within the chapter. After you complete the chapter, you should be able to:

7.1 Explain the importance of sedimentary rocks and summarize the part of the rock cycle that pertains to sediments and sedimentary rocks. List the three categories of sedimentary rocks.

7.2 Describe the primary basis for distinguishing among detrital rocks and discuss how the origin and history of such rocks might be determined.

7.3 Explain the processes involved in the formation of chemical sedimentary rocks and list several examples.

7.4 Outline the successive stages in the formation of coal.

7.5 Describe the processes that convert sediment into sedimentary rock and other changes associated with burial.

7.6 Summarize the criteria used to classify sedimentary rocks.

7.7 Distinguish among three broad categories of sedimentary environments and provide an example of each. List several sedimentary structures and explain why these features are useful to geologists.

7.8 Distinguish between the two broad groups of nonmetallic mineral resources. Discuss the three important fossil fuels associated with sedimentary rocks.

7.9 Relate weathering processes and sedimentary rocks to the carbon cycle.

These eroded sedimentary layers are part of the Vermillion Cliffs, near Badger Creek in northern Arizona. Iron oxide colors some of the layers red. (Photo by Michael Collier)

CHAPTER 6 PROVIDED THE BACKGROUND you need to understand the origin of sedimentary rocks. Recall that weathering of existing rocks begins the process. Next, gravity and agents of erosion such as running water, wind, and glacial ice remove the products of weathering and carry them to a new location, where they are deposited. Usually the particles are broken down further during this transport phase. Following deposition, this material, which is now called *sediment*, becomes lithified (turned to rock). It is from sedimentary rocks that geologists reconstruct many details of Earth's history. Because sediments are deposited in a variety of settings at the surface, the rock layers that they eventually form hold many clues about past surface environments. A layer may represent a desert sand dune, the muddy floor of a swamp, or a tropical coral reef. There are many possibilities. Many sedimentary rocks are associated with important energy and mineral resources and are therefore important economically as well.

7.1 An Introduction to Sedimentary Rocks

Explain the importance of sedimentary rocks and summarize the part of the rock cycle that pertains to sediments and sedimentary rocks. List the three categories of sedimentary rocks.

Most of the solid Earth consists of igneous and metamorphic rocks. Geologists estimate that these two categories represent 90 to 95 percent of the outer 16 kilometers (10 miles) of the crust. Nevertheless, most of Earth's solid surface consists of either sediment or sedimentary rock.

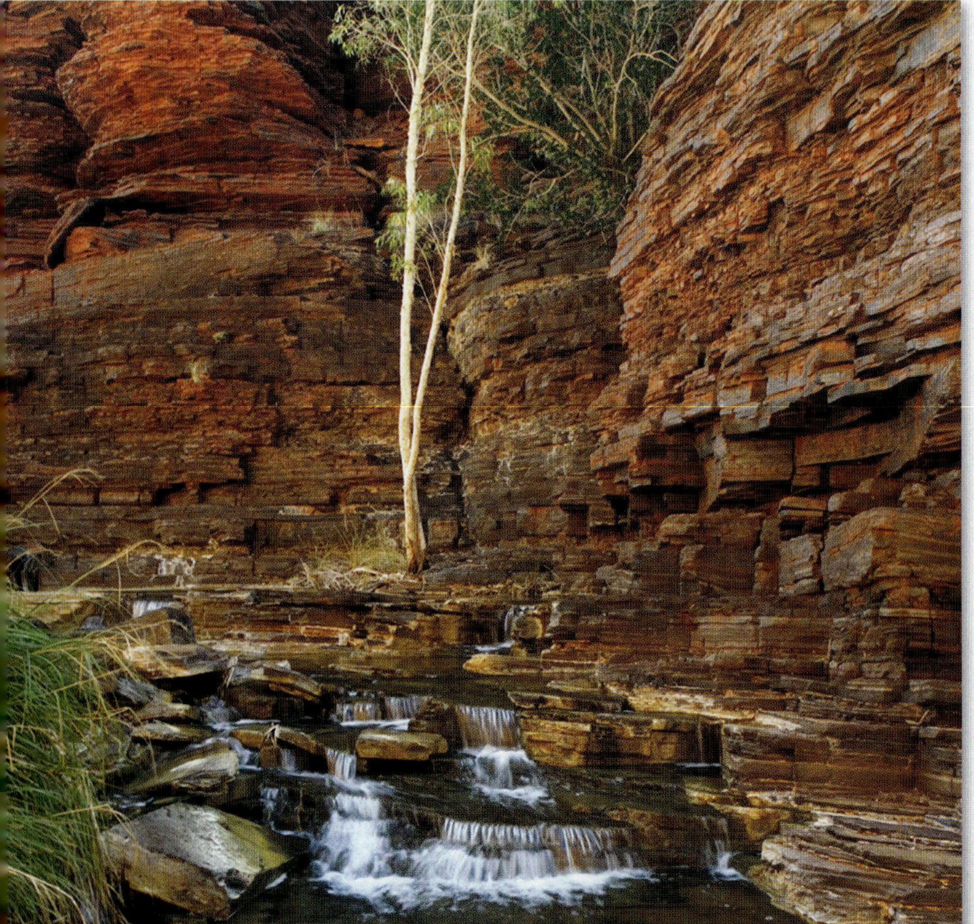

Importance

About 75 percent of land areas are covered by sediments and sedimentary rocks. Across the ocean floor, which represents about 70 percent of Earth's solid surface, virtually everything is covered by sediment. Igneous rocks are exposed only at the crests of mid-ocean ridges and in some volcanic areas. Thus, while sediment and sedimentary rocks make up only a small percentage of Earth's crust, they are concentrated at or near the surface—the interface among the geosphere, hydrosphere, atmosphere, and biosphere. Because of this unique position, sediments and the rock layers that they eventually form contain evidence of past conditions and events at the surface. Based on the compositions, textures, structures, and fossils in sedimentary rocks, experienced geologists can decipher clues that provide insights into past climates, ecosystems, and ocean environments. Furthermore, by studying sedimentary rocks, geologists can reconstruct the configuration of ancient landmasses and the locations and compositions of long-vanished mountain systems. In short, this group of rocks provides geologists with much of the basic information needed to reconstruct the details of Earth history (**Figure 7.1**).

The study of sedimentary rocks has economic significance as well. Coal, which provides a significant portion of our electrical energy, is classified as a sedimentary rock.

◄ **Figure 7.1 Sedimentary rocks record change** Because they contain fossils and other clues about the geologic past, sedimentary rocks are important in the study of Earth history. Vertical changes in rock types represent environmental changes through time. These strata are exposed at Karijini National Park, Western Australia. (Photo by S. Sailer/A. Sailer/AGE Fotostock)

Moreover, other major energy sources—including oil, natural gas, and uranium—are derived from sedimentary rocks. So are major sources of iron, aluminum, manganese, and phosphate fertilizer, plus numerous materials that are essential to the construction industry, such as cement and aggregate. Sediments and sedimentary rocks are also the primary reservoir of groundwater. Thus, an understanding of this group of rocks and the processes that form and modify them is basic to locating and maintaining supplies of many important resources.

Origins

Like other rocks, the **sedimentary rocks** that we see around us and use in so many different ways have their origin in the rock cycle. Figure 7.2 illustrates the portion of the rock cycle that occurs near Earth's surface—the part that pertains to sediments and sedimentary rocks. A brief overview of these processes provides a useful perspective:

- Weathering begins the process. It involves the physical disintegration and chemical decomposition of preexisting igneous, metamorphic, and sedimentary rocks. Weathering generates a variety of products subject to erosion, including various solid particles and ions in solution. These are the raw materials for sedimentary rocks.
- Soluble constituents are dissolved and carried away by runoff and groundwater. Solid particles are frequently moved downslope by gravity, a process termed *mass wasting*, before running water, groundwater,

▼ **SmartFigure 7.2 The big picture** Beginning with weathering, this diagram outlines the portion of the rock cycle that pertains to the formation of sedimentary rocks.

TUTORIAL
https://goo.gl/ZuDYza

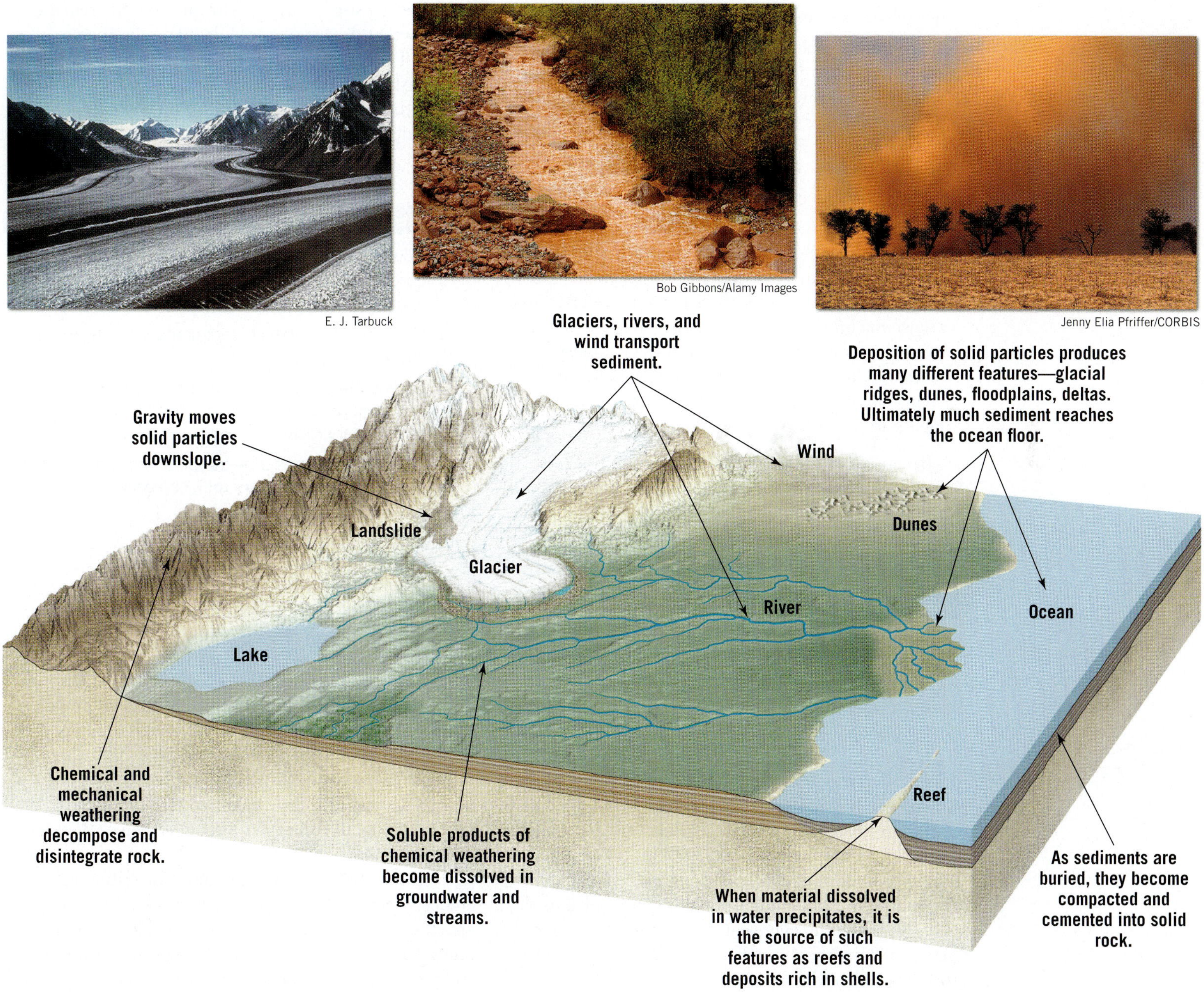

E. J. Tarbuck

Bob Gibbons/Alamy Images

Jenny Elia Pfriffer/CORBIS

Glaciers, rivers, and wind transport sediment.
Deposition of solid particles produces many different features—glacial ridges, dunes, floodplains, deltas. Ultimately much sediment reaches the ocean floor.
Gravity moves solid particles downslope.
Wind
Landslide
Dunes
Glacier
River
Ocean
Lake
Reef
Chemical and mechanical weathering decompose and disintegrate rock.
Soluble products of chemical weathering become dissolved in groundwater and streams.
When material dissolved in water precipitates, it is the source of such features as reefs and deposits rich in shells.
As sediments are buried, they become compacted and cemented into solid rock.

wave activity, wind, and glacial ice remove them. These agents of transport, covered in detail in later chapters, move these materials from the sites where they originated to locations where they accumulate. The transport of sediment is usually intermittent. For example, during a flood, a rapidly moving river moves large quantities of sand and gravel. As the floodwaters recede, particles are temporarily deposited, only to be moved again by a subsequent flood.

- Deposition of solid particles occurs when wind and water currents slow down and as glacial ice melts. The word *sedimentary* actually refers to this process. It is derived from the Latin *sedimentum*, which means "to settle," a reference to solid material settling out of a fluid (water or air). The mud on the floor of a lake, a delta at the mouth of a river, a gravel bar in a streambed, the particles in a desert sand dune, and even household dust are examples.
- The deposition of material dissolved in water is not related to the strength of water currents. Rather, ions in solution are removed when chemical or temperature changes cause material to crystallize and precipitate (solidify out of a liquid solution) or when organisms remove dissolved material to build hard parts such as shells.
- As deposition continues, older sediments are buried beneath younger layers and gradually converted to sedimentary rock (lithified) by compaction and cementation. This and other changes are referred to as *diagenesis* (*dia* = change; *genesis* = origin), a collective term for all the changes (short of metamorphism, discussed in Chapter 8) that take place in texture, composition, and other physical properties after sediments are deposited.

Because the products of weathering are transported, deposited, and transformed into solid rock in a variety of ways, geologists recognize three categories of sedimentary rocks. As this overview reminds us, sediment has two principal sources. First, it may be an accumulation of material that originates and is transported as solid particles derived from both mechanical and chemical weathering. Deposits of this type are termed *detrital*, and the sedimentary rocks they form are called **detrital sedimentary rocks**.

The second major source of sediment is soluble material produced largely by chemical weathering. When these ions in solution are precipitated by either inorganic or biological processes, the material is known as *chemical sediment*, and the rocks formed from it are called **chemical sedimentary rocks**.

The third category is **organic sedimentary rocks**. These rocks form from the carbon-rich remains of organisms. The primary example is coal; this black combustible rock consists of organic carbon from the remains of plants that died and accumulated on the floor of a swamp. The bits and pieces of undecayed plant material that constitute the "sediments" in coal are quite unlike the weathering products that make up detrital and chemical sedimentary rocks.

CONCEPT CHECKS 7.1

1. How does the volume of sedimentary rocks in Earth's crust compare to the volume of igneous and metamorphic rocks?
2. List two ways in which sedimentary rocks are important.
3. Outline the steps that would transform an exposure of granite in the mountains into various sedimentary rocks.
4. List and briefly describe the differences among the three basic sedimentary rock categories.

7.2 Detrital Sedimentary Rocks

Describe the primary basis for distinguishing among detrital rocks and discuss how the origin and history of such rocks might be determined.

Though a wide variety of minerals and rock fragments (*clasts*) may be found in detrital rocks, clay minerals and quartz are the chief constituents of most sedimentary rocks in this category. Recall from Chapter 6 that clay minerals are the most abundant product of the chemical weathering of silicate minerals, especially the feldspars. Clays are fine-grained minerals with sheetlike crystalline structures similar to the micas. The other common mineral, quartz, is abundant because it is extremely durable and very resistant to chemical weathering. Thus, when igneous rocks such as granite are attacked by weathering processes, individual quartz grains are freed.

Other common minerals in detrital rocks are feldspars and micas. Because chemical weathering rapidly transforms these minerals into new substances, their presence in sedimentary rocks indicates that erosion and deposition occurred fast enough to preserve some of the primary minerals from the source rock before they could be decomposed.

Particle size is the primary basis for distinguishing among various detrital sedimentary rocks. Figure 7.3 presents the size categories for particles making up detrital rocks. Particle size allows us to distinguish detrital rocks, and the sizes of the component grains also provide useful

Size Range (millimeters)	Particle Name	Common Name	Detrital Rock
>256	Boulder	Gravel	Conglomerate, Breccia
64–256	Cobble	Gravel	Conglomerate, Breccia
4–64	Pebble	Gravel	Conglomerate, Breccia
2–4	Granule	Gravel	Conglomerate, Breccia
1/16–2	Sand	Sand	Sandstone
1/256–1/16	Silt	Mud	Siltstone
<1/256	Clay	Mud	Shale or Mudstone

0 10 20 30 40 50 60 70 mm

◀ **Figure 7.3 Particle size categories** Particle size is the primary basis for distinguishing among various detrital sedimentary rocks. (Breccia photo by E. J. Tarbuck; all other photos by Dennis Tasa)

information about environments of deposition. Currents of water or air sort the particles by size; the stronger the current, the larger the particle size that can be carried. Gravels, for example, are moved by swiftly flowing rivers as well as by landslides and glaciers. Less energy is required to transport sand; thus, sand is found in such features as windblown dunes and some river deposits and beaches. Very little energy is needed to transport clay, so it settles very slowly. Accumulation of these tiny particles is generally associated with the quiet water of a lake, lagoon, swamp, or certain marine environments.

In order of increasing particle size, common detrital sedimentary rocks include shale, sandstone, and conglomerate or breccia. We will now look at each type and how it forms.

Shale

Shale is a sedimentary rock consisting of silt- and clay-size particles (Figure 7.4). These fine-grained detrital rocks account for well over half of all sedimentary rocks. The particles in these rocks are so small that they cannot be readily identified without great magnification, and this makes shale more difficult to study and analyze than most other sedimentary rocks.

How Does Shale Form? Much of what can be learned about the process that forms shale is related to particle size. The tiny grains in shale indicate that deposition occurs as a result of gradual settling from relatively quiet, nonturbulent currents. Such environments include lakes, river floodplains, lagoons, and portions of the deep-ocean basins. Even in these "quiet" environments, there is usually enough turbulence to keep clay-size particles suspended almost indefinitely. Consequently, much of the clay is deposited only after the individual particles coalesce to form larger aggregates.

Sometimes the chemical composition of rock provides additional information. One example is black shale, which is black because it contains abundant organic matter (carbon). When such a rock is found, it strongly implies that deposition occurred in an oxygen-poor environment such as a swamp, where organic materials do not readily oxidize and decay.

Thin Layers As silt and clay accumulate, they tend to form thin layers, which are commonly referred to as *laminae* (*lamin* = a thin sheet). Initially the particles in the laminae are oriented randomly. This disordered arrangement leaves a high percentage of open space (called *pore space*) that is filled with water. However, this situation usually changes over time, as additional layers of sediment pile up and compact the sediment below.

During this phase, the clay and silt particles take on a more nearly parallel alignment and become tightly packed.

Did You Know?
The term *clay* may be confusing because it has more than one meaning. In the context of detrital particle size, the term *clay* refers to grains less than 1/256 mm, which are thus microscopic (see Figure 7.3). However, the term *clay* is also used as the name of a group of silicate minerals (see Chapter 3, page 84). Although most clay minerals are of clay size, not all clay-size sediment consists of clay minerals!

▼ **Figure 7.4 Shale—the most abundant sedimentary rock** Dark shale containing fossilized plant remains is relatively common. (Photo by E. J. Tarbuck)

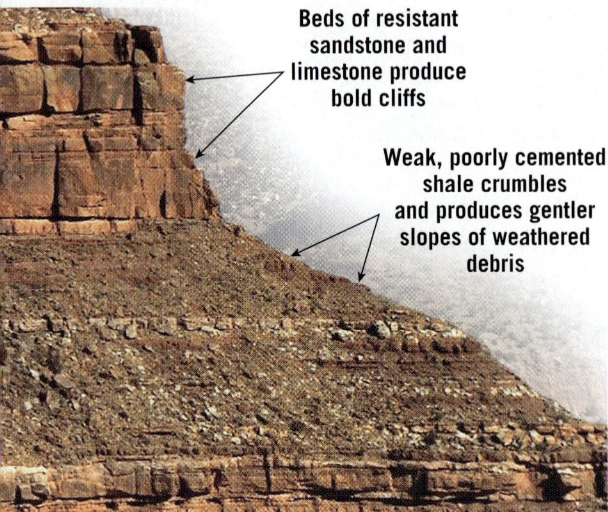
Beds of resistant sandstone and limestone produce bold cliffs
Weak, poorly cemented shale crumbles and produces gentler slopes of weathered debris

▲ **Figure 7.5 Shale crumbles easily** Beds of resistant sandstone and limestone produce bold cliffs. Weaker, poorly cemented shale produces gentler slopes of weathered debris. (Photo by Dennis Tasa)

Did You Know?
The Navajo Sandstone pictured in Figure 7.8 represents a vast area of sand dunes that once covered up to 400,000 sq km (156,000 sq mi), an area the size of California.

This rearrangement of grains reduces the size of the pore spaces and forces out much of the water. Once the grains are pressed closely together, the tiny spaces between particles do not readily permit solutions containing cementing material to circulate. Therefore, geologists often describe shales as being weak because they are poorly cemented and therefore not well lithified.

The inability of water to penetrate shale's microscopic pore spaces explains why this rock often forms barriers to the subsurface movement of water and petroleum. Indeed, rock layers that contain groundwater are commonly underlain by shale beds that block further downward movement. The opposite is true for underground reservoirs of petroleum. They are often capped by shale beds that effectively prevent oil and gas from escaping to the surface.*

Shale, Mudstone, or Siltstone? It is common to apply the term *shale* to all fine-grained sedimentary rocks, especially in a nontechnical context. However, be aware that there is a more restricted use of the term. In this narrower usage, shale must exhibit the ability to split into thin layers along well-developed, closely spaced planes. This property is termed **fissility** (*fissilis* = that which can be cleft or split). If the rock breaks into chunks or blocks, the name *mudstone* is applied. Another fine-grained sedimentary rock that, like mudstone, is often grouped with shale but lacks fissility is *siltstone* (see Figure 7.3). As its name implies, siltstone is composed largely of silt-size particles and contains less clay-size material than shale and mudstone.

Gentle Slopes Although shale is far more common than other sedimentary rocks, it does not usually attract as much notice as other, less abundant, members of this group. The reason is that shale does not form prominent outcrops, as sandstone and limestone often do. Rather, shale crumbles easily and usually forms a cover of soil that hides the unweathered rock below. This is illustrated nicely in the Grand Canyon, where the gentler slopes of weathered shale are quite inconspicuous and overgrown with vegetation, in sharp contrast with the bold cliffs produced by more durable rocks (Figure 7.5).

*The relationship between impermeable beds and the occurrence and movement of groundwater is examined in Chapter 14. Shale beds can be cap rocks in oil traps and are discussed later in the chapter.

Although shale beds may not form striking cliffs and prominent outcrops, some deposits have economic value. Certain shales are quarried to obtain raw material for pottery, brick, tile, and china. Moreover, when mixed with limestone, shale is used to make Portland cement. Some shales contain substantial quantities of oil and natural gas that can be obtained by a process called *hydraulic fracturing*. There is more about this process in the section on energy resources later in the chapter.

Sandstone

Sandstone is the name given to rocks in which sand-size grains predominate (Figure 7.6). After shale, sandstone is the next most abundant sedimentary rock, accounting for approximately 20 percent of the entire group. Sandstones form in a variety of environments, and clues such as the composition, shape, and sorting of the grains may provide information about the environment in which the sediment was deposited.

Sorting All the particles in sandstone are not necessarily identical in size. **Sorting** refers to the degree of similarity in particle size in a sedimentary rock. For example, if all the grains in a sample of sandstone are about the same size, the sand is considered *well sorted*. Conversely, if the rock contains mixed large and small particles, the sand is said to be *poorly sorted* (Figure 7.7). By studying the degree of sorting, we can learn much about the depositing current. Deposits of wind-blown sand are usually better sorted than deposits sorted by wave activity (Figure 7.8). Particles washed by waves are commonly better sorted than materials deposited by streams. Sediment accumulations that exhibit poor sorting usually result when particles are transported for only a relatively short time and then rapidly deposited. For example,

Close up

▲ **Figure 7.6 Quartz sandstone** After shale, sandstone is the next most abundant sedimentary rock. (Photos by Dennis Tasa)

when a turbulent stream reaches the gentler slopes at the base of a steep mountain, its velocity is quickly reduced, and poorly sorted sands and gravels are deposited.

Particle Shape The shapes of sand grains can also help decipher the history of a sandstone (see Figure 7.7). When streams, winds, or waves move sand and other larger sedimentary particles, the grains lose their sharp edges and corners and become more rounded as they collide with other particles during transport. Thus, rounded grains likely have been airborne or waterborne. Further, the degree of rounding indicates the distance or time involved in the transportation of sediment by currents of air or water. Highly rounded grains indicate that a great deal of abrasion and hence a great deal of transport has occurred.

Very angular grains, on the other hand, can imply either of two things: that the materials were transported only a short distance before they were deposited or that they were carried by some medium other than water or wind. For example, when glaciers move sediment, the particles are usually made more irregular by the crushing and grinding action of the ice.

Transport Affects Mineral Composition In addition to affecting the degree of rounding and the amount of sorting that particles undergo, the length of transport by turbulent air and water currents also influences the mineral composition of a sedimentary deposit. Substantial weathering and long transport lead to the gradual destruction of weaker and less stable minerals, including the feldspars and ferromagnesians. Because quartz is very durable, it is the mineral that best survives a long trip in a turbulent environment.

To summarize, the origin and history of sandstone can often be deduced by examining the sorting, roundness, and mineral composition of its constituent grains. Knowing this information allows us to infer that a well-sorted, quartz-rich sandstone consisting of highly rounded grains must be the result of a great deal of transport. Such a rock, in fact, may represent several cycles of weathering, transport, and deposition. We may also conclude that a sandstone containing significant amounts of feldspar and angular grains of ferromagnesian minerals underwent little chemical weathering and transport and was probably deposited close to the source area of the rock particles.

Sorting

Very poorly sorted | Poorly sorted | Well sorted | Very well sorted

Sediments are "very poorly sorted" when there is a wide range of different sizes.

Rocks with particles that are nearly all the same size are "well sorted."

Angularity and Sphericity

Angular | Subangular | Subrounded | Rounded

High sphericity

Low sphericity

Transportation reduces the size and angularity of particles but does not change their general shape.

▲ **SmartFigure 7.7 Sorting and particle shape** *Sorting* refers to the range of particle sizes present in a rock. Geologists describe a particle's shape in terms of its *angularity* (the degree to which edges and corners are rounded) and *sphericity* (how close the shape is to a sphere).

TUTORIAL
https://goo.gl/wG2hAf

▼ **Figure 7.8 Sand dunes consist of well-sorted sediment A.** The Navajo Sandstone represents a vast area of ancient sand dunes that once covered an area the size of California. **B.** These modern dunes are among the highest in North America. (Photo A by Dennis Tasa; photo B by George H. H. Huey/Alamy Images)

A. The orange and yellow cliffs of Utah's Zion National Park expose thousands of feet of Jurassic-age Navajo Sandstone.

B. The quartz grains composing the Navajo Sandstone were deposited by wind as dunes similar to these in Colorado's Great Sand Dunes National Park. The sand is well sorted because all of the particles are practically the same size.

▲ **Figure 7.9**
Conglomerate The gravel-size particles in this rock are rounded. (Photo by E. J. Tarbuck)

Varieties of Sandstone Due to its durability, quartz is the predominant mineral in most sandstones. Such rock is often simply called *quartz sandstone*. When a sandstone contains appreciable quantities of feldspar (25 percent or more), the rock is called *arkose*. In addition to feldspar, arkose usually contains quartz and sparkling bits of mica. The mineral composition of arkose indicates that the grains were derived from granitic source rocks. The particles are generally poorly sorted and angular, which suggests short-distance transport, minimal chemical weathering in a relatively dry climate, and rapid deposition and burial.

A third variety of sandstone is known as *graywacke*. Along with quartz and feldspar, this dark-colored rock contains abundant rock fragments and matrix—finer-grained material in which the fragments are embedded. More than 15 percent of graywacke's volume is matrix. The poor sorting and angular grains characteristic of graywacke suggest that the particles were transported only a relatively short distance from their source area and were then rapidly deposited. Before the sediment could be reworked and sorted further, it was buried by additional layers of material. Graywacke is frequently associated with submarine deposits made by dense sediment-choked torrents called *turbidity currents*.

Did You Know?
The most important and common material used for making glass is silica, which is usually obtained from the quartz in "clean," well-sorted sandstones.

Conglomerate & Breccia

Conglomerate consists largely of gravels (**Figure 7.9**). As Figure 7.3 indicates, these particles can range in size from large boulders to particles as small as peas. The particles are often large enough to be identified as distinctive rock types; thus, they can be valuable in identifying the source areas of sediments. More often than not, conglomerates are poorly sorted, and the spaces between the large gravel particles contain sand or mud (**Figure 7.10**).

Gravels accumulate in a variety of environments and usually indicate the existence of steep slopes or very turbulent currents. The coarse particles in a conglomerate may reflect the action of energetic mountain streams or result from strong wave activity along a rapidly eroding coast. Some glacial and landslide deposits also contain plentiful gravel.

▲ **Figure 7.10** **Poorly sorted sediments** Gravel deposits along Carbon Creek in Grand Canyon National Park are poorly sorted. (Photo by Michael Collier)

If the large particles are angular rather than rounded, the rock is called **breccia** (see Figure 7.3 to compare breccia and conglomerate). Because large particles abrade and become rounded very rapidly during transport, the pebbles and cobbles in a breccia indicate that they did not travel far from their source area before they were deposited. Thus, as with many other sedimentary rocks, conglomerates and breccias contain clues to their history. Their particle sizes reveal the strength of the currents that transported them, and the degree of rounding indicates how far the particles traveled. The fragments within a sample identify the source rocks that supplied them.

CONCEPT CHECKS 7.2

1. What minerals are most abundant in detrital sedimentary rocks? In which rocks do these minerals predominate?
2. What is the primary basis for distinguishing among detrital rocks?
3. Describe how sediments become sorted. What would cause sediments to be poorly sorted?
4. Distinguish between conglomerate and breccia.

7.3 Chemical Sedimentary Rocks

Explain the processes involved in the formation of chemical sedimentary rocks and list several examples.

In contrast to detrital rocks, which form from the solid products of weathering, chemical sediments derive from ions that are carried *in solution* to lakes and seas. This material does not remain dissolved in the water indefinitely, however. Some of it precipitates to form chemical sediments. These become rocks such as limestone, chert, and rock salt.

This precipitation of material occurs in two ways. *Inorganic* (*in* = not, *organicus* = life) processes such as evaporation and chemical activity can produce chemical

Delicate calcite crystals forming in a drop of water at the tip of soda straw stalactite. The formation of crystals is triggered when some carbon dioxide escapes from the water drop.

▲ **Figure 7.11 Cave deposits** An example of a chemical sedimentary rock with an inorganic origin. (Photo by Guillen Photography/Alamy Images; inset photo by Dante Fenolio/Science Source)

sediments. *Organic* (*organicus* = life) processes of water-dwelling organisms also form chemical sediments, said to be of **biochemical** origin.

One example of a deposit resulting from inorganic chemical processes is the dripstone that decorates many caves (Figure 7.11). Another is the salt left behind as a body of seawater evaporates. In contrast, many water-dwelling animals and plants extract dissolved mineral matter to form shells and other hard parts. After the organisms die, their skeletons collect by the millions on the floor of a lake or an ocean as biochemical sediment (Figure 7.12).

Limestone

Representing about 10 percent of the total volume of all sedimentary rocks, **limestone** is the most abundant chemical sedimentary rock. It is composed chiefly of the mineral calcite ($CaCO_3$) and forms either by inorganic means or as a result of biochemical processes. Although the mineral composition of all limestone is similar, many different types exist, reflecting the varied conditions under which limestone originates. Forms that have a marine biochemical origin are by far the most common.

Carbonate Reefs Corals are one important example of organisms that are capable of creating large quantities of marine limestone. These relatively simple invertebrate animals secrete a calcareous (calcium carbonate) external skeleton. Although they are small, corals are capable of creating massive structures called *reefs* (Figure 7.13). Reefs consist of coral colonies made up of great numbers of individuals that live side by side on a calcite structure secreted by the animals. In addition, calcium carbonate–secreting algae live with the corals and help cement the entire structure into a solid mass. A wide variety of other organisms also live in and near the reefs.

Certainly the best-known modern reef is Australia's 2600-kilometer-long (1600-mile-long) Great Barrier Reef, but many lesser reefs also exist. They develop in the shallow, warm waters of the tropics and subtropics equatorward of about 30° latitude. Striking examples exist in The Bahamas, Hawaii, and the Florida Keys.

Modern corals were not the first reef builders. Earth's first reef-building organisms were photosynthesizing bacteria that lived during Precambrian time, more than 2 billion years ago. From fossil remains, it is known that a variety of organisms have constructed reefs, including bivalves (clams and oysters), bryozoans (coral-like animals), and sponges. Corals have been found in fossil reefs as ancient as 500 million years old, but corals similar to the modern colonial varieties have constructed reefs only during the past 60 million years.

In the United States, reefs of Silurian age (416 to 444 million years ago) are prominent features in Wisconsin, Illinois, and Indiana. In western Texas and adjacent southeastern New Mexico,

▼ **Figure 7.12 Coquina** This variety of limestone consists of shell fragments; therefore, it has a biochemical origin. (Rock sample photo by E. J. Tarbuck; beach photo by Donald R. Frazier Photolibrary, Inc./Alamy Images)

▶ **Figure 7.13 Carbonate reefs** Large quantities of biochemical limestone are created by reef-building organisms. (Australia photo by JC Photo/Shutterstock; Texas photo by Michael Collier)

Great Barrier Reef
Australia
This image in Guadalupe Mountains National Park, Texas, shows a small portion of an ancient (Permian-age) reef complex that once formed a 600-kilometer loop around the margin of the Delaware Basin.

Aerial view showing a small portion of Australia's Great Barrier Reef. Located off the coast of Queensland, it extends for 2600 kilometers and consists of more than 2900 individual reefs.

▼ **Figure 7.14 The white chalk cliffs** This prominent deposit underlies large portions of southern England as well as parts of northern France. (Photos by David Wall/Alamy Images)

View of a group of plankton called *coccolithophores* from a scanning electron microscope. Individual plates shaped like hubcaps are only three one-thousandths of a millimeter in diameter; so tiny they could pass through the eye of a needle.

a massive reef complex that formed during the Permian period (251 to 299 million years ago) is strikingly exposed in Guadalupe Mountains National Park (see Figure 7.13).

Coquina & Chalk Although a great deal of limestone is produced by biological processes, this origin is not always evident because shells and skeletons may undergo considerable change before becoming lithified into rock. However, one easily identified biochemical limestone is *coquina*, a coarse rock composed of poorly cemented shells and shell fragments (see Figure 7.12). Another less obvious but nevertheless familiar example is *chalk*, a soft, porous rock made up almost entirely of the hard parts of microscopic marine organisms. Among the most famous chalk deposits are those exposed along the southeast coast of England (**Figure 7.14**).

Inorganic Limestones Limestones having an inorganic origin form when chemical changes or high water temperatures cause calcium carbonate to precipitate out of the water. *Travertine*, the type of limestone commonly seen in caves, is an example (see Figure 7.11). When travertine is deposited in caves, groundwater is the source of the calcium carbonate. As water droplets become exposed to the air in a cavern, some of the carbon dioxide dissolved in the water escapes, causing calcium carbonate to precipitate.

Another variety of inorganic limestone is *oolitic limestone*, a rock composed of small spherical grains called *ooids*. Ooids form in very shallow marine waters as tiny "seed" particles (commonly small shell fragments) are moved back and forth with currents. As the grains are rolled about in the warm water, which is supersaturated with calcium carbonate, they become coated with layer upon layer of the chemical precipitate (**Figure 7.15**).

Small spherical grains called ooids are formed by chemical precipitation of calcium carbonate around a tiny nucleus and are the raw material for oolitic limestone.

▲ **Figure 7.15 Oolitic limestone** Oolitic limestone is an inorganic limestone composed of ooids. Note the dime coin for scale. (Photos by Marli Miller)

Dolostone

Closely related to limestone is **dolostone**, a rock composed of the calcium—magnesium carbonate mineral dolomite [$CaMg(CO_3)_2$]. Although dolostone and limestone sometimes closely resemble one another, they can be easily distinguished by observing their reaction to dilute hydrochloric acid. When a drop of acid is placed on limestone, the reaction (fizzing) is obvious. However, unless dolostone is powdered, it does not visibly react to the acid.

Dolostone's origins remain a subject of discussion among geologists. No marine organisms produce hard parts of dolomite, and the chemical precipitation of dolomite from seawater occurs only under conditions of unusual water chemistry in certain near-shore sites. Yet dolostone is abundant in many ancient sedimentary rock successions. It appears that a significant proportion of dolostone is produced when magnesium-rich waters circulate through limestone and replace some calcium ions with magnesium ions (a process called *dolomitization*). However, not all dolostones appear to be formed by such a process, and their origin remains uncertain.

Chert

Chert is a name used for a number of very compact and hard rocks made of microcrystalline quartz (SiO_2). It can vary in color from nearly white to black; chert often appears gray, brown, or brownish gray and can even appear red or greenish. Color variations are related to trace elements present in the rock. **Figure 7.16** shows some varieties. One well-known form is *flint*, whose dark color results from the organic matter it contains. *Jasper*, a red variety, gets its bright color from iron oxide. *Petrified wood* is chert that is made when silica-rich material such as volcanic ash buries trees. As groundwater rich in dissolved silica from the ash penetrates the wood, silica precipitates, gradually replacing the wood. Fine structures such as growth rings are often preserved. Like glass, most chert has a conchoidal fracture. Its hardness, ease of chipping, and ability to hold a sharp edge made chert a favorite of Native Americans for fashioning points for spears and arrows. Because of chert's durability and extensive use, arrowheads are found in many parts of North America.

Chert most commonly occurs either as layered deposits, called *bedded cherts*, or as somewhat spherical masses called *nodules*, which form within other rocks (commonly limestone) and range in diameter from pea sized to several centimeters. Most bedded cherts are believed to originate from tiny marine organisms (diatoms and radiolarians) that produce hard parts from silica rather than calcium carbonate. These organisms extract silica from seawater (where it is present in only tiny quantities); when they die, their hard parts may accumulate on the seafloor. Some bedded cherts occur in association with lava flows and layers of volcanic ash. For these occurrences, it is probable that the silica was derived from the decomposition of the volcanic ash and not

Did You Know?

In the western and southwestern United States, sedimentary rocks often exhibit a brilliant array of colors. (For examples, see Figure 6.1, Figure 6.8B, this chapter's opening photo, and Figure 7.1.) The walls of Arizona's Grand Canyon feature layers that are red, orange, purple, gray, brown, and buff. Some of the sedimentary rocks in Utah's Bryce Canyon are a delicate pink color. Sedimentary rocks in more humid places are also colorful, but they are usually covered by soil and vegetation.

The most important "pigments" are iron oxides, and only very small amounts are needed to color a rock. Hematite tints rocks red or pink, whereas limonite produces shades of yellow and brown. When sedimentary rocks contain organic matter, it often colors the rocks black or gray (see Figure 7.4).

◄ **Figure 7.16 Colorful chert** Chert is the name applied to a number of dense, hard chemical sedimentary rocks made of microcrystalline quartz. (Petrified wood photo by gracious_tiger/Shutterstock; flint and jasper photos by E. J. Tarbuck; arrowhead photo by Daniel Sambraus/Science Source)

Did You Know?
Each year about 30 percent of the world's supply of salt is extracted from seawater. The seawater is pumped into ponds and allowed to evaporate, leaving behind "artificial evaporites," which are harvested.

from biochemical sources. Chert nodules are sometimes referred to as *secondary cherts*, or *replacement cherts*. They form when silica, originally deposited in one place, dissolves, migrates, and then chemically precipitates elsewhere, replacing older material.

Evaporites

In the geologic past, many areas that are now dry land were shallow marine bays with only a narrow connection to the open ocean. Under these conditions, seawater continually moved into the bay to replace water lost by evaporation. Eventually the waters of the bay became saturated, and salt deposition began. Such deposits are called **evaporites**.

Minerals commonly precipitated in this fashion include halite (sodium chloride, NaCl), which is the chief component of *rock salt*, and gypsum (hydrous calcium sulfate, $CaSO_4 \cdot 2H_2O$), which is the main ingredient in *rock gypsum*. Both have economic importance. In addition to its familiar role as table salt, halite has many other uses, from melting ice on roads to making hydrochloric acid. People have sought, traded, and fought over it for much of human history. Gypsum is the basic ingredient in plaster of Paris and is used in making wallboard and interior plaster.

When a body of seawater evaporates, the minerals that precipitate do so in a sequence that is determined by their solubility. Less soluble minerals precipitate first, and more soluble minerals precipitate later, as salinity increases. For example, gypsum precipitates when about 80 percent of the seawater has evaporated, and halite crystallizes when 90 percent of the water has been removed. During the last stages of this process, potassium and magnesium salts precipitate. One of these last-formed salts, the mineral *sylvite*, is mined as a significant source of potassium ("potash") for fertilizer.

On a smaller scale, evaporite deposits can be seen in such places as Death Valley, California (see Figure 16.8, page 433). Here, following rains or periods of snowmelt in the mountains, streams flow from the surrounding mountains into an enclosed basin. As the water evaporates, dissolved minerals precipitate on the surface, forming a **salt flat**. Utah's extensive Bonneville salt flats, pictured in Figure 7.17, are another well-known example.

CONCEPT CHECKS 7.3

1. Explain how the formation of biochemical sediments differs from the formation of sediments formed by inorganic processes. Use examples as part of your explanation.
2. Distinguish among limestone, dolostone, and chert. Describe several varieties of each.
3. How do evaporites form? What are some examples?

This extensive evaporite deposit is a 30,000-acre expanse of hard white salt that in places is nearly 2 meters thick.

▶ **SmartFigure 7.17 Bonneville salt flats** This well-known Utah site was once a large salt lake. (Photo by Stock Connection/Glow Images; satellite image by NASA)

TUTORIAL
https://goo.gl/eZ5W9p

7.4 Coal: An Organic Sedimentary Rock

Outline the successive stages in the formation of coal.

Coal is different from other sedimentary rocks. Unlike limestone and chert, which are rich in calcite and silica, coal is made of organic matter. Close examination of coal under a magnifying glass often reveals plant structures such as leaves, bark, and wood that have been chemically altered but are still identifiable. This supports the conclusion that coal is the end product of large amounts of plant material being buried for millions of years (Figure 7.18). The formation of coal involves these stages:

1. **Accumulation of plant remains.** The initial stage in coal formation is the accumulation of large quantities of plant remains. Such accumulations result from special conditions because dead plants readily decompose when exposed to the atmosphere or other oxygen-rich environments. One important environment that allows for the buildup of plant material is a swamp. Stagnant swamp water is oxygen deficient, so complete decay (oxidation) of the plant material is not possible. Instead, the plants are attacked by certain bacteria that partly decompose the organic material and liberate oxygen and hydrogen. As these elements escape, the percentage of carbon in the plant matter gradually increases. The bacteria are not able to finish the job of decomposition because their growth is impeded by acids liberated from the plants.
2. **Formation of peat.** The partial decomposition of plant remains in an oxygen-poor swamp creates a layer of *peat*, a soft brown material in which plant structures are still easily recognized. With shallow burial, peat slowly changes to *lignite*, a soft brown coal. Burial increases the temperature of sediments as well as the pressure on them.
3. **Formation of lignite and bituminous coal.** Higher temperatures bring about chemical reactions within the plant materials and yield water and organic gases (volatiles). As the load increases from more sediment on top of the developing coal, the water and volatiles are pressed out, and the proportion of *fixed carbon* (the remaining solid combustible material) increases. The greater the carbon content, the greater the coal's energy ranking as a fuel. During burial, the coal also becomes increasingly compact. For example, deeper burial transforms lignite into a harder, more compacted black rock called *bituminous* coal. A bed of bituminous coal may be only one-tenth as thick as the peat bed from which it formed.
4. **Formation of anthracite coal.** Lignite and bituminous coals are sedimentary rocks. However, when sedimentary layers are subjected to the folding and deformation associated with mountain building, the heat and pressure cause a further loss of volatiles and water, thus increasing the concentration of fixed carbon. This metamorphoses bituminous coal into *anthracite*, a very hard, shiny, black *metamorphic* rock. Although anthracite is a clean-burning fuel, only a relatively small amount of it is mined.

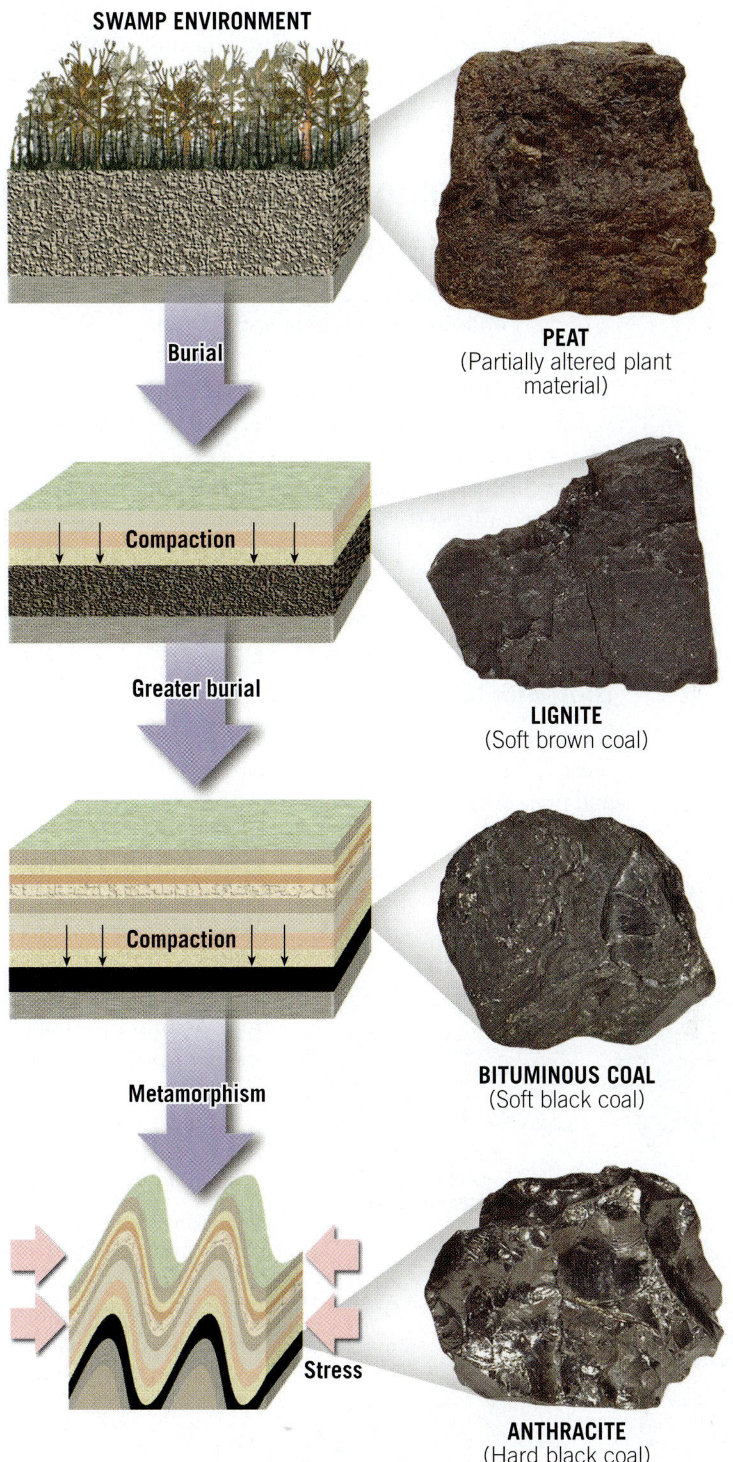

▲ **SmartFigure 7.18 From plants to coal** Successive stages in the formation of coal. (Photos by E. J. Tarbuck)

TUTORIAL
https://goo.gl/KPKKPW

Anthracite is not widespread and is more difficult and expensive to extract than the relatively flat-lying layers of bituminous coal.

Coal is a major energy resource. Its role as a fuel and some of the problems associated with burning coal are discussed later in the chapter.

CONCEPT CHECKS 7.4

1. What is the "raw material" for coal? Under what circumstances does it accumulate?
2. Outline the successive stages in the formation of coal.

7.5 Turning Sediment into Sedimentary Rock: Diagenesis & Lithification

Describe the processes that convert sediment into sedimentary rock and other changes associated with burial.

Did You Know?
In 2015, about 40 percent of the electricity generated in the United States was generated from coal.

We've seen that sediments change as they are converted to sedimentary rock. The term **diagenesis** (*dia* = change, *genesis* = origin) is a collective term for all the chemical, physical, and biological changes that take place after sediments are deposited and during and after lithification to form sedimentary rock. Diagenesis is distinct from the metamorphic processes that may later convert a sedimentary rock to a metamorphic one.

Diagenesis

Burial promotes diagenesis because as sediments are buried, they are subjected to increasing temperatures and pressures. Diagenesis occurs within the upper few kilometers of Earth's crust, at temperatures that are generally less than 150° to 200°C (300° to 400°F). Beyond this somewhat arbitrary temperature threshold, metamorphism is said to occur.

One example of diagenetic change is *recrystallization*, the development of more stable minerals from less stable ones. For example, the mineral aragonite is the less stable form of calcium carbonate ($CaCO_3$). Aragonite is secreted by many marine organisms to form shells and other hard parts, such as the skeletal structures produced by corals. In some environments, large quantities of these solid materials accumulate as sediment. As burial takes place, aragonite recrystallizes to the more stable form of calcium carbonate, calcite, the main constituent in the sedimentary rock limestone.

Another example of diagenesis is the chemical alterations that take place as plant matter is buried gradually in an oxygen-poor environment, evolving into peat and then into coal.

▼ **Figure 7.19** **Compaction and cementation**

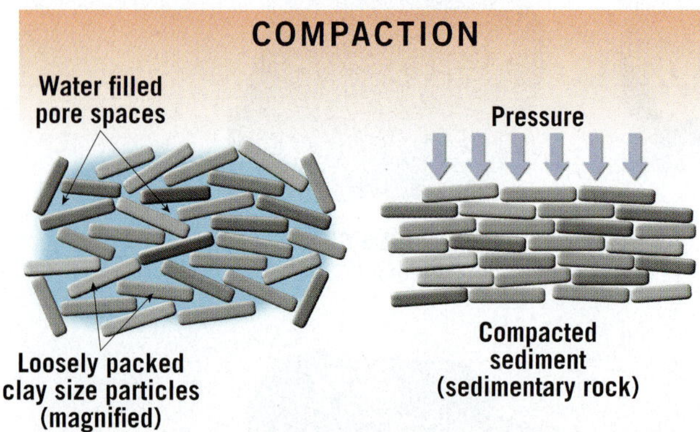

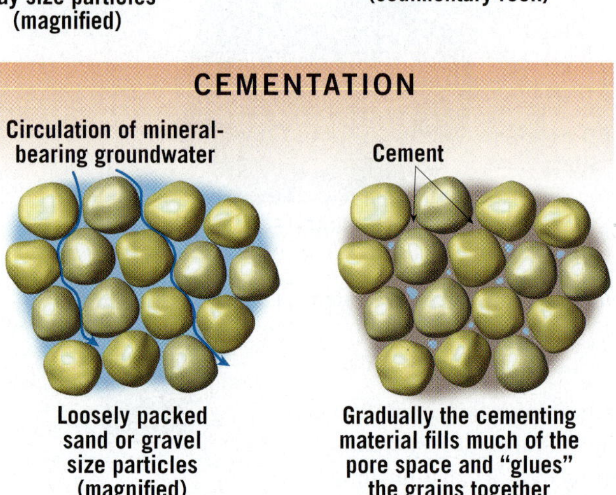

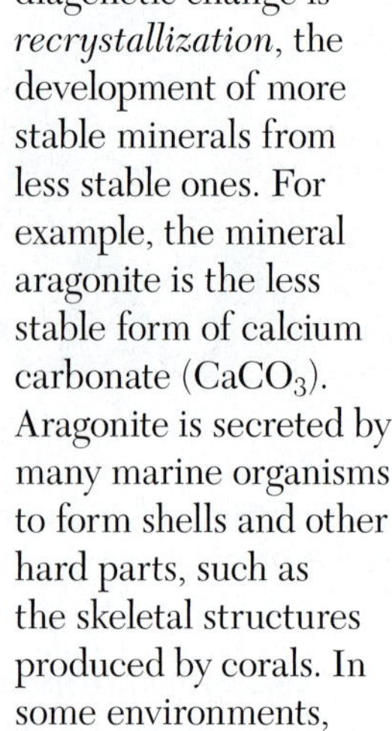

Lithification

Diagenesis includes **lithification**, the processes by which unconsolidated sediments are transformed into solid sedimentary rocks (*lithos* = stone, *fic* = making). Basic lithification processes include compaction and cementation (**Figure 7.19**).

Compaction The most common physical diagenetic change is **compaction**. As described earlier, as sediment accumulates, the weight of overlying material compresses the deeper sediments. The deeper a sediment is buried, the more it is compacted and the firmer it becomes. As the grains are pressed closer and closer, there is considerable reduction in pore space (the open space between particles). For example, when clays are buried beneath several thousand meters of material, the volume of the clay layer may be reduced by as much as 40 percent. As pore space decreases, much of the water that was trapped in the sediments is driven out. Because sands and other coarse sediments are less compressible, compaction is most significant as a lithification process in fine-grained sedimentary rocks such as shale.

Cementation The most important process by which sediments are converted to sedimentary rock is **cementation**, a process in which ions carried in solution by groundwater crystallize in the spaces between sediment grains to form minerals that gradually cement the grains together. Like compaction, this process reduces a rock's porosity.

Calcite, silica, and iron oxide are the most common cements. It is often a relatively simple matter to identify the cementing material. Calcite cement will effervesce in contact with dilute hydrochloric acid. Silica is the hardest cement and thus produces the hardest sedimentary rocks. An orange or dark-red color in a sedimentary rock means that iron oxide is present.

While compaction and cementation are the main lithification processes for most types of sediment, in some cases the original sedimentary particles themselves undergo dissolution and recrystallization. For example, with time and burial, loose sediment consisting of delicate calcium carbonate–rich skeletal debris may be recrystallized into a relatively dense crystalline limestone. Because crystals grow until they fill all the available space, crystalline sedimentary rocks often lack pore spaces. Unless the rocks later develop joints and fractures, they will be relatively impermeable to fluids such as water and oil. Evaporites form initially as solid masses of intergrown crystals and hence can be similarly impermeable.

CONCEPT CHECKS 7.5

1. What is diagenesis?
2. Compaction is most important as a lithification process with which sediment size?
3. List three common cements. How might each be identified?

7.6 Classification of Sedimentary Rocks

Summarize the criteria used to classify sedimentary rocks.

The classification chart in **Figure 7.20** divides sedimentary rocks into major groups: detrital on the left side and chemical/organic on the right. Further, you can see that the main criterion for subdividing the detrital rocks is particle size, whereas the primary basis for distinguishing among different rocks in the chemical group is their mineral composition.

As is the case with many (perhaps most) classifications of natural phenomena, the categories presented in Figure 7.20 are more rigid than the actual state of

Detrital Sedimentary Rocks

ClasticTexture (particle size)	Sediment Name	Rock Name
Coarse (over 2 mm)	Gravel (Rounded particles)	Conglomerate
	Gravel (Angular particles)	Breccia
Medium (1/16 to 2 mm)	Sand	Sandstone (Arkose)*
Fine (1/16 to 1/256 mm)	Mud	Siltstone
Very fine (less than 1/256 mm)	Mud	Shale or Mudstone

*If abundant feldspar is present the rock is called Arkose.

Chemical and Organic Sedimentary Rocks

Composition	Texture	Rock Name	
Calcite, $CaCO_3$	Nonclastic: Fine to coarse crystalline	Crystalline Limestone	
		Travertine	
	Clastic: Visible shells and shell fragments loosely cemented	Coquina	Biochemical Limestone
	Clastic: Various size shells and shell fragments cemented with calcite cement	Fossiliferous Limestone	Biochemical Limestone
	Clastic: Microscopic shells and clay	Chalk	Biochemical Limestone
Quartz, SiO_2	Nonclastic: Very fine crystalline	Chert (light colored) Flint (dark colored) Jasper (red)	
Gypsum $CaSO_4 \cdot 2H_2O$	Nonclastic: Fine to coarse crystalline	Rock Gypsum	
Halite, NaCl	Nonclastic: Fine to coarse crystalline	Rock Salt	
Altered plant fragments	Nonclastic: Fine-grained organic matter	Bituminous Coal	

▶ **Figure 7.20 Identification of sedimentary rocks** The main criterion for naming detrital rocks is particle size. The primary basis for naming chemical and organic sedimentary rocks is their composition.

nature. In reality, many of the sedimentary rocks classified into the chemical group also contain at least small quantities of detrital sediment. Many limestones, for example, contain varying amounts of mud or sand, giving them a "sandy" or "shaly" quality. Conversely, because practically all detrital rocks are cemented with material that was originally dissolved in water, they too are far from being "pure."

▲ **Figure 7.21 Rock salt** Like other evaporites, rock salt has a nonclastic texture because it is composed of intergrown crystals. (Photos by E. J. Tarbuck)

As was the case with the igneous rocks examined in Chapter 5, *texture* is a part of sedimentary rock classification. There are two major textures used in the classification of sedimentary rocks: clastic and nonclastic. The term **clastic** is taken from a Greek word meaning "broken." Rocks that display a clastic texture consist of discrete fragments and particles that are cemented and compacted together. Although cement is present in the spaces between particles, these openings are rarely filled completely. All detrital rocks have a clastic texture. In addition, some chemical sedimentary rocks exhibit this texture. For example, coquina, the limestone composed of shells and shell fragments, is obviously as clastic as a conglomerate or sandstone. The same applies to some varieties of oolitic limestone.

Some chemical sedimentary rocks have a **nonclastic,** or **crystalline** texture, in which the minerals form a pattern of interlocking crystals. The crystals may be microscopically small or large enough to be visible without magnification. Evaporites are of this type (Figure 7.21). As described earlier, some limestones originate as clastic deposits of shell fragments and subsequently undergo recrystallization. Chert can have a similar history.

Because they consist of intergrown crystals, nonclastic sedimentary rocks may resemble igneous rocks. The two rock types are usually easy to distinguish because the minerals contained in nonclastic sedimentary rocks (such as halite, gypsum, and calcite) are quite unlike those found in most igneous rocks.

CONCEPT CHECKS 7.6

1. What is the primary basis for distinguishing (naming) different chemical sedimentary rocks? How is the naming of detrital rocks different?
2. Distinguish between clastic and nonclastic textures. Which texture is associated with all detrital rocks?

7.7 Sedimentary Rocks Represent Past Environments

Distinguish among three broad categories of sedimentary environments and provide an example of each. List several sedimentary structures and explain why these features are useful to geologists.

Sedimentary rocks are important to interpreting Earth's history (Figure 7.22). By understanding the conditions under which sedimentary rocks form, geologists can often deduce the history of a rock, including information about the origin of its component particles, the method of sediment transport, and the nature of the place where the grains eventually came to rest—that is, the environment of deposition.

▶ **SmartFigure 7.22 Utah's Capitol Reef National Park** These tilted sedimentary strata are part of the Waterpocket Fold at Halls Creek. The strata here record changing environments during the Mesozoic era. (Photo by Michael Collier)

MOBILE FIELD TRIP
https://goo.gl/fkQcwr

Importance of Sedimentary Environments

An **environment of deposition,** or **sedimentary environment**, is a geographic setting where sediment is accumulating. Each site is characterized by a particular combination of geologic processes and environmental conditions. Some sediments, such as the chemical sediments that precipitate in water bodies, are solely products of their sedimentary environment. That is, their component minerals originated and were deposited in the same place. Other sediments originate far from the site where they accumulate and are transported from their source by some combination of gravity, water, wind, and ice.

At any given time, the geographic setting and environmental conditions of a sedimentary environment determine the nature of the sediments that accumulate. Geologists carefully study the sediments in present-day depositional environments because the features they find can also be observed in ancient sedimentary rocks.

Geologists apply a thorough knowledge of present-day conditions to reconstruct the ancient environments and geographic relationships of an area at the time particular sedimentary layers were deposited. This process is an excellent example of applying a fundamental principle of modern geology: "The present is the key to the past."* Such analyses often lead to the creation of maps depicting the past geographic distribution of land and sea, mountains and river valleys, deserts and glaciers, and other environments of deposition.

Sedimentary environments are commonly placed into one of three broad categories: continental, marine, or transitional (shoreline). Each category includes many specific subenvironments. **Figure 7.23** includes an idealized diagram illustrating a number of important sedimentary environments associated with each category. Realize that this is just a sampling of the great diversity of depositional environments. Chapters 13 through 17 examine many of these environments in greater detail.

Sedimentary Facies

When we study a series of sedimentary layers, we can see the successive changes in environmental conditions that occurred at a particular place with the passage of time. Changes in past environments may also be seen when a single layer of sedimentary rock is traced laterally. This is true because at any one time, many different depositional environments can exist over a broad area. For example, when sand is accumulating in a beach environment, finer muds are often being deposited in quieter offshore waters. Still farther out, perhaps in a zone where biological activity is high and land-derived sediments are scarce, the deposits consist largely of the calcite-rich remains of small organisms. In this example, different sediments are accumulating adjacent to one another at the same time. Different parts of each layer possess a distinctive set of characteristics that reflect the conditions in a particular environment. The term **facies** is used to describe such sets of sediments. When a sedimentary layer is examined in cross section from one end to the other, each facies grades laterally into another that formed at the same time but that exhibits different characteristics (**Figure 7.24**). The merging of adjacent facies tends to be a gradual transition rather than a sharp boundary, but abrupt changes do sometimes occur.

*For more on this idea, see Section 1.2 in Chapter 1.

Sedimentary Structures

In addition to variations in grain size, mineral composition, and texture, sediments exhibit a variety of structures that are often preserved when they change to sedimentary rock. Some, such as graded beds, are created when sediments are accumulating and are a reflection of the transporting medium. Others, such as *mud cracks*, form after the materials have been deposited and result from processes occurring in the environment. When present, sedimentary structures provide additional information that can be useful in interpreting Earth's history.

Sedimentary rocks form as layer upon layer of sediment accumulates in various depositional environments. These layers, called **strata,** or **beds**, are probably *the single most common and characteristic feature of sedimentary rocks*. Each stratum is unique. It may be a coarse sandstone, a fossil-rich limestone, a black shale, and so on. When you look at **Figure 7.25**, or look back at the chapter-opening photo and at Figures 7.1 and 7.5, you see many such layers, each different from the others. The variations in texture, composition, and thickness reflect the different conditions under which each layer was deposited.

Beds range from microscopically thin to tens of meters thick. Separating these strata are **bedding planes**, relatively flat surfaces along which rocks tend to separate or break. Changes in the grain size or in the composition of the sediment being deposited can create bedding planes. Pauses in deposition can also lead to layering because chances are slight that newly deposited material will be exactly the same as previously deposited sediment. Generally, each bedding plane marks the end of one episode of sedimentation and the beginning of another.

Because sediments usually accumulate as particles that settle from a fluid, most strata are originally deposited as horizontal layers. There are circumstances, however, when sediments do not accumulate in horizontal beds. Sometimes when a bed of sedimentary rock is examined, we see layers within it that are inclined to the horizontal. This is called **cross-bedding**, and it

Did You Know?

These terms are often used when describing the mineral makeup of sedimentary rocks: *siliceous rocks* contain abundant quartz, *argillaceous* rocks are clay rich, and *carbonate* rocks contain calcite or dolomite.

Figure 7.23 Sedimentary environments Each environment is characterized by certain physical, chemical, and biological conditions. Because sediments contain clues about the environment in which they were deposited, sedimentary rocks are important in the interpretation of Earth history. A number of important examples are represented in this idealized diagram.

Beaches that form where wave activity is strong consist mainly of pebbles and cobbles.

Caves that develop in limestone are sites where calcium carbonate is deposited as dripstone.

Sand dunes consist of well-sorted sand grains deposited by the wind.

Beaches, **bars**, and **spits** along low-lying coasts and in sheltered coves are typically composed of well-sorted sand and/or shell fragments.

Tidal flats and **lagoons** are areas where fine clay particles or carbonate-rich muds accumulate.

Deep marine environments adjacent to the continental slope often contain material that was transported by dense underwater currents of suspended sediment. Each layer has coarser particles at the bottom and finer material on top.

Shallow marine environments are sites where sand, clay, and carbonate-rich muds are often deposited. Ripple marks caused by wave activity may be present.

Inland seas and **lakes** in arid environments where evaporation exceeds precipitation, produce evaporite deposits such as rock salt and gypsum.

Alluvial fans consist of coarse sediments that are deposited when mountain streams reach flat lowlands.

Glacial deposits often consist of a poorly-sorted mixture of many different sediment sizes ranging from clay to boulders.

Swamps and **bogs** are quiet-water environments where mud and decayed plant material accumulate.

Streams in mountainous areas erode and deposit a wide variety of sediment, while those in lowlands transport and deposit mostly mud (silt and clay) and sand.

Landslides produce an unsorted jumble of many sediment sizes.

Coral reefs are massive limestone structures that form in warm, shallow clear seas and consist of material secreted by corals and other marine life.

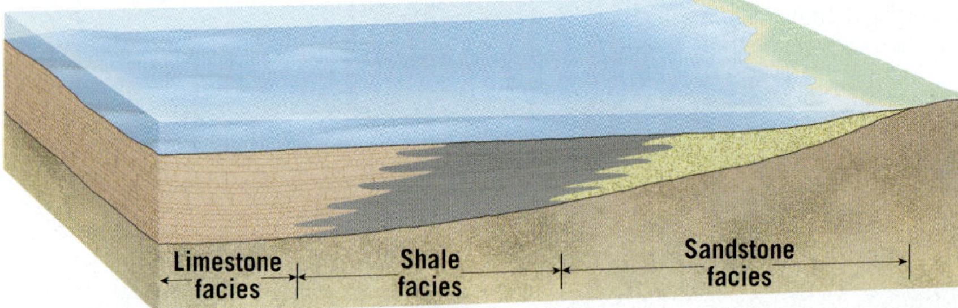

▲ **SmartFigure 7.24 Lateral change** When a sedimentary layer is traced laterally, we may find that it is made up of several different rock types. This occurs because many sedimentary environments can exist at the same time over a broad area. The term *facies* is used to describe such sets of sedimentary rocks. Each facies grades laterally into another that formed at the same time but in a different environment.

TUTORIAL
https://goo.gl/79Abcq

is most characteristic of sand dunes, river deltas, and certain stream channel deposits (see Figure 7.8 and Figure 7.26).

Graded beds are another special type of bedding. In this case, the particles within a single sedimentary layer gradually change from coarse at the bottom to fine at the top. Graded beds are most characteristic of rapid deposition from water containing sediment of varying sizes. When a current experiences a rapid energy loss, the largest particles settle first, followed by successively smaller grains. The deposition of a graded bed is most often associated with a turbidity current, a mass of sediment-choked water that is denser than clear water and that moves downslope along the bottom of a lake or an ocean (Figure 7.27).

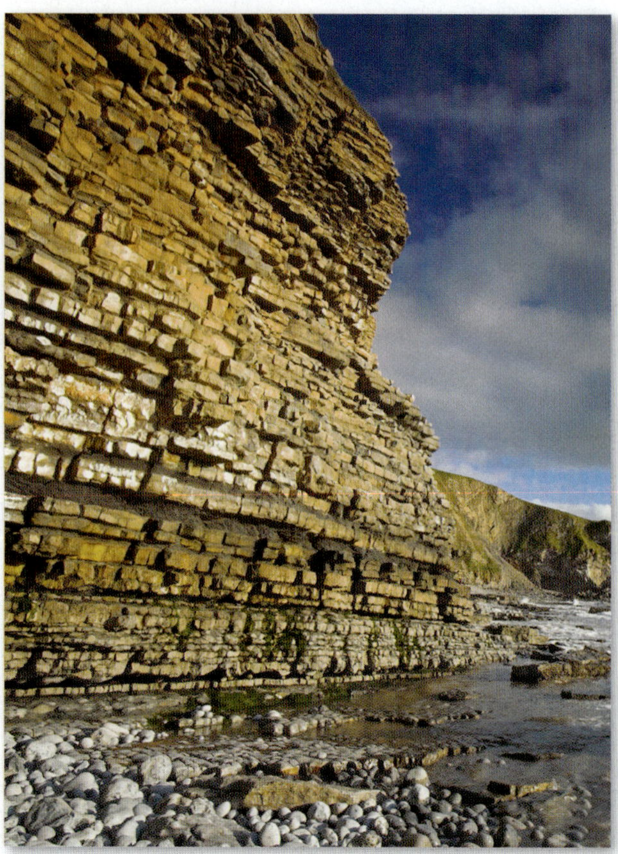

▲ **Figure 7.25 Layers are called strata** This outcrop of sedimentary strata illustrates the characteristic layering of this group of rocks. These strata are exposed at Dunraven Bay along the coast of Wales. (Photo by Adam Burton/Alamy)

► **Figure 7.26 Cross-bedding** Sand dunes typically exhibit thin layers inclined at an angle to the main bedding.

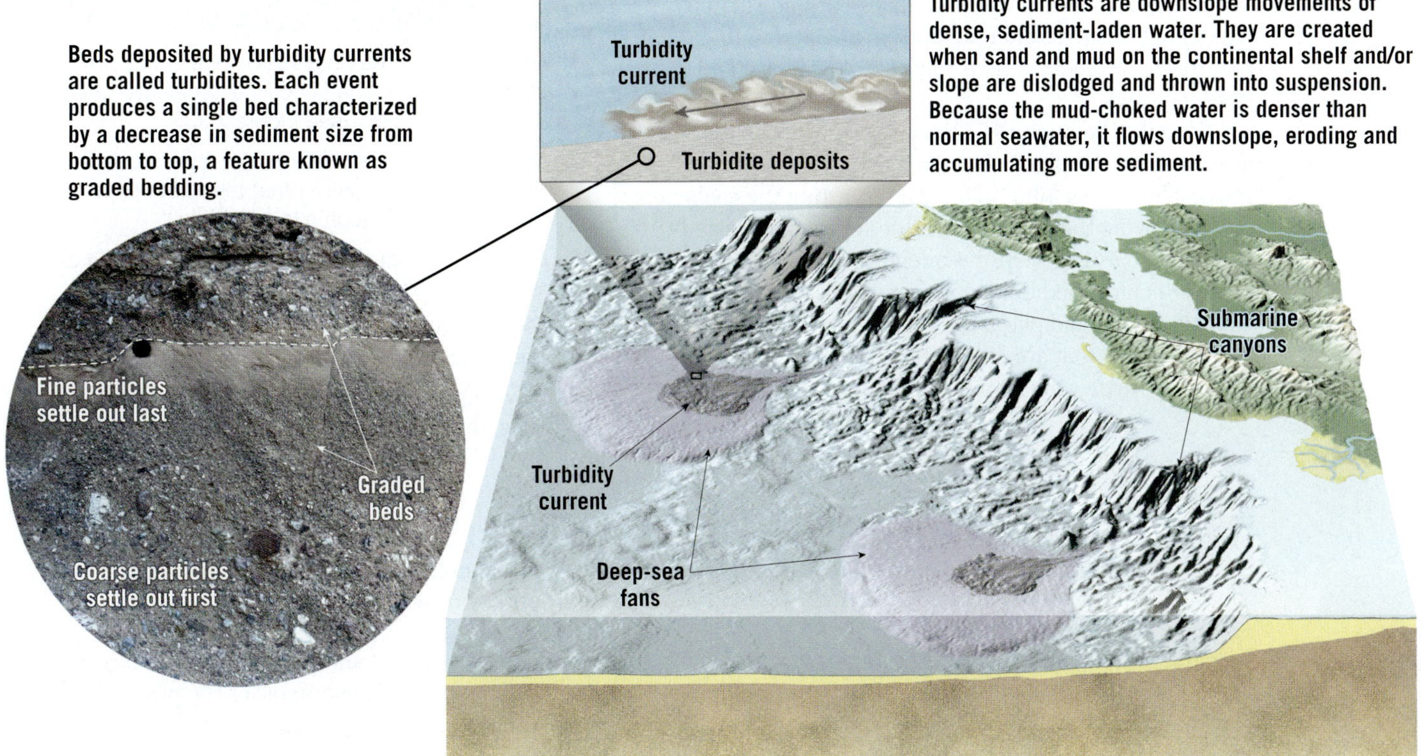

Beds deposited by turbidity currents are called turbidites. Each event produces a single bed characterized by a decrease in sediment size from bottom to top, a feature known as graded bedding.
Turbidity current
Turbidite deposits
Turbidity currents are downslope movements of dense, sediment-laden water. They are created when sand and mud on the continental shelf and/or slope are dislodged and thrown into suspension. Because the mud-choked water is denser than normal seawater, it flows downslope, eroding and accumulating more sediment.
Fine particles settle out last
Graded beds
Coarse particles settle out first
Submarine canyons
Turbidity current
Deep-sea fans

▲ **Figure 7.27 Graded bedding** A graded bed is characterized by a decrease in sediment size from bottom to top. Graded beds are associated with submarine currents known as *turbidity currents.* (Photo by Marli Miller)

Geologists can deduce quite a bit about past environments by examining sedimentary rocks. A conglomerate, for example, may indicate a high-energy environment, such as a surf zone or rushing stream, where only coarse materials remain and finer particles are kept suspended. If the rock is arkose, it may signify a dry climate, where little chemical alteration of feldspar is possible. Carbonaceous shale is a sign of a low-energy, organic-rich environment, such as a swamp or lagoon.

Other features found in some sedimentary rocks also give clues to past environments. Ripple marks are such a feature. **Ripple marks** are small waves of sand that develop on the surface of a sediment layer through the action of moving water or air (Figure 7.28A). The ridges form at right angles to the direction of motion. If the ripple marks are formed by air or water moving in essentially one direction (*current ripple marks*), they will be asymmetrical, with a steeper side in the downcurrent direction and a more gradual slope on the upcurrent side. A stream flowing across a sandy bed and wind blowing over a sand dune produce current ripples. When present in solid rock, they reveal the direction of ancient wind or water currents. Ripple marks that form in response to the back-and-forth movement of waves in near-shore environments are symmetrical and are called *oscillation ripple marks.*

Mud cracks (Figure 7.28B) indicate that the sediment in which they were formed was alternately wet and dry.

▲ **Figure 7.28 Frozen in stone A.** These current ripple marks formed in sandy sediment and are now preserved in rock. (Photo by Tim Graham/Alamy Images) **B.** When muddy sediments dry out, they shrink and create cracks. When the sediment is turned into rock, the mud cracks are preserved. (Photo by Marli Miller)

When exposed to air, wet mud dries out and shrinks, producing cracks. Mud cracks are associated with environments such as tidal flats, shallow lakes, and desert basins.

Fossils, the remains or traces of prehistoric life, are important inclusions in sediment and sedimentary rocks. They are helpful tools for interpreting the geologic past. Knowing the nature of the life-forms that existed at a particular time helps researchers understand past environmental conditions. Further, fossils are important time indicators and play a key role in correlating rocks of similar ages that are from different places. Fossils are discussed further in Chapter 18.

CONCEPT CHECKS 7.7

1. What are the three broad categories of sedimentary environments? List a specific example associated with each category (see Figure 7.23).
2. Why might a single sedimentary layer include different types of rock? What term applies to the different parts of such a layer?
3. What is the single most characteristic feature of sedimentary rocks?
4. What is the difference between cross-bedding and graded bedding?
5. How might mud cracks and ripple marks be useful clues about the geologic past?

7.8 Resources from Sedimentary Rocks

Distinguish between the two broad groups of nonmetallic mineral resources. Discuss the three important fossil fuels associated with sedimentary rocks.

Did You Know?
Producing 1 ton of steel requires about 1/3 ton of limestone and between 2 and 20 lb of fluorspar (fluorite).

In addition to the fact that sedimentary rocks contain critical information about Earth's history, this group is economically important as well. Sedimentary rocks are associated with important mineral and energy resources.

Nonmetallic Mineral Resources

Earth materials that are not used as fuels or processed for the metals they contain are referred to as **nonmetallic mineral resources**. Realize that use of the word *mineral* is very broad in this economic context and is quite different from the geologist's strict definition of mineral found in Chapter 3. Nonmetallic mineral resources are extracted and processed either for the nonmetallic elements they contain or for the physical and chemical properties they possess (Table 7.1). Although these resources have diverse origins, many are sediments or sedimentary rocks. Because a large percentage of these materials are used in manufacturing or to make fertilizers—processes we rarely see—their importance is easy to underestimate.

The quantities of nonmetallic minerals used each year are enormous. Per capita consumption of nonfuel resources in the United States is nearly 11 metric tons, of which more than 94 percent are nonmetallic minerals. Nonmetallic mineral resources are commonly divided into two broad groups: **building materials** and **industrial minerals**. Some substances fall into both groups. Limestone, perhaps the most versatile and widely used rock of all, is the best example (Figure 7.29). As a building material, it is used not only as crushed rock and building

Table 7.1 Uses of Nonmetallic Minerals

Mineral	Uses
Apatite	Phosphorus fertilizers
Asbestos (chrysotile)	Incombustible fibers
Calcite	Aggregate, steelmaking, soil conditioning, chemicals, cement, building stone
Clay minerals (kaolinite)	Ceramics, china
Corundum	Gemstones, abrasives
Diamond	Gemstones, abrasives
Fluorite	Steelmaking, aluminum refining; glass; chemical manufacture
Garnet	Abrasives, gemstones
Graphite	Pencil lead, lubricant, refractories
Gypsum	Plaster of Paris
Halite	Table salt, chemicals, ice control
Muscovite	Insulator in electrical applications
Quartz	Primary ingredient in glass
Sulfur	Chemical and fertilizer manufacture
Sylvite	Potassium fertilizers
Talc	Powder used in paints, cosmetics, etc.

▲ **Figure 7.29 Limestone quarry** Limestone is considered both a building material and an industrial mineral. This quarry is near Amsterdam, Indiana. (Photo by Daniel Dempster/Alamy Images)

stone but also in making cement. As an industrial mineral, limestone is an ingredient in the manufacture of steel and is used in agriculture to neutralize acidic soils.

Building materials also include cut stone, aggregate (sand, gravel, and crushed rock), gypsum for plaster and wallboard, clay for tile and bricks, and cement, which is made from limestone and shale. Cement and aggregate go into the making of concrete, a material that is essential to practically all construction.

A wide variety of resources are classified as industrial minerals. Some are sources of elements or compounds that are used in the manufacture of chemicals and the production of fertilizers. Corundum and garnet are used as abrasives. Deposits of industrial minerals are generally far more restricted in distribution and extent than are deposits of building materials.

Energy Resources

Earth's tremendous industrialization over the past two centuries has been fueled—and is still fueled—by burning coal, petroleum, and natural gas. About 81 percent of the energy consumed in the United States today comes from these basic sources. Figure 7.30 shows where our energy comes from and what it is used for. Our reliance on fossil fuels is obvious. Although we are using greater quantities of energy from alternative sources such as solar, wind, geothermal, and hydroelectric, the U.S. Department of Energy projects that fossil fuels will continue to be very important for decades to come.

Coal In addition to oil and natural gas, coal is commonly called a **fossil fuel**. This designation is appropriate because when coal is burned, energy from the Sun that was stored by plants many millions of years ago is being used, hence the actual burning of "fossils." In the United States, coal fields are widespread and contain supplies that should last for hundreds of years (Figure 7.31).

Coal is plentiful, but its recovery and use present a number of challenges. Surface mining can turn the countryside into a scarred wasteland if careful (and costly) reclamation is not carried out to restore the land. Today all U.S. surface mines are required to reclaim the land. Although underground mining does not scar the landscape to the same degree, it has been costly in terms of human life and health. Strong

Did You Know?
Just 2 km of four-lane highway require more than 85,000 tons of aggregate. The United States produces about 2 billion tons of aggregate per year. This represents about one-half of the entire nonenergy mining volume in the country.

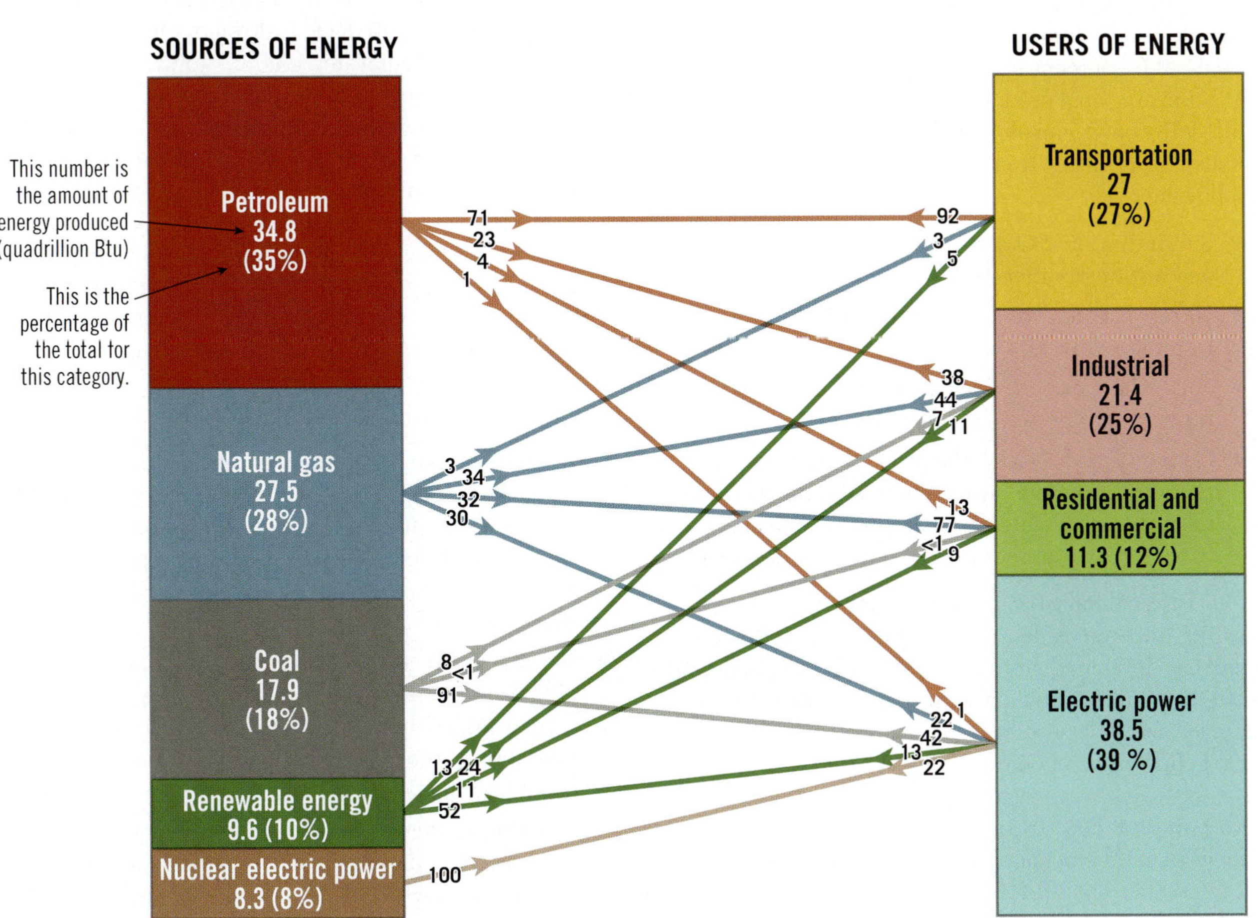

Reading this double graph:
The left side indicates what energy sources we use. The right side shows where we use the energy.
The lines with numbers that connect the graphs provide more details. Use the top line as an example. It shows that 71% of the petroleum is used by the transportation sector. It also indicates that 92% of the energy used by the transportation sector is petroleum.

SmartFigure 7.30 **U.S. energy consumption, 2014** The total was 98.3 quadrillion Btu. A quadrillion is 10 raised to the 15th power, or a million billion. A quadrillion Btu is a convenient unit for referring to U.S. energy use as a whole. (U.S. Department of Energy/Energy Information Agency)

TUTORIAL
https://goo.gl/cbZGI1

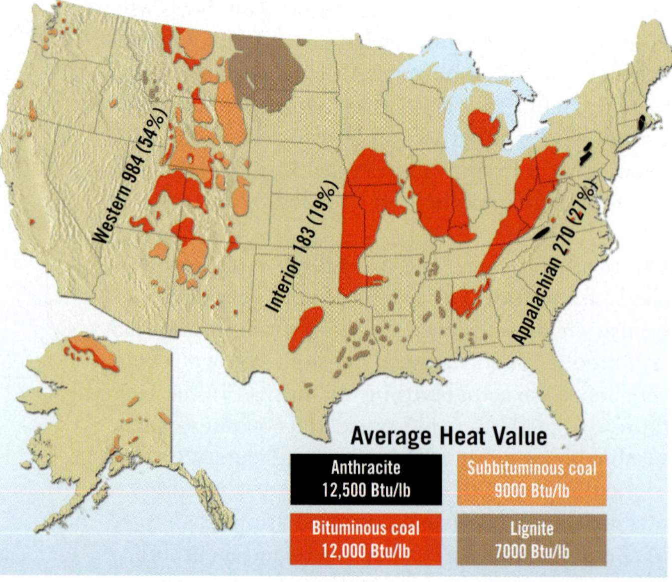

Figure 7.31 Coal fields of the United States Production is shown in millions of tons produced in 2014. The percentage of the total production is indicated in parentheses. Most of the coal mined in the United States is subbituminous (47%) and bituminous (45%). (Data from Energy Information Agency/DOE)

federal safety regulations have made U.S. mining quite safe. However, collapsing roofs, gas explosions, and the required heavy equipment remain hazards. Over the years, the share of coal produced from surface mines has increased significantly, from 51 percent in 1949 to about 65 percent in 2014.

Burning coal produces emissions that adversely affect the environment and human health. Principal emissions resulting from coal combustion include the following:

- Sulfur dioxide (SO_2), which contributes to acid rain and respiratory illnesses
- Nitrogen oxides (NO_x), which contribute to smog and respiratory illnesses
- Particulate matter, which contributes to smog, haze, respiratory illnesses, and lung disease
- Carbon dioxide (CO_2), the primary greenhouse gas produced from the burning of fossil fuels, which plays a significant role in the heating of our atmosphere (Chapter 20 examines this issue in some detail.)

The coal industry has found several ways to reduce sulfur, nitrogen oxides, and other impurities from coal and has developed more effective ways of cleaning coal after it is mined. Coal consumers have shifted toward greater use of less-polluting low-sulfur coal. Yet significant challenges remain.

Did You Know?
Wyoming is far and away the largest producer of coal in the United States. With 39.4 percent of U.S. production in 2014, Wyoming is home to the eight largest mines in the country. More than 90 percent of the coal produced in the United States is used for the generation of electricity. Coal accounted for about 42 percent of U.S. electric power in 2014.

Oil & Natural Gas Together, oil and natural gas provided more than 60 percent of the energy consumed in the United States in 2014. By examining Figure 7.30, you can see that the transportation sector of the U.S. economy relies almost totally on petroleum as an energy source. In 2011, natural gas surpassed coal for the first time in more than 30 years as a source of energy in the United States. An important reason for this is the use of new technologies that have increased production from shale formations. (Hydraulic fracturing is discussed later in this section.) Users of natural gas are almost evenly divided among the three categories other than transportation in Figure 7.30.

Like coal, petroleum and natural gas are biological products derived from the remains of organisms. However, while coal formed mostly from plant material that accumulated in swampy environments on land, as shown in Figure 7.18, oil and natural gas originated mainly from microscopic marine plankton in ancient seas millions of years ago. This organic material accumulated in marine sedimentary basins together with mud, sand, and other sediments that protected it from oxidation. With increased temperature and burial, chemical reactions gradually transformed this organic matter into the liquid and gaseous hydrocarbons we call petroleum and natural gas.

Unlike the organic matter from which they form, petroleum and natural gas are mobile. These fluids are gradually squeezed from the compacting, mud-rich layers where they originate, called the **source rock**, into adjacent permeable beds such as sandstone, where openings between sediment grains are larger. Because this occurs in a marine environment, the rock layers containing the oil and gas are saturated with saltwater. Oil and gas, being less dense than water, migrate upward through the water-filled pore spaces of the enclosing rocks.

A geologic environment that allows for economically significant amounts of oil and gas to accumulate underground is termed an **oil trap**. Several geologic structures can act as oil traps, and all have two basic features: a porous, permeable **reservoir rock** that can yield petroleum and natural gas in sufficient quantities to make drilling worthwhile, and a **cap rock**, such as shale, that is virtually impermeable to oil and gas and hence traps them in the reservoir rock. Figure 7.32 illustrates the following common oil and natural gas traps:

- **Anticline.** One of the simplest traps is an *anticline*, an up-arched series of sedimentary strata (Figure 7.32A). The rising oil and gas collect at the top of the fold. Because of its lower density, natural gas collects above the oil. Both the oil and gas rest on the denser water that saturates the reservoir rock.
- **Fault trap.** When strata are displaced so that a dipping reservoir rock abuts an impermeable bed, a *fault trap* forms, as shown in Figure 7.32B. The upward migration of the oil and gas is halted where they encounter the fault.
- **Salt dome.** In the Gulf coastal plain region of the United States, important accumulations of oil occur in association with *salt domes*. Such areas have thick accumulations of sedimentary strata that include layers of rock salt. Salt occurring at great depths has been forced to rise in columns by the pressure of overlying beds. These rising salt columns gradually deform the overlying strata. Because oil and gas migrate to

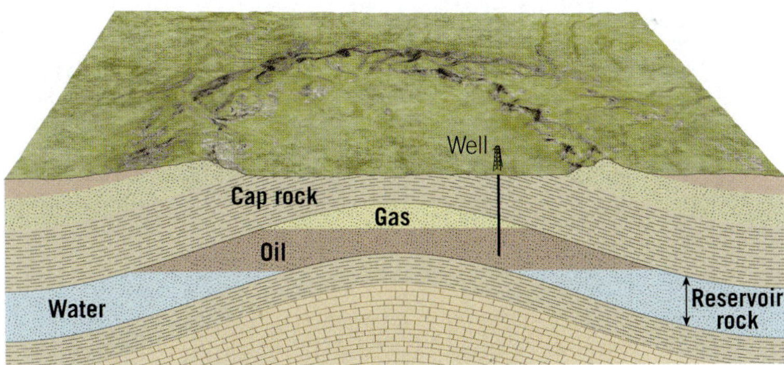

A. Anticline

B. Fault trap

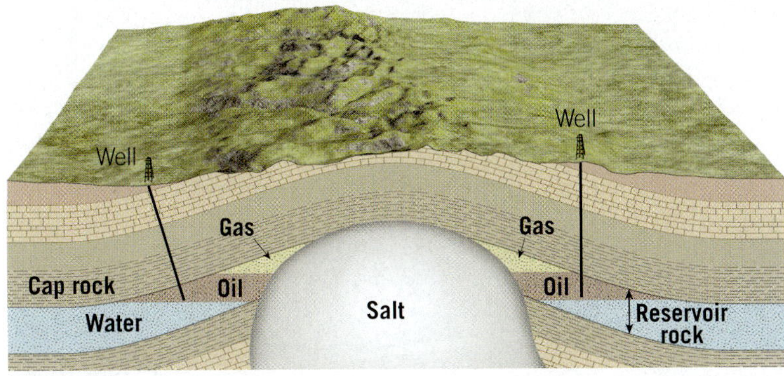

C. Salt dome

D. Stratigraphic (pinchout) trap

▲ SmartFigure 7.32
Common oil traps

the highest level possible, they accumulate in the upturned sandstone beds adjacent to the salt column (Figure 7.32C).

- **Stratigraphic (pinchout) trap.** A *stratigraphic trap* results primarily from the original pattern of sedimentation rather than from structural deformation. The stratigraphic trap illustrated in Figure 7.32D exists because a sloping bed of sandstone thins to the point of disappearance.

Whatever the type of trap, when drilling punctures the lid created by the cap rock, the oil and natural gas, which are under pressure, migrate from the pore spaces of the reservoir rock to the drill pipe. On rare occasions, when fluid pressure is great, it may force oil up the drill hole to the surface, causing a "gusher" at the surface. Usually, however, a pump is required to extract the oil.

Hydraulic Fracturing Some shale deposits contain significant reserves of oil and natural gas that cannot naturally leave because of the rock's low permeability. The practice of **hydraulic fracturing** (often called *fracking*) shatters the shale, opening up cracks through which the oil and natural gas can flow into wells and then be brought to the surface. Figure 7.33 illustrates the process. The fracturing of the shale is initiated by pumping fluids into the rock at very high pressures. The fluid is mostly water but also includes other chemicals that aid in the fracturing process. Some of these chemicals may be toxic, and there are concerns about fracking fluids leaking into freshwater aquifers that supply drinking water. The injection fluid also includes sand, so once fractures open up in the shale, the sand grains can keep them propped open and permit the oil and gas to continue to flow. Once the fracturing has been accomplished, the fracking fluid is brought back to the surface. Often this wastewater is injected into deep disposal wells. In some locations, these injections appear to trigger numerous minor earthquakes. Due to concerns about potential groundwater contamination and induced seismicity, hydraulic fracturing remains a controversial practice. Its environmental effects remain a focus of continuing research.

CONCEPT CHECKS 7.8

1. Nonmetallic mineral resources are commonly divided into what two broad groups? Give some examples of materials that belong to each group.
2. Why are coal, oil, and natural gas called *fossil fuels*? What part of U.S. energy consumption does each represent?
3. Coal has the advantage of being plentiful. What are some of coal's disadvantages?
4. What is an *oil trap*? Sketch and label two examples. What do all oil traps have in common?
5. Describe the circumstances in which hydraulic fracturing is used.

TUTORIAL
https://goo.gl/a2Ki8m

Did You Know?
The first successful oil well was completed by Edwin Drake on August 27, 1859, along Oil Creek near Titusville, Pennsylvania. Oil-bearing strata were encountered at a depth of just 20.7 m (69 ft).

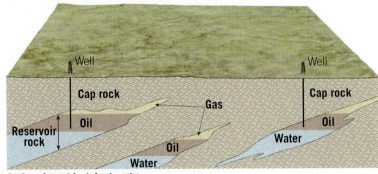

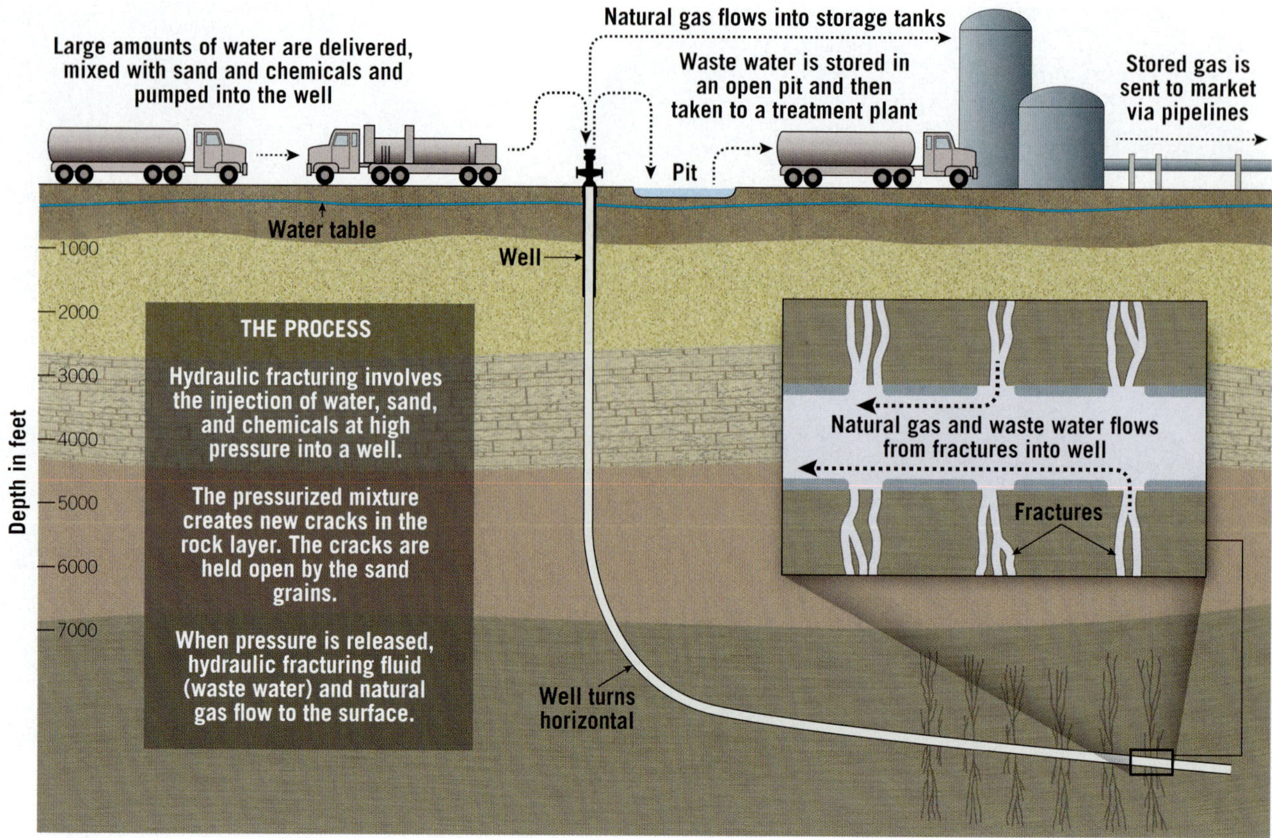
Large amounts of water are delivered, mixed with sand and chemicals and pumped into the well
Natural gas flows into storage tanks
Waste water is stored in an open pit and then taken to a treatment plant
Stored gas is sent to market via pipelines
Pit
Water table
Well
Depth in feet
1000
2000
3000
4000
5000
6000
7000
THE PROCESS
Hydraulic fracturing involves the injection of water, sand, and chemicals at high pressure into a well.
The pressurized mixture creates new cracks in the rock layer. The cracks are held open by the sand grains.
When pressure is released, hydraulic fracturing fluid (waste water) and natural gas flow to the surface.
Natural gas and waste water flows from fractures into well
Fractures
Well turns horizontal

Figure 7.33 Hydraulic fracturing ("fracking") This well-stimulation process is commonly used in low-permeability rocks such as shale to increase oil and/or natural gas production.

7.9 The Carbon Cycle & Sedimentary Rocks

Relate weathering processes and sedimentary rocks to the carbon cycle.

To illustrate the movement of material and energy among the spheres of the Earth system, let us take a brief look at the **carbon cycle** (Figure 7.34). Most carbon is bonded chemically to other elements to form compounds such as carbon dioxide, calcium carbonate, and the hydrocarbons found in coal and petroleum. Carbon is also the basic building block of life, as it readily combines with hydrogen and oxygen to form the fundamental organic compounds that compose living things.

Certainly one of the most active parts of the carbon cycle is the movement of CO_2 from the atmosphere to the biosphere and back again. In the atmosphere, carbon is found mainly as carbon dioxide (CO_2). Atmospheric carbon dioxide is significant because it is a greenhouse gas, which means it is an efficient absorber of energy emitted by Earth and thus influences the heating of the atmosphere, a process described in more detail in Chapter 20. Because many of the processes that operate on Earth involve carbon dioxide, this gas is constantly moving into and out of the atmosphere. Through the process of photosynthesis, plants absorb carbon dioxide from the atmosphere to produce the essential organic compounds needed for growth. Animals that consume these plants (or that consume other animals that eat plants) use these organic compounds as a source of energy and, through the process of respiration, return carbon dioxide to the atmosphere. (Plants also return some CO_2 to the atmosphere via respiration.) Further, when plants die and decay or are burned, this biomass is oxidized, and carbon dioxide is returned to the atmosphere. In the oceans, photosynthetic plankton plays the same role that plants do on land.

Not all plants (and plankton) decay immediately back to carbon dioxide. A small percentage is deposited as sediment. Over long spans of geologic time, considerable biomass is buried with sediment. Under the right conditions, some of these carbon-rich deposits are converted to fossil fuels—coal, petroleum, or natural gas. Some of these fuels are eventually recovered (mined or pumped from a well) and burned to generate electricity and fuel our transportation system. One result of fossil-fuel combustion is the release of huge quantities of carbon dioxide back into the atmosphere.

Carbon also moves from the geosphere and hydrosphere to the atmosphere and back again. For example, volcanic activity early in Earth's history is thought to have been the source of much of our atmosphere's carbon dioxide. One way that carbon dioxide makes its way back to the hydrosphere and then to the solid Earth is by first combining with water to form carbonic acid (H_2CO_3), which then attacks the rocks that compose Earth's crust. One product of this chemical weathering of solid rock is the soluble bicarbonate ion (HCO_3^-), which is carried by groundwater and streams to the ocean. Water-dwelling organisms extract this dissolved material to produce hard parts of calcium carbonate ($CaCO_3$). When the organisms die, these skeletal remains settle to the ocean floor as biochemical sediment and become sedimentary rock. In fact, the crust is by far Earth's largest depository of carbon, where it is a constituent of a variety of rocks, the most abundant being limestone. When the rock is carried to great depths at a subduction zone, where metamorphism and melting can occur, the heated rock recombines into silicate minerals, releasing carbon dioxide. When volcanoes erupt, they vent the gas to the atmosphere and cover the land with fresh silicate rock, beginning the cycle again.

In summary, carbon moves among all four of Earth's major spheres. It is essential to every living thing in the biosphere. In the atmosphere, carbon dioxide is an important greenhouse gas. In the hydrosphere, carbon dioxide is dissolved in lakes, rivers, and the ocean. In the geosphere, carbon is contained in carbonate sediments and sedimentary rocks and is stored as organic matter dispersed through sedimentary rocks and as deposits of coal and petroleum.

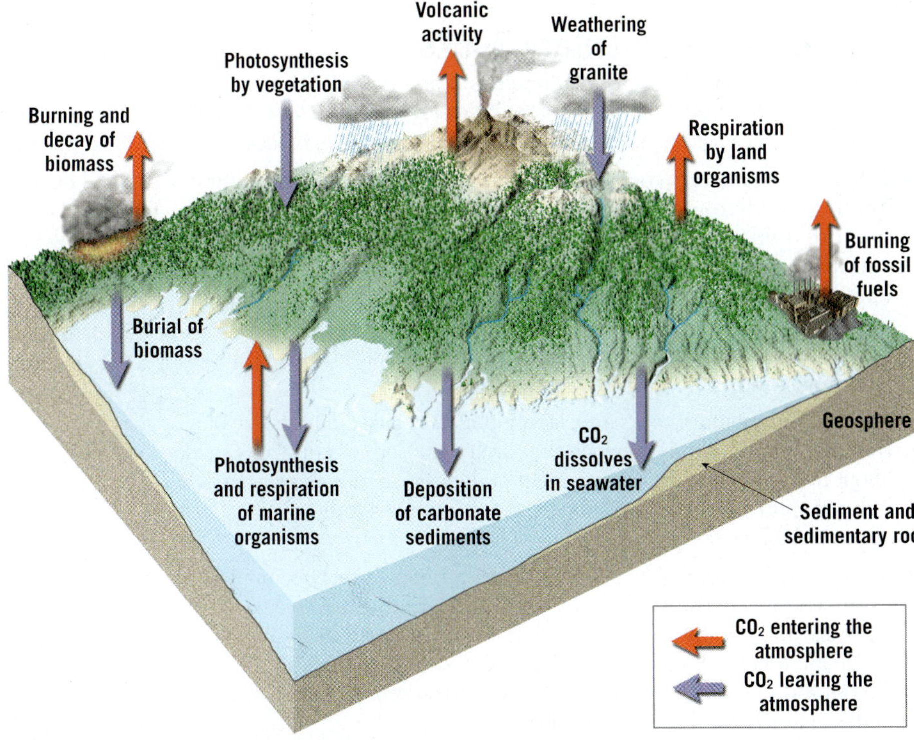

▲ **Figure 7.34 Carbon cycle** This simplified diagram emphasizes the flow of carbon between the atmosphere and the hydrosphere, geosphere, and biosphere. The arrows show whether the flow of carbon is into or out of the atmosphere.

CONCEPT CHECKS 7.9

1. Describe how chemical weathering and the formation of biochemical sediment remove carbon from the atmosphere and store it in the geosphere.
2. Provide an example of carbon moving from the geosphere to the atmosphere.

CONCEPTS IN REVIEW
Sedimentary Rocks

7.1 An Introduction to Sedimentary Rocks

Explain the importance of sedimentary rocks and summarize the part of the rock cycle that pertains to sediments and sedimentary rocks. List the three categories of sedimentary rocks.

KEY TERMS: sedimentary rock, detrital sedimentary rock, chemical sedimentary rock, organic sedimentary rock

- Though igneous and metamorphic rocks make up most of Earth's crust by volume, sediment and sedimentary rocks are concentrated near the surface. There, at the interface of Earth's four spheres, sediments and the rock layers they eventually form record past conditions and events at the surface. Sedimentary rocks contain fossils that show the evolution of life over time.
- Numerous geologic resources are primarily associated with sedimentary rocks. Examples include coal, oil, uranium, and several major ores of metals.
- Sediments are the raw materials from which sedimentary rocks are made. Sediments are produced from the weathering of preexisting rocks. Both solid particles of many sizes and chemical residues, including ions in solution, qualify as sediments.
- Once produced, sediments are transported from their source area by water currents, wind, glacial ice, or simply downhill under the influence of gravity. Eventually, solid particles are deposited at a new site, where they are compacted and/or cemented into a detrital sedimentary rock. Ions in solution are precipitated and become the raw material for chemical sedimentary rocks.
- There are three main varieties of sedimentary rocks: detrital, chemical, and organic.

? Of the two main sources of energy that drive the rock cycle—Earth's internal heat and solar energy—which one is primarily responsible for the formation of sediment and sedimentary rocks? Briefly explain.

7.2 Detrital Sedimentary Rocks

Describe the primary basis for distinguishing among detrital rocks and discuss how the origin and history of such rocks might be determined.

KEY TERMS: shale, fissility, sandstone, sorting, conglomerate, breccia

- Detrital sedimentary rocks are made of solid particles, mostly quartz grains and microscopic clay minerals. Quartz and clay dominate because unlike most other minerals, they are stable at Earth's surface. Feldspars and micas are conspicuous additions to certain detrital sediments, indicating a relatively short time in the chemical weathering environment; these were mechanically weathered, transported a relatively short distance, and deposited with minimal decomposition.
- Detrital sedimentary rocks are classified mainly based on the size of the sedimentary grains of which they are composed. Grain size is a clue about how energetic the environment of deposition was. Bigger particles indicate more powerful transporting currents; finer grains can be deposited only where current energy is relatively low.
- Shale is made mostly of small grains of clay minerals that accumulate in low-energy depositional environments such as the deep sea, lake bottoms, and floodplains adjacent to rivers. Shale exhibits fissility due to the alignment of microscopic clay flakes parallel to bedding. Shale containing a lot of organic material forms in low-oxygen environments and is characterized by a black color.
- Sandstone is dominated by sand-sized grains and may exhibit various degrees of sorting. Sorting is a reflection of how abruptly or gradually the sand was deposited. Rounding of the individual sand grains is another important aspect of a sandstone's texture: Rounder grains signify further transport, while angular grains imply a shorter transport distance. Sandstones also vary in their composition: The presence of minerals that are relatively unstable at the surface (like feldspar) implies that the sand did not undergo much chemical weathering prior to deposition. The greater the proportion of quartz, the more the source sediment was chemically weathered before deposition. Three main varieties of sandstone worth knowing are quartz sandstone, arkose, and graywacke.
- Conglomerate and breccia are characterized by a high proportion of gravel-sized grains. If deposited by water, a conglomerate implies a very energetic current. Grains in a breccia are angular, indicating that the material was deposited closer to its source area. Conglomerate is made of rounded grains, implying a significant amount of transport before deposition.

? Examine these sketches of sediment particles. Which particle—A, B, or C—has traveled farthest from its source? Explain your reasoning.

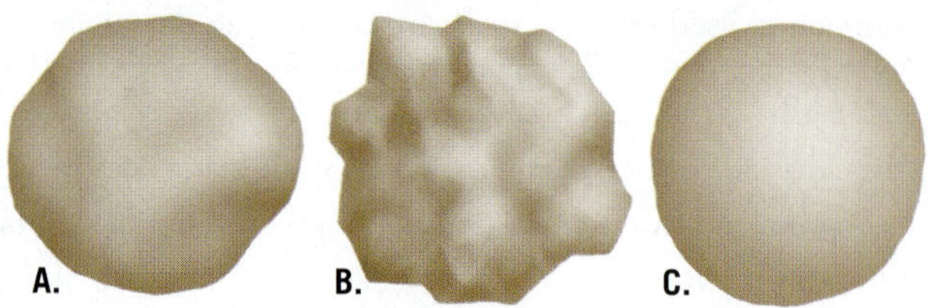

7.3 Chemical Sedimentary Rocks

Explain the processes involved in the formation of chemical sedimentary rocks and list several examples.

KEY TERMS: biochemical, limestone, dolostone, chert, evaporite, salt flat

- Chemical sedimentary rocks are formed when ions dissolved in solution link together to form mineral crystals. Sometimes this happens inorganically (without life being involved), and other times living organisms biochemically extract the ions to precipitate mineral matter as bone or shell.
- Limestone is the most common chemical sedimentary rock. It forms mainly in shallow, warm ocean settings. Limestone is dominated by calcium carbonate. This is the material from which corals construct reefs. Coquina and chalk are also examples of biochemical limestone. Travertine and oolitic limestone are examples of inorganic limestone.
- Dolostone is a chemical sedimentary rock that is dominated by the mineral dolomite. Like calcite, dolomite is a carbonate mineral, but about half of its calcium ions have been replaced by magnesium ions.
- Chert is the general term for rocks made of microcrystalline silica. If the chert is red, it is called jasper. Black chert is flint. Agate is multicolored. When silica replaces plant matter to make petrified wood, often a variety of colors are present.
- Evaporite deposits form when minerals precipitate from an ever-more-concentrated solution of dissolved ions. This is how the vast salt flats of the American West formed. Digging into these deposits reveals rock salt, rock gypsum, and the potassium salt sylvite.

? Does this rock salt more likely have a biochemical or an inorganic origin?

Dennis Tasa

7.4 Coal: An Organic Sedimentary Rock

Outline the successive stages in the formation of coal.

KEY TERM: coal

- Coal forms from large amounts of plant matter buried in low-oxygen depositional environments such as swamps and bogs. Through compression, the peat formed becomes compressed into a low-grade form of coal called lignite. Lignite can be compressed further, driving out volatile components and concentrating carbon to make higher-grade bituminous coal. Metamorphism accompanying mountain building can take this concentration process even further, producing the highest-grade coal, anthracite.

? Considering that it comes from plants, what is the ultimate source of the energy in coal?

7.5 Turning Sediment into Sedimentary Rock: Diagenesis & Lithification

Describe the processes that convert sediment into sedimentary rock and other changes associated with burial.

KEY TERMS: diagenesis, lithification, compaction, cementation

- When sediments are buried to relatively shallow depths (the upper few kilometers of the crust), changes in temperature and pressure trigger a variety of processes called diagenesis, a collective term for all the chemical, physical, and biological changes that occur after sediments are deposited and during and after lithification.
- The transformation of sediment into sedimentary rock is called lithification. The two main processes contributing to lithification are compaction (a reduction in pore space by packing grains more tightly together) and cementation (a reduction in pore space by adding new mineral material that acts as a "glue" to bind the grains to each other).

7.6 Classification of Sedimentary Rocks

Summarize the criteria used to classify sedimentary rocks.

KEY TERMS: clastic, nonclastic (crystalline)

- Sedimentary rocks are classified primarily based on whether they are detrital, chemical, or organic. Detrital rocks are subdivided by grain size, while among the chemical rocks, mineral composition is the key distinguishing characteristic. An additional characteristic is whether the rocks exhibit a clastic or nonclastic (crystalline) texture.
- Crystalline texture is common to nonclastic sedimentary rocks and igneous rocks and is particularly apparent under magnification. However, the minerals involved are totally different and allow the two types of rock to be distinguished.

? Would a limestone made of shell fragments be detrital or chemical? Would its texture be clastic or crystalline (nonclastic)?

7.7 Sedimentary Rocks Represent Past Environments

Distinguish among three broad categories of sedimentary environments and provide an example of each. List several sedimentary structures and explain why these features are useful to geologists.

KEY TERMS: environment of deposition (sedimentary environment), facies, strata (beds), bedding plane, cross-bedding, graded bed, ripple mark, mud crack, fossil

- Different combinations of tectonic, climatic, and biological conditions result in different types of sediment accumulating. The principle of uniformity suggests that the sedimentary record can be interpreted in light of modern depositional environments. Continental, marine, and transitional (shoreline) environments all have distinctive characteristics that allow geologists to identify sedimentary rocks formed in those environments.

(7.7 continued)

- Sedimentary facies are lateral equivalents that represent different depositional conditions operating in adjacent areas at the same time. For instance, a beach today may be depositing sand, while a kilometer or two offshore only mud is being deposited, and further beyond that, carbonate minerals may be precipitating. All are the same age but represent neighboring areas governed by different conditions.
- Sedimentary structures are patterns that form in sedimentary rock at the time of deposition (or shortly thereafter), before the sediments become lithified. They can provide powerful clues to the conditions under which the sediment accumulated.
- Beds (or strata) are sheets of sediment deposited in a more-or-less continuous layer. Sometimes cross-beds are preserved within bedding, allowing geologists to deduce the direction of the depositional current. Ripple marks, small waves created by wind or water on the surface of a sediment layer, also provide useful clues. Graded beds indicate depositional currents that quickly lost their energy. Larger grains settled out first, and the smallest grains settled out last. Mud cracks form when mud contracts upon drying out, and they indicate that the sediment was exposed to air. Fossils serve as useful tools that allow geologists to date and interpret past environmental conditions.

? The accompanying photo shows Los Osos Creek entering California's Morro Bay. Sediment transported by the creek has been deposited as a delta. Which broad category of sedimentary environment is represented in this scene?

Michael Collier

7.8 Resources from Sedimentary Rocks

Distinguish between the two broad groups of nonmetallic mineral resources. Discuss the three important fossil fuels associated with sedimentary rocks.

KEY TERMS: nonmetallic mineral resource, building material, industrial mineral, fossil fuel, source rock, oil trap, reservoir rock, cap rock, hydraulic fracturing

- Earth materials that are not used as fuels or processed for the metals they contain are referred to as nonmetallic resources. Many are sediments or sedimentary rocks. The two broad groups of nonmetallic resources are building materials and industrial minerals. Limestone, perhaps the most versatile and widely used rock of all, is found in both groups.
- Coal, oil, and natural gas are all fossil fuels. In each, the energy of ancient sunlight, captured by photosynthesis, is stored in the hydrocarbons that made up plants or other living things and that were then buried by sediments.
- Coal is formed from compressed plant fragments, originally deposited in ancient swamps. Burning it supplies about 18 percent of U.S. energy use. Coal mining can be risky and environmentally damaging, and burning coal generates several kinds of pollution.
- Oil and natural gas are formed from the heated remains of marine plankton. Together, they account for about 60 percent of U.S. energy use. Both oil and natural gas leave their source rock (typically shale) and migrate to an oil trap made up of other, more porous rocks, called reservoir rock, covered by a suitable impermeable cap rock.
- Hydraulic fracturing (or "fracking") is a method of opening up pore space in otherwise impermeable rocks, permitting natural gas to flow out into wells.

? As you can see here, the Sinclair Oil Company's logo speaks directly to the "fossil" nature of the fuel it sells. However, is it likely that any dinosaur carbon has ended up in Sinclair's oil? Explain.

Vespasian/Alamy

7.9 The Carbon Cycle & Sedimentary Rocks

Relate weathering processes and sedimentary rocks to the carbon cycle.

KEY TERM: carbon cycle

- Carbon is a vital component of the atmosphere, biosphere, geosphere, and hydrosphere. The same atom of carbon that is currently part of your nose may have entered your body in a piece of bread. Prior to that, it may once have been part of a plant, and before that it may have been a carbon dioxide molecule in the atmosphere. How did it get to the atmosphere? Perhaps it escaped into the air after originally being dissolved in the ocean, and perhaps it got to the ocean by being weathered from limestone and then washed down a stream. While individual details vary, the key point is that carbon is a reactive element that is equally at home in rocks, water, the air, and living tissue.

GIVE IT SOME THOUGHT

1. Develop a geologic "life history" of a sedimentary rock. Begin with a mass of igneous bedrock in a mountain area and end with your sedimentary rock being collected by a future geology student. Be as complete as possible.
2. How is the use of the term *clay* in relation to minerals different from the use of the term in Figure 7.3? How are these two uses of the term related?
3. This detrital rock consists of angular grains and is rich in potassium feldspar and quartz. What do the angular grains indicate about the distance the sediment was transported? The source of the sediment in this rock was an igneous mass. Name the likely rock type. Did the sediment in this sample undergo a great deal of chemical weathering? Explain.

E.J. Tarbuck

4. If you hiked to a mountain peak and found limestone at the top, what would that indicate about the likely geologic history of the rock atop the mountain?
5. Why is it not necessary to indicate the texture of detrital rocks on the identification chart for sedimentary rocks (see Figure 7.20)?
6. During a hike in Utah's Zion National Park, you pick up a sedimentary rock sample. When you examine the sample with your hand lens, you see that the rock consists mainly of rounded glassy particles that appear to be quartz. To be sure, you conduct two basic tests. When you check for hardness, the rock easily scratches glass, which is what quartz would do. However, when you place a drop of acid on the sample, it fizzes. Explain how a rock that appears to be rich in quartz could effervesce with acid.

7 Examine the accompanying sketch, which shows three sediment layers on the ocean floor. What term is applied to such layers? What process was responsible for creating these layers? Are these layers more likely part of an offshore lagoon or a deep-sea fan?

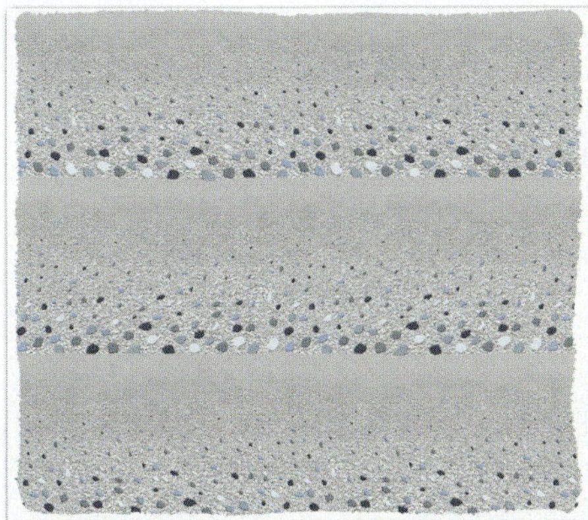

8 This image shows the surface of a sand dune. What term is applied to the wave-like ridges on the surface? Be as specific as possible. Is the prevailing wind direction most likely from the left or from the right? Explain.

Michael Collier

9 This rock sample consists of intergrown crystals. How would you determine whether the rock is sedimentary or igneous? If it is sedimentary, what term describes its texture?

E.J. Tarbuck

10 This chapter includes a discussion of industrial minerals. Are these substances actually minerals? That is, do they meet the definition of *mineral* outlined in Chapter 3? Explain.

11 Follow the history of a single carbon atom as it moves from one part of the Earth system to another. Using any order you choose, include as many of the following sites as possible: a cave, a river, a dinosaur, a coal-fired power plant, a volcanic eruption, a swamp, a can of beer, and a coral reef.

MasteringGeology™

Looking for additional review and test prep materials? Visit the Study Area in MasteringGeology to enhance your understanding of this chapter's content by accessing a variety of resources, including Self-Study Quizzes, Geoscience Animations, SmartFigures, Mobile Field Trips, *Project Condor* Quadcopter videos, *In the News* RSS feeds, flashcards, web links, and an optional Pearson eText.

www.masteringgeology.com

8 Metamorphism & Metamorphic Rocks

FOCUS ON CONCEPTS

Each statement represents the primary learning objective for the corresponding major heading within the chapter. After you complete the chapter, you should be able to:

8.1 Compare and contrast the environments that produce metamorphic, sedimentary, and igneous rocks.

8.2 List and distinguish among the four agents that drive metamorphism.

8.3 Explain how foliated and nonfoliated textures develop.

8.4 List and describe the most common metamorphic rocks.

8.5 Write a description for each of these metamorphic environments: contact metamorphism, hydrothermal metamorphism, subduction zone metamorphism, and regional metamorphism.

8.6 Explain how index minerals are used to establish the metamorphic grade of a rock body.

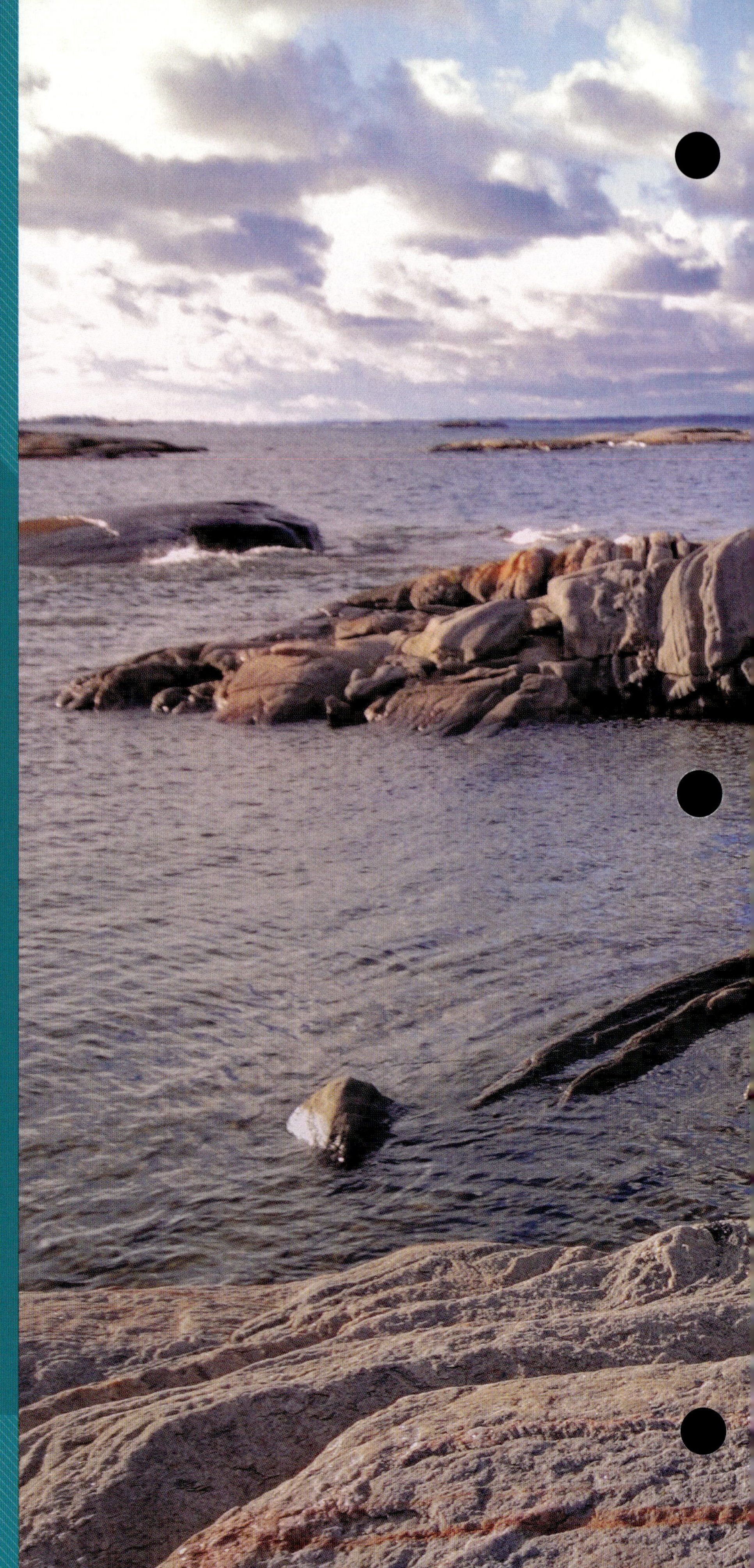

Metamorphic rocks exposed on an island in Georgian Bay, Ontario, Canada. (Photo provided by Charline Xia Ontario Canada Collection/Alamy Stock Photo)

THE FOLDED AND METAMORPHOSED ROCKS shown in the chapter-opening photo were once flat-lying sedimentary strata. Compressional forces of unimaginable magnitude, combined with temperatures hundreds of degrees above surface conditions, prevailed for perhaps thousands or millions of years to produce the deformation displayed by these rocks. Under such extreme conditions, solid rock responds by folding, fracturing, and often flowing. (The processes that deform solid rock are discussed in detail in Chapter 11.) This chapter looks at the tectonic forces that forge metamorphic rocks and how these rocks change in appearance, mineralogy, and sometimes overall chemical composition.

8.1 What Is Metamorphism?

Compare and contrast the environments that produce metamorphic, sedimentary, and igneous rocks.

Unlike some igneous and sedimentary processes that occur in surface or near-surface environments, metamorphism most often occurs deep within Earth, beyond our direct observation. Notwithstanding this significant obstacle, geologists have developed techniques that allow them to learn about the conditions under which metamorphic rocks form. In turn, the study of metamorphic rocks provides important insights into how tectonic processes operate to alter the structure and composition of Earth's crust.

Recall from the discussion of the rock cycle in Chapter 1 that metamorphism is the transformation of one rock type into another rock type. Metamorphic rocks are produced from preexisting sedimentary and igneous rocks, as well as from other metamorphic rocks. Thus, every metamorphic rock has a **parent rock**—the rock from which it was formed.

Metamorphism, which means to "change form," is a process that transforms the mineralogy, texture, and sometimes chemical composition of the parent rock. The **mineralogy** (mineral constituents of a rock) changes because the rock is subjected to new conditions, usually elevated temperatures and pressures, which are significantly different from those in which it initially formed. For example, clay minerals, which are the most common minerals in sedimentary rocks, are stable only at Earth's surface. (Kaolinite is one example of a clay mineral; refer to Figure 3.29, page 85.) When clay minerals are buried to a depth where temperatures exceed 200°C (nearly 400°F), they are transformed into the minerals chlorite and/or muscovite mica. (Chlorite is a mica-like mineral formed by the metamorphism of dark iron- and magnesium-rich silicate minerals.) Under more extreme

Figure 8.1 Metamorphic grade A. Here low-grade metamorphism is illustrated by the transformation of the common sedimentary rock shale to the more compact metamorphic rock slate. **B.** High-grade metamorphic environments obliterate the existing texture and often change the mineralogy of the parent rock. High-grade metamorphism occurs at temperatures that approach those at which rocks melt. (Photos by Dennis Tasa)

conditions, chlorite becomes biotite mica. Metamorphism also alters a rock's texture, producing larger crystals and sometimes a distinct layered or banded appearance.

The degree to which a parent rock changes during metamorphism is called its **metamorphic grade**, and it varies from low grade (low temperatures and pressures) to high grade (high temperatures and pressures). For example, in low-grade metamorphic environments, the common sedimentary rock shale becomes the more compact metamorphic rock slate. During this transformation, the clay minerals in shale are transformed into tiny chlorite and muscovite mica flakes. Hand samples of slate and shale are sometimes difficult to distinguish from one another, illustrating that the transition from sedimentary to metamorphic rock is often gradual and the change subtle (**Figure 8.1A**).

In environments where temperatures and pressures are more extreme, metamorphism causes a transformation so complete that the identity of the parent rock cannot be easily determined. In high-grade metamorphism, such features as bedding planes, fossils, and vesicles that existed in the parent rock are obliterated. Further, when rocks deep in the crust are subjected to compressional stress (like being placed in a giant vise), the entire mass may be deformed, usually by folding (**Figure 8.1B**).

Figure 8.2 illustrates the relationships among metamorphic, sedimentary, and igneous environments. Metamorphism occurs over a range of temperatures that lie between those experienced during the formation of sedimentary rocks (up to about 200°C [400°F]) and temperatures approaching those at which rocks begin to melt (about 700°C [1300°F]). However, *during metamorphism, the rock remains essentially solid*. If significant melting occurs, the rock has entered the realm of igneous activity, as discussed in Chapter 4.

Pressure also plays an important role in metamorphism. Although confining pressure acts to compact sediment to form sedimentary rocks, the pressures involved in metamorphism are even greater—sufficient to convert mineral matter into denser forms having more compact crystalline structures. Thus, metamorphism involves the formation of new minerals from preexisting ones.

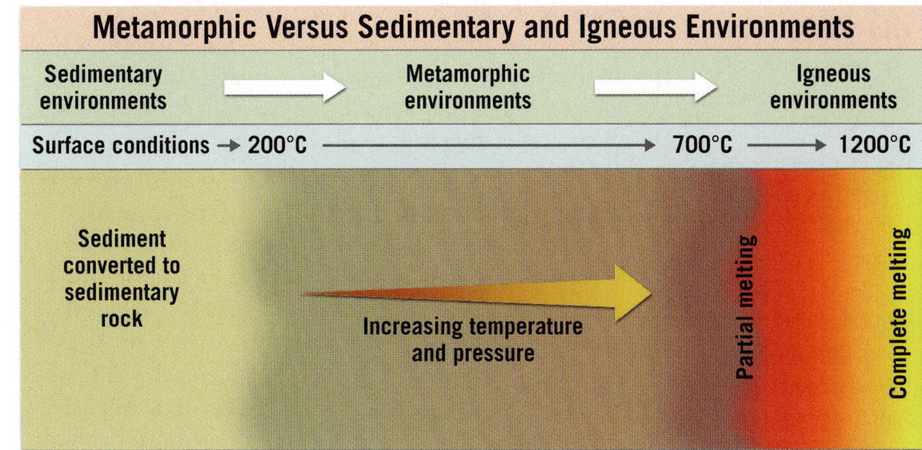

▲ **Figure 8.2**
Metamorphic versus sedimentary and igneous environments Metamorphism occurs over a range of temperatures that lie between those experienced in sedimentary environments and temperatures that approach those at which rocks melt. Pressure, which includes confining pressure and differential stress, also plays a major role in metamorphism.

CONCEPT CHECKS 8.1

1. *Metamorphism* means to "change form." Describe how a rock may change during metamorphism.
2. What is meant by the statement "Every metamorphic rock has a *parent rock*"?
3. Define *metamorphic grade*.

8.2 What Drives Metamorphism?

List and distinguish among the four agents that drive metamorphism.

The agents of metamorphism include *heat*, *pressure*, *differential stress*, and *chemically active fluids*. During metamorphism, rocks may be subjected to all four metamorphic agents simultaneously. However, the degree of metamorphism and the contribution of each agent vary greatly from one environment to another.

Heat as a Metamorphic Agent

The most important factor driving metamorphism is *heat* because it provides the energy needed to produce the chemical reactions that result in the recrystallization of existing minerals. Recall from the discussion of igneous rocks in Chapter 4 that an increase in temperature causes the atoms within a mineral to vibrate more rapidly. Even in a crystalline solid, where atoms are strongly bonded, this elevated level of activity allows individual atoms to migrate more freely between sites in the crystalline structure.

Changes Caused by Heat The formation of new or enlarged mineral grains at the expense of original grains is called **recrystallization**. During this process, the mineralogy of the rock may or may not change, but grains tend to become larger. For example, when quartz sandstone is metamorphosed to form quartzite, the mineralogy does not change; both rocks consist of quartz, but the process results in fewer and larger quartz grains. By contrast, when shale is metamorphosed to slate, the clay minerals are recrystallized and become new minerals—usually chlorite and muscovite.

Although the mineralogy changes in the transition from shale to slate, the overall chemical composition

Did You Know?
The temperature of Earth's crust increases with depth, an idea that can be expressed as *the deeper one goes, the hotter it gets*. This causes considerable problems for underground mining efforts. In the Western Deep Levels mine in South Africa, which is 2.5 mi. deep, the temperature of the rock is hot enough to scorch human skin. Here, the miners work in groups of two: one to mine the rock, and the other to operate the fan that keeps them cool.

remains essentially unchanged. Instead, the existing atoms are rearranged into new crystalline structures that are more stable in the new environment. (In some environments, ions may actually migrate into or out of a rock, thereby changing its overall chemical composition.)

What Is the Source of Heat? There are two primary ways in which a rock may become hotter. First, it may become more deeply buried; Earth's temperature rises with depth. Second, a magma body may be emplaced nearby, heating the surrounding rocks.

Earth's interior is extremely hot, mainly because of heat released from the repeated collision of asteroid-size bodies during the formation of our planet as well as from energy being continually released by the decay of radioactive elements. The rate of increase in temperature with depth is known as the **geothermal gradient**. In the upper crust, this increase in temperature averages about 25°C (45°F) per kilometer (Figure 8.3). Thus, rocks that formed at Earth's surface experience a gradual increase in temperature if they are transported to greater depths. As described earlier, clay minerals tend to become unstable when buried to a depth of about 8 kilometers (5 miles), where temperatures are about 200°C (400°F). The clay minerals begin to recrystallize into new minerals, such as chlorite and muscovite, both of which are stable in this new environment. However, many silicate minerals, particularly those found in crystalline igneous rocks—such as quartz and feldspar—remain stable at these temperatures. Thus, metamorphic changes in quartz and feldspar generally occur at higher temperatures.

Figure 8.3 provides several examples of conditions in which heat drives metamorphism. Environments where rocks may be carried to great depths and heated include convergent plate boundaries, where slabs of sediment-laden oceanic crust are being subducted. Rocks may also become deeply buried in large basins, where gradual subsidence results in thick accumulations of sediment. These basins, exemplified by the Gulf of Mexico, are known to develop low-grade metamorphic conditions near the base of the pile. In addition, continental collisions, which result in mountain building, cause some rocks to be uplifted while others are thrust downward, where elevated temperatures and pressures trigger metamorphism.

Heat may also be transported from the mantle into the shallowest layers of the crust. Rising mantle plumes, upwelling at mid-ocean ridges, and magma generated by partial melting of mantle rock at subduction zones are three such examples (see Figure 8.3). When magma intrudes rocks at shallow depths, the magma cools and releases heat, which "bakes" and transforms the surrounding host rock.

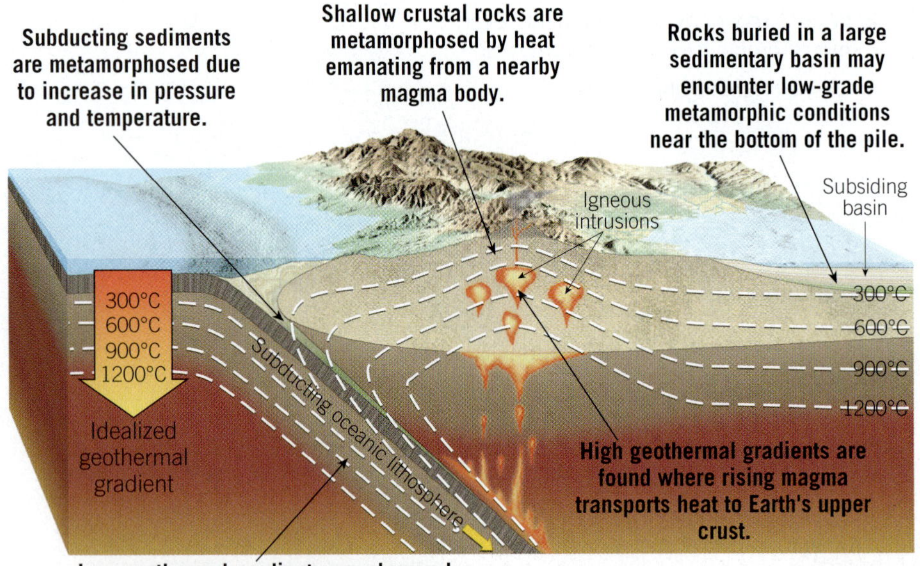
Subducting sediments are metamorphosed due to increase in pressure and temperature.
Shallow crustal rocks are metamorphosed by heat emanating from a nearby magma body.
Rocks buried in a large sedimentary basin may encounter low-grade metamorphic conditions near the bottom of the pile.
Igneous intrusions
Subsiding basin
300°C
600°C
900°C
1200°C
Idealized geothermal gradient
Subducting oceanic lithosphere
High geothermal gradients are found where rising magma transports heat to Earth's upper crust.
Low geothermal gradients are observed in subduction zones because cold oceanic crust and overlying sediments are descending into the mantle.

▲ SmartFigure 8.3 **Sources of heat for metamorphism** The two main sources of heat for metamorphism are the increasing temperature that occurs as we travel deeper into Earth's interior and heat released to the surrounding rocks when a magma body cools.

TUTORIAL
https://goo.gl/fW7IlN

Confining Pressure

Pressure, like temperature, increases with depth because the thickness of the overlying rock increases. Buried rocks are subjected to **confining pressure**, which is analogous to water pressure, in which the forces are applied equally in all directions (Figure 8.4A). Imagine scuba divers—the deeper they dive, the greater the confining pressure.

Confining pressure causes the spaces between mineral grains to close, producing more compact rocks that have greater densities. If the pressure becomes extreme enough, it can cause the atoms in a mineral to pack more closely together to produce a new, denser mineral. Recall from Chapter 3 that the transformation of one mineral (polymorph) to another is called a *phase change*. Confining pressure does *not*, however, fold or fracture rocks.

Differential Stress

In addition to confining pressure, rocks may be subjected to directed pressure. This occurs, for example, at convergent plate boundaries where slabs of lithosphere collide. Here the forces that deform rock are different in different directions and are referred to as **differential stress**. (A discussion of various types of *differential stress* is provided in Chapter 11.) Unlike confining pressure, which "squeezes" rock equally in all directions, differential stresses are greater in one direction than in others.

Differential stress that squeezes a rock mass as if it were placed in a vise is termed **compressional stress**.

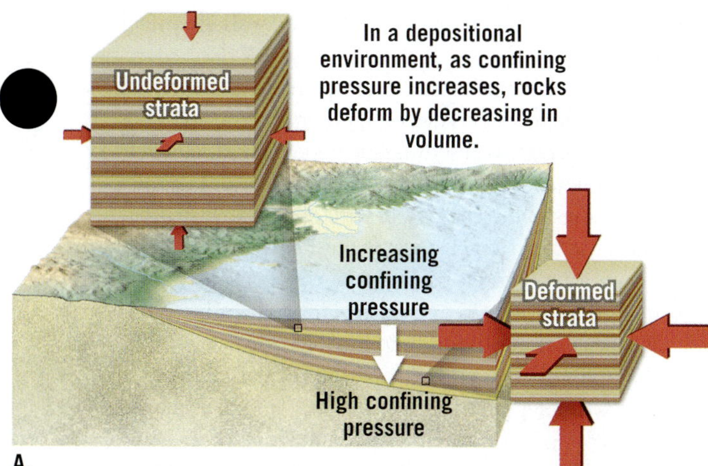

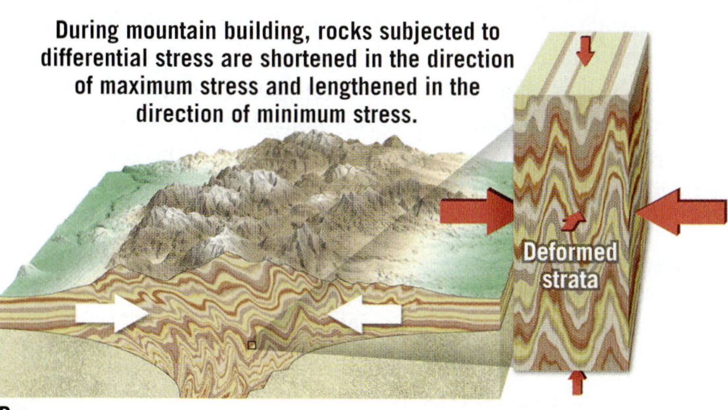

During mountain building, rocks subjected to differential stress are shortened in the direction of maximum stress and lengthened in the direction of minimum stress.
Deformed strata
B.

▲ **SmartFigure 8.4 Confining pressure and differential stress**

https://goo.gl/nBMksQ

As shown in **Figure 8.4B**, rocks subjected to compressional stress are shortened in the direction of greatest stress and elongated, or lengthened, in the direction perpendicular to that stress. Along convergent plate boundaries, the greatest differential stress is directed horizontally in the direction of plate motion. Consequently, in these settings, the crust is greatly shortened (horizontally) and thickened (vertically), resulting in mountainous topography.

In high-temperature, high-pressure environments, rocks are *ductile*, which allows their mineral grains to flatten (like what happens when you step on a piece of clay) when subjected to differential stress. The *metaconglomerate*, also called a stretched pebble conglomerate, shown in **Figure 8.5**, illustrates this tendency. The parent rock, a conglomerate, consisted of nearly spherical pebbles that have been flattened into elongated structures by differential stress. On a larger scale, rocks that are ductile deform by flowing rather than breaking or fracturing. As a result, deeply buried rocks subjected to extremely high temperature and pressure will develop intricate folds when deformed by differential stress (**Figure 8.6**).

By contrast, in near-surface environments where temperatures and pressure are comparatively low, rocks are *brittle* and tend to fracture when subjected to differential stress. Continued deformation grinds and pulverizes the mineral grains into smaller and smaller fragments.

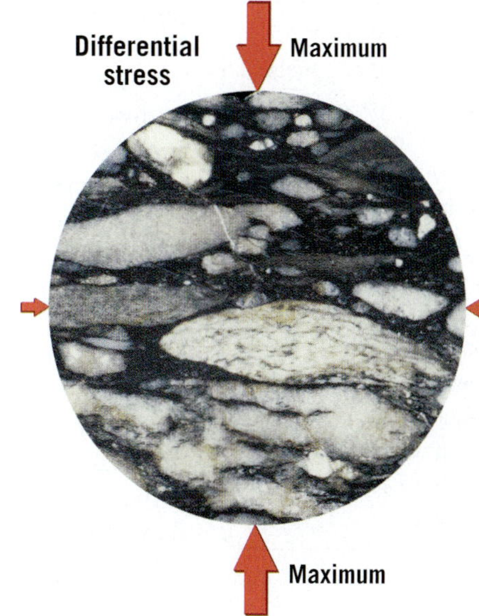

▲ **Figure 8.5 Metaconglomerate, also called stretched pebble conglomerate** This metaconglomerate is made of once nearly spherical pebbles that have been heated and flattened into elongated structures by differential stress. (Photo by E. J. Tarbuck)

Chemically Active Fluids

Water is abundant in Earth's crust. In the upper crust, water occurs as groundwater. At depth, hot water is released into the surrounding rocks when a magma body cools and solidifies. In addition, many minerals, including clays, micas, and amphiboles, are hydrated—which means they contain water in their crystalline structures. Elevated temperatures and pressures cause

◀ **Figure 8.6 Deformed and folded gneiss, Anza-Borrego Desert State Park, California** (Photo by A. P. Trujillo/APT Photos)

Did You Know?
Some low-grade metamorphic rocks contain fossils. These fossils can provide useful clues for determining the original rock type and its depositional environment. In addition, fossils whose shapes have been distorted during metamorphism provide insight into the extent to which the rock has been deformed.

these minerals to dehydrate, expelling hot, mineral-laden water.

Hot, chemically active fluids enhance metamorphism by dissolving and transporting ions from one site in the crystal structure to another, thereby facilitating the process of recrystallization. In increasingly hotter environments, these fluids become correspondingly more reactive.

In some metamorphic environments, hot fluids transport mineral matter over considerable distances. This occurs, for example, when hot, ion-rich fluids, called a *hydrothermal solution*, are expelled from a magma body as it cools and solidifies. If the rocks that surround the pluton differ markedly in chemical composition from the invading fluids, there may be an exchange of ions between these fluids and host rocks. In other words, these fluids bring in new atoms or take out atoms rather than simply reorganizing what is already present. When this occurs, the overall chemical composition of the surrounding rock changes, in a process called *metasomatism*. One example of metasomatism is the formation of the mineral wollastonite ($CaSiO_3$) from calcite ($CaCO_2$) the primary ingredient in limestone. When a silica-rich hydrothermal solution invades limestone, calcite reacts with silica (SiO_2) to generate wollastonite, and the gas carbon dioxide CO_2 is driven off.

The Importance of Parent Rock

Most metamorphic rocks have the same overall chemical composition as the parent rocks from which they formed, except for the possible loss or acquisition of volatiles such as water H_2O and carbon dioxide CO_2 Therefore, when we try to establish from what parent material metamorphic rocks were derived, the most important clue comes from their overall chemical composition.

Consider the large exposures of the metamorphic rock marble found high in the Alps of Southern Europe. Because marble and the common sedimentary rock limestone have the same mineralogy (calcite), it seems reasonable to conclude that limestone is the parent rock of marble. Furthermore, because limestone usually forms in warm, shallow marine environments, we can surmise that considerable deformation must have occurred to convert limy deposits in a shallow sea into marble crags in the lofty Alps.

The mineral makeup of the parent rock also largely determines the degree to which each metamorphic agent will cause change. For example, when magma forces its way into an existing body of rock, high temperatures and hot fluids may alter the host rock. If the host rock is composed of minerals that are comparatively nonreactive, such as quartz, any alterations that may occur will be confined to a narrow zone next to the igneous intrusion. However, when the host rock is limestone, which is highly reactive, the zone of metamorphism may extend far from the intrusion.

CONCEPT CHECKS 8.2

1. List four agents that drive metamorphism.
2. Why is heat considered the most important agent of metamorphism?
3. How is confining pressure different from differential stress?
4. What role do chemically active fluids play in metamorphism?
5. What characteristic of a metamorphic rock is determined primarily by its parent rock?

8.3 Metamorphic Textures

Explain how foliated and nonfoliated textures develop.

Did You Know?
Glacial ice is a metamorphic rock that exhibits ductile flow, much like hot rocks buried deep within Earth's crust. Although we think of glacial ice as being cold, it is in fact "hot" relative to its melting temperature. Therefore, we should not be surprised that the ice within a glacier gradually flows downslope in response to the force of gravity.

The term **texture** is used to describe the size, shape, and arrangement of the mineral grains within a rock. Recall that texture is one characteristic by which igneous and sedimentary rocks are classified. Most igneous and many sedimentary rocks consist of mineral grains or crystals that have a random orientation and thus appear uniform when viewed from any direction. By contrast, metamorphic rocks that contain platy minerals (such as micas) and/or elongated minerals (such as amphiboles) typically display some kind of *preferred orientation*, in which the mineral grains exhibit a parallel to subparallel alignment. Like a fistful of pencils, rocks containing elongated mineral grains that are oriented parallel to each other appear different when viewed from the side than when viewed head-on. A rock that exhibits a preferred orientation of its mineral constituents is said to possess *foliation*.

Foliation

The term **foliation** refers to any planar (nearly flat) arrangement of mineral grains or crystals within a rock. Although foliation may occur in some sedimentary and even a few types of igneous rocks, it is a fundamental characteristic of metamorphosed rocks that have been strongly deformed, mainly by folding. In metamorphic environments, foliation is ultimately driven by compressional stress that shortens rock units, causing mineral grains in preexisting rocks to develop parallel, or nearly parallel, alignments. Examples of foliation include the parallel alignment of platy minerals through *rotation*, *recrystallization*, and *flattening of mineral grains or pebbles*. Foliated textures include *rock cleavage*, in which rocks can be easily split into tabular slabs, and *compositional banding*, in which the separation of dark and light

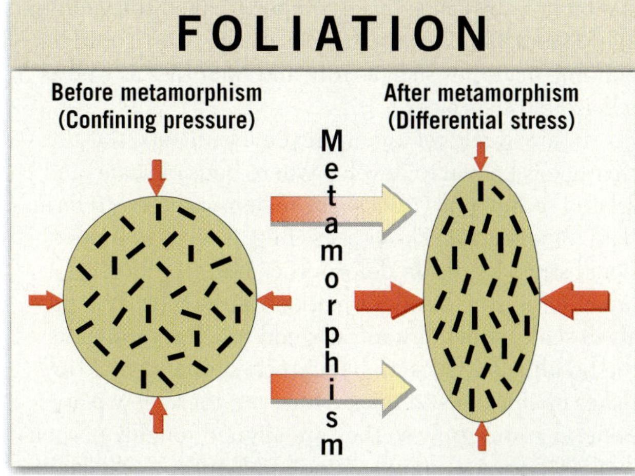

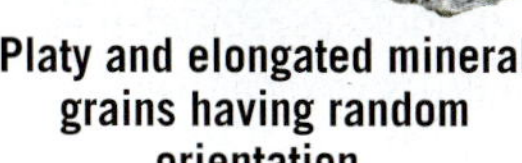

Platy and elongated mineral grains having random orientation.

When differential stress causes rocks to flatten, the mineral grains rotate and align roughly perpendicular to the direction of maximum differential stress.

▲ **SmartFigure 8.7 Mechanical rotation of platy mineral grains to produce foliation** (Photos by E. J. Tarbuck)

ANIMATION
https://goo.gl/ClsjD3

minerals generates a layered appearance. These diverse types of foliation can form in many different ways.

Rotation of Platy Mineral Grains The rotation of existing mineral grains is the easiest of the foliation mechanisms to envision. Figure 8.7 illustrates the mechanics by which platy or elongated mineral grains are rotated. Note that the new alignment of the grains is roughly perpendicular to the direction of maximum stress. Although physical rotation of platy minerals contributes to the development of foliation in low-grade metamorphism, other mechanisms dominate in more extreme environments.

Recrystallization That Produces New Minerals Recall that *recrystallization* is the creation of new mineral grains from preexisting ones. When recrystallization occurs as rock is being subjected to differential stress, any elongated minerals (such as amphiboles) and platy minerals (such as micas) that form tend to recrystallize perpendicular to the direction of maximum stress. Thus, the newly formed mineral grains exhibit a distinct layering, and the metamorphic rocks containing them exhibit foliation.

Flattening Spherically Shaped Grains Mechanisms that flatten existing mineral grains are important in the metamorphism of rocks containing minerals such as quartz, calcite, and olivine. These minerals normally develop roughly spherical crystals and have a rather simple chemical composition.

A change in grain shape can occur as distinct units of a mineral's crystalline structure slide relative to one another along discrete planes, thereby distorting the grain, as shown in Figure 8.8. This type of gradual *solid-state flow* involves slippage that disrupts the crystalline structure as atoms shift positions by breaking existing chemical bonds and forming new ones.

The shape of a mineral may also be altered by a process in which individual atoms move from a location along the margin of the grain that is highly stressed to a less-stressed position on the same grain. This mechanism, called *pressure solution*, is significantly aided by hot, ion-rich water. Mineral matter (ions) dissolves where grains are in contact with each other (areas of high stress) and is deposited in pore spaces (areas of low stress). As a result, the mineral grains tend to become shortened in the direction of maximum stress and

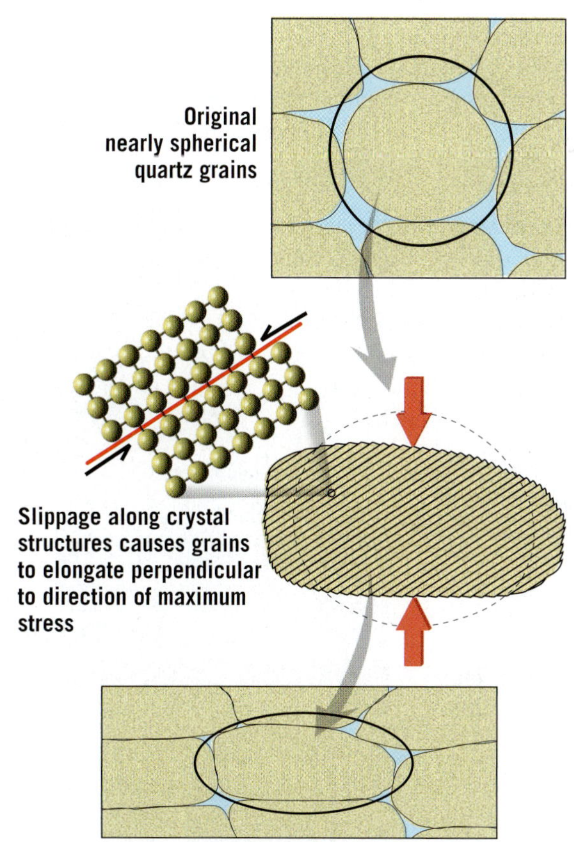

◀ **Figure 8.8 Solid-state flow of mineral grains** Mineral grains can be flattened by solid-state flow when units of a mineral's crystalline structure slide relative to each other. This mechanism involves breaking existing chemical bonds and forming new ones.

▼ **Figure 8.9 Excellent slaty cleavage** Slaty cleavage is exhibited by the rock in this slate quarry. Because slate breaks into flat slabs, it has many uses. (Photo by Fred Bruemmer/Photolibrary) The inset photo shows the use of slate for the roof of a house in Switzerland. (Photo by E. J. Tarbuck)

elongated in the direction of minimum stress. While both of these mechanisms flatten mineral grains, the mineralogy remains the same.

Foliated Textures

Various types of foliation exist, depending largely upon the grade of metamorphism and the mineralogy of the parent rock. We will look at three: *rock*, or *slaty*, *cleavage*; *schistosity*; and *gneissic texture*, or *banding*.

Rock, or Slaty, Cleavage Rocks that split into thin slabs when hit with a hammer exhibit **rock cleavage**. Rock cleavage develops in various metamorphic rocks but is best displayed in slates that exhibit an excellent splitting property called **slaty cleavage** (Figure 8.9). Because it splits easily, slate is used for building materials such as roof and floor tiles as well as billiard table surfaces.

In low-grade metamorphic environments, slaty cleavage is known to develop where beds of shale (and related sedimentary rocks) are metamorphosed to form slate (Figure 8.10). The process begins when compressional stress begins to deform rock units, producing broad folds. Further deformation causes the clay minerals in shale, which initially aligned roughly parallel to the bedding surfaces, to begin recrystallizing into tiny flakes of chlorite and mica. However, these new platy mineral grains grow so they are aligned roughly perpendicular to the maximum differential stress, as shown in Figure 8.10B.

Because slate typically forms during the low-grade metamorphism of shale, evidence of the original sedimentary bedding surfaces is often preserved. However, as Figure 8.10C illustrates, the orientation of slate's cleavage usually develops at an angle to the sedimentary beds. Thus, unlike shale, which splits *along* bedding planes, slate often splits *across* bedding surfaces. Other metamorphic rocks, such as schists and gneisses, sometimes split along planar surfaces and thus also exhibit rock cleavage.

Schistosity At higher temperatures and pressures, the minute mica and chlorite flakes in slate begin to

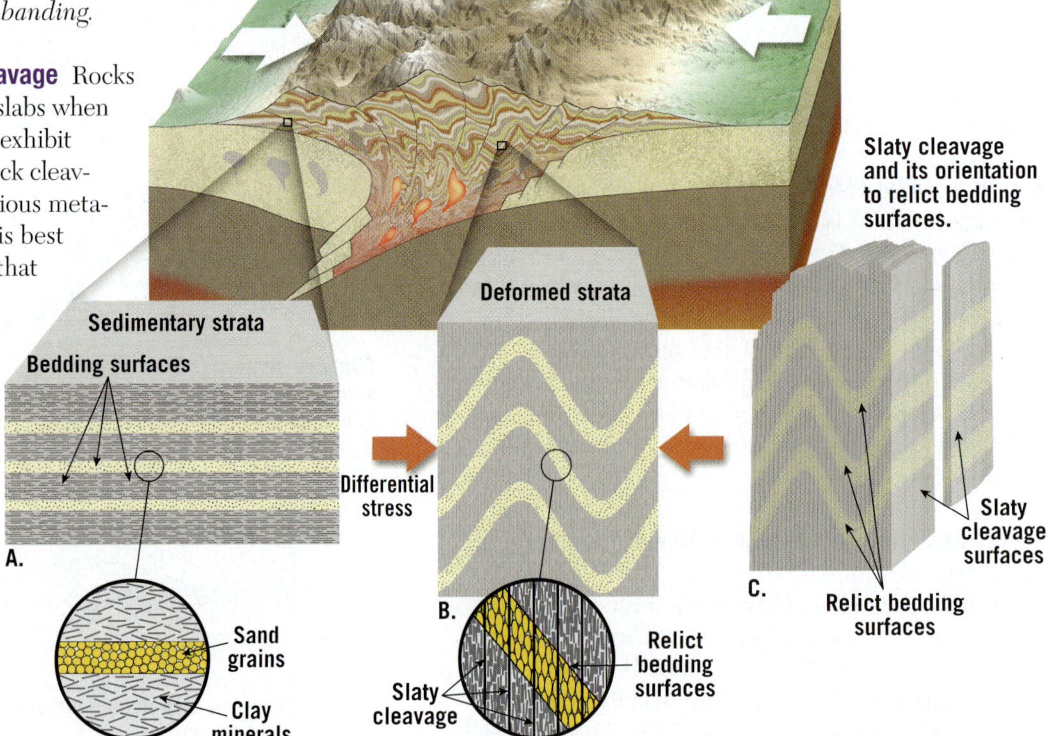

▶ **SmartFigure 8.10 Development of rock cleavage** When shale that is interbedded with sandstone is strongly folded and metamorphosed, the clay minerals begin to recrystallize into tiny flakes of chlorite and mica. These new platy minerals grow so they are aligned roughly perpendicular to the directed stress, which gives slate its foliation.

TUTORIAL
https://goo.gl/CJPGdG

recrystallize into larger muscovite and biotite crystals. When these platy minerals are large enough to be discernible with the unaided eye, they exhibit planar or layered structures called **schistosity**. Rocks that have this type of foliation are termed *schist*. In addition to containing platy minerals, schist often contains deformed quartz and feldspar crystals that appear flattened or lens shaped and are embedded among the mica grains.

Gneissic Texture, or Banding During high-grade metamorphism, ion migration can result in the segregation of minerals, as shown in Figure 8.11. Notice that the dark biotite and amphibole crystals and light silicate minerals (quartz and feldspar) have separated, giving the rock a banded appearance called **gneissic texture**, or **gneissic banding**. Metamorphic rocks with this texture are called *gneiss* (pronounced "nice"). Although they are foliated, gneisses do not usually split as easily as slates and some schists.

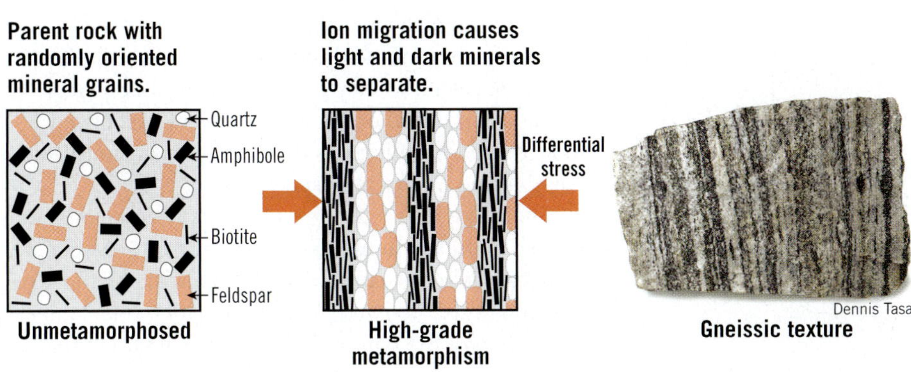

▲ **Figure 8.11 Development of gneissic banding** Gneissic banding develops through the migration of ions that cause felsic and mafic minerals to grow in separate layers.

Other Metamorphic Textures

Metamorphic rocks that *do not* develop a layered or banded appearance as a result of metamorphism are referred to as **nonfoliated**. Nonfoliated metamorphic rocks typically form in metamorphic environments where deformation is minimal or when the parent rock is composed of minerals that develop equidimensional crystals rather than tabular-shaped crystals. For example, when a fine-grained limestone (made of calcite) is metamorphosed, the small calcite grains in limestone recrystallize to form larger uniform-shaped crystals. The resulting metamorphic rock, *marble*, consists of intergrown calcite crystals that lack banding and are similar in appearance to the crystals in coarse-grained igneous rocks. Another common nonfoliated metamorphic rock, called *hornfels*, is generated when clay-rich rocks like shale and mudstone are intruded by a hot magma body. In this environment, where differential stress tends to be absent, the rocks are baked to produce a tough rock that lacks alignment of its platy mineral grains.

Metamorphic rocks may also contain some unusually large grains, called *porphyroblasts*, that are surrounded by a fine-grained matrix of other minerals.

Porphyroblastic textures develop in a wide range of rock types and metamorphic environments when minerals in the parent rock recrystallize to form new minerals. During recrystallization, certain metamorphic minerals, such as garnet, tend to develop *a small number of very large crystals*. By contrast, minerals such as muscovite and biotite typically form *a large number of smaller grains*. As a result, metamorphic rocks that contain large crystals (porphyroblasts) of, for example, garnet embedded in a finer-grained matrix of biotite and muscovite, are relatively common (Figure 8.12).

▲ **Figure 8.12 Garnet–mica schist** The dark red garnet crystals (porphyroblasts) are embedded in a matrix of fine-grained micas. (Photos by E. J. Tarbuck)

CONCEPT CHECKS 8.3

1. Define *foliation*.
2. Briefly describe three ways in which the mineral grains in a rock develop a preferred orientation (that is, foliation).
3. Distinguish among *slaty cleavage*, *schistosity*, and *gneissic* textures.
4. What is meant by *nonfoliated texture*? Name one rock that exhibits this texture.

8.4 Common Metamorphic Rocks

List and describe the most common metamorphic rocks.

Most metamorphic rocks that we observe at Earth's surface were derived from the three most common sedimentary rocks—shale, limestone, and quartz sandstone. Shale is the most likely parent of most slate, phyllite, schist, and gneiss. This sequence of metamorphic rocks reflects an increase in grain size, a change in rock texture, and a change in mineralogy.

Limestone, which is composed of the mineral calcite ($CaCO_3$) is the parent rock of marble, while quartz (SiO_2) sandstone is the parent of quartzite. Because calcite and quartz are simple chemical compounds compared to clay minerals, their mineralogy does not change during metamorphism; calcite usually remains calcite, and quartz remains quartz. Rather, these minerals recrystallize to produce larger fused grains that are the main constituents of marble and quartzite, respectively.

The major characteristics of the most common metamorphic rocks are summarized in **Figure 8.13**. Notice that metamorphic rocks can be broadly classified by the type of foliation exhibited and, to a lesser extent, the chemical composition of the parent rock. It is worth noting that certain rock *names* (slate, schist, and gneiss) are also used to describe rock *texture*.

Metamorphic Rock		Texture	Comments	Parent Rock
Slate	Foliated		Composed of tiny chlorite and mica flakes, breaks in flat slabs called slaty cleavage, smooth dull surfaces	Shale, mudstone, or siltstone
Phyllite	Foliated		Fine-grained, glossy sheen, breaks along wavy surfaces	Shale, mudstone, or siltstone
Schist	Foliated		Medium- to coarse-grained, scaly foliation, micas dominate	Shale, mudstone, or siltstone
Gneiss	Foliated		Coarse-grained, compositional banding due to segregation of light and dark colored minerals	Shale, granite, or volcanic rocks
Marble	Nonfoliated		Medium- to coarse-grained, relatively soft (3 on the Mohs scale), interlocking calcite or dolomite grains	Limestone, dolostone
Quartzite	Nonfoliated		Medium- to coarse-grained, very hard, massive, fused quartz grains	Quartz sandstone
Hornfels	Nonfoliated		Very fine-grained, often exceedingly tough and durable, usually dark colored	Often shale, but can have any composition

▲ **Figure 8.13 Classification of common metamorphic rocks** (Photos by E. J. Tarbuck)

Foliated Metamorphic Rocks

Slate A very fine-grained (less than 0.5 millimeter) foliated rock composed mainly of minute chlorite and mica flakes (too small to be visible to the human eye) is termed **slate**. Slate may also contain tiny quartz and feldspar crystals. Thus, slate generally appears dull and closely resembles shale. A noteworthy characteristic of slate is its excellent rock cleavage, or tendency to break into flat slabs (see Figure 8.9).

Slate is most often generated by the low-grade metamorphism of shale, mudstone, or siltstone. Less frequently it is produced when volcanic ash is metamorphosed. Slate's color depends on its mineral constituents: Black (carbonaceous) slate contains organic material, red slate gets its color from iron oxide, and green slate usually contains a lot of the mineral chlorite.

Phyllite **Phyllite** represents a degree of metamorphism between slate and schist. Its constituent platy minerals are larger than those in slate but not large enough to be readily identifiable with the unaided eye. Although similar to slate in appearance, phyllite can be easily distinguished from slate by its glossy sheen and wavy surface (see Figure 8.13). Phyllite exhibits rock cleavage and is composed mainly of very fine crystals of muscovite, chlorite, or both.

Schist Medium- to coarse-grained metamorphic rocks in which platy minerals are dominant are called **schists**. These flat components commonly include muscovite and biotite with parallel alignments that give the rock its foliated texture (**Figure 8.14**). In addition, schists contain smaller amounts of other minerals, often quartz and feldspar. Some schists are composed mostly of dark minerals

▲ **Figure 8.14 Mica schist** This sample of schist, composed mostly of muscovite and biotite, exhibits foliation. (Photo by E. J. Tarbuck)

(such as amphiboles). As with slate, the parent rock of most schists is shale that has undergone medium- to high-grade metamorphism during a major mountain-building episode.

As you learned in the previous section the term *schist* describes the texture of rocks, and as such it is used to name rocks that have a wide variety of chemical compositions. To indicate composition, mineral names are added. For example, schists composed primarily of muscovite and biotite are called *mica schist* (see Figure 8.14). Mica schists often contain *accessory minerals*, some of which are unique to metamorphic rocks. Some common accessory minerals that occur as porphyroblasts include *garnet*, *staurolite*, and *andalusite*, in which case the rock is called *garnet-mica schist*, *staurolite-mica schist*, or *andalusite-mica schist*.

In addition, schists may be composed largely of the minerals chlorite or talc, in which case they are called *chlorite schist* and *talc schist*, respectively. Both chlorite and talc schists can form when rocks having basaltic compositions undergo metamorphism.

Gneiss **Gneiss** is the term applied to medium- to coarse-grained banded metamorphic rocks in which granular and elongated (as opposed to platy) minerals predominate. The most common minerals in gneiss are quartz, potassium feldspar, and plagioclase feldspar. Most gneisses also contain lesser amounts of biotite, muscovite, and amphibole. Some gneisses split along the layers of platy minerals, but most break in an irregular fashion.

Recall that during high-grade metamorphism, the light and dark components separate, giving gneisses their characteristic banded or layered appearance. Thus, most gneisses consist of alternating bands of white or reddish feldspar-rich zones and layers of dark ferromagnesian minerals (**Figure 8.15**). These banded gneisses often exhibit evidence of deformation, including folds and sometimes faults.

Gneisses having a felsic composition may be derived from granite or its fine-grained equivalent, rhyolite. However, most gneisses are generated through high-grade metamorphism of shale. Therefore, gneiss represents the highest-grade metamorphic rock in the sequence of shale, slate, phyllite, schist, and gneiss. Like schists, gneisses may also include large crystals of accessory minerals such as garnet. Gneisses made up primarily of dark minerals also occur. For example, an amphibole-rich rock that exhibits a gneissic texture is called *amphibolite*.

Did You Know?
Rocks of the Acasta Gneiss Complex, exposed just east of Great Slave Lake in the Northwest Territories of Canada, have yielded radiometric dates of up to 4.03 billion years. These are currently the oldest-known rocks in the world.

Nonfoliated Metamorphic Rocks

Marble The metamorphism of limestone or dolostone produces the crystalline metamorphic rock called **marble** (see Figure 8.13). Pure marble is white and composed essentially of the mineral calcite. Because of its relative softness (3 on the Mohs scale), marble is easy to cut and shape. White marble is particularly prized as a stone from which monuments and statues are carved, such as the Lincoln Memorial in Washington, DC, and the Taj Mahal in India (**Figure 8.16**). Unfortunately, when exposed to acid rain, marble's composition (calcium carbonate) makes it susceptible to chemical weathering.

The parent rocks of most marbles contain impurities that color the stone. Thus, marble can be pink, gray, green, or even black and may contain a variety of accessory minerals (such as chlorite, mica, garnet, and wollastonite). When marble forms from limestone interbedded with shales, it appears banded and exhibits visible foliation. When deformed, these banded marbles may develop highly contorted mica-rich folds that enhance the rocks' artistic appearance. These decorative marbles have been used as building stones since prehistoric times.

Quartzite **Quartzite** is a very hard metamorphic rock formed from quartz sandstone (**Figure 8.17**). Under moderate- to high-grade metamorphism, the quartz

▲ **Figure 8.15 Banded gneiss found in the Adirondacks, New York** (Photo by Michael Collier)

◀ **Figure 8.16 Marble is a widely used building stone** The exterior of India's Taj Mahal is constructed primarily of the metamorphic rock marble. (Photo by Sam DCruz/Shutterstock)

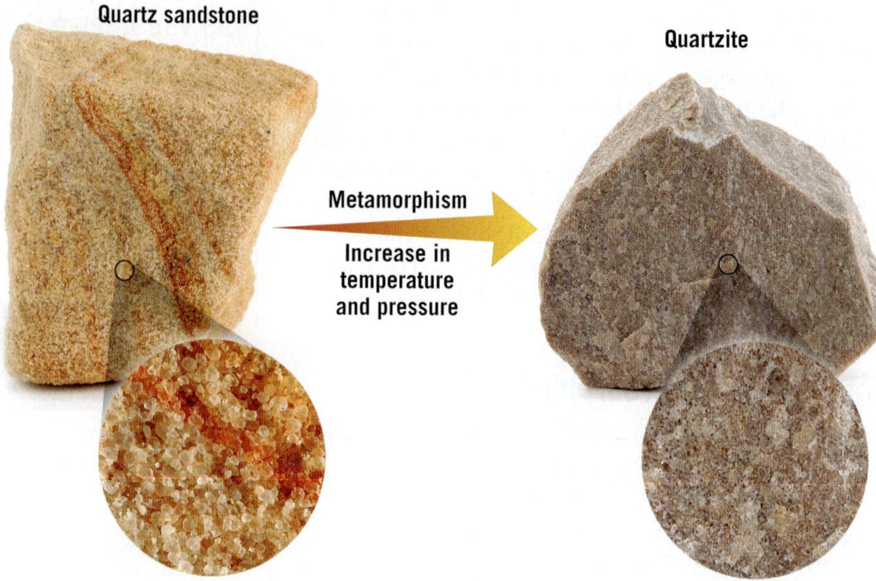

▲ **Figure 8.17 Quartzite** Quartzite is a nonfoliated metamorphic rock formed from quartz sandstone. The close-up images compare the loosely bound grains in a quartz sandstone to the interlocking quartz grains typical of quartzite. (Photos by Dennis Tasa)

grains in sandstone fuse together (inset in Figure 8.17). Recrystallization is often so complete that, when broken, quartzite splits across the original quartz grains rather than along their boundaries. In some instances, sedimentary features such as cross-bedding are preserved and give the rock a banded appearance. Pure quartzite is white, but iron oxide may produce reddish or pinkish stains, while dark mineral grains may impart shades of green or gray.

Hornfels A fine-grained nonfoliated metamorphic rock, **hornfels** is unlike marble and quartzite in that it has a variable mineral composition. The parent rock of most hornfels is shale or another clay-rich rock that has been "baked" by a hot intruding magma body. Hornfels tends to be gray to black in color and quite hard, and it may display conchoidal fracture.

CONCEPT CHECKS 8.4

1. Although slate and phyllite resemble each other, how can they be differentiated?
2. For the rock mica schist, what does *mica* indicate, and what does *schist* indicate?
3. Briefly describe the appearance of the metamorphic rock gneiss and explain how it forms.
4. Describe slate, phyllite, schist, and gneiss in terms of texture and grain size.
5. Compare and contrast marble and quartzite.

8.5 Metamorphic Environments

Write a description for each of these metamorphic environments: contact metamorphism, hydrothermal metamorphism, subduction zone metamorphism, and regional metamorphism.

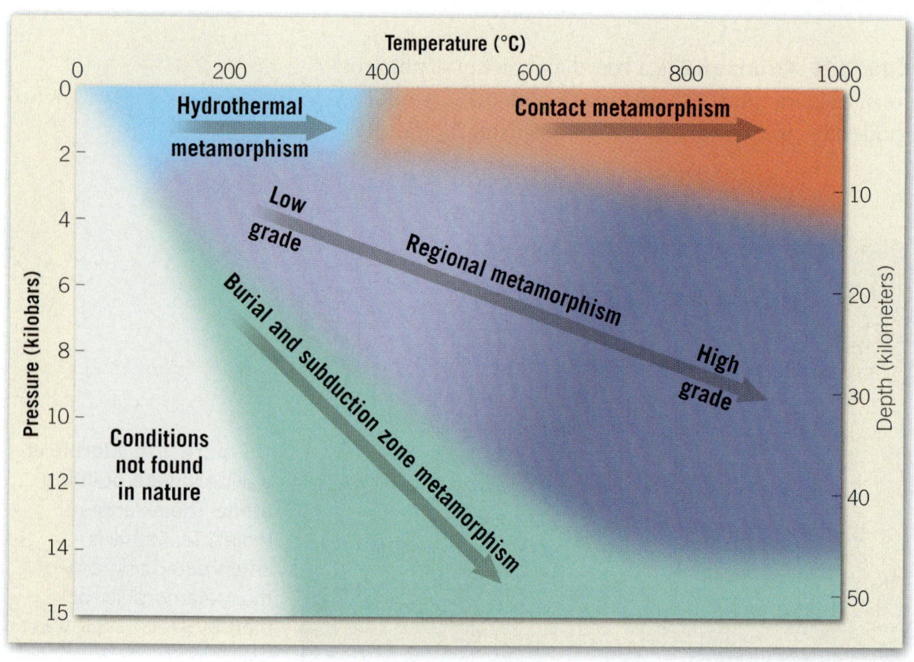

There are many different environments in which metamorphism occurs. Most of these environments occur in the vicinity of plate margins, and several are associated with igneous activity. We will consider the following basic metamorphic environments: *contact*, or *thermal*, *metamorphism*; *hydrothermal metamorphism*; *burial* and *subduction zone metamorphism*; and *regional metamorphism*, as well as a few types of metamorphism that generate relatively small quantities of metamorphic rock.

As Figure 8.18 illustrates, each type of metamorphism occurs over a range of temperatures and pressures. On this graph, temperature, which increases from surface conditions to about 1000°C (1800°F), is displayed along the top, and pressure is shown along the vertical axis. Notice that hydrothermal metamorphism

◄ **Figure 8.18 Metamorphic environments** This graph illustrates the temperatures and pressures typically associated with the major types of metamorphic environments.

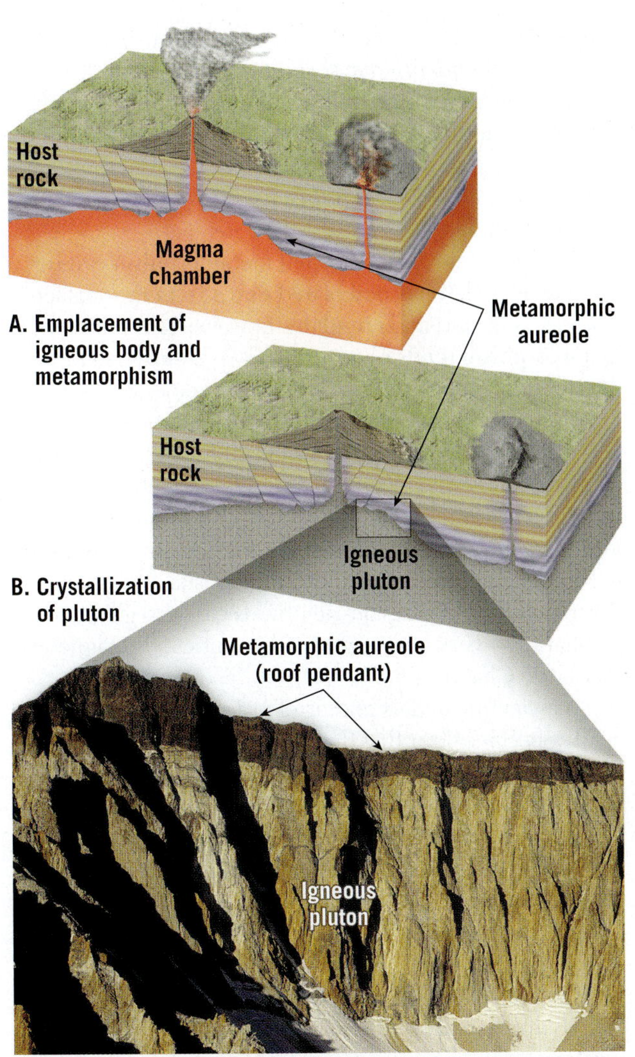

▲ **SmartFigure 8.19 Contact metamorphism** Contact metamorphism produces a zone of alteration called an *aureole* around an intrusive igneous body. In the photo, the dark layer is a type of metamorphic aureole called a *roof pendant*, which consists of metamorphosed host rocks that are in contact with the upper part of the light-colored igneous pluton. The term *roof pendant* implies that the rock was once the roof of a magma chamber. The photo shows the Sierra Nevada, near Bishop, California. (Photo by Michael Collier)

TUTORIAL
https://goo.gl/LVq0Cm

occurs in relatively low-temperature and low-pressure environments, whereas burial and subduction metamorphism occur under relatively low-temperature and high-pressure conditions. Although not shown on this graph, regional and subduction metamorphism involve differential stress, whereas hydrothermal and contact metamorphism do not.

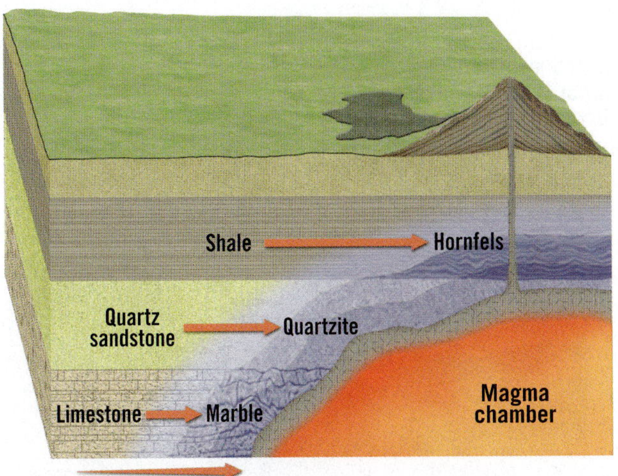

◄ **Figure 8.20 Rocks produced by contact metamorphism** Contact metamorphism of shale yields hornfels, while contact metamorphism of quartz sandstone and limestone produces quartzite and marble, respectively.

Contact, or Thermal, Metamorphism

Contact metamorphism, or **thermal metamorphism**, occurs in Earth's upper crust (low pressure), when rocks immediately surrounding a molten igneous body are "baked" (high temperature). Because contact metamorphism does not involve differential stress, the resulting metamorphic rocks are not foliated.

Contact metamorphism alters rocks in a discrete zone adjacent to the heat source, called an **aureole** (Figure 8.19). The emplacement of small intrusions such as dikes and sills (see Chapter 4) typically form aureoles only a few centimeters thick. By contrast, large molten bodies that eventually cool to form batholiths can produce aureoles that extend outward for several kilometers. These large aureoles often consist of distinct *zones of metamorphism*. Close to the magma body, high-temperature minerals such as garnet may form, whereas farther away, low-grade metamorphism produces minerals such as chlorite.

Depending mainly on the composition of the parent rock, a variety of metamorphic rocks can form in the same setting (Figure 8.20). For example, during contact metamorphism of mudstones and shales, the clay minerals are baked, much as clay is baked in a kiln to make pottery. The result is a very hard, fine-grained metamorphic rock called *hornfels* (see Figure 8.13). Hornfels can also form from a variety of other materials, including volcanic ash and basaltic rocks. Other metamorphic rocks that are produced by contact metamorphism are marble and quartzite (see Figure 8.20). Recall that limestone is the parent of marble and that the metamorphism of quartz sandstone produces quartzite.

Did You Know?
One of America's worst civil-engineering disasters occurred in 1928, when the St. Francis Dam in southern California failed. Huge torrents of water washed down San Francisquito Canyon, destroying 900 buildings and taking nearly 500 lives. The eastern part of the dam was built on highly foliated mica schist that was prone to slippage, as evidenced by an earlier landslide in that area. The immense water pressure at the base of the dam and the weak schist to which it was anchored are thought to have contributed to the failure.

Hydrothermal Metamorphism

When hot, ion-rich water circulates through pore spaces or fractures in rock, a chemical alteration called **hydrothermal metamorphism** may

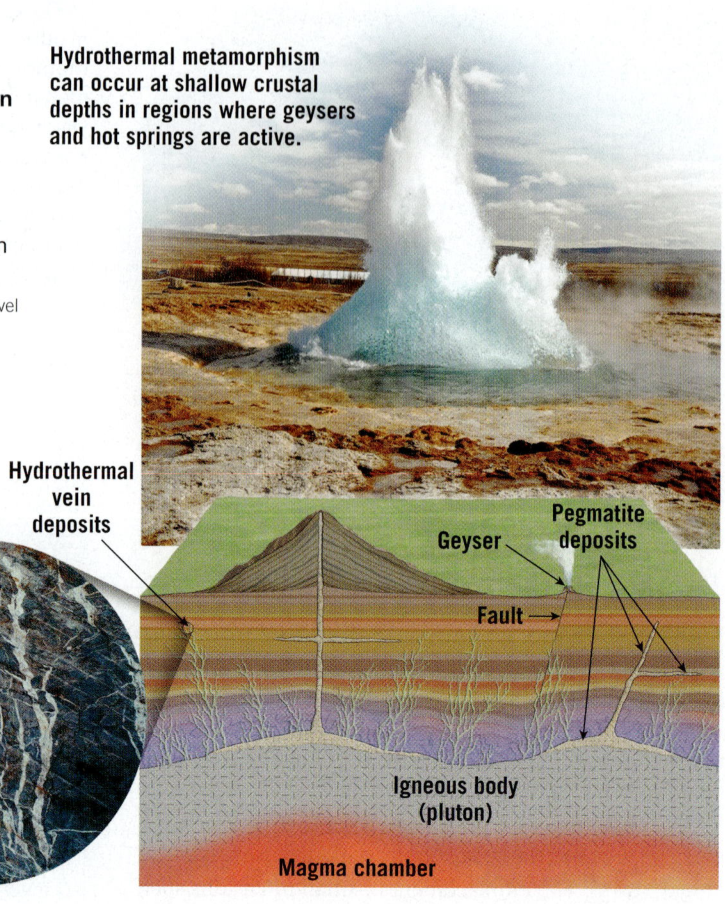

▶ **Figure 8.21 Hydrothermal metamorphism associated with an intrusive igneous body** Pegmatites and hydrothermal mineral deposits form adjacent to an igneous intrusion (pluton). (Photo by Pavel Svofoda/Fotolia)

occur (**Figure 8.21**). Recall that hot mineral-laden fluids called *hydrothermal solutions* contribute to metamorphism by enhancing the recrystallization of existing minerals. In addition, hot ion-rich fluids facilitate the movement of mineral matter into and out of rock bodies, thereby changing their overall chemical composition.

The water for hydrothermal metamorphism can be groundwater that has percolated down from the surface, where it is heated and circulates upward. This type of metamorphism tends to occur at low pressures (shallow depth) and relatively low to moderate temperatures.

Water that drives hydrothermal metamorphism may also arise from igneous activity. As large magma bodies cool and solidify, ion-rich water is released into the surrounding host rocks. When the host rock is porous or highly fractured, mineral matter contained in these fluids may precipitate to form important deposits of copper, silver, and gold. These ion-rich fluids can also generate pegmatites—very coarse-grained granitic (felsic) igneous rocks (see Chapter 4).

As scientific understanding of plate tectonics developed, it became clear that the most widespread occurrence of hydrothermal metamorphism is along the axis of the mid-ocean ridge system (**Figure 8.22**). As plates move apart, upwelling magma from the mantle generates new seafloor. Seawater percolating through the young, hot oceanic crust is heated and chemically reacts with the newly formed basaltic rocks. The result is the conversion of mafic rocks of the oceanic crust and uppermost mantle into the hydrated rocks serpentinite and soapstone (**Figure 8.23**).

Hydrothermal solutions circulating through the seafloor also remove large amounts of metals, such as iron, cobalt, nickel, silver, gold, and copper, from the newly formed crust. These hot, metal-rich fluids eventually rise along fractures and gush from the seafloor at temperatures of about 350°C (660°F), generating particle-filled clouds called *black smokers*. Upon mixing with the cold seawater, sulfides and carbonate minerals containing these heavy metals precipitate to form metallic deposits. Geologists credit this process with the formation of the copper ores mined today on the Mediterranean island of Cyprus.

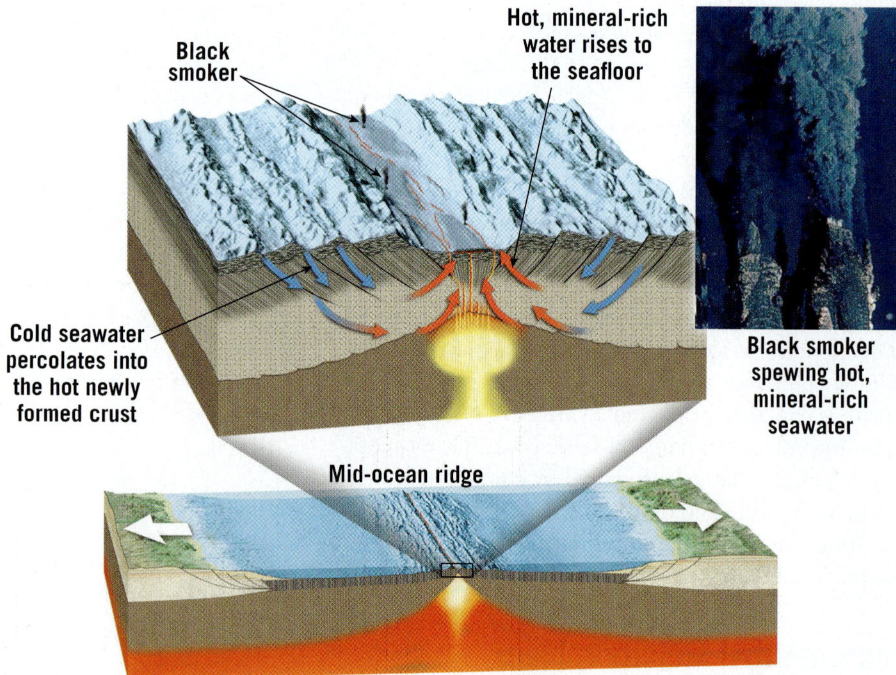

▲ **Figure 8.22 Hydrothermal metamorphism along a mid-ocean ridge** (Photo by Fisheries and Oceans Canada/Uvic-Verena Tunnicliffe/Newscom)

▲ **Figure 8.23 Serpentinite and soapstone** These metamorphic rocks are produced by hydrothermal alteration of mafic rocks along the mid-ocean ridge system. (Photos by Dennis Tasa)

Burial & Subduction Zone Metamorphism

Burial metamorphism tends to occur where massive amounts of sedimentary or volcanic material accumulate in a subsiding basin, such as the Gulf of Mexico (see Figure 8.3). Here, low-grade metamorphic conditions may be reached within the deepest layers. Confining pressure and heat drive the recrystallization of the constituent minerals, changing the texture and/or mineralogy of the rock without appreciable deformation.

Rocks and sediments can also be carried to great depths along convergent boundaries where oceanic lithosphere is being subducted (see Figure 8.3). In this setting, cold, dense oceanic crust and sediments, which are poor conductors of heat, are subducting rapidly enough that pressure increases faster than temperature. This phenomenon, called **subduction zone metamorphism**, differs from burial metamorphism in that differential stress, rather than confining pressure, plays a major role in deforming rock as it is being metamorphosed.

Regional Metamorphism

Regional metamorphism is a common, widespread type of metamorphism typically associated with mountain building, where large segments of Earth's crust are intensely deformed by the collision of two continental crust blocks (**Figure 8.24**). Recall that denser oceanic crust subducts under more buoyant continental crust but that two landmasses will instead collide and deform. Sediments and crustal rocks that form the margins of the colliding continents are folded and faulted and, as a result, shorten and thicken like a rumpled carpet. Continental collisions may also cause crystalline basement rocks lying under sedimentary layers, as well as slices of oceanic crust that once floored the intervening ocean basin, to be uplifted and deformed.

The general thickening of the crust that occurs during mountain building results in buoyant lifting, in which deformed rocks are elevated high above sea level. Crustal thickening also results in the deep burial of large quantities of rock as one crustal block is thrust beneath another. Deep in the roots of mountains, elevated temperatures caused by deep burial are responsible for the most intense metamorphic activity within a mountain belt.

In some settings, deeply buried rocks become heated beyond their melting points, producing magma. When these magma bodies grow large enough to buoyantly rise, they intrude the overlying metamorphic and sedimentary rocks (see Figure 8.24). Consequently, the cores of many mountain belts consist of folded and faulted metamorphic rocks, often intertwined with igneous bodies. Over time, these deformed rock masses are uplifted, and erosion removes the overlying material to expose the igneous and metamorphic core of the mountain range.

Regional metamorphism produces some of the most common metamorphic rocks. In these settings, shale is metamorphosed to produce the sequence of slate, phyllite, schist, and gneiss. In addition, quartz sandstone and limestone are metamorphosed into quartzite and marble.

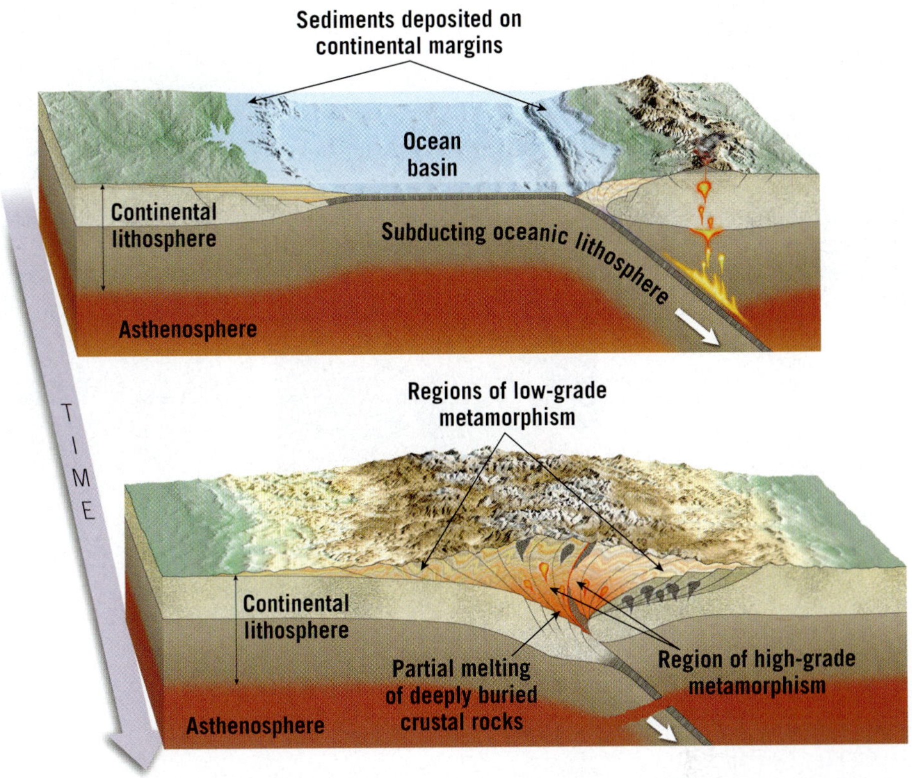

▲ **Figure 8.24 Regional metamorphism** Regional metamorphism is often associated with a continental collision where rocks are squeezed between two converging crustal blocks, resulting in mountain building.

Other Metamorphic Environments

Other, less-common types of metamorphism generate relatively small amounts of metamorphic rock that tends to be geographically localized.

Metamorphism Along Fault Zones Near Earth's surface, rock behaves like a brittle solid. Consequently, movement along a fault zone fractures and pulverizes rock (**Figure 8.25A**). The result is a loosely coherent rock called *fault breccia*, composed of broken and crushed rock fragments (see Figure 8.25A). Displacements along California's San Andreas Fault have created a zone of fault breccia and related rock types more than 1000 kilometers (600 miles) long and up to 3 kilometers (2 miles) wide.

Much of the deformation associated with fault zones occurs at great depth and thus at high temperatures. In this environment, preexisting minerals deform by ductile flow (**Figure 8.25B**). As large slabs of rock move in opposite directions, the minerals in the fault zone between them tend to form elongated grains that give the rock a foliated and/or lineated appearance. Rocks formed in these zones of intense ductile deformation are termed *mylonites* (*mylo* = a mill, *ite* = a stone).

Impact Metamorphism **Impact**, or **shock, metamorphism** occurs when meteorites (fragments of

Did You Know?
Most pigment that gives paint its color and its ability to hide previous coats comes from various natural Earth materials. For example, *umber*, which produces a rich chestnut-brown color, is mined on the island of Cyprus. This material was formed on an ancient seafloor by the hydrothermal alteration of iron- and manganese-rich basaltic lavas.

▶ SmartFigure 8.25
Metamorphism along a fault zone (Photo A by A. P. Trujillo; photo B by Ann Bykerk-Kauffman)

TUTORIAL
https://goo.gl/BbHSb8

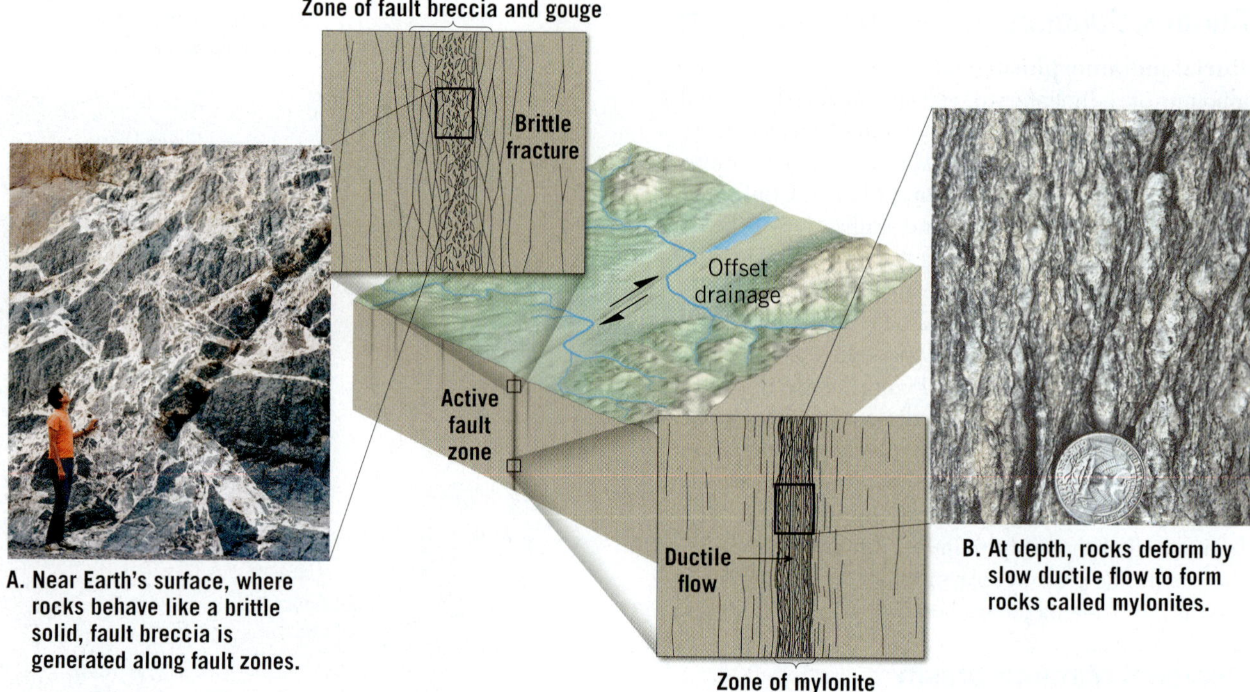

A. Near Earth's surface, where rocks behave like a brittle solid, fault breccia is generated along fault zones.

B. At depth, rocks deform by slow ductile flow to form rocks called mylonites.

comets or asteroids) strike Earth's surface at high speeds. Upon impact, the energy of the once rapidly moving meteorite is transformed into heat energy and shock waves that pass through the surrounding rocks. The result is pulverized, shattered, and sometimes melted rock.

The products of these impacts, called *impactiles*, include mixtures of fused fragmented rock plus glass-rich ejecta that resemble volcanic bombs. In some cases, a very dense form of quartz (*coesite*) and minute diamonds are found. The existence of these high-pressure minerals provides convincing evidence that pressures and temperatures involved in impact metamorphism can be as great as those found in the upper mantle.

CONCEPT CHECKS 8.5

1. In which type of metamorphism does compressional stress play a major role?
2. Name three rocks that are produced by contact metamorphism.
3. What is an *aureole*?
4. What is the agent of hydrothermal metamorphism?
5. Which type of plate boundary is typically associated with regional metamorphism?
6. List the common metamorphic rocks generated by regional metamorphism.

Did You Know?
Although they are rare, metamorphic rocks that contain microscopic diamonds have been discovered at Earth's surface. The existence of diamonds in metamorphic rocks indicates that these rocks formed at very high pressures found at depths of at least 60 mi. How these metamorphic rocks were buried to these great depths and later returned to the surface is still a mystery.

8.6 Metamorphic Zones

Explain how index minerals are used to establish the metamorphic grade of a rock body.

In areas affected by metamorphism, geologists can observe the usually systematic variations in the mineralogy and texture of the altered rocks. These differences are clearly related to variations in the degree of metamorphism that takes place in each metamorphic zone.

Textural Variations

Across areas where regional metamorphism has occurred, rock textures vary based on the intensity of metamorphism. If we begin with a clay-rich sedimentary rock such as shale or mudstone, a gradual increase in metamorphic intensity from low grade to high grade is accompanied by a general coarsening of the grain size. **Figure 8.26** illustrates that as metamorphic intensity increases, shale changes to a fine-grained slate, which then forms phyllite, which, through continued recrystallization, generates a medium-grained schist. Under more intense conditions, a gneissic texture that exhibits layers of dark and light minerals may develop. This systematic transition in metamorphic textures can be observed as we approach the Appalachian Mountains from the west (**Figure 8.27**). Beds of shale, which once extended over large areas of the eastern United States, still occur as nearly flat-lying strata in Ohio. However, in the broadly folded Appalachians of central Pennsylvania, the rocks that once formed flat-lying beds are folded and display a preferred orientation of platy mineral grains, as exhibited by well-developed slaty cleavage. As we move further east, toward the intensely deformed crystalline Appalachians, we find large exposures of schists. Some of the most intense zones of metamorphism are found in Vermont and New Hampshire, where gneissic rocks are exposed at the surface.

Index Minerals & Metamorphic Grade

In addition to textural changes, we encounter corresponding changes in mineralogy as we shift from areas of low-grade metamorphism to areas of high-grade metamorphism. An idealized transition in mineralogy that results from the regional metamorphism of shale is shown in **Figure 8.28**. The first new mineral to form as shale changes to slate is chlorite. At higher temperatures, flakes of muscovite and biotite begin to dominate. Under more extreme conditions, metamorphic rocks may contain garnet and/or staurolite crystals (**Figure 8.29**). At temperatures approaching the melting point of rock, sillimanite forms. Sillimanite is a high-temperature metamorphic mineral used to make porcelains used in extreme environments, such as for spark plugs.

Through the study of metamorphic rocks in their natural settings (called *field studies*) and through experimental studies, researchers have learned that certain minerals, such as those listed in Figure 8.28, are good indicators of the metamorphic environment in which they formed. Using these **index minerals**, geologists distinguish among different zones of regional metamorphism. For example, the mineral chlorite begins to form when temperatures are relatively low—less than 200°C (400°F; see Figure 8.28). Thus, rocks containing chlorite (usually slates) are categorized as *low grade*. By contrast, the mineral sillimanite forms only in extreme environments where temperatures exceed 600°C (1100°F), and rocks containing it

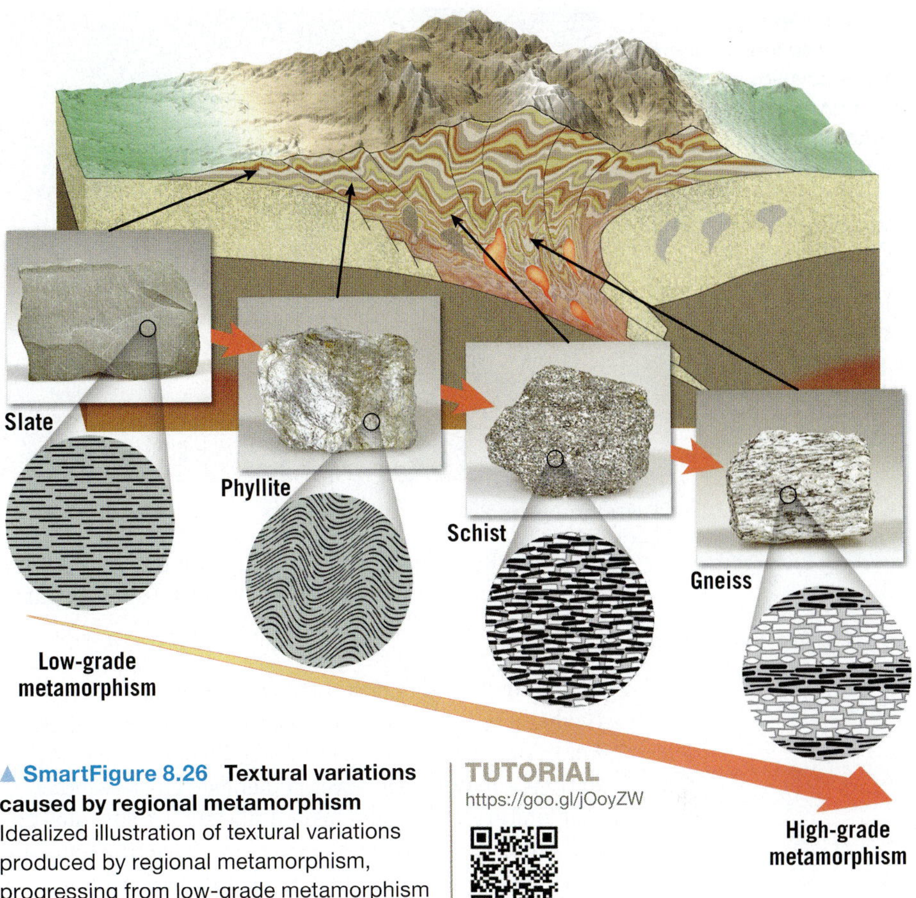

▲ **SmartFigure 8.26 Textural variations caused by regional metamorphism** Idealized illustration of textural variations produced by regional metamorphism, progressing from low-grade metamorphism (slate) to high-grade metamorphism (gneiss). (Photos by E. J. Tarbuck)

TUTORIAL
https://goo.gl/jOoyZW

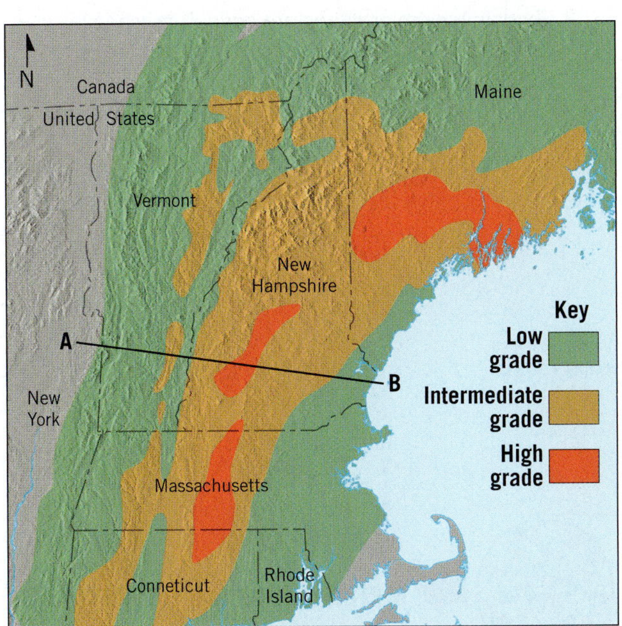

▲ **Figure 8.27 Zones of metamorphic intensities in New England** Highly generalized map that shows areas of low- to high-grade metamorphism in New England.

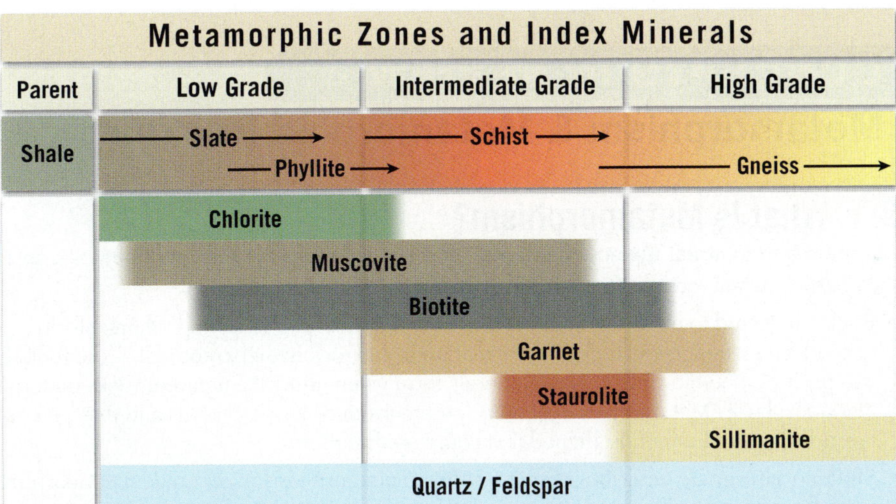

▲ **Figure 8.28 Metamorphic zones and index minerals** This is a typical transition of various index minerals associated with the progression from low-grade to high-grade metamorphism of the rock shale.

▶ SmartFigure 8.29 **Garnet, an index mineral, provides evidence of medium- to high-grade metamorphism** These garnet porphyroblasts are found in a gneiss in the Adirondacks, New York. (Photo by Michael Collier)

TUTORIAL
https://goo.gl/1pgs0t

▲ Figure 8.30 **Migmatite** Under high-grade metamorphism, light-colored (felsic) minerals in a gneiss may begin to melt, while the dark-colored (mafic) minerals remain solid. If this melt solidifies in place, the rock—called a migmatite—will contain light-colored igneous rock intermixed with metamorphic rock composed of dark-colored (mafic) minerals. (Photo by Science Photo Library)

are considered *high grade*. By mapping the occurrences of index minerals, geologists can identify zones of varying metamorphic grades (see Figure 8.27).

Migmatites In the most extreme environments, even the highest-grade metamorphic rocks undergo change. For example, gneissic rocks may be heated sufficiently to trigger melting. However, minerals melt at different temperatures. The light-colored silicates, usually quartz and potassium feldspar, have the lowest melting temperatures and begin to melt first, while the mafic silicates, such as amphibole and biotite, remain solid. When the partially melted rock cools, the light bands will be composed of igneous or igneous-appearing components, while the dark bands will consist of unmelted metamorphic material. Rocks of this type are called **migmatites** (*migma* = mixture, *ite* = a stone) (Figure 8.30). The light color bands in migmatites often form intricate folds and may contain tabular inclusions of the dark components.

Migmatites serve to illustrate the fact that some rocks are considered transitional and do not fit neatly into any of the three basic rock groups.

CONCEPT CHECKS 8.6

1. Describe the different grades of metamorphism that might be encountered moving west to east from Ohio to the crystalline core of the Appalachians.
2. How do geologists use index minerals?
3. Explain why migmatites are difficult to place into any one of the three basic rock groups.

CONCEPTS IN REVIEW
Metamorphism & Metamorphic Rocks

8.1 What Is Metamorphism?

Compare and contrast the environments that produce metamorphic, sedimentary, and igneous rocks.

KEY TERMS: parent rock, metamorphism, mineralogy, metamorphic grade

- Rocks subjected to elevated temperatures and pressures can react and "change form" to produce metamorphic rocks. Every metamorphic rock has a parent rock—the rock it was prior to metamorphism. New minerals form when minerals in parent rocks undergo metamorphosis. Metamorphic reactions tend to produce larger crystals and may generate layers or bands of minerals aligned in a common direction.
- Metamorphic grade describes the intensity of metamorphosis. Low-grade metamorphic rocks strongly resemble their parent rock. High-grade metamorphism destroys textures of parent rocks and features such as fossils.
- Metamorphism takes place in the solid state and in most cases does not involve any melting.

? Compare the processes that produce metamorphic rocks to those that produce igneous or sedimentary rocks.

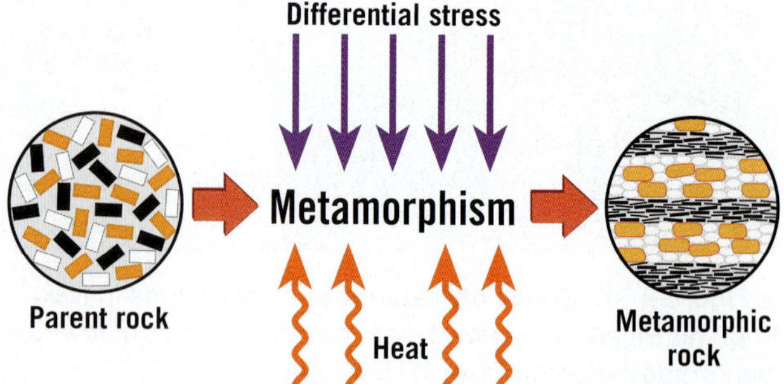

8.2 What Drives Metamorphism?

List and distinguish among the four agents that drive metamorphism.

KEY TERMS: recrystallization, geothermal gradient, confining pressure, differential stress, compressional stress

- Heat, pressure, differential stress, and chemically active fluids are four agents that drive metamorphic reactions. Any one alone may trigger metamorphism, or all four may exert influence simultaneously.
- The burial of rock or the intrusion of a nearby magma body will raise the temperature of a rock. Heat provides energy that drives chemical reactions and results in the recrystallization of the existing minerals. Different minerals have different levels of susceptibility to recrystallization: Some crystals just grow larger, while others react to form new minerals. Quartz is stable across a wide range of temperatures, whereas clay minerals are stable only at low (near-surface) temperatures.
- Confining pressure, which results from burial, is analogous to water pressure in that force is exerted equally in all directions. An increase in confining pressure causes rocks and minerals to compact into more dense configurations.
- Differential stress results from tectonic forces, where the pressure is greater in some directions than in others. Rocks subjected to differential stress deep in Earth's crust tend to respond in a ductile manner, shortening in the direction of greatest stress and elongating in the direction(s) of least stress, producing flattened or stretched grains. If the same differential stress is applied to a rock in the shallow crust, the rock may deform in a brittle fashion instead, breaking into smaller pieces.
- Water is an important chemically active fluid in Earth's crust. Hot water can facilitate numerous chemical reactions and may transport dissolved mineral matter great distances to new locations. This may alter the composition of the metamorphosing rock by introducing or removing certain elements.

? Draw a sketch illustrating the difference between confining pressure and differential stress.

8.3 Metamorphic Textures

Explain how foliated and nonfoliated textures develop.

KEY TERMS: texture, foliation, rock cleavage, slaty cleavage, schistosity, gneissic texture (gneissic banding), nonfoliated, porphyroblastic texture

- Texture can reveal the orientation of stresses that helped form a metamorphic rock. A common metamorphic texture is foliation, the planar arrangement of mineral grains. Foliation forms perpendicular to the direction of maximum differential stress, through a combination of processes: rotation of mineral grains, recrystallization and growth of new mineral grains, and flattening of grains by solid-state flow or pressure solution.
- Several arrangements of mineral grains are classified as varieties of foliation, including slaty (rock) cleavage, schistosity, and gneissic banding.
- Metamorphic rocks that do not exhibit foliation are described as nonfoliated. For example, marble, which is a nonfoliated metamorphic rock, forms when small calcite grains in a parent limestone grow into a mass of large uniformly shaped crystals.
- Porphyroblasts are unusually large crystals of certain minerals, such as garnet, that form in an otherwise finer-grained metamorphic rock. They represent the tendency of some

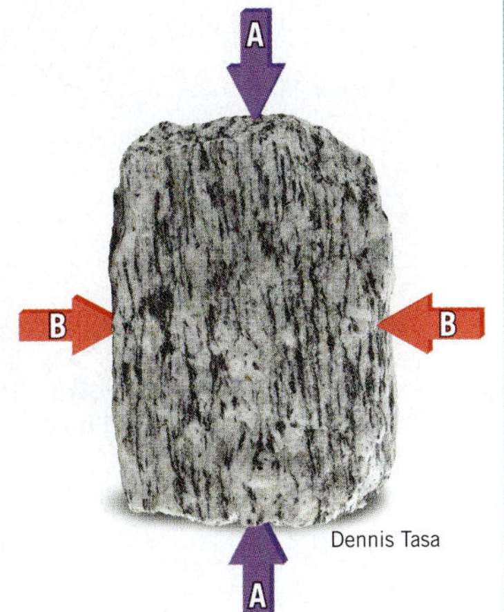

Dennis Tasa

(8.3 continued)

minerals to form a small number of large crystals during recrystallization, while other minerals form a large number of smaller crystals.

? Examine the accompanying photograph. Is this rock foliated or nonfoliated? Which pair of arrows shows the direction of maximum stress?

8.4 Common Metamorphic Rocks

List and describe the most common metamorphic rocks.

KEY TERMS: slate, phyllite, schist, gneiss, marble, quartzite, hornfels

- Common foliated metamorphic rocks include (in order of increasing metamorphic grade) slate, phyllite, schist, and gneiss.
- Common nonfoliated metamorphic rocks include quartzite, marble, and hornfels, which form from quartz sandstone, limestone, and shale, respectively.

? If you were on a class field trip and found an outcrop of nonfoliated light-colored rock, how would you determine whether it is quartzite or marble? How would you distinguish quartzite from quartz sandstone?

Dennis Tasa

8.5 Metamorphic Environments

Write a description for each of these metamorphic environments: contact metamorphism, hydrothermal metamorphism, subduction zone metamorphism, and regional metamorphism.

KEY TERMS: contact (thermal) metamorphism, aureole, hydrothermal metamorphism, burial metamorphism, subduction zone metamorphism, regional metamorphism, impact (shock) metamorphism

- Metamorphism can be triggered by diverse geologic situations. Convergent and divergent plate boundaries are both sites of metamorphism, as are deep stacks of sediments on passive margins.
- Contact metamorphism occurs in Earth's upper crust when heat from a molten igneous body "bakes" the surrounding rocks in a zone called the aureole. Differential stress is typically absent, so the rocks are not foliated.
- Hydrothermal metamorphism occurs when hot, ion-rich water circulates through pore spaces or fractures in rock. These fluids enhance recrystallization and can also import or export elements, changing the rock's chemical composition. Hydrothermal metamorphism produces economically important deposits of metal ore.
- Burial metamorphism occurs when rocks are buried beneath kilometers of overlying rock. Under confining pressure, elevated temperatures cause metamorphic reactions. Subduction zone metamorphism is similar, but with the added influence of differential stress.
- Continental collisions produce regional metamorphism. Rocks at these sites of crustal thickening are subjected to both high temperatures and differential pressure. The resulting belts of metamorphic rock (and associated igneous intrusions) mark zones where continental blocks collided to produce mountains. Long after the mountains have worn away, the collision site is marked by a deformed zone of regional metamorphic rocks that may include slate, schist, marble, and gneiss.
- Other distinctive varieties of metamorphism are associated with faults and meteorite impacts.

8.6 Metamorphic Zones

Explain how index minerals are used to establish the metamorphic grade of a rock body.

KEY TERMS: index mineral, migmatite

- The amount of change in metamorphic rocks increases as the rocks are exposed to higher temperatures and pressures. Metamorphic rocks reveal the degree of metamorphism through their textures and the minerals they contain.
- Grain size increases with higher levels of metamorphism. Under progressively higher temperatures and pressures, shale may transform first to slate, then to phyllite, to schist, and to gneiss.
- Certain minerals can act as indicators of the environment (temperature/pressure) that a metamorphic rock experienced. Chlorite is a mineral associated with low-grade metamorphism, whereas garnet and staurolite are index minerals for intermediate metamorphic grades. Sillimanite indicates a high metamorphic grade.
- In extreme metamorphic environments, the minerals with the lowest melting points (usually quartz and potassium feldspar) may melt while the mafic minerals with higher melting points remain solid. The resulting rocks, called migmatites, consist of light and dark bands that may be intricately folded.

Marli Miller

? When exposed to metamorphic conditions over sufficient time, shale becomes schist. This sample of schist contains porphyroblasts of sillimanite. Do these porphyroblasts indicate low-, intermediate-, or high-grade metamorphism?

GIVE IT SOME THOUGHT

1 Each of the following statements describes one or more characteristics of a particular metamorphic rock. For each statement, identify the metamorphic rock or rocks being described:

- **a.** calcite-rich and nonfoliated
- **b.** loosely coherent and composed of broken fragments that formed along a fault zone
- **c.** represents a grade of metamorphism between slate and schist
- **d.** composed of tiny chlorite and mica grains and displaying excellent rock cleavage
- **e.** foliated and composed predominantly of platy materials
- **f.** composed of alternating bands of light and dark silicate minerals
- **g.** hard and nonfoliated, often produced by contact metamorphism

2 Examine the accompanying close-up images of a conglomerate and a metaconglomerate.

- **a.** Describe how the conglomerate is different from the metaconglomerate.
- **b.** Does this metaconglomerate appear to have been exposed to significant differential stress? Explain.

A. Conglomerate

B. Metaconglomerate

Photos by Dennis Tasa

3 Describe the textural changes that occur as shale goes from low to high metamorphic grade, forming, successively, slate, schist, and gneiss.

4 Examine the accompanying photos that show the geology of the Grand Canyon. Notice that most of the canyon consists of layers of sedimentary rocks, but if you were to hike into the inner gorge, you would encounter the Vishnu Schist, a metamorphic rock.

- **a.** What process might have been responsible for the formation of the Vishnu Schist?
- **b.** What does the Vishnu Schist tell you about the history of the Grand Canyon prior to the formation of the canyon itself?
- **c.** Why is the Vishnu Schist visible at Earth's surface?
- **d.** Is it likely that rocks similar to the Vishnu Schist exist elsewhere but are not exposed at Earth's surface? Explain.

A. Inner Gorge of the Grand Canyon

B. Close up of Vishnu Schist (dark color)

5 One of the rock outcrops in the accompanying photos consists mainly of metamorphic rock. Which do you think it is? Explain why you ruled out the other rock bodies. (Photos by E. J. Tarbuck)

6 Refer to the accompanying diagram and match each labeled area with the appropriate environment listed below:

a. contact metamorphism
b. subduction metamorphism
c. regional metamorphism
d. burial metamorphism
e. hydrothermal metamorphism

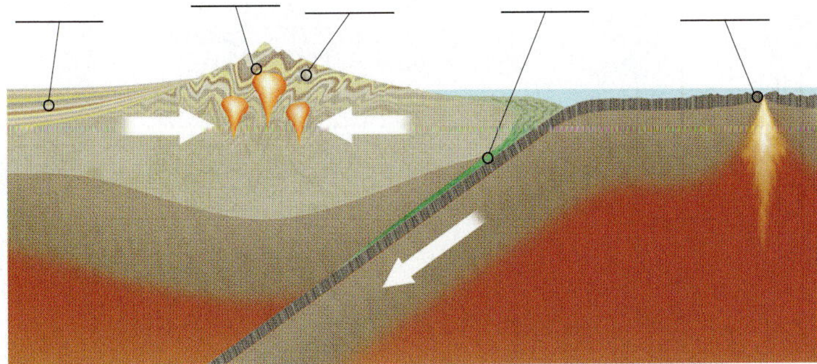

7 Based on the information provided in Figure 8.27, complete the following:

a. Describe how the metamorphic grade changes from west to east across New England along line A–B.
b. How might these metamorphic rocks have formed?
c. Are these zones of metamorphism consistent with the current tectonic setting of New England? Explain.

8 Examine the accompanying close-up images of six different rocks labeled A–F. Classify them as igneous, sedimentary, or metamorphic, based on texture. (*Hint:* There are two of each rock type.)

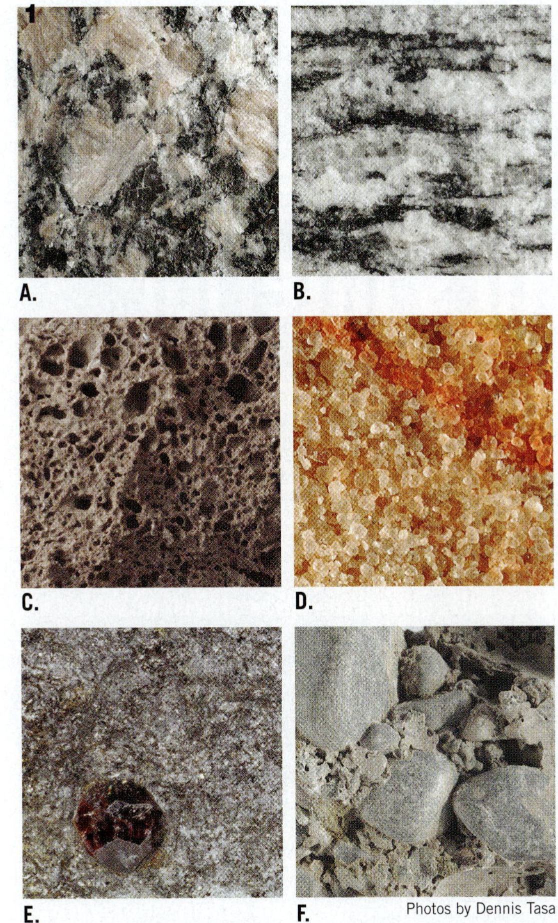

Photos by Dennis Tasa

9 The accompanying image shows a metamorphic rock located in Purgatory Chasm in Newport, Rhode Island, that is made of elongated cobbles composed mainly of quartz.

a. What name would you give to this metamorphic rock?
b. Which set of arrows (red or blue) best represents the direction of maximum differential stress?
c. Is this type of deformation best described as ductile or brittle?

Callan Bentley

MasteringGeology™

Looking for additional review and test prep materials? Visit the Study Area in MasteringGeology to enhance your understanding of this chapter's content by accessing a variety of resources, including Self-Study Quizzes, Geoscience Animations, SmartFigures, Mobile Field Trips, *Project Condor* Quadcopter videos, *In the News* RSS feeds, flashcards, web links, and an optional Pearson eText.

www.masteringgeology.com

9 Earthquakes & Earth's Interior

FOCUS ON CONCEPTS

Each statement represents the primary learning objective for the corresponding major heading within the chapter. After you complete the chapter, you should be able to:

9.1 Sketch and describe the mechanism that generates most earthquakes.

9.2 Compare and contrast the types of seismic waves and describe the principle of the seismograph.

9.3 Explain how seismographs are used to locate the epicenter of an earthquake.

9.4 Distinguish between intensity scales and magnitude scales.

9.5 List and describe the major destructive forces that earthquake vibrations can trigger.

9.6 Locate Earth's major earthquake belts on a world map.

9.7 Compare and contrast the goals of short-range earthquake predictions and long-range forecasts.

9.8 Explain how Earth acquired its layered structure and name and describe each of its major layers.

Tsunami stricking the coast of Minamisoma, Japan on March 11,2011. (Photo by Sadatasgu Tomizawa/AFP/Getty Images)

ON APRIL 25, 2015, Nepal experienced its worst natural disaster in over 80 years, when an earthquake measuring M 7.8* struck this mountainous region of South Asia. Geologists had anticipated a significant earthquake for decades because of Nepal's location high in the Himalayas, above the collisional boundary where India is being thrust into Asia. The quake killed nearly 9000 people and resulted in injuries to almost 23,000 more. The shallow nature of the 50-second quake resulted in widespread destruction and triggered avalanches on Mt. Everest, claiming the lives of 19 people, including international hikers and Sherpa guides. Numerous landslides throughout the region blocked roads and delayed relief efforts, and Nepal's capital, Kathmandu, reportedly shifted 3 meters (10 feet) to the south during the course of the event.

9.1 What Is an Earthquake?

Sketch and describe the mechanism that generates most earthquakes.

An **earthquake** is ground shaking caused by the sudden and rapid movement of one block of rock slipping past another along fractures in Earth's crust, called **faults**. Most faults are *locked*, except for brief, abrupt movements when sudden slippage produces an earthquake. Faults are locked because the confining pressure exerted by the overlying crust is enormous, causing these fractures in the crust to be "squeezed shut."

Earthquakes tend to occur along preexisting faults where internal stresses cause the crustal rocks to rupture or break into two or more units. The location where slippage begins is called the **hypocenter**, or **focus**. The point on Earth's surface directly above the hypocenter is called the **epicenter** (**Figure 9.1**).

Large earthquakes release huge amounts of stored-up energy as **seismic waves**—a form of energy that travels through the lithosphere and Earth's interior.

*An earthquake's *magnitude* (abbreviation M) is a measure of earthquake strength and is discussed later in this chapter.

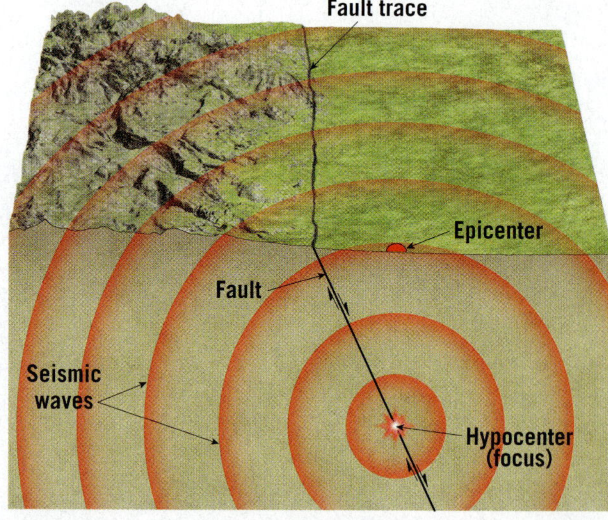

Figure 9.1 Earthquakes hypocenter and epicenter The *hypocenter* is the zone at depth where the initial displacement occurs. The *epicenter* is the surface location directly above the hypocenter.

The energy carried by these waves causes the material that transmits them to shake. Seismic waves are analogous to waves produced when a stone is dropped into a calm pond. Just as the impact of the stone creates a pattern of circular waves, an earthquake generates waves that radiate outward in all directions from the zone of slippage. Although seismic energy dissipates rapidly as it moves away from the source of an earthquake, sensitive instruments can detect earthquakes even when they occur on the opposite side of Earth.

Thousands of earthquakes occur around the world every day. Fortunately, most are small enough that people cannot detect them. Only about 15 strong earthquakes (magnitude 7 or greater) are recorded each year, many of them occurring in remote regions. The occasional large earthquakes that are triggered near a major population centers are among the most destructive natural events on Earth. The shaking of the ground, coupled with the liquefaction of soils, wreaks havoc on buildings, roadways, and other structures. In addition, a quake occurring in a populated area can rupture power and gas lines, causing numerous fires. In the famous 1906 San Francisco earthquake, much of the damage was caused by fires that became uncontrollable when broken water mains left firefighters with only trickles of water (**Figure 9.2**).

Discovering the Causes of Earthquakes

The energy released by volcanic eruptions, massive landslides, and meteorite impacts can generate earthquake-like waves, but these events are usually weak. What mechanism produces a destructive earthquake? As you have learned, Earth is not a static planet. Because fossils of marine organisms have been discovered thousands of meters above sea level, we know that large sections of Earth's crust have been thrust upward. Other regions, such as California's Death Valley, exhibit

Fire triggered when a gas line ruptured during the Northridge earthquake in southern California in 1994.

Figure 9.2 Earthquakes can trigger fires (Reproduced from the collection of the Library of Congress; inset photo by Hal Garb/AFP/Getty Images)

evidence of extensive subsidence. In addition to these vertical displacements, offsets in fences, roads, and other structures indicate that horizontal movements between blocks of Earth's crust are also common (**Figure 9.3**).

The actual mechanism of earthquake generation eluded geologists until H. F. Reid conducted a landmark study following the 1906 San Francisco earthquake. This earthquake was accompanied by horizontal surface displacements of several meters along the northern portion of the San Andreas Fault. Field studies determined that during this single earthquake, the Pacific plate lurched as much as 9.7 meters (32 feet) northward past the adjacent North American plate. To better visualize this, imagine standing on one side of the fault and watching a person on the other side, originally directly in front of you, suddenly slide horizontally 32 feet to your right.

What Reid concluded from his investigations is illustrated in **Figure 9.4**. Over tens to hundreds of years, differential stress slowly bends the crustal rocks on both sides of a fault. This is much like a person bending a limber wooden stick, as shown in Figure 9.4A,B. Frictional resistance keeps the fault from rupturing and slipping. (Friction inhibits slippage and is enhanced by irregularities that occur along the fault surface.) At some point, stress along the fault overcomes frictional resistance, and slippage occurs. Slippage allows the deformed (bent) rock to "snap

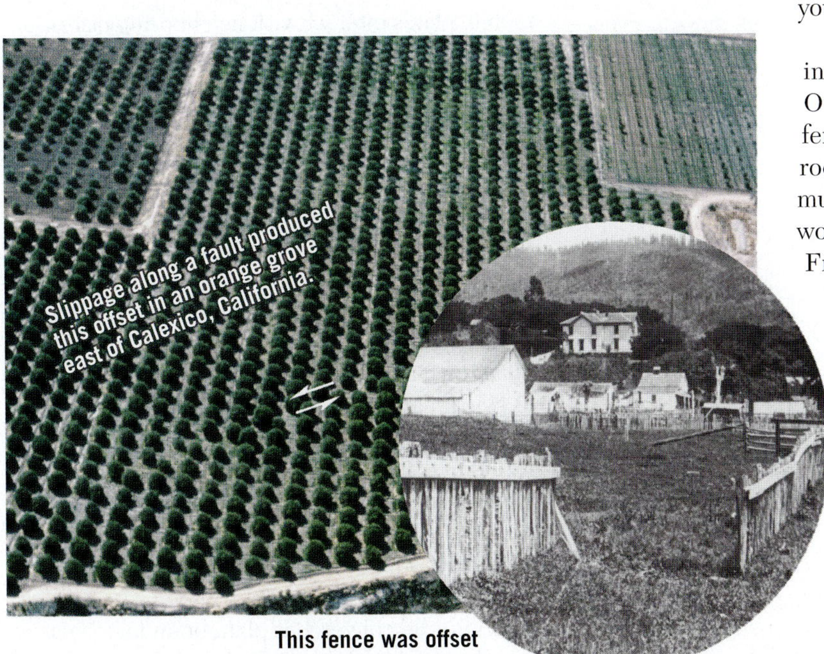

Slippage along a fault produced this offset in an orange grove east of Calexico, California.
This fence was offset 2.5 meters (8.5 feet) during the1906 San Francisco earthquake.

Figure 9.3 Displacement of human-made structures along a fault (Color photo by John S. Shelton/ University of Washington Libraries; inset photo by G. K. Gilbert/USGS)

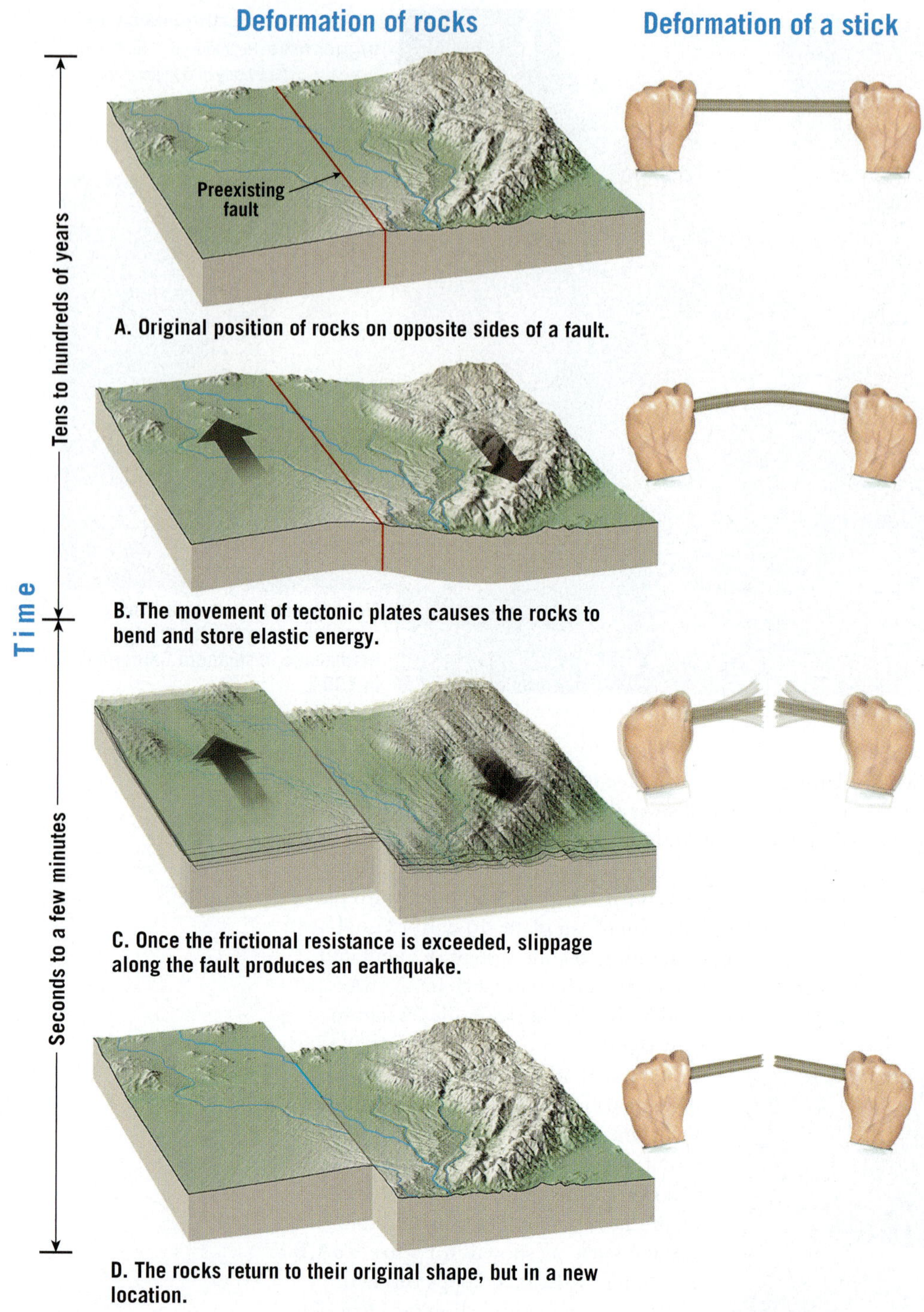

Deformation of rocks

Deformation of a stick

Tens to hundreds of years

Time

Seconds to a few minutes

Preexisting fault

A. Original position of rocks on opposite sides of a fault.

B. The movement of tectonic plates causes the rocks to bend and store elastic energy.

C. Once the frictional resistance is exceeded, slippage along the fault produces an earthquake.

D. The rocks return to their original shape, but in a new location.

▲ **SmartFigure 9.4**
Elastic rebound

TUTORIAL
https://goo.gl/s1w1z3

back" to its original, stress-free shape; a series of earthquake waves radiate outward as it slides (see Figure 9.4C,D). Reid termed this "springing back" **elastic rebound** because the rock behaves elastically, much as a stretched rubber band does when it is released.

Aftershocks & Foreshocks

Strong earthquakes are followed by numerous earthquakes of lesser magnitude, called **aftershocks**, which result from crust along the fault surface adjusting to the displacement caused by the main shock. Aftershocks gradually diminish in frequency and intensity over a period of several months following an earthquake. In a little more than a month following the M 7.0 earthquake that devastated Haiti in 2010, the U.S. Geological Survey detected nearly 60 aftershocks with magnitudes of 4.5 or greater. The two largest aftershocks had magnitudes of 6.0 and 5.9, both large enough to inflict further damage. Hundreds of minor tremors were also felt.

Although aftershocks are weaker than the main earthquake, they often trigger the destruction of already weakened structures. For example, in northwestern Armenia in 1988, where many people lived in large apartment buildings constructed of brick and concrete slabs, a moderate earthquake of magnitude 6.9 weakened many structures, and a strong aftershock of magnitude 5.8 completed the demolition.

In contrast to aftershocks, small earthquakes called **foreshocks** often, but not always, precede major earthquakes by days or, in some cases, several years. Monitoring of foreshocks to predict forthcoming earthquakes has been attempted with only limited success.

Faults & Large Earthquakes

The slippage that occurs along faults can be explained by the plate tectonics theory, which states that large slabs of Earth's lithosphere plates are continually grinding past one another. These mobile plates interact with neighboring plates, straining and deforming the rocks along their margins. Faults associated with convergent and transform plate boundaries are the source of most large earthquakes.

Convergent Plate Boundaries Most of Earth's strongest earthquakes occur along large faults associated with convergent plate boundaries. Compressional forces associated with continental collisions slice Earth's crust along numerous large *thrust faults*, discussed in more detail in Section 11.3. The 2015 Nepal earthquake is one example of an earthquake generated along a thrust fault. The epicenter of the quake was located about 80 kilometers (50 miles) north of Kathmandu, where the Indian plate is advancing into the Eurasian plate at a rate of 4.5 centimeters (about 2 inches) per year, driving the uplift of the Himalayas.

In addition, the convergent plate boundary separating a subducting slab of oceanic lithosphere and the

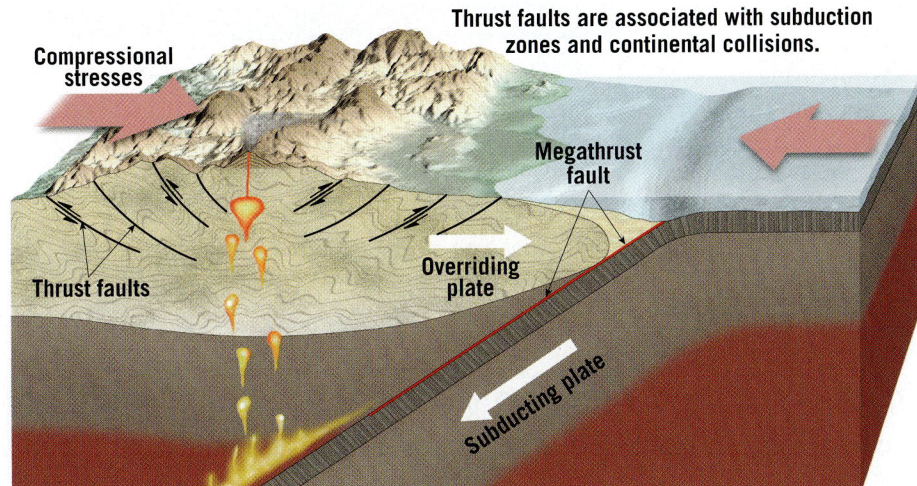

◀ **Figure 9.5 Megathrust faults are the sites of Earth's largest earthquakes** A convergent plate boundary is a site where one plate is subducting beneath another, and the megathrust faults that separate these plates generate most of Earth's largest earthquakes.

overlying plate forms an extensive fault zone, called a **megathrust fault**, that can be several thousand kilometers long (**Figure 9.5**). Along most subduction zones these megathrust faults remain locked for decades or even centuries. As the subducting plate slowly descends, it drags and slowly bends the leading edge of overlying plate, sometimes producing a bulge on the ocean floor (see Figure 9.26, page 254). Once the frictional forces between the two stuck plates is exceeded, the overriding plate snaps back to its original shape. This snapping back generates an earthquake, whose magnitude depends largely on the size of the zone of slippage.

Megathrust faults have produced the majority of Earth's most powerful and destructive earthquakes, including the 2011 Japan quake (M 9.0), the 2004 Indian Ocean (Sumatra) quake (M 9.1), the 1964 Alaska quake (M 9.2), and the largest earthquake yet recorded, the 1960 Chile quake (M 9.5).

Transform Plate Boundaries Faults in which the dominant displacement is horizontal and parallel to the direction of the fault trace (the line where the fault intersects Earth's surface) are called *strike-slip faults*, (discussed in more detail in Section 11.3). Recall from Chapter 2 that *transform plate boundaries* or simply *transform faults* accommodate motion between two tectonic plates. For example, the San Andreas Fault is a large transform fault that separates the North American plate and the Pacific plate (**Figure 9.6**). Most large transform faults are not perfectly straight or continuous; instead, they consist of numerous branches and smaller fractures that display kinks and offsets (see Figure 9.6). Earthquakes can occur along any of these branches.

Fault Rupture & Propagation

By studying earthquakes around the globe, geologists have learned that displacement along large faults occurs along discrete fault segments that often behave differently from one another. Some sections of the San Andreas, for example, exhibit slow, gradual displacement known as **fault creep** and produce little seismic shaking. Other segments slip at relatively closely spaced intervals, producing numerous small to moderate earthquakes. Still other segments remain locked and store elastic energy for a few hundred years before they break loose. Ruptures on segments that have been locked for a hundred years or longer usually result in major earthquakes.

Geologists have also discovered that slippage along large faults, such as the San Andreas, does not occur instantaneously. The initial slip occurs at the hypocenter and propagates (travels) along the fault surface; as each section slips, it puts strain on the next section, causing it to slip, too. Slippage propagates at 2 to 4 kilometers per second—faster than a rifle shot. Rupture

Did You Know?
Literally thousands of earthquakes occur daily! Fortunately, the majority of them are too small for people to feel, and the majority of larger ones occur in remote regions. Their existence is known only because of sensitive seismographs.

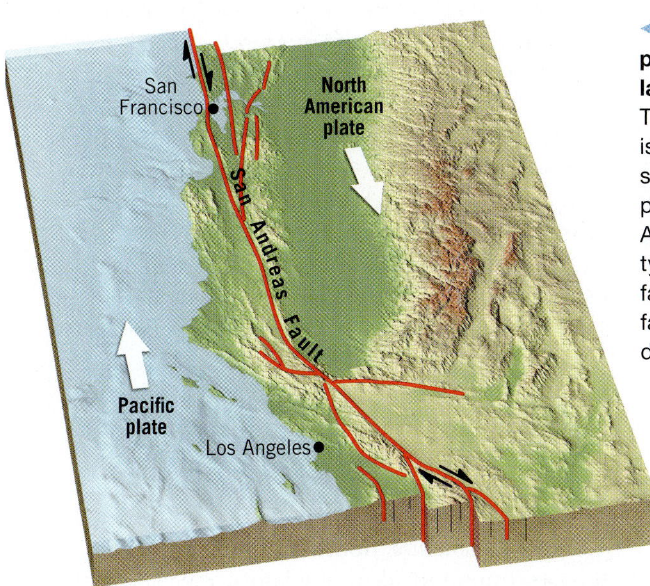

◀ **Figure 9.6 Transform plate boundaries and large earthquakes** The San Andreas Fault is a large fault system separating the Pacific plate from the North American plate. This type of large strike-slip fault is called a transform fault and can generate destructive earthquakes.

of a 100-kilometer (60-mile) fault segment takes about 30 seconds, and rupture of a 300-kilometer (200-mile) segment takes about 90 seconds. As rupturing progresses, it can slow down, speed up, or even jump to a nearby fault segment. Earthquake waves are generated at every point along the fault as that portion of the fault begins to slip.

CONCEPT CHECKS 9.1

1. What is an earthquake? Under what circumstances do most large earthquakes occur?
2. How are faults, hypocenters, and epicenters related?
3. Explain what is meant by *elastic rebound*.
4. What type of fault tends to produce the most destructive earthquakes?

9.2 Seismology: The Study of Earthquake Waves

Compare and contrast the types of seismic waves and describe the principle of the seismograph.

The study of earthquake waves, **seismology**, dates back to attempts made in China almost 2000 years ago to determine the direction from which these waves originated. The earliest known instrument, invented by Zhang Heng, was a large hollow jar containing a weight suspended from the top (Figure 9.7). The suspended weight (similar to a clock pendulum) was connected to the jaws of several large dragon figurines that encircled the container. The jaws of each dragon held a metal ball. When earthquake waves reached the instrument, the relative motion between the suspended mass and the jar would dislodge some of the metal balls into the waiting mouths of frogs directly below.

▲ **Figure 9.7 Ancient Chinese seismograph** During an Earth tremor, the dragons located in the direction of the main vibrations would drop a ball into the mouth of a frog below. (Photo by James E. Patterson Collection)

Instruments That Record Earthquakes

In principle, modern **seismographs**, or **seismometers**, are similar to the instruments used in ancient China. A seismograph has a weight freely suspended from a support that is securely attached to bedrock (Figure 9.8). When vibrations from an earthquake reach the instrument, the **inertia** of the weight keeps it relatively stationary, while Earth and the support move. Inertia can be simply described by this statement: *Objects at rest tend to stay at rest, and objects in motion tend to remain in motion, unless acted upon by an outside force.* You have experienced inertia when you have tried to stop your automobile quickly and your body continued to move forward.

Most seismographs are designed to amplify ground motion in order to detect very weak earthquakes or a great earthquake that has occurred in another part of the world. Instruments used in earthquake-prone areas are designed to withstand the violent shaking that can occur near a quake's epicenter.

Seismic Waves

The records obtained from seismographs, called **seismograms**, provide useful information about the nature of seismic waves. Seismograms reveal that two main types of seismic waves are generated by the slippage of a rock mass. One of these wave types, called **body waves**, travel through Earth's interior. The other type, called **surface waves**, travel in the rock layers just below Earth's surface (Figure 9.9).

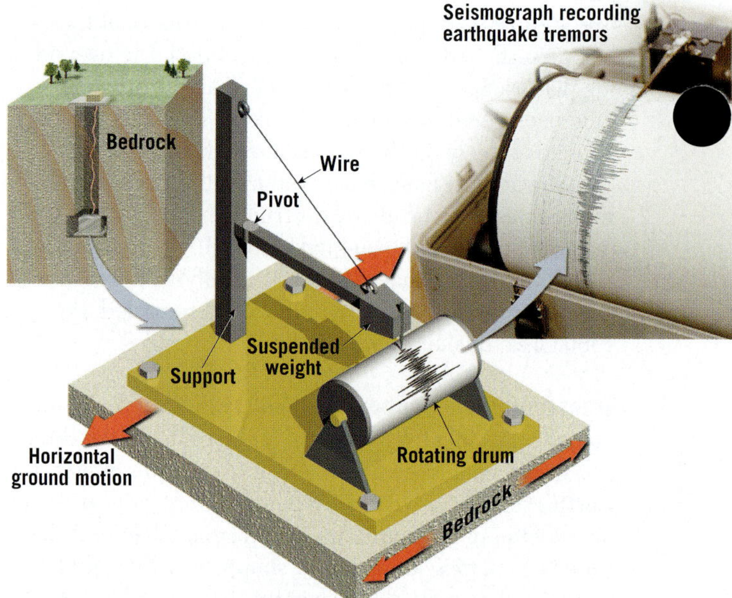

▲ **SmartFigure 9.8 Principle of the seismograph** The inertia of the suspended weight tends to keep it motionless while the recording drum, which is anchored to bedrock, vibrates in response to seismic waves. The stationary weight provides a reference point from which to measure the amount of displacement occurring as a seismic wave passes through the ground. (Photo courtesy of Zephyr/Science Source)

ANIMATION
https://goo.gl/13acww

Body Waves Body waves are further divided into two types—**primary waves**, or **P waves**, and **secondary waves**, or **S waves**—and are identified by their mode of travel through intervening materials. P waves are "push/pull" waves; they momentarily push (compress) and pull (stretch) rocks in the direction the waves are traveling (Figure 9.10A). This wave motion is similar to that generated by striking a drum, which moves air back and forth to create sound. Solids, liquids, and gases resist stresses that change their volume when compressed and, therefore, elastically spring back once the stress is removed. Therefore, P waves can travel through all these materials.

By contrast, S waves "shake" the particles at right angles to their direction of travel. This can be illustrated by fastening one end of a rope and shaking the other end, as shown in Figure 9.10B. Unlike P waves, which temporarily change the *volume* of intervening material by alternately squeezing and stretching it, S waves change the *shape* of the material that transmits them. Because fluids (gases and liquids) do not resist stresses that cause changes in shape—meaning fluids do not return to their original shape once the stress is removed—liquids and gases do not transmit S waves.

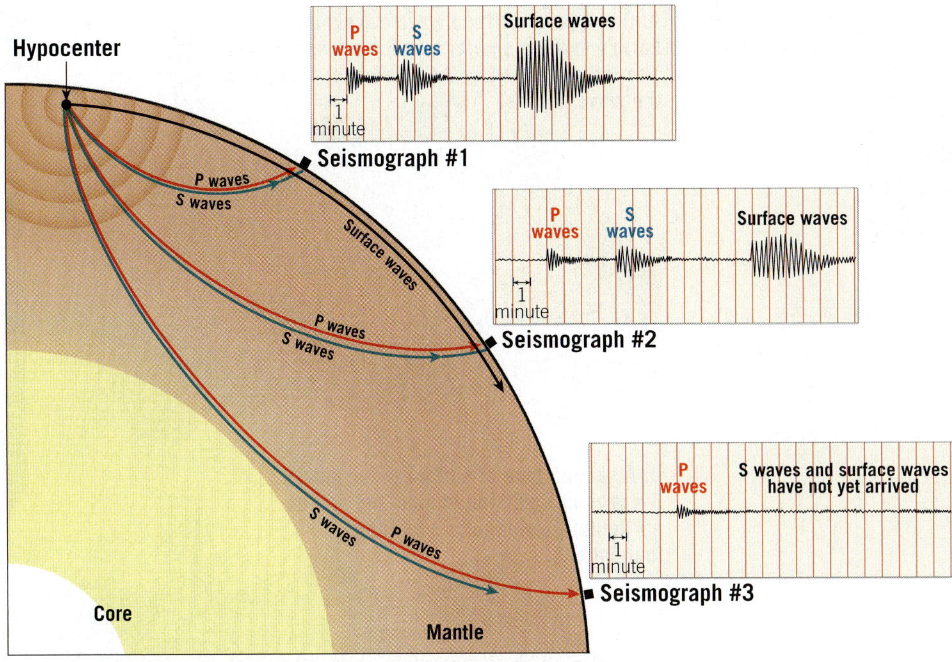

▲ SmartFigure 9.9 **Body waves (P and S waves) versus surface waves** P and S waves travel through Earth's interior, while surface waves travel in the layer directly below the surface. P waves are the first to arrive at a seismic station, followed by S waves and then surface waves.

TUTORIAL
https://goo.gl/PZIJrY

Surface Waves There are two types of surface waves. One type causes Earth's surface and anything resting on it to move up and down, much as ocean swells toss a ship (Figure 9.11A). The second type of surface wave causes Earth's surface to move from side to side. This motion is particularly damaging to the foundations of structures (Figure 9.11B).

Comparing the Speed & Size of Seismic Waves By examining the seismogram shown in Figure 9.12 you can see that the major difference among seismic waves is their speed of travel. P waves are the first to arrive at a recording station, then S waves, and finally surface waves. Generally, in any solid Earth material, P waves travel about 70 percent faster than S waves, and S waves are roughly 10 percent faster than surface waves.

In addition to the velocity differences in the waves, notice in Figure 9.12 that the height, or *amplitude*,

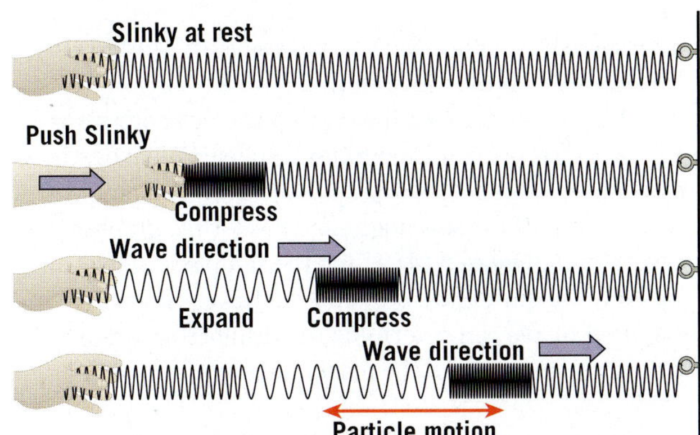

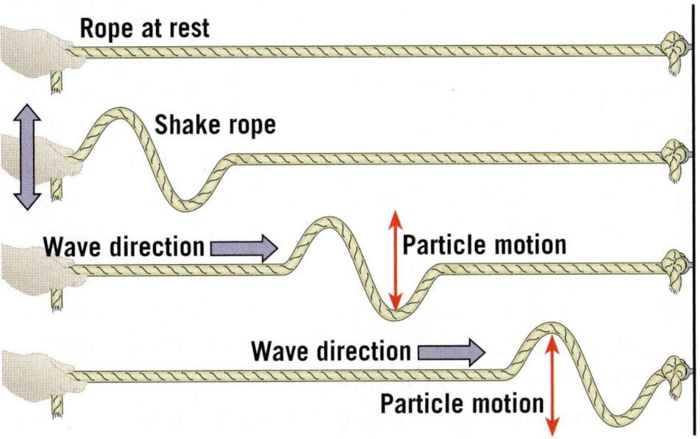

▲ Figure 9.10 **The characteristic motion of P and S waves** During a strong earthquake, ground shaking consists of a combination of various kinds of seismic waves.

▶ **SmartFigure 9.11 Two types of surface waves**

ANIMATION
https://goo.gl/RSV8tN

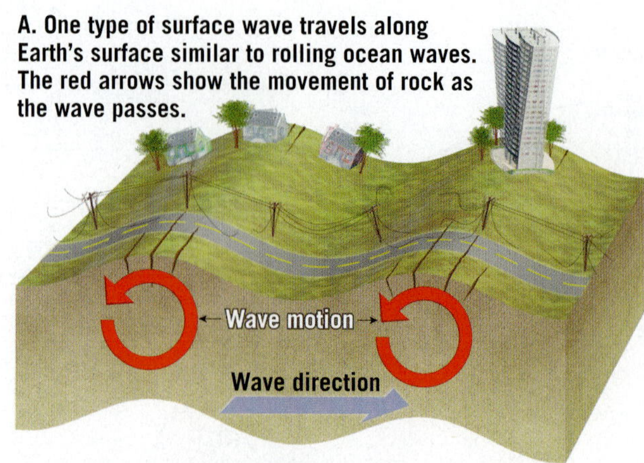

A. One type of surface wave travels along Earth's surface similar to rolling ocean waves. The red arrows show the movement of rock as the wave passes.
Wave motion
Wave direction

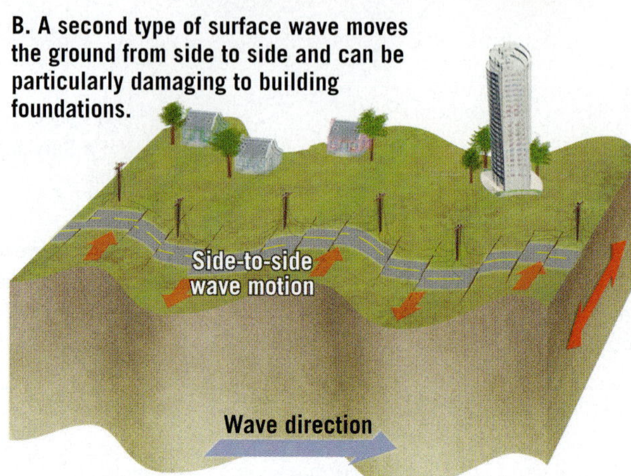

B. A second type of surface wave moves the ground from side to side and can be particularly damaging to building foundations.
Side-to-side wave motion
Wave direction

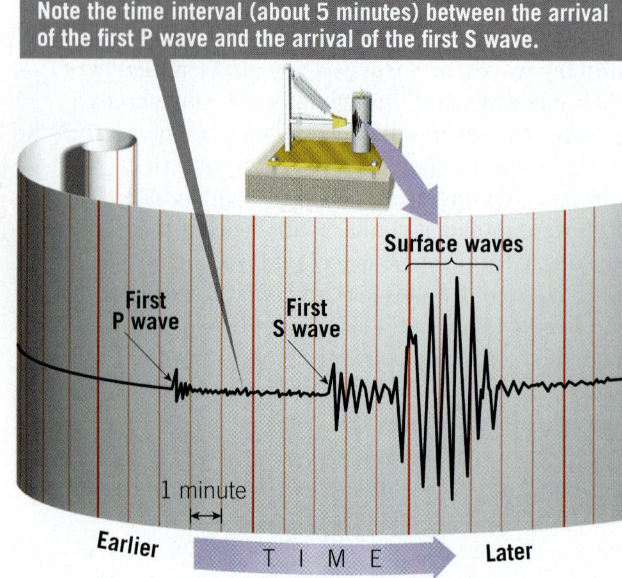

▲ **Figure 9.12 Typical seismogram**

of these wave types also varies. S waves have slightly greater amplitudes than P waves, and surface waves exhibit even greater amplitudes. Surface waves also retain their maximum amplitude longer than P and S waves. As a result, surface waves tend to cause greater ground shaking and, hence, greater property damage, than either P or S waves.

CONCEPT CHECKS 9.2

1. Describe how a seismograph works.
2. List the major differences between P, S, and surface waves.
3. Which type of seismic wave tends to cause the greatest destruction to buildings?

Did You Know?
Some of the most interesting uses of seismographs involve the reconstruction of human catastrophes, such as airline crashes, pipeline explosions, and mining disasters. For example, a seismologist helped with the investigation of Pan Am Flight 103, which was brought down over Lockerbie, Scotland, by a terrorist bomb in 1988. Nearby seismographs recorded six separate impacts, indicating that the plane had broken into that many large pieces before it crashed.

9.3 Locating the Source of an Earthquake

Explain how seismographs are used to locate the epicenter of an earthquake.

When seismologists analyze an earthquake, they first determine its *epicenter*, the point on Earth's surface directly above the hypocenter, or focus (see Figure 9.1). One method used for locating an earthquake's epicenter relies on the fact that P waves travel faster than S waves.

The traveling waves are analogous to two racing automobiles, one faster than the other. The first P wave, like the faster automobile, always wins the race, arriving ahead of the first S wave. The greater the length of the race, the greater the difference in their arrival times at the finish line (the seismic station). Therefore, the longer the interval between the arrival of the first P wave and the arrival of the first S wave, the greater the distance to the epicenter. **Figure 9.13** shows three simplified seismograms for the same earthquake. Based on the P–S interval, which city—New York, Nome, or Mexico City—is farthest from the epicenter?

The system for locating earthquake epicenters was developed by using seismograms from earthquakes whose epicenters could be easily pinpointed based on physical evidence. From these seismograms, travel–time graphs were constructed (**Figure 9.14**). Using the sample seismogram for New York in Figure 9.13 and the travel–time curve in Figure 9.14, we can determine the distance separating the recording station from the earthquake in three steps:

1. Using the seismogram for New York, we determine that the time interval between the arrival of the first P wave and the arrival of the first S wave is 5 minutes.

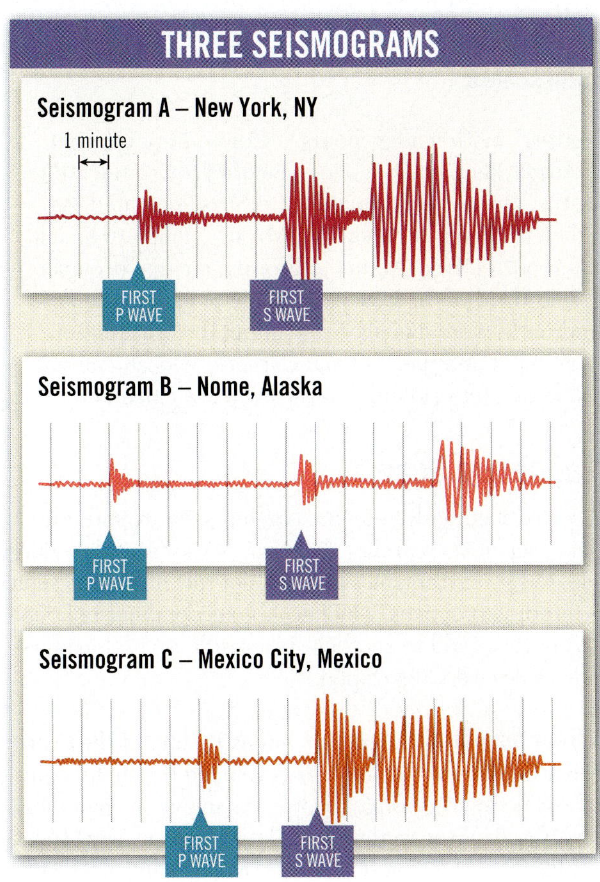

▲ **Figure 9.13 Seismograms of the same earthquake recorded at three different locations**

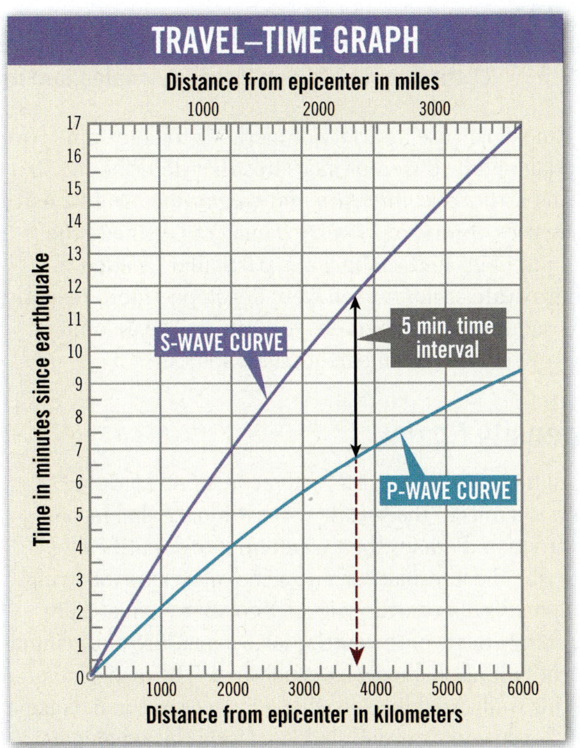

▲ **Figure 9.14 Travel–time graph** A travel–time graph is used to determine the distance to an earthquake's epicenter. The difference in arrival times of the first P and S waves in the example shown is 5 minutes.

2. Using the travel–time graph, we find the location where the vertical separation between P and S curves is equal to the P–S time interval (5 minutes in this example).
3. From the position in step 2, we draw a vertical line to the horizontal axes and read the distance to the epicenter.

Using these steps we determine that the earthquake occurred 3700 kilometers (2300 miles) from the recording instrument in New York City.

Now we know the *distance*, but what about *direction*? The epicenter could be in any direction from the seismic station. Using a method called *triangulation*, we can determine the location of an epicenter if we know the distance to it from two or more additional seismic stations (**Figure 9.15**). On a map or globe, we draw a circle around each seismic station with a radius equal to the distance from that station to the epicenter. The point where the three circles intersect is the approximate epicenter of the quake.

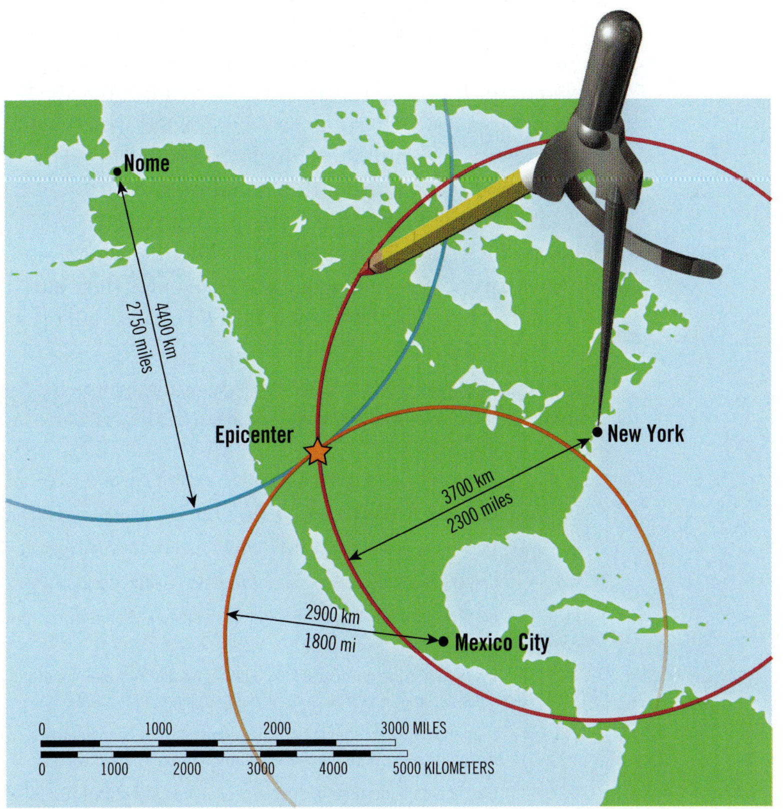

▲ **Figure 9.15 Triangulation to locate an earthquake** This method involves using the distance obtained from three or more seismic stations to establish the location of an earthquake.

CONCEPT CHECKS 9.3

1. What information does a travel–time graph provide?
2. Briefly describe the *triangulation* method used to locate the epicenter of an earthquake.

9.4 Determining the Size of an Earthquake

Distinguish between intensity scales and magnitude scales.

Seismologists use a variety of methods to determine two fundamentally different measures that describe the size of an earthquake: *intensity* and *magnitude*. An **intensity** scale uses observed property damage to estimate the amount of ground shaking at a particular location. **Magnitude** scales, which were developed more recently, use data from seismographs to estimate the amount of energy released at an earthquake's source.

Intensity Scales

Until the mid-1800s, historical records provided the only accounts of the severity of earthquake shaking and destruction. Perhaps the first attempt to scientifically describe the aftermath of an earthquake came following the great Italian earthquake of 1857. By systematically mapping effects of the earthquake, a measure of the intensity of ground shaking was established. The map generated by this study used lines to connect places of equal damage and hence equal ground shaking. Using this technique, zones of intensity were identified, with the zone of highest intensity located near the center of maximum ground shaking and often (but not always) the earthquake epicenter.

In 1902, Giuseppe Mercalli developed a more reliable intensity scale, which is still used today in a modified form. The **Modified Mercalli Intensity scale**, shown in Table 9.1, was developed using California buildings as its standard. Based on the 12-point Mercalli Intensity scale, an area in which some well-built wood structures and most masonry buildings are destroyed by an earthquake would be assigned a Roman numeral X (10).

More recently, the U.S. Geological Survey has developed a website called "Did You Feel It," where Internet users experiencing a quake can enter their zip code and answer questions such as "Did objects fall off shelves?" Within a few hours, a Community Internet Intensity Map, like the one in Figure 9.16 for the 2011 central Virginia earthquake (M 5.8), is generated. As shown in Figure 9.16, shaking strong enough to be felt was reported from Maine to Florida, an area occupied by one-third of the U.S. population. Several national landmarks were damaged, including the Washington Monument and the National Cathedral, located about 130 kilometers (80 miles) away from the epicenter.

Magnitude Scales

To more accurately compare earthquakes around the globe, scientists searched for a way to describe the energy released by earthquakes that did not rely on factors such as building practices, which vary considerably from one part of the world to another. As a result, several magnitude scales were developed.

Richter Magnitude In 1935 Charles Richter of the California Institute of Technology developed the first magnitude scale to use seismic records. As shown in Figure 9.17 (top), the **Richter scale** is calculated by measuring the amplitude of the largest seismic wave (usually an S wave or a surface wave) recorded on a seismogram. Because seismic waves weaken as the distance between the hypocenter and the seismograph increases, Richter developed a method that accounts for the decrease in wave amplitude with increasing distance. Theoretically, as long as equivalent instruments are used, monitoring stations at different locations will obtain the same Richter magnitude for each recorded earthquake. In practice, however, different recording stations often obtain slightly different magnitudes for the same earthquake—a result of the variations in the rock types through which the waves travel.

Table 9.1 Modified Mercalli Intensity Scale

I	Not felt except by a very few under especially favorable circumstances.
II	Felt only by a few persons at rest, especially on upper floors of buildings.
III	Felt quite noticeably indoors, especially on upper floors of buildings, but many people do not recognize it as an earthquake.
IV	During the day felt indoors by many, outdoors by few. Sensation like heavy truck striking building.
V	Felt by nearly everyone, many awakened. Disturbances of trees, poles, and other tall objects sometimes noticed.
VI	Felt by all; many frightened and run outdoors. Some heavy furniture moved; few instances of fallen plaster or damaged chimneys. Damage slight.
VII	Everybody runs outdoors. Damage negligible in buildings of good design and construction; slight to moderate in well-built ordinary structures; considerable in poorly built or badly designed structures.
VIII	Damage slight in specially designed structures; considerable, with partial collapse, in ordinary buildings; great in poorly built structures (falling chimneys, factory stacks, columns, monuments, walls).
IX	Damage considerable in specially designed structures. Buildings shifted off foundations. Ground cracked conspicuously.
X	Most masonry and frame structures destroyed. Some well-built wooden structures destroyed. Ground badly cracked.
XI	Few, if any, (masonry) structures remain standing. Bridges destroyed. Broad fissures in ground.
XII	Damage total. Waves seen on ground surfaces. Objects thrown upward into air.

The green dots on the national map show locations of people who reported feeling earthquakes of similar magnitude, one that occurred in California versus one that occurred in Virginia. The difference is attributable to the rigidity of the bedrock.

USGS Community Internet Intensity Map

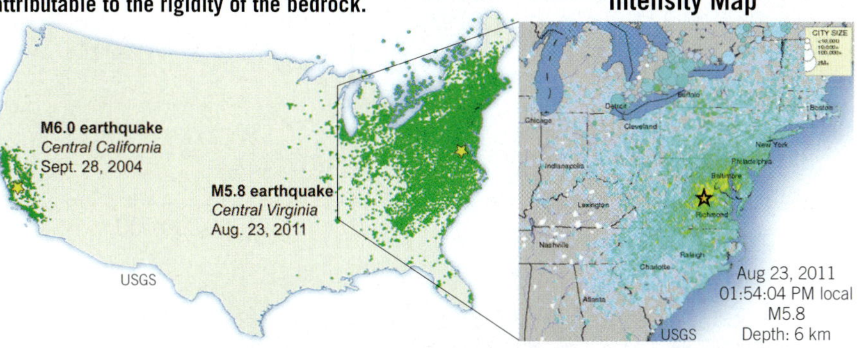

Key for USGS Community Internet Intensity Map

INTENSITY	I	II-III	IV	V	VI	VII	VIII	IX	X
SHAKING	Not felt	Weak	Light	Moderate	Strong	Very strong	Severe	Violent	Extreme
DAMAGE	none	none	none	Very light	Light	Moderate	Moderate/Heavy	Heavy	Very Heavy

▲ **SmartFigure 9.16 USGS Community Internet Intensity Map** Maps like this one are prepared using data collected on the Internet from people responding to questions such as, "Did objects fall off shelves?"

TUTORIAL
https://goo.gl/NANACy

Earthquakes vary enormously in strength, and great earthquakes produce wave amplitudes thousands of times larger than those generated by weak tremors. To accommodate this wide variation, Richter used a *logarithmic scale* to express magnitude, in which a *10-fold* increase in wave amplitude corresponds to an increase of 1 on the magnitude scale. Thus, the intensity of ground shaking for a magnitude 5 earthquake is 10 times greater than that produced by an earthquake having a Richter magnitude (M_L) of 4 (**Figure 9.18**).

In addition, each unit of increase in Richter magnitude equates to roughly a *32-fold increase in the energy released*. Thus, an earthquake with a magnitude of 6.5 releases 32 times more energy than one with a magnitude of 5.5 and roughly 1000 times (32 × 32) more energy than a magnitude 4.5 quake. A major earthquake with a magnitude of 8.5 releases millions of times more energy than the smallest earthquakes felt by humans (**Figure 9.19**).

The convenience of describing the size of an earthquake by a single number that can be calculated quickly from seismograms makes the Richter scale a powerful tool. Seismologists have since modified Richter's work and developed other Richter-like magnitude scales.

Despite its usefulness, the Richter scale is not adequate for describing very large earthquakes. For example, the 1906 San Francisco earthquake and the 1964 Alaska earthquake have roughly the same Richter magnitudes. However, based on the relative size of the affected areas and the associated tectonic changes, the Alaska earthquake released considerably more energy than the San Francisco quake. Thus, the Richter scale is considered *saturated* for major earthquakes because it cannot distinguish among them. Despite this shortcoming, Richter-like scales are still used because they allow for quick calculations.

Moment Magnitude For measuring medium and large earthquakes, seismologists now favor a newer scale called **moment magnitude** (M_W), which estimates the total energy released during an earthquake. Moment magnitude is calculated by determining the average amount of slip on the fault, the area of the fault surface that slipped, and the strength of the faulted rock.

Moment magnitude can also be calculated by modeling data obtained from seismograms. The results are converted to a magnitude number, similar to other magnitude scales. Like the Richter scale, each unit increase

1. Measure the height (amplitude) of the largest wave on the seismogram (23 mm) and plot it on the amplitude scale (right).

2. Determine the distance to the earthquake using the time interval separating the arrival of the first P wave and the arrival of the first S wave (24 seconds) and plot it on the distance scale (left).

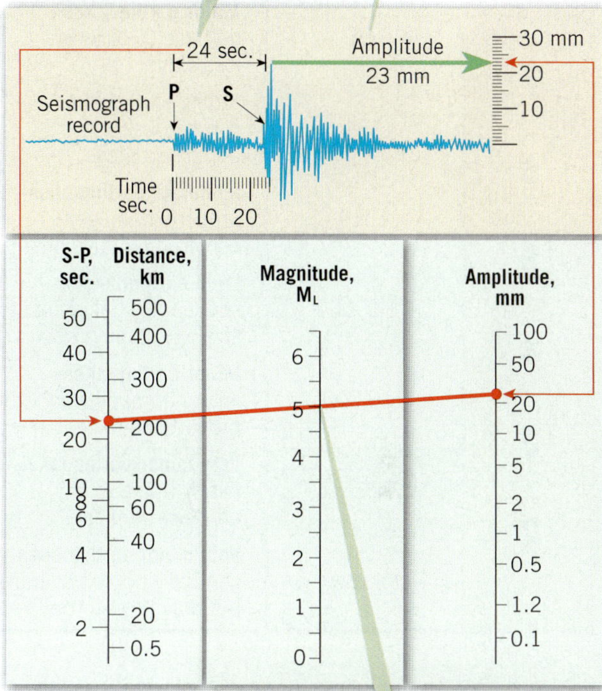

3. Draw a line connecting the two plots and read the Richter magnitude (M_L 5) from the magnitude scale (center).

▲ **Figure 9.17 Determining the Richter magnitude of an earthquake**

Magnitude vs. Ground Motion and Energy

Difference in Magnitude	Difference in Ground Motion (amplitude)	Difference in Energy Release (approximate)
4.0	10,000 times	1,000,000 times
3.0	1000 times	32,000 times
2.0	100 times	1000 times
1.0	10 times	32 times
0.5	3.2 times	5.5 times
0.1	1.3 times	1.4 times

◀ **Figure 9.18 Magnitude versus ground motion and energy released** An earthquake that is 1 magnitude stronger than another (M 6 versus M 5) produces seismic waves that have a maximum amplitude 10 times greater and releases about 32 times more energy than the weaker quake.

Frequency and Energy Released by Earthquakes of Different Magnitudes

Magnitude (Mw)	Average Per Year	Description	Examples	Energy Release (equivalent kilograms of explosive)
9	<1	**Largest recorded earthquakes–** destruction over vast area, massive loss of life possible	Chile, 1960 (M 9.5); Alaska, 1964 (M 9.2); Sumatra, 2004 (M 9.1); Japan, 2011 (M 9.0)	56,000,000,000,000
8	1	**Great earthquakes–** severe economic impact, large loss of life	Chile, 2010 (M 8.8); Mexico City, 1980 (M 8.1)	1,800,000,000,000
7	15	**Major earthquakes–** damage ($ billions), loss of life	San Francisco, California, 1906 (M 7.9); Haiti, 2012 (M 7.0); Ecuador, 2016 (M 7.8)	56,000,000,000
6	134	**Strong earthquakes–** can be destructive in populated areas, loss of life	Kobe, Japan, 1995 (M 6.9); Loma Prieta, California, 1989 (M 6.9); Northridge, California, 1994 (M 6.7)	1,800,000,000
5	1319	**Moderate earthquakes–** property damage to poorly constructed buildings	Mineral, Virginia, 2011 (M 5.8); Northern New York, 1994 (M 5.8); East of Oklahoma City, Oklahoma, 2011 (M 5.6)	56,000,000
4	13,000	**Light earthquakes–** noticeable shaking of items indoors, some property damage	Western Minnesota, 1975 (M 4.6); Arkansas, 2011 (M 4.7)	1,800,000
3	130,000	**Minor earthquakes–** felt by humans, very light property damage, if any	New Jersey, 2009 (M 3.0); Maine, 2006 (M 3.8); Texas, 2015 (M 3.6)	56,000
2	1,300,000	**Very minor earthquakes–** felt by humans, no property damage		1,800
	Unknown	**Very minor earthquakes–** generally not felt by humans, but may be recorded		56

Data from USGS

▲ **Figure 9.19 Annual occurrence of earthquakes with various magnitudes**

Did You Know?
It is a commonly held belief that moderate earthquakes decrease the chances of a major earthquake in the same region, but this is not the case. When you compare the amount of energy released by earthquakes of different magnitudes, it turns out that thousands of moderate tremors would have to occur to equal the huge amount of energy released during one "great" earthquake.

on the moment magnitude scale equates to roughly a 32-fold increase in the energy released.

Because the moment magnitude scale is better at estimating the relative size of very large earthquakes, seismologists have used the moment magnitude scale to recalculate the magnitudes of older strong earthquakes. For example, the 1964 Alaska earthquake, given a Richter magnitude of 8.3, has since been recalculated using the moment magnitude scale, resulting in an upgrade to M_W 9.2. Conversely, the 1906 San Francisco earthquake's Richter magnitude of 8.3 was downgraded to M_W 7.9. The strongest earthquake on record is the 1960 Chilean subduction zone earthquake, at a moment magnitude of 9.5.

CONCEPT CHECKS 9.4

1. What does the Modified Mercalli Intensity scale tell us about an earthquake?
2. What information is used to establish the lower numbers on the Mercalli scale?
3. How much more energy does a magnitude 7.0 earthquake release than a magnitude 6.0 earthquake?
4. Why is the moment magnitude scale favored over the Richter scale for large earthquakes?

9.5 Earthquake Destruction

List and describe the major destructive forces that earthquake vibrations can trigger.

The most violent earthquake ever recorded in North America—the 1964 Alaska earthquake—occurred at 5:36 P.M. on March 27, 1964. Felt over most of the state, the earthquake had a moment magnitude of 9.2 and lasted 3 to 4 minutes. This event left 128 people dead and thousands homeless, and it badly disrupted the state's economy. Within 24 hours of the initial shock, 28 aftershocks were recorded, 10 of them exceeding magnitude 6. The

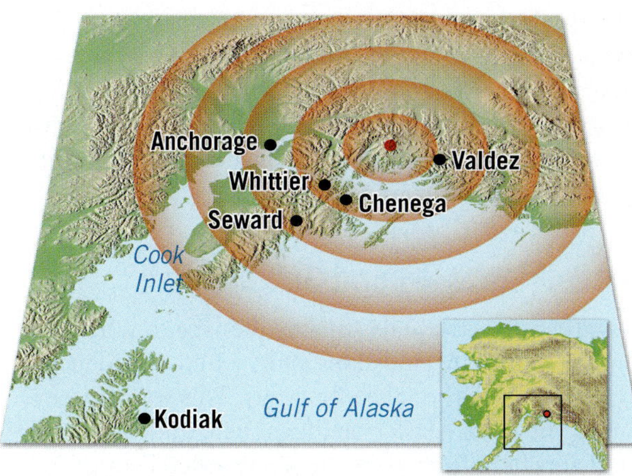

▲ **Figure 9.20 Region most affected by the Alaska earthquake, 1964**

▲ **Figure 9.21 Comparing damage to structures** The poorly designed five-story JCPenney building in Anchorage, Alaska, sustained extensive damage. The steel-frame adjacent building incurred very little structural damage. (Courtesy of NOAA/Seattle)

epicenter and towns hardest hit by the quake are shown in Figure 9.20.

Many factors determine the degree of destruction that accompanies an earthquake. The most obvious is the *magnitude of the earthquake and its proximity to a populated area*. During an earthquake, the region within 20 to 50 kilometers (12 to 30 miles) of the epicenter tends to experience roughly the same degree of ground shaking, and beyond that limit, vibrations usually diminish rapidly. As described earlier, earthquakes that occur in the stable continental interior, such as the New Madrid, Missouri, earthquakes of 1811–1812, are generally felt more over a much larger area than those in earthquake-prone areas such as California.

Destruction from Seismic Vibrations

The 1964 Alaska earthquake provided geologists with insights into the role of ground shaking as a destructive force. As the energy released by an earthquake travels along Earth's surface, it causes the ground to vibrate in a complex manner involving up-and-down as well side-to-side motion. The amount of damage to human-made structures attributable to the vibrations depends on several factors, including (1) the *intensity* and (2) *duration of the vibrations*, (3) the *nature of the material on which structures rest*, and (4) the *building materials and construction practices of the region*.

All the multistory structures in Anchorage were damaged by the vibrations. The more flexible wood-frame residential buildings fared best. A striking example of how construction variations affect earthquake damage is shown in Figure 9.21. You can see that the steel-frame building on the left withstood the vibrations, whereas the poorly designed JCPenney building was badly damaged. Engineers have learned that buildings constructed of blocks and bricks that are not reinforced with steel rods are the most serious safety threats in earthquakes. Unfortunately, most of the structures in the developing world are constructed of unreinforced concrete slabs and bricks made of dried mud—a primary reason the death toll in poor countries such as Haiti and Nepal is usually higher than for earthquakes of similar size in Japan and the United States.

The 1964 Alaska earthquake damaged most large structures in Anchorage, even though they were built according to the earthquake provisions of the Uniform Building Code. Perhaps some of that destruction can be attributed to the unusually long duration of the earthquake. Most quakes involve tremors that last less than a minute. For example, the 1994 Northridge earthquake was felt for about 40 seconds, and the strong vibrations of the 1989 Loma Prieta earthquake lasted less than 15 seconds. But the Alaska quake, which was substantually larger than both, reverberated for 3 to 4 minutes.

Amplification of Seismic Waves Although the whole region near the epicenter experiences about the same intensity of ground shaking, destruction may vary considerably in this area due to the nature of the ground on which the structures are built. Soft sediments, for example, generally amplify the vibrations more than solid bedrock. Thus, the buildings in Anchorage that were situated on unconsolidated sediments experienced heavy structural damage (Figure 9.22). In contrast, most of the town of Whittier, though much nearer the epicenter, rested on a firm foundation of solid bedrock and, therefore, suffered much less damage from seismic vibrations.

▶ **Figure 9.22 Ground failure caused this street in Anchorage, Alaska, to collapse** (Photo by USGS)

Liquefaction The intense shaking of an earthquake can cause loosely packed water-logged materials, such as sandy stream deposits or fill, to be transformed into a substance that acts like a fluid. The phenomenon of transforming a somewhat stable soil into mobile material capable of rising toward Earth's surface is known as **liquefaction**. When liquefaction occurs, the ground may not be capable of supporting buildings, and underground storage tanks and sewer lines may literally float toward the surface (**Figure 9.23**).

During the 1989 Loma Prieta earthquake, in San Francisco's Marina District, foundations failed, and geysers of sand and water shot from the ground, evidence that liquefaction had occurred (**Figure 9.24**). Liquefaction also contributed to the damage inflicted on San Francisco's water system during the 1906 earthquake. During the 2011 Japan earthquake, liquefaction caused entire buildings to sink several feet.

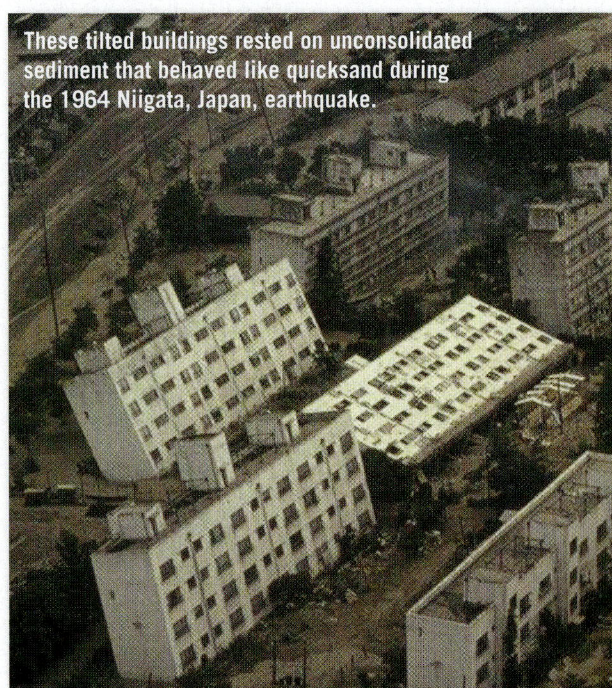

These tilted buildings rested on unconsolidated sediment that behaved like quicksand during the 1964 Niigata, Japan, earthquake.

▶ **Figure 9.23 Effects of liquefaction on buildings** (Photo courtesy of USGS)

Landslides & Ground Subsidence

The greatest earthquake-related damage to structures is often caused by landslides and ground subsidence triggered by earthquake vibrations. This was the case during the 1964 Alaska earthquake in Valdez and Seward, where the violent shaking caused coastal sediments to slump, carrying away both waterfronts. In Valdez, 31 people died when a dock slid into the sea. Because of the threat of recurrence, the entire town of Valdez was relocated to more stable ground about 7 kilometers away.

Much of the damage in Anchorage was attributed to landslides. Homes in Turnagain Heights were destroyed when a layer of clay lost its strength and over 200 acres of land slid toward the ocean (**Figure 9.25**). A portion of this spectacular landslide was left in its natural condition as a reminder of this destructive event. The site was appropriately named "Earthquake Park." Downtown Anchorage was also disrupted as sections of the main business district dropped by as much as 3 meters.

Fire

More than a century ago, San Francisco was the economic center of the western United States, largely because of gold and silver mining. Then, at dawn on April 18, 1906, a violent earthquake struck, triggering an enormous firestorm (see Figure 9.2). Much of the city was reduced to ashes and ruins. It is estimated that 3000 people died and more than half of the city's 400,000 residents were left homeless.

The historic San Francisco earthquake reminds us of the formidable threat of fire, which started when the quake severed gas and electrical lines. The initial ground shaking broke the city's water lines into hundreds of disconnected pieces, which made controlling the fires virtually impossible. The fires, which raged out of control for 3 days, were finally contained when expensive houses along Van Ness Avenue were dynamited to provide a fire break, similar to the strategy used in fighting forest fires.

While few deaths were attributed to the San Francisco fires, other earthquake-initiated fires have been more destructive, claiming many more lives. For example, the 1923 earthquake in Japan triggered an estimated 250 fires, devastating the city of Yokohama and destroying more than half the homes in Tokyo. More than 100,000 deaths were attributed to the fires, which were driven by unusually high winds.

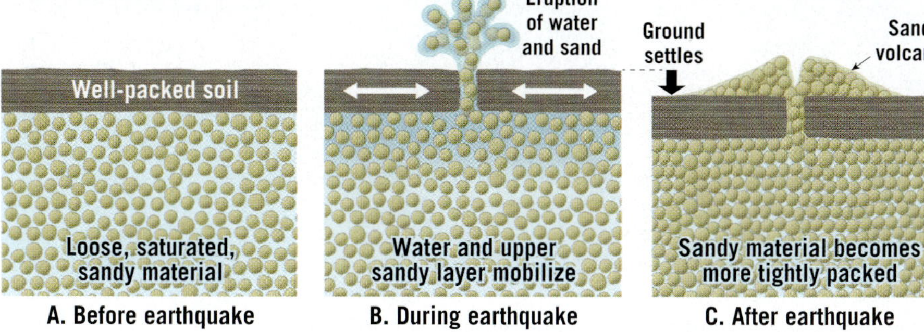

▲ **Figure 9.24 Liquefaction** These "sand volcanoes," produced by the Christchurch, New Zealand, earthquake of 2011, formed when "geysers" of sand and water shot from the ground, an indication that liquefaction occurred. (Photo by Alamy)

A. Vibrations from the Alaska earthquake caused cracks to appear near the edge of the Turnagain Heights bluff.

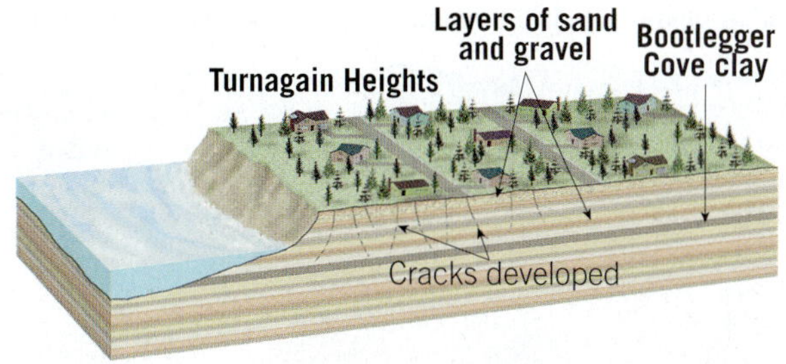

B. Blocks of land began to slide toward the sea on a weak layer called the Bootlegger Cove clay and in less than 5 minutes, as much as 200 meters of the Turnagain Heights bluff area had been destroyed.

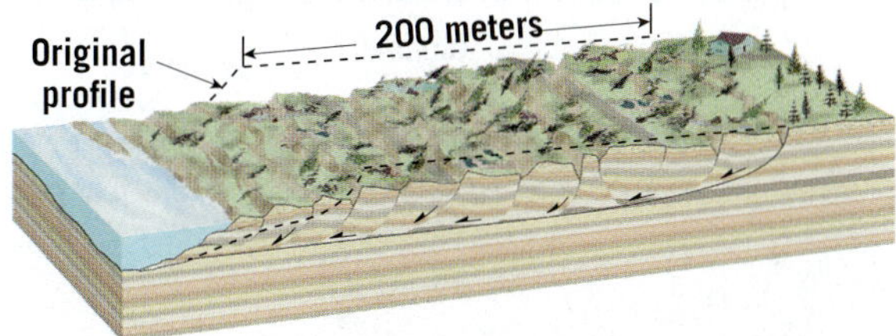

C. Photo of a small area of destruction caused by the Turnagain Heights slide.

▲ **SmartFigure 9.25 Turnagain Heights slide caused by the 1964 Alaska earthquake** (Photo courtesy of USGS)

TUTORIAL
https://goo.gl/q3Pw2q

Tsunamis

Major undersea earthquakes may set in motion a series of large ocean waves that are known by the Japanese name **tsunami** ("harbor wave"). Most tsunamis are generated by displacement along a megathrust fault that suddenly lifts a large slab of seafloor (**Figure 9.26**). Once generated, a tsunami resembles a series of ripples formed when a pebble is dropped into a pond. In contrast to ripples, however, tsunamis advance across the ocean at amazing speeds, about 800 kilometers (500 miles) per hour—equivalent to the speed of a commercial airliner. Despite this striking characteristic, a tsunami in the open ocean can pass undetected because its height (amplitude) is usually less than 1 meter (3 feet), and the distance separating wave crests ranges from 100 to 700 kilometers (60 to 425 miles). However, upon entering shallow coastal waters, these destructive waves "feel bottom" and slow, causing the water to pile up (see Figure 9.26). A few exceptional tsunamis have approached 20 meters (65 feet) in height. As the crest of a tsunami approaches the shore, it appears as a rapid rise in sea level with a turbulent and chaotic surface; it usually does not resemble a breaking wave (**Figure 9.27**).

The first warning of an approaching tsunami is often the rapid withdrawal of water from beaches, the result of the trough of the first large wave preceding the crest. Some inhabitants of the Pacific basin have learned to heed this warning and quickly move to higher ground. Approximately 5 to 30 minutes after the retreat of water, a surge capable of extending several kilometers inland occurs. In a successive fashion, each surge is followed by a rapid oceanward retreat of the sea. Therefore, people experiencing a tsunami should not return to the shore when the first surge of water retreats.

Tsunami Damage from the 2004 Indonesia Earthquake

A massive undersea earthquake of M_W 9.1 occurred near the island of Sumatra on December 26, 2004, sending waves of water racing across the Indian Ocean and Bay of Bengal. It was one of the deadliest natural disasters of any kind in modern times, claiming more than 230,000 lives. As water surged several kilometers inland, cars and trucks were flung around like toys in a bathtub, and fishing boats were rammed into homes.

A. Before the Earthquake: Along subduction zones the megathrust fault located between the subducting plate and the overriding plate can remain locked for decades or even centuries. As the subducting plate slowly descends it drags and gradually bends the leading edge of overlying plate, sometimes producing a bulge on the ocean floor.

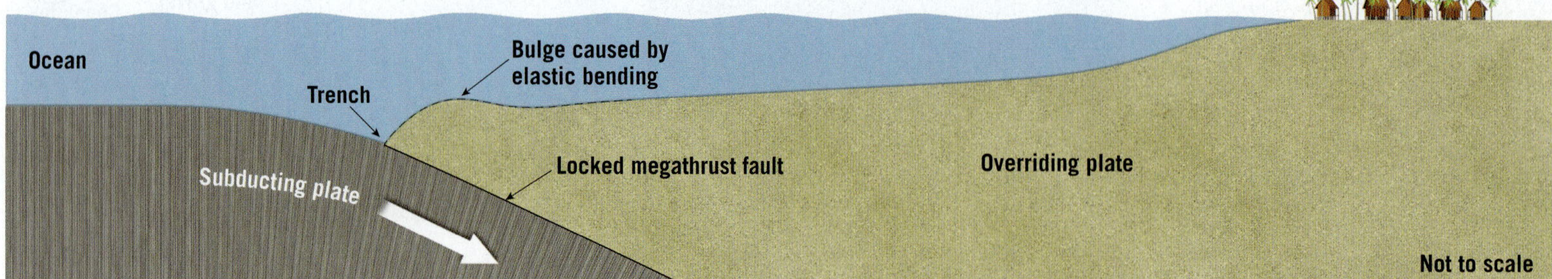

B. During the Earthquake: When the frictional force between the stuck plates is exceeded, the overlying plate lurches seaward and upward. At the same time, the seafloor landward of the zone of displacement stretches and subsides. These rapid up and down vertical motions trigger a tsunami consisting of a train of wave crests and troughs.

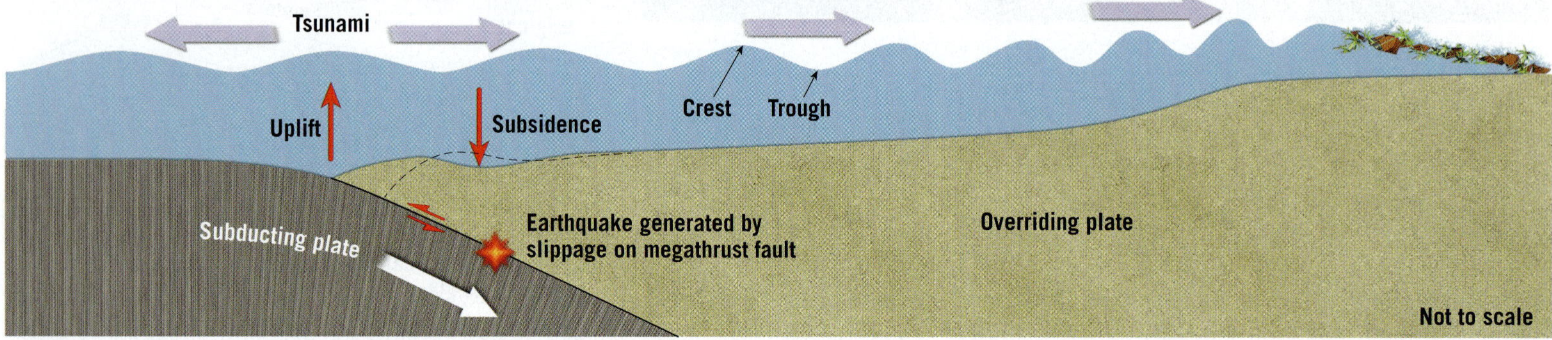

▲ SmartFigure 9.26 **How a tsunami is generated by displacement of the ocean floor during an earthquake** The speed of a tsunami wave correlates with ocean depth. In deep water, these waves can advance at speeds exceeding 800 kilometers (500 miles) per hour. When they enter coastal waters and begin to "feel bottom," they slow down and grow in height. They are still very fast moving, with a speed of 50 kilometers (30 miles) per hour at a depth of 20 meters (65 feet). The size and spacing of the swells in this figure are not to scale.

TUTORIAL
https://goo.gl/glpMvu

In some locations, the backwash of water dragged bodies and huge amounts of debris out to sea.

The destruction was indiscriminate, destroying luxury resorts as well as poor fishing hamlets along the Indian Ocean. Damage was reported as far away as the coast of Somalia in Africa, 4100 kilometers (2500 miles) west of the earthquake epicenter.

Japan Tsunami Because of Japan's location along the circum-Pacific belt and its extensive coastline, it is especially vulnerable to tsunami destruction. The most powerful earthquake to strike Japan in the age of modern seismology was the 2011 Tohoku earthquake (M_W 9.0). This historic earthquake and devastating tsunami resulted in at least 15,890 deaths, more than 3000 people missing, and 6107 injured. Nearly 400,000 buildings, 56 bridges, and 26 railways were destroyed or damaged.

The majority of human casualties and damage after the Tohoku earthquake were caused by a Pacific-wide tsunami that reached a maximum height of about 8.5 meters (28 feet) and traveled inland 10 kilometers (6 miles) in the region of Sendai, Japan (Figure 9.28). Based on physical evidence the run-up height, which is the elevation that a place on land is inundated by water from a tsunami, was almost twice as high—15 meters (50 feet) above sea level. In addition, meltdowns occurred at three inundated nuclear reactors in Japan's Fukushima Daiichi Nuclear Complex. Across the Pacific in California, Oregon, Peru, and Chile, some loss of life occurred, and several houses, boats, and docks were destroyed. The tsunami was

▼ SmartFigure 9.27 **Tsunami generated off the coast of Sumatra, 2004** (AFP/Getty Images, Inc.)

ANIMATION
https://goo.gl/WXeRi6

▲ **Figure 9.28 Japan tsunami, March 2011** (JIJI PRESS/AFP/Getty Images)

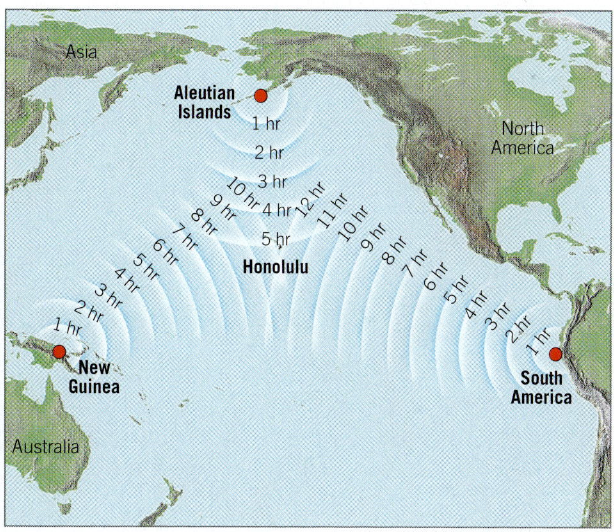

◄ **Figure 9.29 Tsunami travel times** Travel times to Honolulu, Hawaii, from selected locations throughout the Pacific. (Data from NOAA)

generated when a slab of seafloor located 60 kilometers (37 miles) off the east coast of Japan was suddenly "thrust up" an estimated 5 to 8 meters (16 to 26 feet).

Tsunami Warning System In 1946, a large tsunami struck the Hawaiian Islands without warning. A wave more than 15 meters (50 feet) high left several coastal villages in shambles. This destruction motivated the U.S. Coast and Geodetic Survey to establish a tsunami warning system for coastal areas of the Pacific that today includes 26 countries. Seismic observatories throughout the region report large earthquakes to the Tsunami Warning Center in Honolulu. Scientists at the center use deep-sea buoys equipped with pressure sensors to detect energy released by an earthquake. In addition, tidal gauges measure the rise and fall in sea level that accompany tsunamis, and warnings are issued within an hour. Although tsunamis travel very rapidly, there is sufficient time to warn all except those in the areas nearest the epicenter. For example, a tsunami generated near the Aleutian Islands would take 5 hours to reach Hawaii, and one generated near the coast of Chile would travel 15 hours before reaching the shores of Hawaii (Figure 9.29).

CONCEPT CHECKS 9.5

1. List four factors that influence the amount of destruction that seismic vibrations cause to human-made structures.
2. In addition to the destruction created directly by seismic vibrations, list three other types of destruction associated with earthquakes.
3. What is a tsunami? How are tsunamis generated?
4. List at least three reasons an earthquake with a magnitude of 7.0 might result in more death and destruction than a quake with a magnitude of 8.0.

Did You Know?
Although there are no historical records of tsunamis produced by meteorite impacts in the oceans, such events have occurred. Geologic evidence indicates that the most recent occurrence was a megatsunami that devastated portions of the Australian coast sometime around 1500. A meteorite impact 65 million years ago near Mexico's Yucatan Peninsula created one of the largest impact-induced tsunamis ever. The giant wave swept hundreds of kilometers inland around the Gulf of Mexico coastline.

9.6 Where Do Most Earthquakes Occur?

Locate Earth's major earthquake belts on a world map.

About 95 percent of the energy released by earthquakes originates in a few relatively narrow zones, shown in Figure 9.30. These zones of earthquake activity are located along fault surfaces where tectonic plates interact along one of the three types of plate boundaries—convergent, divergent, and transform plate boundaries.

Earthquakes Associated with Plate Boundaries

The previous section described the relationship between plate tectonics and seismic activity. The zone of greatest seismic activity, called the **circum-Pacific belt**, encompasses the coastal regions of Chile, Central America, Indonesia, Japan, and Alaska, including the Aleutian Islands (see Figure 9.30). Most earthquakes in the circum-Pacific belt occur along convergent plate boundaries at which one plate subducts beneath another at a low angle. Recall that the contacts between the subducting and overlying plates are called *megathrust faults* (see Figure 9.5). There are more than 40,000 kilometers (25,000 miles) of subduction boundaries in the circum-Pacific belt where displacement is dominated by thrust faulting. Ruptures occasionally occur along segments that

Did You Know?
The greatest loss of life attributed to an earthquake occurred in northern China. The Chinese made their homes there by carving numerous structures into a powdery, windblown material called *loess*. Early on the morning of January 23, 1556, an earthquake struck the region, collapsing the cavelike structures and killing about 830,000 people. Sadly, in 1920, a similar event collapsed loess structures in much the same way, taking about 200,000 lives.

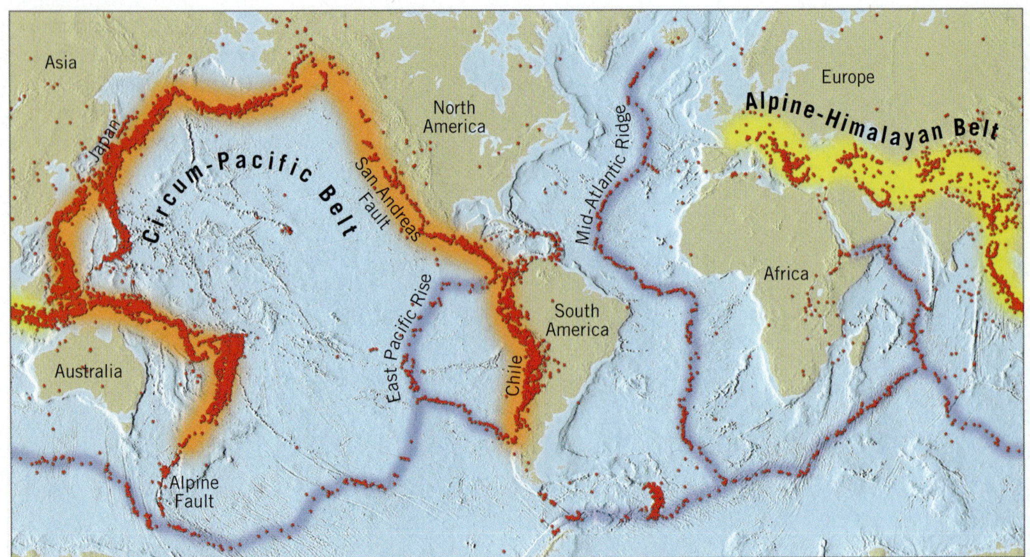

▲ **Figure 9.30 Global earthquake belts** Distribution of nearly 15,000 earthquakes with magnitudes equal to or greater than 5 for a 10-year period. (Data from USGS)

are nearly 1000 kilometers (600 miles) long, generating catastrophic earthquakes with magnitudes of 8 or greater.

Another major concentration of strong seismic activity, referred to as the *Alpine–Himalayan belt*, runs through the mountainous regions that flank the Mediterranean Sea and extends past the Himalayan Mountains (see Figure 9.30). Tectonic activity in this region is mainly attributed to collisions of the African plate and the Indian subcontinent with the vast Eurasian plate (see Figure 2.18, page 47). These plate interactions created many thrust and strike-slip faults that remain active.

Transform faults that run through continental crust are also a source of some large earthquakes. Examples include California's San Andreas Fault, New Zealand's Alpine Fault, and Turkey's North Anatolian Fault, which produced a deadly earthquake in 1999.

Figure 9.30 shows another continuous earthquake belt that extends thousands of kilometers through the world's oceans. This zone coincides with the oceanic ridge system—a divergent plate boundary—an area of frequent but weak seismic activity.

Damaging Earthquakes East of the Rockies

When you think "earthquake," you probably think of the western United States or Japan. However, six major earthquakes and several others that inflicted considerable damage have occurred in the central and eastern United States since colonial times (**Figure 9.31**).

Three of these quakes, which occurred as a cluster, had estimated magnitudes of 7.0 and destroyed what was then the frontier town of New Madrid, Missouri, located in the Mississippi River valley. It has been estimated that if an earthquake the size of the 1811–1812 New Madrid event were to strike in the same location in the next decade, it would result in casualties in the thousands and damages in the tens of billions of dollars.

The greatest historical earthquake in the eastern states occurred on August 31, 1886, in Charleston, South Carolina. This earthquake resulted in 60 deaths, numerous injuries, and great economic loss over an area extending 200 kilometers (120 miles) from Charleston. Within 8 minutes, effects were felt as far away as Chicago and St. Louis, where strong vibrations shook

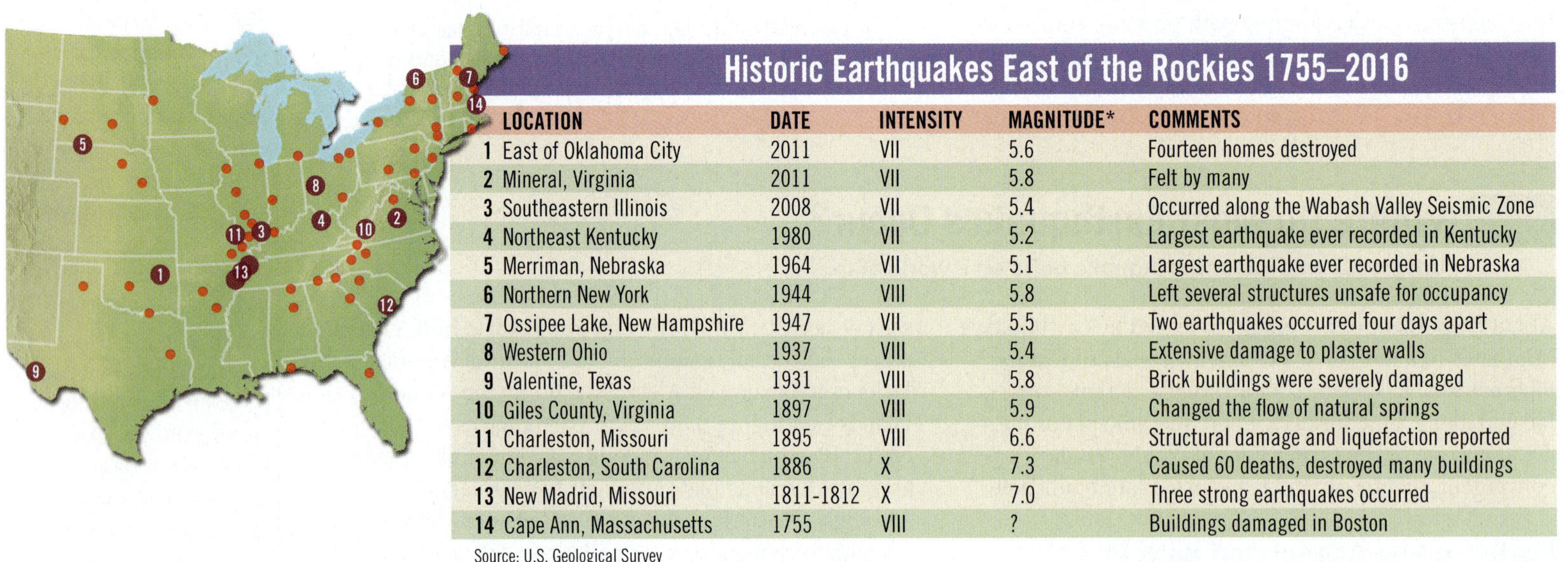

Historic Earthquakes East of the Rockies 1755–2016

LOCATION	DATE	INTENSITY	MAGNITUDE*	COMMENTS
1 East of Oklahoma City	2011	VII	5.6	Fourteen homes destroyed
2 Mineral, Virginia	2011	VII	5.8	Felt by many
3 Southeastern Illinois	2008	VII	5.4	Occurred along the Wabash Valley Seismic Zone
4 Northeast Kentucky	1980	VII	5.2	Largest earthquake ever recorded in Kentucky
5 Merriman, Nebraska	1964	VII	5.1	Largest earthquake ever recorded in Nebraska
6 Northern New York	1944	VIII	5.8	Left several structures unsafe for occupancy
7 Ossipee Lake, New Hampshire	1947	VII	5.5	Two earthquakes occurred four days apart
8 Western Ohio	1937	VIII	5.4	Extensive damage to plaster walls
9 Valentine, Texas	1931	VIII	5.8	Brick buildings were severely damaged
10 Giles County, Virginia	1897	VIII	5.9	Changed the flow of natural springs
11 Charleston, Missouri	1895	VIII	6.6	Structural damage and liquefaction reported
12 Charleston, South Carolina	1886	X	7.3	Caused 60 deaths, destroyed many buildings
13 New Madrid, Missouri	1811-1812	X	7.0	Three strong earthquakes occurred
14 Cape Ann, Massachusetts	1755	VIII	?	Buildings damaged in Boston

Source: U.S. Geological Survey
*Intensity and magnitudes have been estimated for many of these events.

▲ **Figure 9.31 Historical earthquakes east of the Rockies** Large earthquakes are uncommon in the middle of continents, far from the places where plates collide or grind past one another, or where one plate slides beneath another. Nevertheless, several damaging earthquakes have occurred in the central and eastern United States since colonial times.

Figure 9.32 Damage to Charleston, South Carolina, caused by the August 31, 1886, earthquake (Photo courtesy of USGS)

the upper floors of buildings, causing people to rush outdoors. In Charleston alone, more than 100 buildings were destroyed, and 90 percent of the remaining structures were damaged (Figure 9.32).

Earthquakes in the central and eastern United States occur far less frequently than in California, yet history indicates that the East is vulnerable. Further, these shocks east of the Rockies have generally produced structural damage over a larger area than earthquakes of similar magnitude in California. This is because the underlying bedrock in the central and eastern United States is older and more rigid. As a result, seismic waves can travel greater distances with less attenuation (loss of strength) than in the western United States.

Earthquakes that occur away from plate boundaries are called *intraplate earthquakes*. Intraplate earthquakes can be caused by a variety of factors. For example, stress can rejuvenate ancient fault systems that formed when crustal fragments collided to generate the continents billions of years ago (see Figure 12.12, page 335). In addition, the process called fracking, in which a solution is injected into the ground under high pressure to enhance oil and gas production, has contributed to the recent increase in earthquake activity east of the Rockies. Most of these intraplate earthquakes are weak.

CONCEPT CHECKS 9.6

1. What zone on Earth has the greatest amount of seismic activity?
2. What type of plate boundary is associated with Earth's largest earthquakes?
3. Explain why an earthquake east of the Rockies may produce damage over a larger area than one of similar magnitude in California.

Did You Know?
During the 1811–1812 New Madrid earthquake, some areas subsided as much as 4.5 m (15 ft). This subsidence created Lake St. Francis west of the Mississippi and enlarged Reelfoot Lake to the east of the river. Other regions rose, creating temporary waterfalls in the channel of the Mississippi River.

9.7 Can Earthquakes Be Predicted?

Compare and contrast the goals of short-range earthquake predictions and long-range forecasts.

The vibrations that shook the San Francisco area in 1989 caused 63 deaths, heavily damaged the Marina District, and caused the collapse of a double-decked section of I-880 in Oakland, California (Figure 9.33). This level of destruction was the result of an earthquake of moderate magnitude (M_W 6.9). Seismologists warn that other earthquakes of comparable or greater strength can be expected along the San Andreas system, which cuts a nearly 1300-kilometer (800-mile) path through the western one-third of the state. An obvious question is: Can these earthquakes be predicted?

▲ **Figure 9.33 Collapse of the double-decked section of I-880** This section of a double-decked highway, known as the Cypress Viaduct, collapsed during the 1989 Loma Prieta earthquake. (Photo by Paul Sakuma/AP Photo)

Short-Range Predictions

The goal of short-range earthquake prediction is to provide a warning of the location and magnitude of a large earthquake within a narrow time frame (Table 9.2). Substantial efforts to achieve this objective have been attempted in Japan, the United States, China, and Russia—countries where earthquake risks are high. This research has concentrated on monitoring possible **precursors**—events or changes that precede a forthcoming earthquake and thus may provide warning. In California, for example, seismologists monitor changes in ground elevation and variations in strain levels near active faults. Other researchers measure changes in groundwater levels, while still others try to predict earthquakes based on an increase in the frequency of foreshocks that precede some, but not all, earthquakes.

Table 9.2 Some Notable Earthquakes

Year	Location	Deaths (est.)	Magnitude*	Comments
856	Iran	200,000		
893	Iran	150,000		
1138	Syria	230,000		
1268	Asia Minor	60,000		
1290	China	100,000		
1556	Shensi, China	830,000		Possibly the greatest natural disaster
1667	Caucasia	80,000		
1727	Iran	77,000		
1755	Lisbon, Portugal	70,000		Tsunami damage extensive
1783	Italy	50,000		
1908	Messina, Italy	120,000		
1920	China	200,000	7.5	Landslide buried a village
1923	Tokyo, Japan	143,000	7.9	Fire caused extensive destruction
1948	Turkmenistan	110,000	7.3	Almost all brick buildings near epicenter collapsed
1960	Southern Chile	5700	9.5	The largest-magnitude earthquake ever recorded
1964	Alaska	131	9.2	Greatest-magnitude North American earthquake
1970	Peru	70,000	7.9	Great rockslide
1976	Tangshan, China	242,000	7.5	Estimates for the death toll are as high as 655,000
1985	Mexico City	9500	8.1	Major damage occurred 400 km from epicenter
1988	Armenia	25,000	6.9	Poor construction practices increased the death toll
1990	Iran	50,000	7.4	Landslides and poor construction practices led to great damage
1993	Latur, India	10,000	6.4	Located in stable continental interior
1995	Kobe, Japan	5472	6.9	Damages estimated to exceed $100 billion
1999	Izmit, Turkey	17,127	7.4	Nearly 44,000 injured and more than 250,000 displaced
2001	Gujarat, India	20,000	7.9	Millions homeless
2003	Bam, Iran	31,000	6.6	Ancient city with poor construction
2004	Indian Ocean (Sumatra)	230,000	9.1	Devastating tsunami damage
2005	Pakistan/Kashmir	86,000	7.6	Many landslides; 4 million homeless
2008	Sichuan, China	87,000	7.9	Millions homeless, some towns will not be rebuilt
2010	Port-au-Prince, Haiti	250,000–316,000	7.0	More than 300,000 injured and 1.3 million homeless
2011	Japan	16,000	9.0	Majority of the casualties due to a tsunami
2015	Nepal	8000	7.8	About 21,000 injured
2016	Ecuador	700	7.8	More than 28,000 injuries

*Widely differing magnitudes have been estimated for some of these earthquakes. When available, moment magnitudes are used.
Source: U.S. Geological Survey.

Japanese and Chinese scientists have tried to monitor anomalous animal behavior. A few days before the May 12, 2008, earthquake in China's Sichuan Province, the streets of a village near the fault were filled with toads migrating from the mountains. Was this a warning? Perhaps. Walter Mooney, a USGS seismologist, put it best: "Everyone hopes that animals can tell us something we don't know . . . but animal behavior is way too unreliable." Although precursors may exist, we have yet to determine effective ways to interpret and utilize the information.

One claim of a successful short-range prediction, based on an increase in foreshocks, was made by the Chinese government after the February 4, 1975, earthquake in Liaoning Province. According to reports, very few people were killed—even though more than 1 million lived near the epicenter—because the earthquake was "predicted," and the residents were evacuated. Some Western seismologists have questioned this claim and suggest instead that an intense swarm of foreshocks, which began 24 hours before the main earthquake, may have caused many people to evacuate of their own accord.

One year after the Liaoning earthquake, an estimated 240,000 people perished in the Tangshan, China, earthquake, which was *not* predicted. There were no foreshocks. Predictions can also lead to false alarms. In a province near Hong Kong, people reportedly evacuated their dwellings for over a month, but no earthquake followed.

In order for a short-range prediction scheme to be generally accepted, it must be both accurate and reliable. Thus, *it must have a small range of uncertainty in regard to location and timing, and it must produce few failures or false alarms.* Can you imagine the debate that would precede an order to evacuate a large U.S. city, such as Los Angeles or San Francisco? The cost of evacuating millions of people, arranging for living accommodations, and providing for their lost wages would be staggering.

Currently, no reliable method exists for making short-range earthquake predictions. In fact, leading seismologists in the past 100 years have generally concluded that short-range earthquake prediction is not feasible.

Long-Range Forecasts

In contrast to short-range predictions, which aim to predict earthquakes within a time frame of hours or days, long-range forecasts are estimates of how likely it is for an earthquake of a certain magnitude to occur on a time scale of 30 to 100 years or more. These forecasts give statistical estimates of the expected intensity of ground motion for a given area over a specified time frame. Although long-range forecasts are not as informative as we might like, these data provide important guides for building codes so that buildings, dams, and roadways are constructed to withstand expected levels of ground shaking.

Most long-range forecasting strategies are based on evidence that many large faults break in a cyclical manner, producing similar quakes at roughly similar intervals. In other words, as soon as a section of a fault ruptures, the continuing motions of Earth's plates begin to build strain in the rocks again until they fail (rupture) once more. Seismologists have therefore studied historical records of earthquakes to see if there are any discernible patterns so that they can establish the probability of recurrence.

Seismic Gaps Seismologists began to plot the distribution of rupture zones associated with great earthquakes around the globe. The maps revealed that individual rupture zones tend to occur adjacent to one another, without appreciable overlap, thereby tracing out a plate boundary. Because plates are moving at known velocities, the rate at which strain builds can also be estimated.

When these researchers studied historical records, they discovered that some seismic zones had not produced a large earthquake in more than a century or, in some locations, for several centuries. These quiet zones, called **seismic gaps**, are believed to be zones that are storing strain that will be released during a future earthquake. **Figure 9.34** shows a patch (seismic gap) of the megathrust fault that lies offshore of Padang, a low-lying Sumatran city of 800,000 people that has not ruptured since 1797. Scientists are particularly concerned about this seismic gap because rupture of an adjacent segment of the fault caused the 2004 Indian Ocean (Sumatra) earthquake and tsunami that claimed 230,000 lives.

Paleoseismology Another method of long-term forecasting involves **paleoseismology** (*paleo* = ancient, *seismos* = shake), the study of the timing, location, and size of prehistoric earthquakes. Paleoseismology studies are often conducted by digging a trench across a suspected fault zone and then looking for evidence of ancient faulting, such as offset sedimentary

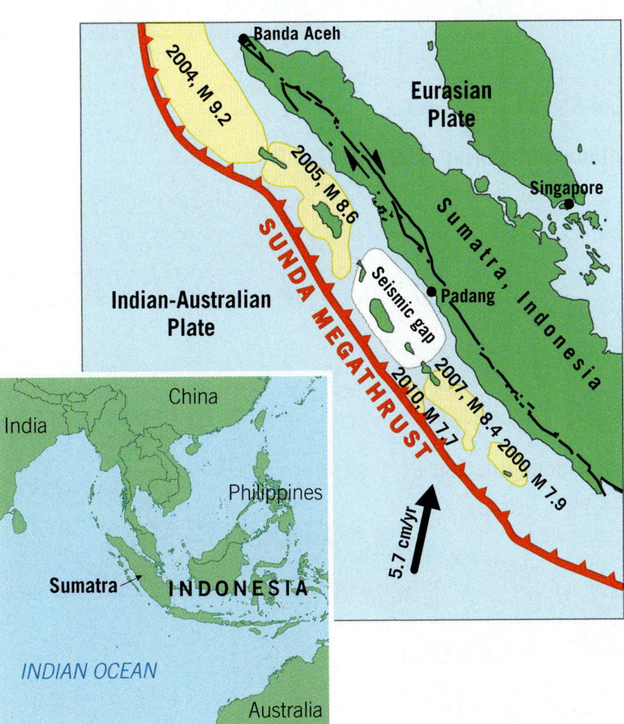

◄ **Figure 9.34 Seismic gaps: Tools for forecasting earthquakes** Seismic gaps are "quiet zones," thought to be inactive, that are storing elastic strain that will eventually produce major earthquakes. This seismic gap occurs along a patch of the megathrust fault where oceanic lithosphere is being subducted beneath Sumatra, near Padang, a low-lying coastal city with a population of 800,000 people.

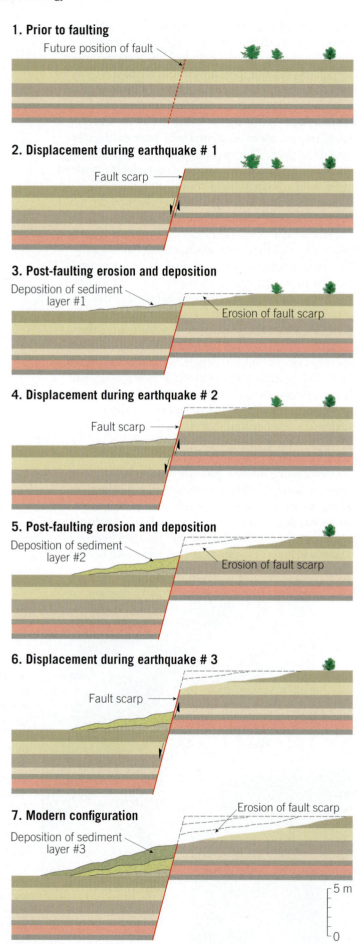

1. Prior to faulting

Future position of fault

2. Displacement during earthquake # 1

Fault scarp

3. Post-faulting erosion and deposition

Deposition of sediment layer #1

Erosion of fault scarp

4. Displacement during earthquake # 2

Fault scarp

5. Post-faulting erosion and deposition

Deposition of sediment layer #2

Erosion of fault scarp

6. Displacement during earthquake # 3

Fault scarp

7. Modern configuration

Erosion of fault scarp

Deposition of sediment layer #3

5 m

0

A.

B. The events depicted in the accompanying diagrams were deciphered by digging a trench (shown here) across the fault zone and studying the displaced sedimentary beds.

◀ **Figure 9.35 Paleoseismology: The study of prehistoric earthquakes** This study was conducted in the Pallet Creek area by digging a trench across a branch of the San Andreas Fault and then looking for evidence of ancient displacements, such as offset sedimentary strata. This simplified diagram shows that vertical displacement occurred on this fault three different times, with each event producing an earthquake. Based on the size of the vertical displacement, these ancient earthquakes had estimated magnitudes of between 6.8 and 7.4. (Photo courtesy of USGS)

strata or mud volcanoes. A large vertical offset of the layers of sediments indicates a large earthquake. Sometimes buried plant debris can be carbon dated, allowing for the timing of recurrence to be established.

One investigation that used this method focused on a segment of the San Andreas Fault that lies north and east of Los Angeles. At this site, the drainage of Pallet Creek has been repeatedly disturbed by successive ruptures along the fault zone (**Figure 9.35**). Trenches excavated across the creek bed have exposed sediments that have been displaced by several large earthquakes over a span of 1500 years. From these data, it was determined that strong earthquakes occur an average of once every 135 years. The last major event, the Fort Tejon earthquake, occurred on this segment of the San Andreas Fault in 1857, roughly 150 years ago. Because earthquakes occur on a cyclical basis, a major event in southern California may be imminent.

Using other paleoseismology techniques, researchers determined that several powerful earthquakes (magnitude 8 or larger) have repeatedly struck the coastal Pacific Northwest over the past several thousand years.

The most recent event, which occurred about 300 years ago, generated a destructive tsunami. As a result of these findings, public officials have taken steps to strengthen some of the region's existing buildings, dams, bridges, and water systems. Even the private sector responded. The U.S. Bancorp building in Portland, Oregon, was strengthened at a cost of $8 million.

CONCEPT CHECKS 9.7

1. Are accurate, short-range earthquake predictions currently possible using modern seismic instruments? Explain.
2. What is the value of long-range earthquake forecasts?

9.8 Earth's Interior

Explain how Earth acquired its layered structure and name and describe each of its major layers.

The detailed studies of seismic waves described earlier help scientists locate and measure earthquakes. Knowing how seismic waves behave has also given us a better understanding of the nature of Earth's interior.

If we could slice Earth in half, the first thing we would notice is that it has distinct layers. The heaviest materials (metals) are in the center. Less dense solids (rocks) are in the middle, and lighter liquids (mainly water) and gases are on top. Within Earth we know these layers as the iron-rich core, the rocky mantle and crust, the liquid ocean, and the gaseous atmosphere.

Probing Earth's Interior: "Seeing" Seismic Waves

How do we know about the structure and properties of Earth's deep interior? Light does not travel through rock, so we must find other ways to "see" into our planet. The best way to learn about Earth's interior is to dig or drill a hole and examine it directly. Unfortunately, this is possible only at shallow depths. The deepest a drilling rig has ever penetrated is only 12.3 kilometers (7.6 miles), which is about 1/500 of the way to Earth's center! Even this was an extraordinary accomplishment because temperature and pressure increase rapidly with depth.

About 3000 earthquakes occur each year that are large enough (about M_W 5.5) to travel all the way through Earth and be recorded by seismographs on the other side of the globe (**Figure 9.36**). The P and S waves from these large earthquakes can be used to "see" into our planet in much the same way that doctors use ultrasound waves to generate sonograms.

Using the waves recorded on seismograms to visualize Earth's interior structure is challenging. Seismic waves do not travel along straight paths; instead, they are *reflected*, *refracted*, and *diffracted* as they pass through our planet. They reflect off boundaries between different layers, they refract (change direction) when passing from one layer to another layer, and they diffract (follow a curved path) around obstacles they encounter. These different wave behaviors have been used to identify the boundaries that exist within Earth.

One of the most noticeable behaviors of seismic waves is that they follow strongly curved paths (see Figure 9.36) because the velocity of seismic waves generally increases with depth. Seismic waves also travel faster when rock is stiffer or less compressible. These properties of stiffness and compressibility can be used to interpret the composition and temperature of the rock. For instance, when rock is hotter, it becomes less stiff (imagine a chocolate bar left out in the sun), and waves travel more slowly. Waves also travel at different speeds through rocks of different compositions. Thus, the speed at which seismic waves travel can help determine both the kinds of rocks that are inside Earth and how hot they are.

Earth's Layered Structure

According to the *nebular theory* (see Section 1.6), Earth and the solar system began to form nearly 5 billion years ago due to the gravitational collapse of a huge cloud of dust and gases called a *nebula*. As material accumulated to form Earth (and for a short period afterward), the high-velocity impact of nebular debris and the decay of radioactive elements caused the temperature of our planet to increase steadily. During this time of intense heating, Earth became hot enough to melt iron and nickel.

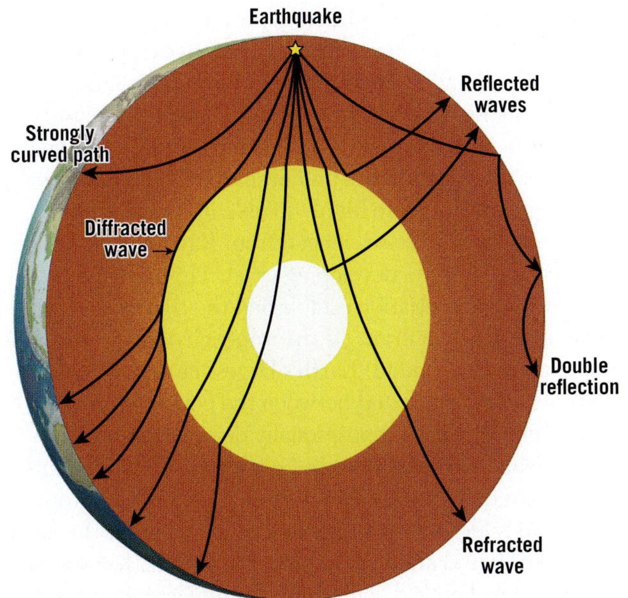

◄ **Figure 9.36 Possible paths that earthquake waves take** Notice that in the mantle, the seismic waves follow curved (refracting) paths rather than straight paths because the seismic velocity of rocks increases with depth due to increasing pressure.

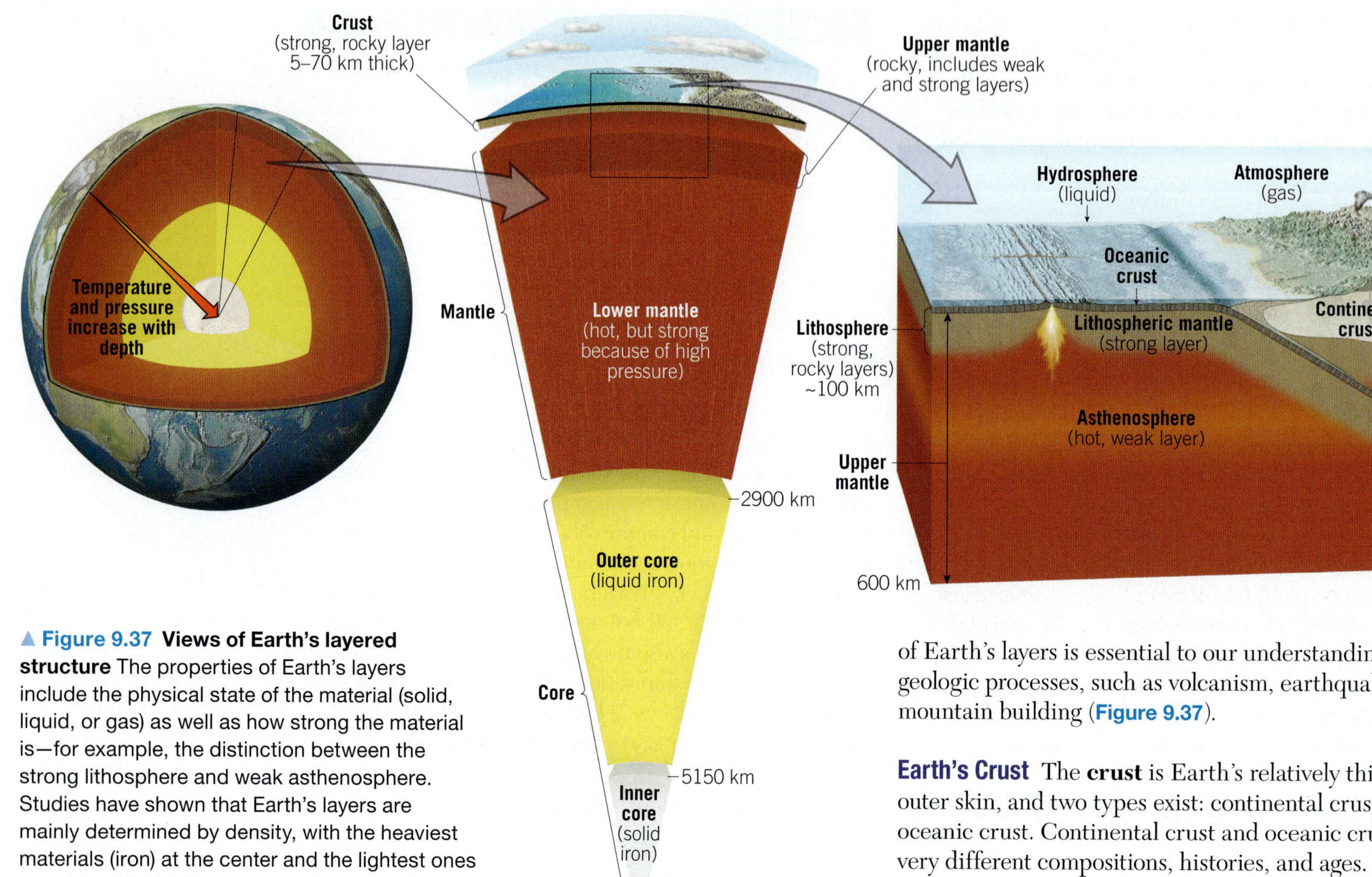

▲ **Figure 9.37 Views of Earth's layered structure** The properties of Earth's layers include the physical state of the material (solid, liquid, or gas) as well as how strong the material is—for example, the distinction between the strong lithosphere and weak asthenosphere. Studies have shown that Earth's layers are mainly determined by density, with the heaviest materials (iron) at the center and the lightest ones (gases and liquids) on the outside.

Melting produced liquid blobs of heavy metal that sank toward the center of the planet. This process occurred rapidly on the scale of geologic time and produced Earth's dense iron-rich core.

The early period of heating resulted in another process, called *chemical differentiation*, whereby melting formed buoyant masses of molten rock that rose toward Earth's surface and solidified to produce a primitive crust. These rocky materials were rich in oxygen and "oxygen-seeking" elements, particularly silicon and aluminum, along with lesser amounts of calcium, sodium, potassium, iron, and magnesium. In addition, some heavy metals such as gold, lead, and uranium, which have low melting points or were highly soluble in the ascending molten masses, were scavenged from Earth's interior and concentrated in the developing crust. This early period of chemical differentiation established the three basic divisions of Earth's interior: (1) the iron-rich *core*, (2) the thin *primitive crust*, and (3) Earth's largest layer, called the *mantle*, which is located between the core and crust.

Earth's three compositionally distinct layers—the crust, mantle, and core—can be further subdivided into zones based on *physical* properties. The physical properties used to define such regions include whether the layer is solid or liquid and how weak or strong it is. Knowledge of both the chemical composition and physical properties of Earth's layers is essential to our understanding of basic geologic processes, such as volcanism, earthquakes, and mountain building (Figure 9.37).

Earth's Crust The **crust** is Earth's relatively thin, rocky outer skin, and two types exist: continental crust and oceanic crust. Continental crust and oceanic crust have very different compositions, histories, and ages. In fact, oceanic crust is compositionally more similar to Earth's mantle than to continental crust.

The ocean crust is about 7 kilometers (4 miles) thick and forms along the mid-ocean ridge system. The rocks of the oceanic crust are younger (180 million years old or less) and denser than continental rocks. Ocean crust has a density of about 3.0 g/cm^3 and is composed of the dark igneous rocks *basalt* and *gabbro*.

Unlike oceanic crust, which has a relatively homogeneous chemical composition, continental crust consists of many rock types. Although the upper crust has an average composition of a granitic rock called *granodiorite*, its composition and structure varies considerably from place to place.

Continental crust averages about 40 kilometers (25 miles) thick but can be more than 70 kilometers (40 miles) thick in mountainous regions such as the Himalayas and the Andes. Continental crust has an average density of about 2.7 g/cm^3, which is much lower than the density of mantle rock. The low density of the continents relative to the mantle explains why continents are buoyant—acting like giant rafts, floating atop the underlying mantle—and why they cannot be readily subducted into the mantle. Because continental rocks cannot be easily recycled into the mantle, continental rocks older than 4 billion years have been found.

Earth's Mantle More than 82 percent of Earth's volume is contained in the **mantle**, a solid, rocky shell that extends to a depth of about 2900 kilometers (1800 miles) beneath Earth's crust. The boundary between the crust and mantle represents a marked change in chemical composition. The dominant rock type in the uppermost mantle is *peridotite*, which is richer in the metals iron and magnesium than are the rocks found in either the continental or oceanic crust.

The upper mantle extends from the crust–mantle boundary down to a depth of about 660 kilometers (410 miles). Recall that the upper mantle is divided into two different parts. The top portion of the upper mantle is part of the stiff *lithosphere*, and beneath that is the weaker *asthenosphere*. The **lithosphere** ("sphere of rock") consists of the crust and uppermost mantle and forms Earth's relatively cool, rigid outer shell. Averaging about 100 kilometers (62 miles) thick, the lithosphere is more than 250 kilometers (155 miles) thick below the oldest portions of the continents (see Figure 9.37).

Beneath Earth's stiff outer layer lies a solid but comparatively weak layer termed the **asthenosphere** ("weak sphere"). Because the asthenosphere and lithosphere are mechanically detached from each other, the lithosphere is able to move independently of the asthenosphere.

From 660 kilometers (410 miles) deep to the top of the core, at a depth of 2900 kilometers (1800 miles), is the lower mantle. Because pressure in the mantle increases with depth (due to the weight of the overlying rock), the mantle becomes stronger with depth. Despite their strength, however, rocks in the lower mantle are very hot and capable of very gradual flow.

Inner & Outer Core The **core** is thought to consist mainly of iron combined with an unknown quantity of nickel, as well as minor amounts of oxygen, silicon, and sulfur—elements that readily form compounds with iron. Because of the extreme pressure found in the core, this iron-rich material has an average density of more than 10 times the density of water, or 10 g/cm^3. The density at Earth's center is about 13 times that of water, or 13 g/cm^3.

The **outer core** is a liquid iron-rich layer 2270 kilometers (1410 miles) thick. The liquid nature of the outer core was discovered when researchers found that S waves do not penetrate the core. Because S waves do not pass through liquids, their inability to pass though the outer core indicates that it must be a liquid.

At Earth's center lies the **inner core**, a solid dense metallic sphere with a radius of 1216 kilometers (754 miles). Despite its higher temperature, the inner core is solid, not molten like the outer core, due to the immense pressures that exist at the center of our planet. Because the inner core is a sphere, whereas Earth's other layers are shells, drawings make the inner core appear much larger than it really is (see Figure 9.37).

CONCEPT CHECKS 9.8

1. How do seismic waves help scientists describe Earth's interior?
2. How did Earth acquire its layered structure?
3. How do continental crust and oceanic crust differ?
4. Contrast the physical characteristics of the asthenosphere and the lithosphere.
5. How are Earth's inner and outer cores different? How are they similar?

CONCEPTS IN REVIEW
Earthquakes & Earth's Interior

9.1 What Is an Earthquake?

Sketch and describe the mechanism that generates most earthquakes.

KEY TERMS: earthquake, fault, hypocenter (focus), epicenter, seismic wave, elastic rebound, aftershock, foreshock, megathrust fault, fault creep

- The sudden movements of large blocks of rock on opposite sides of faults cause most earthquakes. The location where the rock begins to slip is called the hypocenter, or focus. During an earthquake, seismic waves radiate outward from the hypocenter into the surrounding rock. The point on Earth's surface directly above the hypocenter is the epicenter.
- Over tens to hundreds of years, differential stresses gradually bend Earth's crust. Frictional resistance keeps the rock from rupturing and slipping. At some point, the stress overcomes the frictional resistance, and slippage allows the deformed (bent) rock to "spring back" to its original shape, generating an earthquake. The springing back is called elastic rebound.
- Convergent plate boundaries and associated subduction zones are marked by megathrust faults. These large faults are responsible for most of the largest earthquakes in recorded history. Megathrust faults are also capable of generating tsunamis.
- The San Andreas Fault in California is an example of a large strike-slip fault that forms a transform plate boundary capable of generating destructive earthquakes.

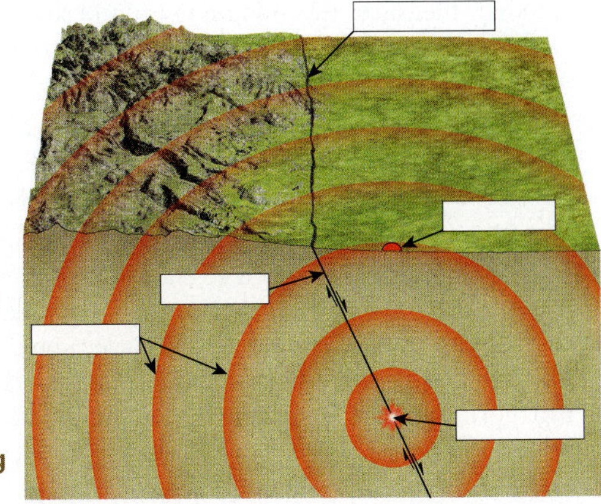

? Label the blanks on the diagram to show the relationship between earthquakes and faults using the following terms: epicenter, seismic waves, fault, fault trace, and hypocenter.

9.2 Seismology: The Study of Earthquake Waves

Compare and contrast the types of seismic waves and describe the principle of the seismograph.

KEY TERMS: seismology, seismograph (seismometer), inertia, seismogram, body waves, surface waves, P waves (primary waves), S waves (secondary waves)

- Seismology is the study of seismic waves. A seismograph measures these waves, using the principle of inertia. While the body of the instrument moves with the waves, the inertia of a suspended weight keeps a sensor stationary to record the displacement between the two.
- A seismogram, a record of seismic waves, reveals two main categories of earthquake waves: body waves (P waves and S waves), which are capable of moving through Earth's interior, and surface waves, which travel only along the upper layers of the crust. P waves are the fastest, S waves are intermediate in speed, and surface waves are the slowest. However, surface waves tend to have the greatest amplitude, S waves are intermediate, and P waves have the lowest amplitude. Large-amplitude waves produce the most shaking, so surface waves usually account for most damage during earthquakes.
- P waves and S waves exhibit different kinds of motion. P waves momentarily push (compress) and pull (stretch) rocks as they travel through a rock body, thereby changing the volume of the rock. S waves impart a shaking motion as they pass through rock, changing the rock's shape but not its volume. Because fluids do not resist forces that change their shape, S waves cannot travel through fluids, whereas P waves can.

? How could you physically demonstrate the difference between P waves and S waves to a friend who hasn't taken a geology course? (*Caution*: Don't hurt your friend!)

9.3 Locating the Source of an Earthquake

Explain how seismographs are used to locate the epicenter of an earthquake.

- Using the difference in arrival times between P and S waves, the distance separating a recording station from an earthquake's epicenter can be determined. When the distances are known from three or more seismic stations, the epicenter can be located using a method called triangulation.

9.4 Determining the Size of an Earthquake

Distinguish between intensity scales and magnitude scales.

KEY TERMS: intensity, magnitude, Modified Mercalli Intensity scale, Richter scale, moment magnitude

- Intensity and magnitude are different measures of earthquake strength. Intensity measures the amount of ground shaking at a location due to an earthquake, and magnitude is an estimate of the actual amount of energy released during an earthquake.
- The Modified Mercalli Intensity scale is a tool for measuring an earthquake's intensity at different locations. The scale is based on verifiable physical evidence that is used to quantify intensity on a 12-point scale.
- The Richter scale takes into account both the maximum amplitude of the seismic waves measured at a given seismograph and that seismograph's distance from the earthquake. The Richter scale is logarithmic, meaning that the next higher number on the scale represents seismic amplitudes that are 10 times greater than those represented by the number below. Furthermore, each larger number on the Richter scale represents the release of about 32 times more energy than the number below it.

(9.4 continued)

- Because the Richter scale does not effectively differentiate between very large earthquakes, the moment magnitude scale was devised. This scale measures the total energy released from an earthquake by considering the strength of the faulted rock, the amount of slippage, and the area of the fault that slipped. Moment magnitude is the modern standard for measuring the size of earthquakes.

9.5 Earthquake Destruction

List and describe the major destructive forces that earthquake vibrations can trigger.

KEY TERMS: liquefaction, tsunami

- Factors influencing how much destruction an earthquake might inflict on a human-made structure include (1) intensity of the shaking, (2) how long shaking persists, (3) the nature of the ground that underlies the structure, and (4) building construction. Buildings constructed of unreinforced bricks and blocks are more likely than other types of structures to be severely damaged in a quake.
- In general, bedrock-supported buildings fare best in an earthquake, as loose sediments amplify seismic shaking.
- Liquefaction may occur when water-logged sediment or soil is severely shaken during an earthquake. Liquefaction can reduce the strength of the ground to the point that it may not support buildings.
- Earthquakes may also trigger landslides or ground subsidence, and they may break gas lines, which can initiate devastating fires.
- Tsunamis are large ocean waves that form when water is displaced, usually by a megathrust fault rupturing on the seafloor. Traveling at the speed of a jet aircraft, a tsunami is hardly noticeable in deep water. However, upon arrival in shallower coastal waters, the tsunami slows down and piles up, producing a wall of water sometimes more than 30 meters (100 feet) in height. Tsunamis cause major destruction in coastal areas if they strike the shoreline. Tsunami warning systems have been established in most of the large ocean basins.

? What geologic phenomenon acts in a similar manner to the water-saturated sandy soil that these students are "playing" in?

Marli Miller

9.6 Where Do Most Earthquakes Occur?

Locate Earth's major earthquake belts on a world map.

KEY TERM: circum-Pacific belt

- Most earthquake energy is released in the circum-Pacific belt, the ring of megathrust faults rimming the Pacific Ocean. Another earthquake belt is the Alpine–Himalayan belt, which runs along the zone where the Eurasian plate collides with the Indian–Australian and African plates.
- Earth's oceanic ridge system produces another belt of earthquake activity. Here seafloor spreading and active transform faults that separate ridge segments generate many frequent small-magnitude quakes. Transform faults in the continental crust, including the San Andreas Fault, can produce large earthquakes.
- Although most destructive earthquakes are produced along plate boundaries, some occur at considerable distances from plate boundaries. Examples include the 1811–1812 New Madrid, Missouri, earthquakes and the 1886 Charleston, South Carolina, earthquake.

? Outline the circum-Pacific earthquake belt on the accompanying map that has plate boundaries drawn in red. Do the same for the Alpine–Himalayan belt.

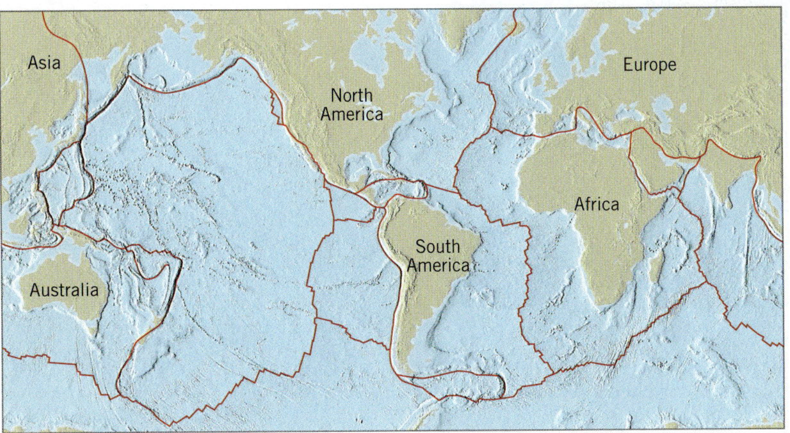

9.7 Can Earthquakes Be Predicted?

Compare and contrast the goals of short-range earthquake predictions and long-range forecasts.

KEY TERMS: precursor, seismic gap, paleoseismology

- Successful earthquake prediction has been an elusive goal of seismology for many years. Shorter-range predictions (for hours or days) are based on precursor events such as changes in ground elevation or in strain levels near a fault. Unfortunately, this type of monitoring is not reliable.
- Long-range forecasts (for time scales of 30 to 100 years) are statistical estimates of the likelihood that an earthquake of a given magnitude will occur. Long-range forecasts are useful because they can guide development of building codes and infrastructure.
- Scientists have identified seismic gaps, portions of faults that have been storing strain for a long time, meaning these sites have great potential for experiencing an earthquake in the not-too-distant future. Paleoseismology is another tool used to make long-range forecasts. Because earthquakes occur on a cyclical basis, determining how frequently they have occurred in the past can give some insight into when they are most likely to occur again.

? Of the earthquake hazards discussed earlier, which is (are) the greatest concern in the region where you live? Why?

9.8 Earth's Interior

Explain how Earth acquired its layered structure and name and describe each of its major layers.

KEY TERMS: crust, mantle, lithosphere, asthenosphere, core, outer core, inner core

- The layered internal structure of Earth developed due to gravitational sorting of Earth materials early in the history of the planet. The densest material settled to form Earth's core, while the least dense material rose to form Earth's crust, oceans, and atmosphere.
- Seismic waves allow geoscientists to "look into" Earth's interior. Like sonograms used to image human organs, seismic waves generated by large earthquakes reveal details about Earth's layered structure.
- Earth has two distinct kinds of crust: oceanic and continental. Oceanic crust is thinner, denser, and younger than continental crust. Oceanic crust also readily subducts, whereas the less dense continental crust does not.
- The uppermost mantle and crust make up Earth's rigid outer shell, called the lithosphere, which overlies the asthenosphere—a solid but relatively weak layer. The lower mantle is a strong solid layer but capable of very gradual flow.
- Earth's core is very dense and composed of a mixture of iron and nickel, with minor amounts of lighter elements. The outer core is liquid, whereas the inner core is solid.

GIVE IT SOME THOUGHT

1 Describe the concept of elastic rebound. Develop an analogy other than a rubber band to illustrate this concept.

2 The accompanying map shows the locations of many of the largest earthquakes in the world since 1900. Refer to the map of Earth's plate boundaries in Figure 2.10, page XXX, and determine which type of plate boundary is most often associated with these destructive events.

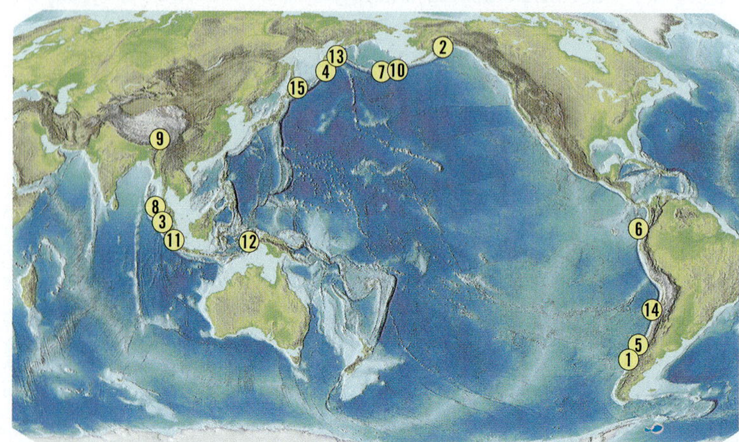

3 Use the accompanying seismogram to answer the following questions:

a. Which of the three types of seismic waves reached the seismograph first?

b. What is the time interval between the arrival of the first P wave and the arrival of the first S wave?

c. Use your answer from Question b and the travel–time graph in Figure 9.14 to determine the distance from the seismic station to the earthquake.

d. Which of the three types of seismic waves had the highest amplitude when they reached the seismic station?

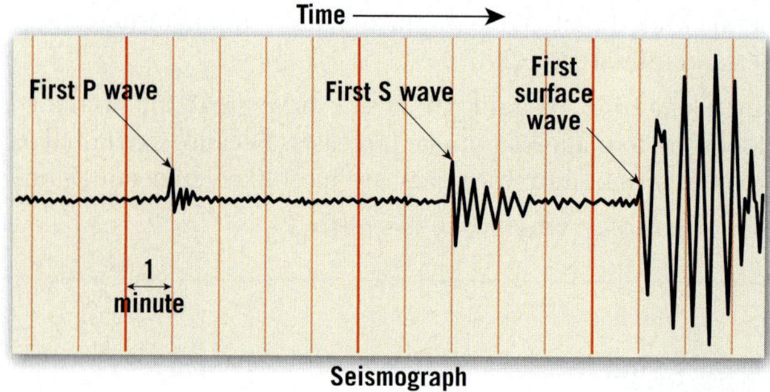

4 You go for a jog on a beach and choose to run near the water, where the sand is well packed and solid under your feet. With each step, you notice that your footprint quickly fills with water, but not water coming in from the ocean. What is this water's source? For what earthquake-related hazard is this phenomenon a good analogy?

5 The accompanying image shows a double-decked section of Interstate 880 (the Nimitz Freeway) that collapsed during the 1989 Loma Prieta earthquake and caused 42 deaths. About 1.4 kilometers of this freeway section, called the Cypress Viaduct, collapsed, while a similar section survived the vibration. Both sections were subsequently demolished and rebuilt as a single-level structure, at a cost of $1.2 billion. Examine the map and seismograms from an aftershock that shows the intensity of shaking observed at three nearby locations to answer the following questions:

a. What type of ground material experienced the least amount of shaking during the aftershock?

b. What type of ground materials experienced the greatest amount of ground shaking during the same event?

c. Which of the two sections of the Cypress Viaduct shown on the map do you think collapsed? Explain.

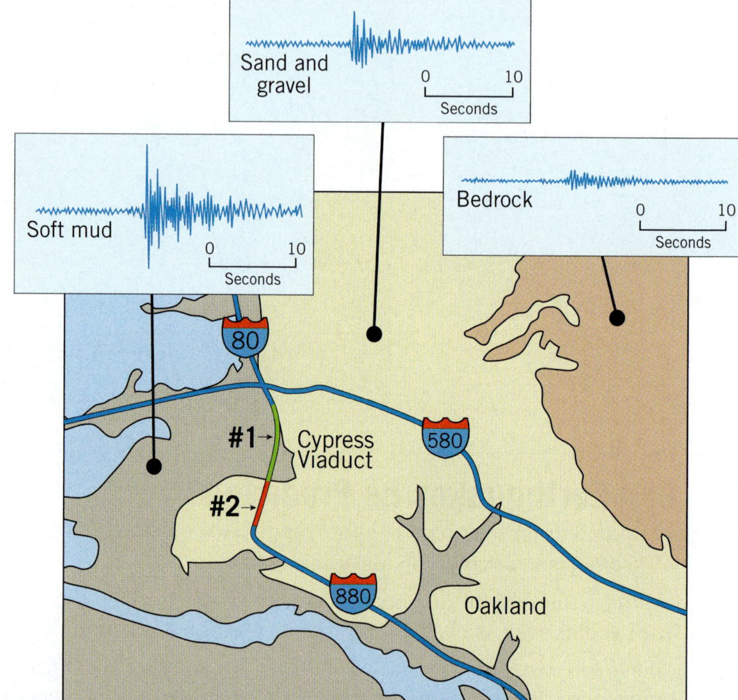

6 Using the accompanying map of the San Andreas Fault, answer the following questions:

a. Which of the four segments (1–4) of the San Andreas Fault do you think is experiencing fault creep?

b. Paleoseismology studies have found that the section of the San Andreas Fault that failed during the Fort Tejon quake (segment 3) produces a major earthquake every 135 years, on average. Based on this information, how would you rate the chances of a major earthquake occurring along this section in the next 30 years? Explain.

c. Do you think San Francisco or Los Angeles has the greater risk of experiencing a major earthquake in the near future? Defend your selection.

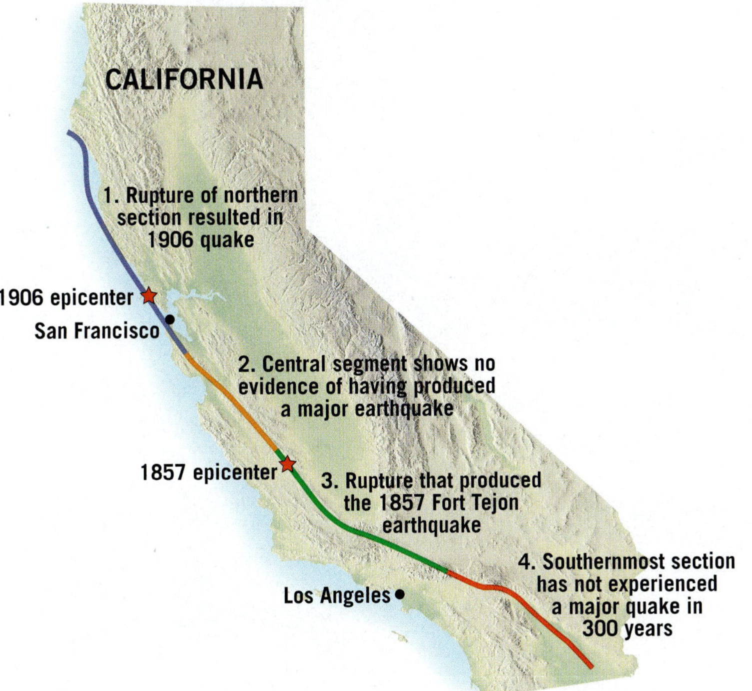

7 On the accompanying Richter scale diagram, first determine the Richter magnitude (M_L) for an earthquake at 400 kilometers distance, with a maximum amplitude of 0.5 millimeter. Second, for this same earthquake (same M_L), determine the amplitude of the biggest waves for a seismograph 40 kilometers from the hypocenter.

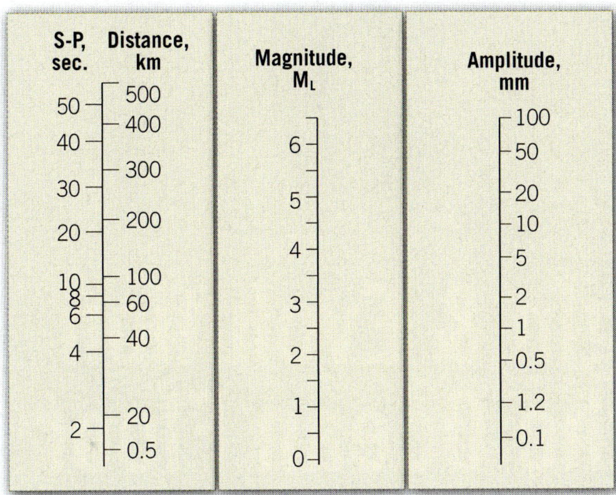

8 Earthquakes below the Yellowstone caldera originate at very shallow depths, about 4 kilometers on average. Below this depth, the rocks are at about 400°C, too hot and weak to store elastic energy. Based on this data, answer the following questions:

a. Calculate the average geothermal gradient in the first 4 kilometers beneath the Yellowstone caldera, assuming an average surface temperature of 0°C (32°F) and a temperature at 4 kilometers of 400°C (752°F). (For this example, the geothermal gradient, which is the increase in temperature with depth, should be measured in degrees centigrade per 100 meters.)

b. At about what depth is the groundwater below the Yellowstone caldera hot enough to "boil" and therefore capable of generating a geyser?

(Photo by Quasarphoto/Fotolia.com)

MasteringGeology™

Looking for additional review and test prep materials? Visit the Study Area in MasteringGeology to enhance your understanding of this chapter's content by accessing a variety of resources, including Self-Study Quizzes, Geoscience Animations, SmartFigures, Mobile Field Trips, *Project Condor* Quadcopter videos, *In the News* RSS feeds, flashcards, web links, and an optional Pearson eText.

www.masteringgeology.com

10

Origin & Evolution of the Ocean Floor

FOCUS ON CONCEPTS

Each statement represents the primary learning objective for the corresponding major heading within the chapter. After you complete the chapter, you should be able to:

10.1 Define *bathymetry* and describe the various bathymetric techniques used to map the ocean floor.

10.2 Compare and contrast a passive continental margin with an active continental margin and list the major features of each.

10.3 List and describe the major features of the deep-ocean basin.

10.4 Sketch and label a cross-sectional view of the Mid-Atlantic Ridge. Explain how a cross section of the East Pacific Rise would look different.

10.5 Write a statement describing how spreading rates affect ridge topography.

10.6 List the four layers of oceanic crust and explain how oceanic crust forms and how it differs from continental crust.

10.7 Outline the steps by which continental rifting results in the formation of new ocean basins.

10.8 Compare and contrast spontaneous subduction and forced subduction.

Crashing waves along the Pacific coast, Soberanes Point, California. (Photo by Jamie Pham/Zoonar/AGE Fotostock)

THE OCEAN IS EARTH'S MOST PROMINENT FEATURE, covering more than 70 percent of its surface. Yet, prior to the 1950s, information about the ocean floor was extremely limited. With the development of modern instruments, our understanding of the diverse topography of the ocean floor improved dramatically. Particularly significant was the discovery of the global oceanic ridge system, a broad elevated landform that stands 2 to 3 kilometers higher than the adjacent deep-ocean basins and is the longest topographic feature on Earth. In this chapter we will examine the topography of the ocean floor and look at the processes that generate its varied features.

10.1 An Emerging Picture of the Ocean Floor

Define *bathymetry* and describe the various bathymetric techniques used to map the ocean floor.

If all water could be drained from the ocean basins, a great variety of features would be observed, including volcanic peaks, deep trenches, extensive plains, linear ridges, and large plateaus. In fact, the topography would be nearly as diverse as that on the continents.

Mapping the Seafloor

The complex nature of ocean-floor topography did not unfold until the historic 3½-year voyage of the HMS *Challenger* (**Figure 10.1**). From December 1872 to May 1876, the *Challenger* expedition made the first comprehensive study of the global ocean ever attempted. During the 127,500-kilometer (79,200-mile) voyage, the ship and its crew of scientists traveled to every ocean except the Arctic. Throughout the voyage, they sampled a multitude of ocean properties, including water depth, which was accomplished by laboriously lowering long weighted lines overboard. The HMS *Challenger* measured the depth of the deepest spot on the seafloor in 1875. This spot, now called the Challenger Deep, is about 10,994 meters (36,060 feet) deep.

Modern Bathymetric Techniques The measurement of ocean depths and the charting of the shape or topography of the ocean floor is known as **bathymetry** (*bathos* = depth, *metry* = measurement). Today, sound energy is often used to measure water depths. The basic approach employs **sonar**, an acronym for *so*und *na*vigation and *r*anging. The first devices that used sound to measure water depth, called **echo sounders**, were developed early in the twentieth century. Echo sounders work by transmitting a sound wave (called a *ping*) into the water in order to produce an echo when it bounces off any object, such as a large marine organism or the ocean floor (**Figure 10.2**). A sensitive receiver

▼ SmartFigure 10.1
HMS *Challenger* The first systematic bathymetric survey of the ocean was made aboard the HMS *Challenger* during its historic 3½-year voyage. (Photo courtesy of the Library of Congress)

TUTORIAL
http://goo.gl/ydE3f7

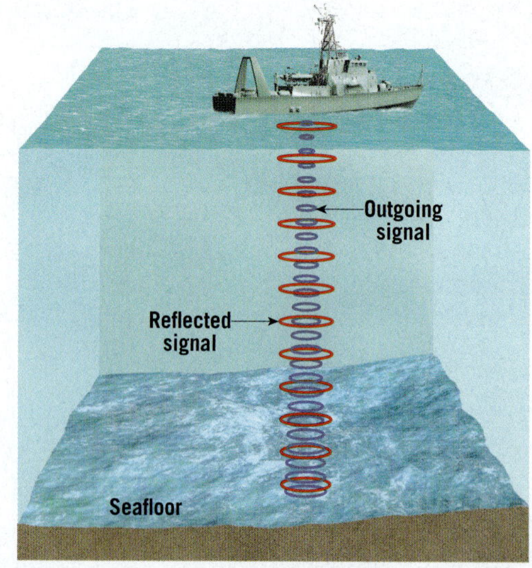

▲ Figure 10.2 **Echo sounder** An echo sounder determines the water depth by measuring the time interval required for an acoustic wave to travel from a ship to the seafloor and back.

A. Sidescan sonar and multibeam sonar operating from the same research vessel.

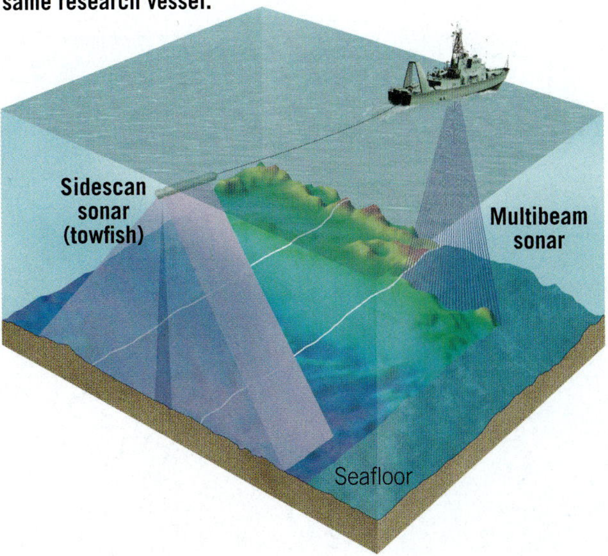

B. Color-enhanced perspective map of the seafloor and coastal landforms in the Los Angeles area of California.

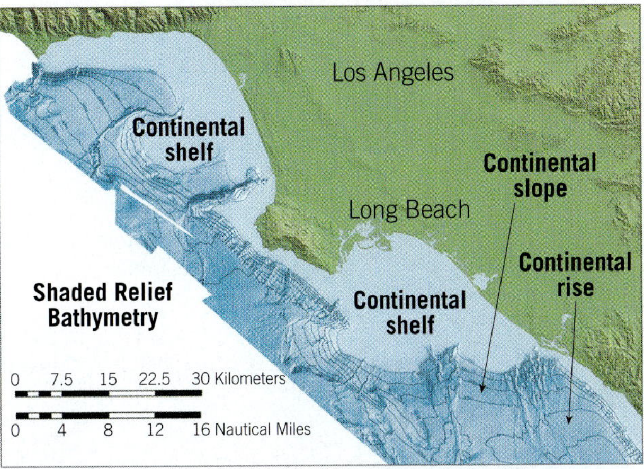

▲ **Figure 10.3 Mapping using sidescan and multibeam sonar.**

intercepts the reflected echo, and a clock precisely records the travel time to fractions of a second. Using the speed of sound waves in water—about 1500 meters (4900 feet) per second—and measuring the time required for the energy pulse to reach the ocean floor and return, the depth can be calculated: Depth = ½ (1500 m/sec × Echo travel time). Bathymetry determined from continuous monitoring of these echoes is plotted to obtain a profile of the ocean floor. By laboriously combining numerous profiles, a detailed chart of the seafloor has been produced.

Following World War II, the U.S. Navy developed *sidescan sonar* to look for explosive devices that had been deployed in shipping lanes (**Figure 10.3A**). Torpedo-shaped instruments towed behind ships send out a fan of sound extending on either side of the ship's path. By combining swaths of sidescan sonar data, oceanographers produced the first photograph-like images of the seafloor (**Figure 10.4**, overleaf).

Although sidescan sonar provides valuable views of the seafloor, it does not provide bathymetric (water depth) data. This drawback was resolved with the development of *high-resolution multibeam* instruments (see Figure 10.3A). These systems use hull-mounted sound sources that send out a fan of sound and then record reflections from the seafloor through a set of narrowly focused receivers aimed at different angles. Rather than obtain the depth of a single point every few seconds, this technique allows a survey ship to map a swath of ocean floor tens of kilometers wide. In addition, these systems collect bathymetric data of such high resolution that they can distinguish depths that differ by less than 1 meter (3 feet). When multibeam sonar is used to map sections of seafloor, the ship travels in a regularly spaced back-and-forth pattern known as "mowing the lawn."

Despite their greater efficiency and enhanced detail, research vessels equipped with multibeam sonar travel at a mere 10 to 20 kilometers (6 to 12 miles) per hour. It would take at least 100 vessels outfitted with this equipment hundreds of years to map the entire seafloor. This explains why only about 5 percent of the seafloor has been mapped in detail (**Figure 10.3B**).

Mapping the Ocean Floor from Space Another technological breakthrough that led to an enhanced understanding of the seafloor involves measuring the shape of the ocean surface from space. After compensating for waves, tides, currents, and atmospheric effects, scientists discovered that the water's surface is not perfectly "flat." Because massive structures such as seamounts and ridges exert stronger-than-average gravitational attraction, they produce elevated areas on the ocean surface. Conversely, canyons and trenches create slight depressions.

Satellites equipped with *radar altimeters* are able to measure subtle differences in sea level by bouncing microwaves off the sea surface (**Figure 10.5**). These devices can measure variations as small as a few centimeters.

▼ **Figure 10.5 Satellite altimeter** A satellite altimeter measures the variation in sea-surface elevation, which is caused by gravitational attraction and mimics the shape of the seafloor. The sea-surface anomaly is the difference between the measured ocean surface and the theoretical ocean surface.

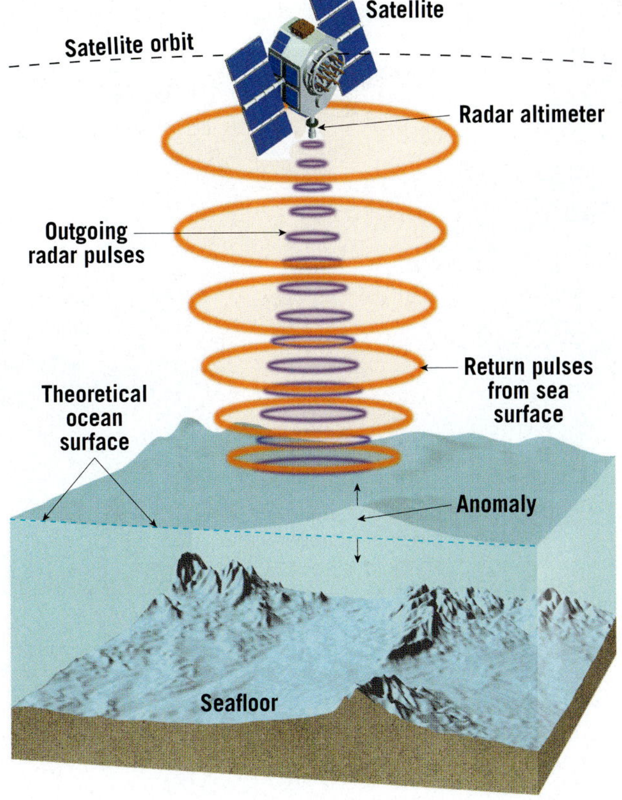

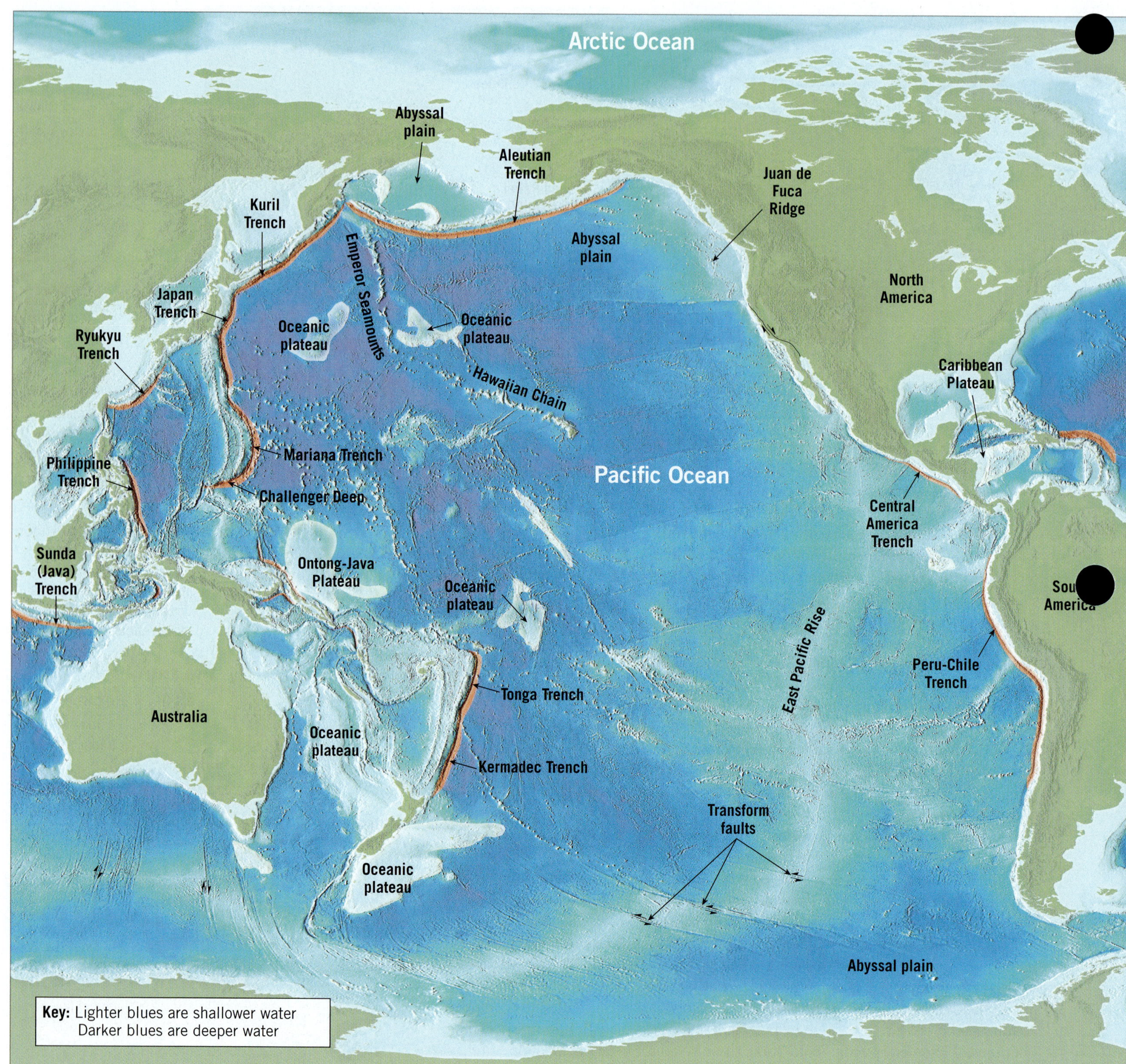

▲ **Figure 10.4 Major features of the seafloor.**

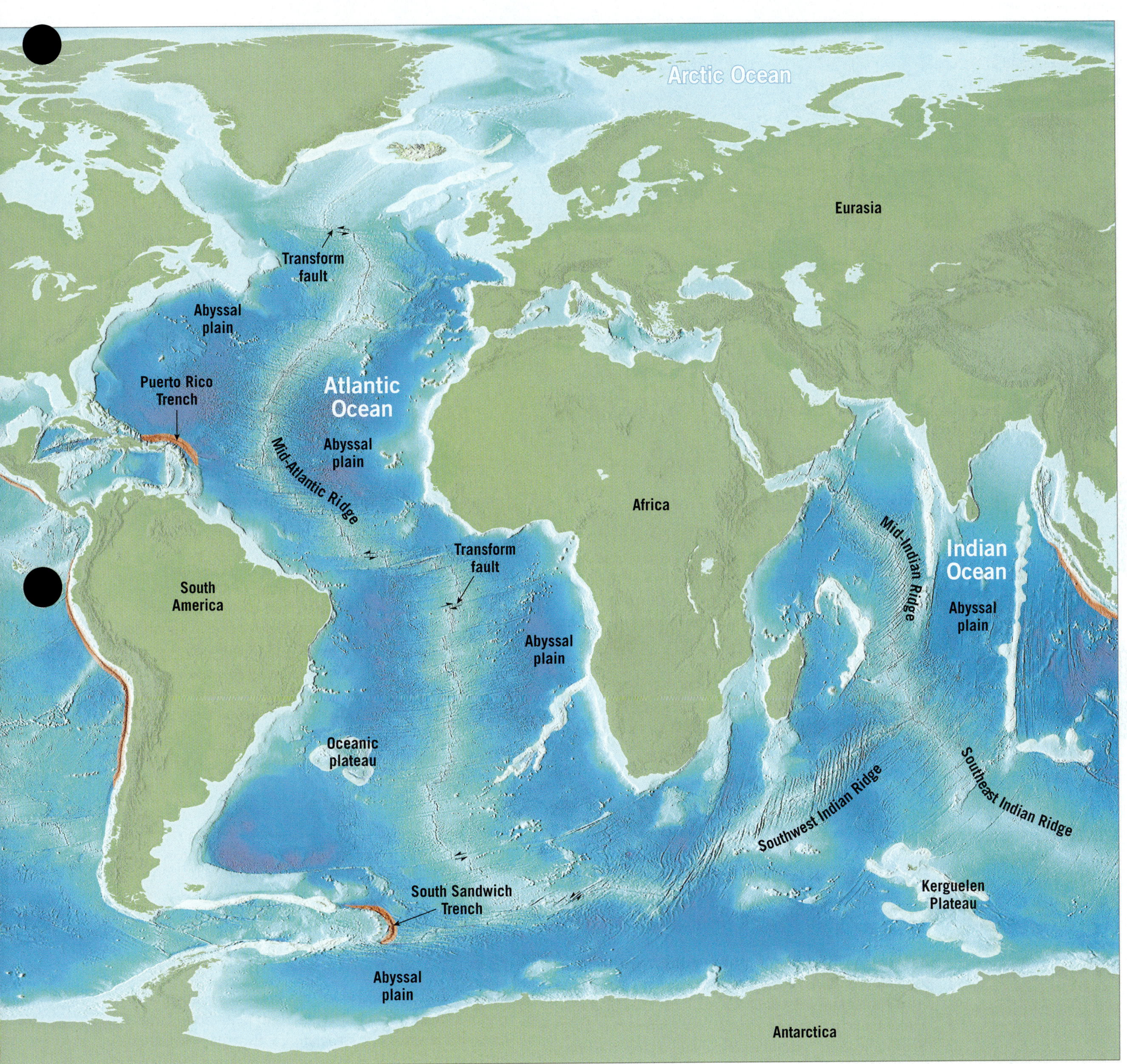
Arctic Ocean
Eurasia
Transform fault
Abyssal plain
Puerto Rico Trench
Atlantic Ocean
Abyssal plain
Mid-Atlantic Ridge
Africa
Transform fault
South America
Abyssal plain
Mid-Indian Ridge
Indian Ocean
Abyssal plain
Oceanic plateau
Southwest Indian Ridge
Southeast Indian Ridge
South Sandwich Trench
Kerguelen Plateau
Abyssal plain
Antarctica

▶ **Figure 10.6 Major topographic divisions of the North Atlantic** The accompanying profile extends from New England to the coast of North Africa and provides a glimpse of the various structures found across the Atlantic basin.

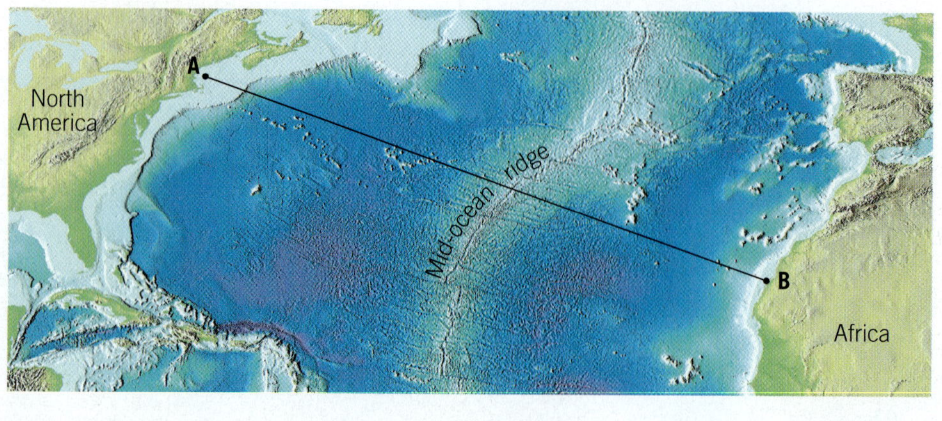

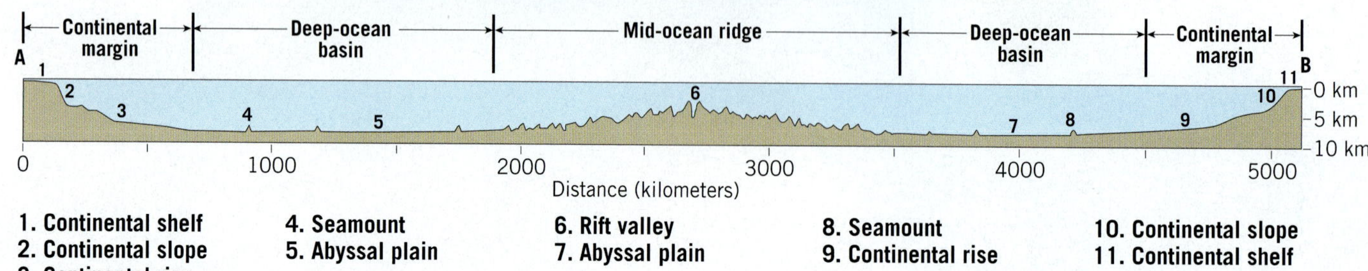

Provinces of the Ocean Floor

Oceanographers studying the topography of the ocean floor identify three major areas: *continental margins, deep-ocean basins*, and *oceanic (mid-ocean) ridges*. The map in **Figure 10.6** outlines these provinces for the North Atlantic Ocean, and the profile at the bottom shows the varied topography. Profiles of this type usually have their vertical dimension exaggerated many times (40 times, in this case) to make topographic features more conspicuous. Vertical exaggeration, however, makes slopes appear *much* steeper than they actually are.

CONCEPT CHECKS 10.1

1. Define *bathymetry*.
2. Assuming that the average speed of sound waves in water is 1500 meters (4900 feet) per second, determine the water depth if the signal sent out by an echo sounder requires 6 seconds to strike bottom and return to the recorder.
3. Describe how satellites orbiting Earth can determine features on the seafloor without being able to directly observe them beneath several kilometers of seawater.
4. What are the three major topographic provinces of the ocean floor?

Did You Know?
The U.S. Navy utilizes the biological sonar of bottlenose dolphins to help defend its ships and facilities. Specifically, dolphins have been trained to detect mines intended to blow up ships. When Navy ships enter mine-infested waters, the trained dolphins are released and use their sonar to locate the deadly mines.

10.2 Continental Margins

Compare and contrast a passive continental margin with an active continental margin and list the major features of each.

As the name implies, the **continental margins** are the outer margins of the continents, where continental crust transitions to oceanic crust. Two types of continental margin have been identified—*passive* and *active*. Nearly the entire Atlantic Ocean and a large portion of the Indian Ocean are surrounded by passive continental margins (see Figure 10.4). By contrast, most of the Pacific Ocean is bordered by active continental margins (subduction zones), as shown in **Figure 10.7**. Notice that some of those active subduction zones lie far beyond the margins of the continents.

Passive Continental Margins

Passive continental margins are tectonically inactive regions located some distance from plate boundaries. As a result, they are not associated with strong earthquakes or volcanic activity. Passive continental margins develop when continental blocks rift apart and are separated by continued seafloor spreading. As a result, the continental blocks are firmly attached to the adjacent oceanic crust.

Most passive margins are relatively wide and are sites where large quantities of sediments are deposited. The features comprising passive continental margins

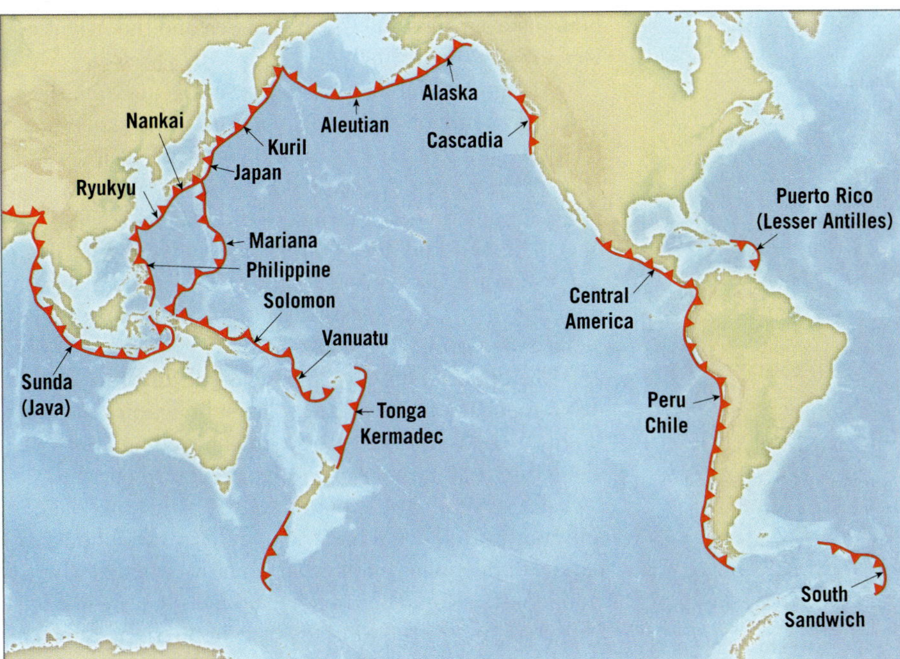

▲ Figure 10.7 **Distribution of active continental margins (subduction zones) surrounding the Pacific basin.**

include the continental shelf, the continental slope, and the continental rise (Figure 10.8).

Continental Shelf The gently sloping, submerged surface that extends from the shoreline toward the deep-ocean basin is called the **continental shelf**. It consists mainly of continental crust, capped with sedimentary rocks and sediments eroded from adjacent landmasses.

The continental shelf varies greatly in width. The shelf extends seaward more than 1500 kilometers (930 miles) along some continental margins yet is almost nonexistent along others. The average inclination of the continental shelf is only about one-tenth of 1 degree, a slope so slight that it would appear to an observer to be a horizontal surface.

The continental shelf tends to be relatively featureless; however, some areas are mantled by extensive glacial deposits and thus are quite rugged. In addition, some continental shelves are dissected by large valleys running from the coastline into deeper waters. Many of these *shelf valleys* are the seaward extensions of river valleys on the adjacent landmass. They were eroded during the last Ice Age (Quaternary period), when enormous quantities of water were stored in vast ice sheets on the continents rather than in the world's oceans, lowering sea level by at least 100 meters (330 feet). Because of this sea-level drop, rivers extended their courses, and land-dwelling plants and animals migrated to the newly exposed portions of the continents. Dredging off the coast of North America has retrieved the ancient remains of numerous land dwellers, including mammoths, mastodons, and horses—providing further evidence that portions of the continental shelves were once above sea level.

Although continental shelves represent only 7.5 percent of the total ocean area, they have economic and political significance because they contain important reservoirs of oil and natural gas (discussed in Section 7.8, page 206) and support important fishing grounds.

Continental Slope Marking the seaward edge of the continental shelf is the **continental slope**, a relatively steep structure that forms the boundary between continental crust and oceanic crust. Although the inclination of the continental slope varies greatly from place to place, it averages about 5 degrees and in places exceeds 25 degrees.

Continental Rise The continental slope merges into a more gradual incline known as the **continental rise**, which may extend seaward for hundreds of kilometers. The continental rise consists of a thick accumulation of sediment that has moved down the continental slope and onto deep-ocean floor. Most of the sediments are delivered to the seafloor by **turbidity currents**, mixtures of sediment and water that periodically flow down **submarine canyons** (see Figure 7.27, page 205). When these muddy slurries emerge from the mouth of a canyon onto the relatively flat ocean floor, they deposit sediment that forms a **deep-sea fan** (see Figure 10.8). As fans from adjacent submarine canyons grow, they merge to produce a continuous wedge of sediment at the base of the continental slope, forming the continental rise.

Active Continental Margins

Active continental margins are located along convergent plate boundaries, where oceanic lithosphere is being subducted beneath the leading edge of a continent

Did You Know?
Faced with a shortage of land along its heavily populated coast, Japan has constructed artificial islands on the adjacent continental shelf. One of these islands is located about 3 mi off the coast of Osaka, where the Kansai Airport was built.

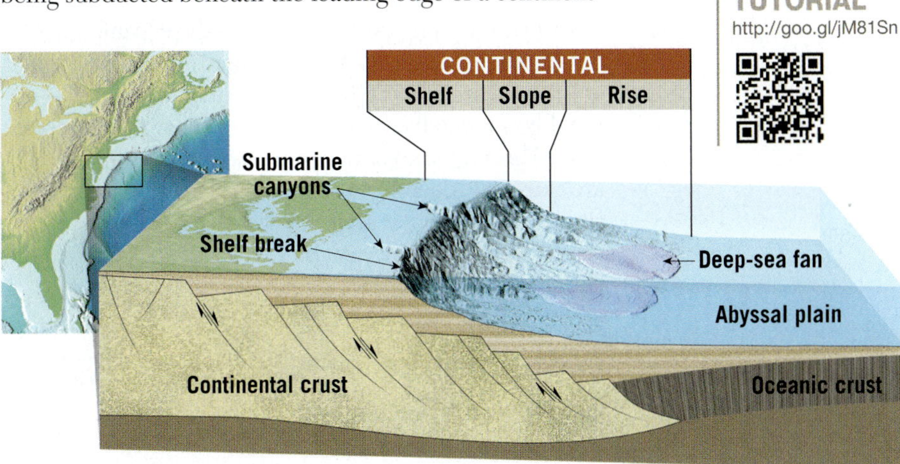

▼ SmartFigure 10.8
Passive continental margins The slopes shown for the continental shelf and continental slope are greatly exaggerated. The continental shelf has an average slope of one-tenth of 1 degree, while the continental slope has an average slope of about 5 degrees.

TUTORIAL
http://goo.gl/jM81Sn

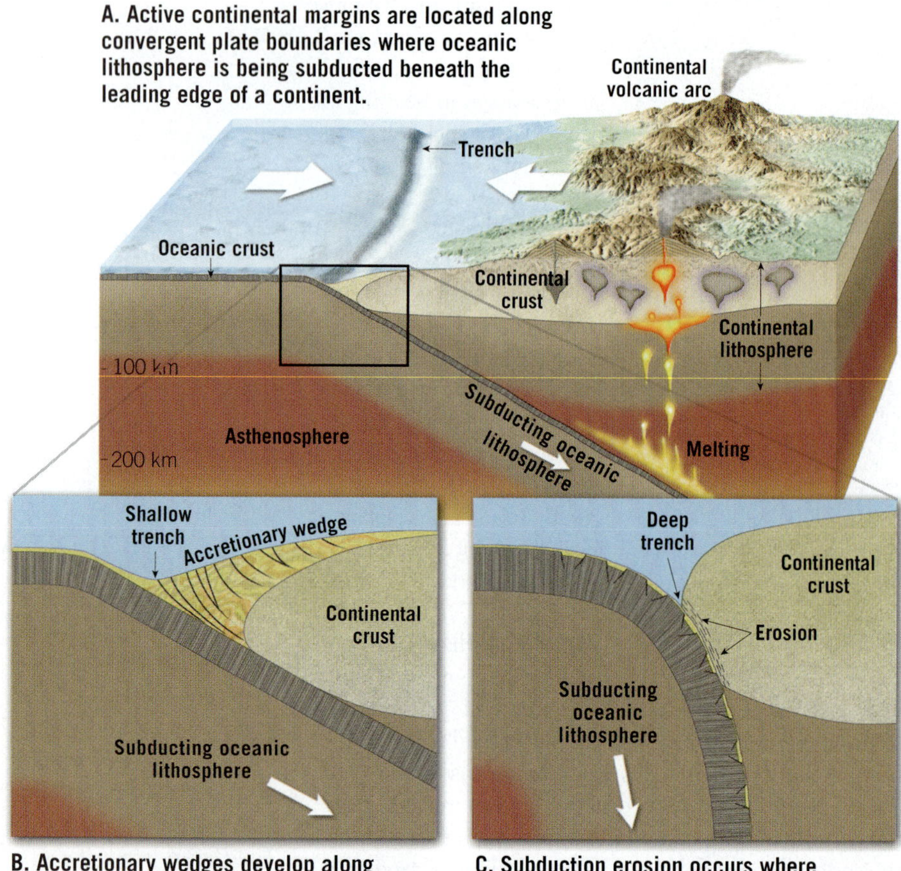

B. Accretionary wedges develop along subduction zones where sediments from the ocean floor are scraped from the descending oceanic plate and pressed against the edge of the overriding plate.

C. Subduction erosion occurs where sediment and rock scraped off the bottom of the overriding plate are carried into the mantle by the subducting plate.

▲ **Figure 10.9 Active continental margins.**

(**Figure 10.9A**). Deep-ocean trenches are the major topographic expression at convergent plate boundaries. Most of these deep, narrow furrows surround the Pacific basin. One exception is the Puerto Rico trench, which forms the boundary between the Caribbean Sea and the Atlantic Ocean (see Figure 10.4).

Along some subduction zones, sediments from the ocean floor and pieces of oceanic crust are scraped from the descending oceanic plate and plastered against the edge of the overriding plate (**Figure 10.9B**). This chaotic accumulation of deformed sediment and scraps of oceanic crust is called an **accretionary wedge** (*ad* = toward, *crescere* = to grow). Prolonged plate subduction can produce massive accumulations of sediment along active continental margins.

The opposite process, known as **subduction erosion**, characterizes many other subduction zones (Figure 10.9C). Rather than sediment accumulating along the front of the overriding plate, sediment and rock are scraped off the bottom of the overriding plate and transported down into the mantle by the subducting plate. Subduction erosion is particularly effective where cold, dense oceanic lithosphere subducts at a steep angle, as exemplified by the Mariana trench. Sharp bending of the subducting plate causes faulting in the ocean crust and a rough surface, as shown in **Figure 10.9C**.

CONCEPT CHECKS 10.2

1. List the three major features of a passive continental margin. Which of these features is considered a flooded extension of the continent? Which one has the steepest slope?
2. Describe the differences between active and passive continental margins and give a geographic example of each.
3. How are active continental margins related to plate tectonics?
4. Briefly explain how an accretionary wedge forms.
5. What is meant by *subduction erosion*?

10.3 Features of Deep-Ocean Basins

List and describe the major features of the deep-ocean basin.

Between the continental margin and the oceanic ridge lies the **deep-ocean basin** (see Figure 10.6). The size of this region—almost 30 percent of Earth's surface—is roughly comparable to the percentage of land above sea level. This region contains *deep-ocean trenches*, which are extremely deep linear depressions in the ocean floor; remarkably flat areas known as *abyssal plains*; tall volcanic peaks called *seamounts* and *guyots*; and large, elevated flood basalt provinces called *oceanic plateaus*.

Deep-Ocean Trenches

Deep-ocean trenches are long, narrow creases in the seafloor that are the deepest parts of the ocean floor. Most trenches are located along the margins of the Pacific Ocean, where many exceed 10 kilometers (6 miles) in depth (see Figure 10.4). The Challenger Deep, located in the Mariana trench, has been measured at 10,994 meters (36,060 feet) below sea level, making it the deepest known part of the world ocean (**Figure 10.10**). Only two trenches are located in the Atlantic—the Puerto Rico trench adjacent to the Lesser Antilles arc (see Figure 10.4) and the South Sandwich trench.

Although deep-ocean trenches represent a very small portion of the area of the ocean floor, they are nevertheless significant geologic features. Trenches are sites of plate convergence where slabs of oceanic lithosphere subduct and plunge back into the mantle. In addition to generating earthquakes as one plate "scrapes" against another, plate subduction also triggers volcanic activity.

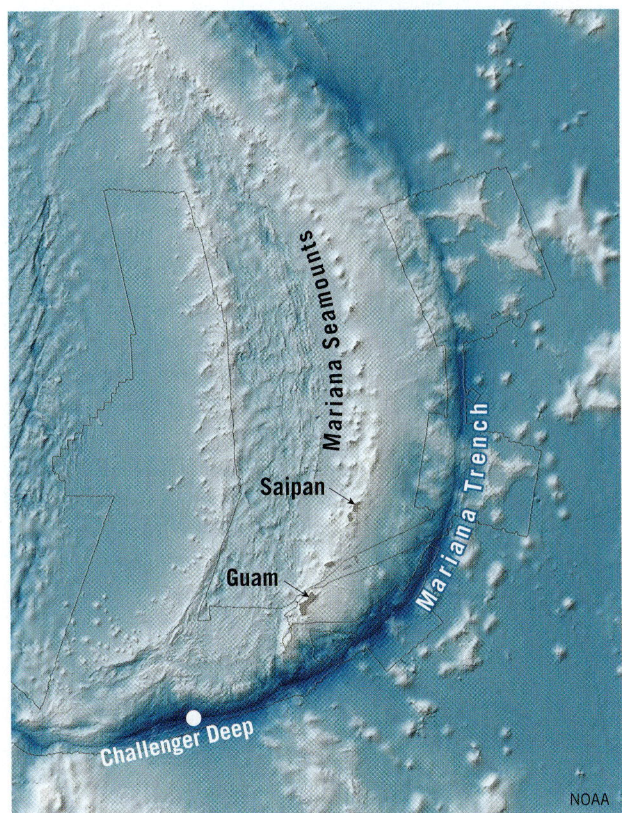

▲ **Figure 10.10 The Challenger Deep** Located near the southern end of the Mariana Trench, the Challenger Deep is the deepest place in the global ocean, about 10,994 meters (36,070 feet) deep. Film director James Cameron (*Titanic* and *Avatar*) made news in March 2012 as the first person to dive to the bottom of the Challenger Deep in more than 50 years.

As a result, a trench tends to run parallel to an arc-shaped row of active volcanoes—a **volcanic island arc** when the overriding plate is oceanic and a **continental volcanic arc** when the oceanic plate subducts under a continental margin (see Figure 2.15A,B, page 45). The volcanic activity associated with the trenches that surround the Pacific Ocean explains why this region is called the *Ring of Fire.*

Abyssal Plains

Abyssal plains (*a* = without, *byssus* = bottom) are flat features of the deep-ocean floor; in fact, they are likely the most level places on Earth (see Figure 10.4). The abyssal plain found off the coast of Argentina, for example, has less than 3 meters (10 feet) of relief over a distance exceeding 1300 kilometers (800 miles). The monotonous topography of abyssal plains is occasionally interrupted by the protruding summit of a partially buried volcanic peak (seamount).

Using **seismic reflection profilers**, researchers have determined that the relatively featureless topography of abyssal plains is due to thick accumulations of sediment that have buried an otherwise rugged ocean floor (**Figure 10.11**). The nature of the sediment indicates that these plains consist primarily of four materials: (1) fine sediments transported far out to sea by turbidity currents, (2) mineral matter that has precipitated out of seawater, (3) shells and skeletons of microscopic marine organisms, and (4) wind-blown clay particales dirived from the land.

Abyssal plains are found in all oceans. However, the Atlantic Ocean has the most extensive abyssal plains because it has few trenches to act as traps for sediment carried down the continental slope.

Did You Know?
Today, coral reefs cover about 230,000 mi^2 of the world's marine areas. Many different stresses can destroy coral reefs, including human activity and rising ocean temperatures due to global warming.

Volcanic Structures on the Ocean Floor

Dotting the seafloor are numerous volcanic structures of various sizes. Many occur as isolated features that resemble volcanic cones on land. Others occur in nearly linear chains that stretch for thousands of kilometers, while still others are massive structures that cover areas the size of Texas.

Seamounts & Volcanic Islands Submarine volcanoes, called **seamounts**, may rise hundreds of meters above the surrounding topography. It is estimated that more than a million seamounts exist. Some grow large enough to become oceanic islands, but most do not have a sufficiently long eruptive history to build a structure above sea level. Although seamounts are found on the floors of all the oceans, they are most common in the Pacific.

Some, like the Hawaiian Island–Emperor Seamount chain, which stretches from the Hawaiian Islands to the Aleutian trench, form over volcanic hot spots (see Figure 2.26, page 54). Others are born near oceanic ridges.

If a volcano grows large enough before plate motion carries it away from its magma source, the structure may emerge as a **volcanic island**. Examples of volcanic

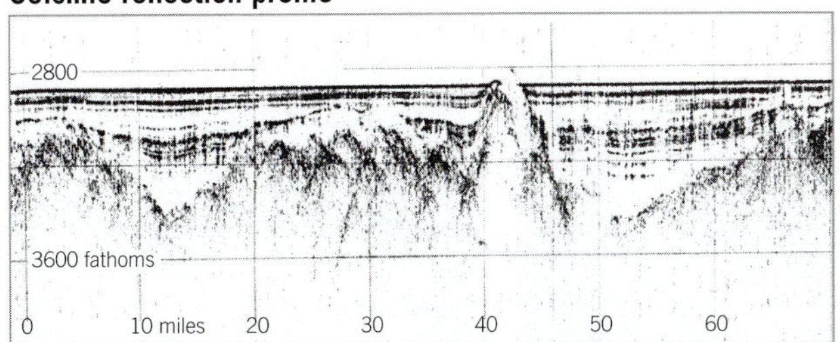

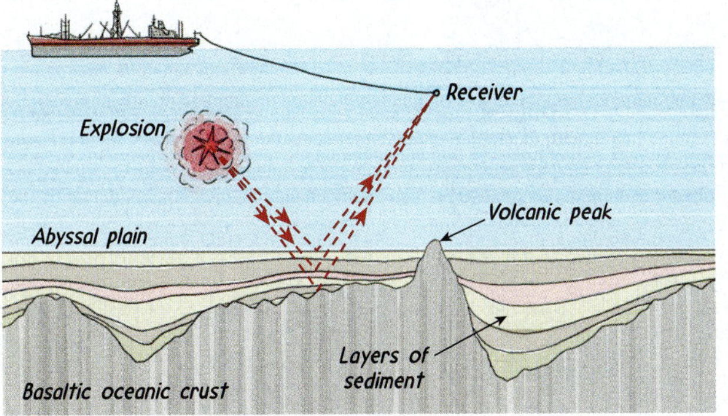

◀ **Figure 10.11 Seismic profile of the ocean floor** This seismic cross section and matching sketch across a portion of the Madeira abyssal plain in the eastern Atlantic Ocean show the irregular oceanic crust buried by sediments. (Image courtesy of Charles Hollister, Woods Hole Oceanographic Institution)

islands include Easter Island, Tahiti, Bora Bora, the Galapagos Islands, and the Canary Islands.

Guyots Inactive volcanic islands are gradually but inevitably lowered to near sea level by the forces of weathering and erosion. As a moving plate slowly carries volcanic islands away from the elevated oceanic ridge or hot spot over which they formed, they gradually sink and disappear below the water surface. Submerged, flat-topped seamounts that formed in this manner are called **guyots**.*

Oceanic Plateaus The ocean floor contains several massive **oceanic plateaus**. Resembling the flood basalt provinces found on the continents, oceanic plateaus are thought to form when the bulbous head of a rising mantle plume melts, producing vast outpourings of fluid basaltic lavas (see Figure 5.33, page 155).

Some oceanic plateaus appear to have formed quickly in geologic terms. Examples include the Ontong Java Plateau, which formed in less than 3 million years, and the Kerguelen Plateau, which formed in 4.5 million years (see Figure 10.4).

Explaining Coral Atolls—Darwin's Hypothesis

Coral **atolls** are ring-shaped structures that extend from slightly above sea level to depths of several thousand meters. What causes atolls to form, and how do they attain such great thicknesses?

*The term *guyot* is named after Arnold Guyot, Princeton University's first geology professor. It is pronounced "GEE-oh" with a hard g, as in "give."

Corals are tiny animals that generally cluster in large numbers and form colonies when linked. Most corals create a hard external skeleton made of calcium carbonate. Some build large calcium carbonate structures, called *reefs*, where new colonies grow atop the strong skeletons of previous colonies. Sponges and algae may attach to the reef, enlarging it further.

Reef-building corals grow best in waters with an average annual temperature of about 24°C (75°F). They cannot survive prolonged exposure to temperatures below 18°C (64°F) or above 30°C (86°F). In addition, reef-builders require clear, sunlit water. Consequently, the depth of most active reef growth is limited to no more than about 45 meters (150 feet).

The strict environmental conditions required for coral growth create an interesting paradox: How can corals—which require warm, shallow, sunlit water no deeper than a few dozen meters—create thick structures such as coral atolls that extend to great depths?

The naturalist Charles Darwin was one of the first to formulate a hypothesis on the origin of ringed-shaped atolls. He did this during a 5-year period (1831–1836) when he sailed aboard the British ship HMS *Beagle* on its famous global circumnavigation. On this journey, Darwin visited many volcanic islands surrounded by coral reefs. A keen observer, Darwin noticed a progression in coral reef development from (1) a *fringing reef* along the margins of a volcano to (2) a *barrier reef* with a volcano in the middle to (3) an *atoll*, consisting of a continuous or broken ring of coral reef surrounding a central lagoon (**Figure 10.12**).

The essence of Darwin's hypothesis is that, in addition to being lowered by erosional forces, many volcanic islands gradually sink. Darwin also

▼ Figure 10.12 Darwin's hypothesis Darwin's hypothesis asserts that, in addition to being lowered by erosional forces, many volcanic islands gradually sink. Darwin also suggested that corals respond to the gradual change in water depth caused by the subsiding volcano by building the reef complex upward. Atolls may eventually become guyots, when the rate of subsidence exceeds the ability of corals to build up the reef, or when the reef-building organisms die out.

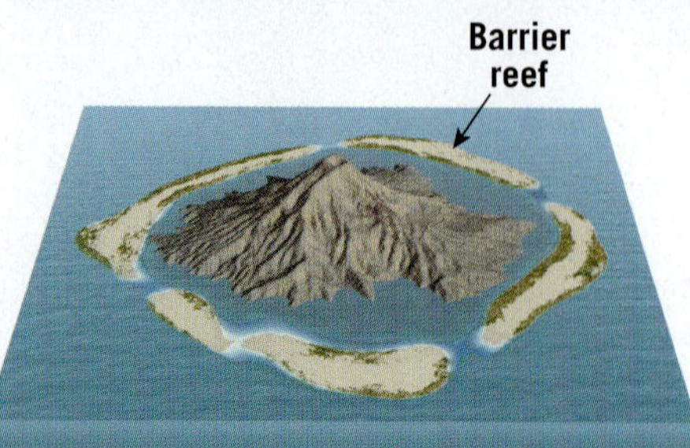

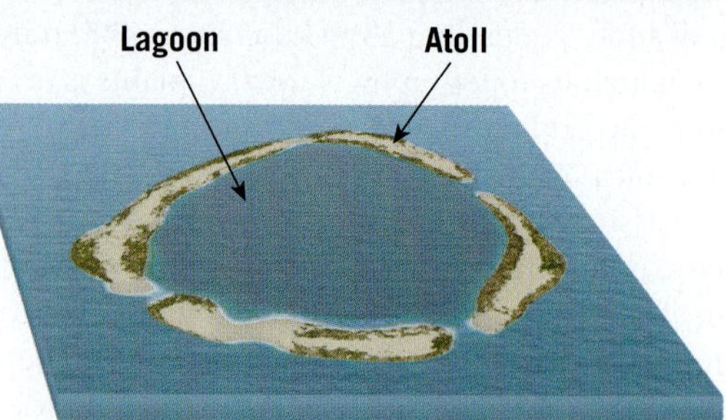

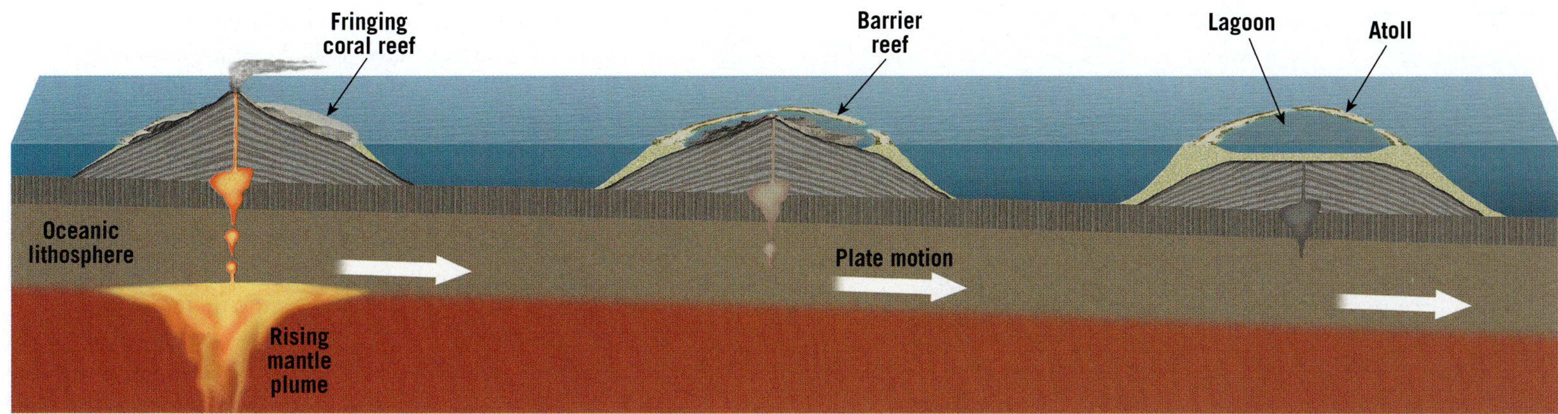

▲ **Figure 10.13 Plate tectonics and coral atolls** The plate tectonics theory provides the most current scientific explanation regarding how volcanic islands become extinct and sink to great depths over long periods of time. Some volcanic islands form over relatively stationary mantle plumes, causing the lithosphere to be buoyantly uplifted. Over spans of millions of years, these volcanic islands gradually sink as the moving plates carry them away from the region of hot-spot volcanism.

hypothesized that corals responded to the gradual change in water depth caused by the subsiding volcano by building the reef complex upward. Eventually, the volcano would submerge beneath the sea, and its remnant would be covered by a type of reef we call an atoll. During Darwin's time, however, there was no plausible mechanism to account for how or why so many volcanic islands sink.

The plate tectonics theory provides the most current scientific explanation of how volcanic islands become extinct and sink to great depths over long periods of time. Some volcanic islands form over relatively stationary mantle plumes, causing the lithosphere to be warmed and buoyantly uplifted. Over spans of millions of years, these volcanic islands gradually sink as the moving plates carry them away from the region of hot-spot volcanism because the lithosphere cools, becomes denser, and sinks (Figure 10.13). Volcanic islands that form on spreading centers also sink when they move away from these areas of upwelling.

CONCEPT CHECKS 10.3

1. Explain how deep-ocean trenches are related to plate boundaries.
2. Why are abyssal plains more extensive on the floor of the Atlantic than on the floor of the Pacific?
3. How does a flat-topped seamount, called a *guyot*, form?
4. What features on the ocean floor most resemble flood basalt provinces on the continents?
5. Using Darwin's hypothesis, place these coral reefs in order from youngest to oldest: *barrier reef*, *atoll*, and *fringing reef*.

10.4 Anatomy of the Oceanic Ridge

Sketch and label a cross-sectional view of the Mid-Atlantic Ridge. Explain how a cross section of the East Pacific Rise would look different.

Along well-developed divergent plate boundaries, the seafloor is elevated, forming a broad linear swell called the **oceanic ridge,** or **mid-ocean ridge** or **rise**. Our knowledge of the oceanic ridge system comes from soundings of the ocean floor, core samples from deep-sea drilling, visual inspection using deep-diving submersibles (Figure 10.14), and firsthand inspection of slices of ocean floor that have been thrust onto dry land during continental collisions.

The oceanic ridge system winds through all major oceans in a manner similar to the seam on a baseball, and it is the longest topographic feature on Earth, exceeding 70,000 kilometers (43,000 miles) in length (Figure 10.15). The crest of the ridge typically stands 2 to 3 kilometers above the adjacent deep-ocean basins and marks the plate boundary, where new oceanic crust is created.

▼ **Figure 10.14 The deep-diving submersible *Alvin*** This submersible is 7.6 meters long, weighs 16 tons, has a cruising speed of 1 knot, and can reach depths of 4000 meters. A pilot and two scientific observers are along during a normal 6- to 10-hour dive. (Photo by Rod Catanach KRT/Newscom)

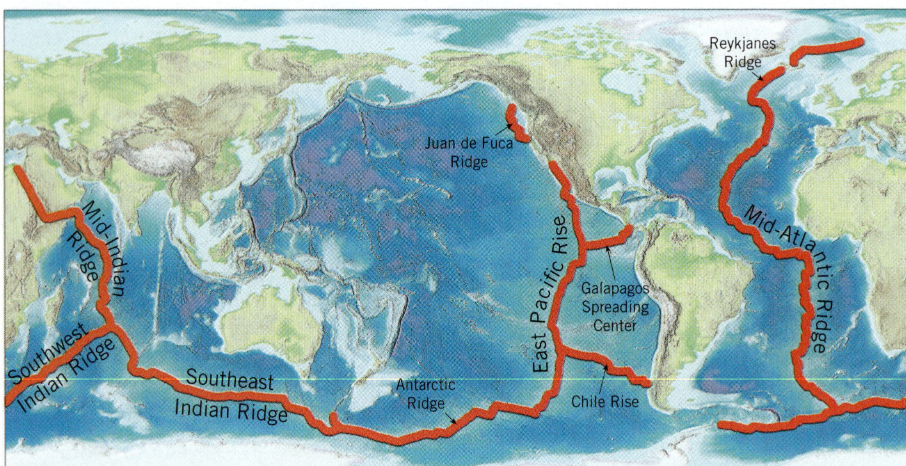

▲ **Figure 10.15**
The oceanic ridge system.

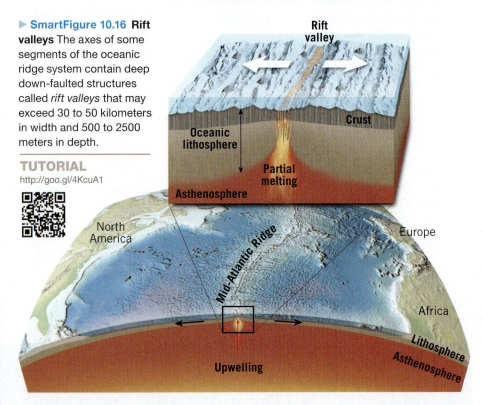

Notice in Figure 10.15 that large sections of the oceanic ridge system have been named based on their locations within the various ocean basins. A ridge may run through the middle of an ocean basin, where it is appropriately called a *mid-ocean ridge*. The Mid-Atlantic Ridge and the Mid-Indian Ridge are examples. By contrast, the East Pacific Rise is *not* a "mid-ocean" feature. Rather, as its name implies, it is located in the eastern Pacific, far from the center of the ocean.

The term *ridge* is somewhat misleading because these features are not narrow and steep, as the term implies, but have widths from 1000 to 4000 kilometers (600 to 2500 miles) and the appearance of broad, elongated swells that exhibit varying degrees of ruggedness. Furthermore, the ridge system is broken into segments that range from a few tens of kilometers to hundreds of kilometers in length. Each ridge segment is offset from the adjacent segment by a transform fault.

Oceanic ridges are as high as some mountains on the continents, but the similarities end there. Whereas most mountain ranges on land form when the compressional forces associated with continental collisions fold and metamorphose thick sequences of sedimentary rocks, oceanic ridges form where upwelling from the mantle generates new oceanic crust. Oceanic ridges consist of layers and piles of newly formed basaltic rocks that are buoyantly uplifted by the hot mantle rocks from which they formed.

Along the axis of some segments of the oceanic ridge system are deep, down-faulted structures called **rift valleys** because of their striking similarity to the continental rift valleys found in East Africa (**Figure 10.16**). Some rift valleys, including those along the rugged Mid-Atlantic Ridge, are 30 to 50 kilometers (20 to 30 miles) wide and have walls that tower 500 to 2500 meters (1600 to 8200 feet) above the valley floor. This makes them comparable to the deepest and widest part of Arizona's Grand Canyon.

▶ **SmartFigure 10.16 Rift valleys** The axes of some segments of the oceanic ridge system contain deep down-faulted structures called *rift valleys* that may exceed 30 to 50 kilometers in width and 500 to 2500 meters in depth.

TUTORIAL
http://goo.gl/4KcuA1

CONCEPT CHECKS 10.4

1. Briefly describe the oceanic ridge system.
2. Although oceanic ridges can be as tall as some mountains found on the continents, list some ways in which the features are different.

10.5 Oceanic Ridges & Seafloor Spreading

Write a statement describing how spreading rates affect ridge topography.

The greatest volume of magma (more than 60 percent of Earth's total yearly output) is produced along the oceanic ridge system in association with seafloor spreading. As plates diverge, fractures created in the oceanic crust fill with molten rock that gradually wells up from the hot mantle below. This molten material slowly cools and crystallizes, producing new slivers of seafloor. This process repeats in episodic bursts, generating new lithosphere that moves away from the ridge crest in a conveyor belt fashion.

Seafloor Spreading

Harry Hess of Princeton University formulated the concept of seafloor spreading in the early 1960s. Later, geologists were able to verify Hess's view that seafloor spreading occurs along the crests of oceanic ridges, where hot mantle rock rises to replace the material that has shifted horizontally (see Figure 10.16). Recall from Chapter 4 that as rock rises, it experiences a decrease in confining pressure that may lead to *decompression melting.*

Partial melting of ultramafic mantle rock produces basaltic magma that has a surprisingly consistent chemical composition. The newly formed melt separates from the mantle rock and rises toward the surface. Along some ridge segments, the melt collects in small, elongated reservoirs located just beneath the ridge crest. Eventually, about 10 to 20 percent of the melt migrates upward along fissures and erupts as lava flows on the ocean floor, while the remainder crystallizes at depth to form the lower crust. This activity continuously adds new basaltic rock to diverging plate margins, temporarily welding them together; they are then broken as spreading continues. Along some ridges, outpourings of pillow lavas build submerged shield volcanoes (seamounts) as well as elongated lava ridges. At other locations, more voluminous lava flows pave the surface to create a relatively subdued topography.

Why Are Oceanic Ridges Elevated?

The primary reason for the elevated position of the ridge system is that newly created oceanic lithosphere is hot and therefore less dense than cooler rocks of the deep-ocean basin. As the newly formed basaltic crust travels away from the ridge crest due to seafloor spreading, it is cooled from above as seawater circulates through the pore spaces and fractures in the rock. Further cooling occurs as this new rock gets farther and farther from the zone of hot mantle upwelling. As a result, the lithosphere gradually cools, contracts, and becomes denser. This thermal contraction accounts for the greater ocean depths that occur away from the ridge. After about 80 million years of cooling and contraction, rock that was once part of an elevated ocean-ridge system becomes part of the deep-ocean basin.

As lithosphere is displaced away from the ridge crest, cooling also causes a gradual increase in lithospheric thickness. This happens because the boundary between the lithosphere and asthenosphere is a thermal (temperature) boundary. Recall from Chapter 9 that the lithosphere is Earth's cool, stiff outer layer, whereas the asthenosphere is a comparatively hot and weak layer. As material in the uppermost asthenosphere ages (cools), it becomes stiff and rigid. Thus, the upper portion of the asthenosphere is gradually converted to lithosphere simply by cooling. Oceanic lithosphere continues to thicken until it is about 80 to 100 kilometers (50 to 60 miles) thick. Thereafter, its thickness remains relatively unchanged until it is subducted.

Spreading Rates & Ridge Topography

When researchers studied various segments of the oceanic ridge system, it became clear that there were topographic differences. These differences appear to have resulted from differences in spreading rates—which largely determine the amount of melt generated at a rift zone. More magma wells up from the mantle at fast spreading centers than at slow spreading centers. This difference in output causes differences in the structure and topography of various ridge segments.

Oceanic ridges that exhibit slow spreading rates from 1 to 5 centimeters per year have prominent rift valleys and rugged topography (**Figure 10.17A**). The Mid-Atlantic and Mid-Indian Ridges are examples. The vertical displacement of large slabs of oceanic crust along normal faults is responsible for the steep walls of the rift valleys. Furthermore, volcanism produces numerous cones in the rift valley, which enhance the rugged topography of the ridge crest.

By contrast, along the Galapagos Ridge, an intermediate spreading rate of 5 to 9 centimeters per year is the norm. As a result, the rift valleys that develop are

Did You Know?

If you were thinking about buying an island hideaway, the Indian Ocean might be the place to look. Numerous small coral islands, including the Maldives, are scattered about the Indian Ocean. The Maldives include more than 1000 islands, most of which are uninhabited. Rising sea levels, however, threaten many of these low-lying coral masses.

A. At slow spreading rates, a prominent central rift valley develops along the ridge crest, and the topography of the ridge is typically rugged.

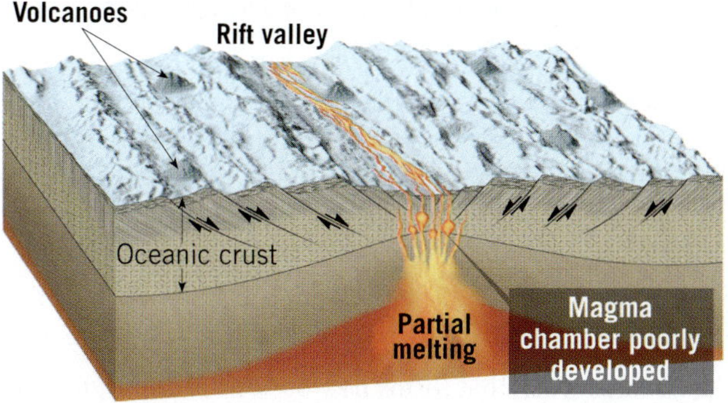

B. Along fast spreading centers, medial rift valleys do not develop, and the topography is comparatively smooth.

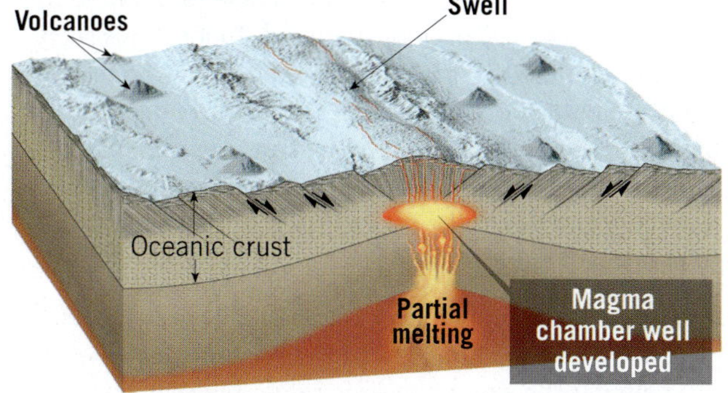

Figure 10.17
Topography of slow and fast spreading centers.

Ridges that have fast spreading rates are characterized by smooth topography, gently sloping flanks, and a central swell.

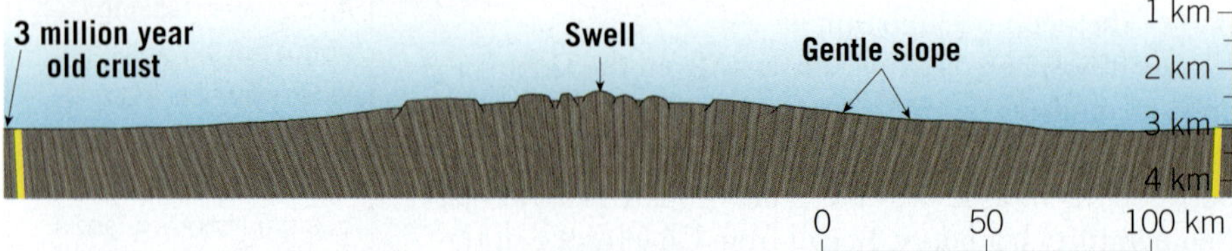

Ridges that have intermediate spreading rates are characterized by small rift valleys (less than 500 meters deep) and moderately sloping flanks.

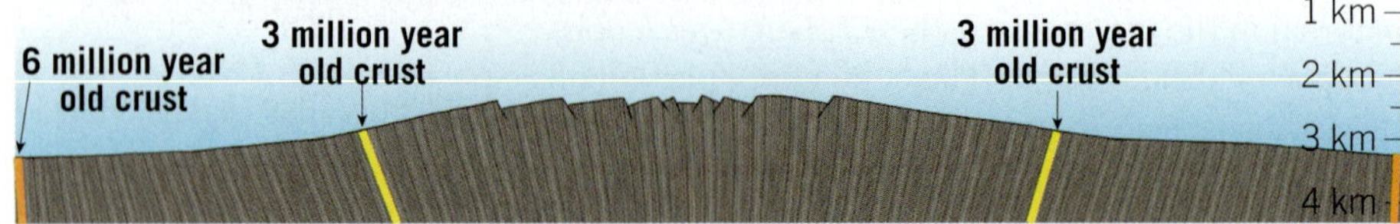

Ridges that have slow spreading rates are characterized by rugged topography, well-developed rift valleys (500-2500 meters deep), and steeply sloping flanks.

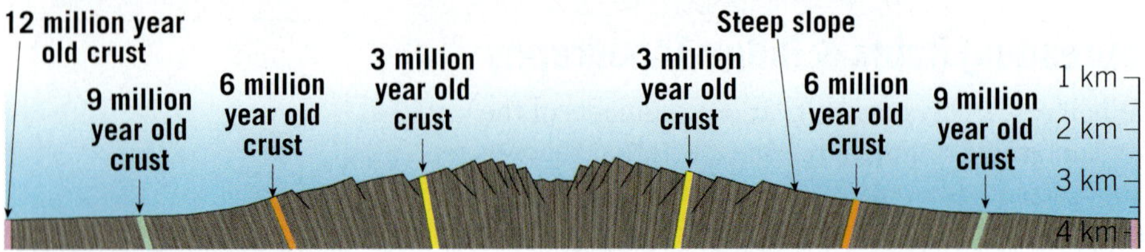

▲ **Figure 10.18 Ridge segments that exhibit fast, intermediate, and slow spreading rates.**

relatively shallow—often less than 200 meters (660 feet) deep. In addition, their topography is more subdued compared to ridges that have slower spreading rates.

At fast spreading centers (greater than 9 centimeters per year), such as along much of the East Pacific Rise, rift valleys are generally absent (Figure 10.17B). Instead, the ridge axis is elevated. These elevated structures, called *swells*, are built from lava flows up to 10 meters (30 feet) thick that have incrementally paved the ridge crest with volcanic rocks (see Figure 10.17B). In addition, because the depth of the ocean depends largely on the age of the seafloor, ridge segments that exhibit faster spreading rates tend to have more gradual profiles than ridges that have slower spreading rates (**Figure 10.18**). Because of these differences in topography, the gently sloping, less rugged portions of fast-spreading ridges are called *rises*.

CONCEPT CHECKS 10.5

1. What is the source of magma for seafloor spreading?
2. What is the primary reason for the elevated position of the oceanic ridge system?
3. Why does the lithosphere thicken as it moves away from the ridge as a result of seafloor spreading?
4. Compare and contrast the topography of a slow spreading center such as the Mid-Atlantic Ridge with one that exhibits a faster spreading rate, such as the East Pacific Rise.

10.6 The Nature of Oceanic Crust

List the four layers of oceanic crust and explain how oceanic crust forms and how it differs from continental crust.

An interesting aspect of oceanic crust is that its thickness and structure are remarkably consistent throughout the entire ocean basin. Seismic soundings indicate that its thickness averages only about 7 kilometers (4.5 miles). Furthermore, it is composed almost entirely of mafic rocks that are underlain by a layer of the ultramafic rock peridotite, which forms the lithospheric mantle.

Although most oceanic crust forms out of view, far below sea level, geologists have been able to examine the structure of the ocean floor firsthand. In locations such as Newfoundland, Cyprus, Oman, and California, slivers of oceanic crust and underlying mantle have been thrust high above sea level. From these exposures and core samples collected by deep-sea drilling ships, researchers have concluded that the ocean crust consists of four distinct layers (**Figure 10.19**):

- **Layer 1.** The upper layer is a sequence of deep-sea sediments or sedimentary rocks. Sediments are very thin near the axes of oceanic ridges but may be several kilometers thick next to continents.
- **Layer 2.** Below the layer of sediments is a rock unit composed mainly of basaltic lavas that contain abundant pillowlike structures called *pillow lavas*.
- **Layer 3.** The middle, rocky layer is made up of numerous interconnected dikes that have a nearly vertical orientation, called a *sheeted dike complex*. These dikes are former pathways where magma rose to feed pillow basalts on the ocean floor.
- **Layer 4.** The lowest layer is mainly gabbro, the coarse-grained equivalent of basalt, which crystallized deeper in the crust without erupting.

When fragments of oceanic crust and the underlying mantle are discovered on land, they are called an **ophiolite complex**. From studies of various ophiolite complexes around the globe and related data, geologists have pieced together a scenario for the formation of the ocean floor.

How Does Oceanic Crust Form?

The molten rock that forms new oceanic crust originates from partial melting of the ultramafic mantle rock.

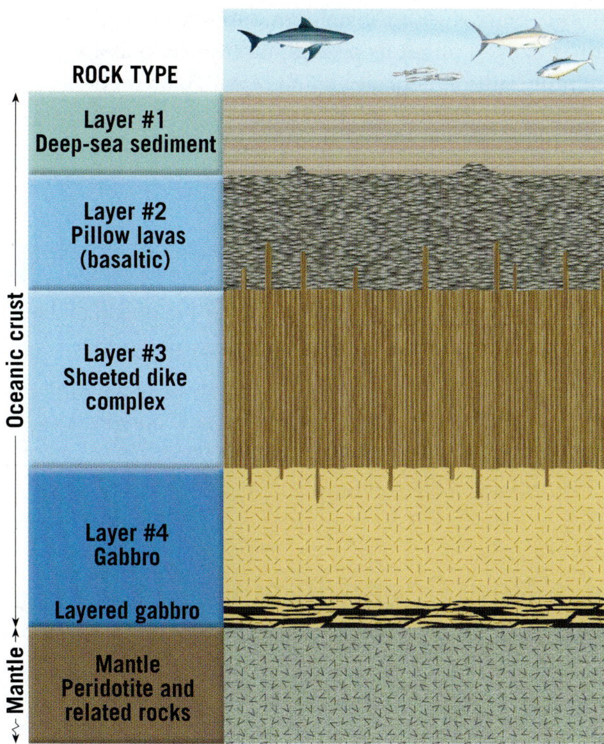

▲ **Figure 10.19 Structure of the oceanic crust** This view of the layered structure of oceanic crust is based on data obtained from ophiolite complexes, seismic profiling, and core samples obtained from deep-sea drilling expeditions. The relative thickness of the layers is not shown to scale.

▲ **Figure 10.20 Cross-sectional view of pillow lava** This pillow lava is exposed along a sea cliff at Cape Wanbrow, New Zealand. Notice that each "pillow" shows an outer dark glassy layer created by rapid cooling, which encloses a dark gray basalt interior. (Photo by GR Roberts/Photo Researchers, Inc.)

This process generates basaltic melt that is less dense than the surrounding solid rock. The newly formed melt rises through the upper mantle along thousands of tiny conduits that feed into a few dozen larger, elongated channels, perhaps 100 meters (300 feet) or more wide. These structures, in turn, feed lens-shaped magma chambers located directly beneath the ridge crest. The addition of melt from below steadily increases the pressure inside the magma chambers. As a result, the rocks above these reservoirs periodically fracture, allowing the melt to ascend along numerous vertical fractures that develop in the ocean crust. Some of the melt cools and solidifies to form dikes. New dikes intrude older dikes, which are still warm and weak, to form a **sheeted dike complex**. This portion of the oceanic crust is usually 1 to 2 kilometers thick.

Roughly 10 to 20 percent of the melt eventually erupts on the ocean floor. Because the surface of these submarine lava flows is chilled quickly by seawater, these flows generally travel no more than a few kilometers before completely solidifying. The forward motion occurs as lava accumulates behind the congealed margin and then breaks through. This process occurs repeatedly, as molten basalt is extruded—like toothpaste from a tightly squeezed tube (see Figure 5.8, page 134). The result is protuberances resembling large bed pillows stacked one atop the other, hence the name **pillow lavas** (Figure 10.20).

In some settings, pillow lavas may build volcano-size mounds that resemble shield volcanoes, whereas in other situations they form elongated ridges tens of kilometers long. These structures are eventually separated from their supply of magma as they are carried away from the ridge crest by seafloor spreading.

The lowest unit of the ocean crust develops from crystallization within the central magma chamber. The first minerals to crystallize are olivine, pyroxene, and occasionally chromite (chromium oxide), which settle through the magma to form a layered zone near the floor of the magma reservoir. The remaining melt tends to cool along the walls of these magma chambers to form massive amounts of coarse-grained gabbro. This portion accounts for up to 5 of the 7 kilometers (3 to 4.5 miles) of ocean crust thickness.

Although molten rock rises continuously from the mantle toward the surface, seafloor spreading occurs in pulse-like bursts. As the melt begins to accumulate in the lens-shaped reservoirs, it is blocked from continuing upward by the stiff overlying rocks. As the amount of melt entering the magma reservoirs increases, pressure rises. Periodically, the pressure exceeds the strength of the overlying rocks, which fracture and initiate a short episode of seafloor spreading.

Interactions Between Seawater & Oceanic Crust

In addition to serving as a mechanism for dissipating Earth's internal heat, the interaction between seawater and the newly formed basaltic crust alters both the seawater and the crust. The permeable and highly fractured lava of the upper oceanic crust allows seawater to penetrate to depths of 2 to 3 kilometers (1 to 2 miles). Seawater

Did You Know?
The circulation of seawater through the oceanic crust extracts heat from the volcanically produced material and is the main method by which the crust is cooled.

circulating through the hot crust is heated and chemically reacts with the basaltic rock in a process called *hydrothermal* (hot water) *metamorphism* (see Figure 8.22, page 231). This alteration causes the dark silicates (olivine and pyroxene) in basalt to form new metamorphic minerals such as chlorite and serpentine. Simultaneously, the hot seawater dissolves ions of silica, iron, copper, and occasionally silver and gold from the hot basalts. When the water temperature reaches a few hundred degrees Celsius, these mineral-rich fluids buoyantly rise along fractures and eventually spew out on the ocean floor.

Studies conducted by submersibles along several segments of the oceanic ridge have photographed these metallic-rich solutions gushing from the seafloor to form particle-filled clouds called **black smokers** (see Figure 8.22, page 231). As the hot liquid (up to 400°C [750°F]) mixes with the cold, mineral-laden seawater, the dissolved minerals precipitate to form massive metallic sulfide deposits, some of which are economically important. Occasionally these deposits grow upward to form underwater chimney-like structures equivalent in height to skyscrapers.

CONCEPT CHECKS 10.6

1. Briefly describe the four layers of the ocean crust.
2. How does a *sheeted dike complex* form?
3. How does hydrothermal metamorphism alter the basaltic rocks that make up the seafloor? How is seawater changed during this process?
4. What is a black smoker?

10.7 Continental Rifting: The Birth of a New Ocean Basin

Outline the steps by which continental rifting results in the formation of new ocean basins.

The breakup of Pangaea nearly 200 million years ago opened a new ocean basin—the Atlantic. Although geoscientists still debate what initiated this event, the breakup of Pangaea illustrates that ocean basins originate when large landmasses break apart.

Evolution of an Ocean Basin

The opening of a new ocean basin begins with the formation of a **continental rift**, an elongated depression along which the entire lithosphere is stretched and thinned. Where the lithosphere is thick, cool, and strong, rifts tend to be narrow—often less than a few hundred kilometers wide. Modern examples of narrow continental rifts include the East African Rift, the Rio Grande Rift (southwestern United States), the Baikal Rift (south-central Siberia), and the Rhine Valley (northwestern Europe). By contrast, where the crust is thin, hot, and weak, rifts can be more than 1000 kilometers (600 miles) wide, as exemplified by the Basin and Range region in the western United States.

In settings where rifting continues, the rift system evolves into a young, narrow ocean basin, such as the present-day Red Sea. Continued seafloor spreading eventually results in the formation of a mature ocean basin bordered by rifted continental margins. The Atlantic Ocean is such a feature.

What follows is an overview of ocean basin evolution, using modern examples to represent the various stages of rifting.

East African Rift The East African Rift is a continental rift that extends through eastern Africa for approximately 3000 kilometers (2000 miles). It consists of several interconnected rift valleys that split into eastern and western sections around Lake Victoria (**Figure 10.21**).

▼ **SmartFigure 10.21 East African Rift Valley** The map shows the extent of the East African Rift Valley, as well as the rifted valley that is now occupied by the Red Sea.

TUTORIAL
http://goo.gl/J08Epg

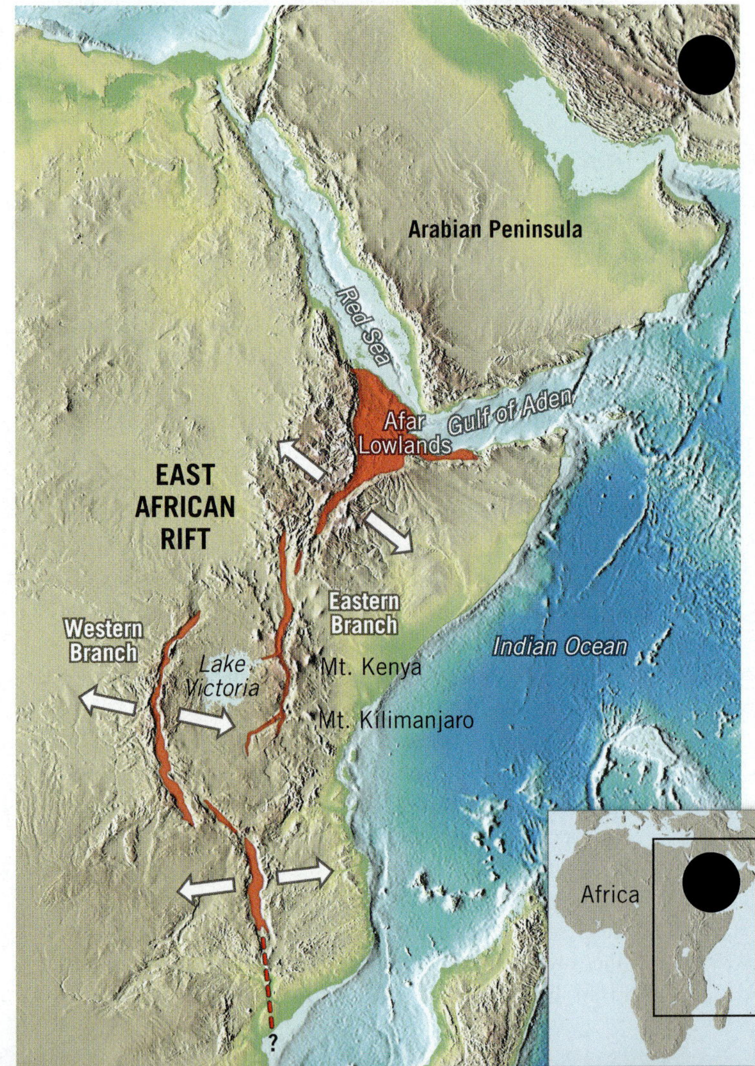

A. Tensional forces and buoyant uplifting of the heated lithosphere cause the upper crust to be broken along normal faults, while the lower crust deforms by ductile stretching.

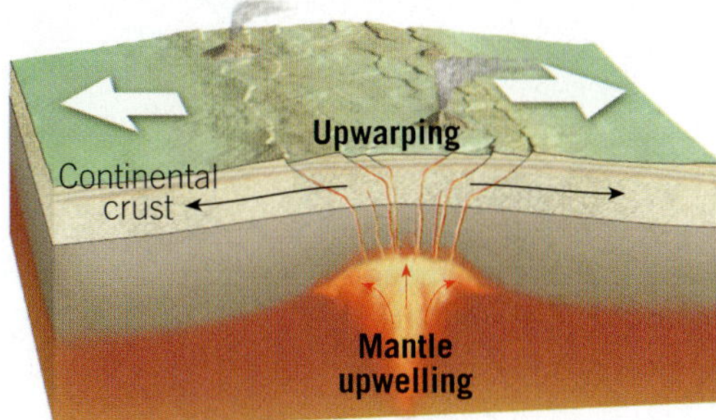

B. As the crust is pulled apart, large slabs of rock sink, generating a rift valley.

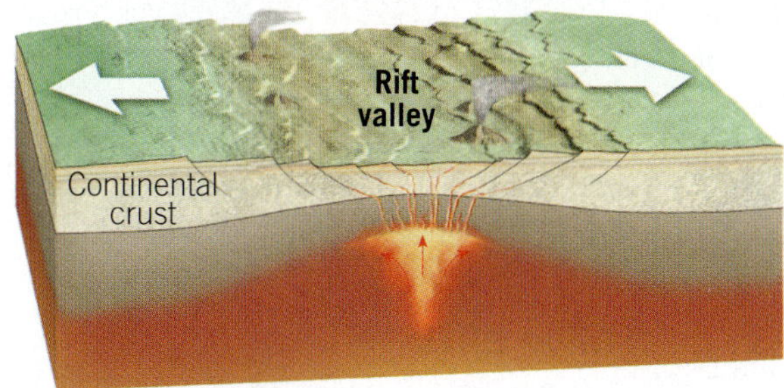

C. Further spreading generates a narrow sea.

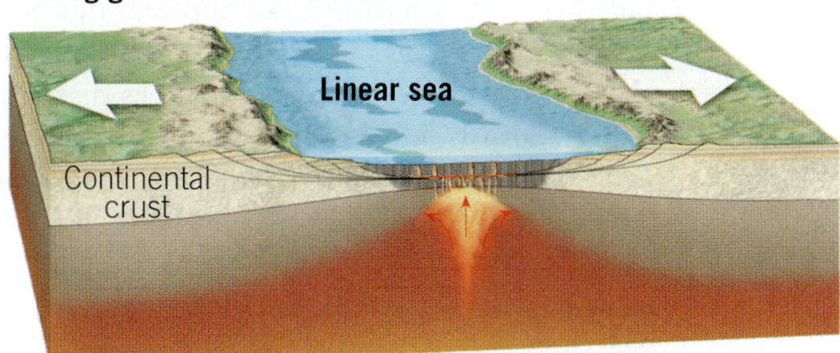

D. Eventually, an expansive ocean basin and ridge system are created.

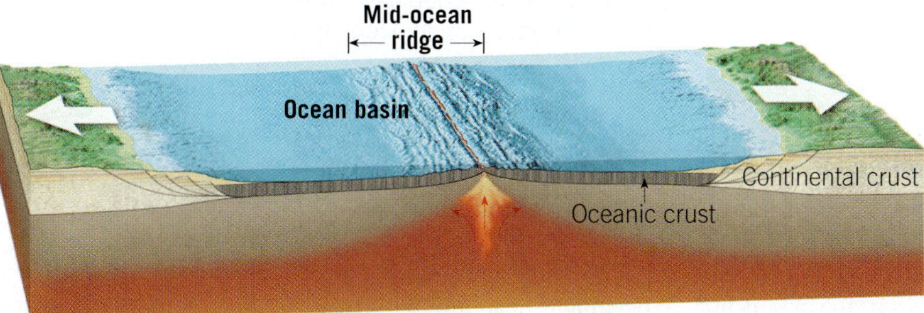

Illustration of the separation of South America and Africa to form the South Atlantic.

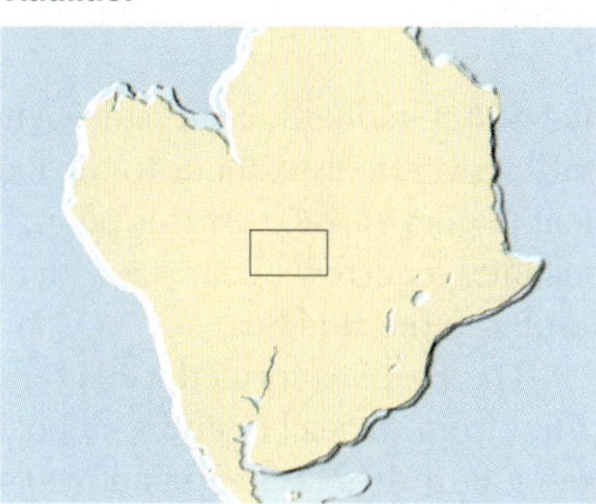

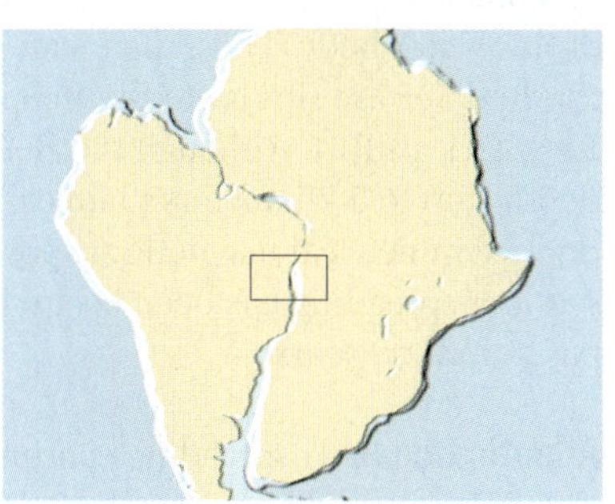

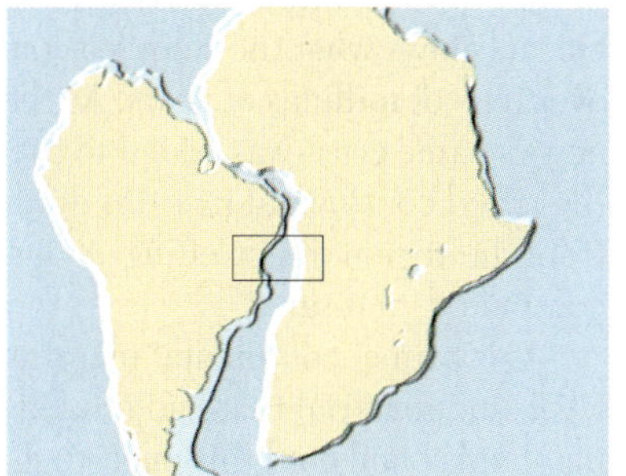

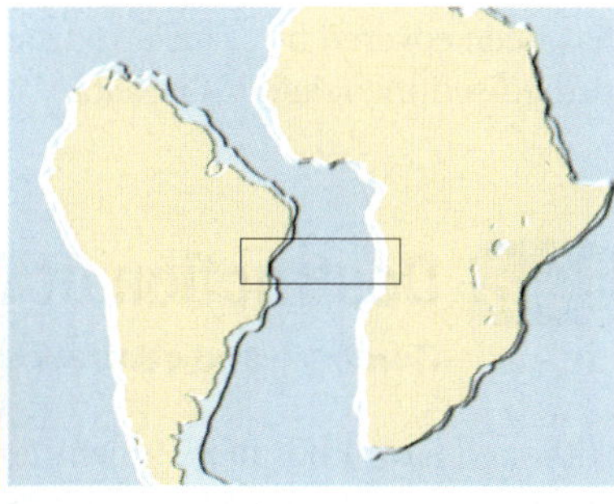

◀ SmartFigure 10.22
Formation of an ocean basin

ANIMATION
http://goo.gl/PeZrDy

Whether this rift will eventually develop into a spreading center, with the Somali subplate separating from the rest of the continent of Africa, is uncertain.

The most recent period of rifting began about 20 million years ago, as upwelling in the mantle intruded the base of the lithosphere (**Figure 10.22A**). Buoyant uplifting of the heated lithosphere led to doming and stretching of the crust. Consequently, the upper crust was broken along high-angle normal faults, producing down-faulted blocks, or *grabens*, while the lower crust deformed by ductile stretching (**Figure 10.22B**). (Faulting and associated structures are examined in detail in the Chapter 11.)

In the early stages of rifting, magma generated by decompression melting of the rising mantle rocks intruded the crust. Occasionally, some of the magma migrated upward along fractures and erupted at the surface. This activity produced extensive basaltic flows within the rift as well as volcanic cones—some forming more than 100 kilometers (60 miles) from the rift axis.

Did You Know?

Communities of organisms reside around hydrothermal vents (black smokers) in dark, hot, sulfur-rich environments where photosynthesis cannot occur. The base of the food web is provided by microbes that use a process called *chemosynthesis* to grow, which allows them and many other organisms to live in this extreme environment.

Examples include Mount Kenya and Mount Kilimanjaro, the highest point in Africa, which rises almost 6000 meters (20,000 feet) above the Serengeti Plain.

Red Sea Gradually, a rift valley lengthens and deepens, until it eventually extends to the margin of the continent (Figure 10.22C). At this point, the continental rift becomes a narrow linear sea with an outlet to the ocean, similar to the Red Sea.

The Red Sea formed when the Arabian Peninsula rifted from Africa beginning about 30 million years ago (see Figure 10.21). Steep fault scarps that rise as much as 3 kilometers (2 miles) above sea level flank the margins of this water body. Thus, the escarpments surrounding the Red Sea are similar to the steep cliffs that border the East African Rift. Although the Red Sea reaches oceanic depths (up to 5 kilometers [3 miles]) in only a few locations, symmetrical magnetic stripes indicate that typical seafloor spreading has been occurring for at least the past 5 million years.

Atlantic Ocean If spreading continues, the Red Sea will grow wider and develop an elevated oceanic ridge similar to the Mid-Atlantic Ridge (Figure 10.22D). The Atlantic Ocean shows what the Red Sea could eventually become over tens of millions of years. As the Atlantic formed, new oceanic crust was added to the diverging plates, and the rifted continental margins gradually moved away from the region of upwelling. As a result, they cooled, contracted, and sank.

Over time, continental margins subsided below sea level, and material that had eroded from the adjacent highlands blanketed this once-rugged topography. The result was a *passive continental margin* on both sides of the Atlantic, consisting of rifted continental crust that has been covered by a thick wedge of relatively undisturbed sediment and sedimentary rock.

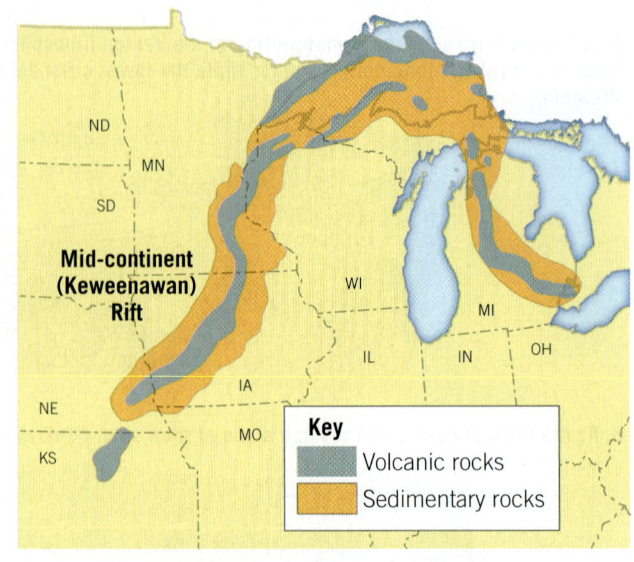

▲ **Figure 10.23 Failed rift** Location of a failed rift extending from the Great Lakes region to Kansas.

Failed Rifts

Not all continental rift valleys develop into full-fledged spreading centers. In the central United States, a failed rift extends from Lake Superior into Kansas (Figure 10.23). This once-active rift valley is filled with clastic sedimentary and basaltic rocks that were extruded onto the crust more than a billion years ago. Why one rift valley develops into a full-fledged active spreading center while others fail to develop is not fully understood.

CONCEPT CHECKS 10.7

1. Name a modern example of a continental rift.
2. Briefly describe each of the four stages in the formation of an ocean basin.

Did You Know? Sea turtles, salmon, and other animals travel great distances across the open ocean so that they can lay their eggs in the same location where they were hatched. Recent research suggests that these animals may use Earth's magnetic field for navigation. Interestingly, many animals that have "homing" abilities also contain minuscule amounts of the magnetic mineral magnetite.

10.8 Destruction of Oceanic Lithosphere

Compare and contrast spontaneous subduction and forced subduction.

Although new lithosphere is continually being produced at divergent plate boundaries, Earth's surface area is not growing larger. In order to balance the amount of newly created lithosphere, there must be a process whereby oceanic lithosphere is destroyed.

Why Oceanic Lithosphere Subducts

The process of plate subduction is complex, and the ultimate fate of oceanic lithosphere is still being debated. Does subducted oceanic crust pile up at the boundary between the upper mantle and the lower mantle, where it is assimilated into the mantle? Or do most plates descend to the core–mantle boundary, where they are heated and eventually rise back toward Earth's surface as mantle plumes?

What scientists do know with some certainty is that oceanic lithosphere will resist subduction unless its overall density is greater than that of the underlying asthenosphere. It takes at least 15 million years for a young slab of oceanic lithosphere to cool sufficiently to become denser than the supporting asthenosphere.

Spontaneous Subduction Subduction zones can be divided into two basic types, based on the nature of the subducting plate. The first type, called a *Mariana-type subduction zone*, is characterized by old, dense lithosphere sinking into the mantle due to its own weight. The lithosphere entering the Mariana trench is about 185 million years old, some of the oldest and densest lithosphere in today's oceans. Along this trench, the subducting slab descends into the mantle at a steep angle that approaches 90 degrees (**Figure 10.24A**). Steep subduction angles produces deep trenches, which account in part for the depth of the Challenger Deep, located at the southern limb of the Mariana trench. The Mariana and most of the other subduction zones in the western Pacific involve cold, dense lithosphere and therefore exhibit **spontaneous subduction**.

It is important to note that the *lithospheric mantle*, which makes up about 80 percent of the descending oceanic slab, drives subduction. Even when the *oceanic crust* is quite old, its density is still less than that of the underlying asthenosphere. Subduction, therefore, depends on lithospheric mantle that is colder and therefore denser than the underlying asthenosphere.

When an oceanic slab descends to about 400 kilometers (250 miles), mineral phase changes (the transitions from low-density mineral to high-density mineral) enhance subduction. At this depth, the transition of olivine to its compact, much denser spinel structure increases the density of the slab, which helps pull the plate into the subduction zone.

Forced Subduction The second type of subduction zone, called the *Peru–Chile–type subduction zone*, is characterized by younger, hotter, and less-dense lithosphere that dips at shallower angles (**Figure 10.24B**). Along Peru–Chile–type boundaries, the lithosphere is too buoyant to subduct spontaneously; rather, it is *forced* beneath the overlying plate by compressional forces.

In areas where **forced subduction** occurs, a strong coupling develops between the overlying plate and the subducting plate, which can result in particularly strong and frequent earthquakes. Stated another way, plate motion generates horizontal compressional forces that cause the upper plate and underlying plate to grind against each other. The result can be folding and thickening of the upper plate and sometimes the formation of mountainous terrains such as those we see today in the Andes. Shallow subduction and strong coupling have also been observed in the past decade along the Sunda subduction zone, off the coast of Sumatra, another region that has experienced several major earthquakes.

Researchers have also determined that unusually thick units of oceanic crust, those that approach 30 kilometers (20 miles) in thickness, are likely to resist subduction. The Ontong Java Plateau, for example, is a thick oceanic plateau, about the size of Alaska, located in the western Pacific (see Figure 10.4). About 20 million years ago, this plateau reached the trench that forms the boundary between the subducting Pacific plate and the overriding Australian–Indian plate. Apparently too buoyant to subduct, the Ontong Java Plateau clogged the trench. We will consider the fate of crustal fragments that are too buoyant to subduct in the next chapter.

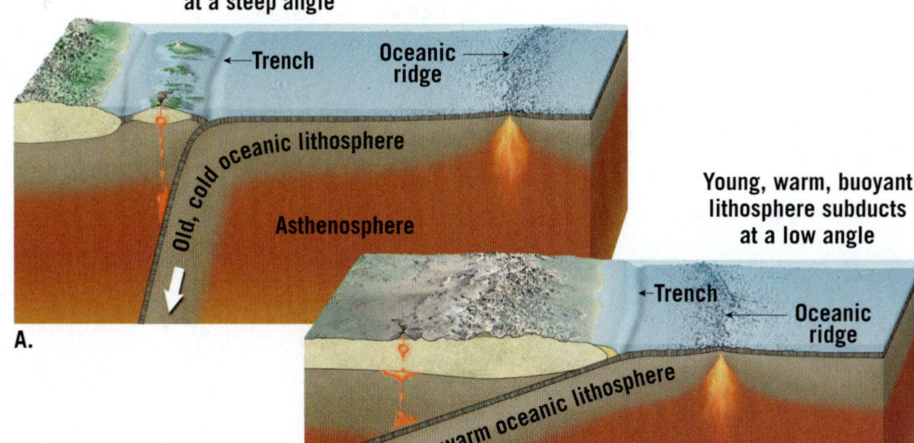

▲ **Figure 10.24 The angle of plate subduction depends on the lithosphere's density**
A. In parts of the Pacific, some oceanic lithosphere exceeds 180 million years in age and descends into the mantle at angles approaching 90 degrees. **B.** Young oceanic lithosphere is warm and buoyant, and it tends to subduct at a low angle.

Subducting Plates: The Demise of Ocean Basins

In the 1970s, geologists began using magnetic stripes and fracture zones on the ocean floor to reconstruct the past 200 million years of plate movement. This research showed that parts of, or even entire, ocean basins have been destroyed along subduction zones. For example, during the breakup of Pangaea shown in Figure 2.22 on page 51, notice that the African plate moved northward, eventually colliding with Eurasia. During this event, the floor of the intervening Tethys Ocean was almost entirely consumed into the mantle, leaving behind a few small remnants—the Eastern Mediterranean Sea and the Black Sea.

Reconstructions of the breakup of Pangaea also helped investigators understand the demise of the Farallon plate—a large oceanic plate that once occupied much of the eastern Pacific basin. At the time of the breakup of Pangaea, the Farallon plate was situated on the eastern side of a spreading center, as shown in **Figure 10.25A**. The spreading center that generated both the Farallon and Pacific plates is the East Pacific Rise.

Beginning about 180 million years ago, the Americas were propelled westward as Pangaea broke up and the Atlantic Ocean started to open. As a result, the Farallon plate began subducting beneath the Americas faster than it was being generated, causing it to decrease in size (**Figure 10.25B**). The three remaining fragments of the once-extensive Farallon plate include the modern Juan de Fuca, Cocos, and Nazca plates.

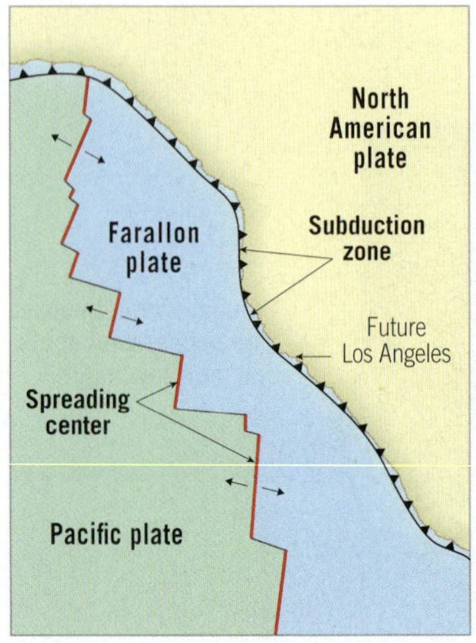

A. 56 million years ago

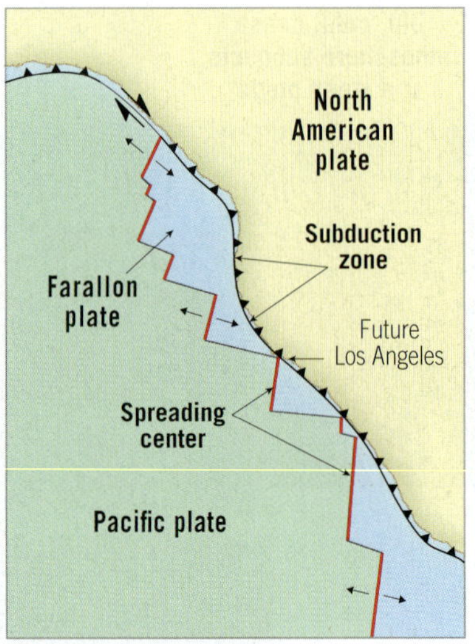

B. 30 million years ago

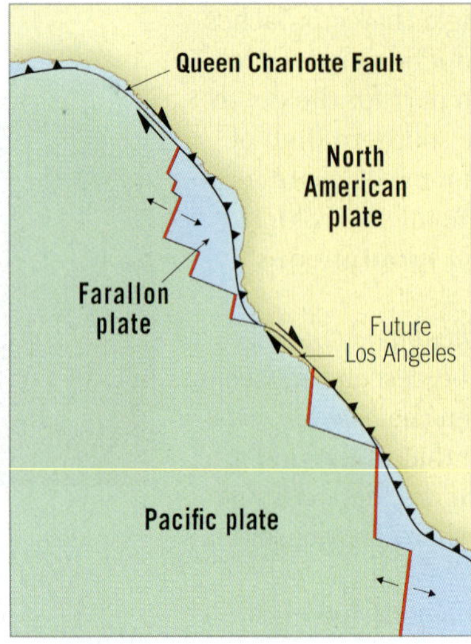

C. 20 million years ago

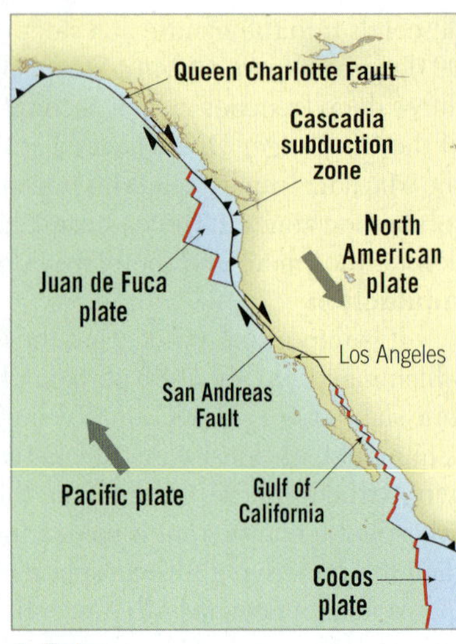

D. Today

▲ **SmartFigure 10.25 The demise of the Farallon plate** Because the Farallon plate was subducting faster than it was being generated, it became steadily smaller. The remaining fragments of the once-mighty Farallon plate are the Juan de Fuca, Cocos, and Nazca plates.

The westward migration of North America also caused a section of the East Pacific Rise to enter the subduction zone that once lay off the coast of California (see Figure 10.25B). As this spreading center subducted, it was destroyed and replaced by a transform fault system that currently accommodates the differential motion between the North American and Pacific plates. This change in plate geometry caused the Pacific plate to capture a sliver of North America (the Baja Peninsula and a portion of southern California) that it carries northwestward toward Alaska at an average rate of about 6 centimeters (2.5 inches) per year.

As more of the ridge subducted, the transform fault system, which we now call the San Andreas Fault, increased in length (**Figure 10.25C**). Today, the southern end of the San Andreas Fault connects to a young spreading center that is generating the Gulf of California (**Figure 10.26**). A similar event generated the Queen Charlotte transform fault, located off the west coast of Canada and southeastern Alaska.

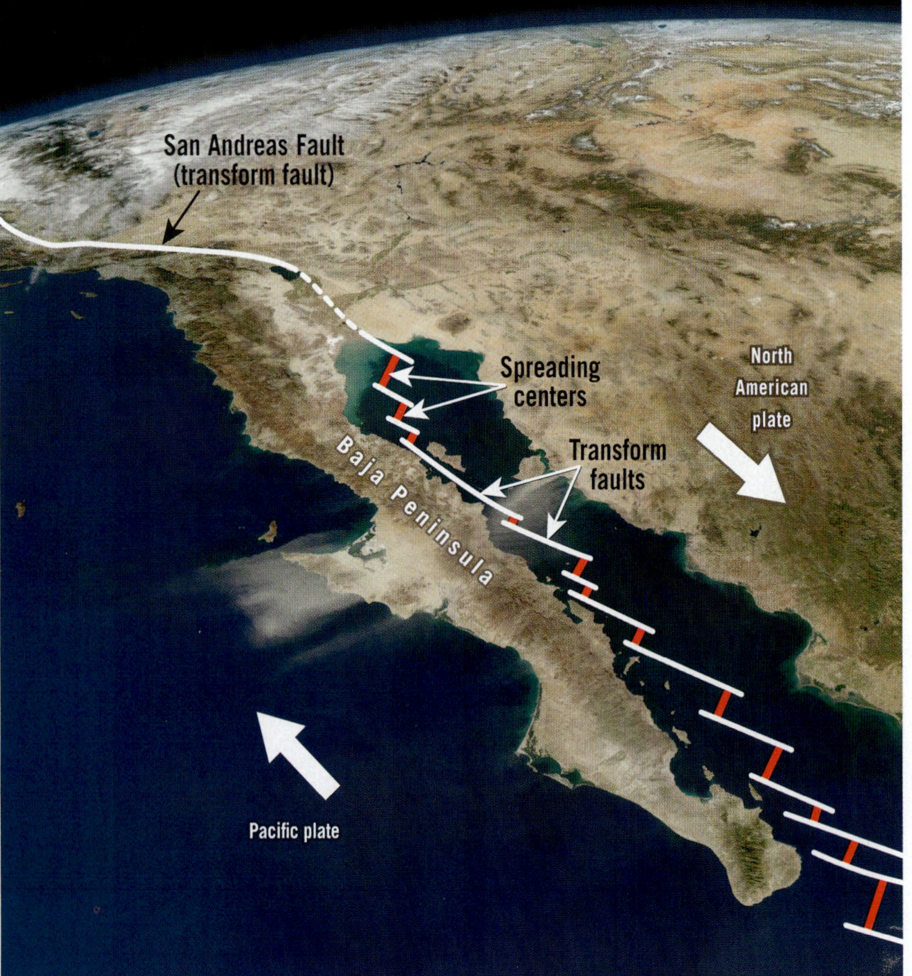

▲ **Figure 10.26 The separation between the Baja Peninsula and North America.** (SeaWiFS Project/ORBIGMAGE/NASA/Goddard Space Flight Center)

CONCEPT CHECKS 10.8

1. Explain why oceanic lithosphere subducts even though the oceanic crust is less dense than the underlying asthenosphere.
2. Compare spontaneous subduction and forced subduction. Provide examples of places where each operates.
3. What role do mineral phase changes play in plate subduction?
4. Explain what happened when the spreading center that generated the Farallon plate collided with the North American plate.

CONCEPTS IN REVIEW

Origin & Evolution of the Ocean Floor

10.1 An Emerging Picture of the Ocean Floor

Define *bathymetry* and describe the various bathymetric techniques used to map the ocean floor.

KEY TERMS: bathymetry, sonar, echo sounder

- Oceanographers map the seafloor with sonar—shipboard instruments that emit pulses of sound that "echo" off the bottom. Satellites are also used to map the ocean floor. Their instruments measure slight variations in sea level that result from differences in the gravitational pull of features on the seafloor.
- Mapping efforts have revealed three major areas of the ocean floor: continental margins, deep-ocean basins, and oceanic ridges.

10.2 Continental Margins

Compare and contrast a passive continental margin with an active continental margin and list the major features of each.

KEY TERMS: continental margin, passive continental margin, continental shelf, continental slope, continental rise, turbidity current, submarine canyon, deep-sea fan, active continental margin, accretionary wedge, subduction erosion

- Continental margins are transitional zones between continental and oceanic crust. Active margins occur along continental plate boundaries where an oceanic slab is subducting beneath a continent. Passive continental margins are on the trailing edge of continents, far from plate boundaries.

(10.2 continued)

- Heading offshore from the shoreline of a passive margin, a submarine traveler first encounters the gently sloping continental shelf and then the steeper continental slope, which marks the end of the continental crust and the beginning of the oceanic crust. Beyond the continental slope is the continental rise, which is made of sediment transported by turbidity currents.
- At an active continental margin, material may be added to the leading edge of a continent in the form of an accretionary wedge (common at shallow-angle subduction zones), or material may be scraped off the edge of a continent by subduction erosion (common at steeply dipping subduction zones).

? What type of continental margin is depicted in this diagram? Name the feature indicated by the question mark.

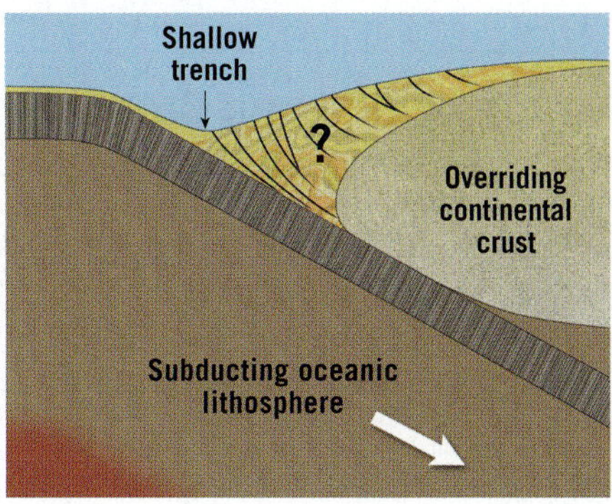

10.3 Features of Deep-Ocean Basins

List and describe the major features of the deep-ocean basin.

KEY TERMS: deep-ocean basin, deep-ocean trench, volcanic island arc, continental volcanic arc, abyssal plain, seismic reflection profiler, seamount, volcanic island, guyot, oceanic plateau, atoll

- The deep-ocean basin makes up about half of the ocean floor's area. Much of it is abyssal plain (deep, featureless, sediment-draped crust). Subduction zones and deep-ocean trenches also occur in deep-ocean basins. Paralleling trenches are volcanic island arcs (if an oceanic slab is subducted under oceanic lithosphere) or continental volcanic arcs (if an oceanic slab is subducted under an overriding continental plate).
- A variety of volcanic structures poke up from the abyssal plain. Seamounts are submarine volcanoes; if they emerge at the ocean's surface, we call them volcanic islands. Guyots are old volcanic islands that have had their tops eroded off before sinking below the sea. Oceanic plateaus are submarine flood basalt provinces—unusually thick sections of oceanic crust formed by massive underwater eruptions of lava.
- Ring-shaped coral atolls are features formed from biological rather than geological processes around volcanic islands. As these islands erode and sink below the surface, corals build upward on existing coral skeletons to take advantage of sunlight and warm water temperatures.

? On this cross-sectional view of a deep-ocean basin, label the following features: seamount, guyot, volcanic island, oceanic plateau, and abyssal plain.

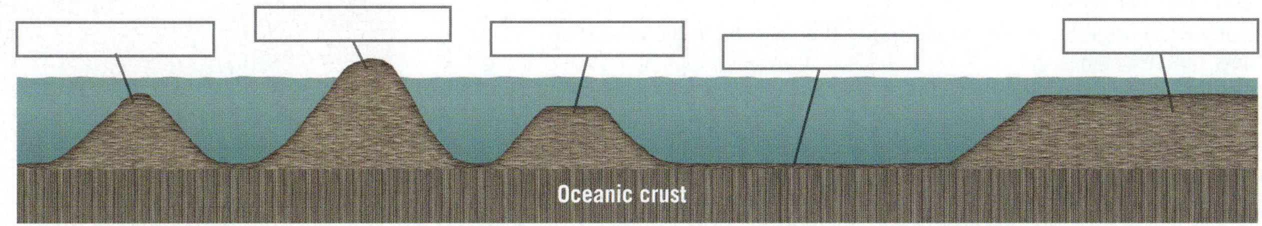

10.4 Anatomy of the Oceanic Ridge

Sketch and label a cross-sectional view of the Mid-Atlantic Ridge. Explain how a cross section of the East Pacific Rise would look different.

KEY TERMS: oceanic ridge (mid-ocean ridge or rise), rift valley

- The oceanic ridge system is the longest topographic feature on Earth, wrapping around the world through all major ocean basins. It is a few kilometers tall, a few thousand kilometers wide, and a few tens of thousands of kilometers long. The ridge summit is where new oceanic crust is generated, and it may be marked by a rift valley.

10.5 Oceanic Ridges & Seafloor Spreading

Write a statement describing how spreading rates affect ridge topography.

- Oceanic ridges form due to seafloor spreading: As plates of oceanic lithosphere move apart, the warm mantle beneath rises and undergoes decompression melting. The resulting magma is basaltic in composition. Some of it erupts along the ridge axis, but much of it crystallizes at depth. The freshly cooled basalt or gabbro makes up new oceanic crust.
- Oceanic ridges are elevated features because they are warm and therefore less dense than older, colder oceanic lithosphere that makes up the deep-ocean basins. Heat loss causes the oceanic crust to subside and be buried. After 80 million years, crust that was once an oceanic ridge can become an abyssal plain.
- The rate at which seafloor spreading occurs determines the shape of an oceanic ridge. Ridges with slow spreading rates (1 to 5 centimeters per year) have prominent rift valleys and rugged topography. Those with fast spreading rates (greater than 9 centimeters per year) lack rift valleys and show a smoother, more subdued topography.

10.6 The Nature of Oceanic Crust

List the four layers of oceanic crust and explain how oceanic crust forms and how it differs from continental crust.

KEY TERMS: ophiolite complex, sheeted dike complex, pillow lava, black smoker

- Ophiolite complexes contain slices of oceanic crust that have been thrust above sea level. They have four distinct layers (from the top down): (1) deep-sea sediment, (2) pillow lava flows, (3) the sheeted dike complex, and (4) the gabbro layer.
- As two divergent plates move apart, fractures open perpendicular to the stretching direction, and lava moves up through these cracks, toward the seafloor. Once the lava has cooled and sealed these fractures shut, they become preserved as dikes of the sheeted dike complex. Magma that cools at depth crystallizes to become gabbro. Any lava that makes it to the seafloor is erupted as pillow lavas, which are gradually buried by deep-sea sedimentation.
- Along mid-ocean ridges, seawater flows through fissures in the oceanic crust and is heated by the hot surrounding rock. The warmer the seawater, the more chemically active it becomes. The hot water causes hydrothermal metamorphism and dissolves metal ions. These hot solutions may spew out of the crust as black smokers.

10.7 Continental Rifting: The Birth of a New Ocean Basin

Outline the steps by which continental rifting results in the formation of new ocean basins.

KEY TERM: continental rift

- When continents rift apart, new ocean basins may form. The East African Rift is an example of the initial stages of continental breakup, with rift valleys that are sites of basaltic volcanism and deposition of detrital sediment. The Red Sea is an ongoing example of more advanced rifting, in which seafloor spreading is occurring and the rift is submerged below sea level. Over time, the rift may widen due to the seafloor spreading, forming an ocean basin flanked by passive continental margins. A modern example of this stage is the Atlantic Ocean.

? Consider the Baja Peninsula of Mexico (see Figure 10.26). Which stage of continental rifting does the Gulf of California most closely match?

10.8 Destruction of Oceanic Lithosphere

Compare and contrast spontaneous subduction and forced subduction.

KEY TERMS: spontaneous subduction, forced subduction

- The production of oceanic lithosphere by seafloor spreading is balanced by subduction, which carries oceanic lithosphere into the mantle. The fate of subducted lithosphere in the mantle is still uncertain.
- When oceanic lithosphere is sufficiently old (and therefore cold), it may begin to sink because of its increased density. The resulting spontaneous subduction zone plunges at an angle near 90 degrees and is marked by a deep trench. The Mariana trench is an example.
- Forced subduction shoves relatively buoyant oceanic lithosphere underneath another plate. This results in shallow subduction angles and large earthquakes along megathrust faults. The Peru–Chile trench is an example.
- Regardless of its initial cause, subduction serves to close ocean basins. This may eventually bring once widely separated landmasses into contact with one another.

? Which type of subduction is illustrated in this diagram? Describe the age of the oceanic lithosphere at the deep-ocean trench compared to that at the oceanic ridge.

GIVE IT SOME THOUGHT

1 How many seconds would it take an echo sounder's ping to make the trip from a ship to the Challenger Deep (10,994 meters [36,060 feet]) and back? Recall that depth = ½ (1500 m/sec × Echo travel time).

2 Referring to Figure 10.17, compare and contrast the topography of the crest of an oceanic ridge that exhibits a slow spreading rate with one that exhibits a fast spreading rate. Give examples of each type of ridge.

3 Mataiva is a small atoll in the Tuamotu Archipelago, French Polynesia. Using the accompanying map, determine the approximate number of atolls in this small region of the western Pacific.

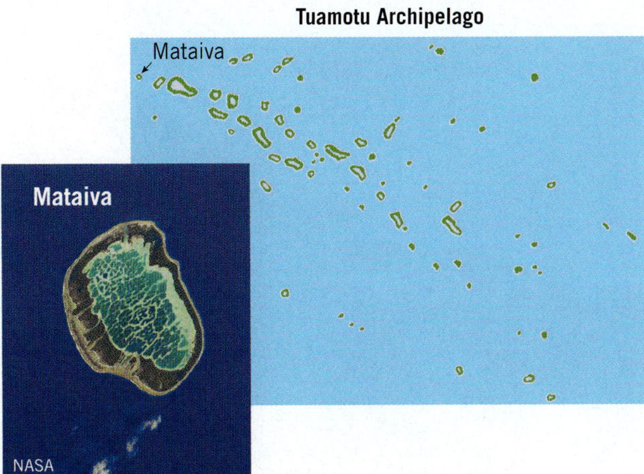

4 Refer to the accompanying map, which shows the eastern seaboard of the United States, to complete the following:

a. Which letter is associated with each of the following: continental shelf, continental rise, and continental slope?

b. How does the size of the continental shelf that surrounds the state of Florida compare with the size of the Florida peninsula?

c. Why are there no deep-ocean trenches on this map?

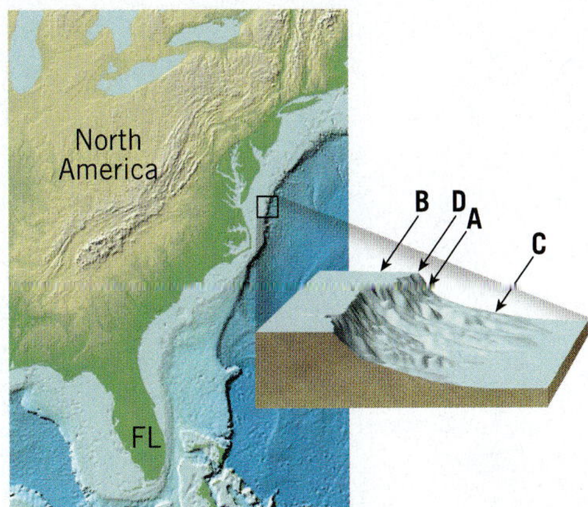

5 Briefly explain why the ocean floor generally gets deeper the farther one travels from the ridge crest.

6 The accompanying photo is a false-color sonar image that shows the ridge crest (linear pink area) of a section of a spreading center.

a. Is the structure along the ridge crest characteristic of a fast spreading center or a slow spreading center? Explain.

b. What name is given to the submerged cone-shaped structure in the lower-left portion of this image?

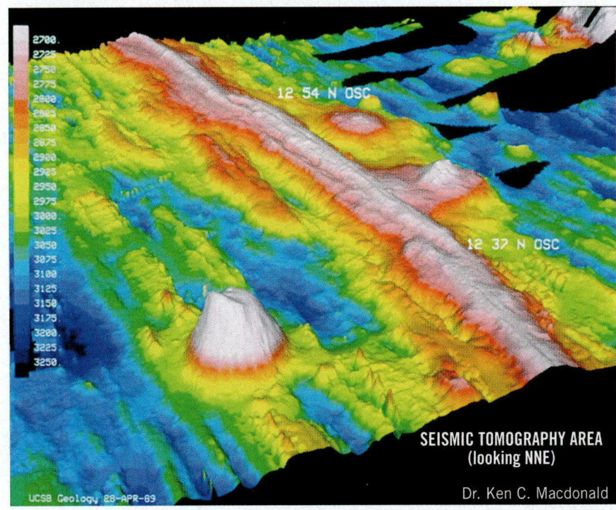

7 Name a location where a new ocean basin may develop in the future. Explain why you chose that site.

8 Refer to Figure 10.4. At which of the following locations would you expect to have the highest angle of subduction: the Puerto Rico subduction zone or the Cascadia subduction zone? Explain.

9 Refer to Figure 10.25. Predict the fate of the Juan de Fuca plate. What type of boundary might the Cascadia subduction zone become in the future? Explain.

10 Explain this statement: Oceans have existed on Earth for over 4 billion years, but the oldest seafloor that has been dated is only about 200 million years old.

11 This image shows lava at a temperature of about 1200°C erupting on the seafloor west of the Tonga trench. What name is given to lava flows like these that erupt underwater?

MasteringGeology™

Looking for additional review and test prep materials? Visit the Study Area in MasteringGeology to enhance your understanding of this chapter's content by accessing a variety of resources, including Self-Study Quizzes, Geoscience Animations, SmartFigures, Mobile Field Trips, *Project Condor* Quadcopter videos, *In the News* RSS feeds, flashcards, web links, and an optional Pearson eText.

www.masteringgeology.com

Crustal Deformation & Mountain Building*

FOCUS ON CONCEPTS

Each statement represents the primary learning objective for the corresponding major heading within the chapter. After you complete the chapter, you should be able to:

11.1 Describe the three types of differential stress and identify the tectonic setting most commonly associated with each. Differentiate stress from strain and brittle from ductile deformation.

11.2 List and describe five types of folds.

11.3 Sketch and briefly describe the relative motion of rock bodies located on opposite sides of normal, reverse, and thrust faults as well as both types of strike-slip faults.

11.4 Locate and name Earth's major mountain belts on a world map.

11.5 Sketch a cross section of an Andean-type mountain belt and describe how its major features are generated.

11.6 Summarize the stages in the development of an Alpine-type mountain belt such as the Appalachians.

11.7 Explain the principle of isostasy and how it contributes to the elevated topography of mountain belts.

*This chapter was revised with the assistance of Professor Callan Bentley.

MountKidd, Alberta Canada, with Kananaskies River in foreground (Photo by Tom Nevesely)

MOUNTAINS PROVIDE SOME OF THE most spectacular scenery on our planet. Poets, painters, and songwriters have captured their splendor. Geologists understand that at some time, all continental regions were mountainous masses and that continents grow over time through the addition of mountains to their flanks. Where lithospheric plates collide, the rocks in the collision zone are crumpled and transformed. Even continental interiors hold the remnants of mountain ranges formed from ancient collisions. This chapter explains the forces that create mountains, while the following chapters focus on the processes that weather and carry away rock material to create Earth's varied topography.

11.1 Crustal Deformation

Describe the three types of differential stress and identify the tectonic setting most commonly associated with each. Differentiate stress from strain and brittle from ductile deformation.

Young mountain belts tower above the surrounding landscapes, exhibiting steep slopes and high relief. Once the tectonic forces that raised these mountains cease, weathering and erosion will, over long spans of geologic time, grind them down until their roots lie flush with the surrounding topography. Even then their rocks hold clues to their former grandeur and to the forces that created them. Folded and faulted rocks like those in **Figure 11.1** attest to the compressional forces that deformed them.

What Causes Rocks to Deform?

Tectonic forces can cause rocks to *move*, *tilt*, and/or *change shape*. For instance, colliding plates can uplift a flat-lying bed of marine limestone so that it is exposed at the surface, rotate the rock at a steep angle, and crumple it into folds. Collectively, all these types of change are called **deformation**. Because rock deformation is caused mainly by tectonic forces, it occurs mostly along plate

▲ **SmartFigure 11.1 Deformed sedimentary strata** These deformed strata are exposed in a road cut near Palmdale, California. In addition to the obvious folding, light-colored beds are offset along a fault located on the right side of the photograph. (Photo by E. J. Tarbuck)

TUTORIAL
https://goo.gl/BpfvcZ

boundaries—the places where lithospheric plates push together, pull apart, or scrape past each other.

Rocks deform in characteristic ways that can be observed when the rocks are exposed on Earth's surface as an outcrop. For example, the folds in Figure 11.1 represent a typical response to compressional forces at depth. Geologists use the term **tectonic structures,** or **geologic structures**, for the structural features that we can observe and that reflect a rock's tectonic history. This chapter will look particularly at three types of tectonic structures: folds, faults, and joints (cracks in the rock).

To understand rock deformation, we first need to look more closely at the concepts of *stress* and *strain*.

Stress: The Cause of Deformation

So far, we've said that tectonic *forces* cause deformation. More precisely, rocks respond to **stress**, which takes into account the area over which a force acts (Stress = Force/Area). Recall from Chapter 8 that a force applied equally in all directions is called **confining pressure**—a force that compacts mineral grains to reduce the volume of rock bodies (Figure 11.2A). However, confining pressure does not cause deformation; instead, most deformation is caused by **differential stress**, in which the force is stronger in one direction and weaker in another.

Does stress acting on a rock always deform it? No, obviously not. When you stand on a rock, you press down on it; when you step off, you don't leave a dent. A rock's *strength* is its resistance to being deformed permanently by a stress. A rock will resist deformation until the stress exceeds its strength. A rock's strength depends on factors such as temperature and confining pressure, as well as the length of time over which the stress is exerted.

We will consider three types of differential stress: *compressional*, *tensional*, and *shear*. On a large scale, each type of stress tends to be associated with one type of plate boundary.

- **Compression.** Differential stress that squeezes a rock mass as if it is in a vise is known as **compressional stress** (Figure 11.2B). Compressional stresses are most often associated with convergent plate boundaries. When plates collide, Earth's crust is generally shortened horizontally and thickened vertically. Over millions of years, this deformation produces mountain belts.
- **Tension.** Differential stress that pulls apart rock bodies is known as **tensional stress** (Figure 11.2C). Along divergent plate boundaries, where plates are moving apart, tensional stresses stretch and lengthen rock bodies horizontally and thin them vertically. For example, in the Basin and Range Province in western North America, tensional forces have fractured and stretched the crust to as much as twice its original width.
- **Shear.** Differential stress can cause rock to **shear**, which involves the movement of one part of a rock body past another (Figure 11.2D). An everyday example of shear is the slippage that occurs between individual playing cards when the top of the deck is moved relative to the bottom. Shear is important at transform plate boundaries, such as the San Andreas Fault, where large segments of Earth's crust slip horizontally past one another.

Strain: A Change in Shape Caused by Stress

Earlier, we said that differential stress can deform a rock body by causing it to *move*, to *tilt*, and/or to *change shape*. When differential stress changes a rock's shape, the resulting deformation (distortion) is called **strain**. By observing and measuring the strain imprinted on a rock body, we can infer the type of stress that deformed the rock.

How does a rigid object like a rock change its shape? On a microscopic scale, this shape change (strain) is often accomplished through shearing. Note that deformation on a large scale often involves shearing on a small scale.

There are two main ways in which a solid rock can change its shape by undergoing small-scale shearing. First, a rock can undergo slippage along parallel surfaces of weakness, such as foliation surfaces or microscopic fractures. Like the deck of cards in Figure 11.3, many tiny slips can add up to a significant change in shape.

Second, shearing can take place within intact mineral grains. This occurs in part because mineral crystals are often far from perfect; the crystal structure may be missing atoms, or the atoms of one element may substitute for those of another. When a crystal undergoes differential stress, these defects are places where bonds can easily break and re-form. The result is slippage along microscopic surfaces within the crystal structure, and the mineral grain as a whole gradually changes shape (see Figure 8.8, page 224). If mineral crystals were perfect, rocks would be much stronger than they actually are.

Mineral grains can also change shape in response to differential stress that does not involve slippage along zones of weakness. This process, called *recrystallization*, involves the movement of atoms from a location that is highly stressed to a less-stressed position on the same grain (see Section 8.3, page 220).

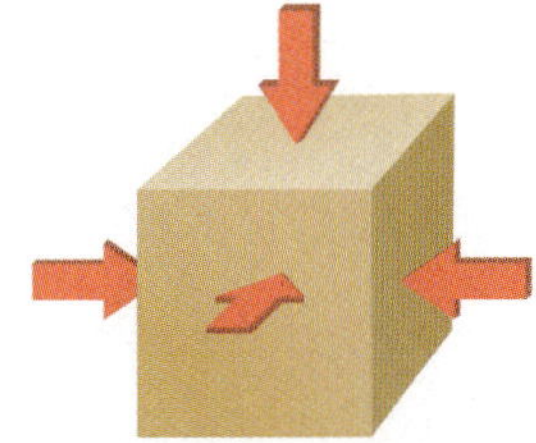
A. Confining pressure

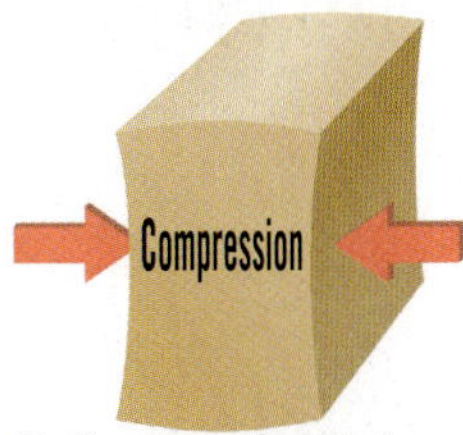

B. Compressional stress (shortening)

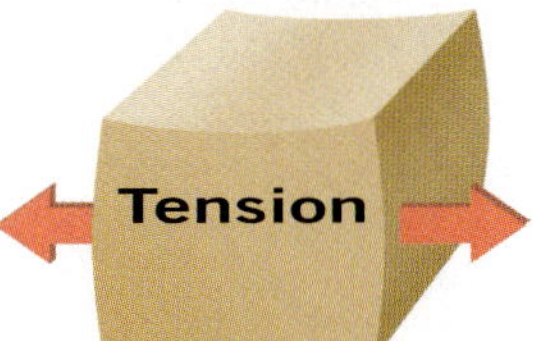

C. Tensional stress (stretching)

D. Shear stress (sliding and tearing)

▲ **Figure 11.2 Three aspects of differential stress: Compression, tension, and shear.**

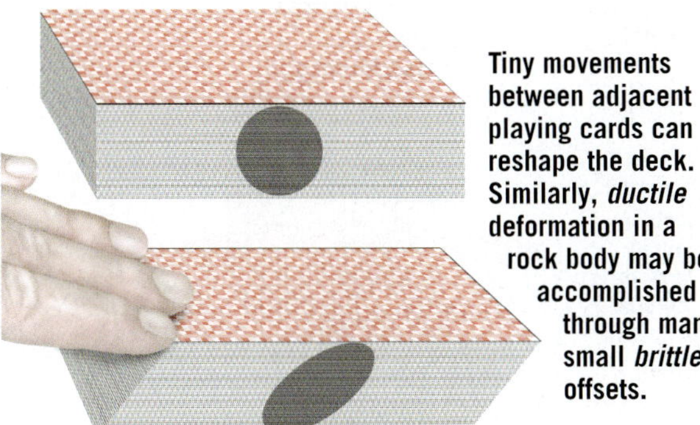

Tiny movements between adjacent playing cards can reshape the deck. Similarly, *ductile* deformation in a rock body may be accomplished through many small *brittle* offsets.

◄ **Figure 11.3 Shearing and the resulting deformation (strain)** An ordinary deck of playing cards with a circle embossed on its side illustrates shearing and the resulting strain.

Types of Deformation

Rocks experience three types of deformation that lead to shape changes (strain): *elastic*, *brittle*, and *ductile*.

Elastic Deformation If you have ever opened a door that has a spring attached, you may have noticed that when you let go of the door, the spring pulls the door shut as it returns to its original shape. We say the spring undergoes **elastic deformation**: It deforms temporarily in response to a stress (the force used to open the door) and returns to its original configuration when the stress is removed. Chemical bonds in a mineral grain act like a spring: When they are elastically deformed, they stretch but do not break, and when the stress is removed, the chemical bonds snap back to their original length. As you saw in Chapter 9, the energy released by most strong earthquakes comes from stored energy that is suddenly released as rock elastically snaps back (elastic rebound) to its original shape.

Brittle Deformation When a rock's strength is exceeded—that is, when it is deformed beyond its ability to respond elastically—it will either break or bend. Rocks that break into smaller pieces exhibit **brittle deformation**. From our everyday experience, we know that glass objects, wooden pencils, ceramic plates, and even our bones exhibit brittle deformation when their strength is surpassed. Brittle deformation occurs when stress breaks the chemical bonds that hold a material together.

Ductile Deformation When an object changes shape without breaking, we say that it has undergone **ductile deformation**. When you knead clay or taffy, you are deforming it in a ductile way. Ductile deformation in rocks takes place largely through the mechanisms described earlier—slippage along surfaces of weakness within the rock and the gradual reshaping of mineral grains. These processes enable rock to flow very slowly, even though it remains in a solid state. Intricate folds are an example of ductile deformation.

Factors That Affect How Rocks Deform

Figure 11.4 compares the types of brittle and ductile deformation. As the figure indicates, rock deformation tends to be brittle at shallow depths and ductile at greater depths. Four factors influence how a rock deforms: (1) temperature, (2) confining pressure, (3) the type of rock, and (4) time.

The Role of Temperature Temperature plays a major role in the behavior of rocks. Where temperatures are high (deep in Earth's crust or adjacent to a heat source such as a magma chamber), rocks are nearer their melting temperatures and are therefore weaker and more capable of ductile deformation (see Figure 11.4). Near the surface or in a comparatively cool environment such as a subduction zone, rocks are more brittle and prone to fracture. This behavior is completely familiar: A cold chocolate bar snaps between your teeth, whereas a warm one bends in your hand.

The Role of Confining Pressure Recall that the confining pressure on rocks increases with depth as the thickness of the overlying rock increases. Because confining pressure squeezes rocks equally from all directions, it tends to make them harder to break and hence less brittle. Thus, at depth, the increase in temperature and

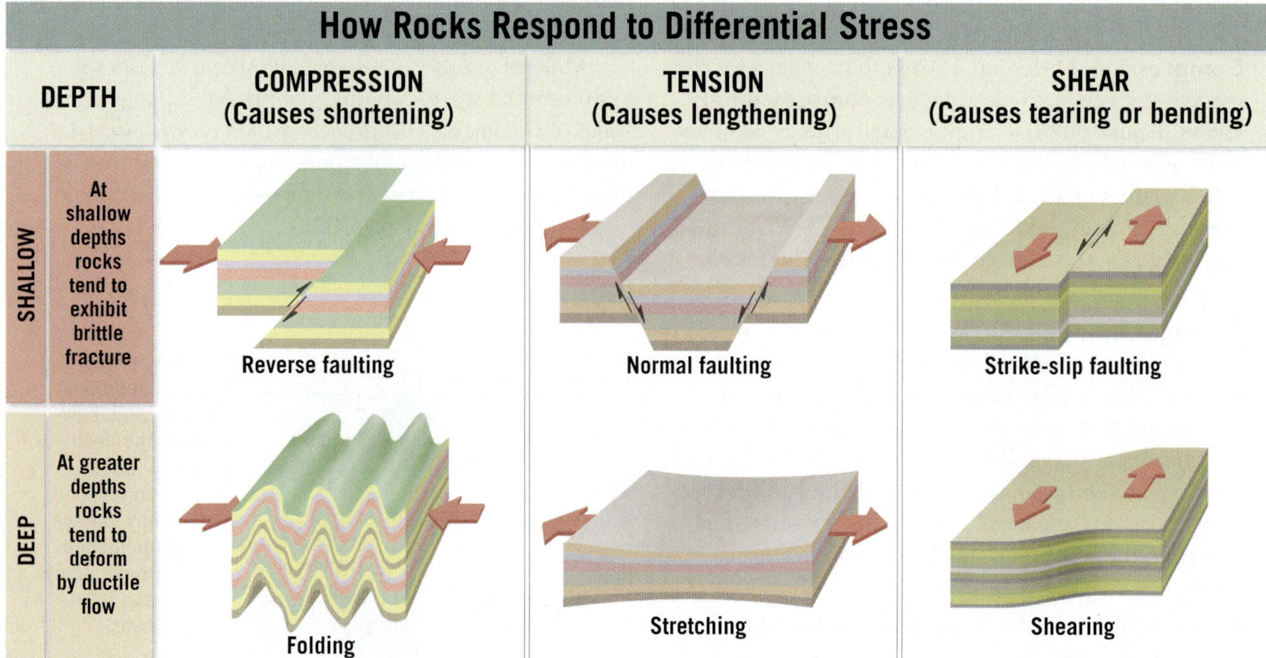

How Rocks Respond to Differential Stress
DEPTH
COMPRESSION (Causes shortening)
TENSION (Causes lengthening)
SHEAR (Causes tearing or bending)
SHALLOW
At shallow depths rocks tend to exhibit brittle fracture
Reverse faulting
Normal faulting
Strike-slip faulting
DEEP
At greater depths rocks tend to deform by ductile flow
Folding
Stretching
Shearing

Figure 11.4 Deformation caused by three types of stress Brittle deformation (fracturing and faulting) dominates in the upper crust, where the temperatures are comparatively cool. By contrast, at depths greater than about 10 kilometers (6 miles), where temperatures are high, rock deforms by ductile flow and folding.

the increase in pressure have complementary effects: The increase in temperature enhances ductile behavior, and the increase in pressure tends to keep the rock intact and thus more likely to bend than to fracture.

The Influence of Rock Type The manner in which a particular rock type responds to stress is greatly influenced by its mineral composition and texture. Granite, basalt, and well-cemented quartz sandstones are examples of strong, brittle rocks that tend to fail by breaking (brittle deformation) when subjected to stresses that exceed their strength. By contrast, clay-rich or weakly cemented sedimentary rocks and foliated metamorphic rocks more readily exhibit ductile deformation. Weak rocks that are most likely to behave in a ductile manner (bend or flow) when subjected to differential stress include rock salt, shale, limestone, and schist.

Rock salt consists of intergrown crystals, so you might think that it would resist ductile deformation. In fact, the opposite is true. Salt crystals readily change shape as the mineral grains continually recrystallize in response to differential stresses, which is why salt layers can rise up through other strata to form salt domes. Glacial ice also consists of crystals that deform easily, accommodating internal flow.

Some outcrops consist of a sequence of alternating layers of weak and strong rocks that have been deformed. Figure 11.5 shows an example in which adjacent metamorphic rock layers responded very differently to the same differential stress. Notice that the central nonfoliated layer broke into chunks, while the flanking foliated layers deformed in a ductile manner, flowing into the gaps between the chunks. The same thing happens when you bite into a s'more: The graham crackers break, while the marshmallow and chocolate layers ooze.

Time as a Factor The folded rocks of mountain belts show that tens of kilometers of compressional strain (shortening) can be accommodated by ductile deformation. For this to happen, however, stress has to be applied slowly enough that the sluggish processes of ductile deformation can keep up. If stress is applied to a rock unit relatively quickly, the rock will deform elastically until its strength is exceeded, and then it will fracture. Taffy exhibits the same behavior on a more familiar timescale: If you hit a bar of taffy against the edge of a table, it will break, but if you put a weight on it and leave it overnight, it will gradually spread and flatten. In practice, rocks that are near Earth's surface tend to accommodate even gradual strain by fracturing; ductile behavior happens mainly at depth.

▲ **Figure 11.5 How rock type influences the type of deformation** This structure is an example of *boudinage*, which forms when some portions of a rock body deform in a ductile fashion and others act as brittle units. The central greenish layer has been broken into a series of chunks, and the surrounding gray material has flowed into the gaps between the chunks. (Photo by Callan Bentley)

CONCEPT CHECKS 11.1

1. What is deformation? List several ways in which a rock body might change during deformation.
2. List the three types of differential stress and briefly describe the changes they can impart to rock bodies.
3. What type of plate boundary is most commonly associated with compressional stress?
4. How is strain different from stress?
5. How is brittle deformation different from ductile deformation?
6. List and describe the four factors that affect whether a rock deforms in a brittle or ductile manner.

Did You Know?

The combination of brittle and ductile behavior seen in the rock in Figure 11.5 resembles links of old-fashioned sausages in a string. Because of this resemblance, the structure is called "boudinage," from the French term *boudin*, meaning sausage.

11.2 Folds: Rock Structures Formed by Ductile Deformation

List and describe five types of folds.

Along convergent plate boundaries, rock strata are often bent into a series of wavelike undulations called **folds**. Folds in flat-lying strata are much like those that would form if you were to hold the ends of a sheet of paper on a flat surface and then push them together. Such folds come in a wide variety of sizes and configurations. Some are broad flexures in which strata hundreds of meters thick have been slightly warped. Others are very tight, even microscopic structures found in metamorphic rocks. Most folds result from *compressional stresses* that result in a lateral shortening and vertical thickening of the crust.

▶ **SmartFigure 11.6 Features associated with symmetrical folds** The axial plane divides a fold as symmetrically as possible, while the hinge line traces the points of maximum curvature of any layer.

CONDOR VIDEO
https://goo.gl/zWIRuV

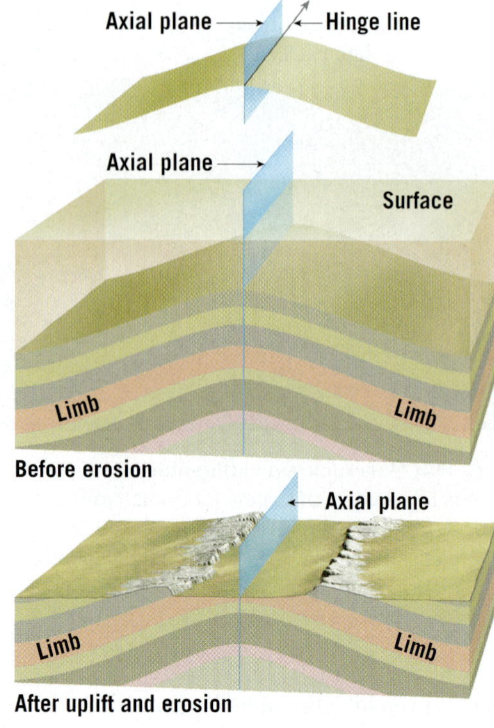

Folds are geologic structures that result when originally planar ("flat") surfaces, such as sedimentary strata, are permanently bent as a result of deformation. Each layer is bent around an imaginary axis called a *hinge line*, or simply a *hinge* (**Figure 11.6**).

Folds are also described by their *axial plane*, which is a surface that connects all the hinge lines of the folded strata. In simple folds, the axial plane is vertical and divides the fold into two roughly symmetrical *limbs*. However, the axial plane often leans to one side so that one limb is steeper than the other.

Anticlines & Synclines

The two most common types of folds are anticlines and synclines (**Figure 11.7**). **Anticlines** usually form by the upfolding, or arching, of sedimentary layers.* Typically found in association with anticlines are downfolds, or troughs, called **synclines**. Notice in Figure 11.7 that the limb of an anticline is also a limb of an adjacent syncline.

Depending on their orientation, these basic folds are described as *symmetrical* when the limbs are mirror images of each other and *asymmetrical* when they are not. The limbs of a symmetrical fold dip in opposite directions but at the same angle. The limbs of an asymmetrical fold dip in opposite directions but at different angles. An asymmetrical fold is said to be *overturned* if both limbs dip in the same direction, with one limb tilted beyond the vertical (see Figure 11.7). An overturned fold can also "lie on its side" so that the axial

*By strict definition, an anticline is a structure in which the oldest strata are found in the center, and a syncline is a structure where the youngest strata are found in the center.

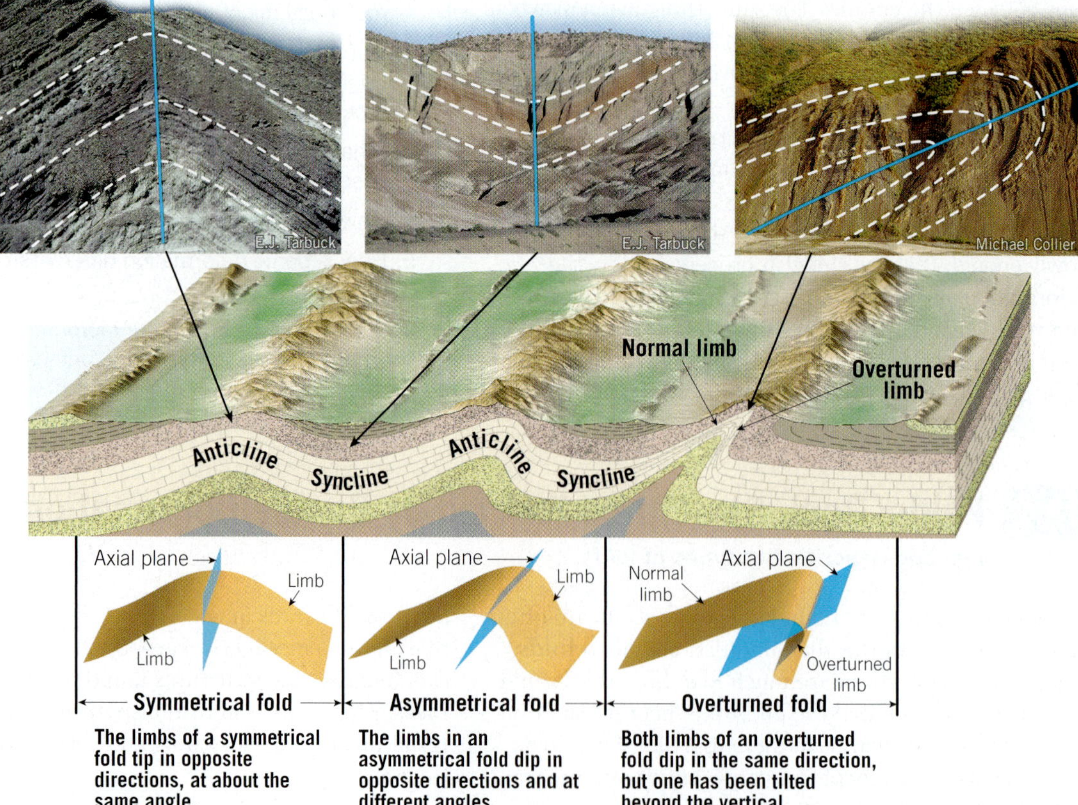

▶ **SmartFigure 11.7 Common types of folds** The arched structures—folds that go up in the middle—are *anticlines*. Their limbs dip away from each another. Folds that go down in the middle (limbs dip toward each other) are *synclines*. Notice that a limb of an anticline is also the limb of the adjacent syncline.

TUTORIAL
https://goo.gl/EwDKUK

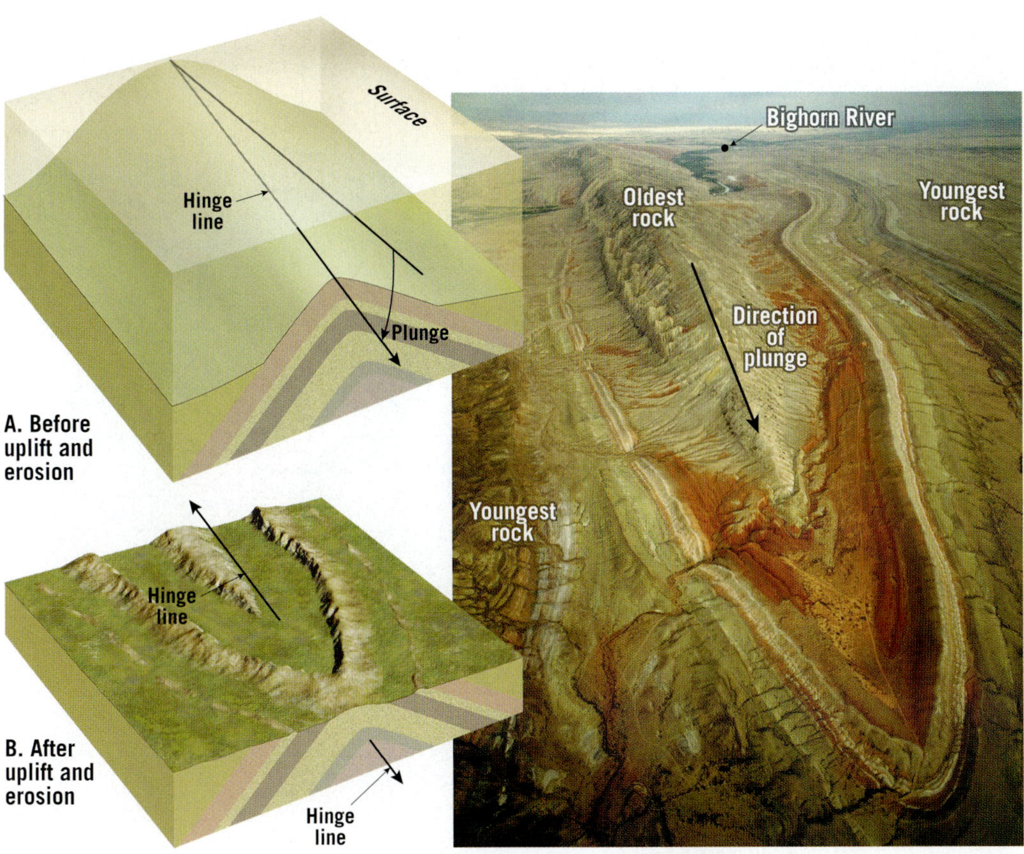

◀ **SmartFigure 11.8 Sheep Mountain, Wyoming** Eroded plunging anticlines, like Sheep Mountain, have an outcrop pattern that "points" in the direction of plunge. (Photo by Michael Collier)

MOBILE FIELD TRIP
https://goo.gl/7LyhfN

plane is horizontal. These *recumbent* folds are common in highly deformed mountain belts such as the Alps.

Folds can also be tilted by tectonic forces so their hinge lines slope downward (**Figure 11.8A**). Folds of this type are said to *plunge* because the hinge lines penetrate Earth's surface. Sheep Mountain, Wyoming, is an example of a plunging anticline (**Figure 11.8C**). Notice in **Figure 11.8B** that a V-shaped pattern is produced when erosion removes the upper layers of a plunging fold and exposes its interior. In the case of an anticline, such as Sheep Mountain, the tip of the V points in the direction of plunge (see Figure 11.8C); the opposite is true for a syncline.

A good example of the topography that can result when erosional forces attack folded sedimentary strata is found in the Valley and Ridge Province of the Appalachians (see Figure 11.32). It is important to realize that anticlines typically do not show up as ridges, nor do synclines show up as valleys. Rather, ridges and valleys result from differential weathering and erosion. For example, in the Valley and Ridge Province, resistant sandstone beds remain as imposing ridges separated by valleys that are cut into more easily eroded shale or limestone beds.

Domes & Basins

When a broad upwarping of basement rock deforms the overlying cover of sedimentary strata to produce a circular or slightly elongated bulge, the feature is called a **dome** (**Figure 11.9A**). The Black Hills of western South Dakota represent a large structural dome generated by upwarping. Here erosion has stripped away the highest portions of the overlying sedimentary beds, exposing older igneous and metamorphic rocks in the center (**Figure 11.10**).

Structural domes can also be formed by the intrusion of magma (laccoliths), as shown in Figure 4.26, page 114. In addition, the upward migration of buried salt deposits can produce salt domes like those beneath the Gulf of Mexico. Salt domes are economically important rock structures because when salt migrates upward, the surrounding oil-bearing sedimentary strata deform to form oil reservoirs (see Figure 7.32, page 209).

The inverse of a dome is a downwarped structure termed a **basin** (**Figure 11.9B**). Several large structural

▼ **SmartFigure 11.9 Domes versus basins** Gentle upwarping and downwarping of crustal rocks produce (**A**) domes and (**B**) basins. Erosion of these structures results in an outcrop pattern that is roughly circular or a more elongated ellipse.

TUTORIAL
https://goo.gl/tD2CXw

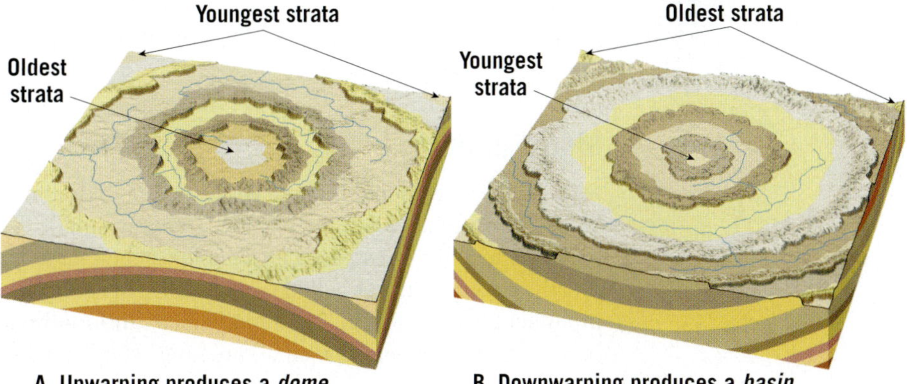

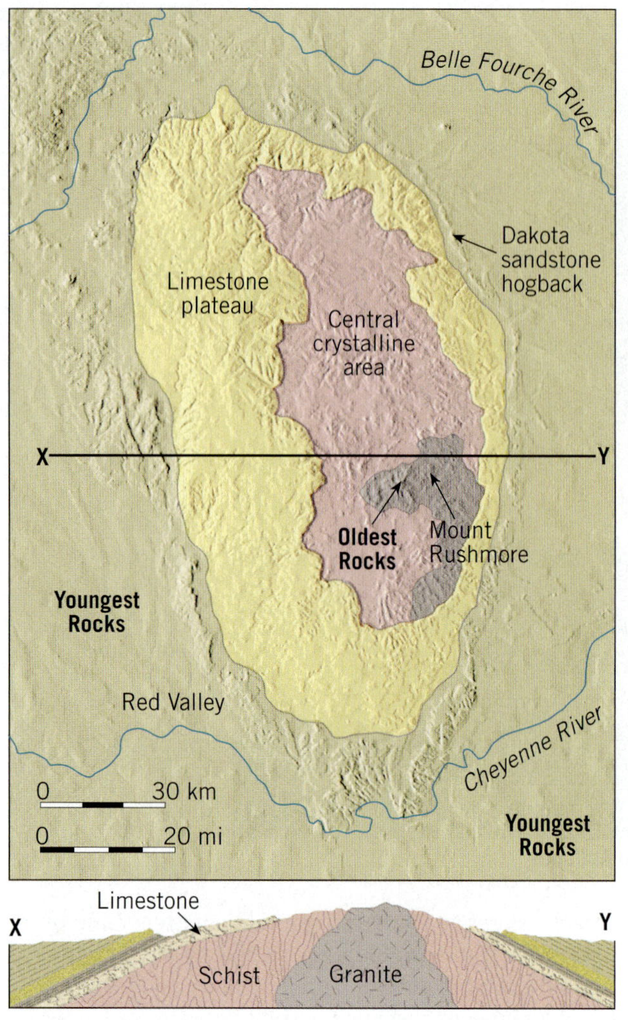

▶ **Figure 11.10 The Black Hills of South Dakota, a large structural dome** The central core of the Black Hills is composed of resistant Precambrian-age igneous and metamorphic rocks. The surrounding rocks are mainly younger limestones and sandstones.

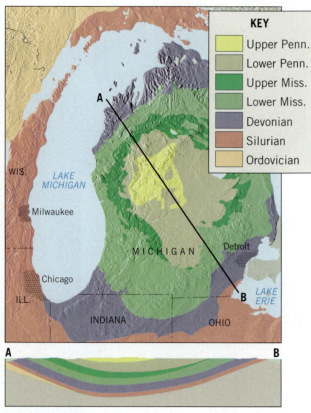

▲ **Figure 11.11 The bedrock geology of the Michigan basin** The youngest rocks are centrally located, and the oldest beds flank this structure.

basins exist in the United States (**Figure 11.11**). The basins of Michigan and Illinois have gently sloping beds similar to saucers. These basins are thought to have resulted from large accumulations of sediment, whose weight caused the crust to subside (see the section "The Principle of Isostasy" later in this chapter). A few structural basins may have resulted from giant meteorite impacts.

Because structural basins usually contain sedimentary beds sloping at low angles, they are identified by the age of the rocks composing them. The youngest rocks are found near the center, and the oldest rocks are at the flanks. By contrast, in a dome, such as the Black Hills, the oldest rocks form the core.

Monoclines

Although we have separated our discussion of folds and faults, they often actually occur together as a result of the same tectonic stresses. One example can be found in broad, regional features called *monoclines*. Particularly prominent features of the Colorado Plateau, **monoclines** (*mono* = one, *kleinen* = incline) are large, steplike folds in otherwise horizontal sedimentary strata (**Figure 11.12**). These folds appear to have resulted from the reactivation of ancient, steep-dipping reverse faults located in basement rocks beneath the plateau. As large blocks of basement rock were displaced upward, the comparatively ductile sedimentary strata above responded by draping over the fault like clothes hanging over a bench. Displacement along these reactivated faults can exceed 1 kilometer (0.6 miles).

CONCEPT CHECKS 11.2

1. Distinguish between anticlines and synclines, between domes and basins, and between anticlines and domes.
2. Draw a cross-sectional view of a symmetrical anticline. Include a line to represent the axial plane and label both limbs.
3. The Black Hills of South Dakota is a good example of what type of geologic structure?
4. Where are the youngest rocks in a structural basin found: near the center or near the margin?
5. Describe how a monocline forms.

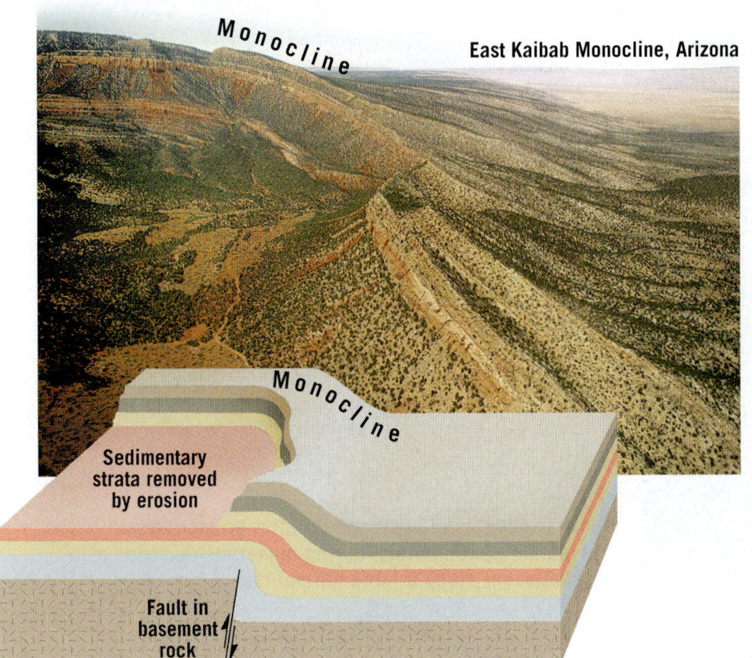

◀ **SmartFigure 11.12 The East Kaibab Monocline, Arizona** This monocline consists of bent sedimentary beds that were deformed by faulting in the bedrock below. The thrust fault does not reach the surface. Other examples of monoclines found on the Colorado Plateau include the Waterpocket Fold and the San Rafael Swell. The inclined strata once extended over the sedimentary layers now exposed at the surface—evidence that a tremendous volume of rock has been eroded from this area.

CONDOR VIDEO
https://goo.gl/dPwXf4

11.3 Faults & Joints: Rock Structures Formed by Brittle Deformation

Sketch and briefly describe the relative motion of rock bodies located on opposite sides of normal, reverse, and thrust faults as well as both types of strike-slip faults.

Faults and joints are both structures that form where brittle deformation leads to fracturing of Earth's crust. A *joint* is simply a fracture, whereas a **fault** is a fracture along which motion has occurred, so that the rocks on either side are offset from each other. Figure 11.13 shows a small fault revealed in a road cut, where the sedimentary beds have been offset by a few meters. Faults of this scale usually occur as single discrete breaks. By contrast, large faults, like the San Andreas Fault in California, have displacements of hundreds of kilometers and consist of many interconnecting fault surfaces. These structures, described as *fault zones*, can be several kilometers wide and are often easier to identify from aerial photographs than at ground level. Sudden movements along faults cause most earthquakes. However, the vast majority of faults are remnants of past deformation and are inactive.

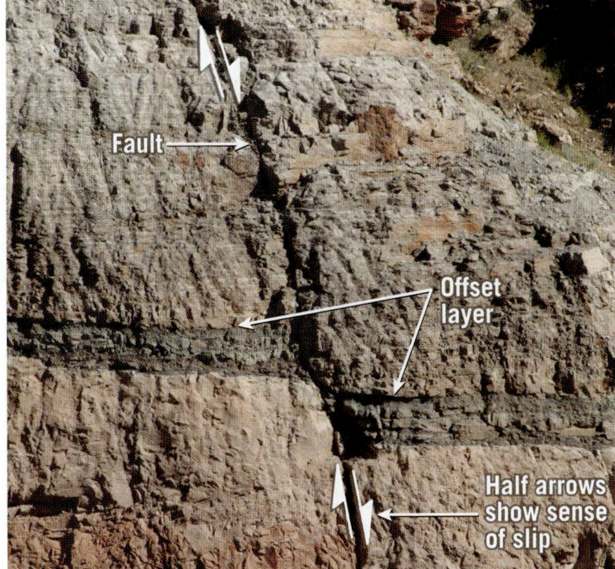

▲ **SmartFigure 11.13 Faults are fractures where slip has occurred** (Photo by E. J. Tarbuck)

CONDOR VIDEO
https://goo.gl/rCweMh

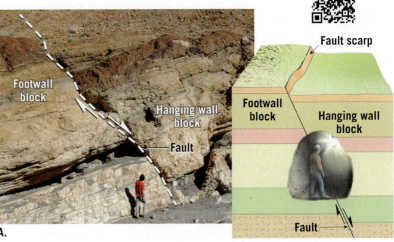

Dip-Slip Faults

Faults in which movement is primarily parallel to the slope of the fault surface are called **dip-slip faults**. ("Dip" refers to the angle at which the fault surface is inclined relative to the horizontal.) Geologists identify the rock surface immediately above the fault as the **hanging wall block** and the rock surface below as the **footwall block** (Figure 11.14). These names were first

▼ **SmartFigure 11.14 Hanging wall block and footwall block** The rock immediately above a fault surface is the *hanging wall block*, and the one below is called the *footwall block*. These terms were coined by miners who excavated ore deposits that formed along fault zones. The miners hung their lanterns on the rocks above the fault trace (hanging wall block) and walked on the rocks below the fault trace (footwall block).

ANIMATION
https://goo.gl/qkD1uL

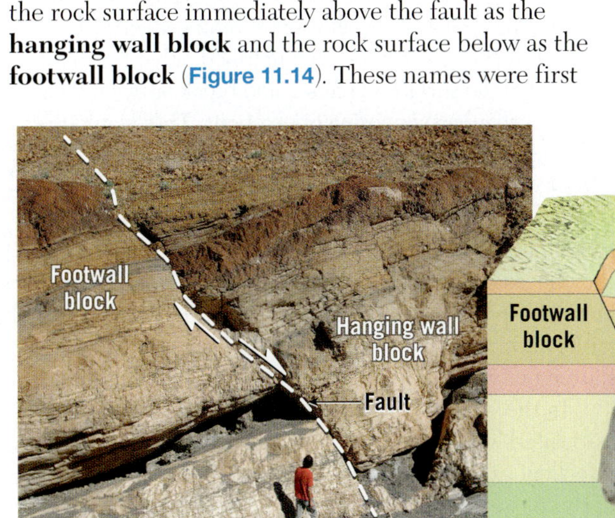

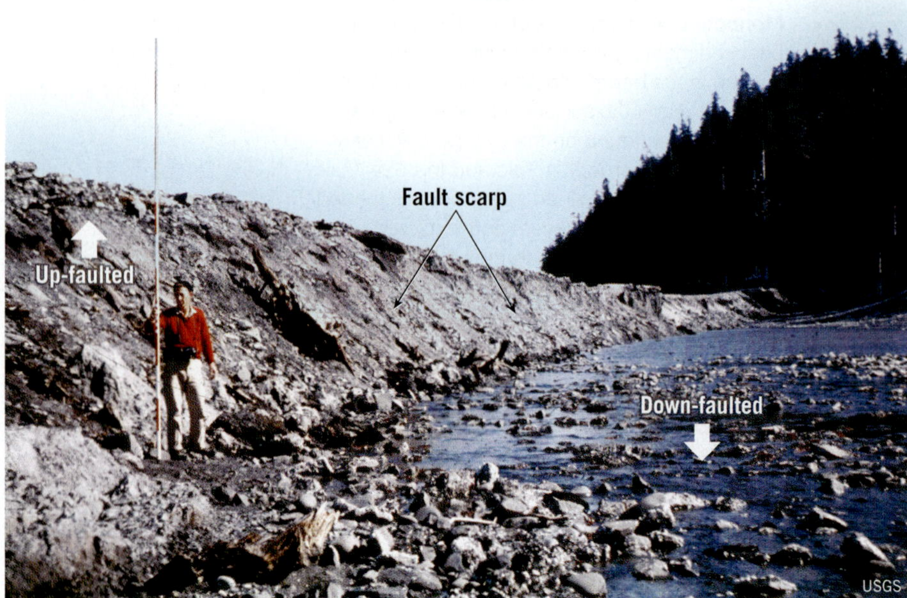

▲ **Figure 11.15 Fault scarp** This fault scarp was created during the Alaska earthquake of 1964. (Photo courtesy of the USGS)

used by prospectors and miners who excavated metallic ore deposits such as gold that had precipitated from hydrothermal solutions along inactive fault zones. The miners would walk on the rocks below the mineralized fault zone (*the footwall block*) and hang their lanterns on the rocks above (*the hanging wall block*).

Vertical displacements along dip-slip faults tend to produce long, low cliffs called **fault scarps**. Fault scarps, such as the one shown in Figure 11.15, are produced by rapid vertical slips that generate earthquakes.

There are two kinds of dip-slip faults: *normal faults* and *reverse faults*.

A. Idealized block diagram showing a normal fault in which the hanging wall block has moved down relative to the footwall block.

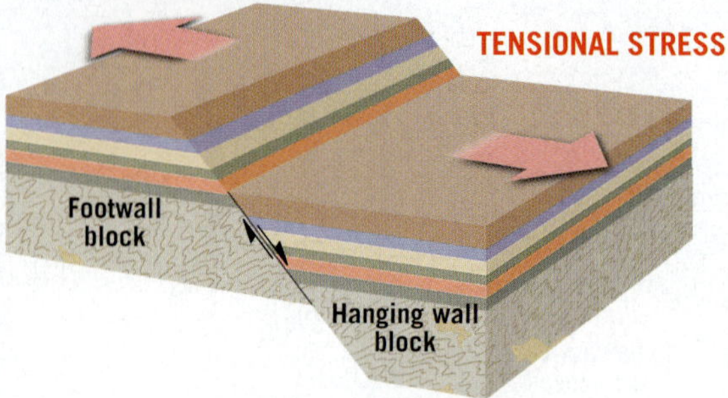

B. Landscape as it might appear after erosion has carved the uplifted block during numerous episodes of displacement along a large normal fault.

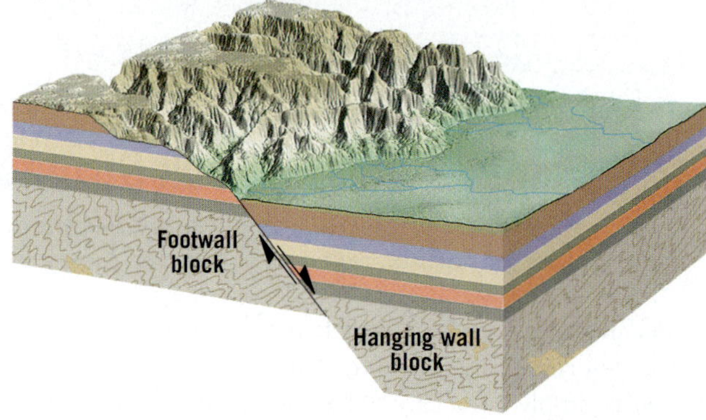

▲ **SmartFigure 11.16 Normal dip-slip fault** The upper diagram illustrates the relative displacement that occurs between the blocks on either side of a fault. The lower diagram shows how erosion may alter the up-faulted blocks.

TUTORIAL
https://goo.gl/eFkg3y

Normal Faults When the hanging wall block moves down relative to the footwall block, dip-slip faults are classified as **normal faults** (Figure 11.16). Normal faults are associated with tensional stresses that pull rock units apart, thereby lengthening the crust laterally and thinning it vertically. This "pulling apart" can be accomplished either by uplift that causes the surface to stretch and break or by horizontal forces that have opposing orientations. These faults are called normal faults because it is "normal" for gravity to pull a block of rock down an inclined plane (the fault surface).

Normal faults occur in a variety of sizes. Some are small, having displacements of only a meter or so, like the one shown in the road cut in Figure 11.13. Others, however, extend for tens of kilometers. Most large normal faults have relatively steep dips at the surface but tend to flatten out with depth.

In the western United States, large normal faults are associated with structures called **fault-block mountains**. Excellent examples of fault-block mountains are found in the Basin and Range Province, a region that encompasses Nevada and portions of the surrounding states (Figure 11.17). Here the crust has been elongated and broken to create more than 200 relatively small mountain ranges. Averaging about 80 kilometers (50 miles) in length, the ranges rise 900 to 1500 meters (3000 to 5000 feet) above the adjacent down-faulted topographic basins.

The topography of the Basin and Range Province evolved in association with a system of normal faults trending roughly north–south. Movements along these faults produced alternating uplifted fault blocks called **horsts** (*horst* = hill) and down-dropped blocks called **grabens** (*graben* = ditch). Horsts form the ranges and are the source of sediments that have accumulated in the topographic basins created by the grabens. As Figure 11.17 illustrates, structures called **half-grabens**, which are tilted fault blocks, also contribute to the alternating topographic highs and lows in the Basin and Range Province.

Also notice in Figure 11.17 that the slopes of many of the large normal faults in the Basin and Range Province decrease with depth and eventually join to form a low-angle, nearly horizontal fault called a **detachment fault**. These faults represent a major boundary between the rocks below, which exhibit ductile deformation, and the rocks above, which exhibit mainly brittle deformation.

Did You Know?
Many faults do not reach the surface and hence do not produce a fault scarp during an earthquake. Faults of this type are termed *blind*, or *hidden faults*. The 1994 Northridge earthquake in California occurred along a blind thrust fault, as did the 2011 Mineral, Virginia, earthquake.

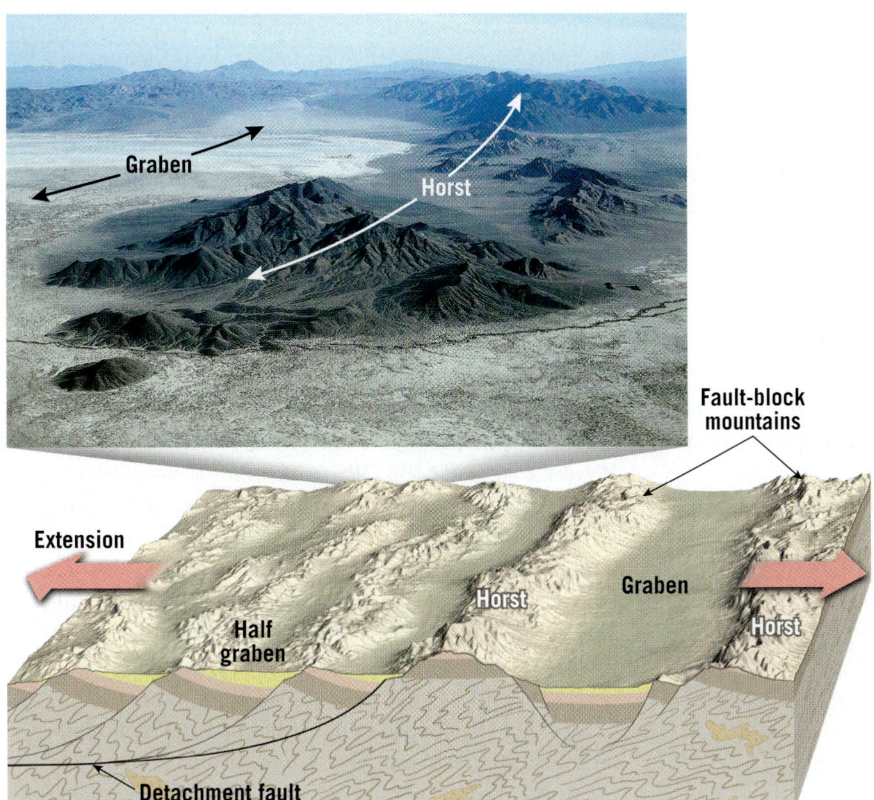

◀ **SmartFigure 11.17 Normal faulting in the Basin and Range Province** Here, tensional stresses have elongated and fractured the crust into numerous blocks. Movement along these faults has tilted the blocks, producing parallel mountain ranges called *fault-block mountains*. The down-faulted blocks (*grabens*) form basins, whereas the up-faulted blocks (*horsts*) erode to form rugged mountainous topography. In addition, numerous tilted blocks (*half-grabens*) form both basins and mountains. (Photo by Michael Collier)

MOBILE FIELD TRIP
https://goo.gl/N7B9bZ

Reverse & Thrust Faults Dip-slip faults in which the hanging wall block moves up relative to the footwall block are called **reverse faults** (Figure 11.18). A **thrust fault** is a type of reverse fault in which the dip is less than 45 degrees. Reverse and thrust faults result from compressional stresses that produce horizontal shortening of the crust.

Most high-angle reverse faults are small and accommodate local displacements in regions dominated by other types of faulting. Thrust faults, on the other hand, exist at all scales, with some large thrust faults having displacements ranging from tens to hundreds of kilometers. Movement along a thrust fault can cause the hanging wall block to be thrust nearly horizontally over the footwall block, as shown in Figure 11.19.

Thrust faulting is most pronounced along convergent plate boundaries. Compressional forces associated with colliding plates generally create folds as well as thrust faults that thicken and shorten the crust to produce mountainous topography (see Figure 11.29). Examples of mountainous belts produced by this type of compressional tectonics include the Alps, Northern Rockies, Himalayas, and Appalachians.

Strike-Slip Faults

A fault in which the dominant displacement is horizontal and parallel to the *trend* (direction) of the fault surface

A. Idealized block diagram showing a reverse fault in which the hanging wall block has moved up relative to the footwall block.

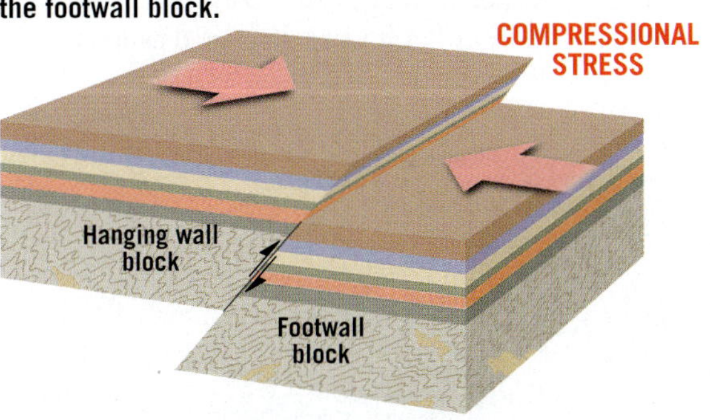

B. Landscape as it might appear after erosion has carved the uplifted block during numerous episodes of displacement along a large reverse fault.

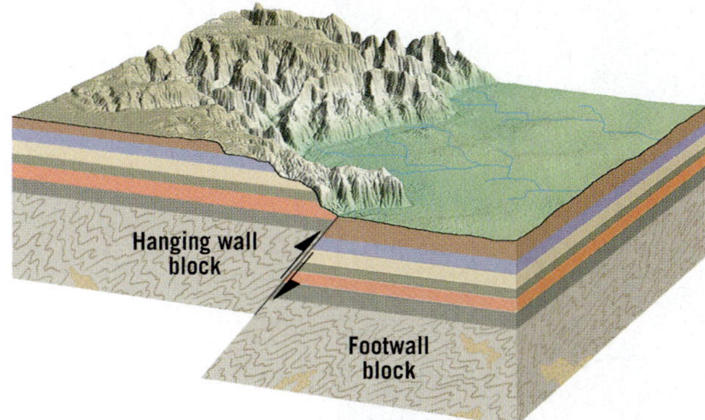

◀ **SmartFigure 11.18 Reverse faults** Reverse faults are generated by compressional stresses that force one block of rock over another.

ANIMATION
https://goo.gl/OvmqP2

▶ **SmartFigure 11.19**
Thrust fault Thrust faults are a type of reverse fault having dips less than 45 degrees.

ANIMATION
https://goo.gl/lejdNe

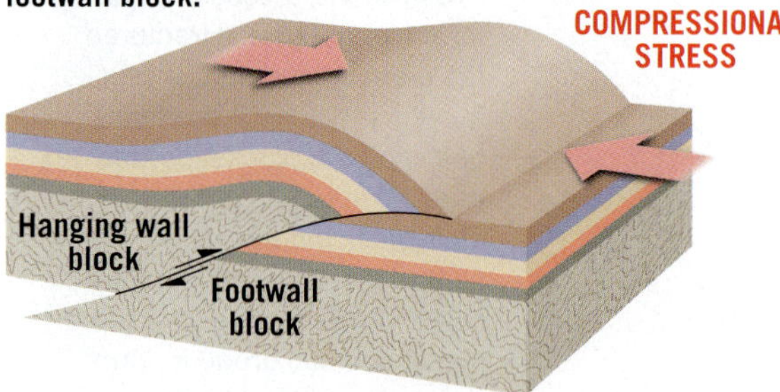

A. Idealized block diagram showing a thrust fault in which the hanging wall block has moved up relative to the footwall block.

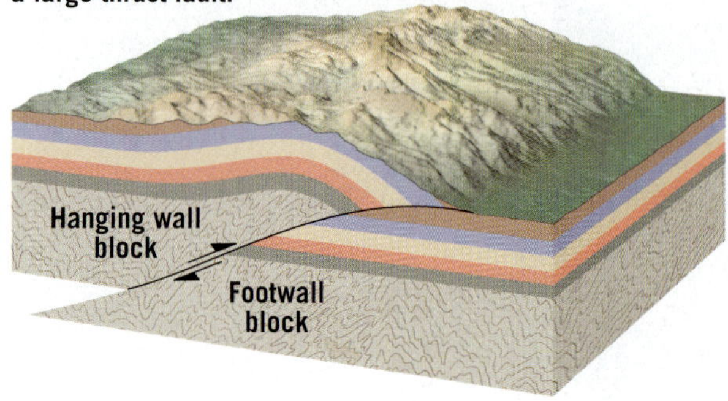

B. Landscape as it might appear after erosion has carved the uplifted block during numerous episodes of displacement along a large thrust fault.

Did You Know?
The lake that is the home of the legendary Loch Ness Monster is located along the Great Glen Fault that divides northern Scotland. Strike-slip movement along this fault zone shattered a wide zone of rock, which glacial erosion subsequently removed, to produce the elongated valley in which Loch Ness is situated. Broken rock is easily eroded, a fact that accounts for the great depth of Loch Ness—as much as 600 ft below sea level.

is called a **strike-slip fault** (Figure 11.20). The earliest scientific records of strike-slip faulting were made following surface ruptures that produced large earthquakes. One of the most noteworthy of these was the great San Francisco earthquake of 1906. During this strong earthquake, structures such as fences and roads that were built across the San Andreas Fault were displaced as much as 4.7 meters (15 feet). Because movement along the San Andreas Fault causes the crustal block on the opposite side of the fault to move to the right as you face the fault, it is called a *right-lateral* strike-slip fault.

The Great Glen Fault in Scotland, which exhibits displacement in the opposite direction, is a well-known example of a *left-lateral* strike-slip fault. Associated with the crushed-up rock along the Great Glen Fault trace are numerous lakes, including the legendary Loch Ness.

As discussed in Chapter 2, **transform faults** are a unique class of strike-slip faults that slice through Earth's lithosphere and accommodate motion between two tectonic plates. Most transform faults cut the oceanic lithosphere and link spreading oceanic ridges (see Figure 2.19, page 48). Others accommodate displacement between continental blocks that slip horizontally past each other. Some of the best-known transform faults include California's San Andreas Fault, New Zealand's Alpine Fault, the Middle East's Dead Sea Fault, and Turkey's North Anatolian Fault (see Figure 11.20). Large transform faults like these accommodate relative displacements of up to several hundred kilometers.

Rather than being a single fracture, most continental transform faults consist of a zone of roughly parallel fractures. While this zone may be up to several kilometers wide, the most recent movement is often along a strand only a few meters wide, which may offset features such as stream channels. Crushed and broken rocks produced during faulting are more easily eroded, so depressions such as linear valleys often mark the locations of large strike-slip faults.

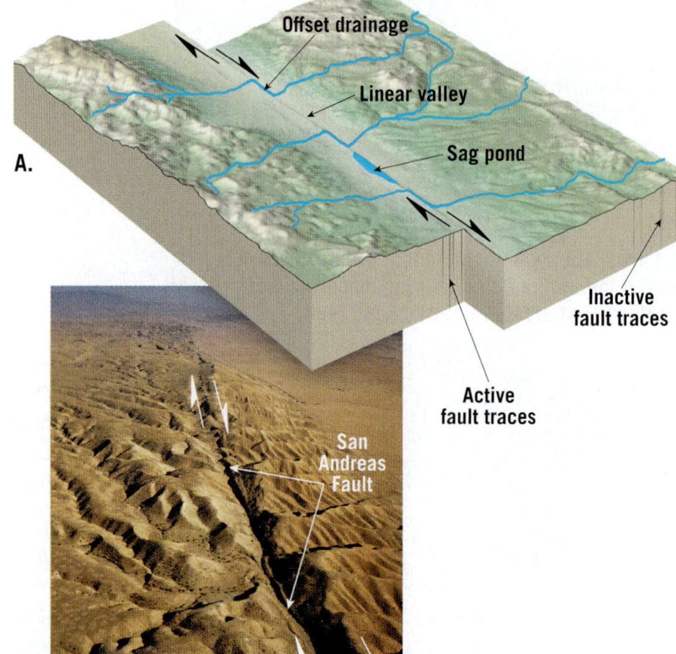

▶ **Figure 11.20**
Strike-slip faults
A. The block diagram illustrates the features associated with large strike-slip faults. Notice how the stream channels have been offset by fault movement. **B.** Aerial view of the San Andreas Fault. (Photo by D. Parker/Science Source)

Joints

Among the most common rock structures are fractures called **joints**. As mentioned earlier, joints differ from faults because no appreciable displacement has occurred along the fracture. Although some joints have a random orientation, most occur in roughly parallel groups.

We have already considered two types of joints. Recall from Chapter 4 that *columnar joints* form when igneous rocks cool and develop shrinkage fractures that produce elongated, pillarlike columns (see Figure 4.29, page 116). In addition, recall from Chapter 6 that sheeting produces a pattern of gently curved joints that develop more or less parallel to the surface of large exposed igneous bodies such as batholiths. Here the jointing results from the gradual expansion that occurs when erosion removes the overlying load and reduces pressure on the rock (see Figure 6.5, page 164).

Most joints are produced when rocks in the outermost crust are deformed, causing the rock to fail by brittle fracture. Extensive joint patterns often develop in response to relatively subtle and often barely perceptible regional upwarping and downwarping of the crust.

Many rocks are broken by two or even three sets of intersecting joints that slice the rock into numerous regularly shaped blocks. These joint sets often exert a strong

◀ **Figure 11.21 Joint control of landforms** This sea arch in northern Scotland displays a “stair-step” weathering pattern because it breaks more or less equally along planes of bedding and two sets of parallel joints. One of the joint sets is vertical and forms the side of the arch facing the camera; the other is perpendicular to that, highlighted with the red lines. (Photo by Iain Sarjeant/Getty Images)

influence on other geologic processes (**Figure 11.21**). For example, chemical weathering tends to be concentrated along joints, and joint patterns influence how groundwater moves through the crust (**Figure 11.22**). Because jointing weakens rocks, highly jointed rocks present a risk to construction projects, including bridges, highways, and dams.

On June 5, 1976, the Teton Dam in Idaho failed, taking 14 lives and causing nearly $1 billion in property damage. This earthen dam, constructed of easily eroded clays and silts, was situated on highly fractured volcanic rocks. Although attempts were made to fill the voids in the jointed rock, water gradually penetrated the subsurface fractures and undermined the dam's foundation. Eventually the moving water cut a tunnel into the easily erodible clays and silts. Within minutes the dam failed, sending a 20-meter- (65-foot-) high wall of water down the Teton and Snake Rivers.

Jointed rocks can be beneficial to humans. Some of the world's largest and most important mineral deposits are located along joint systems. Hydrothermal solutions (mineralized fluids) can migrate into fractured host rocks and precipitate economically significant amounts of copper, silver, gold, zinc, lead, and uranium.

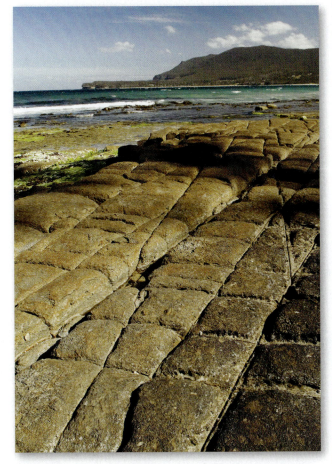

▲ **Figure 11.22 Joint patterns influence how groundwater moves through the crust** As a result, chemical weathering tends to be concentrated along joints. This image shows the joint pattern called Tessellated Pavement, located at Eaglehawk Neck, Tasmania, Australia. (Photo provided by THPStock/Shutterstock)

CONCEPT CHECKS 11.3

1. Contrast the movements that occur along normal and reverse faults. What type of stress is responsible for each kind of fault?
2. What type of fault is associated with fault-block mountains?
3. How are reverse faults different from thrust faults? In what way are they similar?
4. Describe the relative movement along a strike-slip fault.
5. How are joints different from faults?

11.4 Mountain Building

Locate and name Earth's major mountain belts on a world map.

Mountain building has occurred in the recent geologic past at several locations around the world. Young mountain belts include the American Cordillera (*cordillera* means "spine" or "backbone"), which runs along the western margin of the Americas from southernmost South America to Alaska and includes both the Rockies and the Andes; the Alpine–Himalaya chain, which extends along the margin of the Mediterranean, through Iran to northern India and into Indochina; and the mountainous terrains of the western Pacific, which include volcanic island arcs like Japan, the Philippines, and much of Indonesia. Most of these young mountain belts have formed in the past 100 million years (Figure 11.23). Some, including the Himalayas, began their growth as recently as 50 million years ago.

In addition to these young mountain belts, there are several chains of Paleozoic-age mountains on Earth. Although these older mountain belts are deeply eroded and topographically less prominent, they exhibit the same structural features found in younger mountains. The Appalachians in the eastern United States and the Urals in Russia are classic examples of this group of older, well-worn mountain belts.

The term for the processes that collectively produce a mountain belt is **orogenesis** (*oros* = mountain, *genesis* = to come into being). An episode of mountain building is called an **orogeny**. Most major mountain belts display striking visual evidence of great compressional forces that have shortened the crust horizontally while thickening it vertically. These **collisional mountains** usually result from the collision of one or more small crustal fragments to a continental margin or from the closure of an ocean basin that results in the collision of two major landmasses. As a consequence, collisional mountains contain large quantities of preexisting sediments and sedimentary rocks that at one time lay along the margin of a continent and were subsequently faulted and contorted into a series of folds. Although folding and thrust faulting are often the most conspicuous signs of orogenesis, varying degrees of metamorphism and igneous activity are always present.

The theory of plate tectonics provides a model for orogenesis with excellent explanatory power; it accounts for the origin of virtually all the present mountain belts and most of the ancient ones. According to this model, the tectonic processes that generate Earth's major mountainous terrains occur along convergent plate boundaries. We will next revisit the nature of convergent plate boundaries and then examine how the process of subduction has driven mountain building around the globe.

Did You Know?
During Charles Darwin's famous voyage on the H.M.S. *Beagle* (1831–1836), he experienced a strong earthquake that accompanied sudden uplift of the land around the Bay of Concepciòn in Chile. From this and other observations, Darwin concluded that the uplift he witnessed "marked one step in the elevation of a mountain chain."

CONCEPT CHECKS 11.4

1. Define *orogenesis*.
2. Which type of plate boundary is most directly associated with Earth's major mountain belts?

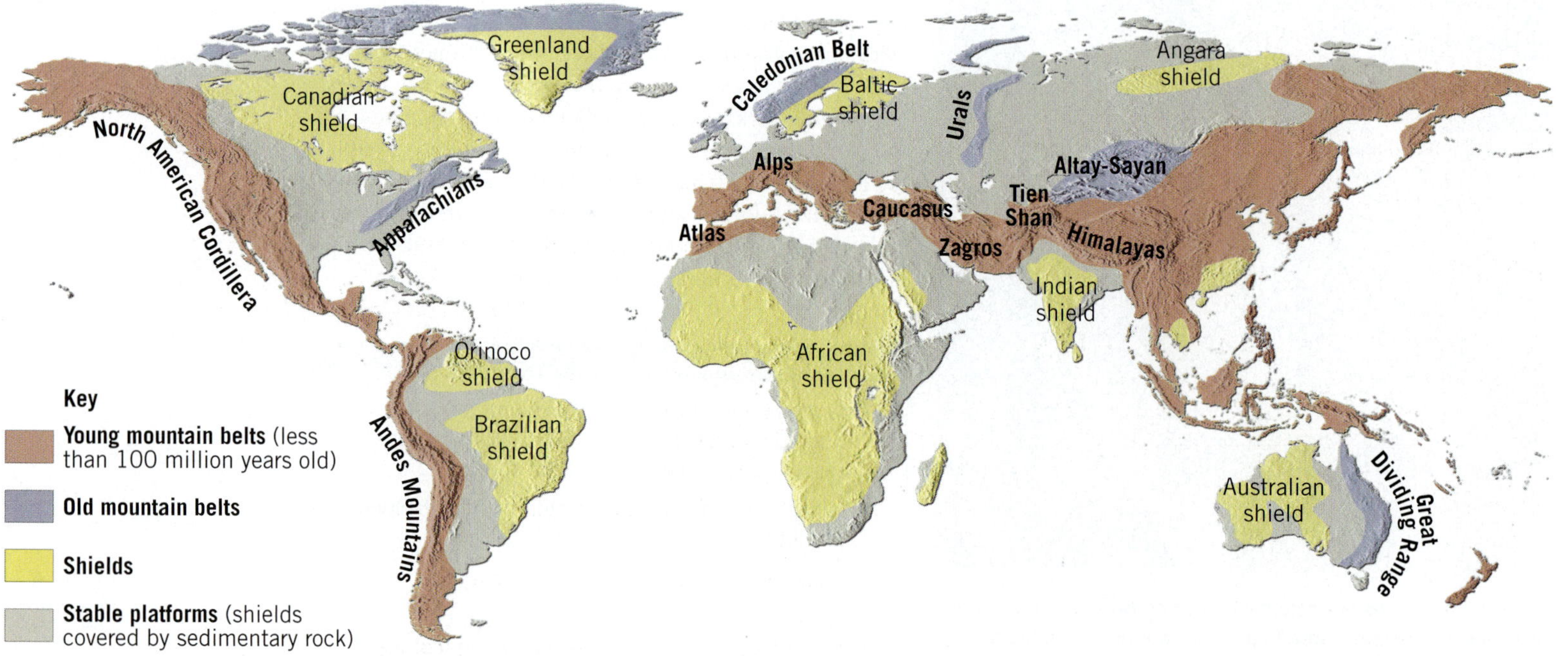

▲ Figure 11.23 **Earth's major mountain belts** Notice the east–west trend of major mountain belts in Eurasia in contrast to the north–south trend of the North and South American Cordilleras.

11.5 Subduction & Mountain Building

Sketch a cross section of an Andean-type mountain belt and describe how its major features are generated.

Where oceanic lithosphere subducts beneath oceanic lithosphere, a *volcanic island arc* and related tectonic features develop. Subduction of oceanic beneath continental lithosphere, on the other hand, results in the formation of a *continental volcanic arc* and mountainous topography along the margin of a continent. Acting like a conveyor belt, oceanic lithosphere may also bring volcanic island arcs and other crustal fragments to a subduction zone. These crustal elements are generally too buoyant to subduct, so they become welded to the overriding plate, which may be another small crustal fragment or a continent. If subduction continues long enough, it can ultimately lead to the closure of an ocean basin and the ensuing collision of two continents.

Island Arc–Type Mountain Building

Island arcs result from the steady subduction of oceanic lithosphere under oceanic lithosphere, which may continue for 200 million years or more. Periodic volcanic activity, the emplacement of igneous plutons at depth, and the accumulation of sediment that is scraped from the subducting plate gradually increase the volume of crustal material capping the upper plate (**Figure 11.24**). Some large volcanic island arcs, such as Japan, owe their size to having been built on fragments of continental crust that have rifted from a large landmass or to the joining of multiple island arcs over time.

The continued growth of a volcanic island arc can generate mountainous topography consisting of nearly parallel belts of igneous and metamorphic rocks. This activity, however, is viewed as just one phase in the development of Earth's major mountain belts. As you will see later, some volcanic arcs are carried by subducting plates to the margin of large continental blocks, where they become involved in large-scale mountain-building episodes.

Continuous subduction of oceanic lithosphere results in the development of thick units of continental-like crust on the overlying plate.

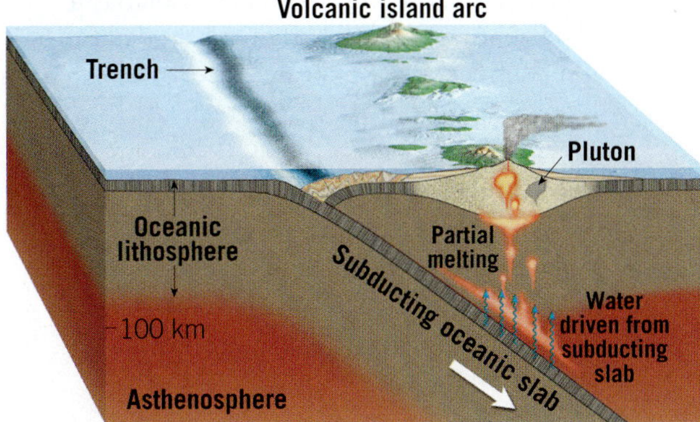

▲ **Figure 11.24 Development of a volcanic island arc** A volcanic island arc forms where one slab of oceanic lithosphere is subducted beneath another slab of the same material.

Andean-Type Mountain Building

Andean-type mountain building is characterized by subduction beneath a continent rather than oceanic lithosphere, as in the Andes Mountains of South America. Subduction along these active continental margins is associated with long-lasting magmatic activity that builds continental volcanic arcs. The result is crustal thickening, with the crust reaching thicknesses of more than 70 kilometers (45 miles).

The first stage in the development of Andean-type mountain belts occurs along *passive continental margins*. The east coast of the United States provides a modern example of a passive continental margin where sedimentation has produced a thick platform of shallow-water sandstones, limestones, and shales (**Figure 11.25A**). At some point, the forces that drive plate motions change, and a subduction zone develops along the margin of the continent. This subduction zone may form because the oceanic lithosphere here is so old and dense that it begins to sink of its own accord. Alternatively, strong compressional forces may help to initiate subduction.

Building Volcanic Arcs Recall that as oceanic lithosphere descends into the mantle, increasing temperatures and pressures drive volatiles (mostly water and carbon dioxide) from the crustal rocks. These mobile fluids migrate upward into the wedge-shaped region of mantle between the subducting slab and upper plate. At a depth of about 100 kilometers (60 miles), these fluids reduce the melting point of hot mantle rock sufficiently to trigger partial melting (**Figure 11.25B**).

Partial melting of the ultramafic mantle rock peridotite generates magmas with basaltic (mafic) compositions. Because these newly formed basaltic magmas are less dense than the rocks from which they originated, they will buoyantly rise. In continental settings, basaltic magma often ponds beneath the less dense rocks of the crust. The hot basaltic magma may heat these overlying crustal rocks sufficiently to generate a silica-rich magma of intermediate and/or felsic (granitic) composition that can rise to form the continental volcanic arc characteristic of an Andean-type subduction zone.

Emplacement of Batholiths Because of its low density and great thickness, continental crust significantly impedes the ascent of molten rock. Consequently, much of the magma that intrudes Earth's crust never reaches

Did You Know?
The highest city in the world, La Paz, Bolivia, is located 3,640 m (11,910 ft) above sea level, more than twice as high as Denver, Colorado (the highest major city in the United States). Bolivia is still rising. Estimates for uplift rates in the central Andes over the past 5 million years range between 0.2 and 3 mm per year.

A. Passive continental margin with an extensive platform of sediments and sedimentary rocks.

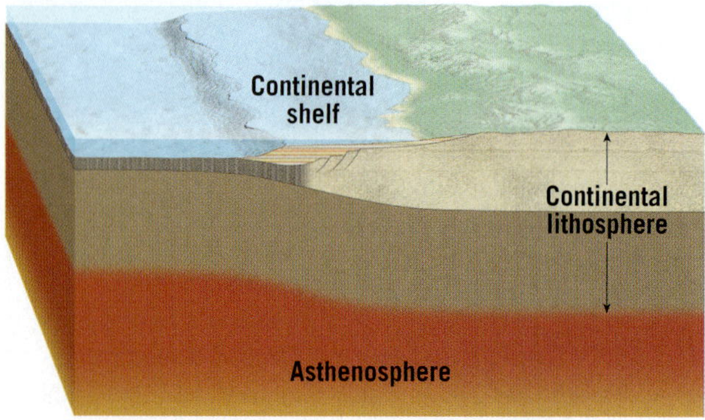

B. Plate convergence generates a subduction zone, and partial melting produces a continental volcanic arc. Compressional forces and igneous activity further deform and thicken the crust, elevating the mountain belt.

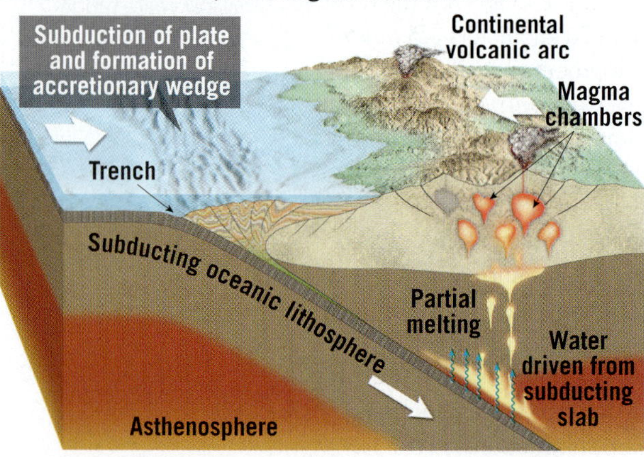

C. Subduction ends and is followed by a period of uplift.

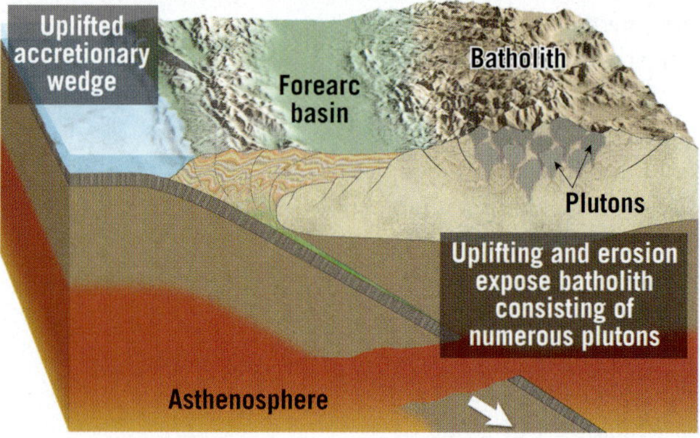

▲ **Figure 11.25 Andean-type mountain building**

the surface; instead, it crystallizes at depth to form massive collections of igneous plutons called *batholiths*. The result of this activity is thickening of Earth's crust.

Eventually, uplift and erosion exhume the batholiths. The American Cordillera contains several large batholiths, including the Sierra Nevada batholith of California, the Coast Range batholith of western Canada, and several large igneous bodies in the Andes. Most batholiths consist of intrusive igneous rocks that range in composition from granite to diorite.

Development of an Accretionary Wedge During the development of volcanic arcs, unconsolidated sediments that are carried on the subducting plate, as well as fragments of oceanic crust, may be scraped off and plastered against the edge of the overriding plate, like a wedge of soil scraped up by an advancing bulldozer. The resulting chaotic accumulation of deformed and thrust-faulted sediments and scraps of ocean crust is called an **accretionary wedge** (see Figure 11.25B).

Some of the sediments in an accretionary wedge are muds that accumulated on the ocean floor and were carried to the subduction zone by plate motion. Additional materials are derived from an adjacent continental volcanic arc and consist of volcanic debris and products of weathering and erosion.

Where sediment is plentiful, prolonged subduction may thicken a developing accretionary wedge so that it protrudes above sea level. This has occurred along the southern end of the Puerto Rico trench, where the Orinoco River basin of Venezuela is a major source of sediments. The resulting wedge emerges to form the island of Barbados.

Forearc Basins As an accretionary wedge thickens, it acts as a barrier to the movement of sediment from the volcanic arc to the trench. As a result, sediments begin to collect between the accretionary wedge and the volcanic arc. This region, which is composed of relatively undeformed layers of sediment and sedimentary rocks, is called a **forearc basin** (see Figure 11.25B,C). Subsidence and continued sedimentation in forearc basins can generate a sequence of nearly horizontal sedimentary strata that can attain thicknesses of several kilometers.

Sierra Nevada, Coast Ranges, & Great Valley

California's Sierra Nevada, Coast Ranges, and Great Valley are excellent examples of the tectonic structures that are typically generated along an Andean-type subduction zone. These structures were produced by the subduction of a portion of the Pacific basin (the Farallon plate) under the western margin of California (see Figure 11.25B). The Sierra Nevada batholith is a remnant of the continental volcanic arc that was produced by many intrusions of magma over a span of more than 100 million years. The Coast Ranges were built from the vast accumulation of sediments (accretionary wedge) that collected along the continental margin.

Beginning about 30 million years ago, subduction gradually ceased along much of the margin of North America, as the spreading center that produced the Farallon plate entered the California trench. The uplifting and erosion that followed removed most of the evidence of past volcanic activity and exposed the core of crystalline igneous and associated metamorphic rocks that make up the Sierra Nevada (Figure 11.25C). The Coast Ranges were uplifted only recently, as evidenced by the young, unconsolidated sediments that currently blanket portions of these highlands.

California's Great Valley is a remnant of the forearc basin that formed between the Sierra Nevada and the accretionary wedge and trench that lay offshore. Throughout much of its history, portions of the Great Valley lay below sea level. This sediment-laden basin contains thick marine deposits and debris eroded from the adjacent continental volcanic arc.

CONCEPT CHECKS 11.5

1. Compare and contrast mountain building at a volcanic island arc with mountain building at an Andean-style continental margin.
2. Describe and give an example of a passive continental margin.
3. In what ways are the Sierra Nevada and the Andes similar?
4. What is an accretionary wedge? Briefly describe its formation.
5. What is a batholith? In what tectonic setting are batholiths being generated?

11.6 Collisional Mountain Belts

Summarize the stages in the development of an Alpine-type mountain belt such as the Appalachians.

Most major mountain belts are generated when one or more buoyant crustal fragments collide with a continental margin as a result of subduction. Whereas oceanic lithosphere, which is relatively dense, readily subducts, continental lithosphere contains significant amounts of low-density crustal rocks and is therefore too buoyant to be subducted deeply or permanently. Consequently, the arrival of a crustal fragment at a trench results in a collision between the two continental blocks.

Cordilleran-Type Mountain Building

A Cordilleran-type orogeny, named after the North American Cordillera, is associated with a Pacific-like ocean—in that, unlike the Atlantic, the Pacific may never close. The rapid rate of seafloor spreading in the Pacific basin is balanced by a high rate of subduction. In this setting, island arcs and small crustal fragments are often carried along until they collide with an active continental margin and accrete (join) onto it. This process of collision and accretion has generated many of the mountainous regions that rim the Pacific. These accreted blocks of crust are called **terranes**. Geologists use this term to describe any crustal fragment that consists of a distinct and recognizable series of rock formations and has been transported and accreted by plate tectonic processes. Notice that "terrane" is a different word from "terrain"; the two are pronounced the same, but "terrain" refers to the shape of the surface topography, or "lay of the land."

The Nature of Terranes What is the nature of the crustal fragments that have become terranes? Some may have been **microcontinents** similar to the modern-day island of Madagascar, located east of Africa in the Indian Ocean. Many others were island arcs similar to Japan, the Philippines, and the Aleutian Islands. Still others may have been submerged oceanic plateaus created by massive submarine outpourings of basaltic lavas (Figure 11.26). More than 100 of these relatively small crustal fragments exist in the modern world.

Accretion & Orogenesis Small structures such as seamounts are generally subducted along with the descending oceanic slab. However, thick sections of oceanic crust, such as the Ontong Java Plateau (which is almost as big as Alaska) or an island arc dominated by low-density andesitic igneous rocks, are too buoyant to subduct. In these situations, a collision between the crustal fragment and the continental margin occurs.

The sequence of events that happen when small crustal fragments reach a Cordilleran-type margin is shown in Figure 11.27. The upper crustal layers are "peeled" from the descending plate and thrust in relatively thin sheets onto the adjacent continental block. Convergence does not generally end with the accretion of a crustal fragment. Rather, new subduction zones typically form seaward of the accreted terrane, and they can carry other island arcs or microcontinents toward a collision with the continental margin. Each collision displaces earlier accreted terranes further inland, adding to the zone of deformation as well as to the thickness and lateral extent of the continental margin.

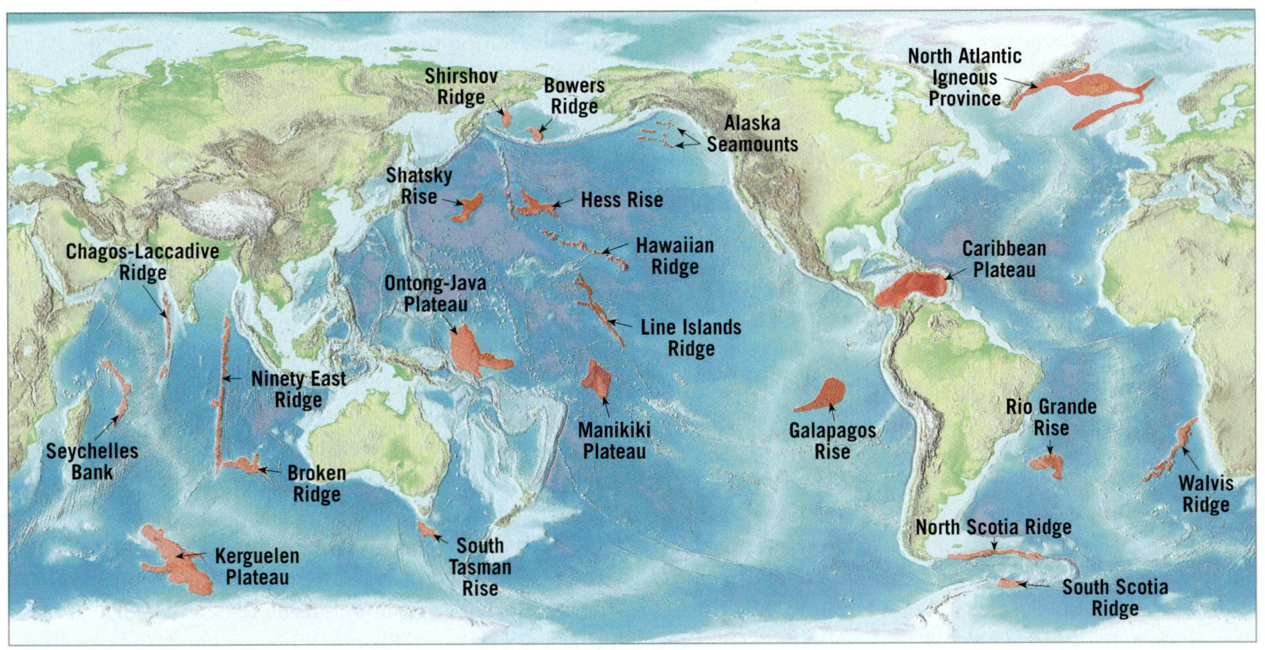

Figure 11.26 **Distribution of present-day oceanic plateaus and other submerged crustal fragments** These blocks of crust, shaded orange, could someday be accreted to continents as new terranes. (Data from Zvi Ben-Avraham and others)

A. A microcontinent and a volcanic island arc are being carried toward a subduction zone.

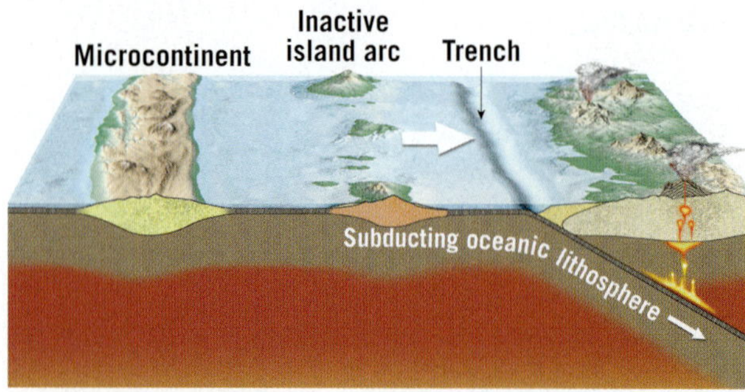

B. The volcanic island arc is sliced off the subducting plate and thrust onto the continent.

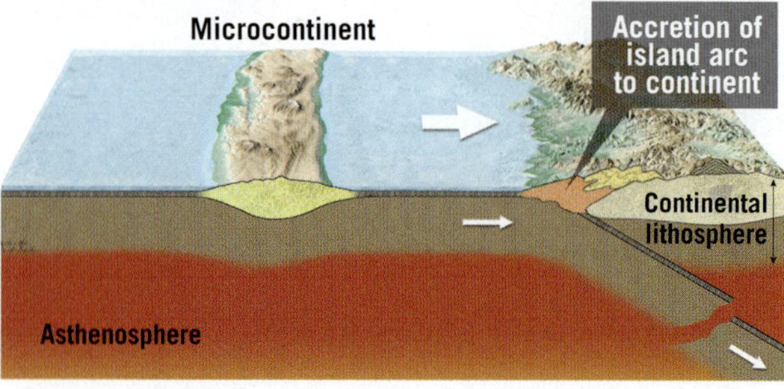

C. A new subduction zone forms seaward of the old subduction zone.

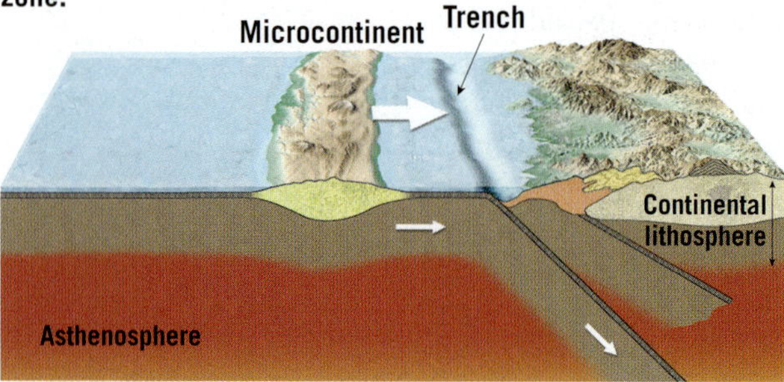

D. The accretion of the microcontinent to the continental margin shoves the remnant island arc further inland and grows the continental margin seaward.

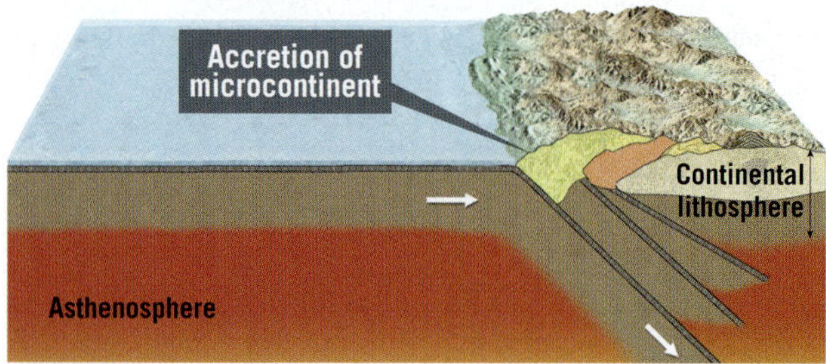

▲ SmartFigure 11.27 **Collision and accretion of small crustal fragments to a continental margin**

TUTORIAL
https://goo.gl/t8gK91

The North American Cordillera The correlation between mountain building and the accretion of crustal fragments was first developed in studies of the North American Cordillera (**Figure 11.28**). Researchers determined that some of the rocks in the orogenic belts of Alaska and British Columbia contained fossil and paleomagnetic evidence indicating that these strata previously lay much closer to the equator.

It is now known that many of the terranes that make up the North American Cordillera were scattered throughout the Pacific, like the island arcs and oceanic plateaus currently distributed in the western Pacific. During the breakup of Pangaea, the eastern portion of the Pacific basin (the Farallon plate) began to subduct under the western margin of North America. This activity resulted in many additions of crustal fragments along the entire Pacific margin of the continent—from Mexico's Baja Peninsula to northern Alaska (see Figure 11.27). Geologists expect that many modern microcontinents will likewise be accreted to active continental margins surrounding the Pacific, producing new orogenic belts.

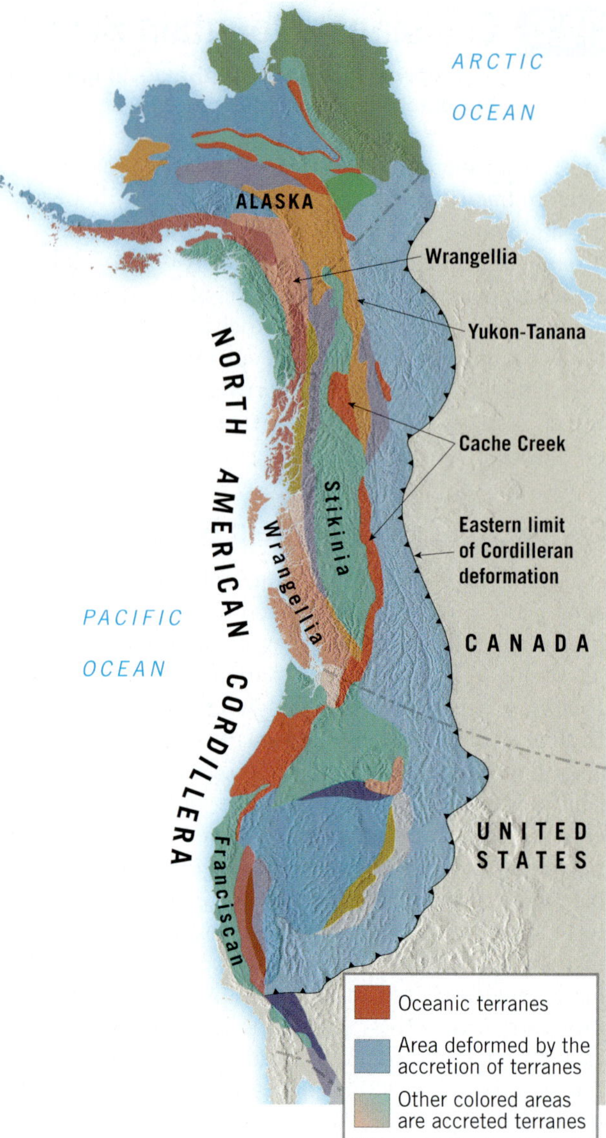

▲ SmartFigure 11.28 **Terranes that have been added to western North America during the past 200 million years** Paleomagnetic studies and fossil evidence indicate that some of these terranes originated thousands of kilometers to the south and west of their present locations. (After D. R. Hutchinson and others)

ANIMATION
https://goo.gl/GjDgFr

Alpine-Type Mountain Building: Continental Collisions

Alpine-type orogenies are episodes of mountain building that occur where two continental masses collide. They are named after the Alps, which have been intensively studied for more than 200 years. Mountain belts formed by the closure of major ocean basins include the Himalayas, Appalachians, Urals, and Alps. Continental collisions result in the development of mountains characterized by laterally shortened and vertically thickened crust, achieved through deformation such as folding and large-scale thrust faulting. Prior to the collision of the two large landmasses, this type of orogeny may also involve the accretion of smaller continental fragments or island arcs that occupied the ocean basin that once separated the two continental blocks.

The zone where two continents collide and are "welded" together is a **suture**. The same term can be used to describe the boundary between two adjacent accreted terranes. This portion of a mountain belt often preserves slivers of oceanic lithosphere that were trapped between the colliding plates. The unique structure of these pieces of oceanic lithosphere, called *ophiolites*, help identify the collision boundary.

Next, we will take a closer look at two examples of collisional mountains: the Himalayas and the Appalachians. The Himalayas, Earth's youngest collisional mountains, are still rising. By contrast, the Appalachians are a much older mountain belt, in which active mountain building ceased about 250 million years ago.

The Himalayas

The mountain-building episode that created the Himalayas began between 50 and 30 million years ago, when India began to collide with Asia. Prior to the breakup of Pangaea, India was located between Africa and Antarctica in the Southern Hemisphere. As Pangaea fragmented, India moved rapidly, geologically speaking, a few thousand kilometers in a northward direction.

The subduction zone that facilitated India's northward migration was near the southern margin of Asia (**Figure 11.29A**). Continued subduction along Asia's margin created an Andean-type plate margin that contained a well-developed continental volcanic arc and an accretionary wedge. India's northern margin, on the other hand, was a passive continental margin consisting of a thick platform of shallow-water sediments and sedimentary rocks.

Geologists have determined that two or perhaps more small crustal fragments were positioned on the subducting plate somewhere between India and Asia. During the closing of the intervening ocean basin, a small crustal fragment, which now forms southern Tibet, reached the trench and was accreted to Asia. This event was followed by the docking of India itself.

As the intervening ocean basin was closing up, the more deformable materials on the continental margins of these landmasses became highly folded and faulted (**Figure 11.29B**). Two major thrust faults and many smaller ones sliced through the Indian crust. Subsequent motion along these thrust faults caused slices of the Indian crust to be stacked one upon the other. Today, these slices make up the bulk of the highest peaks in the Himalayas—many of which are capped by tropical marine limestones that formed along what was once the continental shelf.

The formation of the Himalayas was followed by a period of uplift that raised the Tibetan Plateau. Seismic evidence suggests that a portion of the Indian subcontinent was thrust beneath Tibet—a distance of perhaps

Did You Know?
During the assembly of Pangaea, the landmasses of Europe and Siberia collided to produce the Ural Mountains. Long before the discovery of plate tectonics, this extensively eroded mountain chain was regarded as the boundary between Europe and Asia.

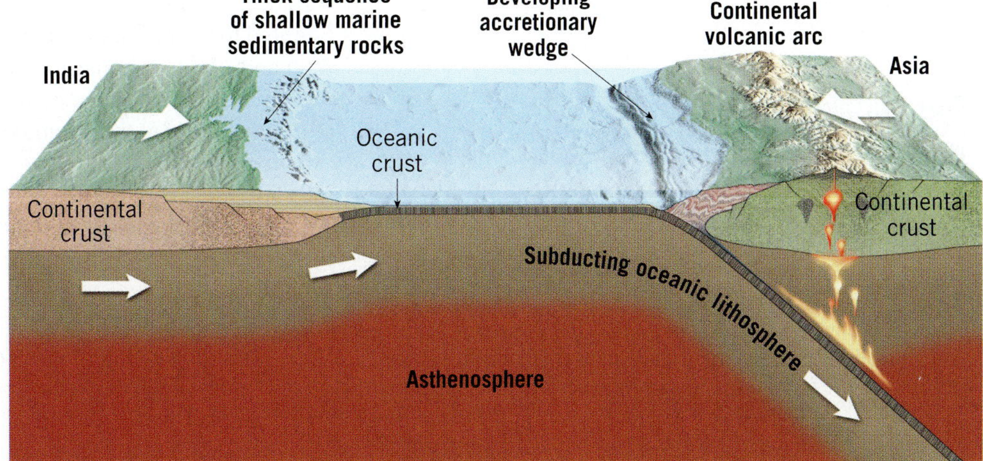

A. Prior to the collision of India and Asia, India's northern margin consisted of a thick platform of continental shelf sediments, whereas Asia's was an active continental margin with a well developed accretionary wedge and volcanic arc.

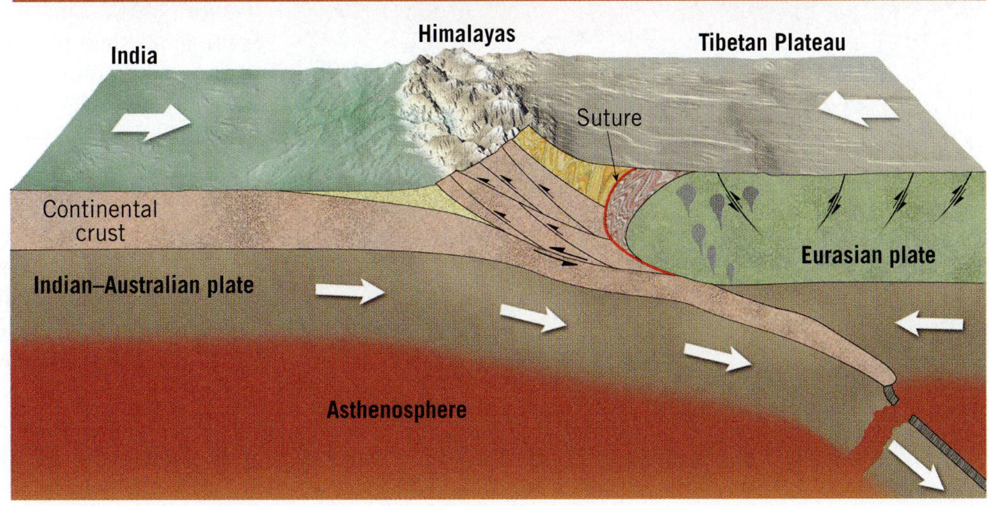

B. The continental collision folded and faulted crustal rocks along the margins of these continents to form the Himalayas. This event was followed by the gradual uplift of the Tibetan Plateau as the subcontinent of India was shoved under Asia.

◀ **SmartFigure 11.29**
Continental collision: The formation of the Himalayas These diagrams illustrate the collision of India with the Eurasian plate that produced the spectacular Himalayas.

ANIMATION
https://goo.gl/x8XSxr

Did You Know?
Edmund Hillary of New Zealand and Tenzing Norgay of Nepal were the first to reach the summit of Mount Everest on May 29, 1953. On their way up, they crossed a major Himalayan detachment fault, recognizable in movies like *Everest* (2015) by the prominent yellow band its footwall traces across the mountainside.

400 kilometers (250 miles). If this occurred, the added crustal thickness would account for the lofty landscape of southern Tibet, which has an average elevation of more than 4,500 meters (14,800 feet), higher than the tallest mountain in the contiguous United States.

The collision with Asia slowed but did not stop the northward movement of India, which has since penetrated at least 2000 kilometers (1200 miles) into the mainland of Asia. Crustal shortening and thickening accommodated some of this motion. Much of the remaining penetration into Asia caused lateral displacement of large blocks of the Asian crust by a mechanism described as *escape tectonics*. As shown in **Figure 11.30**, when India continued its northward trek, parts of Asia were "squeezed" eastward, out of the collision zone. These displaced crustal blocks include much of Southeast Asia (the region between India and China) and sections of China.

Why was the interior of Asia deformed to such a large extent, while India has remained essentially intact? The answer lies in the nature of these diverse crustal blocks. Much of India is a continental shield composed mainly of old Precambrian rocks (see Figure 11.23). This thick, cold slab of crustal material has been intact for more than 2 billion years and is mechanically strong as a result. By contrast, Southeast Asia was assembled more recently, from the collision of several smaller crustal fragments. Consequently, it is still relatively "warm and weak" from recent periods of mountain building (see Figure 11.30).

The Appalachians

The Appalachian Mountains provide great scenic beauty near the eastern margin of North America, from Alabama to Newfoundland. Mountain belts of similar origin that formed during the same period and were once contiguous are found in the British Isles, Scandinavia, northwestern Africa, and Greenland (see Figure 2.6, page 37). The orogenies that generated this extensive mountain system lasted a few hundred million years and resulted in the assembly of the supercontinent Pangaea. Detailed studies of the Appalachians indicate that this mountain belt was the result of three distinct episodes of mountain building.

Our simplified overview begins roughly 750 million years ago, with the breakup of a supercontinent called Rodinia that predates Pangaea. Similar to the breakup of Pangaea, this episode of continental rifting and seafloor spreading generated a new ocean between the rifted continental blocks. Located within this widening ocean basin was a microcontinent near the edge of ancestral Africa.

About 600 million years ago, for reasons geologists do not completely understand, plate motion changed dramatically, and this ancient ocean basin began to close. This led to the development of multiple subduction zones, and the stage was now set for the three orogenic events that would lead to the collision of North America and Africa (**Figure 11.31A**).

Taconic Orogeny Around 450 million years ago, the marginal sea between the volcanic island arc and ancestral North America began to close. The collision that ensued, called the *Taconic Orogeny*, caused the volcanic arc along with ocean sediments located on the upper plate to be accreted to the edge of the larger continental block. The remnants of this volcanic arc and oceanic sediments are recognized today as the metamorphic rocks through much of the Appalachian mountain belt (**Figure 11.31B**). For example, schists beneath New York City and Washington, DC, formed at this time. In addition to this pervasive regional metamorphism, numerous magma bodies intruded the crustal rocks along the entire continental margin.

Map view showing the southeastward displacement of China and the mainland of Southeast Asia as India plowed into Asia.

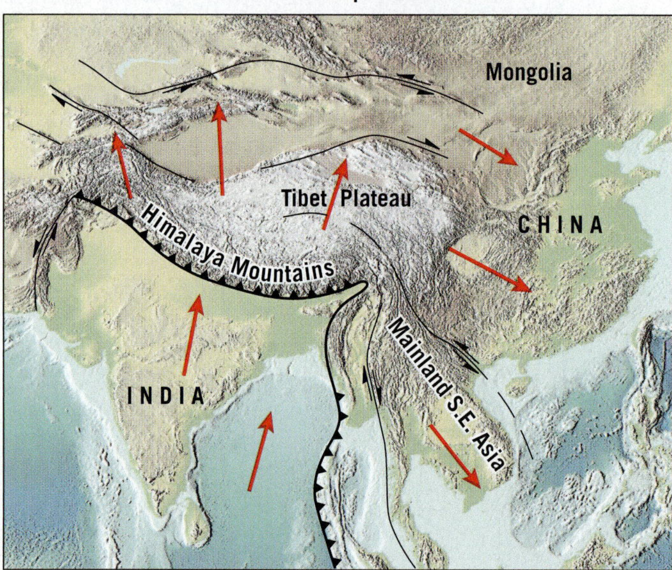

SmartFigure 11.30
India's continued northward migration severely deformed much of China and Southeast Asia

TUTORIAL
http://goo.gl/6SHx6Y

Acadian Orogeny A second episode of mountain building, called the *Acadian Orogeny*, occurred about 350 million years ago. The continued closing of this ancient ocean basin resulted in the collision of a microcontinent with North America (**Figure 11.31C**). This orogeny involved thrust faulting, metamorphism, and the intrusion of many large granite bodies. This event also added substantially to the width of North America, particularly in eastern New England.

Alleghanian Orogeny The final orogeny, called the *Alleghanian Orogeny*, occurred between 250 and 300 million years ago, when Africa collided with North America. This collision displaced material that was accreted earlier by as much as 250 kilometers (155 miles) toward the interior of North America. This event also displaced and further deformed the continental shelf sediments and sedimentary rocks that had once flanked the

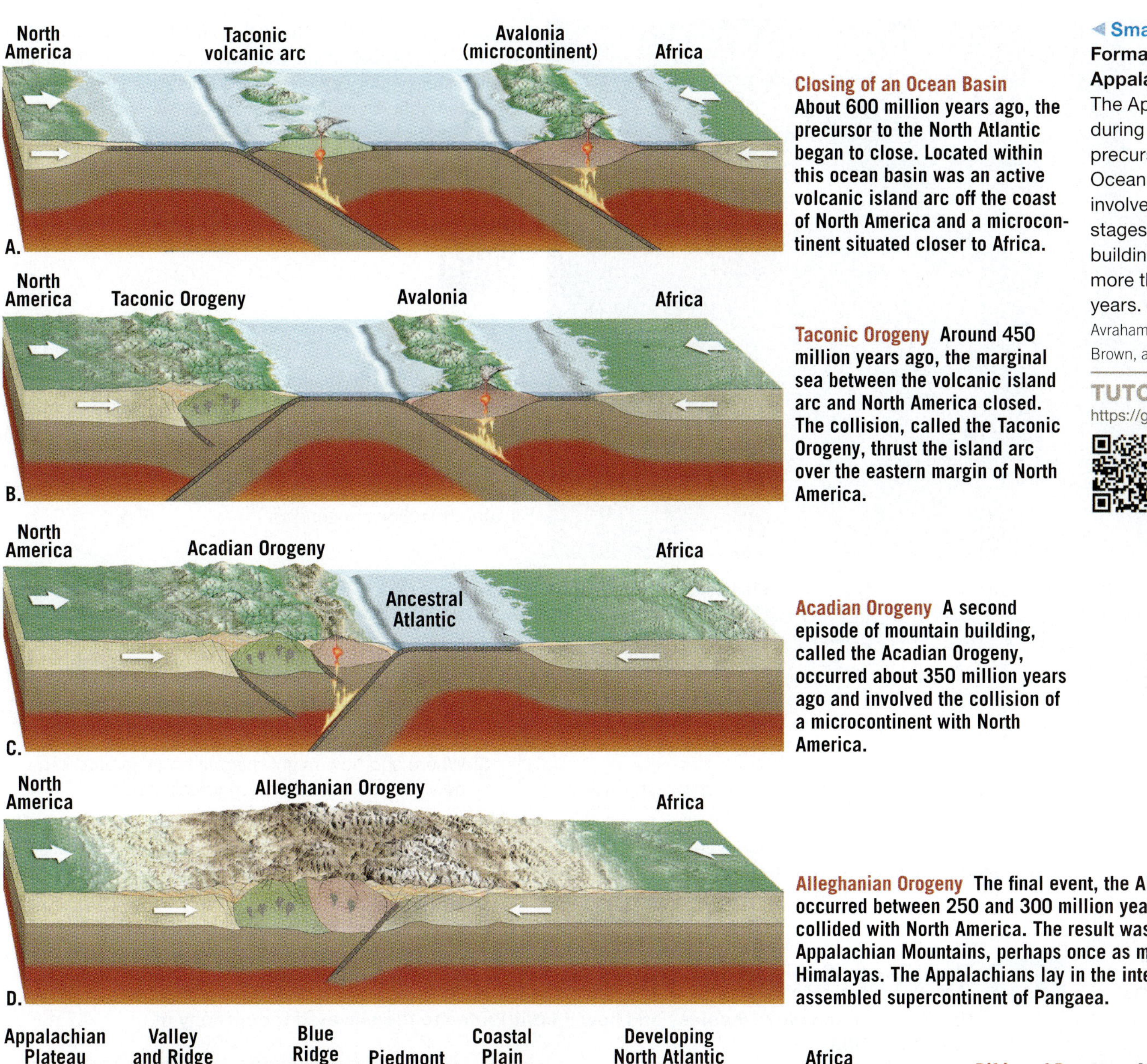

North America
Taconic volcanic arc
Avalonia (microcontinent)
Africa
A.
Closing of an Ocean Basin About 600 million years ago, the precursor to the North Atlantic began to close. Located within this ocean basin was an active volcanic island arc off the coast of North America and a microcontinent situated closer to Africa.
North America
Taconic Orogeny
Avalonia
Africa
B.
Taconic Orogeny Around 450 million years ago, the marginal sea between the volcanic island arc and North America closed. The collision, called the Taconic Orogeny, thrust the island arc over the eastern margin of North America.
North America
Acadian Orogeny
Ancestral Atlantic
Africa
C.
Acadian Orogeny A second episode of mountain building, called the Acadian Orogeny, occurred about 350 million years ago and involved the collision of a microcontinent with North America.
North America
Alleghanian Orogeny
Africa
D.
Alleghanian Orogeny The final event, the Alleghanian Orogeny, occurred between 250 and 300 million years ago, when Africa collided with North America. The result was the formation of the Appalachian Mountains, perhaps once as majestic as the Himalayas. The Appalachians lay in the interior of the newly assembled supercontinent of Pangaea.
Appalachian Plateau
Valley and Ridge
Blue Ridge
Piedmont
Coastal Plain
Developing North Atlantic
Africa
Remnant of Africa
E.
Rifting of Pangaea About 180 million years ago, Pangaea began to break into smaller fragments, a process that ultimately created the modern Atlantic Ocean. Because this new zone of rifting occurred east of the suture that formed when Africa and North America collided, remnants of African crust remain "welded" to the North American plate.

◀ **SmartFigure 11.31**
Formation of the Appalachian Mountains The Appalachians formed during the closing of a precursor to the Atlantic Ocean. This event involved three separate stages of mountain building that spanned more than 300 million years. (Based on Zvi Ben-Avraham, Jack Oliver, Larry Brown, and Frederick Cook)

TUTORIAL
https://goo.gl/YZaDyd

eastern margin of North America (**Figure 11.31D**). Today these folded and thrust-faulted sandstones, limestones, and shales make up the largely unmetamorphosed rocks of the Valley and Ridge Province (**Figure 11.32**). This structural signature of mountain building can be found as far inland as central Pennsylvania and West Virginia.

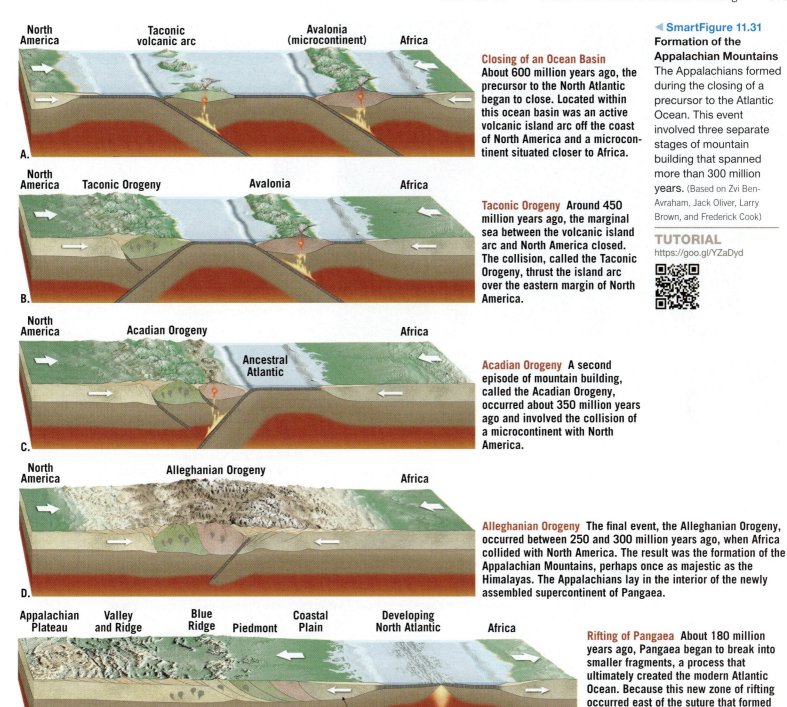

With the collision of Africa and North America, the young Appalachians, perhaps as majestic as the Himalayas, lay along the suture, in the interior of Pangaea. The tectonic forces that built the mountains ceased to drive them upward. Then, about 180 million years ago, the new supercontinent began to break into smaller fragments, a process that ultimately created the modern Atlantic Ocean. Because this new zone of rifting occurred east of the suture that formed when Africa and North America collided, remnants of Africa remain stuck

SmartFigure 11.32 The Valley and Ridge Province This region of the Appalachian Mountains consists of folded and faulted sedimentary strata that were displaced landward along thrust faults as a result of the collision of Africa with North America. (NASA/GSFC/JPL, MISR Science Team)

MOBILE FIELD TRIP
https://goo.gl/Ihn2PI

to the North American plate (**Figure 11.31E**). The crust underlying Florida is an example.

Other mountain ranges built from continental collisions include the Alps and the Urals. The Alps formed as Africa and several smaller crustal fragments collided with Europe during the closing of the Tethys Sea. Similarly, the Urals were deformed and uplifted during the assembly of Pangaea, when northern Europe and northern Asia collided, forming a major portion of Eurasia. Unlike the Appalachian belt, however, the Urals did not break apart again after their orogenesis.

CONCEPT CHECKS 11.6

1. Differentiate between *terrane* and *terrain*.
2. Explain why the continental crust of Asia was deformed more than that of the Indian subcontinent during the formation of the Himalayas.
3. Where and how might magma be generated in a newly formed collisional mountain belt?
4. How does the plate tectonics theory help explain the existence of fossil marine life in rocks atop collisional mountains?

Did You Know?
Which is the tallest mountain on Earth? The answer depends on how you define "tallest." Mount Everest, in the Himalayas, is of course the point on Earth's surface highest above sea level. However, Mauna Kea, a shield volcano on the Big Island of Hawaii, is the tallest peak as measured from the base of the mountain (the Pacific seafloor) to the summit. Because Earth bulges outward along the Equator, the peak of Chimborazo, in Ecuador, is the point on Earth's surface furthest from the center of the planet.

11.7 Vertical Motions of the Crust

Explain the principle of isostasy and how it contributes to the elevated topography of mountain belts.

Beyond the tectonic forces that move rocks laterally and thicken them vertically to produce mountainous topography, additional processes help to shape Earth's surface. As weathering and erosion work to lower mountains, a compensating process called *isostasy* causes them to rise, so they remain mountainous long after the tectonic processes that initially created them have ceased. Also, if tectonic processes raise a mountain belt "too high," the rock at its core will become too weak to support the load, allowing the mountain to spread.

The Principle of Isostasy

During the 1840s, researchers discovered that Earth's low-density crust "floats" on top of the high-density rocks of the mantle, much as wood floats in water. We will use the floating wooden blocks in **Figure 11.33** to explore this idea. You can think of these blocks as a model mountain belt, floating in the mantle. Notice that only about one-quarter of each block projects above the water, and three-quarters is submerged. This is because wood is about three-quarters as dense as water. Similarly, most of the vertical thickness of a mountain belt forms a buoyant root that is "submerged" in the mantle; the remainder projects above the surrounding crust. This concept—that the crust floats in gravitational balance in the mantle—is called **isostasy**.

Notice that the tallest of the floating blocks in Figure 11.33 stands the highest above the water surface and also sits the deepest. Similarly, the greater the crustal thickness of a mountain belt, the higher it will stand above sea level, and also the deeper its roots will be. Thus, the Himalayas, as the tallest range on Earth, also have the deepest roots.

Now, visualize what would happen if you placed a second small block on top of one of the blocks in

▲ SmartFigure 11.33 **The principle of isostasy** This drawing shows how wooden blocks of different thicknesses float in water. In a similar manner, thick sections of crustal material float higher than thinner crustal slabs.

ANIMATION
https://goo.gl/zaZa7C

Figure 11.33. The combined block would sink until it reached a new isostatic (gravitational) balance, at which point its top would be higher than before, and its bottom would be lower. This process of establishing a new gravitational balance in response to loading or unloading is called **isostatic adjustment**. Notice also that as a block rises or sinks, the surrounding water flows to accommodate it. Similarly, the highly viscous mantle will flow, albeit at an excruciatingly slow rate, when weight is added to or subtracted from the overlying crustal blocks.

Applying the concept of isostatic adjustment, we should expect that when weight is added to the crust, the crust will respond by subsiding, causing the underlying mantle rocks to flow away. When the weight is removed, the crust will rebound, and the mantle rock will flow back underneath. (Visualize what happens when a ship's cargo is loaded or unloaded.) Scientists have found evidence for crustal subsidence followed by isostatic rebound in areas formerly overlain by Ice Age glaciers. When continental ice sheets covered portions of North America during the Pleistocene epoch, ice masses that averaged about 3 kilometers (2 miles) thick added weight to the crust and caused downwarping by hundreds of meters. In the 8000 years since this last ice sheet melted, gradual uplift of as much as 330 meters (1000 feet) has occurred in Canada's Hudson Bay region, where the thickest ice had accumulated.

One of the consequences of isostatic adjustment is that, as erosion cuts into a mountain range, removing mass, the range rises in response to the reduced load (Figure 11.34). In fact, because erosion removes material mainly by carving canyons and valleys rather than by uniformly wearing down mountain peaks, isostasy may actually "push" the peaks higher than their original height.

The processes of uplift and erosion continue until the mountain block reaches average crustal thickness. When this occurs, these once-elevated structures are near sea level, and the once-deeply buried interior of the mountain is exposed at the surface. In addition, as mountains are worn down, the eroded sediment is deposited on adjacent landscapes, causing these areas to subside (see Figure 11.34).

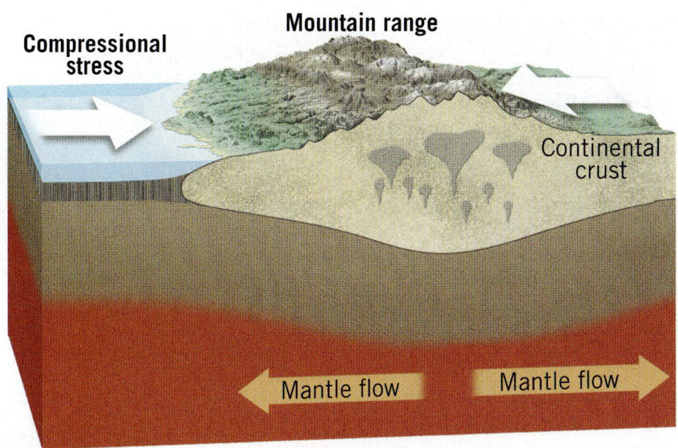
Compressional stress
Mountain range
When compressional mountains are young they are composed of thick, low density crustal rocks that float on the denser mantle below.
Continental crust
Mantle flow
Mantle flow

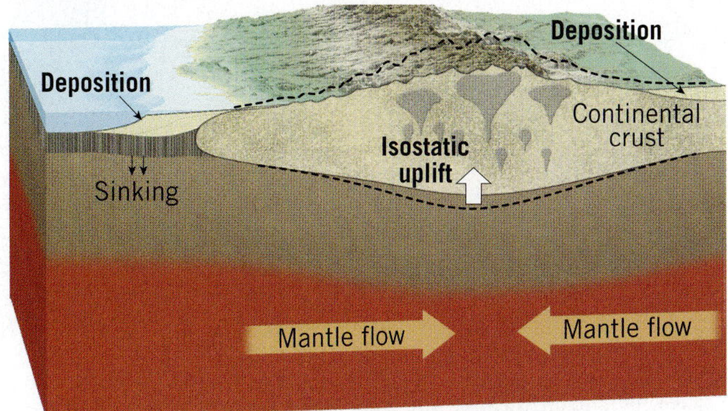
Erosion
As erosion lowers the mountains, the crust rises in response to reduced load in order to maintain isostatic balance.
Deposition
Deposition
Continental crust
Isostatic uplift
Sinking
Mantle flow
Mantle flow

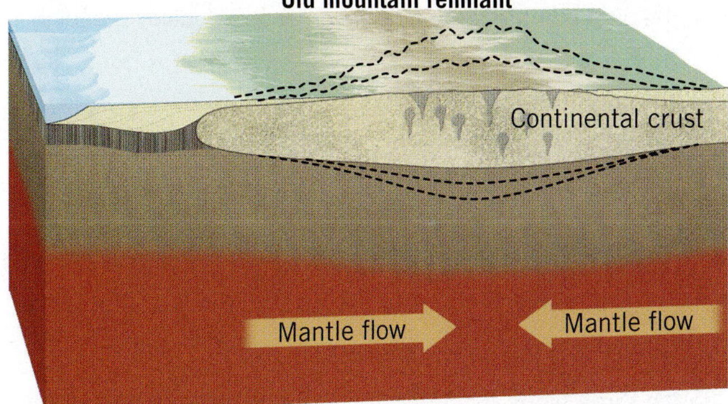
Old mountain remnant
Erosion and uplift continue until the mountains reach "normal" crustal thickness.
Continental crust
Mantle flow
Mantle flow

▲ SmartFigure 11.34 **The effects of isostatic adjustment and erosion on mountainous topography** This sequence illustrates how the combined effects of erosion and isostatic adjustment result in a thinning of the crust in mountainous regions.

TUTORIAL
https://goo.gl/kMchrw

How High Is Too High?

Where compressional forces are great, such as those driving India into Asia, lofty mountains such as the Himalayas result. Is there a limit on how high a mountain can rise? As mountaintops are elevated, gravity-driven processes such as erosion and mass wasting accelerate, carving the deformed strata into rugged landscapes. Equally important, however, is the fact that gravity also acts on the rocks within the mountain belt. The higher the mountain, the greater the downward force on rocks near

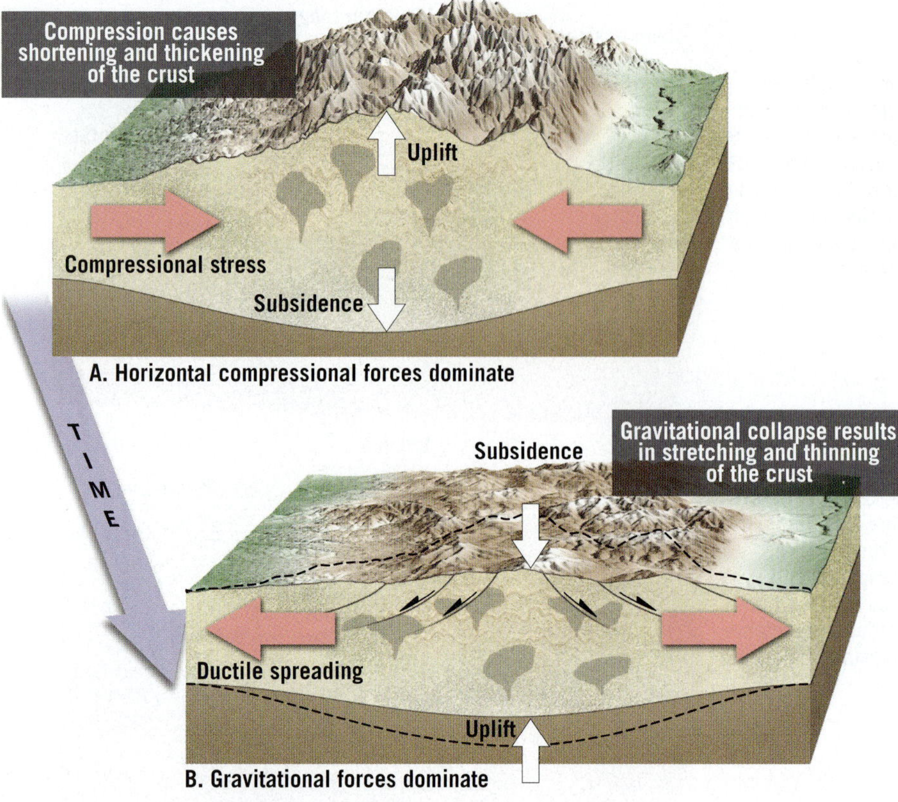

▲ **Figure 11.35 Gravitational collapse** Without compressional forces to support them, mountains gradually collapse under their own weight. Gravitational collapse involves normal faulting in the upper, brittle portion of the crust and ductile spreading in the warm, weak rocks at depth.

the base. Eventually, the rocks deep within the developing mountain, which are relatively warm and weak, begin to flow laterally, as shown in Figure 11.35. This process of **gravitational collapse** is analogous to what happens when a ladle of very thick pancake batter is poured onto a hot griddle. In addition to ductile spreading at depth, this process causes normal faulting and subsidence in the upper, brittle portion of Earth's crust.

Considering these factors, what keeps the Himalayas standing? Simply, the horizontal compressional forces that are driving India into Asia are greater than the vertical force of gravity. However, when India's northward trek ends, the downward pull of gravity, weathering, and erosion will become the dominant forces acting on this mountainous region.

CONCEPT CHECKS 11.7

1. Define *isostasy*.
2. Give one kind of evidence that supports the concept of isostatic uplift.
3. What happens to a floating object when weight is added? Subtracted?
4. Briefly describe how the principle of isostatic adjustment applies to changes in the elevations of mountains.
5. Explain the process whereby mountainous regions experience gravitational collapse.

CONCEPTS IN REVIEW
Crustal Deformation & Mountain Building

11.1 Crustal Deformation

Describe the three types of differential stress and identify the tectonic setting most commonly associated with each. Differentiate stress from strain and brittle from ductile deformation.

KEY TERMS: deformation, tectonic structure (geologic structure), stress, confining pressure, differential stress, compressional stress, tensional stress, shear, strain, elastic deformation, brittle deformation, ductile deformation

- Tectonic (geologic) structures are structures generated when rocks are deformed by bending or breaking; they include folds, faults, and joints.
- Stress is the force that drives rock deformation. When stress acts equally from all directions, we call it confining pressure. When the stress is greatest in one direction we call it differential stress. There are three main types of differential stress: compressional, tensional, and shear.
- A rock's strength is its ability to resist permanent deformation. When the stresses on a rock exceed its strength, the rock deforms, usually by folding or faulting.
- Elastic deformation is caused by a temporary stretching of the chemical bonds in a rock. When the stress is released, the rock returns to its original shape. When the rock's strength is exceeded, bonds break, and the rock deforms in either a brittle or ductile fashion. Brittle deformation fractures rocks, whereas ductile deformation changes a rock's shape.
- Whether a rock deforms in a brittle or ductile manner depends on its temperature and its confining pressure. The hotter a rock, the more likely it is to experience ductile deformation. Greater confining pressure makes a rock stronger and less likely to break. Thus, rock deformation tends to be brittle in the shallow crust and ductile at deeper levels.
- Whether deformation is brittle or ductile also depends on the type of rock. For example, shale is weaker than granite, so it is more prone to ductile deformation. If a rock is forced to deform more quickly than can be accommodated by the slow processes of ductile deformation, it will break.

? Did the quarter on the left in the accompanying image experience brittle or ductile deformation?

Anthony Pleva/Alamy

11.2 Folds: Rock Structures Formed by Ductile Deformation

List and describe five types of folds.

KEY TERMS: fold, anticline, syncline, dome, basin, monocline

- Folds are wavelike undulations in layered rocks that develop through ductile deformation caused by compressional stresses.
- Anticlines usually arise by upfolding, or arching, of sedimentary layers, whereas synclines are downfolds, or troughs. Anticlines and synclines may be symmetrical, asymmetrical, overturned, or recumbent.
- When folded rocks erode to form a series of ridges and valleys, the ridges represent resistant beds (not anticlines), and the valleys represent softer beds (not synclines). A fold is said to plunge when its axis penetrates the ground at an angle. This results in a V-shaped outcrop pattern.
- Domes and basins are large bowl- or saucer-shaped folds that produce roughly circular outcrop patterns. When eroded, a dome has the oldest beds in the middle, and a basin has the oldest beds around the margin.
- Monoclines are large steplike folds in otherwise horizontal strata that develop when beds drape over a vertical offset produced by subsurface faulting.

? What name is given to the geologic structure shown in the accompanying image?

11.3 Faults & Joints: Structures Formed by Brittle Deformation

Sketch and briefly describe the relative motion of rock bodies located on opposite sides of normal, reverse, and thrust faults as well as both types of strike-slip faults.

KEY TERMS: fault, dip-slip fault, hanging wall block, footwall block, fault scarp, normal fault, fault-block mountain, horst, graben, half-graben, detachment fault, reverse fault, thrust fault, strike-slip fault, transform fault, joint

- Faults and joints are fractures in rock that form through brittle deformation.
- A fault is a fracture along which motion occurs, offsetting the rocks on either side. If the movement is in the direction of the fault's dip (or inclination), the rock above the fault plane is the hanging wall block, and the rock below the fault is the footwall block. If the hanging wall moves *down* relative to the footwall, the fault is a normal fault. If the hanging wall moves *up* relative to the footwall, the fault is a reverse fault. Large normal faults with low dip angles are called detachment faults. Large reverse faults with low dip angles are thrust faults.
- Faults that intersect Earth's surface may produce a "step" in the land known as a fault scarp. Areas of tectonic extension, such as the Basin and Range Province, produce fault-block mountains—horsts separated by neighboring grabens or half-grabens.

(11.3 continued)

- Areas of tectonic compression, such as mountain belts, are dominated by reverse faults that shorten the crust horizontally while thickening it vertically.
- Strike-slip faults have most of their movement in a horizontal direction, along the trend of the fault trace. Transform faults are strike-slip faults that serve as tectonic boundaries between lithospheric plates.
- Joints form in the shallow crust when rocks are stressed under brittle conditions. They facilitate groundwater movement and mineralization of economic resources, and they may result in hazards to humans.

? What type of rock structure is shown in each of the accompanying images: faults or joints? Explain how you arrived at your answer.

11.4 Mountain Building

Locate and name Earth's major mountain belts on a world map.

KEY TERMS: orogenesis, orogeny, collisional mountain

- Orogenesis is the making of mountains. An episode of orogenesis is an orogeny. Most orogenesis occurs along convergent plate boundaries, where compressional forces cause folding and faulting, thickening the crust vertically and shortening it horizontally.

? Look at the South American plate on the map. Explain why the Andes Mountains are located on the western margin of South America rather than the eastern margin.

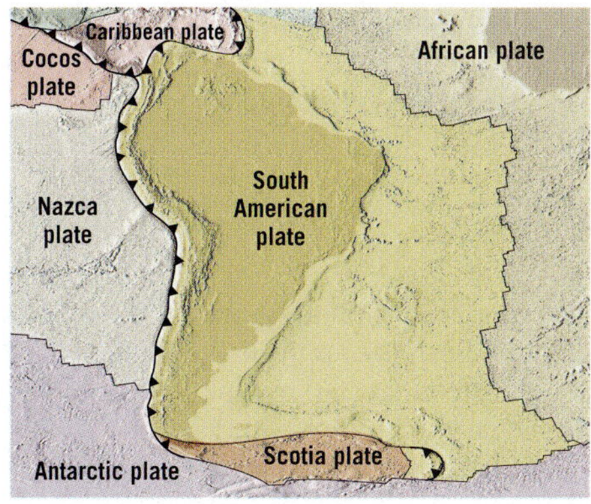

11.5 Subduction & Mountain Building

Sketch a cross section of an Andean-type mountain belt and describe how its major features are generated.

KEY TERMS: accretionary wedge, forearc basin

- The type of convergent margin determines the type of mountains that form. Where one oceanic plate overrides another, a volcanic island arc forms. Where an oceanic plate subducts under a continent, Andean-type mountain building occurs.
- In either case, release of water from the subducted slab triggers melting in the overlying mangle wedge, generating basaltic magmas that rise to the base of the continental crust, where they often pond. The hot basaltic magma may heat the overlying crustal rocks sufficiently to generate a silica-rich magma of intermediate or felsic (granitic) composition.
- Sediment scraped off the subducting plate builds an accretionary wedge. Between the accretionary wedge and the volcanic arc is a relatively calm site of sedimentary deposition, the forearc basin.
- The geography of central California preserves an accretionary wedge (Coast Ranges), a forearc basin (Great Valley), and the roots of an Andean-style mountain belt (Sierra Nevada).

11.6 Collisional Mountain Belts

Summarize the stages in the development of an Alpine-type mountain belt such as the Appalachians.

KEY TERMS: terrane, microcontinent, suture

- A terrane is a relatively small crustal fragment (microcontinent, volcanic island arc, or oceanic plateau) that has been carried by an oceanic plate to a continental subduction zone and then accreted onto the continental margin. The North American Cordillera formed by the accretion of many successive terranes.

(11.6 continued)

- The Himalayas and Appalachians were formed by collisions between continents when the intervening ocean basin subducted completely. The Appalachians were caused by the collision of ancestral North America with ancestral Africa more than 250 million years ago. The Himalayas were formed by the collision of India and Eurasia starting around 50 million years ago and are still rising.

11.7 Vertical Motions of the Crust

Explain the principle of isostasy and how it contributes to the elevated topography of mountain belts.

KEY TERMS: isostasy, isostatic adjustment, gravitational collapse

- Earth's crust floats in the denser material of the mantle the way wood floats in water. If additional weight is placed on the crust (an ice sheet), the crust sinks, and if weight is removed (glacial melting), the crust rebounds. This process of maintaining gravitational equilibrium is called isostatic adjustment. For a mountain belt, isostasy partially offsets the effect of erosion, pushing the mountains up as erosion wears them down.
- When compressional forces raise a mountain belt too high, the rock at the belt's core becomes warm and weak, and the belt spreads, becoming broader and lower.

? Based on the principle of isostasy, predict what will happen to the elevation of the highest peaks in a mountain range if rivers and glaciers erode deep valleys through the mountains, removing large amounts of rock.

GIVE IT SOME THOUGHT

1 Refer to the accompanying diagrams to answer the following:
  a. What type of dip-slip fault is shown in Diagram 1? Were the dominant forces during faulting tensional, compressional, or shear?
  b. What type of dip-slip fault is shown in Diagram 2? Were the dominant forces during faulting tensional, compressional, or shear?
  c. Match the correct pair of arrows in Diagram 3 to the faults in Diagrams 1 and 2.

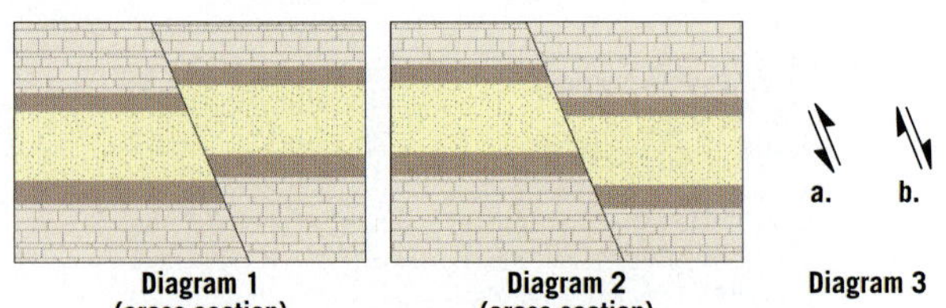

Diagram 1 (cross section) Diagram 2 (cross section) Diagram 3

2 Refer to the accompanying photo to answer the following:
  a. The white line shows the approximate location of a fault that displaced these furrows created by a plow. What type of fault caused the offset shown?
  b. Is this a right-lateral or left-lateral fault? Explain.

USGS

3 Which of the three types of plate boundaries is primarily associated with normal faulting? Thrust faulting? Strike-slip faulting?

4 Suppose that a sliver of oceanic crust were discovered in the interior of a continent. Would this refute the theory of plate tectonics? Explain.

5 The accompanying photo, taken near the bottom of the Grand Canyon, shows a quartz vein that has been deformed.
  a. What type of deformation is exhibited—ductile or brittle?
  b. Did this deformation most likely occur near Earth's surface or at great depth?

6 Examine the diagrams depicting the structure of East Africa and the Canadian Rockies.

a. Characterize the type of faulting found at each location and identify the differential stresses that produced these landforms.

b. Along what type of plate boundary did each of these structures form?

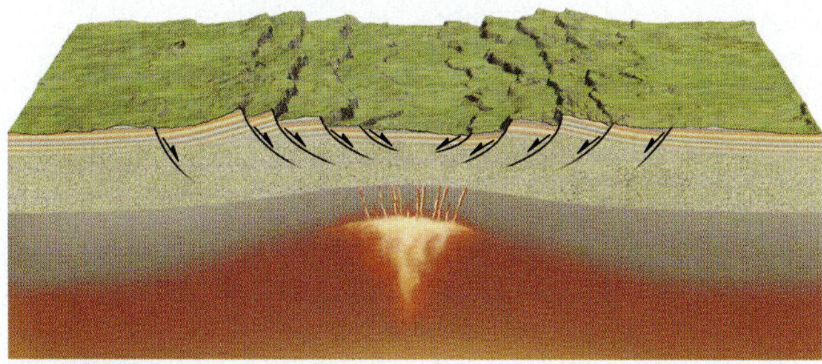

East Africa

Canadian Rockies

7 Refer to the accompanying map, which shows the location of the Galapagos Rise and the Rio Grande Rise to answer the following questions:

a. Name the types of continental margin found on the west and east coasts of South America.

b. Based on your answer to Question a, is the Galapagos Rise or the Rio Grande Rise more likely to end up accreted to a continent? Explain your choice.

c. In the distant future, how might a geologist determine that this accreted landmass is distinct from the continental crust to which it accreted?

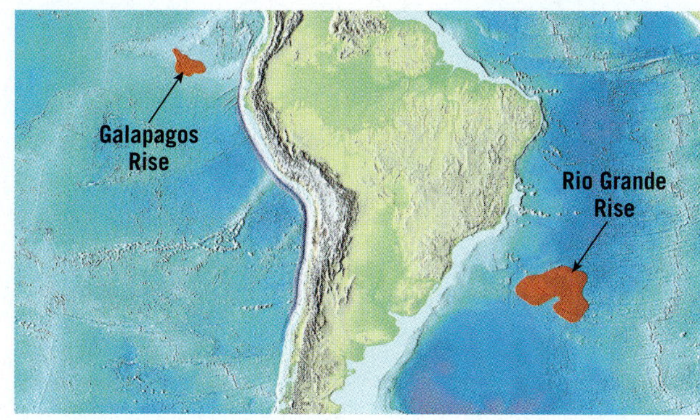

8 The Ural Mountains exhibit a north–south orientation through Eurasia. How does the theory of plate tectonics explain the existence of this mountain belt in the interior of an expansive landmass?

9 Briefly describe the major differences between the evolution of the Appalachian Mountains and the North American Cordillera.

10 Which of the accompanying sketches best illustrates an Andean-type orogeny, a Cordilleran-type orogeny, and an Alpine-type orogeny?

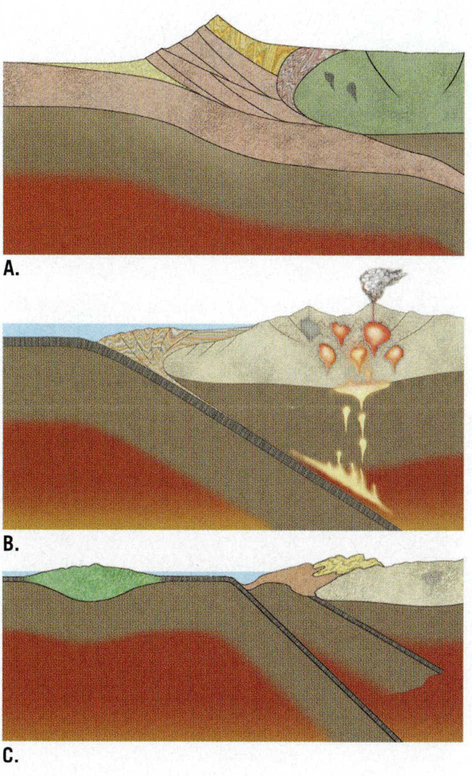

MasteringGeology™

Looking for additional review and test prep materials? Visit the Study Area in MasteringGeology to enhance your understanding of this chapter's content by accessing a variety of resources, including Self-Study Quizzes, Geoscience Animations, SmartFigures, Mobile Field Trips, *Project Condor* Quadcopter videos, *In the News* RSS feeds, flashcards, web links, and an optional Pearson eText.

www.masteringgeology.com

12

Mass Movement on Slopes: The Work of Gravity*

FOCUS ON CONCEPTS

Each statement represents the primary learning objective for the corresponding major heading within the chapter. After you complete the chapter, you should be able to:

12.1 Describe how mass movement processes can cause natural disasters and discuss the role that mass movement plays in the development of landscapes.

12.2 Summarize the factors that control and trigger mass movement processes.

12.3 List and explain the criteria that are commonly used to classify mass movement processes.

12.4 Distinguish among rockfall, rockslide, debris avalanche, debris flow, slump, and earthflow.

12.5 Review the general characteristics of slow mass movement processes and describe the unique issues associated with a permafrost environment.

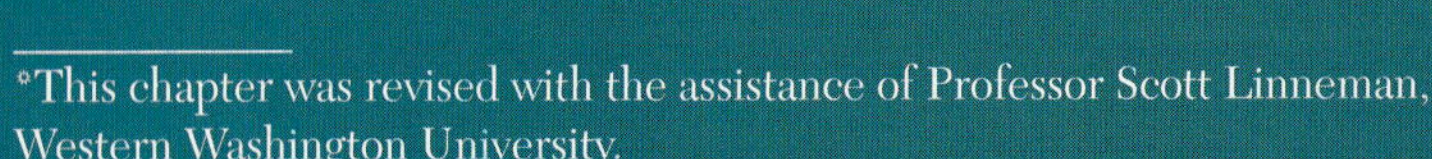

*This chapter was revised with the assistance of Professor Scott Linneman, Western Washington University.

On March 22, 2014, a major landslide occurred near Oso, Washington. An extremely wet winter saturated older landslide deposits until they mobilized on a sunny Saturday morning. A large mass of mud and debris buried a 2.6-square-kilometer (1-square-mile) area, damming a river, engulfing 49 buildings, and blocking a highway. The event claimed 43 lives. (Photo by Michael Collier)

EARTH'S SURFACE IS SELDOM PERFECTLY FLAT but instead consists of slopes of many different varieties. Some are steep and cliff-like; others are moderate or gentle. Slopes can consist of barren rock and rubble or can be mantled with soil and covered by vegetation. Taken together, slopes are the most common elements in our physical landscape. Although most slopes may appear to be stable and unchanging, the force of gravity causes rock and soil to move downslope by the processes we call mass movement. At one extreme, the movement may consist of a roaring debris flow or a thundering rock avalanche. At the other extreme, it may be gradual and practically imperceptible. Landslides are a worldwide natural hazard. When these hazardous processes lead to loss of life and property, they become natural disasters.

12.1 The Importance of Mass Movement

Describe how mass movement processes can cause natural disasters and discuss the role that mass movement plays in the development of landscapes.

Some mass movement processes are dangerous events that represent significant geologic hazards. Perhaps less well known is the fact that mass movement processes play an important role in the development and evolution of many of Earth's varied landscapes. The chapter-opening photo is a striking example of a phenomenon that many people would call a landslide. For most of us, the word *landslide* implies a sudden event in which large quantities of rock and soil plunge down steep slopes. When people and communities are in the way, a natural disaster may result. Landslides constitute major geologic hazards that each year in the United States cause billions of dollars in damages and the loss of dozens of lives. As you will see, many landslides occur in connection with other major natural disasters, including earthquakes, volcanic eruptions, wildfires, and severe storms.

Landslides are spectacular examples of a basic geologic process called mass movement. **Mass movement** refers to the downslope movement of rock, regolith, and soil under the direct influence of gravity. If you were to view several images of events called landslides, you would likely notice that the term seems to refer to several different things—from mudflows to rock avalanches. This variety reflects the fact that although many people, including geologists, frequently use the

A.

B.

► SmartFigure 12.1
Landslides are geologic hazards A. In January 2011, following a period of torrential rains, a thick slurry of mud, appropriately called a mudflow, buried these cars in Nova Friburgo, Brazil. Heavy rains are an important trigger of mass movement processes. (ZumaPress.com/Newscom) **B.** This deadly landslide that buried 300 homes in Santa Tecia, El Salvador, was triggered by an earthquake. (Photo by Ed Harp/U.S. Geological Survey)

MOBILE FIELD TRIP
https://goo.gl/kjQUFs

SmartFigure 12.2
Excavating the Grand Canyon The walls of the canyon extend far from the channel of the Colorado River. This results primarily from the transfer of weathered debris downslope to the river and its tributaries by mass movement processes. (Photo by Bryan Brazil/Shutterstock)

TUTORIAL
https://goo.gl/KIMCNq

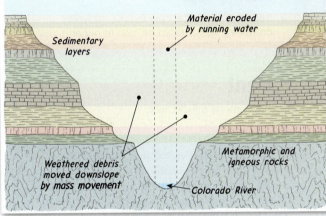

word *landslide*, the term has no specific definition in geology. Rather, it is a popular nontechnical word used to describe any or all relatively rapid forms of mass movement.

Landslides as Geologic Hazards

Though they are often used in the same conversation, "hazard" and "risk" have different meanings. A *geologic hazard* is a process that can cause harm to people or property if they are sufficiently exposed. *Risk* is the probability that exposure to a hazard will cause harm. Thus we can think of Risk = Hazard × Exposure. Landslides are not just events that occur in remote mountains and canyons. People frequently live where rapid but rare mass movement events occur. Consequently, people living in susceptible areas often do not appreciate the risks of living where they do. However, media reports remind us that such events occur with some regularity around the world. Two examples are provided in **Figure 12.1**. In the example from Brazil, we see a city street buried in a mudflow that was triggered by heavy rains. By contrast, in El Salvador steep slopes, weak rock, and the shock of an earthquake combined to produce a deadly event.

The Role of Mass Movement in Landscape Development

In the evolution of most landscapes, mass movement is the step that follows weathering. Landscapes slowly change when rock and soil are removed from the places where they originate. Once weathering weakens and breaks apart rock, mass movement transfers the debris downslope, where a stream or glacier, acting as a conveyor belt, usually carries it away. Although there may be many intermediate stops along the way, most sediment is eventually transported to its ultimate destination: the sea.

The combined effects of mass movement and running water produce stream valleys, which are among the most common and conspicuous of Earth's landforms. If streams alone were responsible for creating the valleys in which they flow, the valleys would be very narrow features. However, the fact that most river valleys are much wider than they are deep is a strong indication of the significance of mass movement processes in supplying material to streams. This is illustrated by the Grand Canyon (**Figure 12.2**). The walls of the canyon extend far from the Colorado River due to the transfer of weathered debris downslope to the river and its tributaries by mass movement processes. In this manner, streams and mass movement combine to modify and sculpt the surface. Of course, glaciers, groundwater, waves, and wind are also important agents in shaping landforms and developing landscapes.

Slopes Change Through Time

It is clear that if mass movement is to occur, there must be slopes that rock, soil, and regolith can move down. Earth's mountain-building and volcanic processes produce these slopes through sporadic changes in the elevations of landmasses and the ocean floor. If dynamic internal processes did not continually produce regions having higher elevations, the system that moves debris to lower elevations would gradually slow and eventually cease.

Most rapid and spectacular mass movement events occur in areas of rugged, geologically young mountains. Newly formed mountains are rapidly eroded by rivers and glaciers into regions characterized by steep and unstable slopes. These areas often include valleys in thick deposits of unconsolidated material. It is in such settings that many massive destructive landslides, such as the ones described at the beginning of the chapter, occur. As mountain building subsides, mass movement and erosional processes lower the land. Through time, steep and rugged mountain slopes give way to gentler, more subdued terrain. Thus, as a landscape ages, massive and rapid mass movement processes give way to smaller, less dramatic downslope movements that are often imperceptibly slow.

CONCEPT CHECKS 12.1

1. Define *mass movement*. How does mass movement differ from erosional agents such as streams, glaciers, and wind?
2. In what sort of landscape are rapid mass movement processes most likely to occur? Describe how these geologic hazards might become geologic risks.
3. Sketch or describe how mass movement combines with stream erosion to expand valleys.

12.2 Controls & Triggers of Mass Movement

Summarize the factors that control and trigger mass movement processes.

Whether a slope fails depends on the relative size of the force pulling material downslope (gravity) and the resisting strength of the rock or soil. While gravity is the driving force of mass movement, several factors play important roles in overcoming friction and material strength to create downslope movements. Long before a landslide occurs, various processes work to weaken slope material, gradually making it more and more susceptible to the pull of gravity. During this span, the slope appears stable but gets closer and closer to being unstable. Eventually, the strength of the slope is sufficiently weakened, or the slope becomes steeper, to the point that something causes it to cross the threshold from stability to instability. Such an event that initiates downslope movement is called a **trigger**. Remember that a trigger is not the sole cause of a mass movement event but just the last of many causes. Among the common factors that trigger mass movement processes are saturation of material with water, oversteepening of slopes, removal of anchoring vegetation, and ground vibrations from earthquakes.

Many rapid mass movement events occur without discernible triggers. Slope materials gradually weaken over time under the influence of long-term weathering, infiltration of water, and other physical processes. Eventually, if the strength falls below what is necessary to maintain slope stability, a landslide will occur. The timing of such events is random, and thus accurate prediction is impossible.

The Role of Water

Mass movement is sometimes triggered when heavy rains or periods of snowmelt saturate surface materials. The water does not transport the material. Rather, it allows gravity to more easily set the material in motion. This was the case in January 2005, when a massive debris flow (popularly called a mudslide) swept through La Conchita, California, a small coastal community northwest of Los Angeles (**Figure 12.3**). Like the Oso landslide shown in the chapter-opening photo, the La Conchita debris flow happened when older landslide deposits on the hillside became saturated during a winter of unusually high precipitation.

When the pores in sediment become filled with water, the cohesion among particles is decreased, and they can move past one another with relative ease. For example, when sand is slightly moist, it sticks together quite well. However, if enough water is added to fill the openings between the grains, the sand will ooze out in all directions (**Figure 12.4**). Thus, saturation reduces the internal resistance of materials, which are then easily set in motion by the force of gravity. When clay is wetted, it becomes very slick—another example of the "lubricating" effect of water. Water also adds considerable weight to a mass of material. The added weight in itself may cause the downslope forces to exceed the resisting forces, which leads to the mass of material sliding or flowing downslope.

Oversteepened Slopes

Oversteepening of slopes is another trigger of many mass movements. Oversteepening takes place in many situations. For example, as a stream (or a bulldozer) cuts into a valley wall, it removes material from the base of the wall. The steeper slope increases the downslope forces and reduces the resisting forces, causing the material to fall or slide

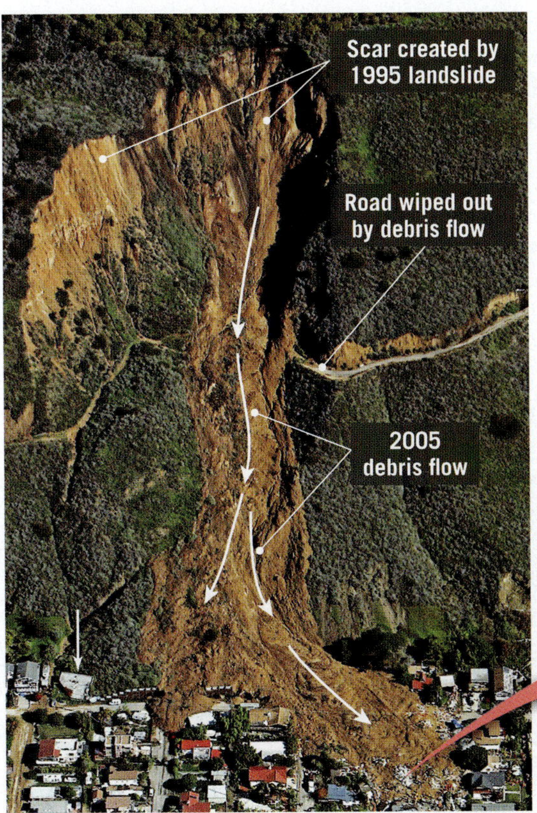
Scar created by 1995 landslide
Road wiped out by debris flow
2005 debris flow

Houses buried in mud and debris. The thick debris moved some houses instead of flowing around them.
January 10, 2005–the debris flow occurred after the heaviest rainfall of an exceptionally wet rainy season
Rainfall (centimeters)
10
9
8
7
6
5
4
3
2
1
0
11/1/04
11/10/04
11/20/04
11/30/04
12/10/04
12/20/04
12/30/04
1/9/05
1/19/05

◀ **Figure 12.3 Heavy rains trigger debris flows** On January 10, 2005, a massive debris flow, originating from the saturated deposits of a 1995 landslide, swept through La Conchita, California, a small coastal town situated on a narrow coastal strip between the shoreline and a steep bluff. The event, which occurred after a span of near record rainfall, killed 10 people and destroyed many homes. (Photos by Kevork Djansezian/AP Images)

into the stream. Furthermore, through their activities, people often create oversteepened and unstable slopes that become prime sites for mass movement (**Figure 12.5**).

Loose, granular particles (sand-size or coarser) assume a stable slope called the **angle of repose** (*reposen* = to be at rest). This is the steepest angle at which loose material remains stable (**Figure 12.6**). Depending on the size and shape of the particles, the angle varies from 25 to 40 degrees. Larger, more angular particles maintain the steepest slopes. If the angle is increased, the rock debris adjusts by moving downslope.

Oversteepening is not just important because it triggers movements of unconsolidated granular materials. Oversteepening also produces unstable slopes and mass movements in cohesive

▼ **Figure 12.4 Saturation reduces friction** When water saturates sediment, friction among particles is reduced, allowing material to move downslope.

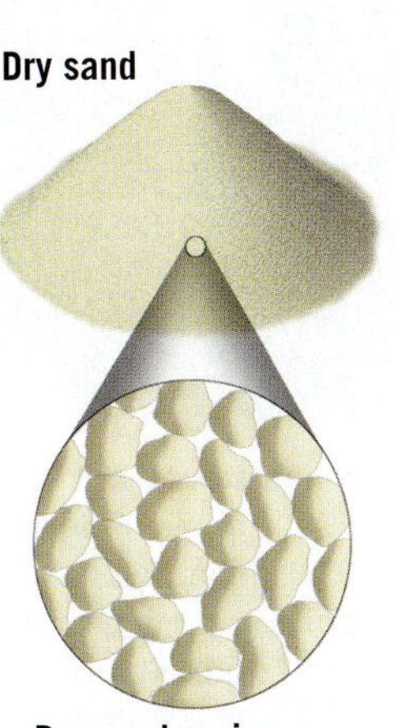
Dry sand
Dry sand grains are bound mainly by friction with one another

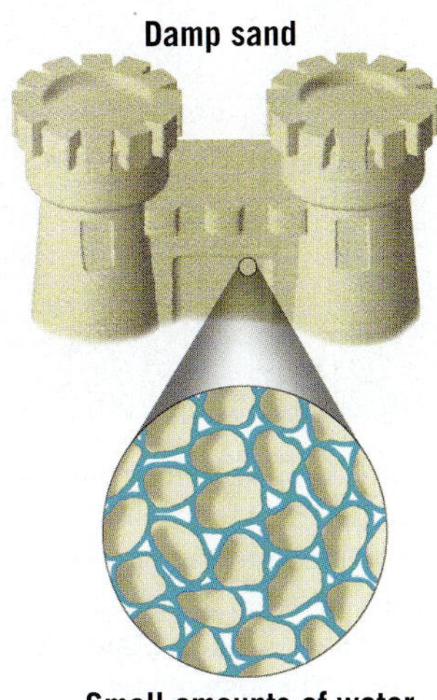
Damp sand
Small amounts of water increase the cohesion among sand grains

Wet sand
Saturation reduces friction and causes the sand to flow

▼ **Figure 12.5 Unstable slopes** Natural processes such as stream and wave erosion can oversteepen slopes. Changing the slope to accommodate a new house or road can also lead to instability and a destructive mass movement event.

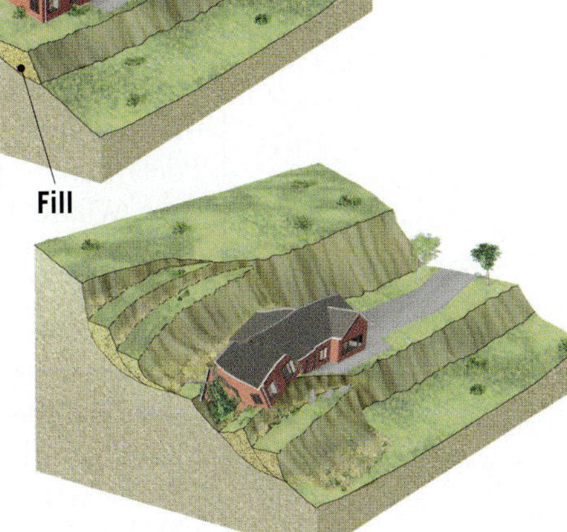
Heavy rainfall
Oversteepened hillslope
Fill

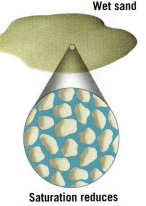

▶ **Figure 12.6 Angle of repose** The angle of repose is the steepest angle at which an accumulation of granular particles remains stable. Larger, more angular particles maintain the steepest slopes. (Photo by G. Leavens/Science Source)

soils, regolith, and bedrock because the steeper angle increases gravity's effect. The response is not immediate, as with loose, granular material, but sooner or later, one or more mass movement processes will eliminate the oversteepening and restore stability to the slope.

Removal of Vegetation

Plants protect against erosion and contribute to the stability of slopes because their root systems bind together soil and regolith. In addition, plants shield the soil surface from the erosional effects of raindrop impact. Where plants are lacking, mass movement is enhanced, especially if slopes are steep and water is plentiful. When anchoring vegetation is removed by forest fires or by people (for timber, farming, or development), surface materials frequently move downslope.

An example illustrating the anchoring effect of plants occurred several decades ago, on steep slopes near Menton, France. Farmers replaced olive trees, which have deep roots, with a more profitable but shallow-rooted crop: carnations. When the less stable slope failed, the landslide took 11 lives.

In July 1994 a severe wildfire swept Storm King Mountain, west of Glenwood Springs, Colorado, denuding the slopes of vegetation. Two months later, heavy rains resulted in numerous debris flows—rapid mass movement events involving water-saturated rock and soil. One debris flow blocked Interstate 70 and threatened to dam the Colorado River. A 5-kilometer (3-mile) length of the highway was inundated with tons of rock, mud, and burned trees. The closure of Interstate 70 imposed costly delays on users of this major highway.

Wildfires are inevitable in the western United States, and fast-moving, highly destructive debris flows triggered by intense rainfall are some of the most dangerous postfire hazards (Figure 12.7). Such events are particularly dangerous because they tend to occur with little warning. Their mass and speed make them particularly destructive. Postfire debris flows are most common in the 2 years after a fire. The very first intense rain event following a wildfire is likely to trigger a debris-flow event. It takes much less rain to trigger debris flows in burned areas than in unburned areas. In southern California, as little as 7 millimeters (0.3 inch) of rain in 30 minutes has triggered debris flows.

In addition to eliminating plants that anchor the soil, fire can promote mass movement in other ways. Following a wildfire, the upper part of the soil may become dry and loose. As a result, even in dry weather, the soil tends to move down steep slopes. Moreover, fire can "bake" the ground, creating a water-repellant layer at a shallow depth. This nearly impermeable barrier prevents or slows the infiltration of water, resulting in increased surface runoff during rains. The result can be dangerous torrents of viscous mud and rock debris.

Did You Know?
One of the most expensive landslides in history occurred April 10, 2013, in the Bingham Canyon open-pit copper mine in Utah. Careful monitoring of the oversteepened slope allowed the event to be predicted, which meant that the mining operations could be stopped and the mine evacuated, preventing loss of life. However, the loss of productivity cost the mine's owner an estimated $770 million. The landslide was the largest non-volcanic event in North America in modern times, moving 65 million m^3—enough rock to cover all of New York City's Central Park with 20 m of debris.

▲ **Figure 12.7 Wildfires contribute to mass movement** In September 2014, this wildfire consumed 9000 acres of dense forest on steep terrain near Fresh Pond, California. During the summer, wildfires are becoming more common in many parts of the western United States. Millions of acres are burned each year. The loss of anchoring vegetation sets the stage for increased mass movements and more sediment delivered to stream systems. (Photo by Raymond Gehman/National Geographic Stock)

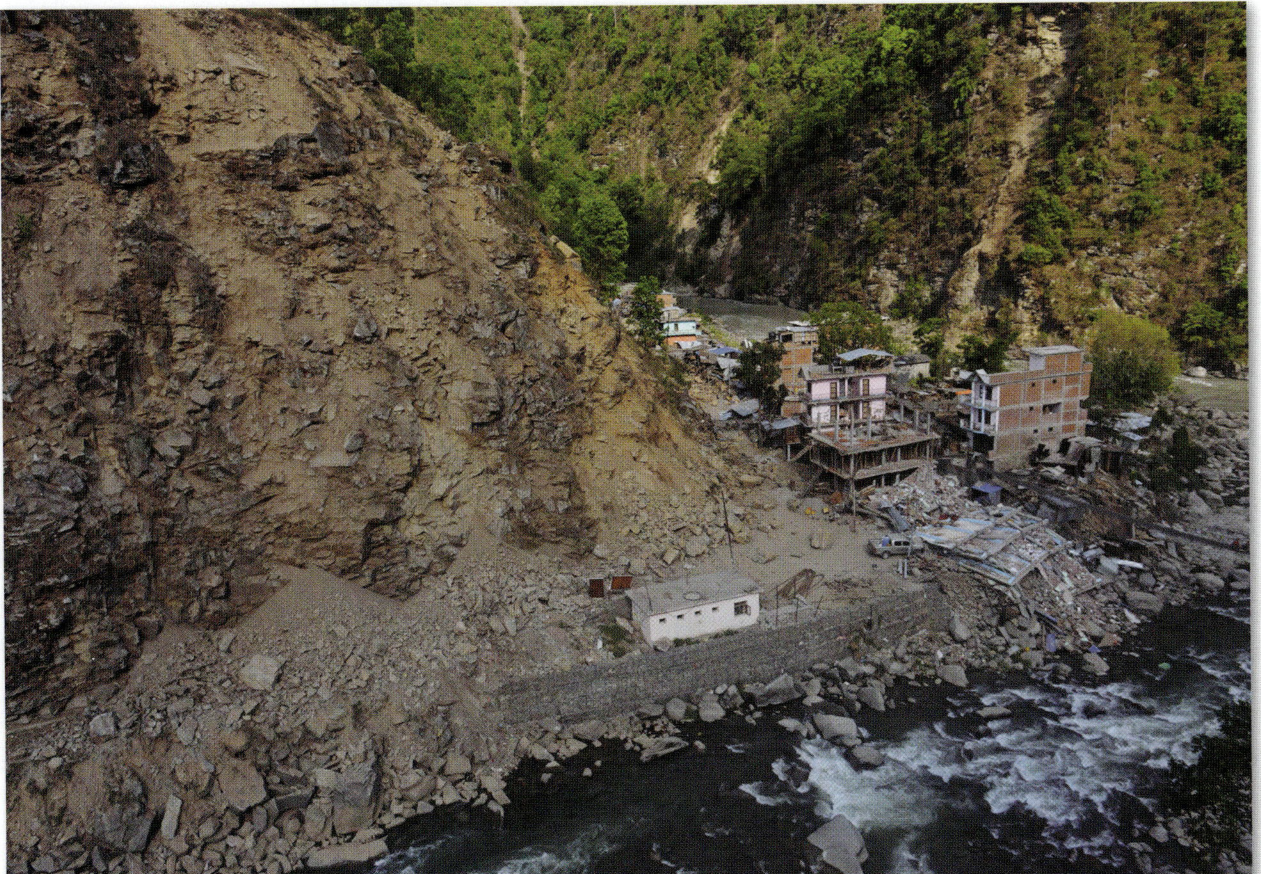

◀ **Figure 12.8**
Earthquakes as triggers A major earthquake in Nepal in April 2015 triggered hundreds of landslides, which destroyed roads and blocked rivers. The damage shown here occurred in the village of Singati in northeastern Nepal. (Photo by Prakash Mathema/AFP/Getty Images)

Earthquakes as Triggers

Conditions that favor mass movement may exist in an area for a long time without movement occurring. An additional factor is sometimes necessary to trigger the movement. Among the most important and dramatic triggers are earthquakes. An earthquake and its aftershocks can dislodge enormous volumes of rock and unconsolidated material (**Figure 12.8**).

Examples from Plate Boundaries: California and Nepal A memorable U.S. example of an earthquake triggering mass movement occurred in January 1994, when a quake struck the Los Angeles region of southern California. Named for its epicenter in the town of Northridge, the magnitude 6.7 event produced estimated losses of $20 billion. Some of the losses resulted from more than 11,000 landslides in an area of about 10,000 square kilometers (3900 square miles) that were set in motion by the quake. Most were shallow rockfalls and slides, but some were much larger and filled canyon bottoms with jumbles of soil, rock, and plant debris. The debris in canyon bottoms creates a secondary threat because it can mobilize during rainstorms, producing debris flows. Such flows are common and often disastrous in southern California.

On April 25, 2015, a magnitude 7.8 earthquake shook much of central Nepal. The earthquake and its aftershocks triggered thousands of landslides in the rugged Himalayas and caused nearly 8900 fatalities. Rock avalanches and debris slides thundered down steep mountain slopes, burying buildings and blocking roads and rail lines. The landslides also dammed rivers, creating many lakes. Earthquake-created lakes present a dual danger. Apart from the upstream floods that occur as a lake builds behind the natural dam, the piles of rubble that form the dam may be unstable. Another quake, or simply the pressure of water behind it, could burst the dam, sending a wall of water downstream. Such floods may also occur when water begins to cascade over the top of the dam.

Liquefaction Intense ground shaking during earthquakes can cause water-saturated surface materials to lose their strength and behave as fluidlike masses that flow (see Figure 9.24, page 253). This process, called *liquefaction*, was a major cause of property damage in Anchorage, Alaska, during the massive 1964 Good Friday earthquake—the largest quake to strike North America in the twentieth century.

Did You Know?
According to the U.S. Geological Survey, studies of debris flows (often called *mudslides* in the media) from burned areas in southern California and other western states have ranged in volume from as small as 800 yd^3 to as large as 400,000 yd^3. This larger volume is equivalent to 50,000 dump truck loads!

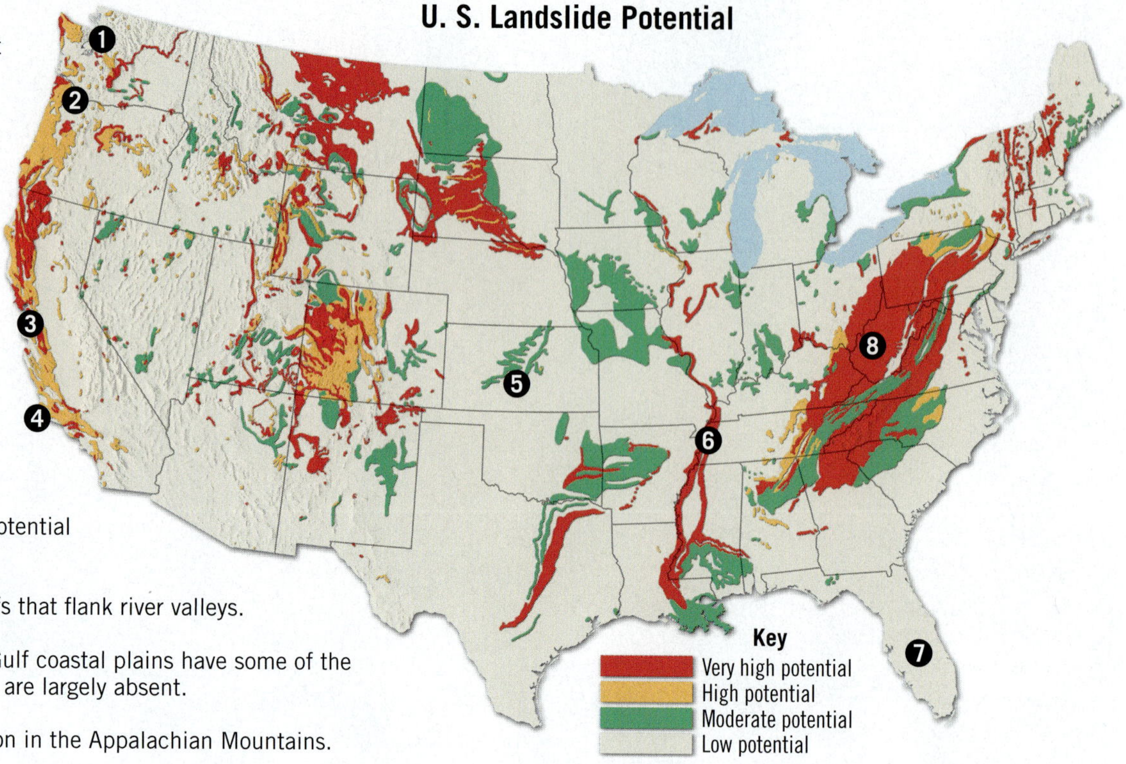

U. S. Landslide Potential

1 In the **Seattle area**, potential threats include volcanic mudflows from Mount Rainier and earthquake-triggered landslides in glacial sediments.

2 In the mountainous parts of the **Pacific Northwest**, heavy rains and melting snow often trigger rapid forms of mass movement.

3 **Coastal California's** steep slopes have a high landslide potential often triggered by winter storms or ground shaking associated with earthquakes.

4 Strong wave activity undercuts and oversteepens coastal cliffs.

5 In the **center of the country**, the plains states are relatively flat, so landslide potential is mostly low-to-moderate.

6 High potential occurs along steep bluffs that flank river valleys.

7 **Florida** and the adjacent Atlantic and Gulf coastal plains have some of the lowest potentials because steep slopes are largely absent.

8 In the **East**, landslides are most common in the Appalachian Mountains.

Key
Very high potential
High potential
Moderate potential
Low potential

▲ **Figure 12.9 Landslide potential in the contiguous 48 states** (From the U.S. Geological Survey)

The Potential for Landslides

Figure 12.9 shows the landslide potential for the contiguous United States. All states experience some damage from rapid mass movement processes, but it is obvious that not all areas of the country have the same landslide hazard potential. As you might expect, there are greater landslide risks in mountain areas. In the East, landslides are most common in the Appalachian Mountains. In the Pacific Northwest, water from winter rains and melting snow often trigger rapid forms of mass movement, including slumps in the glacial sediments that underlie Seattle's and Portland's hilly neighborhoods. Coastal California's steep slopes have a high landslide potential. Here mass movement events may be triggered by winter storms or by the ground shaking associated with earthquakes. Landslides are also triggered when strong wave activity undercuts and oversteepens coastal cliffs.

A glance at the map shows that Florida and the adjacent Atlantic and Gulf coastal plains have some of the lowest landslide potentials because steep slopes are largely absent. In the center of the country, the Plains states are relatively flat, so landslide potential is mostly low to moderate. High-potential areas are along the steep bluffs that flank river valleys.

Did You Know?
Landslides are expensive. The U.S. Geological Survey estimates that the total losses from landslides each year in the United States are between $2 billion and $4 billion (2009 dollars). This is a conservative estimate, as there is no uniform method or overall agency that keeps track of or reports landslide losses. Landslides result in extremely high monetary losses in other countries, too, but there are no reliable estimates of the amounts.

CONCEPT CHECKS 12.2

1. How does water affect mass movement processes? Compare the circumstances that led to the killer landslides at Oso, Washington, and La Conchita, California.
2. Describe the significance of the angle of repose.
3. How might a wildfire influence the risk of mass movement?
4. Describe the relationship between earthquakes and landslides.
5. What factors increase the risks of landslides?

12.3 Classification of Mass Movement Processes

List and explain the criteria that are commonly used to classify mass movement processes.

Geologists use the term *mass movement* to describe a broad array of different processes. Generally, the different types are classified based on the type of material involved, the kind of motion displayed, and the velocity of the movement.

Type of Material

The classification of mass movement processes on the basis of the material involved in the movement depends on whether the descending mass began as unconsolidated material or as bedrock. If soil and regolith dominate, terms such as *debris*, *mud*, or *earth* are used in the description. In contrast, when a mass of bedrock breaks loose and moves downslope, the term *rock* may be part of the description.

Type of Motion

In addition to characterizing the type of material involved in a mass movement event, the way in which the material moves may also be important. Generally, the kind of motion is described as either a fall, a slide, or a flow.

Fall When the movement in a mass movement event involves the free fall of detached individual pieces of any size, it is termed a **fall**. Fall is a common form of movement on slopes that are so steep that loose material cannot remain on the surface. The rock may fall directly to the base of the slope or move in a series of leaps and bounds over other rocks along the way. Rockfalls are the primary way in which **talus slopes** are built and maintained (see Figure 6.4, page 164). Many falls result when freeze and thaw cycles and/or the action of plant roots reduce rock strength to the point that gravity overwhelms the resisting forces. Although signs along bedrock cuts on highways warn of falling rock, few of us have actually witnessed such an event in progress. However, as Figure 12.10 illustrates, they do indeed occur.

When large masses of rock plunge from great heights, they hit the ground with enormous force and often trigger additional mass movement events. One deadly example occurred in Peru. In May 1970, an earthquake caused a huge mass of rock and ice to break free from the precipitous north face of Nevados Huascarán, the loftiest peak in the Peruvian Andes. The material plunged nearly a kilometer and pulverized on impact. The rock avalanche that followed rushed down the mountainside, made fluid by trapped air and ice. Along the way, it ripped loose millions of tons of additional debris that ultimately and tragically buried more than 20,000 people in the towns of Yungay and Ranrahirca.

A different effect triggered by a rockfall occurred in Yosemite National Park on July 10, 1996. When two large rock masses broke loose from steep cliffs and fell about 500 meters (1640 feet) to the floor of Yosemite Valley, the impacts were great enough to be recorded at seismic stations 200 kilometers (125 miles) from the site. As the dislodged rock masses struck the ground, they generated atmospheric pressure waves that were comparable in velocity to a tornado or hurricane. The force of the air blasts uprooted and snapped more than 1000 trees, including some that were 40 meters (130 feet) tall.

Slide Many mass movement processes are described as **slides**. The term refers to mass movements in which there is a distinct zone of weakness separating the slide material from the more stable underlying material. Two basic types of slides are recognized. In *rotational slides*, the surface of rupture is a concave-upward curve that resembles the shape of a spoon, and the descending material exhibits a downward and outward rotation. By contrast, in a *translational slide*, a mass of material moves along a relatively flat surface such as a joint, fault, or bedding plane. Such slides exhibit little rotation or backward tilting.

Flow The third type of movement common to mass movement processes is termed **flow**. Flow occurs when material moves downslope as a viscous, often turbulent, fluid. Most flows are saturated with water and typically move as lobes or tongues.

Rate of Movement

Some of the events that have been described so far involved very rapid rates of movement. For example, it is estimated that the debris that rushed down the slopes of Peru's Nevados Huascarán moved at speeds in excess of 200 kilometers (125 miles) per hour. This most rapid type of mass movement is termed a **rock avalanche**. Geologists used to think that rock avalanches, such as the one that produced the scene in Figure 12.11A, must literally "float on air" as they move downslope. That is, high velocities might result when air becomes trapped and compressed beneath the falling mass of debris, allowing it to move as a buoyant, flexible sheet across the surface. But more recent studies of landslide deposits on other

▼ **SmartFigure 12.10 Watch out for falling rock!** This rockfall blocked a highway. (Main photo by Europics/Newscom)

ANIMATION
https://goo.gl/eeKKoS

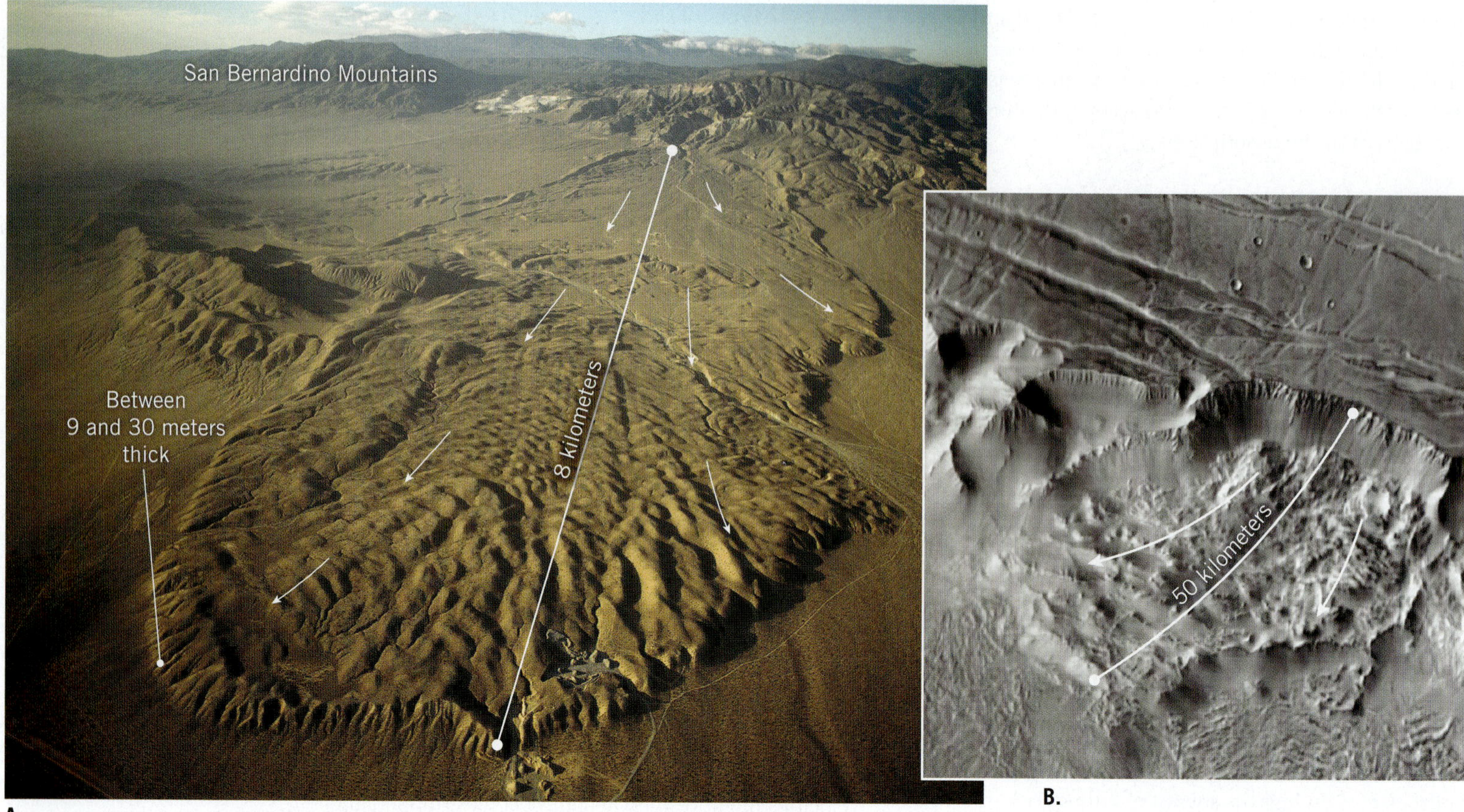

▲ **Figure 12.11**

Landslides on Earth and Mars A. The prehistoric Blackhawk rock avalanche is considered one of the largest non-volcanic landslides in North America. (Photo by Michael Collier) **B.** The Blackhawk event is just a fraction the size of the landslides in the Valles Marineris canyons of Mars. Water probably played an important role in both situations, allowing the deposits to extend so far from their sources. (NASA)

planetary bodies have made us question this hypothesis. Similar deposits from long runout landslides on Mars seem to have been lubricated by interactions of rock and water (**Figure 12.11B**).

Most mass movements, however, do not move with the speed of a rock avalanche. In fact, a great deal of mass movement is imperceptibly slow. One process that we will examine later, termed *creep*, results in particle movements that are usually measured in millimeters or centimeters per year. Thus, as you can see, rates of movement can be spectacularly sudden or exceptionally gradual. Although various types of mass movement are often classified as either rapid or slow, such a distinction is highly subjective because a wide range of rates exists between the two extremes. Even the velocity of a single process at a particular site can vary considerably.

CONCEPT CHECKS 12.3

1. List and sketch three ways material can move during mass movement.
2. How has scientific thinking changed about how rock avalanches move at such great speeds?

12.4 Common Forms of Mass Movement: Rapid to Slow

Distinguish among rockfall, rockslide, debris avalanche, debris flow, slump, and earthflow.

Did You Know?

The U.S. Geological Survey estimates that between 25 and 50 people are killed by landslides annually in the United States The worldwide death toll is much higher.

Classifying kinds of mass movement can be tricky because of the variable factors of speed, three-dimensional geometry, and type of material. The common mass movement processes discussed in this section are slump, rockslide, debris flow, and earthflow. They are most common where slopes are steep. The speed of movement varies from barely perceptible to very rapid.

Slump refers to the downward sliding of a mass of rock or unconsolidated material moving as a unit along a curved surface (**Figure 12.12**). Usually the slumped material does not travel spectacularly fast, and it does not travel very far. Slump is a common form of mass movement, especially in thick accumulations of cohesive materials such as clay. The ruptured surface is characteristically spoon shaped and concave upward or outward. As the movement occurs, a crescent-shaped scarp is created at the head, and the block's upper surface is sometimes tilted backward. Although slump may involve

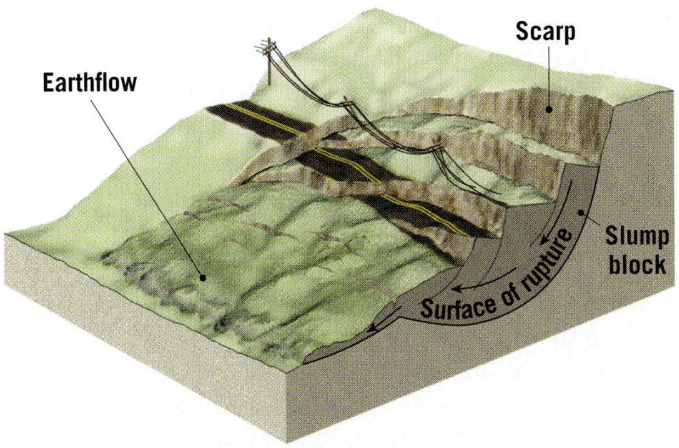

▲ **Figure 12.12 Slump** Slump occurs when material slips downslope en masse along a curved surface of rupture. It is an example of a rotational slide. Earthflows frequently form at the base of the slump.

a single mass, it often consists of multiple blocks. Sometimes water collects between the base of the scarp and the top of the tilted block. As this water percolates downward along the ruptured surface, it may promote further instability and additional movement.

Slump commonly occurs because a slope has been oversteepened. The material on the upper portion of a slope is held in place by the material at the bottom of the slope. As this anchoring material at the base is removed, the material above is made unstable and reacts to the pull of gravity. One relatively common example is a valley wall that becomes oversteepened by a meandering river. Figure 12.13 provides an example in which a coastal cliff has been undercut by wave action at its base. Slumping may also occur when a slope is overloaded, causing internal stress on the material below. This type of slump often occurs where weak, clay-rich material underlies layers of stronger, more resistant rock such as sandstone. The seepage of water through the upper layers reduces the strength of the clay below, resulting in slope failure.

Rockslide & Debris Avalanche

Rockslides occur when blocks of bedrock break loose and slide down a slope (Figure 12.14). If the material involved is largely unconsolidated, the term **debris slide** is used instead. After the initial sliding stage, the material can break up into a chaotic **debris avalanche**. Such events are among the fastest and most destructive mass movements. Usually rockslides take place in a geologic setting where the rock strata are inclined, or where joints and fractures exist parallel to the slope. When such a rock unit is undercut at the base of the slope, it loses support, and the rock eventually gives way. Sometimes a rockslide is triggered when rain or melting snow lubricates the underlying surface to the point that friction is no longer sufficient to hold the rock unit in place. As a result, rockslides tend to be most common during spring, when heavy rains and melting snow are most prevalent.

As mentioned earlier, earthquakes can trigger rockslides and other mass movements. There are many well-known examples. On August 17, 1959, a severe earthquake west of Yellowstone National Park triggered a massive slide in the canyon of the Madison River in southwestern Montana. In a matter of moments, an estimated 27 million cubic meters of rock, soil, and trees slid into the canyon. The debris dammed the river and buried a campground and highway. More than 20 unsuspecting campers perished.

Heavy rains and melting snow, rather than an earthquake, triggered another major rockslide—a debris avalanche in the Yellowstone region. The legendary Gros Ventre rockslide occurred not far from the site of the Madison Canyon slide, 34 years earlier. The Gros Ventre River flows west from the northernmost part of the Wind

▶ **Figure 12.13 Slump at Point Fermin, California** Waves undercut the base of the steep slope, making it unstable. (Photo by John S. Shelton/University of Washington Libraries)

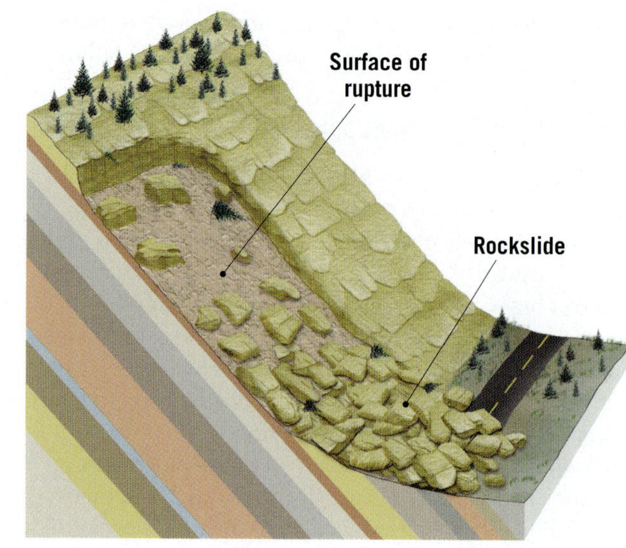

Figure 12.14 Rockslide These rapid movements are classified as translational slides, in which the material moves along a relatively flat surface with little or no rotation or backward tilting.

SmartFigure 12.15 Gros Ventre rockslide This massive slide occurred on June 23, 1925, just east of the small town of Kelly, Wyoming. (Photo by Michael Collier)

TUTORIAL
https://goo.gl/ek79Ry

The side of the mountain gave way when the tilted sandstone bed, that had been cut through by the river, could no longer maintain its position atop the saturated bed of clay.

Gros Ventre landslide debris
Former land surface
Scar
Lake
Clay bed
Sandstone
Limestone
0 0.5 1
Kilometer
Rupture surface

Even though the Gros Ventre rockslide occurred in 1925, the scar left on the side of Sheep Mountain is still a prominent feature.

River Range in northwestern Wyoming, through Grand Teton National Park, and eventually empties into the Snake River. On June 23, 1925, a massive rockslide took place in its valley, just east of the small town of Kelly. In the span of just a few minutes, a great mass of sandstone, shale, and soil crashed down the south side of the valley, carrying with it a dense pine forest. The volume of debris, estimated at 38 million cubic meters (50 million cubic yards), created a 70-meter- (230-foot-) high dam on the Gros Ventre River. Because the river was completely blocked, a lake was formed. It filled so quickly that a house that had been 18 meters (60 feet) above the river floated off its foundation 18 hours after the slide. In 1927, the lake overflowed the dam, partially draining the lake and resulting in a devastating flood downstream.

Why did the Gros Ventre rockslide take place? Figure 12.15 shows a diagrammatic cross-sectional view of the geology of the valley. You will notice that (1) the sedimentary strata in this area dip (tilt) 15 to 21 degrees; (2) underlying the bed of sandstone is a relatively thin layer of clay; and (3) at the bottom of the valley, the river had cut through much of the sandstone layer. During the spring of 1925, water from heavy rains and melting snow seeped through the sandstone, saturating the clay below. Because much of the sandstone layer had been cut through by the Gros Ventre River, the layer had virtually no support at the bottom of the slope. Eventually the sandstone could no longer hold its position on the wetted clay, and gravity pulled the mass down the side of the valley. The circumstances at this location were such that the event was inevitable.

Debris Flow

Debris flow is a relatively rapid type of mass movement that involves a flow of soil and regolith containing a large amount of water (Figure 12.16). The La Conchita debris flow (see Figure 12.3) is one example. Debris flows are sometimes called **mudflows** when the material is primarily fine-grained. As you will see, debris flows are a particular problem in semiarid mountainous regions and on steep volcanic slopes, where they are called *lahars*. Because of their fluid properties, debris flows frequently follow canyons and stream channels. In populated areas, debris flows can pose a significant hazard to life and property.

Debris Flows in Semiarid Regions

When a cloudburst or rapidly melting mountain snows create a sudden flood in a semiarid region, large quantities of soil and regolith are washed into nearby stream channels because there is usually little vegetation to anchor the surface material. The end product is a flowing tongue of well-mixed mud, soil, rock, and water. Its consistency may range from that of wet concrete to a soupy mixture not much thicker than muddy water. The rate of flow, therefore, depends not only on the slope but also on the water content. Dense debris flows are capable of carrying or pushing large boulders, trees, and even houses with relative ease.

Debris flows pose a serious hazard to development in relatively dry mountainous areas such as southern California. The construction of homes on canyon hillsides and the removal of native vegetation by brush fires and other means have increased the frequency of these destructive events. Moreover, when a debris flow reaches

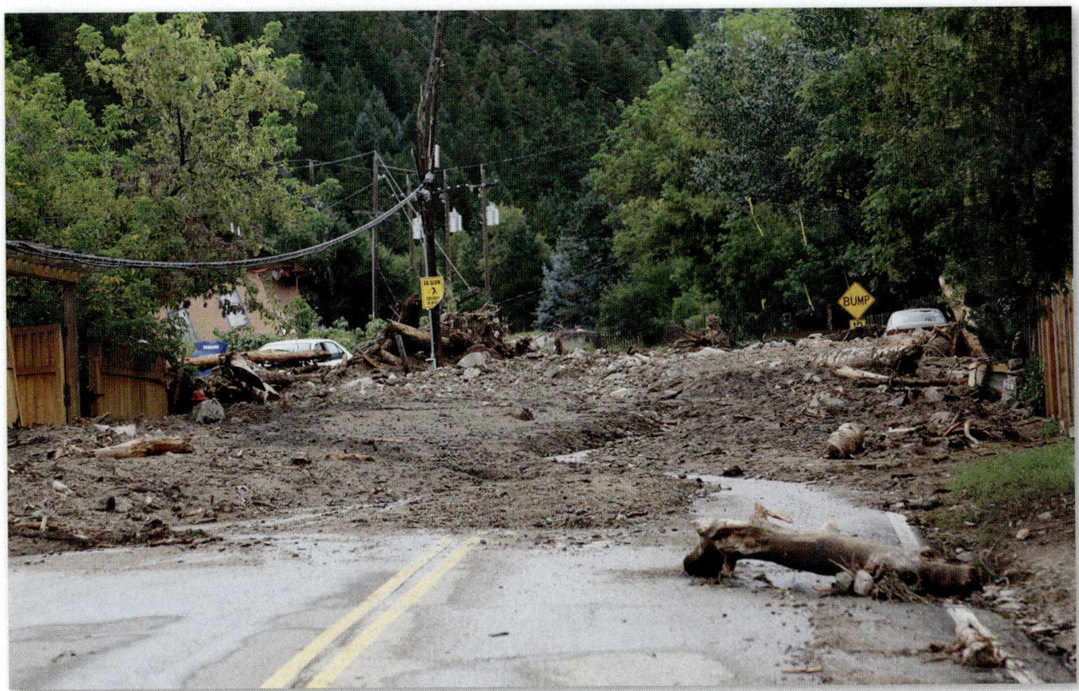

◄ **Figure 12.16 Debris flow in the Colorado Front Range** During the week of September 9 to 13, 2013, residents in and near Boulder, Colorado, received a harsh reminder of the dangers posed by debris flows. During that 5-day span, nearly continuous rainfall triggered numerous flash floods and more than 1100 debris flows in an area covering more than 3400 square kilometers (1300 square miles). (Photo by Rick Wilking/Reuters)

the end of a steep, narrow canyon, it spreads out, covering the area beyond the mouth of the canyon with a mixture of wet debris. This material contributes to the buildup of fanlike deposits, called *alluvial fans*, at canyon mouths. The fans are relatively easy to build on, often have nice views, and are close to the mountains; in fact, like the nearby canyons, many have become preferred sites for development. Because debris flows occur only sporadically, the public is often unaware of the potential hazard of such sites.

Lahars Debris flows composed mostly of volcanic materials on the flanks of volcanoes are called **lahars**. The word originated in Indonesia, a volcanic region that has experienced many of these often-destructive events. Historically, lahars have been some of the deadliest volcano-related hazards. They can occur either during an eruption or when a volcano is quiet. They take place when highly unstable layers of ash and debris become saturated with water and flow down steep volcanic slopes, generally following existing stream channels (Figure 12.17). Heavy rainfalls often trigger these flows. Others are initiated when large volumes of ice and snow are melted by heat flowing to the surface from within the volcano or by the hot gases and near-molten debris emitted during a violent eruption.

When Mount St. Helens erupted in May 1980, several massive lahars were created. The flows and accompanying floods raced down the valleys of the north and south forks of the Toutle River at speeds often in excess of 30 kilometers (20 miles) per hour. Fortunately, the affected area was not densely settled. Nevertheless, more than 200 homes were destroyed or severely damaged. Most bridges met a similar fate.

In November 1985, lahars were produced during the eruption of Nevado del Ruiz, a 5300-meter (17,400-foot) volcano in the Andes Mountains of Colombia.

▼ **Figure 12.17 Lahar at Mt. Redoubt** An eruption in April 2009 of this volcano on Alaska's Kenai Peninsula sent lahars down the valley of the Drift River. The dark color of the lahar contrasts sharply with the surrounding snow-covered landscape. (NASA)

Lahar following valley of Drift River

Mt. Redoubt erupting ash column

Ash-covered snow

Ash plume spreading downwind

Cook Inlet

N

Geologist's Sketch

▲ **Figure 12.18 Earthflow** This small tongue-shaped earthflow occurred on a newly formed slope along a recently constructed highway in central Illinois. It formed in clay-rich material following a period of heavy rain. Notice the small slump at the head of the earthflow. (Photo by E. J. Tarbuck)

The eruption melted much of the snow and ice that capped the uppermost 600 meters (2000 feet) of the peak, producing torrents of hot, viscous mud, ash, and debris. The lahars moved outward from the volcano, following the valleys of three rain-swollen rivers that radiate from the peak. The flow that moved down the valley of the Lagunillas River was the most destructive. It devastated the town of Armero, 48 kilometers (30 miles) from the mountain, causing more than 25,000 deaths.

Earthflow

We have seen that debris flows are frequently confined to channels in semiarid regions. In contrast, **earthflows** most often form on hillsides in humid areas during times of heavy precipitation or snowmelt. When water saturates the soil and regolith on a hillside, the material may break away, leaving a scar on the slope and forming a tongue- or teardrop-shaped mass that flows downslope (Figure 12.18).

The materials most commonly involved are rich in clay and silt and contain only small proportions of sand and coarser particles. Earthflows range in size from bodies a few meters long, a few meters wide, and less than a meter deep to masses more than 1 kilometer long, several hundred meters wide, and more than 10 meters deep. Because earthflows are quite viscous, they generally move at slower rates than the more fluid debris flows described in the preceding section. They are characterized by a gradual movement that may go on for periods ranging from days to years. Depending on the steepness of the slope and the material's consistency, measured velocities range from less than a millimeter a day up to several meters a day. Over the time span that earthflows are active, movement is typically faster during wet periods than during drier times. In addition to occurring as isolated hillside phenomena, earthflows commonly take place in association with large slumps. In this situation, they may be seen as tongue-like flows at the base of the slump block.

CONCEPT CHECKS 12.4

1. Without looking at Figures 12.12 and 12.13, sketch and label a simple cross section (side view) of a slump.
2. Both slumps and rockslides move by sliding. How do these processes differ?
3. What factors led to the massive rockslide at Gros Ventre, Wyoming?
4. How is a lahar different from a debris flow that might occur in southern California?
5. Contrast earthflows and debris flows.

12.5 Very Slow Mass Movements

Review the general characteristics of slow mass movement processes and describe the unique issues associated with a permafrost environment.

Movements such as rockslides, rock avalanches, and lahars are certainly the most spectacular and catastrophic forms of mass movement. These dangerous events deserve intensive study to enable more effective prediction, timely warnings, and better controls to save lives. However, because of their large size and spectacular nature, they give us a false impression of their importance as mass movement processes. Indeed, sudden movements are responsible for moving less material than the slower and far more subtle action of creep. Whereas rapid types of mass movement are characteristic of mountains and steep hillsides, creep takes place on both steep and gentle slopes and is thus much more widespread.

Creep

Creep is a type of mass movement that involves the gradual downhill movement of soil and regolith. One factor that contributes to creep is the alternating expansion

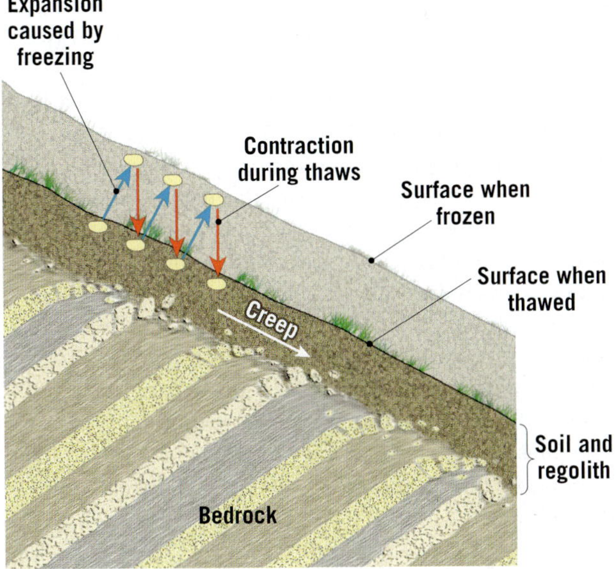

▲ **SmartFigure 12.19 Creep** The repeated expansion and contraction of the surface material causes a net downslope migration of soil and rock particles.

TUTORIAL
https://goo.gl/DVmxZQ

and contraction of surface material caused by freezing and thawing or wetting and drying. As shown in Figure 12.19, freezing or wetting lifts particles at right angles to the slope, and thawing or drying allows the particles to fall back to a slightly lower level. Each cycle therefore moves the material a tiny distance downslope. Creep is aided by anything that disturbs the soil, such as raindrop impact and disturbance by plant roots and burrowing animals. Creep is also promoted when the ground becomes saturated with water. Following a heavy rain or snowmelt, a water-logged soil may lose its internal cohesion, allowing gravity to pull the material downslope. Because creep is imperceptibly slow, the process cannot be observed in action. However, the effects of creep can be observed. Creep causes fences and utility poles to tilt and retaining walls to be displaced.

Solifluction

When soil is saturated with water, the soggy mass may flow downslope at a rate of a few millimeters or a few centimeters per day or per year. Such a process is called **solifluction** (literally "soil flow"). It is a type of mass movement that is common wherever water cannot escape from the saturated surface layer by infiltrating to deeper levels. A dense clay hardpan in soil or an impermeable bedrock layer can promote solifluction.

Solifluction is also common in regions underlain by permafrost. *Permafrost* refers to the permanently frozen ground that occurs in association with Earth's harsh tundra and subarctic climates. (There is more about permafrost in the next section.) Solifluction occurs in a zone above the permafrost called the *active layer*, which thaws to a depth of about a meter during the brief high-latitude summer and then refreezes in winter. During the summer season, water is unable to percolate into the impervious permafrost layer below. As a result, the active layer becomes saturated and slowly flows. The process can occur on slopes as gentle as 2 to 3 degrees. Where there is a well-developed mat of vegetation, a solifluction sheet may move in a series of well-defined lobes or as a series of partially overriding folds (Figure 12.20).

The Sensitive Permafrost Landscape

Many of the mass movement disasters described in this chapter had sudden and disastrous impacts on people. When the activities of people cause ice contained in permanently frozen ground to melt, the impact is more gradual and less deadly. Nevertheless, because permafrost regions are sensitive and fragile landscapes, the scars resulting from poorly planned actions can remain for generations.

Permanently frozen ground, known as **permafrost**, occurs where summers are too short and cool to melt more than a shallow surface layer. Deeper ground remains frozen year-round. Permafrost is extensive in the lands surrounding the Arctic Ocean. Strictly speaking, permafrost is defined only on the basis of temperature; that is, it is ground with temperatures that have remained below 0°C (32°F) continuously for 2 years or more. The degree to which ice is present in the ground strongly affects the behavior of the surface material.

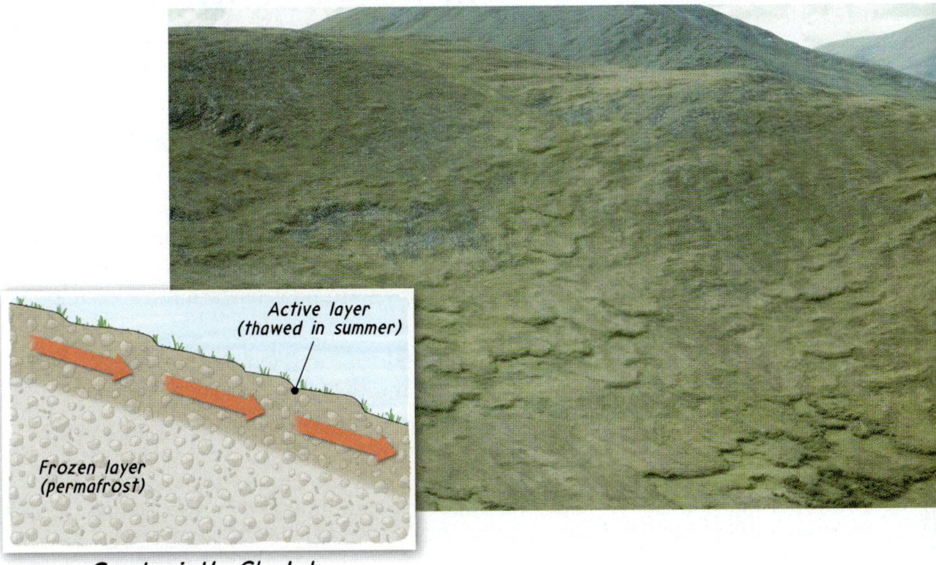

▼ **Figure 12.20 Solifluction lobes near the Arctic Circle in Alaska** Solifluction occurs in permafrost regions when the active layer thaws in summer. (Photo by James E. Patterson, courtesy of F. K. Lutgens)

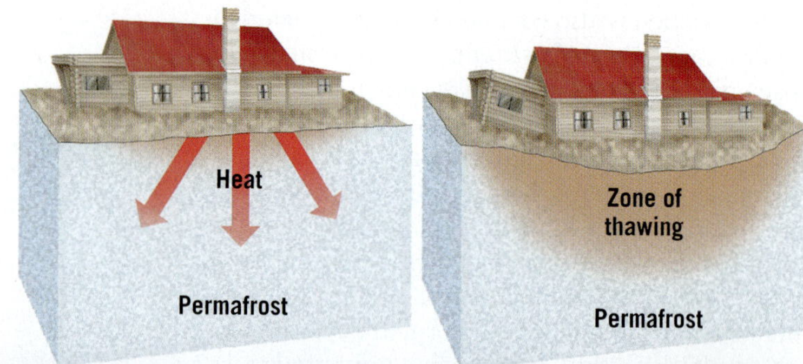

▲ **SmartFigure 12.21 When permafrost thaws** This building, located south of Fairbanks, Alaska, subsided because of thawing permafrost. Notice that the right side, which was heated, settled much more than the unheated porch on the left. (Photo by Steve McCutcheon, U.S. Geological Survey)

TUTORIAL
https://goo.gl/rvlQNg

Knowing how much ice is present and where it is located is very important when it comes to constructing roads, buildings, and other projects in areas underlain by permafrost.

When people disturb the surface, such as by removing the insulating vegetation mat or by constructing roads and buildings, the delicate thermal balance is disturbed, and the permafrost can thaw. Thawing produces unstable ground that may slide, slump, subside, and undergo severe frost heaving. When a heated structure is built directly on permafrost that contains a high proportion of ice, thawing creates soggy material into which a building can sink (**Figure 12.21**). One solution is to place buildings and other structures on piles, like stilts. The piles allow subfreezing air to circulate between the floor of the building and the soil and thereby keep the ground frozen.

CONCEPT CHECKS 12.5

1. Describe the basic mechanisms that contribute to creep. How might you recognize that creep is occurring?
2. During what season does solifluction in the Arctic occur? Explain why it occurs only during that season.
3. What is permafrost? How might disturbing permafrost lead to unstable ground that may slide, flow, or subside?

CONCEPTS IN REVIEW

Mass Movement on Slopes: The Work of Gravity

12.1 The Importance of Mass Movement

Describe how mass movement processes can cause natural disasters and discuss the role that mass movement plays in the development of landscapes.

KEY TERM: mass movement

- After weathering breaks apart rock, gravity moves the debris downslope, in a process called mass movement. Sometimes this occurs rapidly as a landslide, and at other times the movement is slower. Landslides are a significant geologic hazard, taking many lives and destroying property.
- Mass movement serves an important role in landscape development. It widens stream valleys and helps tear down the mountains thrust up by plate tectonics.

? The Oso landslide shown in the chapter-opening photo surprised many people. In this detailed topographic map of the valley (derived using lidar technology), the deposit of the 2014 Oso landslide is shown by the cross-hatch pattern. Colored areas show older landslide deposits, distinguished by age. Study the map and explain whether you think the 2014 event should have been anticipated.

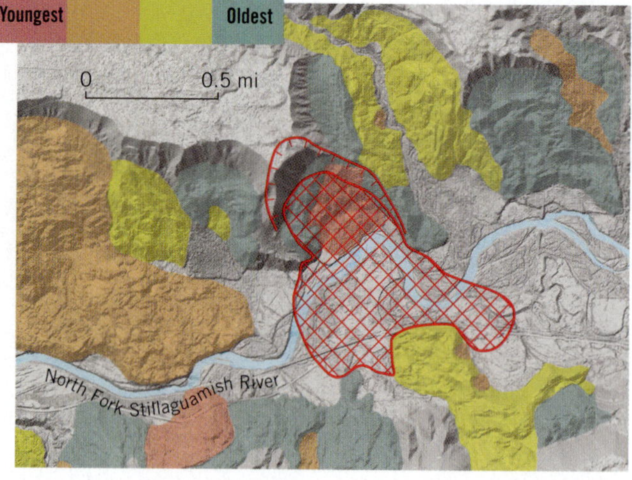

12.2 Controls & Triggers of Mass Movement

Summarize the factors that control and trigger mass movement processes.

KEY TERMS: trigger, angle of repose

- An event that initiates a mass movement process is referred to as a trigger. The addition of water, oversteepening of the slope, removal of vegetation, and shaking due to an earthquake are four important examples of triggers. Not all landslides are triggered by one of these four processes, but many are.
- Water added to a slope can expand the pores, separating grains and causing them to lose their cohesion. Water also lubricates the contacts between the grains and adds a significant amount of mass to a wetted slope.
- Granular materials can pile up to a certain angle of slope, but granular piles steeper than that critical angle will spontaneously collapse outward to form a gentler slope. For most geologic materials, this angle of repose varies between 25 and 40 degrees from horizontal. Oversteepened slopes are likely to fail with a landslide.
- The roots of plants (especially plants with deep roots) act as a three-dimensional "net" that holds soil and regolith particles in place. When the plants die, the soil loses an important support structure. Plants may be removed naturally (through wildfires, for instance) or by humans seeking to harvest timber, plant crops, or build structures.
- Earthquakes are significant triggers that deliver an energetic jolt to slopes poised on the brink of failure.

? Do all mass movement events have triggers? Explain.

12.3 Classification of Mass Movement Processes

List and explain the criteria that are commonly used to classify mass movement processes.

KEY TERMS: fall, talus slope, slide, flow, rock avalanche

- There are a variety of Earth materials (rock, soil, and regolith) and a variety of rates of mass movement. The type of material and the nature of the motion are combined into classifying terms for the different types of mass movement.
- Rockfalls occur when pieces of bedrock detach and fall freely through the air, slamming into the ground below with tremendous force. Repeated rockfalls generate a talus slope, the characteristic "apron" of angular rock debris that accumulates below mountain cliffs. Slides occur when discrete blocks of rock or unconsolidated material slip downslope on a planar or curved surface. Unlike in a fall, the material in a slide does not drop through the air. Flows occur when individual grains or particles move randomly in a slurry, a viscous mixture of water-saturated materials.
- Rock and debris avalanches on Earth and Mars move incredibly rapidly over surprising distances (as much as tens of kilometers in a few minutes). Other forms of mass movement are much slower, moving perhaps only millimeters per year.

12.4 Common Forms of Mass Movement: Rapid to Slow

Distinguish among rockfall, rockslide, debris avalanche, debris flow, slump, and earthflow.

KEY TERMS: slump, rockslide, debris slide, debris avalanche, debris flow, mudflow, lahar, earthflow

- A slump is a distinctive and common form of mass movement in which coherent blocks of material move downhill on a spoon-shaped slip surface. Slumps are often marked by curved scarps that open up at their tops. They are frequently triggered by oversteepening, such as that caused by stream erosion of a valley wall.
- Rockslides are rapid mass movement events in which a coherent block of rock slides downhill along a planar surface. Often this is a preexisting structure such as a joint or a bedding plane. Situations where these surfaces dip into a valley at an angle are especially dangerous when the slide transforms into a chaotic debris avalanche.
- Debris flows occur when unconsolidated soil or regolith becomes saturated with water and moves downhill in a slurry, picking up other objects (trees, houses, livestock) along the way. Varieties of debris flow include mudflows, which are dominated by small particle sizes, and lahars, which involve volcanic materials. Debris flows can move quickly—up to 30 kilometers (20 miles) per hour.
- Earthflow is characterized by a similar loss of coherence between grains in unconsolidated material, but it is much slower than a debris flow. Typically, sites of earthflow show an uphill scarp and a lobe of viscous soil on the downhill side.

? This rockslide occurred in the rugged Himalayas of northern India. Identify a feature in the photo that may have been a factor that contributed to the slide. Speculate about what might have triggered the event.

Sajjad Hussain/AFP/Getty Images

12.5 Very Slow Mass Movements

Review the general characteristics of slow mass movement processes and describe the unique issues associated with a permafrost environment.

KEY TERMS: creep, solifluction, permafrost

- Creep is a widespread and important form of mass movement that is very slow. It occurs when freezing (or wetting) causes soil particles to be pushed out away from the slope, only to drop down to a lower position following thawing (or drying). In contrast, solifluction is the gradual flow of a saturated surface layer that is underlain by an impermeable zone. In arctic regions, the impermeable zone is permafrost.
- Permafrost, permanently frozen ground, covers large portions of North America and Siberia. Constructing buildings and other infrastructure in such regions requires special planning. Leaking heat can melt permafrost, causing a loss of volume and triggering flow in the formerly frozen soil. The resulting subsidence can be devastating.

? This pipeline carries heated oil from Alaska's North Slope to a port along Alaska's south coast. Notice that the pipeline is not buried but rather suspended above ground. Suggest a reason the pipeline in this image is not buried.

Michael Collier

GIVE IT SOME THOUGHT

1. Describe a type of mass movement that might occur in your home area. Remember to consider characteristics such as climate, surface materials, and steepness of slopes. Does your example have a trigger?
2. Rivers, groundwater, glaciers, wind, and waves can all move and deposit sediment. Geologists refer to these phenomena as *agents of erosion*. Mass movement also involves the movement and deposition of sediment, yet it is *not* classified as an agent of erosion. How is mass movement different?
3. Describe at least one situation in which an internal process might cause or contribute to a mass movement event.
4. Do you think it is likely that landslides frequently occur on the Moon? Explain why or why not.
5. Heavy rains in late July 2010 triggered the mass movement that occurred in this mountain valley near Durango, Colorado. Heavy equipment is clearing away material that blocked railroad tracks and significantly narrowed the adjacent stream channel. Was this event more likely a rockfall, creep, or a debris flow? Most of us are familiar with the phrase "One thing leads to another." It certainly applies to the Earth system. Suppose the material from the mass movement event shown here had completely filled the stream. What other natural hazard might have developed?

Soaring Tree Adventures

6 When the rail line in the accompanying photo was built in rural Alaska in the 1930s, the terrain was relatively level. Not long after the railroad was completed, a great deal of subsidence and shifting of the ground occurred, turning the tracks into the "roller coaster" shown here. As a result, the rail line had to be abandoned. Suggest a reason the ground became unstable and shifted.

L.A. Yewle/O.J. Ferrians, Jr./USGS

7 This photo shows landslide debris atop Buckskin Glacier in Denali National Park in the rugged Alaska Range. The glacier feeds a river that flows into Cook Inlet, just west of the city of Anchorage. Cook Inlet is an arm of the North Pacific.

a. Based on what you learned about sorting of sediment in Chapter 7 (Figure 7.7, page 191), would you expect the material deposited by the landslide to be well sorted? Why or why not?

b. Also referring to Chapter 7 and Figure 7.7, would you expect the particles in the landslide debris to be rounded or angular? Explain.

c. These mountains are clearly being sculpted by glaciers. However, other processes also have important roles. Identify these processes and briefly describe the role that each plays in the evolution of this mountainous landscape.

Michael Collier

8 Mass movement is influenced by many processes associated with all four spheres of the Earth system. Select two items from the list below. For each, outline a series of events that relate the item to various spheres and to a mass movement process. The following example does this for the term *frost wedging*: *Frost wedging involves rock (geosphere) being broken when water (hydrosphere) freezes. Freeze–thaw cycles (atmosphere) promote frost wedging. When frost wedging loosens a rock on a cliff, the fragment tumbles to the base of the cliff. This event, rockfall, is an example of mass movement.* Now you give it a try. Use your imagination.

- Deforestation
- Spring thaw/melting snow
- Highway road cut
- Crashing waves
- Cavern formation (see Figure 14.30, page 388)

9 The Swift Creek landslide is a large, complex slump in northwest Washington State that transitions into an earthflow and also generates debris flows during heavy rain events. Explore the Swift Creek Landslide Observatory website (http://landslide.geol.wwu.edu). The landslide moves on average 3 to 4 meters per year. That's about 1 centimeter per day! To see this slow movement over an extended period, click TimeLapse (upper right) to try out the video generator, using these settings:

Start Date: 01/01/2007
End Date: 01/01/2014
Time of Day: Between 12-3pm
Camera: Upper
Framerate: Medium (20 frames/second)
Video Length: 10 seconds

Then click Generate. (The site may take about 30 seconds to generate your custom time-lapse video.)

After you try this, use the video generator to determine whether the Swift Creek Landslide moves more during the wet season (November–April) or during the dry season (May–October).

MasteringGeology™

Looking for additional review and test prep materials? Visit the Study Area in MasteringGeology to enhance your understanding of this chapter's content by accessing a variety of resources, including Self-Study Quizzes, Geoscience Animations, SmartFigures, Mobile Field Trips, *Project Condor* Quadcopter videos, *In the News* RSS feeds, flashcards, web links, and an optional Pearson eText.

www.masteringgeology.com

13

Running Water

FOCUS ON CONCEPTS

Each statement represents the primary learning objective for the corresponding major heading within the chapter. After you complete the chapter, you should be able to:

13.1 List the hydrosphere's major reservoirs and describe the different paths that water takes through the hydrologic cycle.

13.2 Describe the nature of drainage basins and river systems. Sketch and briefly explain four basic drainage patterns.

13.3 Discuss streamflow and the factors that cause it to change.

13.4 Outline the ways in which streams erode, transport, and deposit sediment.

13.5 Contrast bedrock and alluvial stream channels. Distinguish between two types of alluvial channels.

13.6 Contrast narrow V-shaped valleys, broad valleys with floodplains, and valleys that display incised meanders or stream terraces.

13.7 List the major depositional landforms associated with streams and describe the formation of these features.

13.8 Summarize the various categories of floods and the common measures of flood control.

The Colorado River winding through Canyonlands National Park in southern Utah. When this meandering path was established, the river flowed across a relatively flat landscape. Subsequently, the region was gradually lifted upward, while downward erosion lowered the riverbed. The meandering pattern persists, but the loops, locked within confining walls, are now referred to as incised meanders. (Photo by Michael Collier)

CONSIDER THE BITTERSWEET RELATIONSHIP we have with rivers. They are vital economic tools—used as highways to move goods and as sources of water for irrigation and energy—as well as prime locations for recreation. When considered as part of the Earth system, rivers and streams represent a fundamental link in the constant cycling of our planet's water. As streams move across the land, they erode more terrain and transport more sediment than any other process. Soils and other materials can be washed away, only to be deposited elsewhere. Because human populations concentrate along rivers, flooding represents enormous potential for destruction and loss of life.

This chapter provides an overview of the hydrologic cycle, the nature of river systems, the types of river channels and the factors that produce them, the influences of running water on our planet's landscapes, and the nature of floods and their impact on people.

13.1 Earth as a System: The Hydrologic Cycle

List the hydrosphere's major reservoirs and describe the different paths that water takes through the hydrologic cycle.

Water is constantly moving among Earth's different spheres—the *hydrosphere*, the *atmosphere*, the *geosphere*, and the *biosphere*. This unending circulation of water is called the **hydrologic cycle**. Earth is the only planet in the solar system that has a global ocean and a hydrologic cycle.

Earth's Water

Water is almost everywhere on Earth—in the oceans, glaciers, rivers, lakes, air, soil, and living tissue. All these "reservoirs" constitute Earth's hydrosphere. In all, the water content of the hydrosphere is an estimated 1.36 billion cubic kilometers (326 million cubic miles).

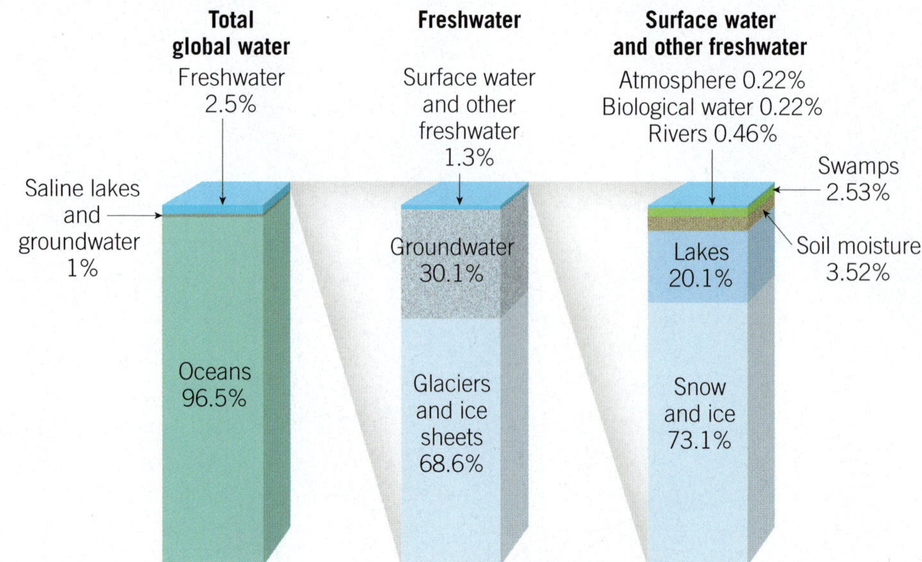

▼ **Figure 13.1 Distribution of Earth's water**

The vast bulk of it, about 96.5 percent, is stored in the global ocean. Ice sheets and glaciers account for an additional 1.76 percent, leaving just slightly more than 2 percent to be divided among lakes, streams, groundwater, and the atmosphere (**Figure 13.1**). Although the percentage of Earth's total water found in each of the latter sources is just a small fraction of the total inventory, the absolute quantities are great.

Water's Paths

The hydrologic cycle is a gigantic, worldwide system powered by energy from the Sun, in which the atmosphere provides a vital link between the oceans and continents (**Figure 13.2**). **Evaporation**, the process by which liquid water changes into water vapor (gas), is how water enters the atmosphere from the ocean and, to a much lesser extent, from the continents. Winds transport this moisture-laden air, often great distances. Complex processes of cloud formation eventually result in precipitation. The precipitation that falls into the ocean has completed its cycle and is ready to begin another. The water that falls on the continents, however, must make its way back to the ocean.

What happens to precipitation once it has fallen on land? A portion of the water soaks into the ground (called **infiltration**), slowly moving downward, then moving laterally, and finally seeping into lakes, streams, or directly into the ocean. When the rate of rainfall exceeds the ground's ability to absorb it, the surplus water flows over the surface into lakes and streams, a process called **runoff**. Much of the water that infiltrates or runs off eventually returns to the atmosphere because of evaporation from the soil, lakes, and streams.

Also, some of the water that soaks into the ground is absorbed by plants, which then release it into the atmosphere. This process is called **transpiration**. Because both evaporation and transpiration involve the transfer of water from the surface directly to the atmosphere, they are often considered together as the combined process of **evapotranspiration**.

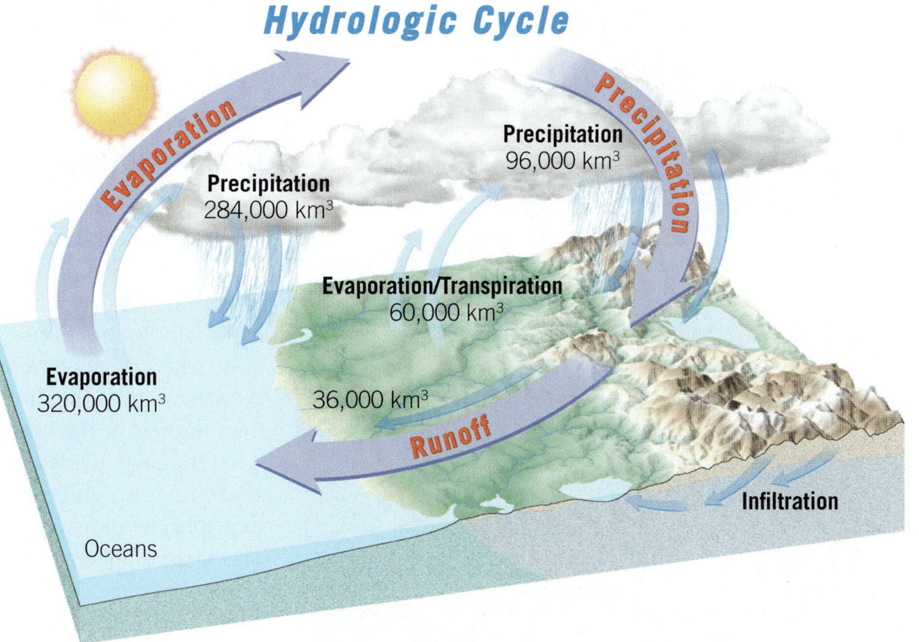

◀ SmartFigure 13.2
The hydrologic cycle The primary movement of water through the cycle is shown by the large arrows. A number refers to the annual amount of water taking a particular path.

TUTORIAL
https://goo.gl/ZYRTcf

Storage in Glaciers

When precipitation falls in very cold places—at high elevations or high latitudes—the water may not immediately soak in, run off, or evaporate. Instead, it may become part of a snowfield or a glacier. In this way, glaciers store large quantities of water on land. If present-day glaciers were to melt and release all their water, sea level would rise by several dozen meters. Such a rise would submerge many heavily populated coastal areas. As you will see in Chapter 15, over the past 2 million years, huge ice sheets have formed and melted on several occasions, each time affecting the balance of the hydrologic cycle.

Water Balance

Figure 13.2 also shows Earth's overall *water balance*, or the volume that passes through each part of the cycle annually. The amount of water vapor in the air at any one time is just a tiny fraction of Earth's total water supply. But the *absolute* quantities that are cycled through the atmosphere over a 1-year period are immense—some 380,000 cubic kilometers (91,000 cubic miles)—enough to cover Earth's entire surface to a depth of about 1 meter (39 inches).

It is important to know that the hydrologic cycle is *balanced*. Because the total amount of water vapor in the atmosphere remains about the same, the average annual precipitation worldwide must be equal to the quantity of water evaporated. However, for all the continents taken together, precipitation exceeds evaporation. Conversely, over the oceans, evaporation exceeds precipitation. Because the level of the world ocean is not dropping, the system must be in balance. In Figure 13.2, the 36,000 cubic kilometers (8600 cubic miles) of water that annually makes its way from the land to the ocean causes enormous erosion. In fact, this immense volume of moving water is *the single most important agent sculpting Earth's land surface*. In the rest of this chapter, we will examine stream characteristics, erosion and deposition by running water, and floods, a significant natural hazard.

CONCEPT CHECKS 13.1

1. Describe or sketch the movement of water through the hydrologic cycle. Once precipitation has fallen on land, what paths might the water take?
2. What is meant by the term *evapotranspiration*?
3. Over the oceans, evaporation exceeds precipitation, yet sea level does not drop. Explain why.

Did You Know?
Each year a field of crops may transpire the equivalent of a water layer 2 ft deep over the entire field. The same area of trees may pump twice this amount into the atmosphere.

13.2 Running Water

Describe the nature of drainage basins and river systems. Sketch and briefly explain four basic drainage patterns.

Much of the precipitation that falls on land either soaks into the ground (infiltration) or remains at the surface, moving downslope as runoff. The amount of water that runs off compared to the amount that infiltrates depends on several factors: (1) the intensity and duration of rainfall, (2) the amount of water already in the soil, (3) the nature of the surface material, (4) the slope of the land, and (5) the extent and type of vegetation. When the surface material is impermeable or when it becomes saturated, runoff is the dominant process. Runoff is also

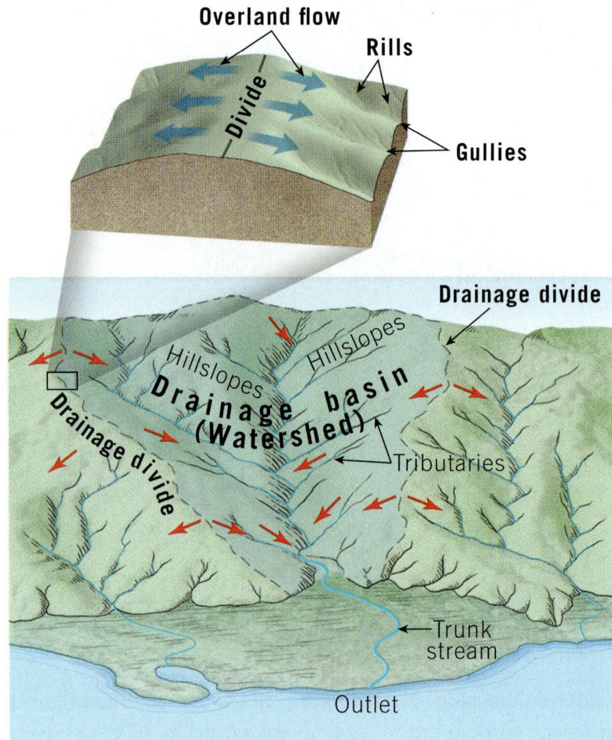

▶ **Figure 13.3 Drainage basin and divide** A drainage basin or watershed is the area drained by a stream and its tributaries. Boundaries between basins are called divides.

high where slopes are steep and in cities where large areas are covered with impermeable buildings, roads, and parking lots.

Runoff initially flows in broad, thin sheets across slopes in a process called *sheet flow*. This unconfined flow eventually develops threads of current that form tiny channels called *rills*. Rills meet to form *gullies*, which join to form streams. Then, when they reach an undefined size, they are called rivers. Although the terms *river* and *stream* are often used interchangeably, geologists define **stream** as water that flows in a channel, regardless of size. **River**, on the other hand, is a general term for streams that carry substantial amounts of water and have numerous tributaries.

Drainage Basins

Every stream drains an area of land called a **drainage basin,** or **watershed** (**Figure 13.3**). Each drainage basin is bounded by an imaginary line called a **divide**, something that is clearly visible as a sharp ridge in some mountainous areas but can be more difficult to determine when the topography is subdued. The outlet, where the stream exits the drainage basin, is at a lower elevation than the rest of the basin.

Drainage divides range in scale from a small ridge separating two gullies on a hillside to a *continental divide* that splits an entire continent into enormous watersheds. The Mississippi River has the largest drainage basin in North America, collecting and carrying 40 percent of the flow in the United States (**Figure 13.4**).

By looking at the drainage basin in Figure 13.3, it should be obvious that slopes cover most of the area. Water erosion on the hillsides is aided by the impact of raindrops and by sheet flow, moving downslope as sheets or in rills toward a stream channel (see Figures 6.21 and 6.22). Hillslope erosion is the main source of fine particles (clays and fine sand) carried in stream channels.

If you could observe streams in an area similar to that depicted in Figure 13.3 over several years, you would see many of them lengthen by **headward erosion**—that is, by extending the heads of their channels upslope. Headward erosion occurs when the surface flow converging at the head of a channel has enough power to cut the channel deeper (*downcut*). This lowering of the channel creates steeper headward

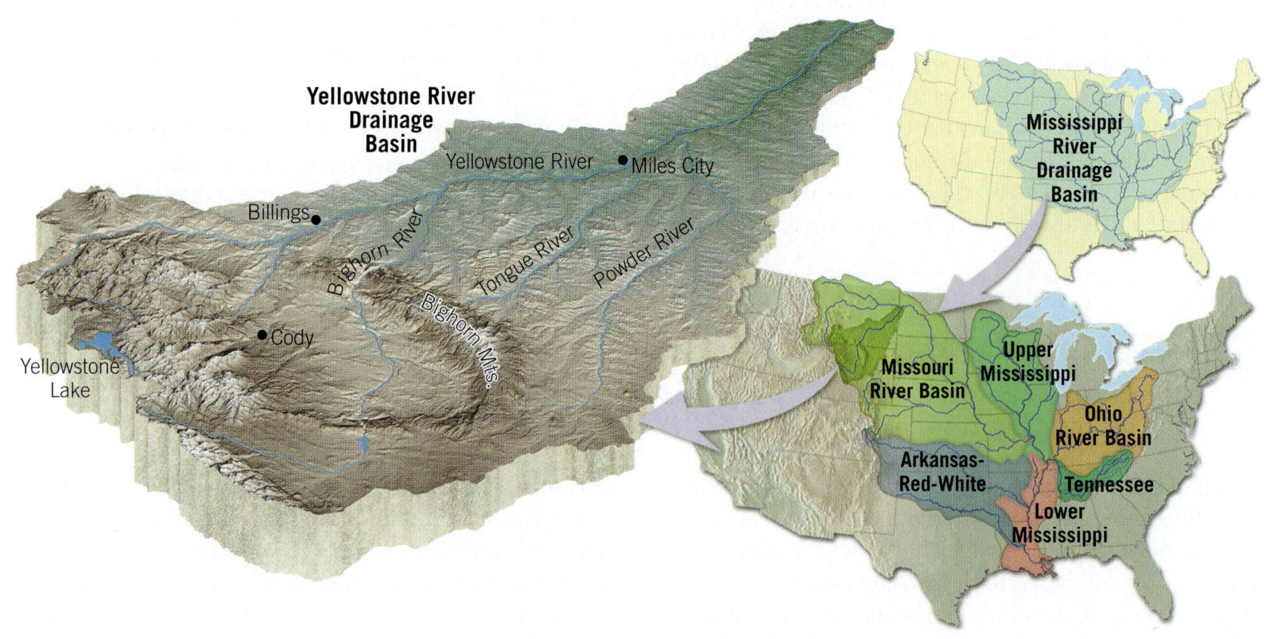

▶ **SmartFigure 13.4 Mississippi River drainage basin** The drainage basin of the Mississippi River forms a funnel that stretches from Montana and southern Canada in the west to New York State in the east that runs down to a spout in Louisiana. It consists of many smaller drainage basins. The drainage basin of the Yellowstone River is one of many that contribute water to the Missouri River, which, in turn, is one of many that make up the drainage basin of the Mississippi River.

TUTORIAL
https://goo.gl/HU34IU

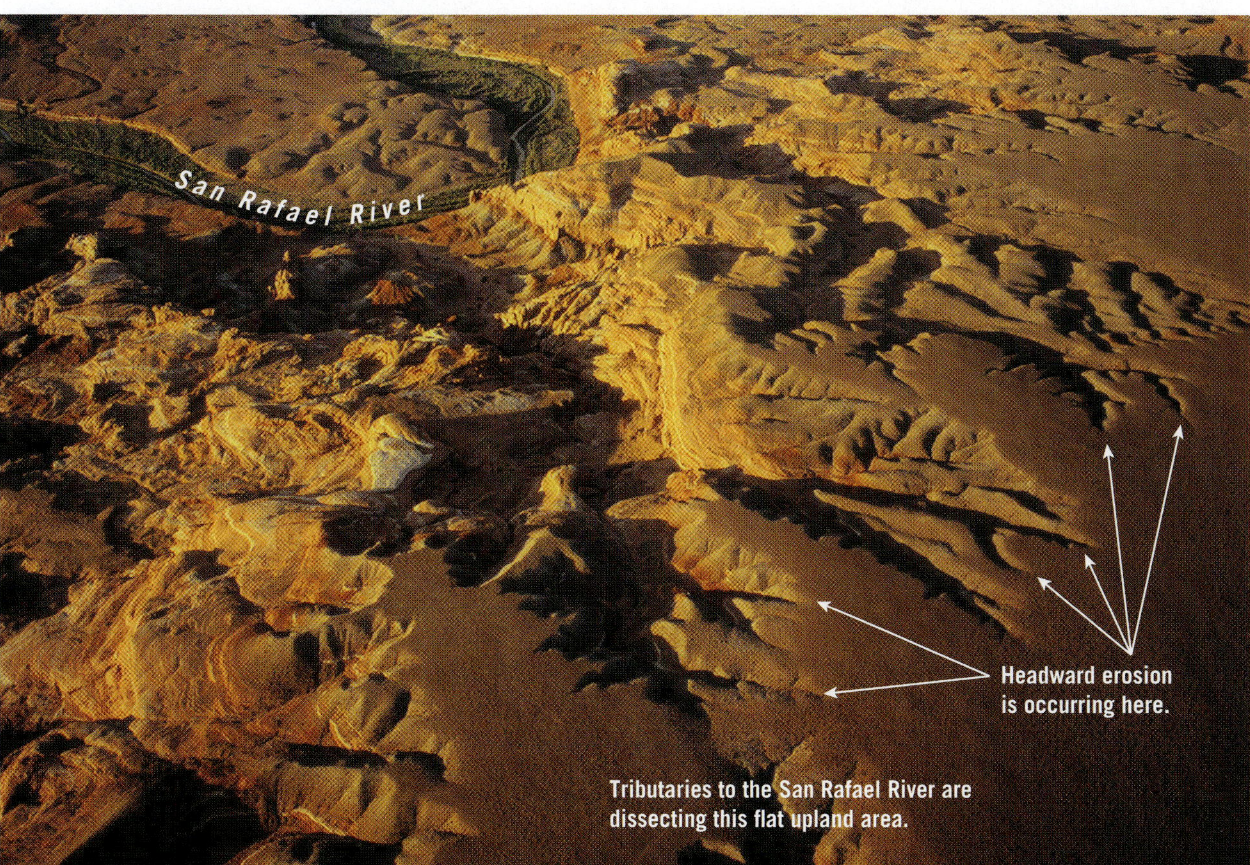

San Rafael River
Headward erosion is occurring here.
Tributaries to the San Rafael River are dissecting this flat upland area.

SmartFigure 13.5 Headward erosion A stream lengthens its course by extending the head of its valley upslope into previously undissected terrain. (Photo by Michael Collier)

TUTORIAL
https://goo.gl/J3ulnM

slopes that erode more quickly. Thus, through headward erosion, a valley extends into previously undissected terrain (**Figure 13.5**).

River Systems

A river system includes not only its network of stream channels but its entire drainage basin. It can be divided into three zones based on the process that dominates in each. These are the zones of *sediment production* (where erosion dominates), *sediment transport*, and *sediment deposition* (**Figure 13.6**). It is important to recognize that sediment is being eroded, transported, and deposited along the entire length of a stream, regardless of which process is dominant within each zone.

Sediment Production The zone of *sediment production*, where most of the sediment is derived, is located in the headwaters region of the river system. Much of the sediment carried by streams begins as bedrock that is subsequently broken down by weathering and then moved downslope by mass wasting and by way of sheet flow and rills. Bank erosion can also contribute significant amounts of sediment. In addition, scouring of the channel bed deepens the channel and adds to the stream's sediment load.

Sediment Transport Downstream from the zone of sediment production, sediment acquired by a stream is transported through the channel network along sections referred to as *trunk streams*. When trunk streams are in balance, the amount of sediment eroded from their banks equals the amount deposited elsewhere in the channel. Although trunk streams rework their channels over time, they are not a source of sediment, nor do they accumulate or store it.

Sediment Deposition When a river approaches the ocean or another large body of water, it slows, and the energy to transport sediment is greatly reduced. Most of the sediments either accumulate at the mouth of the river to form a delta, are reconfigured by wave action to form a variety of coastal features, or are moved far offshore

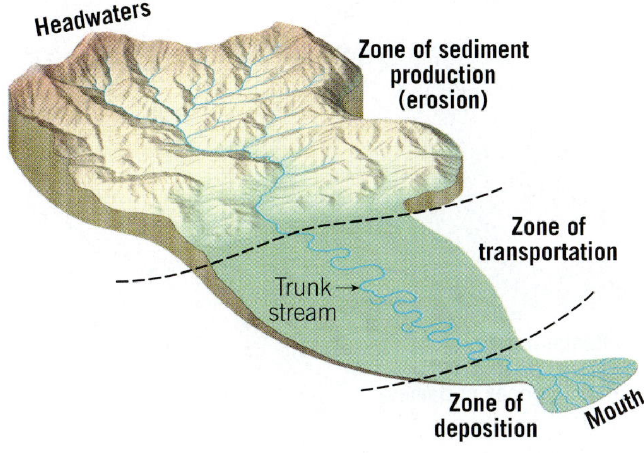

Figure 13.6 Zones of a river Each of the three zones is based on the process that dominates in that part of the river system.

Did You Know?
North America's largest river is the Mississippi. South of Cairo, Illinois, where the Ohio River joins it, the Mississippi is more than 1 mi wide. Each year the "Mighty Mississippi" carries about half a billion tons of sediment to the Gulf of Mexico.

by ocean currents. Because coarse sediments tend to be deposited upstream, it is primarily the fine sediments (clay, silt, and fine sand) that eventually reach the ocean. Taken together, erosion, transportation, and deposition are the processes by which rivers move Earth's surface materials and sculpt landscapes.

Drainage Patterns

Drainage systems, which are interconnected networks of streams, can exhibit a variety of patterns. The pattern that develops depends primarily on the kind of rock present and/or the structural pattern of joints, faults, and folds. Figure 13.7 illustrates four drainage patterns.

Dendritic Pattern The most common drainage pattern is the **dendritic pattern** (see Figure 13.7A), so called because it resembles the branching pattern of a deciduous tree (*dendritic* means "treelike"). A dendritic pattern forms where the underlying material is relatively uniform in its resistance to erosion and hence does not control the pattern of streamflow. The pattern is therefore determined chiefly by the direction of slope of the land.

Radial Pattern When streams diverge from a central area like spokes from the hub of a wheel, the pattern is said to be **radial** (see Figure 13.7B). This pattern typically develops on isolated volcanic cones and domal uplifts.

Rectangular Pattern A **rectangular pattern** exhibits many right-angle bends (see Figure 13.7C). This pattern develops when the bedrock is crisscrossed by a series of joints and/or faults. Because fractured rock tends to weather and erode more easily than unbroken rock, the geometric pattern of joints guides the paths of streams as they carve their valleys.

Trellis Pattern A **trellis pattern** is a rectangular drainage pattern in which tributary streams are nearly parallel to one another and have the appearance of a garden trellis (see Figure 13.7D). This pattern forms in areas underlain by alternating bands of resistant and less-resistant rock and is particularly well displayed in the folded Appalachian Mountains, where both weak and strong strata outcrop in nearly parallel belts. Water gaps occur where a stream cuts across ridges of resistant rock.

Formation of a Water Gap To fully understand the drainage pattern displayed by a stream, it is often useful to consider a stream's entire history. For example, river valleys occasionally cut through a ridge or mountainous topography that lies across their path. This situation is illustrated by the trellis pattern in Figure 13.7D. The steep-walled notch followed by the river is called a **water gap**.

Why do streams cut *across* such structures and not flow *around* them? One possibility is that a **superposed stream** eroded its channel into an existing

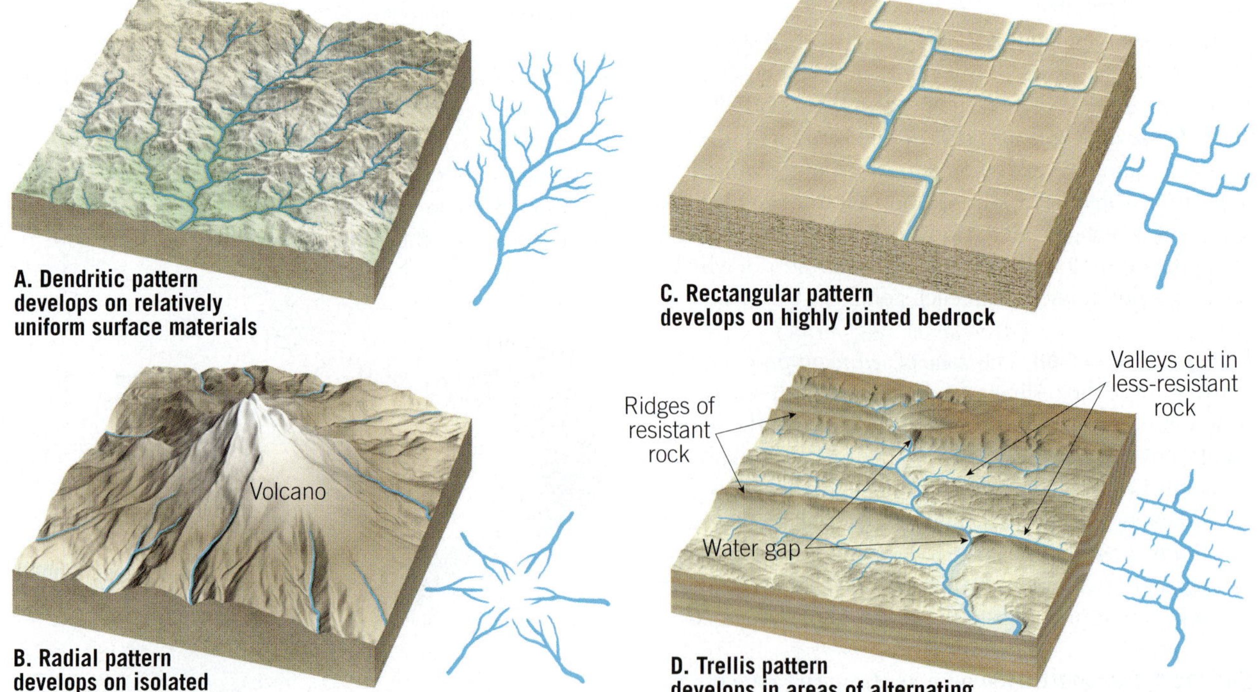

A. Dendritic pattern develops on relatively uniform surface materials
B. Radial pattern develops on isolated volcanic cones or domes
Volcano
C. Rectangular pattern develops on highly jointed bedrock
D. Trellis pattern develops in areas of alternating weak and resistant bedrock
Ridges of resistant rock
Valleys cut in less-resistant rock
Water gap

▶ SmartFigure 13.7
Drainage patterns
Networks of streams form a variety of patterns. The mobile field trip shows how water gaps form, such as those in part D of this figure.

MOBILE FIELD TRIP
http://goo.gl/YAa9Ny

structure. This can occur when folded beds or resistant rocks are buried beneath layers of relatively flat-lying sediments or sedimentary strata. Streams originating on the overlying strata establish their courses without regard to the structures below. Then, as the valley deepens, the river continues to cut its valley into the underlying structure.

CONCEPT CHECKS 13.2

1. List several factors that cause infiltration and runoff to vary from place to place and time to time.
2. Draw a simple sketch of a drainage basin and a divide and label each.
3. What are the three main zones of a river system?
4. Prepare a simple sketch of the four drainage patterns discussed in this section. Briefly describe why streams exhibit each pattern.

13.3 Streamflow Characteristics

Discuss streamflow and the factors that cause it to change.

The water in stream and river channels moves under the influence of gravity. In very slowly flowing streams, water moves in nearly straight-line paths parallel to the stream channel; this is called **laminar flow** (Figure 13.8A). However, streams typically exhibit **turbulent flow** (Figure 13.8B). Strong turbulent behavior occurs in whirlpools and eddies, as well as in roiling whitewater rapids. Even streams that appear smooth on the surface often exhibit turbulent flow near the bottom and sides of the channel, where flow resistance is greatest. Turbulence contributes to a stream's ability to erode its channel because it acts to lift sediment from the streambed.

An important factor influencing stream turbulence is the water's flow velocity. As the velocity of a stream increases, the flow becomes more turbulent. Flow velocities can vary significantly from place to place along a stream channel, as well as over time, in response to variations in the amount and intensity of precipitation. If you have ever waded into a stream, you may have noticed that the strength of the current increased as you moved into deeper parts of the channel. This is related to the fact that frictional resistance is greatest near the banks and bed of a stream channel.

Flow velocities are determined at gaging stations by averaging measurements taken at various locations across the stream's channel. Some sluggish streams have flow velocities of less than 1 kilometer (0.62 mile) per hour, whereas stretches of some fast-flowing rivers may exceed 30 kilometers (19 miles) per hour.

Factors Affecting Flow Velocity

The ability of a stream to erode and transport material is directly related to its flow velocity. Even slight variations in flow rate can lead to significant changes in the sediment load transported by a stream. Factors that influence flow velocities and, therefore, control a stream's potential to do "work" include (1) channel slope, or gradient, (2) channel cross-sectional shape, (3) channel size and roughness, and (4) discharge, or the amount of water flowing in the channel.

Gradient The slope of a stream channel, expressed as the vertical drop of a stream over a specified distance, is called **gradient**. Portions of the lower Mississippi River have very low gradients, about 10 centimeters or less per kilometer (6.3 inches per mile). By contrast, some mountain streams have channels that drop at a rate of more than 40 meters per kilometer (210 feet per mile)—a gradient 400 times steeper than that of the lower Mississippi. Gradient also varies along the length of a particular channel. When the gradient is steeper, more gravitational energy is available to drive channel flow.

Channel Shape As water in a stream channel moves, it encounters a significant amount of frictional resistance. The *cross-sectional shape* (a slice taken across the channel) determines, to a large extent, the amount of

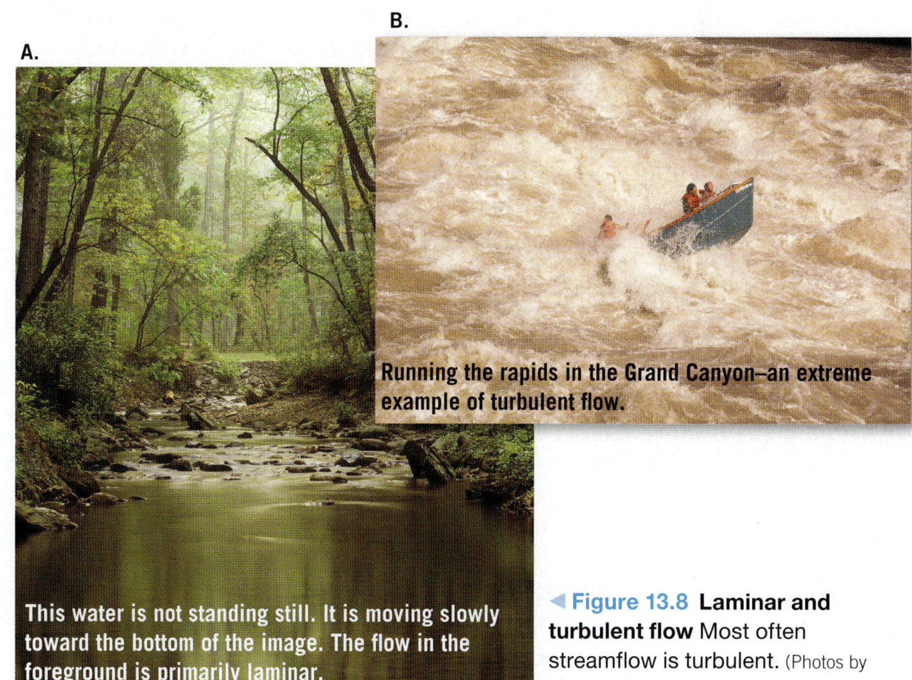

A.
B.
Running the rapids in the Grand Canyon–an extreme example of turbulent flow.
This water is not standing still. It is moving slowly toward the bottom of the image. The flow in the foreground is primarily laminar.

Figure 13.8 Laminar and turbulent flow Most often streamflow is turbulent. (Photos by Michael Collier)

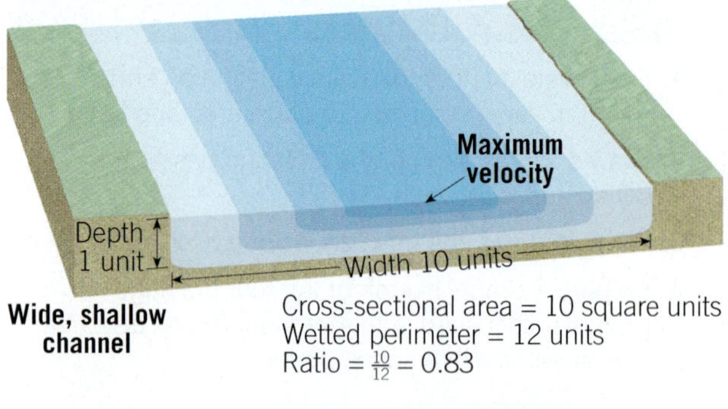

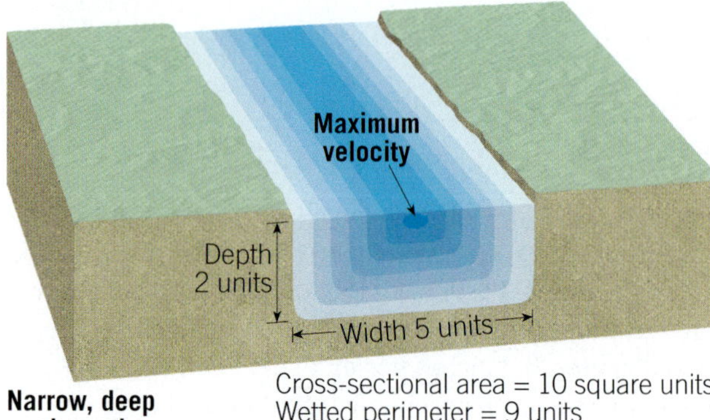

▶ **Figure 13.9 Influence of channel shape on velocity** The stream with the smaller wetted perimeter has less frictional drag and will flow more rapidly, all else being equal.

flow in contact with the banks and bed of the channel. This measure is referred to as the **wetted perimeter**. The most efficient channel is one with the least wetted perimeter for its cross-sectional area. Figure 13.9 compares two channels that differ only in shape: One is wide and shallow, the other is narrow and deep. Although the cross-sectional areas of the two are identical, the narrow and deep one has less water in contact with the channel (a smaller wetted perimeter) and therefore less frictional drag. As a result, if all other factors are equal, the water will flow more efficiently and at a higher velocity in a channel that is roughly as wide as it is deep than in one that is wide and shallow.

Channel Size & Roughness All other factors being equal, flow velocities are higher in large channels than in small channels. The depth of the water in a given channel also affects the frictional resistance that the channel exerts on flow. Maximum flow velocity occurs when a stream is *bankfull*, before water starts to inundate the floodplain. At this stage, the channel's ratio of the cross-sectional area to wetted perimeter is highest, and streamflow is most efficient. A final factor is how *rough* a channel is. Elements such as boulders, irregularities in the channel bed, and woody debris create turbulence that significantly reduces flow velocity.

Did You Know?
The Amazon River is responsible for about 20 percent of all the water reaching the ocean via rivers. Its nearest rival, Africa's Congo River, delivers about 4 percent of the total.

Discharge Streams vary in size from small headwater creeks less than a meter wide to large rivers with widths of several kilometers. The size of a stream channel is largely determined by the amount of water supplied from the drainage basin. The measure most often used to compare the size of streams is **discharge**—the volume of water flowing past a certain point in a given unit of time. Discharge, usually measured in cubic meters per second or cubic feet per second, is determined by multiplying a stream's cross-sectional area by its velocity. The largest river in North America, the Mississippi, has an average discharge at its mouth of about 16,800 cubic meters (593,000 cubic feet) per second (Figure 13.10). That figure is dwarfed by South America's mighty Amazon River, which discharges 13 times more water than the Mississippi.

The discharge of a river system changes over time because of variations in the amount of precipitation

▲ **SmartFigure 13.10 The Mighty Mississippi near Helena, Arkansas** The Mississippi is North America's largest river. From head to mouth, it is nearly 3900 kilometers (2400 miles) long. Its watershed encompasses about 40 percent of the lower 48 states and includes all or parts of 31 states and 2 Canadian provinces. Average discharge at its mouth is about 16,800 cubic meters (593,000 cubic feet) per second. (Photo by Michael Collier)

MOBILE FIELD TRIP
https://goo.gl/UyqOHJ

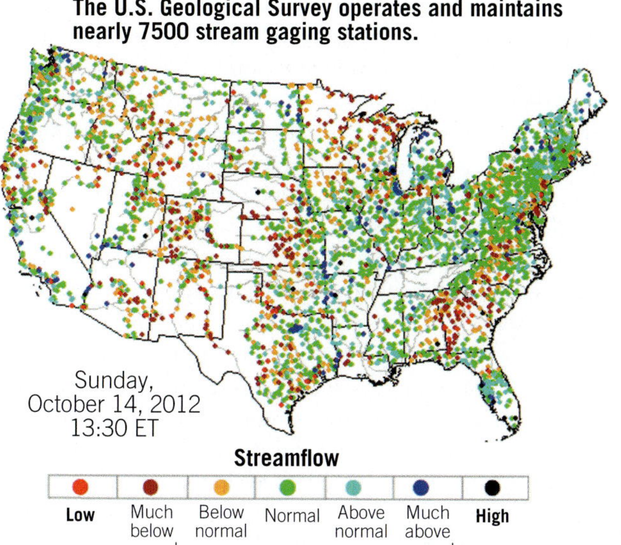

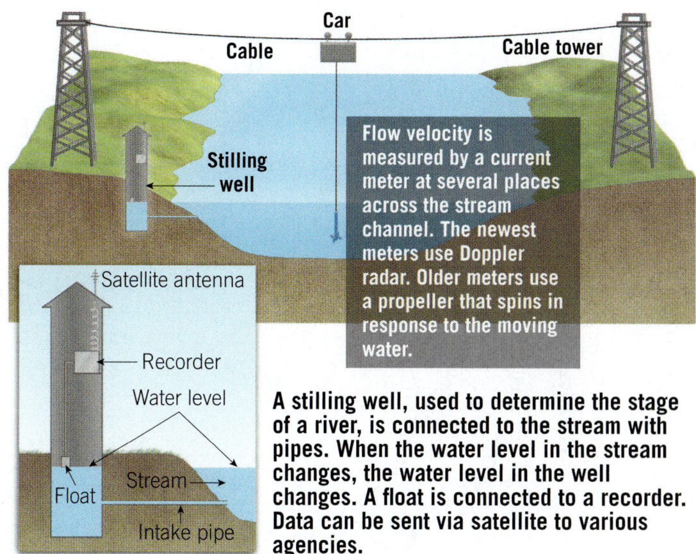

Car
Cable
Cable tower
Stilling well
Flow velocity is measured by a current meter at several places across the stream channel. The newest meters use Doppler radar. Older meters use a propeller that spins in response to the moving water.
Satellite antenna
Recorder
Water level
Float
Stream
Intake pipe
A stilling well, used to determine the stage of a river, is connected to the stream with pipes. When the water level in the stream changes, the water level in the well changes. A float is connected to a recorder. Data can be sent via satellite to various agencies.

▲ **Figure 13.11 Stream gaging stations** The United States has a dense network of stream gaging stations. To determine a channel's discharge, it is necessary to survey the channel to determine its shape and calculate its area. A stilling well measures the stage, and a current meter determines flow velocity.

received by the watershed. Studies show that when discharge increases, the width, depth, and flow velocity of the channel all increase predictably. As we saw earlier, when the size of the channel increases, proportionally less water is in contact with the bed and banks of the channel. Thus, friction, which acts to retard the flow, is reduced, resulting in an increase in the rate of flow.

Monitoring Streamflow The U.S. Geological Survey maintains a network of about 7500 stream gaging stations that collect basic data about the country's surface-water resources (**Figure 13.11**). Data collected include flow velocity, discharge, and river stage. *Stage* is the height of the water surface relative to a fixed reference point. This measurement is frequently reported by the media, especially when a river approaches or surpasses *flood stage*. Streamflow measurements are essential components of river models used to make flood forecasts and issue warnings. There are other applications as well. The data are used in making decisions related to water supply allocations, operation of wastewater treatment plants, design of highway bridges, and recreation activities.

Changes Downstream

One useful way of studying a stream is to examine its **longitudinal profile**. Such a profile is simply a cross-sectional view of a stream from its source area (called the **head,** or **headwaters**) to its **mouth**, the point downstream where it empties into another water body—a river, a lake, or the ocean. **Figure 13.12** shows that the most obvious feature of a typical longitudinal profile is its concave shape—a result of the decrease in slope that occurs from the headwaters to the mouth. In addition, local irregularities exist in the profiles of most streams; the flatter sections may be associated with lakes or reservoirs, and the steeper sections are sites of rapids or waterfalls.

The change in slope observed on most stream profiles is usually accompanied by an increase in discharge and channel size and a reduction in sediment particle

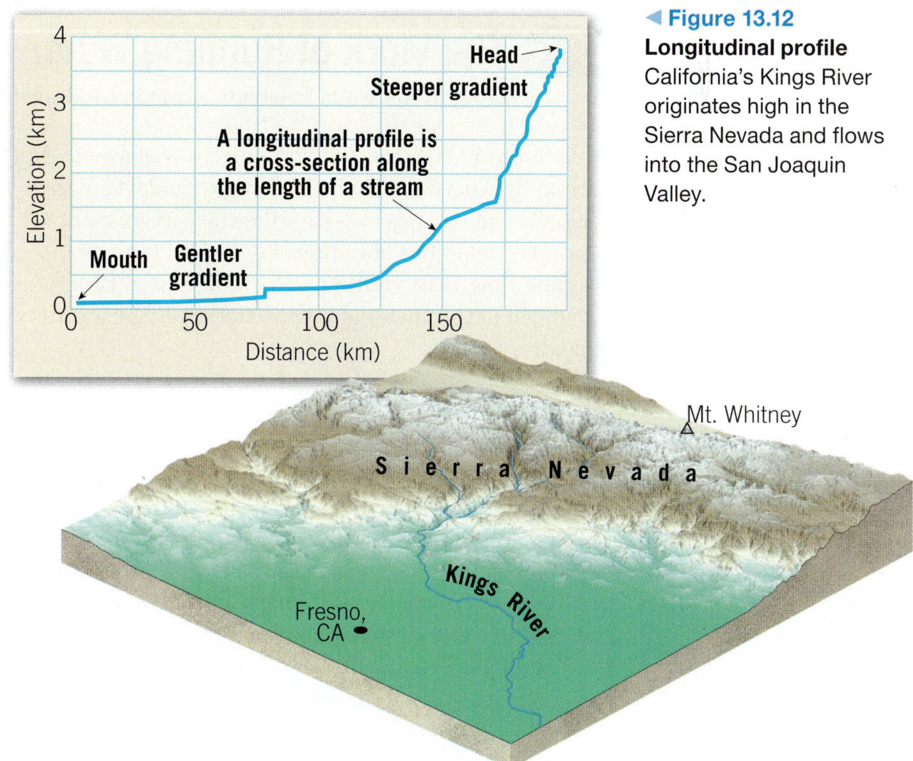

◀ **Figure 13.12**
Longitudinal profile
California's Kings River originates high in the Sierra Nevada and flows into the San Joaquin Valley.

▶ SmartFigure 13.13
Channel changes from head to mouth Although the gradient decreases toward the mouth of a stream, increases in discharge and channel size and decreases in roughness more than offset the decrease in slope. Consequently, flow velocity usually increases toward the mouth.

TUTORIAL
https://goo.gl/mFXfG4

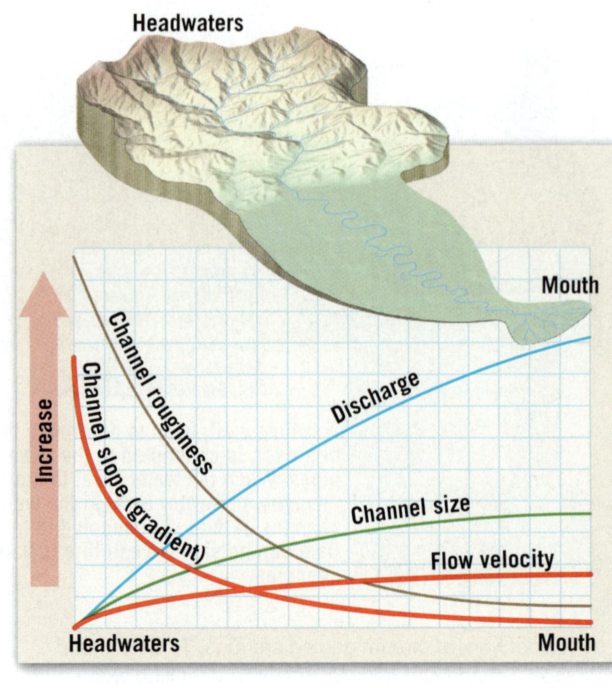

size (**Figure 13.13**). For example, data from successive gaging stations along most rivers show that, in humid regions, discharge increases toward the mouth. This occurs because, as we move downstream, more and more tributaries contribute water to the main channel. In the case of the Amazon, for example, about 1000 tributaries join the main river along its 6500-kilometer (4000-mile) course across South America.

In order to accommodate the growing volume of water, channel size typically increases downstream as well. Recall that flow velocities are higher in large channels than in small channels. Furthermore, observations show a general decline in sediment size downstream, making the channel smoother and more efficient.

Although the channel slope decreases—the gradient becomes less steep—toward a stream's mouth, the flow velocity generally increases. This fact contradicts our intuitive assumption that narrow headwater streams are swift whereas a broad river flowing across subtle topography will be slow. Actually, the increase in channel size and discharge and decrease in channel roughness that occur downstream compensate for the decrease in slope, making the stream more efficient (see Figure 13.13). Thus, the average flow velocity is typically lower in headwater streams than in wide, placid-appearing rivers.

CONCEPT CHECKS 13.3

1. Contrast laminar flow and turbulent flow.
2. Summarize the factors that influence flow velocity.
3. What is a longitudinal profile?
4. What typically happens to channel width, channel depth, flow velocity, and discharge between the headwaters and the mouth of a stream? Briefly explain why these changes occur.

13.4 The Work of Running Water

Outline the ways in which streams erode, transport, and deposit sediment.

Streams are Earth's most important erosional agents. Not only do they have the ability to downcut and widen their channels, but streams also have the capacity to transport enormous quantities of sediment delivered to them by overland flow, mass wasting, and groundwater. Eventually, much of this material is deposited to create a variety of landforms.

Stream Erosion

A stream's ability to accumulate and transport soil and weathered rock is aided by the work of raindrops, which knock sediment particles loose (see Figure 6.21, page 177). When the ground is saturated, rainwater cannot infiltrate, so it flows downslope, transporting some of the material it dislodges. On barren slopes, the flow of muddy water (sheet flow) often produces small channels (rills), which in time can evolve into larger gullies (see Figure 6.22, page 177).

Once flow is confined in a channel, the erosional power of a stream is related to its slope and discharge. The rate of erosion is also dependent on the relative resistance of the bank and bed material. In general, channels composed of unconsolidated materials are more easily eroded than channels cut into bedrock.

When a channel is sandy, the particles are easily dislodged from the bed and banks, and they are lifted into the moving water. Moreover, banks consisting of sandy material are often undercut, dumping even more loose debris into the water to be carried downstream. Banks that consist of coarse gravels or cohesive clay and silt particles tend to be relatively resistant to erosion. Thus, channels with cohesive silty banks are generally narrower than comparable ones with sandy banks.

Streams cut channels into bedrock through three main processes: *quarrying*, *abrasion*, and *corrosion*.

Quarrying **Quarrying** involves the removal of blocks from the bed of a stream channel. This process is aided by fracturing and weathering that loosen the blocks sufficiently so they are movable during times of high flow rates. Quarrying is mainly a result of the impact forces exerted by flowing water.

Abrasion The process by which the bed and banks of a bedrock channel are ceaselessly bombarded by particles carried into the flow is termed **abrasion**. The individual sediment grains are also abraded by their many impacts with the channel and with one another. Thus, by scraping, rubbing, and bumping, abrasion erodes a bedrock channel and simultaneously smooths and rounds the abrading particles. That is why smooth, rounded cobbles and pebbles are found in streams. Abrasion also results in a reduction in the size of the sediments transported by streams.

Features common to some bedrock channels are circular depressions known as **potholes**, which are created by the abrasive action of particles swirling in fast-moving eddies (Figure 13.14). The rotational motion of sand and pebbles acts like a drill that bores the holes. As the particles wear down to nothing, they are replaced by new ones that continue to drill the streambed. Eventually, smooth depressions several meters across and deep may result.

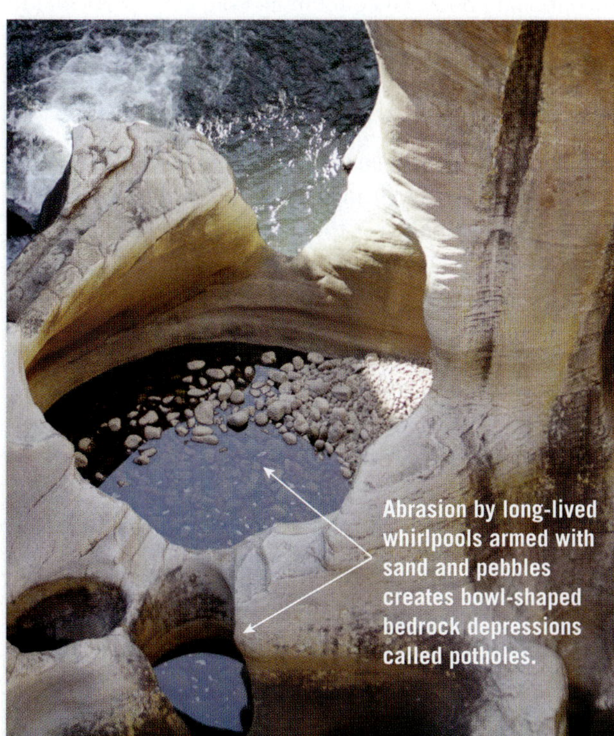

Abrasion by long-lived whirlpools armed with sand and pebbles creates bowl-shaped bedrock depressions called potholes.

▲ **Figure 13.14 Potholes** The rotational motion of swirling pebbles acts like a drill to create potholes. (Photo by Elmari Joubert/Alamy Images)

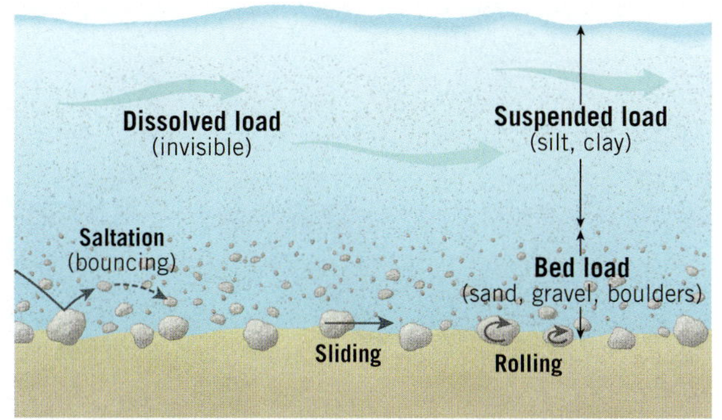

◀ **SmartFigure 13.15**
Transport of sediment Streams transport their load of sediment in three ways. The dissolved and suspended loads are carried in the general flow. The bed load includes coarse sand, gravel, and boulders that move by rolling, sliding, and saltation.

ANIMATION
https://goo.gl/SMpTzA

Corrosion Bedrock channels formed in soluble rock such as limestone are susceptible to **corrosion**—a process in which rock is gradually dissolved by the flowing water. Corrosion is a type of chemical weathering between the solutions in the water and the mineral matter that makes up the bedrock.

Transport of Sediment by Streams

All streams, regardless of size, transport some weathered rock material (Figure 13.15). Streams also sort the solid sediment they transport because finer, lighter material is carried more readily than larger, heavier particles. Streams transport their load of sediment in three ways: (1) in solution (*dissolved load*), (2) in suspension (*suspended load*), and (3) by sliding, skipping, or rolling along the bottom (*bed load*).

Dissolved Load Most of the **dissolved load** is brought to a stream by groundwater and is dispersed throughout the flow. When water percolates through the ground, it acquires soluble soil compounds. Then it seeps through cracks and pores in bedrock, dissolving additional mineral matter. Eventually much of this mineral-rich water finds its way into streams.

The velocity of streamflow has essentially no effect on a stream's ability to carry its dissolved load; material in the solution goes wherever the stream goes. Precipitation of the dissolved mineral matter occurs when the chemistry of the water changes, when organisms extract mineral matter to create hard parts, or when the water enters an inland "sea," located in an arid climate where the rate of evaporation is high.

Suspended Load Usually streams transport the greatest amount of sediment as the **suspended load**. Indeed, the muddy appearance created by suspended sediment is the most obvious portion of a stream's load (Figure 13.16). Usually only very fine sand, silt, and clay particles are carried this way, but during flood stage, larger particles can also be transported in suspension. During flood stage, the total quantity

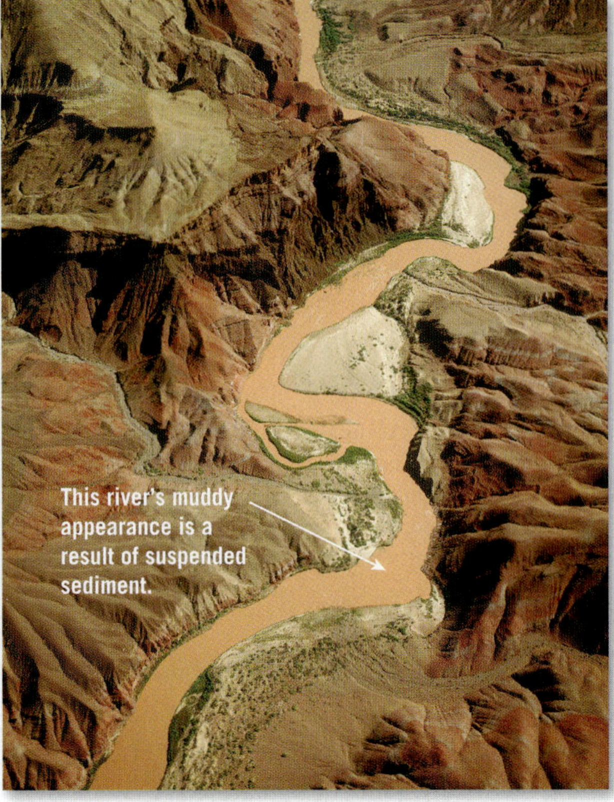

This river's muddy appearance is a result of suspended sediment.

▶ **Figure 13.16** **Suspended load** An aerial view of the Colorado River in the Grand Canyon. Heavy rains washed sediment into the river. (Photo by Michael Collier)

of material carried in suspension also increases dramatically; people whose homes have become sites for the deposition of this material can verify this.

The type and amount of material carried in suspension are controlled by two factors: the flow velocity and the settling velocity of each sediment grain. **Settling velocity** is defined as the speed at which a particle falls through a still fluid. The larger the particle, the more rapidly it settles toward the streambed (**Figure 13.17**). In addition to size, the shape and specific gravity of particles also influence settling velocity. Flat grains sink through water more slowly than spherical grains, and dense particles fall toward the bottom more rapidly than less dense particles. As long as flow velocity exceeds settling velocity, sediment remains suspended and is transported downstream. Deposition occurs when flow velocity falls below the settling velocity of a particle.

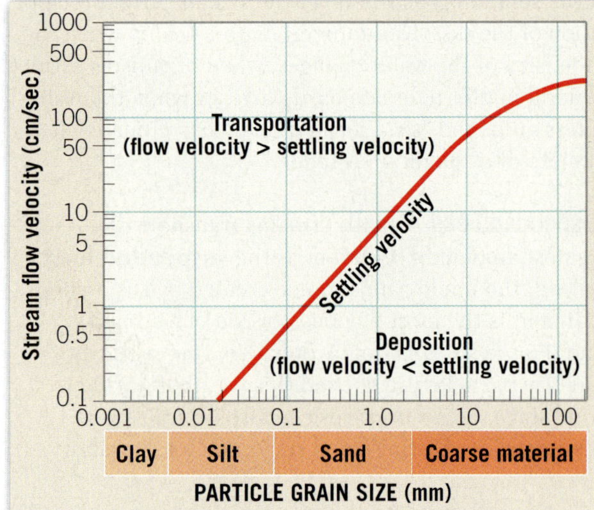

▶ **Figure 13.17** **Settling velocity** The speed at which a particle falls through still water is its settling velocity. Deposition of suspended particles occurs when their settling velocities are greater than the stream's flow velocity.

Bed Load Coarse materials, including coarse sands, gravels, and even boulders, typically move along the bed of a channel as **bed load**. The particles that make up the bed load move by rolling, sliding, and saltation. Sediment moving by **saltation** (*saltare* = to leap) appears to jump or skip along the streambed (see Figure 13.15). This occurs as particles are propelled upward by collisions or lifted by the current and then carried downstream a short distance until gravity pulls them back to the bed of the stream. Particles that are too large or heavy to move by saltation either roll or slide along the bottom, depending on their shape.

The movement of bed load through a stream network tends to be less rapid and more localized than the movement of suspended load. A study conducted on a glacially fed river in Norway determined that suspended sediments took only a day to exit the drainage basin, while the bed load required several decades to travel the same distance. Depending on the discharge and slope of the channel, coarse gravels may only be moved during times of high flow, while boulders move only during exceptional floods. Once set in motion, large particles are usually carried short distances. Along some stretches of a stream, bed load cannot be carried at all until it is broken into smaller particles.

Capacity & Competence A stream's ability to carry solid particles is described using two criteria: *capacity* and *competence*. **Capacity** is the maximum load of solid particles a stream can transport per unit time. The greater the discharge, the greater the stream's capacity for hauling sediment. Consequently, large rivers with high flow velocities have large capacities.

Competence is a measure of a stream's ability to transport particles based on size rather than quantity. Flow velocity is key: Swift streams have greater competencies than slow streams, regardless of channel size. A stream's competence increases proportionately to the square of its velocity. Thus, if the velocity of a stream doubles, the impact force of the water increases four times; if the velocity triples, the force increases nine times, and so forth. Hence, large boulders that are often visible in low water and seem immovable can, in fact, be transported during exceptional floods because of the stream's increased competence.

By now it should be clear why the greatest erosion and transportation of sediment occur during periods of high water associated with floods. The increase in discharge results in greater capacity; the increased velocity produces greater competency. Rising velocity makes the water more turbulent, and larger particles are set in motion. In the course of a few days, or perhaps a few hours, a stream at flood stage can erode and transport more sediment than it does during several months of normal flow.

Deposition of Sediment by Streams

Deposition occurs whenever a stream slows, causing a reduction in competence. Put another way, particles are deposited when the flow velocity is less than the settling velocity. As a stream's flow velocity decreases, sediment begins to settle, largest particles first. In this manner, stream transport provides a mechanism by which solid particles of various sizes are separated. This process, called **sorting**, explains why particles of similar size are deposited together.

The general term for sediment deposited by streams is **alluvium**. Many different depositional features are composed of alluvium. Some occur within stream channels, some occur on the valley floor adjacent to a channel, and some are found at the mouth of a stream. We will consider the nature of these features later in the chapter.

CONCEPT CHECKS 13.4

1. Describe two processes by which streams cut channels in bedrock.
2. In what three ways does a stream transport its load? Which part of the load moves most slowly?
3. Explain the difference between capacity and competency.
4. What is settling velocity? What factors influence settling velocity? Does settling velocity affect the dissolved load?

Did You Know?
It is estimated that each year the world's rivers transport about 2.4 mi^3 of weathered material to lakes and oceans either as detrital sediment or as dissolved ions. If erosion occurred evenly over Earth's land surface, this quantity of sediment would represent a lowering of nearly 0.3 in over a 100-year span.

13.5 Stream Channels

Contrast bedrock and alluvial stream channels. Distinguish between two types of alluvial channels.

A basic characteristic that distinguishes streamflow from overland flow is that it is confined in a channel. A stream channel can be thought of as an open conduit consisting of the streambed and banks that act to confine flow except, of course, during floods.

Although this is somewhat oversimplified, we can divide stream channels into two basic types. *Bedrock channels* are channels in which the streams are actively cutting into solid rock. In contrast, when the bed and banks are composed mainly of unconsolidated sediment or alluvium, the channel is called an *alluvial channel*.

Bedrock Channels

As the name suggests, **bedrock channels** are cut into the underlying strata and typically form in the headwaters of river systems where streams have steep gradients. The energetic flow tends to transport coarse particles that actively abrade the bedrock channel. Potholes are often visible evidence of the erosional forces at work.

Steep bedrock channels often develop a sequence of *steps* and *pools*, relatively flat segments (pools) where alluvium tends to accumulate, and steep segments (steps) where bedrock is exposed. The steep areas contain rapids or, occasionally, waterfalls.

The channel pattern exhibited by streams cutting into bedrock is controlled by the underlying geologic structure. Even when flowing over rather uniform bedrock, streams tend to exhibit winding or irregular patterns rather than flowing in straight channels. Anyone who has gone white-water rafting has observed the steep, winding nature of a stream flowing in a bedrock channel.

Alluvial Channels

Alluvial channels form in sediment that was previously deposited in the valley. When the valley floor reaches sufficient width, material deposited by the stream can form a *floodplain* that borders the channel. Because the banks and beds of alluvial channels are composed of unconsolidated sediment (alluvium), they can undergo significant changes in shape as material is continually being eroded, transported, and redeposited. The major factors affecting the shapes of these channels are the average size of the sediment being transported, the channel's gradient, and discharge.

Alluvial channel patterns reflect a stream's ability to transport its load at a uniform rate while expending the least amount of energy. Thus, the size and type of sediment being carried help determine the nature of the stream channel. Two common types of alluvial channels are *meandering channels* and *braided channels*.

Meandering Channels Streams that transport much of their load in suspension generally move in sweeping bends called **meanders**. These streams flow in relatively deep, smooth channels and primarily transport mud (silt and clay), sand, and occasionally fine gravel. The lower Mississippi River exhibits this type of channel.

Did You Know?
In 2014, renewable energy represented about 10 percent of U.S. energy consumption. Hydroelectric power plants contribute 26 percent of the country's renewable energy. Most of this energy is produced at large dams.

▶ **SmartFigure 13.18 Formation of cut banks and point bars** By eroding its outer bank and depositing material on the inside of the bend, a stream is able to shift its channel.

TUTORIAL
https://goo.gl/eFrBNv

Maximum velocity
Maximum velocity
Erosion of a cut bank along the Newaukum River in southwestern Washington State.
USGS
Deposition of point bar
White River near Vernal, Utah
The cut bank forms on the outside of a meander where flow velocity and turbulence are greatest.
Point bar
Erosion of cut bank
As water slows on the inside of a meander, coarser material is deposited as a point bar.
Michael Collier
Maximum velocity

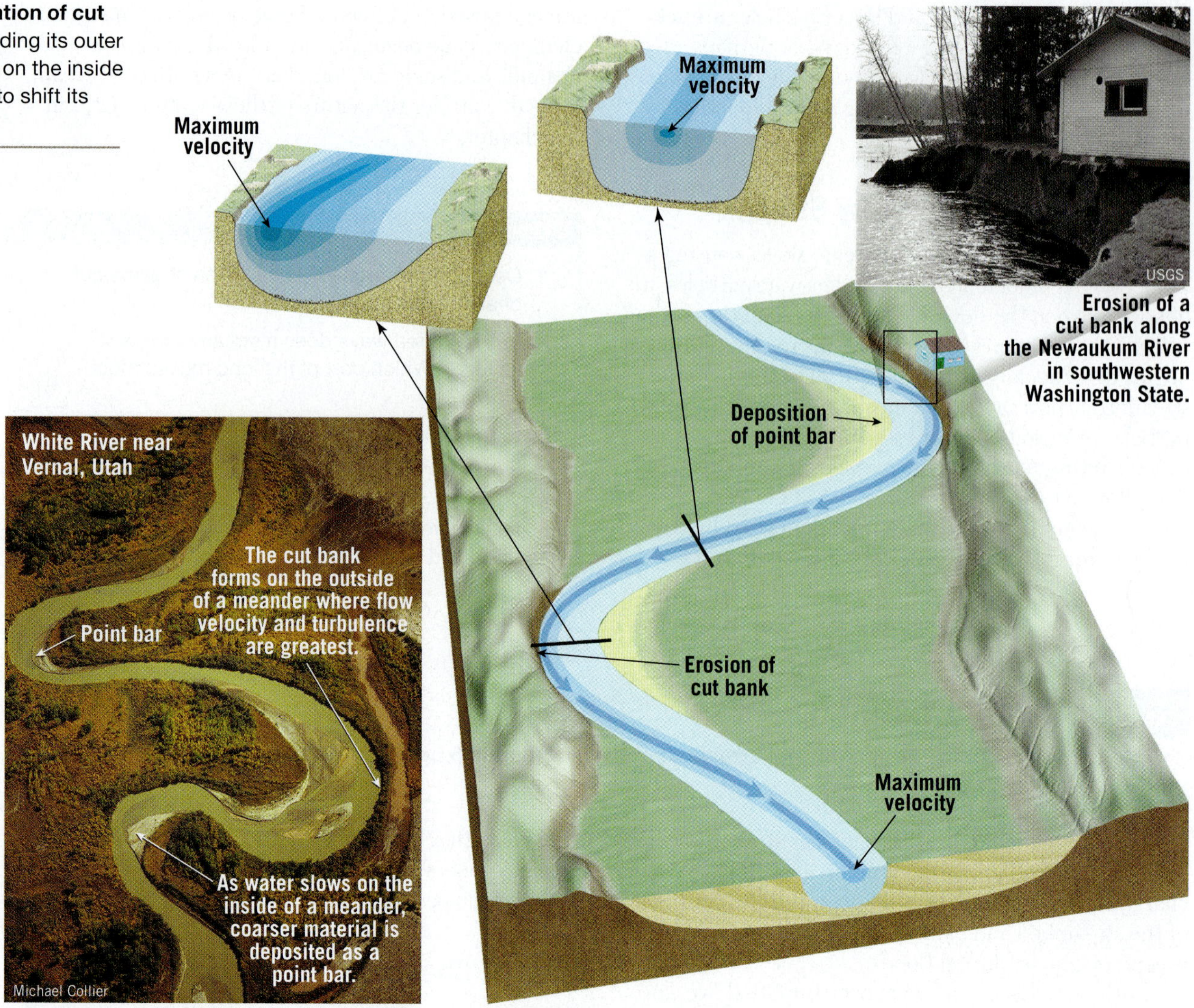

Meandering channels evolve over time as individual bends migrate across the floodplain. Most of the erosion is focused at the outside of the meander, where velocity and turbulence are greatest. In time, the outside bank is undermined, especially during periods of high water. Because the outside of a meander is a zone of active erosion, it is referred to as a **cut bank** (Figure 13.18). Debris acquired by the stream at the cut bank moves downstream, where the coarser material is generally deposited as **point bars** on the insides of meanders. In this manner, meanders migrate laterally by eroding the outside of the bends and depositing sediment on the inside, without appreciably changing their shape.

In addition to migrating laterally, the bends in a channel also migrate down the valley. This occurs because erosion is more effective on the downstream (downslope) side of the meander. Sometimes the downstream migration of a meander is slowed when it reaches a more resistant bank material. This allows the next meander upstream to gradually erode the material between the two meanders, as shown in Figure 13.19. Eventually, the river may erode through the narrow neck of land, forming a new, shorter channel segment called a **cutoff**. Because of its shape, the abandoned bend is called an **oxbow lake**.

Braided Channels Some streams consist of a complex network of converging and diverging channels that thread their way among numerous islands or gravel bars (Figure 13.20). Because these channels have an interwoven appearance, they are called **braided channels**. Braided channels form where a large portion of a stream's load consists of coarse sediment (sand and gravel) and the stream has a highly variable discharge. Because the bank material is readily erodible, braided channels are wide and shallow.

One setting in which braided streams form is at the end of glaciers, where there is a large seasonal variation in discharge. During the summer, large amounts of ice-eroded sediment are dumped into the meltwater streams flowing away from the glacier. However, when flow is sluggish, the stream deposits the coarsest material as

Did You Know?
In 2014 about 29 percent of the total U.S. hydroelectric power was generated in the state of Washington, which also has the nation's largest hydroelectric facility—Grand Coulee Dam on the Columbia River.

▲ **SmartFigure 13.19 Formation of an oxbow lake** Oxbow lakes occupy abandoned meanders. Aerial view of an oxbow lake created by the meandering Green River near Bronx, Wyoming. (Photo by Michael Collier)

ANIMATION
https://goo.gl/0lWPw5

▲ **Figure 13.20 Braided stream** The Knik River is a classic braided stream with multiple channels separated by migrating gravel bars. The Knik is choked with sediment from four melting glaciers in the Chugach Mountains north of Anchorage, Alaska. (Photo by Michael Collier)

elongated structures called *bars*. This process causes the flow to split into several paths around the bars. During the next period of high flow, the laterally shifting channels erode and redeposit much of this coarse sediment, thereby transforming the entire streambed. In some braided streams, the bars have built semipermanent islands anchored by vegetation.

CONCEPT CHECKS 13.5

1. Are bedrock channels more likely to be found near the head or the mouth of a stream?
2. Describe or sketch the evolution of a meander, including how an oxbow lake forms.
3. Describe a situation that might cause a stream channel to become braided.

13.6 Shaping Stream Valleys

Contrast narrow V-shaped valleys, broad valleys with floodplains, and valleys that display incised meanders or stream terraces.

A **stream valley** consists of a channel and the surrounding terrain that directs water to the stream. It includes the *valley floor*, which is the lower, flatter area that is partially or totally occupied by the stream channel, and the sloping *valley walls* that rise above the valley floor on both sides. Alluvial channels often flow in valleys that have wide valley floors consisting of sand and gravel deposited in the channel and clay and silt deposited by floods. Bedrock channels, on the other hand, tend to be located in narrow V-shaped valleys. In some arid regions, where weathering is slow and rock is particularly resistant, narrow valleys having nearly vertical walls are also found. Such features are called *slot canyons* (**Figure 13.21**). Stream valleys exist on a continuum from narrow, steep-sided valleys to valleys that are so flat and wide that the valley walls are not discernible.

Streams, with the aid of weathering and mass wasting, shape the landscape through which they flow. As a result, streams continuously modify the valleys they occupy.

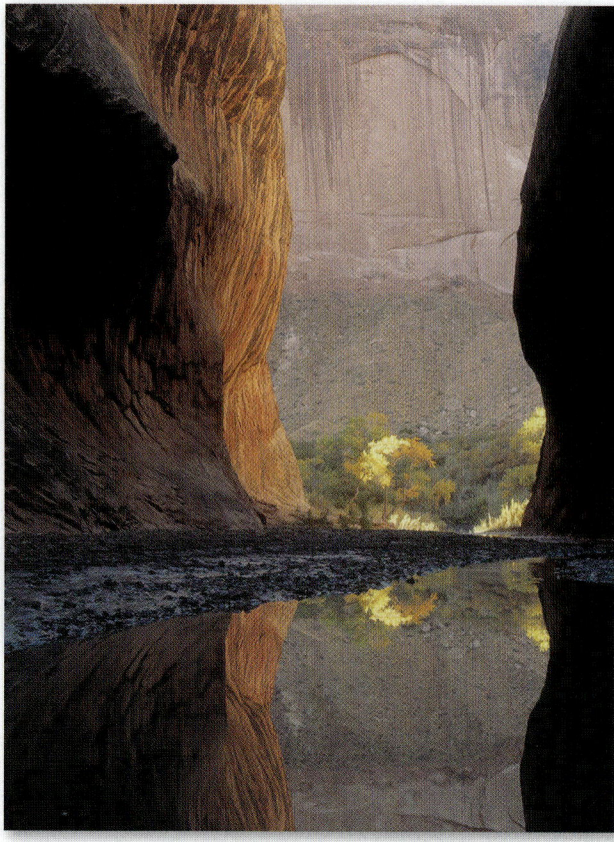

▶ **Figure 13.21 Halls Creek Narrows** This is a classic slot canyon in Utah's Capitol Reef National Park. (Photo by Michael Collier)

Base Level & Graded Streams

In 1875 John Wesley Powell, the pioneering geologist who led the first expedition down the Colorado River and through the Grand Canyon in 1869 and later headed the U.S. Geological Survey, introduced the concept of a downward limit to stream erosion, which he called **base level**. A fundamental concept in the study of stream activity, base level is defined as the lowest elevation to which a stream can erode its channel. Essentially, it is the level at which the mouth of a stream enters the ocean, a lake, or a trunk stream. Powell identified two types of base level: "We may consider the level of the sea to be a grand base level, below which the dry lands cannot be eroded; but we may also have, for local and temporary purposes, other base levels of erosion."*

Sea level, which Powell called "grand base level," is now referred to as **ultimate base level**. **Local** (or **temporary**) **base levels** include lakes, resistant layers of rock, and rivers that act as base levels for their tributaries. All of these base levels limit a stream's ability to downcut its channel.

Changes in base level cause corresponding adjustments in the "work" that streams perform. When a dam is built across a stream, the reservoir that forms behind it raises the base level of the stream (Figure 13.22). Upstream from the reservoir, the stream gradient is reduced, lowering its velocity and, hence, its sediment-transporting ability. As a result, the stream deposits material, thereby building up its channel. This process continues until the stream again has a gradient sufficient to carry its load. The profile of the new channel will be similar to the old but somewhat higher.

If, on the other hand, base level is lowered by a drop in sea level, the stream will have excess energy and downcut its channel to establish a balance with its new base level. Erosion first occurs near the mouth and then progresses upstream, creating a new stream profile.

Observing streams that adjust their profiles to changes in base level led to the concept of a graded stream. A **graded stream** has the necessary slope and other channel characteristics to maintain the minimum velocity required to transport the material supplied to it. On average, a graded system is neither eroding nor depositing material but simply transporting it. When a stream reaches equilibrium, it becomes a self-regulating system in which a change in one characteristic causes a change in the others to counteract the effect.

▼ **Figure 13.22 Building a dam** The base level upstream from the reservoir is raised, which reduces the stream's flow velocity and leads to deposition and a reduced gradient.

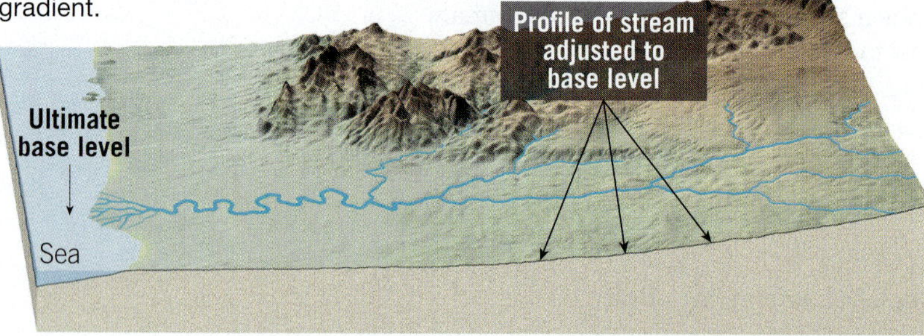

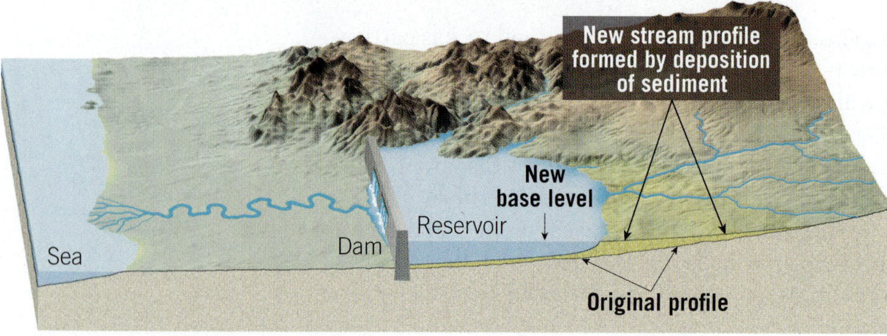

Valley Deepening

When a stream's gradient is steep, the channel is well above base level, and the dominant erosional process is **downcutting**, the lowering of the streambed toward base level. Abrasion caused by the bed load sliding and rolling along the bottom, as well as the hydraulic power of fast-moving water, slowly lowers the streambed. The result is usually a V-shaped valley with steep sides. A classic example of a V-shaped valley is the section of the Yellowstone River shown in Figure 13.23.

The most prominent features of V-shaped valleys are *rapids* and *waterfalls*. Both occur where the stream's gradient increases significantly, a situation usually caused by

**Exploration of the Colorado River of the West* (Washington, DC: Smithsonian Institution, 1875), p. 203.

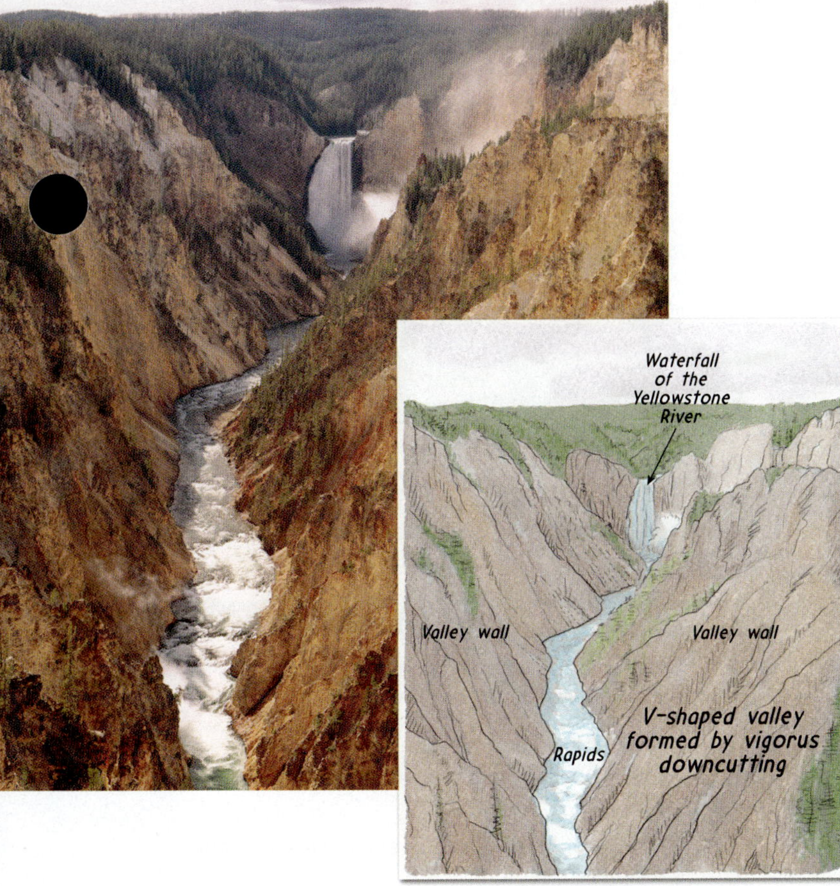

▲ **Figure 13.23 Yellowstone River** The V-shaped valley, rapids, and waterfalls indicate that the river is downcutting vigorously. (Photo by Charles A. Blakeslee/AGE Fotostock)

variations in the erodibility of the bedrock into which a stream channel is cutting. Resistant beds create rapids by acting as a temporary base level upstream while allowing downcutting to continue downstream. Over time erosion usually eliminates the resistant rock. Waterfalls occur where streams make vertical drops.

Valley Widening

As a stream approaches a graded condition—in which it primarily works to transport sediment—downcutting becomes less dominant. At this point, the stream's channel takes on a meandering pattern, and more of its energy is directed from side to side. As a result, the valley widens as the river cuts away at one bank and then the other. The continuous lateral erosion caused by shifting meanders gradually produces a broad, flat valley floor covered with alluvium (Figure 13.24). This feature is called a **floodplain** because it becomes inundated when the river overflows its banks during flood stage. Over time the floodplain will widen to a point where the stream is actively eroding the valley walls in only a few places. In the case of the lower Mississippi River, for example, the distance from one valley wall to the opposite wall sometimes exceeds 160 kilometers (100 miles).

When a river erodes laterally and creates a floodplain as described, it is called an *erosional floodplain*. Floodplains can be depositional as well. *Depositional floodplains* are produced by major fluctuations in conditions, such as changes in base level or climate. The floodplain in California's Yosemite Valley is one such feature; it was produced when a glacier gouged the valley floor about 300 meters (1000 feet) deeper than its former level. After the glacial ice melted, running water refilled the valley with alluvium. The Merced River currently winds across a relatively flat floodplain that forms much of the floor of Yosemite Valley.

Did You Know?
The world's highest uninterrupted waterfall is Venezuela's Angel Falls. Named for American aviator Jimmie Angel, who first sighted the falls from the air in 1933, the water plunges 3212 ft.

Incised Meanders & Stream Terraces

We usually find streams with highly meandering courses on floodplains in wide valleys. However, some rivers have meandering channels that flow in steep, narrow bedrock valleys. Such meanders are called **incised meanders** (*incisum* = to cut into); a classic example is pictured in the chapter-opening photo.

How do these features form? Originally the meanders probably developed on the floodplain of a stream that was in balance with its base level (Figure 13.25A). Then, a change in base level caused the stream to begin downcutting. Such a change can be caused either by a drop in a downstream base level or by uplift of the land on which the river is flowing.

▼ SmartFigure 13.24
Development of an erosional floodplain
Continuous side-to-side erosion by shifting meanders gradually produces a broad, flat valley floor. Alluvium deposited during floods covers the valley floor.

CONDOR VIDEO
https://goo.gl/0sxaWY

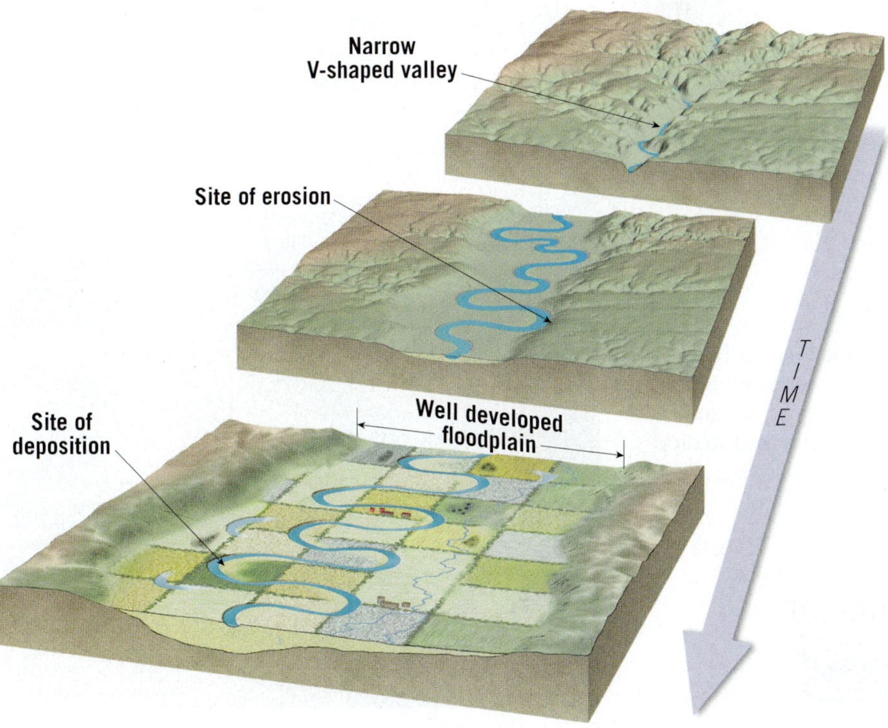

▶ **SmartFigure 13.25**
Incised meanders These diagrams illustrate the development of incised meanders such as those shown in the chapter-opening photo.

TUTORIAL
https://goo.gl/P3ZIz9

A. Before uplift of the Colorado Plateau, the river was meandering on a floodplain.

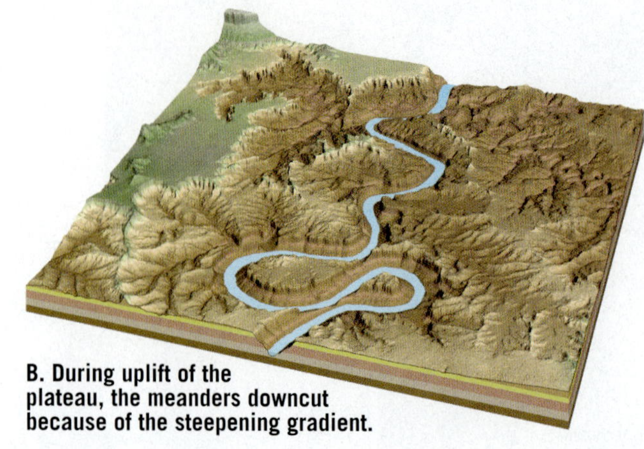

B. During uplift of the plateau, the meanders downcut because of the steepening gradient.

For example, regional uplifting of the Colorado Plateau in the southwestern United States generated incised meanders on several rivers (**Figure 13.25B**). As the plateau gradually rose, meandering rivers began downcutting because of their steepening gradient.

Other features associated with a relative drop in base level are **stream terraces**. After a river that had been flowing on a floodplain has adjusted to a relative drop in base level, it may once again produce a floodplain at a level below the old one. As shown in **Figure 13.26**, remnants of the former floodplain are left as relatively flat surfaces above the newly forming floodplain.

CONCEPT CHECKS 13.6

1. Define *base level* and distinguish between ultimate base level and local, or temporary, base level.
2. What is a graded stream?
3. Explain why V-shaped valleys often contain rapids and/or waterfalls.
4. Describe or sketch how an erosional floodplain develops.
5. Describe two situations that would trigger the formation of incised meanders.

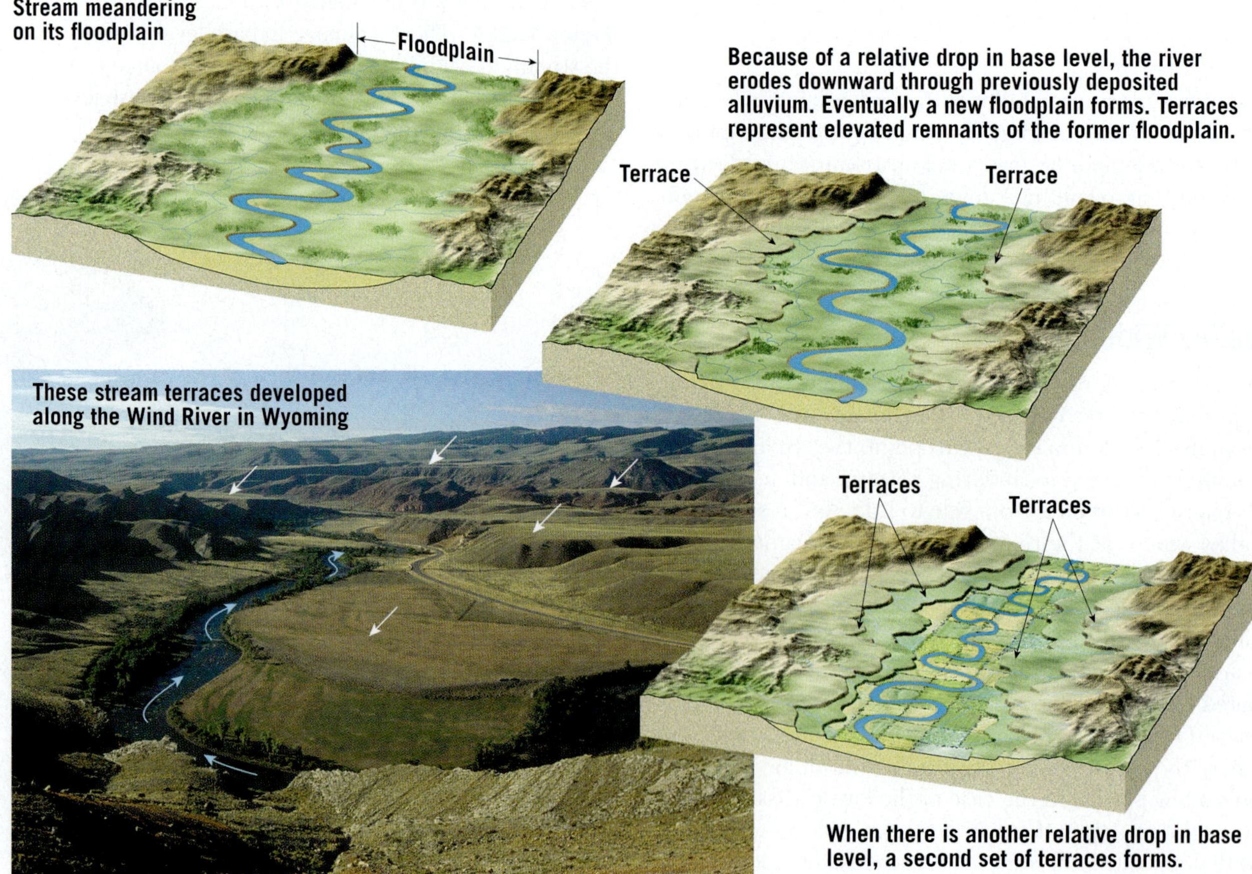

These stream terraces developed along the Wind River in Wyoming

▶ **SmartFigure 13.26**
Stream terraces Terraces result when a stream adjusts to a relative drop in base level. (Photo by Greg Hancock)

CONDOR VIDEO
https://goo.gl/YSP8UV

13.7 Depositional Landforms

List the major depositional landforms associated with streams and describe the formation of these features.

Recall that streams continually pick up sediment in one part of their channel and deposit it downstream. These small-scale channel deposits are called **bars**. Such features, however, are only temporary, as the material will be picked up again and eventually carried to the ocean. In addition to sand and gravel bars, streams also create depositional features that have longer life spans. These include *deltas*, *natural levees*, and *alluvial fans*.

Deltas

Deltas form where sediment-laden streams enter the relatively still waters of a lake, an inland sea, or the ocean (**Figure 13.27**). As the stream's forward motion is slowed, sediments are deposited by the dying current, producing three types of beds. *Foreset beds* are composed of coarse particles that drop almost immediately upon entering the water to form layers that slope downcurrent from the delta front. The foreset beds are subsequently covered by thin, horizontal *topset beds* deposited during floods. The finer silts and clays settle away from the mouth in nearly horizontal layers called *bottomset beds*.

As a delta grows outward from the shoreline, streamflow becomes sluggish because the gradient is decreasing. This eventually causes the channel to become choked with sediment. As a consequence, the river seeks shorter, higher-gradient routes to base level. Figure 13.27 shows the main channel dividing into several smaller ones, called **distributaries**, that carry water away from the main channel in varying paths to base level. After numerous shifts in the main flow from one distributary to the next, a delta may grow into the triangular shape of the Greek letter delta (Δ), although several other shapes exist. Differences in the configurations of shorelines and variations in the nature and strength of wave activity are responsible for the shape and structure of each delta. Many of the world's great rivers have created massive deltas, each with its own peculiarities and typically more complex than the one illustrated in Figure 13.27.

Not all rivers have deltas. A river that transports a large sediment load may lack a delta because ocean waves and powerful currents quickly redistribute the material soon after it is deposited. The Columbia River in the Pacific Northwest is an example. In other cases, rivers do not carry sufficient quantities of sediment to build a delta. The St. Lawrence River, for example, has little opportunity to pick up much sediment between its head in Lake Ontario and its mouth in the Gulf of St. Lawrence.

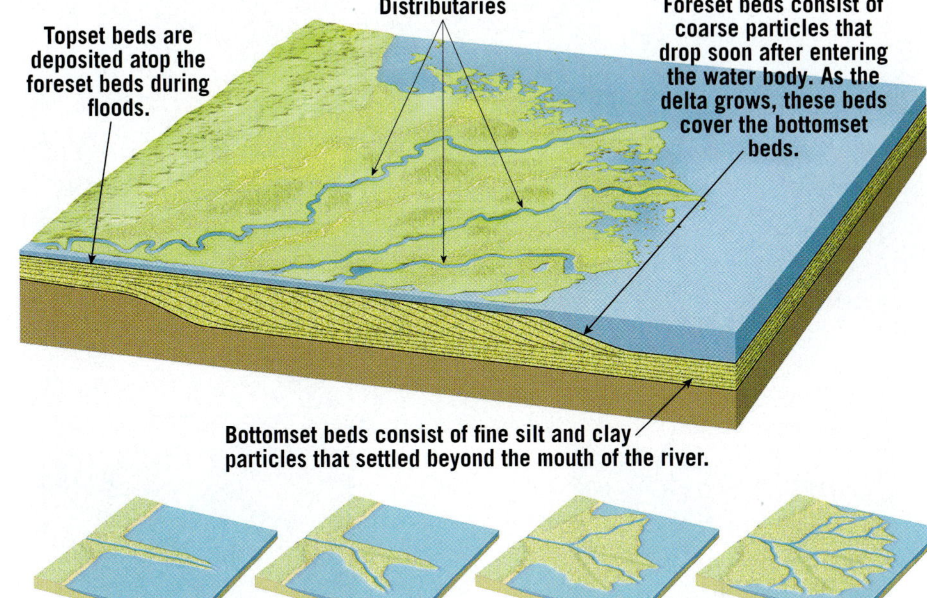

Distributaries
Topset beds are deposited atop the foreset beds during floods.
Foreset beds consist of coarse particles that drop soon after entering the water body. As the delta grows, these beds cover the bottomset beds.
Bottomset beds consist of fine silt and clay particles that settled beyond the mouth of the river.
As the stream extends its channel, the gradient is reduced. During flood stage some of the flow is diverted to a shorter, higher-gradient route forming a new distributary.

▲ **Figure 13.27**
Formation of a simple delta Structure and growth of a simple delta that forms in relatively quiet waters.

The Mississippi River Delta

The Mississippi River delta resulted from the accumulation of huge quantities of sediment derived from the vast region drained by the river and its tributaries. New Orleans currently rests where there was once ocean.

History & Structure The portion of the Mississippi delta that has formed during the past 6000 years is shown in **Figure 13.28**. As the figure illustrates, the delta is actually a series of seven coalescing subdeltas. Each subdelta formed when the main flow was diverted from one channel to a shorter, more direct path to the Gulf of Mexico. The individual subdeltas intertwine and partially cover each other to produce a complex structure. It is also apparent from Figure 13.28 that after each channel was abandoned, coastal erosion modified the newly formed subdelta. The present subdelta (number 7 in Figure 13.28), called a *bird-foot* delta because of the configuration of its distributaries, has been built by the Mississippi River over the past 500 years.

At present, this active bird-foot delta has extended seaward almost as far as natural forces will allow. In fact, for many years the river has been "struggling" to cut through a narrow neck of land and shift its course to the Atchafalaya River (see inset in Figure 13.28). If this were to happen, the Mississippi would abandon the lowermost 500 kilometers (300 miles) of its channel in favor of the Atchafalaya's much shorter 225-kilometer (140-mile) route to the Gulf.

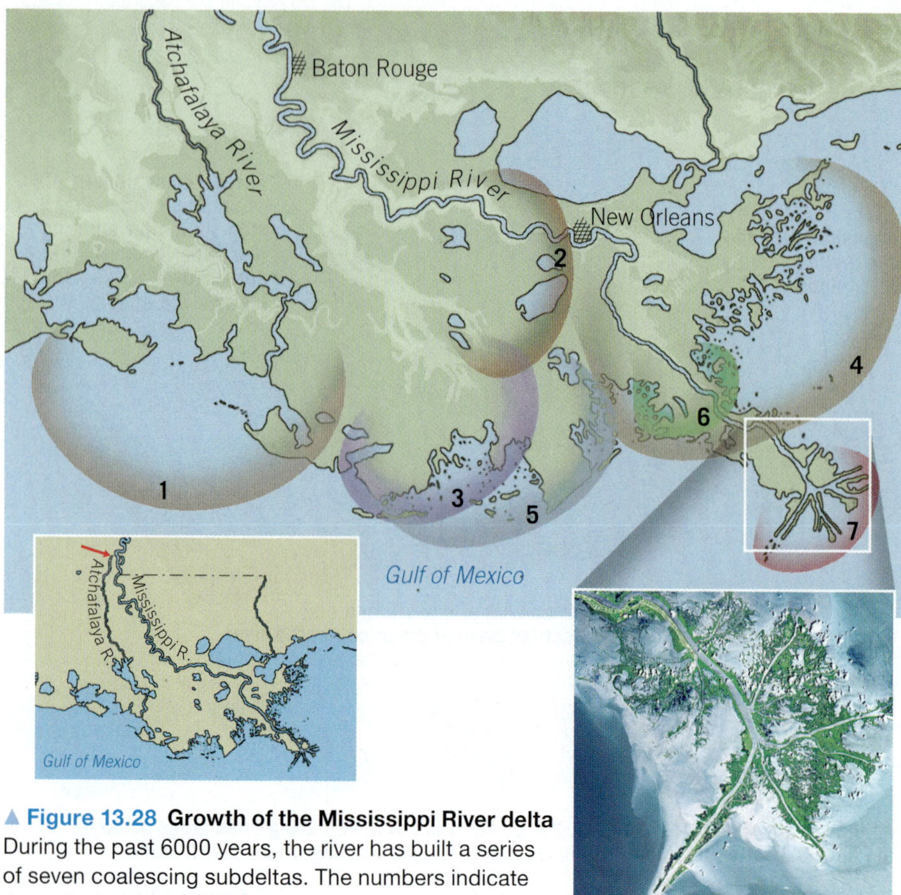

▲ Figure 13.28 Growth of the Mississippi River delta During the past 6000 years, the river has built a series of seven coalescing subdeltas. The numbers indicate the order in which the subdeltas were deposited. The present bird-foot delta (number 7) represents the activity of the past 500 years. The left inset shows the point where the Mississippi may sometime break through (arrow) and the shorter path it would take to the Gulf of Mexico. (Image courtesy of JPL/Cal tech/NASA)

In a concerted effort to keep the Mississippi on its present course, a dam-like structure was erected at the site where the channel was trying to break through. A flood in 1973 weakened the control structure, and the river again threatened to shift. This event caused the U.S. Army Corps of Engineers to construct a massive auxiliary dam that was completed in the mid-1980s. For the time being, at least, the inevitable has been avoided, and the Mississippi River continues to flow past Baton Rouge and New Orleans on its way to the Gulf of Mexico.

Vanishing Wetlands The delta of the Mississippi River in Louisiana is a biologically significant region that includes about 12,000 square kilometers (3 million acres) of coastal wetlands—40 percent of all coastal wetlands in the contiguous United States (**Figure 13.29**). These flat areas that are only slightly above sea level are sheltered from the wave action of hurricanes and winter storms by low-lying offshore barrier islands. Both the wetlands and offshore islands were formed and maintained by sediments carried to the Gulf of Mexico by the Mississippi River.

The coastal wetlands of Louisiana are disappearing at an alarming rate—accounting for 80 percent of the wetland loss in the lower 48 states. According to the U.S. Geological Survey, Louisiana has lost more than 5000 square kilometers (1900 square miles) of coastal land since the early 1930s. If wetland loss continues at this rate, another 3000 square kilometers (1170 square miles) will vanish by the year 2050. Why is this occurring?

Before Europeans settled the delta, the Mississippi River regularly overflowed its banks in seasonal floods. Huge quantities of sediment were deposited atop the delta, which kept the land elevated above sea level. However, with settlement came flood-control efforts and the desire to maintain and improve navigation on the river. Artificial levees (earthen mounds built parallel to the river) were constructed to contain the rising river during flood stage. Over time the levees were extended all the way to the mouth of the Mississippi to keep the channel open to navigation.

The effects have been straightforward: The levees prevent sediment from being dispersed onto the wetlands. As a result, sediment is not added in sufficient quantities to offset compaction, subsidence, and wave erosion. Thus, the size of the delta and the extent of the wetlands are shrinking. The problem has been exacerbated by a decline in the sediment load of the Mississippi, which has decreased approximately 50 percent over the past 100 years. A substantial portion of the reduction is due to sediment being trapped upstream by dams on the river's many tributaries.

Natural Levees

Meandering rivers that occupy valleys with broad floodplains tend to build **natural levees** (*lever* = to raise)

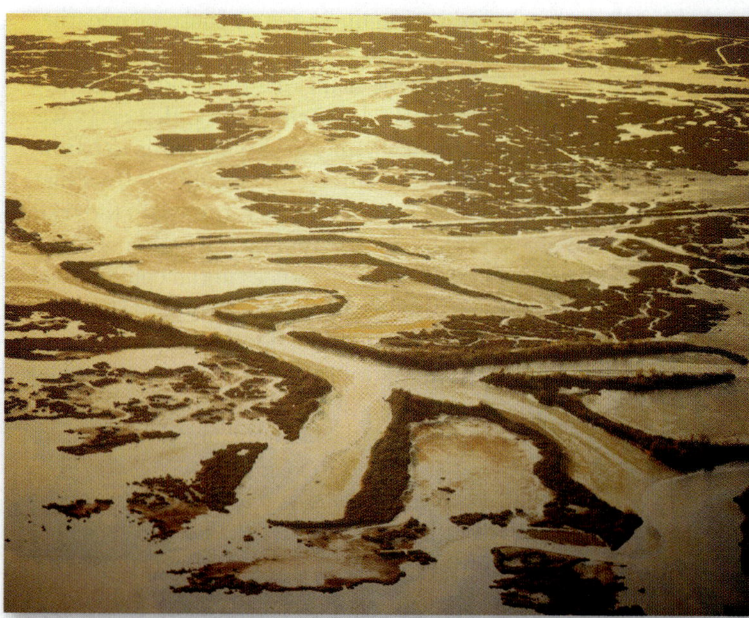

▲ SmartFigure 13.29 Coastal wetlands These low-lying, sediment-starved swamps, marshes, and bayous are disappearing at an alarming rate. (Photo by Michael Collier)

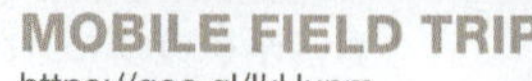

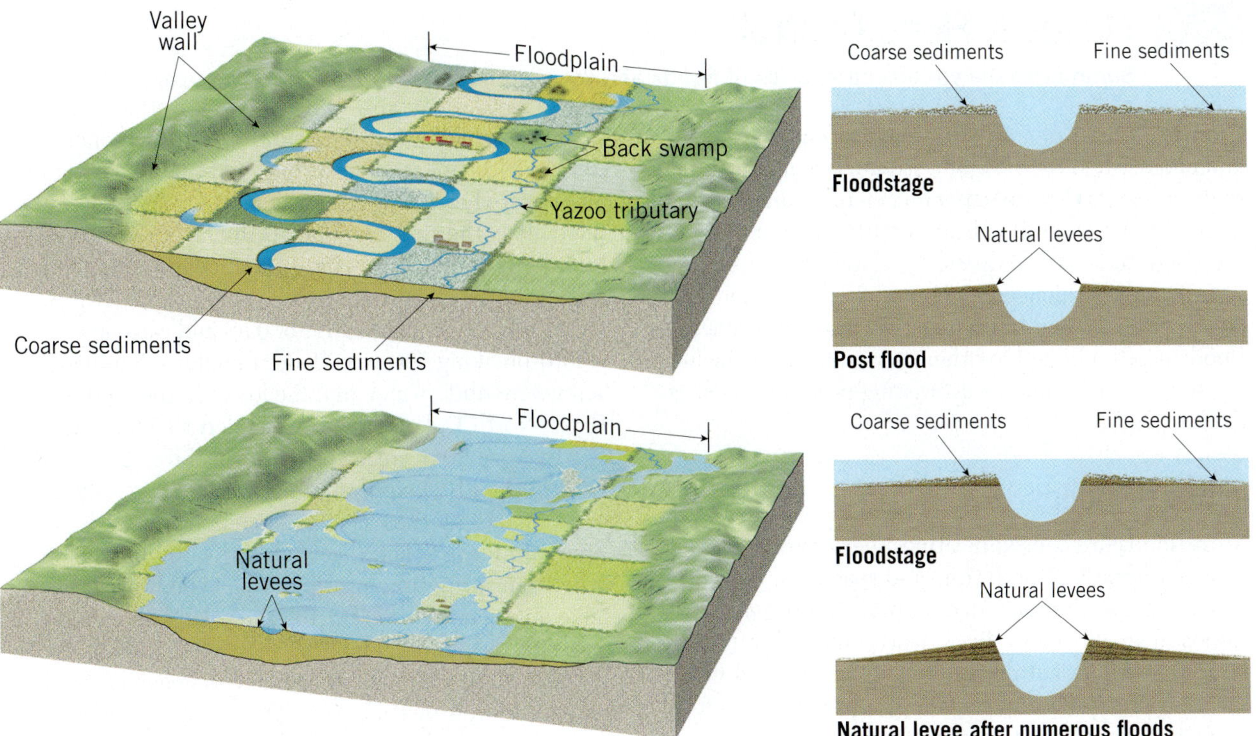

SmartFigure 13.30
Formation of a natural levee These gently sloping structures that parallel a river channel are created by repeated floods. Because the ground next to the channel is higher than the adjacent floodplain, back swamps and yazoo tributaries may develop.

ANIMATION
https://goo.gl/UdXGH3

that parallel their channels on both banks (**Figure 13.30**). Natural levees are built by years of successive floods. When a stream overflows onto the floodplain, the water flows over the surface as a broad sheet. Because flow velocity drops rapidly, the coarser portion of the suspended load is immediately deposited adjacent to the channel. As the water spreads across the floodplain, a thin layer of fine sediment is laid down over the valley floor. This uneven distribution of material produces the gentle, almost imperceptible, slope of the natural levee (see Figure 13.30).

The natural levees of the lower Mississippi River rise 6 meters (20 feet) above the adjacent valley floor. The area behind a levee is characteristically poorly drained for the obvious reason that water cannot flow over the levee and into the river. Marshes called **back swamps** often result. When a tributary stream enters a river valley that has a substantial natural levee, it often flows for many kilometers through the back swamps before finding an opening where it enters the main river. Such streams are called **yazoo tributaries**, after the Yazoo River, which parallels the lower Mississippi River for more than 300 kilometers (185 miles).

Alluvial Fans

Alluvial fans are fan-shaped deposits that accumulate along steep mountain fronts (see Figure 16.8, page 433). When a mountain stream emerges onto a relatively flat lowland, its gradient drops, and it deposits a large portion of its sediment load. Although alluvial fans are more prevalent in arid climates, they are occasionally found in humid regions.

Mountain streams, because of their steep gradients, carry much of their sediment load as coarse sand and gravel. Because alluvial fans are composed of these same coarse materials, water that flows across them readily soaks in. When a stream emerges from its valley onto an alluvial fan, its flow divides itself into several distributary channels. The fan shape is produced because the main flow swings back and forth between the distributaries from a fixed point where the stream exits the mountain.

Between rainy periods in deserts, little or no water flows across an alluvial fan, which is evident in the many dry channels that cross its surface. Thus, fans in dry regions grow intermittently, receiving considerable water and sediment only during wet periods.

Steep canyons in dry regions are prime locations for debris flows. Therefore, as would be expected, many alluvial fans have debris-flow deposits interbedded with the coarse alluvium.

CONCEPT CHECKS 13.7

1. Sketch a cross section of a simple delta and distinguish among the three types of beds that compose it.
2. Explain why the Mississippi delta consists of seven coalescing subdeltas.
3. Briefly describe the formation of a natural levee. How is this feature related to back swamps and yazoo tributaries?
4. Describe the formation of an alluvial fan.

13.8 Floods & Flood Control

Summarize the various categories of floods and the common measures of flood control.

Floods are part of the natural behavior of a stream and occur when the flow of a stream becomes so great that it exceeds the capacity of its channel and overflows its banks. Although floods are natural phenomena, the magnitude and frequency of flooding is often significantly influenced by human activities such as clearing forests and building cities. The occurrence of floods is often linked to other natural hazards, including severe storms and mass-wasting processes such as debris flows.

Types of Floods

Most floods are caused by atmospheric processes that can vary greatly in both time and space. An hour or less of intense thunderstorm rainfall can trigger flash floods in small valleys. By contrast, major floods in large river valleys often result from an extraordinary series of precipitation events over a broad region for many days or weeks. Common flood types include *regional floods*, *flash floods*, *ice-jam floods*, and *dam-failure floods*.

A.

▼ **Figure 13.31 Floods are dangerous natural hazards A.** Although short lived, flash floods can be powerful and often occur with little advance warning. The event shown here occurred in an urban setting. (AP Photo/Sue Ogrocki) **B.** In most years floods are responsible for the greatest number of storm-related deaths. The average number of hurricane deaths was dramatically affected by Hurricane Katrina (more than 1000). For all other years on this graph, hurricane fatalities numbered fewer than 20. (AP Photo/Sue Ogrocki)

Regional Floods Most regional floods are seasonal. Rapid melting of snow in spring and/or heavy spring rains often overwhelm rivers. For example, the extensive 1997 flood along the Red River in North Dakota was preceded by an especially snowy winter and an early spring blizzard. Early April brought rapidly rising temperatures, melting the snow in a matter of days and causing a record-breaking 500-year flood. Roughly 4.5 million acres were underwater, and the losses in the Grand Forks, North Dakota, region exceeded $3.5 billion.*

In April 2011, unrelenting storms brought record rains to the Mississippi watershed. The Ohio Valley, which makes up the eastern portion of the Mississippi's drainage basin, received nearly 300 percent of its normal springtime precipitation. When that rainfall combined with water from the past winter's large and extensive, rapidly melting snowpack, the Mississippi River and many of its tributaries began to swell to record levels by early May. The resulting floods were among the largest and most damaging in nearly a century. Like most other regional floods, these were associated with weather phenomena that could be forecast with a good deal of accuracy. This allowed adequate time to warn and evacuate thousands of people who were in harm's way. Although economic losses approached $4 billion, loss of life was small.

Flash Floods Unlike extensive regional floods, *flash floods* are more limited in extent. Flash floods occur with little warning and can be deadly because they produce a rapid rise in water levels and can have a devastating flow velocity (**Figure 13.31**). Several factors influence flash flooding. Among them are rainfall intensity and duration, surface conditions, and topography. Urban areas are susceptible to flash floods because a high percentage of the surface area is composed of impervious roofs, streets, and parking lots, where runoff is very rapid (**Figure 13.32**). Mountainous areas are susceptible because steep slopes can quickly funnel runoff into narrow canyons.

Ice-Jam Floods Frozen rivers are especially susceptible to ice-jam floods. As the level of a stream rises, it breaks up ice and creates ice floes that can accumulate on channel obstructions. Jams of this nature create temporary ice dams across the channel. Water trapped upstream can rise rapidly and overflow the channel banks. When an ice dam fails, water

*Ice jams also contribute to floods on the Red River of the North. See the section "Ice-Jam Floods."

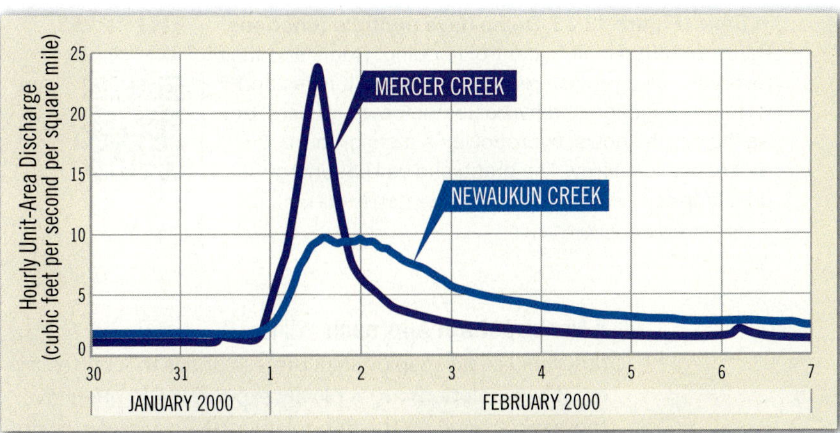

▲ **Figure 13.32 Effect of urban development on flooding** Streamflow in Mercer Creek, an *urban stream* in western Washington, increased more quickly, reached a higher peak discharge, and had a larger volume during a 1-day storm on February 1, 2000, than streamflow in Newaukum Creek, a nearby *rural stream*. Streamflow during the following week, however, was greater in Newaukum Creek.

behind the dam is often released with sufficient force to inflict considerable damage downstream.

These floods are often associated with northward-flowing rivers in the Northern Hemisphere. The Red River of the North, mentioned earlier, is an example. Siberia has several northward-flowing rivers, including the Ob, Lena, and Yenisei, that frequently experience these floods. When spring arrives, ice melts in the warmer southern portions of a river and its drainage basin, while farther north the river remains frozen. Water flowing from the south "backs up" behind the frozen northern portion of the river.

Dam-Failure Floods Human interference with stream systems can cause floods. A prime example is the failure of a dam or an artificial levee designed to contain small or moderate floods. When larger floods occur, the dam or levee may fail, resulting in the water behind it being released as a flash flood. The bursting of a dam in 1889 on the Little Conemaugh River caused the devastating Johnstown, Pennsylvania, flood that took more than 2200 lives.

Flood Control

Several strategies have been devised to eliminate or lessen the catastrophic impact of floods on our lives and the environment. Engineering efforts include the construction of artificial levees, river channelization, and the building of flood-control dams.

Artificial Levees *Artificial levees* are earthen mounds built on river banks to increase the volume of water the channel can hold. Levees, used since ancient times, are the most commonly used stream-containment structures. In some locations, concrete floodwalls are constructed to function as artificial levees.

Many artificial levees were not built to withstand periods of extreme flooding. For example, numerous levees failed during the summer of 1993, when the upper Mississippi and many of its tributaries experienced record flooding. During that event, floodwalls at St. Louis, Missouri, created a bottleneck for the river that led to increased flooding upstream of the city.

Channelization *Channelization* involves altering a stream channel in order to make the flow more efficient. This may simply involve clearing a channel of obstructions or dredging a channel to make it wider and deeper.

A more radical alteration involves straightening a channel by creating *artificial cutoffs*. The idea is that by shortening the stream, the gradient and hence the flow velocity are increased. By increasing velocity, the larger discharge associated with flooding can be dispersed more rapidly.

Beginning in the early 1930s, the U.S. Army Corps of Engineers created many artificial cutoffs on the Mississippi for the purpose of increasing the efficiency of the channel and reducing the threat of flooding. In all, the river was shortened more than 240 kilometers (150 miles). The program has been somewhat successful in reducing the height of the river in flood stage. However, because the river's tendency toward meandering still exists, preventing the river from returning to its previous course has been difficult.

Flood-Control Dams *Flood-control dams* are built to store floodwater and then release it slowly, in a controlled manner. Since the 1920s, thousands of dams have been built on rivers; nearly every major river in the United States has at least one dam. Many dams have significant non-flood-related functions, such as providing water for irrigated agriculture and for hydroelectric power generation (Figure 13.33). Many reservoirs are also major regional recreational facilities.

Although dams are effective in reducing flooding and provide other benefits, their construction and maintenance also have significant costs and consequences. For example, reservoirs created by dams may cover valuable farmland, forests, historic sites, and scenic valleys. Large dams can also cause significant damage to river ecosystems that have developed over thousands of years.

Furthermore, building dams is not a permanent solution to flooding. Sedimentation behind a dam gradually diminishes the volume of its reservoir, reducing the long-term effectiveness of this flood-control measure.

Did You Know?
Just 6 in of fast-moving water can knock a person down. Many cars will float and be swept away in only 2 ft of water. More than half of all U.S. flash-flood fatalities are automobile related.

Did You Know?
Urban development increases runoff. As a result, peak discharge and flood frequency increase in urban areas. Humans have covered an amazing amount of land with buildings, parking lots, and roads. A recent study indicated that the area of such impervious surfaces in the United States (excluding Alaska and Hawaii) amounts to more than nearly 44,000 mi^2, which is slightly less than the area of the state of Ohio.

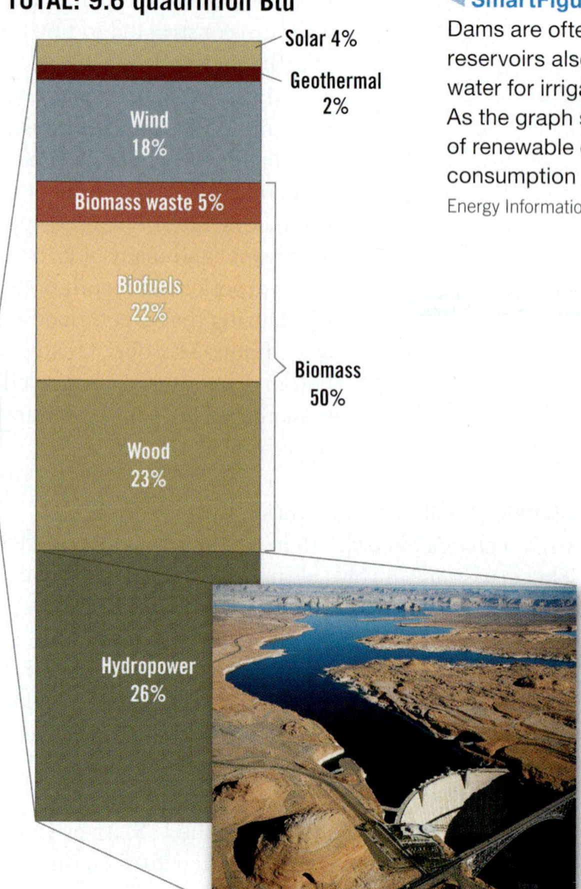

SmartFigure 13.33 Dams have multiple functions Dams are often built to control flooding. Dams and their reservoirs also provide recreational opportunities and water for irrigation and hydroelectric power generation. As the graph shows, hydropower is a major form of renewable energy. The graph shows U.S. energy consumption in 2014. (Photo by Michael Collier; data from U.S. Energy Information Administration)

TUTORIAL
https://goo.gl/zseNGR

A Nonstructural Approach All the flood-control measures described so far employ structural solutions to "control" rivers. These solutions are typically expensive and often give those who reside on the floodplain a false sense of security.

Many scientists and engineers advocate a nonstructural approach to flood control. They suggest that sound floodplain management is an alternative to artificial levees, dams, and channelization. By identifying high-risk flood areas, appropriate zoning regulations can be implemented to minimize development and promote safer, more appropriate land use.

CONCEPT CHECKS 13.8

1. List and distinguish among four types of floods.
2. Describe three basic flood-control strategies. What are some drawbacks of each?
3. What is meant by a *nonstructural approach* to flood control?

CONCEPTS IN REVIEW
Running Water

13.1 Earth as a System: The Hydrologic Cycle

List the hydrosphere's major reservoirs and describe the different paths that water takes through the hydrologic cycle.

KEY TERMS: hydrologic cycle, evaporation, infiltration, runoff, transpiration, evapotranspiration

- Water moves through the hydrosphere's different reservoirs by evaporating, condensing into clouds, and falling as precipitation. Once it reaches the ground surface, rain can either soak into the soil, evaporate, be returned to the atmosphere by plant transpiration, or run off over the surface.
- At any given time, running water represents a small fraction of the total water on Earth. Nevertheless, huge quantities are cycled through this part of the hydrologic system, making running water the most important erosional agent sculpting Earth's surface.

13.2 Running Water

Describe the nature of drainage basins and river systems. Sketch and briefly explain four basic drainage patterns.

KEY TERMS: stream, river, drainage basin (watershed), divide, headward erosion, dendritic pattern, radial pattern, rectangular pattern, trellis pattern, water gap, superposed stream

- The land area that contributes water to a stream is its drainage basin. Drainage basins are separated by imaginary lines called divides.
- River systems tend to erode at the upstream end, transport sediment through the middle section, and deposit sediment at the downstream end.
- A stream effectively erodes in a headward direction, thereby lengthening its course.
- A water gap is a steep-walled notch through which a stream flows. Water gaps are often associated with superposed streams.

? Identify each of the drainage patterns depicted in the accompanying sketch.

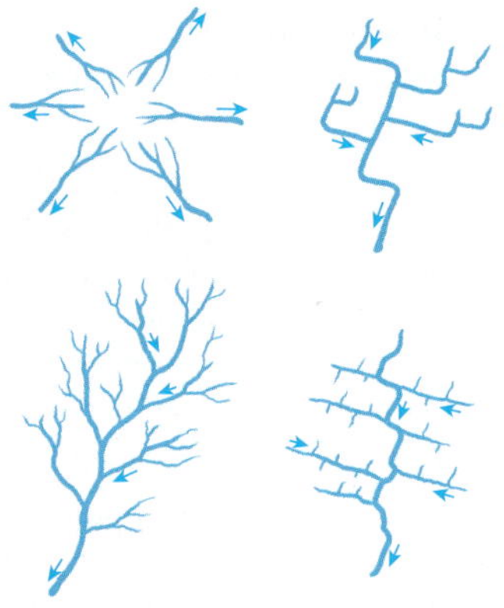

13.3 Streamflow Characteristics

Discuss streamflow and the factors that cause it to change.

KEY TERMS: laminar flow, turbulent flow, gradient, wetted perimeter, discharge, longitudinal profile, head (headwaters), mouth

- The flow of water in a stream may be laminar or turbulent. A stream's flow velocity is influenced by channel gradient; size, shape, and roughness of the channel; and discharge.
- A cross-sectional view of a stream from head to mouth is a longitudinal profile. Usually the gradient and roughness of the stream channel decrease going downstream, whereas the size of the channel, stream discharge, and flow velocity increase in the downstream direction.

? Sketch a typical longitudinal profile. Where does most erosion happen? Where is sediment transport the dominant process?

13.4 The Work of Running Water

Outline the ways in which streams erode, transport, and deposit sediment.

KEY TERMS: quarrying, abrasion, pothole, corrosion, dissolved load, suspended load, bed load, settling velocity, saltation, capacity, competence, sorting, alluvium

- Streams are powerful agents of erosion that carve solid rock through quarrying, abrasion, and the focused "drilling" that results in potholes. Turbulent water also lifts loose particles from the streambed. In areas of soluble bedrock such as limestone, stream water can also corrode the landscape by dissolving the rock.

(13.4 continued)

- Sediment is transported in a stream either in solution, suspended in the water, or rolling or saltating along the bottom of the stream. Compared to slow-moving rivers, fast-moving rivers can carry a greater total amount of sediment (capacity) and larger individual particles (competence). Flooding increases both capacity and competence, which is why rivers do most of their work during short-lived times of peak flow.
- Sediment deposited by streams is called alluvium. Typically, streams are efficient agents of sorting, meaning that they deposit similarly sized grains in the same area.

13.5 Stream Channels

Contrast bedrock and alluvial stream channels. Distinguish between two types of alluvial channels.

KEY TERMS: bedrock channel, alluvial channel, meander, cut bank, point bar, cutoff, oxbow lake, braided channel

- Bedrock channels are cut into solid rock. They typically show steps (highlighted by waterfalls or rapids) and pools (segments that are relatively flat).
- Alluvial channels are dominated by streamflow moving through sediment previously deposited by the river. A floodplain usually covers much of a valley floor, with the river itself moving through a channel that may meander or exhibit a braided pattern.
- Meanders are enhanced through erosion at the cut bank (outside edge of the meander) and deposition of sediment on the point bar (inside edge of the meander). The shape of the meander may grow more and more exaggerated until it loops back on itself. Once a cutoff is formed, the main current abandons the old meander loop, which becomes an oxbow lake.
- Braided channels occur in streams that experience a high degree of variability in discharge. During times of low flow, the river moves in an interwoven network of channels between bars of coarse-grained alluvium.

? The town of Carter Lake is the *only* portion of the state of Iowa that lies on the west side of the Missouri River. It is bounded on the north by its namesake, Carter Lake, on the south by the Missouri River, and on the east and west by Nebraska. After examining the map, prepare a hypothesis that explains how this unusual situation could have developed.

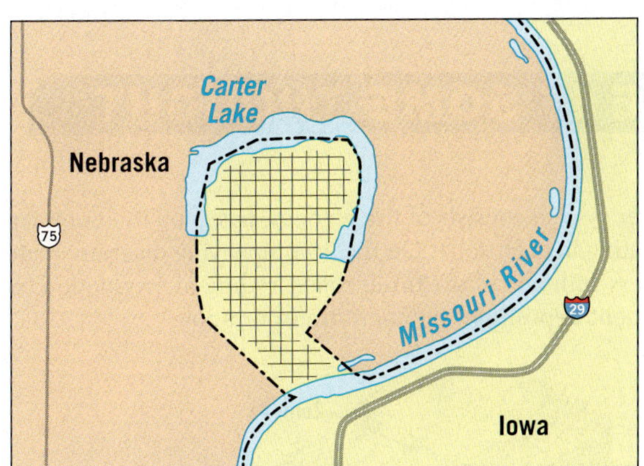

13.6 Shaping Stream Valleys

Contrast narrow V-shaped valleys, broad valleys with floodplains, and valleys that display incised meanders or stream terraces.

KEY TERMS: stream valley, base level, ultimate base level, local (temporary) base level, graded stream, downcutting, floodplain, incised meander, stream terrace

- A stream valley includes the channel itself, the adjacent floodplain, and relatively steep valley walls. Stream valley width and shape vary a lot from V-shaped with narrow floodplains to broad floodplains of alluvium and relatively subdued valley walls.
- Streams erode downward until they approach base level, which is usually the level at which the stream enters another stream, a lake, or the ocean. A river flowing toward the sea (ultimate base level) may encounter several local base levels along its route. These could be lakes or resistant rock units that retard the downcutting of the stream. A graded stream has reached equilibrium with its base level and primarily works to transport sediment.
- Stream valleys are widened through the meandering action of the stream, which erodes the valley walls and widens the floodplain. If the base level drops, the stream downcuts. If it is underlain by bedrock, the stream may then develop incised meanders. Streams underlain by deep alluvium are likely to develop terraces.

? Meanders are associated with a river that is eroding from side to side, whereas narrow canyons are associated with rivers that are vigorously downcutting. The river in this image is confined to a narrow canyon but is also meandering. Explain.

Michael Collier

13.7 Depositional Landforms

List the major depositional landforms associated with streams and describe the formation of these features.

KEY TERMS: bar, delta, distributary, natural levee, back swamp, yazoo tributary, alluvial fan

- A delta may form where a river deposits sediment in another water body at its mouth. The partitioning of streamflow into multiple distributaries spreads sediment in different directions. In the United States, the Mississippi River is an example of a major river with a dynamic delta system.
- Natural levees result from sediment deposited along the margins of a river channel by many years of successive floods. Because they slope gently away from the channel, the surrounding land is poorly drained, resulting in back swamps.
- Alluvial fans are fan-shaped deposits of alluvium that form where steep mountain fronts drop down into adjacent valleys.

13.8 Floods & Flood Control

Summarize the various categories of floods and the common measures of flood control.

KEY TERMS: flood

- When a stream receives more discharge than its channel can hold, a flood occurs. Four major factors that can cause a flood are large amounts of precipitation (or melting) across a region, sudden pulses of precipitation in areas of high runoff, temporary dams made of floating ice (which then break up, releasing impounded water), and failure of human-built dams, causing the sudden escape of water from a reservoir.
- Three main structural strategies exist for coping with floods: construction of artificial levees to constrain streamflow to the channel, alterations to make a stream channel's flow more efficient, and building of dams so that a sudden influx of water will be temporarily stored and released slowly to the river system. A nonstructural approach involves sound floodplain management. Here, a solid scientific understanding of flood dynamics informs policy and regulation of areas that are subject to flooding.

? Artificial levees are constructed to protect property from floods. Sometimes artificial levees in rural areas are intentionally opened up to protect a city from experiencing flooding. How does this work?

GIVE IT SOME THOUGHT

1 A river system consists of three zones, based on the dominant process operating in each zone. On the accompanying diagram, match each process with one of the three zones: sediment production (erosion), sediment deposition, sediment transportation.

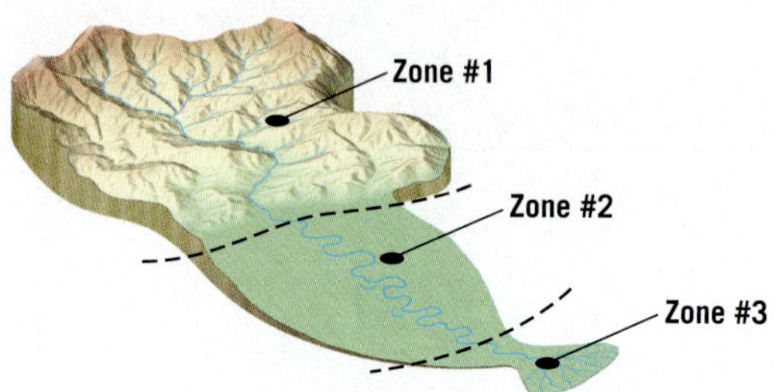

2 If you collect a jar of water from a stream, what part of its load will settle to the bottom of the jar? What portion will remain in the water indefinitely? What part of the stream's load would probably not be represented in your sample?

3 Streamflow is affected by several variables, including discharge, gradient, and channel roughness, size, and shape. Develop a scenario in which a mass wasting event influences a stream's flow. Explain what led up to, or triggered, the event and describe how the mass wasting process influenced the stream's flow.

4 The meandering White River in Arkansas is a tributary of the Mississippi River. In this aerial view, the color of the river is brown. What part of the stream's load gives it this color? If a channel were created across the narrow neck of land shown by the arrow, how would the river's gradient change? How would the flow velocity be affected by the formation of such a channel?

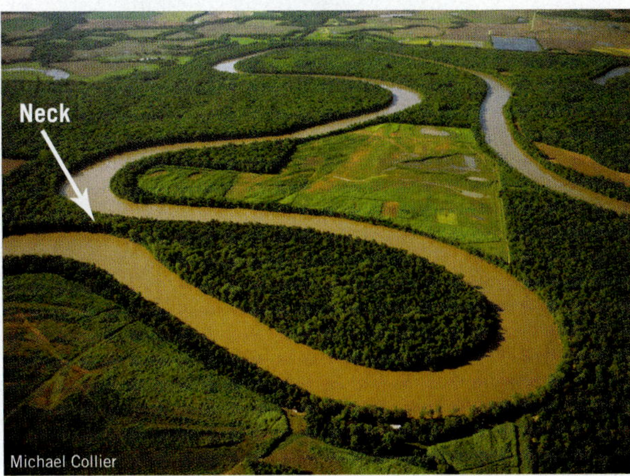

Michael Collier

5 This image, taken from the International Space Station, shows a plateau and canyons in central Saudi Arabia. Over a long time span, how will the canyons likely change? How will the plateau be affected? Is there a term that describes the process?

NASA

6 This satellite image shows the delta of the Yukon River in the Bering Sea along the coast of Alaska. Explain why the river breaks into numerous channels as it crosses the delta. What term is applied to these channels? Notice the cloud of sediment in the water surrounding the delta. Are these sediments more likely sand and gravel or silt and clay? Explain. After the cloud of material settles to the seafloor and the delta expands farther into the Bering Sea, which beds of the delta will these sediments become?

NASA

7 Several times during the past 2.5 million years, huge ice sheets (continental-size glaciers) formed and spread across large parts of Northern Hemisphere landmasses and then gradually melted away.

a. How do you think the formation of ice sheets affected sea level?

b. How would rivers flowing into the ocean have been affected as the ice sheets expanded?

c. What kind of adjustments would these rivers have made as the glacial ice melted?

8 Building a dam is one method of regulating the flow of a river to control flooding. Dams and their reservoirs may also provide recreational opportunities and water for irrigation and hydroelectric power generation.

a. When a dam is built, how will the behavior of the stream likely change upstream from the reservoir?

b. Given enough time, how might the reservoir change?

c. What are some possible environmental impacts of building a dam on a large river?

9 The accompanying images show lag times between rainfall and peak flow (flooding) for an urban area and a rural area. Which image (A or B) represents the rural area? Explain your choice.

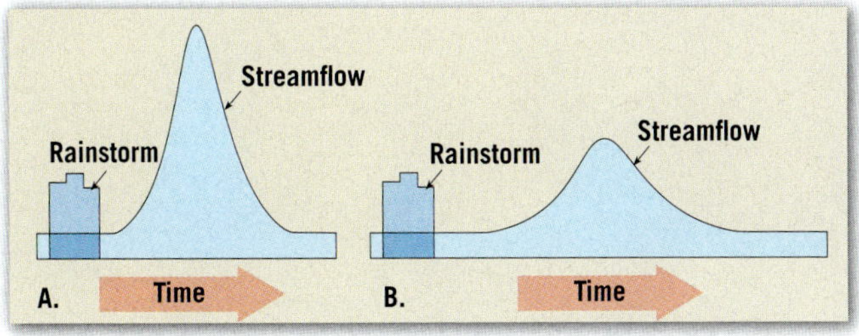

MasteringGeology™

Looking for additional review and test prep materials? Visit the Study Area in MasteringGeology to enhance your understanding of this chapter's content by accessing a variety of resources, including Self-Study Quizzes, Geoscience Animations, SmartFigures, Mobile Field Trips, *Project Condor* Quadcopter videos, *In the News* RSS feeds, flashcards, web links, and an optional Pearson eText.

www.masteringgeology.com

14 Groundwater

FOCUS ON CONCEPTS

Each statement represents the primary learning objective for the corresponding major heading within the chapter. After you complete the chapter, you should be able to:

14.1 Describe the importance of groundwater as a source of freshwater and groundwater's roles as a geologic agent.

14.2 Prepare a sketch with labels that summarizes the distribution of water beneath Earth's surface. Discuss the factors that cause variations in the water table and describe the interactions between groundwater and streams.

14.3 Summarize the factors that influence the storage and movement of groundwater. Discuss how groundwater movement is measured and the different scales of movement.

14.4 Discuss water wells and their relationship to the water table. Sketch and label a simple artesian system.

14.5 Distinguish among springs, hot springs, and geysers. List the geologic factors that favor geothermal energy development.

14.6 List and discuss important environmental problems associated with groundwater.

14.7 Explain the formation of caverns and the development of karst topography.

The dissolving action of acidic groundwater created these caverns. Later, groundwater deposited the dripstone decorations. (Photo by Images & Stories / Alamy Stock Photo)

ALTHOUGH IT IS HIDDEN FROM VIEW, vast quantities of water exist in the cracks, crevices, and pore spaces of rock and soil. Groundwater can be found almost everywhere beneath Earth's surface and is a major source of water worldwide. Groundwater is a valuable natural resource that provides about half of our drinking water and is essential to the vitality of agriculture and industry. In addition to its utility, groundwater plays a crucial role in sustaining streamflow between precipitation events—especially during protracted dry periods. Many ecosystems depend on groundwater discharge into streams, lakes, and wetlands. In some regions, large-scale development has caused groundwater levels to decline, resulting in water shortages, streamflow depletion, land subsidence, and increased pumping costs. Groundwater pollution is also a serious issue in some places.

14.1 The Importance of Groundwater

Describe the importance of groundwater as a source of freshwater and groundwater's roles as a geologic agent.

Did You Know?
The term "acre-foot" is often used to express large quantities of water such as the volume of a reservoir or amounts allocated for irrigation. As the term implies, an acre-foot is the volume of water needed to cover 1 acre of land to a depth of 1 ft. An acre-foot is equal to 326,000 gal.

Groundwater is one of our most important and widely available resources, yet people's perceptions of the subsurface environment from which it comes are often unclear and incorrect. This is because the groundwater environment is largely hidden from view except in caves and mines, and the impressions people gain from these subsurface openings are misleading. Observations on the land surface give an impression that Earth is "solid." This view remains when we enter a cave and see water flowing in a channel that appears to have been cut into solid rock.

Because of such observations, many people believe that groundwater occurs only in underground "rivers." In reality, most of the subsurface environment is not "solid" at all. It includes countless tiny *pore spaces* between grains of soil and sediment, plus narrow joints and fractures in bedrock. Together, these spaces add up to an immense volume. Where these subsurface pore spaces are saturated with water, that stored water is called **groundwater**. It is in these small openings that groundwater collects and moves.

Groundwater & the Hydrosphere

When we consider the entire hydrosphere, or all of Earth's water, only about six-tenths of 1 percent is located underground. Nevertheless, this small percentage, stored in the rocks and sediments beneath Earth's surface, is a vast quantity. When the oceans are excluded and only sources of freshwater are considered, the significance of groundwater becomes more apparent.

Figure 14.1 gives estimates of the distribution of freshwater in the hydrosphere. Clearly the largest volume occurs as glacial ice. Groundwater is ranked second, with slightly more than 30 percent of the total. However, when ice is excluded and just liquid water is considered, 96 percent of all freshwater is groundwater. Without question, *groundwater represents the largest reservoir of freshwater that is readily available to humans.* Its value in terms of economics and human well-being is incalculable.

Geologic Importance of Groundwater

Geologically, groundwater is important as an erosional agent. The dissolving action of groundwater slowly removes soluble rock such as limestone, allowing surface depressions known as *sinkholes* to form as well as creating subterranean caverns such as those shown in the chapter-opening photo. The final section of this chapter describes the landforms associated with subsurface water. Groundwater is also an equalizer of streamflow. Much of the water that flows in rivers is not direct runoff from rain and snowmelt. Rather, a large percentage of precipitation soaks into the ground and then moves slowly under the surface to stream channels. Groundwater is thus a form of storage that sustains streams during periods when rain does not fall. Therefore, when we see water flowing in a river during

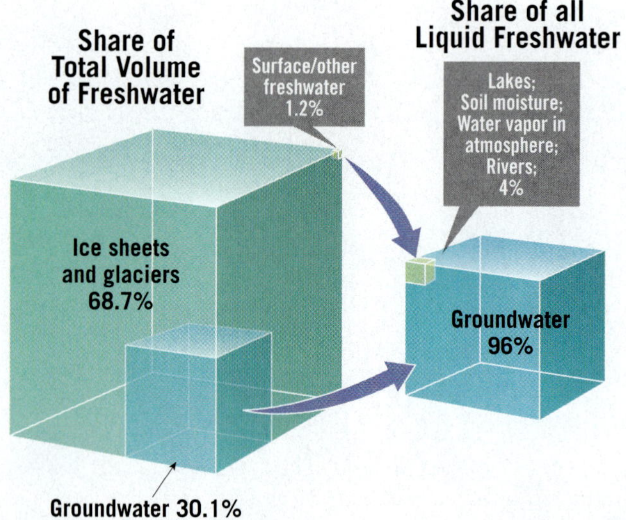

▲ **Figure 14.1 Earth's freshwater** Groundwater is the major reservoir of liquid freshwater.

a dry period, it is rain that fell at some earlier time and was stored underground.

Groundwater: A Basic Resource

Water is basic to life. It has been called the "bloodstream" of both the biosphere and society. Each day in the United States we use about 306 billion gallons of freshwater. According to the U.S. Geological Survey, about 75 percent of our freshwater comes from surface sources. Groundwater provides the remaining 25 percent (**Figure 14.2**). One of the advantages of groundwater is that it exists almost everywhere across the country and, thus, is often available in places that lack reliable surface sources such as lakes and rivers. Water in a groundwater system is stored in subsurface pore spaces and fractures. As water is withdrawn from a well, the connected pore spaces and fractures act as a "pipeline" that allows water to gradually move from one part of the hydrologic system to where it is being withdrawn.

What are the primary ways that we use groundwater? The U.S. Geological Survey identifies several categories that are shown in Figure 14.2. More groundwater is used for irrigation than for all other uses combined. There are nearly 60 million acres (nearly 243,000 square kilometers [about 93,700 square miles]) of irrigated land in the United States. That is an area nearly the size of the state of Wyoming. The vast majority (75 percent) of the irrigated land is in the 17 conterminous western states, where annual precipitation is typically less than 20 inches (50 centimeters). About 42 percent of the water used for irrigation is groundwater.

Public and domestic uses include water for indoor and outdoor household purposes as well as water used for commercial purposes. Common indoor uses include drinking, cooking, bathing, washing clothes and dishes, and flushing toilets. If you are curious about how much water an average American uses each day for indoor domestic purposes, look at **Figure 14.3**. Major outdoor uses are watering lawns and gardens. Water for domestic use may come from a public supply or may be self-supplied.* Practically all (98 percent) those whose water is self-supplied rely on groundwater.

Another category, aquaculture, involves water used for fish hatcheries, fish farms, and shellfish farms. Many mining operations require significant quantities of water, as do industrial processes such as petroleum refining and the manufacture of chemicals, plastics, paper, steel, and concrete.

Did You Know?
It takes about 6 gal of water to grow a head of lettuce and nearly 49 gal of water to produce just one 8-oz glass of milk. Producing a single serving of steak requires more than 2600 gal of water.

CONCEPT CHECKS 14.1

1. What percentage of Earth's *total freshwater supply* is groundwater?
2. What share of Earth's *liquid freshwater* is groundwater?
3. List two geologic roles that groundwater plays.
4. What share of U.S. freshwater is provided by groundwater? What is most groundwater used for?

*According to the U.S. Geological Survey, *public supply* refers to water withdrawn by suppliers that furnish water to at least 25 people or have at least 15 connections.

▲ **Figure 14.2 Sources and uses of freshwater** Each day in the United States, we use 306 billion gallons of freshwater. Groundwater is the source of nearly one-quarter of the total. More groundwater is used for irrigation than for all other uses combined. (Data from U.S. Geological Survey)

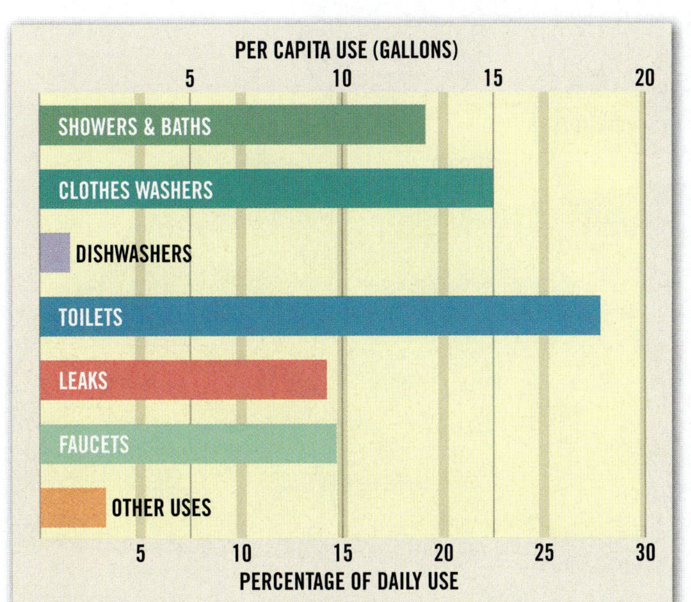

◄ **Figure 14.3 How we use water** According to the American Water Works Association, daily indoor per capita use in an average single-family home is 69.3 gallons. By installing more efficient water fixtures and regularly checking for leaks, households could reduce this figure by about one-third.

Did You Know?
Groundwater pumping to support rapid population growth in south-central Arizona (including the Tucson and Phoenix areas) has resulted in water level declines (drawdown) of between 300 and 500 ft.

14.2 Groundwater & the Water Table

Prepare a sketch with labels that summarizes the distribution of water beneath Earth's surface. Discuss the factors that cause variations in the water table and describe the interactions between groundwater and streams.

When rain falls on Earth's land surface, some of the water runs off, some returns to the atmosphere by evaporation and transpiration, and the remainder soaks into the ground. This last path is the primary source of practically all groundwater. The amount of water that takes each of these paths, however, varies greatly both in time and space. Influential factors include steepness of slope, nature of the surface material, intensity of rainfall, and type and amount of vegetation. For example, heavy rains falling on steep slopes underlain by impervious materials will obviously result in a high percentage of the water running off. Conversely, if rain falls steadily and gently on more gradual slopes composed of materials that are easily penetrated by the water, a much larger percentage of water soaks into the ground.

▼ **SmartFigure 14.4 Water beneath Earth's surface** The shape of the water table is usually a subdued replica of the surface topography. During periods of drought, the water table falls, reducing streamflow and drying up some wells.

https://goo.gl/llzbEM

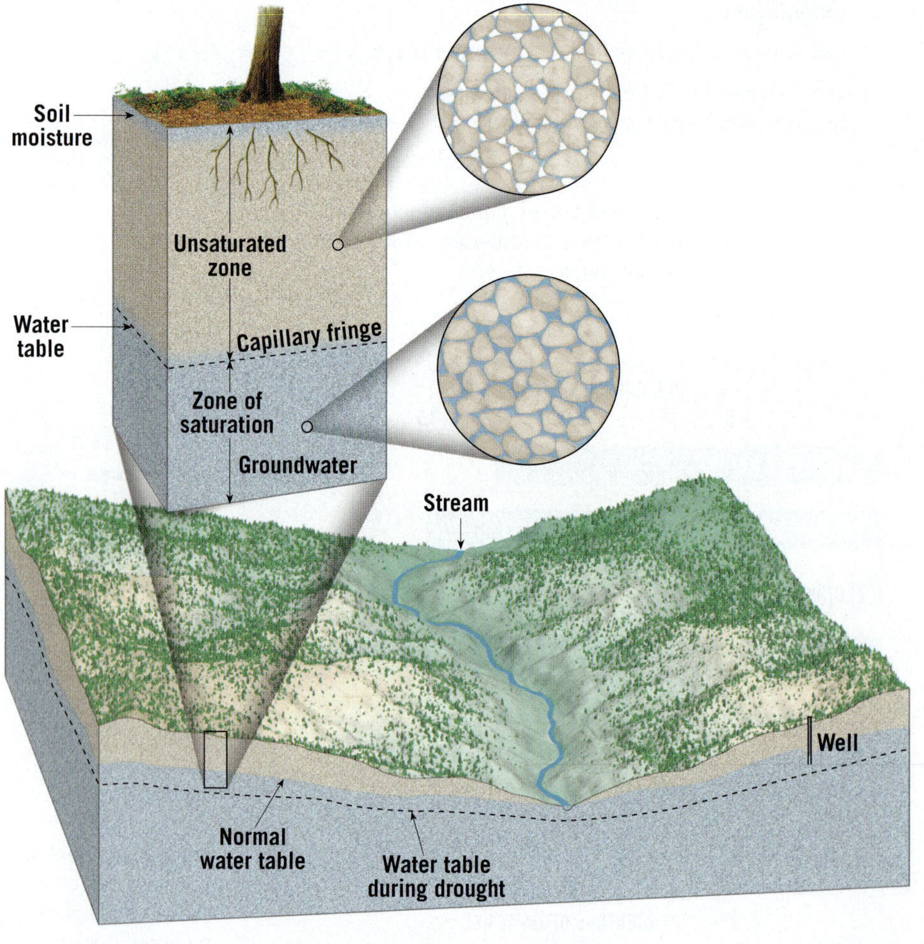

Distribution of Groundwater

Some of the water that soaks into the ground does not travel far because it is held by molecular attraction as a surface film on soil particles. This near-surface zone is called the **zone of soil moisture**. It is crisscrossed by roots, voids left by decayed roots, and animal and worm burrows that enhance the infiltration of rainwater into the soil. Soil water is used by plants in life functions and transpiration. Some soil water also evaporates directly back into the atmosphere.

Water that is not held as soil moisture percolates downward until it reaches a zone where all the open spaces in sediment and rock are completely filled with water (Figure 14.4). This is the **zone of saturation** (also called the **phreatic zone**). Water in the zone of saturation is called *groundwater*. The upper limit of this zone is known as the **water table**. Extending upward from the water table is the **capillary fringe** (*capillus* = hair). Here groundwater is held by surface tension in tiny passages between grains of soil or sediment. The area above the water table that includes the capillary fringe and the zone of soil moisture is called the **unsaturated zone** (also known as the **vadose zone**). The pore spaces in this zone contain both air and water. Although a considerable amount of water can be present in the unsaturated zone, this water cannot be pumped by wells because it clings too tightly to rock and soil particles. By contrast, below the water table, the water pressure is great enough to allow water to enter wells, thus permitting groundwater to be withdrawn for use. We will examine wells more closely later in the chapter.

Variations in the Water Table

The water table, the upper limit of the zone of saturation, is a very significant feature of the groundwater system. Knowing the water table level is important in predicting the productivity of wells, explaining the changes in the flow of springs and streams, and accounting for fluctuations in the levels of lakes.

The water table level is highly variable and can range from zero, when the zone of saturation is at the surface, to hundreds of meters below the surface in some places. An important characteristic of the water table is that its configuration varies seasonally and from year to year because the addition of water to the groundwater system is closely related to the quantity, distribution, and timing of precipitation. Except where the water table is at the surface, we cannot observe it directly.

A groundwater network is an array of wells where water levels are routinely measured. The U.S. Geological Survey in conjunction with state agencies maintains an extensive network of about 20,000 observation wells. This map shows the network of wells in Missouri.

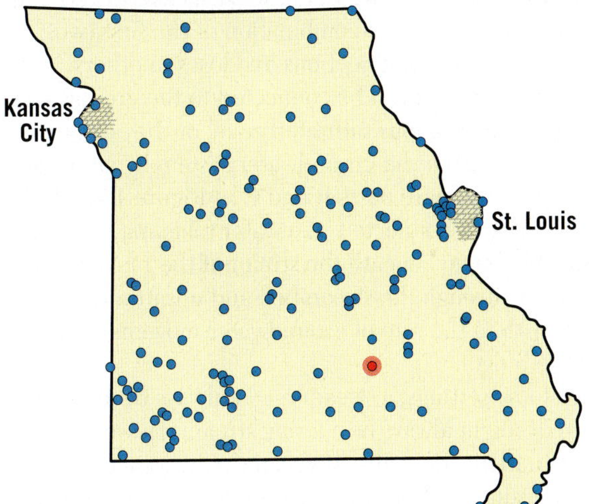

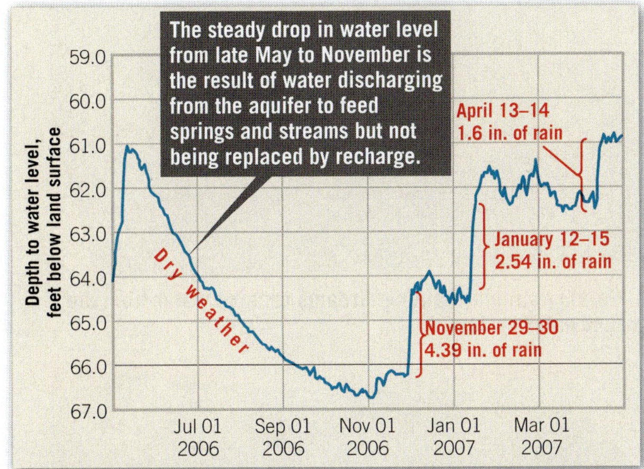

Groundwater recharge and discharge at the Akers observation well in Shannon County, Missouri. The height of the water table in this well and many others in this region shows a steady decline between spring and fall. This occurs because recharge is low during these months. However, water continues to move through the aquifer to supply springs and streams in the area, maintaining their flow even in dry years.

Water level data can be accessed remotely.

▲ **Figure 14.5 Monitoring the water table** Water level measurements from observation wells are a basic and important source of data. The well location highlighted in red on the map is the well whose data are shown above. (Photo by Missouri Department of Natural Resources)

However, the elevation of the water table is mapped and studied in detail by examining water levels in wells (**Figure 14.5**). The U.S. Geological Survey and state agencies maintain and monitor an extensive network of observation wells to provide statistics about groundwater levels. Such data are the basis for maps that reveal that the water table is rarely level, as we might expect a table to be (**Figure 14.6**). Instead, its shape is usually a subdued replica of the surface topography, reaching its highest elevations beneath hills and descending toward valleys (see Figure 14.4). In a wetland (swamp), the water table is right at the surface. Lakes and streams generally occupy areas low enough that the water table is above the land surface.

Several factors contribute to the irregular surface of the water table. One important influence is the fact that groundwater moves very slowly and at varying rates under different conditions. Because of this, water tends to "pile up" beneath high areas between stream valleys. If rainfall were to cease completely, these water table "hills" would slowly subside and gradually approach the level of the valleys. However, new supplies of rainwater are usually added frequently enough to prevent this. Nevertheless, in times of extended drought, the water table may drop enough to dry up shallow wells (see Figure 14.4). Other causes for the uneven water table are variations in precipitation and differences in the permeability of surface and subsurface materials from place to place.

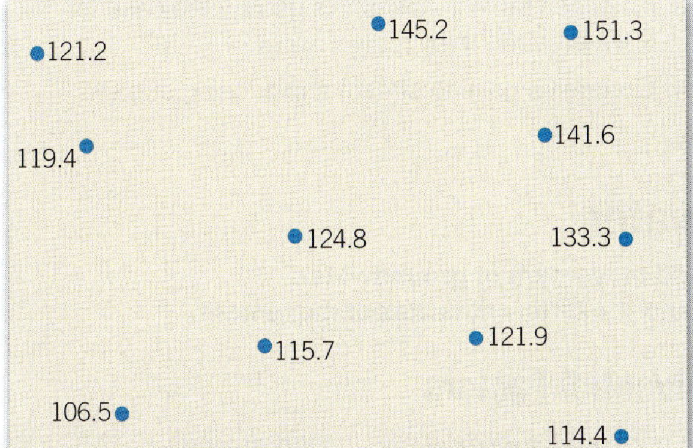

Step 1 The locations of observation wells and the elevations of the water table above sea level are plotted on the map.

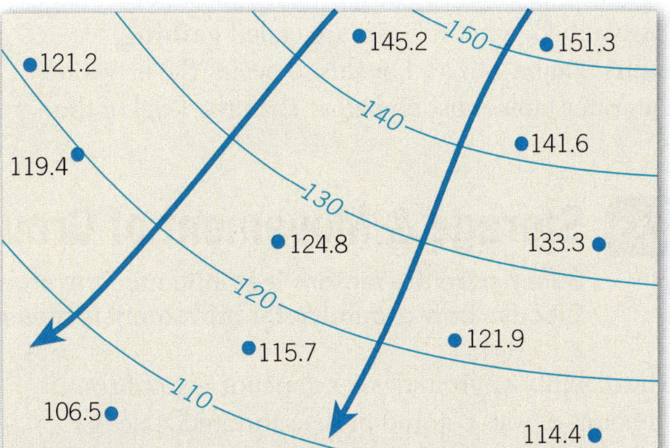

Step 2 Data points are used to guide the drawing of water-table contour lines. Groundwater flow lines can be added to show water movement in the upper portion of the zone of saturation. Groundwater moves perpendicular to the contours and down the slope of the water table.

◀ **Figure 14.6 Mapping the water table** The water table coincides with the water level in observation wells. (Based on U.S Geological Survey)

▶ **Figure 14.7**
Interactions between the groundwater system and streams (Based on U.S Geological Survey)

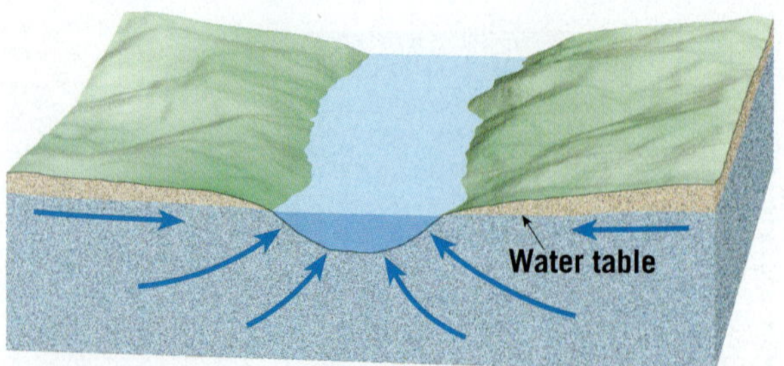

A. Gaining stream **Gaining streams receive water from the groundwater system.**

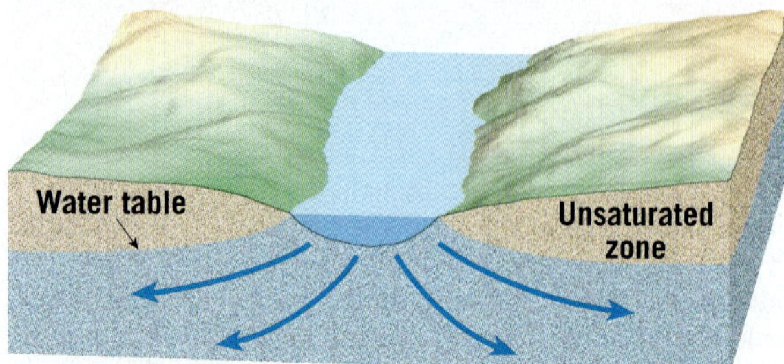

B. Losing stream (connected) **Losing streams provide water to the groundwater system.**

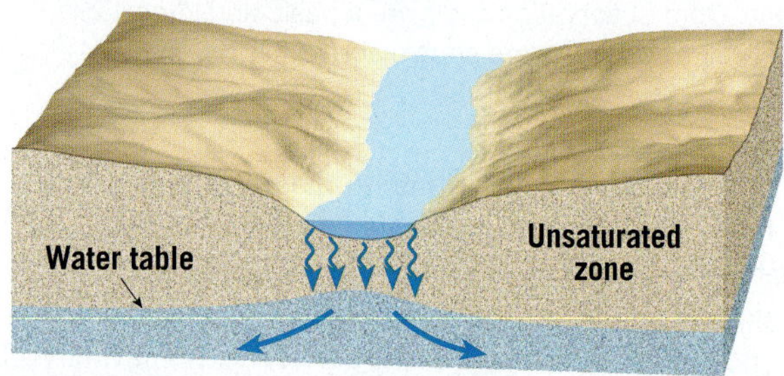

C. Losing stream (disconnected) **When losing streams are separated from the groundwater system by the unsaturated zone, a bulge may form in the water table.**

Interactions Between Groundwater & Streams

The interaction between the groundwater system and streams is a basic link in the hydrologic cycle. This interaction can take place in one of three ways. Streams may gain water from the inflow of groundwater through the streambed. Such streams are called **gaining streams** (Figure 14.7A). For this to occur, the elevation of the water table must be higher than the level of the surface of the stream. Streams may lose water to the groundwater system by outflow through the streambed. The term **losing stream** is applied to this situation (Figure 14.7B,C). When this happens, the elevation of the water table must be lower than the surface of the stream. The third possibility is a combination of the first two: A stream gains in some sections and loses in others.

Losing streams can be connected to the groundwater system by a continuous saturated zone, or they can be disconnected from the groundwater system by an unsaturated zone. Compare parts B and C in Figure 14.7. When the stream is disconnected, the water table may have a discernible bulge beneath the stream if the rate of water movement through the streambed and unsaturated zone is greater than the rate of groundwater movement away from the bulge.

In some settings, a stream might always be a gaining stream or might always be a losing stream. However, in many situations, flow direction can vary a great deal along a stream; some sections receive groundwater, and other sections lose water to the groundwater system. Moreover, the direction of flow can change over a short time span due to storms adding water near the stream bank or when temporary flood peaks move down the channel.

Groundwater contributes to streams in most geologic and climatic settings. Even where streams are primarily losing water to the groundwater system, certain sections may receive groundwater inflow during some seasons. One study of 54 streams in all parts of the United States found that 52 percent of the streamflow was contributed by groundwater. The groundwater contribution ranged from a low of 14 percent to a maximum of 90 percent. Groundwater is also a major source of water for lakes and wetlands.

CONCEPT CHECKS 14.2

1. When rain falls on land, what factors influence the amount of water that soaks in?
2. Define *groundwater* and relate it to the water table.
3. A kitchen table is flat. Is this usually the case for a water table? Why?
4. Contrast a gaining stream and a losing stream.

14.3 Storage & Movement of Groundwater

Summarize the factors that influence the storage and movement of groundwater.
Discuss how groundwater movement is measured and the different scales of movement.

The availability of groundwater depends not only on the amount of water stored in the saturated zone but also on the ability of groundwater to move through the subsurface environment. What factors influence the storage and movement of groundwater? What is the nature of groundwater movement? This section addresses the answers to these basic questions.

Influential Factors

The nature of subsurface materials strongly influences the rate of groundwater movement and the amount of groundwater that can be stored. Two factors are especially important: porosity and permeability.

Porosity Water soaks into the ground because bedrock, sediment, and soil contain countless voids, or openings. These openings are similar to those of a sponge and are often called *pore spaces*. The quantity of groundwater that can be stored depends on the **porosity** of the material, which is the percentage of the total volume of rock or sediment that consists of pore spaces (Figure 14.8). Voids most often are spaces between sedimentary particles, but also common are joints, faults, cavities formed by the dissolving of soluble rock such as limestone, and vesicles (voids left by gases escaping from lava).

Variations in porosity can be great. Sediment is commonly quite porous, and open spaces may occupy 10 to 50 percent of the sediment's total volume. Pore space depends on the size and shape of the grains, how they are packed together, the degree of sorting, and, in sedimentary rocks, the amount of cementing material. For example, clay may have a porosity as high as 50 percent, whereas some gravels may have only 20 percent voids.

Where sediments are poorly sorted, the porosity is reduced because the finer particles tend to fill the openings among the larger grains. Most igneous and metamorphic rocks, as well as some sedimentary rocks, are composed of tightly interlocking crystals such that the voids between the grains may be negligible. In these rocks, fractures must provide the porosity.

Permeability Porosity alone cannot measure a material's capacity to yield groundwater. Rock or sediment may be very porous yet still not allow water to move through it. The pores must be *connected* to allow water flow, and they must be *large enough* to allow flow. Thus, the **permeability** (*permeare* = to penetrate) of a material—its ability to *transmit* a fluid—is also very important.

Groundwater moves by twisting and turning through small interconnected openings. The smaller the pore spaces, the more slowly the water moves. For example, the ability of a clay deposit to store water may be great, due to high porosity, but its pore spaces are so small that water is unable to move through it. Thus, we say the clay is *impermeable*.

Aquitards & Aquifers Impermeable layers that hinder or prevent water movement are termed **aquitards** (*aqua* = water, *tard* = slow). Clay is a good example. On the other hand, larger particles, such as sand or gravel, have larger pore spaces. Therefore, the water moves with relative ease. Permeable rock strata or sediment that transmit groundwater freely are called **aquifers** (*aqua* = water, *fer* = carry). Sands and gravels are common examples.

In summary, porosity is not always a reliable guide to the amount of groundwater that can be produced, and permeability is significant in determining the rate of groundwater movement and the quantity of water that might be pumped from a well.

The beaker on the left is filled with 1000 ml of sediment. The beaker on the right is filled with 1000 ml of water.
1000 ml
500 ml
1000 ml
500 ml

The sediment-filled beaker now contains 500 ml of water. Pore spaces (porosity) must represent 50 percent of the volume of the sediment.
1000 ml
500 ml
1000 ml
500 ml

▲ **Figure 14.8 Porosity demonstration** Porosity is the percentage of the total volume of rock or sediment that consists of pore spaces.

How Groundwater Moves

The movement of water in the atmosphere and on the land surface is relatively easy to visualize, but the movement of groundwater is not. Near the beginning of the chapter, we mentioned the common misconception that groundwater occurs in underground rivers that resemble surface streams. Although subsurface streams do exist, they are *not* common. Rather, as you learned in the preceding sections, groundwater exists in the pore spaces and fractures in rock and sediment. Thus, contrary to any impressions of rapid flow that an underground river might evoke, the movement of most groundwater is exceedingly slow, from pore to pore.

A Simple Groundwater Flow System Figure 14.9 depicts a simple example of a *groundwater flow system*—a three-dimensional body of Earth material saturated with moving groundwater. It shows groundwater moving along flow paths from **recharge areas**, where groundwater is being replenished, to **discharge areas** along streams, where groundwater is flowing back to the surface. Discharge also occurs at springs, lakes, or wetlands, and in coastal areas, as groundwater seeps into bays or the ocean. Transpiration by plants whose roots extend to near the water table represents another form of groundwater discharge. Wells, where groundwater is being pumped to the surface, are artificial discharge areas.

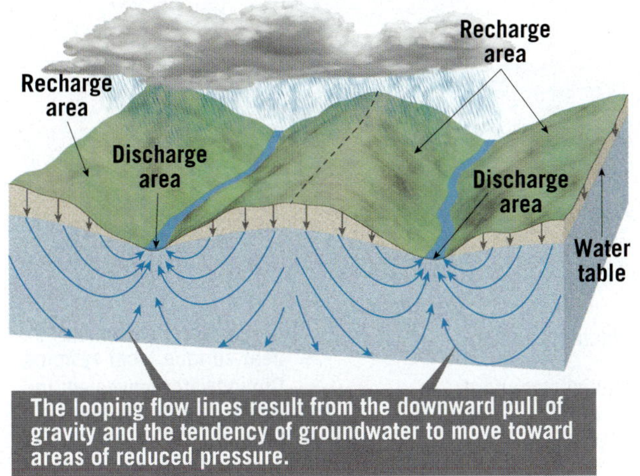
Recharge area
Recharge area
Discharge area
Discharge area
Water table
The looping flow lines result from the downward pull of gravity and the tendency of groundwater to move toward areas of reduced pressure.

◀ **Figure 14.9 Groundwater movement** Arrows show paths of groundwater movement through uniformly permeable material.

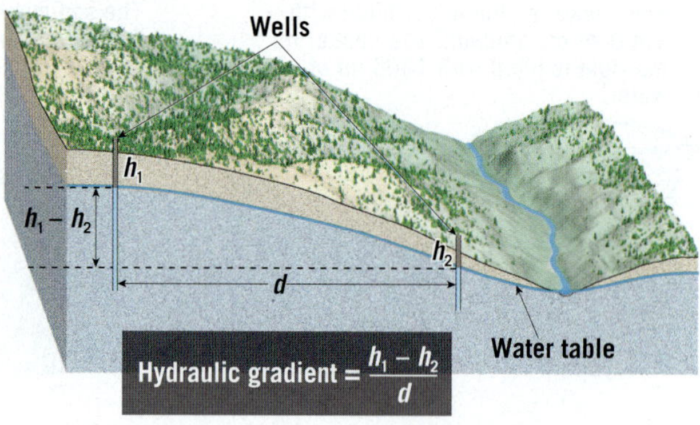

Figure 14.10 Hydraulic gradient The hydraulic gradient is determined by measuring the difference in elevation between two points on the water table ($h_1 - h_2$) divided by the distance between them, *d*. Wells are used to determine the height of the water table.

The energy that makes groundwater move is provided by the force of gravity. In response to gravity, water moves from areas where the water table is high to zones where the water table is lower. Although some water takes the most direct path down the slope of the water table, much of the water follows long, curving paths.

Figure 14.9 shows water percolating into a stream from all possible directions. Some paths clearly turn upward, apparently against the force of gravity, and enter through the bottom of the channel. This is easily explained: The deeper you go into the zone of saturation, the greater is the water pressure. Thus, the looping curves followed by water in the saturated zone may be thought of as a compromise between the downward pull of gravity and the tendency of water to move toward areas of reduced pressure. As a result, water at any given height is under greater pressure beneath a hill than beneath a stream channel, and the water tends to migrate toward points of lower pressure.

Measuring Groundwater Movement Our modern understanding of groundwater movement was founded on the mid-nineteenth century work of the French scientist-engineer Henri Darcy. Among the experiments Darcy carried out was one that showed that the velocity of groundwater flow is proportional to the slope of the water table: The steeper the slope, the faster the water moves (because the steeper the slope, the greater the pressure difference between two points). The water table slope is known as the **hydraulic gradient** and can be expressed as follows:

$$\text{Hydraulic gradient} = \frac{h_1 - h_2}{d}$$

where h_1 is the elevation of one point on the water table, h_2 is the elevation of a second point, and d is the horizontal distance between the two points (Figure 14.10).

Darcy also experimented with different materials, such as coarse sand and fine sand, by measuring the rate of flow through sediment-filled tubes that were tilted at varying angles. He found that flow velocity varied with the permeability of the sediment: Groundwater flows more rapidly through sediments having greater permeability than through materials having lower permeability. This factor is known as **hydraulic conductivity** and is a coefficient that takes into account the permeability of the aquifer and the viscosity of the fluid.

To determine discharge (Q)—that is, the actual volume of water that flows through an aquifer in a specified time—the following equation is used:

$$Q = \frac{KA(h_1 - h_2)}{d}$$

where $\frac{h_1 - h_2}{d}$ is the hydraulic gradient, K is the hydraulic conductivity, and A is the cross-sectional area of the aquifer. This expression has come to be called **Darcy's law**, in honor of the pioneering French scientist-engineer. Using this equation, if you know an aquifer's hydraulic gradient, conductivity, and cross-sectional area, you can calculate its discharge.

Different Scales of Movement The geographic extent of groundwater flow systems varies from a few square kilometers or less to tens of thousands of square kilometers. The length of flow paths ranges from a few meters to tens and sometimes hundreds of kilometers. Figure 14.11 is a cross section of a hypothetical region in which a deep groundwater flow system is overlain by and connected to several more shallow local flow systems. The subsurface geology exhibits a complicated arrangement of aquifer units with high hydraulic conductivity and aquitard units with low hydraulic conductivity.

Starting near the top of Figure 14.11, the blue arrows represent water movement in several local groundwater systems that occur in the upper water table aquifer. These groundwater systems are

SmartFigure 14.11 Hypothetical groundwater flow system The diagram includes subsystems at three different scales. Variations in surface topography and subsurface geology can produce a complex situation. The horizontal scale of the figure could range from tens to hundreds of kilometers.

TUTORIAL
https://goo.gl/oy6YtD

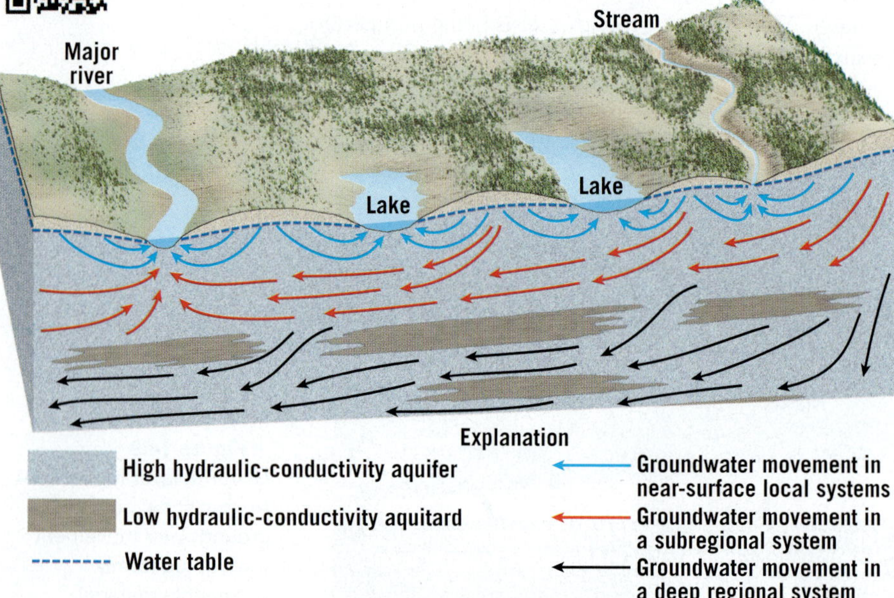

separated by groundwater divides at the center of the hills and discharge into the nearest surface-water body. Beneath these shallow systems, red arrows show water movement in a somewhat deeper system in which groundwater does not discharge into the nearest surface-water body but into a more distant one. Finally, the black arrows show groundwater movement in a deep regional system that lies beneath the more shallow ones and is connected to them. The horizontal scale of the figure could range from tens to hundreds of kilometers.

CONCEPT CHECKS 14.3

1. Distinguish between porosity and permeability.
2. What is the difference between an aquifer and an aquitard?
3. What factors cause water to follow the paths shown in Figure 14.9?
4. Relate groundwater movement to hydraulic gradient and hydraulic conductivity.
5. Contrast groundwater movement in a near-surface local system with that in a deep regional system.

Did You Know?
The rate of groundwater movement is highly variable. One way to measure it is to introduce dye into a well and observe long it takes for the dye to appear in another well at a known distance from the first. A typical rate for many aquifers is about 50 ft per year (about 1.7 in per day).

14.4 Wells & Artesian Systems

Discuss water wells and their relationship to the water table. Sketch and label a simple artesian system.

According to the National Groundwater Association, there are about 16 million water wells for all purposes in the United States. Private household wells constitute the largest share—more than 13 million. About 500,000 new residential wells are drilled each year.

Wells

The most common method for removing groundwater is to use a **well**, a hole bored into the zone of saturation. Wells serve as small reservoirs into which groundwater migrates and from which it can be pumped to the surface. The use of wells dates back many centuries and continues to be an important method of obtaining water today. Groundwater is the principal source of drinking water for about 50 percent of the U.S. population and provides about 96 percent of the water used for rural domestic supplies.

The water table level may fluctuate considerably during the course of a year, dropping during dry periods and rising following periods of rain. Therefore, to ensure a continuous supply of water, a well must penetrate below the water table. Whenever water is withdrawn from a well, the water table around the well is lowered. This effect, termed **drawdown**, decreases with increasing distance from the well. The result is a depression in the water table, roughly conical in shape, known as a **cone of depression** (Figure 14.12). Because the cone of depression increases the hydraulic gradient near the well, groundwater flows more rapidly toward the opening. For most smaller domestic wells, the cone of depression is negligible. However, when wells are heavily pumped for irrigation or for industrial purposes, drawdown can be great enough to create a very wide and steep cone of depression. This may substantially lower the water table in an area and cause nearby shallow wells to become dry. Figure 14.12 illustrates this situation.

Drilling a successful well is a familiar challenge in areas where groundwater is the primary source of supply. One well may be successful at a depth of 10 meters (33 feet), whereas a neighbor may have to go twice as deep to find an adequate supply. Still others may be forced to go deeper or try a different site altogether. When subsurface materials are heterogeneous, the amount of water that a well can provide may vary a great deal over short distances. For example, when two nearby wells are drilled to the same level and only one is successful, it may be because there is a perched water table beneath one of them.

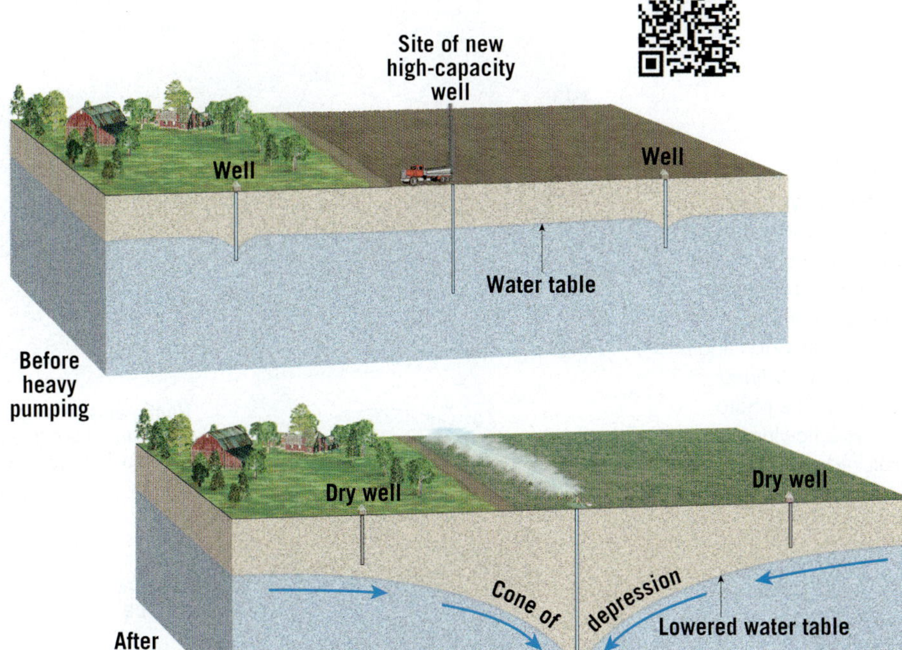

SmartFigure 14.12
Cone of depression For most small domestic wells, the cone of depression is negligible. When wells are heavily pumped, the cone of depression can be large and may lower the water table such that nearby shallower wells may be left dry.

ANIMATION
https://goo.gl/QidrnV

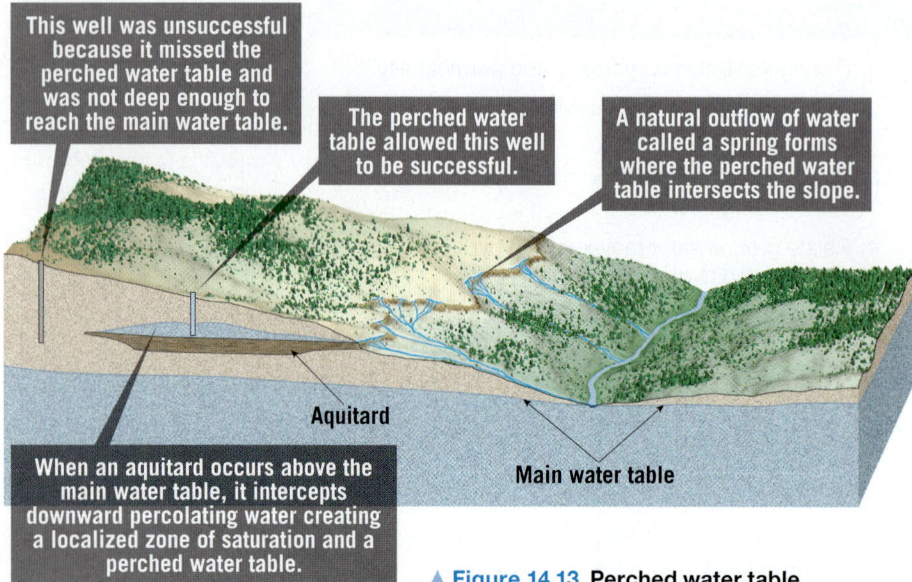

This well was unsuccessful because it missed the perched water table and was not deep enough to reach the main water table.
The perched water table allowed this well to be successful.
A natural outflow of water called a spring forms where the perched water table intersects the slope.
Aquitard
Main water table
When an aquitard occurs above the main water table, it intercepts downward percolating water creating a localized zone of saturation and a perched water table.

▲ **Figure 14.13 Perched water table**

As Figure 14.13 illustrates, a **perched water table** forms where an aquitard is situated above the main water table. Massive igneous and metamorphic rocks provide a second example. These crystalline rocks are usually not very permeable, except where they are cut by many intersecting joints and fractures. Therefore, when a well drilled into such rock does not intersect an adequate network of fractures, it is likely to be unproductive.

Artesian Systems

In most wells, water cannot rise on its own. If water is first encountered at a depth of 30 meters (100 feet), it remains at that level, fluctuating perhaps 1 or 2 meters (3 to 6 feet) with seasonal wet and dry periods. However, in some wells, water rises without being pumped, sometimes overflowing at the surface. Such wells are abundant in the Artois region of northern France, and so we call these self-rising wells *artesian*.

The term **artesian** is applied to *any* situation in which groundwater under pressure rises above the level of the aquifer. For an artesian system to exist, two conditions usually are met (Figure 14.14): (1) Water is confined to an aquifer that is inclined so that one end can receive water, and (2) aquitards, both above and below the aquifer, must be present to prevent the water from escaping. Such an aquifer is called a **confined aquifer**. When such a layer is tapped, the pressure created by the weight of the water above forces the water to rise. If there were no friction, the water in the well would rise to the level of the water at the top of the aquifer. However, friction reduces the height of the pressure surface. The greater the distance from the recharge area (where water enters the inclined aquifer), the greater the friction and the less the rise of water.

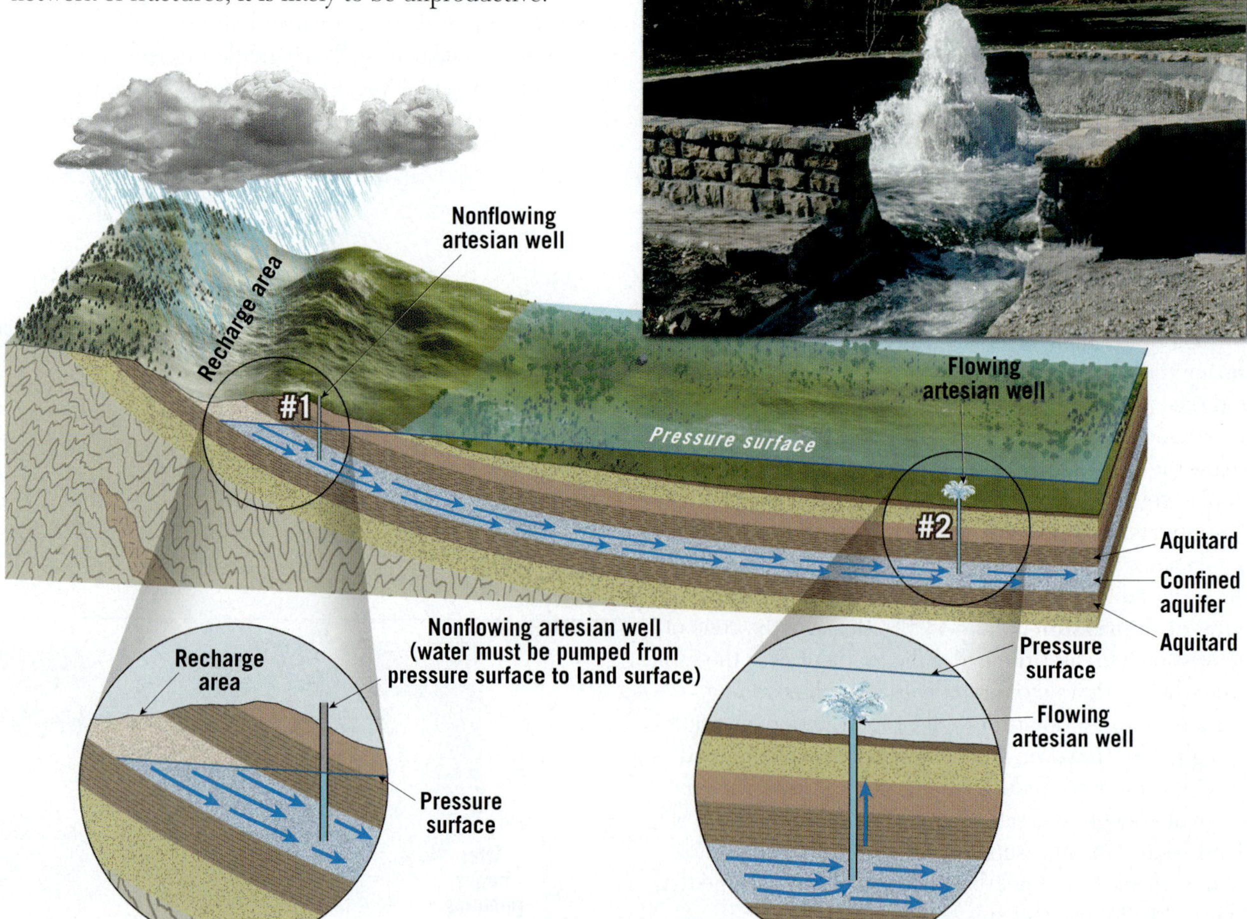

▶ **SmartFigure 14.14**
Artesian systems These groundwater systems occur where an inclined aquifer is surrounded by impermeable beds (aquitards). Such aquifers are called *confined aquifers*. The photo shows a flowing artesian well. (Photo from the James E. Patterson Collection, courtesy of F. K. Lutgens)

TUTORIAL
https://goo.gl/JkOALo

In Figure 14.14, well 1 is a **nonflowing artesian well** because at this location, the pressure surface is below ground level. When the pressure surface is above the ground and a well is drilled into the aquifer, a **flowing artesian well** is created (well 2 in Figure 14.14). Not all artesian systems are wells. *Artesian springs* exist where groundwater reaches the surface by rising along a natural fracture such as a fault rather than through an artificially produced hole. In deserts, artesian springs are sometimes responsible for creating oases.

Artesian systems act as conduits, often transmitting water great distances from remote areas of recharge to points of discharge. A well-known artesian system in South Dakota provides a good example (Figure 14.15). In the western part of the state, the edges of a series of sedimentary layers have been bent up to the surface along the flanks of the Black Hills. One of these beds, the permeable Dakota Sandstone, is sandwiched between impermeable strata and gradually dips into the ground toward the east. When the aquifer was first tapped, water poured from the ground surface, creating fountains many meters high. In some places the force of the water was sufficient to power waterwheels. Such scenes no longer occur because thousands of additional wells now tap the same aquifer. This has depleted the reservoir and lowered the water table in the recharge area. As a consequence, the pressure has dropped to the point where many wells have stopped flowing altogether and now have to be pumped.

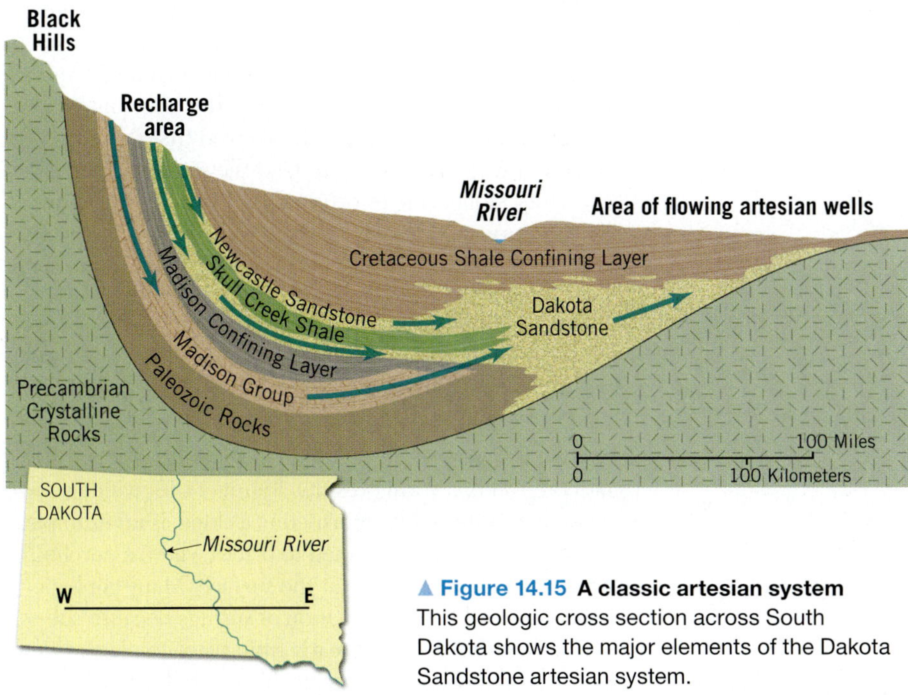

▲ **Figure 14.15 A classic artesian system** This geologic cross section across South Dakota shows the major elements of the Dakota Sandstone artesian system.

On a different scale, city water systems can be considered examples of artificial artesian systems (Figure 14.16). The water tower, into which water is pumped, represents the area of recharge, the pipes the confined aquifer, and the faucets in homes the flowing artesian wells.

CONCEPT CHECKS 14.4

1. Define *drawdown* and relate this to *cone of depression*.
2. In Figure 14.13, two wells were drilled to the same depth. Why was one successful and the other not?
3. Sketch a simple cross section of an artesian system with a flowing well. Label aquitards, aquifers, and the pressure surface.
4. Why do some artesian wells not flow at Earth's surface?

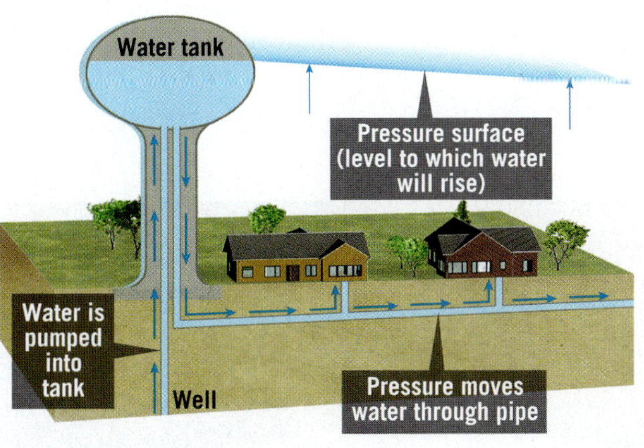

◀ **Figure 14.16 City water systems** City water systems can be considered to be artificial artesian systems.

14.5 Springs, Geysers, & Geothermal Energy

Distinguish among springs, hot springs, and geysers. List the geologic factors that favor geothermal energy development.

The phenomena described in this section often arouse people's curiosity and wonder. The fact that springs, hot springs, and geysers seem rather mysterious is not difficult to understand, for here is water (sometimes very hot water) flowing or spewing from the ground in all kinds of weather, in seemingly inexhaustible supply, but with no obvious source.

Springs

Not until the middle of the seventeenth century did the French physicist Pierre Perrault invalidate the age-old assumption that precipitation could not adequately account for the amount of water emanating from springs and flowing in rivers. Over several years, Perrault computed the quantity of water that fell on France's Seine River basin. He then calculated the mean annual runoff by measuring the river's discharge. After allowing for the loss of water by evaporation, he showed that there was sufficient water remaining to feed the springs. Thanks to Perrault's pioneering efforts and the measurements by many afterward, we now know that the source of springs is water from the zone of saturation and that the ultimate source of this water is precipitation.

Whenever the water table intersects Earth's surface, a natural outflow of groundwater results, and we call this a **spring** (**Figure 14.17**). Springs often form when an aquitard blocks the downward movement of groundwater and causes the water to move laterally. Where the permeable bed crops out, a spring results. Figure 14.13, which shows a perched water table intersecting a slope, is an example.

Springs are not confined to places where a perched water table creates a flow at the surface. Many geologic situations lead to the formation of springs because subsurface conditions vary greatly from place to place. Even in areas underlain by impermeable crystalline rocks, permeable zones may exist in the form of fractures or solution channels. If these openings fill with water and intersect the ground surface along a slope, a spring results.

▼ **Figure 14.17 Thunder Spring** A spring is a natural outflow of groundwater that occurs when the water table intersects the surface. Thunder Spring emerges from a deep joint and cave system in the Muav Limestone along the North Rim of the Grand Canyon. (Photo by Michael Collier)

Hot Springs

There is no universally accepted definition of *hot spring*. One frequently used definition is that the water in a **hot spring** is 6° to 9°C (10° to 15°F) warmer than the mean annual air temperature for the locality where it occurs. In the United States alone, there are more than 1000 such springs.

Temperatures in deep mines and oil wells usually rise with increasing depth, an average of about 25°C (45°F) per kilometer. You learned in Chapter 4 that this is called the *geothermal gradient*. Therefore, when groundwater circulates at great depths, it becomes heated. If the hot water rises rapidly to the surface, it may emerge as a hot spring. The water of some hot springs in the eastern United States is heated in this manner. The springs at Warm Springs, Georgia, the presidential retreat of Franklin Roosevelt, are one example. The temperature of these hot springs is always near 32°C (90°F). At Hot Springs National Park, Arkansas, water temperatures average about 60°C (140°F).

The great majority (more than 95 percent) of the hot springs (and geysers) in the United States are found in the West. This is because the sources of heat for most hot springs are magma bodies and hot igneous rocks, and igneous activity has occurred more recently in the West than elsewhere. The hot springs and geysers of the Yellowstone region are well-known examples.

Geysers

Geysers are intermittent hot springs or fountains in which columns of water are ejected with great force at various intervals, often rising 30 to 60 meters (100 to 200 feet) into the air. After the jet of water ceases, a column of steam rushes out, usually with a thunderous roar. Perhaps the most famous geyser in the world is Old Faithful in Yellowstone National Park (**Figure 14.18**). The great abundance, diversity, and spectacular nature of Yellowstone's geysers and other thermal features undoubtedly was the primary reason for its becoming the first national park in the United States. Geysers are also found in other parts of the world, notably New Zealand and Iceland. In fact, the Icelandic word *geysa*, meaning "to gush," gives us the name *geyser*.

▼ **Figure 14.18 Old Faithful** This geyser in Wyoming's Yellowstone National Park is one of the most famous in the world. (Photo by Jeff Vanuga/Corbis)

How Geysers Work Geysers occur where extensive underground chambers exist within hot igneous rocks. How they operate is shown in Figure 14.19. As relatively cool groundwater enters the chambers, it is heated by the surrounding rock. At the bottom of the chambers, the water is under great pressure because of the weight of the overlying water. This great pressure prevents the water from boiling at the normal surface boiling point of 100°C (212°F). For example, water at the bottom of a 300-meter (1000-foot) water-filled chamber must attain nearly 230°C (450°F) before it will boil. The heating causes the water to expand, with the result that some is forced out at the surface. This loss of water reduces the pressure on the remaining water in the chamber, which lowers the boiling point. A portion of the water deep within the chamber quickly turns to steam, and the geyser erupts. Following eruption, cool groundwater again seeps into the chamber, and the cycle begins anew.

Geyser Deposits When groundwater from hot springs and geysers flows out at the surface, material in solution is often precipitated, producing an accumulation of chemical sedimentary rock. The material deposited at any given place commonly reflects the chemical makeup of the rock through which the water circulated. When the water contains dissolved silica, a material called *siliceous sinter*, or *geyserite*, is deposited around the spring. When the water contains dissolved calcium carbonate, a form of limestone called *travertine*, or *calcareous tufa*, is deposited. The latter term is used if the material is spongy and porous.

The deposits at Mammoth Hot Springs in Yellowstone National Park are more spectacular than most others (Figure 14.20). As the hot water flows upward through a series of channels and then out at the surface, the reduced pressure allows carbon dioxide to separate and escape from the water. The loss of carbon dioxide causes the water to become supersaturated with calcium carbonate, which then precipitates. In addition to containing dissolved silica and calcium carbonate, some hot springs contain sulfur, which gives water a poor taste and unpleasant odor. This is undoubtedly the case at Rotten Egg Spring, Nevada.

Geothermal Energy

Geothermal energy is harnessed by tapping natural underground reservoirs of steam and hot water. These occur where subsurface temperatures are high, due to relatively recent volcanic activity. Geothermal energy is put to use in two ways: The steam and hot water are used for space heating and to generate electricity.

Iceland, a large volcanic island astride the Mid-Atlantic Ridge, provides a good example. In Iceland's capital, Reykjavik, steam and hot water are pumped into buildings throughout the city for space heating. They also warm greenhouses, where fruits and vegetables are grown throughout the year. In the United States,

A. Water near the bottom is heated to near its boiling point. The boiling point is higher at the bottom because pressure is high due to the weight of all the water above.

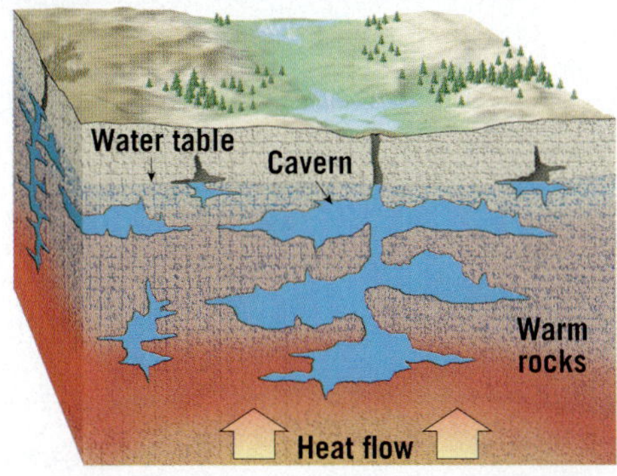

B. Higher up in the geyser the water is also heated, therefore it expands causing some to flow out at the top. This outflow reduces the pressure on the water at the bottom.

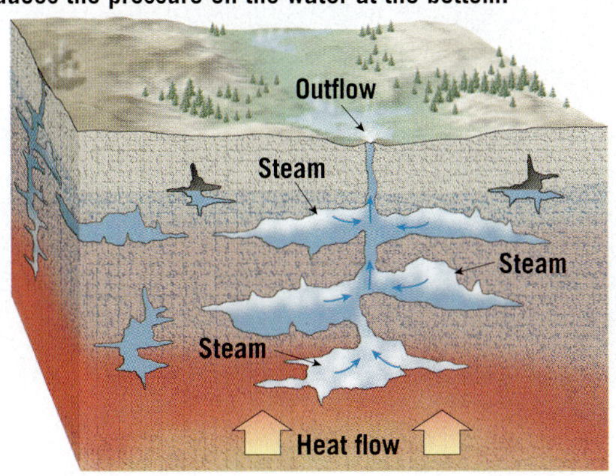

C. When pressure is reduced at the bottom, boiling occurs. Some of the bottom water flashes into steam. The expanding steam triggers an eruption. Then the water flows back in and the whole process begins anew.

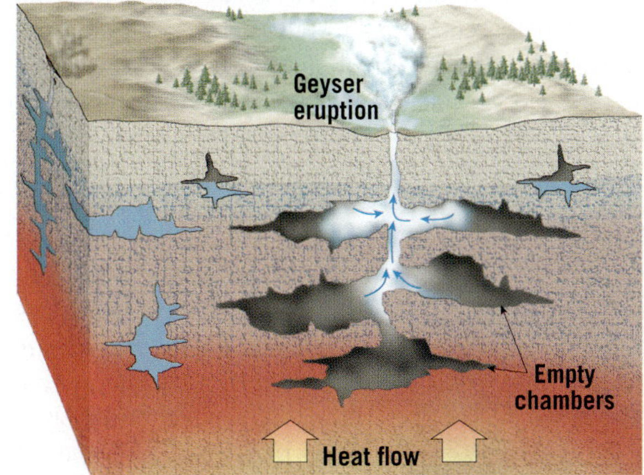

▲ **SmartFigure 14.19 How a geyser works** A geyser can form if the underground plumbing does not allow heat to be readily distributed by convection.

TUTORIAL
https://goo.gl/erKVt3

Did You Know?
Many people think that Old Faithful erupts so reliably—every hour on the hour—that you can set your watch by it. So goes the legend, but it's not true. Time spans between eruptions vary from about 65 minutes to more than 90 minutes and have generally increased over the years, thanks to changes in the geyser's plumbing.

▶ **Figure 14.20 Yellowstone's Mammoth Hot Springs** Although most of the deposits associated with geysers and hot springs in Yellowstone National Park are silica-rich geyserite, the deposits here consist of a form of limestone called travertine. (Photo by Jamie and Judy Wild/Danita Delimont/Alamy)

▼ **Figure 14.21 The Geysers** This facility, near the city of Santa Rosa in northern California, is the world's largest electricity-generating geothermal development. Most of the steam wells are about 3000 meters (10,000 feet) deep. California's 35 geothermal power plants were responsible for 80 percent of U.S. geothermal power production in 2014. (AP Photo/Calpine)

localities in several western states use hot water from geothermal sources for space heating.

The first commercial geothermal power plant in the United States was built in 1960 at The Geysers, north of San Francisco (**Figure 14.21**). These facilities remain the world's largest geothermal installation, generating nearly 1000 megawatts of electrical power. In addition to The Geysers, geothermal development is occurring elsewhere in the western United States, including Nevada, Utah, and the Imperial Valley in southern California. The U.S. geothermal generating capacity in 2014 of 3525 megawatts accounted for about 2 percent of U.S. renewable energy consumption, enough to supply more than 3 million homes.

The following geologic factors favor a geothermal reservoir of commercial value:

1. *A potent source of heat*, such as a large magma chamber deep enough to ensure adequate pressure and slow cooling but not so deep that the natural water circulation is inhibited. Such magma chambers are most likely in regions of recent volcanic activity.
2. *Large and porous reservoirs with channels connected to the heat source*, near which water can circulate and then be stored in the reservoir.
3. A *cap of low-permeability rocks* that inhibits the flow of water and heat to the surface. A deep, well-insulated reservoir contains much more stored energy than a similar but uninsulated reservoir.

Geothermal sources are not expected to provide a high percentage of the world's growing energy needs. Nevertheless, in regions where its potential can be developed, the use of geothermal energy will continue to grow.

CONCEPT CHECKS 14.5

1. Describe some circumstances that lead to the formation of a spring.
2. What warms the waters that flow at Hot Springs National Park, Arkansas, and at Warm Springs, Georgia?
3. What is the source of heat for most hot springs and geysers? How is this reflected in the distribution of these features?
4. Describe what occurs to cause a geyser to erupt.

14.6 Environmental Problems

List and discuss important environmental problems associated with groundwater.

Like many of our other valuable natural resources, groundwater is being exploited at an increasing rate. In some areas, overuse threatens the groundwater supply. In other places, groundwater withdrawal has caused the ground and everything resting on it to sink. Still other localities are concerned with possible contamination of the groundwater supply.

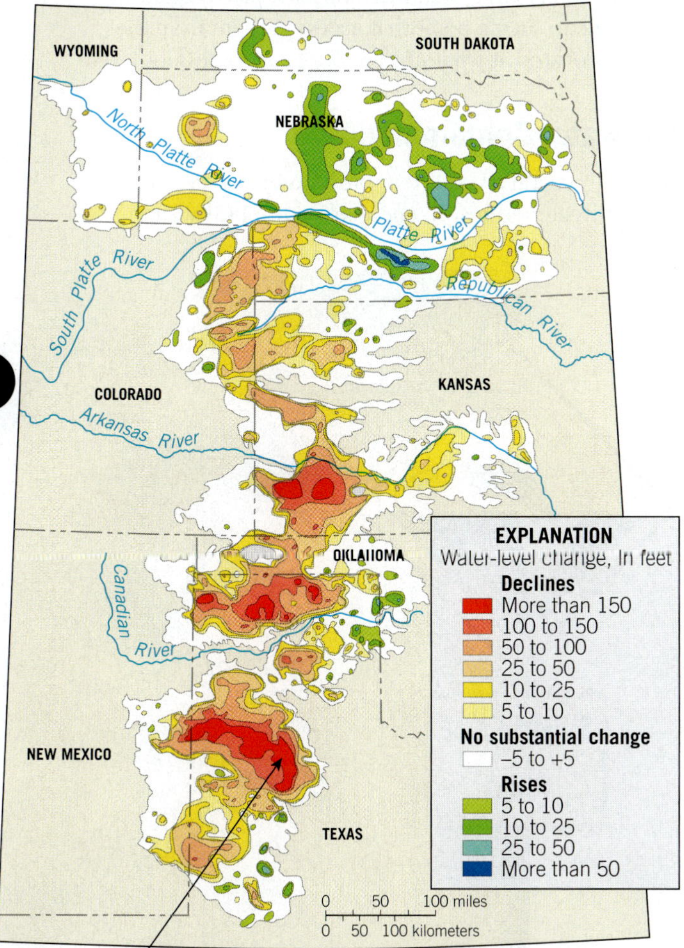

▲ **Figure 14.22 High Plains aquifer** The map shows changes in groundwater levels from predevelopment (about 1950) to 2013. Extensive pumping for irrigation has led to water level declines in excess of 30 meters (100 feet) in parts of four states. Water level rises have occurred where surface water is used for irrigation, such as along the Platte River in Nebraska. (Based on U.S. Geological Survey)

Treating Groundwater as a Nonrenewable Resource

Many think that groundwater is an endlessly renewable resource, for it is continually replenished by rainfall and melting snow. But in some regions, groundwater must be treated as a *nonrenewable* resource because the amount of water available to recharge the aquifer is significantly less than the amount being withdrawn. In such situations, groundwater is essentially being mined.

High Plains Aquifer The High Plains aquifer occupies a large, relatively dry region that extends from South Dakota to western Texas (Figure 14.22). The extensive agricultural economy in the region is heavily dependent on groundwater for irrigation. In fact, this aquifer accounts for about 30 percent of all groundwater withdrawn for irrigation in the country. Because evaporation rates are high and precipitation is modest, there is little rainwater to recharge the aquifer. Thus, in some parts of the region, where intense irrigation has been practiced for an extended period, depletion of groundwater has been severe. Figure 14.22 bears this out.

Did You Know?
Because of its high porosity, excellent permeability, and great size, the High Plains aquifer, the largest aquifer in the United States, accumulated huge amounts of groundwater—enough freshwater to fill Lake Huron.

Impact of Prolonged Drought The map in Figure 14.23 shows the extent and severity of drought conditions in California in late January 2016. Entering its fifth year of drought, nearly 64 percent of the state was experiencing extreme or exceptional drought. Because of the long span

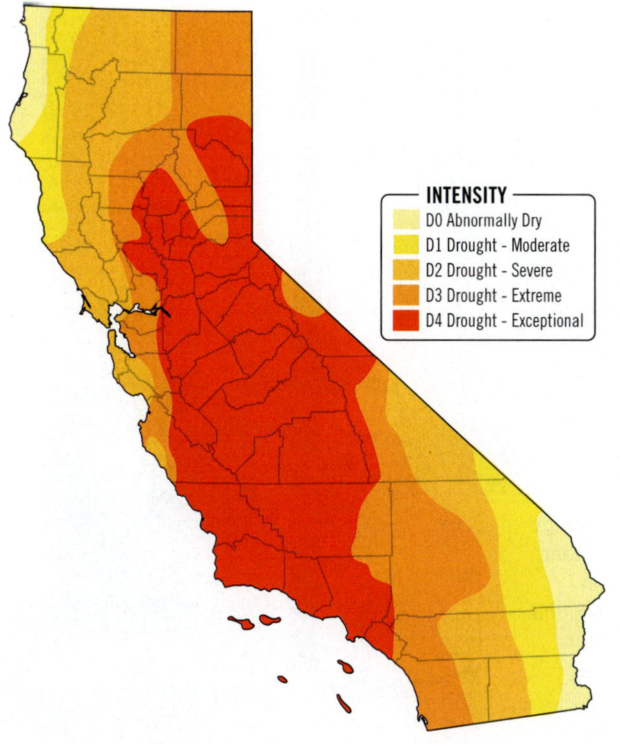

◀ **Figure 14.23 Drought status map for California on January 26, 2016** Much of the West was experiencing drought at this time. California was suffering most. More than 95 percent of the state had at least *severe drought* conditions. The prolonged dry spell led to accelerated groundwater pumping and a significant lowering of the water table. To examine current and archived drought maps, go to http://droughtmonitor.unl.edu. (Data from National Drought Mitigation Center)

Did You Know?
In the contiguous United States, the size of the area affected by land subsidence due to groundwater withdrawal amounts to an estimated 10,000 mi^2—an area about the size of Massachusetts!

of abnormally dry weather, the amount of water stored in lakes and reservoirs was severely depleted. To make up for the shortfall, groundwater use soared. Many areas experienced a dramatic increase in well drilling—not only more wells but deeper wells to reach the dropping water table. Nearly 60 percent of the state's water needs were being met by groundwater, up from 40 percent in years when rain and snow were normal.

Land Subsidence Caused by Groundwater Withdrawal

As you will see later in this chapter, surface subsidence can result from natural processes related to groundwater. However, the ground may also sink when water is pumped from wells faster than natural recharge processes can replace it. This effect is particularly pronounced in areas underlain by thick layers of unconsolidated sediments. As the water is withdrawn, the water pressure drops, and the weight of the overburden is transferred to the sediment. The greater pressure packs the sediment grains tightly together, and the ground subsides.

Many areas may be used to illustrate land subsidence caused by excessive pumping of groundwater from relatively loose sediment. A classic example in the United States occurred in the San Joaquin Valley of California, where subsidence reached 9 meters (30 feet) in some areas (**Figure 14.24**). Many other cases of land subsidence due to groundwater pumping exist in the United States, including portions of southern Arizona (**Figure 14.25**); Las Vegas, Nevada; New Orleans and Baton Rouge, Louisiana; and the Houston–Galveston area of Texas. In the low-lying coastal area between Houston and Galveston, land subsidence of 1.5 to 3 meters (5 to 9 feet) has occurred. The result is that about 78 square kilometers (30 square miles) are permanently flooded.

Outside the United States, one of the most spectacular examples of subsidence occurred in Mexico City, a portion of which is built on a former lake bed. In the first half of the twentieth century, thousands of wells were sunk into the water-saturated sediments beneath the city. As water was withdrawn, portions of the city subsided by as much as 6 to 7 meters (20 to 23 feet). In some places buildings have sunk to such a point that access to them from the street is located at what used to be the second-floor level!

Saltwater Contamination

In many coastal areas, the groundwater resource is being threatened by the encroachment of saltwater.

▶ **Figure 14.24 That sinking feeling!** The San Joaquin Valley, an important agricultural area, relies heavily on irrigation. Between 1925 and 1975, this part of the valley subsided 9 meters (30 feet) because of the withdrawal of groundwater and the resulting compaction of sediments. (Photo courtesy of U.S. Geological Survey)

▲ **Figure 14.25 Land subsidence in south-central Arizona** In southern Arizona, heavy pumping has led to water table declines of up to 180 meters (600 feet). This has triggered extensive and uneven permanent compaction of sediments and the formation of large fissures (cracks) in the ground around the margins of subsiding basins. Some rural roads have signs that warn of the potential hazard. (Photo by Todd Shipman/Arizona Geological Survey)

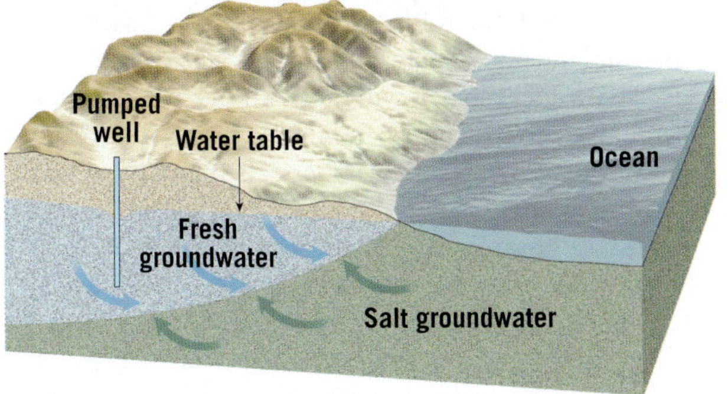
Because freshwater is less dense than saltwater, it floats on the saltwater and forms a lens-shaped body that may extend to considerable depths below sea level.
Pumped well
Water table
Ocean
Fresh groundwater
Salt groundwater

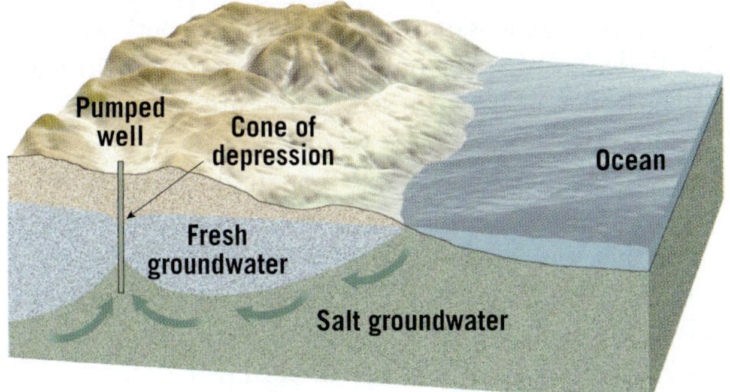
If excessive pumping lowers the water table, the base of the freshwater zone will rise 40 times that amount. The result may be saltwater contamination of wells.
Pumped well
Cone of depression
Ocean
Fresh groundwater
Salt groundwater

Figure 14.26
Saltwater contamination Heavy pumping in coastal areas can cause encroachment of saltwater and threaten the supply of fresh groundwater.

To understand this problem, let us examine the relationship between fresh groundwater and salty groundwater. **Figure 14.26** shows a cross section that illustrates this relationship in a coastal area underlain by permeable homogeneous materials. Freshwater is less dense than saltwater, so it floats on the saltwater and forms a large lens-shaped body that may extend to considerable depths below sea level. In such a situation, if the water table is 1 meter (3 feet) above sea level, the base of the freshwater body will extend to a depth of about 40 meters (130 feet) below sea level. Stated another way, the depth of the freshwater below sea level is about 40 times greater than the elevation of the water table above sea level. Thus, when excessive pumping lowers the water table by a certain amount, the bottom of the freshwater zone rises by 40 times that amount. Therefore, if groundwater withdrawal continues to exceed recharge, there will come a time when the elevation of the saltwater will be sufficiently high to be drawn into wells, thus contaminating the freshwater supply. This is called *saltwater intrusion*, or *saltwater contamination*. Deep wells and wells near the shore are usually the first to be affected.

In urbanized coastal areas, the problems created by excessive pumping are compounded by a decrease in the rate of natural recharge. As more and more of the surface is covered by streets, parking lots, and buildings, surface runoff increases and infiltration into the soil is diminished.

One way to correct the problem of saltwater intrusion of groundwater resources is a network of recharge wells. These wells allow wastewater to be pumped back into the groundwater system. A second method of correction is accomplished by building large recharge basins. These basins collect surface drainage and allow it to seep into the ground. On New York's Long Island, where the problem of saltwater intrusion was recognized more than 60 years ago, both of these methods have been employed with considerable success (**Figure 14.27**).

Groundwater Contamination

The pollution of groundwater is a serious matter, particularly in areas where aquifers provide a large part of the water supply. One common source of groundwater pollution is sewage. Its sources include an ever-increasing number of septic tanks, as well as inadequate or broken sewer systems and farm wastes.

If sewage water that is contaminated with bacteria enters the groundwater system, it may become purified through natural processes. The harmful bacteria may be mechanically filtered by the sediment through which the water percolates, destroyed by chemical oxidation, and/or assimilated by other organisms. For purification to occur, however, the aquifer must be of the correct composition. For example, extremely permeable aquifers (such as highly fractured crystalline rock, coarse gravel, or cavernous limestone) have such large openings that contaminated groundwater may travel long distances without being filtered and cleansed. In this case, the water flows too rapidly and is not in contact with the surrounding material long enough for

Figure 14.27
Recharge basins Much of Long Island, New York, is completely dependent on groundwater. To help maintain the water table and prevent saltwater intrusion, more than 2000 recharge basins have been constructed. Recharge basins are used in many places, not just coastal areas. (Photo by Alan Cressler)

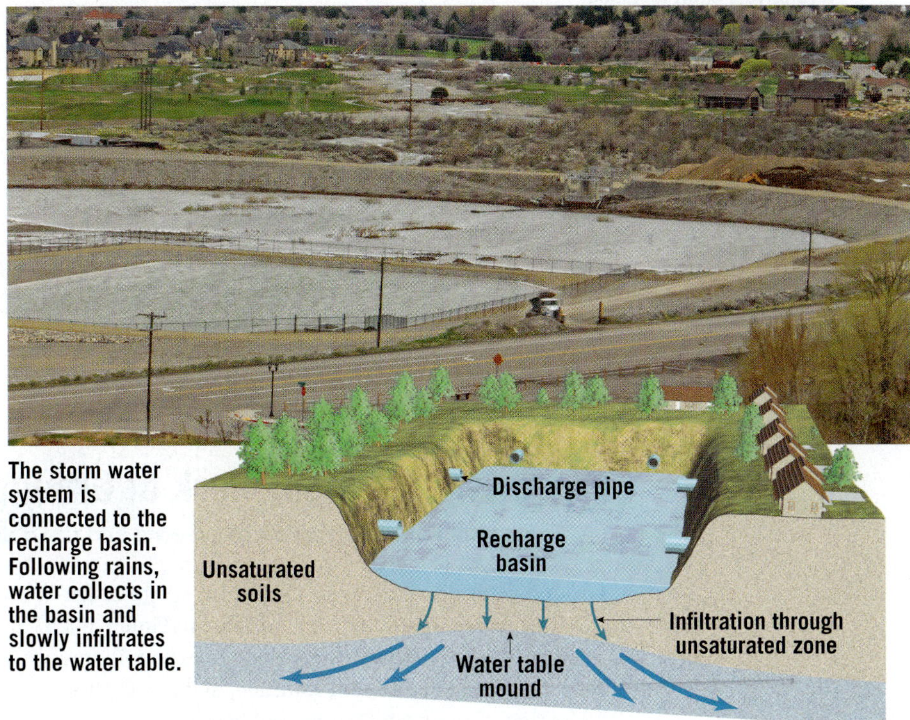
The storm water system is connected to the recharge basin. Following rains, water collects in the basin and slowly infiltrates to the water table.
Discharge pipe
Recharge basin
Unsaturated soils
Infiltration through unsaturated zone
Water table mound

purification to occur. This is the problem at well 1 in Figure 14.28A.

On the other hand, when the aquifer is composed of sand or permeable sandstone, water can sometimes be purified after traveling only a few dozen meters through it. The openings between sand grains are large enough to permit water movement, yet the movement of the water is slow enough to allow ample time for its purification (well 2, Figure 14.28B).

Other sources and types of contamination also threaten groundwater supplies. These include widely used substances such as highway salt, fertilizers that are spread across the land surface, and pesticides. In addition, a wide array of chemicals and industrial materials may leak from pipelines, storage tanks, landfills, and holding ponds. Some of these pollutants are classified as *hazardous*, meaning that they are either flammable, corrosive, explosive, or toxic. In landfills, potential contaminants are heaped onto mounds or spread directly over the ground. As rainwater oozes through the refuse, it may dissolve a variety of organic and inorganic materials. If the leached material reaches the water table, it will mix with the groundwater and contaminate the supply. Similar problems may result from leakage of shallow excavations called *holding ponds* into which a variety of liquid wastes are disposed.

Because groundwater movement is usually slow, polluted water can go undetected for a long time. In fact, contamination is sometimes discovered only after drinking water has been affected and people become ill. By this time, the volume of polluted water may be very large, and even if the source of contamination is removed immediately, the problem is not solved. Although the sources of groundwater contamination are numerous, there are relatively few solutions.

Once the source of the problem has been identified and eliminated, the most common practice is simply to abandon the water supply and allow the pollutants to be flushed away gradually. This is the least costly and easiest solution, but the aquifer must remain unused for many years. To accelerate this process, polluted water is sometimes pumped out and treated. Following removal of the tainted water, the aquifer is allowed to recharge naturally, or in some cases the treated water or other freshwater is pumped back in. This process is costly, time-consuming, and may be risky because there is no way to be certain that all of the contamination has been removed. Clearly, the most effective solution to groundwater contamination is prevention.

▼ **Figure 14.28**
Comparing two aquifers In this example, the limestone aquifer allowed the contamination to reach a well, but the sandstone aquifer did not.

Although the contaminated water has traveled more than 100 meters before reaching Well 1, the water moves too rapidly through the cavernous limestone to be purified.

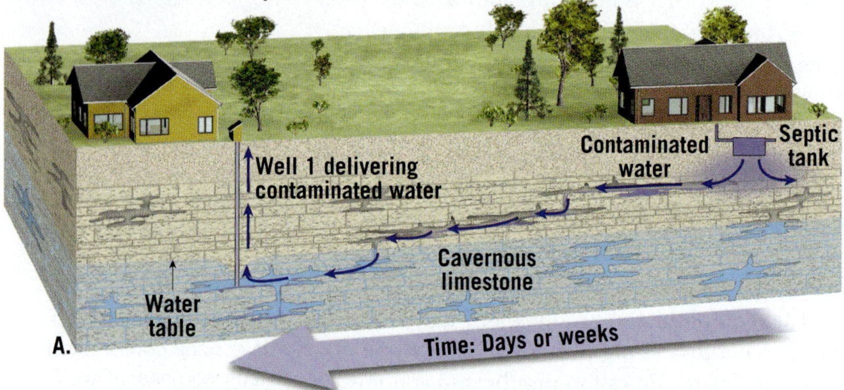

As the discharge from the septic tank percolates through the permeable sandstone, it moves more slowly and is purified in a relatively short distance.

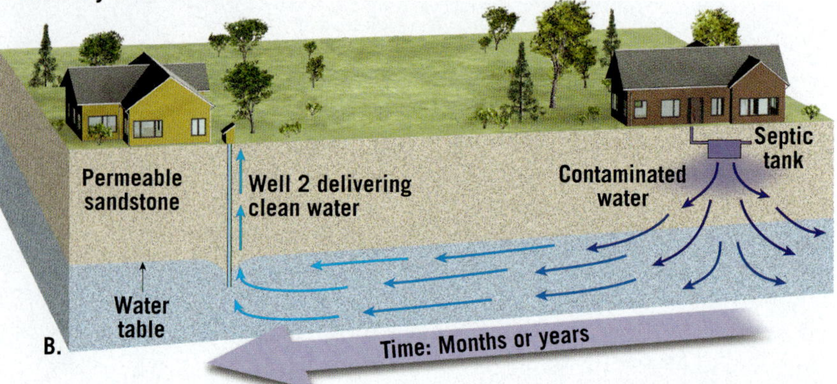

CONCEPT CHECKS 14.6

1. Describe the problem associated with pumping groundwater for irrigation in the southern High Plains and in areas experiencing prolonged drought, such as California in 2016.
2. Explain why ground may subside after groundwater is pumped to the surface.
3. Which type of aquifer material would be most effective in purifying polluted groundwater: coarse gravel, sand, or cavernous limestone?
4. Describe a significant problem that may arise when groundwater is heavily pumped at a coastal site.

14.7 The Geologic Work of Groundwater

Explain the formation of caverns and the development of karst topography.

Groundwater dissolves rock. This fact is key to understanding how caverns and sinkholes form (Figure 14.29). Soluble rocks, especially limestone, underlie millions of square kilometers of Earth's surface, and it is in these rocks that groundwater carries on its important role as an erosional agent. Limestone is nearly insoluble in pure water but is quite easily dissolved by water containing small quantities of carbonic acid, and most groundwater contains this acid. It forms because rainwater readily dissolves carbon dioxide from the air and from decaying

▲ SmartFigure 14.29 **Kentucky's Mammoth Cave area** Portions of Kentucky are underlain by limestone. Dissolution by groundwater has created a landscape characterized by caves and sinkholes. (Photo by Michael Collier)

MOBILE FIELD TRIP
https://goo.gl/jt13CE

plants. Therefore, when groundwater comes in contact with limestone, the carbonic acid reacts with the calcite (calcium carbonate) in the rocks to form calcium bicarbonate, a soluble material that is then carried away in solution.

Caverns

The most spectacular results of groundwater's erosional handiwork are limestone **caverns**. In the United States alone, about 17,000 caves have been discovered, and more are being found every year. Although most are relatively small, some have spectacular dimensions. Mammoth Cave in Kentucky and Carlsbad Caverns in southeastern New Mexico are famous examples. The Mammoth Cave system is the most extensive in the world, with more than 540 kilometers (335 miles) of interconnected passages. The dimensions at Carlsbad Caverns are impressive in a different way. Here we find the largest and perhaps most spectacular single chamber. The Big Room at Carlsbad Caverns has an area equivalent to 14 football fields and enough height to accommodate the U.S. Capitol building.

Cavern Development Most caverns are created at or just below the water table, in the zone of saturation. Here acidic groundwater follows lines of weakness in the rock, such as joints and bedding planes. As time passes, the dissolving process slowly creates cavities and gradually enlarges them into caverns. Material that is dissolved by the groundwater is eventually discharged into streams and carried to the ocean.

In many caves, development has occurred at several levels, with the current cavern-forming activity occurring at the lowest elevation. This situation reflects the close relationship between the formation of major subterranean passages and the river valleys into which they drain. As streams cut their valleys deeper, the water table drops as the elevation of the river drops. Consequently, during periods when surface streams are rapidly downcutting, surrounding groundwater levels drop rapidly, and cave passages are abandoned by the water while the passages are still relatively small in cross-sectional area. Conversely, when the entrenchment of streams is slow or negligible, there is time for large cave passages to form.

How Dripstone Forms Certainly the features that arouse the greatest curiosity for most cavern visitors are the stone formations that give some caverns a wonderland appearance. These are not erosional features, like the cavern itself, but depositional features created by the seemingly endless dripping of water over great spans of time. Recall from our discussion of hot springs that the calcium carbonate left behind produces the limestone we call travertine. These cave deposits, however, are also commonly called *dripstone*, an obvious reference to their mode of origin. Although the formation of caverns takes place in the zone of saturation, the deposition of dripstone is not possible until the caverns are above the water table in the unsaturated zone. As soon as the chamber is filled with air, the stage is set for the decoration phase of cavern building to begin.

Dripstone Features—Speleothems The various dripstone features found in caverns are collectively called **speleothems** (*spelaion* = cave, *thema* = put), no two of which are exactly alike. Perhaps the most familiar speleothems are **stalactites** (*stalaktos* = trickling). These icicle-like pendants hang from the ceiling of the cavern and form where water seeps through cracks above. When the water reaches air in the cave, some of the dissolved carbon dioxide escapes from the drop, and calcite precipitates. Deposition occurs as a ring around the edge of the water drop. As drop after drop follows, each leaves an infinitesimal trace of calcite behind, and a hollow limestone tube is created. Water then moves through the tube, remains suspended momentarily at the end, contributes a tiny ring of calcite, and falls to the cavern floor. The stalactite just described is appropriately called a *soda straw* (Figure 14.30A). Often the hollow tube of the soda straw becomes plugged or its supply of water increases. In either case, the water is forced to flow and hence deposit along the outside of the tube. As deposition continues, the stalactite takes on the more common conical shape.

Speleothems that form on the floor of a cavern and reach upward toward the ceiling are called **stalagmites** (*stalagmos* = dropping). The water supplying the calcite for stalagmite growth falls from the ceiling and splatters over the surface. As a result, stalagmites do not have a central tube and are usually more massive in appearance and rounded on their upper ends than stalactites. Given enough time, a downward-growing stalactite and an upward-growing stalagmite may join to form a *column* (Figure 14.30B).

Did You Know?
Although most caves and sinkholes are associated with regions underlain by limestone, these features can also form in gypsum and rock salt (halite) because these rocks are highly soluble and readily dissolved.

A.

B.

◀ Figure 14.30 **Cave decorations** Speleothems are of many types, including stalactites, stalagmites, and columns. **A.** Close-up of a delicate live soda-straw stalactite in Chinn Springs Cave, Independence County, Arkansas. (Photo by Dante Fenolio/Science Source) **B.** Stalagmites and stalactites in New Mexico's Carlsbad Caverns National Park. (Photo by Fritz Poelking/Glow Images)

Karst Topography

Many areas of the world have landscapes that, to a large extent, have been shaped by the dissolving power of groundwater. Such areas are said to exhibit **karst topography**, named for the Krs Plateau in Slovenia, located along the northeastern shore of the Adriatic Sea, where such topography is strikingly developed. In the United States, karst landscapes occur in many areas that are underlain by limestone, including portions of Kentucky, Tennessee, Alabama, southern Indiana, and central and northern Florida (Figure 14.31). Generally, arid and semiarid areas are too dry to develop karst topography. When solution features exist in such regions, they are likely to be remnants of a time when rainier conditions prevailed.

Sinkholes Karst areas typically have irregular terrain punctuated with many depressions, called **sinkholes**, or **sinks** (Figure 14.32). In the limestone areas of Florida, Kentucky, and southern Indiana, there are tens of thousands of these depressions, varying in depth from just 1 or 2 meters (3 or 6 feet) to more than 50 meters (165 feet).

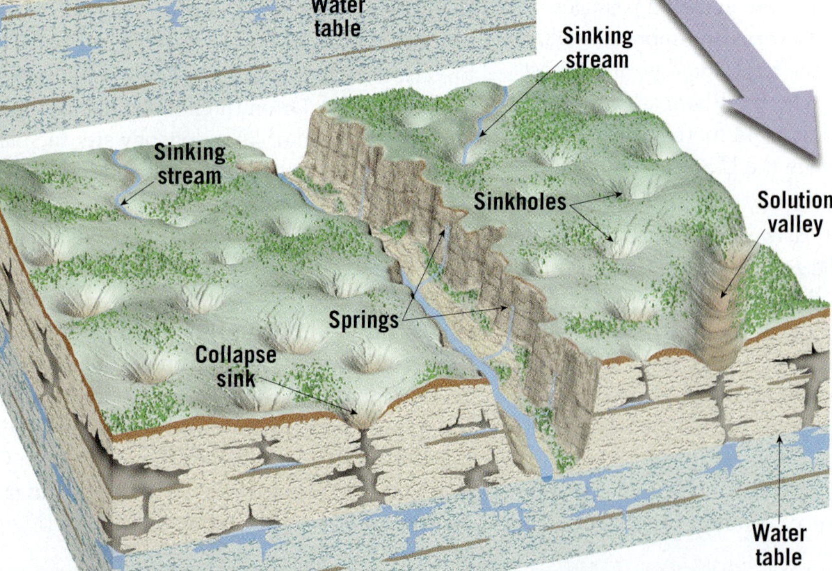
During early stages, groundwater percolates through limestone along joints and bedding planes. Solution activity creates and enlarges caverns at and below the water table.
Water table
Limestone
Sinkholes
TIME
Sinking stream
Sinkholes
With time, caverns grow larger and the number and size of sinkholes increase. Surface drainage is frequently funneled below ground.
Water table
Sinking stream
Sinking stream
Sinkholes
Solution valley
Collapse of caverns and coalescence of sinkholes form larger, flat-floored depressions. Eventually solution activity may remove most of the limestone from the area, leaving isolated remnants as in Figure 14.33.
Springs
Collapse sink
Water table

◀ Figure 14.31 **Development of a karst landscape**

Groundwater was responsible for creating these sinkholes west of Timaru on New Zealand's South Island.

This small sinkhole formed suddenly when the roof of a cavern collapsed, eliminating the backyard of this house in Lake City, Florida.

◀ **Figure 14.32 Sinkholes** Karst landscapes are typically punctuated with these depressions. The white spots in the New Zealand photo are grazing sheep. (Left photo by David Wall/Alamy Images; right photo by AP Photo/The Florida Times-Union, Jon M. Fletcher)

Did You Know?
America's largest bat colonies are found in caves. For example, Bracken Cave in central Texas is the summer home of 20 million Mexican free-tail bats. They spend their days in total darkness more than 2 mi. inside the cave. Each night when they leave the cave to feed, they consume more than 220 tons of insects.

Sinkholes commonly form in two ways. Some develop gradually over many years, without any physical disturbance to the rock. In these situations, the limestone immediately below the soil is dissolved by downward-sweeping rainwater that is freshly charged with carbon dioxide. With time, the bedrock surface is lowered, and the fractures into which the water seeps are enlarged. As the fractures grow in size, soil subsides into the widening voids, from which it is removed by groundwater flowing in the passages below. These depressions are usually shallow and have gentle slopes.

In contrast, sinkholes can also form abruptly and without warning when the roof of a cavern collapses under its own weight. Typically, the depressions created in this manner are steep-sided and deep. When they form in populous areas, they may represent a serious geologic hazard. Such a situation is shown in the lower photo of Figure 14.32.

In addition to a surface pockmarked by sinkholes, karst regions characteristically show a striking lack of surface drainage (streams). Following rainfall, the runoff is quickly funneled belowground through sinks. It then flows through caverns until it finally reaches the water table. Where streams do exist at the surface, their paths are usually short. The names of such streams often give a clue to their fate. The Mammoth Cave area of Kentucky, for example, is home to Sinking Creek, Little Sinking Creek, and Sinking Branch. Some sinkholes become plugged with clay and debris, creating small lakes or ponds.

Tower Karst Some regions of karst development exhibit landscapes that look very different from the sinkhole-studded terrain depicted in Figure 14.31. One striking example is an extensive region in southern China that is described as exhibiting **tower karst**. As Figure 14.33 shows, the term *tower* is appropriate because the landscape consists of a maze of isolated steep-sided hills that rise abruptly from the ground. Each is riddled with interconnected caves and passageways. This type of karst topography forms in wet tropical and subtropical regions having thick beds of highly jointed

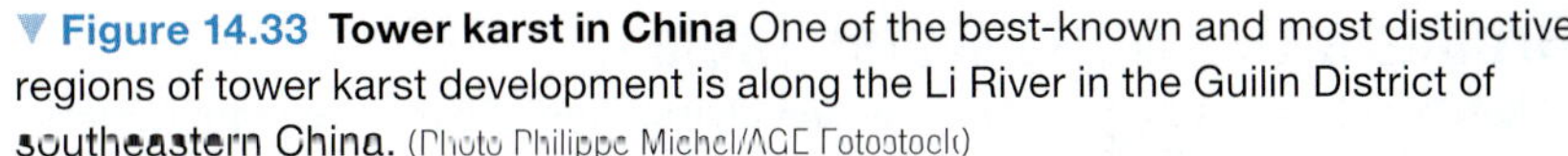
▼ **Figure 14.33 Tower karst in China** One of the best-known and most distinctive regions of tower karst development is along the Li River in the Guilin District of southeastern China. (Photo Philippe Michel/AGE Fotostock)

limestone. Here groundwater has dissolved large volumes of limestone, leaving only these residual towers. Karst development occurs more rapidly in tropical climates due to the abundant rainfall and the greater availability of carbon dioxide from the decay of lush tropical vegetation. The extra carbon dioxide in the soil means there is more carbonic acid for dissolving limestone. Other tropical areas of advanced karst development include portions of Puerto Rico, western Cuba, and northern Vietnam.

CONCEPT CHECKS 14.7

1. How does groundwater create caverns?
2. What causes cavern formation to stop at one level (depth) but continue or begin at a lower level?
3. How do stalactites and stalagmites form?
4. Describe two ways in which sinkholes form.

CONCEPTS IN REVIEW
Groundwater

14.1 The Importance of Groundwater

Describe the importance of groundwater as a source of freshwater and groundwater's roles as a geologic agent.

KEY TERM: groundwater

- Groundwater is water stored below Earth's surface, mainly in tiny pore spaces between rock or sediment grains. Groundwater represents the largest reservoir of freshwater that is readily available to humans and is a critical resource for human civilization.
- Groundwater is important geologically because it dissolves rock to make sinkholes and caverns and supplies surface streams with additional water.
- Each day in the United States we use about 306 billion gallons of freshwater. Groundwater provides about 76 billion gallons, or 25 percent of the total. More groundwater is used for irrigation than for all other uses combined.

? Examine Figure 14.1 and answer these questions: How much of Earth's freshwater is groundwater? How much of Earth's liquid freshwater is groundwater?

14.2 Groundwater & the Water Table

Prepare a sketch with labels that summarizes the distribution of water beneath Earth's surface. Discuss the factors that cause variations in the water table and describe the interactions between groundwater and streams.

KEY TERMS: zone of soil moisture, zone of saturation (phreatic zone), water table, capillary fringe, unsaturated zone (vadose zone), gaining stream, losing stream

- Some of the rain that falls on land soaks into the ground. Typically, a hole dug into the ground penetrates this zone of soil moisture and then crosses the unsaturated zone where pore spaces contain both water and air. The soil grows moist again just above the water table, in the capillary fringe. Crossing the water table, the boundary between the groundwater below and the unsaturated zone above, the hole begins to fill—to the height of the water table—with water that flows in from the zone of saturation.
- Streams and groundwater interact in one of three ways: Streams gain water from the inflow of groundwater (gaining stream); they lose water through the streambed to the groundwater system (losing stream); or they do both, gaining in some sections and losing in others.

? Examine this cross section, which shows the distribution of water in loose sediments. Provide the correct term for each lettered feature.

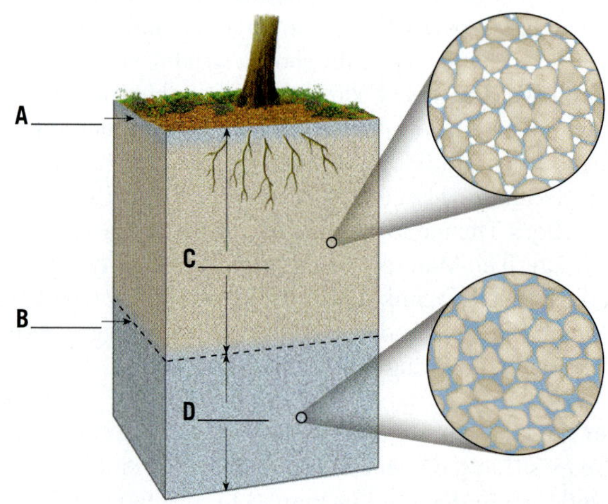

14.3 Storage & Movement of Groundwater

Summarize the factors that influence the storage and movement of groundwater. Discuss how groundwater movement is measured and the different scales of movement.

KEY TERMS: porosity, permeability, aquitard, aquifer, recharge area, discharge area, hydraulic gradient, hydraulic conductivity, Darcy's law

- The quantity of water that can be stored in a material depends on the material's porosity (the volume of open spaces). The permeability (the ability to transmit a fluid through interconnected pore spaces) of a material is a very important factor controlling the movement of groundwater.
- Materials with very small pore spaces (such as clay) hinder or prevent groundwater movement and are called aquitards. Aquifers consist of materials with larger pore spaces (such as sand) that are permeable and transmit groundwater freely.
- Groundwater flows slowly through underground pore spaces, moving on average only a few centimeters per day. Driven by gravity and pressure, it moves as a three-dimensional mass from areas of recharge (where water is added) to areas of discharge (where water leaves the groundwater system), such as springs, gaining streams, or wells drilled by people.
- French scientist-engineer Henri Darcy pioneered the quantification of groundwater flow by measuring the slope of the water table (hydraulic gradient) and the permeability of the sediment or rock (hydraulic conductivity). Darcy's law combines these in an equation to estimate an aquifer's discharge.
- Groundwater flows both short and long distances at both shallow and deep levels. Closer to the surface, the flow is more local in scale, while at greater depths, the flow occurs over regional scales.

14.4 Wells & Artesian Systems

Discuss water wells and their relationship to the water table. Sketch and label a simple artesian system.

KEY TERMS: well, drawdown, cone of depression, perched water table, artesian, confined aquifer, nonflowing artesian well, flowing artesian well

- For centuries, humans have been obtaining groundwater by drilling wells. As water is pumped out, the water table immediately adjacent to the well drops. This drawdown results in a "dimple" in the surface of the water table called the cone of depression. If there is sufficient drawdown, the cone of depression might encompass a large enough area that neighboring wells might go dry.
- A perched water table results from groundwater "piled up" atop an aquitard that is above the main body of groundwater. The shape of the water table is complex, which results in challenges for people trying to drill productive wells.
- Artesian wells tap into inclined aquifers bounded above and below by aquitards. For a system to qualify as artesian, the water in the well must be under sufficient pressure that it can rise above the top of the confined aquifer. Artesian wells may be flowing or nonflowing, depending on whether the pressure surface is above or below the ground surface.

? In 1900, when this well was drilled near Woonsocket in eastern South Dakota, a "gusher" of water nearly 30 meters (100 feet) high resulted. Describe or sketch the subsurface geologic situation that was responsible for this fountain of water. What term is applied to a well such as this?

N.H. Darton/USGS

14.5 Springs, Geysers, & Geothermal Energy

Distinguish among springs, hot springs, and geysers. List the geologic factors that favor geothermal energy development.

KEY TERMS: spring, hot spring, geyser, geothermal energy

- Springs are naturally occurring spots where groundwater leaves the ground and flows out onto the surface. They may be due to the intersection of a perched water table and Earth's surface.
- Hot springs are like regular springs but hot. They transfer heat from the deeper crust to the surface. Most often, this heat comes from relatively shallow bodies of magma.
- Geysers are intermittent hot springs that "erupt" hot water periodically. They are fed by underground chambers that fill with water that warms past the surface-level boiling point. Expansion of the heated water causes some of it to flow out of the surface vent, reducing the pressure below. As a result, water in the heated chambers flashes to vapor and expands rapidly, forcing its way to the surface in an eruption. Geysers can precipitate silica or calcium carbonate around the geyser vent, producing the rocks siliceous sinter (geyserite) or travertine (tufa).
- Geothermal energy uses heat in Earth's subsurface to produce hot water and steam for space heating and to produce electricity. A geothermal site requires (1) a potent source of heat, (2) large subsurface reservoirs, and (3) a cap of low-permeability rock.

? This photo from the 1930s shows Franklin Roosevelt enjoying the hot springs at the presidential retreat at Warm Springs, Georgia. The temperature of this water is always near 32°C (90°F). This area has no history of recent igneous activity. What is the likely reason these springs are so warm?

New York Daily News/Getty Images

14.6 Environmental Problems

List and discuss important environmental problems associated with groundwater.

- Groundwater can be "mined" by being extracted at a rate that is greater than the rate of replenishment. When groundwater is a nonrenewable resource, the water table drops. In parts of the High Plains aquifer, the water table has fallen by more than 45 meters (150 feet).
- The extraction of groundwater can cause pore spaces to decrease in volume and the grains of loose Earth materials to pack more closely together. This overall compaction of sediment volume results in subsidence of the land surface.
- Saltwater contamination is a common environmental problem in coastal areas. Fresh groundwater "floats" on salty groundwater due to its lower density. If sufficient freshwater is pumped out to lower the water table by some amount, the base of the freshwater lens will rise about 40 times that amount. Deep wells may begin to access the deeper, salty water instead.
- Contamination of groundwater with sewage, highway salt, fertilizer, and industrial chemicals is another issue of critical concern. Once groundwater is contaminated, the problem is very difficult to solve, requiring expensive remediation or abandonment of the aquifer.

14.7 The Geologic Work of Groundwater

Explain the formation of caverns and the development of karst topography.

KEY TERMS: cavern, speleothem, stalactite, stalagmite, karst topography, sinkhole (sink), tower karst

- Groundwater dissolves rock, in particular limestone, leaving behind void spaces in the rock. Caverns form at the zone of saturation, but later dropping of the water table may leave them open and dry—and available for people to explore.
- Dripstone is rock deposited by dripping of water that contains dissolved calcium carbonate inside caverns. Speleothems are features made of dripstone and include stalactites, stalagmites, and columns.
- Karst topography is a distinctive type of landscape dominated by the dissolving of limestone near Earth's surface. Collapsing caverns show up as sinkholes. Streams flowing on the surface may "sink" into the subterranean cavern system, and in other places the same water may reemerge as a spring. If enough limestone is dissolved, only isolated pinnacles of limestone will remain, towering over the landscape as tower karst.

? Identify the three speleothems shown in this photograph. Would these speleothems have formed in the saturated zone or unsaturated zone? Why?

Miroslav/AGE Fotostock

GIVE IT SOME THOUGHT

1 The cemetery in the photo is located in New Orleans, Louisiana. In this area, all burial plots are aboveground. Based on what you have learned in this chapter, suggest a reason for this rather unusual practice.

Caitlin Mirra/Shutterstock

2 Imagine a water molecule that is part of a groundwater system in an area of gently rolling hills in the eastern United States. Describe some possible paths the molecule might take through the hydrologic cycle in the following scenarios:

a. If it were pumped from the ground to irrigate a farm field
b. If there were a long period of heavy rainfall
c. If the water table in the vicinity of the molecule developed a steep cone of depression due to heavy pumping from a nearby well

Combine your understanding of the hydrologic cycle with your imagination and include possible short-term and long-term destinations and information about how the molecule gets to these places via evaporation, transpiration, condensation, precipitation, infiltration, and/or runoff. Remember to consider possible interactions with streams, lakes, groundwater, the ocean, and the atmosphere.

3 The drainage basin of the Republican River occupies portions of Colorado, Nebraska, and Kansas. A significant part of the basin is considered semiarid. In 1943, the three states made a legal agreement regarding sharing the river's water. In 1998, Kansas went to court to force farmers in southern Nebraska to substantially reduce the amount of groundwater used for irrigation. Nebraska officials claimed that the farmers were not taking water from the Republican River and thus were not violating the 1943 agreement. The court ruled in favor of Kansas.

a. Explain why the court ruled that groundwater in southern Nebraska should be considered part of the Republican River system.

b. How might heavy irrigation in a drainage basin influence the flow of a river?

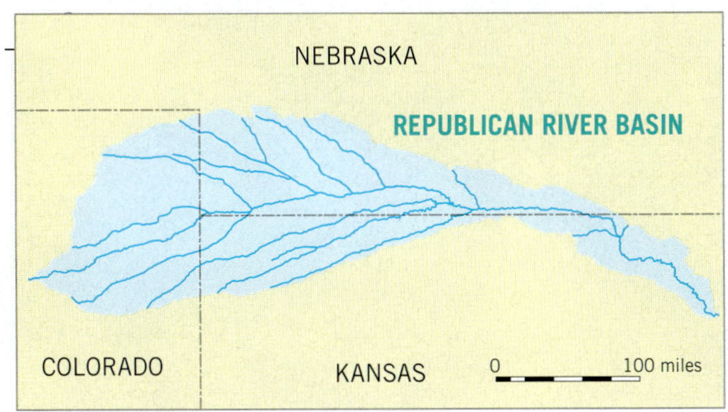

4 During a trip to a grocery store, your friend wants to buy some bottled water. Some brands promote the fact that their product is artesian. Other brands boast that their water comes from a spring. Your friend asks, "Is artesian water or spring water necessarily better than water from other sources?" How would you answer?

5 Imagine that you are an environmental scientist who has been hired to solve a groundwater contamination problem. Several homeowners have noticed that their well water has a funny smell and taste. Some think the contamination is coming from a landfill, but others think it might be a nearby cattle feedlot or chemical plant. Your first step is to gather data from wells in the area and prepare the map of the water table shown here.

a. Based on your map, can any of the three potential sources of contamination be eliminated? If so, explain.

b. What other steps would you take to determine the source of the contamination?

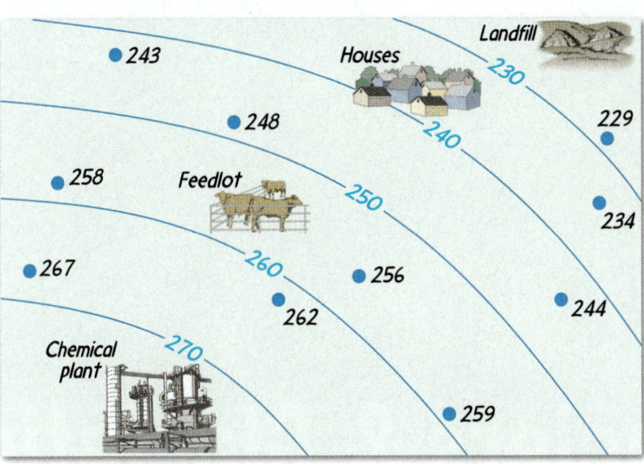

6 An acquaintance is considering purchasing a large tract of productive irrigated farmland in western Texas. His intention is to continue growing crops on the land for years to come. If he asked your opinion about the area he selected, what figure in this chapter would you consult before you responded? How would this figure help this person evaluate his potential purchase?

7 Sinkholes commonly form in one of two ways. Examine the accompanying photo, which shows a sinkhole in Winter Park, Florida, and describe how it likely formed.

AP Photo

MasteringGeology™

Looking for additional review and test prep materials? Visit the Study Area in MasteringGeology to enhance your understanding of this chapter's content by accessing a variety of resources, including Self-Study Quizzes, Geoscience Animations, SmartFigures, Mobile Field Trips, *Project Condor* Quadcopter videos, *In the News* RSS feeds, flashcards, web links, and an optional Pearson eText.

www.masteringgeology.com

15

Glaciers & Glaciation

FOCUS ON CONCEPTS

Each statement represents the primary learning objective for the corresponding major heading within the chapter. After you complete the chapter, you should be able to:

15.1 Explain the role of glaciers in the hydrologic and rock cycles and describe the different types of glaciers, their characteristics, and their present-day distribution.

15.2 Describe how glaciers move, the rates at which they move, and the significance of the glacial budget.

15.3 Discuss the processes of glacial erosion. Identify and describe the major topographic features sculpted by glacial erosion.

15.4 Distinguish between the two basic types of glacial drift. List and describe the major depositional features associated with glacial landscapes.

15.5 Describe and explain several important effects of Ice Age glaciers other than erosional and depositional landforms.

15.6 Briefly discuss the development of glacial theory and summarize current ideas on the causes of ice ages.

Hiker at the terminus of Exit Glacier in Kenai Fiords National Park near Seward, Alaska. (Photo by Michael Collier)

CLIMATE HAS A STRONG INFLUENCE on the nature and intensity of Earth's external processes. This fact is dramatically illustrated in this chapter because the existence and extent of glaciers are largely controlled by Earth's changing climate.

Like the running water and groundwater that were the focus of the preceding two chapters, glaciers represent a significant erosional process. These moving masses of ice are responsible for creating many unique landforms and are part of an important link in the rock cycle in which the products of weathering are transported and deposited as sediment.

Today glaciers cover nearly 10 percent of Earth's land surface; however, in the recent geologic past, ice sheets were three times more extensive, covering vast areas with ice thousands of meters thick. Many regions still bear the marks of these glaciers. The landscapes of such diverse places as the Alps, Cape Cod, and Yosemite Valley were fashioned by now-vanished masses of glacial ice. Moreover, Long Island, the Great Lakes, and the fiords of Norway and Alaska all owe their existence to glaciers. Glaciers, of course, are not just a phenomenon of the geologic past. As you will see, they are still sculpting the landscape and depositing debris in many regions today.

15.1 Glaciers: A Part of Two Basic Cycles

Explain the role of glaciers in the hydrologic and rock cycles and describe the different types of glaciers, their characteristics, and their present-day distribution.

Glaciers are a part of two fundamental cycles in the Earth system: the hydrologic cycle and the rock cycle. Earlier you learned that the water of the hydrosphere is constantly cycled through the atmosphere, biosphere, and geosphere. Time and time again, water evaporates from the oceans into the atmosphere, precipitates on the land, and flows in rivers and underground back to the sea. However, when precipitation falls at high elevations or high latitudes, the water may not immediately make its way toward the sea. Instead, it may become part of a glacier. Although the ice will eventually melt, allowing the water to continue its path to the sea, water can be stored as glacial ice for many tens, hundreds, or even thousands of years.

A **glacier** is a thick ice mass that forms over hundreds or thousands of years. It originates on land from the accumulation, compaction, and recrystallization of snow. A glacier appears to be motionless, but it is not; glaciers move very slowly. Like running water, groundwater, wind, and waves, glaciers are dynamic erosional agents that accumulate, transport, and deposit sediment. As such, glaciers are among the agents that perform a basic function in the rock cycle. Although glaciers are found in many parts of the world today, most are located in remote areas, either near Earth's poles or in high mountains.

Valley (Alpine) Glaciers

Literally thousands of relatively small glaciers exist in lofty mountain areas, where they usually follow valleys that were originally occupied by streams. Unlike the rivers that previously flowed in these valleys, the glaciers advance slowly, perhaps only a few centimeters per day. Because of their setting, these moving ice masses are termed **valley glaciers,** or **alpine glaciers** (Figure 15.1). Each glacier actually is a stream of ice, bounded by precipitous rock walls, that flows downvalley from an accumulation center near its head. Like rivers, valley glaciers can be long or short, wide or narrow, single or with branching tributaries. Generally, alpine glaciers are longer than they are wide. Some extend for just a fraction of a kilometer, whereas others go on for many tens of kilometers. The west branch of the Hubbard Glacier, for example, runs through 112 kilometers (nearly 70 miles) of mountainous terrain in Alaska and the Yukon Territory.

Ice Sheets

In contrast to valley glaciers, **ice sheets** exist on a much larger scale. The low total annual solar radiation reaching the poles makes these regions hospitable to great ice accumulations. Presently each of Earth's polar regions supports an ice sheet: on Greenland in the Northern Hemisphere and on Antarctica in the Southern Hemisphere (Figure 15.2).

Ice Age Ice Sheets About 18,000 years ago, glacial ice covered not only Greenland and Antarctica but also large portions of North America, Europe, and Siberia. That period in Earth history is appropriately known as the *Last Glacial Maximum*. The term implies that there

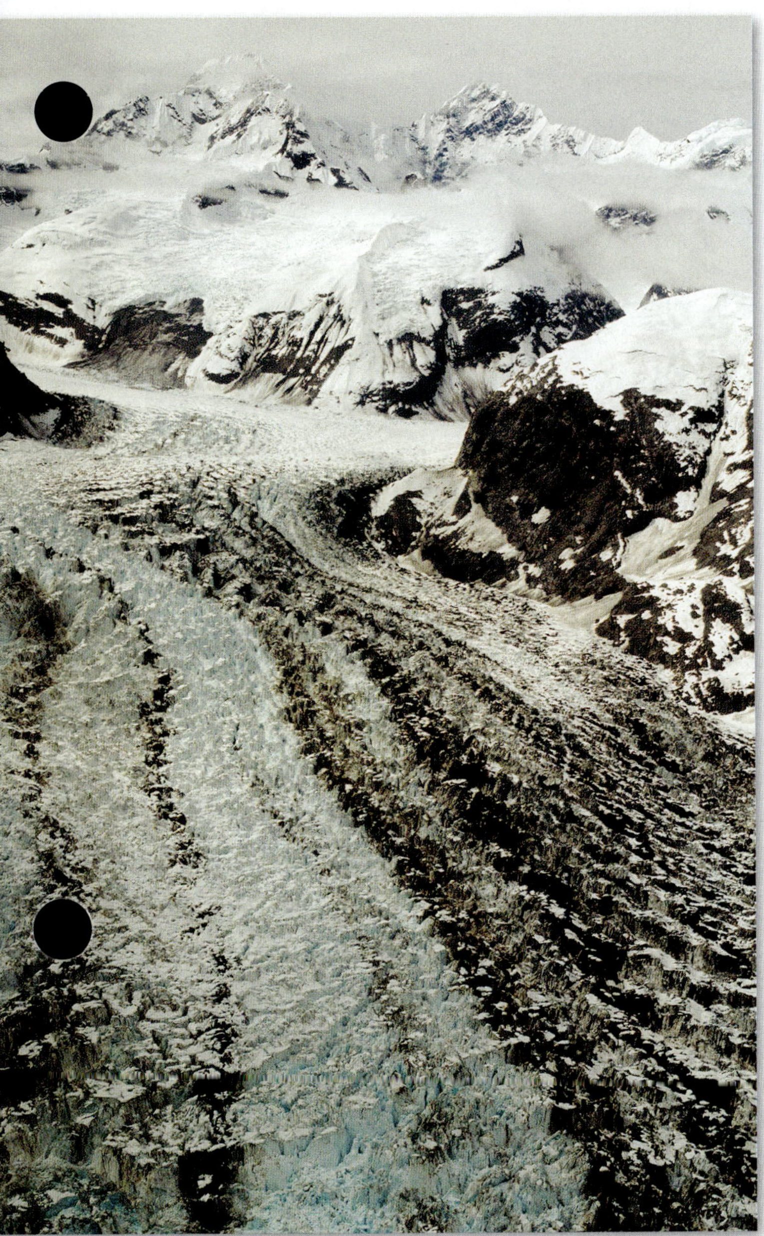

▲ **Figure 15.1 Valley glacier** This tongue of ice, also called an *alpine glacier*, is still eroding the Alaskan landscape. Dark stripes of sediment within the glacier are called medial moraines. This is Johns Hopkins Glacier in Alaska's Glacier Bay National Park. (Photo by Michael Collier)

were other glacial maximums, which is indeed the case. Throughout the Quaternary period, which began about 2.6 million years ago and extends to the present, ice sheets have formed, advanced over broad areas, and then wasted away. These alternating glacial and interglacial periods have occurred over and over again.

Greenland & Antarctica Some people mistakenly think that the North Pole is covered by glacial ice, but this is not the case. The ice that covers the Arctic Ocean is **sea ice**—frozen seawater. Sea ice floats because ice is less dense than liquid water. Although sea ice never completely disappears from the Arctic, the area covered with sea ice expands and contracts with the seasons. The thickness of sea ice ranges from a few centimeters for new ice to 4 meters (13 feet) for sea ice that has survived for years. By contrast, glaciers can be hundreds or thousands of meters thick.

Glaciers form on land, and in the Northern Hemisphere, Greenland supports an ice sheet. Greenland extends between about 60° and 80° north latitude. This largest island on Earth is covered by an imposing ice sheet that occupies 1.7 million square kilometers (more than 660,000 square miles), or about 80 percent of the island. Averaging nearly 1500 meters (5000 feet) thick, the ice extends 3000 meters (10,000 feet) above the island's bedrock floor in some places.

In the Southern Hemisphere, the huge Antarctic Ice Sheet attains a maximum thickness of about 4300 meters (14,000 feet) and covers nearly the entire continent, an area of more than 13.9 million square kilometers (5.4 million square miles). Because of the proportions of these huge features, they are often called *continental ice sheets*. Indeed, the combined areas of present-day continental ice sheets represent almost 10 percent of Earth's land area.

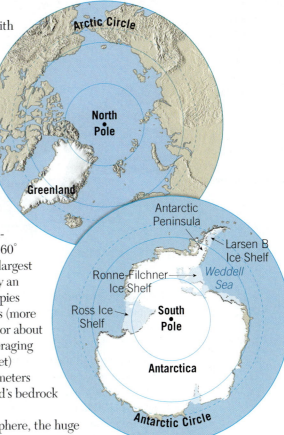

Greenland's ice sheet occupies 1.7 million square kilometers (663,000 square miles), about 80 percent of the island.

Arctic Circle
North Pole
Greenland
Antarctic Peninsula
Larsen B Ice Shelf
Ronne-Filchner Ice Shelf
Weddell Sea
Ross Ice Shelf
South Pole
Antarctica
Antarctic Circle

The area of the Antarctic Ice Sheet is almost 14 million square kilometers (5,460,000 square miles). Ice shelves occupy an additional 1.4 million square kilometers (546,000 square miles).

▲ **SmartFigure 15.2 Ice sheets** The only present-day ice sheets are those covering Greenland and Antarctica. Their combined areas represent almost 10 percent of Earth's land area.

VIDEO
https://goo.gl/x7UVyL

These enormous masses flow out in all directions from one or more snow-accumulation centers and completely obscure all but the highest areas of underlying terrain. Even sharp variations in the topography beneath a glacier usually appear as relatively subdued undulations on the surface of the ice. Such topographic differences, however, do affect the behavior of the ice sheets, especially near their margins, by guiding flow in certain directions and creating zones of faster and slower movement.

Ice Shelves Along portions of the Antarctic coast, glacial ice flows into the adjacent ocean, creating features called **ice shelves**. These large, relatively flat masses of glacial ice extend seaward from the coast but remain attached to the land along one or more sides. About 80 percent of the ice lies below the surface of the ocean, so in shallow water, the ice shelf "touches bottom" and is

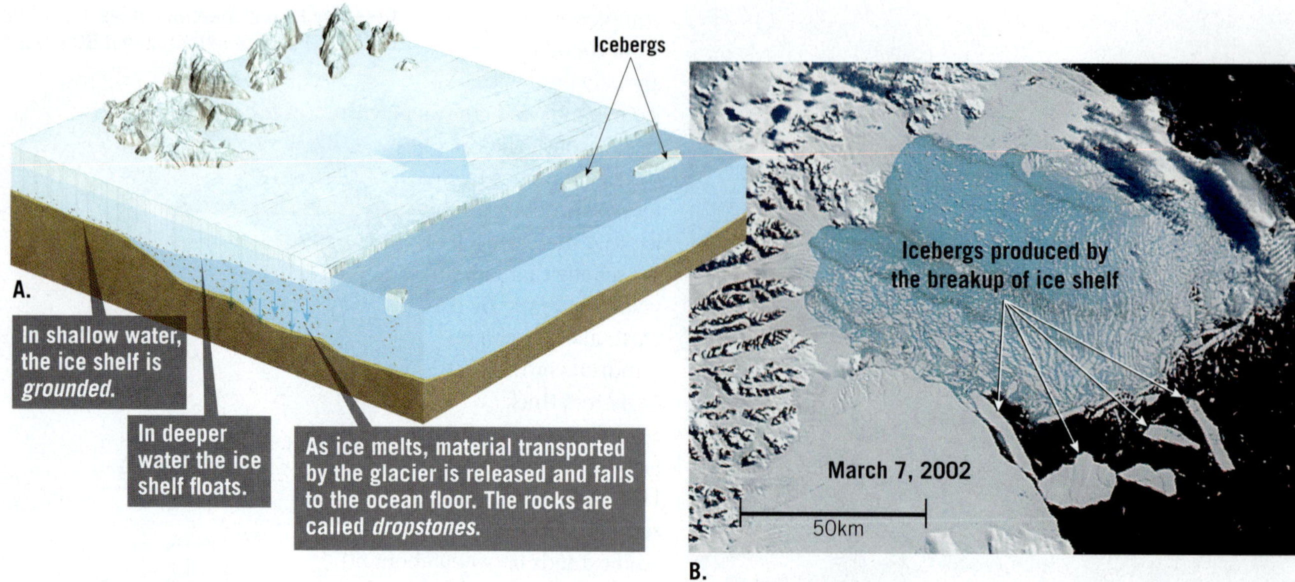
Icebergs
A.
In shallow water, the ice shelf is *grounded.*
In deeper water the ice shelf floats.
As ice melts, material transported by the glacier is released and falls to the ocean floor. The rocks are called *dropstones.*
Icebergs produced by the breakup of ice shelf
March 7, 2002
50km
B.

▶ **Figure 15.3**
Ice shelves A. An ice shelf forms when a glacier or an ice sheet flows into the adjacent ocean. **B.** This satellite image documents the breakup of the Larsen B Ice Shelf adjacent to the Antarctic Peninsula in 2002. Thousands of icebergs were created in the process. (NASA)

said to be *grounded*. In deeper water, the ice shelf floats (**Figure 15.3A**). There are ice shelves along more than half of the Antarctic coast but very few in Greenland.

The shelves are thickest on their landward sides, and they become thinner seaward. They are sustained by ice from the adjacent ice sheet, and they are also nourished by snowfall and the freezing of seawater to their bases. Antarctica's ice shelves extend over approximately 1.4 million square kilometers (0.6 million square miles). The Ross and Ronne-Filchner Ice Shelves are the largest, with the Ross Ice Shelf alone covering an area approximately the size of Texas (see Figure 15.2). In recent years, satellite monitoring has shown that some ice shelves are unstable and breaking apart. For example, during a 35-day span in February and March 2002, an ice shelf on the eastern side of the Antarctic Peninsula, known as the Larsen B Ice Shelf, broke apart and separated from the continent. Thousands of icebergs were set adrift in the adjacent Weddell Sea. The event was captured in satellite imagery (Figure 15.3B). This was not an isolated happening but part of a trend related to accelerated climate change. In fact, a 2015 NASA study predicts that the last remnant of the once-vast Larsen B Ice Shelf will break apart by 2020.

Did You Know?
Sea ice is sensitive to climate change. As our planet warms, the area of the Arctic Ocean that is covered by sea ice in summer has been declining for more than 3 decades. In addition, the sea ice that remains through the summer has been getting thinner and thus more susceptible to loss in subsequent summers. For more about this phenomenon, see Figure 20.35 and the section "The Changing Arctic" in Chapter 20.

Other Types of Glaciers

In addition to valley glaciers and ice sheets, scientists have identified other types of glaciers. Covering some uplands and plateaus are masses of glacial ice called **ice caps**. Like ice sheets, ice caps completely bury the underlying landscape, but they are much smaller than the continental-scale features. Ice caps occur in many places, including Iceland and several of the large islands in the Arctic Ocean (**Figure 15.4**).

Often ice caps and ice sheets feed **outlet glaciers** that flow down valleys around the margin of the cap or sheet. In essence, these are valley glaciers by which the ice cap or sheet drains through mountainous terrain to the sea. Where they encounter the ocean, some outlet glaciers spread out as floating ice shelves. Often large numbers of icebergs are produced.

ICELAND
Reykjavik
Vatnajökull ice cap
Ice caps completely bury the underlying terrain but are much smaller than ice sheets.

▲ **SmartFigure 15.4 Iceland's Vatnajökull ice cap** In 1996 the Grímsvötn Volcano erupted beneath this ice cap, an event that triggered melting and floods. (NASA)

MOBILE FIELD TRIP
https://goo.gl/CAZDCb

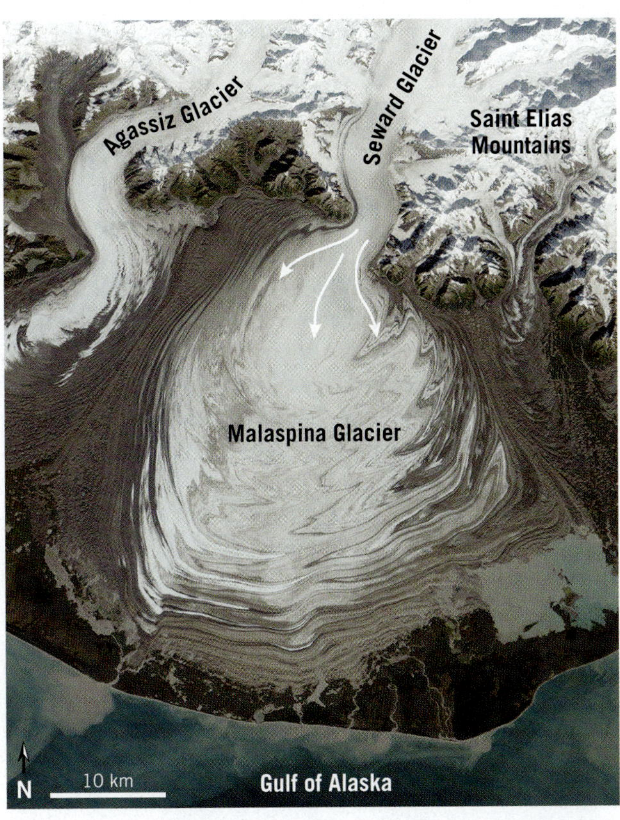

Piedmont glaciers occupy broad lowlands at the bases of steep mountains and form when one or more alpine glaciers emerge from the confining walls of mountain valleys. Here the advancing ice spreads out to form a broad lobe (Figure 15.5). The size of individual piedmont glaciers varies greatly. Among the largest is the broad Malaspina Glacier along the coast of southern Alaska. It covers thousands of square kilometers of the flat coastal plain at the foot of the lofty St. Elias Range.

CONCEPT CHECKS 15.1

1. Where are glaciers found today? What percentage of Earth's land surface do they cover?
2. Describe how glaciers fit into the hydrologic cycle. What role do they play in the rock cycle?
3. List and briefly distinguish among four types of glaciers.
4. What is the difference between an ice sheet and an ice shelf? How are they related?

◀ **Figure 15.5 Piedmont glacier** The ice of a piedmont glacier spills from a steep valley onto a relatively flat plain, where it spreads out. Malaspina Glacier, in southeastern Alaska, fills most of this image. Covering roughly 3880 square kilometers (1500 square miles), it extends nearly 45 kilometers (28 miles) from the mountain front nearly to the sea. (NASA)

Did You Know?
It's hard to imagine how large the Greenland Ice Sheet really is, but try this. The ice sheet is long enough to extend from Key West, Florida, to 100 miles north of Portland, Maine. Its width could reach from Washington, DC, to Indianapolis, Indiana. Put another way, the ice sheet is 80 percent as big as the entire United States east of the Mississippi River. The area of Antarctica's ice sheet is more than eight times as great!

15.2 Formation & Movement of Glacial Ice

Describe how glaciers move, the rates at which they move, and the significance of the glacial budget.

Snow is the raw material from which glacial ice originates; therefore, glaciers form in areas where more snow falls in winter than melts during the summer. In the high-latitude polar realm, where annual snowfall is modest, glaciers form because temperatures are so low that little of the snow melts. Glaciers can form in mountains because temperatures drop with an increase in altitude. So even near the equator, glaciers may form at elevations above about 5000 meters (16,400 feet). For example, Tanzania's Mount Kilimanjaro, which is located practically astride the equator at an altitude of 5895 meters (19,336 feet), has glaciers at its summit. The elevation above which snow remains throughout the year is called the snowline. As you would expect, the elevation of the snowline varies with latitude. Near the equator, it occurs high in the mountains, whereas in the vicinity of the 60th parallel, it is at or near sea level. Before a glacier is created, however, snow must be converted into glacial ice.

Glacial Ice Formation

When temperatures remain below freezing following a snowfall, the fluffy accumulation of delicate hexagonal crystals soon changes. First, the extremities of the crystals evaporate, and the water vapor recondenses near the centers of the crystals. This process of recrystallization makes the snowflakes smaller, thicker, and more spherical, giving them the consistency of coarse sand. The snow also compacts, reducing the pore spaces between grains. The resulting granular recrystallized snow is called **firn** and is commonly found making up old snow banks near the end of winter. As more snow is added, the pressure on the lower layers gradually increases, compacting the ice grains at depth. Once the thickness of ice and snow exceeds 50 meters (165 feet), the weight is sufficient to fuse firn into a solid mass of interlocking ice crystals. Glacial ice has now been formed.

The rate at which this transformation occurs varies. In regions where the annual snow accumulation is great, burial is relatively rapid, and snow may turn to glacial ice in a matter of a decade or less. Where the yearly addition of snow is less abundant, burial is slow, and the transformation of snow to glacial ice may take hundreds of years.

How Glaciers Move

The way in which ice moves is complex and of two basic types. The first of these, *plastic flow*, involves movement

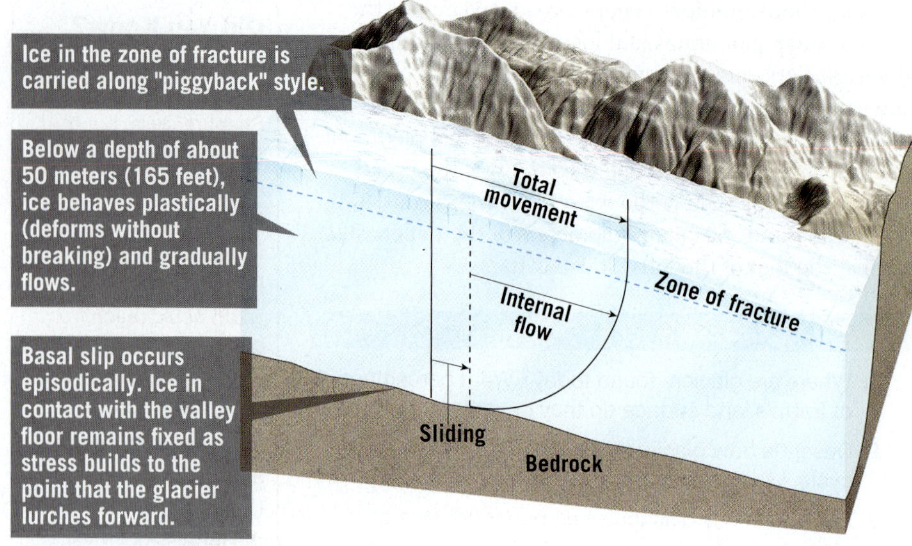

Ice in the zone of fracture is carried along "piggyback" style.
Below a depth of about 50 meters (165 feet), ice behaves plastically (deforms without breaking) and gradually flows.
Basal slip occurs episodically. Ice in contact with the valley floor remains fixed as stress builds to the point that the glacier lurches forward.
Total movement
Zone of fracture
Internal flow
Sliding
Bedrock

▲ SmartFigure 15.6
Movement of a glacier This vertical cross section through a glacier shows that movement is divided into two components. Also notice that the rate of movement is slowest at the base of the glacier, where frictional drag is greatest.

TUTORIAL
https://goo.gl/UAyBCE

within the ice. Ice behaves as a brittle solid until the pressure on it is equivalent to the weight of about 50 meters (165 feet) of ice. Once that load is surpassed, ice behaves as a plastic material, and flow begins. Such flow occurs because of the molecular structure of ice. Glacial ice consists of layers of molecules stacked one upon the other. The bonds between layers are weaker than those within each layer. Therefore, when a stress exceeds the strength of the bonds between the layers, the layers remain intact and slide over one another.

A second and often equally important mechanism of glacial movement occurs when an entire ice mass slips along the ground. The lowest portions of most glaciers probably move by this sliding process, called *basal slip*. **Figure 15.6** illustrates the effects of the two basic types of glacial motion. This vertical profile through a glacier also shows that not all the ice flows forward at the same rate. Frictional drag with the bedrock floor causes the lower portions of the glacier to move more slowly.

The uppermost 50 meters (165 feet) of a glacier is appropriately referred to as the **zone of fracture**. Because there is not enough overlying ice to cause plastic flow, this upper part of the glacier consists of brittle ice. Consequently, the ice in this zone is carried along piggyback-style by the ice below. When the glacier moves over irregular terrain, the zone of fracture is subjected to tension, resulting in cracks called **crevasses** (**Figure 15.7**). These gaping cracks, which often make travel across glaciers dangerous, may extend to depths of 50 meters (165 feet). Below this depth, plastic flow seals them off.

Did You Know?
Glaciers are found on all continents except Australia. Surprisingly, tropical Africa has a small area of glacial ice atop its highest mountain, Mount Kilimanjaro. However, studies show that the mountain has lost more than 80 percent of its ice since 1912 and that all of it may be gone by 2020.

Observing & Measuring Movement

Unlike the movement of water in streams, the movement of glacial ice is not obvious. If we could watch a valley glacier move, we would see that, as with the water in a river, all of the ice does not move downstream at the same rate. Flow is greatest in the center of the glacier because the drag created by the walls and floor of the valley slow the base and sides.

Early in the nineteenth century, the first experiments on glacier movement were designed and carried out in the Alps. Markers were placed in a straight line across an alpine glacier, and the line's position was marked on the valley walls so that if the ice moved, the change in position could be detected. Periodically the positions of the markers were recorded, showing the movement just described. Although most glaciers move too slowly for direct visual detection, the experiments successfully demonstrated that movement nevertheless occurs. The experiment illustrated in **Figure 15.8** was carried out at Switzerland's Rhône Glacier later in the nineteenth century. It not only traced the movement of markers within the ice but also mapped the position of the glacier's terminus.

▲ Figure 15.7 **Crevasses** As a glacier moves, internal stresses cause large cracks to develop in the brittle upper portion of the glacier, called the zone of fracture. Crevasses can extend to depths of 50 meters (165 feet) and can make travel across glaciers dangerous. (Photo by Wave/Glow Images)

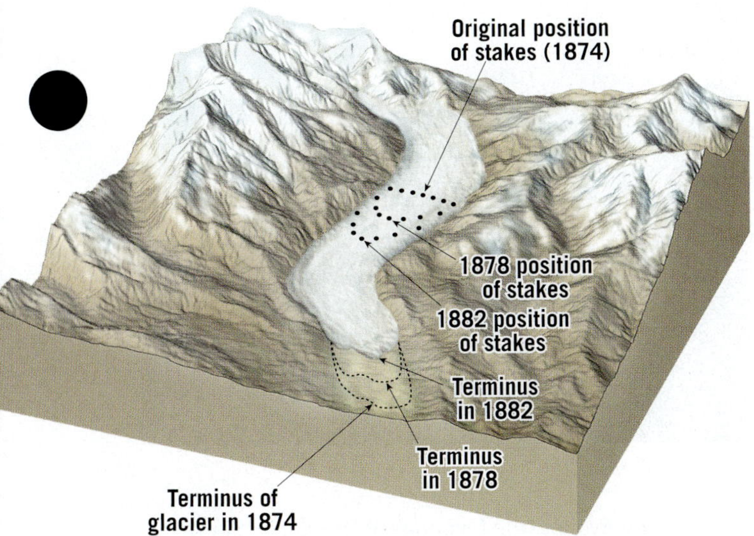

▲ **Figure 15.8 Measuring the movement of a glacier** Ice movement and changes in the terminus of Rhône Glacier, Switzerland. In this classic study of a valley glacier, the movement of stakes clearly shows that glacial ice moves and that movement along the sides of the glacier is slower than movement in the center. Also notice that even though the ice front was retreating, the ice within the glacier was advancing. (NASA)

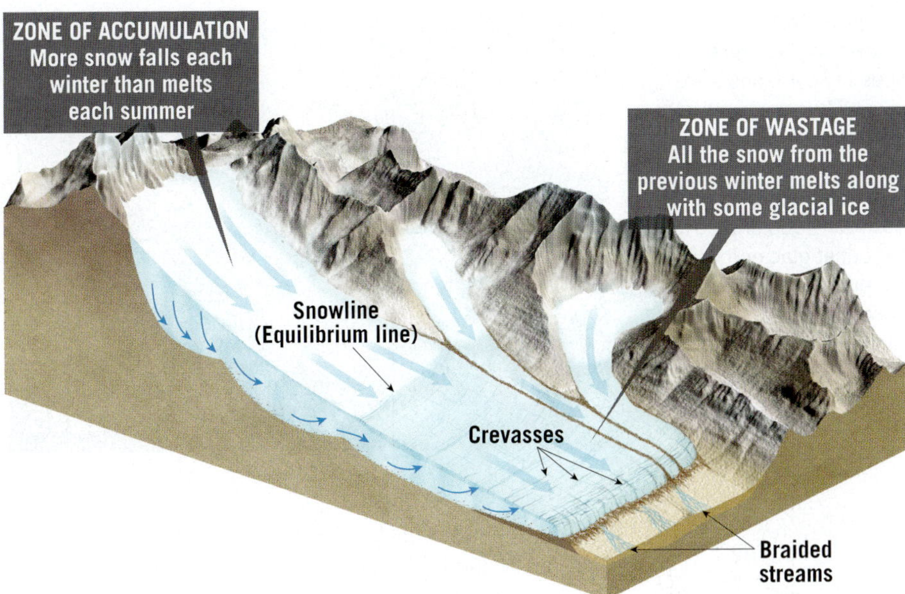

▲ **SmartFigure 15.9 Zones of a glacier** The snowline separates the zone of accumulation and the zone of wastage. Whether the ice front advances, retreats, or remains stationary depends on the balance or lack of balance between accumulation and wastage (ablation).

TUTORIAL
https://goo.gl/y3g5ig

How rapidly does glacial ice move? Average rates vary considerably from one glacier to another. Some glaciers move so slowly that trees and other vegetation may become well established in the debris that accumulates on the glacier's surface. Others advance up to several meters each day. Recent satellite imaging provided insights into movements within the Antarctic Ice Sheet (see Figure 1.8). Portions of some outlet glaciers move at rates greater than 800 meters (2600 feet) per year; on the other hand, ice in some interior regions creeps along at less than 2 meters (6.5 feet) per year. Movement of some glaciers is characterized by occasional periods of extremely rapid advance called *surges*, followed by periods of much slower movement.

Budget of a Glacier: Accumulation Versus Wastage

Snow is the raw material from which glacial ice originates; therefore, glaciers form in areas where more snow falls in winter than melts during the summer. Glaciers are constantly gaining and losing ice.

Glacial Zones Snow accumulation and ice formation occur in the **zone of accumulation**. Its outer limits are defined by the **snowline** or **equilibrium line**—the elevation at which the accumulation and wasting of glacial ice is equal. As noted earlier, the elevation of this boundary varies greatly, from sea level in polar regions to altitudes approaching 5000 meters (16,000 feet) near the equator. Above the snowline, in the zone of accumulation, the addition of snow thickens the glacier and promotes movement. Below the snowline is the **zone of wastage**. Here there is a net loss to the glacier as all of the snow from the previous winter melts, along with some of the glacial ice (Figure 15.9).

The loss of ice by a glacier is termed **ablation**. In addition to melting, glaciers waste away as large pieces of ice break off the front of the glacier in a process called **calving**. Calving creates **icebergs** in places where the glacier has reached the sea or a lake (Figure 15.10). Because icebergs are just slightly less dense than seawater, they float very low in the water, with more than 80 percent of their mass submerged. Along the margins of Antarctica's ice shelves, calving is the primary means by which ice is lost. The relatively flat icebergs produced here can be several kilometers across and up to about 600 meters (2000 feet) thick (see the icebergs in Figures 15.3). By comparison, thousands of irregularly shaped icebergs are produced by outlet glaciers flowing

Geologist's Sketch
Only about 20 percent or less of an iceberg protrudes above the waterline.

◀ **Figure 15.10 Icebergs** Icebergs form when large masses of ice break off from the front of a glacier after it reaches a water body, in a process known as calving. (Photo by Radius Images/Photolibrary)

▶ **Figure 15.11**
Retreating glaciers
These two images were taken 78 years apart from about the same vantage point along the southwest coast of Greenland. Between 1935 and 2013 the outlet glacier that is the primary focus of these photos retreated about 3 kilometers (about 2 miles).
(NASA/Earth Observatory)

from the margins of the Greenland Ice Sheet. Many drift southward and find their way into the North Atlantic, where they can be hazardous to navigation.

Glacial Budget Whether the margin of a glacier is advancing, retreating, or remaining stationary depends on the budget of the glacier. The **glacial budget** is the balance, or lack of balance, between accumulation at the upper end of the glacier and ablation at the lower end. If ice accumulation exceeds ablation, the glacial front advances until the two factors balance. When this happens, the terminus of the glacier is stationary.

If a warming trend increases ablation and/or if a drop in snowfall decreases accumulation, the ice front retreats. As the terminus of the glacier retreats, the extent of the zone of wastage diminishes. Therefore, in time a new balance will be reached between accumulation and wastage, and the ice front will again become stationary.

Whether the margin of a glacier is advancing, retreating, or stationary, the ice within the glacier continues to flow forward. In the case of a receding glacier, the ice still flows forward but not rapidly enough to offset ablation. This point is illustrated well in Figure 15.8. As the line of stakes within the Rhone Glacier continued to move downvalley, the terminus of the glacier slowly retreated upvalley.

Glaciers in Retreat: Unbalanced Glacial Budgets Because glaciers are sensitive to changes in temperature and precipitation, they provide clues about changes in climate. With few exceptions, valley glaciers around the world have been retreating at unprecedented rates over the past century. Many valley glaciers have disappeared altogether. For example, 150 years ago, there were 147 glaciers in Montana's Glacier National Park. Today only 37 remain. Greenland's ice sheet and portions of Antarctica's ice are also shrinking. The photos in **Figure 15.11** provide an example.

CONCEPT CHECKS 15.2

1. Describe two components of glacial movement.
2. How rapidly does glacial ice move? Provide some examples.
3. What are crevasses, and where do they form?
4. Relate the glacial budget to the two zones of a glacier.
5. Under what circumstances does the front of a glacier advance? Retreat? Remain stationary?

15.3 Glacial Erosion

Discuss the processes of glacial erosion. Identify and describe the major topographic features sculpted by glacial erosion.

Glaciers are capable of great erosion. For anyone who has observed the terminus of an alpine glacier, the evidence of its erosive force is clear (**Figure 15.12**). You can witness firsthand the release of rock material of various sizes from the ice as it melts. All signs lead to the conclusion that the ice has scraped, scoured, and torn rock from the floor and walls of the valley and carried it downvalley. It should be pointed out, however, that in mountainous regions, processes of mass movement also make substantial contributions to the sediment load of a glacier.

Once rock debris is acquired by a glacier, the enormous competence of ice does not allow the debris to settle out like the load carried by a stream or by the wind. Indeed, as a medium of sediment transport, ice has no equal. Consequently, glaciers can carry huge blocks that no other erosional agent could possibly budge. Although today's glaciers are of limited importance as erosional

▲ **Figure 15.12 Evidence of glacial erosion** As the terminus of this glacier wastes away, it deposits large quantities of sediment. This image near the terminus of Exit Glacier in Alaska's Kenai Fjords National Park shows that the rock debris dropped by the melting ice is a jumbled mixture of different-size sediments. (Photo by Michael Collier)

agents, many landscapes that were modified by the widespread glaciers of the Ice Age still reflect, to a high degree, the work of ice.

How Glaciers Erode

Glaciers erode the land primarily in two ways: plucking and abrasion. First, as a glacier flows over a fractured bedrock surface, it loosens and lifts blocks of rock and incorporates them into the ice. This process, known as **plucking**, occurs when meltwater penetrates the cracks and joints of bedrock beneath a glacier and freezes. Because water expands when it freezes, it exerts tremendous leverage that pries the rock loose. In this manner, sediment of all sizes becomes part of the glacier's load.

The second major erosional process is **abrasion** (**Figure 15.13**). As the ice and its load of rock fragments slide over bedrock, they function like sandpaper, smoothing and polishing the surface below. The pulverized rock produced by the glacial "grist mill" is appropriately called **rock flour**. So much rock flour may be produced that meltwater streams flowing out of a glacier often have the cloudy appearance of skim milk and offer visible evidence of the grinding power of ice. Lakes fed by such streams frequently have a distinctive turquoise color.

When the ice at the bottom of a glacier contains large rock fragments, long scratches and grooves called **glacial striations** may even be gouged into the bedrock (see Figure 15.13A). These linear grooves provide clues to the direction of ice flow. By mapping the striations over large areas, patterns of glacial flow can often be reconstructed. Ice that carries fine particles can also polish rock. The broad expanses of smoothly polished granite in Yosemite National Park provide an excellent example (see Figure 15.13B).

As is the case with other agents of erosion, the rate of glacial erosion is highly variable. This rate is largely

Glacial abrasion created the scratches and grooves in this bedrock.
A.
Glacially polished granite in California's Yosemite National Park.
B.

◄ **Figure 15.13 Glacial abrasion** Moving glacial ice, armed with sediment, acts like sandpaper, scratching and polishing rock. (Photos by Michael Collier)

controlled by four factors: (1) speed of glacier movement; (2) ice thickness; (3) shape, abundance, and hardness of the rock fragments in the ice at the base of the glacier; and (4) erodibility of the surface beneath the glacier. These factors can all vary from place to place and from one time to another, with resulting variation in the degree of landscape modification.

Landforms Created by Glacial Erosion

The erosional effects of valley glaciers and ice sheets are quite different. A visitor to a glaciated mountain region is likely to see sharp and angular topography. This is because alpine glaciers tend to accentuate the irregularities of the mountain landscape by creating steeper canyon walls and making bold peaks even more jagged. By contrast, continental ice sheets generally override the terrain and hence subdue rather than accentuate the irregularities they encounter. Although the erosional potential of ice sheets is enormous, landforms carved by these huge ice masses usually do not inspire the same awe as do the erosional features created by valley glaciers. **Figure 15.14** shows a hypothetical mountain area before, during, and after glaciation. You will refer to this figure often in the following discussion.

Glaciated Valleys A hike up a glaciated valley reveals a number of striking ice-created features. The valley itself is often a dramatic sight. Unlike streams, which create their own valleys, glaciers take the path of least resistance and follow the paths of existing stream valleys. Prior to glaciation, mountain valleys are characteristically

▶ **SmartFigure 15.14**
Erosional landforms created by alpine glaciers The unglaciated landscape (**A**) is modified by valley glaciers (**B**). After the ice recedes (**C**), the terrain looks very different than it looked before glaciation. (Arête photo by James E. Patterson; cirque photo by Marli Miller; hanging valley photo by John Warden/SuperStock)

MOBILE FIELD TRIP
https://goo.gl/SYujm3

◀ **SmartFigure 15.15**
A U-shaped glacial trough Prior to glaciation, a mountain valley is typically narrow and V-shaped. During glaciation, an alpine glacier widens, deepens, and straightens the valley, creating the classic U-shape shown here. This glacial trough is in the Sierra Nevada, west of Bishop, California. (Photo by Michael Collier)

ANIMATION
https://goo.gl/mzSk4k

narrow and V-shaped because streams are well above base level and are therefore downcutting. However, during glaciation a narrow valley is transformed as the glacier widens and deepens it, creating a U-shaped **glacial trough** (see Figure 15.14C and **Figure 15.15**). In addition to producing a broader and deeper valley, the glacier also straightens the valley. As ice flows around sharp curves, its great erosional force removes the spurs of land that extend into the valley.

The amount of glacial erosion that takes place will vary in different valleys in a mountainous area. Prior to glaciation, the mouths of tributary streams join the main valley (or *trunk* valley) at the elevation of the stream in that valley. During glaciation, the amount of ice flowing through the main valley can be much greater than the amount advancing down each tributary. Consequently, the valley containing the main glacier (or *trunk* glacier) is eroded deeper than the smaller valleys that feed it. Thus, after the ice has receded, the valleys of tributary glaciers are left standing above the main glacial trough and are termed **hanging valleys** (see Figure 15.14C). Rivers flowing through hanging valleys can produce spectacular waterfalls, such as those in Yosemite National Park, California.

Cirques At the head of a glacial valley is a characteristic and often imposing feature associated with an alpine glacier, called a **cirque**. As the photo in Figure 15.14 illustrates, these bowl-shaped depressions have precipitous walls on three sides and are open on the downvalley side. The cirque is the focal point of the glacier's growth because it is the area of snow accumulation and ice formation. Cirques begin as irregularities in the mountainside that are subsequently enlarged by frost wedging and plucking along the sides and bottom of the glacier. The glacier in turn acts as a conveyor belt that carries away the debris. After the glacier has melted away, the cirque basin is sometimes occupied by a small lake called a **tarn** (see Figure 15.14C).

Arêtes & Horns The Alps, Northern Rockies, and many other mountain landscapes sculpted by valley glaciers reveal more than glacial troughs and cirques. In addition, sinuous, knife-edged ridges called **arêtes** and sharp, pyramid-like peaks termed **horns** project above the surroundings (see Figure 15.14C). Both features can originate from the same basic process: the enlargement of cirques produced by plucking and frost action. Several cirques around a single high mountain create the spires of rock called horns. As the cirques enlarge and converge,

Figure 15.16 The Matterhorn Horns are sharp, pyramid-like peaks that were shaped by alpine glaciers. The Matterhorn, in the Swiss Alps, is a famous example. (Photo by Andy Selinger/AGE fotostock)

an isolated horn is produced. A famous example is the Matterhorn in the Swiss Alps (**Figure 15.16**).

Arêtes can form in a similar manner except that the cirques are not clustered around a point but rather exist on opposite sides of a divide. As the cirques grow, the divide separating them is reduced to a very narrow, knifelike partition. An arête can also be created when glaciers that flow in parallel valleys narrow the intervening ridge as they scour and widen their valleys.

Roche Moutonnée In many glaciated landscapes, but most frequently where continental ice sheets have modified the terrain, the ice carves small streamlined hills from protruding bedrock knobs. Such an asymmetrical knob of bedrock is called a **roche moutonnée** (French for "sheep rock"). They are formed when glacial abrasion smooths the gentle slope facing the oncoming ice and plucking steepens the opposite side as the ice rides over the knob (**Figure 15.17**). Roches moutonnées indicate the direction of glacial flow because the gentler slope is generally on the side from which the ice advanced.

Fiords **Fiords** are deep, often spectacular steep-sided inlets of the sea that are present at high latitudes where mountains are adjacent to the ocean (**Figure 15.18**).

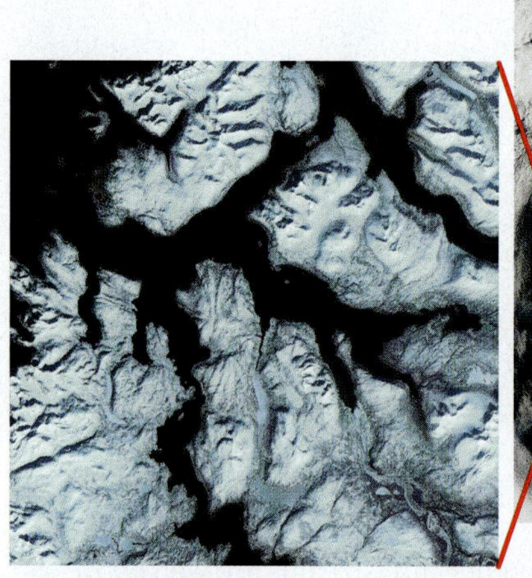

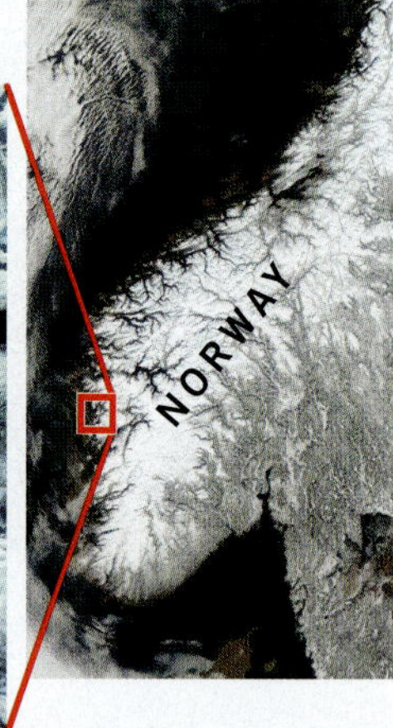

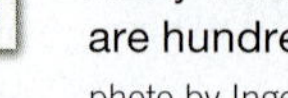

Figure 15.17 Roche mountonnée This classic example is in Yosemite National Park, California. The gentle slope was abraded, and the steep slope was plucked. The glacier moved from right to left. (Photo by E. J. Tarbuck)

Figure 15.18 Fiords The coast of Norway is known for its many fiords. Frequently these ice-sculpted inlets of the sea are hundreds of meters deep. (Satellite images courtesy of NASA; photo by Inger Yoshio Tomii/SuperStock)

They are drowned glacial troughs that became submerged as the ice left the valleys and sea level rose following the Ice Age. The depths of fiords may exceed 1000 meters (3300 feet). However, the great depths of these flooded troughs are only partly explained by the post–Ice Age rise in sea level. Unlike the situation governing the downward erosional work of rivers, sea level does not act as base level for glaciers. As a consequence, glaciers are capable of eroding their beds far below the surface of the sea. For example, a 300-meter- (1000-foot-) thick glacier can carve its valley floor more than 250 meters (820 feet) below sea level before downward erosion ceases and the ice begins to float. Norway, British Columbia, Greenland, New Zealand, Chile, and Alaska all have coastlines characterized by fiords.

CONCEPT CHECKS 15.3

1. How do glaciers acquire their load of sediment?
2. What are some visible effects of glacial erosion?
3. What factors influence a glacier's ability to erode?
4. How does a glaciated mountain valley differ in appearance from a mountain valley that was not glaciated? Describe the erosional features created by valley glaciers.
5. Relate fiords to glacial troughs.

15.4 Glacial Deposits

Distinguish between the two basic types of glacial drift. List and describe the major depositional features associated with glacial landscapes.

A glacier picks up and transports a huge load of rock debris as it slowly advances across the land. Where the ice melts, these materials are deposited. Such deposits can play a significant role in forming the physical landscape. For example, in many areas once covered by the continental ice sheets of the recent Ice Age, the bedrock is rarely exposed because the terrain is completely mantled by glacial deposits tens or even hundreds of meters thick. The general effect is to level the topography. Indeed, rural country scenes that are familiar to many of us—rocky pastures in New England, wheat fields in the Dakotas, rolling farmland in the Midwest—result directly from glacial deposition.

Glacial Drift

Long before the theory of an extensive ice age was ever proposed, much of the soil and rock debris covering portions of Europe was recognized as having come from somewhere else. At the time, these "foreign" materials were believed to have been "drifted" into their present positions by floating ice during an ancient flood. As a consequence, the term *drift* was applied to this sediment. Although rooted in an incorrect concept, this term was so well established by the time the true glacial origin of the debris became widely recognized that it remained part of the basic glacial vocabulary. Today **glacial drift** is an all-embracing term for sediments of glacial origin, no matter how, where, or in what shape they were deposited.

Geologists divide glacial drift into two distinct types: (1) materials deposited directly by the glacier, which are known as *till*, and (2) sediments laid down by glacial meltwater, called *stratified drift*.

Glacial Till As glacial ice melts and drops its load of rock fragments, **till** is deposited. Unlike moving water and wind, ice cannot sort the sediment it carries; therefore, deposits of till are characteristically unsorted mixtures of many particle sizes (Figure 15.19). A close examination of this sediment shows that many of the pieces are scratched and polished as a result of being dragged along by the glacier. Such pieces help distinguish till from other deposits that are a mixture of different sediment sizes, such as material from a debris flow or a rockslide.

Boulders found in the till or lying free on the surface are called **glacial erratics** if they are different

Glacial till is an unsorted mixture of many different sediment sizes.
A close examination of glacial till often reveals cobbles that have been scratched as they were dragged along by the ice.

Figure 15.19 **Glacial till** Unlike sediment deposited by running water and wind, material deposited directly by a glacier is not sorted. Figure 15.12 provides another good example of till. (Top photo by Michael Collier; bottom photo by E. J. Tarbuck)

▲ **Figure 15.20 Glacial erratics** Large glacially transported boulders are relatively common features on Cape Cod, where this photo was taken, and in many other places where glacial deposits are prominent. Such boulders are called *glacial erratics*. (Photo by Michael Collier)

from the bedrock below (**Figure 15.20**). Of course, this means that they must have been derived from a source outside the area where they are found. Although the source for most erratics is unknown, the origin of some can be determined. In many cases, boulders were transported as far as 500 kilometers (300 miles) from their source area and, in a few instances, more than 1000 kilometers (600 miles). Therefore, by studying glacial erratics as well as the mineral composition of the remaining till, geologists are sometimes able to trace the path of a lobe of ice.

In portions of New England and other areas, erratics dot pastures and farm fields. In fact, in some places, these large rocks were cleared from fields and piled to make fences and walls. Keeping the fields clear, however, is an ongoing chore because each spring, newly exposed erratics appear as wintertime frost heaving lifts them to the surface.

Stratified Drift As the name implies, **stratified drift** is sorted according to the size and weight of the particles. Ice is not capable of sorting the way running water can, and therefore these materials are not deposited directly by the glacier as till is but instead reflect the sorting action of glacial meltwater.

Some deposits of stratified drift are made by streams issuing directly from the glacier. Other stratified deposits involve sediment that was originally laid down as till and was later picked up, transported, and redeposited by meltwater beyond the margin of the ice. Accumulations of stratified drift often consist largely of sand and gravel because the meltwater is not capable of moving larger material and because the finer rock flour remains suspended and is commonly carried far from the glacier. In consequence, these deposits may be actively mined as aggregate for road work and other construction projects.

Moraines, Outwash Plains, & Kettles

Perhaps the most widespread features created by glacial deposition are *moraines*, which are simply layers or ridges of till. Lateral and medial moraines are found only in mountain valleys, whereas end moraines and ground moraines are associated with areas affected by either ice sheets or valley glaciers.

Lateral & Medial Moraines The sides of a valley glacier accumulate large quantities of debris from the valley walls. When the glacier wastes away, these materials are left as ridges, called **lateral moraines**, along the sides of the valley. **Medial moraines** are formed when two valley glaciers coalesce to form a single ice stream (**Figure 15.21**). The till that was once carried along the edges of each glacier joins to form a single dark stripe of debris within the newly enlarged glacier. The creation of these dark stripes within the ice stream is one obvious proof that glacial ice moves because the medial moraine could not form if the ice did not flow downvalley. A large alpine glacier may have several medial moraines, each forming where a tributary glacier joins the main valley.

Geologist's Sketch

◀ **SmartFigure 15.21 Formation of a medial moraine** Kennicott Glacier is a 43-kilometer- (27-mile-) long valley glacier that is sculpting the mountains in Alaska's Wrangell–St. Elias National Park. The dark stripes of sediment are medial moraines. (Photo by Michael Collier)

MOBILE FIELD TRIP
https://goo.gl/CWZNtK

End & Ground Moraines Sometimes a glacier is compared to a conveyor belt. No matter whether the front of a glacier or ice sheet is advancing, retreating, or stationary, the glacier or sheet is constantly moving sediment forward and dropping it at its terminus.

An **end moraine** is a ridge of till that forms at the terminus of a glacier or ice sheet whenever the terminus is stationary. That is, the end moraine forms when the ice is wasting away near the end of the glacier at a rate equal to the forward advance of the glacier. Although the terminus of the glacier is stationary, the ice continues to flow forward, delivering a continuous supply of sediment in the same manner a conveyor belt delivers goods to the end of a production line. As the ice melts, the till is dropped, and the end moraine grows. The longer the ice front remains stationary, the larger the ridge of till becomes. End moraines are common features of both valley glaciers and ice sheets. Figure 15.12 shows a portion of an end moraine at the terminus of Alaska's Exit Glacier.

Eventually, ablation exceeds nourishment. At this point, the front of the glacier begins to recede in the direction from which it originally advanced. However, as the ice front retreats, the conveyor-belt action of the glacier continues to provide fresh supplies of sediment to the terminus. In this manner, a large quantity of till is deposited as the ice melts away, creating a rock-strewn, undulating plain. This gently rolling layer of till deposited as the ice front recedes is termed **ground moraine**. It has a leveling effect, filling in low spots and clogging old stream channels, often leading to a derangement of the existing drainage system. In areas where this layer of till is still relatively fresh, such as the northern Great Lakes region, poorly drained swampy lands are quite common.

During an overall retreat, a glacier will often pause for a time, with ablation and nourishment in temporary balance. When this happens, a new end moraine forms. The pattern of end moraine formation and ground moraine deposition may be repeated many times before the glacier has completely vanished. Such a pattern is illustrated in **Figure 15.22**. The very first end moraine to form signifies the farthest advance of the glacier and is called the *terminal end moraine*. End moraines that form as the ice front occasionally stabilizes during retreat are termed *recessional end moraines*. Terminal and recessional moraines are essentially alike; the only difference between them is their relative positions.

End moraines deposited by the most recent stage of Ice Age glaciation are prominent features in many parts of the Midwest and Northeast. In Wisconsin, the wooded, hilly terrain of the Kettle Moraine near Milwaukee is a particularly picturesque example. A well-known example in the Northeast is Long Island. This linear strip of glacial sediment that extends northeastward from New York City is part of an end moraine complex that stretches from eastern Pennsylvania to Cape Cod, Massachusetts (**Figure 15.23**).

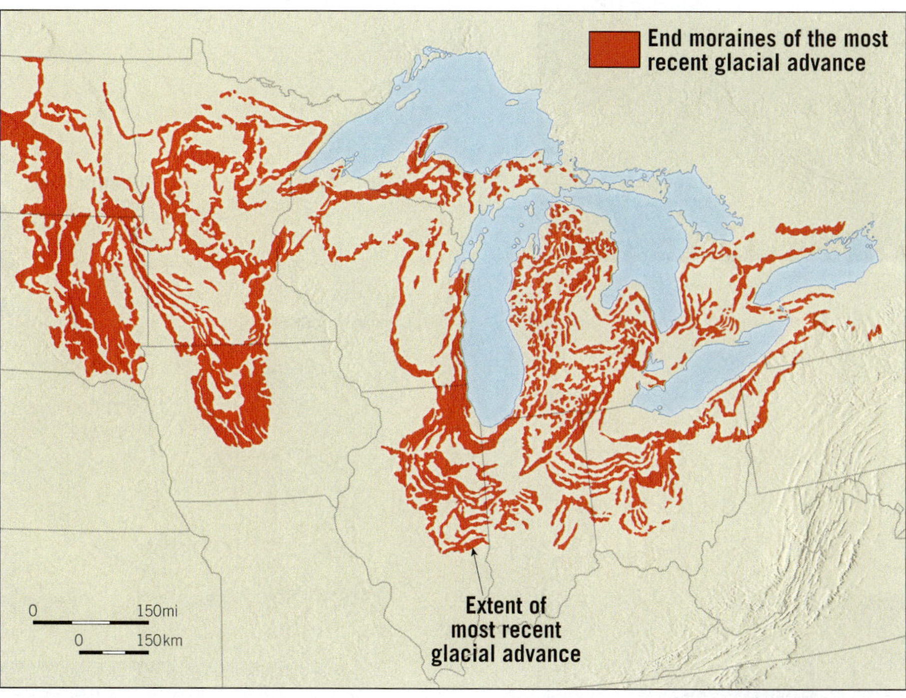

▲ **Figure 15.22 End moraines of the Great Lakes region** End moraines deposited during the most recent stage of glaciation are the most prominent features in many parts of this region.

Figure 15.24 represents a hypothetical area during glaciation and after the retreat of ice sheets. This figure depicts landscape features, such as the end moraines just described, as well as depositional landforms similar to what you might encounter if you were traveling in the upper Midwest or New England. You will be referred to this figure several times as you read the following paragraphs on glacial deposits.

Outwash Plains & Valley Trains At the same time that an end moraine is forming, water from the melting glacier cascades over and through the till, sweeping some of it out in front of the growing ridge of unsorted debris. Meltwater generally emerges from the ice in rapidly

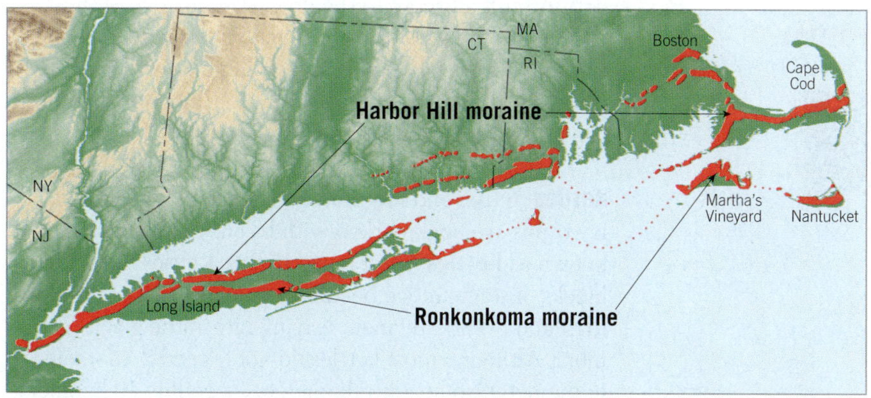

▲ **Figure 15.23 Two significant end moraines in the Northeast** The Ronkonkoma moraine, which was deposited about 20,000 years ago, extends through central Long Island, Martha's Vineyard, and Nantucket. The Harbor Hill moraine formed about 14,000 years ago and extends along the north shore of Long Island, through southern Rhode Island and Cape Cod.

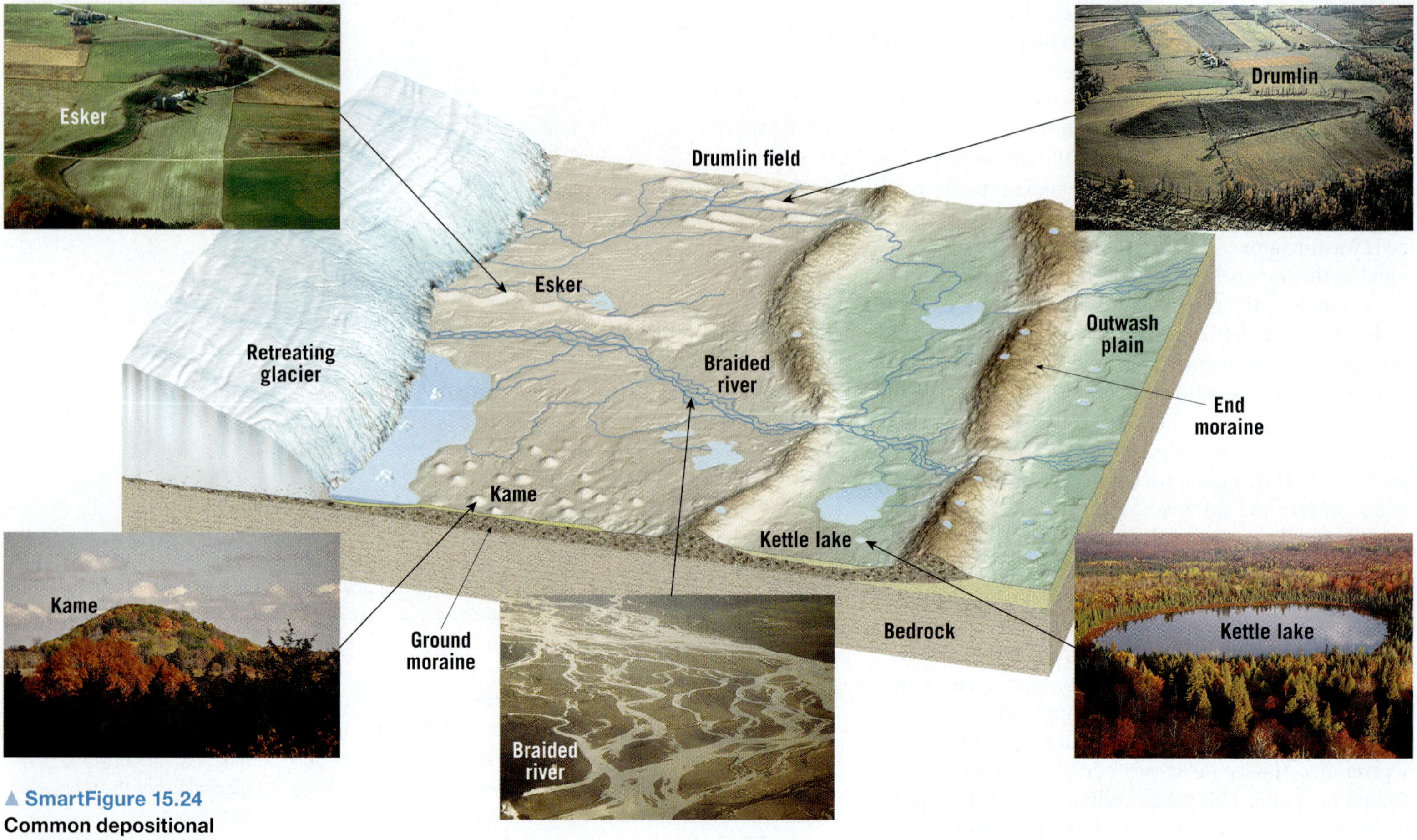

▲ SmartFigure 15.24
Common depositional landforms This diagram depicts a hypothetical area affected by ice sheets in the recent geologic past.
(Drumlin photo courtesy of Ward's Natural Science Establishment; esker photo by Richard P. Jacobs/ JLM Visuals; kame photo by John Dankwardt; kettle lake photo by Carlyn Iverson/Science Source; braided river photo by Michael Collier)

TUTORIAL
https://goo.gl/76mWl1

moving streams that are often choked with suspended material and carry a substantial bed load as well. Water leaving the glacier moves onto the relatively flat surface beyond and rapidly loses velocity. As a consequence, much of its bed load is dropped, and the meltwater begins weaving a complex pattern of braided channels. Such a situation is shown in Figure 13.20. In this way, a broad, ramplike surface composed of stratified drift is built adjacent to the downstream edge of most end moraines (see Figure 15.24). When the feature is formed in association with an ice sheet, it is termed an **outwash plain**; when it is largely confined to a mountain valley, it is usually called a **valley train**.

Kettles Often end moraines, outwash plains, and valley trains are pockmarked with basins or depressions known as **kettles** (see Figure 15.24). Kettles form when blocks of stagnant ice become wholly or partly buried in drift and eventually melt, leaving pits in the glacial sediment. Although most kettles do not exceed 2 kilometers in diameter, some with diameters exceeding 10 kilometers occur in Minnesota. Likewise, the typical depth of most kettles is less than 10 meters, although the vertical dimensions of some approach 50 meters (165 feet). In many cases, water eventually fills the depression and forms a pond or lake.

Drumlins, Eskers, & Kames

Moraines are not the only landforms deposited by glaciers. Some landscapes are characterized by numerous elongate parallel hills made of till. Other areas exhibit conical hills and relatively narrow winding ridges composed largely of stratified drift.

Drumlins Streamlined, asymmetrical hills composed of till are called **drumlins** (see Figure 15.24). They range in height from 15 to 60 meters (50 to 200 feet) and average 0.4 to 0.8 kilometer (0.25 to 0.50 mile) in length. The steep side of the hill faces the direction *from* which the ice advanced, whereas the gentler slope points in the direction the ice moved. Drumlins are not found singly but rather occur in clusters, called *drumlin fields*. One such cluster, east of Rochester, New York, is estimated to contain about 10,000 drumlins. Their streamlined shape indicates that they were molded in the zone of flow within an active glacier. It is thought that drumlins originate when glaciers advance over previously deposited drift and reshape the material.

Eskers & Kames In some areas that were once occupied by glaciers, sinuous ridges composed largely of sand and gravel can be found. Known as **eskers**, these ridges are deposited by meltwater rivers flowing within, on top of, and beneath a mass of motionless, stagnant glacial ice (see Figure 15.24). Many sediment sizes are carried by the torrents of meltwater in the ice-banked channels, but only the coarser material can settle out of the turbulent stream. In some areas they are mined for sand and gravel, and for this reason, eskers are disappearing in some localities.

Kames are steep-sided hills that, like eskers, are composed of sand and gravel (see Figure 15.24). Kames originate when glacial meltwater washes sediment into openings and depressions in the stagnant wasting terminus of a glacier. When the ice eventually melts away, the stratified drift is left behind as mounds or hills.

CONCEPT CHECKS 15.4

1. What term can be applied to any glacial deposit? Distinguish between till and stratified drift.
2. How are medial moraines and lateral moraines related? In what kind of setting are these features found?
3. Contrast end moraines and ground moraines. Relate their formation to the budget of a glacier.
4. List and briefly describe four depositional features other than moraines. Describe the formation of a kettle.

Did You Know?
One well-known example of a water-filled kettle is Walden Pond near Concord, Massachusetts. It is here that the noted transcendentalist Henry David Thoreau lived alone for 2 years in the 1840s and about which he wrote his famous book *Walden; or Life in the Woods*.

15.5 Other Effects of Ice Age Glaciers

Describe and explain several important effects of Ice Age glaciers other than erosional and depositional landforms.

In addition to their massive erosional and depositional work, the ice sheets and glaciers of the Ice Age had other effects, sometimes profound, on the landscape. For example, as the ice advanced and retreated, animals and plants were forced to migrate or perish; a number of plants and animals became extinct. Other effects of Ice Age glaciers described in this section involve adjustments in Earth's crust due to the addition and removal of ice and sea-level changes associated with the formation and melting of ice sheets. The advance and retreat of ice sheets also led to significant changes in the routes taken by rivers. In some regions, glaciers acted as dams that created large lakes. When these ice dams failed, the effects on the landscape were profound. In areas that today are deserts, lakes of another type, called pluvial lakes, formed.

Crustal Subsidence & Rebound

In areas that were major centers of ice accumulation, such as Scandinavia and the Canadian Shield, the land has been slowly rising over the past several thousand years. Uplifting of almost 300 meters (1000 feet) has occurred in the Hudson Bay region. This, too, is the result of the continental ice sheets. But how can glacial ice cause such vertical crustal movement? We now understand that the land is rising because the added weight of the 3-kilometer- (2-mile-) thick mass of ice caused downwarping of Earth's crust. Following the removal of this immense load, the crust has been adjusting by gradually rebounding upward ever since.

Sea-Level Changes

One of the most interesting and perhaps dramatic effects of the Ice Age was the fall and rise of sea level that accompanied the advance and retreat of the glaciers. Although the total volume of glacial ice today is great, exceeding 25 million cubic kilometers, during the Last Glacial Maximum the volume of glacial ice amounted to about 70 million cubic kilometers, or 45 million cubic kilometers more than at present. Because we know that the snow from which glaciers are made ultimately comes from the evaporation of ocean water, the growth of ice sheets must have caused a worldwide drop in sea level (**Figure 15.25**). Indeed, estimates suggest that sea level was as much as 100 meters (330 feet) lower than it is today. Thus, land that is presently flooded by the oceans was dry. The Atlantic coast of the United States lay more

Did You Know?
Studies have shown that the retreat of glaciers may reduce the stability of faults and hasten earthquake activity. When the weight of the glacier is removed, the ground rebounds. In tectonically active areas experiencing postglacial rebound, earthquakes may occur sooner and/or be stronger than if the ice were present.

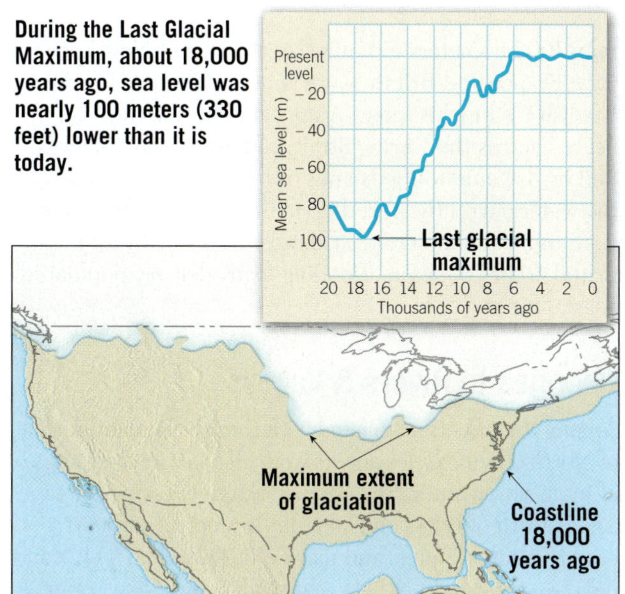

During the Last Glacial Maximum, the shoreline extended out onto the present-day continental shelf.

◀ **SmartFigure 15.25**
Changing sea level As ice sheets form and then melt away, sea level falls and rises, causing the shoreline to shift.

ANIMATION
https://goo.gl/QL4wnv

Figure 15.26 Changing rivers The advance and retreat of ice sheets caused major changes in the routes followed by rivers in the central United States.

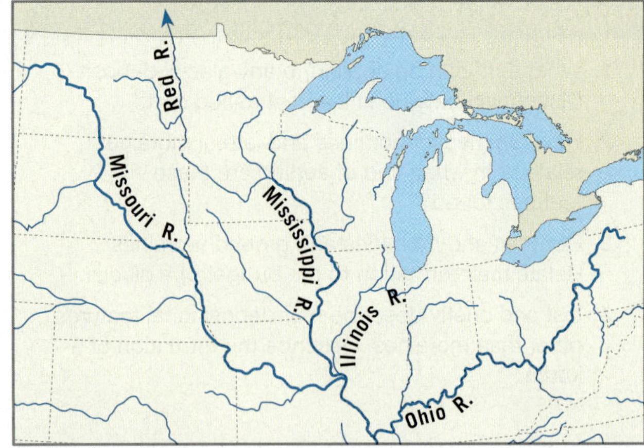

A. This map shows the Great Lakes and the familiar present-day pattern of rivers. Quaternary ice sheets played a major role in creating this pattern.

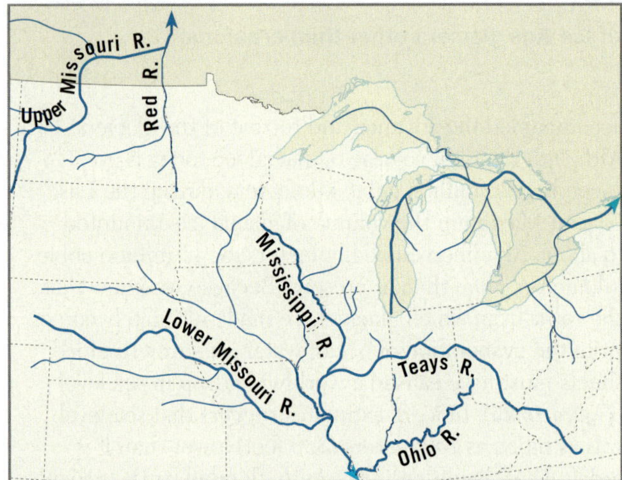

B. Reconstruction of drainage systems prior to the Ice Age. The pattern was very different from today, and the Great Lakes did not exist.

than 100 kilometers (60 miles) to the east of New York City, France and Britain were joined where the famous English Channel is today, Alaska and Siberia were connected across the Bering Strait, and Southeast Asia was tied by dry land to the islands of Indonesia. Conversely, if the water currently locked up in the Antarctic Ice Sheet were to melt completely, sea level would rise by an estimated 60 or 70 meters, flooding many densely populated coastal areas.

Changes to Rivers & Valleys

Among the effects associated with the advance and retreat of North American ice sheets were changes in the routes of many rivers and in the size and shape of many valleys. The pattern of rivers and lakes in the central and northeastern United States (and many other places as well) can only be understood in light of glacial history. The upper Mississippi River drainage basin provides a good example.

Figure 15.26A shows the familiar present-day pattern of rivers in the central United States, with the Missouri, Ohio, and Illinois Rivers as major tributaries to the Mississippi. Figure 15.26B depicts drainage systems in this region prior to the Ice Age. The pattern is *very* different from the present. This remarkable transformation of river systems resulted from the advance and retreat of the Quaternary ice sheets.

Notice that prior to the Ice Age, a significant part of the Missouri River drained north toward Hudson Bay. Moreover, the Mississippi River did not follow the present Iowa–Illinois boundary but rather flowed across west-central Illinois, where the lower Illinois River flows today. The preglacial Ohio River barely reached to the present-day state of Ohio, and the rivers that today feed the Ohio in western Pennsylvania flowed north and drained into the North Atlantic. The Great Lakes were created by glacial erosion during the Ice Age. Prior to the Pleistocene epoch, the basins occupied by these huge lakes were lowlands with rivers that ran eastward to the Gulf of St. Lawrence.

The large Teays River was a significant feature prior to the Ice Age (see Figure 15.26B). The Teays flowed from West Virginia across Ohio, Indiana, and Illinois, and it discharged into the Mississippi River not far from present-day Peoria. This river valley, which would have rivaled the Mississippi in size, was completely obliterated during the Pleistocene, buried by glacial deposits hundreds of feet thick. Today the sands and gravels in the buried Teays valley make it an important aquifer.

Ice Dams Create Proglacial Lakes

Ice sheets and alpine glaciers can act as dams to create lakes by trapping glacial meltwater and blocking the flow of rivers. Some of these lakes are relatively small, short-lived impoundments. Others can be large and exist for hundreds or thousands of years.

Figure 15.27 is a map of Lake Agassiz—the largest lake to form during the Ice Age in North America. It came into existence about 12,000 years ago and lasted for about 4500 years. With the retreat of the ice sheet came enormous volumes of meltwater. The Great Plains generally slope upward to the west. As the terminus of the ice sheet receded northeastward, meltwater was trapped between the ice on one side and the sloping land on the other, causing Lake Agassiz to deepen and spread across the landscape. Such water bodies are termed **proglacial lakes**, referring to their position just beyond the outer limits of a glacier or ice sheet. The history of Lake Agassiz is complicated by the dynamics of the ice sheet, which, at various times, readvanced and affected lake levels and drainage systems. Where drainage occurred depended on the water level of the lake and the position of the ice sheet.

Did You Know?
Antarctica's ice sheet weighs so much that some parts of the continent are depressed by an estimated 900 meters (3000 feet) or more.

▲ **Figure 15.27** **Glacial Lake Agassiz** This lake was an immense feature—bigger than all of the present-day Great Lakes combined. Modern-day remnants of this proglacial water body are still major landscape features.

Lake Agassiz left marks over a broad region. Former beaches, many kilometers from any water, mark former shorelines. Several modern river valleys, including the Red River and the Minnesota River, were originally cut by water entering or leaving the lake. Present-day remnants of Lake Agassiz include Lake Winnipeg, Lake Manitoba, Lake Winnipegosis, and Lake of the Woods. The sediments of the former lake basin are now fertile agricultural land.

Research shows that the shifting of glaciers and the failure of ice dams can cause the rapid release of huge volumes of water. Such events occurred during the history of Lake Agassiz. A dramatic example of such glacial outbursts occurred in the Pacific Northwest between about 15,000 and 13,000 years ago and is briefly described in Figure 15.28.

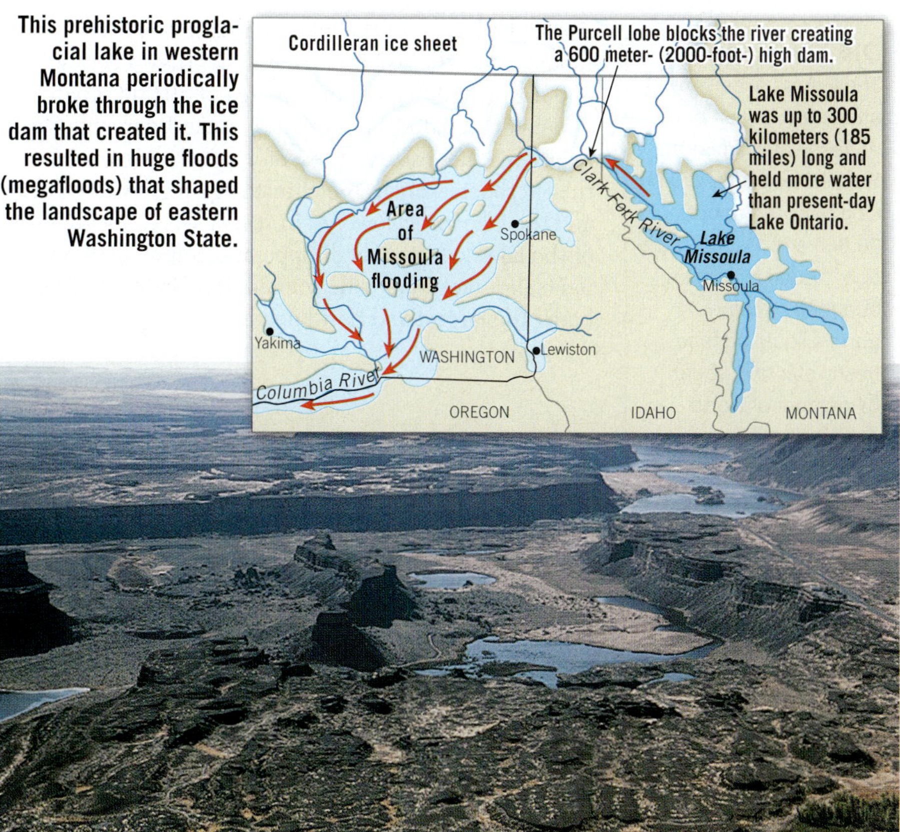

▲ **Figure 15.28** **Lake Missoula and the Channeled Scablands** During a span of 1500 years, more than 40 megafloods from Lake Missoula carved the Channeled Scablands. (Photo by John S. Shelton/ University of Washington Libraries)

Pluvial Lakes

While the formation and growth of ice sheets was an obvious response to significant changes in climate, the existence of the glaciers themselves triggered important climatic changes in the regions beyond their margins. In arid and semiarid areas on all the continents, temperatures were lower and thus evaporation rates were lower, but at the same time, precipitation totals were moderate. This cooler, wetter climate formed many **pluvial lakes** (*pluvia* = rain).. In North America the greatest concentration of pluvial lakes occurred in the vast Basin and Range region of Nevada and Utah (Figure 15.29).

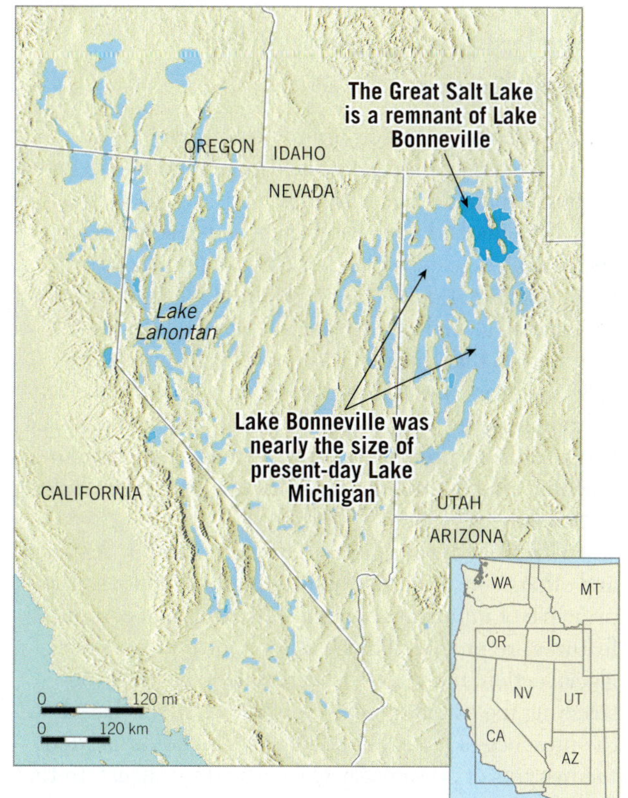

◄ **Figure 15.29** **Pluvial lakes** During the Ice Age, the Basin and Range region experienced a wetter climate than it has today. Many basins turned into large lakes.

Did You Know?
The Great Lakes constitute the largest body of freshwater on Earth. Formed by glacial erosion between about 10,000 and 12,000 years ago, the lakes contain about 20 percent of Earth's liquid surface freshwater and about 84 percent of North America's.

By far the largest of the lakes in this region was Lake Bonneville. With maximum depths exceeding 300 meters (1000 feet) and an area of 50,000 square kilometers (20,000 square miles), Lake Bonneville was nearly the same size as present-day Lake Michigan. As the ice sheets waned, the climate again grew more arid, and the lake levels lowered in response. Although most of the lakes completely disappeared, a few small remnants of Lake Bonneville remain, the Great Salt Lake being the largest and best known.

CONCEPT CHECKS 15.5

1. List and briefly describe five effects of Ice Age glaciers aside from the formation of major erosional and depositional features.
2. Examine Figure 15.25 and determine how much sea level has changed since the Last Glacial Maximum.
3. Compare the two parts of Figure 15.26 and identify three major changes to the flow of rivers in the central United States during the Ice Age.
4. Contrast proglacial lakes and pluvial lakes. Give an example of each.

15.6 The Ice Age

Briefly discuss the development of glacial theory and summarize current ideas on the causes of ice ages.

In the preceding pages, we mentioned the Ice Age, a time when ice sheets and alpine glaciers were far more extensive than they are today. As noted, there was a time when the most popular explanation for what we now know to be glacial deposits was that the materials had been drifted in by means of icebergs or perhaps simply swept across the landscape by a catastrophic flood. What convinced geologists that an extensive ice age was responsible for these deposits and many other glacial features?

Historical Development of the Glacial Theory

In 1821 a Swiss engineer, Ignaz Venetz, presented a paper suggesting that glacial landscape features occurred at considerable distances from the existing glaciers in the Alps. This implied that the glaciers had once been larger and occupied positions farther downvalley. Another Swiss scientist, Louis Agassiz, doubted the proposal of widespread glacial activity put forth by Venetz. He set out to prove that the idea was not valid. However, his 1836 fieldwork in the Alps convinced him of the merits of his colleague's hypothesis. In fact, a year later Agassiz hypothesized a great ice age that had extensive and far-reaching effects—an idea that was to give Agassiz widespread fame.

The proof of the glacial theory proposed by Agassiz and others constitutes a classic example of applying the principle of uniformitarianism. Realizing that certain features are produced by no other known process but glacial action, the scientists were able to begin reconstructing the extent of now-vanished ice sheets based on the presence of features and deposits found far beyond the margins of present-day glaciers and ice sheets. In this manner, the development and verification of the glacial theory continued during the nineteenth century, and through the efforts of many scientists, a knowledge of the nature and extent of former ice sheets became clear.

By the beginning of the twentieth century, geologists had largely determined the extent of the Ice Age glaciation. Further, during the course of their investigations, they discovered that many glaciated regions had not one but several layers of drift. Moreover, close examination of these older deposits showed well-developed zones of chemical weathering and soil formation, as well as the remains of plants that require warm temperatures. The evidence was clear: There had been not just one glacial advance but many, each separated by an extended period when climates were as warm as or warmer than the present. The Ice Age had not simply been a time when the ice advanced over the land, lingered for a while, and then receded. Rather, the period was a very complex event, characterized by a number of advances and withdrawals of glacial ice.

By the early twentieth century, a fourfold division of the Ice Age had been established for both North America and Europe. The divisions were based largely on studies of glacial deposits. In North America each of the four major stages was named for the midwestern state where deposits of that stage were well exposed and/or were first studied. These are, in order of occurrence, the Nebraskan, Kansan, Illinoian, and Wisconsinan. These traditional divisions remained in place for many years, until it was learned that sediment cores from the ocean floor contain a much more complete record of climate change during the Ice Age. Unlike the glacial record on land, which is punctuated by many unconformities, seafloor sediments provide an uninterrupted record of climatic cycles for this period. Studies of these seafloor sediments showed that glacial/interglacial cycles had occurred about every 100,000 years. About 20 such cycles of cooling and warming were identified for the span we call the Ice Age.

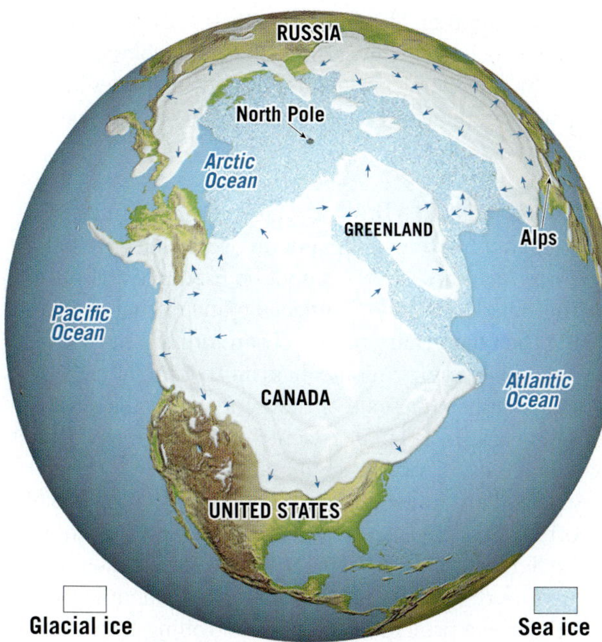

▲ **Figure 15.30 Where was the ice?** This map shows the maximum extent of ice sheets in the Northern Hemisphere during the Ice Age.

During the Ice Age, ice left its imprint on almost 30 percent of Earth's land area, including about 10 million square kilometers of North America, 5 million square kilometers of Europe, and 4 million square kilometers of Siberia (**Figure 15.30**). The amount of glacial ice in the Northern Hemisphere was roughly twice that in the Southern Hemisphere. The primary reason is that the southern polar ice could not spread far beyond the margins of Antarctica. By contrast, North America and Eurasia provided great expanses of land for the spread of ice sheets.

Today we know that the Ice Age began between 2 million and 3 million years ago. This means that most of the major glacial stages occurred during a division of the geologic time scale called the **Quaternary period**. However, this period does not encompass all of the last glacial period. Antarctica's ice sheets, for example, probably formed at least 30 million years ago.

Causes of Ice Ages

A great deal is known about glaciers and glaciation. Much has been learned about glacier formation and movement, the extent of glaciers past and present, and the features created by glaciers, both erosional and depositional. However, the causes of ice ages are not completely understood.

Although widespread glaciation has been rare in Earth's history, the Ice Age that encompassed most of the Quaternary period is not the only glacial period for which records exist. Earlier glaciations are indicated by deposits called **tillite**, a sedimentary rock formed when glacial till becomes lithified (**Figure 15.31**). Such deposits, found in strata of several different ages, usually contain striated rock fragments, and some overlie grooved and polished bedrock surfaces or are associated with sandstones and conglomerates that show features of outwash deposits. For example, our Chapter 2 discussion of evidence supporting the continental drift hypothesis mentioned a glacial period that occurred in late Paleozoic time (see Figure 2.7). In addition, two Precambrian glacial episodes have been identified in the geologic record, the first approximately 2 billion years ago and the second about 600 million years ago.

Any theory that attempts to explain the causes of ice ages must successfully answer two basic questions:

- *What causes the onset of glacial conditions?* For continental ice sheets to have formed, average temperature must have been somewhat lower than at present and perhaps substantially lower than throughout much of geologic time. Thus, a successful theory would have to account for the cooling that finally leads to glacial conditions.
- *What caused the alternating glacial and interglacial stages that have been documented for the Quaternary period?* Whereas the first question deals with long-term trends in temperature on a scale of millions of years, this question relates to much shorter-term changes.

Although the scientific literature contains many hypotheses related to the possible causes of glacial periods, we will discuss only a few major ideas to summarize current thought.

Plate Tectonics Probably the most attractive proposal for explaining the fact that extensive glaciations have occurred only a few times in the geologic past comes from the theory of plate tectonics. Because glaciers can form only on land, we know that landmasses must exist somewhere in the higher latitudes before an ice age can

◀ **Figure 15.31 Tillite** When glacial till is lithified, it becomes the sedimentary rock known as *tillite*. Tillite strata are evidence for ice ages that occurred prior to the Quaternary period. (Photo by Brian Roman)

commence. Many scientists suggest that ice ages have occurred only when Earth's shifting crustal plates have carried the continents from tropical latitudes to more poleward positions.

Glacial features in present-day Africa, Australia, South America, and India indicate that these regions, which are now tropical or subtropical, experienced an ice age near the end of the Paleozoic era, about 250 million years ago. However, there is no evidence that ice sheets existed during this same period in what are today the higher latitudes of North America and Eurasia. For many years this puzzled scientists. Was the climate in these relatively tropical latitudes once like it is today in Greenland and Antarctica? Why did glaciers not form in North America and Eurasia? Until the plate tectonics theory was formulated, there had been no reasonable explanation.

Today scientists understand that the areas containing these ancient glacial features were joined together as a single supercontinent, called Pangaea, located at latitudes far to the south of their present positions. Later this landmass broke apart, and its pieces, each moving on a different plate, migrated toward their present locations (**Figure 15.32**). Now we know that during the geologic past, plate movements accounted for many dramatic climate changes as landmasses shifted in relation to one another and moved to different latitudinal positions. Changes in oceanic circulation also must have occurred, altering the transport of heat and moisture and consequently the climate as well. Because the rate of plate movement is very slow—a few centimeters annually—appreciable changes in the positions of the continents occur only over great spans of geologic time. Thus, climate changes triggered by shifting plates are extremely gradual and happen on a scale of millions of years.

The supercontinent Pangaea showing the area covered by glacial ice near the end of the Paleozoic era.

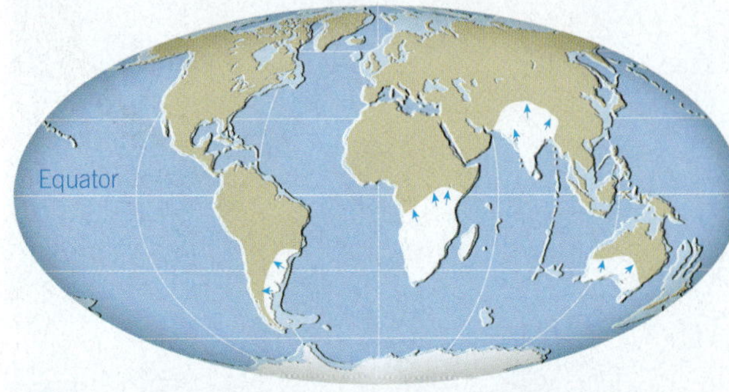

The continents as they appear today. The white areas indicate where evidence of the late Paleozoic ice sheets exists.

Figure 15.32 A late Paleozoic ice age Shifting tectonic plates sometimes move landmasses to high latitudes, where the formation of ice sheets is possible.

Variations in Earth's Orbit Because climatic changes brought about by moving plates are extremely gradual, the plate tectonics theory cannot be used to explain the alternating glacial and interglacial climates that occurred during the Quaternary period. Therefore, we must look to some other triggering mechanism that might cause climate change on a scale of thousands rather than millions of years. Today many scientists strongly suspect that the climate oscillations that characterized the Quaternary period are linked to variations in Earth's orbit. This hypothesis was first developed and strongly advocated by the Serbian astrophysicist Milutin Milankovitch and is based on the premise that variations in incoming solar radiation are a principal factor in controlling Earth's climate.

Milankovitch formulated a comprehensive mathematical model based on the following elements (**Figure 15.33**):

- Variations in the shape (*eccentricity*) of Earth's orbit about the Sun
- Changes in *obliquity*—that is, changes in the angle that Earth's axis makes with the plane of our planet's orbit
- The wobbling of Earth's axis, called *precession*

Using these factors, Milankovitch calculated variations in the receipt of solar energy and the corresponding surface temperature of Earth back into time, in an attempt to correlate these changes with the climate fluctuations of the Quaternary. In explaining climate changes that result from these three variables, note that they cause little or no variation in the *total* solar energy reaching the ground. Instead, their impact is felt because they change the degree of contrast between the seasons. Somewhat milder winters in the middle to high latitudes mean greater snowfall totals, whereas cooler summers bring a reduction in snowmelt.

Among the studies that have added credibility to the astronomical hypothesis of Milankovitch is one in which J. D. Hays and colleagues analyzed deep-sea sediments containing certain climatically sensitive microorganisms to establish a chronology of temperature changes going back nearly a half million years.* This time scale of climate change was then compared to astronomical calculations of eccentricity, obliquity, and precession to determine whether a correlation existed. Although the study was very involved and mathematically complex, the

*J. D. Hays, John Imbrie, and N. J. Shackelton, "Variations in the Earth's Orbit: Pacemaker of the Ice Ages," *Science*, 194 (1976): 1121–1132.

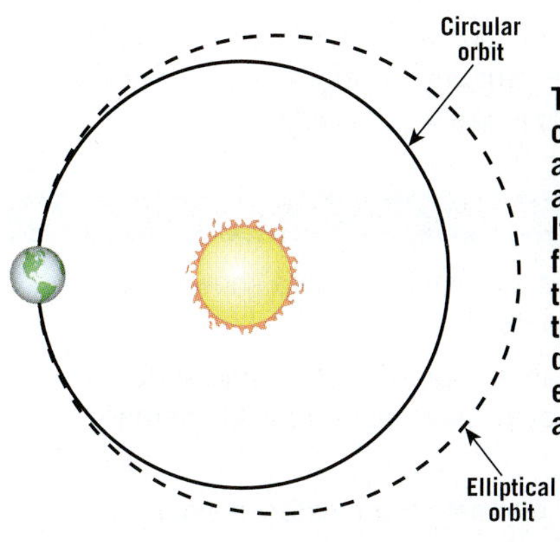

Circular orbit
The shape of Earth's orbit changes during a cycle that spans about 100,000 years. It gradually changes from nearly circular to more elliptical and then back again. This diagram greatly exaggerates the amount of change.
Elliptical orbit

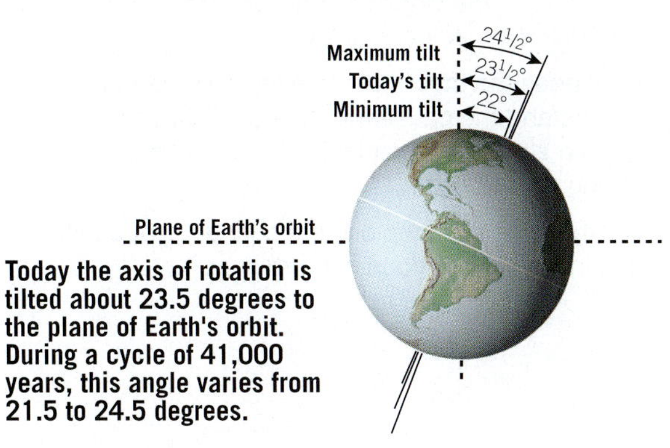

Maximum tilt
Today's tilt
Minimum tilt
24½°
23½°
22°
Plane of Earth's orbit
Today the axis of rotation is tilted about 23.5 degrees to the plane of Earth's orbit. During a cycle of 41,000 years, this angle varies from 21.5 to 24.5 degrees.

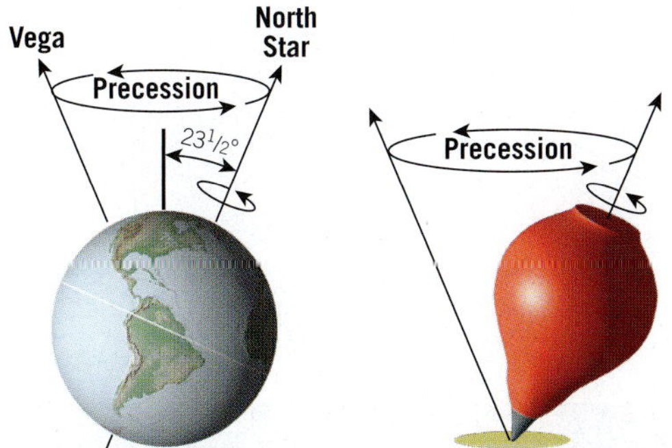

Vega
North Star
Precession
23½°
Precession
Earth's axis wobbles like a spinning top. Consequently, the axis points to different spots in the sky during a cycle of about 26,000 years.

▲ **SmartFigure 15.33 Orbital variations** Periodic variations in Earth's orbit are linked to alternating glacial and interglacial conditions during the Ice Age.

conclusions were straightforward: The researchers found that major variations in climate over the past several hundred thousand years were closely associated with changes in the geometry of Earth's orbit; that is, cycles of climate change were shown to correspond closely with the periods of obliquity, precession, and orbital eccentricity. More specifically, the authors stated: "It is concluded that changes in the earth's orbital geometry are the fundamental cause of the succession of Quaternary ice ages."*

*J. D. Hays et al., p. 1131.

Let us briefly summarize the ideas just described: Plate tectonics theory provides an explanation for the widely spaced and nonperiodic onset of glacial conditions at various times in the geologic past, whereas the theory proposed by Milankovitch and supported by the work of J. D. Hays and his colleagues furnishes an explanation for the alternating glacial and interglacial episodes of the Quaternary.

Other Factors Variations in Earth's orbit correlate closely with the timing of glacial–interglacial cycles. However, the variations in solar energy reaching Earth's surface caused by these orbital changes do not adequately explain the magnitude of the temperature changes that occurred during the most recent Ice Age. Other factors must also have contributed. One factor involves variations in the composition of the atmosphere. Other influences are related to changes in the reflectivity of Earth's surface and in ocean circulation. Let's take a brief look at these factors.

Chemical analyses of air bubbles that become trapped in glacial ice at the time of ice formation indicate that the Ice Age atmosphere contained less of the gases carbon dioxide and methane than the post–Ice Age atmosphere (**Figure 15.34**). Carbon dioxide and methane are important "greenhouse" gases, which means they trap radiation emitted by Earth and contribute to the heating of the atmosphere. When the amount of carbon dioxide and methane in the atmosphere increases, global temperatures rise, and when there is a reduction in these gases, as occurred during glacial cycles, temperatures fall. Therefore, reductions in the concentrations of greenhouse gases help explain the magnitude of the temperature drop that occurred during glacial times. Although scientists know that concentrations of carbon dioxide and methane dropped, they do not know what caused the drop. As often occurs in science, observations gathered during one investigation yield information and raise questions that require further analysis and explanation.

Did You Know?
Antarctica is the fifth-largest continent and is about twice as large as Australia. It also has the distinction of being the coldest, driest, and windiest continent and also has the highest average elevation.

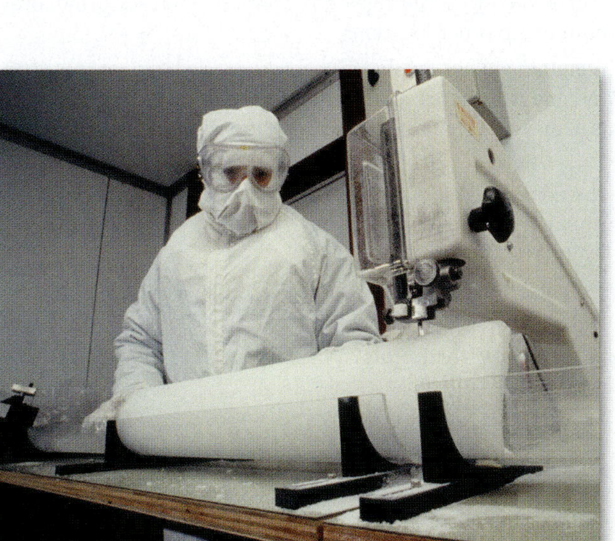

◀ **Figure 15.34 Ice cores contain clues to shifts in climate** This scientist is slicing an ice core from Antarctica for analysis. He is wearing protective clothing and a mask to minimize contamination of the sample. Chemical analyses of ice cores can provide important data about past climates. (Photo by British Antarctic Survey/Science Source)

Did You Know?
McMurdo Station is the main U.S. scientific research station on Antarctica. It is the largest installation on the continent, capable of supporting more than 1200 residents. The total population at all research stations on Antarctica is about 4000 in summer and 1000 in winter. There are no indigenous human residents on the continent.

Obviously, whenever Earth enters an ice age, extensive areas of land that were once ice free are covered with ice and snow. In addition, a colder climate causes the area covered by sea ice (frozen surface seawater) to expand. Ice and snow reflect a large portion of incoming solar energy back to space. Thus, energy that would have warmed Earth's surface and the air above is lost, and global cooling is reinforced.

Yet another factor that influences climate during glacial times relates to ocean currents. Research has shown that ocean circulation changes during ice ages. For example, studies suggest that the warm current that transports large amounts of heat from the tropics toward higher latitudes in the North Atlantic was significantly weaker during the Ice Age. This would lead to a colder climate in Europe, amplifying the cooling attributable to orbital variations.

In conclusion, we emphasize that the ideas just discussed do not represent the only possible explanations for ice ages. Although interesting and attractive, these proposals are certainly not without critics, nor are they the only possibilities currently under study. Other factors may be—and probably are—involved.

CONCEPT CHECKS 15.6

1. What was the best source of data showing Ice Age climate cycles?
2. About what percentage of Earth's land surface was affected by glaciers during the Quaternary period?
3. Where were ice sheets more extensive during the Ice Age: the Northern Hemisphere or the Southern Hemisphere? Why?
4. How does the theory of plate tectonics help us understand the causes of ice ages? Does this explain alternating glacial/interglacial climates during the Ice Age?
5. Briefly summarize the climate change hypothesis that involves variations in Earth's orbit.

CONCEPTS IN REVIEW
Glaciers & Glaciation

15.1 Glaciers: A Part of Two Basic Cycles

Explain the role of glaciers in the hydrologic and rock cycles and describe the different types of glaciers, their characteristics, and their present-day distribution.

KEY TERMS: glacier, valley glacier (alpine glacier), ice sheet, sea ice, ice shelf, ice cap, outlet glacier, piedmont glacier

- A glacier is a thick mass of ice originating on land from the compaction and recrystallization of snow, and it shows evidence of past or present flow. Glaciers are part of both the hydrologic cycle and the rock cycle because they store and release freshwater, and they grind away at rock, distributing the resulting sediment to other locations.
- Valley glaciers flow down mountain valleys, while ice sheets are very large masses of ice, such as those that cover Greenland and Antarctica. During the Last Glacial Maximum, around 18,000 years ago, Earth was in an Ice Age that covered large areas of the land surface with glacial ice.
- When valley glaciers leave the confining mountains, they may spread out into broad lobes called piedmont glaciers. Similarly, ice shelves form when glaciers flow into the ocean and spread out to form a wide layer of floating ice.
- Ice caps are like smaller ice sheets. Both ice sheets and ice caps may be drained by outlet glaciers, which often resemble valley glaciers and piedmont glaciers.

? What term is applied to the ice at the North Pole? What term best describes Greenland's ice? Are both considered glaciers? Explain.

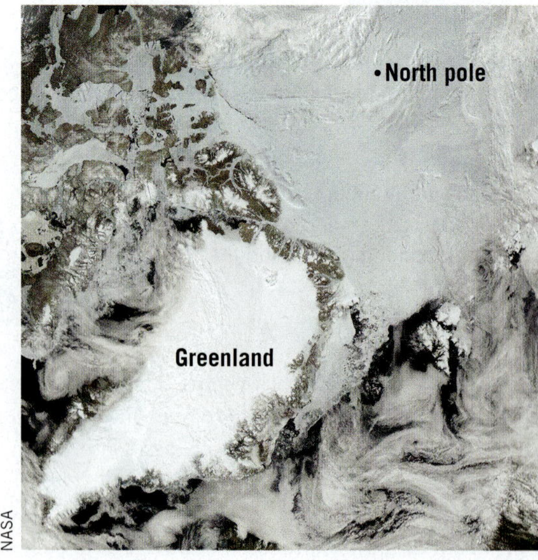

15.2 Formation & Movement of Glacial Ice

Describe how glaciers move, the rates at which they move, and the significance of the glacial budget.

KEY TERMS: firn, zone of fracture, crevasse, zone of accumulation, snowline, equilibrium line, zone of wastage, ablation, calving, iceberg, glacial budget

- When snow piles up sufficiently, it recrystallizes to dense granules of firn, which can then pack together even more tightly to become glacial ice.
- When ice is put under pressure, it flows very slowly. The uppermost 50 meters (165 feet) of a glacier are under too little pressure to flow, so they constitute the zone of fracture, breaking open into cracks called crevasses. Most glaciers also move by sliding along their bed (basal slip).
- Fast glaciers may move 800 meters (2600 feet) per year, while slow glaciers may move only 2 meters (6.5 feet) per year. Some glaciers experience periodic surges of sudden movement.
- A glacier's budget is the balance between gain of snow in the zone of accumulation and loss of ice (ablation) in the downstream zone of wastage. Glaciers waste away mainly through melting and iceberg calving. When ablation exceeds accumulation, the terminus retreats. Ice continues to flow downstream even when the terminus is retreating.

? This image shows that melting is one way that glacial ice wastes away. What is another way that ice is lost from a glacier? What is the general term for loss of ice from a glacier?

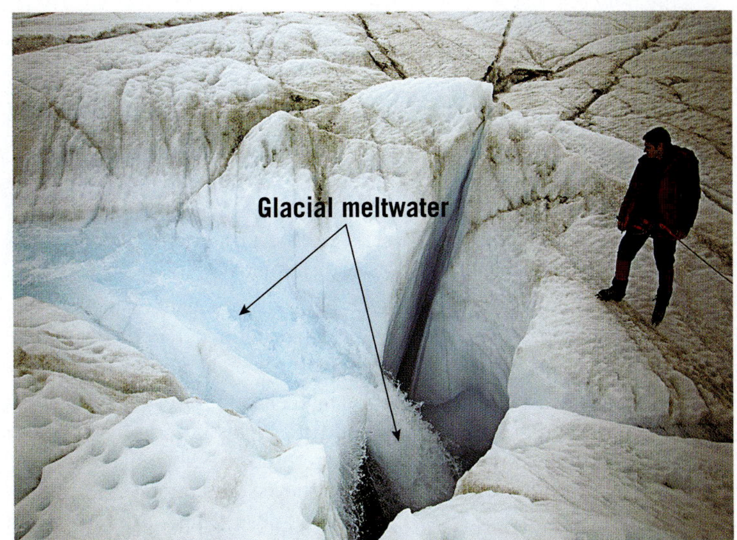

Robbie Shone/Photo Researchers, Inc

15.3 Glacial Erosion

Discuss the processes of glacial erosion. Identify and describe the major topographic features sculpted by glacial erosion.

KEY TERMS: plucking, abrasion, rock flour, glacial striation, glacial trough, hanging valley, cirque, tarn, arête, horn, roche moutonnée, fiord

- Glaciers acquire sediment through plucking from the bedrock beneath the glacier, by abrasion of the bedrock using sediment already in the ice, and by mass-movement processes dropping debris on top of the glacier. Grinding of the bedrock produces grooves and scratches called glacial striations.
- Erosional features produced by valley glaciers include glacial troughs, hanging valleys, cirques, arêtes, horns, and fiords.

? Examine the illustration of a mountainous landscape after glaciation. Identify the landforms that resulted from glacial erosion.

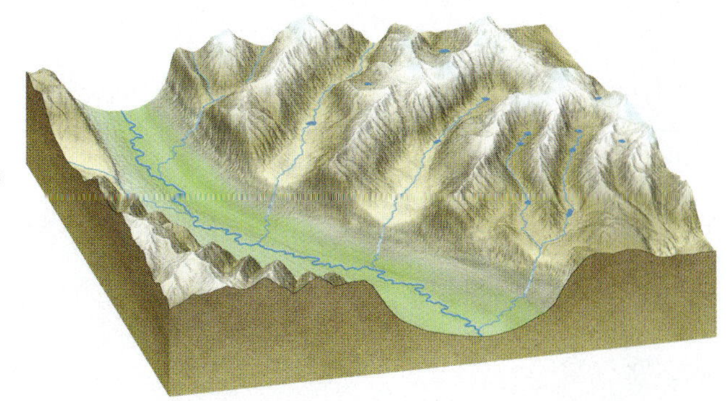

15.4 Glacial Deposits

Distinguish between the two basic types of glacial drift. List and describe the major depositional features associated with glacial landscapes.

KEY TERMS: glacial drift, till, glacial erratic, stratified drift, lateral moraine, medial moraine, end moraine, ground moraine, outwash plain, valley train, kettle, drumlin, esker, kame

- Any sediment of glacial origin is called drift. The two distinct types of glacial drift are till, which is unsorted material deposited directly by the ice, and stratified drift, which is sediment sorted and deposited by meltwater from a glacier.
- The most widespread features created by glacial deposition are layers or ridges of till, called moraines. Associated with valley glaciers are lateral moraines, formed along the sides of the valley, and medial moraines, formed between two valley glaciers that have merged. End moraines, which mark the former position of the front of a glacier, and ground moraines, undulating layers of till deposited as the ice front retreats, are common to both valley glaciers and ice sheets.

? Examine the illustration of depositional features formed in the wake of a retreating ice sheet. Name the features and indicate which landforms are composed of till and which are composed of stratified drift.

15.5 Other Effects of Ice Age Glaciers

Describe and explain several important effects of Ice Age glaciers other than erosional and depositional landforms.

KEY TERMS: proglacial lake, pluvial lake

- Ice sheets are so heavy that they can cause the crust to flex downward under their tremendous load. After the glaciers melt off, that weight is released, and the crust slowly rebounds vertically upward.
- Ice sheets are nourished by water that ultimately comes from the ocean, so when ice sheets grow, sea level falls, and when they melt, sea level rises. At the Last Glacial Maximum, global sea level was about 100 meters (330 feet) lower than it is today. At that time, the coastlines of the modern continents were vastly different than today's coastlines.
- The advance and retreat of ice sheets caused significant changes to the paths that rivers follow. In addition, glaciers deepened and widened stream valleys and lowlands to create features such as the Great Lakes.
- An ice sheet can act as a dam by trapping meltwater or blocking the flow of rivers to create proglacial lakes. Glacial Lakes Agassiz and Missoula both impounded tremendous quantities of water. In Lake Missoula's case, the water drained out in huge torrents when the ice dam periodically broke.
- Pluvial lakes existed during the height of the Ice Age but occurred far from the actual glaciers, in a climate that was cooler and wetter than today's. Lake Bonneville is a classic example that existed in the area that is now Utah and Nevada. A shrunken remnant of it survives as the Great Salt Lake.

15.6 The Ice Age

Briefly discuss the development of glacial theory and summarize current ideas on the causes of ice ages.

KEY TERMS: Quaternary period, tillite

- The idea of a geologically recent Ice Age was born in the early 1800s in Switzerland. By applying a uniformitarian approach to field study of glacial landforms, Louis Agassiz and others established that only the former presence of tremendous quantities of glacial ice could explain the landscape of Europe (and later North America and Siberia). As additional research accumulated, especially data from the study of seafloor sediments, it was revealed that the Quaternary period was marked by many advances and retreats of glacial ice.
- While rare, glacial episodes have occurred in Earth history prior to the recent glaciations we call the Ice Age. Lithified till, called tillite, is a major line of evidence for these ancient ice ages. There are several reasons that glacial ice might accumulate globally, including the position of the continents, which is driven by plate tectonics. Antarctica's position over the South Pole is doubtless a key reason for its massive ice sheets, for instance.
- The Quaternary period is marked by not only glacial advances but also intervening episodes of glacial retreat. One way to explain these oscillations is through variations in Earth's orbit, which lead to seasonal variations in the distribution of solar radiation. The orbit's shape varies (eccentricity), the tilt of the planet's rotational axis varies (obliquity), and the axis slowly "wobbles" over time (precession). These three effects, which occur on different time scales, collectively do a good job of accounting for alternating colder and warmer periods during the Quaternary.
- Additional factors that may be important for initiating or ending glaciations include rising or falling levels of greenhouse gases, changes in the reflectivity of Earth's surface, and variations in the ocean currents that redistribute heat energy from warmer to colder regions.

? About 250 million years ago, parts of India, Africa, and Australia were covered by ice sheets, while Greenland, Siberia, and Canada were ice free. Explain why this occurred.

GIVE IT SOME THOUGHT

1 The accompanying diagram shows the results of a classic experiment used to determine how glacial ice moves in a mountain valley. The experiment occurred over an 8-year span. Refer to the diagram and answer the following:

a. What was the average yearly rate at which ice in the center of the glacier advanced?

b. About how fast was the center of the glacier advancing *per day*?

c. Calculate the average rate at which ice along the sides of the glacier moved forward.

d. Why was the rate at the center different than along the sides?

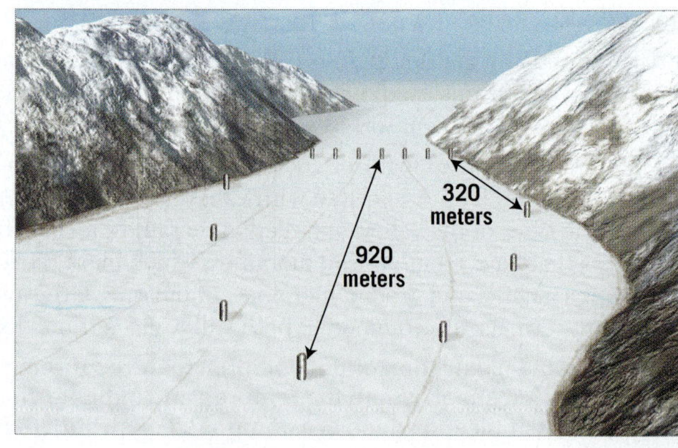

2 Studies have shown that during the Ice Age, the margins of some ice sheets advanced southward from the Hudson Bay region at rates ranging from about 50 to 320 meters per year.

a. Determine the maximum amount of time required for an ice sheet to move from the southern end of Hudson Bay to the south shore of present day Lake Erie, a distance of 1600 kilometers.

b. Calculate the minimum number of years required for an ice sheet to move this distance.

3 This iceberg is floating in the ocean near the coast of Greenland.

a. How do icebergs form? What term applies to this process?

b. Using the knowledge you have gained about these features, explain the common phrase "It's only the tip of the iceberg."

c. Is an iceberg the same as sea ice? Explain.

d. If this iceberg were to melt, how would sea level be affected?

Andrzej Gibasiewicz/Shutterstock

4 If Earth were to experience another Ice Age, one hemisphere would have substantially more expansive ice sheets than the other. Would it be the Northern Hemisphere or the Southern Hemisphere? What is the reason for the large disparity?

5 Is the glacial deposit shown here an example of till or stratified drift? Is it more likely part of an end moraine or an esker?

E. J. Tarbuck

6 While taking a break from a hike in the Northern Rockies with a fellow geology enthusiast, you notice that the boulder you are sitting on is part of a deposit that consists of a jumbled mixture of many different sediment sizes. Since you are in an area that once had extensive valley glaciers, your colleague suggests that the deposit must be glacial till. Although you know this is certainly a good possibility, you remind your companion that other processes in mountain areas also produce unsorted deposits. What might such a process be? How might you and your friend determine whether this deposit is actually glacial till?

7 This is a small portion of a topographic map of an area in upstate New York that was affected by ice sheets during the Quaternary period. The area is characterized by numerous hills composed of till. What term is applied to these hills? Did the ice advance from the top or from the bottom of the map area? Explain.

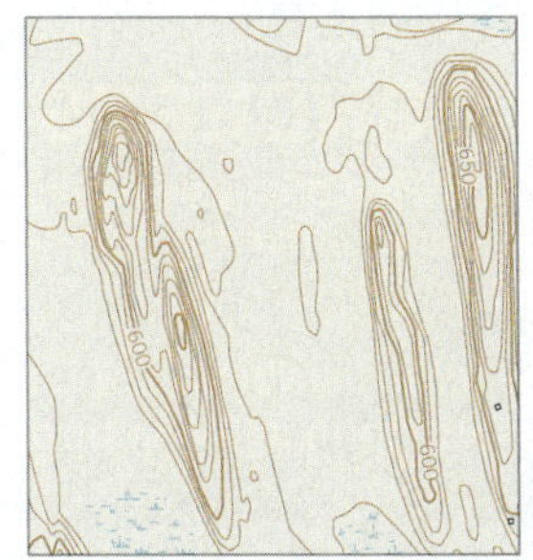

8 If the budget of a valley glacier were balanced for an extended time span, what feature would you expect to find at the terminus of the glacier? Is it composed of till or stratified drift? Now assume that the glacier's budget changes so that ablation exceeds accumulation. How would the terminus of the glacier change? Describe the deposit you would expect to form under these conditions.

9 Assume that you and a nongeologist friend are visiting Alaska's Hubbard Glacier, shown here. After studying the glacier for quite a long time, your friend asks, "Do these things really move?" How would you convince your friend that this glacier does indeed move, using evidence that is clearly visible in this image?

Michael Collier

MasteringGeology™

Looking for additional review and test prep materials? Visit the Study Area in MasteringGeology to enhance your understanding of this chapter's content by accessing a variety of resources, including Self-Study Quizzes, Geoscience Animations, SmartFigures, Mobile Field Trips, *Project Condor* Quadcopter videos, *In the News* RSS feeds, flashcards, web links, and an optional Pearson eText.

www.masteringgeology.com

16

Deserts & Wind

FOCUS ON CONCEPTS

Each statement represents the primary learning objective for the corresponding major heading within the chapter. After you complete the chapter, you should be able to:

16.1 Describe the general distribution of Earth's dry lands and explain why deserts form in the subtropics and middle latitudes.

16.2 Summarize the geologic roles of weathering, running water, and wind in arid and semiarid climates.

16.3 Discuss the stages of landscape evolution in the Basin and Range region of the western United States.

16.4 Describe the ways that wind transports sediment and the processes and features associated with wind erosion.

16.5 Discuss the formation and movement of dunes and distinguish among different dune types. Explain how loess deposits differ from deposits of sand.

When rains occur in deserts, they are often heavy and brief. Because rainfall intensity is high, not all of the water can soak in, and rapid runoff and erosion result. (Photo by Thomas Weber / Alamy Stock Photo)

CLIMATE HAS A STRONG INFLUENCE ON THE NATURE and intensity of Earth's external processes. This was clearly demonstrated in the preceding chapter on glaciers. Desert landscapes and their development provide another excellent example of the strong link between climate and geology. As you will see, arid regions are not dominated by a single geologic process. Rather, the effects of tectonic forces, running water, and wind are all apparent. Because these processes combine in different ways from place to place, the appearance of desert landscapes varies a great deal as well.

Did You Know?
The dry regions of the world encompass about 16.4 million mi^2, a surprising 30 percent of Earth's land surface. No other climate group covers so large a land area.

16.1 Distribution & Causes of Dry Lands

Describe the general distribution of Earth's dry lands and explain why deserts form in the subtropics and middle latitudes.

Desert landscapes frequently appear stark. Their profiles are not softened by a continuous carpet of soil and abundant plant life. Instead, barren rocky outcrops with steep, angular slopes are common. At some places the rocks are tinted orange and red. At others they are gray and brown and streaked with black. For many visitors, desert scenery exhibits a striking beauty; to others, the terrain seems bleak. No matter which feeling is elicited, it is clear that deserts are very different from the more humid places where most people live.

What Is Meant by *Dry*?

We all recognize that deserts are dry places, but just what is meant by the term *dry*? That is, how much rain defines the boundary between humid and dry regions? Sometimes it is arbitrarily defined by a single rainfall figure, such as 25 centimeters (10 inches) per year of precipitation. However, the concept of dryness is a relative one that refers to any situation in which an ongoing water deficiency exists.

Climatologists define a **dry climate** as a climate in which yearly precipitation is not as great as the potential loss of water by evaporation. Therefore, dryness is not only related to annual rainfall totals but is also a function of evaporation, which in turn is dependent upon temperature. As temperatures climb, potential evaporation also increases. As little as 15 to 25 centimeters (6 to 10 inches) of precipitation per year may be sufficient to support coniferous forests in northern Scandinavia or Siberia, where evaporation into the cool, humid air is slight and a surplus of water remains in the soil. However, the same amount of rain falling on New Mexico or Iran supports only a sparse vegetative cover because evaporation into the hot, dry air is great. So, clearly, no specific amount of precipitation can serve as a universal boundary for dry climates.

Within water-deficient regions, two climatic types are commonly recognized: **desert**, or arid, and **steppe**, or semiarid. The two share many features; their differences are primarily a matter of degree. The steppe is a marginal and more humid variant of the desert and is a transition zone that surrounds the desert and separates it from bordering humid climates. The world map of the distribution of desert and steppe regions shows that dry lands are concentrated in the subtropics and in the middle latitudes (**Figure 16.1**).

Did You Know?
Most deserts are not lifeless and barren places. Although reduced in amount and different in character, both plant and animal life are usually present. All forms of desert life have developed adaptations that make them highly tolerant of drought. Plants are widely dispersed and provide little ground cover, but many kinds flourish in the desert.

Subtropical Deserts & Steppes

The heart of the subtropical dry climates lies in the vicinities of the Tropics of Cancer and Capricorn. Figure 16.1 shows a virtually unbroken desert environment stretching for more than 9300 kilometers (5800 miles) from the Atlantic coast of North Africa to the dry lands of northwestern India. In addition to this single great expanse, the Northern Hemisphere contains another, much smaller area of subtropical desert and steppe in northern Mexico and the southwestern United States.

In the Southern Hemisphere, dry climates dominate Australia. Almost 40 percent of the continent is desert, and much of the remainder is steppe. In addition, arid and semiarid areas occur in southern Africa and make a limited appearance in coastal Chile and Peru.

What causes these bands of low-latitude desert? The answer is the global distribution of air pressure and winds. **Figure 16.2A**, an idealized diagram of Earth's atmospheric circulation, helps visualize the relationship. Heated air in the pressure belt known as the *equatorial low* rises to great heights (usually between 15 and 20 kilometers) and then spreads out. As the upper-level flow reaches 20° to 30° latitude, north or south, it sinks toward the surface. Air that rises through the atmosphere expands and cools, a process that leads to the development of clouds and precipitation. For this reason, the areas under the influence of the equatorial low are among the rainiest on Earth. Just the opposite is true for the regions in the vicinity of 30° north and south latitude, where high pressure predominates. In these zones, known as the *subtropical highs*, air is subsiding. When air sinks, it is compressed and warmed. Such conditions are just opposite of what is needed to

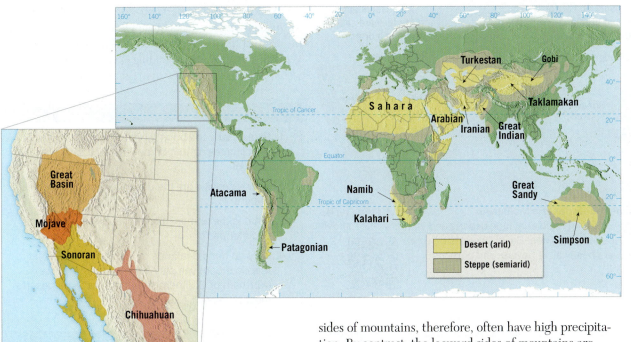

◀ **SmartFigure 16.1 Dry climates** Arid and semiarid climates cover about 30 percent of Earth's land surface. The dry region of the American West is commonly divided into four deserts, two of which extend into Mexico.

TUTORIAL
https://goo.gl/mCl6C8

produce clouds and precipitation. Consequently, these regions are known for their clear skies, sunshine, and ongoing dryness (**Figure 16.2B**).

Middle-Latitude Deserts & Steppes

Unlike their low-latitude counterparts, middle-latitude deserts and steppes are not controlled by the subsiding air masses associated with zones of high pressure. Instead, many of these dry lands exist because they are sheltered in the deep interiors of large landmasses. They are far removed from the ocean, which is the ultimate source of moisture for cloud formation and precipitation. One well-known example is the Gobi Desert of central Asia (see Figure 16.1).

The presence of high mountains across the paths of prevailing winds is another factor that separates middle-latitude arid and semiarid areas from water-bearing, maritime air masses. Mountains force the air to lose much of its water. The mechanism is straightforward: As prevailing winds meet mountain barriers, air is forced to ascend. When air rises, it expands and cools, a process that can produce clouds and precipitation. The windward sides of mountains, therefore, often have high precipitation. By contrast, the leeward sides of mountains are usually much drier (**Figure 16.3**). This situation exists because air reaching the leeward side has lost much of its moisture, and if the air descends, it is compressed and warmed, making cloud formation even less likely. The dry region that results is referred to as a **rainshadow**.

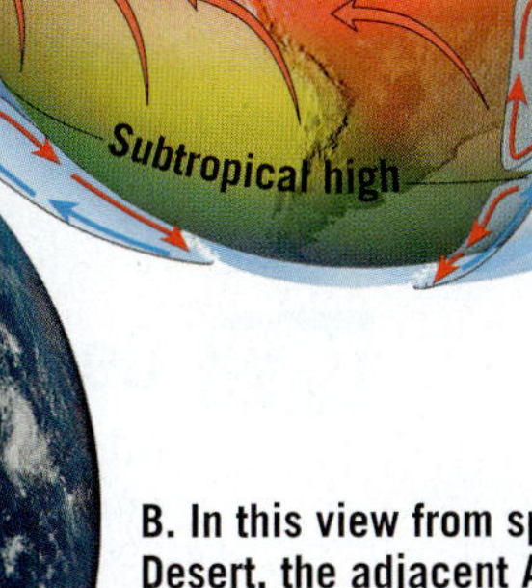

A. Subtropical deserts and steppes are centered between 20° and 30° north and south latitude in association with the subtropical high-pressure belts. Dry subsiding air inhibits cloud formation and precipitation.

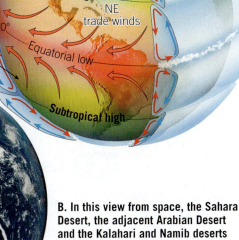

B. In this view from space, the Sahara Desert, the adjacent Arabian Desert and the Kalahari and Namib deserts are clearly visible as tan-colored, cloud-free zones. The band of clouds across central Africa and the adjacent oceans coincides with the equatorial low-pressure belt.

▶ **SmartFigure 16.2 Subtropical deserts** The distribution of subtropical deserts and steppes is closely related to the global distribution of air pressure. (NASA)

ANIMATION
https://goo.gl/a9ZZIT

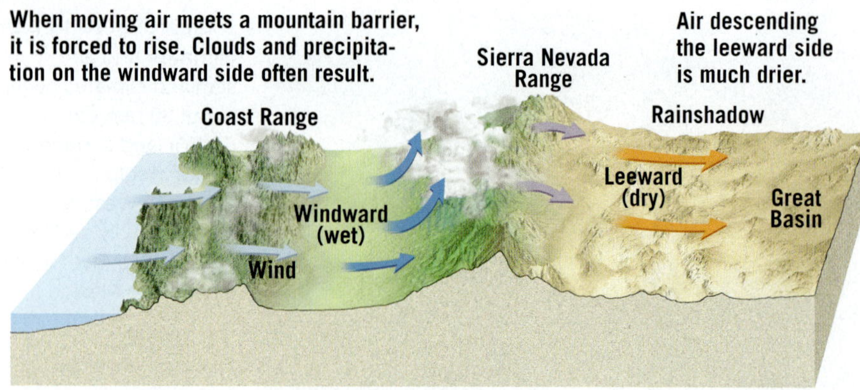
When moving air meets a mountain barrier, it is forced to rise. Clouds and precipitation on the windward side often result.
Sierra Nevada Range
Air descending the leeward side is much drier.
Coast Range
Rainshadow
Leeward (dry)
Great Basin
Windward (wet)
Wind

▲ **SmartFigure 16.3 Rainshadow deserts** Mountains frequently contribute to the aridity of middle-latitude deserts and steppes by creating a rainshadow. The Great Basin Desert is a rainshadow desert that covers nearly all of Nevada and portions of adjacent states.

ANIMATION
https://goo.gl/4ak8IN

Middle-latitude deserts provide an example of how tectonic processes affect climate. Rainshadow deserts exist by virtue of the mountains produced when plates collide. Without such mountain-building episodes, wetter climates would prevail where many dry regions exist today.

CONCEPT CHECKS 16.1

1. Explain why the boundary between humid and dry climates cannot be defined by a single rainfall amount.
2. How extensive are the desert and steppe regions of Earth?
3. What is the primary cause of subtropical deserts and steppes?
4. Why do middle-latitude dry regions exist? What role do mountains play? In which hemisphere (Northern or Southern) are middle-latitude deserts most common? Explain.

Did You Know?
The Sahara of North Africa is the world's largest desert. Extending from the Atlantic Ocean to the Red Sea, it covers about 3.5 million mi^2, an area about the size of the United States. By comparison, the largest desert in the United States, Nevada's Great Basin Desert, has an area that is less than 5 percent of the Sahara.

Figure 16.4, which shows the distribution of precipitation in western Washington State, is a good example. When prevailing winds from the Pacific Ocean to the west (left) meet the mountains, rainfall totals are high. By comparison, precipitation on the leeward (eastern) side of the mountains is relatively meager. In North America, the foremost mountain barriers to moisture from the Pacific are the Coast Ranges, Sierra Nevada, and Cascades. In Asia, the great Himalayan chain prevents the summertime monsoon flow of moist Indian Ocean air from reaching the interior of the continent.

Because the Southern Hemisphere lacks extensive land areas in the middle latitudes, only small areas of desert and steppe occur in this latitude range, existing primarily near the southern tip of South America in the rainshadow of the towering Andes.

ANNUAL PRECIPITATION

cm	in.
>500	>200
405–499	160–199
250–404	100–159
150–249	60–99
50–149	20–59
25–49	10–19
<25	<10

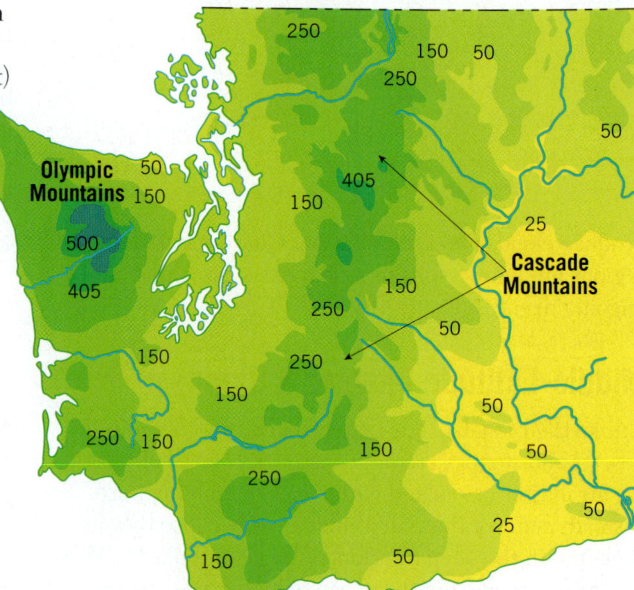
Olympic Mountains
Cascade Mountains
250
150
50
405
500
25

▲ **Figure 16.4 Distribution of precipitation in western Washington State** The Olympic and Cascade Mountains receive abundant rainfall. The semiarid eastern portion of the area shown here is in a rainshadow.

16.2 Geologic Processes in Arid Climates

Summarize the geologic roles of weathering, running water, and wind in arid and semiarid climates.

The angular hills, the sheer canyon walls, and the desert surface of pebbles or sand contrast sharply with the rounded hills and curving slopes of more humid places. To a visitor from a humid region, a desert landscape may seem to have been shaped by forces different from those operating in well-watered areas. However, although the contrasts may be striking, they do not reflect different processes. They merely disclose the differing effects of the same processes that operate under contrasting climatic conditions.

Dry-Region Weathering

Recall from Chapter 6 that water plays an important role in chemical weathering. Consequently, chemical weathering processes are not as prominent in dry regions as in humid regions. In humid regions, relatively fine-textured soils support an almost continuous cover of vegetation. The slopes and rock edges are rounded, reflecting the strong influence of chemical weathering in a humid climate. By contrast, much of the weathered debris in deserts consists of unaltered rock and mineral fragments—the result of mechanical weathering processes. In dry lands, rock weathering of any type is greatly reduced because of the lack of moisture and the scarcity of organic acids from decaying plants. However, chemical weathering is not completely lacking in deserts. Over long spans of time, clays and thin soils do form, and many iron-bearing silicate minerals oxidize, producing the rust-colored stain that tints some desert landscapes.

The Role of Water

Deserts have scant precipitation and few major rivers. Nevertheless, water plays an important role in shaping landscapes in dry regions. Permanent streams are normal in humid regions, but practically all desert streambeds are dry most of the time. Deserts have intermittent or **ephemeral streams** (*ephemero* = short-lived), which means they carry water only in response to specific episodes of rainfall (**Figure 16.5**).

A typical ephemeral stream might flow only a few days or perhaps just a few hours during the year, following sporadic rains. In some years the channel might carry no water at all. This fact is obvious even to the casual traveler who notices numerous bridges with no streams beneath them or numerous dips in the road where dry channels cross. When the rare heavy showers do come, as in the chapter-opening photo, so much rain falls in such a short time that all of it cannot soak in. Because desert vegetative cover is sparse, runoff is largely unhindered and consequently rapid, often creating flash floods along valley floors. These floods are quite unlike floods in humid regions. A flood on a river like the Mississippi may take several days to reach its crest and then subside. But desert floods arrive suddenly and subside quickly. Because much surface material in a desert is not anchored by vegetation, the amount of erosional work that occurs during a single short-lived rain event is impressive.

The wadi in its usual dry state.

Following a rainy period, freshly sprouted vegetation turns the wadi green.

▲ **Figure 16.6 Wadi in North Africa** These two satellite images show how rain transformed a wadi in Niger. (NASA)

In the dry western United States, different names are used for ephemeral streams, including *wash* and *arroyo*. In other parts of the world, a dry desert stream may be a *wadi* (Arabian Peninsula and North Africa), a *donga* (South America), or a *nullah* (India). The satellite images in **Figure 16.6** show a wadi in the Sahara Desert.

Humid regions are notable for their integrated drainage systems. But in arid regions, streams usually lack an extensive system of tributaries. In fact, a basic characteristic of desert streams is that they are small and die out before reaching the sea. Because the water table is usually far below the surface, few desert streams can draw upon it as streams do in humid regions (see Figure 14.7, page 374). Without a steady supply of water, the combination of evaporation and infiltration soon depletes the stream.

The few permanent streams that do cross arid regions, such as the

Most of the time desert stream channels are dry.
An ephemeral stream shortly after a heavy shower. Although such floods are short-lived, they cause large amounts of erosion.
A familiar sign in desert areas. Roads dip into washes which can rapidly fill with water following a heavy rain.
POTENTIAL FLASH FLOOD AREAS
NEXT 21 MILES

▲ **Figure 16.5 Ephemeral stream** This example is near Arches National Park in southern Utah. (Photos by E. J. Tarbuck)

Did You Know?
Some of the highest temperatures ever recorded have been in deserts. The highest authentically recorded temperature in the United States is also the extreme for the world. On July 10, 1913, the temperature in California's Death Valley reached 134°F.

Did You Know?
Not all deserts are hot. Cold temperatures are experienced in middle-latitude deserts. For example, at Ulaanbaatar in Mongolia's Gobi Desert, the average *high* temperature on January days is only −2°F!

Colorado and Nile Rivers, originate *outside* the desert, often in well-watered mountains. Here the water supply must be great, or the stream will lose all its water as it crosses the desert. For example, after the Nile leaves its headwaters in the lakes and mountains of central Africa, it traverses almost 3000 kilometers (1900 miles) of the Sahara without a single tributary. By contrast, in humid regions the discharge of a river grows as it flows downstream because tributaries and groundwater contribute additional water along the way.

It should be emphasized that, although annual rainfall totals are low and rain events are infrequent, *running water does most of the erosional work in deserts.* This is contrary to the common misconception that wind is the most important erosional agent sculpting desert landscapes. Although wind erosion is indeed more significant in dry areas than elsewhere, most desert landforms are carved by running water. As you will see later in this chapter, in Section 16.4, the main role of wind is in the transportation and deposition of sediment, which creates and shapes the ridges and mounds we call dunes.

CONCEPT CHECKS 16.2

1. How does the rate of rock weathering in dry climates compare to the rate in humid regions?
2. What is an ephemeral stream?
3. When a permanent stream such as the Nile River crosses a desert, does the river's discharge increase or decrease? How does this compare to a river in a humid area?
4. What is the most important erosional agent in deserts?

Did You Know?
Not all low-latitude deserts are hot, sunny places with low humidity and cloudless skies. In west coast subtropical deserts, such as the Atacama and Namib, cold ocean currents are responsible for cool temperatures, frequent fogs, and gloomy periods of dense (but rainless) low clouds.

16.3 Basin & Range: The Evolution of a Desert Landscape

Discuss the stages of landscape evolution in the Basin and Range region of the western United States.

As discussed earlier, arid regions typically lack permanent streams and often have **interior drainage**. This means that they exhibit a discontinuous pattern of intermittent streams that do not flow out of the desert to the ocean. In the United States, the dry Basin and Range region provides an excellent example. The region includes southern Oregon, all of Nevada, western Utah, southeastern California, southern Arizona, and southern New Mexico. The name *Basin and Range* is an apt description for this almost 800,000-square-kilometer (300,000-square-mile) region because it is characterized by more than 200 relatively small mountain ranges that rise 900 to 1500 meters (3000 to 5000 feet) above the basins that separate them. The origin of these fault-block mountains is examined in Chapter 11. In this discussion we look at how surface processes change the landscape.

In this region, as in others like it around the world, erosion mostly occurs without reference to the ocean (ultimate base level) because the interior drainage never reaches the sea. Even where permanent streams flow to the ocean, few tributaries exist, and thus only a narrow strip of land adjacent to the stream has sea level as its ultimate level of land reduction.

The block diagrams in Figure 16.7 depict how the landscape has evolved in the Basin and Range region and illustrate the landforms described in the following paragraphs. During and following uplift of the mountains, mass wasting and running water carve the elevated masses and deposit large quantities of sediment in adjacent basins. Relief is greatest during the early stage because elevation differences gradually diminish as the mountains are lowered and sediment fills the basins.

When the occasional torrents of water produced by sporadic rains move down the mountain canyons, they are heavily loaded with sediment. Emerging from the confines of the canyon, the runoff spreads over the gentler slopes at the base of the mountains and quickly loses velocity. Consequently, most of its load is dumped within a short distance. The result is a cone of debris at the mouth of a canyon known as an **alluvial fan**. Because the coarsest (heaviest) material is dropped first, the head of the fan is steepest, having a slope of perhaps 10 to 15 degrees. Moving down the fan, the size of the sediment and the steepness of the slope decrease, and the fan merges imperceptibly with the basin floor. An examination of the fan's surface would likely reveal a braided channel pattern because of the water shifting its course as successive channels became choked with sediment. Over the years a fan enlarges, eventually coalescing with fans from adjacent canyons to produce an apron of sediment called a **bajada** along the mountain front.

On the rare occasions of abundant rainfall, streams may flow across the bajada to the center of the basin, converting the basin floor into a shallow **playa lake**. Playa lakes are temporary features that last only a few days or at best a few weeks before evaporation and infiltration remove the water. The dry, flat lake bed that remains is called a **playa**. Playas are typically composed of fine silts and clays and are occasionally encrusted with salts precipitated during evaporation (see Figure 7.17, page 196). These precipitated salts may be unusual. A case in point is the sodium borate (better known as borax) mined from ancient playa lake deposits in Death Valley, California.

With the ongoing erosion of the mountain mass and the accompanying sedimentation, the local relief continues to diminish. Eventually, nearly the entire mountain mass is gone. Thus, by the late stages of erosion, the

▶ **SmartFigure 16.7 Landscape evolution in the Basin and Range region** As erosion of the mountains and deposition in the basins continue, relief diminishes.

TUTORIAL
https://goo.gl/FNO7OS

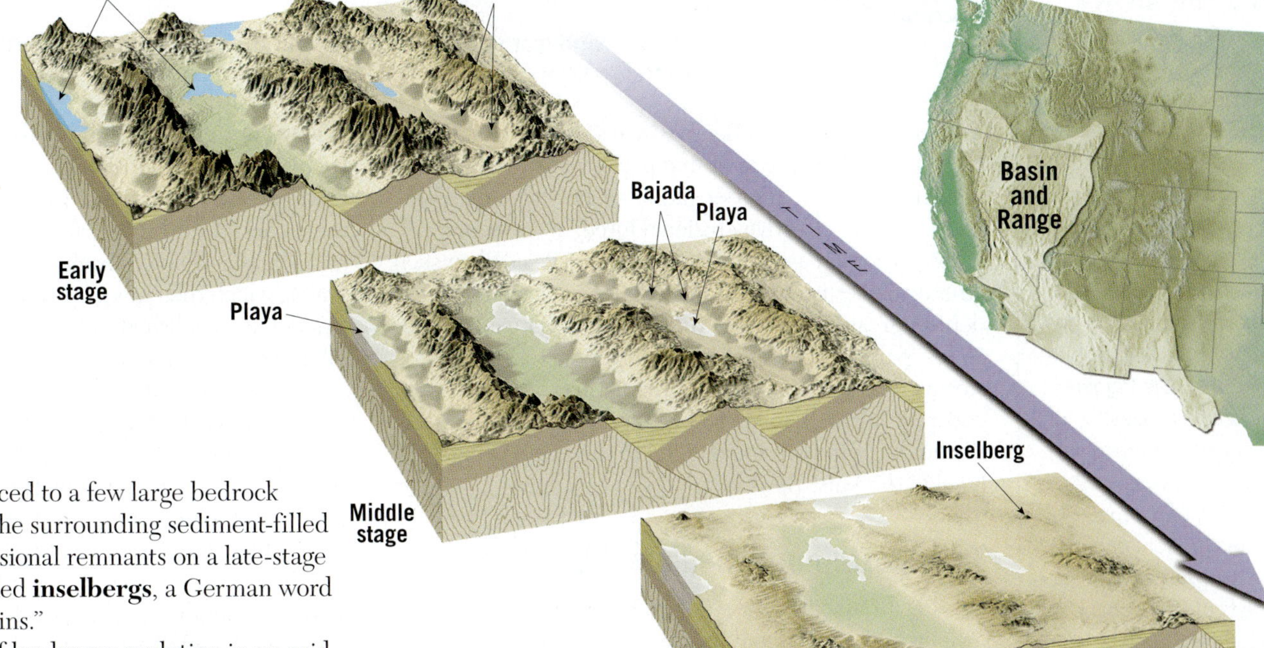

mountain areas are reduced to a few large bedrock knobs projecting above the surrounding sediment-filled basin. These isolated erosional remnants on a late-stage desert landscape are called **inselbergs**, a German word meaning "island mountains."

Each of the stages of landscape evolution in an arid climate depicted in Figure 16.7 can be observed in the Basin and Range region. Recently uplifted mountains in an early stage of erosion are found in southern Oregon and northern Nevada. Death Valley, California, and southern Nevada fit into the more advanced middle stage, whereas the late stage, with its inselbergs, can be seen in southern Arizona.

Figure 16.8 shows images of Death Valley, in which many of the features just described are visible. The February 2005 satellite image shows the area shortly after a rare heavy rain. As has occurred many times over thousands of years, a wide, shallow playa lake formed in the lowest spot. By May 2005, only 3 months after the storm, the valley floor had returned to being a dry, salt-encrusted playa.

CONCEPT CHECKS 16.3

1. What is meant by *interior drainage*?
2. Describe the features and characteristics associated with each stage in the evolution of a mountainous desert.
3. Where in the United States can each stage of desert landscape evolution be observed?

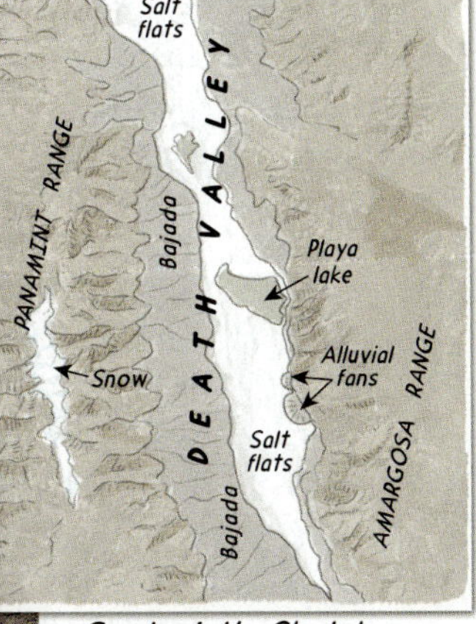

▶ **SmartFigure 16.8 Death Valley: A classic Basin and Range landscape** Shortly before the satellite image (left) was taken in February 2005, heavy rains led to the formation of a small playa lake—the pool of greenish water on the basin floor. By May 2005, the lake had reverted to a salt-covered playa. (NASA) The small photo is a closer view of one of Death Valley's many alluvial fans. The Condor Video associated with this figure takes a closer look at alluvial fans. (Photo by Michael Collier)

CONDOR VIDEO
https://goo.gl/j5dALU

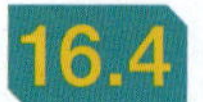

Did You Know?
In some regions deserts are expanding. The problem is called *desertification*, a term that refers to the alteration of land to desertlike conditions as a result of human activities. It commonly takes place on the margins of deserts and results primarily from inappropriate land use. It is triggered when the modest natural vegetation in marginal areas is removed by plowing or overgrazing. When drought occurs, as it inevitably does, and the vegetative cover has been destroyed beyond the minimum to hold the soil against erosion, the destruction becomes irreversible. Desertification is occurring in many places but is particularly serious in the region south of the Sahara Desert known as the Sahel.

16.4 Wind Erosion

Describe the ways that wind transports sediment and the processes and features associated with wind erosion.

Wind erosion is more effective in arid lands than in humid areas because in humid places moisture binds particles together, and vegetation anchors the soil. For wind to be an effective erosional force, dryness and scant vegetation are important prerequisites. When such circumstances exist, wind may pick up, transport, and deposit great quantities of fine sediment. During the 1930s, parts of the Great Plains experienced vast dust storms. The plowing under of the natural vegetative cover for farming, followed by severe drought, exposed the land to wind erosion and led to the area's being labeled the Dust Bowl.

Transportation of Sediment by Wind

Moving air, like moving water, is turbulent and able to pick up loose debris and transport it to other locations. Just as in a stream, the velocity of wind increases with height above the surface. Also as with a stream, wind transports fine particles in suspension, while heavier ones are carried as bed load. However, the transport of sediment by wind differs from that of running water in two significant ways. First, wind's lower density compared to water renders it less capable of picking up and transporting coarse materials. Second, because wind is not confined to channels, it can spread sediment over large areas, as well as high into the atmosphere.

Bed Load The **bed load** that wind carries consists of sand grains. Observations in the field and experiments using wind tunnels indicate that windblown sand moves mainly by skipping and bouncing along the surface—a process termed **saltation** (Latin for "to jump").

The movement of sand grains begins when wind reaches a velocity sufficient to overcome the inertia of the resting particles. At first the sand rolls along the surface. When a moving sand grain strikes another grain, one or both of them may jump into the air. Once in the air, the grains are carried forward by the wind until gravity pulls them back toward the surface. When the sand hits the surface, it either bounces back into the air or dislodges other grains, which then jump upward. In this manner, a chain reaction is established, filling the air near the ground with saltating sand grains in a short period of time (Figure 16.9).

Bouncing sand grains never travel far from the surface. Even when winds are very strong, the height of the saltating sand seldom exceeds 1 meter (3 feet) and usually is no greater than 0.5 meter (1.5 feet). Some sand grains are too large to be thrown into the air by impact from other particles. When this is the case, the energy provided by the impact of the smaller saltating grains drives the larger grains forward. Estimates indicate that between 20 and 25 percent of the sand transported in a sandstorm is moved in this way.

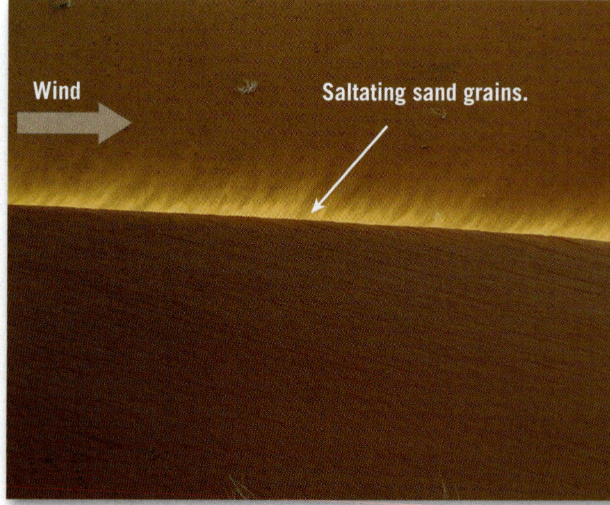

▶ SmartFigure 16.9
Transporting sand The bed load that wind carries consists of sand grains that move by bouncing along the surface. Sand never travels far from the surface, even when winds are very strong. (Photo by Bernd Zoller/Photolibrary)

ANIMATION
https://goo.gl/wnc9t6

Suspended Load Unlike sand, finer particles of dust can be swept high into the atmosphere by the wind. Because dust is often composed of rather flat particles that have large surface areas compared to their weight, it is relatively easy for turbulent air to counterbalance the pull of gravity and keep these fine particles airborne for hours or even days. Although both silt and clay can be carried in suspension, silt commonly makes up the bulk of the **suspended load** because the reduced level of chemical weathering in deserts provides only small amounts of clay.

Fine particles are easily carried by the wind, but they are not so easily picked up to begin with. This is because wind velocity is practically zero within a very thin layer close to the ground. Thus, the wind cannot lift the sediment by itself. Instead, the dust must be ejected or spattered into the moving air by bouncing sand grains or other disturbances. This idea is illustrated nicely by a dry, unpaved country road on a windy day. When the road is undisturbed, the wind raises little dust. However, as a car or truck moves over the road, the layer of silt is kicked up, creating a thick cloud of dust.

Although the suspended load is usually deposited relatively near its source, high winds are capable of carrying large quantities of dust great distances (Figure 16.10). In the 1930s, silt that was picked up in Kansas was transported to New England and beyond, into the North Atlantic. Similarly, dust blown from the Sahara has been traced as far as the Caribbean.

Erosional Features

Compared to running water and glaciers, wind is a relatively modest force in sculpting landforms. Recall that even in deserts, most erosion is performed by intermittent running

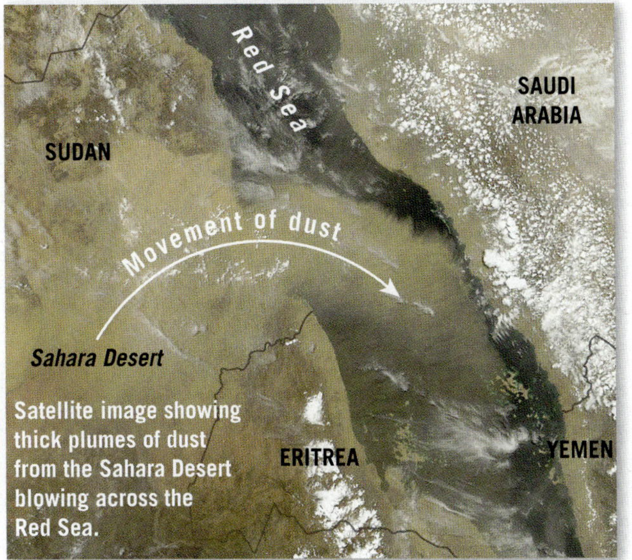

A dust storm blackens the Colorado sky in this historic image from the Dust Bowl of the 1930's.

◀ **SmartFigure 16.10 Wind's suspended load** These are two dramatic examples of suspended load. Dust storms can cover huge areas, and dust can be transported great distances. (Black-and-white image courtesy of U.S.D.A./Natural Resources Conservation Service; satellite image courtesy of NASA)

VIDEO
https://goo.gl/P4bXft

water, not by wind. Nevertheless, features created by wind erosion are significant elements of some landscapes.

Deflation & Blowouts One way that wind erodes is by **deflation** (*de* = out, *flat* = blow), the lifting and removal of loose material. Deflation sometimes is difficult to notice because the entire surface is being lowered at the same time, but it can be significant. In portions of the 1930s Dust Bowl, vast areas of land were lowered by as much as 1 meter in only a few years.

The most noticeable results of deflation in some places are shallow depressions appropriately called **blowouts** (Figure 16.11). In the Great Plains region, from Texas north to Montana, thousands of blowouts are visible on the landscape. They range from small dimples less than 1 meter deep and 3 meters wide to depressions that approach 50 meters deep and several kilometers across. The factor that controls the depths of these basins (that is, acts as base level) is the local water table. When blowouts are lowered to the water table, damp ground and vegetation prevent further deflation.

Did You Know?
The driest place on Earth is in Chile's Atacama Desert, a narrow belt of dry land that extends along the Pacific Coast of South America. The town of Arica has an average rainfall of just 0.03 in. per year. Over a span of 59 years, this region received a *total* of less than 2 in. of rain.

Blowout
In this example, deflation has removed about 1.2 meters (4 feet) of soil–the distance from the man's outstretched arm to his feet.

▲ **Figure 16.11 Blowouts** Deflation is especially effective in creating these depressions when the land is dry and largely unprotected by anchoring vegetation. (Photo courtesy of U.S.D.A./Natural Resources Conservation Service)

Desert Pavement In portions of many deserts, the surface consists of a closely packed layer of coarse particles. This veneer of pebbles and cobbles, called **desert pavement**, is only one or two stones thick (Figure 16.12). Beneath is a layer containing a significant proportion of silt and sand. When desert pavement is present, it is an important control on wind erosion because pavement stones are too large for deflation to remove. When this armor is disturbed, wind can easily erode the exposed fine silt.

◀ **Figure 16.12 Desert pavement** This closely packed veneer of pebbles and cobbles is only one or two stones thick. It is underlain by material containing a significant proportion of finer particles. (Photo by Michael Collier)

For many years, the most common explanation for the formation of desert pavement was that it develops when wind removes sand and silt from poorly sorted surface deposits. As **Figure 16.13A** illustrates, the concentration of larger particles at the surface gradually increases as the finer particles are blown away. Eventually the surface is completely covered with pebbles and cobbles too large to be moved by the wind.

Studies have shown that the process depicted in Figure 16.13A is not an adequate explanation for all environments in which desert pavement exists. For example, in many places, desert pavement is underlain by a relatively thick layer of silt that contains few if any pebbles and cobbles. In such a setting, deflation of fine sediment could not leave behind a layer of coarse particles. Geologists have also found that in some areas, the pebbles and cobbles composing desert pavement have all been exposed at the surface for about the same length of time. This would not be the case for the process shown in Figure 16.13A. Here, the coarse particles that make up the pavement reach the surface over an extended time span, as deflation gradually removes the fine material.

As a result, an alternate explanation for desert pavement was formulated (**Figure 16.13B**). This hypothesis suggests that pavement develops on a surface that initially consists of coarse particles. Over time, protruding cobbles trap fine, windblown grains that settle and sift downward through the spaces between the larger surface stones. The process is aided by infiltrating rainwater. In this model, the cobbles composing the pavement were never buried. Moreover, it successfully explains the lack of coarse particles beneath the desert pavement.

▼ **SmartFigure 16.13 Formation of desert pavement A.** This model shows an area with poorly sorted surface deposits. Over time, deflation lowers the surface, and coarse particles become concentrated. **B.** In this model, the surface is initially covered with cobbles and pebbles. Over time, windblown dust accumulates at the surface and gradually sifts downward.

TUTORIAL
https://goo.gl/Q9CZw8

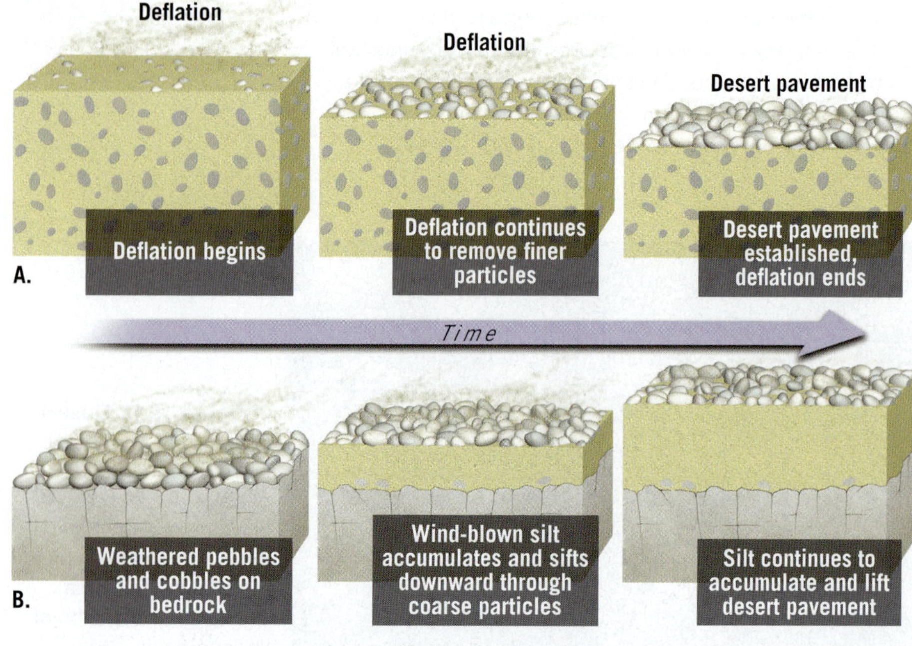

Ventifacts & Yardangs Like glaciers and streams, wind also erodes by **abrasion** (*ab* = away, *radere* = to scrape). In dry regions as well as along some beaches, windblown sand cuts and polishes exposed rock surfaces. Abrasion sometimes creates interestingly shaped stones called **ventifacts** (**Figure 16.14A**). The side of such a stone that is exposed to the prevailing wind is abraded, leaving it polished, pitted, and with sharp edges. If the wind is not consistently from one direction or if the pebble becomes reoriented, the stone may have several faceted surfaces.

Unfortunately, abrasion often gets credit for accomplishments beyond its capabilities. Such features as balanced rocks that stand high atop narrow pedestals and intricate detailing on tall pinnacles are not the results of abrasion. Sand seldom travels more than 1 meter above the surface, so the wind's sandblasting effect is obviously limited in vertical extent.

▲ **Figure 16.14 Shaped by the wind** The sandblasting effect of the wind creates ventifacts (**A**) and yardangs (**B**). (Photo A by Richard M. Busch; photo B by Mike P. Shepherd/Alamy Images)

In addition to creating ventifacts, wind erosion is responsible for creating much larger features, called yardangs (from the Turkistani word *yar*, meaning "steep bank"). A **yardang** is a streamlined, wind-sculpted landform that is oriented parallel to the prevailing wind (Figure 16.14B). Individual yardangs are generally small features that stand less than 5 meters (16 feet) high and no more than about 10 meters (32 feet) long. Because the sand-blasting effect of wind is greatest near the ground, these abraded bedrock remnants are usually narrower at their base. Sometimes yardangs are large features. Peru's Inca Valley contains yardangs that approach 100 meters (330 feet) in height and several kilometers in length. Some yardangs in the desert of Iran reach 150 meters (nearly 500 feet) in height.

CONCEPT CHECKS 16.4

1. Describe the way in which wind transports sand. When winds are strong, how high above the surface can sand be carried?
2. How does wind's suspended load differ from its bed load?
3. Why is wind erosion relatively more important in dry regions than in humid areas?
4. What factor limits the depths of blowouts?
5. Briefly describe two hypotheses used to explain the formation of desert pavement.
6. What are yardangs and ventifacts?

16.5 Wind Deposits

Discuss the formation and movement of dunes and distinguish among different dune types. Explain how loess deposits differ from deposits of sand.

Although wind is relatively unimportant in producing *erosional* landforms, significant *depositional* landforms are created by the wind in some regions. Accumulations of windblown sediment are particularly conspicuous in the world's dry lands and along many sandy coasts. Wind deposits are of two distinctive types: (1) mounds and ridges of sand from the wind's bed load, which we call *dunes*, and (2) extensive blankets of silt, called *loess*, that once were carried in suspension.

Sand Deposits

As is the case with running water, wind drops its load of sediment when velocity falls and the energy available for transport diminishes. Thus, sand begins to accumulate wherever an obstruction across the path of the wind slows its movement. Unlike many deposits of silt, which form blanketlike layers over large areas, winds commonly deposit sand in mounds or ridges called **dunes** (Figure 16.15).

Moving air encountering an object, such as a clump of vegetation or a rock, sweeps around and over the object, leaving a "shadow" of slower-moving air behind the obstacle and a smaller zone of quieter air just in front of the obstacle. Some of the saltating sand grains moving with the wind come to rest in these wind shadows. As the accumulation of sand continues, it becomes a more imposing barrier to the wind and thus a more efficient trap for even more sand. If there is a sufficient supply of sand and the wind blows steadily for a long enough time, the mound of sand grows into a dune.

Many dunes have an asymmetrical profile, with the leeward (sheltered) slope being steep and the windward slope more gently inclined. The dunes in Figure 16.15 are a good example. Sand moves up the gentler slope on the windward side by saltation. Just beyond the crest of the dune, where the wind velocity is reduced, the sand

▼ SmartFigure 16.15
White Sands National Monument The dunes at this landmark in southeastern New Mexico are composed of gypsum. The dunes slowly migrate with the wind. (Photos by Michael Collier)

MOBILE FIELD TRIP
https://goo.gl/UFKrAJ

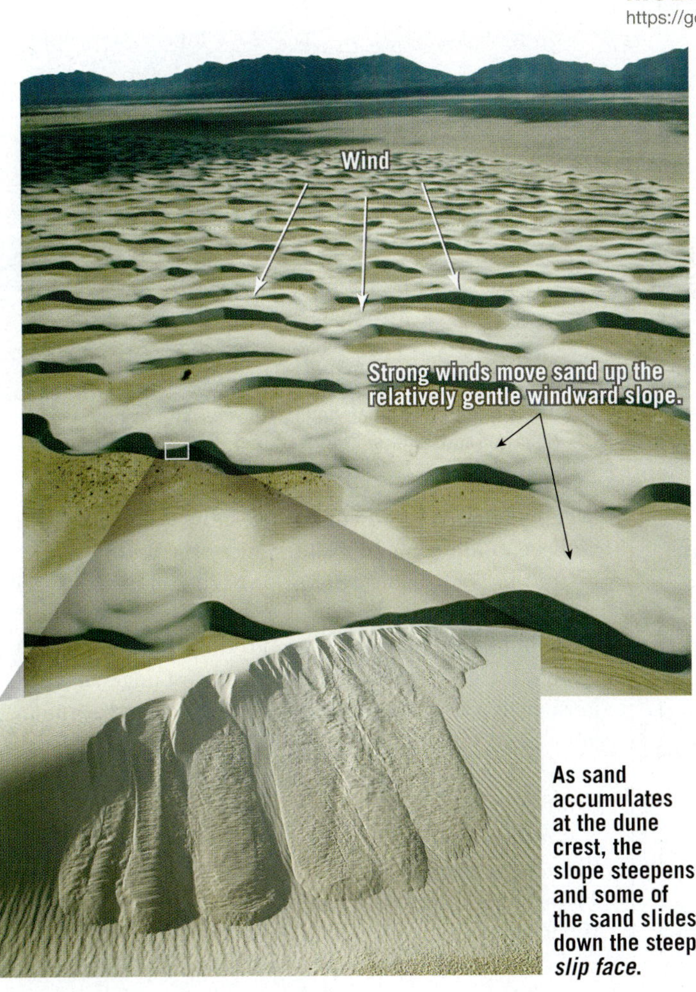
Wind
Strong winds move sand up the relatively gentle windward slope.
As sand accumulates at the dune crest, the slope steepens and some of the sand slides down the steep *slip face*.

▶ SmartFigure 16.16 **Cross-bedding** As sand is deposited on the slip face, layers form that are inclined in the direction the wind is blowing. With time, complex patterns develop in response to changes in wind direction. (Photo by Dennis Tasa)

TUTORIAL
https://goo.gl/ltqAKo

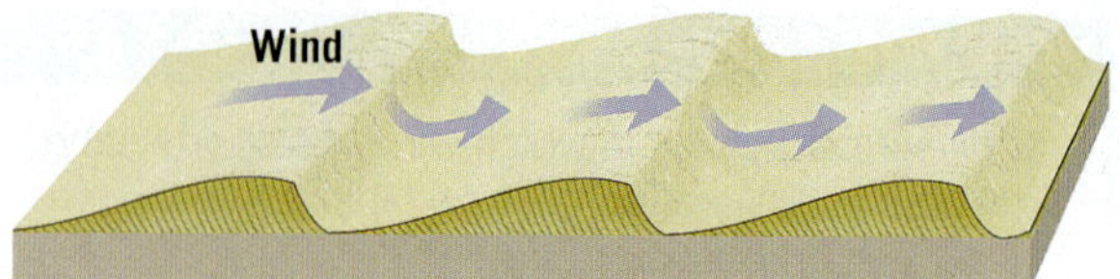

Dunes commonly have an asymmetrical shape and migrate with the wind.

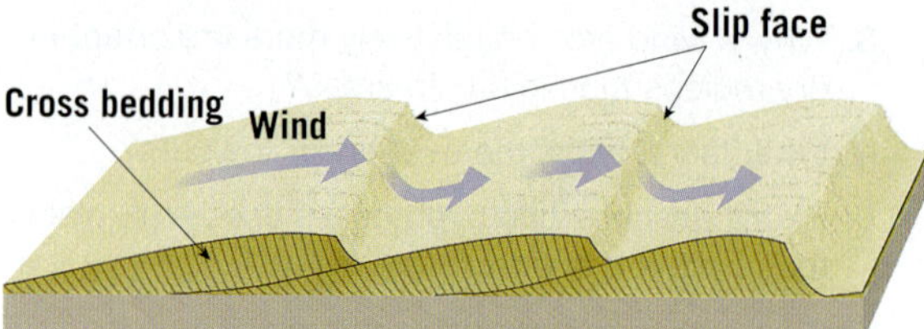

Sand grains deposited on the slip face at the angle of repose create the cross bedding of dunes.

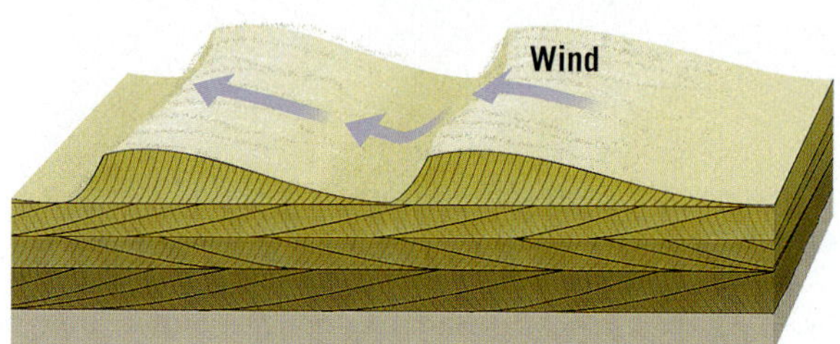

When dunes are buried and become part of the sedimentary rock record, the cross bedding is preserved.

Cross bedding is an obvious characteristic of the Navajo Sandstone in Zion National Park, Utah.

Did You Know?
Deserts don't necessarily consist of mile after mile of drifting sand dunes. Surprisingly, sand accumulations represent only a small percentage of the total desert area. For example, dunes cover only one-tenth of the Sahara's area. The sandiest of all deserts, the Arabian, is one-third sand covered.

accumulates. As more sand collects, the slope steepens, and eventually some of it slides under the pull of gravity. In this way, the leeward slope of the dune, called the **slip face**, maintains an angle of about 34 degrees, the angle of repose for loose dry sand. (Recall from Chapter 12 that the angle of repose is the steepest angle at which loose material remains stable.) Continued sand accumulation, coupled with periodic slides down the slip face, results in the slow migration of the dune in the direction of air movement.

As sand is deposited on the slip face, layers form that are inclined in the direction the wind is blowing. These sloping layers are called **cross-beds** (Figure 16.16). When the dunes are eventually buried under other layers of sediment and become part of the sedimentary rock record, their asymmetrical shape is destroyed, but the cross-beds remain as testimony to their origin. Nowhere is cross-bedding more prominent than in the sandstone walls of Zion Canyon in southern Utah (see Figure 16.16).

Types of Sand Dunes

Dunes are not just random heaps of windblown sediment. Rather, they are accumulations that usually assume patterns that are surprisingly consistent. Addressing this point, a leading early investigator of dunes, the British engineer R. A. Bagnold, observed: "Instead of finding chaos and disorder, the observer never fails to be amazed at a simplicity of form, an exactitude of repetition, and a geometric order."

Dunes come in a wide assortment of forms, and some irregular dunes do not fit easily into any category. For simplicity, we will consider just the six basic dune types shown in Figure 16.17. Several factors influence the form and size that dunes ultimately assume. These include wind direction and velocity, availability of sand, and the amount of vegetation present.

Barchan Dunes Solitary sand dunes shaped like crescents and with their tips pointing downwind are called **barchan dunes** (see Figure 16.17A). These dunes form where supplies of sand are limited and the surface is relatively flat, hard, and lacking vegetation. They migrate slowly with the wind at a rate of up to 15 meters (50 feet) per year. Their size is usually modest, with the largest barchans reaching heights of about 30 meters (100 feet), while the maximum spread between their tips approaches 300 meters (1000 feet). When the wind direction is nearly constant, the crescent form of these dunes is nearly symmetrical. However, when the wind direction is not perfectly fixed, one tip becomes larger than the other.

Transverse Dunes In regions where the prevailing winds are steady, sand is plentiful, and vegetation is sparse or absent, the dunes form a series of long ridges that are separated by troughs and oriented at right angles to the prevailing wind. Because of this orientation, they are termed **transverse dunes** (see Figure 16.17B). Many coastal dunes are of this type. In addition, transverse dunes are common in many arid regions where the extensive surface of wavy sand is sometimes called a *sand sea*. In some parts of the Sahara and Arabian Deserts, transverse dunes reach heights of 200 meters, are 1 to 3 kilometers across, and can extend for distances of 100 kilometers or more.

There is a relatively common dune form that is intermediate between isolated barchans and extensive waves of transverse dunes. Such dunes, called **barchanoid dunes**, form scalloped rows of sand oriented at right angles to the wind (see Figure 16.17C). The rows resemble a series of barchans that have been positioned side by side.

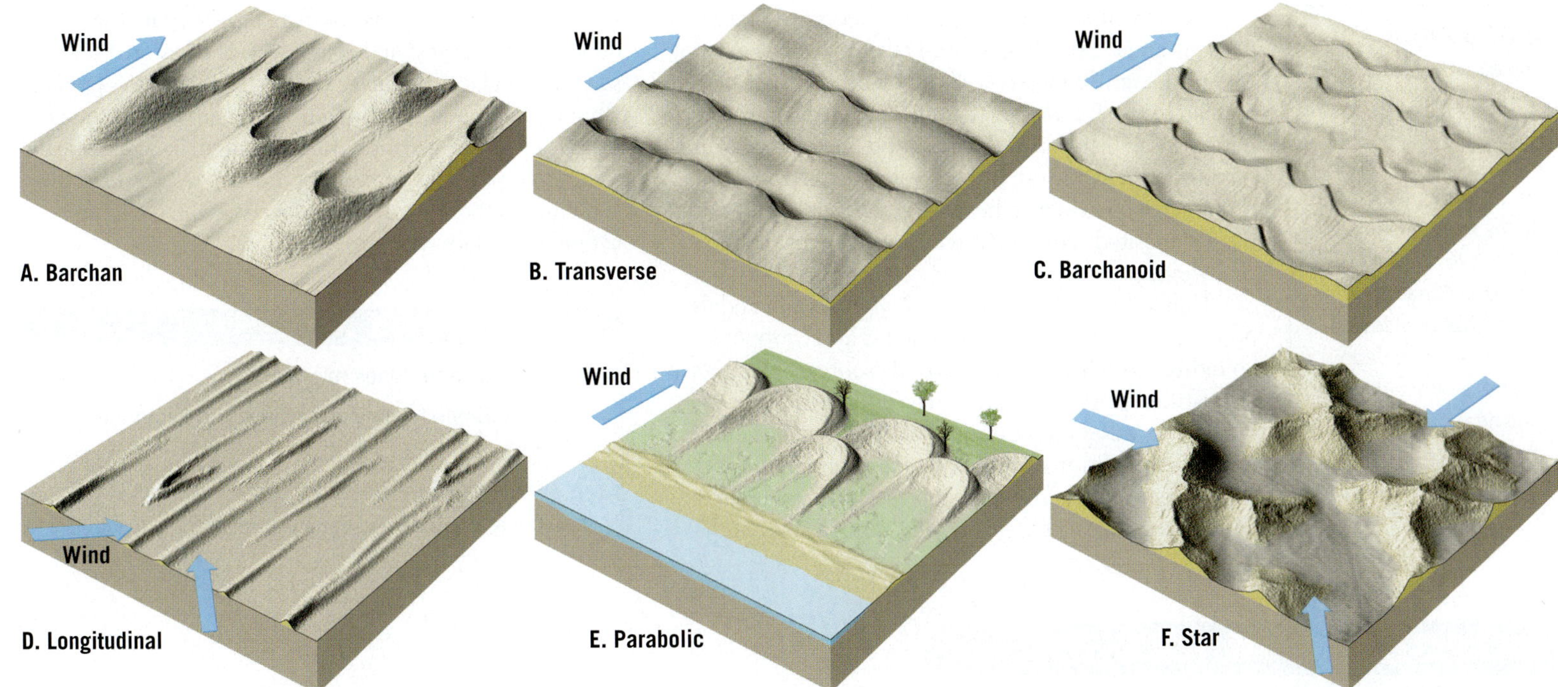

▲ SmartFigure 16.17
Types of sand dunes Factors that influence the form and size of dunes include wind direction and velocity, the availability of sand, and the amount of vegetation.

TUTORIAL
https://goo.gl/zeh51r

Longitudinal Dunes Long ridges of sand that form more or less parallel to the prevailing wind where sand supplies are moderate are called **longitudinal dunes** (see Figure 16.17D). Apparently, the prevailing wind direction must vary somewhat but still remain in the same quadrant of the compass. Although the smaller types are only 3 or 4 meters high and several dozens of meters long, in some large deserts, longitudinal dunes can reach great size. For example, in portions of North Africa, Arabia, and central Australia, these dunes may approach a height of 100 meters (330 feet) and extend for distances of more than 100 kilometers (62 miles).

Parabolic Dunes Unlike the other dune types described thus far, **parabolic dunes** form where vegetation partially covers the sand. The shape of these dunes resembles the shape of barchans except that their tips point into the wind rather than downwind (see Figure 16.17E). Parabolic dunes often form along coasts where there are strong onshore winds and abundant sand. If the sand's sparse vegetative cover is disturbed at some spot, deflation creates a blowout. Sand is then transported out of the depression and deposited as a curved rim, which grows higher as deflation enlarges the blowout.

Star Dunes Confined largely to parts of the Sahara and Arabian Deserts, **star dunes** are isolated hills of sand that exhibit a complex form (see Figure 16.17F). Their name is derived from the fact that the bases of these dunes resemble multipointed stars. Usually three or four sharp-crested ridges diverge from a central high point that in some cases may approach a height of 90 meters (300 feet). As their form suggests, star dunes develop where wind directions are variable.

Loess (Silt) Deposits

In some parts of the world, the surface topography is mantled with deposits of windblown silt, called **loess**. Over periods of perhaps thousands of years, dust storms deposited this material. When loess is breached by streams or road cuts, it tends to maintain vertical cliffs and lacks any visible layers, as you can see in Figure 16.18.

The distribution of loess worldwide indicates that there are two primary sources for this sediment: deserts and glacial outwash deposits. The thickest and most extensive deposits of loess on Earth occur in western and northern China. They were blown there from the extensive desert basins of central Asia. Accumulations of 30 meters (100 feet) are common, and thicknesses of more than 100 meters (330 feet) have been measured. It is this fine, buff-colored sediment that gives the Yellow River (Huang Ho) its name.

This vertical bluff near the Mississippi River in southern Illinois is about 3 meters (10 feet) high.

◀ Figure 16.18
Loess In some regions, the surface is mantled with deposits of windblown silt. (Photo from the James E. Patterson Collection, courtesy of F. K. Lutgens)

Did You Know?
The highest dunes in the world are located along the southwest coast of Africa, in the Namib Desert. In places, these huge dunes reach heights of 1000 to 1167 ft. The dunes at Great Sand Dunes National Park in southern Colorado are the highest in North America, rising more than 700 ft above the surrounding terrain.

In the United States, deposits of loess are significant in many areas, including South Dakota, Nebraska, Iowa, Missouri, and Illinois, as well as portions of the Columbia Plateau in the Pacific Northwest. The correlation between the distribution of loess and important farming regions in the Midwest and eastern Washington State is not just a coincidence because soils derived from this wind-deposited sediment are among the most fertile in the world.

Unlike the deposits in China, which originated in deserts, the loess in the United States and Europe is an indirect product of glaciation. Its source is deposits of stratified drift. During the retreat of the ice sheets, many river valleys were choked with sediment deposited by meltwater. Strong westerly winds sweeping across the barren floodplains picked up the finer sediment and dropped it as a blanket on the eastern sides of the valleys. Such an origin is confirmed by the fact that loess deposits are thickest and coarsest on the leeward side of such major glacial drainage outlets as the Mississippi and Illinois Rivers and rapidly thin with increasing distance from the valleys. Furthermore, the angular, mechanically weathered particles composing the loess are essentially the same as the rock flour produced by the grinding action of glaciers.

CONCEPT CHECKS 16.5

1. How do sand dunes migrate?
2. List and briefly distinguish among basic dune types.
3. How is loess different from sand?
4. How are some loess deposits related to glaciers?

CONCEPTS IN REVIEW
Deserts & Wind

16.1 Distribution & Causes of Dry Lands

Describe the general distribution of Earth's dry lands and explain why deserts form in the subtropics and middle latitudes.

KEY TERMS: dry climate, desert, steppe, rainshadow

- Dry climates cover about 30 percent of Earth's land area. These regions have yearly precipitation totals that are less than the potential loss of water through evaporation. Evaporation depends on temperature, and deserts may occur in hot or cold climates. Deserts are drier than steppes, but both climatic types are considered water deficient.
- Dry climates in subtropical latitudes are associated with the global distribution of air pressure and winds. Near the equator, warm, moist air rises (causing lots of rain) and then moves to 20° or 30° latitude before sinking back toward Earth's surface. The subsiding air brings clear skies, copious sunshine, and dry conditions to these zones of subtropical high pressure.
- Deserts also occur in continental interiors of the middle latitudes. Most are subject to the rainshadow effect, in which moist air moving inland from oceans is intercepted by mountainous obstacles. As the air is forced to rise, it cools, producing clouds and precipitation on the windward slopes. By contrast, the leeward side, called the rainshadow, tends to by quite dry.

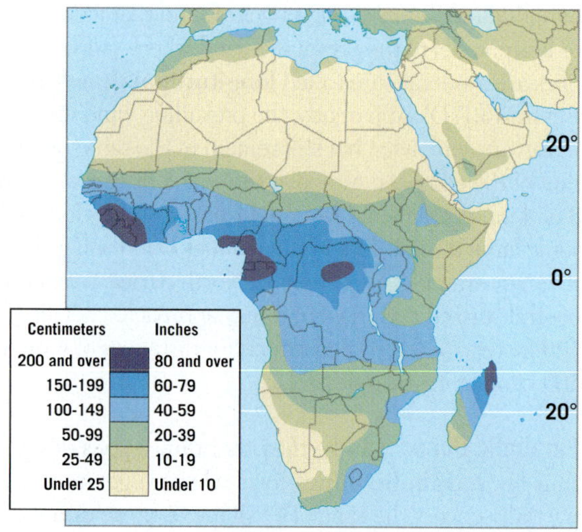

? This map shows annual precipitation across the continent of Africa. At which latitudes is the atmosphere rising? At which latitudes is it sinking? How does this atmospheric circulation influence the continent's climates?

16.2 Geologic Processes in Arid Climates

Summarize the geologic roles of weathering, running water, and wind in arid and semiarid climates.

KEY TERM: ephemeral stream

- In dry lands rock weathering of any type is greatly reduced because of the lack of water and the scarcity of organic acids from decaying plants.
- Practically all desert streams are dry most of the time and are said to be ephemeral. Their channels are carved out largely by flash floods that occur during sporadic storm events.
- Permanent desert streams originate in wetter climates. They must carry a great volume of water to keep from losing all their water as they cross the desert.
- Running water is responsible for most of the erosional work in a desert. Although wind erosion is more significant in dry areas than in other environments, it still cannot match running water as an agent of erosion in deserts.

Leon Werdinger / Alamy Stock Photo

? This is a typical desert stream shortly after a heavy rain. What term is applied to such a stream? How will this scene likely change in the near future?

16.3 Basin & Range: The Evolution of a Desert Landscape

Discuss the stages of landscape evolution in the Basin and Range region of the western United States.

KEY TERMS: interior drainage, alluvial fan, bajada, playa lake, playa, inselberg

- The Basin and Range region is a distinctive area of western North America that demonstrates some key principles of mountainous desert landscapes. The mountain ranges and intervening valleys were produced through normal faulting, but then the rock mass of the mountain ranges was redistributed by weathering, erosion, transportation, and deposition. With sufficient time, the overall effect is to reduce topographic relief by grinding down the mountains and depositing the sedimentary debris in the low-lying basins.
- Alluvial fans form during the early stage of arid landscape development, when the relief is highest. Over time, these fans grow and merge, so that by the middle stage, continuous bajadas of sediment cover the line where the ranges meet the basins. On the basin floor, playa lakes may form during wet periods and then dry up to produce salty playas.
- When almost all of the mountains have been worn down and most of the low-lying basins have been filled in with sediment, the landscape has reached the late stage of development. It is then marked by isolated inselbergs of bedrock poking up through a sea of sediment.

? Identify the lettered features in this photo. How did they form?

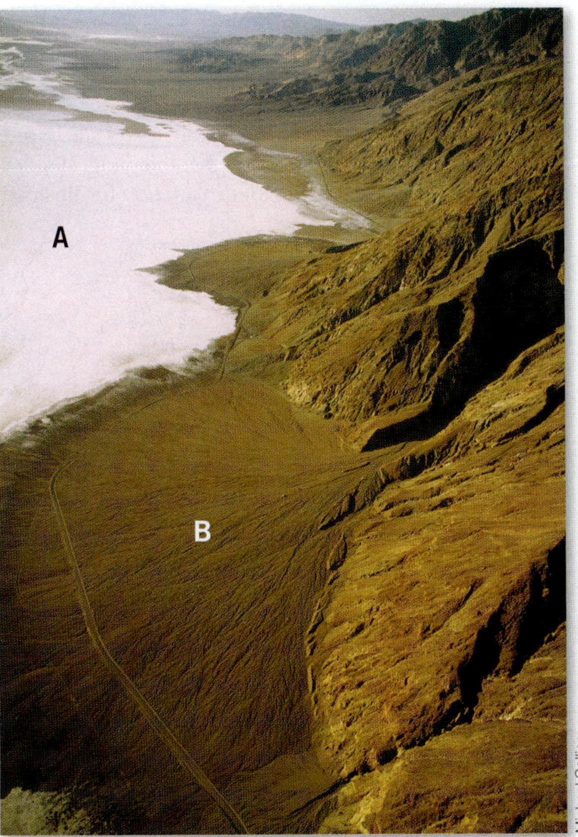

Michael Collier

16.4 Wind Erosion

Describe the ways that wind transports sediment and the processes and features associated with wind erosion.

KEY TERMS: bed load, saltation, suspended load, deflation, blowout, desert pavement, abrasion, ventifact, yardang

- A current of air can pick up and transport sediment, though with a lower competence than a current of water or glacial ice. Wind cannot pick up the coarser particles that water can, but it is capable of transporting sediment over vast areas and even high into the atmosphere.
- A portion of the sediment transported by wind is bed load that bounces along Earth's surface. Generally, these saltating particles of sand never get more than 0.5 meter above the ground.
- Other sediment is fine grained enough to stay suspended in the air. Clay and silt are both small enough particles that they can be carried as suspended load. Once aloft, they can be carried tremendous distances, across entire continents and ocean basins.
- Wind is capable of erosion, though it should be emphasized that water is the most important agent of erosion in desert regions. The Dust Bowl is a classic example of a massive episode of soil erosion by wind during the 1930s. Localized deflation may produce shallow depressions known as blowouts.
- Desert pavement is a thin layer of coarse pebbles and cobbles that covers some desert surfaces. Two models have been proposed to explain the formation of desert pavement: (1) a model in which deflation strips the finer grains from a deposit of poorly sorted sediments and (2) an additive model in which pebbles and cobbles trap smaller windblown grains in the spaces between them.
- Ventifacts are produced when individual rocks become abraded by windblown sediment, giving the rocks a polished, pitted surface. This effect is most pronounced on the side that faces the prevailing wind. Similar "sand-blasting" of the land surface may sculpt rock outcrops into streamlined yardangs that are longest parallel to the prevailing wind direction.

? This satellite image shows a large plume of windblown sediment over portions of Iran, Afghanistan, and Pakistan in March 2012. Is the wind-transported material in the image more likely bed load or suspended load? What term is applied to the erosional process that was responsible for producing this plume?

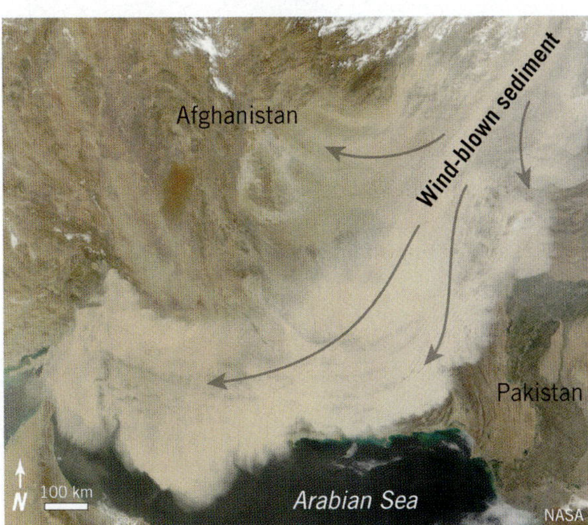

16.5 Wind Deposits

Discuss the formation and movement of dunes and distinguish among different dune types. Explain how loess deposits differ from deposits of sand.

KEY TERMS: dune, slip face, cross-bed, barchan dune, transverse dune, barchanoid dune, longitudinal dune, parabolic dune, star dune, loess

- Wind deposits are of two distinct types: (1) mounds and ridges of sand, called dunes, which are formed from sediment that is carried as part of the wind's bed load, and (2) extensive blankets of silt, called loess, carried by wind in suspension.
- Dunes accumulate due to the differences in wind energy on the upwind and downwind sides of some obstacle. Wind moves sand up the more gently sloping upwind side, across the crest of the dune, where it settles out in the calmer air on the downwind side of the dune, the steeply sloped slip face. When the sand on the slip face piles up past the angle of repose, it spontaneously collapses in small "avalanches" of sand. Over time, this causes a dune to slowly move in the direction of the prevailing wind, with the slip face "leading the way." Inside the dune, the buried slip faces may be preserved as cross-beds.
- Loess is windblown silt, deposited over large areas sometimes in thick blankets. Most loess is derived from either (1) deserts or (2) areas that have recently been glaciated. Winds blow across stratified drift and pick up silt-size grains, carrying them in suspension to the site of deposition.

(16.5 continued)

- There are six major kinds of dunes. Their shapes result from the pattern of prevailing winds, the amount of available sand, and the presence of vegetation.

? This is an aerial view of a dune in northern Arizona. Which one of the basic dune types is it? Sketch a simple profile (side view) of this dune. Include an arrow to show the prevailing wind direction and label the dune's slip face.

Michael Collier

GIVE IT SOME THOUGHT

1 Albuquerque, New Mexico, receives an average of 20.7 centimeters (8.07 inches) of rainfall annually. Albuquerque is considered a desert under the commonly used Köppen climate classification. The Russian city of Verkhoyansk is located near the Arctic Circle in Siberia. Yearly precipitation at Verkhoyansk averages 15.5 centimeters (6.05 inches), about 5 centimeters (2 inches) less than Albuquerque, yet it is classified as a humid climate. Explain why this is the case.

2 Compare and contrast the sediment deposited by a stream, wind, and a glacier. Which deposit should have the most uniform grain size? Which one would exhibit the poorest sorting? Explain your choices.

3 Is either of the following statements true? Are they both true? Explain your answer.

a. Wind is more effective as an agent of erosion in dry places than in humid places.

b. Wind is the most important agent of erosion in deserts.

4 Examine the precipitation map for the state of Nevada. Notice that the areas receiving the most precipitation resemble long, slender "islands" scattered across the state. Provide an explanation for this pattern.

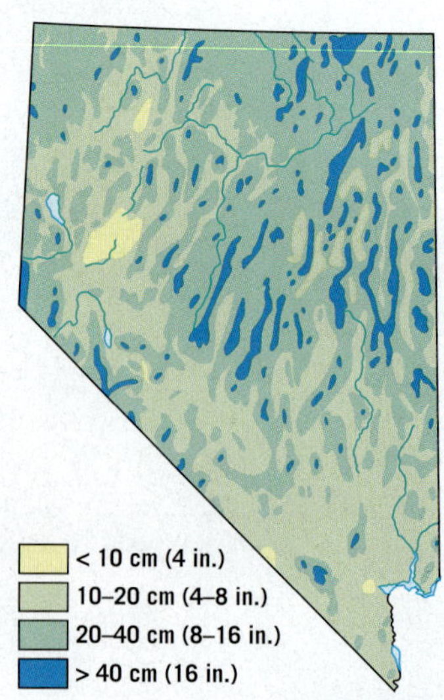

5 This satellite image shows a small portion of the Zagros Mountains in dry southern Iran. Streams in this region flow only occasionally. The green tones on the image identify productive agricultural areas.

a. Identify the large feature that is labeled with a question mark.

b. Explain how the feature named in Question a formed.

c. What term is used to describe streams like the ones that occur in this region?

d. Speculate on the likely source of water for the agricultural areas in this image.

NASA

6 Bryce Canyon National Park, shown in this photo, is in dry southern Utah. It is carved into the eastern edge of the Paunsaugunt Plateau. Erosion has sculpted the colorful limestone into bizarre shapes, including spires called "hoodoos." As you and a nongeologist companion are viewing the scenery in Bryce Canyon, your friend says, "It's amazing how wind has created this incredible scenery!" Now that you have studied arid landscapes, how would you respond to your companion's statement?

Ozoptimes/Shutterstock

MasteringGeology™

Looking for additional review and test prep materials? Visit the Study Area in MasteringGeology to enhance your understanding of this chapter's content by accessing a variety of resources, including Self-Study Quizzes, Geoscience Animations, SmartFigures, Mobile Field Trips, *Project Condor* Quadcopter videos, *In the News* RSS feeds, flashcards, web links, and an optional Pearson eText.

www.masteringgeology.com

17

Shorelines

FOCUS ON CONCEPTS

Each statement represents the primary learning objective for the corresponding major heading within the chapter. After you complete the chapter, you should be able to:

17.1 Explain why the shoreline is considered a dynamic interface. List the factors that influence the height, length, and period of a wave and describe the motion of water within a wave.

17.2 Explain how waves erode and move sediment along the shore.

17.3 Describe the features typically created by wave erosion and those resulting from sediment deposited by longshore transport processes.

17.4 Distinguish between emergent and submergent coasts. Contrast the erosion problems faced on the Atlantic and Gulf coasts with those along the Pacific coast.

17.5 Describe the basic structure and characteristics of a hurricane and the three broad categories of hurricane destruction.

17.6 Summarize the ways in which people deal with shoreline erosion problems.

17.7 Explain the cause of tides and their monthly cycles. Describe the horizontal flow of water that accompanies the rise and fall of tides.

In contrast to the gently sloping coastal plains of the Atlantic and Gulf coasts, the Pacific coast is characterized by relatively narrow beaches that are often backed by steep cliffs and mountain ranges. These crashing waves and sea stacks are at Soberanes Point along the California coast. (Photo by Jamie Pham/Zoonar/AGE Fotostock)

THE RESTLESS WATERS OF THE OCEAN are constantly in motion. Winds generate surface currents, the gravity of the Moon and Sun produces tides, and density differences create deep-ocean circulation. Further, waves carry the energy from storms to distant shores, where their impact erodes the land.

Shorelines are dynamic environments. Their topography, geologic makeup, and climate vary greatly from place to place. Continental and oceanic processes converge along the shore to create landscapes that frequently undergo rapid change. When it comes to the deposition of sediment, shore areas are transition zones between marine and continental environments.

17.1 The Shoreline & Ocean Waves

Explain why the shoreline is considered a dynamic interface. List the factors that influence the height, length, and period of a wave and describe the motion of water within a wave.

The **shoreline** is the line that marks the contact between land and sea. Each day, as tides rise and fall, the position of the shoreline migrates. Over longer time spans, as sea level rises or falls, the average position of the shoreline gradually shifts.

A Dynamic Interface

Nowhere is the restless nature of the ocean more noticeable than along the shore—the dynamic interface among air, land, and sea. An **interface** is a common boundary where different parts of a system interact. This is certainly an appropriate designation for the coastal zone. Here we can see the rhythmic rise and fall of tides and observe waves constantly rolling in and breaking.

▼ **Figure 17.1 Teetering on the edge** Bluff failure caused by storm waves in March 2016 resulted in these apartments in Pacifica, California, being condemned. When these buildings were erected in the 1970s, they were safely away from the cliffs. Over the years, several measures were attempted to reduce erosion of the sandstone cliffs. All proved to be inadequate. (Photo by Terry Chea/AP Photo)

Sometimes the waves are low and gentle. At other times they pound the shore with awesome fury.

Although it may not be obvious, the shoreline is constantly being modified by waves. Crashing surf erodes the land. Wave activity also moves sediment toward and away from the shore, as well as along it. Such activity sometimes produces narrow sandbars and fragile offshore islands that frequently change size and shape as storm waves come and go.

The nature of present-day shorelines is not just the result of the relentless attack of the land by the sea. Rather, the shore has a complex character that results from multiple geologic processes. For example, practically all coastal areas were affected by the worldwide rise in sea level that accompanied the melting of ice sheets following the Last Glacial Maximum (see Figure 15.25, page 411). As the sea encroached landward, the shoreline retreated, becoming superimposed upon existing landscapes that had resulted from such diverse processes as stream erosion, glaciation, volcanic activity, and the forces of mountain building.

Today the coastal zone is experiencing intensive human activity (**Figure 17.1**). Unfortunately, people often treat the shoreline as if it were a stable platform on which structures can safely be built. This attitude inevitably leads to conflicts between people and nature. As you will see, many coastal landforms, especially beaches and barrier islands, are relatively fragile, short-lived features that are inappropriate sites for development. The image of the New Jersey shoreline in **Figure 17.2** is an example.

Ocean Waves

Ocean waves travel along the interface between ocean and atmosphere. They can carry energy from a storm far out at sea over distances of several thousand kilometers. That's why even on calm days, the ocean still has waves

▲ **Figure 17.2 Hurricane Sandy** A portion of the New Jersey shoreline shortly after this huge storm struck in late October 2012. The extraordinary storm surge caused much of the damage pictured here. Many shoreline areas are intensively developed. Often the shifting shoreline sands and the desire of people to occupy these areas are in conflict. (Photo by Mario Tama/Getty Images)

that travel across its surface. When observing waves, always remember that you are watching *energy* travel through a medium (water). If you make waves by tossing a pebble into a pond, or by splashing in a pool, or by blowing across the surface of a cup of coffee, you are imparting *energy* to the liquid, and the waves you see are the visible evidence of the energy passing through.

Wind-generated waves provide most of the energy that shapes and modifies shorelines. Where the land and sea meet, waves that may have traveled unimpeded for hundreds or thousands of kilometers suddenly encounter a barrier that will not allow them to advance farther and must absorb their energy. Stated another way, the shore is the location where a practically irresistible force confronts an almost immovable object. The conflict that results is never-ending and sometimes dramatic.

Wave Characteristics

Most ocean waves derive their energy and motion from the wind. When a breeze is less than 3 kilometers (2 miles) per hour, only small wavelets appear. At greater wind speeds, more stable waves gradually form and advance with the wind.

Characteristics of ocean waves are illustrated in Figure 17.3, which shows a simple, nonbreaking wave form. The tops of the waves are the *crests*, which are separated by *troughs*. Halfway between the crests and troughs is the *still water level*, which is the level the water would occupy if there were no waves. The vertical distance between trough and crest is called the **wave height**, and the horizontal distance between successive crests (or troughs) is the **wavelength**. The time it takes one full wave—one wavelength—to pass a fixed position is the **wave period**.

The height, length, and period that are eventually achieved by a wave depend on three factors: (1) the wind speed, (2) the length of time the wind has blown, and (3) the **fetch**, or the distance the wind has traveled across open water. As the quantity of energy transferred from the wind to the water increases, the height and steepness of the waves increase as well. Eventually a critical point is reached where waves grow so tall that they topple over, forming ocean breakers called *whitecaps*.

For a particular wind speed, there is a maximum fetch and duration of wind beyond which waves will no longer increase in size. When the maximum fetch and duration are reached for a given wind velocity, the waves are said to be "fully developed." The reason that waves can grow no further is that they are losing as much energy through the breaking of whitecaps as they are receiving from the wind.

When wind stops or changes direction or when waves leave the stormy area where they were created, the waves continue on without relation to local winds. The waves also undergo a gradual change to *swells*, which are lower and longer and may carry a storm's energy to distant shores. Because many independent wave systems exist at the same time, the sea surface acquires a complex, irregular pattern. Hence, the sea waves we watch from the shore are often a mixture of swells from faraway storms and waves created by local winds.

Circular Orbital Motion

Waves can travel great distances across ocean basins. In one study, waves generated near Antarctica were tracked as they traveled through the Pacific Ocean basin. After

Did You Know?
About half of the world's human population lives on or within 60 mi of a coast. The proportion of the U.S. population residing within 45 mi of a coast is well in excess of 50 percent. The concentration of such large numbers of people so near the shore means that hurricanes and tsunamis place millions at risk.

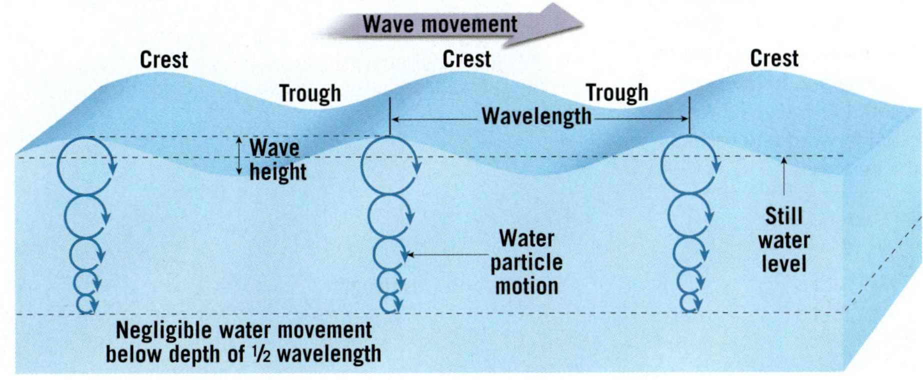

▼ **SmartFigure 17.3**
Wave basics An idealized nonbreaking wave, showing its basic parts and the movement of water with increasing depth.

ANIMATION
https://goo.gl/h0NxLb

▶ **SmartFigure 17.4**
Passage of a wave The movements of the toy boat show that the wave form advances, but the water does not advance appreciably from the original position. In this sequence, the wave moves from left to right as the boat (and the water in which it is floating) rotates in an approximate circle.

TUTORIAL
https://goo.gl/uH2qhW

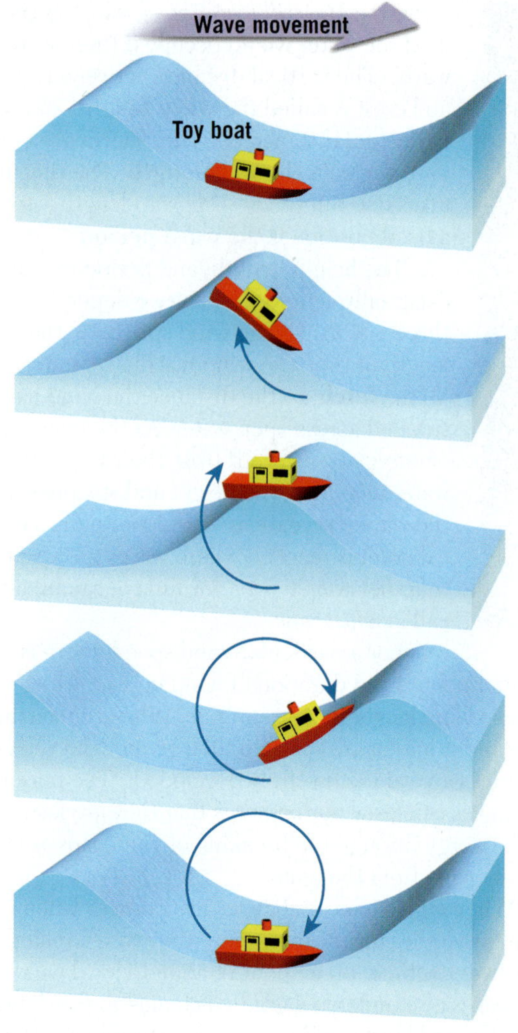

more than 10,000 kilometers (over 6000 miles), the waves finally expended their energy a week later, along the shoreline of the Aleutian Islands of Alaska. The water itself doesn't travel this distance, but the wave form does.

As the wave travels, the water passes the energy along by moving in a circle. This movement is called *circular orbital motion*.

Observation of an object floating in waves reveals that it moves not only up and down but also slightly forward and backward with each successive wave. When the movement of the toy boat shown in **Figure 17.4** is traced as a wave passes, it can be seen that the boat moves in a circle and returns to essentially the same place. Circular orbital motion allows a wave form (the wave's shape) to move forward *through the water*, while the individual water particles that transmit the wave move in a circle. Wind moving across a field of wheat causes a similar phenomenon: The wheat itself doesn't travel across the field, but the waves do.

The energy contributed by the wind to the water is transmitted not only along the surface of the sea but also downward. However, beneath the surface, the circular motion rapidly diminishes until, at a depth equal to one-half the wavelength measured from the still water level, the movement of water particles becomes negligible. This depth is known as the *wave base*. The dramatic decrease of wave energy with depth is shown by the rapidly diminishing diameters of water-particle orbits in Figure 17.3.

Waves in the Surf Zone

As long as a wave is in deep water, it is unaffected by water depth (**Figure 17.5**, *left*). However, when a wave approaches the shore, the water becomes shallower and influences wave behavior. The wave begins to "feel bottom" at a water depth equal to its wave base. Such depths interfere with water movement at the base of the wave and slow its advance (see Figure 17.5, *center*).

As a wave advances toward the shore, the slightly faster waves farther out to sea catch up, decreasing

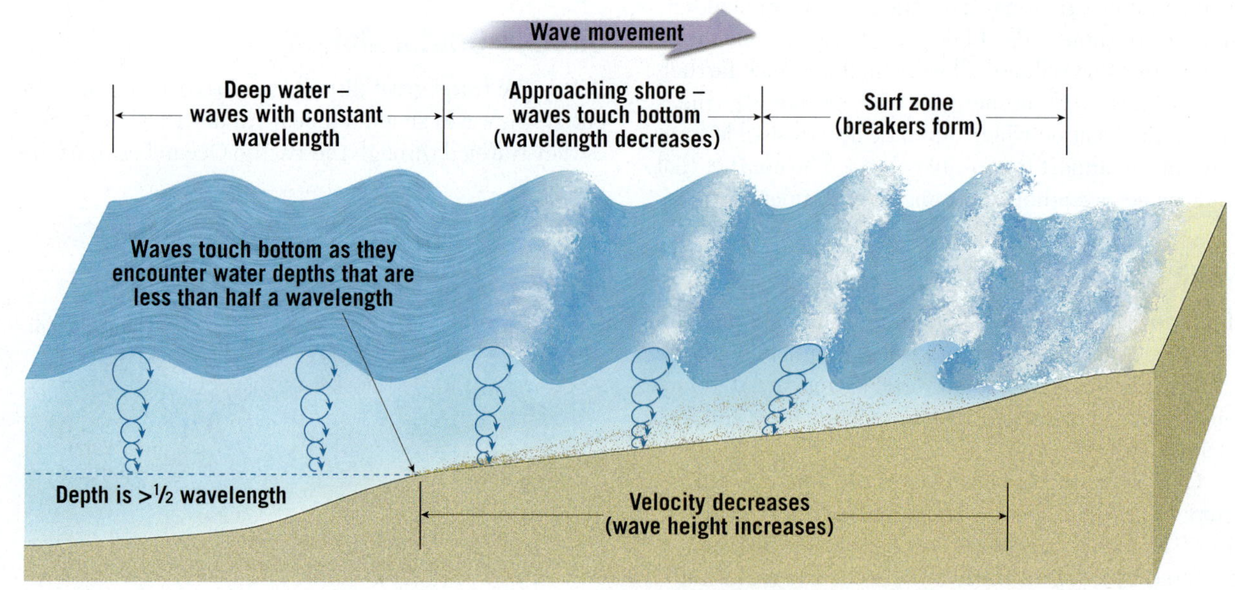

▶ **SmartFigure 17.5**
Waves approaching the shore Waves touch bottom as they encounter water depths that are less than half a wavelength. As a result, the wave speed decreases, and the faster-moving waves farther from shore begin to catch up, which causes the distance between waves (the wavelength) to decrease. This causes an increase in wave height, to the point where the waves finally pitch forward and break in the surf zone.

ANIMATION
https://goo.gl/4PiqxA

the wavelength. As the speed and length of the wave diminish, the wave steadily grows higher. Finally, a critical point is reached when the wave is too steep to support itself and the wave front collapses, or *breaks* (see Figure 17.5, *right*), causing water to advance up the shore.

The turbulent water created by breaking waves is called **surf**. On the landward margin of the surf zone, the *swash*—the turbulent sheet of water from collapsing breakers—moves up the slope of the beach. When the energy of the swash has been expended, the water flows back down the beach toward the surf zone as *backwash*.

CONCEPT CHECKS 17.1

1. Why is the shoreline described as being an interface?
2. Aside from ocean waves, what other factors influence the nature of present-day shorelines?
3. List three factors that determine the height, length, and period of a wave.
4. Describe the motion of a floating object as a wave passes.
5. How do a wave's speed, wavelength, and height change as the wave moves into shallow water and breaks?

17.2 Beaches & Shoreline Processes

Explain how waves erode and move sediment along the shore.

For many, a beach is the sandy area where people lie in the sun and walk along the water's edge. Technically, a **beach** is an accumulation of sediment found along the landward margin of a water body. Along straight coasts, beaches may extend for tens or hundreds of kilometers. Where coasts are irregular, beach formation may be confined to the relatively quiet waters of bays.

Beaches are composed of whatever material is locally abundant. The sediment for some beaches is derived from the erosion of adjacent cliffs or nearby coastal mountains. Other beaches are built from sediment delivered to the coast by rivers. Although the mineral makeup of many beaches is dominated by durable quartz grains, other minerals may be dominant. For example, in areas such as southern Florida, where there are no mountains or other sources of rock-forming minerals nearby, most beaches are composed of shell fragments and the remains of organisms that live in coastal waters (**Figure 17.6A**). Some beaches on volcanic islands in the open ocean are composed of

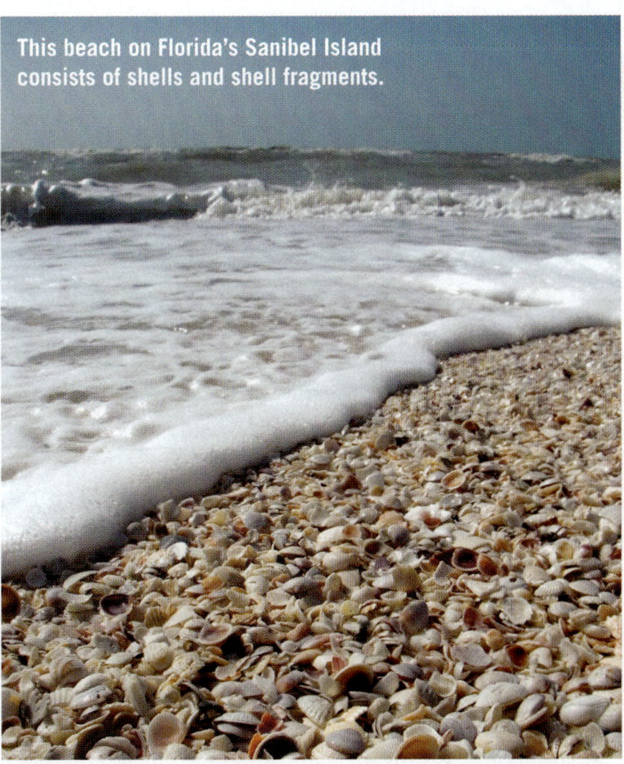

This beach on Florida's Sanibel Island consists of shells and shell fragments.

A.

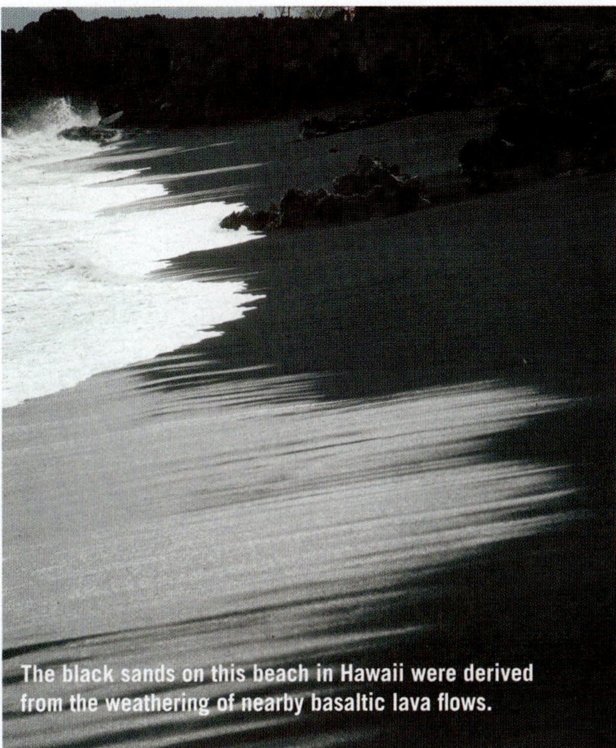

The black sands on this beach in Hawaii were derived from the weathering of nearby basaltic lava flows.

B.

Figure 17.6 Beaches A beach is an accumulation of sediment on the landward margin of an ocean or a lake and can be thought of as material in transit along the shore. Beaches are composed of whatever material is locally available. (Photo A by David R. Frazier/Photo Library/Alamy Images; photo B by E. J. Tarbuck)

▲ **Figure 17.7 Storm waves** When large waves break against the shore, the force of the water can be powerful and the erosional work that is accomplished can be great. These storm waves are breaking along the coast of Wales. (The Photo Library/Alamy Images)

weathered grains of basaltic lava or of coarse debris eroded from coral reefs that develop around islands in low latitudes (Figure 17.6B).

Regardless of the composition, the material that comprises a beach does not stay in one place. Instead, crashing waves are constantly moving it. Thus, beaches can be thought of as material in transit along the shore.

Wave Erosion

During calm weather, wave action is minimal. However, just as streams do most of their work during floods, so too do waves accomplish most of their work during storms. The impact of storm-induced waves against the shore can be awesome in its violence (Figure 17.7). Each breaking wave may hurl thousands of tons of water against the land, sometimes causing the ground to literally tremble. It is no wonder that cracks and crevices are quickly opened in cliffs, seawalls, breakwaters, and anything else that is subjected to these enormous shocks. Water is forced into every opening, causing air in the cracks to become highly compressed by the thrust of crashing waves. When the wave subsides, the air expands rapidly, dislodging rock fragments and enlarging and extending fractures.

In addition to the erosion caused by wave impact and pressure, **abrasion**—the sawing and grinding action of the water armed with rock fragments—is also important. In fact, abrasion is probably more intense in the surf zone than in any other environment. Smooth, rounded stones and pebbles along the shore are obvious reminders of the relentless grinding action of rock against rock in the surf zone. The chapter-opening photo and Figure 17.8A are good examples. Further, the waves use such fragments as "tools" as they cut horizontally into the land (Figure 17.8B).

Sand Movement on the Beach

Beaches are sometimes called "rivers of sand." The energy from breaking waves often causes large quantities of sand to move roughly parallel to the shoreline, both along the beach face and in the surf zone. Wave energy

A.

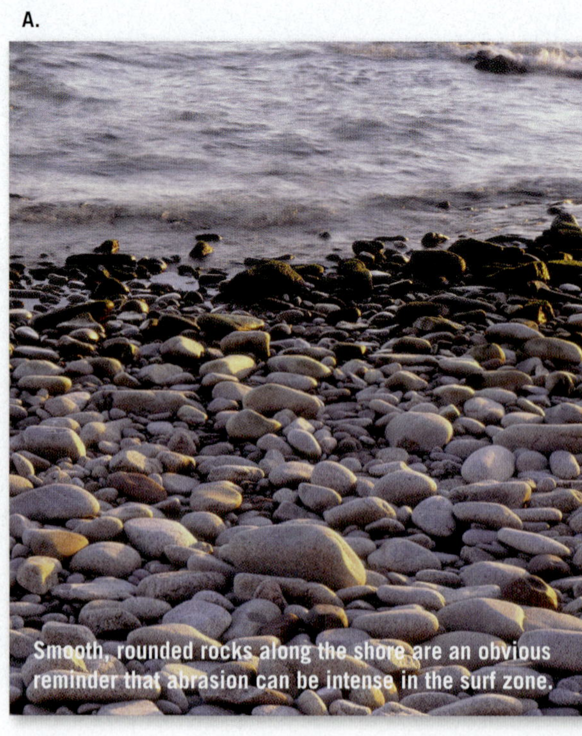

Smooth, rounded rocks along the shore are an obvious reminder that abrasion can be intense in the surf zone.

B.

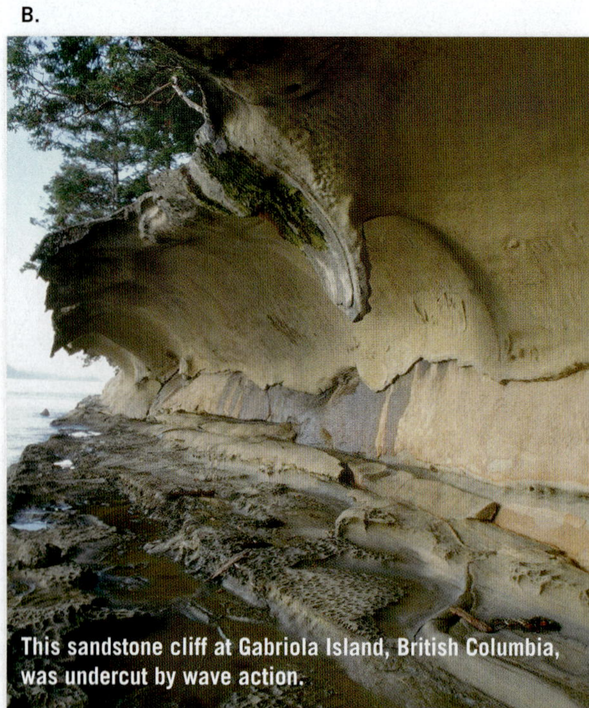

This sandstone cliff at Gabriola Island, British Columbia, was undercut by wave action.

▶ **Figure 17.8 Abrasion: Sawing and grinding** Breaking waves armed with rock debris can do a great deal of erosional work. (Photo A by Michael Collier; photo B by Fletcher and Baylis/Science Source)

also causes sand to move perpendicular to (toward and away from) the shoreline.

Movement Perpendicular to the Shoreline If you stand ankle deep in water at the beach, you will see that swash and backwash move sand toward and away from the shoreline. Whether there is a net loss or addition of sand depends on the level of wave activity. When wave activity is relatively light (less energetic waves), much of the swash soaks into the beach, which reduces the backwash. Consequently, the swash dominates and causes a net movement of sand up the beach face.

When high-energy waves prevail, the beach is saturated from previous waves, so much less of the swash soaks in. As a result, erosion occurs because backwash is strong and causes a net movement of sand down the beach face.

Along many beaches, light wave activity is the rule during the summer. Therefore, a wide sandy beach gradually develops. During winter, when storms are frequent and more powerful, strong wave activity erodes and narrows the beach. A wide beach that may have taken months to build can be dramatically narrowed in just a few hours by the high-energy waves created by a strong winter storm.

Wave Refraction The bending of waves, called **wave refraction**, plays an important part in shoreline processes (**Figure 17.9**). It affects the distribution of energy along the shore and thus strongly influences where and to what degree erosion, sediment transport, and deposition will take place.

The shore is seldom oriented exactly parallel to approaching ocean waves. Rather, most waves move toward the shore at an angle. However, when they reach the shallow water of a smoothly sloping bottom, they bend and tend to become parallel to the shore. Such bending occurs because the part of the wave nearest the shore reaches shallow water and slows first, whereas the end that is still in deep water continues forward at its full speed. The net result is a wave front that may approach nearly parallel to the shore, regardless of the original direction of the wave.

Because of refraction, wave impact is concentrated against the sides and ends of headlands that project into the water, whereas wave attack is weakened in bays. This differential wave attack along irregular coastlines is illustrated in Figure 17.9. The shallow water near a headland causes waves to refract toward the headland, focusing their energy and also causing the waves to attack the headland from all three sides. By contrast, refraction in the bays causes waves to diverge and expend less energy. In these zones of weakened wave activity, sediments can accumulate and form sandy beaches. Over a long period, erosion of the headlands and deposition in the bays will straighten an irregular shoreline.

Longshore Transport Although waves are refracted, most still reach the shore at some angle, however slight. Consequently, the uprush of water from each breaking wave (the swash) is at an oblique angle to the shoreline. However, the backwash is straight down the slope of the beach. The effect of this pattern of water movement is to transport sediment in a zigzag pattern along the beach

Did You Know?
During a storm that struck the coast of Scotland, a 1350-ton portion of a steel-and-concrete breakwater was torn from the rest of the structure and moved toward shore. Five years later the 2600-ton unit that replaced the first met a similar fate.

SmartFigure 17.9
Wave refraction As waves first touch bottom in the shallows along an irregular coast, they are slowed; they then bend (refract) and align nearly parallel to the shoreline. (Photo by Rich Reid/National Geographic/Getty Images)

TUTORIAL
https://goo.gl/3m2GHC

As these waves approach nearly straight on, refraction causes the wave energy to be concentrated at headlands (resulting in erosion) and dispersed in bays (resulting in deposition).

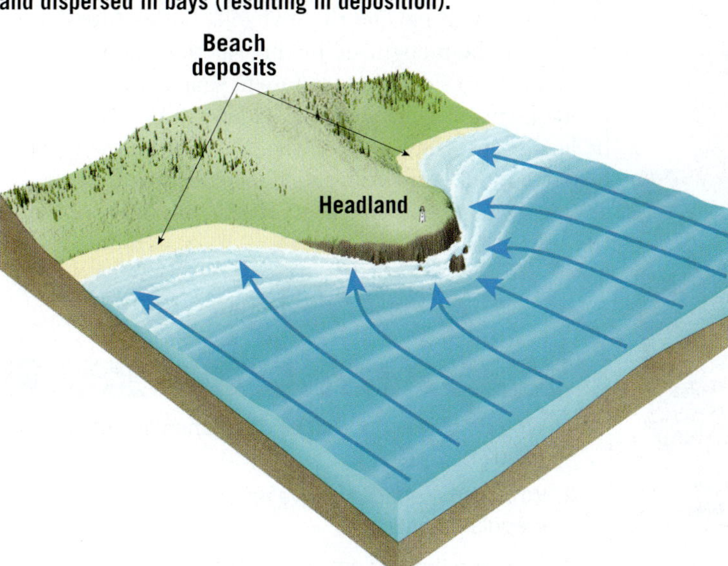
Beach deposits
Headland

Waves travel at original speed in deep water
Waves "feel bottom" and slow down in surf zone
Shoreline
Result: waves bend so that they strike the shore more directly

Wave refraction at Rincon Point, California

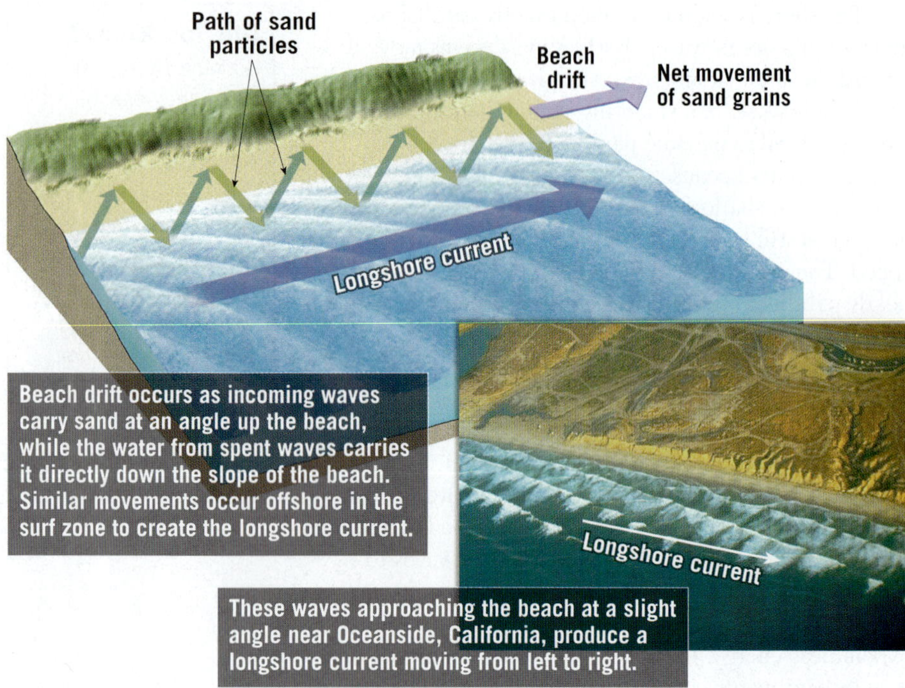

Path of sand particles
Beach drift
Net movement of sand grains
Longshore current
Beach drift occurs as incoming waves carry sand at an angle up the beach, while the water from spent waves carries it directly down the slope of the beach. Similar movements occur offshore in the surf zone to create the longshore current.
Longshore current
These waves approaching the beach at a slight angle near Oceanside, California, produce a longshore current moving from left to right.

▲ **SmartFigure 17.10 The longshore transport system** The two components of the transport system, beach drift and longshore currents, are created by breaking waves that approach the beach at an angle. These processes transport large quantities of material along the beach and in the surf zone. (Photo by John S. Shelton/University of Washington Libraries)

TUTORIAL
https://goo.gl/uE0I0E

face (**Figure 17.10**). This movement is called **beach drift**, and it can transport sand and pebbles hundreds or even thousands of meters each day. However, a more typical rate is 5 to 10 meters per day.

Waves that approach the shore at an angle also produce currents within the surf zone that flow parallel to the shore and move substantially more sediment than beach drift. Because the water here is turbulent, these **longshore currents** easily move the fine suspended sand and roll larger sand and gravel along the bottom. When the sediment transported by longshore currents is added to the quantity moved by beach drift, the total amount can be very large. At Sandy Hook, New Jersey, for example, the quantity of sand transported along the shore over a 48-year period averaged almost 750,000 tons annually. For a 10-year period in Oxnard, California, more than 1.5 million tons of sediment moved along the shore each year.

Both rivers and coastal zones move water and sediment from one area (*upstream*) to another (*downstream*). As a result, the beach has often been characterized as a "river of sand." Beach drift and longshore currents, however, move in a zigzag pattern, whereas rivers flow mostly in a turbulent, swirling fashion. In addition, the direction of flow of longshore currents along a shoreline can change if the direction from which waves approach the beach changes, whereas rivers always flow in the same direction (downhill). Nevertheless, longshore currents generally flow southward along both the Atlantic and Pacific shores of the United States.

Rip Currents **Rip currents** are concentrated movements of water that flow *opposite* the direction of breaking waves. (Sometimes rip currents are incorrectly called *rip tides*, although they are unrelated to tidal phenomena.) Most of the backwash from spent waves finds its way back to the open ocean as an unconfined flow across the ocean bottom called *sheet flow*. However, sometimes a portion of the returning water moves seaward in the form of surface rip currents. These currents do not travel far beyond the surf zone before breaking up and can be recognized by the way they interfere with incoming waves or by the sediment that is often suspended within the rip current (**Figure 17.11**). They can be hazardous to swimmers, who, if caught in them, can be carried out away from shore. The best strategy for exiting a rip current is to swim *parallel* to the shore for a few tens of meters.

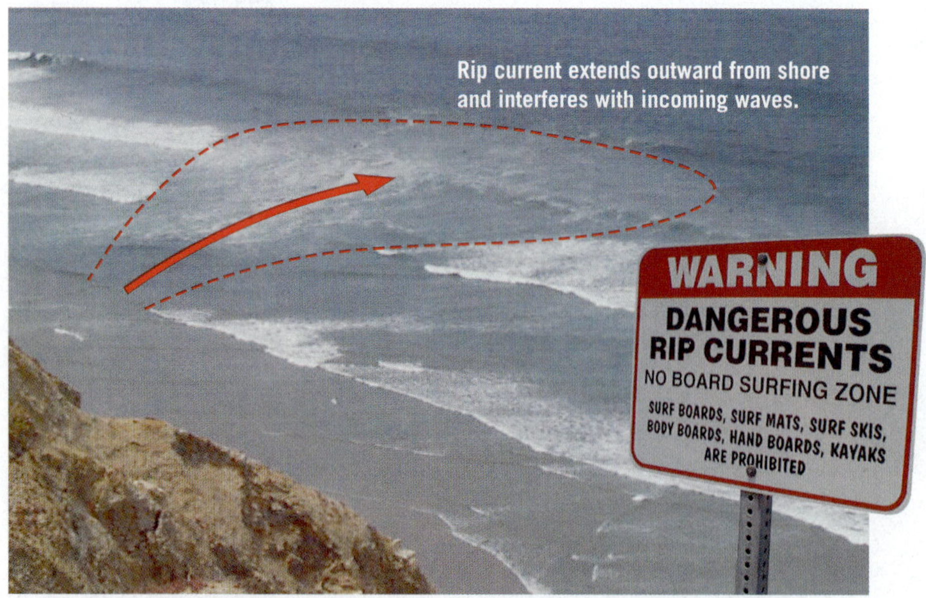

NO BOARD SURFING ZONE
SURF BOARDS, SURF MATS, SURF SKIS, BODY BOARDS, HAND BOARDS, KAYAKS ARE PROHIBITED

▲ **Figure 17.11 Rip current** These concentrated movements of water flow opposite the direction of breaking waves. (Photo by A. P. Trujillo/APT Photos)

CONCEPT CHECKS 17.2

1. Why do waves approaching the shoreline often bend?
2. What is the effect of wave refraction along an irregular coastline?
3. Describe the two processes that contribute to longshore transport.

17.3 Shoreline Features

Describe the features typically created by wave erosion and those resulting from sediment deposited by longshore transport processes.

A fascinating assortment of shoreline features can be observed along the world's coastal regions. Although the same processes cause change along every coast, not all coasts respond in the same way. Interactions among different processes and the relative importance of each process depend on local factors, including (1) the proximity of a coast to sediment-laden rivers, (2) the degree of tectonic activity, (3) the topography and composition of the land, (4) prevailing winds and weather patterns, and (5) the configuration of the coastline and near-shore areas. Features that originate primarily because of erosion are called *erosional features*, whereas accumulations of sediment produce *depositional features*.

Erosional Features

Many coastal landforms owe their origin to erosional processes. Such erosional features are common along the rugged and irregular New England coast and along the steep shorelines of the west coast of the United States.

Wave-Cut Cliffs, Wave-Cut Platforms, and Marine Terraces As the name implies, **wave-cut cliffs** originate in the cutting action of the surf against the base of coastal land. As erosion progresses, rocks overhanging the notch at the base of the cliff crumble into the surf, and the cliff retreats. A relatively flat, benchlike surface, called a **wave-cut platform**, is left behind by the receding cliff (Figure 17.12, *left*). The platform broadens as wave attack continues. Some debris produced by the breaking waves remains along the water's edge as sediment on the beach, and the remainder is transported farther seaward. If a wave-cut platform is uplifted above sea level by tectonic forces, it becomes a **marine terrace** (see Figure 17.12, *right*). Marine terraces are easily recognized by their gentle seaward-sloping shape and are often desirable sites for coastal roads, buildings, or agriculture.

Sea Arches & Sea Stacks Because of refraction, waves vigorously attack headlands that extend into the sea. The surf erodes the rock selectively, wearing away the softer or more highly fractured rock at the fastest rate. At first, sea caves may form. When two caves on opposite sides of a headland unite, a **sea arch** results (Figure 17.13). Eventually the arch falls in, leaving an isolated remnant, or **sea stack**, on the wave-cut platform (see Figure 17.13). In time, it too will be consumed by the action of the waves.

Depositional Features

Sediment that is transported along the shore tends to be deposited in areas where wave energy is low. Such processes produce a variety of depositional features.

Spits, Bars, and Tombolos Where beach drift and longshore currents are active, several features related to the movement of sediment along the shore may develop.

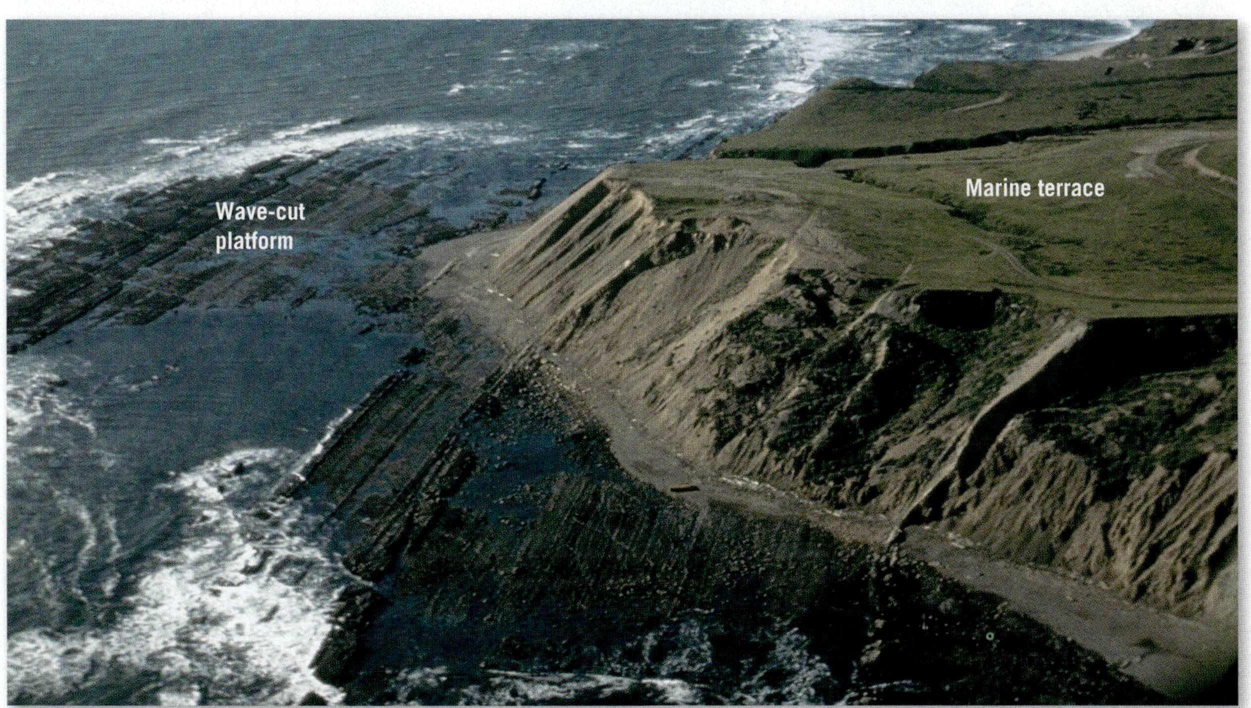

Figure 17.12 Wave-cut platform and marine terrace This wave-cut platform is exposed at low tide along the California coast at Bolinas Point near San Francisco. A wave-cut platform was uplifted to create the marine terrace. (Photo by John S. Shelton/University of Washington Libraries)

▲ **Figure 17.13 Sea stack and sea arch** These features at the tip of Mexico's Baja Peninsula resulted from vigorous wave attack on a headland. (Photo by Lew Robertson/Getty Images)

A **spit** is an elongated ridge of sand that projects from the land into the mouth of an adjacent bay. Often the end of a spit that is in the water hooks landward in response to the dominant direction of the longshore current (Figure 17.14). The term **baymouth bar** is applied to a sandbar that completely crosses a bay, sealing it off from the open ocean (see Figure 17.14). Such a feature tends to form across a bay where currents are weak, allowing a spit to extend to the other side. A **tombolo** (*tombolo* = mound), a ridge of sand that connects an island to the mainland or to another island, forms in much the same manner as a spit.

Barrier Islands The Atlantic and Gulf coastal plains are relatively flat and slope gently seaward. The shore zone is characterized by **barrier islands**. These low ridges of land parallel the coast at distances from 3 to 30 kilometers offshore. From Cape Cod, Massachusetts, to Padre Island, Texas, nearly 300 barrier islands rim the coast (Figure 17.15).

Most barrier islands are 1 to 5 kilometers wide and 15 to 30 kilometers long. The tallest features are sand dunes, which usually reach heights of 5 to 10 meters; in a few areas, unvegetated dunes are more than 30 meters high. The lagoons separating these narrow islands from the shore represent zones of relatively quiet water that allow small craft traveling between New York and northern Florida to avoid the rough waters of the North Atlantic.

Barrier islands form in several ways. Some originated as spits that were severed from the mainland by wave erosion or by the general rise in sea level after the last episode of glaciation. Others are created when turbulent waters in the line of breakers heap up sand scoured from the ocean floor. Because these sand barriers rise above normal sea level, the sand likely piles up as a result of the work of storm waves at high tide. Finally, some barrier islands may be former sand dune ridges that originated along the shore during the last glacial period, when sea level was lower. When the ice sheets melted, sea level rose and flooded the area behind the beach–dune complex.

▶ SmartFigure 17.14
Coastal Massachusetts
A. High-altitude image of a well-developed spit and baymouth bar along the coast of Martha's Vineyard, Massachusetts. (Image courtesy of USDA-ASCS)
B. This photograph, taken from the International Space Station, shows Provincetown Spit at the tip of Cape Cod. (NASA image)

MOBILE FIELD TRIP
https://goo.gl/fz5L6k

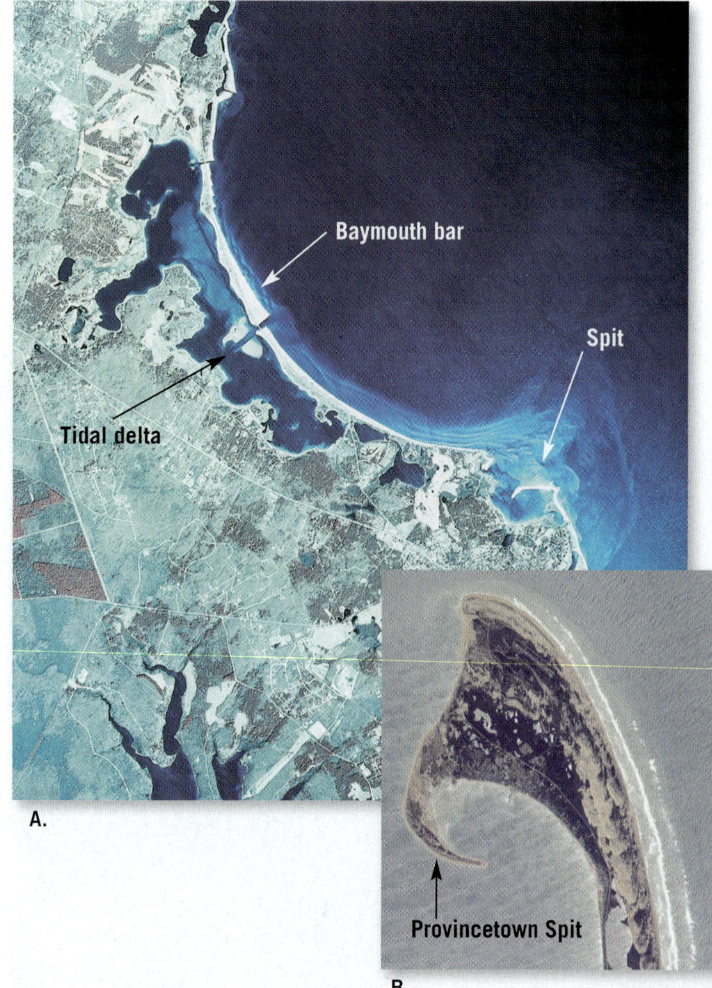

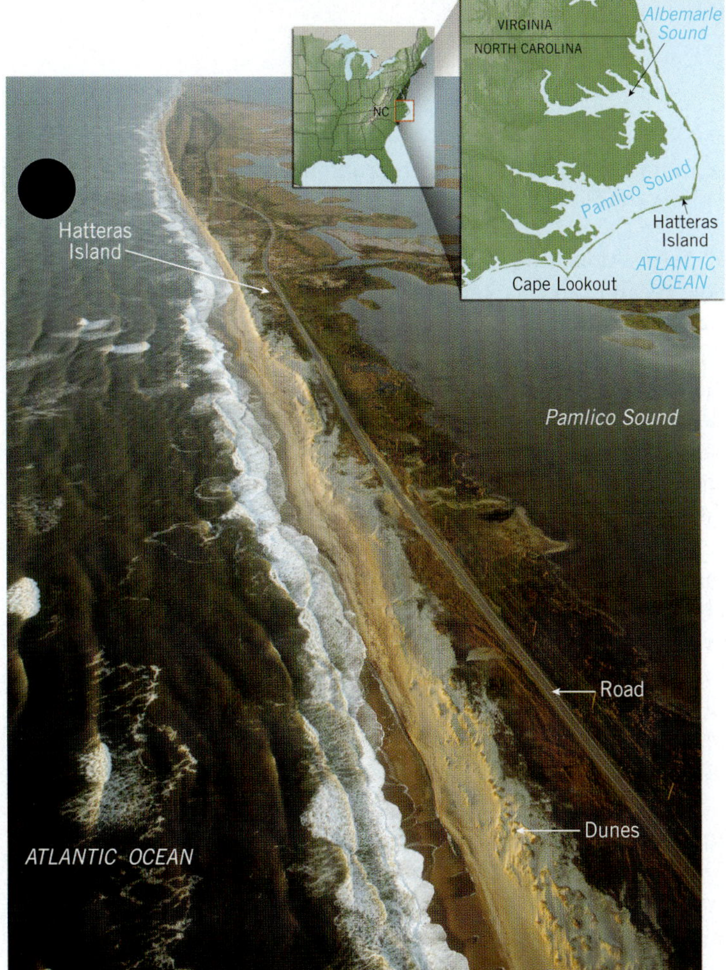

▲ **Figure 17.15 Barrier islands** Nearly 300 barrier islands rim the Gulf and Atlantic coasts. The islands along the coast of North Carolina are excellent examples. (Photo by Michael Collier)

The Evolving Shore

A shoreline continually undergoes modification, regardless of its initial configuration. Along a coastline that is characterized by varied geology, the pounding surf may at first increase its irregularity because the waves will erode the weaker rocks more easily than the stronger ones. However, if a shoreline remains tectonically stable, marine erosion and deposition will eventually produce a straighter, more regular coast. **Figure 17.16** illustrates the evolution of an initially irregular coast. As waves erode the headlands, creating cliffs and a wave-cut platform, sediment is carried along the shore. Some material is deposited in the bays, while other debris is formed into spits and baymouth bars. At the same time, rivers fill the bays with sediment. Ultimately, a generally straight, smooth coast results.

CONCEPT CHECKS 17.3

1. How is a marine terrace related to a wave-cut platform?
2. Describe the formation of each labeled feature in Figure 17.16.
3. List three ways that a barrier island may form.

Did You Know?

Along shorelines composed of unconsolidated material rather than solid rock, the rate of erosion by breaking waves can be extraordinary. In parts of Britain, where waves have the easy task of eroding glacial deposits of sand, gravel, and clay, the coast has been worn back 3 to 5 km (2 to 3 mi) since Roman times (2000 years ago). Waves have swept away many villages and ancient landmarks.

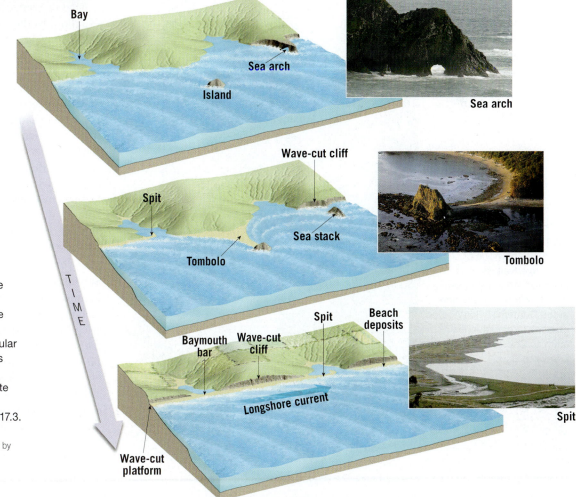

Bay
Sea arch
Island
Sea arch
Wave-cut cliff
Spit
Sea stack
Tombolo
Tombolo
TIME
Baymouth bar
Wave-cut cliff
Spit
Beach deposits
Longshore current
Wave-cut platform
Spit

► **Figure 17.16 The evolving shore** These diagrams illustrate changes that can take place through time along an initially irregular coastline that remains relatively stable. The diagrams also illustrate many of the features described in Section 17.3. (Top and bottom photos by E. J. Tarbuck; middle photo by Michael Collier)

17.4 Contrasting America's Coasts

Distinguish between emergent and submergent coasts. Contrast the erosion problems faced on the Atlantic and Gulf coasts with those along the Pacific coast.

The shoreline along the Pacific coast of the United States is strikingly different from that of the Atlantic and Gulf coast regions. Some of the differences are related to plate tectonics. The west coast represents the leading edge of the North American plate, and thus, it experiences active uplift and deformation. By contrast, the east coast is far from any active plate boundary and is relatively quiet tectonically. Because of this basic geologic difference, the types of shoreline erosion problems along America's opposite coasts are different.

Coastal Classification

The great variety of shorelines demonstrates their complexity. Indeed, to understand any particular coastal area, many factors must be considered, including rock types, size and direction of waves, frequency of storms, tidal range, and offshore topography. In addition, recall from Chapter 15 that practically all coastal areas were affected by the worldwide rise in sea level that accompanied the melting of ice sheets following the Last Glacial Maximum. Finally, tectonic events that elevate or drop the land or change the volume of ocean basins must be taken into account. The numerous factors that influence coastal areas make shoreline classification difficult.

Many geologists classify coasts based on the changes that have occurred with respect to sea level. This commonly used classification divides coasts into two general categories: emergent and submergent. **Emergent coasts** develop because an area experiences either uplift of the land or a drop in sea level. Conversely, **submergent coasts** are created when sea level rises or the land subsides.

▶ SmartFigure 17.17
East coast estuaries The lower portions of many river valleys were flooded by the rise in sea level that followed the end of the Quaternary Ice Age, creating large estuaries such as Chesapeake and Delaware Bays.

TUTORIAL
https://goo.gl/CkLRFL

Emergent Coasts In some areas, the coast is clearly emergent because rising land or a falling sea level exposes wave-cut cliffs and platforms. Excellent examples include portions of coastal California, where uplift has occurred in the recent geologic past. The marine terrace shown in Figure 17.12 illustrates this situation. In the case of the Palos Verdes Hills, south of Los Angeles, seven different terrace levels exist, indicating seven episodes of uplift. The ever-persistent sea is now cutting a new platform at the base of the cliff. If uplift follows, it too will become an elevated marine terrace.

Other examples of emergent coasts include regions that were once buried beneath great ice sheets. When glaciers were present, their weight depressed the crust, and when the ice melted, the crust began gradually to spring back. Consequently, prehistoric shoreline features may now be found high above sea level. The Hudson Bay region of Canada is such an area; portions of it are still rising at a rate of more than 1 centimeter per year.

Submergent Coasts In contrast to the preceding examples, other coastal areas show definite signs of submergence. Shorelines that have been submerged in the relatively recent past are often highly irregular because the sea typically floods the lower reaches of river valleys flowing into the ocean. The ridges separating the valleys, however, remain above sea level and project into the sea as headlands. These drowned river mouths, which are called **estuaries**, characterize many coasts today. Along the Atlantic coastline, the Chesapeake and Delaware Bays are examples of large estuaries created by submergence (Figure 17.17). The picturesque coast of Maine,

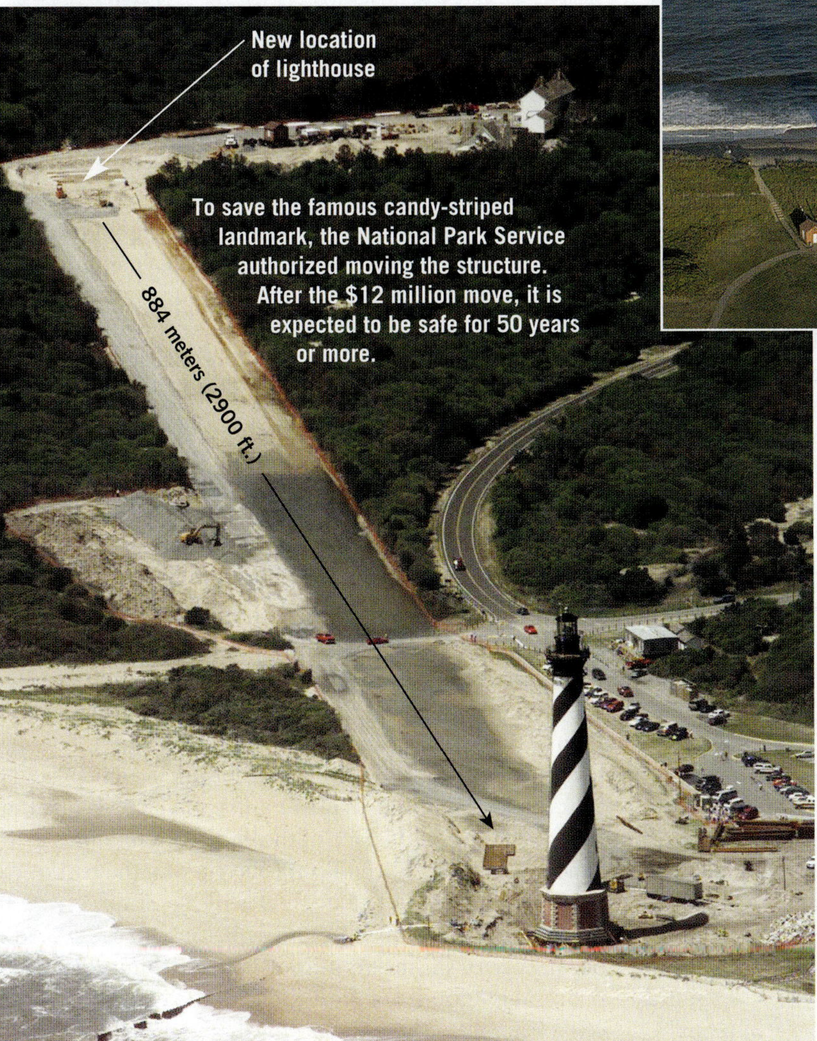

Various attempts to protect the lighthouse failed. They included building groins and beach nourishment. By 1999, when this photo was taken, the lighthouse was only 36 meters (120 ft.) from the water.
New location of lighthouse
To save the famous candy-striped landmark, the National Park Service authorized moving the structure. After the $12 million move, it is expected to be safe for 50 years or more.
884 meters (2900 ft.)

Figure 17.18 Relocating the Cape Hatteras lighthouse After the failure of a number of efforts to protect this 21-story lighthouse, the nation's tallest lighthouse, from being destroyed due to a receding shoreline, the structure finally had to be moved. (Top photo by Don Smetzer/PhotoEdit Inc.; bottom photo by Drew C. Wilson/ Virginian-Pilot/AP Photo)

particularly in the vicinity of Acadia National Park, is another excellent example of an area that was flooded by the postglacial rise in sea level and transformed into a highly irregular coastline.

Keep in mind that most coasts have complicated geologic histories. With respect to sea level, at various times many coasts have emerged and then submerged. Each time, they may retain some of the features created during the previous situation.

Atlantic & Gulf Coasts

Much of the coastal development along the Atlantic and Gulf coasts has occurred on barrier islands. Typically, a barrier island, also termed a *barrier beach* or *coastal barrier*, consists of a wide beach that is backed by dunes and separated from the mainland by marshy lagoons. The broad expanses of sand and exposure to the ocean have made barrier islands exceedingly attractive sites for development. Unfortunately, development has taken place more rapidly than our understanding of barrier island dynamics has increased.

Because barrier islands face the open ocean, they receive the full force of major storms that strike the coast. When a storm occurs, the barriers absorb the energy of the waves primarily through the movement of sand. Figure 17.18, which shows changes at Cape Hatteras National Seashore, reinforces this point. The process and problems that result were recognized years ago and accurately described as follows:

> Waves may move sand from the beach to offshore areas or, conversely, into the dunes; they may erode the dunes, depositing sand onto the beach or carrying

it out to sea; or they may carry sand from the beach and the dunes into the marshes behind the barrier, a process known as overwash. The common factor is movement. Just as a flexible reed may survive a wind that destroys an oak tree, so the barriers survive hurricanes and nor'easters not through unyielding strength but by giving before the storm.

This picture changes when a barrier is developed for homes or as a resort. Storm waves that previously rushed harmlessly through gaps between the dunes now encounter buildings and roadways. Moreover, since the dynamic nature of the barriers is readily perceived only during storms, homeowners tend to attribute damage to a particular storm, rather than to the basic mobility of coastal barriers. With their homes or investments at stake, local residents are more likely to seek to hold the sand in place and the waves at bay than to admit that development was improperly placed to begin with.*

*Frank Lowenstein, "Beaches or Bedrooms—The Choice as Sea Level Rises," *Oceanus* 28 (no. 3, Fall 1985): p. 22 © Woods Hole Oceanographic Institute.

▶ **Figure 17.19 Pacoima Dam and Reservoir** Dams such as this one in the San Gabriel Mountains near Los Angeles trap sediment that otherwise would have nourished beaches along the nearby coast. (Photo by Michael Collier)

Pacific Coast

In contrast to the broad, gently sloping coastal plains of the Atlantic and Gulf coasts, much of the Pacific coast is characterized by relatively narrow beaches that are backed by steep cliffs and mountain ranges. The chapter-opening photo provides a good example. Recall that America's western margin is a more rugged and tectonically active region than the eastern margin. Because uplift continues, a rise in sea level in the West is not so readily apparent. Nevertheless, like the shoreline erosion problems facing the Atlantic coast's barrier islands, west coast difficulties also stem largely from the human alteration of a natural system.

A major problem facing the Pacific shoreline, and especially portions of southern California, is a significant narrowing of many beaches. The bulk of the sand on many of these beaches is supplied by rivers that transport it from the mountains to the coast. Over the years, this natural flow of material to the coast has been interrupted by dams built for irrigation and flood control. The reservoirs effectively trap the sand that would otherwise nourish the beach environment (**Figure 17.19**). When the beaches were wider, they protected the cliffs behind them from the force of storm waves. Now, however, the waves move across the narrowed beaches without losing much energy and cause more rapid erosion of the sea cliffs. Figure 17.1 provides an example. In efforts over the years to halt cliff retreat at Pacifica, California, the city piled rocks along the beach, drilled reinforcement rods into the bluffs, and coated the face of the cliffs with reinforced concrete. Ultimately, the Pacific Ocean won the battle.

Shoreline erosion along the Pacific coast varies considerably from year to year, largely because of the sporadic occurrence of storms. As a result, when the infrequent but serious episodes of erosion occur, the damage is often blamed on the unusual storms and not on coastal development or the sediment-trapping dams that may be great distances away. As sea level rises at an increasing rate in the years to come, increased shoreline erosion and sea-cliff retreat should be expected along many parts of the Pacific coast. Coastal vulnerability to sea-level rise is examined in more detail as part of a discussion of the possible consequences of global warming in Chapter 20.

CONCEPT CHECKS 17.4

1. Are estuaries associated with submergent or emergent coasts? Explain.
2. What observable features would lead you to classify a coastal area as emergent?
3. Briefly describe what happens when storm waves strike an undeveloped barrier island.
4. How might building a dam on a river that flows to the sea affect a coastal beach?

17.5 Hurricanes: The Ultimate Coastal Hazard

Describe the basic structure and characteristics of a hurricane and the three broad categories of hurricane destruction.

Whirling tropical cyclones—the greatest storms on Earth—occasionally have wind speeds exceeding 300 kilometers (185 miles) per hour. In the United States they are known as **hurricanes**, in the western Pacific they are called *typhoons*, and in the Indian Ocean they are simply called *cyclones*. No matter which name is used, these storms are among the most destructive of natural disasters (Figure 17.20).

The vast majority of hurricane-related deaths and damage are caused by relatively infrequent yet powerful storms. Of course, the deadliest and most costly storm in recent memory occurred in August 2005, when Hurricane Katrina devastated the Gulf coast of Louisiana, Mississippi, and Alabama. Although hundreds of thousands of people fled before the storm made landfall, thousands of others were caught by the storm. In addition to the human suffering and tragic loss of life that were left in the wake of Hurricane Katrina, the financial losses caused by the storm are practically incalculable.

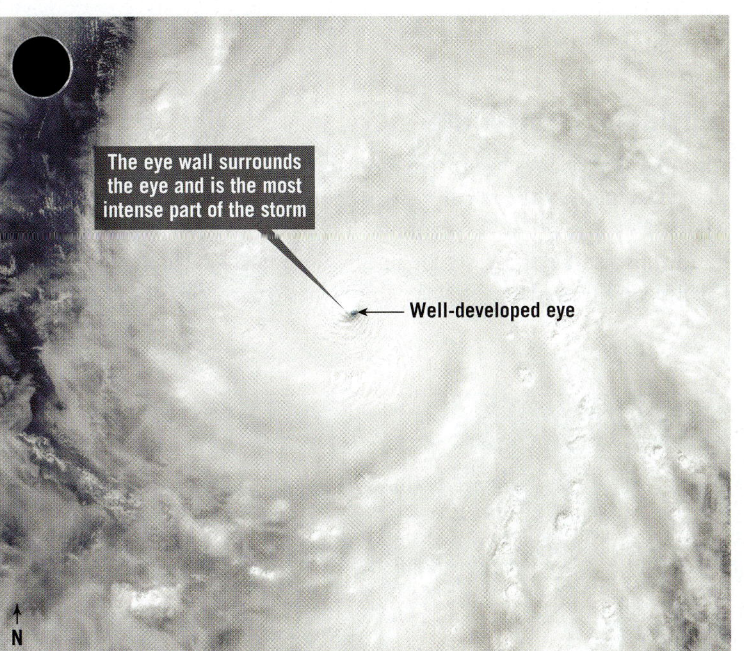

▲ **Figure 17.20 Hurricane Patricia** This satellite image from October 23, 2015, shows a hurricane over the eastern Pacific Ocean near the west coast of Mexico. With sustained winds of 352 kilometers (200 miles) per hour, it was the strongest hurricane ever recorded in the Western Hemisphere. Fortunately, the storm quickly weakened after making landfall. The counterclockwise spiral of the clouds indicates that it is a Northern Hemisphere storm. In the Southern Hemisphere, the spiral of a cyclone is clockwise. (NASA)

Our coasts are vulnerable. People are flocking to live near the ocean. The concentration of large numbers of people near the shoreline means that hurricanes place millions at risk. Moreover, the potential costs of property damage are incredible. As sea level continues to rise in coming decades, low-lying, densely populated coastal areas will become even more vulnerable to the destructive effects of major storms.

Did You Know?
Hurricane season is different in different parts of the world. People in the United States are usually most interested in Atlantic storms. The Atlantic hurricane season officially extends from June 1 through November 30. More than 97 percent of tropical activity in that region occurs during this 6-month span. Statistically, the "heart" of the season is August through October, and peak activity is in early to mid-September.

Profile of a Hurricane

A hurricane is a heat engine that is fueled by the energy liberated when huge quantities of water vapor condense. The amount of energy produced by a typical hurricane in just a single day is truly immense. To get the engine started, a large quantity of warm, moist air is required, and a continuous supply is needed to keep it going.

Hurricane Formation As the graph in Figure 17.21 illustrates, hurricanes most often form in late summer and early fall. It is during this span that sea-surface temperatures reach 27°C (80°F) or higher and are thus able to provide the necessary heat and moisture to the air (Figure 17.22). This ocean-water temperature requirement explains why hurricane formation over the relatively cool waters of the South Atlantic and eastern South Pacific is extremely rare. For the same reason, few hurricanes form poleward of 20° latitude. Although water temperatures are sufficiently high, hurricanes do not develop within about 5° of the equator because the Coriolis effect (the force related to Earth's rotation that gives storms their "spin") is too

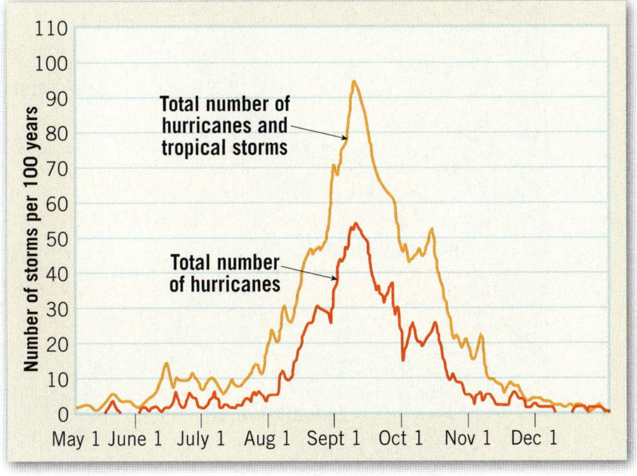

◀ **Figure 17.21 When do Atlantic hurricanes occur?** Frequency of tropical storms and hurricanes from May 1 through December 31 in the Atlantic basin. The graph shows the number of storms to be expected over a span of 100 years. The period from late August through October is clearly the most active. (Data from National Hurricane Center/NOAA)

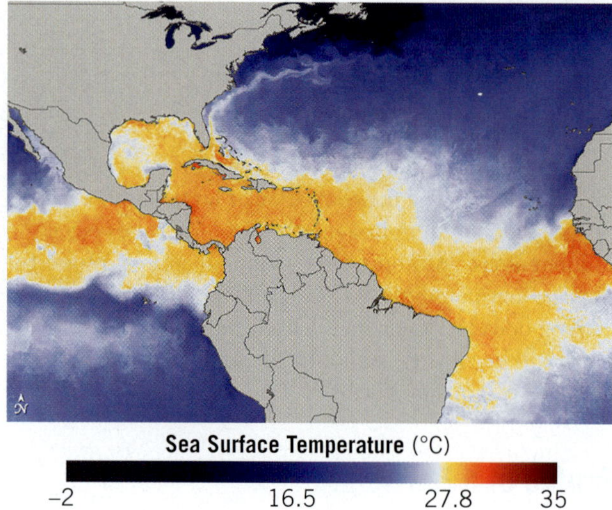

▶ **Figure 17.22 Sea-surface temperatures** Among the necessary ingredients for a hurricane is warm ocean temperatures above 27°C (80°F). This color-coded satellite image from June 1, 2010, shows sea-surface temperatures at the beginning of hurricane season. (NASA)

weak there. Figure 17.23 shows the regions where most hurricanes form.

Pressure Gradient Hurricanes are intense low-pressure centers, which means that as you move toward the center of the storm, air pressure gets lower and lower. Such storms are said to have a very steep pressure gradient. *Pressure gradient* refers to how rapidly the pressure changes per unit distance and is shown on a map with *isobars*, lines of equal pressure. Just as the spacing of contour lines on a topographic map indicates how steep or gentle a slope is, the spacing of isobars on a weather chart shows how rapidly air pressure is changing. Closely spaced isobars indicate a steep pressure gradient and stronger winds. A steep pressure gradient generates the rapid, inward-spiraling winds of a hurricane. As the air rushes toward the center of the storm, its velocity increases. This is similar to skaters with their arms extended spinning faster as they pull their arms in close to their bodies.

Storm Structure As the inward rush of warm, moist surface air approaches the core of a storm, it turns upward and ascends in a ring of cumulonimbus cloud towers (Figure 17.24). This doughnut-shaped wall of intense convective activity surrounding the center of the storm is called the **eye wall**. It is here that the greatest wind speeds and heaviest rainfall occur. Surrounding the eye wall are curved bands of clouds that trail away in a spiral fashion. Near the top of the hurricane, the airflow is outward, carrying the rising air away from the storm center, thereby providing room for more inward flow at the surface.

At the very center of the storm is the **eye** of the hurricane (see Figure 17.24). This well-known feature is a zone about 20 kilometers (12.5 miles) in diameter where precipitation ceases and winds subside. It offers a brief but deceptive break from the extreme weather in the enormous curving wall clouds that surround it. The air within the eye gradually descends and heats by compression, making it the warmest part of the storm. Although many people believe that the eye is characterized by clear blue skies, this is usually not the case because the subsidence in the eye is seldom strong enough to produce cloudless conditions. Although the sky appears much brighter in this region, scattered clouds at various levels are common.

Hurricane Destruction

Although hurricanes are tropical or subtropical in origin, their destructive effects can be experienced far from where they originate. For example, in 2012 Hurricane Sandy (also called Superstorm Sandy) originated in the Caribbean Sea and affected the entire eastern seaboard

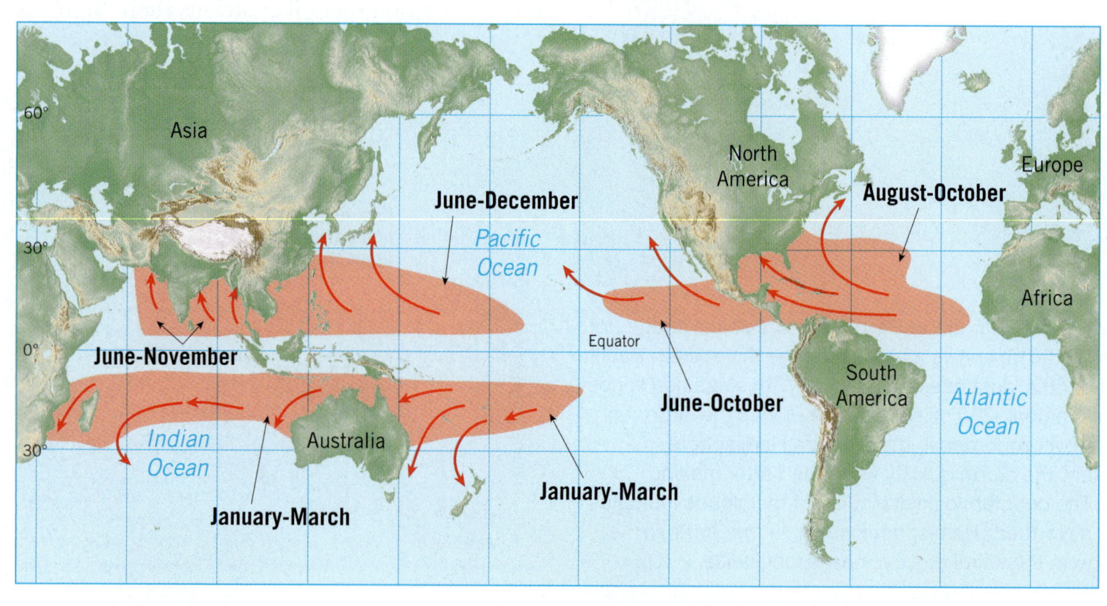

▶ **SmartFigure 17.23 Hurricane source regions and paths** The map shows the regions where most hurricanes form as well as their principal months of occurrence and the tracks they most commonly follow. Hurricanes do not develop within about 5° of the equator because the Coriolis effect (a force related to Earth's rotation that gives storms their "spin") there is too weak. Because warm ocean-surface temperatures are necessary for hurricane formation, hurricanes seldom form poleward of 20° latitude or over the cool waters of the south Atlantic and the eastern south Pacific.

TUTORIAL
https://goo.gl/7sDNGV

from Florida to Maine. Destruction was especially great in New Jersey and New York, even though Sandy was downgraded from hurricane status by that time (see Figure 17.2).

The amount of damage caused by a hurricane depends on several factors, including the size and population density of the area affected and the shape of the ocean bottom near the shore. The most significant factor, of course, is the strength of the storm itself. By studying past storms, a scale called the *Saffir–Simpson hurricane scale* was established to rank the relative intensities of hurricanes. As Table 17.1 indicates, a *category 5* storm is the worst possible, and a *category 1* hurricane is least severe.

During the hurricane season, it is common to hear scientists and reporters alike use the numbers from the Saffir–Simpson scale. When Hurricane Katrina made landfall, sustained winds were 200 kilometers (125 miles) per hour, making it a strong category 3 storm. Storms that fall into category 5 are rare. Hurricane Camille, a 1969 storm that caused catastrophic damage along the coast of Mississippi, is one well-known example.

Once a hurricane makes landfall, it loses energy because it is cut off from its energy source—warm ocean water—and is usually downgraded to a lower category. However, these storms are so large and violent that their effects are often felt far inland. Damage caused by hurricanes can be divided into three categories: storm surge, wind damage, and inland flooding.

Storm Surge Without question, the most devastating damage in the coastal zone is caused by storm surge. It not only accounts for a large share of coastal property losses but is also responsible for a high percentage of all hurricane-caused deaths. A **storm surge** is a dome of water 65 to 80 kilometers (40 to 50 miles) wide that sweeps across the coast near the point where the eye makes landfall. If all wave activity were smoothed out, the storm surge would be the height of the water above normal tide level. In addition, tremendous wave activity is superimposed on the surge. This surge of water can inflict immense damage on low-lying coastal areas. Figure 17.25 and Figure 17.2 are both good examples. The worst surges occur in places like the Gulf of Mexico, where the continental shelf is very shallow and gently sloping. In addition, local features such as bays and rivers can cause the surge to double in height and increase in speed.

As a hurricane advances toward the coast in the Northern Hemisphere, storm surge is always most intense on the right side of the eye, where winds are

Outflow of air at the top of the hurricane is important because it prevents the convergent flow at lower levels from "filling in" the storm.

Eye

Sinking air in the eye warms by compression.

Eye wall, the zone where winds and rain are most intense.

Tropical moisture spiraling inward creates rain bands that pinwheel around the storm center.

Cross section of a hurricane. Note that the vertical dimension is greatly exaggerated. (After NOAA)

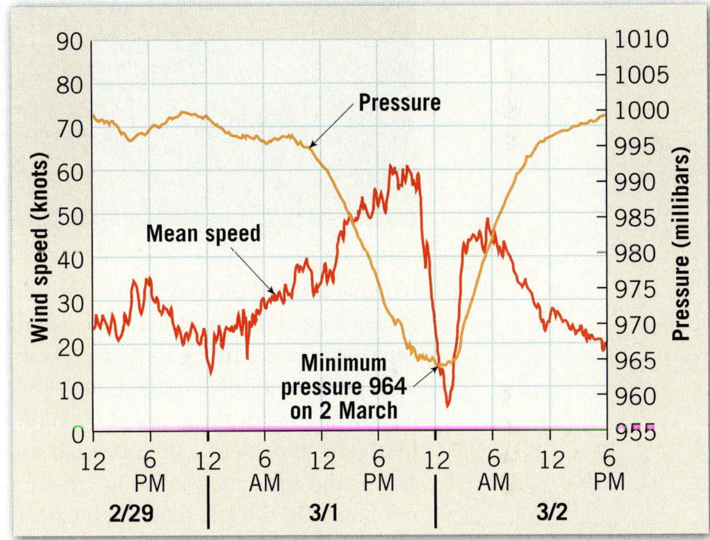

Measurements of surface pressure and wind speed during the passage of Cyclone Monty at Mardie Station, Western Australia, between February 29 and March 2, 2004. (Hurricanes are called "cyclones" in this part of the world.)

▲ SmartFigure 17.24 **Cross section of a hurricane** (Data from World Meteorological Organization)

VIDEO
https://goo.gl/mGbcwY

Table 17.1 Saffir–Simpson Hurricane Scale

Scale Number (category)	Central Pressure (millibars)	Wind Speed (kph)	Wind Speed (mph)	Storm Surge (meters)	Storm Surge (feet)	Damage
1	≥980	119–153	74–95	1.2–1.5	4–5	*Minimal*
2	965–979	154–177	96–110	1.6–2.4	6–8	*Moderate*
3	945–964	178–209	111–130	2.5–3.6	9–12	*Extensive*
4	920–944	210–250	131–155	3.7–5.4	13–18	*Extreme*
5	<920	>250	>155	>5.4	>18	*Catastrophic*

▶ **Figure 17.25 Storm surge destruction** This is Crystal Beach, Texas, on September 16, 2008, 3 days after Hurricane Ike came ashore. At landfall the storm had sustained winds of 165 kilometers (105 miles) per hour. The extraordinary storm surge caused most of the damage shown here. (Photo by REUTERS/Smiley N. Pool/Pool)

blowing *toward* the shore. On this side of the storm, the forward movement of the hurricane also contributes to the storm surge. In Figure 17.26, assume that a hurricane with peak winds of 175 kilometers (109 miles) per hour is moving toward the shore at 50 kilometers (31 miles) per hour. In this case, the net wind speed on the right side of the advancing storm is 225 kilometers (140 miles) per hour. On the left side, the hurricane's winds are blowing opposite the direction of storm movement, so the net winds are *away* from the coast at 125 kilometers (78 miles) per hour. Along the shore facing the left side of the oncoming hurricane, the water level may actually decrease as the storm makes landfall.

Wind Damage Destruction caused by wind is perhaps the most obvious of the classes of hurricane damage. Debris such as signs, roofing materials, and small items left outside become dangerous flying missiles during hurricanes. For some structures, the force of the wind is sufficient to cause total ruin. Mobile homes are particularly vulnerable. High-rise buildings are also susceptible to hurricane-force winds. Upper floors are most vulnerable because wind speeds usually increase with height. Recent research suggests that people should stay below the 10th floor of a building but remain above any floors at risk for flooding. In regions with good building codes, wind damage is usually not as catastrophic as storm-surge damage. However, hurricane-force winds affect a much larger area than storm surge and can cause huge

Did You Know?

The difference between a *hurricane* and a *tropical storm* is related to intensity. Both are tropical cyclones—circular zones of low pressure with strong, inward-spiraling winds. When sustained winds are between 37 and 74 mph, the cyclone is called a tropical storm. It is during this phase that a name is given (Andrew, Fran, Rita, and so on). When the cyclone's sustained winds exceed 74 mph, it has reached hurricane status.

▲ **Figure 17.26 An approaching hurricane** This hypothetical Northern Hemisphere hurricane, with peak winds of 175 kilometers (109 miles) per hour, is moving toward the coast at 50 kilometers (31 miles) per hour. On the right side of the advancing storm, the 175-kilometer-per-hour winds are in the same direction as the movement of the storm (50 kilometers per hour). Therefore the *net* wind speed on the right side of the storm is 225 kilometers (140 miles) per hour *toward* the coast. On the left side, the hurricane's winds are blowing opposite the direction of storm movement, so the *net* winds of 125 kilometers (78 miles) per hour are *away* from the coast. Storm surge will be greatest along the part of the coast hit by the right side of the advancing hurricane.

economic losses. For example, in 1992 it was largely the winds associated with Hurricane Andrew that produced more than $25 billion of damage in southern Florida and Louisiana.

Hurricanes sometimes produce tornadoes that contribute to the storm's destructive power. Studies have shown that more than half of the hurricanes that make landfall produce at least one tornado. In 2004 the number of tornadoes associated with tropical storms and hurricanes was extraordinary. Tropical Storm Bonnie and five landfalling hurricanes—Charley, Frances, Gaston, Ivan, and Jeanne—produced nearly 300 tornadoes that affected the southeastern and mid-Atlantic states.

Heavy Rains & Inland Flooding The torrential rains that accompany most hurricanes bring a third significant threat: flooding. Whereas the effects of storm surge and strong winds are concentrated in coastal areas, heavy rains may affect places hundreds of kilometers from the coast for up to several days after the storm has lost its hurricane-force winds.

In September 1999, Hurricane Floyd brought flooding rains, high winds, and rough seas to a large portion of the Atlantic seaboard. More than 2.5 million people evacuated their homes from Florida north to the Carolinas and beyond. It was the largest peacetime evacuation in U.S. history up to that time. Torrential rains falling on already saturated ground created devastating inland flooding. Altogether Floyd dumped more than 48 centimeters (19 inches) of rain on Wilmington, North Carolina, with 33.98 centimeters (13.38 inches) falling in a single 24-hour span.

Another well-known example is Hurricane Camille (1969). Although this storm is best known for its exceptional storm surge and the devastation it brought to coastal areas, the greatest number of deaths associated with this storm occurred in the Blue Ridge Mountains of Virginia 2 days after Camille's landfall. Many places received more than 25 centimeters (10 inches) of rain.

CONCEPT CHECKS 17.5

1. What factors influence where and when hurricane formation takes place?
2. Distinguish between the eye and eye wall of a hurricane.
3. What are the three broad categories of hurricane damage? Which one is responsible for the greatest number of hurricane-related deaths?
4. Which side of an advancing hurricane in the Northern Hemisphere has the strongest winds and highest storm surge—right or left? Explain.

17.6 Stabilizing the Shore

Summarize the ways in which people deal with shoreline erosion problems.

Compared with natural hazards such as earthquakes, volcanic eruptions, and landslides, shoreline erosion is often perceived to be a more continuous and predictable process that appears to cause relatively modest damage to limited areas. In reality, the shoreline is a dynamic place that can change rapidly in response to natural forces. Exceptional storms are capable of eroding beaches and cliffs at rates that greatly exceed the long-term average. Such bursts of accelerated erosion not only significantly affect the natural evolution of a coast but also can have a profound impact on people who reside in the coastal zone. Erosion along our coasts causes significant property damage. Huge sums are spent annually not only to repair damage but also to prevent or control erosion. Already a problem at many sites, shoreline erosion is certain to become an increasingly serious problem as extensive coastal development continues.

During the past 100 years, growing affluence and increasing demands for recreation have brought unprecedented development to many coastal areas. As both the number and the value of buildings have increased, so too have efforts to protect property from storm waves by stabilizing the shore. Also, controlling the natural migration of sand is an ongoing struggle in many coastal areas. Such interference can result in unwanted changes that are difficult and expensive to correct.

Hard Stabilization

Structures built to protect a coast from erosion or to prevent the movement of sand along a beach are collectively known as **hard stabilization**. Hard stabilization can take many forms and often results in predictable yet unwanted outcomes. Hard stabilization includes jetties, groins, breakwaters, and seawalls.

Jetties Since relatively early in America's history, a principal goal in coastal areas has been the development and maintenance of harbors. In many cases, this has involved the construction of jetty systems. **Jetties** are usually built in pairs and extend into the ocean at the entrances to rivers and harbors. With the flow of water confined to a narrow zone, the ebb and flow caused by the rise and fall of the tides keep the sand in motion and prevent deposition in the channel. However, as illustrated in Figure 17.27, a jetty may act as a dam against which the

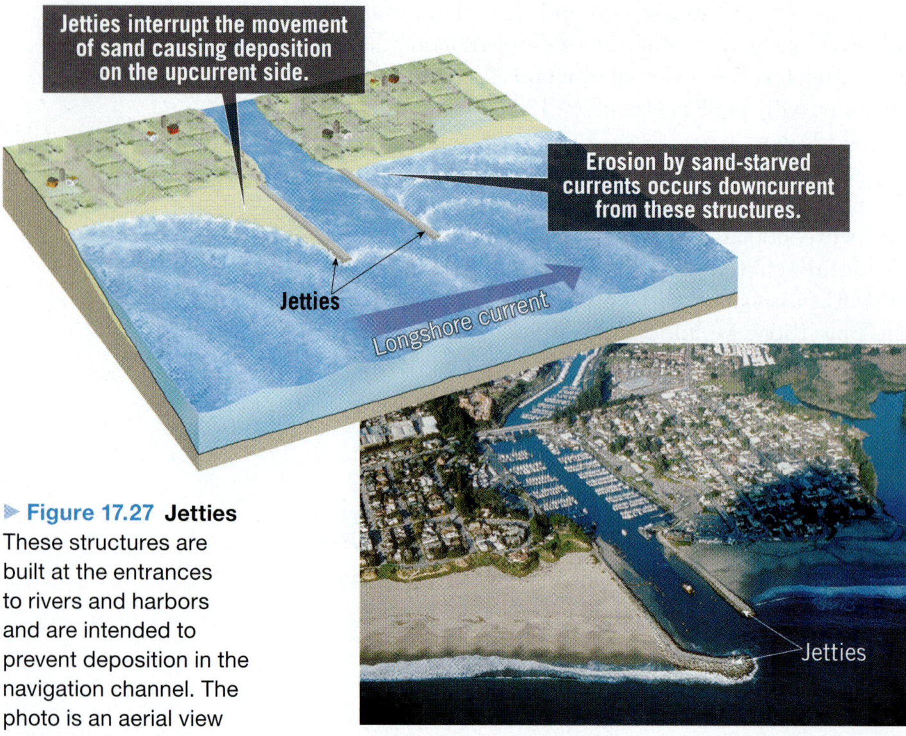
Jetties interrupt the movement of sand causing deposition on the upcurrent side.
Erosion by sand-starved currents occurs downcurrent from these structures.
Jetties
Longshore current
Jetties

► **Figure 17.27 Jetties** These structures are built at the entrances to rivers and harbors and are intended to prevent deposition in the navigation channel. The photo is an aerial view at Santa Cruz Harbor, California. (Photo by U.S. Army Corps of Engineers)

▲ **Figure 17.28 Groins** These wall-like structures trap sand that is moving parallel to the shore. This series of groins is along the shoreline near Chichester, Sussex, England. (Photo by Sandy Stockwell/London Aerial Photo Library/CORBIS)

longshore current and beach drift deposit sand. At the same time, wave activity removes sand on the other side. Because the other side is not receiving any new sand, there is soon no beach at all.

Groins To maintain or widen beaches that are losing sand, groins are sometimes constructed. A **groin** (*groin* = ground) is a barrier built at a right angle to the beach to trap sand that is moving parallel to the shore. Groins are usually constructed of large rocks but may also be composed of wood. These structures often do their job so effectively that the longshore current beyond the groin becomes sand starved. As a result, the current erodes sand from the beach on the downstream side of the groin.

To offset this effect, property owners downstream from the structure may erect a groin on their property. In this manner, the number of groins multiplies, resulting in a *groin field* (Figure 17.28). The New Jersey shoreline is a good example of groin proliferation, where hundreds of these structures have been built. Because it has been shown that groins often do not provide a satisfactory solution, they are no longer the preferred choice for keeping beach erosion in check.

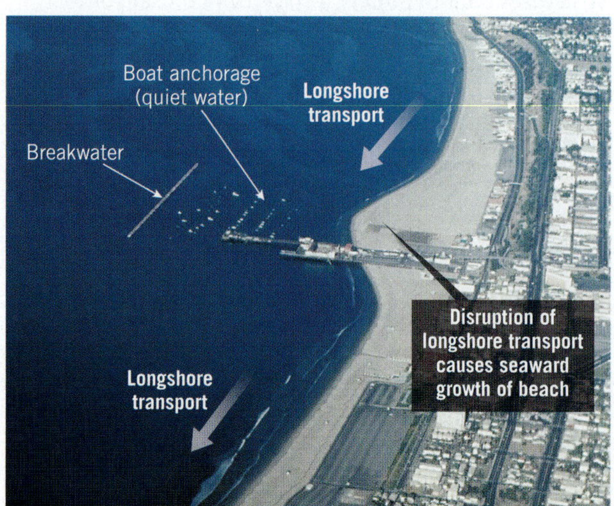
Boat anchorage (quiet water)
Longshore transport
Breakwater
Disruption of longshore transport causes seaward growth of beach
Longshore transport

► **Figure 17.29 Breakwater** Aerial view of a breakwater at Santa Monica, California. The structure appears as a line in the water behind which many boats are anchored. The construction of the breakwater disrupted longshore transport and caused the seaward growth of the beach. (Photo by John S. Shelton/University of Washington Libraries)

Breakwaters & Seawalls Hard stabilization can be built parallel to the shoreline. One such structure is a **breakwater**, which protects boats from the force of large breaking waves by creating a quiet water zone near the shoreline. However, when a breakwater is constructed, the reduced wave activity along the shore behind the structure may allow sand to accumulate. If this happens, the marina will eventually fill with sand, while the downstream beach erodes and retreats. At Santa Monica, California, where the building of a breakwater has created such a problem, the city uses a dredge to remove sand from the protected quiet water zone and deposit it downstream, where longshore currents and beach drift continue to move the sand down the coast (Figure 17.29).

Another type of hard stabilization built parallel to the shoreline is a **seawall**, which is designed to armor the coast and defend property from the force of breaking waves. Waves expend much of their energy as they move across an open beach. Seawalls cut this process short by reflecting the force of unspent waves seaward. As a consequence, the beach to the seaward side of the seawall experiences significant erosion and may in some instances be eliminated entirely (Figure 17.30). Once the width of the beach is reduced, the seawall is subjected to even greater pounding by the waves. Eventually this battering causes the wall to fail, and a larger, more expensive wall must be built to take its place.

The wisdom of building temporary protective structures along shorelines is increasingly questioned. Many

Figure 17.30 Seawall Seabright in northern New Jersey once had a broad, sandy beach. A seawall 5 to 6 meters (16 to 18 feet) high and 8 kilometers (5 miles) long was built to protect the town and the railroad that brought tourists to the beach. After the wall was built, the beach narrowed dramatically. (Photo by Rafael Macia/Science Source)

Did You Know?

Communities along the Atlantic coast of southern Florida have been replenishing their beaches with dredged-up sand for decades. The result is that many areas have exhausted their supply of offshore sand that is environmentally sound and easily accessible. One proposed solution that is under consideration is to grind up recycled glass and transform it into beach sand.

coastal scientists and engineers are of the opinion that halting an eroding shoreline with protective structures benefits only a few and seriously degrades or destroys the natural beach and the value it holds for the majority. Protective structures divert the ocean's energy temporarily from private properties but usually refocus that energy on the adjacent beaches. Many structures interrupt the natural sand flow in coastal currents, robbing affected beaches of vital sand replacement.

Alternatives to Hard Stabilization

Armoring the coast with hard stabilization has several potential drawbacks, including the cost of the structure and the loss of sand on the beach. Alternatives to hard stabilization include beach nourishment and relocation.

Beach Nourishment One approach to stabilizing shoreline sands without hard stabilization is **beach nourishment**. As the term implies, this practice involves adding large quantities of sand to the beach system (Figure 17.31). Extending beaches seaward makes buildings along the shoreline less vulnerable to destruction by storm waves and enhances recreational uses. Without sandy beaches, tourism suffers.

The process of beach nourishment is straightforward. Sand is pumped by dredges from offshore or trucked from inland locations. The "new" beach, however, will not be the same as the former beach. Because replenishment sand is from somewhere else, typically not from another beach, it is new to the beach environment. The new sand is often different in size, shape, sorting, and composition. Such differences pose problems in terms of erodibility and the kinds of life the new sand will support.

Beach nourishment is not a permanent solution to the problem of shrinking beaches. The same processes that removed the sand in the first place eventually remove the replacement sand as well. Nevertheless, the number of nourishment projects has increased in recent years, and many beaches, especially along the Atlantic coast, have had their sand replenished many times. Virginia Beach, Virginia, has been nourished more than 50 times.

Beach nourishment is costly. For example, a modest project might involve 38,000 cubic meters (50,000 cubic yards) of sand distributed across about 1 kilometer (0.6 mile) of shoreline. A good-sized dump truck holds about 7.6 cubic meters (10 cubic yards) of sand. So this small project would require about 5000 dump-truck loads. Many projects extend for many miles. Nourishing beaches typically costs millions of dollars per mile.

Changing Land Use Instead of building structures such as groins and seawalls to hold the beach in place or adding sand to replenish eroding beaches, another option is available. Many coastal scientists and planners are calling for a policy shift from defending and rebuilding beaches and coastal property in high-hazard areas to *relocating* storm-damaged buildings in those places and letting nature reclaim the beach. This option is similar to an approach the federal government adopted for river floodplains following the devastating 1993 Mississippi

Figure 17.31 Beach nourishment If you visit a beach along the Atlantic coast, it is more and more likely that you will walk into the surf zone atop an artificial beach. (Photo by Michael Weber/imagebroker/Alamy Images)

Did You Know?
Tides can be used to generate electrical power when a dam is constructed across the mouth of a bay or an estuary in a coastal area that has a large tidal range. The narrow opening between the bay and the open ocean magnifies the variations in water level as tides rise and fall, and the strong in-and-out flow is used to drive turbines. The largest plant yet constructed is along the coast of France.

River floods, in which vulnerable structures are either abandoned or relocated on higher, safer ground.

A recent example of changing land use occurred on New York's Staten Island following Hurricane Sandy in 2012. The state turned some vulnerable shoreline areas of the island into waterfront parks. The parks act as buffers to protect inland homes and businesses from strong storms while providing the community with needed open space and access to recreational opportunities.

Land use changes can be controversial. People with significant near-shore investments want to rebuild and defend coastal developments from the erosional wrath of the sea. Others, however, argue that with sea level rising, the impact of coastal storms will get worse in the decades to come, and oft-damaged structures should be abandoned or relocated to improve personal safety and reduce costs. Such ideas will no doubt be the focus of much study and debate as states and communities evaluate and revise coastal land-use policies.

CONCEPT CHECKS 17.6

1. List at least three examples of hard stabilization and describe what each is intended to do. How does each affect distribution of sand on a beach?
2. What are two alternatives to hard stabilization, and what are the potential problems associated with each?

17.7 Tides

Explain the cause of tides and their monthly cycles. Describe the horizontal flow of water that accompanies the rise and fall of tides.

Tides are daily changes in the elevation of the ocean surface caused by gravitational interactions of Earth with the Moon and Sun. Their rhythmic rise and fall along coastlines have been known since antiquity. Other than waves, they are the easiest ocean movements to observe (Figure 17.32).

Although known for centuries, tides were not explained satisfactorily until Sir Isaac Newton applied the law of gravitation to them. Newton showed that there is a mutual attractive force between two bodies and that because oceans are free to move, they are deformed by this force. Hence, ocean tides result from the gravitational attraction exerted upon Earth by the Moon and, to a lesser extent, by the Sun.

Causes of Tides

It is easy to see how the Moon's gravitational force can cause the water to bulge on the side of Earth nearest the Moon. In addition, however, an equally large tidal bulge is produced on the side of Earth directly opposite the Moon (Figure 17.33).

Both tidal bulges are caused, as Newton discovered, by the pull of gravity. Gravity is inversely proportional to the

High tide

▼ **Figure 17.32 Bay of Fundy tides** High tide and low tide at Hopewell Rocks on the Bay of Fundy. Tidal flats are exposed during low tide. (High tide photo by Ray Coleman/Science Source; low tide photo by Jeffrey Greenberg/Science Source)

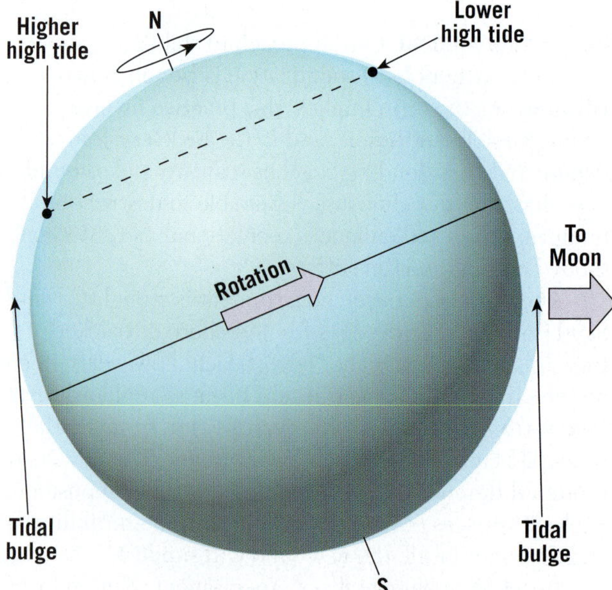

▲ **Figure 17.33 Idealized tidal bulges caused by the Moon** If Earth were covered to a uniform depth with water, there would be two tidal bulges: one on the side of Earth facing the Moon (right) and the other on the opposite side of Earth (left). Depending on the Moon's position, tidal bulges may be inclined relative to Earth's equator. In this situation, Earth's rotation causes an observer to experience two unequal high tides during a day.

square of the distance between two objects, meaning simply that it quickly weakens with distance. In this case, the two objects are the Moon and Earth. Because the force of gravity decreases with distance, the Moon's gravitational pull on Earth is slightly greater on the near side of Earth than on the far side. The result of this differential pulling is to stretch (elongate) the "solid" Earth very slightly. In contrast, the world ocean, which is mobile, is deformed quite dramatically by this effect, producing the two opposing tidal bulges.

Because the position of the Moon changes only moderately in a single day, the tidal bulges remain in place while Earth rotates "through" them. For this reason, if you stand on the seashore for 24 hours, Earth will rotate you through alternating areas of deeper and shallower water. As you are carried into each tidal bulge, the tide rises, and as you are carried into the intervening troughs between the tidal bulges, the tide falls. Therefore, most places on Earth experience two high tides and two low tides each day.

Further, the tidal bulges migrate as the Moon revolves around Earth about every 29 days. As a result, the tides, like the time of moonrise, shift about 50 minutes later each day. After 29 days the cycle is complete, and a new one begins.

In many locations, there may be an inequality between the high tides during a given day. Depending on the position of the Moon, the tidal bulges may be inclined to the equator, as in Figure 17.33. This figure illustrates that one high tide experienced by an observer in the Northern Hemisphere is considerably higher than the high tide half a day later. In contrast, a Southern Hemisphere observer would experience the opposite effect.

Monthly Tidal Cycle

The primary body that influences the tides is the Moon, which makes one complete revolution around Earth every 29.5 days. The Sun, however, also influences the tides. It is far larger than the Moon, but because it is much farther away, its effect is considerably less. In fact, the Sun's tide-generating effect is only about 46 percent that of the Moon.

Near the times of new and full moons, the Sun and Moon are aligned, and their forces on tides are added together (**Figure 17.34A**). The combined gravity of these two tide-producing bodies causes larger tidal bulges (higher high tides) and deeper tidal troughs (lower low tides), producing a large tidal range. These are called the **spring tides** (*springen* = to rise up), which have no connection with the spring season but occur twice a month, during the time when the Earth–Moon–Sun system is aligned. Conversely, at about the time of the first and third quarters of the Moon, the gravitational forces of the Moon and Sun act on Earth at right angles, and each partially offsets the influence of the other (**Figure 17.34B**). As a result, the daily tidal range is less. These are called **neap tides** (*nep* = scarcely or barely touching), and they also occur twice each month. Each month, then, there are two spring tides and two neap tides, each about 1 week apart.

Tidal Currents

Tidal current is the term used to describe the *horizontal* flow of water that accompanies the rise and fall of the tide. These water movements induced by tidal forces can be important in some coastal areas. Tidal currents flow in one direction during a portion of the tidal cycle and reverse their flow during the remainder. Tidal currents that advance into the coastal zone as the tide rises are called **flood currents**. As the tide falls, seaward-moving water generates **ebb currents**. Periods of little or no current, called *slack water*, separate flood and ebb. The areas covered and uncovered by these alternating tidal currents are **tidal flats** (see Figure 17.32). Depending on the nature of the coastal zone, tidal flats vary from narrow strips seaward of the beach to zones that may extend for several kilometers.

Although tidal currents are generally not important in the open sea, they can be rapid in bays, river estuaries, straits, and other narrow places. Off the coast of Brittany in France, for example, tidal currents that accompany a high tide of 12 meters (40 feet) may attain a speed of 20 kilometers (12 miles) per hour. While tidal currents are not generally major agents of erosion and sediment transport,

Did You Know?
The world's largest tidal range (difference between successive high and low tides) is found in the northern end of Nova Scotia's Bay of Fundy. Here the maximum spring tidal range is about 17 meters (56 feet). This leaves boats "high and dry" during low tide.

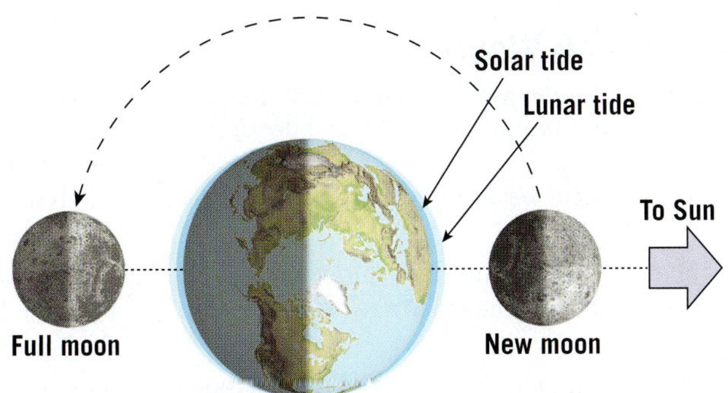

A. Spring Tide When the Moon is in the full or new position, the tidal bulges created by the Sun and Moon are aligned and there is a large tidal range.

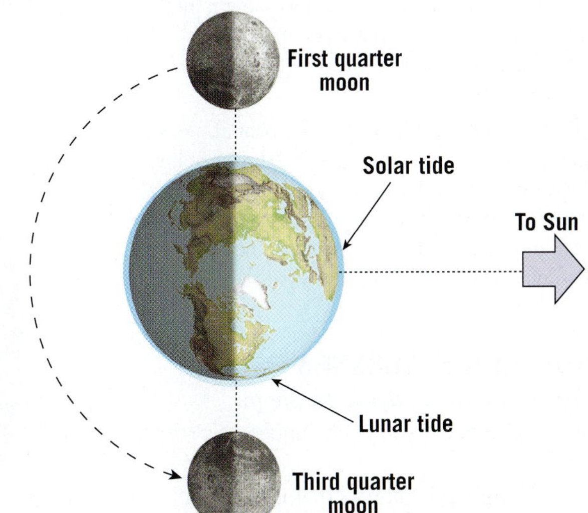

B. Neap Tide When the Moon is in the first-or third-quarter position, the tidal bulges produced by the Moon are at right angles to the bulges created by the Sun and the tidal range is smaller.

◀ **SmartFigure 17.34**
Spring and neap tides
Earth–Moon–Sun positions influence the tides.

ANIMATION
http://goo.gl/4rUIGf

notable exceptions occur where tides move through narrow inlets. Here they constantly scour the small entrances to many good harbors that would otherwise be blocked.

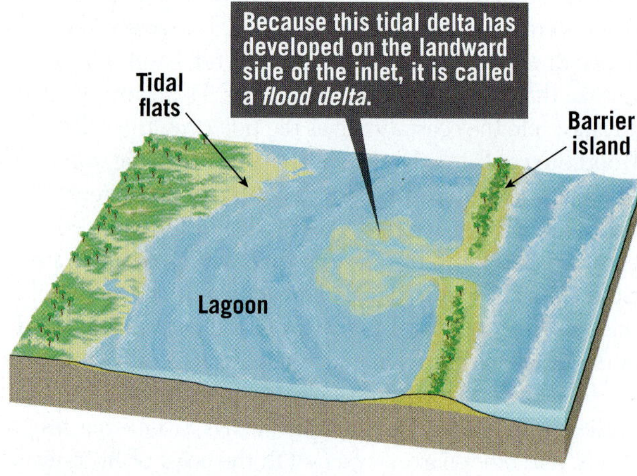
Because this tidal delta has developed on the landward side of the inlet, it is called a *flood delta.*
Tidal flats
Barrier island
Lagoon

▶ **Figure 17.35 Tidal deltas** As a rapidly moving tidal current (flood current) moves through a barrier island's inlet into the quiet waters of the lagoon, the current slows and deposits sediment, creating a tidal delta. Because this tidal delta has developed on the landward side of the inlet, it is called a *flood delta*. Such a tidal delta is shown in Figure 17.14A.

Sometimes deposits called **tidal deltas** are created by tidal currents (**Figure 17.35**). They may develop either as *flood deltas* landward of an inlet or as *ebb deltas* on the seaward side of an inlet. Because wave activity and longshore currents are reduced on the sheltered landward side, flood deltas are more common and more prominent (see Figure 17.14A). They form after the tidal current moves rapidly through an inlet. As the current emerges from the narrow passage into more open waters, it slows and deposits its load of sediment.

CONCEPT CHECKS 17.7

1. Explain why an observer can experience two unequal high tides during a single day.
2. Distinguish between *neap tides* and *spring tides*.
3. Contrast *flood current* and *ebb current*.

CONCEPTS IN REVIEW

Shorelines

17.1 The Shoreline & Ocean Waves

Explain why the shoreline is considered a dynamic interface. List the factors that influence the height, length, and period of a wave and describe the motion of water within a wave.

KEY TERMS: shoreline, interface, wave height, wavelength, wave period, fetch, surf

- The shoreline is a transition zone between marine and continental environments. It is a dynamic interface, a boundary where land, sea, and air meet and interact.
- Energy from waves plays an important role in shaping the shoreline, but many factors contribute to the character of particular shorelines.
- Waves are moving energy, and most ocean waves are initiated by wind. The three factors that influence the height, wavelength, and period of a wave are (1) wind speed, (2) length of time the wind has blown, and (3) fetch, the distance that the wind has traveled across open water. Once waves leave a storm area, they are termed *swells*, which are symmetrical, longer-wavelength waves.
- As waves travel, water particles transmit energy by circular orbital motion, which extends to a depth equal to one-half the wavelength (the wave base). When a wave enters water that is shallower than the wave base, it slows down, which allows waves farther from shore to catch up. As a result, wavelength decreases and wave height increases. Eventually the wave breaks, creating turbulent surf in which water rushes toward the shore.

17.2 Beaches & Shoreline Processes

Explain how waves erode and move sediment along the shore.

KEY TERMS: beach, abrasion, wave refraction, beach drift, longshore current, rip current

- A beach is composed of any locally derived material that is in transit along the shore.
- Waves provide most of the energy that modifies shorelines. Wave erosion is caused by wave impact pressure and abrasion (the sawing and grinding action of water armed with rock fragments).

(17.2 continued)

- As they approach the shore, waves refract (bend) to align nearly parallel to the shore. Refraction occurs because a wave travels more slowly in shallower water, allowing the part still in deeper water to catch up. Wave refraction causes wave erosion to be concentrated against the sides and ends of headlands and dispersed in bays.
- Waves that approach the shore at an angle transport sediment parallel to the shoreline. On the beach face, this longshore transport is called beach drift, and it is due to the fact that the incoming swash pushes sediment obliquely upward, whereas the backwash pulls it directly downhill. Longshore currents are a similar phenomenon in the surf zone, capable of transporting very large quantities of sediment parallel to a shoreline.

? What process is causing wave energy to be concentrated on the headland? Predict how this area will appear in the future.

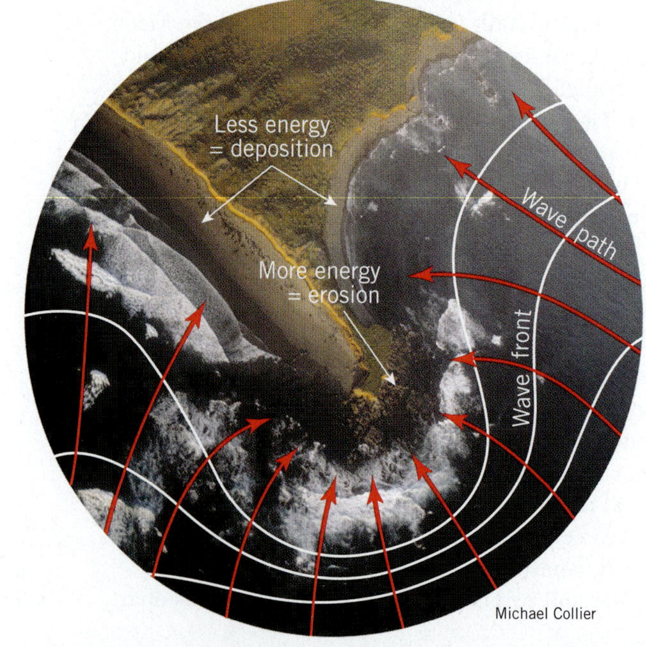
Less energy
= deposition
More energy
= erosion
Wave path
Wave front

Michael Collier

17.3 Shoreline Features

Describe the features typically created by wave erosion and those resulting from sediment deposited by longshore transport processes.

KEY TERMS: wave-cut cliff, wave-cut platform, marine terrace, sea arch, sea stack, spit, baymouth bar, tombolo, barrier island

- Erosional features include wave-cut cliffs (which originate from the cutting action of the surf against the base of coastal land), wave-cut platforms (relatively flat, bench-like surfaces left behind by receding cliffs), and marine terraces (uplifted wave-cut platforms). Erosional features also include sea arches (formed when a headland is eroded and two sea caves from opposite sides unite) and sea stacks (formed when the roof of a sea arch collapses).
- Some of the depositional features that form when sediment is moved by beach drift and longshore currents are spits (elongated ridges of sand that project from the land into the mouth of an adjacent bay), baymouth bars (sandbars that completely cross a bay), and tombolos (ridges of sand that connect an island to the mainland or to another island). Along the Atlantic and Gulf coastal plains, the coastal region is characterized by offshore barrier islands, which are low ridges of sand that parallel the coast.

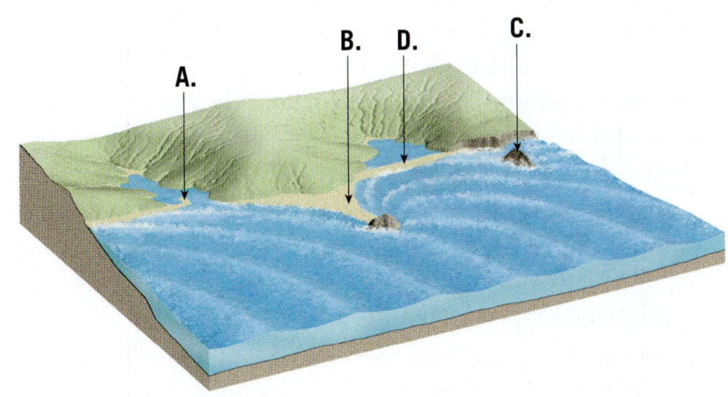

? Identify the lettered features in this diagram.

17.4 Contrasting America's Coasts

Distinguish between emergent and submergent coasts. Contrast the erosion problems faced on the Atlantic and Gulf coasts with those along the Pacific coast.

KEY TERMS: emergent coast, submergent coast, estuary

- Coasts can be classified by their changes relative to sea level. Emergent coasts are sites of either land uplift or sea-level fall. Marine terraces are features of emergent coasts. Submergent coasts are sites of land subsidence or sea-level rise. One characteristic of submergent coasts is drowned river valleys called estuaries. In the United States, the Pacific coast is emergent, and the Atlantic and Gulf coasts are submergent.
- The Atlantic and Gulf coasts of the United States are lined in many places by barrier islands—dynamic expanses of sand that see a lot of change during storm events. Many of these low and narrow islands have also been prime sites for real estate development.
- The Pacific coast's big issue is the narrowing of beaches due to sediment starvation. Rivers that drain to the coast (bringing sand to it) have been dammed, resulting in reservoirs that trap sand and prevent it from reaching the coast. Narrower beaches offer less resistance to incoming waves, often leading to erosion of bluffs behind the beach.

Michael Collier

? What term is applied to the masses of rock protruding from the water in this photo? How did they form? Is the location more likely along the Gulf Coast or the coast of California? Explain.

17.5 Hurricanes: The Ultimate Coastal Hazard

Describe the basic structure and characteristics of a hurricane and the three broad categories of hurricane destruction.

KEY TERMS: hurricane, eye wall, eye, storm surge

- Hurricanes are fueled by warm, moist air and usually form in the late summer when sea-surface temperatures are highest. Water vapor in rising warm air condenses, releasing heat and triggering the formation of dense clouds and heavy rain. Because of a steep pressure gradient, air rushes into the center of the storm. The Coriolis effect and ocean-water temperatures strongly influence where hurricanes form.
- The eye at the center of a hurricane has the lowest pressure, is relatively calm, and lacks rain. The surrounding eye wall has the strongest winds and most intense rainfall. The Saffir–Simpson scale classifies storms based on their air pressure and wind speed.
- Most hurricane damage comes from one or a combination of three causes: storm surge, wind damage, or inland flooding due to heavy rains. Storm surge is ocean water that gets pushed up above the normal water level by the strong winds. In the Northern Hemisphere hurricanes rotate counterclockwise, and storm surge is greatest on the right side of an advancing hurricane. This is due to the combination of the storm's forward movement and strong winds blowing toward the shore.

Lucas Jackson/Reuters

? This coastal scene shows hurricane destruction. Which one of the three basic classes of damage was most likely responsible for this destruction? What is your reasoning?

17.6 Stabilizing the Shore

Summarize the ways in which people deal with shoreline erosion problems.

KEY TERMS: hard stabilization, jetty, groin, breakwater, seawall, beach nourishment

- Hard stabilization refers to any structures built along the coastline to prevent movement of sand or shoreline erosion. Jetties project out from the coast with the goal of keeping inlets open. Groins are also oriented perpendicular to the coast, but their goal is to slow beach erosion by longshore currents. Breakwaters are parallel to the coast but located some distance offshore. Their goal is to blunt the force of incoming ocean waves, often to protect boats. Like breakwaters, seawalls are parallel to the coast, but they are built on the shoreline itself. Often the installation of hard stabilization results in increased erosion elsewhere.
- Beach nourishment is an expensive alternative to hard stabilization. Sand is pumped onto a beach from some other area, temporarily replenishing the sediment supply. Another possibility is relocating buildings away from high-risk areas and leaving the beach to be shaped by natural processes.

? Based on their position and orientation, identify the four kinds of hard stabilization illustrated in this diagram.

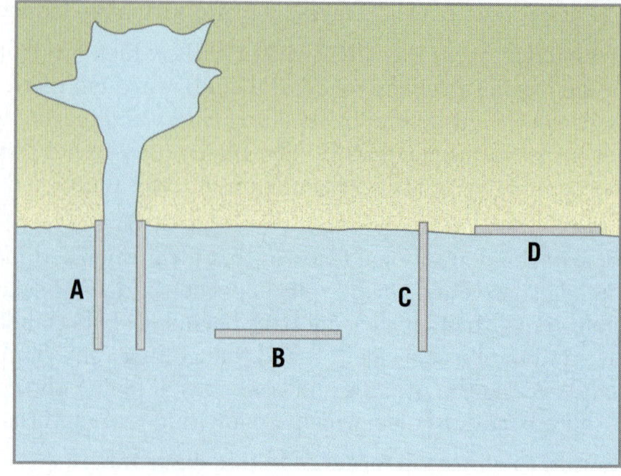

17.7 Tides

Explain the cause of tides and their monthly cycles. Describe the horizontal flow of water that accompanies the rise and fall of tides.

KEY TERMS: tide, spring tide, neap tide, tidal current, flood current, ebb current, tidal flat, tidal delta

- Tides are daily changes in ocean-surface elevation. They are caused by gravitational pull on ocean water by the Moon and, to a lesser extent, the Sun. When the Sun, Earth, and Moon all line up about every 2 weeks (full moon and new moon), the tides are most exaggerated. When a quarter Moon is in the sky, it indicates that the Moon is pulling on Earth's water at a right angle relative to the Sun, and the daily tidal range is minimized, as the two forces partially counteract one another.
- A flood current is the landward movement of water during the shift between low tide and high tide. As high tide transitions to low tide again, the movement of water away from the land is an ebb current. Ebb currents may expose tidal flats to the air. If a tide passes through an inlet, the current may carry sediment that gets deposited as a tidal delta.

GIVE IT SOME THOUGHT

1. During a visit to the beach, you and a friend get in a rubber raft and paddle out into deep water *beyond* the surf zone. Tiring, you stop and take a rest. Describe the movement of the raft during your rest. How does this movement differ, if at all, from what you have experienced if you had stopped paddling while *in* the surf zone?
2. Examine the aerial photo that shows a portion of the New Jersey coast. What term is applied to the wall-like structures that extend into the water? What is their purpose? In what direction are beach drift and longshore currents moving sand: toward the top of the photo or toward the bottom?
3. You and a friend set up an umbrella and chairs at a beach. Your friend then goes into the surf zone to play Frisbee with another person. Several minutes later, your friend looks back toward the beach and is surprised to see that she is no longer near where the umbrella and chairs were set up. Although she is still in the surf zone, she is 30 yards away from where she started. How would you explain to your friend why she moved along the shore?

John S. Shelton/University of Washington Libraries

4 This surfer is enjoying a ride on a large wave along the coast of Maui.

a. What was the source of energy that created this wave?

b. How was the wavelength changing just prior to the time when this photo was taken?

c. Why was the wavelength changing?

d. Many ocean waves exhibit circular orbital motion. Is that true of the wave in this photo? Explain.

Ron Dahlquist/Getty Images

5 Assume that it is late September 2018, and Hurricane Gordon, a category 5 storm, is projected to follow the path shown on the accompanying map. The path of the arrow represents the path of the hurricane's eye. Answer the following questions:

a. Should the city of Houston expect to experience Gordon's fastest winds and greatest storm surge? Explain why or why not.

b. What is the greatest threat to life and property if this storm approaches the Dallas–Fort Worth area? Explain your reasoning.

6 A friend wants to purchase a vacation home on a barrier island. If consulted, what advice would you give your friend?

7 Hurricane Rita was a major storm that struck the Gulf coast in late September 2005, less than a month after Hurricane Katrina. The accompanying graph shows changes in air pressure and wind speed from the storm's beginning as an unnamed tropical disturbance north of the Dominican Republic on September 18 until its last remnants faded away in Illinois on September 26. Use the graph to answer these questions:

a. Which line represents air pressure, and which line represents wind speed? How did you figure this out?

b. What was the storm's maximum wind speed, in knots? Convert this answer to kilometers per hour by multiplying by 1.85.

c. What was the lowest pressure attained by Hurricane Rita?

d. Using wind speed as your guide, what was the highest category reached on the Saffir–Simpson scale? On what day was this status reached?

e. When landfall occurred, what was the category of Hurricane Rita?

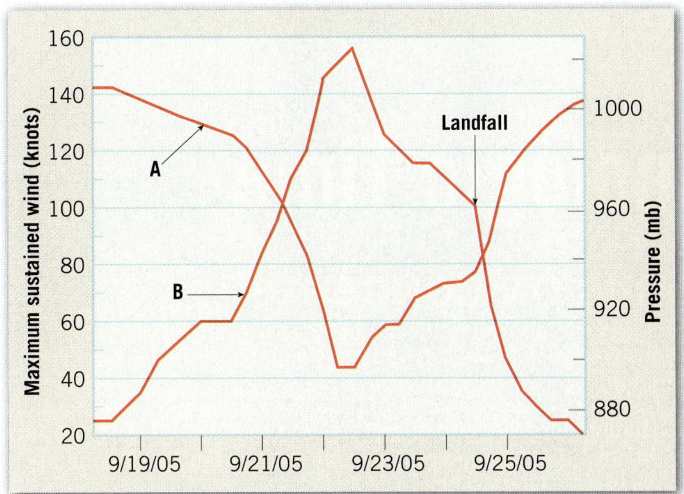

8 The force of gravity plays a critical role in creating ocean tides. The more massive an object, the stronger its gravitational pull. Explain why the Sun's influence is only about half that of the Moon, even though the Sun is much more massive than the Moon.

9 This photo shows a portion of the Maine coast. The brown muddy area in the foreground is influenced by tidal currents. What term is applied to this muddy area? Name the type of tidal current this area will experience in the hours to come.

Marli Miller

MasteringGeology™

Looking for additional review and test prep materials? Visit the Study Area in MasteringGeology to enhance your understanding of this chapter's content by accessing a variety of resources, including Self-Study Quizzes, Geoscience Animations, SmartFigures, Mobile Field Trips, *Project Condor* Quadcopter videos, *In the News* RSS feeds, flashcards, web links, and an optional Pearson eText.

www.masteringgeology.com

18

Geologic Time

FOCUS ON CONCEPTS

Each statement represents the primary learning objective for the corresponding major heading within the chapter. After you complete the chapter, you should be able to:

18.1 Distinguish between numerical and relative dating and apply relative dating principles to determine a time sequence of geologic events.

18.2 Define *fossil* and discuss the conditions that favor the preservation of organisms as fossils. List and describe various types of fossils.

18.3 Explain how rocks of similar age that are in different places can be matched up.

18.4 Discuss three ways that atomic nuclei change and explain how unstable isotopes are used to determine numerical dates.

18.5 Explain how reliable numerical dates are determined for layers of sedimentary rock.

18.6 Distinguish among the four basic time units that make up the geologic time scale and explain why the time scale is considered to be a dynamic tool.

The Colorado River above Havasu Creek in Grand Canyon National Park. Millions of years of Earth history are exposed in the canyon's rock walls. (Photo by Michael Collier)

IN THE LATE EIGHTEENTH CENTURY, James Hutton recognized the immensity of Earth history and the importance of time as a component in all geologic processes. In the nineteenth century, Sir Charles Lyell and others effectively demonstrated that Earth had experienced many episodes of mountain building and erosion, which must have required great spans of geologic time. Although these pioneering scientists understood that Earth was very old, they had no way of determining its age in years. Was it tens of millions, hundreds of millions, or even billions of years old? Long before geologists could establish a geologic calendar that included numerical dates in years, they gradually assembled a time scale using relative dating principles. What are these principles? What part do fossils play? With the discovery of radioactivity and radiometric dating techniques, geologists can now assign quite accurate dates to many of the events in Earth history. What is radioactivity? Why is it a good "clock" for dating the geologic past?

18.1 Creating a Time Scale: Relative Dating Principles

Distinguish between numerical and relative dating and apply relative dating principles to determine a time sequence of geologic events.

Figure 18.1 shows a hiker resting atop the Permian-age Kaibab Formation at Cape Royal on the North Rim of the Grand Canyon. Beneath him are thousands of meters of sedimentary strata that go as far back as Cambrian time, more than 540 million years ago. These strata rest atop even older sedimentary, metamorphic, and igneous rocks from a span known as the Precambrian. Some of these rocks are 2 billion years old. Although the Grand Canyon's rock record has numerous interruptions, the rocks beneath the hiker contain clues to great spans of Earth history. Earth's long and complicated history is recorded in the structure, sequence, and properties of its rocks, sediments, and fossils.

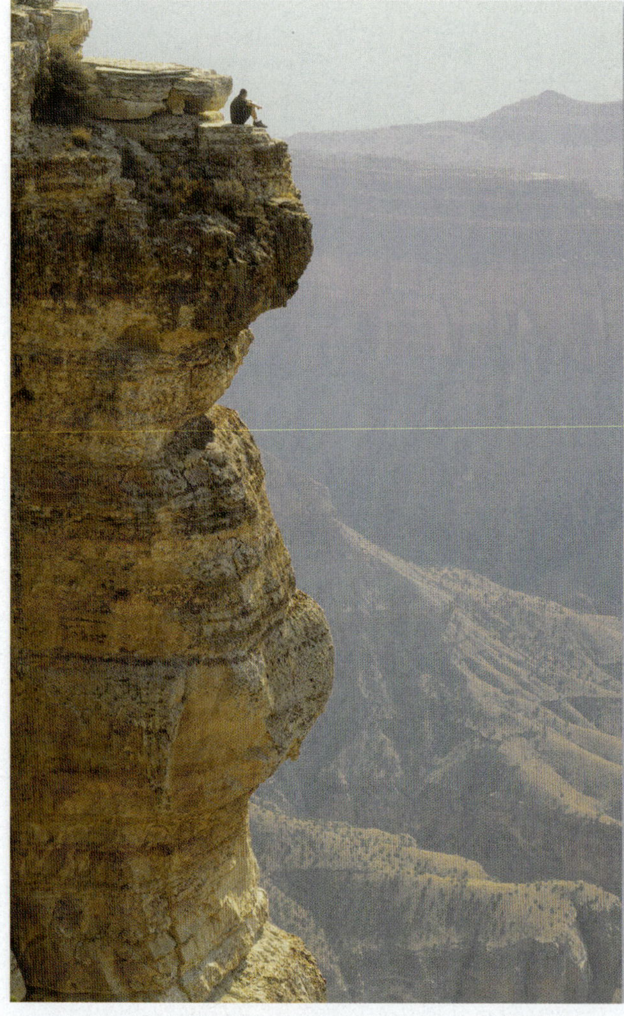

▲ **Figure 18.1 Contemplating geologic time** This hiker is resting atop the Kaibab Formation, the uppermost layer in the Grand Canyon. (Photo by Michael Collier)

The Importance of a Time Scale

Like the pages in a long and complicated history book, rocks record the geologic events and changing life-forms of the past. The book, however, is not complete. Many pages, especially in the early chapters, are missing. Others are tattered, torn, or smudged. Yet enough of the book remains to allow much of the story to be deciphered.

Interpreting Earth history is a prime goal of the science of geology. Like a modern-day sleuth, a geologist must interpret the clues found preserved in the rocks. By studying rocks, especially sedimentary rocks, and the features they contain, geologists can unravel the complexities of the past.

Geologic events by themselves, however, have little meaning until they are put into a time perspective. Studying history, whether it is the Civil War or the age of dinosaurs, requires a calendar. Among geology's major contributions to human knowledge are the *geologic time scale* and the discovery that Earth history is exceedingly long.

Numerical & Relative Dates

The geologists who developed the geologic time scale revolutionized the way people think about time and perceive our planet. They learned that Earth is much older than anyone had previously imagined and that its surface and interior have been changed over and over again by the same geologic processes that operate today.

Numerical Dates During the late 1800s and early 1900s, attempts were made to determine Earth's age. Although some of the methods appeared promising at the time, none of those early efforts proved to be reliable. What those scientists were seeking was a **numerical date**. Such dates specify the actual number of years that have passed since an event occurred. Today, our understanding of radioactivity allows us to accurately determine numerical dates for rocks that represent important events in Earth's distant past. We will study radioactivity later in this chapter. Prior to the discovery of radioactivity, geologists had no reliable method of numerical dating and had to rely solely on relative dating.

Relative Dates When we place rocks in their proper *sequence of formation*—indicating which formed first, second, third, and so on—we are establishing **relative dates**. Such dates cannot tell us how long ago something took place, only that it followed one event and preceded another. The relative dating techniques that were developed are valuable and still widely used. Numerical dating methods did not replace these techniques; they simply supplemented them. To establish a relative time scale, a few basic principles or rules had to be discovered and applied. They were major breakthroughs in thinking at the time, and their discovery was an important scientific achievement.

Principle of Superposition

Nicolas Steno, a Danish anatomist, geologist, and priest (1638–1686), was the first to recognize a sequence of historical events in an outcrop of sedimentary rock layers. Working in the mountains of western Italy, Steno applied a very simple rule that has become the most basic principle of relative dating—the **principle of superposition** (*super* = above; *positum* = to place). This principle simply states that in an undeformed sequence of sedimentary rocks, each bed is older than the one above and younger than the one below. Although it may seem obvious that a rock layer could not be deposited with nothing beneath it for support, it was not until 1669 that Steno clearly stated this principle.

This rule also applies to other surface-deposited materials, such as lava flows and beds of ash from volcanic eruptions. Applying the principle of superposition to the beds exposed in the upper portion of the Grand Canyon, we can easily place the layers in their proper order. Among those that are pictured in Figure 18.2, the sedimentary rocks in the Supai Group are the oldest, followed in order by the Hermit Shale, Coconino Sandstone, Toroweap Formation, and Kaibab Limestone.

Principle of Original Horizontality

Steno is also credited with recognizing the importance of another basic rule, the **principle of original horizontality**, which states that layers of sediment are generally deposited in a horizontal position. Thus, if we observe rock layers that are flat, it means they have not been disturbed and still have their *original* horizontality. The layers in the Grand Canyon illustrate this in Figures 18.1

Figure 18.2 Superposition According to the principle of superposition, the Supai Group is oldest of these layers in the upper portion of the Grand Canyon, and the Kaibab Limestone is youngest.

▲ **Figure 18.3 Original horizontality** Most layers of sediment are deposited in a nearly horizontal position. When we see strata that are folded or tilted, we can assume that they were moved into that position by crustal disturbances *after* their deposition. (Photo by Marco Simoni/Robert Harding World Imagery)

and 18.2. But if they are folded or inclined at a steep angle, they must have been moved into that position by crustal disturbances sometime *after* their deposition (Figure 18.3).

Principle of Lateral Continuity

The **principle of lateral continuity** refers to the fact that sedimentary beds originate as continuous layers that extend in all directions until they eventually grade into a different type of sediment or until they thin out at the edge of the basin of deposition (Figure 18.4). For example, when a river creates a canyon, we can assume that identical or similar strata on opposite sides once spanned the canyon. Although rock outcrops may be separated by a considerable distance, the principle of lateral continuity tells us that those outcrops once formed a continuous layer. This principle allows geologists to relate rocks in isolated outcrops to one another. Combining the principles of lateral continuity and superposition lets us extend relative age relationships over broad areas. This process, called *correlation*, is examined in Section 18.3.

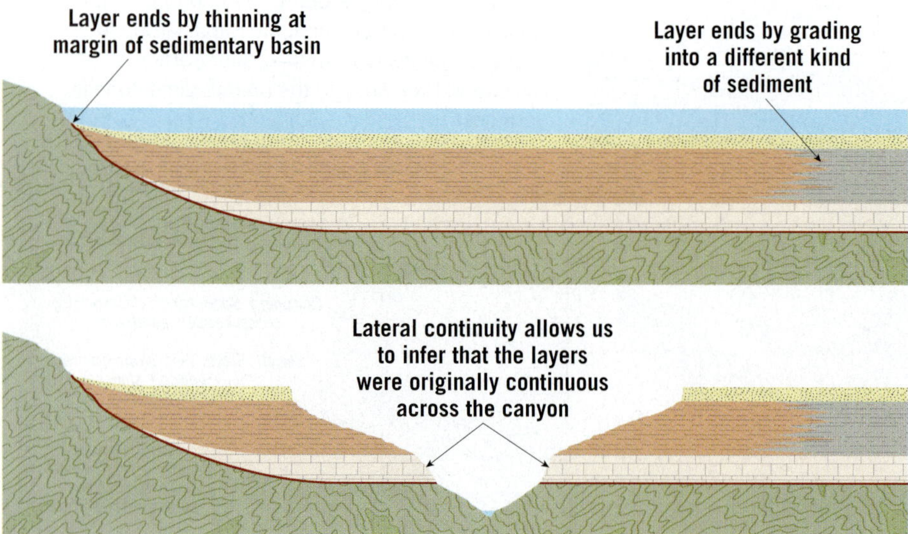
Layer ends by thinning at margin of sedimentary basin
Layer ends by grading into a different kind of sediment
Lateral continuity allows us to infer that the layers were originally continuous across the canyon

▲ **Figure 18.4 Lateral continuity** Sediments are deposited over a large area in a continuous sheet. Sedimentary strata extend continuously in all directions until they thin out at the edge of a depositional basin or grade into a different type of sediment.

Principle of Cross-Cutting Relationships

Figure 18.5 shows a mass of rock that is offset by a fault, a fracture in rock along which displacement occurs. It is clear that the rocks must be older than the fault that broke them. The **principle of cross-cutting relationships** states that geologic features that cut across rocks must have formed *after* the rocks they cut through. Igneous intrusions provide another example. The dike shown in Figure 18.6 is a tabular mass of igneous rock that cuts through the surrounding rocks. The magmatic heat from igneous intrusions often creates a narrow "baked" zone of contact metamorphism on the adjacent rock, also indicating that the intrusion occurred after the surrounding rocks were in place.

Principle of Inclusions

Sometimes inclusions can aid in the relative dating process. *Inclusions* are fragments of one rock unit that have been enclosed within another. The **principle of inclusions** is logical and straightforward: The rock mass

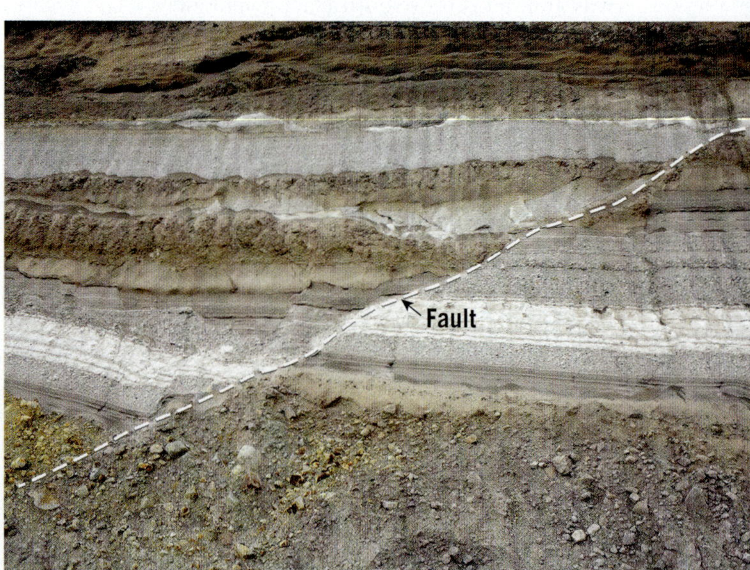
Fault

▲ **SmartFigure 18.5 Cross-cutting fault** The rocks are older than the fault that displaced them. (Morley Read/Alamy)

VIDEO
https://goo.gl/gLu1IR

▲ **Figure 18.6 Cross-cutting dike** An igneous intrusion is younger than the rocks that are intruded. (Photo by Jonathan.s.kt)

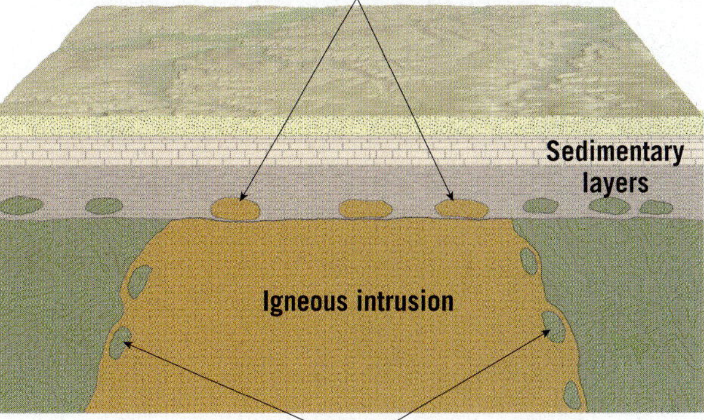

These inclusions of igneous rock contained in the adjacent sedimentary layer indicate that the sediments were deposited atop the weathered igneous mass and thus are younger.
Sedimentary layers
Igneous intrusion
Xenoliths are inclusions in an igneous intrusion that form when pieces of surrounding rock are incorporated into magma.

◀ **SmartFigure 18.7 Inclusions** The rock containing inclusions is younger than the inclusions.

TUTORIAL
https://goo.gl/s8USDC

adjacent to the one containing the inclusions must have been there first in order to provide the rock fragments. Therefore, the rock mass that contains inclusions is the younger of the two. For example, when magma intrudes into surrounding rock, blocks of the surrounding rock may become dislodged and incorporated into the magma. If these pieces do not melt, they remain as inclusions, known as *xenoliths*. In another example, when sediment is deposited atop a weathered mass of bedrock, pieces of the weathered rock become incorporated into the younger sedimentary layer (**Figure 18.7**).

Unconformities

When we observe layers of rock that have been deposited essentially without interruption, we call them **conformable**. Particular sites exhibit conformable beds representing certain spans of geologic time. However, no place on Earth has a complete set of conformable strata.

Throughout Earth history, the deposition of sediment has been interrupted over and over again. All such breaks in the rock record are termed *unconformities*. An **unconformity** represents a long period during which deposition ceased, erosion removed previously formed rocks, and then deposition resumed. In each case, uplift and erosion are followed by subsidence and renewed sedimentation. Unconformities are important features because they represent significant geologic events in Earth history. There are three basic types of unconformities, and their recognition helps geologists identify what intervals of time are not represented by strata and thus are missing from the geologic record.

Angular Unconformity Perhaps the most easily recognized unconformity is an **angular unconformity**. It consists of tilted or folded sedimentary rocks that are overlain by younger, more flat-lying strata. An angular unconformity indicates that during a pause in deposition, a period of deformation (folding or tilting) and erosion occurred (**Figure 18.8**).

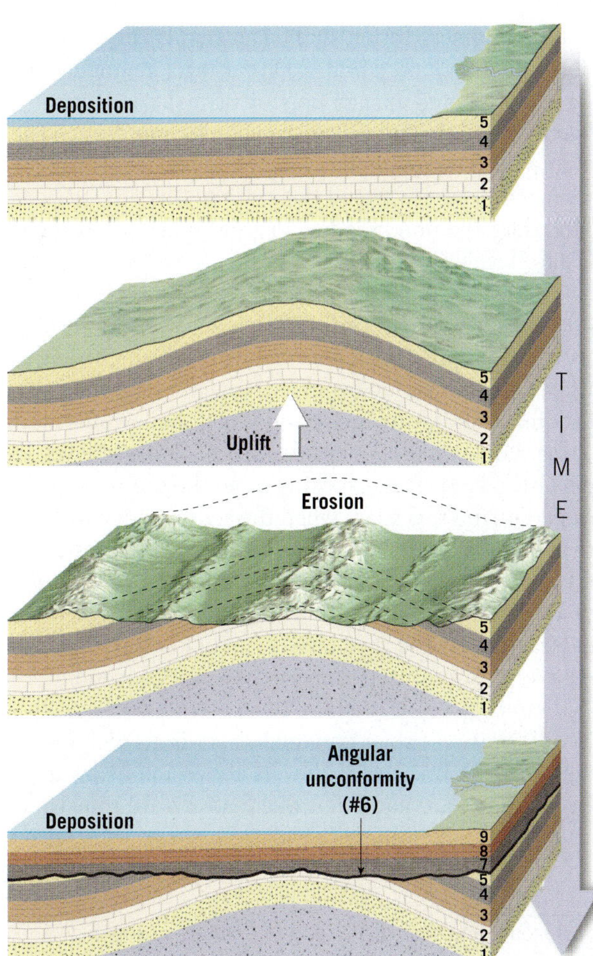

◀ **SmartFigure 18.8 Formation of an angular unconformity** An angular unconformity represents an extended period during which deformation and erosion occurred. Numbers on the diagram indicate the order in which events occurred.

TUTORIAL
https://goo.gl/fkUmV5

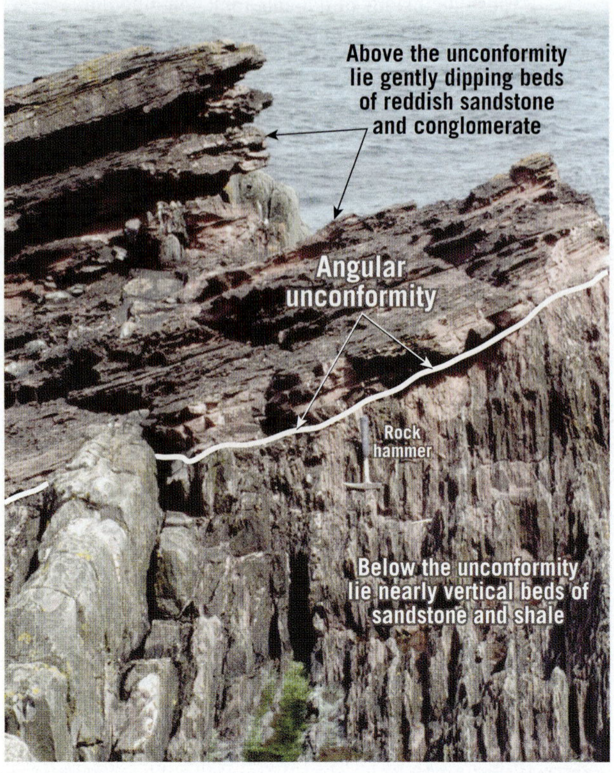
Above the unconformity lie gently dipping beds of reddish sandstone and conglomerate
Angular unconformity
Rock hammer
Below the unconformity lie nearly vertical beds of sandstone and shale

Figure 18.9 Siccar Point, Scotland James Hutton studied this famous unconformity in the late 1700s. (Photo by Marli Miller)

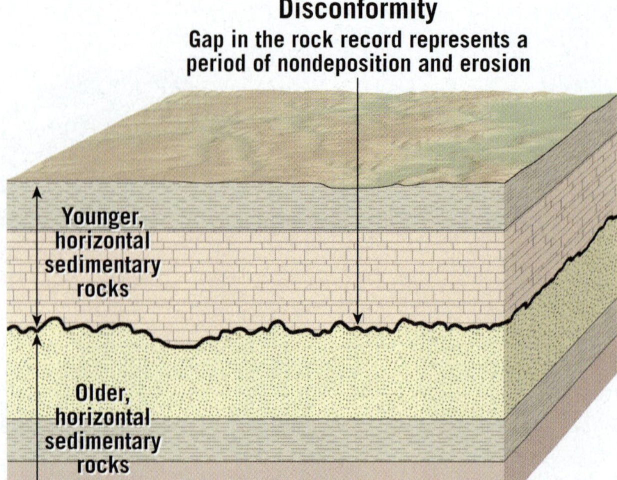
Disconformity
Gap in the rock record represents a period of nondeposition and erosion
Younger, horizontal sedimentary rocks
Older, horizontal sedimentary rocks

Figure 18.10 Disconformity The layers on both sides of this gap in the rock record are essentially parallel.

When James Hutton studied an angular unconformity in Scotland more than 200 years ago, he understood that it represented a major episode of geologic activity (Figure 18.9).* He and his colleagues also appreciated the immense time span implied by such relationships. When a companion later wrote of their visit to this site, he stated that "the mind seemed to grow giddy by looking so far into the abyss of time."

Did You Know?
Early attempts at determining Earth's age proved to be unreliable. One method reasoned that if the rate at which sediment accumulates could be determined, as well as the total thickness of sedimentary rock that had been deposited during Earth history, an estimate of Earth's age could be made. All that was necessary was to divide the rate of sediment accumulation into the total thickness of sedimentary rock. This method was riddled with difficulties. Can you think of some of them?

Disconformity A **disconformity** is a gap in the rock record that represents a period during which erosion rather than deposition occurred. Imagine that a series of sedimentary layers is deposited in a shallow marine setting. Following this period of deposition, sea level falls or the land rises, exposing some the sedimentary layers. During this span, when the sedimentary beds are above sea level, no new sediment accumulates, and some of the existing layers are eroded away. Later, sea level rises or the land subsides, submerging the landscape. Now the surface is again below sea level, and a new series of sedimentary beds is deposited. The boundary separating the two sets of beds is a disconformity—a span for which there is no rock record (Figure 18.10). Because the layers above and below a disconformity are parallel, these features are sometimes difficult to identify unless you notice evidence of erosion such as a buried stream channel.

Nonconformity The third basic type of unconformity is a **nonconformity**, in which younger sedimentary strata overlie older metamorphic or intrusive igneous rocks (Figure 18.11). Just as angular unconformities and some disconformities imply crustal movements, so too do nonconformities. Intrusive igneous masses and metamorphic rocks originate far below the surface. Thus, for a nonconformity to develop, there must be a period of uplift and erosion of overlying rocks. Once exposed at the surface, the igneous or metamorphic rocks are subjected to weathering and erosion and then undergo subsidence and renewed sedimentation.

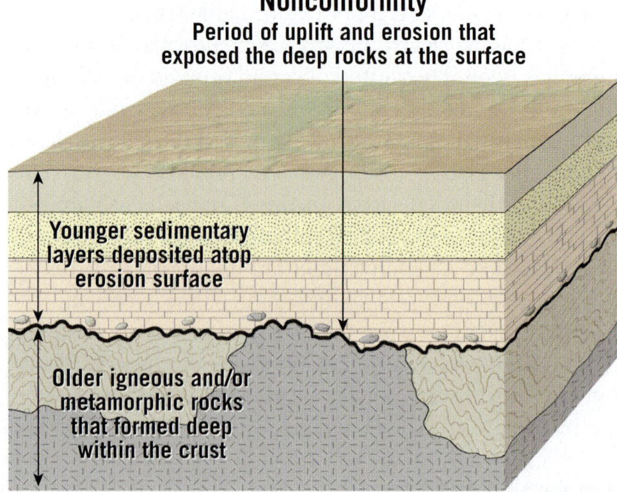
Nonconformity
Period of uplift and erosion that exposed the deep rocks at the surface
Younger sedimentary layers deposited atop erosion surface
Older igneous and/or metamorphic rocks that formed deep within the crust

Figure 18.11 Nonconformity Younger sedimentary rocks rest atop older metamorphic or igneous rocks.

*This pioneering geologist is discussed in the section on the birth of modern geology in Chapter 1.

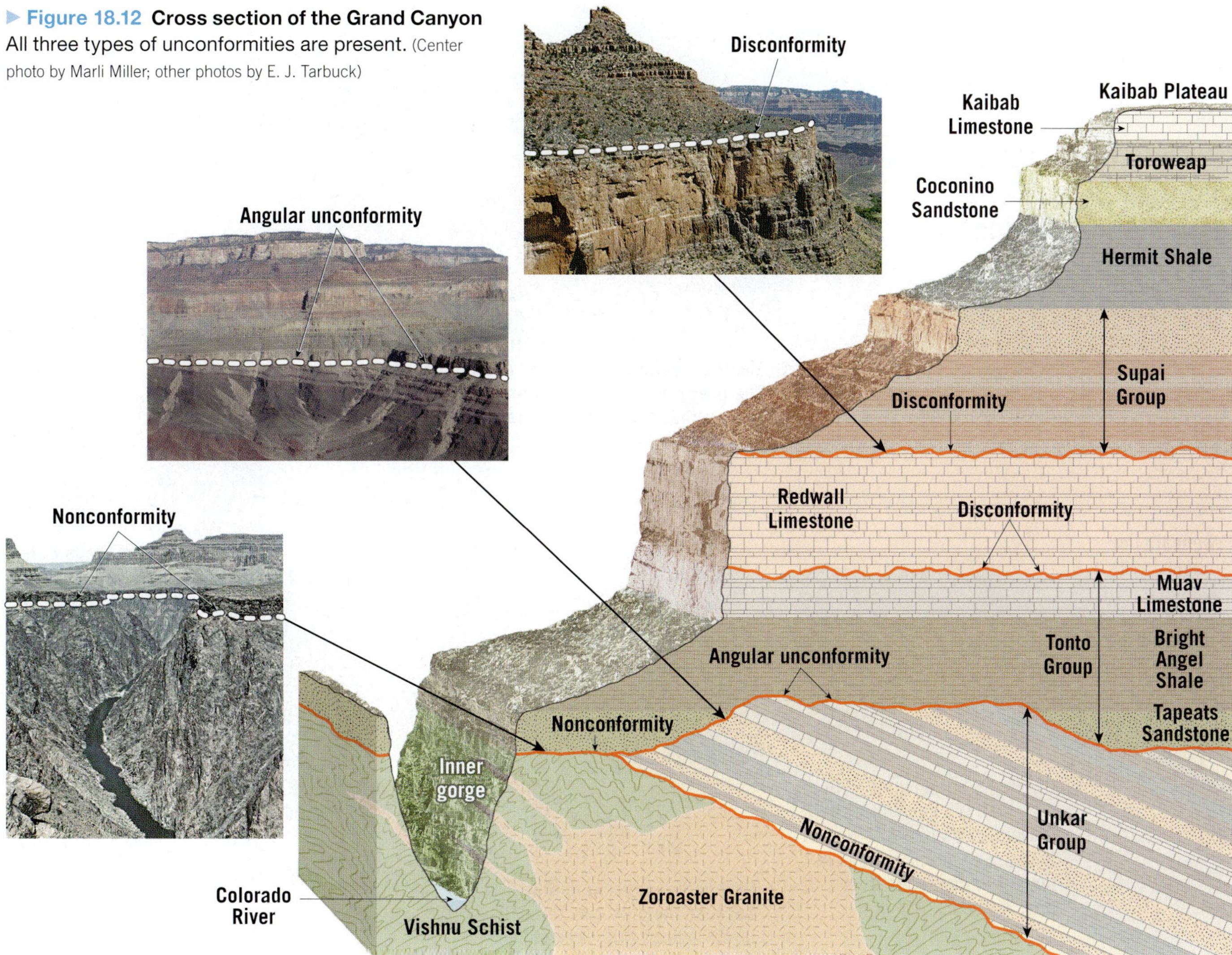

Figure 18.12 Cross section of the Grand Canyon All three types of unconformities are present. (Center photo by Marli Miller; other photos by E. J. Tarbuck)

Unconformities in the Grand Canyon The rocks exposed in the Grand Canyon of the Colorado River represent a tremendous span of geologic history. It is a wonderful place to take a trip through time. The canyon's colorful strata record a long history of sedimentation in a variety of environments—advancing seas, rivers and deltas, tidal flats and sand dunes. But the record is not continuous. Unconformities represent vast amounts of time that have not been recorded in the canyon's layers. Figure 18.12 is a geologic cross section of the Grand Canyon. All three types of unconformities can be seen in the canyon walls.

Applying Relative Dating Principles

If you apply the principles of relative dating to the hypothetical geologic cross section in Figure 18.13, you can place in proper sequence the rocks and the events they represent. The statements within the figure summarize the logic used to interpret the cross section.

In this example, we establish a relative time scale for the rocks and events in the area of the cross section. Remember that this method gives us no idea how many years of Earth history are represented, for we have no numerical dates. Nor do we know how this area compares to any other.

CONCEPT CHECKS 18.1

1. Distinguish between numerical dates and relative dates.
2. Sketch and label four simple diagrams that illustrate each of the following: superposition, original horizontality, lateral continuity, and cross-cutting relationships.
3. What is the significance of an unconformity?
4. Distinguish among angular unconformity, disconformity, and nonconformity.

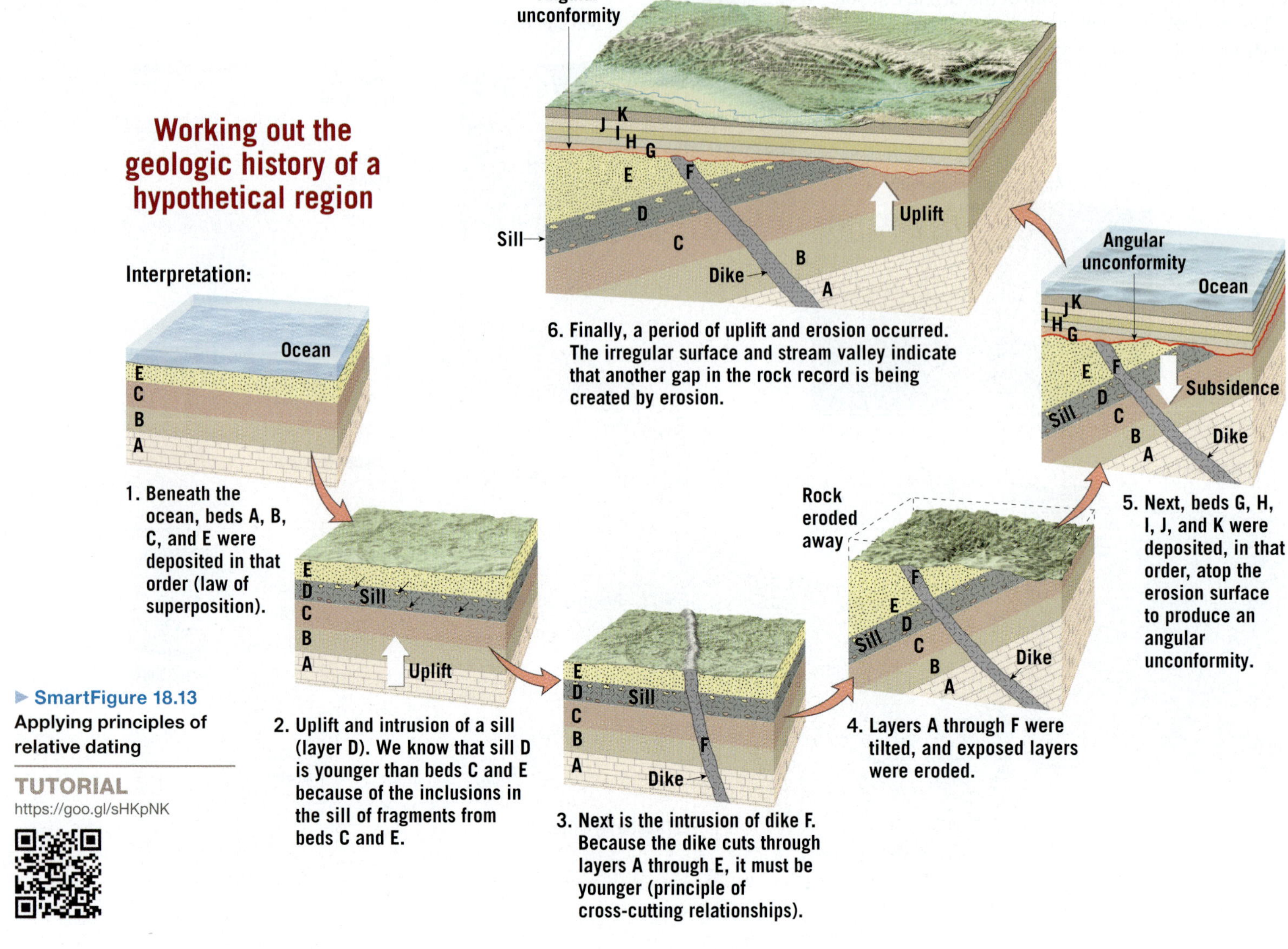
Working out the geologic history of a hypothetical region
Interpretation:
Ocean
E
C
B
A
1. Beneath the ocean, beds A, B, C, and E were deposited in that order (law of superposition).
E
D
Sill
C
B
A
Uplift
2. Uplift and intrusion of a sill (layer D). We know that sill D is younger than beds C and E because of the inclusions in the sill of fragments from beds C and E.
E
D
Sill
C
B
A
F
Dike
3. Next is the intrusion of dike F. Because the dike cuts through layers A through E, it must be younger (principle of cross-cutting relationships).
Rock eroded away
F
E
Sill
D
C
B
A
Dike
4. Layers A through F were tilted, and exposed layers were eroded.
Angular unconformity
Ocean
K
J
I
H
G
E
F
Sill
D
C
B
A
Subsidence
Dike
5. Next, beds G, H, I, J, and K were deposited, in that order, atop the erosion surface to produce an angular unconformity.
Angular unconformity
K
J
I
H
G
E
F
D
Sill
C
Uplift
B
Dike
A
6. Finally, a period of uplift and erosion occurred. The irregular surface and stream valley indicate that another gap in the rock record is being created by erosion.

▶ SmartFigure 18.13
Applying principles of relative dating

TUTORIAL
https://goo.gl/sHKpNK

18.2 Fossils: Evidence of Past Life

Define *fossil* and discuss the conditions that favor the preservation of organisms as fossils. List and describe various types of fossils.

Did You Know?
The word *fossil* comes from the Latin *fossilium*, which means "dug up from the ground." As originally used by medieval writers, a fossil was *any* stone, ore, or gem that came from an underground source. In fact, many early books on mineralogy are called books of fossils. The current meaning of *fossil* came about during the 1700s.

Fossils, the remains or traces of prehistoric life, are important inclusions in sediment and sedimentary rocks. They are basic and important tools for interpreting the geologic past. The scientific study of fossils is called **paleontology**. It is an interdisciplinary science that blends geology and biology in an attempt to understand all aspects of the evolution of life over the vast expanse of geologic time. Knowing the nature of the life-forms that existed at a particular time helps researchers understand past environmental conditions. Further, fossils are important time indicators and play a key role in correlating rocks of similar ages that are from different places.

Types of Fossils

Fossils are of many types. The remains of relatively recent organisms may not have been altered at all. Objects such as teeth, bones, and shells are common examples (Figure 18.14). Far less common are entire animals, flesh included, that have been preserved because of rather unusual circumstances. Remains of prehistoric elephants called mammoths that were frozen in the Arctic tundra of Siberia and Alaska are examples, as are the mummified remains of sloths preserved in a dry cave in Nevada.

Permineralization When mineral-rich groundwater permeates porous tissue such as bone or wood, minerals precipitate out of solution and fill pores and empty spaces, a process called *permineralization*. The formation of *petrified wood* involves permineralization with silica, often from a volcanic source such as a surrounding layer of volcanic ash. The wood is gradually transformed into chert, sometimes with colorful bands from impurities

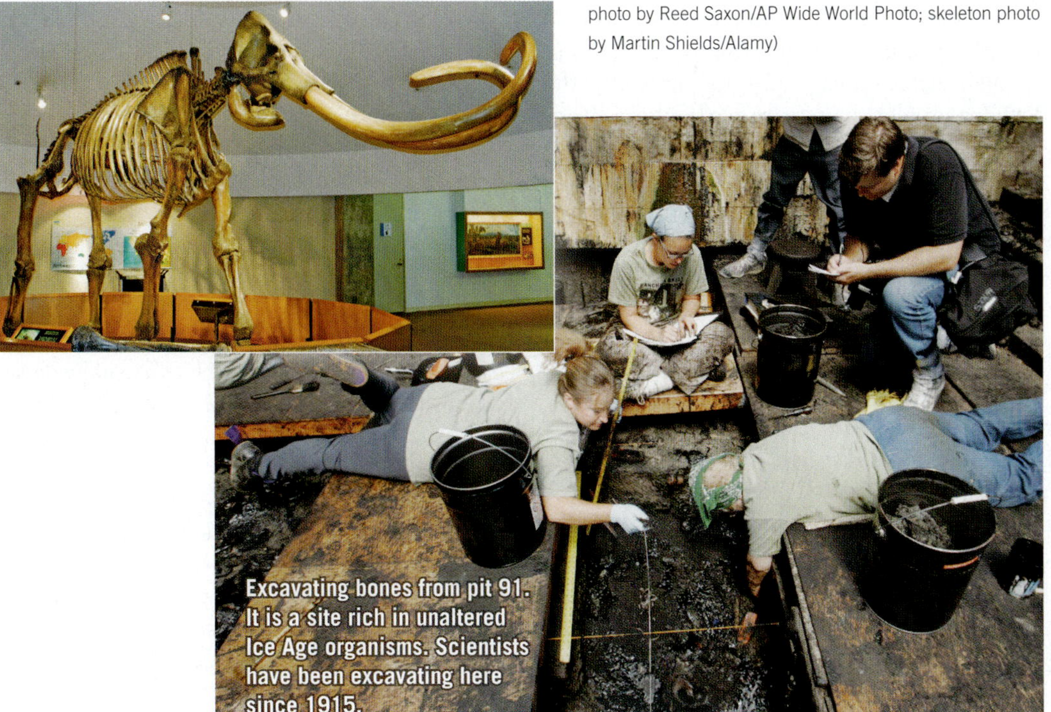

Skeleton of a mammoth, a prehistoric relative of modern elephant, from the La Brea tar pits.
Excavating bones from pit 91. It is a site rich in unaltered Ice Age organisms. Scientists have been excavating here since 1915.

◀ **Figure 18.14 La Brea tar pits** The fossils here are actual (unaltered) remains. (Excavation photo by Reed Saxon/AP Wide World Photo; skeleton photo by Martin Shields/Alamy)

such as iron or carbon (**Figure 18.15A**). The word *petrified* literally means "turned into stone." Sometimes the microscopic details of the petrified structure are faithfully retained.

Molds & Casts Another common class of fossils is *molds* and *casts*. When a shell or another structure is buried in sediment and then dissolved by underground water, a *mold* is created. The mold faithfully reflects the shape and surface marking of the organism; however, it does not reveal any information concerning its internal structure. If these hollow spaces are subsequently filled with mineral matter, *casts* are created (**Figure 18.15B**).

Carbonization & Impressions A type of fossilization called *carbonization* is particularly effective at preserving leaves and delicate animal forms. It occurs when fine sediment encases the remains of an organism. As time passes, pressure squeezes out the liquid and gaseous components and leaves behind a thin residue of carbon (**Figure 18.15C**). Black shale deposited as organic-rich mud in oxygen-poor environments often contains abundant carbonized remains. If the film of carbon is lost from a fossil preserved in fine-grained sediment, a replica of the surface, called an *impression*, may still show considerable detail (**Figure 18.15D**).

Amber Delicate organisms, such as insects, are difficult to preserve, and consequently they are relatively rare in the fossil record. However, *amber*—the hardened

▼ **Figure 18.15 Types of fossils** (Photo A by Bernhard Edmaier/Science Source; photo B by E. J. Tarbuck; photo C by Florissant Fossil Beds National Monument; photo D by E. J. Tarbuck; photo E by Colin Keates/Dorling Kindersley Media Library; photo F by E. J. Tarbuck)

A. Petrified wood preserved by permineralization

B. A trilobite preserved as a mold and cast

C. A fossil bee preserved as a thin carbon film

D. Fishes preserved as detailed impressions

E. Spider preserved in amber

F. A coprolite (fossil dung)—an example of a trace fossil

Did You Know?
People frequently confuse paleontology and archaeology. Paleontologists study fossils and are concerned with *all* life-forms in the geologic past. By contrast, archaeologists focus on the material remains of past human life. These remains include both the objects used by people long ago, called *artifacts*, and the buildings and other structures associated with where people lived, called *sites*.

resin of ancient trees—can preserve them in exquisite three-dimensional detail. The spider in Figure 18.15E was preserved after being trapped in a drop of sticky resin. Resin sealed off the insect from the atmosphere and protected the remains from damage by water and air. As the resin hardened, a protective pressure-resistant case was formed.

Trace Fossils In addition to the fossils already mentioned, there are numerous other types, many of them only traces of prehistoric life. Examples of such indirect evidence include:

- Tracks—animal footprints made in soft sediment that later turned into sedimentary rock.
- Burrows—tubes in sediment, wood, or rock made by an animal. These holes may later become filled with mineral matter and preserved. Some of the oldest-known fossils are believed to be worm burrows.
- Coprolites—fossil dung and stomach contents that can provide useful information pertaining to the size and food habits of organisms (Figure 18.15F).
- Gastroliths—highly polished stomach stones that were used in the grinding of food by some dinosaurs and other organisms.

Conditions Favoring Preservation

Only a tiny fraction of the organisms that have lived during the geologic past have been preserved as fossils. Normally, the remains of an animal or a plant are destroyed. Under what circumstances are they preserved? Two special conditions appear to be necessary: rapid burial and the possession of hard parts.

When an organism perishes, its soft parts usually are quickly eaten by scavengers or decomposed by bacteria. Occasionally, however, the remains are buried by sediment. When this occurs, the remains are protected from the surface environment, where destructive processes operate. Rapid burial, therefore, is an important condition favoring preservation.

In addition, animals and plants have a much better chance of being preserved as part of the fossil record if they have hard parts. Although traces and imprints of soft-bodied animals such as jellyfish, worms, and insects exist, they are not common. Flesh usually decays so rapidly that preservation is exceedingly unlikely. Hard parts such as shells, bones, and teeth predominate in the record of past life.

Because preservation is contingent on special conditions, the record of life in the geologic past is biased. The fossil record of those organisms with hard parts that lived in areas of sedimentation is quite abundant. However, we get only an occasional glimpse of the vast array of other life-forms that did not meet the special conditions favoring preservation.

CONCEPT CHECKS 18.2

1. Describe several ways that an animal or a plant can be preserved as a fossil.
2. List three examples of trace fossils.
3. What conditions favor the preservation of an organism as a fossil?

18.3 Correlation of Rock Layers

Explain how rocks of similar age that are in different places can be matched up.

Did You Know?
Even when organisms die and their tissues decay, the organic compounds (hydrocarbons) of which they were made may survive in sediments. These are called *chemical fossils*. Commonly, these hydrocarbons form oil and gas, but some residues can persist in the rock record and can be analyzed to determine the kind of organisms from which they are derived.

To develop a geologic time scale that is applicable to the entire Earth, rocks of similar age in different regions must be matched up. Such a task is called **correlation**. Correlating the rocks from one place to another makes possible a more comprehensive view of the geologic history of a region. Figure 18.16, for example, shows the correlation of strata at three sites on the Colorado Plateau in southern Utah and northern Arizona. No single locale exhibits the entire sequence, but correlation reveals a more complete picture of the sedimentary rock record.

Correlation Within Limited Areas

Within a limited area, geologists can correlate rocks of one locality with those of another by simply walking along the outcropping edges, but this may not be possible when the rocks are mostly concealed by soil and vegetation. Correlation over short distances is often achieved by noting the position of a bed in a sequence of strata. Or a layer may be identified in another location if it is composed of distinctive or uncommon minerals. However, when correlation between widely separated areas or between continents is the objective, geologists must rely on fossils.

Fossils & Correlation

The existence of fossils had been known for centuries, yet it was not until the late 1700s and early 1800s that their significance as geologic tools was made evident. During this period, an English engineer and canal

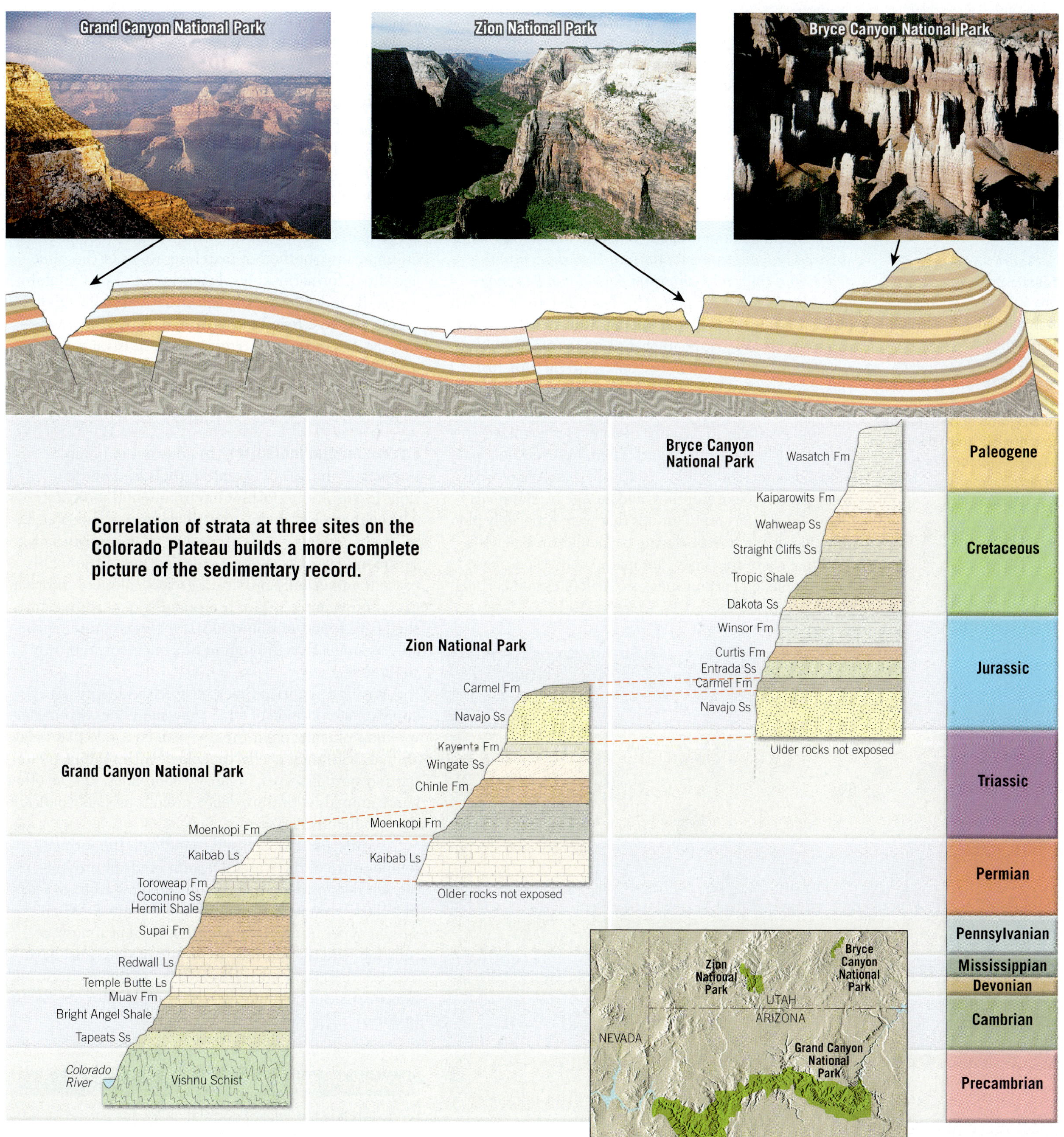

▲ **Figure 18.16 Correlation** Matching strata at three locations on the Colorado Plateau. (Photos by E. J. Tarbuck)

▲ **Figure 18.17 Index fossils** Since microfossils are often very abundant, widespread, and quick to appear and become extinct, they constitute ideal index fossils. This scanning electron micrograph shows marine microfossils from the Miocene epoch. (Photo by Biophoto Associates/ Science Source)

builder, William Smith, discovered that each rock formation in the canals he worked on contained fossils unlike those in the beds either above or below. Further, he noted that sedimentary strata in widely separated areas could be identified—and correlated—based on their distinctive fossil content.

Principle of Fossil Succession Based on Smith's classic observations and the findings of many later geologists, one of the most important and basic principles in historical geology was formulated: *Fossil organisms succeed one another in a definite and determinable order, and therefore any time period can be recognized by its fossil content.* This has come to be known as the **principle of fossil succession**. In other words, when fossils are arranged according to their age, they do not present a random or haphazard picture. To the contrary, fossils document the evolution of life through time.

For example, an Age of Trilobites is recognized quite early in the fossil record. Then, in succession, paleontologists recognize an Age of Fishes, an Age of Coal Swamps, an Age of Reptiles, and an Age of Mammals. These "ages" pertain to groups that were especially plentiful and characteristic during particular time periods. Within each of the "ages" are many subdivisions, based, for example, on certain species of trilobites and certain types of fish, reptiles, and so on. This same succession of dominant organisms, never out of order, is found on every continent.

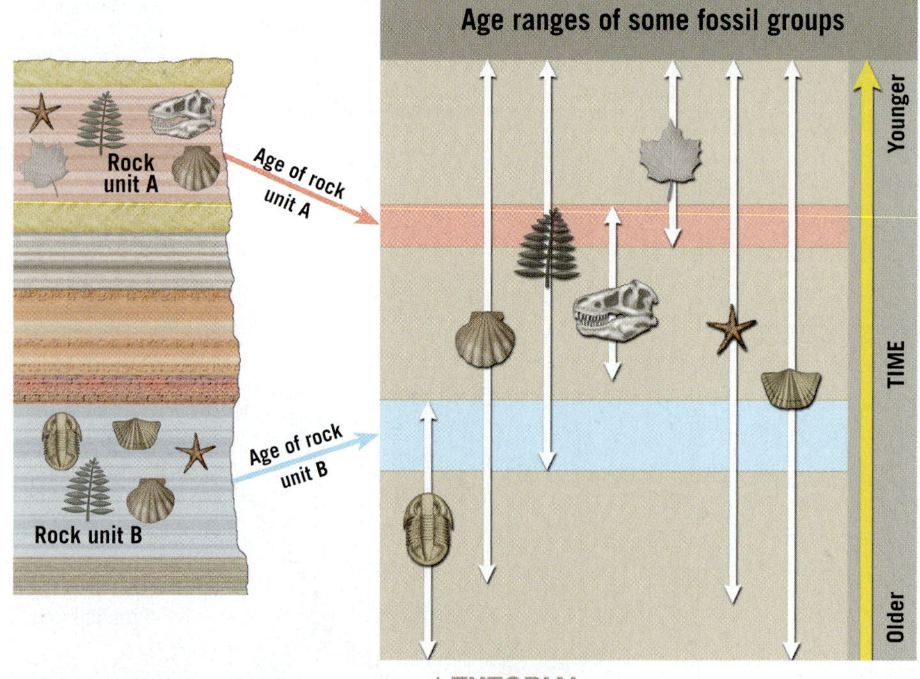

▲ **SmartFigure 18.18 Fossil assemblage** Overlapping ranges of fossils help date rocks more exactly than using a single fossil.

TUTORIAL
https://goo.gl/IpkEBT

Index Fossils & Fossil Assemblages When fossils were found to be time indicators, they became the most useful means of correlating rocks of similar age in different regions. Geologists pay particular attention to certain fossils called **index fossils** (Figure 18.17). These fossils are widespread geographically but limited to a short span of geologic time, so their presence provides an important method of matching rocks of the same age. Rock formations, however, do not always contain a specific index fossil. In such situations, a group of fossils, called a **fossil assemblage**, is used to establish the age of the bed. Figure 18.18 illustrates how an assemblage of fossils may be used to date rocks more precisely than could be accomplished by the use of any single fossil.

Environmental Indicators In addition to being important, and often essential, tools for correlation, fossils are important environmental indicators. Although we can deduce much about past environments by studying the nature and characteristics of sedimentary rocks, a close examination of the fossils present can usually provide a great deal more information. For example, when the remains of certain clam shells are found in limestone, a geologist quite reasonably assumes that the region was once covered by a shallow sea.

Fossils can also at times be used to identify the approximate position of an ancient shoreline. Given what we know of living organisms, we can conclude that fossil animals with thick shells, capable of withstanding pounding and surging waves, inhabited shorelines. On the other hand, animals with thin, delicate shells probably indicate deep, calm offshore waters.

Fossils also can be used to indicate the former temperature of the water. Certain kinds of present-day corals must live in warm and shallow tropical seas like those around Florida and The Bahamas. When similar types of coral are found in ancient limestones, they indicate the marine environment that must have existed when they were alive. These examples illustrate how fossils can help unravel the complex story of Earth history.

CONCEPT CHECKS 18.3

1. What is the goal of correlation?
2. State the principle of fossil succession in your own words.
3. Contrast index fossil and fossil assemblage.
4. Along with their value for correlation, how else are fossils useful to geologists?

18.4 Numerical Dating with Nuclear Decay

Discuss three ways that atomic nuclei change and explain how unstable isotopes are used to determine numerical dates.

In addition to establishing relative dates by using the principles described in the preceding sections, scientists can also obtain reliable numerical dates for events in the geologic past. For example, we know that Earth is about 4.6 billion years old and that the dinosaurs became extinct about 66 million years ago. Dates that are expressed in millions and billions of years truly stretch our imagination because our personal calendars involve time measured in hours, weeks, and years. In this section you will learn about radioactivity and its application in radiometric dating. Our understanding of changes in the nuclei of atoms has allowed us to determine that geologic time is vast. This immense span is often referred to as *deep time*. Radiometric dating allows us to measure it quantitatively.

Reviewing Basic Atomic Structure

Recall from Chapter 3 that each atom has a *nucleus* that contains protons and neutrons and that the nucleus is orbited by electrons. *Electrons* have a negative electrical charge, and *protons* have a positive charge. A *neutron* has no charge (it is electrically neutral), but it can be converted to a positively charged proton plus a negatively charged electron.

The *atomic number* (each element's identifying number) is the number of protons in the nucleus. Every element has a different number of protons and thus a different identifying atomic number (hydrogen = 1, carbon = 6, oxygen = 8, uranium = 92, etc.). Atoms of the same element always have the same number of protons, so the atomic number stays constant.

Practically all of an atom's mass (99.9 percent) is in the nucleus, indicating that electrons have virtually no mass at all. So, by adding the protons and neutrons in an atom's nucleus, we derive the atom's *mass number*. The number of neutrons can vary, and these variants, or *isotopes*, have different mass numbers.

To summarize with an example, uranium's nucleus always has 92 protons, so its atomic number is always 92. But its neutron population varies, so uranium has three isotopes: uranium-234 (protons + neutrons = 234), uranium-235, and uranium-238. All three isotopes are mixed in nature. They look the same and behave the same in chemical reactions.

Changes to Atomic Nuclei

Usually, the forces that stabilize atomic nuclei are strong. However, in some isotopes, the forces that bind protons and neutrons are not strong enough to keep them together forever. Such nuclei are *unstable* and spontaneously break apart in a process called **nuclear decay** (also called **radioactive decay**). As time goes by, more and more of the unstable atoms decay, producing an ever-growing number of stable isotopes. Not all isotopes are unstable—there are stable isotopes, too—but here we focus on the unstable isotopes and the stable isotopes they produce.

What happens when unstable atoms break apart? Three common types of nuclear decay are illustrated in Figure 18.19:

- Alpha particles (α particles) may be emitted from the nucleus. An alpha particle is composed of 2 protons

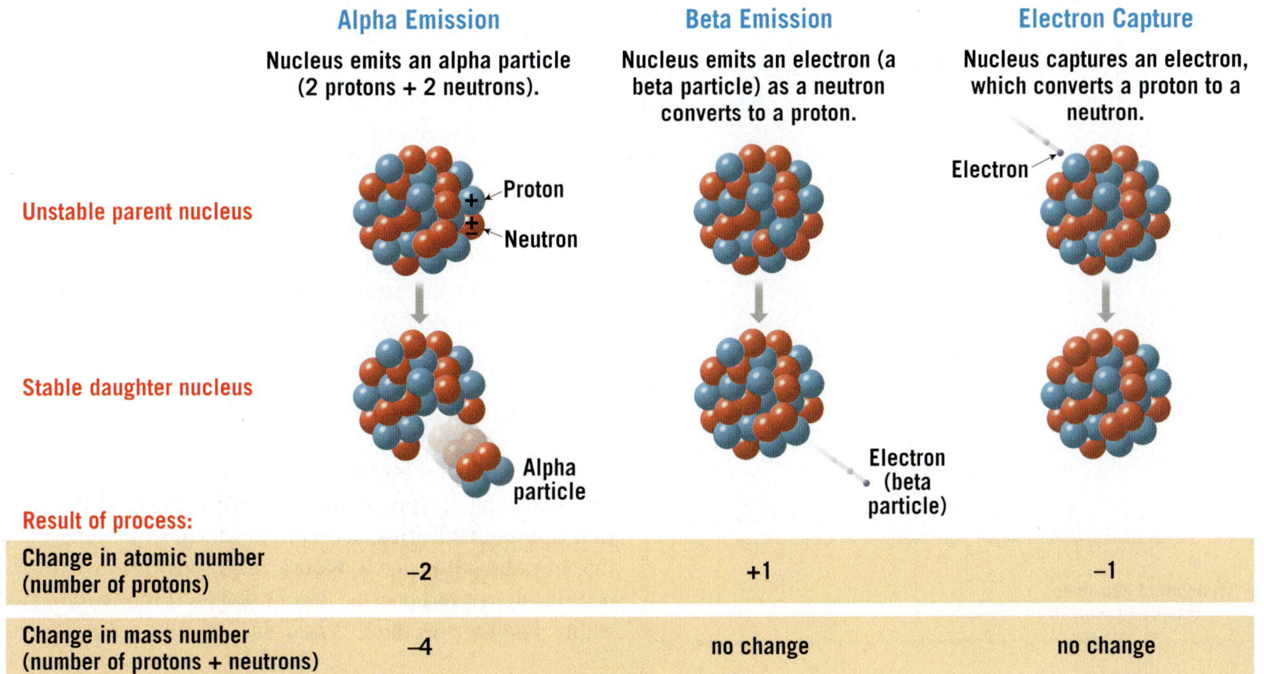
Alpha Emission
Nucleus emits an alpha particle (2 protons + 2 neutrons).
Beta Emission
Nucleus emits an electron (a beta particle) as a neutron converts to a proton.
Electron Capture
Nucleus captures an electron, which converts a proton to a neutron.
Unstable parent nucleus
Proton
Neutron
Electron
Stable daughter nucleus
Alpha particle
Electron (beta particle)
Result of process:

Result of process:	Alpha Emission	Beta Emission	Electron Capture
Change in atomic number (number of protons)	–2	+1	–1
Change in mass number (number of protons + neutrons)	–4	no change	no change

Figure 18.19 Changing atomic nuclei Notice that in each example, the number of protons (atomic number) in the nucleus changes, thus producing a different element.

▶ **Figure 18.20 Decay of U-238** Uranium-238 is an example of a nuclear decay series. Before the stable end product (Pb-206) is reached, many different isotopes are produced as intermediate steps.

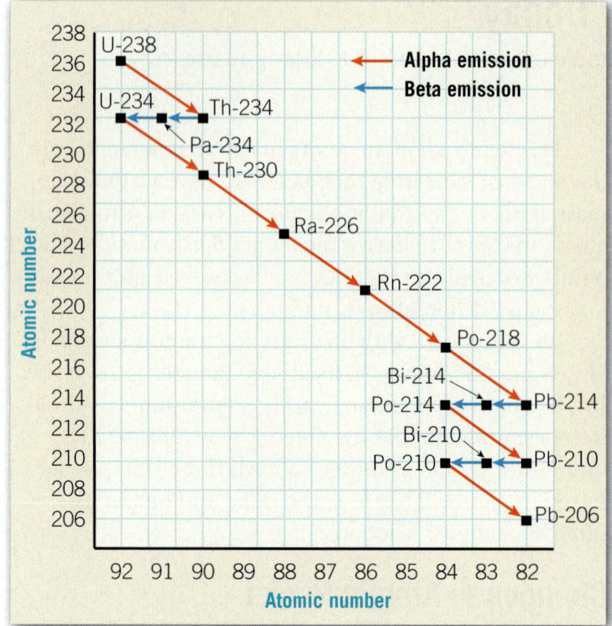

and 2 neutrons. Thus, the emission of an alpha particle means that the mass number of the isotope is reduced by 4, and the atomic number is lowered by 2.

- When an electron (often confusingly referred to as a "beta particle," or β particle), is emitted from a nucleus, the mass number remains unchanged because electrons have practically no mass. However, the electron is produced when a neutron (which has no charge) decays to produce the electron plus a proton. Because the nucleus now contains one more proton than before, the atomic number increases by 1. It's no longer the same element!
- Sometimes an electron is captured by the nucleus. The electron combines with a proton and forms an additional neutron. As in the last example, the mass number remains unchanged. However, because the nucleus now contains one fewer proton, the atomic number decreases by 1.

An unstable (radioactive) isotope is referred to as the *parent*, and the isotopes resulting from the decay of the parent are termed the *daughter products*. But the path from parent to daughter isn't always direct. Uranium-238, one of the most important isotopes for geologic dating, provides an example of the complexity (**Figure 18.20**). When the radioactive parent, uranium-238 (atomic number 92, mass number 238) decays, it follows a number of steps, emitting a total of 8 alpha particles and 6 electrons before finally becoming the stable daughter product lead-206 (atomic number 82, mass number 206).

Radiometric Dating

Nuclear decay provides a reliable way of calculating the ages of rocks and minerals that contain particular unstable isotopes. The procedure is called **radiometric dating**. Radiometric dating is reliable because the rates of decay for many isotopes have been precisely measured and do not vary under the physical conditions that exist in Earth's outer layers. Therefore, each unstable isotope used for dating has been decaying at a fixed rate since the formation of the mineral crystals in which we find it, and the products of its decay have been accumulating in that crystal at a corresponding rate. For example, some minerals are able to incorporate uranium atoms in their crystal lattice. When such a mineral crystallizes from magma, it contains no lead (the stable daughter product) from previous decay. The radiometric "clock" starts at this point. As the uranium in this newly formed mineral decays, atoms of the daughter product accumulate, trapped in the crystal, and eventually build up to measurable levels. Similarly, when a crystal of feldspar forms, some of the potassium atoms incorporated into its lattice will be the unstable isotope potassium-40. These atoms will decay at a steady rate by electron capture to produce the daughter argon-40. Over time, there is less and less of the parent potassium and more and more of the daughter argon.

Half-Life

The time required for half of the nuclei in a sample of a given unstable isotope to decay is called the **half-life** of that isotope. Half-life is a common way of expressing the rate of radioactive decay. **Figure 18.21** illustrates what occurs when a radioactive parent decays directly into its stable daughter product. When the quantities of parent

▼ **SmartFigure 18.21 Changing parent/daughter ratios** Change is exponential. Half of the unstable parent atoms remain after one half-life. After a second half-life, one-quarter of the parent atoms remain, and so forth.

TUTORIAL
https://goo.gl/o58WCb

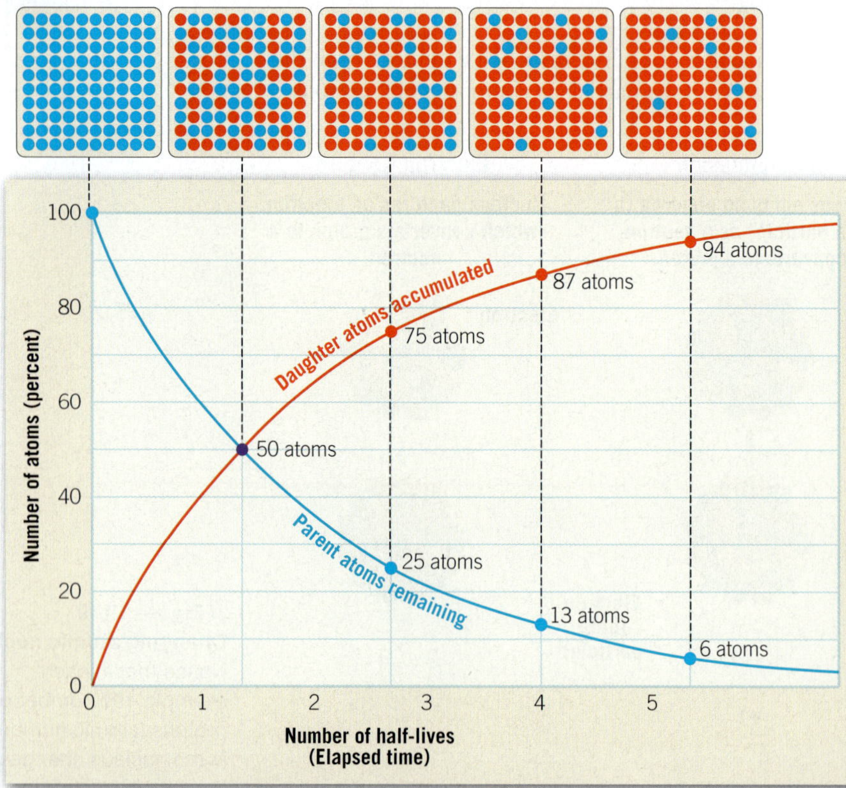

and daughter are equal (ratio 1:1), we know that one half-life has transpired. When one-quarter of the original parent atoms remain and three-quarters have decayed to the daughter product, the parent/daughter ratio is 1:3, and we know that two half-lives have passed. After three half-lives, the ratio of parent atoms to daughter atoms is 1:7 (one parent atom for every seven daughter atoms).

If the half-life of a radioactive isotope is known and the parent/daughter ratio can be determined, the age of the sample can be calculated. For example, assume that the half-life of a hypothetical unstable isotope is 1 million years, and the parent/daughter ratio in a sample is 1:15. This ratio indicates that four half-lives have passed and that the sample must be 4 million years old.

Notice that the *percentage* of radioactive atoms that decay during one half-life is always the same: 50 percent. However, the *actual number* of atoms that decay with the passing of each half-life continually decreases. Thus, as the percentage of radioactive parent atoms declines, the proportion of stable daughter atoms rises, with the increase in daughter atoms just matching the drop in parent atoms. This fact is the key to radiometric dating.

Using Unstable Isotopes

Of the many radioactive isotopes that exist in nature, five have proved particularly useful in providing radiometric ages for ancient rocks (Table 18.1). Rubidium-87, thorium-232, and the two listed isotopes of uranium are used only for dating rocks that are millions of years old, but potassium-40 is more versatile. Although the half-life of potassium-40 is 1.3 billion years, analytical techniques make it possible to detect tiny amounts of its stable daughter product, argon-40, in some rocks that are younger than 100,000 years. Another important reason for its frequent use is that potassium is an abundant constituent of many common minerals, particularly micas and feldspars.

Table 18.1 Isotopes Frequently Used in Radiometric Dating

Radioactive Parent	Stable Daughter Product	Currently Accepted Half-Life Values
Uranium-238	Lead-206	4.5 billion years
Uranium-235	Lead-207	704 million years
Thorium-232	Lead-208	14.1 billion years
Rubidium-87	Strontium-87	47.0 billion years
Potassium-40	Argon-40	1.3 billion years

A Complex Process Although the basic principle of radiometric dating is simple, the actual procedure is quite complex. The chemical analysis that determines the quantities of parent and daughter must be painstakingly precise. In addition, some radioactive materials do not decay directly into the stable daughter product, and this fact may further complicate the analysis. In the case of uranium-238, there are 13 intermediate unstable daughter products formed before the 14th and last daughter product, the stable isotope lead-206, is produced (see Figure 18.20).

Sources of Error It is important to understand that an accurate radiometric date can be obtained only if there has been no leakage of parent or daughter isotopes between the mineral crystal and its surroundings in the time since the mineral formed. This is not always the case. In fact, a limitation of the potassium-argon method arises from the fact that argon is a gas, and it may leak from minerals, resulting in a radiometric age that is lower than the actual age. Indeed, losses can be significant if the rock is subjected to high temperatures. If the rock is heated to the point where *all* of the argon in its minerals escapes, then its radiometric clock will be reset, and radiometric dating will give the time of thermal resetting, not the true age of the rock.

For other radiometric clocks, a loss of daughter atoms can occur if the rock has been subjected to weathering or leaching. To avoid such a problem, one simple safeguard is to use only fresh, unweathered material and not samples that exhibit signs of chemical alteration.

To guard against error in radiometric dating, scientists often use cross-checks, subjecting a sample to two different methods. If the results agree, the likelihood is high that the date is reliable. If the results are appreciably different, other cross-checks must be employed to determine which, if either, is correct.

Earth's Oldest Rocks Radiometric dating has produced literally thousands of dates for events in Earth history. Rocks exceeding 3.5 billion years in age are found on all of the continents. Earth's oldest rocks (so far) may be as old as 4.28 billion years (b.y.). Discovered in northern Quebec, Canada, on the shores of Hudson Bay, these rocks may be remnants of Earth's earliest crust. Rocks from western Greenland have been dated at 3.7 to 3.8 b.y., and rocks nearly as old are found in the Minnesota River valley and northern Michigan (3.5 to 3.7 b.y.), in southern Africa (3.4 to 3.5 b.y.), and in western Australia (3.4 to 3.6 b.y.). Tiny crystals of the mineral zircon having radiometric ages as old as 4.3 b.y. have been found in younger sedimentary rocks in western Australia. The source rocks for these tiny durable grains either no longer exist or have not yet been found.

Radiometric dating has vindicated the ideas of Hutton, Darwin, and others, who more than 150 years ago inferred that geologic time must be immense. Indeed, modern dating methods have proved that there has been enough time for the processes we observe to have accomplished tremendous tasks.

Dating with Carbon-14

To date relatively recent events, carbon-14 is used. Carbon-14 is the radioactive isotope of carbon. The process is often called **radiocarbon dating**. Because the half-life of carbon-14 is only 5730 years, radiocarbon dating can be

Did You Know?
Although movies and cartoons have depicted humans and dinosaurs living side by side, this was never the case. Dinosaurs flourished during the Mesozoic era and became extinct about 65 million years ago. Humans and their close ancestors did not appear on the scene until the late Cenozoic, more than 60 million years *after* the demise of dinosaurs.

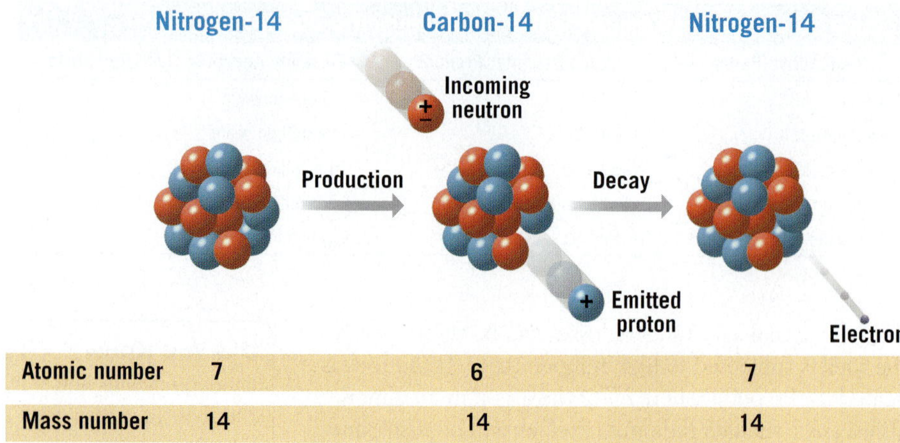

▲ **Figure 18.22**
Carbon-14 Production and decay of radiocarbon. These sketches represent the nuclei of the respective atoms.

used for dating events from the historic past as well as those from very recent geologic history. In some cases carbon-14 can be used to date events as far back as 70,000 years.

Carbon-14 (^{14}C) is continuously produced in the upper atmosphere as a result of cosmic-ray bombardment. Cosmic rays (high-energy particles) shatter the nuclei of gas atoms, releasing neutrons. Some of the neutrons are absorbed by nitrogen atoms (atomic number 7, mass number 14), causing each nucleus to emit a proton. As a result, the atomic number decreases by 1 (to 6), and a different element, carbon-14, is created (**Figure 18.22**). This isotope of carbon quickly becomes incorporated into carbon dioxide, which circulates in the atmosphere and is absorbed by living matter. As a result, all organisms—including you—contain a small amount of carbon-14. You "top off" your ^{14}C levels every time you eat something.

As long as an organism is alive, the decaying radiocarbon is continually replaced, and the proportions of carbon-14 and carbon-12 remain constant. Carbon-12 is the stable and most common isotope of carbon. However, when any plant or animal dies, the amount of carbon-14 gradually decreases as it decays to nitrogen-14 by beta emission. By comparing the proportions of carbon-14 and carbon-12 in a sample, radiocarbon dates can be determined. It is important to emphasize that carbon-14 can only be used to date organic materials, such as wood, charcoal, bones, flesh, and cloth.

Did You Know?

Dating with carbon-14 is useful to archaeologists and historians as well as geologists. For example, University of Arizona researchers used carbon-14 dating to determine the age of the Dead Sea Scrolls, considered to be some of the greatest archaeological discoveries of the twentieth century. Parchment from the scrolls dates between 150 B.C.E. and 5 B.C.E. Portions of the scrolls contain dates that match those determined by the carbon-14 measurements.

▲ **Figure 18.23** **Cave art** Chauvet Cave in southern France, discovered in 1994, contains some of the earliest-known cave paintings. Radiocarbon dating indicates that most of the images were drawn between 30,000 and 32,000 years ago. (Photo by Javier Trueba/MSF/Science Source)

Although carbon-14 is only useful in dating the last small fraction of geologic time, it is a valuable tool for anthropologists, archaeologists, and historians, as well as for geologists who study very recent Earth history (**Figure 18.23**). In fact, the development of radiocarbon dating was considered so important that the chemist who discovered this application, Willard F. Libby, received a Nobel Prize in 1960.

CONCEPT CHECKS 18.4

1. List four ways that unstable nuclei change. For each type, describe how the atomic number and atomic mass change.
2. Sketch a simple diagram that explains the idea of half-life.
3. Why is radiometric dating a reliable method for determining numerical dates?
4. For what time span does radiocarbon dating apply?

18.5 Determining Numerical Dates for Sedimentary Strata

Explain how reliable numerical dates are determined for layers of sedimentary rock.

Although reasonably accurate numerical dates have been worked out for the periods of the geologic time scale, the task is not without difficulty. The primary difficulty in assigning numerical dates to units of time is that not all rocks can be dated by using radiometric methods. For a radiometric date to be useful, all the minerals in the rock must have formed at about the same time. For this reason, unstable isotopes can be used to determine when minerals in an igneous rock crystallized and when pressure and heat created new minerals in a metamorphic rock.

However, samples of sedimentary rock can only rarely be dated directly by radiometric means. Although a detrital sedimentary rock may include particles that contain unstable isotopes, the rock's age cannot be accurately determined because the grains composing the rock are not the same age as the rock in which they occur. Rather, the sediments have been weathered from rocks of diverse ages.

Radiometric dates obtained from metamorphic rocks may also be difficult to interpret because the age of a particular mineral in a metamorphic rock does not necessarily represent the time when the rock initially formed. Instead, the date might indicate any one of a number of subsequent metamorphic phases.

If samples of sedimentary rocks rarely yield reliable radiometric ages, how can numerical dates be assigned to sedimentary layers? Usually geologists must relate the strata to datable igneous masses, as in Figure 18.24. In this example, radiometric dating has determined the ages of the volcanic ash bed in the Morrison Formation and the dike cutting the Mancos Shale and Mesaverde Formation. The sedimentary beds below the ash are obviously older than the ash, and all the layers above the ash are younger (based on the principle of superposition). The dike is younger than the Mancos Shale and the Mesaverde Formation but older than the Wasatch Formation because the dike does not intrude this topmost layer (based on the principle of cross-cutting relationships).

From this kind of evidence, geologists estimate that the Morrison Formation was deposited more than 160 million years ago, as indicated by the ash bed. Further, they conclude that deposition of the Wasatch Formation began after the intrusion of the dike, 66 million years ago. This is one example of literally thousands that illustrate how datable materials are used to "bracket" the various episodes in Earth history within specific time periods. It shows the necessity of combining laboratory dating methods with field observations of rocks.

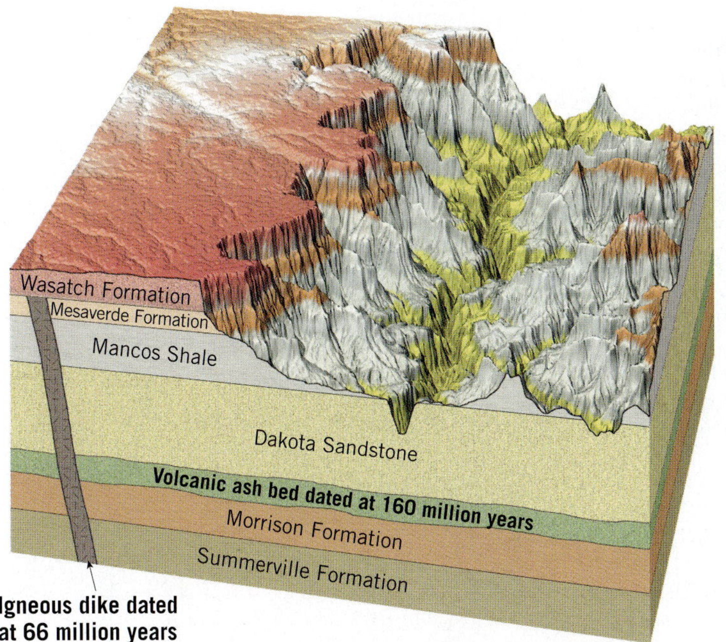

◀ **Figure 18.24 Dating sedimentary strata** Numerical dates for sedimentary layers are usually determined by examining their relationship to igneous rocks.

CONCEPT CHECKS 18.5

1. Briefly explain why it is often difficult to assign a reliable numerical date to a sample of sedimentary rock.
2. How might a numerical date for a layer of sedimentary rock be determined?

18.6 The Geologic Time Scale

Distinguish among the four basic time units that make up the geologic time scale and explain why the time scale is considered to be a dynamic tool.

Geologists have divided the whole of geologic history into units of varying length. Together, they compose the **geologic time scale** of Earth history (Figure 18.25). The major units of the time scale were delineated during the nineteenth century, principally by scientists in Western Europe and Great Britain. Because radiometric dating was unavailable at that time, the entire time scale was created using methods of relative dating. It was only in the twentieth century that radiometric methods permitted numerical dates to be added.

Structure of the Time Scale

The geologic time scale subdivides the 4.6-billion-year history of Earth into many different units and provides a meaningful time frame within which the events of the geologic past are arranged. As shown in Figure 18.25, **eons** represent the greatest expanses of time. The eon that began about 542 million years ago is the **Phanerozoic**, a term derived from Greek words meaning "visible life." It is an appropriate description because the rocks and deposits of the Phanerozoic eon contain abundant fossils that document major evolutionary trends.

Another glance at the time scale reveals that eons are divided into **eras**. The Phanerozoic eon consists of the **Paleozoic era** (*paleo* = ancient, *zoe* = life), the **Mesozoic era** (*meso* = middle, *zoe* = life), and the **Cenozoic era** (*ceno* = recent, *zoe* = life). As the names imply, these eras are bounded by profound worldwide changes in life-forms.*

*Major changes in life-forms are discussed in Chapter 19.

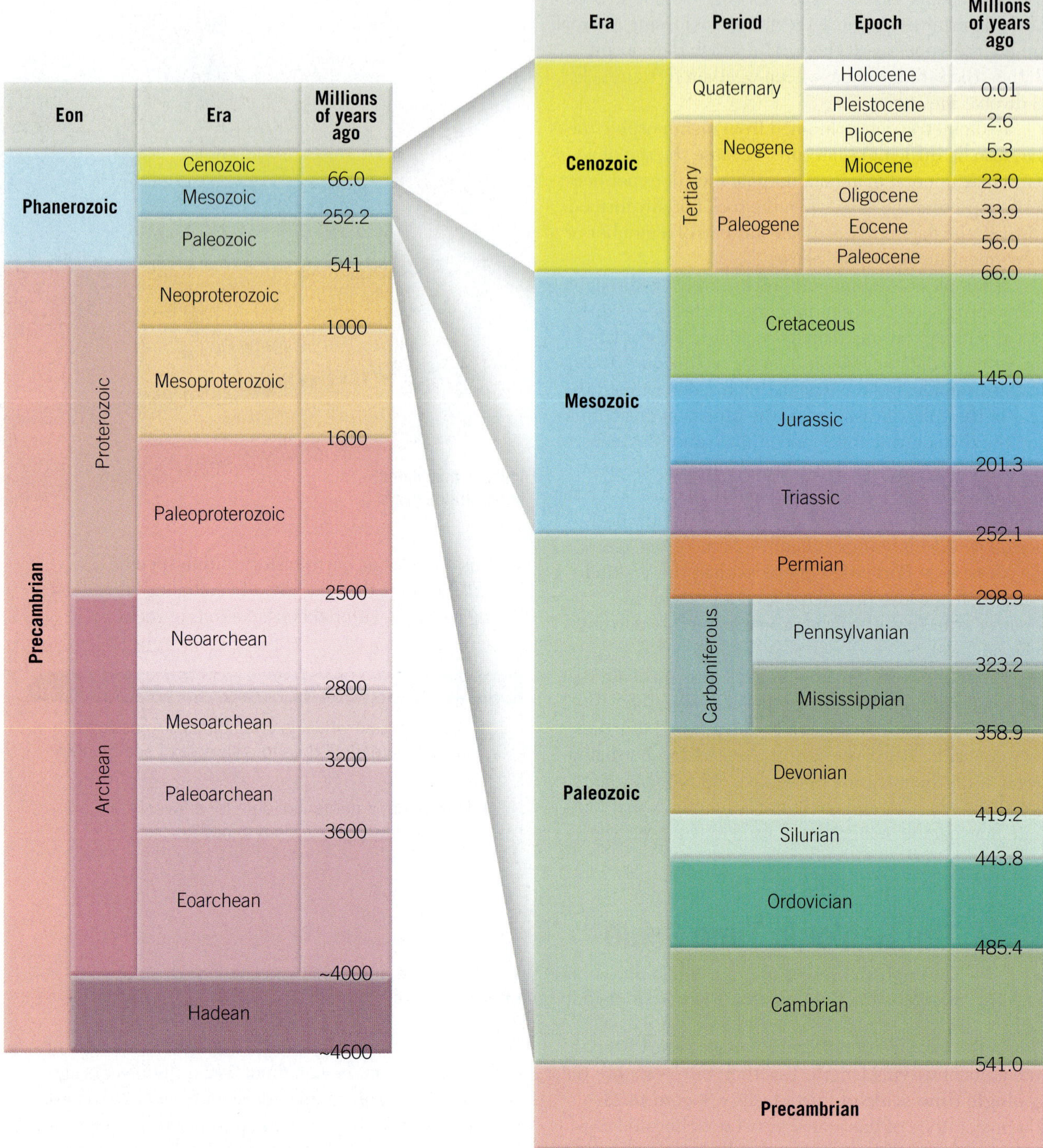

▶ **Figure 18.25**
Geologic time scale: A basic reference The time scale divides the vast 4.6-billion-year history of Earth into eons, eras, periods, and epochs. Numbers on the time scale represent time in millions of years before the present. The Precambrian accounts for more than 88 percent of geologic time. Numerical dates were added long after the time scale was established using relative dating techniques. The dates appearing on this time scale are those currently accepted by the International Commission on Stratigraphy (ICS) in 2015. The color scheme used on this chart was selected because it is similar to that used by the ICS.

Each era of the Phaneroic eon is further divided into time units known as **periods**. The Paleozoic has seven, and the Mesozoic and Cenozoic each have three. Each of these periods is characterized by a somewhat less profound change in life-forms compared with the eras.

Each of the periods is divided into still smaller units called **epochs**. As you can see in Figure 18.25, seven epochs have been named for the periods of the Cenozoic. The epochs of other periods usually are simply termed *early*, *middle*, and *late*.

Precambrian Time

Notice that the geologic time scale is considerably less detailed prior to the beginning of the Cambrian period, 541 million years ago. The nearly 4 billion years that precede the Cambrian are divided into two eons, the **Archean** (*archaios* = ancient) and the **Proterozoic** (*proteros* = before, *zoe* = life). It is also common for this vast expanse of time to simply be referred to as the **Precambrian**.

Why is the huge expanse of Precambrian time, which represents about 88 percent of Earth history, not divided into numerous eras, periods, and epochs? The reason is that Precambrian history is not known in great enough detail. In geology, as in human history, the farther back we go, the less we know. We know much more about the past decade than about the first century C.E. Equally, in Earth history, the more recent past has the freshest, least disturbed, and most observable record. The further back in time a geologist goes, the more fragmented the record and clues become. There are also other reasons to explain our lack of a detailed time scale for this vast segment of Earth history:

- The first abundant fossil evidence does not appear in the geologic record until the beginning of the Cambrian period. Prior to the Cambrian, simple life-forms such as algae, bacteria, fungi, and worms predominated. All of these organisms lack hard parts, an important condition favoring preservation. For this reason, there is only a meager Precambrian fossil record. Many exposures of Precambrian rocks have been studied in some detail, but correlation is often difficult when fossils are lacking.
- Because Precambrian rocks are very old, most have been subjected to a great many changes. Much of the Precambrian rock record is composed of highly distorted metamorphic rocks. This makes the interpretation of past environments difficult because many of the clues present in the original sedimentary rocks have been destroyed.

Radiometric dating has provided a partial solution to the troublesome task of dating and correlating Precambrian rocks. But untangling the complex Precambrian record still remains a daunting task.

Terminology & the Geologic Time Scale

Some terms are associated with the geologic time scale but are not officially recognized as being a part of it. The best known, and most common, example is *Precambrian*—the informal name for the eons that came before the current Phanerozoic eon. Although the term *Precambrian* has no formal status on the geologic time scale, it has been traditionally used as though it does.

Hadean is another informal term that is found on some versions of the geologic time scale and is used by many geologists. It refers to the earliest interval (eon) of Earth history—before the oldest-known rocks. When the term was coined in 1972, the age of Earth's oldest-known rocks was about 3.8 billion years. Today that number stands at slightly greater than 4 billion, and, of course, is subject to revision. The name *Hadean* derives from *Hades*, Greek for "underworld"—a reference to the "hellish" conditions that prevailed on Earth early in its history.

Effective communication in the geosciences requires that the geologic time scale consist of standardized divisions and dates. So, who determines which names and dates on the geologic time scale are "official"? The organization that is largely responsible for maintaining and updating this important document is the International Commission on Stratigraphy (ICS), a committee of the International Union of Geological Sciences. Advances in the geosciences require that the scale be periodically updated to include changes in unit names and boundary age estimates.

For example, the geologic time scale shown in Figure 18.25 was updated in 2015. After considerable dialogue among geologists who focus on very recent Earth history, the ICS changed the date for the start of the Quaternary period and the Pleistocene epoch from 1.8 million to 2.6 million years ago. Perhaps by the time you read this, other changes will have been made.

If you were to examine a geologic time scale from just a few years ago, it is quite possible that you would see the Cenozoic era divided into the Tertiary and Quaternary periods. However, on more recent versions, the space formerly designated as Tertiary is divided into the Paleogene and Neogene periods. As our understanding of this time span has changed, so too has its designation on the geologic time scale. Today, the Tertiary period is considered a "historic" name and is given no official status on the ICS version of the time scale. Many time scales still contain references to the Tertiary period, though, including Figure 18.25. One reason for this is that a great deal of past (and some current) geologic literature uses this name.

For those who study historical geology, it is important to realize that the geologic time scale is a dynamic tool that continues to be refined as our knowledge and understanding of Earth history evolve.

CONCEPT CHECKS 18.6

1. List the four basic units that make up the geologic time scale.
2. Why is *zoic* part of so many names on the geologic time scale?
3. What term applies to *all* of geologic time prior to the Phanerozoic eon? Why is this span *not* divided into as many smaller time units as the Phanerozoic eon?
4. To what does the term *Hadean* apply? Is it an "official" part of the geologic time scale?

Did You Know?

Some scientists have suggested that the Holocene epoch has ended and that we have entered a new epoch called the *Anthropocene*. It is considered to be the span (beginning in the early 1800s) in which the global environmental effects of increased human population and economic development have dramatically transformed Earth's surface. Although this term is used currently as an informal metaphor for human-caused global environmental change, a number of scientists feel that there is merit in recognizing the Anthropocene as a new "official" geologic epoch.

CONCEPTS IN REVIEW
Geologic Time

18.1 Creating a Time Scale: Relative Dating Principles

Distinguish between numerical and relative dating and apply relative dating principles to determine a time sequence of geologic events.

KEY TERMS: numerical date, relative date, principle of superposition, principle of original horizontality, principle of lateral continuity, principle of cross-cutting relationships, principle of inclusions, conformable, unconformity, angular unconformity, disconformity, nonconformity

- The two types of dates that geologists use to interpret Earth history are (1) relative dates, which put events in their proper sequence of formation, and (2) numerical dates, which pinpoint the time in years when an event took place.
- Relative dates can be established using the principles of superposition, original horizontality, cross-cutting relationships, and inclusions. Unconformities, gaps in the geologic record, may be identified during the relative dating process.

? The accompanying photo shows four features. Place the features in the proper sequence, from oldest to youngest. Explain your reasoning.

18.2 Fossils: Evidence of Past Life

Define *fossil* and discuss the conditions that favor the preservation of organisms as fossils. List and describe various types of fossils.

KEY TERMS: fossil, paleontology

- Fossils are remains or traces of ancient life. Paleontology is the branch of science that studies fossils.
- Fossils can form through many processes. For an organism to be preserved as a fossil, it usually needs to be buried rapidly. Also, an organism's hard parts are most likely to be preserved because soft tissue decomposes rapidly in most circumstances.

? What term is used to describe the type of fossil that is shown here? Briefly describe how it formed.

18.3 Correlation of Rock Layers

Explain how rocks of similar age that are in different places can be matched up.

KEY TERMS: correlation, principle of fossil succession, index fossil, fossil assemblage

- Matching up exposures of rock that are the same age but are in different places is called correlation. By correlating rocks from around the world, geologists developed the geologic time scale and obtained a fuller perspective on Earth history.
- Fossils can be used to correlate sedimentary rocks in widely separated places by using the rocks' distinctive fossil content and applying the principle of fossil succession. The principle states that fossil organisms succeed one another in a definite and determinable order, and, therefore, a time period can be recognized by examining its fossil content.
- Index fossils are particularly useful in correlation because they are widespread and associated with a relatively narrow time span. The overlapping ranges of fossils in an assemblage may be used to establish an age for a rock layer that contains multiple fossils.
- Fossils may be used to establish ancient environmental conditions that existed when sediment was deposited.

18.4 Numerical Dating with Nuclear Decay

Discuss three ways that atomic nuclei change and explain how unstable isotopes are used to determine numerical dates.

KEY TERMS: nuclear (radioactive) decay, radiometric dating, half-life, radiocarbon dating

- Nuclear decay is the spontaneous breaking apart of certain unstable atomic nuclei. Three common forms of nuclear decay are (1) emission of an alpha particle from the nucleus, (2) emission of a beta particle (electron) from the nucleus, and (3) capture of an electron by the nucleus.
- Radiometric dating refers to the procedure by which unstable isotopes are used to determine numerical ages of rocks and minerals. It is reliable because the rates of decay for the isotopes that are used have been precisely measured and do not vary.
- The length of time it takes for one-half of the nuclei of an unstable parent isotope to change into its stable daughter product is called the half-life of that isotope. If the half-life is known, and the parent/daughter ratio can be measured, the age of the sample can be calculated.

? Measurements of zircon crystals containing trace amounts of uranium from a specimen of granite yield parent/daughter ratios of 25 percent parent (uranium-235) and 75 percent daughter (lead-206). The half-life of uranium-235 is 704 million years. How old is the granite?

18.5 Determining Numerical Dates for Sedimentary Strata

Explain how reliable numerical dates are determined for layers of sedimentary rock.

- Sedimentary strata are usually not directly datable using radiometric techniques because they consist of the material produced by the weathering of other rocks. A particle in a sedimentary rock comes from some older source rock. If you were to date the particle using unstable isotopes, you would get the age of the source rock, not the age of the sedimentary rock.
- One way geologists assign numerical dates to sedimentary rocks is to use relative dating principles to relate them to datable igneous masses, such as dikes and volcanic ash beds. A layer may be older than one igneous feature and younger than another.

? Express the numerical age of the sandstone layer in the diagram as accurately as possible.

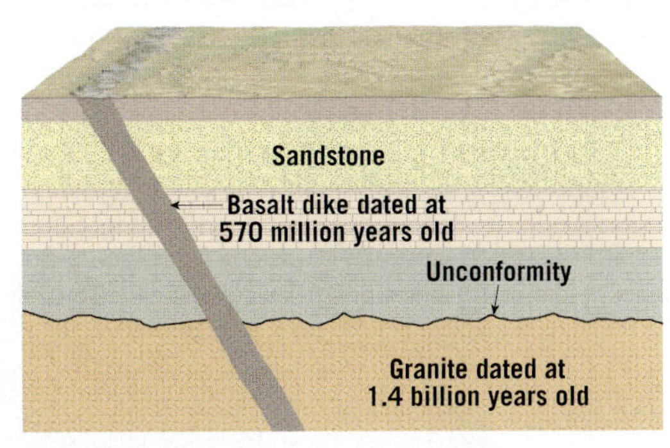

18.6 The Geologic Time Scale

Distinguish among the four basic time units that make up the geologic time scale and explain why the time scale is considered to be a dynamic tool.

KEY TERMS: geologic time scale, eon, Phanerozoic eon, era, Paleozoic era, Mesozoic era, Cenozoic era, period, epoch, Archean, Proterozoic, Precambrian

- Earth history is divided into units of time on the geologic time scale. Eons are divided into eras, which each contain multiple periods. Periods are divided into epochs.
- Precambrian time includes the Archean and Proterozoic eons. It is followed by the Phanerozoic eon, which is well documented by abundant fossil evidence, resulting in many subdivisions.
- The geologic time scale is a work in progress, continually being refined as new information becomes available.

? Is the Mesozoic an example of an eon, an era, a period, or an epoch? What about the Jurassic?

GIVE IT SOME THOUGHT

1 The accompanying image shows the metamorphic rock gneiss, a basaltic dike, and a fault. Place these three features in their proper sequence (which came first, second, and third) and explain your logic.

Marli Miller

2 A mass of granite is in contact with a layer of sandstone. Using a principle described in this chapter, explain how you might determine whether the sandstone was deposited on top of the granite or whether the magma that formed the granite was intruded after the sandstone was deposited.

3 This scenic image is from Monument Valley in the northeastern corner of Arizona. The bedrock in this region consists of layers of sedimentary rocks. Although the prominent rock exposures ("monuments") in this photo are widely separated, we can infer that they represent a once-continuous layer. Discuss the principle that allows us to make this inference.

Michael Collier

4 The accompanying photo shows two layers of sedimentary rock. The lower layer is shale from the late Mesozoic era. Note the old river channel that was carved into the shale after it was deposited. Above is a younger layer of boulder-rich breccia. Are these layers conformable? Explain why or why not. What term from relative dating applies to the line separating the two layers?

Callan Bentley

5 These polished stones are called *gastroliths*. Explain how such objects can be considered fossils. What category of fossil are they? Name another example of a fossil in this category.

Francois Gohier/Photo Researchers, Inc.

6 If an unstable isotope of thorium (atomic number 90, mass number 232) emits 6 alpha particles and 4 beta particles during the course of radioactive decay, what are the atomic number and mass number of the stable daughter product?

7 A hypothetical unstable isotope has a half-life of 10,000 years. If the ratio of radioactive parent to stable daughter product is 1:3, how old is the rock that contains the radioactive material?

8 Solve the problems below that relate to the magnitude of Earth history. To make calculations easier, round Earth's age to 5 billion years.

a. What percentage of geologic time is represented by recorded history? (Assume 5000 years for the length of recorded history.)

b. Humanlike ancestors (hominins) have been around for roughly 5 million years. What percentage of geologic time is represented by these ancestors?

c. The first abundant fossil evidence for multicellular organisms does not appear until the beginning of the Cambrian period, about 540 million years ago. What percentage of geologic time is represented by this abundant fossil evidence?

9 A portion of a popular college text in historical geology includes 10 chapters (281 pages) in a unit titled "The Story of Earth." Two chapters (49 pages) are devoted to Precambrian time. By contrast, the last two chapters (67 pages) focus on the most recent 23 million years, with 25 of those pages devoted to the Holocene Epoch, which began 10,000 years ago.

a. Compare the percentage of pages devoted to the Precambrian to the percentage of geologic time that this span represents.

b. How does the number of pages about the Holocene compare to its percentage of geologic time?

c. Suggest some reasons the text seems to have such an unequal treatment of Earth history.

10 This scene in Montana's Glacier National Park shows layers of Precambrian sedimentary rocks. The darker layer contained within the sedimentary layers is igneous. The narrow, light-colored areas adjacent to the igneous rock were created when molten material that formed the igneous rock baked the adjacent rock.

a. Is the igneous layer more likely a lava flow that was laid down at the surface prior to the deposition of the layers above it or a sill that was intruded after all the sedimentary layers were deposited? Explain.

b. Is it likely that the igneous layer will exhibit a vesicular texture? Explain.

c. To which group (igneous, sedimentary, or metamorphic) does the light-colored rock belong? Relate your explanation to the rock cycle.

Marli Miller

11 The accompanying diagram is a cross-section of a hypothetical area. Place the lettered features in the proper sequence from oldest to youngest. Where in the sequence can you identify an unconformity?

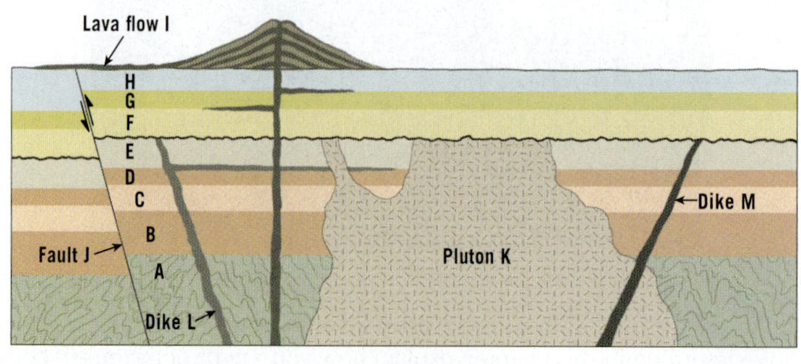

12 This is a close-up view of the detrital sedimentary rock conglomerate. Assume that this rock contains radioactive isotopes that will yield numerical dates.

a. Although radioactive isotopes are present, a reliable numerical date for this conglomerate cannot be *accurately* determined. Explain.

b. How might a numerical age range be established for the conglomerate layer?

E.J. Tarbuck

MasteringGeology™

Looking for additional review and test prep materials? Visit the Study Area in MasteringGeology to enhance your understanding of this chapter's content by accessing a variety of resources, including Self-Study Quizzes, Geoscience Animations, SmartFigures, Mobile Field Trips, *Project Condor* Quadcopter videos, *In the News* RSS feeds, flashcards, web links, and an optional Pearson eText.

www.masteringgeology.com

19 Earth's Evolution Through Geologic Time

FOCUS ON CONCEPTS

Each statement represents the primary learning objective for the corresponding major heading within the chapter. After you complete the chapter, you should be able to:

19.1 List the principal characteristics that make Earth unique among the planets in the solar system.

19.2 Outline the major stages in Earth's evolution, from the Big Bang to the formation of our planet's layered internal structure.

19.3 Describe how Earth's atmosphere and oceans formed and evolved through time.

19.4 Explain the formation of continental crust, how continental crust becomes assembled into continents, and the role that the supercontinent cycle has played in this process.

19.5 List and discuss the major geologic events in the Paleozoic, Mesozoic, and Cenozoic eras.

19.6 Describe some of the hypotheses on the origin of life and the characteristics of early prokaryotes, eukaryotes, and multicellular organisms.

19.7 List the major developments in the history of life during the Paleozoic era.

19.8 Briefly explain the major developments in the history of life during the Mesozoic era.

19.9 Discuss the major developments in the history of life during the Cenozoic era.

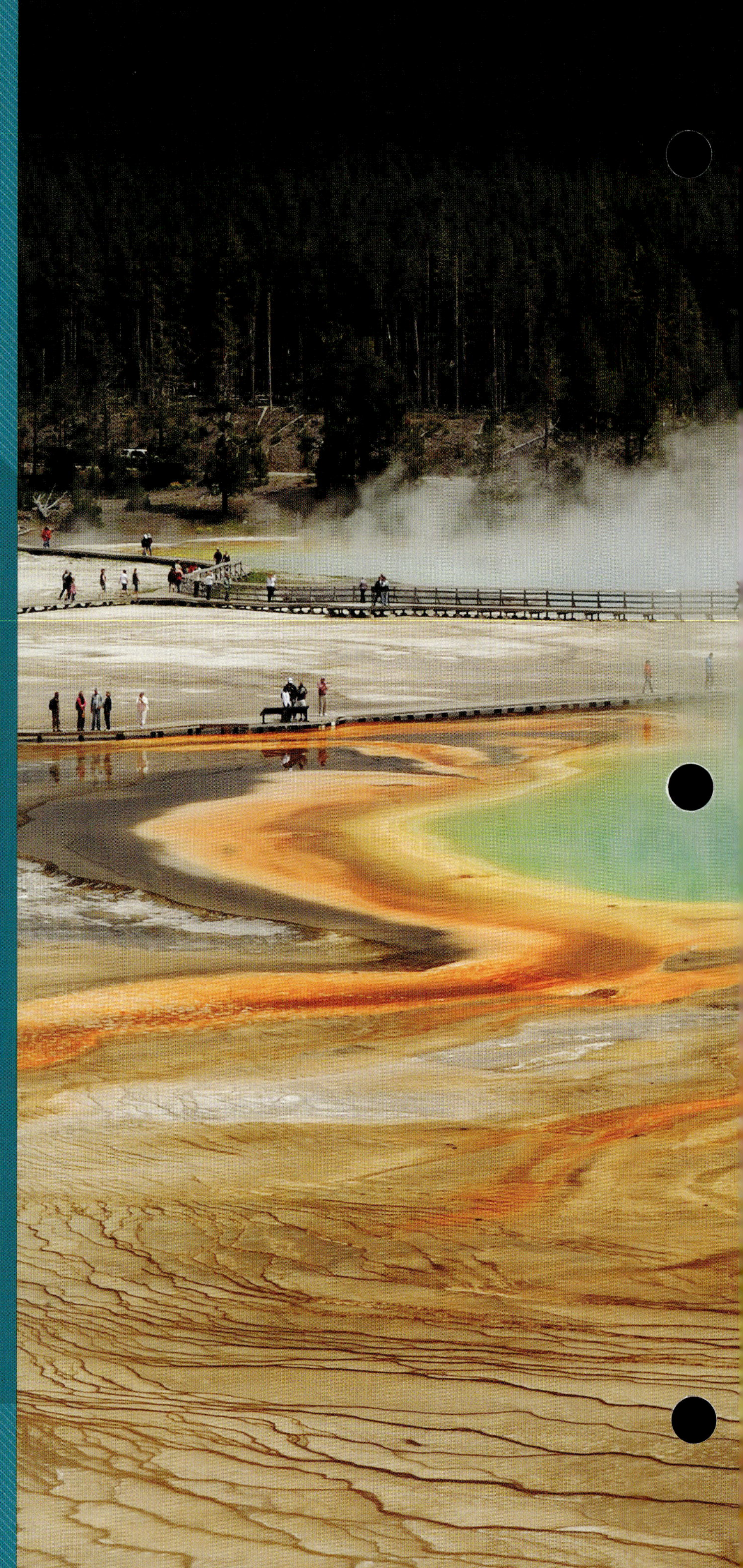

The Grand Prismatic Pool in Yellowstone National Park is a hot-water pool that gets its blue color from several species of heat-tolerant cyanobacteria. Microscopic fossils of organisms similar to modern cyanobacteria are among Earth's oldest fossils. (Photo by Don Johnston/Glow Images)

EARTH HAS A LONG AND COMPLEX HISTORY. Time and again, the splitting and colliding of continents has led to the formation of new ocean basins and the creation of great mountain ranges. Furthermore, the nature of life on our planet has undergone dramatic changes through time.

19.1 Is Earth Unique?

List the principal characteristics that make Earth unique among the planets in the solar system.

Although we have now sent spacecraft to every planet in our solar system and have identified several thousand planets that orbit other stars, we know of only one place that supports life—our own modest-sized planet Earth. Life on Earth is ubiquitous; it is found in boiling mudpots and hot springs, in the deep abyss of the ocean, and even under the Antarctic Ice Sheet. Living space on our planet, however, is significantly limited when we consider the needs of individual organisms, particularly humans. The global ocean covers 71 percent of Earth's surface, but only a few hundred meters below the water's surface, pressures are so intense that humans cannot survive without an atmospheric diving suit or submersible. In addition, many continental areas are too steep, too high, or too cold for us to inhabit (**Figure 19.1**).

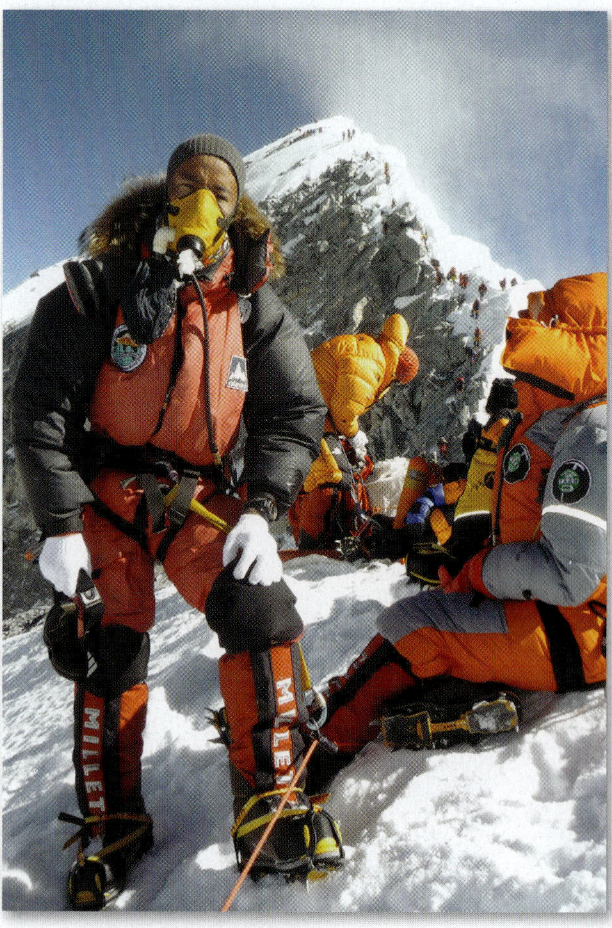

Figure 19.1 Climbers near the top of Mount Everest Much of Earth's surface is uninhabitable by humans. At this altitude, the level of oxygen is only one-third the amount available at sea level. (Photo by STR/AFP/Getty Images)

What fortuitous events produced a planet so hospitable to life? Earth was not always as we find it today. During its formative years, our planet became hot enough to support a magma ocean. It also survived a several-hundred-million-year period of extreme bombardment by asteroids, to which the heavily cratered surfaces of Mars and the Moon testify. The oxygen-rich atmosphere that makes higher life-forms possible developed relatively recently. Serendipitously, Earth seems to be the right planet, in the right location, at the right time.

The Right Planet

What are some of the characteristics that make Earth unique among the planets? Consider the following:

- If Earth were considerably larger (more massive), its force of gravity would be proportionately greater. Like the giant planets, Earth might have retained a thick, hostile atmosphere consisting of ammonia and methane, and possibly hydrogen and helium.
- If Earth were much smaller, oxygen, water vapor, and other volatiles would escape into space and be lost forever. Thus, like the Moon and Mercury, both of which lack appreciable atmospheres, Earth would be devoid of life.
- If Earth did not have a rigid lithosphere overlaying a weak asthenosphere, plate tectonics would not operate. The continental crust (Earth's "highlands") would not have formed without the recycling of plates. Consequently, the entire planet would likely be covered by an ocean a few kilometers deep. As author Bill Bryson so aptly stated, "There might be life in that lonesome ocean, but there certainly wouldn't be baseball."*
- Most surprising, perhaps, is the fact that if our planet did not have a molten metallic outer core, most of the life-forms on Earth would not exist. Fundamentally, without the flow of iron in the core, Earth could not support a magnetic field. It is the magnetic field that prevents lethal cosmic rays from showering Earth's surface and stripping away our atmosphere.

*A *Short History of Nearly Everything* (Broadway Books, 2003).

The Right Location

One primary factor that determines whether a planet is suitable for higher life-forms is its location in the solar system. The following scenarios substantiate Earth's advantageous position:

- If Earth were about 10 percent closer to the Sun, our atmosphere would be more like that of Venus and consist mainly of the greenhouse gas carbon dioxide. Earth's surface temperature would then be too hot to support higher life-forms.
- If Earth were about 10 percent farther from the Sun, the problem would be reversed: It would be too cold. The oceans would freeze over, and Earth's active water cycle would not exist. Without liquid water, all life would perish.
- Earth is near a star of modest size. Stars like the Sun have a life span of roughly 10 billion years and emit radiant energy at a fairly constant level during most of this time. Giant stars, on the other hand, consume their nuclear fuel at very high rates and "burn out" in a few hundred million years. Therefore, Earth's proximity to a modest-sized star allowed enough time for the evolution of humans, who first appeared on this planet only a few million years ago.

The Right Time

The last, but certainly not the least, fortuitous factor for Earth is timing. The first organisms to inhabit Earth were extremely primitive and came into existence roughly 3.8 billion years ago. From that point in Earth's history, innumerable changes occurred: Life-forms came and went, and the physical environment of our planet was transformed in many ways. Consider two of the many timely Earth-altering events:

- Earth's atmosphere has developed over time. Earth's primitive atmosphere is thought to have been composed mostly of nitrogen, water vapor, methane, and carbon dioxide—but no free oxygen, that is, oxygen not combined with other elements. Fortunately, microorganisms evolved that released oxygen into the atmosphere through the process of *photosynthesis*. About 2.5 billion years ago, an atmosphere with free oxygen came into existence. The result was the evolution of the ancestors of the vast array of multicellular organisms that we find on Earth today.
- About 66 million years ago, our planet was struck by an asteroid 10 kilometers (6 miles) in diameter. This impact likely caused a mass extinction during which nearly three-quarters of all plant and animal species were obliterated—including dinosaurs.* Although this may not seem lucky, the extinction of dinosaurs opened new habitats for small mammals that survived the impact. These habitats, along with evolutionary forces, led to the development of the many large mammals that occupy our modern world (Figure 19.2). Without this event, mammals might have remained mostly small and inconspicuous.

▲ **Figure 19.2**
Paleontologists uncover the remains of a 10-million-year-old rhinoceros at a dig site near Orchard, Nebraska
(Photo by JOSE MENDEZ/EPA/Newscom)

As various observers have noted, Earth developed under "just right" conditions to support higher life-forms. Astronomers refer to this as the *Goldilocks scenario*. Like the classic "Goldilocks and the Three Bears" fable, Venus is too hot (Papa Bear's porridge), Mars is too cold (Mama Bear's porridge), but Earth is just right (Baby Bear's porridge).

Viewing Earth's History

The remainder of this chapter focuses on the origin and evolution of planet Earth—the one place in the universe we know fosters life. As you learned in Chapter 18, researchers utilize many tools to interpret clues about Earth's past. Using these tools, as well as clues contained in the rock record, scientists continue to unravel many complex events of the geologic past. This chapter provides a brief overview of the history of our planet and its life-forms—a journey that takes us back about 4.6 billion years, to the formation of Earth. Later, we will consider how our physical world assumed its present state and how Earth's inhabitants changed through time. As you read this chapter, refer to the *geologic time scale* presented in Figure 19.3.

CONCEPT CHECKS 19.1

1. In what way is Earth unique among the planets of our solar system?
2. Explain why Earth is just the right size.
3. Why is Earth's molten, metallic core important to humans living today?
4. Why is Earth's location in the solar system ideal for the development of higher life-forms?

Did You Know?

The names of several periods on the geologic time scale refer to places that have prominent strata of that age. For example, the Cambrian period is taken from the Roman name for Wales (Cambria). The Permian is named for the province of Perm in Russia, while the Jurassic period gets its name from the Jura Mountains located between France and Switzerland.

* We use the term *dinosaurs* to refer to all members of this group except birds.

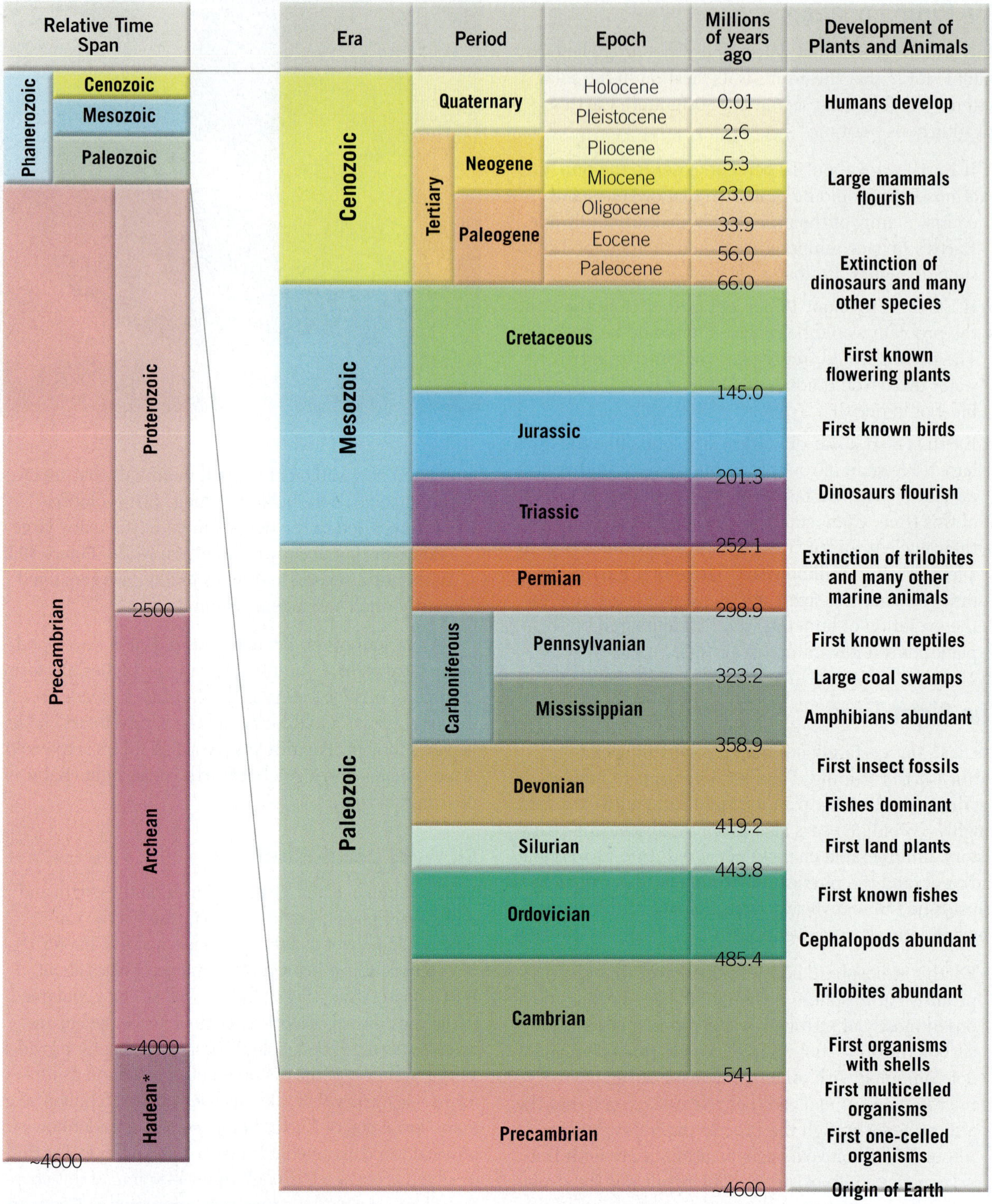

Relative Time Span		Millions of years ago
Phanerozoic	Cenozoic	
	Mesozoic	
	Paleozoic	
Precambrian	Proterozoic	2500
	Archean	~4000
	Hadean*	~4600

Era	Period	Epoch	Millions of years ago	Development of Plants and Animals
Cenozoic	Quaternary	Holocene	0.01	Humans develop
		Pleistocene	2.6	
	Tertiary – Neogene	Pliocene	5.3	
		Miocene	23.0	Large mammals flourish
	Tertiary – Paleogene	Oligocene	33.9	
		Eocene	56.0	
		Paleocene	66.0	Extinction of dinosaurs and many other species
Mesozoic	Cretaceous		145.0	First known flowering plants
	Jurassic		201.3	First known birds
	Triassic		252.1	Dinosaurs flourish
Paleozoic	Permian		298.9	Extinction of trilobites and many other marine animals
	Carboniferous – Pennsylvanian		323.2	First known reptiles Large coal swamps
	Carboniferous – Mississippian		358.9	Amphibians abundant
	Devonian		419.2	First insect fossils Fishes dominant
	Silurian		443.8	First land plants
	Ordovician		485.4	First known fishes Cephalopods abundant
	Cambrian		541	Trilobites abundant First organisms with shells
Precambrian			~4600	First multicelled organisms First one-celled organisms Origin of Earth

* Hadean is the informal name for the span that begins at Earth's formation and ends with Earth's earliest-known rocks.

▲ **Figure 19.3 The geologic time scale** Numbers represent time in millions of years before the present. The Precambrian accounts for about 88 percent of geologic time.

19.2 Birth of a Planet

Outline the major stages in Earth's evolution, from the Big Bang to the formation of our planet's layered internal structure.

The universe began about 13.8 billion years ago with the *Big Bang*, when all matter and space came into existence. Shortly thereafter, the two simplest elements, hydrogen and helium, formed. These basic elements were the ingredients for the first star systems. Several billion years later, our home galaxy, the Milky Way, came into existence. It was within a band of stars and nebular debris in an arm of this spiral galaxy that the Sun and planets took form nearly 4.6 billion years ago.

From the Big Bang to Heavy Elements

One of the products of the Big Bang was an array of subatomic particles, including protons, neutrons, and electrons (**Figure 19.4**). Later, as this debris cooled, these subatomic particles combined to generate atoms of hydrogen and helium, the two lightest elements. Within a few hundred million years, clouds of these gases condensed and coalesced into billions of stars that formed the first galactic systems.

As these gases contracted to become the first stars, heating triggered the process of *nuclear fusion*. Within the interiors of stars, hydrogen nuclei convert to helium nuclei, releasing enormous amounts of radiant energy (heat, light, and cosmic rays). Astronomers have determined that in stars more massive than our Sun, other thermonuclear reactions occur, generating all the elements on the periodic table up to number 26, iron. The heaviest elements (beyond number 26) are created only at extreme temperatures during the explosive death of a star eight or more times as massive as the Sun. During these cataclysmic **supernova** events, exploding stars produce all the elements heavier than iron and spew them into interstellar space. It is from such debris, as well as pre-existing gases, that our Sun and solar system formed. Based on the Big Bang scenario, all the atoms in your body except for hydrogen were produced billions of years ago, in the hot interior of now-defunct stars, and the gold in your jewelry was produced during a supernova explosion that occurred in some distant place.

From Planetesimals to Protoplanets

Recall that the solar system, including Earth, formed about 4.6 billion years ago from the **solar nebula**, a large rotating cloud of interstellar dust and gas (see Figure 19.4E). As the solar nebula contracted, most of the matter collected in the center to create the hot *protosun*. The remaining materials formed a thick, flattened, rotating disk, within which matter gradually cooled and condensed into grains and clumps of icy, rocky, and metallic material. Repeated collisions resulted in most of the material eventually collecting into asteroid-sized objects called **planetesimals**.

The composition of planetesimals was largely determined by their proximity to the protosun. As you might expect, temperatures were highest in the inner solar system and decreased toward the outer edge of the disk. Therefore, between the present orbits of Mercury and Mars, the planetesimals were composed mainly of materials with high melting temperatures—metals and rocky substances. The planetesimals that formed beyond the orbit of Mars, where temperatures are low, contained high percentages of ices—water, carbon dioxide, ammonia, and methane—as well as smaller amounts of rocky and metallic debris.

Through repeated collisions and accretion (sticking together), these planetesimals grew into eight **protoplanets**, as well as dwarf planets and some larger moons (see Figure 19.4G). During this process, the same amount of matter was concentrated into fewer and fewer bodies, each having greater and greater masses.

At some point in Earth's early evolution, a giant impact occurred between a Mars-sized object and a young, semimolten Earth. This collision ejected huge amounts of debris into space, some of which coalesced to form the Moon (see Figure 19.4J,K,L).

Earth's Early Evolution

As material continued to collide and accumulate, the high-velocity impacts of interplanetary debris (planetesimals) and the decay of radioactive elements caused the temperature of our planet to steadily increase. This early period of heating resulted in a magma ocean that was perhaps several hundred kilometers deep. Within the magma ocean, buoyant masses of molten rock rose toward the surface and eventually solidified to produce thin rafts of crustal rocks. Geologists call this early period of Earth's history the **Hadean**, which began with Earth's formation about 4.6 billion years ago and ended roughly 4 billion years ago (**Figure 19.5**). The name *Hadean* is derived from the Greek word *Hades*, meaning "the underworld," referring to the "hellish" conditions on Earth at the time.

During this period of intense heating, Earth became so hot that iron and nickel began to melt. Melting produced liquid blobs of heavy metal that sank toward the center of Earth under their own weight. This process occurred rapidly on the scale of geologic time and produced Earth's dense iron-rich core. As you learned in Chapter 9, the formation of a molten iron core was the first of many stages of chemical differentiation in which

▶ **SmartFigure 19.4**
Major events that led to the formation of early Earth Ages are in billions of years (Ga).

TUTORIAL
http://goo.gl/aqiQ0Z

A. Big Bang 13.8 Ga
B. Hydrogen and helium atoms created
C. Our galaxy forms 10 Ga
D. Heavy elements synthesized b supernova explosions
E. Solar nebula begins to contract 4.7 Ga
F. As material collects to form the protosun rotation flattens nebula
G. Accretion of planetesimals to form Earth and the other planets
Earth
H. Continual bombardment and the decay of radioactive elements produces magma ocean
I. Chemical differentiation produces Earth's layered structure
J. Mars-size object impacts young Earth 4.6 Ga
K. Debris orbits Earth and accretes
L. Formation of Earth–Moon system 4.5 Ga
M. Outgassing produces Earth's primitive atmosphere and ocean

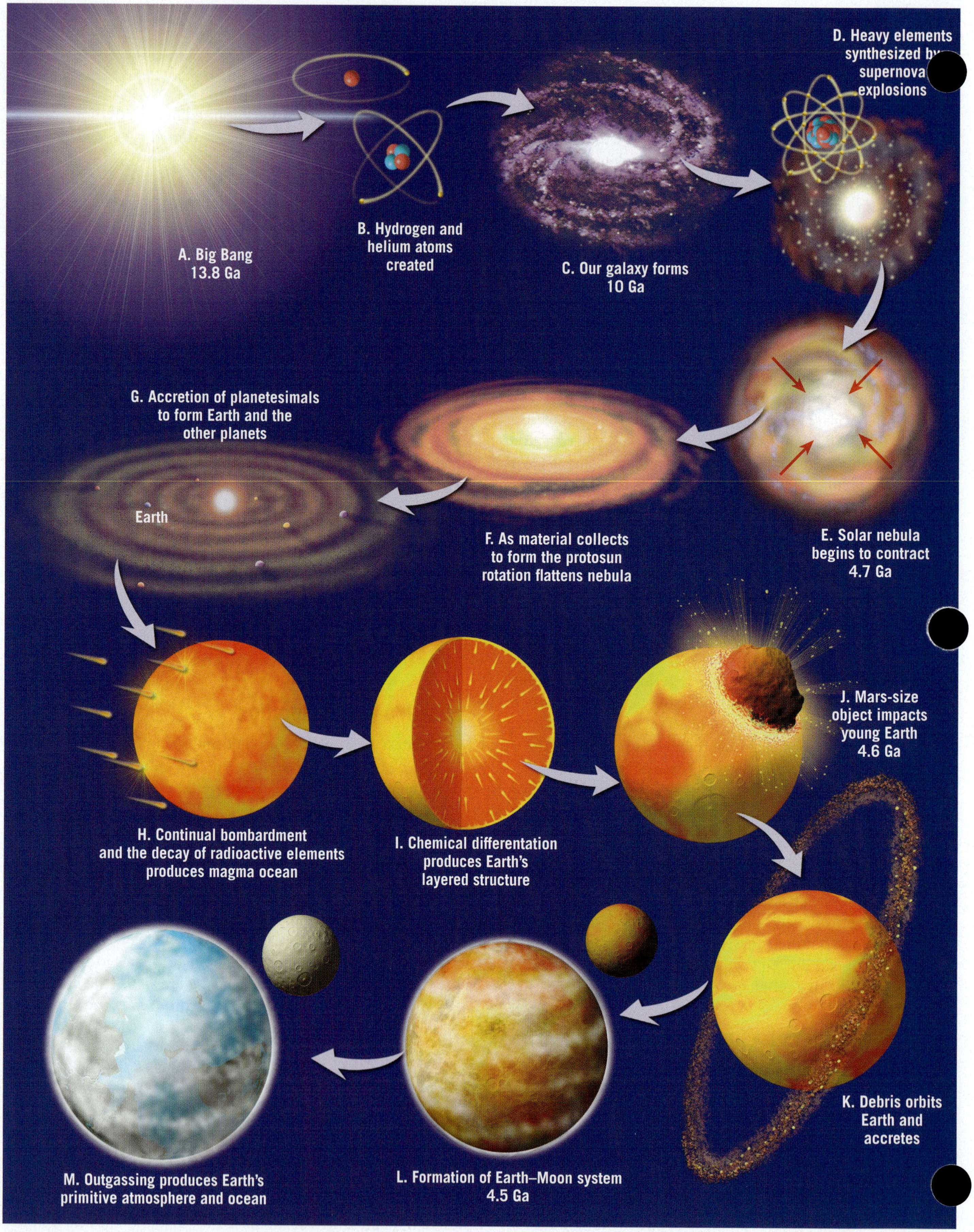

Figure 19.5 Artistic depiction of Earth during the Hadean The Hadean is an unofficial eon of geologic time that occurred before the Archean. Its name refers to the "hellish" conditions on Earth. During the early Hadean, Earth had a magma ocean and experienced intense bombardment by nebular debris.

Earth converted from a homogeneous body, with roughly the same matter at all depths, to a layered planet with material sorted by density (see Figure 19.4I).

This period of chemical differentiation established the three major divisions of Earth's interior: the iron-rich *core*, the thin *primitive crust*, and Earth's thickest layer, the *mantle*, located between the core and the crust. In addition, the lightest materials—including water vapor, carbon dioxide, and other gases—escaped to form a primitive atmosphere and, shortly thereafter, the oceans.

CONCEPT CHECKS 19.2

1. What two elements made up most of the very early universe?
2. Name the cataclysmic event in which an exploding star produces all the elements heavier than iron.
3. Briefly describe the formation of the planets from the solar nebula.
4. Describe the conditions on Earth during the Hadean.

19.3 Origin and Evolution of the Atmosphere and Oceans

Describe how Earth's atmosphere and oceans formed and evolved through time.

We can be thankful for our atmosphere; without it, there would be no greenhouse effect, and Earth would be nearly 60°F colder. Earth's water bodies would be frozen nearly solid, making the hydrologic cycle nonexistent.

The air we breathe is a relatively stable mixture of 78 percent nitrogen, 21 percent oxygen, about 1 percent argon (an inert gas), and small amounts of other gases such as carbon dioxide and water vapor. However, our planet's original atmosphere was substantially different.

Earth's Primitive Atmosphere

Early in Earth's formation, its atmosphere likely consisted of gases most common in the early solar system: hydrogen, helium, methane, ammonia, carbon dioxide, and water vapor. The lightest of these—hydrogen and helium—most likely escaped into space because Earth's gravity was too weak to hold them. The remaining gases—methane, ammonia, carbon dioxide, and water vapor—contain the basic ingredients of life: carbon, hydrogen, oxygen, and nitrogen. This early atmosphere was enhanced by a process called **outgassing**, through which gases trapped in the planet's interior are released. Outgassing from hundreds of active volcanoes still remains an important planetary function worldwide (Figure 19.6). However, early in Earth's history, when massive heating and fluid-like motion occurred in the mantle, the gas output would likely have been

▲ **Figure 19.6**
Outgassing produced Earth's first enduring atmosphere Outgassing continues today from hundreds of active volcanoes worldwide. (Photo by Lee Frost/Robert Harding)

immense. These early eruptions probably released mainly water vapor, carbon dioxide, and sulfur dioxide, with minor amounts of other gases. Most importantly, free oxygen was not present in Earth's primitive atmosphere.

Oxygen in the Atmosphere

As Earth cooled, water vapor condensed to form clouds, and torrential rains began to fill low-lying areas, which eventually became the oceans. In those oceans, nearly 3.5 billion years ago, photosynthesizing bacteria began to release oxygen into the water. During *photosynthesis*, organisms use the Sun's energy to produce organic material (energetic molecules of sugar containing hydrogen and carbon) from carbon dioxide (CO_2) and water (H_2O). One of the earliest known bacteria, *cyanobacteria* (once called blue-green algae), began to produce oxygen as a by-product of photosynthesis.

Initially, the newly released free oxygen was readily captured by chemical reactions with organic matter and dissolved iron in the ocean. It seems that large quantities of iron were released into the early ocean through submarine volcanism and associated hydrothermal vents. Iron has tremendous affinity for oxygen. When these two elements join, they become iron oxide (rust). These early iron oxide accumulations on the seafloor created alternating layers of iron-rich rocks and chert, called **banded iron formations**. Most banded iron deposits accumulated in the Precambrian eon, between 3.5 and 2 billion years ago, and represent the world's most important reservoir of iron ore.

As the number of oxygen-generating organisms increased, oxygen began to build in the atmosphere. Chemical analysis of rock suggests that oxygen first appeared in significant amounts in the atmosphere around 2.5 billion years ago, a phenomenon termed the **Great Oxygenation Event**. Thereafter, oxygen levels in the atmosphere gradually climbed.

For the next billion years, oxygen levels in the atmosphere probably fluctuated but remained below 10 percent of current levels. Prior to the start of the Cambrian period 541 million years ago, which coincided with the evolution of organisms with skeletal hard parts, the level of free oxygen in the atmosphere began to increase. The availability of abundant oxygen in the atmosphere contributed to the proliferation of aerobic life-forms (oxygen-consuming organisms). On the other hand, it likely wiped out huge portions of Earth's anaerobic organisms (organisms that do not require oxygen), for which oxygen is poisonous. One apparent spike in oxygen levels occurred during the Pennsylvanian period (300 million years ago), when oxygen made up about 35 percent of the atmosphere, compared to today's level of 21 percent.

Another positive benefit of the Great Oxygenation Event is that, when struck by sunlight, oxygen molecules form a compound called *ozone* (O_3), a type of oxygen molecule composed of three oxygen atoms. Ozone, which absorbs much of the Sun's harmful ultraviolet radiation before it reaches Earth's surface, is concentrated between 10 and 50 kilometers (6 to 30 miles) above Earth's surface, in a layer called the *stratosphere*. Thus, as a result of the Great Oxygenation Event, Earth's landmasses were protected from ultraviolet radiation, which is particularly harmful to DNA—the genetic blueprints for living organisms. Marine organisms had always been shielded from harmful ultraviolet radiation by seawater, but the development of the atmosphere's protective ozone layer made the continents more hospitable as well.

Evolution of the Oceans

When Earth cooled sufficiently to allow water vapor to condense, rainwater fell and collected in low-lying areas. By 4 billion years ago, scientists estimate that as much as 90 percent of the current volume of seawater was contained in the developing ocean basins. Because volcanic eruptions released into the atmosphere large quantities of sulfur dioxide, which readily combines with water to form sulfuric acid, the earliest rainwater was highly acidic. The level of acidity was even greater than the acid rain that damaged lakes and streams in eastern North America during the latter part of the twentieth century. Consequently, Earth's rocky surface weathered at an accelerated rate. The products released by chemical weathering included atoms and molecules of

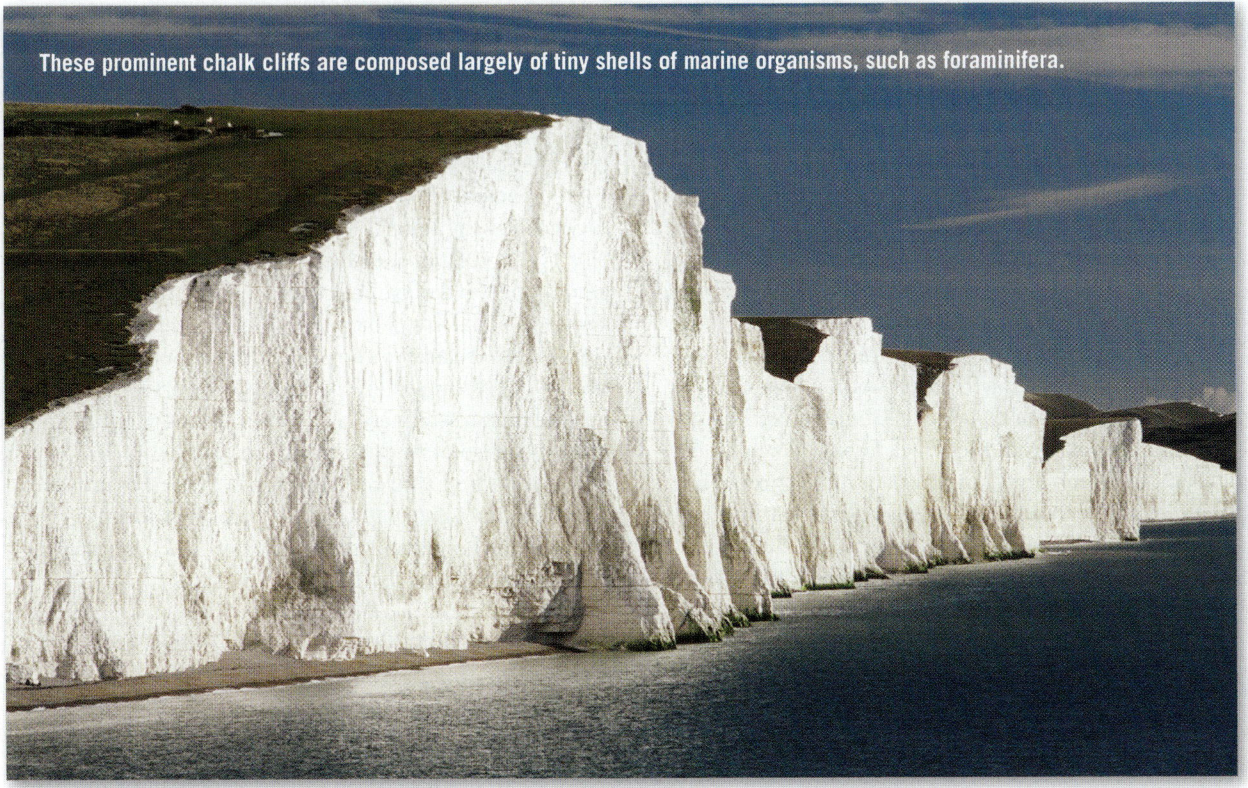

These prominent chalk cliffs are composed largely of tiny shells of marine organisms, such as foraminifera.

▲ **Figure 19.7 White Cliffs of Dover, England** Similar chalk deposits are also found in northern France.
(Photo by Imagesources/Glow Images)

various substances—including sodium, calcium, potassium, and silica—that were carried by running water into the newly formed oceans. Some of these dissolved substances precipitated to become chemical sediment that mantled the ocean floor. Other substances formed soluble salts, which increased the salinity of seawater. Research suggests that the salinity of the oceans initially increased rapidly but has remained relatively constant over the past 2 billion years.

Earth's oceans also serve as a repository for tremendous volumes of carbon dioxide, a major constituent of the primitive atmosphere. This is significant because carbon dioxide is a greenhouse gas that strongly influences the heating of the atmosphere. Venus, once thought to be very similar to Earth, has an atmosphere composed of 97 percent carbon dioxide, which produced an extreme greenhouse effect. As a result, Venus's surface temperature is 475°C (nearly 900°F).

Carbon dioxide is readily soluble in seawater, where it often combines with other atoms or molecules to produce various chemical precipitates. One of the most common compounds generated by mineral precipitation is calcium carbonate ($CaCO_3$), which makes up limestone, the most abundant chemical sedimentary rock. About 541 million years ago, marine organisms began to extract large quantities of calcium carbonate from seawater to make their shells and other hard parts. Trillions of tiny marine organisms, such as foraminifera, deposited their shells on the seafloor at the end of their life cycle. Some of these deposits can be observed today in the chalk beds exposed along the White Cliffs of Dover, England (**Figure 19.7**). By "locking up" carbon dioxide, these limestone deposits store this greenhouse gas so it cannot easily reenter the atmosphere.

CONCEPT CHECKS 19.3

1. What is meant by *outgassing*, and what modern phenomenon serves that role today?
2. List the most abundant gases that were added to Earth's early atmosphere through the process of outgassing.
3. Why was the evolution of photosynthesizing bacteria important for the evolution of large, oxygen-consuming organisms like ourselves?
4. Why was rainwater highly acidic early in Earth's history?
5. How does the ocean remove carbon dioxide from Earth's atmosphere? What role do tiny marine organisms, such as foraminifera, play in the removal of carbon dioxide?

19.4 Precambrian History: The Formation of Earth's Continents

Explain the formation of continental crust, how continental crust becomes assembled into continents, and the role that the supercontinent cycle has played in this process.

Earth's first 4 billion years are encompassed in the time span called the *Precambrian*. Representing nearly 90 percent of Earth's history, the Precambrian is divided into the *Archean eon* ("ancient age") and the *Proterozoic eon* ("early life age"); and an informal time span referred to as the Hadean. Our knowledge of this ancient time is limited because much of the early rock record has been obscured by the very Earth processes you have been studying, especially plate tectonics, erosion, and deposition. Most Precambrian rocks lack fossils, which hinders correlation of rock units. In addition, rocks this old are often metamorphosed and deformed, extensively eroded, and frequently concealed by younger strata. Indeed, Precambrian history is written in scattered, speculative episodes, like a long book with many missing chapters.

Earth's First Continents

Geologists have discovered tiny crystals of the mineral zircon in continental rocks that formed 4.4 billion years ago—evidence that the continents began to form early in Earth's history. By contrast, the oldest rocks found in the ocean basins are generally less than 200 million years old.

What differentiates continental crust from oceanic crust? Recall that oceanic crust is a relatively dense (3.0 g/cm^3) homogeneous layer of basaltic rocks derived from partial melting of the rocky upper mantle. In addition, oceanic crust is thin, averaging only 7 kilometers (4 miles) thick. Continental crust, on the other hand, is composed of a variety of rock types, has an average thickness of nearly 40 kilometers (25 miles), and contains a large percentage of low-density (2.7 g/cm^3), silica-rich rocks such as granite.

The significance of the differences between continental crust and oceanic crust cannot be overstated in a review of Earth's geologic evolution. Oceanic crust, because it is relatively thin and dense, is found several kilometers below sea level—unless of course it has been pushed onto a landmass by tectonic forces. Continental crust, because of its great thickness and lower density, may extend well above sea level. Also, recall that dense oceanic crust of normal thickness readily subducts, whereas thick, buoyant blocks of continental crust resist being recycled into the mantle.

Making Continental Crust The formation of continental crust is a continuation of the gravitational segregation of Earth materials that began during the final stage of our planet's formation. Dense metallic material, mainly iron and nickel, sank to form Earth's core, leaving behind the less dense rocky material that forms the mantle. It is from Earth's rocky mantle that low-density, silica-rich minerals were gradually distilled to form continental crust. This process is analogous to making sour mash whiskeys. In the production of whiskeys, various grains such as corn are fermented to generate alcohol, with sour mash being the by-product. This mixture is then heated or distilled, which drives off the lighter material (alcohol) and leaves behind the sour mash. In a similar manner, partial melting of mantle rocks generates low-density, silica-rich materials that buoyantly rise to the surface to form Earth's crust, leaving behind the dense mantle rocks (see Chapter 4). However, little is known about the details of the mechanisms that generated these silica-rich melts during the Archean eon.

Earth's first crust was probably ultramafic in composition, but because physical evidence no longer exists, we are not certain. The hot, turbulent mantle that most likely existed during the Archean eon recycled most of this crustal material back into the mantle. In fact, it may have been continuously recycled, in much the same way that the "crust" that forms on a lava lake is repeatedly replaced with fresh lava from below (**Figure 19.8**).

The oldest preserved continental rocks occur as small, highly deformed *terranes*, which are incorporated within somewhat younger blocks of continental crust (**Figure 19.9**). One of these is a 3.8-billion-year-old terrane located near Isua, Greenland. Slightly older crustal rocks, called the Acasta Gneiss, have been discovered in Canada's Northwest Territories.

Some geologists think that some type of plate-like motion operated early in Earth's history. In addition,

▼ **Figure 19.8 Earth's early crust was continually recycled** (Photo courtesy of the USGS)

The crust covering this lava lake is continually being replaced with fresh lava from below, much like the way Earth's crust was recycled early in its history.

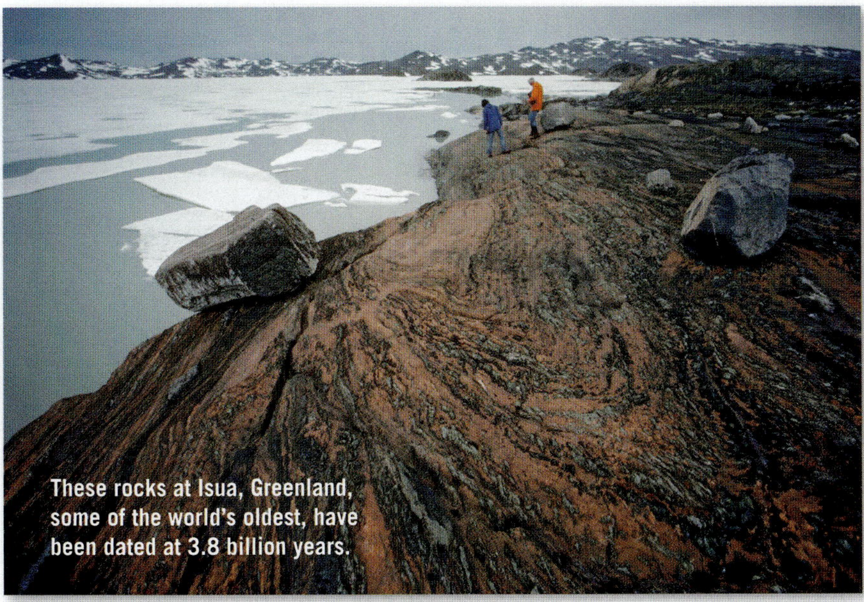

These rocks at Isua, Greenland, some of the world's oldest, have been dated at 3.8 billion years.

▲ **Figure 19.9 Earth's oldest preserved continental rocks are more than 3.8 billion years old** (Photo courtesy of James L. Amos/CORBIS)

hot-spot volcanism was likely active during this time. However, because the mantle was hotter in the Archean than it is today, both of these phenomena would have progressed at faster rates than their modern counterparts. Hot-spot volcanism, due to mantle plumes, is thought to have created immense shield volcanoes as well as oceanic plateaus. Simultaneously, subduction of oceanic crust generated volcanic island arcs. Collectively, these relatively small crustal fragments represent the first phase in creating stable, continent-size landmasses.

From Continental Crust to Continents The growth of larger continental masses was accomplished through collision and accretion of many thin, highly mobile crustal fragments, as illustrated in **Figure 19.10**. This type of collisional tectonics deformed and metamorphosed sediments caught between converging crustal fragments, thereby shortening and thickening the developing crust. In the deepest regions of these collision zones, partial melting of the thickened crust generated silica-rich magmas that ascended and intruded the rocks above. This led to the formation of large crustal provinces that, in turn, accreted with others to form even larger crustal blocks called **cratons**.

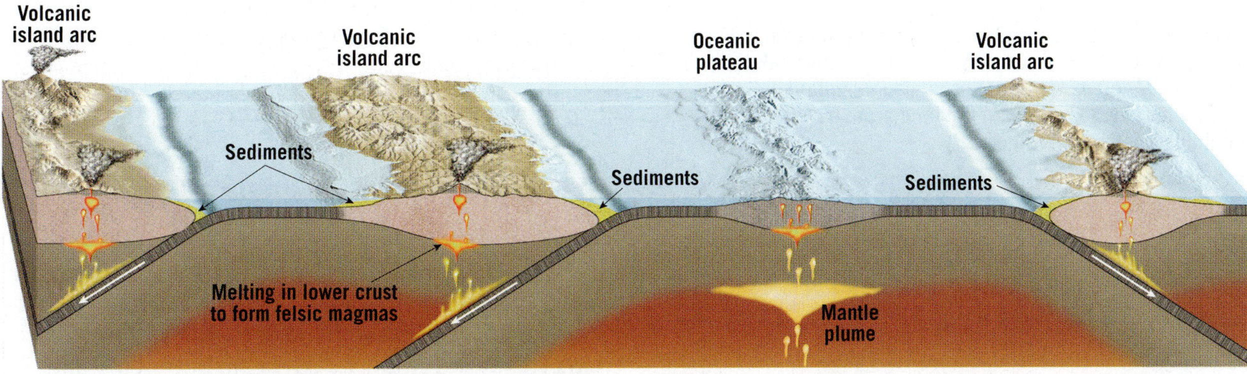

A. Scattered crustal fragments separated by ocean basins

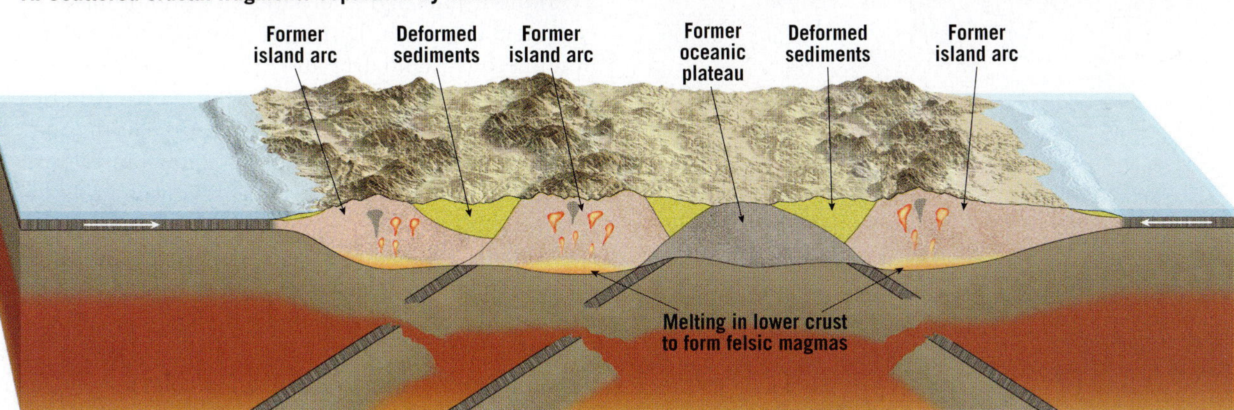

B. Collision of volcanic island arcs and oceanic plateau to form a larger crustal block

◄ **SmartFigure 19.10 The formation of continents** The growth of large continental masses occurs through the collision and accretion of smaller crustal fragments.

TUTORIAL
http://goo.gl/7y4htd

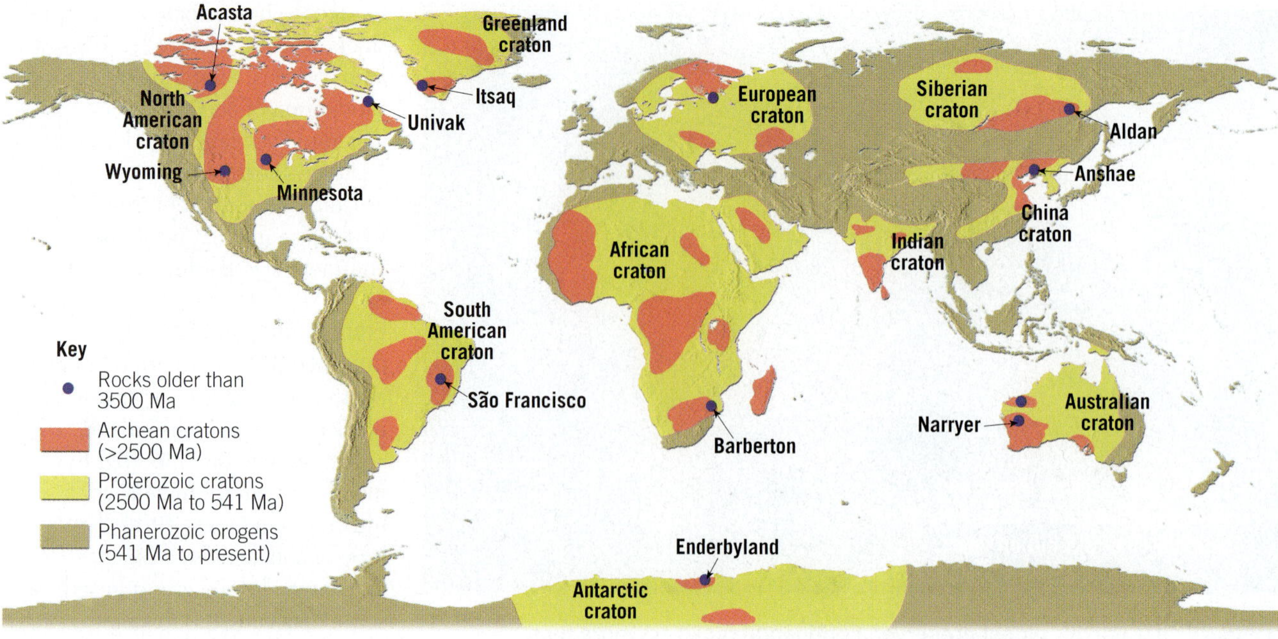

Figure 19.11
Distribution of crustal material remaining from the Archean and Proterozoic eons Ages are in billions of years (Ga).

The assembly of a large craton involves the accretion of several crustal blocks that cause major mountain-building episodes similar to India's collision with Asia. Figure 19.11 shows the extent of crustal material that was produced during the Archean and Proterozoic eons. The regions within a modern continent where these ancient cratons are exposed at the surface are called **shields**.

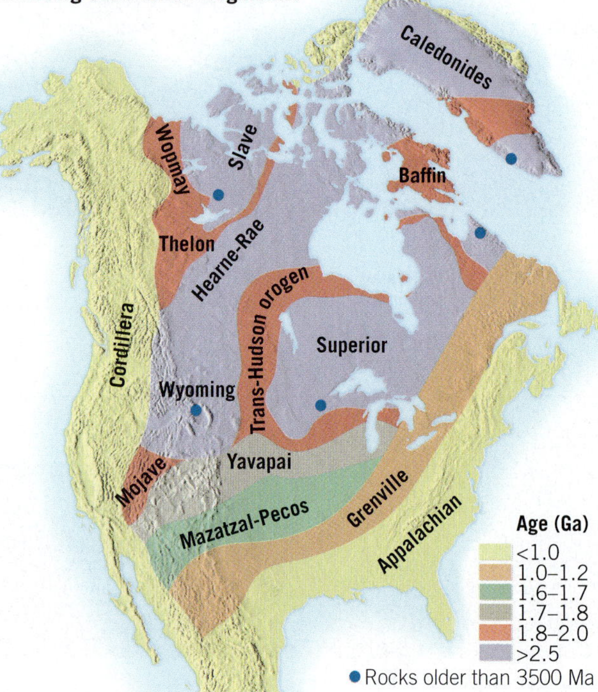

SmartFigure 19.12
The major geologic provinces of North America The age of each province is in billions of years (Ga).

TUTORIAL
http://goo.gl/QGu7RI

Although the Precambrian was a time when much of Earth's continental crust was generated, a substantial amount of crustal material was destroyed as well. Some was lost via weathering and erosion. In addition, during much of the Archean, it appears that thin slabs of continental crust were subducted into the mantle. However, by about 3 billion years ago, cratons grew sufficiently large and thick to resist subduction. After that time, weathering and erosion became the primary processes of crustal destruction. By the close of the Precambrian, an estimated 85 percent of the modern continental crust had formed.

The Making of North America

North America provides an excellent example of the development of continental crust and its piecemeal assembly into a continent. Notice in Figure 19.12 that very little continental crust older than 3.5 billion years remains. In the late Archean, between 3 and 2.5 billion years ago, there was a period of major continental growth. During this span, the accretion of numerous island arcs and other fragments generated several large crustal provinces. North America contains some of these crustal units, including the Superior and Hearne/Rae cratons shown in Figure 19.12, but just where these ancient continental blocks formed is unknown.

About 1.9 billion years ago, these crustal provinces collided to produce the Trans-Hudson mountain belt (see Figure 19.12). (Such mountain-building episodes were not restricted to North America; ancient deformed strata of similar age are also found on other continents.) This event built the North American craton, around which several large and numerous small crustal fragments were

▲ **Figure 19.13 Possible configuration of the supercontinent Rodinia** For clarity, the continents are drawn with somewhat modern shapes, not their actual shapes from 1 billion years ago. (After P. Hoffman, J. Rogers, and others)

later added. One of these late arrivals is the Appalachians province. In addition, several terranes were added to the western margin of North America during the Mesozoic and Cenozoic eras to generate the mountainous North American Cordillera.

Supercontinents of the Precambrian

At different times, parts of what is now North America combined with other continental landmasses to form a supercontinent. **Supercontinents** are large landmasses that contain all, or nearly all, the existing continents. Pangaea was the most recent, but certainly not the only, supercontinent to exist in the geologic past. The earliest well-documented supercontinent, *Rodinia*, formed during the Proterozoic eon, about 1.1 billion years ago (**Figure 19.13**). Although geologists are still studying its construction, it is clear that Rodinia's configuration was quite different from Pangaea's. One obvious distinction is that North America's position near the center of this ancient landmass.

Between 800 and 600 million years ago, Rodinia gradually split apart. By the end of the Precambrian many of the fragments had reassembled, producing a large landmass in the Southern Hemisphere called *Gondwana*, comprised mainly of present-day South America, Africa, India, Australia, and Antarctica (**Figure 19.14**). Other continental fragments also developed—North America, Siberia, and Northern Europe. We consider the fate of these Precambrian landmasses in the next section.

Supercontinent Cycle The **supercontinent cycle** involves rifting and dispersal of one supercontinent followed by a long period during which the fragments are gradually reassembled into a new supercontinent with a different configuration. The assembly and dispersal of supercontinents had a profound impact on the evolution of Earth's continents. In addition, this phenomenon greatly influenced global climates and contributed to periodic episodes of rising and falling sea level.

Supercontinents & Climate The movement of continents changes the patterns of ocean currents and global winds, which influences the global distribution of temperature and precipitation. The formation of the Antarctic's vast ice sheet is one example of how the movement of a continent is thought to have contributed to climate change. Although eastern Antarctica remained over the South Pole for more than 100 million years, Antarctica was not covered by a stable continental-scale ice sheet until about 34 million years ago. Prior to this period of glaciation, South America and Antarctica where connected. As shown in **Figure 19.15A**, this arrangement of landmasses helped maintain a circulation pattern in which warm ocean currents reached the coast of Antarctica and aided in keeping Antarctica mainly ice free. This is similar to the way in which the modern Gulf Stream helps keep Iceland mostly ice free, despite its name. As South America separated from Antarctica and moved northward, a pattern of ocean circulation developed that flowed from west to east around the entire continent of Antarctica (**Figure 19.15B**). This cold current, called the West Wind Drift, effectively isolated the entire Antarctic coast from the warm, poleward-directed currents in the southern oceans. This change in circulation, along with a period of

A. Continent of Gondwana

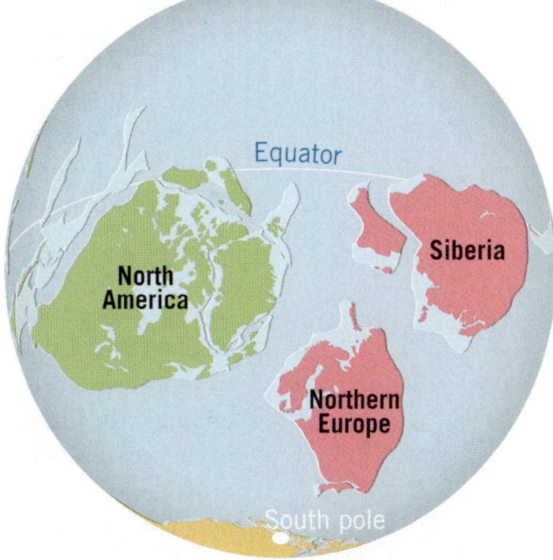

B. Continents not a part of Gondwana

▼ **Figure 19.14 Reconstruction of Earth as it may have appeared in late Precambrian time** The southern continents were joined into a single landmass called Gondwana. Other landmasses that were not part of Gondwana include North America, northwestern Europe, and northern Asia. (After P. Hoffman, J. Rogers, and others)

50 million years ago warm ocean currents kept Antarctica nearly ice free.

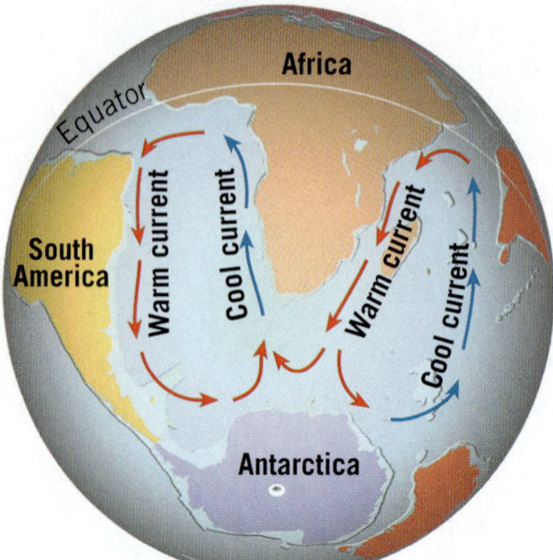

A. Antarctica not extensively glaciated

As South America separated from Antarctica, the West Wind Drift developed. This newly formed ocean current effectively cut Antarctica off from warm currents and contributed to the formation of its vast ice sheets.

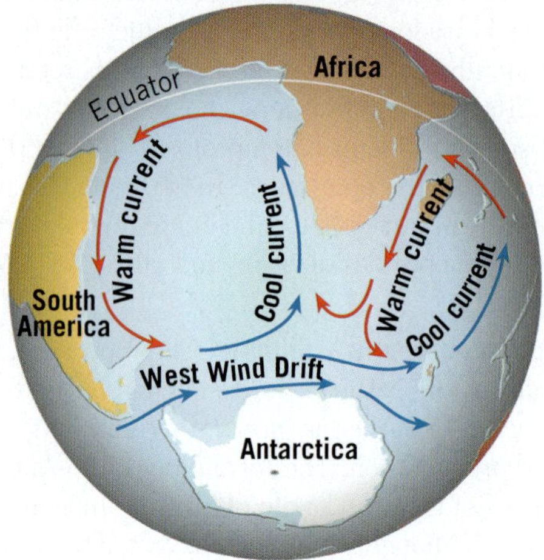

B. Antarctica covered by continental-size ice sheet

▲ SmartFigure 19.15
Connection between ocean circulation and the climate in Antarctica

TUTORIAL
http://goo.gl/0rwB5K

global cooling, probably resulted in the growth of Antarctica's massive ice sheet.

Local and regional climates are also influenced by large mountain systems created by the collision of large cratons. One example is the collision of the subcontinent of India with southern Asia generated the Himalayas. Because of their high elevations, mountains exhibit markedly lower average temperatures than surrounding lowlands. In addition, air rising over these lofty structures promotes condensation and precipitation, leaving the region downwind relatively dry. A modern analogy is the wet, heavily forested western slopes of the Sierra Nevada compared to the dry climate of the Great Basin Desert that lies directly to the east.

Supercontinents & Sea-Level Changes Significant and numerous sea-level changes have been documented in geologic history, and many of them appear to have been related to the assembly and dispersal of supercontinents. If sea level rises, shallow seas advance onto the continents. Evidence for periods when the seas advanced onto the continents includes thick sequences of ancient marine sedimentary rocks that blanket large areas of modern landmasses—including much of the eastern two-thirds of the United States.

The supercontinent cycle and sea-level changes are directly related to rates of *seafloor spreading*. When the rate of spreading is rapid, as it is along the East Pacific Rise today, the production of warm oceanic crust is also high. Because warm oceanic crust is less dense (takes up more space) than cold crust, fast-spreading ridges occupy more volume in the ocean basins than do slow-spreading centers. (Think of getting into a tub filled with water.) As a result, when rates of seafloor spreading increase, more seawater is displaced, which results in the sea level rising. This, in turn, causes shallow seas to advance onto the low-lying portions of the continents.

CONCEPT CHECKS 19.4

1. Briefly explain how low-density continental crust was produced from Earth's rocky mantle.
2. Describe how cratons came into being.
3. What is the supercontinent cycle? What supercontinent preceded Pangaea?
4. Give an example of how the movement of a continent can trigger climate change.
5. Explain how the rate of seafloor spreading is related to changes in sea level.

19.5 Geologic History of the Phanerozoic: The Formation of Earth's Modern Continents

List and discuss the major geologic events in the Paleozoic, Mesozoic, and Cenozoic eras.

The time span since the close of the Precambrian, called the *Phanerozoic eon*, encompasses 541 million years and is divided into three eras: *Paleozoic*, *Mesozoic*, and *Cenozoic*. The beginning of the Phanerozoic is marked by the appearance of the first life-forms with hard parts such as shells, scales, bones, or teeth—all of which greatly enhance the chances for an organism to be preserved in the fossil record. Thus, the study of Phanerozoic crustal history was aided by the availability of fossils, which improved our ability to date and correlate geologic events. Moreover, because every organism is associated with its own particular environmental niche, the greatly improved fossil record provided invaluable information for deciphering ancient environments.

Paleozoic History

As the Paleozoic era opened, what is now North America hosted no plants or animals large enough to be seen—just tiny microorganisms such as bacteria. There were no Appalachian or Rocky Mountains; the continent was largely a barren lowland. Several times during the early Paleozoic, shallow seas moved inland and then receded from the continental interior, leaving

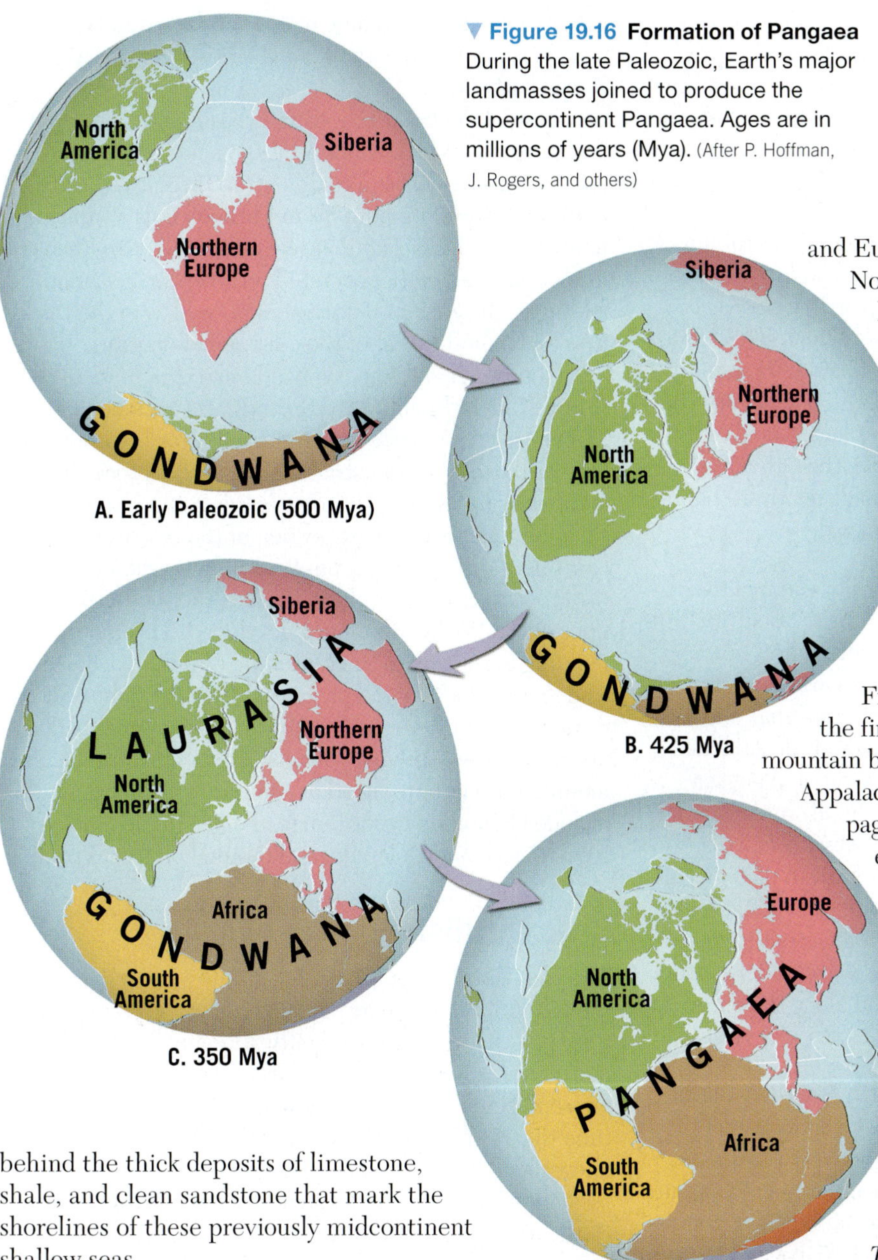

Figure 19.16 Formation of Pangaea During the late Paleozoic, Earth's major landmasses joined to produce the supercontinent Pangaea. Ages are in millions of years (Mya). (After P. Hoffman, J. Rogers, and others)

behind the thick deposits of limestone, shale, and clean sandstone that mark the shorelines of these previously midcontinent shallow seas.

Formation of Pangaea One of the major events of the Paleozoic era was the formation of the supercontinent **Pangaea**. This event began with a series of collisions that over millions of years joined North America, Europe, Siberia, and other smaller crustal fragments to form a large continent called **Laurasia** (Figure 19.16). Located south of Laurasia was the vast southern continent called **Gondwana**, which encompassed five modern landmasses—South America, Africa, Australia, Antarctica, and India—and perhaps portions of China. Evidence of extensive continental glaciation places this landmass near the South Pole. By the late Paleozoic, Gondwana had migrated northward and collided with Laurasia to begin the final stage in Pangaea's assembly.

The accretion of all of Earth's major landmasses to form Pangaea spans more than 300 million years and resulted in the formation of several mountain belts. The collision of northern Europe (mainly Norway) with Greenland produced the Caledonian Mountains, whereas the joining of northern Asia (Siberia) and Europe created the Ural Mountains. Northern China is also thought to have accreted to Asia by the end of the Paleozoic, whereas southern China may not have become part of Asia until after Pangaea had begun to rift. (Recall that India did not begin to accrete to Asia until about 50 million years ago.)

Pangaea reached its maximum size between 300 and 250 million years ago, as Africa collided with North America (see Figure 19.16D). This event marked the final and most intense period of mountain building in the long history of the Appalachian Mountains (see Figure 11.31, page 316). This mountain-building event produced the Central Appalachians of the Atlantic states, as well as New England's northern Appalachians and mountainous structures that extend into Canada (Figure 19.17).

Mesozoic History

Spanning about 186 million years, the Mesozoic era is divided into three periods: the *Triassic*, *Jurassic*, and *Cretaceous*. Major geologic events of the Mesozoic include the breakup of Pangaea and the evolution of our modern ocean basins.

Changes in Sea Levels The Mesozoic era began with much of the world's continents above sea level. The exposed Triassic strata are primarily red sandstones and mudstones that lack marine fossils, features that indicate a terrestrial environment. (The red color in sandstone comes from the oxidation of iron.)

As the Jurassic period opened, the sea invaded western North America. Adjacent to this shallow sea, extensive continental sediments were deposited on what is now the Colorado Plateau. The most prominent is the Navajo Sandstone, a cross-bedded, quartz-rich layer that in some places approaches 300 meters (1000 feet) thick. These remnants of massive dunes indicate that an enormous desert occupied much of the American Southwest during early Jurassic times (**Figure 19.18**). Another well-known Jurassic deposit is the Morrison Formation—one of the world's richest storehouse of dinosaur fossils. Included are the fossilized bones of massive dinosaurs such as *Apatosaurus*, *Brachiosaurus*, and *Stegosaurus*.

Coal Formation in Western North America As the Jurassic period gave way to the Cretaceous, shallow seas again encroached upon much of western North America, as well as the Atlantic and Gulf coastal regions. This led to the formation of "coal swamps"(see Chapter 7) similar to those of the Paleozoic era. Today, the Cretaceous coal deposits in the western United States and Canada are economically important. For example, the Crow Native American reservation in Montana holds nearly 20 billion tons of high-quality, Cretaceous-age coal.

The Breakup of Pangaea Another major event of the Mesozoic era was the breakup of Pangaea. About 185 million years ago, a rift developed between what is now North America and western Africa, marking the birth of the Atlantic Ocean. As Pangaea gradually broke apart, the westward-moving North American plate began to override the Pacific basin. This tectonic event triggered a continuous wave of deformation that moved inland along the entire western margin of North America.

Formation of the North American Cordillera By Jurassic times, subduction of the Pacific basin under the North American plate began to produce the chaotic mixture of rocks that exist today in the Coast Ranges of California (see Figure 11.25, page 314). Further inland, igneous activity was widespread, and for more than 100 million years volcanism was rampant as huge masses of magma rose to within a few kilometers of Earth's surface. The remnants of this activity include the granitic plutons of the Sierra Nevada, as well as the Idaho batholith and British Columbia's Coast Range batholith.

The subduction of the Pacific basin under the western margin of North America also resulted in the piecemeal addition of crustal fragments to the entire Pacific margin of the continent—from Mexico's Baja Peninsula to northern Alaska (see Figure 11.28, page 313). Each collision displaced earlier accreted terranes farther inland, adding to the zone of deformation as well as to the thickness and lateral extent of the continental margin.

Compressional forces moved huge rock units in a shingle-like fashion toward the east. Across much of North America's western margin, older rocks were thrust eastward over younger strata, for distances exceeding 150 kilometers (90 miles). Ultimately, this activity was responsible for generating a vast portion of the North American Cordillera that extends from Wyoming to Alaska.

Toward the end of the Mesozoic, the southern portions of the Rocky Mountains developed. This mountain-building event, called the *Laramide Orogeny*, occurred when large blocks of deeply buried Precambrian rocks were lifted nearly vertically along steeply dipping faults, upwarping the overlying younger sedimentary strata. The mountain ranges produced by the Laramide Orogeny include Colorado's Front Range, the Sangre de Cristo of New Mexico and Colorado, and the Bighorns of Wyoming.

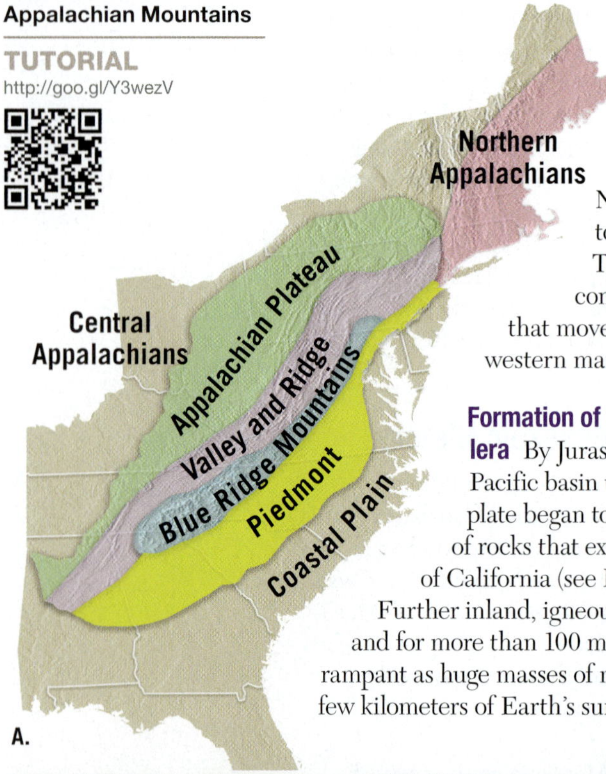

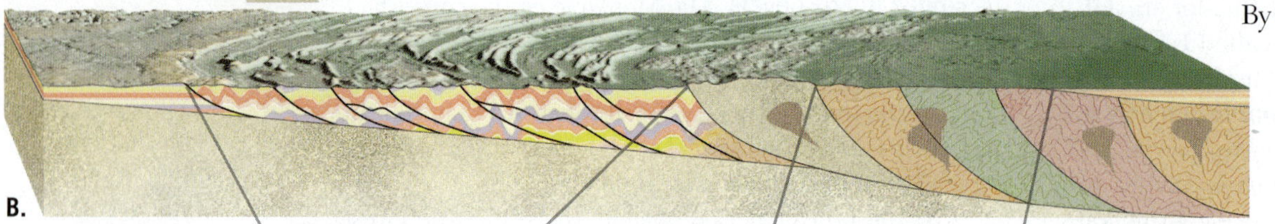

Cenozoic History

The Cenozoic era, or "era of recent life," encompasses the past 66 million years of Earth history. It was during this span that the physical landscapes and life-forms of our modern world came into existence. The Cenozoic era represents a considerably smaller fraction of geologic time than either the Paleozoic or the Mesozoic, but we know much more about this time span because the rock formations are more widespread and less disturbed than those of any preceding era.

Most of North America was above sea level during the Cenozoic era. However, the eastern and western margins of the continent experienced markedly dissimilar events because of their different plate boundary relationships. The Atlantic and Gulf coastal regions, far removed from an active plate boundary, were tectonically stable. By contrast, western North America was the leading edge of the North American plate, and the plate interactions during the Cenozoic account for many events of mountain building, volcanism, and earthquakes.

Eastern North America The stable continental margin of eastern North America was the site of abundant marine sedimentation. The most extensive deposits surrounded the

▼ SmartFigure 19.17 **Major provinces of the Appalachian Mountains**

TUTORIAL http://goo.gl/Y3wezV

A.

B.

Appalachian Plateau (Underlain by nearly flat-lying sedimentary strata of Paleozoic age.)

Valley and Ridge (Highly folded and thrust-faulted sedimentary rocks of Paleozoic age.)

Blue Ridge (Hilly to mountainous terrain consisting of slices of basement rock of Precambrian age.)

Piedmont (Crustal fragments of metamorphosed sedimentary and igneous rocks that were added to North America.)

Coastal Plain (Area of low relief underlain by gradualy sloping sedimentary strata and unlithified sediments.)

Figure 19.18 Massive, cross bedded sandstone cliffs in Zion National Park These sandstone cliffs are the remnants of ancient sand dunes that were part of an enormous desert during the Jurassic period. (Photo by Michael Collier; inset photo by Dennis Tasa)

Close-up view of cross bedding in the Navajo Sandstone, Zion National Park

Gulf of Mexico, from the Yucatan Peninsula to Florida, where a massive buildup of sediment caused the crust to downwarp. In many instances, faulting created structures in which oil and natural gas accumulated. Today, these and other petroleum traps (see Figure 7.32, page 209) are the Gulf Coast's most economically important resource, as evidenced by numerous offshore drilling platforms.

Early in the Cenozoic, the Appalachians had eroded to create a low plain. Later, isostatic adjustments again raised the region and rejuvenated its rivers. Streams eroded with renewed vigor, gradually sculpting the surface into its present-day topography. Sediments from this erosion were deposited along the eastern continental margin, where they accumulated to a thickness of many kilometers. Today, portions of the strata deposited during the Cenozoic are exposed as the gently sloping Atlantic and Gulf coastal plains, where a large percentage of the eastern and southeastern United States population resides (see Figure 19.17).

Western North America In the West, the Laramide Orogeny, responsible for building the southern Rocky Mountains, was coming to an end. As erosional forces lowered the mountains, the basins between uplifted ranges began to fill with sediment. East of the Rockies, a large wedge of sediment from the eroding mountains created the gently sloping Great Plains.

Beginning in the Miocene epoch, about 20 million years ago, a broad region from northern Nevada into Mexico experienced crustal extension that created more than 100 fault-block mountain ranges. Today, they rise abruptly above the adjacent basins, forming the Basin and Range Province (see Figure 11.17, page 305).

During the development of the Basin and Range Province, the entire western interior of the continent gradually uplifted. This event elevated the Rockies and rejuvenated many of the West's major rivers. As the rivers became incised, many spectacular gorges were created, including the Grand Canyon of the Colorado River, the Grand Canyon of the Snake River, and the Black Canyon of the Gunnison River (see Figure 13.25, page 362).

Volcanic activity was also common in the West during much of the Cenozoic. Beginning in the Miocene epoch, great volumes of fluid basaltic lava flowed from fissures in portions of present-day Washington, Oregon, and Idaho. These eruptions built the 3.4-million-square-kilometer (1.3-million-square-mile) Columbia Plateau. Immediately west of the vast Columbia Plateau, volcanic activity was different in character. Here, more viscous magmas with higher silica content erupted explosively, creating the Cascades, a chain of stratovolcanoes extending from northern California into Canada, some of which are still active (**Figure 19.19**).

As the Cenozoic was drawing to a close, the effects of mountain building, volcanic activity, isostatic adjustments, and extensive erosion and sedimentation created the physical landscape we know today. All that remained of the Cenozoic era was the final 2.6-million-year episode called the Quaternary period. During this most recent, and ongoing, phase of Earth's history, humans evolved and the action of glacial ice, wind, and running water added to our planet's long, complex geologic history.

CONCEPT CHECKS 19.5

1. During which period of geologic history did the supercontinent Pangaea come into existence? During which period did it begin to break apart?
2. Describe the climate of the present-day American Southwest during early Jurassic time.
3. Where is most Cretaceous age coal found today in the United States?
4. Compare and contrast eastern and western North America's geology during the Cenozoic era.

▶ **Figure 19.19 Mount Shasta, California** This volcano is one of several large composite cones that comprise the Cascade Range. (Photo by Michael Collier)

19.6 Earth's First Life

Describe some of the hypotheses on the origin of life and the characteristics of early prokaryotes, eukaryotes, and multicellular organisms.

The oldest fossils provide evidence that life on Earth was established at least 3.5 billion years ago (**Figure 19.20**). Microscopic fossils similar to modern cyanobacteria have been found in silica-rich chert deposits worldwide. Notable examples include southern Africa, where rocks date to more than 3.1 billion years ago, and the Lake Superior region of western Ontario and northern Minnesota, where the Gunflint Chert contains some fossils older than 2 billion years. Chemical traces of organic matter in even older rocks have led paleontologists to conclude that life may have existed much earlier.

Origin of Life

How did life begin? This question sparks considerable debate, and hypotheses abound. Requirements for life, in addition to a hospitable environment, include the chemical raw materials that are found in essential molecules such as proteins. Proteins are made from organic compounds called *amino acids*. The first amino acids may have been synthesized from carbon dioxide and nitrogen, both of which were plentiful in Earth's primitive atmosphere. Some scientists suggest that these gases could have been easily reorganized into useful organic molecules by ultraviolet light. Others consider lightning to have been the impetus, as the well-known experiments conducted by biochemists Stanley Miller and Harold Urey attempted to demonstrate.

Still other researchers suggest that amino acids arrived "ready-made," delivered by asteroids or comets that collided with a young Earth. Evidence for this hypothesis comes from a group of meteorites, called *carbonaceous chrondrites*, which contain amino acid–like organic compounds.

Yet another hypothesis proposes that the organic material needed for life came from the methane and hydrogen sulfide that spews from deep-sea hydrothermal vents (black smokers)(see Figure 8.22, page 230). It is also possible that life originated in hot springs similar to those in Yellowstone National Park.

Earth's First Life: Prokaryotes

Regardless of where or how life originated, it is clear that the journey from "then" to "now" involved change (see Figure 19.20). The first known organisms were simple,

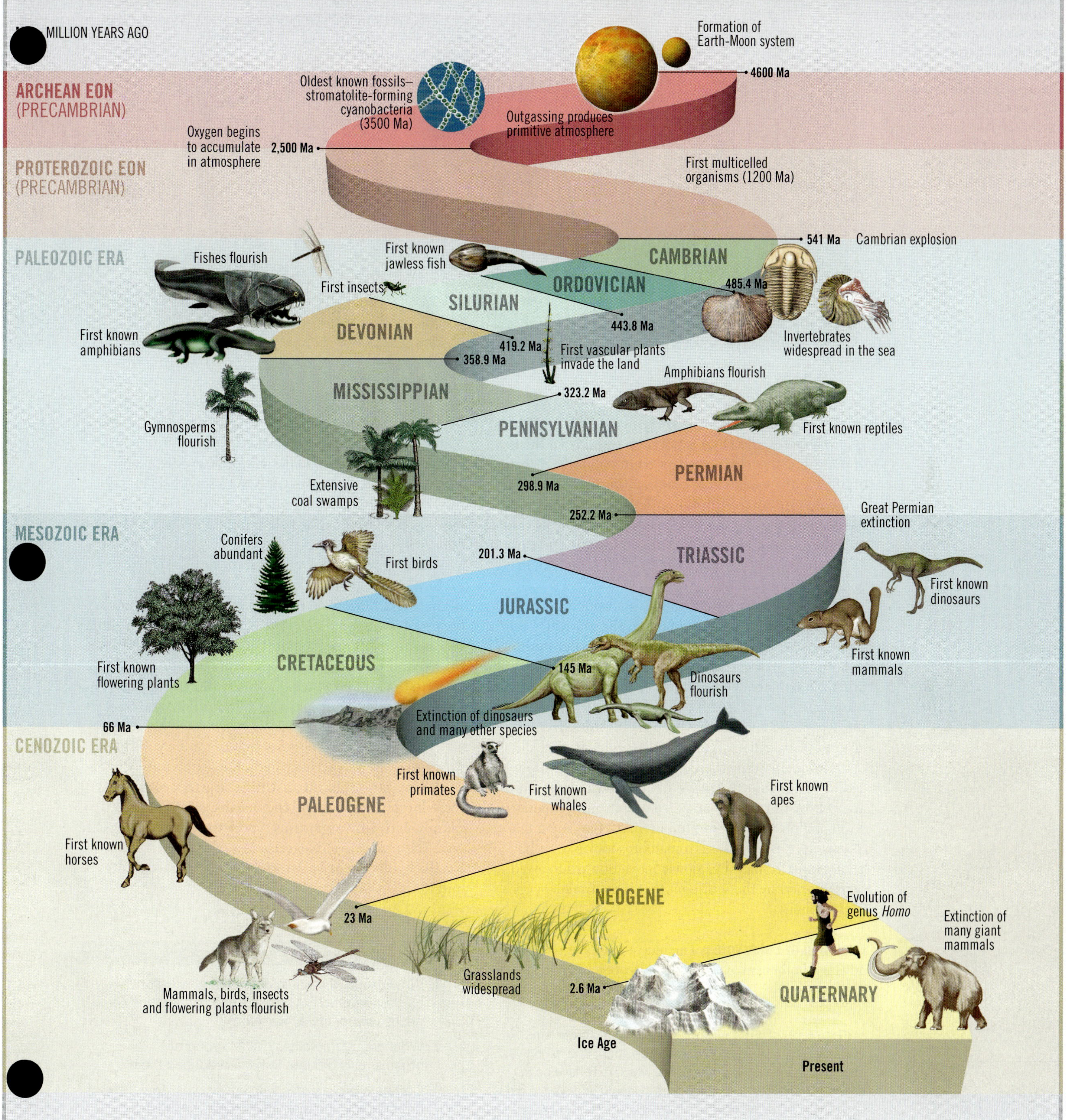

▲ **Figure 19.20** **Evolution of life though geologic time**

Figure 19.21 Stromatolites are among the most common Precambrian fossils A. Cross-section though fossil stromatolites deposited by cyanobacteria. (Photo by Sinclair Stammers/Science Source) **B.** Modern stromatolites exposed at low tide in western Australia. (Photo by Bill Bachman/Science Source)

single-cell **prokaryotes** (bacteria and similar microbes). In prokaryotes, the DNA is not segregated from the rest of the cell in a nucleus. A major triumph in the history of life was the evolution of photosynthesis: the ability to use energy obtained from sunlight to convert carbon dioxide into the organic molecules on which living things are based. These abundant sources of energy and carbon compounds would have enabled life to become more widespread and abundant.

Recall that photosynthesis by ancient cyanobacteria, a type of prokaryote, contributed to the gradual rise in the level of oxygen, first in the ocean and later in the atmosphere. These early organisms, which began to inhabit Earth about 3.5 billion years ago, dramatically transformed our planet. Fossil evidence for the existence of these microscopic bacteria includes distinctively layered mats, called **stromatolites**, composed of slimy material secreted by these organisms, along with trapped sediments (**Figure 19.21A**). What is known about these ancient fossils comes mainly from the study of modern stromatolites like those found in Shark Bay, Australia (**Figure 19.21B**). Today's stromatolites look like stubby pillars built as microbes slowly move upward to avoid being buried by the sediment that is continually deposited on them.

Evolution of Eukaryotes The oldest fossils of more advanced organisms, called **eukaryotes**, are about 2.1 billion years old. Eukaryotic cells have their genetic material segregated into a nucleus, and they are more complex in other ways than their prokaryotic precursors. While the first eukaryotes were single-celled, all the complex multicellular organisms that now inhabit our planet—trees, birds, fish, reptiles, and humans—are eukaryotes.

During much of the Precambrian, life consisted exclusively of single-celled organisms. It wasn't until perhaps 1.2 billion years ago that multicellular eukaryotes evolved. Green algae, one of the first multicellular organisms, contained chloroplasts (used in photosynthesis) and were the likely ancestors of modern plants. The first primitive marine animals did not appear until somewhat later, perhaps 600 million years ago (**Figure 19.22**).

Fossil evidence suggests that organic evolution progressed at an excruciatingly slow pace until the end of the Precambrian. At that time, Earth's continents were largely barren, and the oceans were populated mainly with tiny organisms, many too small to be seen with the naked eye. Nevertheless, the stage was set for the evolution of larger and more complex plants and animals.

Figure 19.22 Ediacaran fossil Ediacaran organisms like this one may have come into existence about 600 million years ago. These soft-bodied organisms, which may have been animals, were up to 1 meter (3 feet) in length and are the oldest large multicellular fossils so far discovered. (Photo Sinclair Stammers/Science Source)

CONCEPT CHECKS 19.6

1. What group of organic compounds is essential for the formation of proteins and is therefore necessary for life as we know it?
2. What are stromatolites? What group of organisms is thought to have produced them?
3. Compare *prokaryotes* with *eukaryotes*. To which group do all complex multicellular organisms belong?

19.7 Paleozoic Era: Life Explodes

List the major developments in the history of life during the Paleozoic era.

The Cambrian period marks the beginning of the Paleozoic era, a time span that saw the emergence of a spectacular variety of new life-forms. All major invertebrate (animals lacking backbones) groups became widespread during the Cambrian, including jellyfish, sponges, worms, mollusks (clams and snails), and arthropods (insects and crabs). This huge expansion in biodiversity is often referred to as the **Cambrian explosion**.

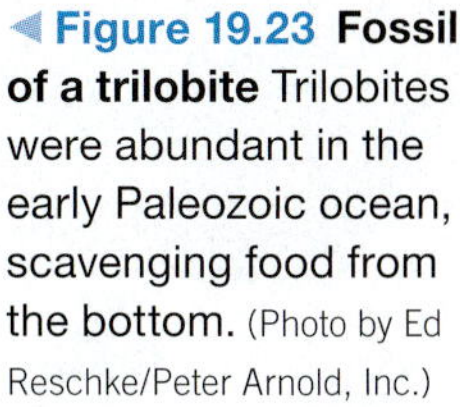

◀ **Figure 19.23 Fossil of a trilobite** Trilobites were abundant in the early Paleozoic ocean, scavenging food from the bottom. (Photo by Ed Reschke/Peter Arnold, Inc.)

Early Paleozoic Life-Forms

The Cambrian period was the golden age of *trilobites* (Figure 19.23). Trilobites developed a flexible exoskeleton of a protein called chitin (similar to a lobster shell), which enabled them to be mobile and search for food by burrowing through soft sediment. More than 600 genera of these mud-burrowing scavengers and grazers flourished worldwide.

The Ordovician period marked the appearance of abundant cephalopods—mobile, highly developed mollusks that became the major predators of their time (Figure 19.24). Descendants of these cephalopods include the squid, octopus, and chambered nautilus that inhabit our modern oceans. Cephalopods were the first truly large organisms on Earth, including one species that reached a length of nearly 10 meters (30 feet).

The early diversification of animals was driven, in part, by the emergence of predatory lifestyles. The larger mobile cephalopods preyed on trilobites that were typically smaller than a child's hand. The evolution of efficient movement was often associated with the development of greater sensory capabilities and more complex nervous systems. These early animals elaborated sensory devices for detecting light, odor, and touch.

Approximately 450 million years ago, green algae that had adapted to survive at the water's edge gave

Did You Know?
Possessing hard parts (shells or skeletons) greatly enhances the likelihood of organisms being preserved as fossils, but there have been rare occasions when large numbers of soft-bodied organisms were preserved. One of the best examples is the Burgess Shale, located in the Canadian Rockies, where more than 100,000 unique fossils have been uncovered.

◀ **Figure 19.24 Artistic depiction of a shallow Ordovician sea** During the Ordovician period (488–444 million years ago), the shallow waters of an inland sea over central North America contained an abundance of marine invertebrates. Shown in this reconstruction are (1) corals, (2) a trilobite, (3) a snail, (4) brachiopods, and (5) a straight-shelled cephalopod. (The Field Museum/Getty Images)

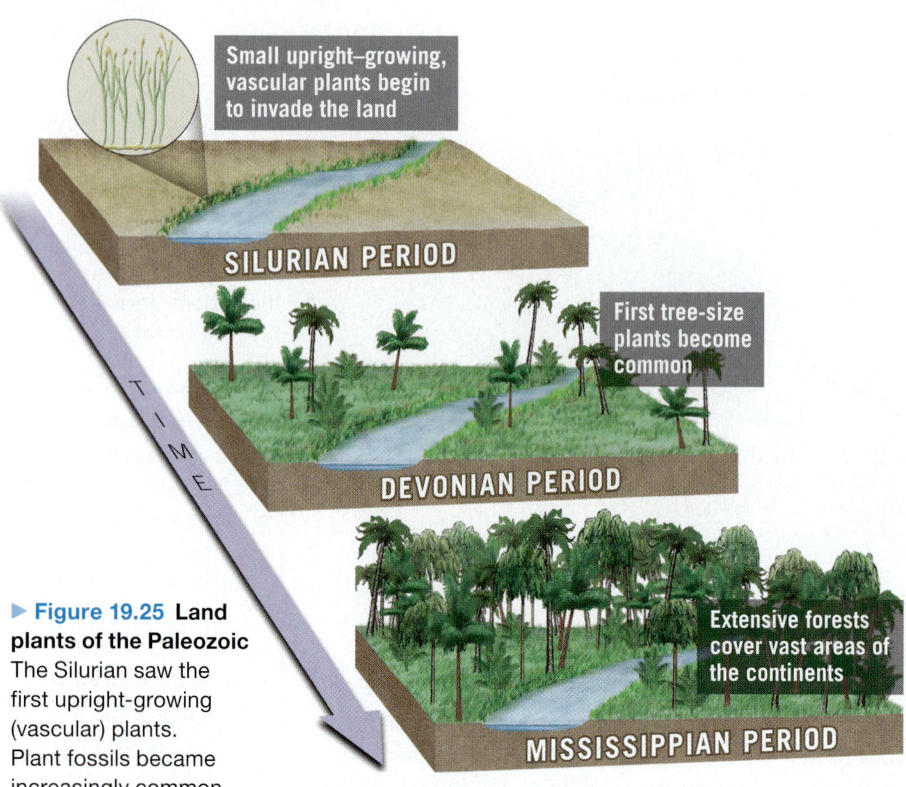

▶ **Figure 19.25 Land plants of the Paleozoic** The Silurian saw the first upright-growing (vascular) plants. Plant fossils became increasingly common from the Devonian onward.

rise to the first multicellular land plants. The primary difficulty in sustaining plant life on land was obtaining water and staying upright, despite gravity and winds. By 410 million years ago, early plants had begun to stand upright, in the form of leafless vertical spikes about the size of a human index finger. By the beginning of the Mississippian period, there were forests with trees tens of meters tall (**Figure 19.25**).

In the ocean, fish perfected an internal skeleton as a new form of support. Armor-plated fish that evolved during the Ordovician continued to adapt. Their armor plates thinned to lightweight scales that increased their speed and mobility. Other fish evolved during the Devonian, including primitive sharks with cartilage skeletons and bony fish—the groups in which many modern fish are classified. Fish, the first large vertebrates, proved to be faster swimmers than invertebrates and possessed more acute senses and larger brains. They became major predators of the sea, which is why the Devonian period is often referred to as the "Age of the Fishes."

Vertebrates Move to Land

During the Devonian, a group of fish called the *lobe-finned fish* began to adapt to terrestrial environments (**Figure 19.26A**). Like many other fish that live in oxygen-poor water, lobe-finned fish used lungs to supplement their gills for breathing. They also had stout fins with an internal skeleton. By the late Devonian, lobe-finned fish had evolved into air-breathing amphibians that had strong legs but still retained a fishlike head and tail (**Figure 19.26B**).

Modern amphibians, such as frogs, toads, and salamanders, are small and occupy limited biological niches. However, conditions during the late Paleozoic were ideal for these newcomers to land. Large tropical swamps teeming with large insects and millipedes extended across North America, Europe, and Siberia (**Figure 19.27**). With virtually no predatory risks, amphibians diversified rapidly. Some even took on lifestyles and forms similar to those of modern reptiles such as crocodiles.

Despite their success, amphibians were not fully adapted to life out of water. In fact, *amphibian* means "double life" because these animals typically need water for reproduction, even if they live mainly on land. Frogs, for instance, lay their eggs in water and develop as aquatic tadpoles with gills and tails before maturing into air-breathing adults with legs.

Reptiles: The First True Terrestrial Vertebrates

During the Mississippian period, some amphibians evolved features that allowed them to be more fully terrestrial and thus became the first reptiles (**Figure 19.28**). These included improved lungs for active lifestyles and "waterproof" skin that prevented the loss of body fluids. Most importantly, reptiles developed shell-covered eggs laid on land. Eliminating the water-dwelling stage (like the tadpole stage in frogs) was an important evolutionary step. Of interest is the fact that the watery fluid within the reptilian egg closely resembles seawater in chemical composition. Because the reptile embryo develops in this

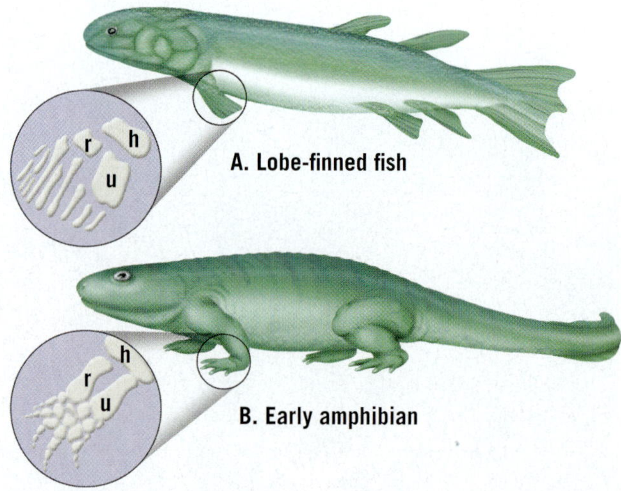

▲ **Figure 19.26 Comparison of the anatomical features of a lobe-finned fish and an early amphibian A.** The fins on the lobe-finned fish contained the same basic elements (*h*, humerus, or upper arm; *r*, radius, and *u*, ulna, which correspond to the lower arm) as those of the amphibians. **B.** This amphibian is shown with the standard five toes, but early amphibians had as many as eight toes. Eventually the amphibians evolved to have a standard toe count of five.

◀ **Figure 19.27 Artistic depiction of a Pennsylvanian-age coal swamp** Shown are scale trees (left), seed ferns (lower left), and horsetails (right). Also note the large dragonfly. (The Field Museum/Getty Images)

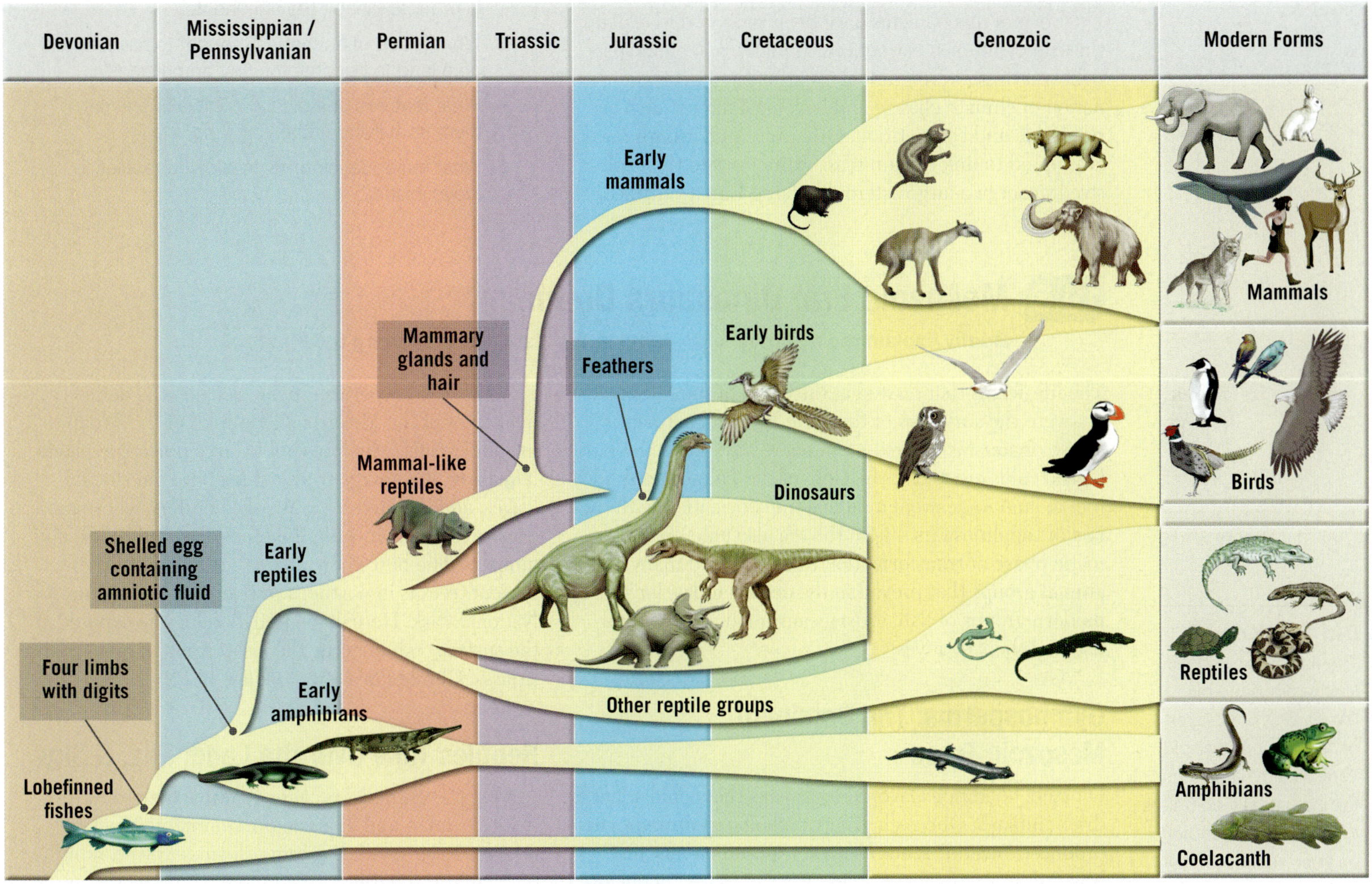

Devonian | Mississippian / Pennsylvanian | Permian | Triassic | Jurassic | Cretaceous | Cenozoic | Modern Forms

Early mammals
Mammals
Mammary glands and hair
Feathers
Early birds
Birds
Mammal-like reptiles
Dinosaurs
Shelled egg containing amniotic fluid
Early reptiles
Reptiles
Four limbs with digits
Early amphibians
Other reptile groups
Amphibians
Lobefinned fishes
Coelacanth

▲ **SmartFigure 19.28 Relationships of major land-dwelling vertebrate groups and their divergence from lobefin fish**

TUTORIAL
http://goo.gl/q6ltN0

watery environment, the shelled egg has been characterized as a "private aquarium" in which the embryos of these land vertebrates spend their water-dwelling stage of life. With this "sturdy egg," the remaining ties to the water were broken, and reptiles moved inland.

The Great Permian Extinction

At the close of the Permian period a **mass extinction** occurred, in which a large number of Earth's species became extinct. During this mass extinction, 70 percent of all land-dwelling vertebrate species and perhaps 90 percent of all marine organisms were obliterated; it was the most severe of five mass extinctions that occurred over the past 500 million years. Each extinction wreaked havoc with the existing biosphere, wiping out large numbers of species. In each case, however, survivors created new biological communities, which were often more diverse. Therefore, mass extinctions can actually invigorate life on Earth, as the few hardy survivors eventually filled the environmental niches left behind by the victims.

Several mechanisms have been proposed to explain these ancient mass extinctions. Initially, paleontologists believed these were gradual events caused by a combination of climate change and biological forces, such as predation and competition. Other research groups have attempted to link certain mass extinctions to the explosive impact of a large asteroid striking Earth's surface.

The most widely held view is that the Permian mass extinction was driven mainly by volcanic activity because it coincided with a period of voluminous eruptions of flood basalts that blanketed about 1.6 million square kilometers (624,000 square miles), an area nearly the size of Alaska. This series of eruptions, which lasted roughly 1 million years, occurred in northern Russia, in an area called the Siberian Traps. It was the largest volcanic eruption in the past 500 million years. The release of huge amounts of carbon dioxide likely generated a period of accelerated greenhouse warming, while the emission of sulfur dioxide is credited with producing copious amounts of acid rain and low-oxygen conditions in the marine environment. These drastic environmental changes likely put excessive stress on many of Earth's life-forms.

CONCEPT CHECKS 19.7

1. What is the Cambrian explosion?
2. Describe the obstacles plants had to overcome in order to inhabit the continents.
3. What group of animals is thought to have moved onto land to become the first amphibians?
4. What features of typical amphibians prevent them from living entirely on land?
5. What major developments allowed reptiles to move inland?

19.8 Mesozoic Era: Dinosaurs Dominant

Briefly explain the major developments in the history of life during the Mesozoic era.

The life-forms that existed at the dawn of the Mesozoic era were the survivors of the great Permian extinction. These organisms diversified in many ways to fill the biological voids created at the close of the Paleozoic. While life on land underwent a radical transformation with the rise of the dinosaurs, life in the sea also entered a dramatic phase of transformation that produced many of the animal groups that prevail in the oceans today, including modern groups of fish, crustaceans, mollusks, and starfish and their relatives.

Did You Know?
The first fossil evidence for the dinosaur–bird link came from a layer of 150-million-year-old Jurassic limestone containing a nearly complete imprint of *Archaeopteryx*, an intermediate animal that had wings and feathers but also teeth and a skeleton similar to that of a small dinosaur (see Figure 19.31).

Gymnosperms: The Dominant Mesozoic Trees

On land, conditions favored organisms that could adapt to drier climates. One such group of plants, **gymnosperms**, produced "naked" seeds that are exposed on modified leaves that usually form cones. The seeds are not enclosed in fruits, as are apple seeds, for example. Unlike the first plants to invade the land, seed-bearing gymnosperms did not depend on free-standing water for fertilization, like the more primitive ferns. Consequently, they could expand easily into drier habitats.

The gymnosperms quickly became the dominant trees of the Mesozoic. Examples of this group include cycads, which resemble large pineapple plants (**Figure 19.29**); ginkgo trees, which have fan-shaped leaves; and the largest plants, the conifers, whose modern descendants include the pines, firs, redwoods, and junipers. The best-known fossil occurrence of these ancient trees is in northern Arizona's Petrified Forest National Park. Here, huge petrified logs lie exposed at the surface, exhumed by the weathering of rocks of the Triassic Chinle Formation (**Figure 19.30**).

Reptiles Take Over the Land, Sea, & Sky

Among the animals, reptiles readily adapted to the drier Permian and Triassic environment. The first reptiles were small, but larger forms evolved rapidly, particularly the dinosaurs. One of the largest was *Ultrasaurus*, which weighed more than 80 tons and measured over 30 meters (100 feet) from head to tail. Some of the largest dinosaurs were carnivorous (for example, *Tyrannosaurus*), whereas others were herbivorous (like ponderous *Apatosaurus*).

▲ **Figure 19.29 A cycad, a type of gymnosperm that was very common in the Mesozoic** These plants have palm-like leaves and large cones. (Photo by Jiri Loun/Science Source)

◀ **Figure 19.30 Petrified logs of Triassic age, Arizona's Petrified Forest National Park** (Photo by Bernd Siering/AGE Fotostock)

Some reptiles evolved specialized characteristics that allowed them to occupy drastically different environments. One group, the pterosaurs, became airborne. How the largest pterosaurs—some of which had wing spans greater than 8 meters (26 feet) and weighed more than 90 kilograms (200 pounds)—took flight is still unknown. Another group, exemplified by the fossil *Archaeopteryx*, evolved from dinosaurs and gave rise to modern birds (**Figure 19.31**). *Archaeopteryx* had feathered wings but retained teeth, clawed digits in its wings, and a long tail with many vertebrae. A recent study concluded that *Archaeopteryx* were unable to use flapping flight. Rather, by running and leaping into the air, these bird-like reptiles escaped predators with glides and downstrokes. Other researchers disagree and envision them as climbing animals that glided down to the ground, following the idea that birds evolved from tree-dwelling gliders. Whether birds took to the air from the ground *up* or from the trees *down* is a question scientists continue to debate.

Other reptiles returned to the sea, including fish-eating *plesiosaurs* and *ichthyosaurs* (**Figure 19.32**). These reptiles became proficient swimmers, breathing by means of lungs rather than gills.

For nearly 160 million years, dinosaurs flourished. However, by the close of the Mesozoic, dinosaurs, like many other reptiles, became extinct. Select reptile groups have survived to recent times, including turtles, snakes, crocodiles, and lizards. The huge, land-dwelling dinosaurs, the marine plesiosaurs, and the flying pterosaurs are known only through the fossil record. What caused this great extinction?

Demise of the Dinosaurs

The boundaries between divisions on the geologic time scale represent times of significant geological and/or biological change. Of special interest is the boundary between the Mesozoic era ("middle life") and the Cenozoic era

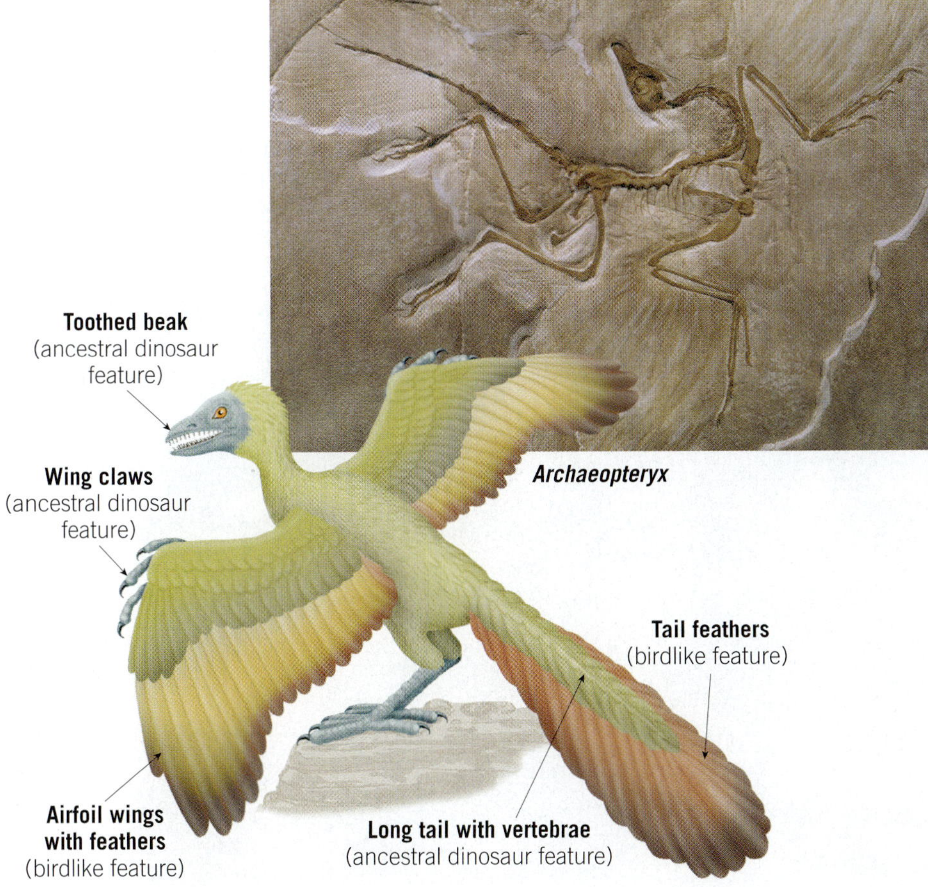

▼ **Figure 19.31** ***Archaeopteryx*, a transitional form related to modern birds** *Archaeopteryx* had feathered wings and a feathered tail that appear to be intended for flight, but in many ways it resembled ancestral dinosaurs. The sketch shows an artist's reconstruction of *Archaeopteryx*. (Photo by Michael Collier)

▲ **Figure 19.32 Reptiles returned to the sea** *Ichthyosaurs* are one of the groups of land animals that returned to the sea during the Mesozoic. (Photo by Chip Clark)

Did You Know?
Dinosaurs were not the only large animals to roam Earth in the distant past. One example is a crocodile known as *Sarcosuchus imperator*. This huge river dweller weighed 8 metric tons (more than 17,600 lb!) and was about 12 m (40 ft) long—as long as *Tyrannosaurus rex* and much heavier. Its jaws were roughly as long as an adult human is tall.

("recent life"), about 66 million years ago. During this transition roughly three-quarters of all plant and animal species died out in another mass extinction. This boundary marks the end of the era in which dinosaurs and other large reptiles dominated the landscape (**Figure 19.33**) and the beginning of the era when mammals assumed that role.

What could have triggered the extinction of one of the most successful groups of land animals? An increasing number of researchers support the view that the dinosaurs fell victim to a "one–two punch." The first blow occurred during the last few million years of the Mesozoic era, when climate data indicates that average temperature over the land increased by more than 20°C (40°F) over a few tens of thousands of years—a blink of the eye in geologic time. This period of global warming coincided with massive basaltic eruptions, which produced the Deccan Plateau, a volcanic area in present-day India. The Deccan eruptions presumably released massive amounts of carbon dioxide, which caused a period of greenhouse warming that resulted in a dramatic rise in temperatures. This period of global warming is thought to have snuffed out some species and hobbled others.

The final blow came about 66 million years ago, when our planet was struck by a stony meteorite, a relic from the formation of the solar system. The errant mass of rock was approximately 10 kilometers (6 miles) in diameter and was traveling at about 90,000 kilometers per hour at the time of impact. It collided with the southern portion of North America in a shallow tropical sea—now Mexico's Yucatan Peninsula (**Figure 19.34**).

Following the impact, suspended dust greatly reduced the amount of sunlight reaching Earth's surface, which resulted in global cooling ("impact winter") and inhibited photosynthesis, disrupting food production. Long after the dust settled, the sulfur compounds added to the atmosphere by the blast remained suspended and reflected solar radiation back to space, perpetuating the unusually cold temperatures.

One piece of evidence that points to a catastrophic collision 66 million years ago is a thin layer of sediment, less than 1 centimeter thick, discovered in several places around the globe. This sediment contains a high level of the element *iridium*, which is rare in Earth's crust but found in high proportions in stony meteorites

▼ **Figure 19.33 Artist's rendering of Allosaurus** *Allosaurus* was a large carnivorous dinosaur that lived in the late Jurassic period (155 to 145 million years ago). (Image by Roger Harris/Science Source)

► **Figure 19.34 Chicxulub crater** The Chicxulub crater is a giant impact crater that formed about 66 million years ago and has since filled with sediment. It is 180 kilometers (110 miles) in diameter and likely contributed to the demise of the dinosaurs.

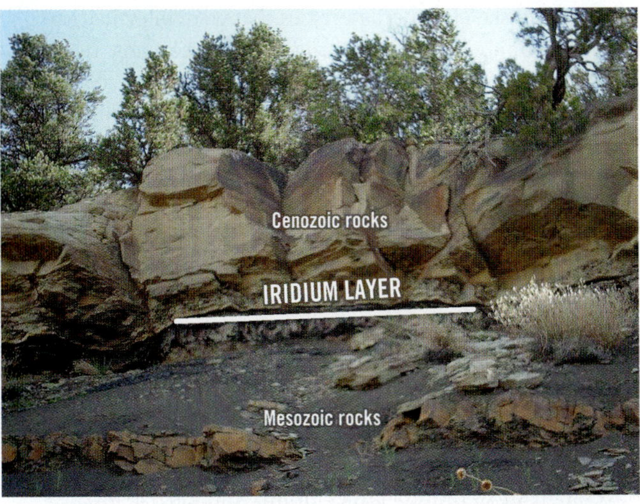

▲ **Figure 19.35 Iridium layer: Evidence for a catastrophic impact 66 million years ago** A thin layer of sediment has been discovered worldwide at Earth's physical boundary separating the Mesozoic and Cenozoic eras. This sediment contains a high level of iridium, an element that is rare in Earth's crust but found in much higher concentrations in stony meteorites. (Photo courtesy of Nationalparks)

(Figure 19.35). This layer presumably contains scattered remains of the meteorite responsible for the environmental changes that provided the second and final blow that led to the demise of many reptile groups.

Regardless of what caused this massive extinction, its outcome provides a valuable lesson in understanding the role that catastrophic events play in shaping our planet's physical landscape and biosphere. The extinction of the dinosaurs opened habitats for the mammals that survived. These new habitats, along with evolutionary forces, led to the development of the great diversity of mammals that occupy our modern world.

CONCEPT CHECKS 19.8

1. What group of plants became the dominant trees during the Mesozoic era? Name a modern descendant of this group.
2. What group of reptiles gave rise to modern birds?
3. What was the dominant reptile group on land during the Mesozoic?
4. Identify two groups of reptiles that returned to life in the sea.

Did You Know?

Both bats and humans are five-fingered mammals, but they have developed very different physical structures and lifestyles. Humans have evolved hands for grasping, whereas bats have proportionally much longer "fingers" that support their wings for flying.

19.9 Cenozoic Era: Mammals Diversify

Discuss the major developments in the history of life during the Cenozoic era.

During the Cenozoic, mammals replaced dinosaurs as the prominent vertebrates on land. Among the plants, the **angiosperms** (flowering plants), which had appeared and begun to diversify during the later Mesozoic, similarly came to dominate most terrestrial environments.

The development of flowering plants strongly influenced the evolution of both birds and mammals that feed on seeds and fruits, as well as many insect groups. During the middle of the Cenozoic, one type of angiosperm, grasses, spread rapidly to form grasslands—a completely novel type of landscape. At the same time, herbivorous mammal groups gave rise to a great variety of grazers, and carnivores evolved to prey on them (Figure 19.36).

During the Cenozoic, the ocean teemed with modern fish such as tuna, swordfish, and barracuda. In addition, some mammals, including seals, whales, and walruses, took up life in the sea.

From Dinosaurs to Mammals

Mammals coexisted with dinosaurs for nearly 100 million years but remained mostly small and probably mostly nocturnal. Then,

▼ **Figure 19.36 Angiosperms became the dominant plants during the Cenozoic** Angiosperms, commonly known as flowering plants, consist of a group of seed plants that have reproductive structures called flowers and fruits. **A.** The most diverse and widespread of modern plants, many angiosperms display easily recognizable flowers. **B.** Some angiosperms, including grasses, have very tiny flowers. The expansion of the grasslands during the Cenozoic era greatly increased the diversity of grazing mammals and the predators that feed on them. (Photo A by Mike Potts/Nature Picture Library; photo B by Torleif/CORBIS)

Did You Know?
The famed La Brea Tar Pits, located in downtown Los Angeles, contain a rich and very well-preserved assemblage of mammals that roamed southern California from 40,000 to 80,000 years ago. Tar pits form when crude oil seeps to the surface and the light portion evaporates, leaving behind sticky pools of heavy tar (see Figure 18.14, page 481).

about 66 million years ago, fate intervened when the large meteorite collided with Earth and dealt the final crashing blow to the reign of the dinosaurs. This transition from one dominant group to another is clearly visible in the fossil record.

Mammals are unique in having hair and in suckling their young with milk. Like birds, they are "warm blooded," maintaining a high and constant body temperature. With the demise of the large Mesozoic reptiles, Cenozoic mammals diversified rapidly. The many forms that exist today evolved from small primitive mammals that were characterized by short legs; flat, five-toed feet; and small brains. Their development and specialization took four principal directions: increase in size, increase in brain capacity, specialization of teeth to accommodate more diverse diets, and specialization of limbs to be better equipped for a particular lifestyle or environment.

Marsupial & Placental Mammals

Two groups of mammals, the marsupials and the placentals, evolved during the Mesozoic. The groups differ principally in their modes of reproduction. Young marsupials are born live at a very early stage of development. At birth, the tiny and immature young enter the mother's pouch to suckle and complete their development. Today, marsupials are found primarily in Australia, where they took a separate evolutionary path largely isolated from placental mammals. Modern marsupials include kangaroos, opossums, and koalas (**Figure 19.37**).

▲ **Figure 19.37 Kangaroos, examples of marsupial mammals** After the breakup of Pangaea, the Australian marsupials evolved independently. (Photo by Martin Harvey/Peter Arnold Inc.)

Placental mammals (eutherians), conversely, develop within the mother's body for a much longer period, so birth occurs when the young are comparatively mature. Members of this group include wolves, elephants, bats, manatees, and monkeys. Most modern mammals, including humans, are placental.

Humans: Mammals with Large Brains & Bipedal Locomotion

Both fossil and genetic evidence suggest that, by 6.5 million years ago, our own group, the hominins, had branched off from the line leading to modern chimpanzees. We have a good record of this evolution in fossils found in several sedimentary basins in Africa, in particular the Rift Valley system in East Africa.

The genus *Australopithecus*, which came into existence about 4.2 million years ago, showed skeletal characteristics that were intermediate between our apelike ancestors and modern humans. In particular, *Australopithecus* walked upright. Evidence for this bipedal stride includes footprints preserved in 3.2-million-year-old ash deposits at Laetoli, Tanzania (**Figure 19.38**). This new way of moving around made it possible for our human ancestors to leave forested habitats and to travel long distances for hunting and gathering food.

The earliest fossils of our genus, *Homo*, include the remains of *Homo habilis*, nicknamed "handy man" because they were often found with sharp stone tools in sedimentary deposits from 2.4 to 1.5 million years ago. *Homo habilis* had a shorter jaw and a larger brain than its predecessor.

During the next 1.3 million years, our ancestors developed substantially larger brains and long slender legs with hip joints adapted for long-distance walking. These species (including *Homo erectus*) ultimately gave rise to our species, *Homo sapiens*, as well as to some extinct related species, including the Neanderthals (*Homo neanderthalis*). Despite having the same-sized

▲ **Figure 19.38 Footprints of *Australopithecus* in ash deposits at Laetoli, Tanzania** (Photo by John Reader/Science Source)

brain as present-day humans and being able to fashion hunting tools from wood and stone, Neanderthals became extinct about 28,000 years ago. At one time, Neanderthals were considered a stage in the evolution of *Homo sapiens*, but that view has largely been abandoned.

Based on our current understanding, *Homo sapiens* originated in Africa about 200,000 years ago and began to spread around the globe. The oldest-known *Homo sapiens* fossils outside Africa were found in the Middle East and date to 115,000 years ago. Humans are known to have coexisted with Neanderthals and other prehistoric populations, with remains found in Siberia, China, and Indonesia. Further, there is mounting genetic evidence that our ancestors interbred with members of some of these groups.

By 36,000 years ago, humans were producing spectacular cave paintings in Europe (**Figure 19.39**). About 28,000 years ago, all prehistoric hominin populations except for *Homo sapiens* died out.

▲ **Figure 19.39 Cave painting of animals by early humans** (Photo courtesy of Sisse Brimberg/National Geographic Society)

Large Mammals & Extinction

During the rapid mammal diversification of the Cenozoic era, some species became very large. For example, by the Oligocene epoch, a hornless rhinoceros evolved that stood nearly 5 meters (16 feet) high. It is the largest land mammal known to have existed. As time passed, many other mammals evolved to large forms—more, in fact, than now exist. Many of these large forms were common as recently as 11,000 years ago. However, a wave of late Pleistocene extinctions rapidly eliminated many of these animals from the landscape.

North America experienced the extinction of mastodons and mammoths, both relatives of the modern elephants (**Figure 19.40**). In addition, saber-toothed cats, giant beavers, large ground sloths, horses, giant bison, and others died out. In Europe, late Pleistocene extinctions included woolly rhinos, large cave bears, and Irish elk. Scientists remain puzzled about the reasons for these more recent extinctions of large mammals. Because these large animals survived several major glacial advances and interglacial periods, it is difficult to ascribe their extinctions to climate change. Some scientists hypothesize that early humans hastened the decline of these mammals by selectively hunting large forms.

▼ **Figure 19.40 Mammoths** These relatives of modern elephants were among the large mammals that became extinct at the close of the Ice Age. (Image courtesy of INTERFOTO/Alamy)

Did You Know?
The first modern humans (*Homo sapiens*) began their migration out of Africa roughly 100,000 years ago. After migrating across Asia, they entered the Americas across the Bering land-bridge about 15,000 years ago.

CONCEPT CHECKS 19.9

1. What animal group became the dominant land animals of the Cenozoic era?
2. Explain how the demise of the dinosaurs impacted the development of mammals.
3. Where did researchers discover most of the evidence for the early evolution of our hominin ancestors been discovered?
4. What two characteristics best separate humans from other mammals?
5. Describe one hypothesis that explains the extinction of large mammals in the late Pleistocene.

CONCEPTS IN REVIEW
Earth's Evolution Through Geologic Time

19.1 Is Earth Unique?

List the principal characteristics that make Earth unique among the planets in the solar system.

- As far as we know, Earth is unique among planets in the fact that it hosts life. The planet's size, composition, and location all contribute to conditions that support life as we know it.

19.2 Birth of a Planet

Outline the major stages in Earth's evolution, from the Big Bang to the formation of our planet's layered internal structure.

KEY TERMS: supernova, solar nebula, planetesimal, protoplanet, Hadean

- The universe is thought to have formed about 13.8 billion years ago, with the Big Bang, which generated space, time, energy, and matter, including the elements hydrogen and helium. Elements heavier than hydrogen and helium were synthesized by nuclear reactions occurring either in the cores of stars or during supernova explosions.
- Earth and the solar system formed around 4.6 billion years ago, with the contraction of a solar nebula. Collisions between clumps of matter in this spinning disc resulted in the growth of planetesimals and then protoplanets. Over time, the matter of the solar nebula was concentrated into a smaller number of larger bodies: the Sun, the rocky inner planets, the icy outer planets, moons, comets, and asteroids.
- Heat production within Earth during its formative years was much higher than it is today, thanks to the kinetic energy of impacting asteroids and planetesimals, as well as the decay of radioactive isotopes. These high temperatures caused rock and iron to melt. This allowed iron to sink to form Earth's core and rocky material to rise to form the mantle and crust.

? The accompanying image shows Comet Shoemaker-Levy 9 impacting Jupiter in 1994. After this event, what happened to Jupiter's total mass? How was the number of objects in the solar system affected?

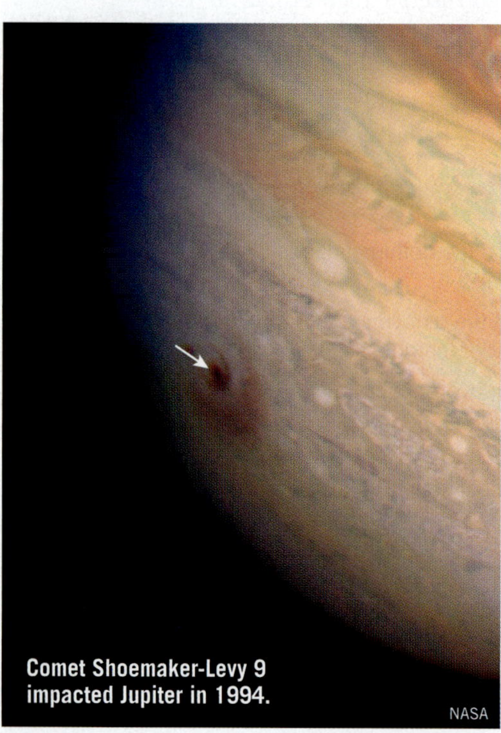

Comet Shoemaker-Levy 9 impacted Jupiter in 1994. NASA

19.3 Origin & Evolution of the Atmosphere and Oceans

Describe how Earth's atmosphere and oceans formed and evolved through time.

KEY TERMS: outgassing, banded iron formation, Great Oxygenation Event

- Earth's atmosphere is essential for life. It evolved as volcanic outgassing added mainly water vapor and carbon dioxide to the primordial atmosphere of gases common in the early solar system: methane and ammonia.
- Free oxygen began to accumulate through photosynthesis by cyanobacteria, which released oxygen as a waste product. Much of this early oxygen immediately reacted with iron dissolved in seawater and settled to the ocean floor as chemical sediments called banded iron formations. The Great Oxygenation Event of 2.5 billion years ago marks the first evidence of significant amounts of free oxygen in the atmosphere.
- Earth's oceans formed after the planet's surface had cooled. Soluble ions weathered from the crust were carried to the ocean, making it salty. The oceans also absorbed tremendous amounts of carbon dioxide from the atmosphere.

19.4 Precambrian History: The Formation of Earth's Continents

Explain the formation of continental crust, how continental crust becomes assembled into continents, and the role that the supercontinent cycle has played in this process.

KEY TERMS: craton, shield, supercontinent, supercontinent cycle

- The Precambrian includes the Archean and Proterozoic eons. However, our knowledge of what occurred during these eons of geologic time is rather limited because erosion destroyed much of the rock record.
- Continental crust was produced over time through the recycling of ultramafic and mafic crust in an early version of plate tectonics. Small crustal fragments formed and amalgamated into large crustal provinces called cratons. Over time, North America and other continents grew through the accretion of new terranes around the edges of this central "nucleus" of crust.
- Early cratons not only merged but sometimes rifted apart. The supercontinent Rodinia formed around 1.1 billion years ago and then rifted apart, opening new ocean basins. Over time, these ocean basins also closed to form a new supercontinent called Pangaea around 250 million years ago. Like Rodinia before it, Pangaea broke up as part of the ongoing supercontinent cycle.
- The formation of elevated oceanic ridges following the breakup of a supercontinent displaced enough water that sea level rose, and shallow seas flooded the continents. The breakup of continents can also influence the direction of ocean currents, with important consequences for climate.

? Consult Figure 19.12 and briefly summarize the history of the assembly of North America over the past 3.5 billion years.

19.5 Geologic History of the Phanerozoic: The Formation of Earth's Modern Continents

List and discuss the major geologic events in the Paleozoic, Mesozoic, and Cenozoic eras.

KEY TERMS: Pangaea, Laurasia, Gondwana

- The Phanerozoic eon encompasses the Paleozoic, Mesozoic, and Cenozoic eras, which cover the past 541 million years of geologic time.
- In the Paleozoic era, North America experienced a series of collisions that resulted in the rise of the young Appalachian mountain belt, as part of the assembly of Pangaea. High sea levels caused the ocean to cover vast areas of the continent and resulted in a thick sequence of sedimentary strata.
- During the Mesozoic, Pangaea broke up, and the Atlantic Ocean began to form. As the North American continent moved westward, the Cordillera began to rise due to subduction and the accretion of terranes along the west coast. In the Southwest, vast deserts deposited thick layers of dune sand, while environments in the East were conducive to the formation and subsequent burial of coal swamps.
- In the Cenozoic era, a thick sequence of sediments was deposited along North America's Atlantic margin and the Gulf of Mexico. Meanwhile, western North America experienced an extraordinary episode of crustal extension; the Basin and Range Province resulted.

? Contrast the tectonics of eastern and western North America during the Mesozoic era.

19.6 Earth's First Life

Describe some of the hypotheses on the origin of life and the characteristics of early prokaryotes, eukaryotes, and multicellular organisms.

KEY TERMS: prokaryote, stromatolite, eukaryote

- Life began from nonlife. Amino acids, a necessary building block for proteins, may have been assembled with energy from ultraviolet light or lightning, or in a hot spring, or were delivered later to Earth via meteorites.
- The first organisms were relatively simple single-celled prokaryotes that thrived in low-oxygen environments. They may have formed by 3.8 billion years ago. The advent of photosynthesis allowed microbial mats to build up and form stromatolites.
- Eukaryotes have larger, more complex cells than prokaryotes. The oldest-known eukaryotic cells date from around 2.1 billion years ago. Eukaryotic cells gave rise to the great diversity of multicellular organisms.

? The accompanying image shows fossil stromatolites. In what way did the ancient cyanobacteria that built these structures contribute to the evolution of Earth's atmosphere?

M.C. Rygel

19.7 Paleozoic Era: Life Explodes

List the major developments in the history of life during the Paleozoic era.

KEY TERMS: Cambrian explosion, mass extinction

- Abundant fossil hard parts appear in sedimentary rocks from the beginning of the Cambrian period. These shells and other skeletal material came from a profusion of new animals, including trilobites and cephalopods.
- Plants colonized the land around 400 million years ago and soon diversified into forests.
- In the Devonian, some lobe-finned fishes gradually evolved into the first amphibians. A subset of the amphibian population evolved waterproof skin and shelled eggs and split off to become the reptile line.
- The Paleozoic era ended with the largest mass extinction in the geologic record. This deadly event may have been related to the eruption of the Siberian Traps flood basalts.

? What advantages do reptiles have over amphibians for life on dry land?

19.8 Mesozoic Era: Dinosaurs Dominant

Briefly explain the major developments in the history of life during the Mesozoic era.

KEY TERM: gymnosperm

- Plants diversified during the Mesozoic. The flora of that time was dominated by gymnosperms, the first plants with seeds that allowed them to migrate beyond the edge of water bodies.
- Reptiles diversified, too. The dinosaurs came to dominate the land, pterosaurs took to the air, and a suite of marine reptiles swam the seas. The first birds evolved during the Mesozoic, exemplified by *Archaeopteryx*, a transitional fossil.
- As with the Paleozoic, the Mesozoic ended with a mass extinction, probably due to a massive meteorite impact in what is now Chicxulub, Mexico.

19.9 Cenozoic Era: Mammals Diversity

Discuss the major developments in the history of life during the Cenozoic era.

KEY TERM: angiosperm

- Flowering plants, called angiosperms, diversified and spread around the world through the Cenozoic era.
- Once the giant Mesozoic reptiles were extinct, mammals were able to diversify on the land, in the air, and in the oceans. Mammals are warm blooded, have hair, and nurse their young with milk. Marsupial mammals are born very young and then move to a pouch on the mother, while placental mammals spend a longer time *in utero* and are born in a relatively mature state compared to marsupials.
- Humans evolved from primate ancestors in Africa over a period of about 6.5 million years. They are distinguished from their ape ancestors by an upright, bipedal posture and large brains, as well as the use of elaborate tools. The oldest anatomically modern human fossils are 200,000 years old. Some of these humans migrated out of Africa and coexisted with Neanderthals and other human-like populations.

GIVE IT SOME THOUGHT

1 Refer to the geologic time scale in Figure 19.3. The Precambrian accounts for nearly 90 percent of geologic time. Why do you think it has fewer divisions than the rest of the time scale?

2 Referring to Figure 19.4, write a brief summary of the events that led to the formation of Earth.

3 The accompanying photograph shows layered iron-rich rocks called banded iron formations. What does the existence of these 2.5-billion-year-old rocks tell us about the evolution of Earth's atmosphere?

Blue Gum Pictures/Alamy

4 Describe how the sudden appearance of oxygen in the atmosphere about 2.5 billion years ago influenced the development of modern life-forms.

5 Five mass extinctions, in which 50 percent or more of Earth's marine species became extinct, are documented in the fossil record. Use the accompanying graph, which depicts the time and extent of each mass extinction, to answer the following:

a. Which of the five mass extinctions was the *most extreme*? Identify this extinction by name and when it occurred.

b. When did the *most recent* mass extinction occur?

c. During the most recent mass extinction, what prominent animal group was eliminated?

d. What group of terrestrial animals experienced a major period of diversification following the most recent mass extinction?

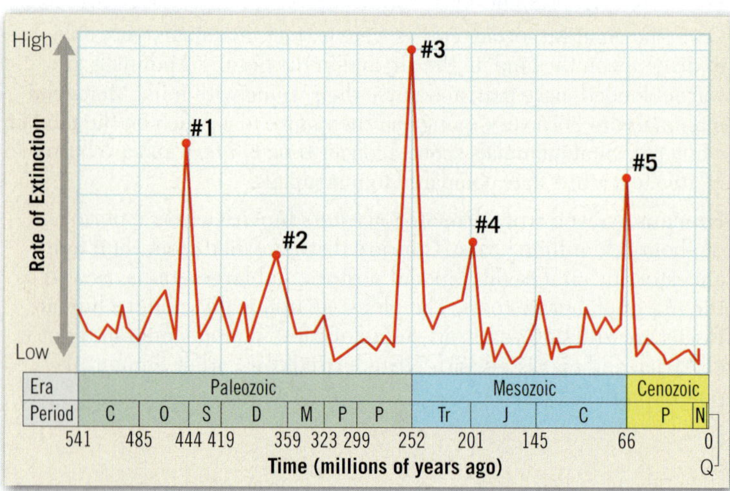

6 Currently oceans cover about 71 percent of Earth's surface. However, early in Earth history the oceans covered a greater percentage of the surface. Explain.

7 Contrast the eastern and western margins of North America during the Cenozoic era in terms of their relationships to plate boundaries.

8 Between 300 and 250 million years ago, plate movement assembled all the previously separated landmasses together to form the supercontinent Pangaea. The formation of Pangaea resulted in deeper ocean basins and a drop in sea level, causing shallow coastal areas to dry up. Thus, in addition to rearranging the geography of our planet, continental drift had a major impact on life on Earth. Use the accompanying diagrams showing the movement of hypothetical landmasses and the information above to answer the following:

a. Which of the following types of habitats would likely diminish in size during the formation of a supercontinent: deep-ocean habitats, wetlands, shallow marine environments, or terrestrial (land) habitats? Explain.

b. During the breakup of a supercontinent, would sea level remain the same, rise, or fall?

c. Explain how and why the development of an extensive oceanic ridge system that forms during the breakup of a supercontinent affects sea level.

Four hypothetical continents

TIME

Newly formed supercontinent

9 Some scientists have proposed that the environments around black smokers is similar to the extreme conditions that existed early in Earth history and study the unusual life that exists for clues about how the earliest life may have survived. Compare and contrast the environment of a black smoker to the environment on Earth approximately 3 to 4 billion years ago. Are there parallels between the two, and if so, do you think black smokers are good examples of the environment that earliest life may have experienced? Explain.

10 Most of the vast North American coal resources located from Pennsylvania to Illinois formed during the Pennsylvanian and Mississippian periods of Earth history. (This time period is also known as the Carboniferous period.) Examine the accompanying diagram, which illustrates the geographic position of North America during the period of coal formation.

a. Where, relative to the equator, was the United States located during the time of coal formation?

b. Why is it unlikely that a similar coal-forming environment will repeat itself in North America in the near future?

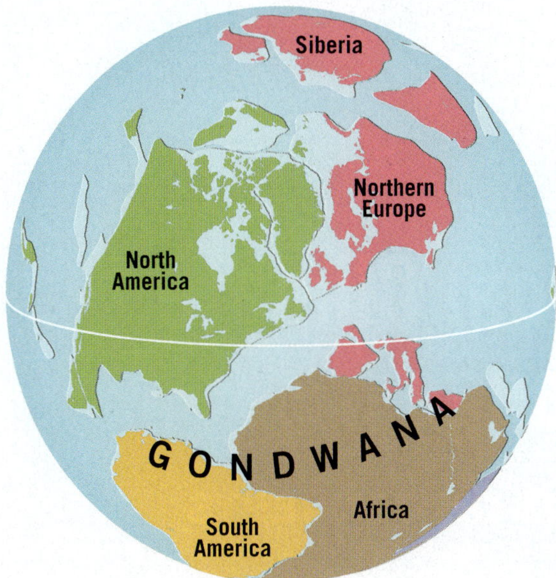

MasteringGeology™

Looking for additional review and test prep materials? Visit the Study Area in MasteringGeology to enhance your understanding of this chapter's content by accessing a variety of resources, including Self-Study Quizzes, Geoscience Animations, SmartFigures, Mobile Field Trips, GEODe Tutorials, RSS feeds, flashcards, web links, and an optional Pearson eText.

www.masteringgeology.com

20 Global Climate Change

FOCUS ON CONCEPTS

Each statement represents the primary learning objective for the corresponding major heading within the chapter. After you complete the chapter, you should be able to:

20.1 List the major parts of the climate system and some connections between climate and geology.

20.2 Explain why unraveling past climate changes is important and discuss several ways in which such changes are detected.

20.3 Describe the composition of the atmosphere and the atmosphere's vertical changes in pressure and temperature.

20.4 Outline the basic processes involved in heating the atmosphere.

20.5 Discuss hypotheses that relate to natural causes of climate change.

20.6 Summarize the nature and cause of the atmosphere's changing composition since about 1750. Describe the climate's response.

20.7 Contrast positive and negative feedback mechanisms and provide examples of each.

20.8 Summarize some of the possible consequences of global warming.

Glaciers are sensitive to changes in temperature and precipitation and therefore provide clues about changes in climate. Like most other glaciers and ice sheets worldwide, Margerie Glacier in Alaska's Glacier Bay National Park is losing mass and retreating. (Photo by Don Paulson/AGE Fotostock)

CLIMATE HAS A SIGNIFICANT IMPACT ON PEOPLE, and we are learning that people also have a strong influence on climate. Today global climate change caused by humans is a major environmental issue. Unlike changes in the geologic past, which were natural variations, modern climate change is dominated by human influences that are sufficiently large that they exceed the bounds of natural variability. Moreover, these changes are likely to continue for many centuries. The effects of this venture into the unknown with climate could be very disruptive not only to humans but to many other life-forms as well. The latter portion of this chapter examines the ways in which humans may be changing global climate.

20.1 Climate & Geology

List the major parts of the climate system and some connections between climate and geology.

The term **weather** refers to the state of the atmosphere at a given time and place. Changes in weather are frequent and sometimes seemingly erratic. In contrast, **climate** is a description of aggregate weather conditions, based on observations over many decades. Climate is often defined simply as "average weather," but this definition is inadequate because variations and extremes are also important parts of a climate description.

The Climate System

Throughout this book you have frequently been reminded that Earth is a complex system that consists of many interacting parts. A change in any one part can produce changes in any or all of the other parts—often in ways that are neither obvious nor immediately apparent. Key to understanding climate change and its causes is the fact that climate is related to all parts of the Earth system. We must recognize that there is a **climate system** that derives its energy from the Sun and includes the atmosphere, hydrosphere, geosphere, biosphere, and cryosphere. The first four were discussed in Chapter 1; the **cryosphere** refers to the portion of Earth's surface where water is in solid form. This includes snow, glaciers, sea ice, freshwater ice, and frozen ground (*permafrost*). The climate system *involves the exchanges of energy and moisture that occur among the five spheres*. These exchanges link the atmosphere to the other spheres so that the whole functions as an extremely complex interactive unit. Changes in the climate system do not occur in isolation. Rather, when one part of the climate system changes, the other components react. The major components of the climate system are shown in **Figure 20.1**.

The climate system provides a framework for the study of climate. The interactions and exchanges among the parts of the climate system create a complex network that links the five spheres. As you will see, because the climate system involves all of Earth's spheres, data from many sources are used to study and decipher climate change.

Climate–Geology Connections

Climate has a profound impact on many geologic processes. When climate changes, these processes respond. A glance back at the rock cycle in Chapter 1 reminds us about many of the connections. Of course, rock weathering has an obvious climate connection, as do processes that are associated with arid, tropical, and glacial landscapes. Phenomena such as debris flows and river flooding are often triggered by atmospheric events such as periods of extraordinary rainfall. Clearly, the atmosphere is a basic link in the hydrologic cycle. Other climate–geology connections involve the impact of internal processes on the atmosphere. For example, the particles and gases emitted by volcanoes can change the composition of the atmosphere, and mountain building can have a significant impact on regional temperature, precipitation, and wind patterns.

The study of sediments, sedimentary rocks, and fossils clearly demonstrates that, through the ages, practically every place on our planet has experienced wide swings in climate, such as from ice ages to conditions associated with subtropical coal swamps or desert dunes. Time scales for climate change vary from decades to millions of years.

CONCEPT CHECKS 20.1

1. Distinguish between *weather* and *climate*.
2. What are the five major parts of the climate system?
3. List at least five connections between climate and geology.

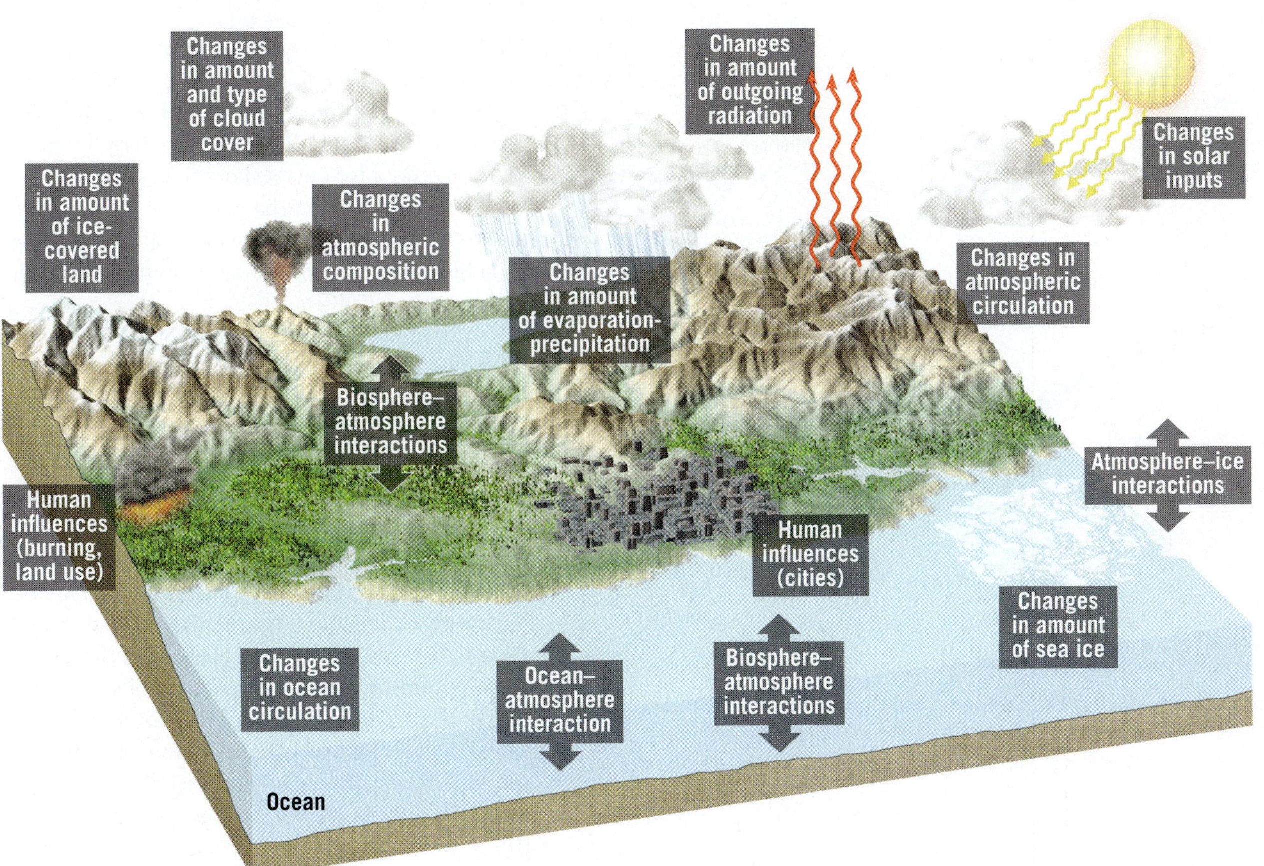

SmartFigure 20.1
Earth's climate system Schematic view showing several components of Earth's climate system. Many interactions occur among the various components on a wide range of space and time scales, making the system extremely complex.

VIDEO
http://goo.gl/La1sq3

20.2 Detecting Climate Change

Explain why unraveling past climate changes is important and discuss several ways in which such changes are detected.

Climate not only varies from place to place, it is also naturally variable over time. During the great expanse of Earth history, and long before humans were roaming the planet, there have been many shifts—from warm to cold and from wet to dry and back again.

Climates Change

By using fossils and many other geologic clues, scientists have reconstructed Earth's climate going back hundreds of millions of years. Chapter 19, which surveys Earth's history, also discusses the geologic evidence for natural climate variability.

Over long time scales (tens to hundreds of millions of years), Earth's climate can be broadly characterized as being a warm "greenhouse" or a cold "icehouse." During greenhouse times, there is little, if any, permanent ice at either pole, and relatively warm temperate climates are found even at high latitudes. During icehouse conditions, global climate is cool enough to support ice sheets at one or both poles. Earth's climate has gradually transitioned between these two categories only a few times over the past 542 million years. During this span, known as the Phanerozoic ("visible life") eon, the environment and climate are relatively well documented because fossils are abundant.

The most recent transition occurred during the Cenozoic era. The early Cenozoic was a time of greenhouse climates, similar to those the dinosaurs experienced during the preceding Mesozoic era. By about 34 million years ago, permanent ice sheets were present at the South Pole, ushering in icehouse conditions (Figure 20.2). Climate warmed during the Miocene epoch about (20 million years ago) as mammal populations reached their greatest diversity. Climate then cooled. In North America, the lush "greenhouse" forests (there were palm trees in Wyoming and banana plants in Oregon) were replaced by open grasslands. Grassland ecosystems are better suited for a cooler, drier "icehouse" climate. By 2.6 million years ago (the start of the Quaternary epoch), Earth's climate was cold enough to support vast ice sheets at both poles. In the Northern Hemisphere, ice advanced nearly as far south as the present-day Ohio River

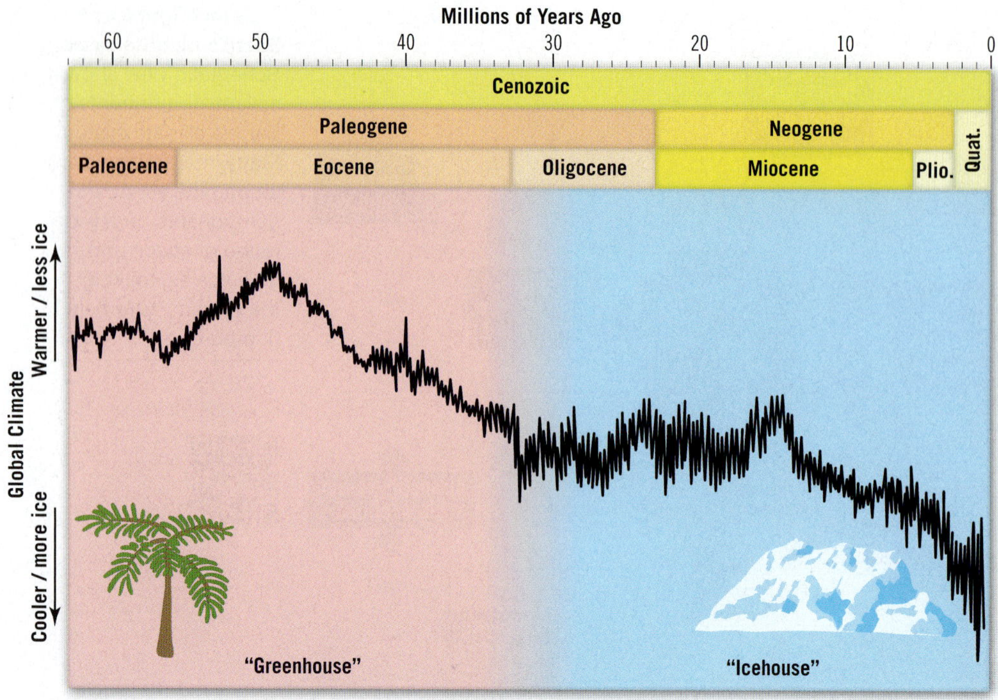

▲ **Figure 20.2 Relative climate change during the Cenozoic era** During the past 65 million years, Earth's climate shifted from being a warm "greenhouse" to being a cool "icehouse." Climate is not stable when viewed over long time spans. Earth has experienced several back-and-forth shifts between warm and cold.

and then subsequently retreated to Greenland. For the past 800,000 years, this cycle of ice advance and retreat has occurred about every 100,000 years. The last major ice sheet advance reached a maximum about 18,000 years ago.

How do we know about these changes? What are the causes? The next sections take a look at how scientists decipher Earth's climate history. Later we will explore some significant natural causes of climate change.

▼ **Figure 20.3 Ancient bristlecone pines** Some of these trees in California's White Mountains are more than 4000 years old. The study of tree growth rings is one way that scientists reconstruct past climates. (Photo by Stock Connection Blue/Alamy)

Proxy Data

High-technology and precision instrumentation are now available to study the composition and dynamics of the atmosphere. But such tools are recent inventions and therefore have been providing data for only a short time span. To understand fully the behavior of the atmosphere and to anticipate future climate change, we must somehow discover how climate has changed over broad expanses of time.

Instrumental records go back only a couple centuries, at best, and the further back we go, the less complete and more unreliable the data become. To overcome this lack of direct measurements, scientists must decipher and reconstruct past climates by using indirect evidence. **Proxy data** come from natural recorders of climate variability, such as seafloor sediments, glacial ice, fossil pollen, and tree growth rings, as well as from historical documents (Figure 20.3). Scientists who analyze proxy data and reconstruct past climates are engaged in the study of **paleoclimatology**. The main goal of such work is to understand the climate of the past in order to assess the current and potential future climate in the context of natural climate variability. In the following discussion, we will briefly examine some of the important sources of proxy data.

Seafloor Sediment: A Storehouse of Climate Data

We know that the parts of the Earth system are linked so that a change in one part can produce changes in any or all of the other parts. In this section, you will see how changes in atmospheric and oceanic temperatures are reflected in the nature of life in the sea.

Most seafloor sediments contain the remains of organisms that once lived near the sea surface (the ocean–atmosphere interface). When such near-surface organisms die, their shells slowly settle to the floor of the ocean, where they become part of the sedimentary record (Figure 20.4). These seafloor sediments are useful recorders of worldwide climate change because the numbers and types of organisms living near the sea surface change with the climate. For this reason, scientists are tapping the huge reservoir of data in seafloor sediments. The sediment cores gathered by drilling ships and other research vessels have provided invaluable data that have significantly expanded our knowledge and understanding of past climates.

One notable example of how seafloor sediments add to our understanding of climate change relates to unraveling the fluctuating atmospheric conditions of the Quaternary Ice Age. The records of temperature changes contained in cores of sediment from the ocean floor have proven critical to our present understanding of this recent span of Earth history. There is more about this topic in the section "Causes of Ice Ages" in Chapter 15.

Oxygen Isotope Analysis

The isotopes of oxygen in water molecules or in the shells of marine organisms are an important source of proxy data on past climate conditions. **Oxygen isotope analysis** is based on precise measurement of the ratio between two isotopes of oxygen: ^{16}O, which is the most common, and the heavier ^{18}O. A molecule of H_2O can form from either ^{16}O or ^{18}O, but water molecules made with the lighter isotope, ^{16}O, evaporate more readily from the oceans. Because of this, precipitation (and hence the glacial ice that it may form) is enriched in ^{16}O. This leaves a greater concentration of the heavier isotope, ^{18}O, in the ocean water. Thus, during periods when glaciers are extensive, more of the lighter ^{16}O is tied up in ice, so the concentration of ^{18}O in seawater increases. Conversely, during warmer interglacial periods, when the amount of glacial ice decreases dramatically, more ^{16}O is returned to the sea, so the proportion of ^{18}O relative to ^{16}O in ocean water also drops. Now, if we had some ancient recording of the changes of the $^{18}O/^{16}O$ ratio, we could determine when there were glacial periods and, therefore, when the climate grew cooler.

Fortunately, we do have such a recording. Certain marine microorganisms secrete shells of calcium carbonate ($CaCO_3$), and the prevailing oceanic $^{18}O/^{16}O$ ratio is reflected in the composition of these hard parts. When the organisms die, their hard parts settle to the ocean floor, becoming part of the sediment layers there. Consequently, periods of glacial activity can be determined from variations in the oxygen isotope ratio found in shells of certain microorganisms buried in deep-sea sediments. A higher ratio of ^{18}O to ^{16}O in shells indicates a time when ice sheets were growing larger.

The $^{18}O/^{16}O$ ratio also varies with temperature. More ^{18}O is evaporated from the oceans when temperatures are high, and less is evaporated when temperatures are low. Therefore, the heavy isotope is more abundant in the precipitation of warm eras and less abundant during colder periods. Using this principle, scientists studying the layers of ice and snow in glaciers have been able to determine past temperature changes.

◀ **Figure 20.4**
Foraminifera These single-celled amoeba-like organisms, also called *forams*, are extremely abundant and found throughout the world's oceans. Although the foram record in ocean sediment goes back farther, the remains of these microscopic organisms are most commonly used to study climate change during the Cenozoic era. The chemical composition of their hard parts depends on water temperature and the presence or absence of large ice sheets. Because of this relationship, scientists analyze foram shells to estimate ocean temperatures and the existence of ice sheets. (Photo by Biophoto Associates/ Science Source)

Climate Change Recorded in Glacial Ice

Ice cores are an indispensable source of data for reconstructing past climates. Research based on vertical cores taken from the Greenland and Antarctic Ice Sheets has changed our basic understanding of how the climate system works.

Scientists collect samples by using a drilling rig, like a small version of an oil drill. A hollow shaft follows the drill head into the ice, and an ice core is extracted. In this way, cores that sometimes exceed 3000 meters (9800 feet) in length and may represent more than 700,000 years of climate history are acquired for study (**Figure 20.5A**).

▼ **SmartFigure 20.5 Ice cores: Important sources of climate data A.** The National Ice Core Laboratory is a physical plant for storing and studying cores of ice taken from glaciers around the world. These cores represent a long-term record of material deposited from the atmosphere. The lab enables scientists to conduct examinations of ice cores, and it preserves the integrity of these samples in a repository for the study of global climate change and past environmental conditions. (Photo by USGS/National Ice Core Laboratory) **B.** This graph, showing temperature variations over the past 40,000 years, is derived from oxygen isotope analysis of ice cores recovered from the Greenland Ice Sheet. (Based on U.S. Geological Survey)

TUTORIAL
http://goo.gl/QHehvC

A.

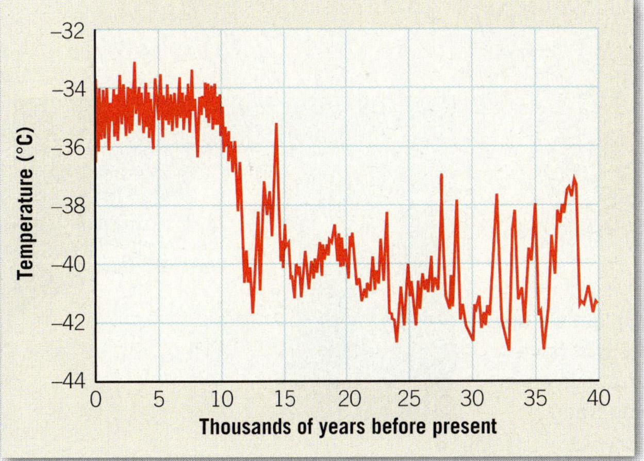

B.

Did You Know?
Bristlecone pines are remarkable for their great age and ability to survive adverse growing conditions. Some of the best examples occur in California's White Mountains, where they grow in scattered groves at elevations between 5600 and 11,200 ft. In 2012 one of the trees there was determined to be 5062 years old! Because they are among the oldest living things on the planet, Bristlecone pines are especially valuable in tree-ring studies.

Scientists are able to produce a record of changing air temperatures and snowfall by means of the *oxygen isotope analysis* described above. Using this technique, scientists are able to produce a record of past temperature changes. A portion of such a record is shown in **Figure 20.5B**.

Air bubbles trapped in the ice also record variations in atmospheric composition. Changes in carbon dioxide and methane are linked to fluctuating temperatures. The cores also include atmospheric fallout, such as wind-blown dust, volcanic ash, pollen, and modern-day pollution.

Tree Rings: Archives of Environmental History

If you look at the end of a log, you will see that it is composed of a series of concentric rings (**Figure 20.6A**). Tree rings can be a very useful source of proxy data on past climates. Every year, in temperate regions, trees add a layer of new wood under the bark. Characteristics of each tree ring, such as thickness and density, reflect the environmental conditions (especially climate) that prevailed during the year when the ring formed. Favorable growth conditions produce a wide ring; unfavorable ones produce a narrow ring. Trees growing at the same time in the same region show similar tree-ring patterns.

Because a single growth ring is usually added each year, the age of the tree when it was cut can be determined by counting the rings. If the year of cutting is known, the age of the tree and the year in which each ring formed can be determined by counting back from the outside ring. Scientists are not limited to working with trees that have been cut down. Small, nondestructive core samples can be taken from living trees (**Figure 20.6B**).

To make the most effective use of tree rings, extended patterns known as *ring chronologies* are established. They are produced by comparing the patterns of rings among trees in an area. If the same pattern can be identified in two samples, one of which has been dated, the second sample can be dated from the first by matching the ring patterns common to both. Tree-ring chronologies extending back thousands of years have been established for some regions. To date a timber sample of unknown age, its ring pattern is matched against the reference chronology.

Tree-ring chronologies are unique archives of environmental history and have important applications in such disciplines as climate, geology, ecology, and archaeology. For example, tree rings are used to reconstruct climate variations within a region for spans of thousands of years prior to human historical records. Knowing such long-term variations is of great value when interpreting the recent record of climate change.

Figure 20.6
Tree rings A. Each year a growing tree produces a layer of new cells beneath the bark. If the tree is cut down and the trunk is examined, each year's growth can be seen as a ring. These rings are useful records of past climate because the amount of growth (the thickness of a ring) depends on precipitation and temperature. (Photo by Victor Zastolskiy/Fotolia) **B.** Scientists are not limited to working with trees that have been cut down. Small, nondestructive core samples can be taken from living trees. (Photo by Gregory K. Scott/Science Source)

Other Types of Proxy Data

In addition to the sources already discussed, other sources of proxy data that are used to gain insight into past climates include fossil pollen, corals, and historical documents.

Fossil Pollen Climate is a major factor influencing the distribution of vegetation, and the nature of the plant community occupying an area is a reflection of the climate. Pollen and spores are parts of the life cycles of many plants, and because they have very resistant walls, they are often the most abundant, easily identifiable, and best-preserved plant remains in sediments (**Figure 20.7**). By analyzing pollen from accurately dated sediments, scientists can obtain high-resolution records of vegetation changes in an area. Past climates can be reconstructed from such information.

Corals Coral reefs consist of colonies of corals, invertebrates that live in warm, shallow waters and form atop the hard material left behind by past corals. Corals build their hard skeletons from calcium carbonate ($CaCO_3$) extracted from seawater. The carbonate contains isotopes of oxygen that can be used to determine the temperature of the water in which the coral grew. The portion of the skeleton that forms in winter has a density that is different from that formed in summer because of variations in growth rates related to temperature and other environmental factors. Thus, corals exhibit seasonal growth bands very much like those observed in trees. The accuracy and reliability of the climate data extracted from corals has been established by comparing recent instrumental records to coral records for the same period. Oxygen isotope analysis of coral growth rings can also serve as a proxy measurement for precipitation, particularly in areas where large variations in annual rainfall occur.

▲ **Figure 20.7 Pollen** This false-color image from an electron microscope shows an assortment of pollen grains. Note how the size, shape, and surface characteristics differ from one species to another. Analysis of the types and abundance of pollen in lake sediments and peat deposits provides information about how climate has changed over time. (Photo by David AMI Images/Science Source)

▲ **Figure 20.8 Harvest dates as climate clues** Historical records can sometimes be helpful in the analysis of past climates. The date for the beginning of the grape harvest in the fall is an integrated measure of temperature and precipitation during the growing season. These dates have been recorded for centuries in Europe and provide a useful record of year-to-year climate variations. (Photo by SGM/AGE Fotostock)

Historical Documents Historical documents sometimes contain helpful information. Although it may seem that such records should readily lend themselves to climate analysis, that is not the case. Most manuscripts were written for purposes other than climate description. Furthermore, writers understandably neglected periods of relatively stable atmospheric conditions and mentioned only droughts, severe storms, memorable blizzards, and other extremes. Nevertheless, records of crops, floods, and the migration of people have furnished useful evidence of the possible influences of changing climate (Figure 20.8).

CONCEPT CHECKS 20.2

1. What are proxy data, and why are they necessary in the study of climate change?
2. Why are seafloor sediments useful in the study of past climates? Aside from seafloor sediments, list four sources of proxy climate data.
3. Explain how past temperatures are determined using oxygen isotope analysis.

20.3 Some Atmospheric Basics

Describe the composition of the atmosphere and the atmosphere's vertical changes in pressure and temperature.

To better understand climate change, it is helpful to possess some basic knowledge about the composition and structure of the atmosphere.

Composition of the Atmosphere

Air is *not* a unique element or compound. Rather, air is a *mixture* of many discrete gases, each with its own physical properties, in which varying quantities of tiny solid and liquid particles are suspended.

Clean, Dry Air As you can see in Figure 20.9, clean, dry air is composed almost entirely of two gases—78 percent nitrogen and 21 percent oxygen. Although these gases are the most plentiful components of air and are of great significance to life on Earth, they have little or no effect on weather phenomena. The remaining 1 percent of dry air is mostly the inert gas argon (0.93 percent) plus tiny quantities of a number of other gases. Carbon dioxide, although present in only minute amounts (0.0400 percent, or 400 parts per million), is nevertheless an important constituent of air because it has the ability to absorb heat energy radiated by Earth and thus influences the heating of the atmosphere.

Air includes many gases and particles that vary significantly from time to time and from place to place. Important examples include water vapor, ozone, and tiny solid and liquid particles.

▶ **SmartFigure 20.9**
Composition of the atmosphere Proportional volume of gases composing dry air. Nitrogen and oxygen obviously dominate.

TUTORIAL
http://goo.gl/EJEV5u

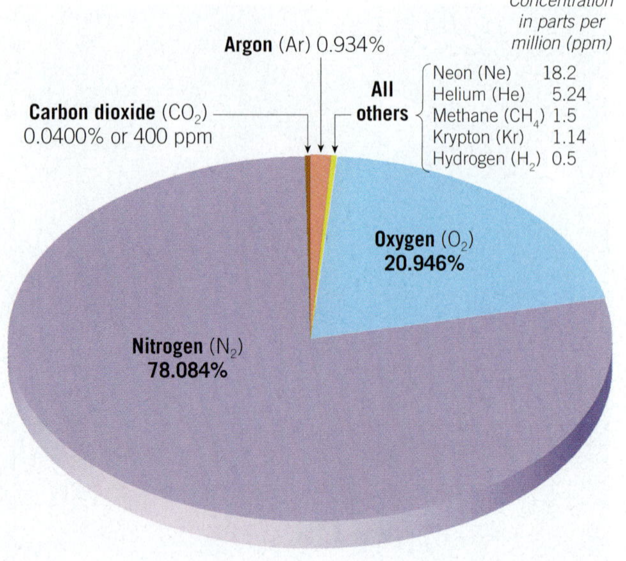

Water Vapor The amount of *water vapor* in the air varies considerably, from practically none at all up to about 4 percent by volume. Why is such a small fraction of the atmosphere so significant? Certainly the fact that water vapor is the source of all clouds and precipitation is enough to explain its importance. However, water vapor has other roles. Like carbon dioxide, it has the ability to absorb heat energy given off by Earth as well as some solar energy. It is, therefore, important when we examine the heating of the atmosphere.

▶ **SmartFigure 20.10**
Aerosols This satellite image shows two examples of aerosols. First, a dust storm is blowing across northeastern China toward the Korean Peninsula. Second, a dense haze toward the south (bottom center) is human-generated air pollution. (NASA)

VIDEO
http://goo.gl/t8anlg

Ozone Another important component of the atmosphere is *ozone*. It is a form of oxygen that combines three oxygen atoms into each molecule (O_3). Ozone is not the same as the oxygen we breathe, which has two atoms per molecule (O_2). There is very little ozone in the atmosphere, and its distribution is not uniform. It is concentrated well above Earth's surface, in a layer called the *stratosphere*, at an altitude of between 10 and 50 kilometers (6 and 31 miles). The presence of the ozone layer in our atmosphere is crucial to those who dwell on Earth: Ozone absorbs the potentially harmful ultraviolet (UV) radiation from the Sun. If ozone did not filter out a great deal of the ultraviolet radiation, and if the Sun's UV rays reached the surface of Earth undiminished, our planet would be uninhabitable for most life as we know it.

Aerosols The movements of the atmosphere are sufficient to keep a large quantity of solid and liquid particles suspended within it. Although visible dust sometimes clouds the sky, these relatively large particles are too heavy to stay in the air very long. Still, many particles are microscopic and remain suspended for considerable periods of time. They may originate from many sources, both natural and human made, and include sea salts from breaking waves, fine soil blown into the air, smoke and soot from fires, pollen and microorganisms lifted by the wind, ash and dust from volcanic eruptions, and more. Collectively, these tiny solid and liquid particles are called **aerosols**.

From a meteorological standpoint, these tiny, often invisible particles can be significant. First, many act as surfaces on which water vapor can condense, an important function in the formation of clouds and fog. Second, aerosols can absorb or reflect incoming solar radiation. Thus, when an air pollution episode is occurring, or when ash fills the sky following a volcanic eruption, the amount of sunlight reaching Earth's surface can be measurably reduced (**Figure 20.10**).

Extent & Structure of the Atmosphere

To say that the atmosphere begins at Earth's surface and extends upward is obvious. But where does the atmosphere end, and where does outer space begin? There is no sharp boundary; the atmosphere rapidly thins as you travel away from Earth, until there are too few gas molecules to detect.

Pressure Changes with Height To understand the vertical extent of the atmosphere, let us examine changes in atmospheric pressure with height. Atmospheric pressure is simply the weight of the air above. At sea level, the average pressure is slightly more than 1000 millibars. This corresponds to a weight of slightly more than 1 kilogram per square centimeter (14.7 pounds per square inch). Obviously, the pressure at higher altitudes is less (**Figure 20.11**).

One-half of the atmosphere lies below an altitude of 5.6 kilometers (3.5 miles). At about 16 kilometers (10 miles), 90 percent of the atmosphere has been traversed, and

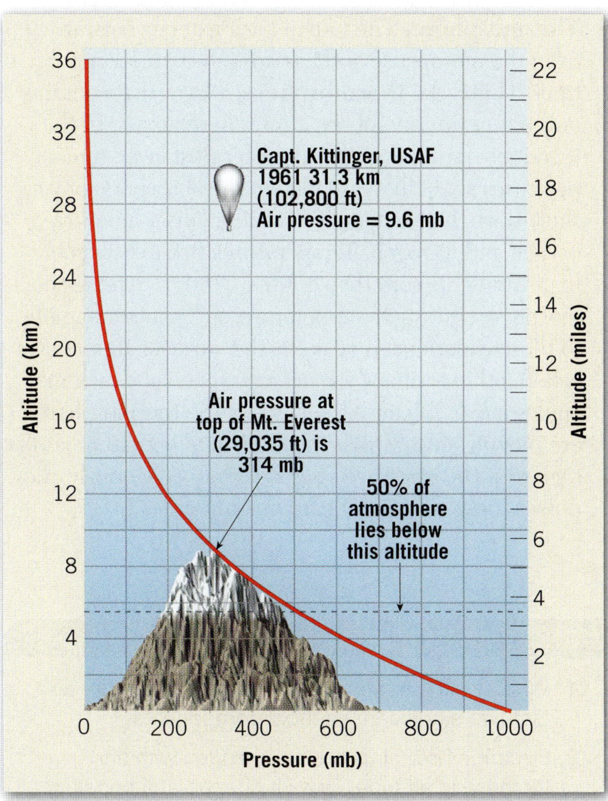

▲ **Figure 20.11 Vertical changes in air pressure** Pressure decreases rapidly near Earth's surface and more gradually at greater heights.

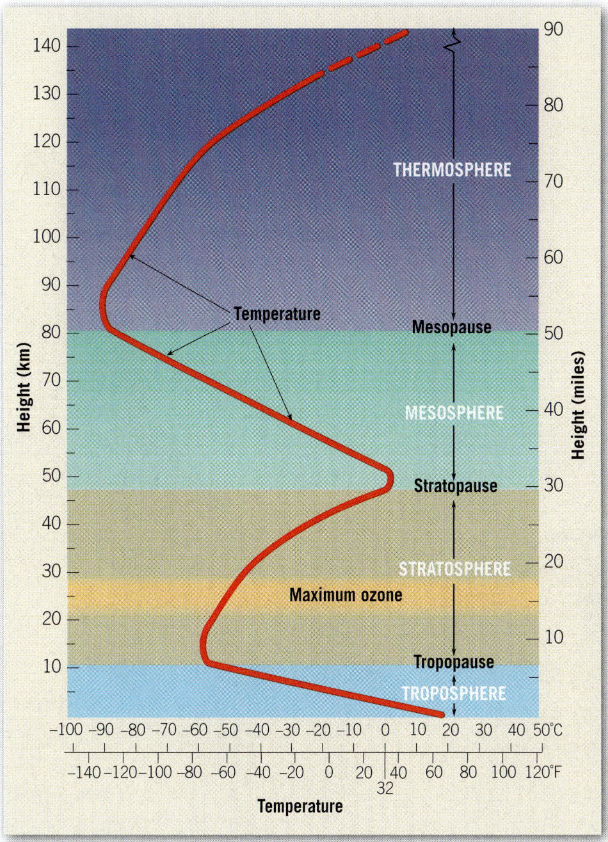

▲ **Figure 20.12 Thermal structure of the atmosphere**

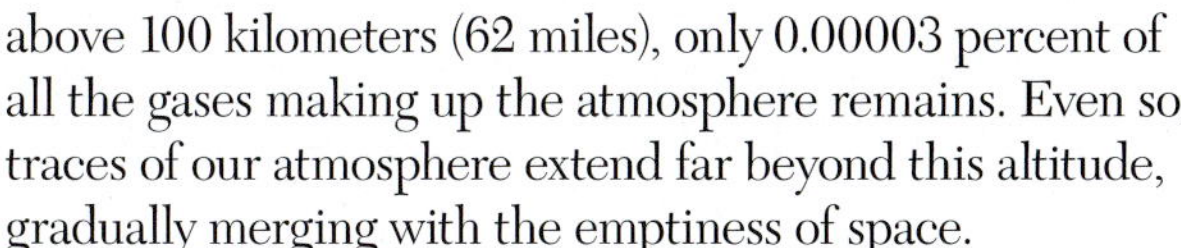

above 100 kilometers (62 miles), only 0.00003 percent of all the gases making up the atmosphere remains. Even so, traces of our atmosphere extend far beyond this altitude, gradually merging with the emptiness of space.

Temperature Changes In addition to vertical changes in air pressure, there are also changes in air temperature as we ascend through the atmosphere. Earth's atmosphere is divided vertically into four layers, on the basis of temperature (**Figure 20.12**):

- **Troposphere.** The bottom layer, in which we live, is characterized by a decrease in temperature with increasing altitude and is called the **troposphere**. The term literally means the region where air "turns over," a reference to the appreciable vertical mixing of air in this lowermost zone. The troposphere is the chief focus of meteorologists because it is in this layer that essentially all important weather phenomena occur. The temperature decrease in the troposphere is called the *environmental lapse rate*. Its average value is 6.5°C per kilometer (3.5°F per 1000 feet), a figure known as the *normal lapse rate*. It should be emphasized, however, that the environmental lapse rate is not a constant; it varies over time and from one place to another and hence must be regularly measured. To determine the actual environmental lapse rate in a particular place and time, as well as to gather information about vertical changes in pressure, wind, and humidity, radiosondes are used. A **radiosonde** is an instrument package that is attached to a weather balloon and transmits data by radio as it ascends through the atmosphere (**Figure 20.13**). The thickness of the troposphere is not the same everywhere; it varies with latitude and the season. On average, the drop in temperature that defines the troposphere continues to a height of about 12 kilometers (7.4 miles). The outer boundary of the troposphere is the *tropopause*.
- **Stratosphere.** Beyond the tropopause is the **stratosphere**. In the stratosphere, the temperature remains constant to a height of about 20 kilometers (12 miles) and then begins a gradual increase that continues until the *stratopause*, at a height of nearly 50 kilometers (30 miles) above Earth's surface. Below the tropopause, atmospheric properties such as temperature and humidity are readily transferred by large-scale turbulence and mixing. Above the tropopause, in the stratosphere, they are not. Temperatures increase in the stratosphere because it is in this layer that the atmosphere's ozone is concentrated. Recall that ozone absorbs ultraviolet radiation from the Sun. As a consequence, the stratosphere is heated.
- **Mesosphere.** In the third layer, the **mesosphere**, temperatures again decrease with height until, at the *mesopause*, more than 80 kilometers (50 miles) above the surface, the temperature approaches −90°C (−130°F). The coldest temperatures anywhere in the atmosphere occur at the mesopause.

Did You Know?

Although naturally occurring ozone in the stratosphere is critical to life on Earth, it is regarded as a pollutant when produced at ground level because it can damage vegetation and be harmful to human health. Ozone is a major component in a noxious mixture of gases and particles called *photochemical smog*. It forms as a result of reactions that are triggered by sunlight and occur among pollutants emitted by motor vehicles and industrial sources.

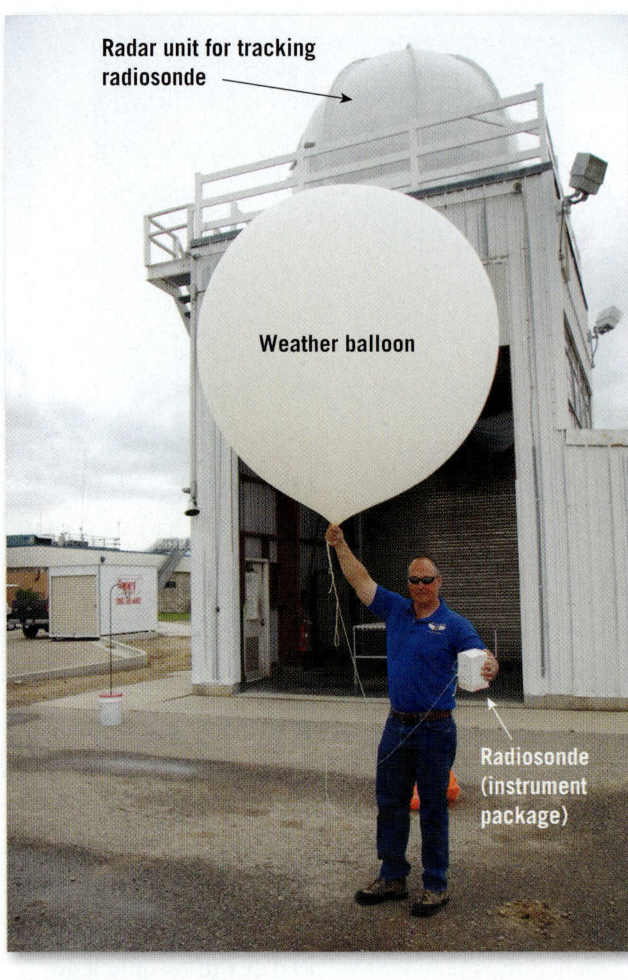

Figure 20.13
Radiosonde A radiosonde is a lightweight package of instruments that is carried aloft by a small weather balloon. It transmits data on vertical changes in temperature, pressure, and humidity in the troposphere. The troposphere is where practically all weather phenomena occur; therefore, it is very important to have frequent measurements. (Photo by David R. Frazier Photolibrary, Inc./Alamy Stock)

- **Thermosphere.** The fourth layer extends outward from the mesopause and has no well-defined upper limit. This is the **thermosphere**, a layer that contains only a *tiny fraction* of the atmosphere's mass. In the extremely rarefied air of this outermost layer, temperatures again increase, due to the absorption of very short-wave, high-energy solar radiation by atoms of oxygen and nitrogen. Temperatures rise to extremely high values of more than 1000°C (1800°F) in the thermosphere. But such temperatures are not comparable to those experienced near Earth's surface. Temperature is defined in terms of the average speed at which molecules move. Because the gases of the thermosphere are moving at very high speeds, the temperature is very high. But the gases are so sparse that, collectively, they possess only an insignificant amount of heat.

CONCEPT CHECKS 20.3

1. What are the major components of clean, dry air? List two significant variable components.
2. Describe how air pressure changes with an increase in altitude. Does it change at a constant rate?
3. The atmosphere is divided vertically into four layers, on the basis of temperature. Name the layers from bottom to top and indicate how temperatures change in each.

20.4 Heating the Atmosphere

Outline the basic processes involved in heating the atmosphere.

Nearly all the energy that drives Earth's variable weather and climate comes from the Sun. Before we can adequately describe how Earth's atmosphere is heated, it is helpful to know something about solar energy and what happens to this energy once it is intercepted by Earth.

Energy from the Sun

From our everyday experience, we know that the Sun emits light and heat as well as the ultraviolet rays that cause suntan. Although these forms of energy comprise a major portion of the total energy that radiates from the Sun, they are only part of a large array of energy called *radiation*, or *electromagnetic radiation*. This array, or spectrum, of electromagnetic energy is shown in Figure 20.14. All radiation—whether x-rays, microwaves, or radio waves—transmits energy through the vacuum of space at 300,000 kilometers (186,000 miles) per second and only slightly slower through our atmosphere. When an object absorbs any form of radiant energy, the result is an increase in molecular motion, which causes a corresponding increase in temperature.

To better understand how the atmosphere is heated, it is useful to have a general understanding of the basic laws governing radiation:

- **All objects, at whatever temperature, emit radiant energy.** Not only hot objects like the Sun but also Earth, including its polar ice caps, continually emit energy.
- **Hotter objects radiate more total energy per unit area than do colder objects.**
- **The hotter the radiating body, the shorter the wavelength of maximum radiation.** The Sun, with a surface temperature of about 5700°C, radiates maximum energy at 0.5 micrometer, which is in the visible range. The maximum radiation for Earth occurs at a wavelength of 10 micrometers, well within the infrared (heat) range. Because the wavelength for maximum Earth radiation is roughly 20 times longer

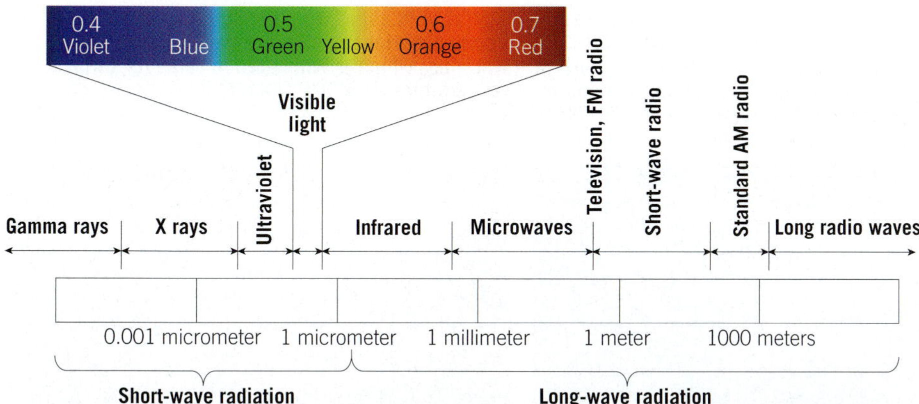

▲ SmartFigure 20.14 **The electromagnetic spectrum** This diagram illustrates the wavelengths and names of various types of radiation. Visible light consists of an array of colors we commonly call the "colors of the rainbow." (Photo by Dennis Tasa)

VIDEO
http://goo.gl/6xl4sR

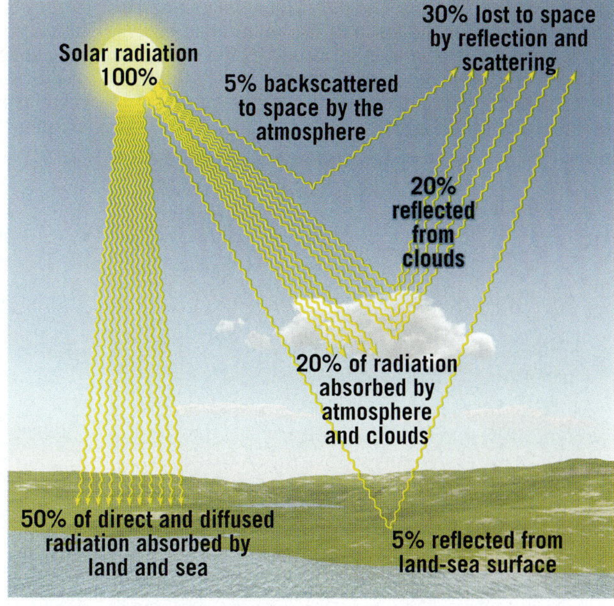

▲ SmartFigure 20.15 **Paths taken by solar radiation** This diagram shows the average distribution of incoming solar radiation by percentage. More solar radiation is absorbed by Earth's surface than by the atmosphere.

TUTORIAL
http://goo.gl/IqkUlj

than the maximum solar radiation, Earth radiation is often called *long-wave radiation*, and solar radiation is called *short-wave radiation*.

- **Objects that are good absorbers of radiation are good emitters as well.** Earth's surface and the Sun approach being perfect radiators because they absorb and radiate with nearly 100 percent efficiency for their respective temperatures. On the other hand, *gases are selective absorbers and emitters of radiation*. For some wavelengths, the atmosphere is nearly transparent (that is, little radiation is absorbed). For other wavelengths, however, the atmosphere is nearly opaque (that is, it is a good absorber). Experience tells us that the atmosphere is transparent to visible light; hence, these wavelengths readily reach Earth's surface. This is not the case for the longer-wavelength radiation emitted by Earth.

The Paths of Incoming Solar Energy

Figure 20.15 shows the paths taken by incoming solar radiation averaged for the entire globe. Notice that the atmosphere is quite transparent to incoming solar radiation. On average, about 50 percent of the solar energy reaching the top of the atmosphere passes through the atmosphere and is absorbed at Earth's surface. Another 20 percent is absorbed directly by clouds and certain atmospheric gases (including oxygen and ozone) before reaching the surface. The remaining 30 percent is reflected back to space by the atmosphere, clouds, and reflective surfaces such as snow and ice. The fraction of the total radiation that is reflected by a surface is called its **albedo** (Figure 20.16). Thus, the albedo for Earth as a whole (the *planetary albedo*) is 30 percent.

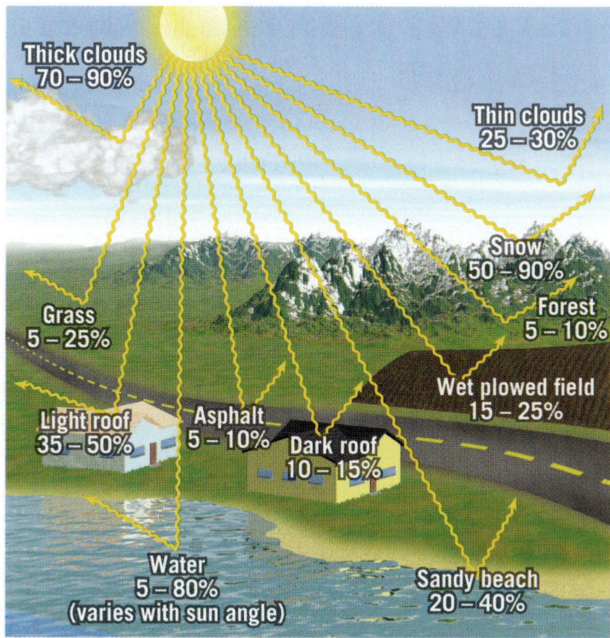

▲ Figure 20.16 **Albedo (reflectivity) of various surfaces** In general, light-colored surfaces tend to be more reflective than dark-colored surfaces and thus have higher albedos.

Airless bodies like the Moon All incoming solar radiation reaches the surface. Some is reflected back to space. The rest is absorbed by the surface and radiated directly back to space. As a result the lunar surface has a much lower average surface temperature than Earth.

All outgoing long-wave energy is reradiated directly back to space

Incoming short-wave solar radiation

David Cole/Alamy

Bodies with modest amounts of greenhouse gases like Earth The atmosphere absorbs some of the longwave radiation emitted by the surface. A portion of this energy is radiated back to the surface and is responsible for keeping Earth's surface 33°C (59°F) warmer than it would otherwise be.

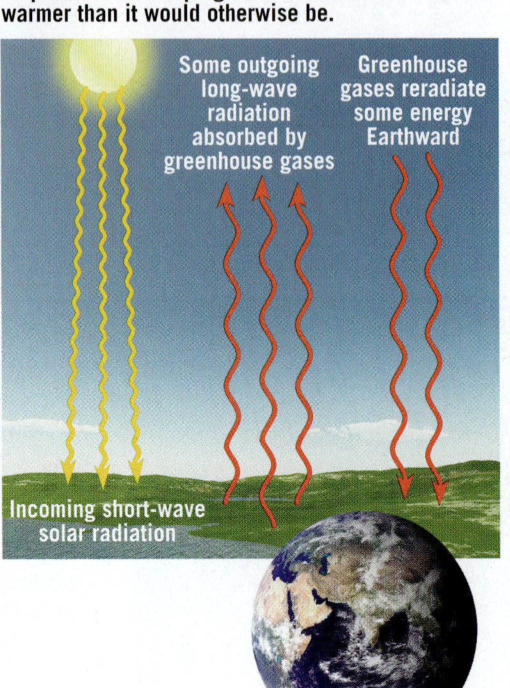
Some outgoing long-wave radiation absorbed by greenhouse gases
Greenhouse gases reradiate some energy Earthward
Incoming short-wave solar radiation
NASA

Bodies with abundant greenhouse gases like Venus Venus experiences extraordinary greenhouse warming, which is estimated to raise its surface temperature by 523°C (941°F).

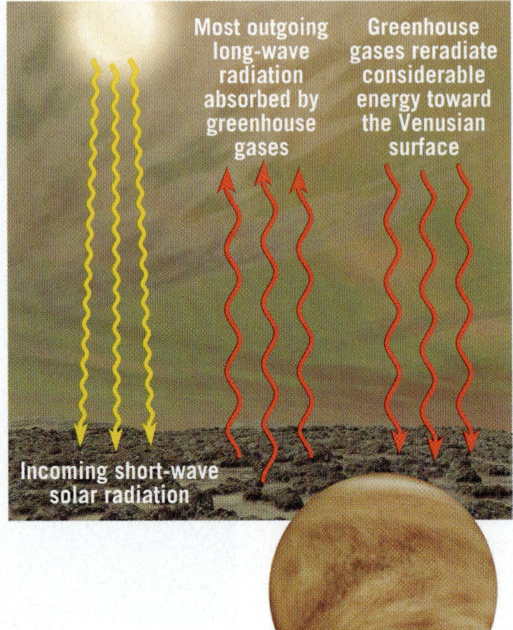
Most outgoing long-wave radiation absorbed by greenhouse gases
Greenhouse gases reradiate considerable energy toward the Venusian surface
Incoming short-wave solar radiation
NASA

▲ SmartFigure 20.17
The greenhouse effect The greenhouse effect of Earth compared with two of our close solar system neighbors.

TUTORIAL
http://goo.gl/2wxon2

What determines whether solar radiation will be transmitted to the surface, scattered, or reflected outward? It depends greatly on the wavelength of the energy being transmitted, as well as on the nature of the intervening material.

The numbers shown in Figure 20.15 represent global averages. The actual percentages at different times and locations vary greatly, primarily due to changes in the percentage of light reflected and scattered back to space. For example, if the sky is overcast, a higher percentage of light is reflected back to space than when the sky is clear.

Heating the Atmosphere: The Greenhouse Effect

If Earth had no atmosphere, it would experience an average surface temperature far below freezing. But the atmosphere warms the planet and makes Earth livable. The extremely important role the atmosphere plays in heating Earth's surface has been named the **greenhouse effect**.

As discussed earlier, cloudless air is largely transparent to incoming short-wave solar radiation and, hence, transmits it to Earth's surface. By contrast, a significant fraction of the long-wave radiation emitted by Earth's land–sea surface is absorbed by water vapor, carbon dioxide, and other trace gases in the atmosphere. This energy heats the air and increases the rate at which it radiates energy, both out to space and back toward Earth's surface. Without this complicated game of "pass the hot potato," Earth's average temperature would be −18°C (−0.4°F), rather than the current average temperature of 15°C (59°F) (Figure 20.17). These absorptive gases in our atmosphere make Earth habitable for humans and other lifeforms. As you will see in the sections that follow, changes in the air's composition, both natural and human caused, impact the greenhouse effect in ways that can cause the atmosphere to either become warmer or cooler.

The greenhouse effect was so named because greenhouses are warmed in part by a similar phenomenon. The glass in a greenhouse allows short-wave solar radiation to enter and be absorbed by the objects inside. These objects radiate energy at longer wavelengths, to which glass is nearly opaque. The heat, therefore, is "trapped" in the greenhouse. In reality, the *major* reason air in a greenhouse warms is because the glass restricts exchange with cooler air outside. Nevertheless, the term *greenhouse effect* remains.

CONCEPT CHECKS 20.4

1. What are the three paths that incoming solar radiation takes? What might cause the percentage taking each path to vary?
2. Explain why the atmosphere is heated chiefly by radiation from Earth's surface.
3. Prepare a sketch with labels that explains the greenhouse effect.

20.5 Natural Causes of Climate Change

Discuss hypotheses that relate to natural causes of climate change.

A great variety of hypotheses have been proposed to explain climate change. Several have gained wide support, only to lose it and then sometimes to regain it. Some explanations are controversial. This is to be expected because planetary atmospheric processes are so large scale and complex that they cannot be reproduced physically in laboratory experiments. Rather, climate and its changes must be simulated mathematically (modeled), using powerful computers.

In this section we examine several current hypotheses that have earned serious consideration from the scientific community. They describe "natural" mechanisms of climatic change, causes that are unrelated to human activities. A later section examines human-induced climate changes, including the effect of rising carbon dioxide levels caused primarily by our burning of fossil fuels.

As you read this section, you will find that more than one hypothesis may explain the same change in climate. In fact, several mechanisms may interact to shift climate. Also, no single hypothesis can explain climate change on all time scales. A proposal that explains variations over millions of years generally cannot explain fluctuations over hundreds of years. If our atmosphere and its changes ever become fully understood, we will probably see that climate change is caused by many of the mechanisms discussed here, plus new ones yet to be proposed.

Plate Movements & Orbital Variations

In Chapter 15, the section "Causes of Ice Ages" describes two natural mechanisms of climate change. Recall that the movement of lithospheric plates gradually moves Earth's continents closer to or farther from the equator. Although these shifts in plates are very slow, they can have a dramatic impact on climate over spans of millions of years. Moving landmasses can also lead to significant shifts in ocean circulation, which influences heat transport around the globe.*

A second natural mechanism of climate change related to the causes of ice ages involves variations in Earth's orbit. Changes in the shape of the orbit (*eccentricity*), variations in the angle that Earth's axis makes with the plane of its orbit (*obliquity*), and the wobbling of the axis (*precession*) cause fluctuations in the seasonal and latitudinal distribution of solar radiation (see Figure 15.33, page 417). These variations, in turn, contributed to the alternating glacial–interglacial episodes of the Ice Age.

Volcanic Activity & Climate Change

The idea that explosive volcanic eruptions might alter Earth's climate was first proposed many years ago. It is still regarded as a plausible explanation for some aspects of climatic variability. Explosive eruptions emit huge quantities of gases and fine-grained debris into the atmosphere (**Figure 20.18**). The greatest eruptions are sufficiently powerful to inject material high into the atmosphere, and strong upper-air winds spread it around the globe, where it remains for many months or even years.

The Effect of Volcanic Aerosols on Climate Suspended volcanic material filters out a portion of the incoming

*For more on this, see the section "Supercontinents, Mountain Building, and Climate" in Chapter 19.

Did You Know?
A meteorite colliding with Earth could trigger climate change. For example, the most strongly supported hypothesis for the extinction of dinosaurs (about 65.5 million years ago) is related to such an event. When a large (about 6 mi in diameter) meteorite struck Earth, huge quantities of debris were blasted high into the atmosphere. For months the encircling dust cloud greatly restricted the amount of light reaching Earth's surface. Without sufficient sunlight for photosynthesis, delicate food chains collapsed. When the sunlight returned, more than half of the species on Earth, including the dinosaurs and many marine organisms, had become extinct. There is more about this in Chapter 19.

This satellite image shows the sulfur dioxide (SO_2) plume in shades of purple and black. Climate may be affected when large quantities of SO_2 are injected into the atmosphere.

This image was taken from the International Space Station and shows a plume of volcanic ash streaming southeastward from the volcano.

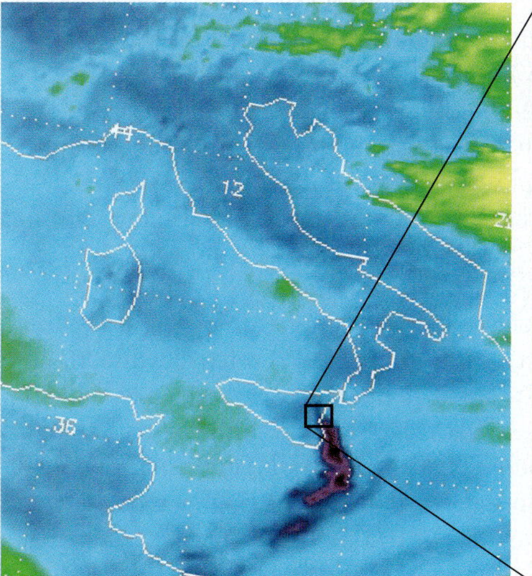

Figure 20.18 Mount Etna erupting in October 2002 This volcano on the island of Sicily is Europe's largest and most active volcano. (NASA images)

solar radiation, which in turn lowers temperatures in the troposphere. More than 200 years ago, Benjamin Franklin used this idea to argue that material from the eruption of a large Icelandic volcano could have reflected sunlight back to space and therefore might have been responsible for the unusually cold winter of 1783–1784.

Perhaps the most notable cool period linked to a volcanic event is the "year without a summer" that followed the 1815 eruption of Mount Tambora in Indonesia. The eruption of Tambora is the largest of modern times. During April 7–12, 1815, this nearly 4000-meter- (13,000-foot-) high volcano violently expelled an estimated 100 cubic kilometers (24 cubic miles) of volcanic debris. The impact of the volcanic aerosols on climate is believed to have been widespread in the Northern Hemisphere. From May through September 1816, an unprecedented series of cold spells affected the northeastern United States and adjacent portions of Canada. There was heavy snow in June and frost in July and August. Abnormal cold was also experienced in much of Western Europe. Similar, although apparently less dramatic, effects have been associated with other great explosive volcanoes, including Indonesia's Krakatoa in 1883.

Three more recent volcanic events have provided considerable data and insight regarding the impact of volcanoes on global temperatures. The eruptions of Washington State's Mount St. Helens in 1980, the Mexican volcano El Chichón in 1982, and Mount Pinatubo in the Philippines in 1991 have given scientists opportunities to study the atmospheric effects of volcanic eruptions with the aid of more sophisticated technology than was previously available. Satellite images and remote-sensing instruments allowed scientists to closely monitor the effects of the clouds of gases and ash that these volcanoes emitted.

▼ **Figure 20.19 Volcanic haze reducing sunlight at Earth's surface** The reflective haze produced by some volcanic eruptions is not volcanic ash but tiny sulfuric acid aerosols. (NASA)

A plume of white haze from Anatahan Volcano blankets a portion of the Philippine Sea in April 2005. The haze consisted of tiny droplets of sulfuric acid formed when sulfur dioxide from the volcano combined with water in the atmosphere. The plume is bright and reflects sunlight back to space.

A.

Net solar radiation at Hawaii's Mauna Loa Observatory relative to 1970 (zero on the graph). The eruptions of El Chichón and Mt. Pinatubo clearly caused temporary drops in solar radiation reaching the surface.

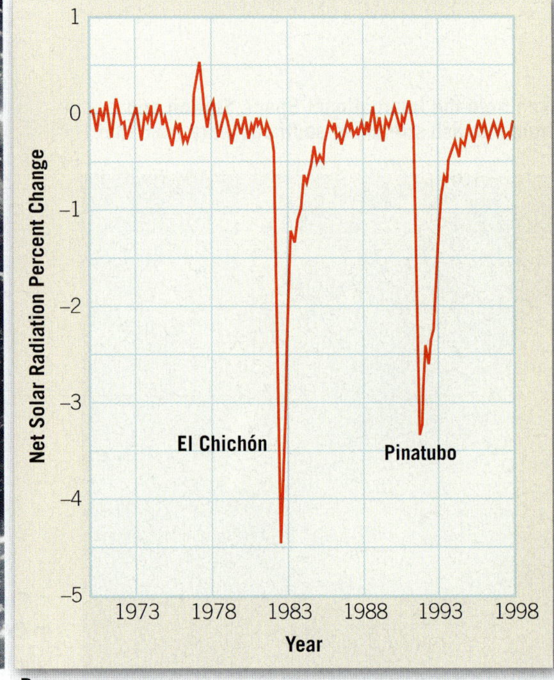

B.

Volcanic Ash & Dust When Mount St. Helens erupted, there was immediate speculation about the possible effects on climate. Could such an eruption cause our climate to change? There is no doubt that the large quantity of volcanic ash emitted by the explosive eruption had significant local and regional effects for a short period. Still, studies indicated that any longer-term lowering of hemispheric temperatures was negligible. The cooling was so slight—probably less than 0.1°C (0.2°F)—that it could not be distinguished from other natural temperature fluctuations.

Sulfuric Acid Droplets Two years of monitoring and studies following the 1982 El Chichón eruption indicated that it had a greater cooling effect on global mean temperature than Mount St. Helens—on the order of 0.3° to 0.5°C (0.5° to 0.9°F). El Chichón's eruption was *less explosive* than the Mount St. Helens blast, so why did it have a greater impact on global temperatures? The reason is that the material emitted by Mount St. Helens was largely fine ash that settled out in a relatively short time. El Chichón, on the other hand, emitted far greater quantities of sulfur dioxide gas (an estimated 40 times more) than Mount St. Helens. This gas combines with water vapor in the stratosphere to produce a dense cloud of tiny sulfuric acid particles (**Figure 20.19A**). These particles take several years to settle out completely. They lower the troposphere's mean temperature because they reflect solar radiation back to space (**Figure 20.19B**).

We now understand that volcanic clouds that remain in the stratosphere for a year or more are composed largely of sulfuric-acid droplets and not of dust, as was once thought. Thus, the volume of fine debris emitted during an explosive event is not an accurate criterion for predicting the global atmospheric effects of an eruption.

Mount Pinatubo in the Philippines erupted explosively in June 1991, injecting 25 to 30 million tons of sulfur dioxide into the stratosphere. The event provided scientists with an opportunity to study the climatic impact of a major explosive volcanic eruption using NASA's spaceborne Earth Radiation Budget Experiment. During the next year, the haze of tiny aerosols increased reflectivity and lowered global temperatures by 0.5°C (0.9°F).

The impact on global temperature of eruptions like El Chichón and Mount Pinatubo is relatively minor, but many scientists agree that the cooling produced could alter the general pattern of atmospheric circulation for a limited period. Such a change could, in turn, influence the weather in some regions. Predicting, or even identifying, specific regional effects still presents a considerable challenge to atmospheric scientists.

The preceding examples illustrate that the impact on climate of a single volcanic eruption, no matter how great, is relatively small and short-lived. The graph in Figure 20.19B reinforces this point. Therefore, if the processes discussed in this section are to have a pronounced impact over an extended period, many great eruptions, closely spaced in time, need to occur. Because no such period of explosive volcanism is known to have occurred in historic times, it is most often mentioned as a possible contributor to prehistoric climatic shifts.

Volcanism & Global Warming The Cretaceous period is the last period of the Mesozoic era, the era of *middle life* that is often called the "age of dinosaurs." It began about 145.5 million years ago and ended about 65.5 million years ago, with the extinction of the dinosaurs (and many other life-forms as well).*

The Cretaceous climate was among the warmest in Earth's long history. Dinosaurs, which are associated with mild temperatures, ranged north of the Arctic Circle. Tropical forests existed in Greenland and Antarctica, and coral reefs grew as much as 15° latitude closer to the poles than at present. Deposits of peat that would eventually form widespread coal beds accumulated at high latitudes. Sea level was as much as 200 meters (650 feet) higher than it is today, consistent with a lack of polar ice sheets as well as with an abundance of active and hence high-standing oceanic ridges.

What caused the unusually warm climates of the Cretaceous period? Among the significant factors that may have contributed was an enhanced greenhouse effect due to an increase in the amount of carbon dioxide in the atmosphere. But where did the additional CO_2 come from?

Many geologists suggest that the probable source was volcanic activity. Carbon dioxide is one of the gases emitted during volcanism, and there is now considerable geologic evidence that the Middle Cretaceous was a time of an unusually high rate of volcanic activity. Several huge oceanic lava plateaus were produced on the floor of the western Pacific during this span. These vast features were associated with hot spots that may have been produced by large mantle plumes. Massive outpourings of lava over millions of years would have been accompanied by the release of huge quantities of CO_2, which in turn would have enhanced the atmospheric greenhouse effect. *Thus, the warmth that characterized the Cretaceous may have had its origins deep in Earth's mantle.*

This example illustrates the interrelationships among parts of the Earth system. Seemingly unrelated materials and processes turn out to be linked. Here you have seen how processes originating deep in Earth's interior are connected directly or indirectly to the atmosphere, the oceans, and the biosphere.

*For more about the end of the Cretaceous, see Chapter 19

Solar Variability & Climate

Among the most persistent hypotheses of climate change have been those based on the idea that the Sun is a variable star and that its energy output varies through time. The effect of such changes would seem direct and easily understood: Increases in solar output would cause the atmosphere to warm, and reductions would result in cooling. This notion is appealing because it can be used to explain climate change of any length or intensity. However, no major *long-term* variations in the total intensity of solar radiation have yet been measured outside the atmosphere. Such measurements were not even possible until satellite technology became available. We can now measure solar output, but we still need many decades of records before we begin to sense how variable (or invariable) energy from the Sun really is.

Some hypotheses for climate change relate to sunspot cycles. The most conspicuous and best-known features on the surface of the Sun are the dark blemishes called **sunspots** (Figure 20.20). Sunspots are huge magnetic storms that extend from the Sun's surface deep into the interior. Moreover, these spots are associated with the Sun's ejection of huge masses of particles that, on reaching Earth's upper atmosphere, interact with gases to produce displays known as the *aurora borealis*, or Northern Lights, in the Northern Hemisphere.

Sunspots occur in cycles, with the number of sunspots reaching a maximum about every 11 years

▼ SmartFigure 20.20
Sunspots Both images show sunspot activity at the same location on the solar disk at the same time on March 5, 2012, using two different instruments from NASA's Solar Dynamics Observatory. (NASA)

VIDEO
http://goo.gl/97bha6

This view shows an approximation of the Sun's surface. The black spots surrounded by deep orange is a sunspot region where magnetic activity is extremely intense.

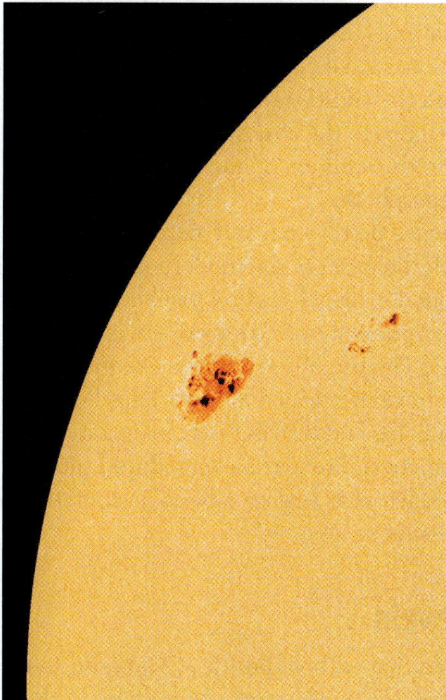

The instrument that produced this image used ultraviolet, radio, and other parts of the electromagnetic spectrum. Looping lines show solar plasma following magnetic field lines.

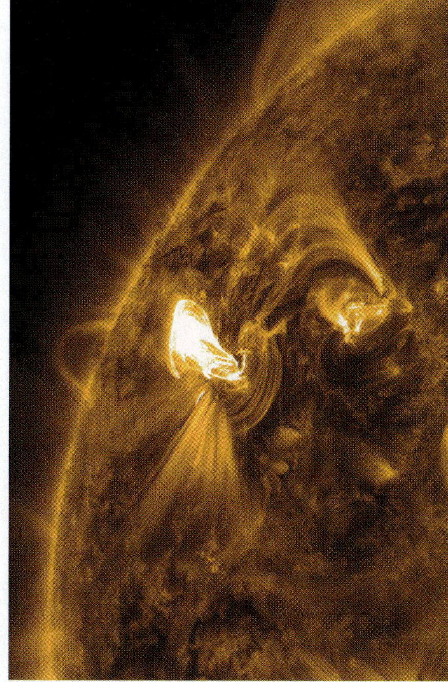

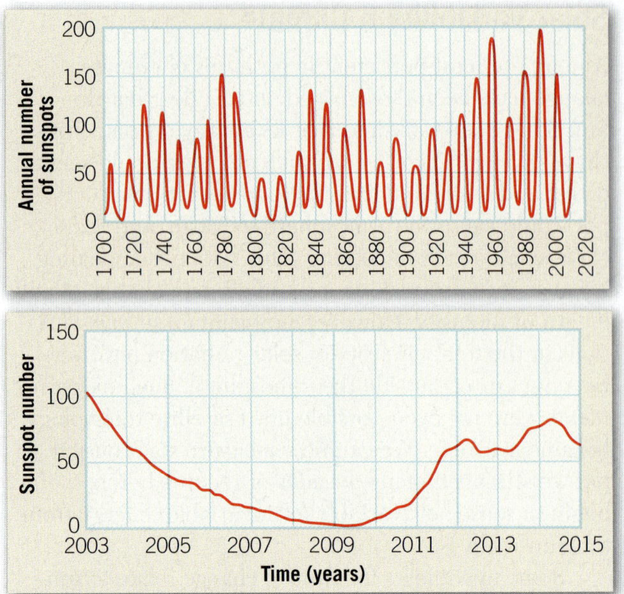

Figure 20.21 Mean annual sunspot numbers The number of sunspots reaches a maximum about every 11 years.

(Figure 20.21). During periods of maximum sunspot activity, the Sun emits slightly more energy than during sunspot minimums. Based on measurements from space that began in 1978, the variation during an 11-year cycle is about 0.1 percent. Although sunspots are dark, they are surrounded by brighter areas, which apparently offsets the effect of the dark spots. It appears that this change in solar output is too small and the cycles are too short to have any appreciable effect on global temperatures. However, there is a possibility that longer-term variations in solar output may affect climates on Earth.

For example, the span between 1645 and 1715 is a period, known as the *Maunder minimum*, during which sunspots were largely absent. This period of missing sunspots closely corresponds with a period in climate history known as the *Little Ice Age*, an especially cold period in Europe. For some scientists, this correlation suggests that a reduction in the Sun's output was likely responsible at least in part for this cold episode. Other scientists seriously question this notion. Their hesitation stems, in part, from subsequent investigations using different climate records from around the world that failed to find a significant correlation between sunspot activity and climate.

Did You Know?
Based on recent research that used satellite data, there appears to be no connection between changes in the Sun's brightness and recent global warming. The scientists who carried out this detailed analysis state that variations in solar brightness measured since 1978, which is as far back as these measurements go, are too small to have contributed appreciably to accelerated global warming over the past 30 years.

CONCEPT CHECKS 20.5

1. Describe and briefly explain the effect on global temperatures of the eruptions of El Chichón and Mount Pinatubo.
2. How might volcanism lead to global warming?
3. What are sunspots? How does solar output change as sunspot numbers change? Is there a solid connection between sunspot numbers and climate change on Earth?

20.6 Human Impact on Global Climate

Summarize the nature and cause of the atmosphere's changing composition since about 1750. Describe the climate's response.

So far we have examined potential causes of climate change that are natural. In this section, we examine how humans contribute to global climate change. One impact largely results from the addition of carbon dioxide and other greenhouse gases to the atmosphere. A second impact is related to the addition of human-generated aerosols to the atmosphere.

Human influence on regional and global climate did not just begin with the onset of the modern industrial period. There is good evidence that people have been modifying the environment over extensive areas for thousands of years. The use of fire and the overgrazing of marginal lands by domesticated animals have both reduced the abundance and distribution of vegetation. By altering ground cover, humans have modified such important climate factors as surface albedo, evaporation rates, and surface winds.

Rising CO_2 Levels

Earlier you learned that carbon dioxide (CO_2) represents only about 0.0400 percent (400 parts per million) of the gases that make up clean, dry air. Nevertheless, it is a very significant component meteorologically. Carbon dioxide is influential because it is transparent to incoming short-wavelength solar radiation, but it is not transparent to some of the longer-wavelength outgoing Earth radiation. A portion of the energy leaving Earth's surface is absorbed by atmospheric CO_2. This energy is subsequently re-emitted, part of it back toward the surface, thereby keeping the air near the ground warmer than it would be without CO_2. Thus, along with water vapor, carbon dioxide is largely responsible for the atmosphere's greenhouse effect.

The tremendous industrialization of the past two centuries has been fueled—and still is fueled—by burning fossil fuels: coal, natural gas, and petroleum (Figure 20.22). Combustion of these fuels has added great quantities of carbon dioxide to the atmosphere. Figure 20.23 shows changes in CO_2 concentrations at Hawaii's Mauna Loa Observatory, where measurements have been made since 1958. The graph shows an annual seasonal cycle and a steady upward trend over the years. The up-and-down of the seasonal cycle is due to the vast

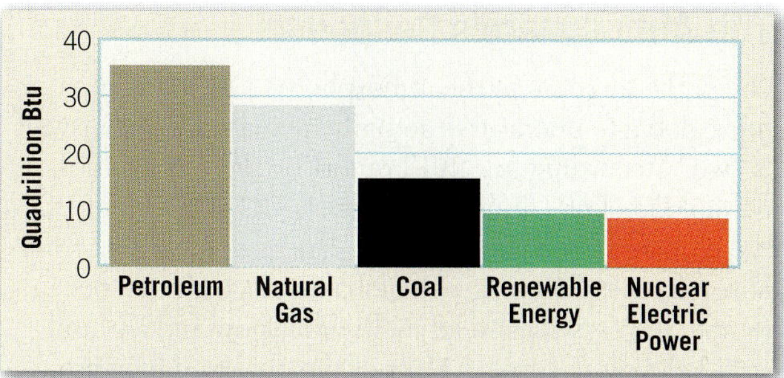

▲ **Figure 20.22 U.S. energy consumption** The graph shows energy consumption in 2015. The total was 97.65 quadrillion Btu. A quadrillion is 10 raised to the 15th power, or a billion million. The burning of fossil fuels represents about 81.3 percent of the total. (Based on data from U.S. Energy Information Administration)

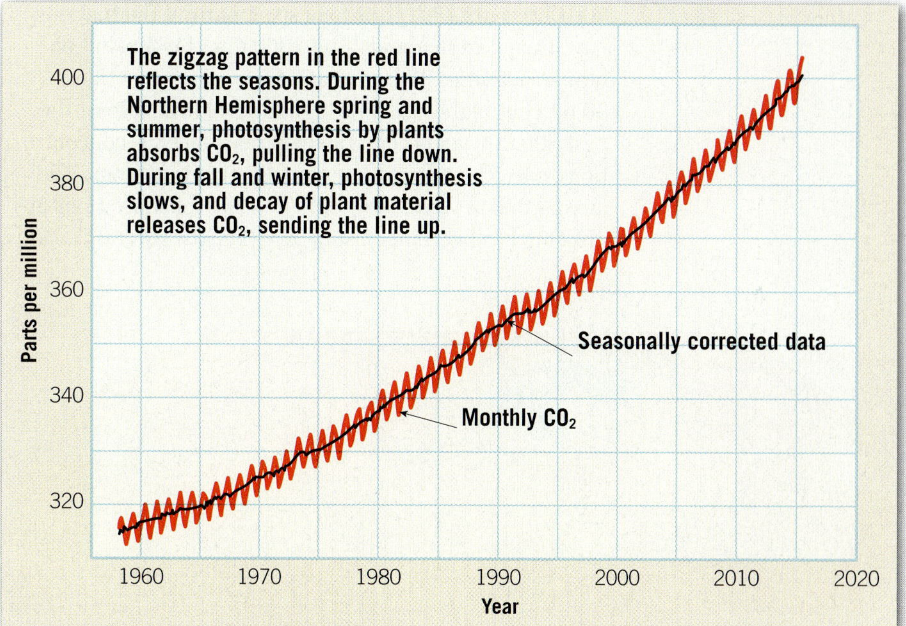

▲ **SmartFigure 20.23 Monthly CO_2 concentrations** Atmospheric CO_2 has been measured at Mauna Loa Observatory, Hawaii, since 1958. There has been a consistent increase since monitoring began. This graphic portrayal is known as the *Keeling Curve*, in honor of the scientist who originated the measurements. (Based on NOAA)

TUTORIAL
http://goo.gl/Y479oq

land area of the Northern Hemisphere, which contains the majority of land-based vegetation. During spring and summer in the Northern Hemisphere, when plants are absorbing CO_2 as part of photosynthesis, concentrations decrease. The annual increase in atmospheric CO_2 during the cold months occurs as vegetation dies and leaves fall and decompose, releasing CO_2 back into the air.

The use of coal and other fuels is the most prominent means by which humans add CO_2 to the atmosphere, but it is not the only way. The clearing of forests also contributes substantially because CO_2 is released as vegetation is burned or decays (**Figure 20.24**). Deforestation is particularly pronounced in the tropics, where vast tracts are cleared for ranching and agriculture, or subjected to inefficient commercial logging operations. All major tropical forests—including those in South America, Africa, Southeast Asia, and Indonesia—are disappearing. According to United Nations estimates, more than 10 million hectares (25 million acres) of tropical forest were permanently destroyed *each year* during the decades of the 1990s and 2000s, although this rate has slowed in recent years.

Some of the excess CO_2 is taken up by plants or is dissolved in the ocean, but an estimated 45 percent remains in the atmosphere. **Figure 20.25** is a graphic record of changes in atmospheric CO_2 extending back

▼ **Figure 20.24 Tropical deforestation** Clearing of the tropical rain forest is a serious environmental issue. In addition to causing a loss of biodiversity, it is a significant source of carbon dioxide. Fires are frequently used to clear the land. This scene is in Brazil's Amazon basin. (Photo by Nigel Dickinson/Alamy)

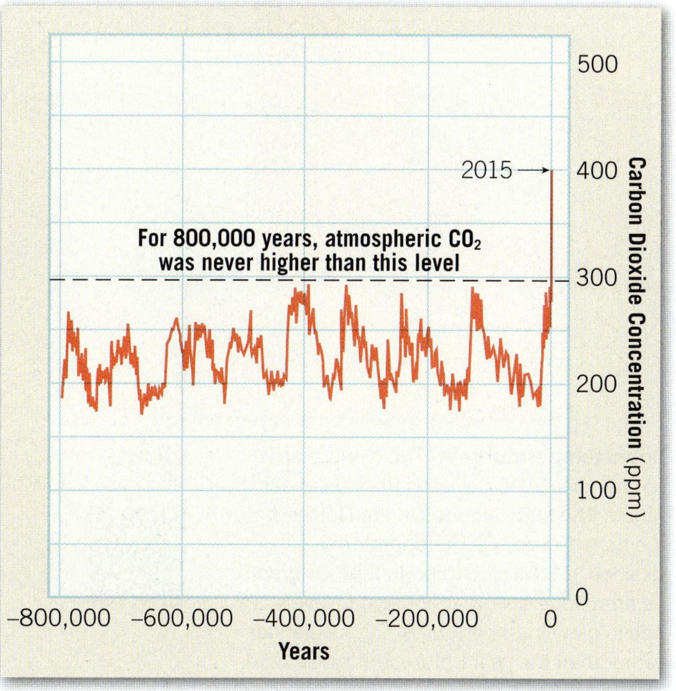

◀ **Figure 20.25 CO_2 concentrations over the past 800,000 years** Most of these data come from the analysis of air bubbles trapped in ice cores. The record since 1958 comes from direct measurements at Mauna Loa Observatory, Hawaii. The rapid increase in CO_2 concentrations since the onset of the Industrial Revolution is obvious. (Based on data from NOAA)

800,000 years. Over this long span, natural fluctuations have varied from about 180 to 300 ppm. As a result of human activities, the present CO_2 level is more than 30 percent higher than its highest level over at least the past 800,000 years. The rapid increase in CO_2 concentrations since the onset of industrialization is obvious. The annual rate at which atmospheric CO_2 concentrations are growing has been increasing over the past several decades.

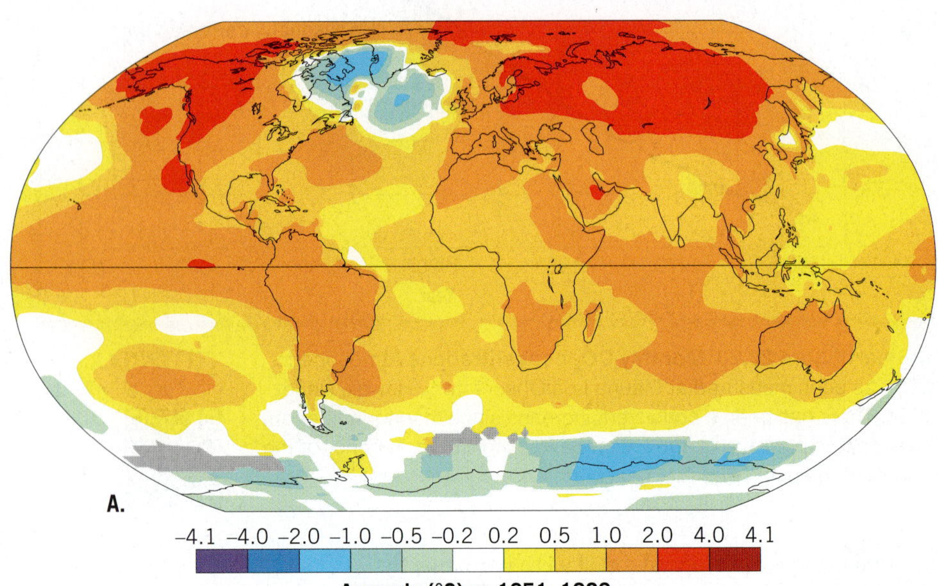

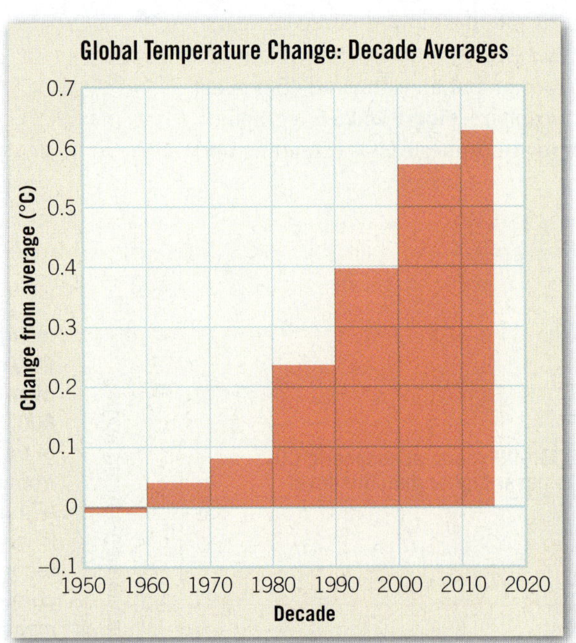

▲ **SmartFigure 20.26 Global temperatures A.** The world map shows how average temperatures in 2015 deviated from the mean for the 1951–1980 base period. The high latitudes in the Northern Hemisphere clearly stand out. (NASA/GISS) **B.** The decade-by-decade temperature trend since 1950 is illustrated in this bar graph. Continued increases in the atmosphere's greenhouse gas levels are driving a long-term increase in global temperatures. Each calendar year is not necessarily warmer than the year before, but since 1950, each decade has been warmer than the previous one. (Data from NASA)

VIDEO
https://goo.gl/zz5VKN

The Atmosphere's Response

Given the increase in the atmosphere's CO_2 content, have global temperatures actually increased? The answer is "yes." According to a 2013 report by the Intergovernmental Panel on Climate Change (IPCC), "Warming of the climate system is unequivocal, as is now evident from observations of increases in global average air and ocean temperatures, widespread melting of snow and ice, and rising global sea level."* Most of the observed increase in global average temperatures since the mid-twentieth century is *extremely likely* due to the observed increase in human-generated greenhouse gas concentrations. As used by the IPCC, *extremely likely* indicates a probability of 95 to 100 percent. The planet's average surface air temperature has risen about 1°C (1.8°F) since the late nineteenth century, with most of the warming occurring in the past 35 years. Fifteen of the 16 warmest years on record have occurred since 2001. Surface air temperatures in 2015 were the warmest since record keeping began in 1880, shattering the mark set in 2014 (**Figure 20.26A**).

Weather patterns and other natural cycles cause fluctuations in average temperatures from year to year. This is especially true on regional and local levels. For example, while the globe experienced notably warm temperatures in 2014, parts of the continental United States were cooler than normal. By contrast, 2014 was the warmest year on record for much of Europe and parts of Russia, and ocean temperatures were at a record high. Regardless of regional differences in any year, increases in greenhouse gas levels are causing a long-term rise in global temperatures. Although each calendar year will not necessarily be warmer than the one before, scientists expect each decade to be warmer than the previous one. An examination of the decade-by-decade temperature trend in **Figure 20.26B** bears this out.

What about the future? Projections for the years ahead depend in part on the quantities of greenhouse gases that are emitted. **Figure 20.27** shows the best estimates of global warming for several different scenarios. The 2013 IPCC report states that if there is a doubling of the pre-industrial level of carbon dioxide (280 ppm) to 560 ppm, the *likely* temperature increase will be in the range of 2° to 4.5°C (3.5° to 8.1°F). The increase is *very unlikely* (1 to 10 percent probability) to be less than 1.5°C (2.7°F), and values higher than 4.5°C (8.1°F) are possible.

The Role of Trace Gases

Carbon dioxide is not the only gas contributing to a global increase in temperature. In recent years atmospheric scientists have come to realize that human industrial and agricultural activities are causing a buildup of

*IPCC, "Summary for Policymakers," in *Climate Change 2013: The Physical Science Basis*. The Intergovernmental Panel on Climate Change is an authoritative group of scientists that provides advice to the world community through periodic reports that assess the state of knowledge of the causes and effects of climate change.

several trace gases that also play significant roles. The substances are called **trace gases** because their concentrations are much lower than the concentration of carbon dioxide. The most important trace gases are methane (CH_4), nitrous oxide (N_2O), and chlorofluorocarbons (CFCs). These gases absorb wavelengths of outgoing radiation from Earth that would otherwise escape into space. Although individually their impact is modest, taken together these trace gases play a significant role in warming the troposphere.

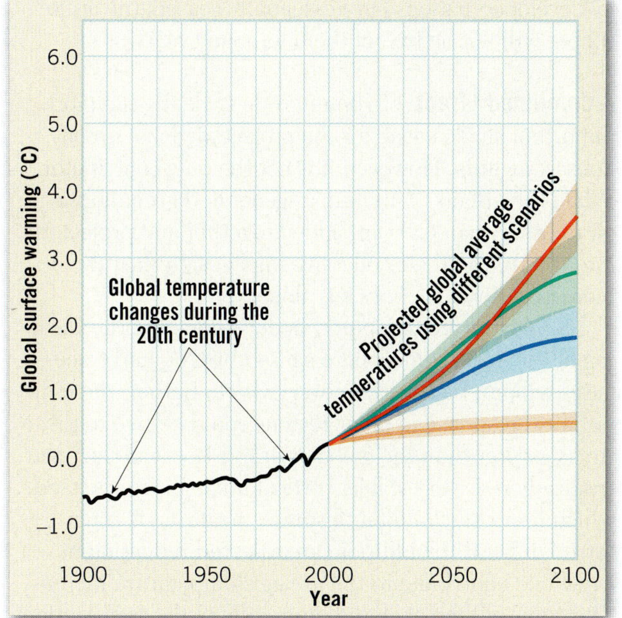

SmartFigure 20.27 **Temperature projections to 2100** The right half of the graph shows projected global warming based on different emissions scenarios. The shaded zone adjacent to each colored line shows the uncertainty range for each scenario. The basis for comparison (0.0 on the vertical axis) is the global average for the period 1980 to 1999. The orange line represents the scenario in which CO_2 concentrations were held constant at values for the year 2000. (Data from NOAA)

VIDEO
http://goo.gl/3xMnoA

Methane Although present in much smaller amounts than CO_2, methane's significance is greater than its relatively small concentration would indicate (**Figure 20.28**). This is because methane is about 20 times more effective than CO_2 at absorbing infrared radiation emitted by Earth.

Methane is produced by *anaerobic* bacteria in wet places where oxygen is scarce. (*Anaerobic* means "without air," specifically oxygen.) Such places include swamps, bogs, wetlands, and the guts of termites and grazing animals such as cattle and sheep. Methane is also generated in flooded paddy fields ("artificial swamps") used for growing rice. Mining of coal and drilling for oil and natural gas are other sources because methane is a product of their formation.

The increase in the concentration of methane in the atmosphere has been in step with the growth in human population. This relationship reflects the close link between methane formation and agriculture. As population increases, so do the number of cattle and rice paddies.

Nitrous Oxide Sometimes called "laughing gas," nitrous oxide is also building in the atmosphere, although not as rapidly as methane (see Figure 20.28). The increase results primarily from agricultural activity. When farmers use nitrogen fertilizers to boost crop yield, some of the nitrogen enters the air as nitrous oxide. This gas is also produced by high-temperature combustion of fossil fuels. Although the annual release into the atmosphere is small, nitrous oxide is about 300 times more effective (by weight) than CO_2 as a greenhouse gas. Also, the average residence time of a nitrous oxide molecule in the atmosphere is about 114 years! If nitrogen fertilizer and fossil fuel use grow at projected rates, nitrous oxide's contribution to greenhouse warming may approach half that of methane.

CFCs Unlike methane and nitrous oxide, chlorofluorocarbons (CFCs) are not naturally present in the atmosphere. CFCs are manufactured chemicals with many uses that have gained notoriety because they are responsible for ozone depletion in the stratosphere. The role of CFCs in global warming is less well known. CFCs are very effective greenhouse gases. They were not developed until the 1920s and were not used in great quantities until the 1950s. Although corrective action has been taken, CFC levels will not drop rapidly. CFCs remain in the atmosphere for decades, so even though CFC emissions

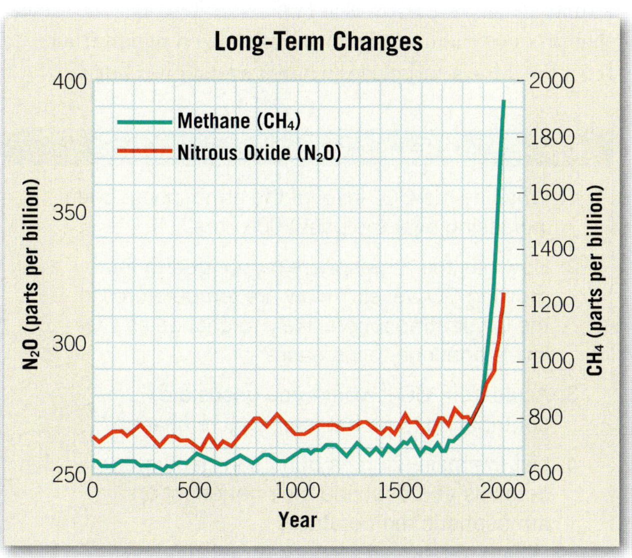

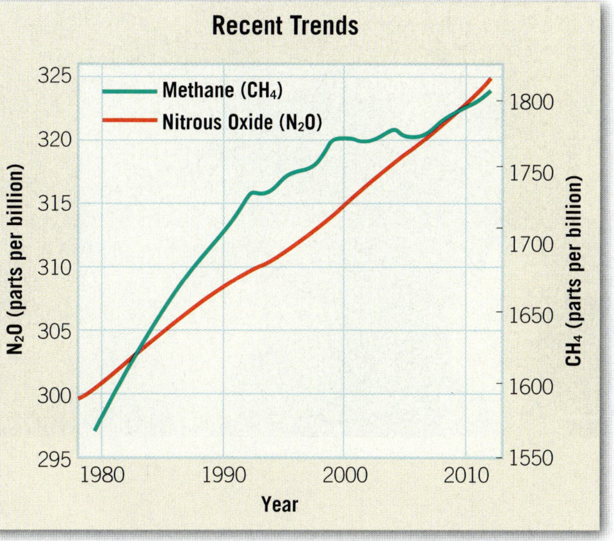

Figure 20.28 Methane and nitrous oxide Although CO_2 is most important, these trace gases also contribute to global warming. Over the 2000-year span shown here, there were relatively minor fluctuations until the industrial era. The graph on the right shows recent trends. (Based on data from U.S. Global Change Research Program and NOAA)

Did You Know?
The Intergovernmental Panel on Climate change (IPCC) referred to in the discussion of global warming was established by the United Nations Environment Programme and the World Meteorological Organization in 1988 to assess the scientific, technical, and socioeconomic information that is relevant to understanding climate change. It is an authoritative group that provides periodic reports regarding the state of knowledge about causes and effects of climate change. More than 250 authors and nearly 1100 scientific reviewers from 55 countries contributed to *Climate Change 2013: The Physical Science Basis*.

by developed nations have essentially ceased, the atmosphere will not be free of them for many years.

A Combined Effect Carbon dioxide is clearly the most important single cause for the projected global greenhouse warming. However, it is not the only contributor. When the effects of all human-generated greenhouse gases other than CO_2 are added together and projected into the future, their collective impact significantly increases the impact of CO_2 alone.

Sophisticated computer models show that the warming of the lower atmosphere caused by CO_2 and trace gases will not be the same everywhere. Rather, the temperature response in polar regions could be two to three times greater than the global average. Because the polar troposphere is very stable, vertical mixing is suppressed, which limits the amount of surface heat that is transferred upward. In addition, an expected reduction in sea ice would contribute to the greater temperature increase. This topic will be explored more fully in the next section.

How Aerosols Influence Climate

Increasing the levels of carbon dioxide and other greenhouse gases in the atmosphere is the most direct human influence on global climate. But it is not the only impact. Global climate is also affected by human activities that contribute to the atmosphere's aerosol content. Recall that aerosols are the tiny, often microscopic, liquid and solid particles that are suspended in the air. Unlike cloud droplets, aerosols are present even in relatively dry air. Atmospheric aerosols are composed of many different materials, including soil, smoke, sea salt, and sulfuric acid. Natural sources are numerous and include such phenomena as dust storms and volcanoes.

Most human-generated aerosols come from two sources: sulfur dioxide emitted during the combustion of fossil fuels and smoke from vegetation burned to clear agricultural land. Chemical reactions in the atmosphere convert the sulfur dioxide into sulfate aerosols, the same material that produces acid precipitation. The satellite images in **Figure 20.29** provide an example.

How do aerosols affect climate? Aerosols act directly by reflecting sunlight back to space and indirectly by making clouds "brighter" reflectors. The second effect relates to the fact that many aerosols (such as those composed of salt or sulfuric acid) attract water and thus are especially effective as cloud condensation nuclei. The large quantities of aerosols produced by human activities (especially industrial emissions) trigger an increase in the number of cloud droplets that form within a cloud. A greater number of small droplets increases the cloud's brightness, causing more sunlight to be reflected back to space.

One category of aerosols, called **black carbon**, is soot generated by combustion processes and fires. Unlike most other aerosols, black carbon warms the atmosphere because it is an effective absorber of incoming solar radiation. In addition, when deposited on snow and ice, black carbon reduces surface albedo, thus increasing the amount of light absorbed. Nevertheless, despite the warming effect of black carbon, the overall effect of atmospheric aerosols is to cool Earth.

Studies indicate that the cooling effect of human-generated aerosols offsets a portion of the global warming caused by the growing quantities of greenhouse gases in the atmosphere. The magnitude and extent of the cooling effect of aerosols is uncertain. This uncertainty is a significant hurdle in advancing our understanding of how humans alter Earth's climate.

It is important to point out some significant differences between global warming by greenhouse gases and aerosol cooling. After being emitted, carbon dioxide and trace gases remain in the atmosphere for many decades. By contrast, aerosols released into the troposphere remain there for only a few days or, at most, a few weeks before they are "washed out" by precipitation, limiting their effects. Because of their short lifetime in the troposphere, aerosols are distributed unevenly over the globe. As expected, human-generated aerosols are concentrated near the areas that produce them—namely industrialized regions that burn fossil fuels and places where vegetation is burned.

▼ **Figure 20.29 Human-generated aerosols** These satellite images show a serious air pollution episode that plagued China on October 8, 2010. (NASA)

This satellite image shows the extremely high levels of aerosols associated with this air pollution episode. At an index value of 4, aerosols are so dense that you would have difficulty seeing the midday Sun.

The source of these pollutants was coal-burning power plants, agricultural burning, and industrial processes.
Beijing
Zhengzhou
N
200 km

CONCEPT CHECKS 20.6

1. Why has the CO_2 level of the atmosphere been increasing over the past 200 years?
2. How has the atmosphere responded to the growing CO_2 levels? How are temperatures in the lower atmosphere likely to change as CO_2 levels continue to increase?
3. Aside from CO_2, what trace gases are contributing to global temperature change?
4. List the main sources of human-generated aerosols and describe their net effect on atmospheric temperatures.

20.7 Climate Feedback Mechanisms

Contrast positive and negative feedback mechanisms and provide examples of each.

Climate is a very complex interactive physical system. Thus, when any component of the climate system is altered, scientists must consider many possible outcomes. These possible outcomes are called **climate feedback mechanisms**. They complicate climate-modeling efforts and add uncertainty to climate predictions.

Types of Feedback Mechanisms

What climate feedback mechanisms are related to carbon dioxide and other greenhouse gases? One important mechanism is that warmer surface temperatures increase evaporation rates. This, in turn, increases the water vapor in the atmosphere. Remember that water vapor is an even more powerful absorber of radiation emitted by Earth than is carbon dioxide. Therefore, with more water vapor in the air, the temperature increase caused by carbon dioxide and trace gases is reinforced.

Scientists who model global climate change indicate that the temperature increase at high latitudes may be two to three times greater than the global average. This assumption is based in part on the likelihood that the area covered by sea ice will decrease as surface temperatures rise. Because ice reflects a much larger percentage of incoming solar radiation than does open water, the melting of sea ice replaces a highly reflective surface with a relatively dark surface (Figure 20.30). The result is a substantial increase in the solar energy absorbed at the surface. This in turn feeds back to the atmosphere and magnifies the initial temperature increase created by higher levels of greenhouse gases.

The climate feedback mechanisms discussed thus far magnify the temperature rise caused by the buildup of greenhouse gases. Because these effects reinforce the initial change, they are called **positive feedback mechanisms**. On the other hand, **negative feedback mechanisms** produce results that are just the opposite of the initial change and tend to offset it.

One probable result of a global temperature rise would be an accompanying increase in cloud cover due to the higher moisture content of the atmosphere. Most clouds are good reflectors of solar radiation. At the same time, however, they are also good absorbers and emitters of radiation emitted by Earth. Consequently, clouds produce two opposite effects. They exert a negative feedback because they increase Earth's albedo and thus reduce the amount of solar energy available to heat the atmosphere. On the other hand, clouds exert a positive feedback by absorbing and emitting radiation that would otherwise be lost from the troposphere. Which effect, if either, is stronger? Observations and modeling demonstrate that the overall effect of clouds is to slightly cool the planet. This means that the warming caused by the buildup of greenhouse gases is slightly offset by the increase in cloud cover.

The problem of global warming caused by human-induced changes in atmospheric composition continues to be one of the most studied aspects of climate change. Although no models yet incorporate the full range of potential factors and feedbacks, there is a strong scientific consensus that the increasing concentrations of atmospheric carbon dioxide and trace gases is creating a warmer planet with a different distribution of climate regimes.

Computer Models of Climate: Important yet Imperfect Tools

Earth's climate system is amazingly complex. Comprehensive state-of-the-science climate simulation models are among the tools used to develop possible climate-change scenarios. Called *general circulation models* (*GCMs*), they are based on fundamental laws of physics and chemistry and incorporate human and biological interactions. GCMs are used to simulate many variables, including temperature, rainfall, snow cover, soil moisture, winds, clouds, sea ice, and ocean circulation, over the entire globe through the seasons and over spans of decades.

In many other fields of study, hypotheses can be tested by direct experimentation in a laboratory or by field observations and measurements. However, this is often not possible in the study of climate. Rather, scientists must construct computer models of how our planet's climate system works. If we understand the climate system correctly and construct the model appropriately, the

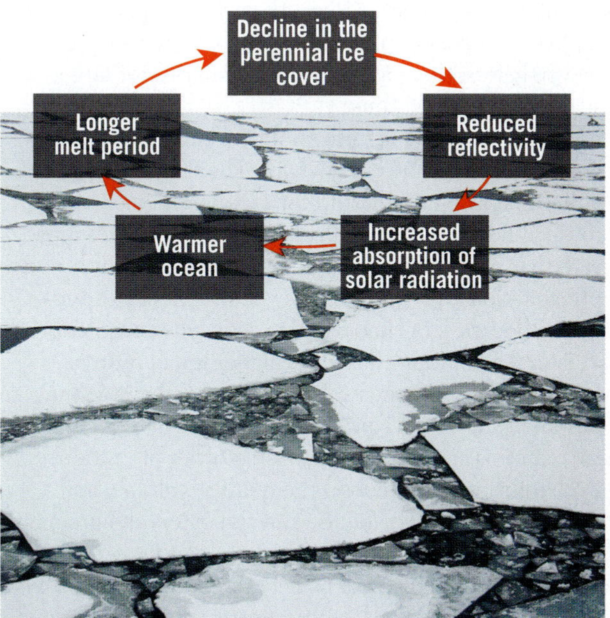

SmartFigure 20.30 Sea ice as a feedback mechanism The image shows the springtime breakup of sea ice near Antarctica. The diagram shows a likely feedback loop. A reduction in sea ice acts as a positive feedback mechanism because surface albedo decreases, and the amount of energy absorbed at the surface increases. (Photo by Radius Images/Alamy)

VIDEO
http://goo.gl/XKhbz8

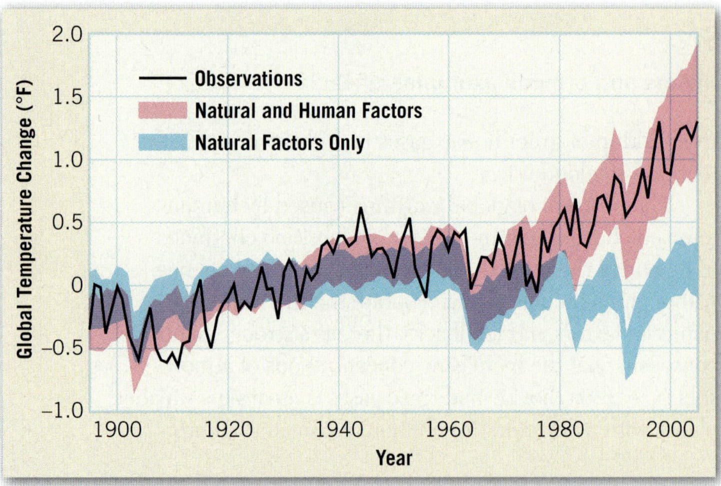

▲ **Figure 20.31 Separating human and natural influences on climate** The blue band shows how global average temperatures would have changed due to natural forces only, as simulated by climate models. The pink band shows model projections of the effects of human and natural forces combined. The black line shows actual observed global average temperatures. As the blue band indicates, without human influences, temperatures over the past century would actually have first warmed and then cooled slightly over recent decades. Bands of color are used to express the range of uncertainty. (U.S. Global Change Research Program)

behavior of the model climate system should mimic the behavior of Earth's climate system (**Figure 20.31**).

What factors influence the accuracy of climate models? Clearly, mathematical models are *simplified* versions of the real Earth and cannot capture its full complexity, especially at smaller geographic scales. Moreover, computer models used to simulate future climate change must make many assumptions that significantly influence predictions. They must consider a wide range of possible changes in population, economic growth, fossil fuel consumption, technological development, improvements in energy efficiency, and more.

Despite many obstacles, our ability to use supercomputers to simulate climate is very good and continues to improve. Although today's models are far from infallible, they are powerful tools for understanding what Earth's future climate might be like.

CONCEPT CHECKS 20.7

1. Distinguish between positive and negative feedback mechanisms.
2. Provide at least one example of each type of feedback mechanism.
3. What factors influence the accuracy of computer models of climate?

20.8 Some Consequences of Global Warming

Summarize some of the possible consequences of global warming.

Did You Know?
A *scenario* is an example of what might happen under a particular set of assumptions. Scenarios are tools that allow us to examine questions about an uncertain future. For example, future trends in fossil-fuel use and other human activities are uncertain. Therefore, scientists have developed a set of scenarios for how the climate may change based on a wide range of possibilities for these variables.

What consequences can be expected as the carbon dioxide content of the atmosphere reaches a level that is twice what it was early in the twentieth century? Because the climate system is complex, predicting the occurrence of specific effects in particular places is speculative. It is not yet possible to pinpoint such changes. Nevertheless, there are plausible scenarios for larger scales of space and time.

As noted, the magnitude of the temperature increase will not be the same everywhere. The temperature rise will probably be smallest in the tropics and increase toward the poles. As for precipitation, the models indicate that some regions will experience significantly more precipitation and runoff. However, others will experience a decrease in runoff due to reduced precipitation or greater evaporation caused by higher temperatures.

Table 20.1 lists possible effects of global warming based on the IPCC's projections for the late twenty-first century, ranked in decreasing order of certainty. Probabilities are based on the quality, volume, and consistency of the evidence and the extent of agreement among scientists. The risk associated with any of the projections in Table 20.1 is a combination of the probability of occurrence and the severity of the damage if it were to occur. This means that even projections labeled "unlikely" should not be ignored because if any of those events were to happen, the consequences would be extremely serious.

Sea-Level Rise

A significant impact of human-induced global warming is a rise in sea level. As this occurs, coastal cities, wetlands, and low-lying islands could be threatened with more frequent flooding, increased shoreline erosion, and saltwater encroachment into coastal rivers and aquifers.

How is a warmer atmosphere related to a rise in sea level? One significant factor is thermal expansion. Higher air temperatures warm the adjacent upper layers of the ocean, which in turn causes the water to expand and sea level to rise.

A second factor contributing to global sea-level rise is melting glaciers. With few exceptions, glaciers around the world have been retreating at unprecedented rates over the past century. Some mountain glaciers have disappeared altogether. A satellite study spanning 20 years showed that the mass of the Greenland and Antarctic Ice Sheets dropped an average of

Table 20.1 IPCC Projections for the Late 21st Century

Projection	Probability
• Cold days and nights will be warmer and less frequent over most land areas • Hot days and nights will be warmer and more frequent over most land areas • The extent of permafrost will decline • Ocean acidification will increase as the atmosphere accumulates CO_2 • Northern Hemisphere glaciation will not initiate before the year 3000 • Global mean sea level will rise and continue to do so for many centuries	**Virtually certain (99–100%)**
• Arctic sea ice cover will continue to shrink and thin, and Northern Hemisphere spring snow cover will decrease • The dissolved oxygen content of the ocean will decrease by a few percent • The rate of increase in atmospheric CO_2, methane, and nitrous oxide will reach levels unprecedented in the past 10,000 years • The frequency of warm spells and heat waves will increase • The frequency of heavy precipitation events will increase • Precipitation amounts will increase in high latitudes • The ocean's conveyor-belt circulation will weaken • The rate of sea-level rise will exceed that of the late twentieth century • Extreme high sea-level events will increase, as will ocean wave heights of midlatitude storms	**Very likely (90–100%)**
• If the atmospheric CO_2 level stabilizes at double the present level, global temperatures will rise by between 1.5°C (2.7°F) and 4.5°C (8.1°F) • Areas affected by drought will increase • Precipitation amounts will decline in the subtropics • The loss of glaciers will accelerate in the next few decades	**Likely (66–100%)**
• Intense tropical cyclone activity will increase • The West Antarctic Ice Sheet will pass the melting point if global warming exceeds 5°C (9°F)—this is relative, not absolute	**About as likely as not (33–66%)**
• Antarctic and Greenland Ice Sheets will collapse due to surface warming	← **Not likely (0–33%)**

0 10 20 30 40 50 60 70 80 90
Probability (%)

475 gigatons per year. (A gigaton is 1 billion metric tons.) That is enough water to raise sea level 1.5 millimeters (0.05 inch) per year. The loss of ice was not steady but was occurring at an accelerating rate during the study period. During the same span, mountain glaciers and ice caps lost an average of slightly more than 400 gigatons per year.

Research indicates that sea level has risen about 25 centimeters (9.75 inches) since 1870, with the rate of sea-level rise accelerating in recent years. What about future changes in sea level? As **Figure 20.32** indicates, the estimates of future sea-level rise are uncertain. The four scenarios depicted on the graph represent estimates based on different degrees of ocean warming and ice sheet loss and range from 0.2 meter (8 inches) to 2 meters (6.6 feet). The lowest scenario is an extrapolation of the annual rate of sea-level rise that occurred between 1870 and 2000 (1.7 millimeter/year). However, when the rate of sea-level rise for the period 1993 to 2012 is examined, the annual change is 3.17 millimeters/year. Such data show that there is a reasonable chance that sea level will rise considerably more than the lowest scenario indicates.

Scientists realize that even modest rises in sea level along a *gently* sloping shoreline, such as the Atlantic and Gulf coasts of the United States, will lead to significant erosion and severe permanent inland flooding

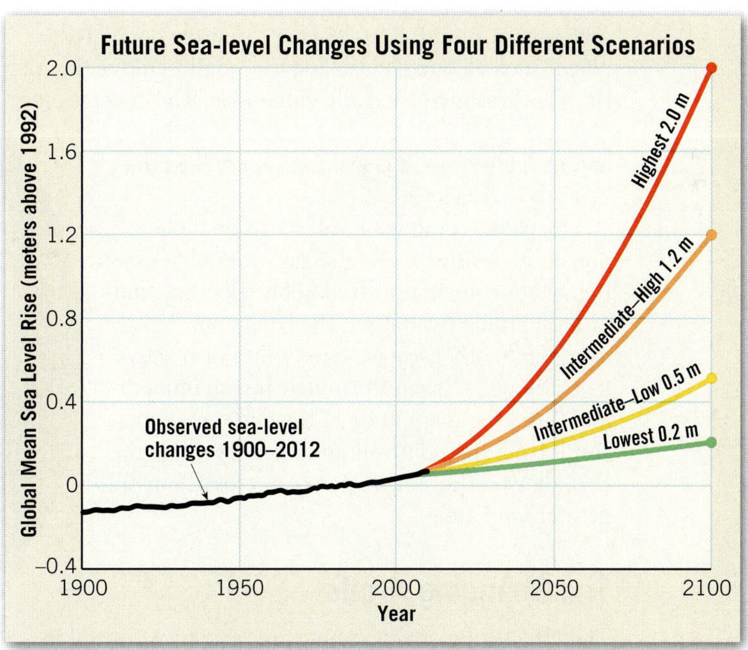

▲ **SmartFigure 20.32 Changing sea level** This graph shows changes in sea level between 1900 and 2012 and projections to 2100, using four different scenarios. Currently the highest and lowest projections are considered to be extremely unlikely. The greatest uncertainty surrounding estimates involves the rate and magnitude of ice sheet loss from Greenland and Antarctica. Zero on the graph represents mean sea level in 1992. (NOAA)

VIDEO
http://goo.gl/BqepKo

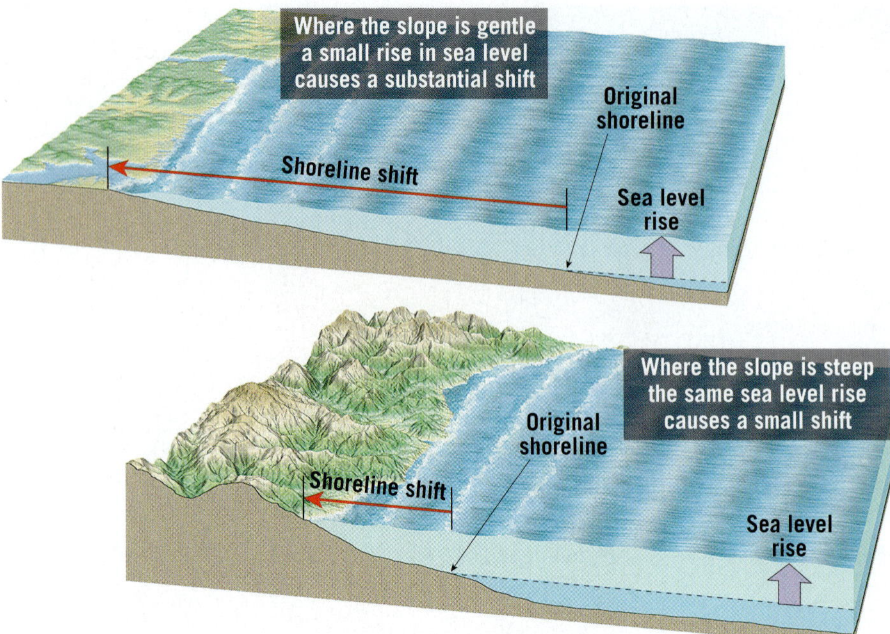

▲ **SmartFigure 20.33 Slope of the shoreline** The slope of the shoreline is critical to determining the degree to which sea-level changes will affect it. As sea level gradually rises, the shoreline retreats, and structures that were once thought to be safe from wave attack become vulnerable.

TUTORIAL
http://goo.gl/u5fVXR

(**Figure 20.33**). If this happens, many beaches and wetlands will disappear, and coastal civilization will be severely disrupted. Low-lying and densely populated places such as Bangladesh and the small island nation of the Maldives are especially vulnerable. The average elevation in the Maldives is 1.5 meters (less than 5 feet), and its highest point is just 2.4 meters (less than 8 feet) above sea level.

Because rising sea level is a gradual phenomenon, coastal residents may overlook it as an important contributor to shoreline flooding and erosion problems. Rather, the blame may be assigned to other forces, especially storm activity. Although a given storm may be the immediate cause, the magnitude of the destruction may result from the relatively small sea-level rise that allowed the storm's power to cross a much greater land area.

The Changing Arctic

The effects of global warming are most pronounced in the high latitudes of the Northern Hemisphere. For more than 30 years, the extent and thickness of sea ice have been rapidly declining. In addition, permafrost temperatures have been rapidly rising, and the area affected by permafrost has been decreasing. Meanwhile, alpine glaciers and the Greenland Ice Sheet have been shrinking. Another sign that the Arctic is rapidly warming is related to plant growth (**Figure 20.34**). A 2013 study showed that vegetation growth at northern latitudes now resembles that which characterized areas 4° to 6° of latitude farther south as recently as 1982. That is a distance of 400 to 700 kilometers (250 to 430 miles). One researcher characterized the finding this way: "It's like Winnipeg, Manitoba, moving to Minneapolis-St. Paul in only 30 years."

Arctic Sea Ice Climate models are in general agreement that one of the strongest signals of global warming should be a loss of sea ice in the Arctic. This is indeed occurring. The map in **Figure 20.35A** compares the average sea ice extent for September 2016 to the long-term average for the period 1981 to 2010. On September 10th the extent was about 4.14 million square kilometers (1.6 million square miles)—the second lowest minimum of the satellite era, which began in 1979. (September represents the end of the melt period, when the area covered by sea ice is at a minimum.) **Figure 20.35B**, which shows

▼ **SmartFigure 20.34 Climate change spurs plant growth beyond 45° north** Of the 26 million square kilometers (10 million square miles) of northern vegetated lands, about 40 percent showed increases in plant growth during the 30-year period ending in 2012 (green and blue on the satellite image). (NASA)

VIDEO
http://goo.gl/rRnMdj

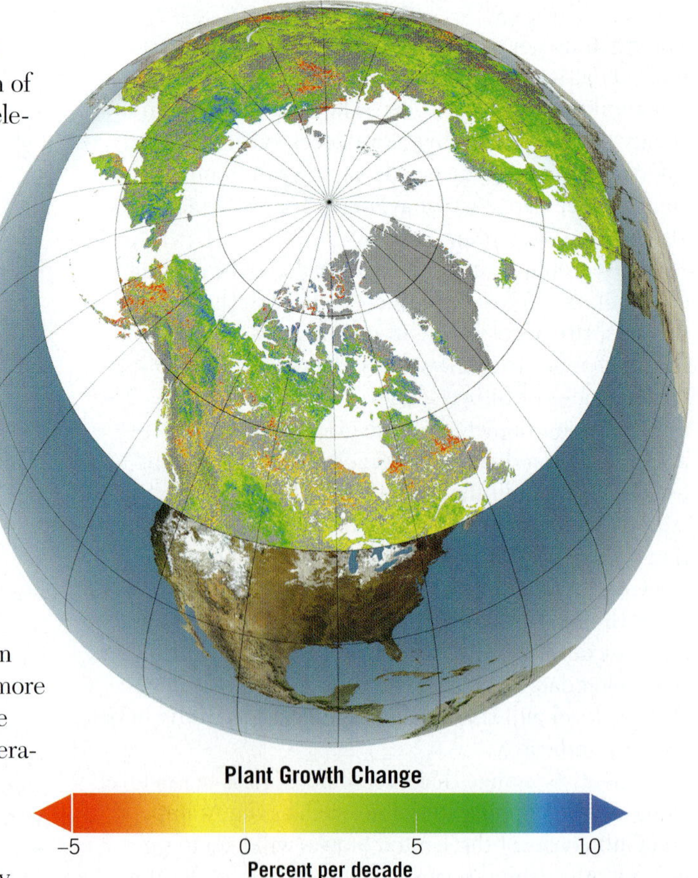

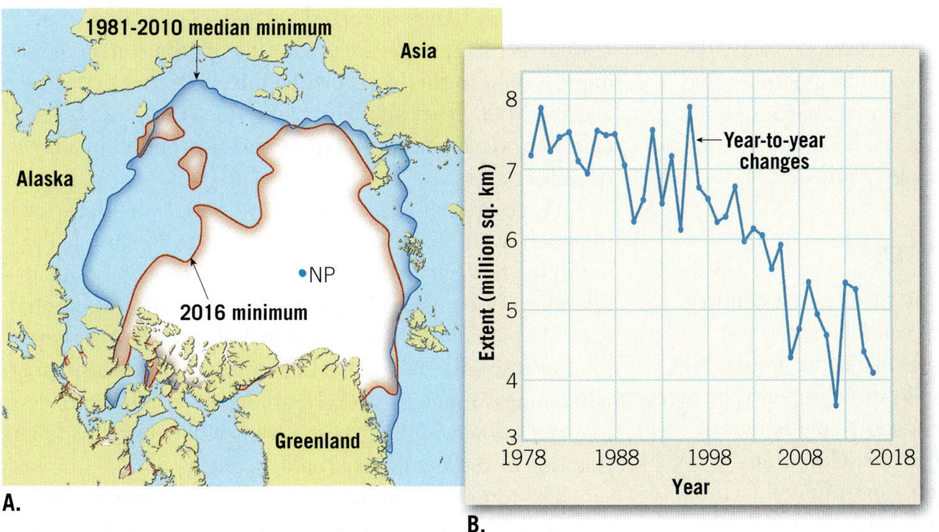

◀ **SmartFigure 20.35 Tracking sea ice changes** Sea ice is frozen seawater. In winter the Arctic Ocean is completely ice covered. In summer, a portion of the ice melts. **A.** This map shows the extent of sea ice in early September 2016 compared to the average extent of the period 1981 to 2010. The sea ice that does not melt in summer is getting thinner. **B.** The graph clearly depicts the trend in the area covered by sea ice at the end of the summer melt period. (Data from National Snow and Ice Data Center)

VIDEO
http://goo.gl/bVornN

year-to-year changes, clearly depicts the trend. Not only is the area covered by sea ice declining, but the remaining sea ice has become thinner and therefore more vulnerable to further melting. Models that best match historical trends project that Arctic waters may be virtually ice free in the late summer by the 2030s. As noted earlier in this chapter, a reduction in sea ice is a positive feedback mechanism that reinforces global warming.

The area covered by Arctic sea ice when it reaches its *maximum* extent in late winter is also declining. On March 24, 2016, the extent of sea ice peaked at 14.5 million square kilometers (5.6 million square miles)—the smallest maximum since satellite measurements started. Indeed, as of 2016, the 13 smallest maximums had all occurred in the preceding 13 years.

Permafrost Chapter 12 included a brief discussion of permanently frozen ground called *permafrost* that occurs in large portions of the high latitudes of the Northern Hemisphere. Mounting evidence indicates that the extent of permafrost in the Northern Hemisphere has decreased over the past decade, as would be expected under long-term warming conditions.

Studies in Alaska show that thawing is occurring in interior and southern parts of the state where permafrost temperatures are near the thaw point. As Arctic temperatures continue to rise, some models project that near-surface permafrost may be lost entirely from large parts of Alaska by the end of the century.

Thawing permafrost represents a potentially significant positive feedback mechanism that may reinforce global warming. When vegetation dies in the Arctic, cold temperatures inhibit its decomposition. As a consequence, over thousands of years, a great deal of organic matter has become stored in the permafrost. When the permafrost thaws, organic matter that may have been frozen for millennia comes out of "cold storage" and decomposes. The result is the release of carbon dioxide and methane—greenhouse gases that contribute to global warming. Thus, like decreasing sea ice, thawing permafrost is a positive feedback mechanism.

Increasing Ocean Acidity

The human-induced increase in the amount of carbon dioxide in the atmosphere has some serious implications for ocean chemistry and for marine life. Recent studies show that about one-third of the human-generated CO_2 currently ends up in the oceans. This additional carbon dioxide lowers the ocean's pH, making seawater more acidic. The pH scale is shown and briefly described in **Figure 20.36**.

When atmospheric CO_2 dissolves in seawater (H_2O), it forms carbonic acid (H_2CO_3). This lowers the oceans' pH and changes the balance of certain chemicals found naturally in seawater. In fact, the oceans have already absorbed enough carbon dioxide for surface waters to have experienced a pH decrease of 0.1 pH unit since preindustrial times, with additional pH decrease likely in the future. Moreover, if the current trend in carbon dioxide emissions continues, the oceans will experience a pH decrease of at least 0.2 pH unit by 2100—a change in ocean chemistry that has not occurred for millions of years. This shift toward acidity and the resulting changes in ocean chemistry make it more difficult for certain

▼ **Figure 20.36 The pH scale** This is the common measure of the degree of acidity or alkalinity of a solution. The scale ranges from 0 to 14, with a value of 7 indicating a solution that is neutral. Values below 7 indicate greater acidity, whereas numbers above 7 indicate greater alkalinity. It is important to note that the pH scale is logarithmic; that is, each whole number increment indicates a tenfold difference. Thus, pH 4 is 10 times more acidic than pH 5 and 100 times (10 × 10) more acidic than pH 6.

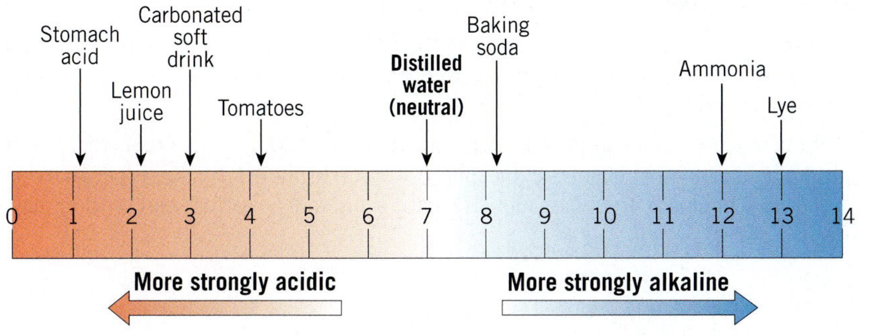

marine creatures to build hard parts out of calcium carbonate. The decline in pH thus threatens a variety of calcite-secreting organisms as diverse as microbes and corals, which concerns marine scientists because of the potential consequences for other sea life that depends on the health and availability of these organisms.

The Potential for Surprises

You have seen that climate in the twenty-first century, unlike during the preceding thousand years, is not expected to be stable. Rather, change is occurring. The amount and rate of future climate shifts depends primarily on current and future human-caused emissions of heat-trapping gases and airborne particles. Many of the changes will probably be gradual environmental shifts, nearly imperceptible from year to year. Nevertheless, the effects, accumulated over decades, will have powerful economic, social, and political consequences.

Despite our best efforts to understand future climate shifts, there is also the potential for surprises: Due to the complexity of Earth's climate system, we might experience relatively sudden, unexpected changes or see some aspects of climate shift in unanticipated ways. Many future climate scenarios predict steadily changing conditions, giving the impression that humanity will have time to adapt. However, the scientific community has indicated that at least some changes will be abrupt, perhaps crossing a threshold, or "tipping point," so quickly that there will be little time to react. This is a reasonable concern because abrupt changes occurring over periods as short as decades or even years have been a natural part of the climate system throughout Earth history. The paleoclimate record described earlier in the chapter contains ample evidence of such abrupt changes. For example, one such abrupt change occurred at the end of a time span known as the *Younger Dryas*, a time of abnormal cold and drought in the Northern Hemisphere that occurred about 12,000 years ago. Following this 1000-year-long cold period, the Younger Dryas abruptly ended in a few decades or less.

There are many examples of potential surprises, each of which would have large consequences. We simply do not know how far the climate system or other systems it affects can be pushed before they respond in unexpected ways. Even if the chance of any particular surprise happening is small, the chance that at least one such surprise will occur is much greater. In other words, although we may not know which of these events will occur, it is likely that one or more will eventually occur.

The impact on climate of an increase in atmospheric CO_2 and trace gases is obscured by some uncertainties. Yet climate scientists continue to improve our understanding of the climate system and the potential impacts and effects of global climate change. Policymakers are confronted with responding to the risks posed by greenhouse gas emissions, knowing that our understanding is imperfect. They must also face the fact that climate-induced environmental changes cannot be reversed quickly, if at all, due to the lengthy time scales associated with the climate system.

CONCEPT CHECKS 20.8

1. Describe the factors that are causing sea level to rise.
2. Is global warming greater near the equator or near the poles? Explain.
3. Based on Table 20.1, what projected changes relate to something other than temperature?

CONCEPTS IN REVIEW
Global Climate Change

20.1 Climate & Geology

List the major parts of the climate system and some connections between climate and geology.

KEY TERMS: weather, climate, climate system, cryosphere

- Climate is the aggregate of weather conditions for a place or region over a long period of time. Over time, if those conditions shift (for instance, to become hotter or cooler, wetter or drier), the climate is said to have changed.
- Earth's climate system is a complex interchange of energy and moisture that occurs among the atmosphere, hydrosphere, geosphere, biosphere, and cryosphere (ice, snow, permafrost). When the climate changes, geologic processes such as weathering, mass wasting, and erosion may change as well.

20.2 Detecting Climate Change

Explain why unraveling past climate changes is important and discuss several ways in which such changes are detected.

KEY TERMS: proxy data, paleoclimatology, oxygen isotope analysis

- The geologic record yields multiple kinds of indirect evidence about past climate. These proxy data are the focus of paleoclimatology. Proxy data can come from seafloor sediment, oxygen isotopes, cores of glacial ice, tree rings, coral growth bands, fossil pollen, and even historical documents.
- Trees grow thicker rings in warmer, wetter years and thinner rings in colder, drier years. The pattern of ring thickness can be matched up between trees of overlapping ages for a long-term record of a region's climate.
- Oxygen isotope analysis is based on the difference between heavier ^{18}O and lighter ^{16}O isotopes of oxygen. Water molecules containing ^{16}O evaporate more readily and thus are enriched in rain, snow, and glacial ice, leaving ocean water enriched in ^{18}O. Consequently, the $^{18}O/^{16}O$ ratio of ocean water rises during times of glaciation. During warm periods, glacial ice melts, and also the difference in evaporation between ^{16}O- and ^{18}O-containing water is smaller, so the $^{18}O/^{16}O$ ratio of ocean water falls. Oxygen isotopes can be measured in the shells of fossil marine organisms and in the water molecules that make up glacial ice. Glacial ice also traps small samples of the atmosphere in air bubbles.

? Why might the sediment core being collected by this oceanographic research vessel be useful to scientists studying climate change?

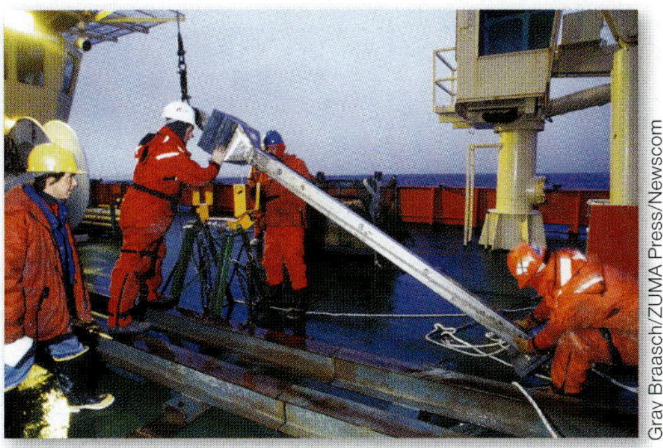

Gray Braasch/ZUMA Press/Newscom

20.3 Some Atmospheric Basics

Describe the composition of the atmosphere and the atmosphere's vertical changes in pressure and temperature.

KEY TERMS: aerosol, troposphere, radiosonde, stratosphere, mesosphere, thermosphere

- Air is a mixture of many discrete gases, and its composition varies from time to time and place to place. Two gases, nitrogen and oxygen, make up 99 percent of the volume of clean, dry air. Carbon dioxide, although present in only minute amounts (0.0400 percent, or 400 parts per million), is an efficient absorber of energy emitted by Earth and thus influences the heating of the atmosphere.
- Two important variable components of air are water vapor and aerosols. Like carbon dioxide, water vapor can absorb heat given off by Earth. Aerosols (tiny solid and liquid particles) are important because these often invisible particles act as surfaces on which water vapor can condense and are also good absorbers and reflectors (depending on the particles) of incoming solar radiation.
- The atmosphere is densest closest to the surface of Earth. It thins rapidly with increasing altitude and gradually fades off into space. Temperature varies through a vertical section of the atmosphere. Generally, temperatures fall with increasing altitude in the troposphere, warm with altitude in the stratosphere, cool with altitude in the mesosphere, and increase again in the very tenuous thermosphere.

(20.3 continued)

? This graph shows changes in an atmospheric element from Earth's surface to a height of about 140 kilometers (90 miles). Which element is being depicted: air pressure, humidity, or temperature? Show how this graph is used to divide the atmosphere into layers.

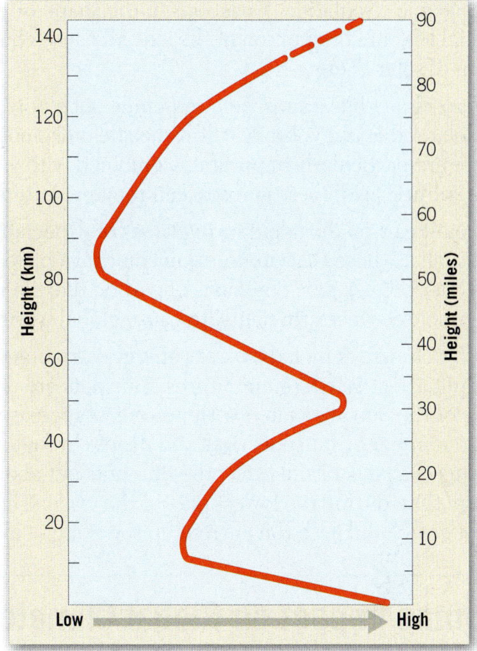

20.4 Heating the Atmosphere

Outline the basic processes involved in heating the atmosphere.

KEY TERMS: albedo, greenhouse effect

- Electromagnetic radiation is energy emitted in the form of rays, or waves, called electromagnetic waves. All radiation can transmit energy through the vacuum of space. Electromagnetic waves have different wavelengths. Visible light is the only portion of the electromagnetic spectrum we can see. Some of the basic laws that govern radiation as it heats the atmosphere are (1) all objects emit radiant energy, (2) hotter objects radiate more total energy than do colder objects, (3) the hotter the radiating object, the shorter the wavelength of maximum radiation, and (4) objects that are good absorbers of radiation are good emitters as well. Gases are selective absorbers, meaning that each specific type of gas absorbs and emits certain wavelengths but not others.
- Approximately 50 percent of the solar energy that strikes the top of the atmosphere reaches Earth's surface. About 30 percent is reflected back to space. The remaining 20 percent that does not reach Earth's surface is absorbed by clouds and the atmosphere's gases. The amount of solar energy that is absorbed (versus reflected or scattered) depends on its wavelengths and on the size and nature of the absorbing or reflecting substance.
- Radiant energy that is absorbed heats Earth and eventually is reradiated skyward. Because Earth has a much lower surface temperature than the Sun, its radiation is in the form of long-wave infrared radiation. Because certain atmospheric gases, primarily water vapor and carbon dioxide, can absorb long-wave radiation, the atmosphere is heated from the ground up. The transmission of short-wave solar radiation by the atmosphere, coupled with the selective absorption of Earth's long-wave radiation by atmospheric gases, results in the warming of the atmosphere and is called the greenhouse effect.

20.5 Natural Causes of Climate Change

Discuss hypotheses that relate to natural causes of climate change.

KEY TERM: sunspot

- The natural functions of the Earth system produce climate change. The position of lithospheric plates can influence the climate of the continents as well as oceanic circulation. Variations in the shape of Earth's orbit, angle of axial tilt, and orientation of the axis all cause changes in the distribution of solar energy.
- Volcanic aerosols act like a sunshade, screening out a portion of incoming solar radiation. Volcanic sulfur dioxide emissions that reach the stratosphere are particularly important. Combined with water to form tiny droplets of sulfuric acid, these aerosols can remain aloft for several years.
- Volcanoes emit carbon dioxide. During times of especially large eruptions, such as those that produced oceanic lava plateaus during the Cretaceous period, volcanic emissions of carbon dioxide may contribute to the greenhouse effect sufficiently to cause global warming.
- Since Earth's climate is fueled by solar energy, variations in the Sun's energy output affect Earth temperatures. Sunspots are dark features on the surface of the Sun associated with periods of increased solar energy output. The number of sunspots rises and drops throughout an 11-year cycle. During the peak of the cycle, the Sun puts out about 0.1 percent more energy than during the lowest part of the cycle. This small cyclical effect is not correlated with the current episode of global warming.

20.6 Human Impact on Global Climate

Summarize the nature and cause of the atmosphere's changing composition since about 1750. Describe the climate's response.

KEY TERMS: trace gases, black carbon

- Humans have been modifying the environment for thousands of years. Clearing or burning ground cover and overgrazing land have changed important climatic factors such as surface albedo, evaporation rates, and surface winds.
- Human activities produce climate change by releasing carbon dioxide (CO_2) and trace gases. Humans release CO_2 when they cut down forests or burn fossil fuels such as coal, oil, and natural gas. A steady rise in atmospheric CO_2 levels has been documented at Mauna Loa, Hawaii, and other locations around the world.
- More than half of the carbon released by humans is absorbed by new plant matter or dissolved in the oceans. About 45 percent remains in the atmosphere, where it can influence climate for decades. Air bubbles trapped in glacial ice reveal that there is currently about 30 percent more CO_2 than the atmosphere has contained in the past 800,000 years.
- As a result of extra heat retained by added CO_2, Earth's atmosphere has warmed by about 1°C (1.8°F) since the late 1800s, most of it since 1980. Temperatures are projected to increase by another 2°C to 4.5°C (3.5°F to 8.1°F) in the future.
- Trace gases such as methane, nitrous oxide, and CFCs also play significant roles in increasing global temperature.
- When emitted due to human activities, tiny liquid and solid particles suspended in the air, called aerosols, have an effect on global climate. Many aerosols reflect a portion of incoming solar radiation back to space and therefore have a cooling effect. Some aerosols, called black carbon, absorb incoming solar radiation and warm the atmosphere. When black carbon is deposited on snow and ice, it reduces surface albedo and increases the amount of light absorbed at the surface.

? Do aerosols spend more or less time in the atmosphere than greenhouse gases such as carbon dioxide? What is the significance of this difference in residence time? Explain.

20.7 Climate Feedback Mechanisms

Contrast positive and negative feedback mechanisms and provide examples of each.

KEY TERMS: climate feedback mechanism, positive feedback mechanism, negative feedback mechanism

- A change in one part of the climate system may trigger changes in other parts of the climate system that amplify or diminish the initial effect. These climate feedback mechanisms are called positive feedback mechanisms if they reinforce the initial change and negative feedback mechanisms if they counteract the initial effect.
- An example of a positive feedback mechanism is the melting of sea ice due to global warming. A decrease in sea ice exposes more low-albedo open water, which increases warming, which causes more sea ice to melt, and so forth. Cloud production can constitute a negative feedback mechanism: Warming increases evaporation, which leads to the formation of more clouds, which causes cooling.
- Computer models of climate give scientists a tool for testing hypotheses about climate change. Although these models are simpler than the real climate system, they are useful tools for predicting the future climate.

? This ice breaker is plowing through sea ice in the Arctic Ocean. What spheres of the climate system are represented in this photo? How has the area covered by summer sea ice been changing since satellite monitoring began in 1979? How does this change influence temperatures in the Arctic?

Wolfgang Bechtold/Imagebroker/Alamy

20.8 Some Consequences of Global Warming

Summarize some of the possible consequences of global warming.

- In the future, Earth's surface temperature is likely to continue to rise. The temperature increase will likely be greatest in the polar regions and least in the tropics. Some areas will get drier, and other areas will get wetter.
- Sea level is predicted to rise for several reasons, including the melting of glacial ice and thermal expansion. (A given mass of seawater takes up more volume when it is warm than when it is cool.) Low-lying, gently sloped coastal areas (which are often highly populated) are most at risk.
- The extent and thickness of sea ice in the Arctic have been declining since satellite observations began in 1979.
- Because of the warming of the Arctic, permafrost is melting, releasing CO_2 and methane to the atmosphere in a positive feedback loop.
- Because the climate system is complicated, dynamic, and imperfectly understood, it could produce sudden, unexpected changes with little warning.

GIVE IT SOME THOUGHT

1 Refer to Figure 20.1, which illustrates various components of Earth's climate system. The boxed labels represent interactions or changes that occur in the climate system. Select three of these boxes and provide an example of an interaction or change associated with each. Explain how these interactions may influence temperature.

2 When this weather balloon was launched, the surface air temperature was 17°C. The balloon is now at an altitude of 1 kilometer. What term is applied to the instrument package being carried aloft by the balloon? In what layer of the atmosphere is the balloon? If average conditions prevail, what is the air temperature at this altitude? How did you figure this out?

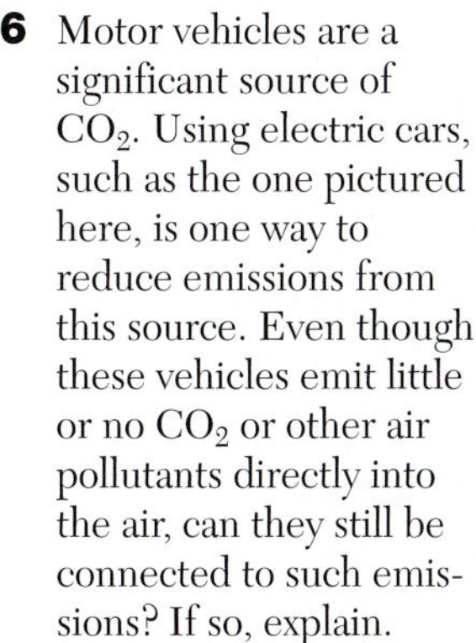

David R. Frazier/Science Source

3 Figure 20.15 shows that about 30 percent of the Sun's energy intercepted by Earth is reflected or scattered back to space. If Earth's albedo were to increase to 50 percent, how would you expect Earth's average surface temperature to change? Explain.

4 Volcanic events, such as the eruptions of El Chichón and Mount Pinatubo, have been associated with drops in global temperatures. During the Cretaceous period, volcanic activity was associated with global warming. Explain the apparent paradox.

5 The accompanying photo is a 2005 view of Athabasca Glacier in the Canadian Rockies. A line of boulders in the foreground marks the outer limit of the glacier in 1992. Is the behavior of Athabasca Glacier shown in this image typical of other glaciers around the world? Describe a significant impact of such behavior.

Hughrocks

6 Motor vehicles are a significant source of CO_2. Using electric cars, such as the one pictured here, is one way to reduce emissions from this source. Even though these vehicles emit little or no CO_2 or other air pollutants directly into the air, can they still be connected to such emissions? If so, explain.

David Pearson/Alamy

7 During a conversation, an acquaintance indicates that he is skeptical about global warming. When you ask him why he feels that way, he says, "The past couple of years in this area have been among the coolest I can remember." While you assure this person that it is useful to question scientific findings, you suggest to him that his reasoning in this case may be flawed. Use your understanding of the definition of *climate* along with one or more graphs in the chapter to persuade this person to reevaluate his reasoning.

8 A 2015 report by the National Research Council recommends that climate intervention research be conducted to explore strategies to help offset global warming. One strategy would involve injecting aerosols into the stratosphere.

a. Explain how such an aerosol strategy might influence global warming.

b. What *natural* cause of climate change operates on the same principle as the one you describe in your answer to Part a?

9 This large cattle feedlot is in the Texas panhandle. How might consuming less beef influence global climate change?

Glow Images

MasteringGeology™

Looking for additional review and test prep materials? Visit the Study Area in MasteringGeology to enhance your understanding of this chapter's content by accessing a variety of resources, including Self-Study Quizzes, Geoscience Animations, SmartFigures, Mobile Field Trips, *Project Condor* Quadcopter videos, *In the News* RSS feeds, flashcards, web links, and an optional Pearson eText.

www.masteringgeology.com

APPENDIX Metric and English Units Compared

Units

1 kilometer (km) = 1000 meters (m)
1 meter (m) = 100 centimeters (cm)
1 centimeter (cm) = 0.39 inch (in.)
1 mile (mi) = 5280 feet (ft)
1 foot (ft) = 12 inches (in.)
1 inch (in.) = 2.54 centimeters (cm)
1 square mile (mi^2) = 640 acres (a)
1 kilogram (kg) = 1000 grams (g)
1 pound (lb) = 16 ounces (oz)
1 fathom = 6 feet (ft)

Conversions

When you want to convert:	multiply by:	to find:
Length		
inches	2.54	centimeters
centimeters	0.39	inches
feet	0.30	meters
meters	3.28	feet
yards	0.91	meters
meters	1.09	yards
miles	1.61	kilometers
kilometers	0.62	miles
Area		
square inches	6.45	square centimeters
square centimeters	0.15	square inches
square feet	0.09	square meters
square meters	10.76	square feet
square miles	2.59	square kilometers
square kilometers	0.39	square miles
Volume		
cubic inches	16.38	cubic centimeters
cubic centimeters	0.06	cubic inches
cubic feet	0.028	cubic meters
cubic meters	35.3	cubic feet
cubic miles	4.17	cubic kilometers
cubic kilometers	0.24	cubic miles
liters	1.06	quarts
liters	0.26	gallons
gallons	3.78	liters
Masses and Weights		
ounces	28.35	grams
grams	0.035	ounces
pounds	0.45	kilograms
kilograms	2.205	pounds

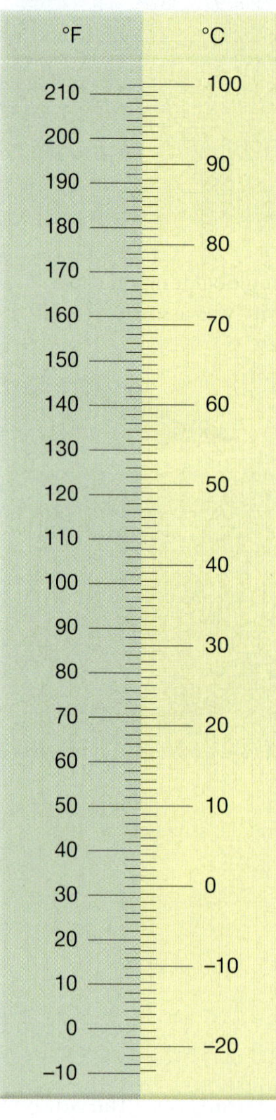

Temperature

When you want to convert degrees Fahrenheit (°F) to degrees Celsius (°C), subtract 32 degrees and divide by 1.8.

When you want to convert degrees Celsius (°C) to degrees Fahrenheit (°F), multiply by 1.8 and add 32 degrees.

When you want to convert degrees Celsius (°C) to Kelvins (K), delete the degree symbol and add 273. When you want to convert Kelvins (K) to degrees Celsius (°C), add the degree symbol and subtract 273.

◀ **Figure A.1** A comparison of Fahrenheit and Celsius temperature scales.

GLOSSARY

A

Aa flow A type of lava flow that has a jagged, blocky surface.

Ablation A general term for the loss of ice and snow from a glacier.

Abrasion The grinding and scraping of a rock surface by the friction and impact of rock particles carried by water, wind, and ice.

Abyssal plain A very level area of the deep-ocean floor, usually lying at the foot of the continental rise.

Accretionary wedge A large wedge-shaped mass of sediment that accumulates in subduction zones. Here sediment is scraped from the subducting oceanic plate and accreted to the overriding crustal block.

Active continental margin A margin that is usually narrow and consists of highly deformed sediments. Such margins occur where oceanic lithosphere is being subducted beneath the margin of a continent.

Active layer The zone above the permafrost that thaws in summer and refreezes in winter.

Aerosol Tiny solid and liquid particles suspended in the atmosphere.

Aftershock A smaller earthquake that follows the main earthquake.

Albedo The reflectivity of a substance, usually expressed as a percentage of the incident radiation reflected.

Alluvial channel A stream channel in which the bed and banks are composed largely of unconsolidated sediment (alluvium) that was previously deposited in the valley.

Alluvial fan A fan-shaped deposit of sediment formed when a stream's slope is abruptly reduced.

Alluvium Unconsolidated sediment deposited by a stream.

Alpine glacier A glacier confined to a mountain valley, which in most instances had previously been a stream valley. Also called a *valley glacier.*

Ambiguous property A property of a mineral that is not diagnostic because it varies among different specimens of the mineral.

Andesite A gray, fine-grained igneous rock, primarily of volcanic origin and commonly exhibiting a porphyritic texture.

Andesitic composition A compositional group of igneous rocks that contains at least 25 percent dark silicate minerals. The other dominant mineral is plagioclase feldspar. Also called *Intermediate composition*.

Angiosperm A flowering plant in which fruits contain the seeds.

Angle of repose The steepest angle at which loose material remains stationary without sliding downslope.

Angular unconformity An unconformity in which the older strata dip at an angle different from that of the younger beds.

Antecedent stream A stream that continued to downcut and maintain its original course as an area along its course was uplifted by faulting or folding.

Anthracite A hard metamorphic form of coal that burns cleanly and hot.

Anticline A fold in sedimentary strata that resembles an arch.

Aphanitic texture A texture of igneous rocks in which the crystals are too small for individual minerals to be distinguished without the aid of a microscope. Also called *fine-grained texture.*

Aquifer Rock or sediment through which groundwater moves easily.

Aquitard An impermeable bed that hinders or prevents groundwater movement.

Archean The first eon of Precambrian time. The eon preceding the Proterozoic. It extends between 4.5 and 2.5 billion years ago.

Arête A narrow, knifelike ridge separating two adjacent glaciated valleys.

Artesian A system in which groundwater under pressure rises above the level of the aquifer.

Arkose A feldspar-rich sandstone.

Assimilation In igneous activity, the process of incorporating country rock into a magma body.

Asteroid One of thousands of small planetlike bodies, ranging in size from a few hundred kilometers to less than 1 kilometer across. Most asteroids' orbits lie between those of Mars and Jupiter.

Asthenosphere A subdivision of the mantle situated below the lithosphere. This zone of weak material exists below a depth of about 100 kilometers (60 miles) and in some regions extends as deep as 700 kilometers (430 miles). The rock within this zone is easily deformed.

Atmosphere The gaseous portion of a planet, the planet's envelope of air. One of the traditional subdivisions of Earth's physical environment.

Atoll A coral island that consists of a nearly continuous ring of coral reef surrounding a central lagoon.

Atom The smallest particle that exists as an element.

Atomic mass unit A mass unit equal to exactly one-twelfth the mass of a carbon-12 atom.

Atomic number The number of protons in the nucleus of an atom.

Augite A black, opaque silicate mineral of the pyroxene group that is a dominant component of basalt.

Atomic weight The average of the atomic masses of isotopes for a given element.

Aureole A zone or halo of contact metamorphism found in the country rock surrounding an igneous intrusion.

B

Backarc basin A basin that forms on the side of a volcanic arc away from the trench.

Back swamp A poorly drained area on a floodplain resulting when natural levees are present.

Bajada An apron of sediment along a mountain front created by the coalescence of alluvial fans.

Banded iron formation A finely layered iron and silica-rich (chert) layer deposited mainly during the Precambrian.

Bar Common term for sand and gravel deposits in a stream channel.

Barchan dune A solitary sand dune shaped like a crescent, with its tips pointing downwind.

Barchanoid dune A type of dune that forms scalloped rows of sand oriented at right angles to the wind. This form is intermediate between isolated barchans and extensive waves of transverse dunes.

Barrier island A low, elongate ridge of sand that parallels the coast.

Basal slip A mechanism of glacial movement in which the ice mass slides over the surface below.

Basalt A fine-grained igneous rock of mafic composition.

Basalt plateau The broad and extensive accumulation of lava from a succession of flows emanating from fissure eruptions.

Basaltic composition A compositional group of igneous rocks indicating that the rock contains substantial dark silicate minerals and calcium-rich plagioclase feldspar. Also called *mafic composition.*

Base level The level below which a stream cannot erode.

Basin A circular downfolded structure.

Batholith A large mass of igneous rock that formed when magma was emplaced at depth, crystallized, and subsequently exposed by erosion.

Bathymetry The measurement of ocean depths and the charting of the topography of the ocean floor.

Baymouth bar A sandbar that completely crosses a bay, sealing it off from the main body of water.

Beach An accumulation of sediment found along the landward margin of the ocean or a lake.

Beach drift The transport of sediment in a zigzag pattern along a beach, caused by the uprush of water from obliquely breaking waves.

Beach nourishment A process in which large quantities of sand are added to the beach system to offset losses caused by wave erosion. Building beaches seaward improves beach quality and storm protection.

Bed load Sediment moved along the bottom of a stream by moving water, or particles moved along the ground surface by wind.

Bedding plane A nearly flat surface that separates two beds of sedimentary rock. Each bedding plane marks the end of one deposit and the beginning of another having different characteristics.

Bedrock channel A channel in which a stream is cutting into solid rock. Such channels typically form in the headwaters or river systems where gradients are high.

Beds Parallel layers of sedimentary rock. Also called *strata*.

Biochemical A type of chemical sediment that forms when material dissolved in water is precipitated by water-dwelling organisms. Shells are common examples.

Biogenous sediment Seafloor sediments consisting of material of marine-organic origin.

Biomass Organic material that is renewable energy derived from trees, crops, and waste. Examples include biofuels such as ethanol and biodiesel, as well as biogas, which is methane recovered from landfills.

Biosphere The totality of life-forms on Earth.

Biotite A dark iron-rich mineral and a member of the mica family with excellent cleavage.

Black carbon Soot generated by combustion processes and fires.

Black smoker A hydrothermal vent on the ocean floor that emits a black cloud of hot, metal-rich water.

Block lava Lava that has a surface of angular blocks associated with material having andesitic and rhyolitic compositions.

Blowout A depression excavated by wind in easily eroded materials.

Body waves Seismic wave that travel through Earth's interior.

Bowen's reaction series A concept proposed by N. L. Bowen that illustrates the relationships between magma and the minerals crystallizing from it during the formation of igneous rocks.

Braided channel Type of alluvial channel consisting of a broad network of diverging and converging channels. Forms where the stream's load includes abundant coarse material and the discharge is highly variable.

Breakwater A structure that protects a near-shore area from breaking waves.

Breccia A sedimentary rock composed of angular fragments that were lithified.

Brittle deformation The loss of strength by a material, usually in the form of sudden fracturing.

Building material Category of nonsilicate mineral resource used in building.

Burial metamorphism Low-grade metamorphism that occurs in the lowest layers of very thick accumulations of sedimentary strata.

C

Calcite Calcium carbonate ($CaCO_3$), one of the two most common carbonate minerals.

Caldera A large depression typically caused by collapse or ejection of the summit area of a volcano.

Calving Wastage of a glacier that occurs when large pieces of ice break into the water.

Cambrian explosion The huge expansion in biodiversity that occurred at the beginning of the Paleozoic era.

Cap rock A necessary part of an oil trap. The cap rock is impermeable and hence keeps upwardly mobile oil and gas from escaping at the surface.

Capacity The total amount of sediment that a stream is able to transport.

Capillary fringe A relatively narrow zone at the base of the zone of aeration. Here water rises from the water table in tiny, threadlike openings between grains of soil or sediment.

Carbon cycle An Earth system in which carbon moves through the atmosphere, hydrosphere, biosphere, and geosphere, in different directions.

Carbonic acid A weak acid formed when carbon dioxide is dissolved in water. It plays an important role in chemical weathering.

Catastrophism The concept that Earth was shaped by catastrophic events of a short-term nature.

Cavern A naturally formed underground chamber or series of chambers most commonly produced by solution activity in limestone.

Cementation One way in which sedimentary rocks are lithified. As material precipitates from water that percolates through the sediment, open spaces are filled and particles are joined into a solid mass.

Cenozoic era A time span on the geologic time scale beginning about 65.5 million years ago, following the Mesozoic era.

Chemical bond A strong attractive force that exists between atoms in a substance. It involves the transfer or sharing of electrons that allows each atom to attain a full valence shell.

Chemical compound A substance formed by the chemical combination of two or more elements in definite proportions and usually having properties different from those of its constituent elements.

Chemical sedimentary rock Sedimentary rock consisting of material that was precipitated from water by either inorganic or organic means.

Chemical weathering The processes by which the internal structure of a mineral is altered by the removal and/or addition of elements.

Chert A durable sedimentary rock formed of microcrystalline quartz.

Cinder cone A rather small volcano built primarily of ejected lava fragments that consist mostly of pea- to walnut-size lapilli. Also called a *scoria cone.*

Circum-Pacific belt An area approximately 40,000 kilometers (24,000 miles) in length surrounding the basin of the Pacific Ocean where oceanic lithosphere is continually subducted beneath the surrounding continental plates causing most of Earth's largest earthquakes.

Cirque An amphitheater-shaped basin at the head of a glaciated valley produced by frost wedging and plucking.

Clastic texture A sedimentary rock texture consisting of broken fragments of preexisting rock.

Clay (Clay mineral) A group of light-colored silicates that typically form as products of chemical weathering of igneous rocks. Major components of soil and sedimentary rocks. Kaolinite is a common clay mineral derived from the weathering of feldspar.

Cleavage The tendency of a mineral to break along planes of weak bonding.

Climate A description of aggregate weather conditions; the sum of all statistical weather information that helps describe a place or region.

Climate feedback mechanism Various outcomes that may result when one of this complex interactive physical system's elements is altered.

Climate system Exchanges of energy and moisture occurring among the atmosphere, hydrosphere, lithosphere, biosphere, and cryosphere.

Coal A sedimentary rock consisting primarily of organic matter, formed in stages from accumulations of large quantities of undecayed plant material. Used as a fossil fuel.

Coarse-grained texture An igneous rock texture in which the crystals are roughly equal in size and large enough so the individual minerals can be identified without the aid of a microscope. Also called *Phaneritic texture*.

Collisional mountains Mountains in which great horizontal forces have shortened and thickened the crust. Most major mountain belts are of this type.

Color A phenomenon of light by which otherwise identical objects may be differentiated.

Column A feature found in caves that is formed when a stalactite and stalagmite join.

Columnar jointing A pattern of cracks that forms during cooling of molten rock to generate columns.

Compaction A type of lithification in which the weight of overlying material compresses more deeply buried sediment. It is most important in the fine-grained sedimentary rocks such as shale.

Competence A measure of the largest particle a stream can transport; a factor dependent on velocity.

Composite volcano A volcano composed of both lava flows and pyroclastic material. Also called a *stratovolcano*.

Compound A substance formed by the chemical combination of two or more elements in definite proportions and usually having properties different from those of its constituent elements.

Compressional mountains Mountains in which great horizontal forces have shortened and thickened the crust. Most major mountain belts are of this type.

Compressional stress Differential stress that shortens a rock body.

Concordant A term used to describe intrusive igneous masses that form parallel to the bedding of the surrounding rock.

Conduction The transfer of heat through matter by molecular activity.

Conduit A pipelike opening through which magma moves toward Earth's surface. It terminates at a surface opening called a vent.

Cone of depression A cone-shaped depression in the water table immediately surrounding a well.

Confined aquifer An aquifer that has impermeable layers (aquitards) both above and below.

Confining pressure Stress that is applied uniformly in all directions.

Conformable A reference to layers of sediment deposited without interruption.

Conglomerate A sedimentary rock composed of rounded, gravel-size particles.

Contact metamorphism Changes in rock caused by the heat from a nearby magma body. Also called *thermal metamorphism.*

Continent Large, continuous areas of land that include the adjacent continental shelf and islands that are structurally connected to the mainland.

Continental drift A hypothesis, credited largely to Alfred Wegener, which suggested that all present continents once existed as a single supercontinent. Further, beginning about 200 million years ago, the supercontinent began breaking into smaller continents, which then "drifted" to their present positions.

Continental margin The portion of the seafloor that is adjacent to the continents. It may include the continental shelf, continental slope, and continental rise.

Continental rift A linear zone along which contin ental lithosphere stretches and pulls apart. Its creation may mark the beginning of a new ocean basin.

Continental rise The gently sloping surface at the base of the continental slope.

Continental shelf The gently sloping submerged portion of the continental margin, extending from the shoreline to the continental slope.

Continental slope The steep gradient that leads to the deep-ocean floor and marks the seaward edge of the continental shelf.

Continental volcanic arc Mountains formed in part by igneous activity associated with the subduction of oceanic lithosphere beneath a continent. Examples include the Andes and the Cascades.

Convection The transfer of heat by the mass movement or circulation of a substance.

Convergent plate boundary A boundary in which two plates move together, resulting in oceanic lithosphere being thrust beneath an overriding plate, eventually to be reabsorbed into the mantle. It can also involve the collision of two continental plates to create a mountain system. See Also called a *subduction zone.*

Coral reef A structure formed in a warm, shallow, sunlit ocean environment that consists primarily of the calcite-rich remains of corals as well as the limy secretions of algae and the hard parts of many other small organisms.

Core The innermost layer of Earth. It is thought to be largely an iron–nickel alloy, with minor amounts of oxygen, silicon, and sulfur.

Correlation The process of establishing the equivalence of rocks of similar age in different areas.

Corrosion The process by which soluble rock is gradually dissolved by flowing water.

Covalent bond A chemical bond produced by the sharing of electrons.

Crater The depression at the summit of a volcano or a depression that is produced by a meteorite impact.

Craton The part of the continental crust that has attained stability; that is, it has not been affected by significant tectonic activity during the Phanerozoic eon. It consists of the shield and the stable platform.

Creep The slow downhill movement of soil and regolith.

Crevasse A deep crack in the brittle surface of a glacier.

Cross bed Thin inclined layers that form as sand is deposited on the slip face of a dune. See also *cross bedding.*

Cross-bedding A structure in which relatively thin layers are inclined at an angle to the main bedding. Cross-bedding is formed by currents of wind or water.

Cross-cutting *See* Principle of cross-cutting relationships.

Crust The very thin, outermost layer of Earth.

Cryosphere The portion of Earth's surface where water is in solid form, including snow, glaciers, sea ice, freshwater ice, and frozen ground. It is one of the spheres of the climate system.

Crystal Any natural solid with an ordered, repetitive atomic structure.

Crystal shape Refers to the common or characteristic shape of a crystal or an aggregate of crystals. Also called *habit*.

Crystalline Term for the texture of sedimentary rocks in which the minerals form a pattern of interlocking crystals. Also called *nonclastic texture*.

Crystallization The formation and growth of a crystalline solid from a liquid or gas.

Crystal settling A process that occurs during the crystallization of magma, in which the earlier-formed minerals are denser than the liquid portion and settle to the bottom of the magma chamber.

Curie point The temperature above which a material loses its magnetization.

Cut bank The area of active erosion on the outside of a meander.

Cutoff A short channel segment created when a river erodes through the narrow neck of land between meanders.

D

Darcy's law An equation which states that groundwater discharge depends on the hydraulic gradient, hydraulic conductivity, and cross-sectional area of an aquifer.

Dark silicate A silicate mineral that contains ions of iron and/or magnesium in its structure. Dark silicates are dark in color and have a higher specific gravity than nonferromagnesian silicates. Also called *ferromagnesian silicate*.

Daughter product An isotope that results from radioactive decay.

Debris avalanche A rapid and chaotic mass movement of material that is largely unconsolidated.

Debris flow A flow of soil and regolith that contains a large amount of water. Most common in semiarid mountainous regions and on the slopes of some volcanoes.

Debris slide See *Rockslide*.

Decompression melting Melting that occurs as rock ascends due to a drop in confining pressure.

Deep-ocean basin The portion of seafloor that lies between the continental margin and the oceanic ridge system. This region comprises almost 30 percent of Earth's surface.

Deep-ocean trench A narrow, elongated depression of the seafloor.

Deep-sea fan A cone-shaped deposit at the base of the continental slope. The sediment is transported to the fan by turbidity currents that follow submarine canyons.

Deflation The lifting and removal of loose material by wind.

Deformation General term for the processes of folding, faulting, shearing, compression, or extension of rocks as the result of various natural forces.

Delta An accumulation of sediment formed where a stream enters a lake or an ocean.

Dendritic pattern A stream system that resembles the pattern of a branching tree.

Density A property of matter defined as mass per unit volume.

Desert One of the two types of dry climate; the driest of the dry climates.

Desert pavement A layer of closely spaced coarse pebbles and gravel that cover barren, rocky deserts to form a relatively smooth surface.

Detachment A nearly horizontal fault that may extend for hundreds of kilometers below the surface. Such a fault represents a boundary between rocks that exhibit ductile deformation and rocks that exhibit brittle deformation.

Detrital sedimentary rock Rocks that form from the accumulation of materials that originate and are transported as solid particles derived from both mechanical and chemical weathering.

Diagenesis A collective term for all the chemical, physical, and biological changes that take place after sediments are deposited and during and after lithification.

Diagnostic property Property of a mineral that aids in mineral identification. Taste or feel, crystal shape, and streak are examples of diagnostic properties.

Differential stress Forces that are unequal in different directions.

Differential weathering The variation in the rate and degree of weathering caused by such factors as mineral makeup, degree of jointing, and climate.

Dike A tabular-shaped intrusive igneous feature that cuts through the surrounding rock.

Diorite A coarse-grained intrusive igneous rock primarily composed of plagioclase feldspar and amphibole minerals.

Dip The angle at which a rock layer or fault is inclined from the horizontal. The direction of dip is at a right angle to the strike.

Dip-slip fault A fault in which the movement is parallel to the dip of the fault.

Discharge The quantity of water in a stream that passes a given point in a period of time.

Discharge area A location, such as a spring or a stream, where groundwater flows back to the surface.

Disconformity A type of unconformity in which the beds above and below are parallel.

Discontinuity A sudden change of depth in one or more of the physical properties of the material making up Earth's interior. The boundary between two dissimilar materials in Earth's interior as determined by the behavior of seismic waves.

Discordant A term used to describe plutons that cut across existing rock structures, such as bedding planes.

Disseminated deposit Any economic mineral deposit in which the desired mineral occurs as scattered particles in the rock but in sufficient quantity to make the deposit an ore.

Dissolution A common form of chemical weathering, it is the process of dissolving into a homogeneous solution, as when an acidic solution dissolves limestone.

Dissolved load The portion of a stream's load that is carried in solution.

Distributary A section of a stream that leaves the main flow.

Divergent plate boundary A boundary in which two plates move apart, resulting in upwelling of material from the mantle to create new seafloor. Also called *spreading center*.

Divide An imaginary line that separates the drainage of two streams, often found along a ridge.

Dolomite Calcium/magnesium carbonate, $CaMg(CO_3)_2$, one of the two most common carbonate minerals.

Dolostone A chemical sedimentary rock formed from dolomite, a calcium–magnesium carbonate mineral.

Dome A roughly circular upfolded structure.

Downcutting The lowering of a streambed toward base level as turbulent water lifts unconsolidated material or when bedrock channels are lowered by means of quarrying, abrasion, and corrosion.

Drainage basin The land area that contributes water to a stream. Also called a *watershed*.

Drawdown The difference in height between the bottom of a cone of depression and the original height of the water table.

Drift See *Glacial drift*.

Drumlin A streamlined symmetrical hill composed of glacial till. The steep side of the hill faces the direction from which the ice advanced.

Dry climate A climate in which yearly precipitation is less than the potential loss of water by evaporation.

Ductile deformation A type of solid-state flow that produces a change in the size and shape of a rock body without fracturing. Occurs at depths where temperatures and confining pressures are high.

Dune A hill or ridge of wind-deposited sand.

E

Earth system science An interdisciplinary study that seeks to examine Earth as a system composed of numerous interacting parts or subsystems.

Earthflow The downslope movement of water-saturated, clay-rich sediment. Most characteristic of humid regions.

Earthquake Vibration of Earth produced by the rapid release of energy.

Ebb current The movement of tidal current away from the shore.

Echo sounder An instrument used to determine the depth of water by measuring the time interval between emission of a sound signal and the return of its echo from the bottom.

Economic mineral A mineral used extensively in the manufacture of products.

Effusive eruption A quiescent eruption that produces mainly outpourings of fluid lava.

Elastic deformation Rock deformation in which the rock will return to nearly its original size and shape when the stress is removed.

Elastic rebound The sudden release of stored strain in rocks that results in movement along a fault.

Electron A negatively charged subatomic particle that has a negligible mass and is found outside an atom's nucleus.

Element A substance that cannot be decomposed into simpler substances by ordinary chemical or physical means.

Eluviation The washing out of fine soil components from the *A* horizon by downward-percolating water.

Emergent coast A coast where land formerly below sea level has been exposed by crustal uplift or a drop in sea level or both.

End moraine A ridge of till marking a former position of the front of a glacier.

Energy levels Spherically shaped, negatively charged zones that surround the nucleus of an atom.

Environment of deposition A geographic setting where sediment accumulates. Each site is characterized by a particular combination of geologic processes and environmental conditions. Also called *sedimentary environment*.

Eon The largest time unit on the geologic time scale, next in order of magnitude above era.

Ephemeral stream A stream that is usually dry because it carries water only in response to specific episodes of rainfall. Most desert streams are of this type.

Epicenter The location on Earth's surface that lies directly above the focus of an earthquake.

Epoch A unit of the geologic time scale that is a subdivision of a period.

Equilibrium line For a glacier, the elevation at which the accumulation and wasting of glacial ice is equal.

Era A major division on the geologic time scale; eras are divided into shorter units called periods.

Erosion The incorporation and transportation of material by a mobile agent, such as water, wind, or ice.

Eruption column Buoyant plumes of hot, ash-laden gases that can extend thousands of meters into the atmosphere.

Escape velocity The initial velocity an object needs to escape from the surface of a celestial body.

Esker A sinuous ridge composed largely of sand and gravel deposited by a stream flowing in a tunnel beneath a glacier near its terminus.

Estuary A funnel-shaped inlet of the sea that formed when a rise in sea level or subsidence of land caused the mouth of a river to be flooded.

Eukaryote An organism whose genetic material is enclosed in a nucleus; plants, animals, and fungi are eukaryotes.

Evaporation The process of converting a liquid to a gas.

Evaporite A sedimentary rock formed of material deposited from solution by evaporation of the water.

Evapotranspiration The combined effect of evaporation and transpiration.

Exfoliation dome A large, dome-shaped structure, usually composed of granite, that is formed by sheeting.

External process A process such as weathering, mass wasting, or erosion that is powered by the Sun and contributes to the transformation of solid rock into sediment.

Extrusive igneous rock Igneous rock formed when magma solidifies at Earth's surface. Also called *volcanic rock.*

Eye A zone of scattered clouds and calm averaging about 20 kilometers (12 miles) in diameter at the center of a hurricane.

Eye wall The doughnut-shaped area of intense cumulonimbus development and very strong winds that surrounds the eye of a hurricane.

F

Facies A portion of a rock unit that possesses a distinctive set of characteristics that distinguishes it from other parts of the same unit.

Fall A type of movement that is common to mass-wasting processes that refers to the free falling of detached individual pieces of any size.

Fault A break in a rock mass along which movement has occurred.

Fault creep Gradual displacement along a fault. Such activity occurs relatively smoothly and with little noticeable seismic activity.

Fault scarp A cliff created by movement along a fault. It represents the exposed surface of the fault prior to modification by weathering and erosion.

Fault-block mountain A mountain that is formed by the displacement of rock along a fault.

Feldspar A group of nonferromagnesian silicate minerals; by far the most plentiful silicate group in Earth's crust.

Felsic composition A compositional group of igneous rocks indicating the rock is composed almost entirely of light-colored silicates. Also called *granitic composition*.

Ferromagnesian silicate A silicate mineral that contains ions of iron and/or magnesium in its structure. Dark silicates are dark in color and have a higher specific gravity than nonferromagnesian silicates. Also called a *dark silicate.*

Fetch The distance that the wind has traveled across the open water.

Fine-grained texture A texture of igneous rocks in which the crystals are too small for individual minerals to be distinguished without the aid of a microscope. Also called *aphanitic texture*.

Fiord A steep-sided inlet of the sea formed when a glacial trough was partially submerged.

Firn Granular, recrystallized snow. A transitional stage between snow and glacial ice.

Fissility The property of splitting easily into thin layers along closely spaced, parallel surfaces, such as bedding planes in shale.

Fissure A crack in rock along which there is a distinct separation.

Fissure eruption An eruption in which lava is extruded from narrow fractures or cracks in the crust.

Flood The overflow of a stream channel that occurs when discharge exceeds the channel's capacity. The most common and destructive geologic hazard.

Flood basalt Flow of basaltic lava that issues from numerous cracks or fissures and commonly covers extensive areas to thicknesses of hundreds of meters.

Flood current The tidal current associated with the increase in the height of the tide.

Floodplain The flat, low-lying portion of a stream valley subject to periodic inundation.

Flow A type of movement common to mass-wasting processes in which water-saturated material moves downslope as a viscous fluid.

Flowing artesian well An artesian well in which water flows freely at Earth's surface because the pressure surface is above ground level.

Focus The zone within Earth where rock displacement produces an earthquake. Also called the *hypocenter.*

Fold A bent layer or series of layers that were originally horizontal and subsequently deformed.

Fold-and-thrust belts Regions within compressional mountain systems where large areas have been shortened and thickened by the processes of folding and thrust faulting, as exemplified by the Valley and Ridge province of the Appalachians.

Foliated texture A texture of metamorphic rocks that gives the rock a layered appearance.

Foliation A term for a linear arrangement of textural features often exhibited by metamorphic rocks.

Footwall block The rock surface below a fault.

Forced subduction A process that occurs at Peru–Chile–type subduction zones in which lithosphere is too buoyant to subduct spontaneously but is forced beneath the overriding plate.

Forearc basin The region located between a volcanic arc and an accretionary wedge where shallow-water marine sediments typically accumulate.

Foreshocks Small earthquakes that often precede a major earthquake.

Fossil The remains or traces of organisms preserved from the geologic past.

Fossil assemblage The overlapping ranges of a group of fossils (assemblage) collected from a layer. By examining such an assemblage, the age of the sedimentary layer can be established.

Fossil fuel General term for any hydrocarbon that may be used as a fuel, including coal, oil, natural gas, bitumen from tar sands, and shale oil.

Fossil magnetism The natural remnant magnetism in rock bodies. The permanent magnetization acquired by rock that can be used to determine the location of the magnetic poles and the latitude of the rock at the time it became magnetized. Also called *paleomagnetism*.

Fossil succession The definite and determinable order in which fossil organisms occur. Fossil succession enables us to identify many time periods by their fossil content.

Fracture Any break or rupture in rock along which no appreciable movement has taken place.

Fracture zone A linear zone of irregular topography on the deep-ocean floor that follows transform faults and their inactive extensions.

Fragmental texture An igneous rock texture resulting from the consolidation of individual rock fragments that are ejected during a violent volcanic eruption. Also called *pyroclastic texture*.

Frost wedging The mechanical breakup of rock caused by the expansion of freezing water in cracks and crevices.

Fumarole A vent in a volcanic area from which fumes or gases escape.

G

Gabbro A dark-green to black intrusive igneous rock composed of dark silicate minerals. Gabbro makes up a significant percentage of oceanic crust.

Gaining stream Streams that gain water from the inflow of groundwater through the streambed.

Garnet Silicate mineral composed of individual silica tetrahedrons. Most often brown to deep red; has a glassy luster, lacks cleavage, and exhibits conchoidal fracture.

Geologic structure All features created by the processes of deformation, from minor fractures in bedrock to major mountain chains. Also called *rock structure, tectonic structure.*

Geologic time The span of time since the formation of Earth, about 4.6 billion years.

Geologic time scale The division of Earth history into blocks of time—eons, eras, periods, and epochs. The time scale was created using relative dating principles.

Geology The science that examines Earth, its form and composition, and the changes that it has undergone and is undergoing.

Geosphere The solid Earth; one of Earth's four basic spheres.

Geothermal energy Natural steam used for power generation.

Geothermal gradient The gradual increase in temperature with depth in the crust. The average is 30°C per kilometer in the upper crust.

Geyser A fountain of hot water ejected periodically from the ground.

Glacial budget The balance, or lack of balance, between ice formation at the upper end of a glacier and ice loss in the zone of wastage.

Glacial drift An all-embracing term for sediments of glacial origin, no matter how, where, or in what shape they were deposited.

Glacial erratic An ice-transported boulder that was not derived from the bedrock near its present site.

Glacial striations Scratches and grooves on bedrock caused by glacial abrasion.

Glacial trough A mountain valley that has been widened, deepened, and straightened by a glacier.

Glacier A thick mass of ice originating on land from the compaction and recrystallization of snow that shows evidence of past or present flow.

Glass (volcanic) Natural glass that is produced when molten lava cools too rapidly to permit recrystallization. Volcanic glass is a solid composed of unordered atoms.

Glassy texture A term used to describe the texture of certain igneous rocks, such as obsidian, that contain no crystals.

Gneiss Medium- to coarse-grained banded metamorphic rocks in which granular and elongated minerals dominate.

Gneissic banding A texture of metamorphic rocks in which dark and light silicate minerals are separated, giving the rock a banded appearance. Also called *gneissic texture.*

Gneissic texture A texture of metamorphic rocks in which dark and light silicate minerals are separated, giving the rock a banded appearance. Also called *gneissic banding.*

Gondwana The southern portion of Pangaea consisting of South America, Africa, Australia, India, and Antarctica.

Graben A valley formed by the downward displacement of a fault-bounded block.

Graded bed A sediment layer characterized by a decrease in sediment size from bottom to top.

Graded stream A stream that has the correct channel characteristics to maintain exactly the velocity required to transport the material supplied to it.

Gradient The slope of a stream, generally expressed as the vertical drop over a fixed distance.

Granite An abundant, coarse-grained igneous rock, composed of about 10–20 percent quartz and 50 percent potassium feldspar. Granite is used as a building material.

Granitic composition A compositional group of igneous rocks indicating the rock is composed almost entirely of light-colored silicates. Also called *felsic composition*.

Gravitational collapse The gradual subsidence of mountains caused by lateral spreading of weak material located deep within these structures.

Great Oxygenation Event A time about 2.5 billion years ago, when a significant amount of oxygen appeared in the atmosphere.

Greenhouse effect The transmission of short-wave solar radiation by the atmosphere coupled with the selective absorption of longer-wavelength terrestrial radiation, especially by water vapor and carbon dioxide, resulting in warming of the atmosphere.

Groin A short wall built at a right angle to the seashore to trap moving sand.

Ground moraine An undulating layer of till deposited as an ice front retreats.

Groundmass The matrix of smaller crystals within an igneous rock that has porphyritic texture.

Groundwater Water in the zone of saturation.

Guyot A submerged, flat-topped seamount.

Gymnosperm A group of seed-bearing plants that includes conifers and Ginkgo. The term means "naked seed," a reference to the unenclosed condition of the seeds.

Gypsum A hydrated calcium sulfate mineral. It is the mineral from which plaster, drywall, and other similar building materials are composed.

H

Habit Refers to the common or characteristic shape of a crystal or an aggregate of crystals. Also called *crystal shape*.

Hadean The earliest time interval (eon) of Earth history. The time before the planet's first rocks.

Half-graben A tilted fault block in which the higher side is associated with mountainous topography and the lower side is a basin that fills with sediment.

Half-life The time required for one-half of the atoms of a radioactive substance to decay.

Halite The mineral name for common table salt, NaCl. Commonly found in thick layers that are the last vestiges of ancient evaporated seas.

Hanging valley A tributary valley that enters a glacial trough at a considerable height above the floor of the trough.

Hanging wall block The rock surface immediately above a fault.

Hard stabilization An artificial structure built to protect a coast or to prevent the movement of sand along a beach. Examples include groins, jetties, breakwaters, and seawalls.

Hardness A mineral's resistance to scratching and abrasion.

Head The beginning or source area for a stream. Also called *headwaters*.

Headward erosion The extension upslope of the head of a valley due to erosion.

Headwaters The beginning or source area for a stream. Also called *head*.

Historical geology A major division of geology that deals with the origin of Earth and its development through time. Usually involves the study of fossils and their sequence in rock beds.

Horizon A layer in a soil profile.

Horn A pyramid-like peak formed by glacial action in three or more cirques surrounding a mountain summit.

Hornblende A dark green to black mineral of the amphibole group, often found in igneous rocks.

Hornfels A fine-grained nonfoliated metamorphic rock formed from various minerals.

Horst An elongate, uplifted block of crust bounded by faults.

Host rock Pre-existing crustal rocks intruded by magma. Host rock may be displaced or assimilated by magmas.

Hot spot A concentration of heat in the mantle, capable of producing magma that, in turn, extrudes onto Earth's surface. The intraplate volcanism that produced the Hawaiian Islands is one example.

Hot spot tracks A chain of volcanic structures produced as a lithospheric plate moves over a mantle plume.

Hot spring A spring in which the water is 6–9°C (10–15°F) warmer than the mean annual air temperature of its locality.

Humus Organic matter in soil that is produced by the decomposition of plants and animals.

Hurricane A tropical cyclonic storm that has winds in excess of 119 kilometers (74 miles) per hour.

Hydraulic conductivity A factor relating to ground-water flow; it is a coefficient that takes into account the permeability of the aquifer and the viscosity of the fluid.

Hydraulic fracturing A method of opening up pore space in otherwise impermeable rocks, permitting natural gas and oil to flow out into wells. Also called *fracking*.

Hydraulic gradient The slope of the water table. It is determined by finding the height difference between two points on the water table and dividing by the horizontal distance between the two points.

Hydroelectric power Electricity generated by falling water that is used to drive turbines.

Hydrologic cycle The unending circulation of Earth's water supply. The cycle is powered by energy from the Sun and is characterized by continuous exchanges of water among the oceans, the atmosphere, and the continents.

Hydrosphere The water portion of our planet; one of the traditional subdivisions of Earth's physical environment.

Hydrothermal metamorphism Chemical alterations that occur as hot, ion-rich water circulates through fractures in rock.

Hydrothermal solution The hot, watery solution that escapes from a mass of magma during the latter stages of crystallization. Such solutions may alter the surrounding country rock and are frequently the source of significant ore deposits.

Hypocenter The zone within Earth where rock displacement produces an earthquake. Also called the *focus*.

Hypothesis A tentative explanation that is then tested to determine if it is valid.

I

Ice cap A mass of glacial ice covering a high upland or plateau and spreading out radially.

Ice sheet A very large, thick mass of glacial ice flowing outward in all directions from one or more accumulation centers.

Ice shelf A large, relatively flat mass of floating ice that forms where glacial ice flows into bays and that extends seaward from the coast but remains attached to the land along one or more sides.

Iceberg A mass of floating ice produced by a calving glacier. Usually 20 percent or less of the iceberg protrudes above the waterline.

Igneous rock Rock formed from the crystallization of magma.

Immature soil A soil that lacks horizons.

Impact metamorphism Metamorphism that occurs when meteorites strike Earth's surface. Also called *shock metamorphism*.

Incised meander A meandering channel that flows in a steep, narrow valley. It forms either when an area is uplifted or when the base level drops.

Inclusion A piece of one rock unit that is contained within another. Inclusions are used in relative dating. The rock mass adjacent to the one containing the inclusion must have been there first in order to provide the fragment.

Index fossil A fossil that is associated with a particular span of geologic time.

Index mineral A mineral that is a good indicator of the metamorphic environment in which it formed. Used to distinguish different zones of regional metamorphism.

Industrial mineral Nonmetallic mineral resources that are used in industry.

Inertia A property by which objects at rest tend to remain at rest, and objects in motion tend to stay in motion unless either is acted upon by an outside force.

Infiltration The movement of surface water into rock or soil through cracks and pore spaces.

Inner core The solid innermost layer of Earth, about 1216 kilometers (754 miles) in radius.

Inner planets The innermost planets of our solar system, which include Mercury, Venus, Earth, and Mars. Also known as the terrestrial planets because of their Earth-like internal structure and composition.

Inselberg An isolated mountain remnant characteristic of the late stage of erosion in a mountainous arid region.

Intensity A measure of the degree of earthquake shaking at a given locale, based on the amount of damage.

Interface A common boundary where different parts of a system interact.

Interior drainage A discontinuous pattern of intermittent streams that do not flow to the ocean.

Intermediate composition A compositional group of igneous rocks that contains at least 25 percent dark silicate minerals. The other dominant mineral is plagioclase feldspar. Also called *andesitic composition*.

Internal process A process such as mountain building or volcanism that derives its energy from Earth's interior and elevates Earth's surface.

Intraplate volcanism Igneous activity that occurs within a tectonic plate, away from plate boundaries.

Intrusion A structure that results from the emplacement and crystallization of magma beneath the surface of Earth. Also called a *pluton*.

Intrusive igneous rock Igneous rock that formed below Earth's surface. Also called *plutonic rock*.

Ion An atom or a molecule that possesses an electrical charge.

Ionic bond A chemical bond between two oppositely charged ions that is formed by the transfer of valence electrons from one atom to the other.

Island arc A chain of volcanic islands generally located a few hundred kilometers from a trench where there is active subduction of one oceanic plate beneath another. Also called a *volcanic island arc*.

Isostasy The concept that Earth's crust is "floating" in gravitational balance upon the material of the mantle.

Isostatic adjustment Compensation of the lithosphere when weight is added or removed. When weight is added, the lithosphere responds by subsiding, and when weight is removed, there is uplift.

Isotopes Varieties of the same element that have different mass numbers; their nuclei contain the same number of protons but different numbers of neutrons.

J

Jetty A structure (typically paired) that extends into the ocean at the entrance to a harbor or river and is built for the purpose of protecting against storm waves and sediment deposition.

Joint A fracture in rock along which there has been no movement.

Jovian planet One of the Jupiter-like planets, Jupiter, Saturn, Uranus, and Neptune. These planets have relatively low densities.

K

Kame A steep-sided hill composed of sand and gravel, originating when sediment collected in openings in stagnant glacial ice.

Karst topography A type of topography formed on soluble rock (especially limestone) primarily by dissolution. It is characterized by sinkholes, caves, and underground drainage.

Kettle Depressions created when blocks of ice become lodged in glacial deposits and subsequently melt.

L

Laccolith A massive igneous body intruded between preexisting strata.

Lahar A debris flow on the slopes of a volcano that results when unstable layers of ash and debris become saturated and flow downslope, usually following stream channels.

Laminar flow The movement of water particles in straight-line paths that are parallel to the channel. The water particles move downstream without mixing.

Large igneous province Voluminous accumulations of lava extruded along fissures that produce broad, flat features that are also referred to as basalt plateaus.

Lateral continuity (principle of) A principle which states that sedimentary beds originate as continuous layers that extend in all directions until they grade into a different type of sediment or thin out at the edge of a sedimentary basin.

Lateral moraine A ridge of till along the sides of a valley glacier composed primarily of debris that fell to the glacier from the valley walls.

Laurasia The northern portion of Pangaea, consisting of North America and Eurasia.

Lava Magma that reaches Earth's surface.

Lava dome A bulbous mass associated with an old-age volcano, produced when thick lava is slowly squeezed from the vent. Lava domes may act as plugs to deflect subsequent gaseous eruptions.

Lava tube A tunnel in hardened lava that acts as a horizontal conduit for lava flowing from a volcanic vent. Lava tubes allow fluid lavas to advance great distances.

Law A formal statement of the regular manner in which a natural phenomenon occurs under given conditions.

Leaching The depletion of soluble materials from the upper soil by downward-percolating water.

Light silicate A silicate mineral that lacks iron and/or magnesium. Light silicates are generally lighter in color and have lower specific gravities than dark silicates. Also called *nonferromagnesian silicate.*

Limestone A chemical sedimentary rock composed chiefly of calcite. Limestone can form by inorganic means or from biochemical processes.

Liquefaction The transformation of a stable soil into a fluid that is often unable to support buildings or other structures.

Lithification The process, generally involving cementation and/or compaction, of converting sediments to solid rock.

Lithosphere The rigid outer layer of Earth, including the crust and upper mantle.

Lithospheric plate A coherent unit of Earth's rigid outer layer that includes the crust and upper unit. Also called a *plate.*

Local base level The level of a lake, resistant rock layer, or any other base level that stands above sea level. Also called the *temporary base level.*

Loess Deposits of windblown silt, lacking visible layers, generally buff-colored, and capable of maintaining a nearly vertical cliff.

Longitudinal dunes Long ridges of sand oriented parallel to the prevailing wind; these dunes form where sand supplies are limited.

Longitudinal profile A cross section of a stream channel along its descending course from the head to the mouth.

Longshore current A near-shore current that flows parallel to the shore.

Losing stream A stream that loses water to the groundwater system by outflow through the streambed.

Lower mantle See *Mesosphere.*

Luster The appearance or quality of light reflected from the surface of a mineral.

M

Mafic composition A compositional group of igneous rocks indicating that the rock contains substantial dark silicate minerals and calcium-rich plagioclase feldspar. Also called *Basaltic composition.*

Magma A body of molten rock found at depth, including any dissolved gases and crystals.

Magma mixing The process of altering the composition of a magma through the mixing of material from another magma body.

Magmatic differentiation The process of generating more than one rock type from a single magma.

Magnetic reversal A change in Earth's magnetic field from normal to reverse or vice versa.

Magnetic time scale A scale that shows the ages of magnetic reversals and is based on the polarity of lava flows of various ages.

Magnetometer A sensitive instrument used to measure the intensity of Earth's magnetic field at various points.

Magnitude An estimate of the total amount of energy released during an earthquake, based on seismic records.

Mantle One of Earth's compositional layers. The solid rocky shell that extends from the base of the crust to a depth of 2900 kilometers (1800 miles).

Mantle plume A mass of hotter-than-typical mantle material that ascends toward the surface, where it may lead to igneous activity. These plumes of solid yet mobile material may originate as deep as the core–mantle boundary.

Marble A relatively soft, nonfoliated metamorphic rock formed from limestone or dolostone. Marble of various colors is used for building stones and monuments.

Marine terrace A wave-cut platform that has been exposed above sea level.

Mass extinction An event in which a large percentage of species become extinct.

Mass movement The downslope movement of rock, regolith, and soil under the direct influence of gravity. Also called *mass wasting.*

Massive An igneous pluton that is not tabular in shape.

Mass number The sum of the number of neutrons and protons in the nucleus of an atom.

Mass wasting *See* mass movement.

Meander A looplike bend in the course of a stream.

Mechanical weathering The physical disintegration of rock, resulting in smaller fragments.

Medial moraine A ridge of till formed when lateral moraines from two coalescing alpine glaciers join.

Megathrust fault The plate boundary separating a subducting slab of oceanic lithosphere and the overlying plate.

Melt The liquid portion of magma excluding the solid crystals.

Mercalli intensity scale *See* Modified Mercalli Intensity scale.

Mesosphere The part of the mantle that extends from the core–mantle boundary to a depth of 660 kilometers (410 miles). Also known as the lower mantle.

Mesozoic era A time span on the geologic time scale between the Paleozoic and Cenozoic eras—from about 248 to 65.5 million years ago.

Metallic bond A chemical bond that is present in all metals that may be characterized as an extreme type of electron sharing in which the electrons move freely from atom to atom.

Metamorphic grade The degree to which a parent rock changes during metamorphism. It varies from low grade (low temperatures and pressures) to high grade (high temperatures and pressures).

Metamorphic rock Rock formed by the alteration of preexisting rock deep within Earth (but still in the solid state) by heat, pressure, and/or chemically active fluids.

Metamorphism The changes in mineral composition and texture of a rock subjected to high temperatures and pressures within Earth.

Microcontinent Relatively small fragment of continental crust; may lie above sea level, such as the island of Madagascar, or may be submerged, as exemplified by the Campbell Plateau located near New Zealand.

Mid-ocean ridge A continuous mountainous ridge on the floor of all the major ocean basins and varying in width from 500 to 5000 kilometers (300 to 3000 miles). The rifts at the crests of these ridges represent divergent plate boundaries. Also called *mid-ocean rise, oceanic ridge.*

Mid-ocean rise A continuous mountainous ridge on the floor of all the major ocean basins and varying in width from 500 to 5000 kilometers (300 to 3000 miles). The rifts at the crests of these ridges represent divergent plate boundaries. Also called *mid-ocean ridge, oceanic ridge.*

Migmatite A rock exhibiting both igneous and metamorphic rock characteristics. Such rocks may form when light-colored silicate minerals melt and then crystallize, while the dark silicate minerals remain solid.

Mineral A naturally occurring, inorganic crystalline material with a unique chemical structure.

Mineralogy The study of minerals.

Mineral resource All discovered and undiscovered deposits of a useful mineral that can be extracted now or at some time in the future.

Modified Mercalli Intensity Scale A 12-point scale developed to evaluate earthquake intensity, based on the amount of damage to various structures.

Mohs scale A series of 10 minerals used as a standard in determining hardness.

Moment magnitude A more precise measure of earthquake magnitude than the Richter scale that is derived from the amount of displacement that occurs along a fault zone.

Monocline A one-limbed flexure in strata. The strata are usually flat-lying or very gently dipping on both sides of the monocline.

Mountain belt A geographic area of roughly parallel and geologically connected mountain ranges developed as a result of plate tectonics.

Mouth The point downstream where a river empties into another stream or water body.

Mud crack A feature in some sedimentary rocks that forms when wet mud dries out, shrinks, and cracks.

Mudflow See *Debris flow.*

Muscovite A common member of the mica family. It is light-colored with a pearly luster and has excellent cleavage.

N

Natural levee An elevated landform composed of alluvium that parallels some streams and acts to confine their waters, except during floodstage.

Neap tide The lowest tidal range, occurring near the times of the first and third quarters of the Moon.

Nebular theory A model for the origin of the solar system that supposes a rotating nebula of dust and gases that contracted to form the Sun and planets.

Negative feedback mechanism As used in climatic change, any effect that is opposite the initial change and tends to offset it.

Neutron A subatomic particle found in the nucleus of an atom. The neutron is electrically neutral, with a mass approximately equal to that of a proton.

Nonclastic A term for the texture of sedimentary rocks in which the minerals form a pattern of interlocking crystals. Also called *crystalline texture*.

Nonconformity An unconformity in which older metamorphic or intrusive igneous rocks are overlain by younger sedimentary strata.

Nonferromagnesian silicate A silicate mineral that lacks iron and/or magnesium. Light silicates are generally lighter in color and have lower specific gravities than dark silicates. Also called *light silicate*.

Nonflowing artesian well An artesian well in which water does not rise to the surface because the pressure surface is below ground level.

Nonfoliated Describes metamorphic rocks that do not exhibit foliation.

Nonmetallic mineral resource A mineral resource that is not a fuel or processed for the metals it contains.

Nonrenewable Refers to a resource that forms or accumulates over such long time spans that it must be considered as fixed in total quantity.

Nonsilicate Refers to mineral groups that lack silica in their structures and account for less than 10 percent of Earth's crust.

Normal fault A fault in which the rock above the fault plane has moved down relative to the rock below.

Normal polarity A magnetic field the same as that which presently exists.

Nuclear decay The spontaneous decay of certain unstable atomic nuclei. Alao *radioactive decay*.

Nucleus The small, heavy core of an atom that contains all of its positive charge and most of its mass.

Nuée ardente Incandescent volcanic debris buoyed up by hot gases that moves downslope in an avalanche fashion. Also called a *pyroclastic flow*.

Numerical date The number of years that have passed since an event occurred.

O

Obsidian A volcanic glass of felsic composition.

Ocean basin A deep submarine region that lies beyond the continental margins.

Oceanic plateau An extensive region on the ocean floor that is composed of thick accumulations of pillow basalts and other mafic rocks that, in some cases, exceed 30 kilometers (20 miles) in thickness.

Oceanic ridge A continuous mountainous ridge on the floor of all the major ocean basins and varying in width from 500 to 5000 kilometers (300 to 3000 miles). The rifts at the crests of these ridges represent divergent plate boundaries. Also called *mid-ocean ridge, mid-ocean rise*.

Oceanic ridge system A continuous elevated zone on the floor of all the major ocean basins and varying in width from 500 to 5000 kilometers (300–3000 miles). The rifts at the crests of ridges represent divergent plate boundaries.

Octet rule A rule which states that atoms combine in order that each may have the electron arrangement of a noble gas (that is, the outer energy level contains eight neutrons).

Oil trap A geologic structure that allows for significant amounts of oil and gas to accumulate.

Olivine A high-temperature dark silicate mineral typically found in basalt.

Ophiolite complex The sequence of rocks that make up the oceanic crust. The three-layer sequence includes an upper layer of pillow basalts, a middle zone of sheeted dikes, and a lower layer of gabbro.

Ore deposit Usually a useful metallic mineral that can be mined at a profit. The term is also applied to certain nonmetallic minerals such as fluorite and sulfur.

Organic sedimentary rock Sedimentary rock composed of organic carbon from the remains of plants that died and accumulated on the floor of a swamp. Coal is the primary example.

Original horizontality See *principle of original horizontality*.

Orogenesis The processes that collectively result in the formation of mountains.

Orogeny The processes that collectively result in the formation of mountains.

Outer core A layer beneath the mantle about 2270 kilometers (1410 miles) thick, which has the properties of a liquid.

Outer planets The outermost planets of our solar system, which include Jupiter, Saturn, Uranus, and Neptune. They are known as the Jovian planets.

Outgassing The escape of dissolved gases from molten rocks.

Outlet glacier A tongue of ice normally flowing rapidly outward from an ice cap or ice sheet, usually through mountainous terrain to the sea.

Outwash plain A relatively flat, gently sloping plain consisting of materials deposited by melt-water streams in front of the margin of an ice sheet.

Oxbow lake A curved lake that is created when a stream cuts off a meander.

Oxidation The removal of one or more electrons from an atom or ion. So named because elements commonly combine with oxygen.

Oxygen isotope analysis A method of deciphering past temperatures based on the precise measurement of the ratio between two isotopes of oxygen, ^{16}O and ^{18}O. Analysis is commonly made of seafloor sediments and cores from ice sheets.

P

P waves The fastest type of earthquake wave, which travels by compression and expansion of the medium. Also called *primary waves*.

Pahoehoe flow A lava flow with a smooth to ropy surface.

Paleoclimatology The study of ancient climates; the study of climate and climate change prior to the period of instrumental records using proxy data.

Paleomagnetism The natural remnant magnetism in rock bodies. The permanent magnetization acquired by rock that can be used to determine the location of the magnetic poles and the latitude of the rock at the time it became magnetized. Also called *fossil magnetism*.

Paleontology The systematic study of fossils and the history of life on Earth.

Paleoseismology The study of the timing, location, and size of prehistoric earthquakes.

Paleozoic era A time span on the geologic time scale between the Precambrian and Mesozoic eras—from about 542 million to 251 million years ago.

Pangaea The proposed supercontinent that 200 million years ago began to break apart and form the present landmasses.

Parabolic dune A sand dune that is similar in shape to a barchan dune except that its tips point into the wind. These dunes often form along coasts that have strong onshore winds, abundant sand, and vegetation that partly covers the sand.

Parasitic cone A volcanic cone that forms on the flank of a larger volcano.

Parent material The material on which a soil develops.

Parent rock The rock from which a metamorphic rock formed.

Partial melting The process by which most igneous rocks melt. Since individual minerals have different melting points, most igneous rocks melt over a temperature range of a few hundred degrees. If the liquid is squeezed out after some melting has occurred, a melt with a higher silica content results.

Passive continental margin A margin that consists of a continental shelf, continental slope, and continental rise. They are not associated with plate boundaries and therefore experience little volcanism and few earthquakes.

Pegmatite A very coarse-grained igneous rock (typically granite) commonly found as a dike associated with a large mass of plutonic rock that has smaller crystals. Crystallization in a water-rich environment is believed to be responsible for the very large crystals.

Perched water table A localized zone of saturation above the main water table, created by an impermeable layer (aquiclude).

Peridotite An igneous rock of ultramafic composition thought to be abundant in the upper mantle.

Period A basic unit of the geologic time scale that is a subdivision of an era. Periods may be divided into smaller units called epochs.

Periodic table An arrangement of the elements in which atomic number increases from the left to right and elements with similar properties appear in columns called families or groups.

Permafrost Any permanently frozen subsoil. Usually found in the subarctic and arctic regions.

Permeability A measure of a material's ability to transmit water.

Phaneritic texture An igneous rock texture in which the crystals are roughly equal in size and large enough so the individual minerals can be identified without the aid of a microscope. Also called *coarse-grained texture*.

Phanerozoic eon The part of geologic time that is represented by rocks containing abundant fossil evidence. The eon extending from the end of the Proterozoic eon (540 million years ago) to the present.

Phenocryst A conspicuously large crystal embedded in a matrix of finer-grained crystals.

Phreatic zone The zone where all open spaces in sediment and rock are completely filled with water. Also called *zone of saturation*.

Phyllite A metamorphic rock composed mainly of fine crystals of muscovite, chlorite, or both.

Physical geology A major division of geology that examines the materials of Earth and seeks to understand the processes and forces acting beneath and upon Earth's surface.

Piedmont glacier A glacier that forms when one or more alpine glaciers emerge from the confining walls of mountain valleys and spread out to create a broad sheet in the lowlands at the base of the mountains.

Pillow lava Basaltic lava that solidifies in an underwater environment and develops a structure that resembles a pile of pillows. Also called *pillow basalts*.

Plagioclase A type of feldspar containing both sodium and calcium ions that freely substitute for one another depending on the crystallization environment.

Planetesimal A solid celestial body that accumulated during the first stages of planetary formation. Planetesimals aggregated into increasingly larger bodies, ultimately forming the planets.

Plastic flow A type of glacial movement that occurs within a glacier, below a depth of approximately 50 meters (165 feet), in which the ice is not fractured.

Plate A coherent unit of Earth's rigid outer layer that includes the crust and upper unit. Also called a *lithospheric plate.*

Plate tectonics See *Theory of plate tectonics.*

Playa The flat central area of an undrained desert basin.

Playa lake A temporary lake in a playa.

Pleistocene epoch An epoch of the Quaternary period that began about 2.6 million years ago and ended about 10,000 years ago. Best known as a time of extensive continental glaciation.

Plucking A process by which pieces of bedrock are lifted out of place by a glacier.

Pluton A structure that results from the emplacement and crystallization of magma beneath the surface of Earth. Also called an *intrusion.*

Plutonic rock Igneous rocks that form below Earth's surface. Named after Pluto, the god of the lower world in classical mythology. Also called *intrusive igneous rock.*

Pluvial lake A lake formed during a period of increased rainfall. For example, this occurred in many nonglaciated areas during periods of ice advance elsewhere.

Point bar A crescent-shaped accumulation of sand and gravel deposited on the inside of a meander.

Polymerization The ability of silicate tetrahedra to link together in a variety of configurations, including chains, sheets, and three-dimensional structures.

Porosity The volume of open spaces in rock or soil.

Porphyritic texture An igneous rock texture characterized by two distinctively different crystal sizes. The larger crystals are called phenocrysts, whereas the matrix of smaller crystals is termed the groundmass.

Porphyroblastic texture A texture of metamorphic rocks in which particularly large grains (porphyroblasts) are surrounded by a fine-grained matrix of other minerals.

Porphyry An igneous rock that has a porphyritic texture.

Positive feedback mechanism As used in climatic change, any effect that acts to reinforce the initial change.

Potassium feldspar An abundant, relatively hard light silicate mineral containing potassium ions in its structure.

Pothole A depression formed in a stream channel by the abrasive action of the water's sediment load.

Precambrian All geologic time prior to the Phanerozoic eon. A term encompassing both the Archean and Proterozoic eons.

Precursor Events or changes that precede an earthquake and may provide a warning.

Primary waves The fastest type of earthquake wave, which travels by compression and expansion of the medium. Also called *P waves.*

Principle of cross-cutting relationships A principle of relative dating which states that a rock or fault is younger than any rock or fault through which it cuts.

Principle of fossil succession A principle by which fossil organisms succeed one another in a definite and determinable order, and any time period can be recognized by its fossil content.

Principle of inclusions A principle that uses pieces of rock contained within another to determine a relative date. According to the principle of inclusions, the rock mass that provided the inclusion is older than the rock mass containing the inclusion.

Principle of lateral continuity See *Lateral continuity (principle of).*

Principle of original horizontality A principle by which layers of sediment are generally deposited in a horizontal or nearly horizontal position.

Principle of superposition A principle which states that in any undeformed sequence of sedimentary rocks, each bed is older than the one above and younger than the one below.

Proglacial lake A lake created when a glacier acts as a dam blocking the flow of a river or trapping glacial meltwater. The term refers to the position of such lakes just beyond the outer limits of a glacier.

Prokaryote Refers to the cells or organisms such as bacteria whose genetic material is not enclosed in a nucleus.

Proterozoic The eon following the Archean and preceding the Phanerozoic. It extends between 2500 and 542 million years ago.

Proton A positively charged subatomic particle found in the nucleus of an atom.

Protoplanet A developing planetary body that grows by the accumulation of planetesimals.

Proxy data Data gathered from natural recorders of climate variability such as tree rings, ice cores, and ocean-floor sediments.

Pumice A light-colored, glassy vesicular rock commonly having a granitic composition.

P wave The fastest earthquake wave, which travels by compression and expansion of the medium. Also called *primary wave.*

Pyroclastic flow Incandescent volcanic debris buoyed up by hot gases that moves downslope in an avalanche fashion. Also called a *nuée ardente.*

Pyroclastic material The volcanic rock ejected during an eruption. Pyroclastic materials include ash, bombs, and blocks. Also called *tephra.*

Pyroclastic rock The volcanic rock ejected during an eruption, including ash, bombs, and blocks.

Pyroclastic texture An igneous rock texture resulting from the consolidation of individual rock fragments that are ejected during a violent volcanic eruption. Also called *fragmental texture.*

Q

Quarrying Removing loosened blocks from the bed of a channel during times of high flow rates.

Quartz A common silicate mineral consisting entirely of silicon and oxygen that resists weathering.

Quartzite A hard, nonfoliated metamorphic rock formed from quartz sandstone.

Quaternary period The most recent period on the geologic time scale. It began about 2.6 million years ago and extends to the present.

R

Radial pattern A system of streams running in all directions, away from a central elevated structure, such as a volcano.

Radioactive decay The spontaneous decay of certain unstable atomic nuclei. Also called *nuclear decay.*

Radioactivity See *Radioactive decay.*

Radiocarbon dating Dating of events from the very recent geologic past (the past few tens of thousands of years) based on the fact that the radioactive isotope of carbon is produced continuously in the atmosphere. Also called *carbon-14 dating.*

Radiometric dating The procedure of calculating the absolute ages of rocks and minerals that contain certain radioactive isotopes.

Radiosonde A lightweight package of weather instruments fitted with a radio transmitter and carried aloft by a balloon.

Rainshadow The situation in which a dry area on the lee side of a mountain range. Many middle-latitude deserts are of this type.

Recessional moraine An end moraine formed as the ice front stagnated during glacial retreat.

Recharge area An area where groundwater is replenished.

Recrystallization The formation of new mineral crystals in a rock that tend to be larger than the original crystals.

Rectangular pattern A drainage pattern characterized by numerous right angle bends that develops on jointed or fractured bedrock.

Reflection (seismic) The redirection of some seismic waves back to the surface when the waves hit a boundary between different Earth materials.

Refraction See *Wave refraction.*

Regional metamorphism Metamorphism associated with large-scale mountain building.

Regolith The layer of rock and mineral fragments that nearly everywhere covers Earth's land surface.

Relative date Rocks placed in their proper sequence or order of formation based on geologic principles.

Renewable Refers to a resource that is virtually inexhaustible or that can be replenished over relatively short time spans.

Reserve Already identified deposits from which minerals can be extracted profitably.

Reservoir rock The porous, permeable portion of an oil trap that yields oil and gas.

Residual soil Soil developed directly from the weathering of the bedrock below.

Reverse fault A fault in which the material above the fault plane moves up in relation to the material below.

Reverse polarity A magnetic field opposite that which presently exists.

Rhyolite The fine-grained equivalent of the igneous rock granite, composed primarily of the light-colored silicates.

Richter scale A scale of earthquake magnitude based on the amplitude of the largest seismic wave.

Ridge push A mechanism that may contribute to plate motion. It involves the oceanic lithosphere sliding down the oceanic ridge under the pull of gravity.

Rift valley A long, narrow trough bounded by normal faults. It represents a region where divergence is taking place.

Rills Tiny channels that develop as unconfined flow begins producing threads of current.

Ring of Fire The zone of active volcanoes surrounding the Pacific Ocean.

Rip current A strong, narrow surface or near-surface current of short duration and high speed that moves seaward through the breaker zone at nearly a right angle to the shore.

Ripple marks Small waves of sand that develop on the surface of a sediment layer by the action of moving water or air.

River A general term for a stream that carries a substantial amount of water and has numerous tributaries.

Roche moutonnée An asymmetrical knob of bedrock that is formed when glacial abrasion smoothes the gentle slope facing the advancing ice sheet and plucking steepens the opposite side as the ice overrides the knob.

Rock A consolidated mixture of minerals.

Rock avalanche Very rapid downslope movement of rock and debris. These rapid movements may be aided by a layer of air trapped beneath the debris, and they have been known to reach speeds of over 200 kilometers (125 miles) per hour.

Rock cleavage The tendency of rocks to split along parallel, closely spaced surfaces. These surfaces are often highly inclined to the bedding planes in the rock.

Rock cycle A model that illustrates the origin of the three basic rock types and the interrelatedness of Earth materials and processes.

Rock flour Ground-up rock produced by the grinding effect of a glacier.

Rock-forming minerals The relatively few minerals that make up most of the rocks in Earth's crust.

Rockslide The rapid slide of a mass of rock downslope, along planes of weakness.

Rock structure All features created by the processes of deformation, from minor fractures in bedrock to a major mountain chain. Also called a *geologic structure*.

Runoff Water that flows over land rather than infiltrating into the ground.

S

S waves A type of seismic wave that involves oscillation perpendicular to the direction of propagation. Also called *secondary waves*.

Salinity The proportion of dissolved salts to pure water, usually expressed in parts per thousand (‰).

Salt flat A white crust on the ground that is produced when water evaporates and leaves behind its dissolved materials.

Saltation Transportation of sediment through a series of leaps or bounces.

Sandstone An abundant, durable detrital sedimentary rock primarily composed of sand-size grains.

Schist Medium- to coarse-grained metamorphic rocks having a foliated texture, in which platy minerals dominate.

Schistosity A type of foliation that is characteristic of coarser-grained metamorphic rocks. Such rocks have a parallel arrangement of platy minerals such as the micas.

Scientific method The process by which researchers raise questions, gather data, and formulate and test scientific hypotheses.

Scoria Vesicular ejecta that is the product of basaltic magma.

Scoria cone A rather small volcano built primarily of ejected lava fragments that consist mostly of pea- to walnut-size lapilli. Also called a *cinder cone*.

Sea arch An arch formed by wave erosion when caves on opposite sides of a headland unite.

Sea ice Frozen seawater that is associated with polar regions. The area covered by sea ice expands in winter and shrinks in summer.

Seafloor spreading A hypothesis, first proposed in the 1960s by Harry Hess, which suggested that new oceanic crust is produced at the crests of mid-ocean ridges, which are the sites of divergence.

Seamount An isolated volcanic peak that rises at least 1000 meters (3300 feet) above the deep-ocean floor.

Sea stack An isolated mass of rock standing just offshore, produced by wave erosion of a headland.

Seawall A barrier constructed to prevent waves from reaching the area behind the wall. Its purpose is to defend property from the force of breaking waves.

Secondary enrichment The concentration of minor amounts of metals that are scattered through unweathered rock into economically valuable concentrations by weathering processes.

Secondary waves A type of seismic wave that involves oscillation perpendicular to the direction of propagation. Also called *S waves*.

Sediment Unconsolidated particles created by the weathering and erosion of rock by chemical precipitation from solution in water, or from the secretions of organisms, and transported by water, wind, or glaciers.

Sedimentary environment A geographic setting where sediment accumulates. Each site is characterized by a particular combination of geologic processes and environmental conditions. Also called an *environment of deposition*.

Sedimentary rock Rock formed from the weathered products of preexisting rocks that have been transported, deposited, and lithified.

Seismic gap A segment of an active fault zone that has not experienced a major earthquake over a span when most other segments have. Such segments are probable sites for future major earthquakes.

Seismic reflection profiler An instrument that is used to view the rock structure beneath a blanket of sediment by means of strong, low-frequency sound waves that penetrate the sediments and reflect off the contacts between rock layers and fault zones.

Seismic wave A form of elastic energy released during an earthquake that causes vibrations in the materials that transmit them.

Seismogram A record made by a seismograph.

Seismograph An instrument that records earthquake waves. Also called a *seismometer*.

Seismology The study of earthquakes and seismic waves.

Seismometer An instrument that records earthquake waves. Also called a *seismograph*.

Settling velocity The speed at which a particle falls through a still fluid. The size, shape, and specific gravity of particles influence settling velocity.

Shale The most common sedimentary rock, consisting of silt- and clay-size particles.

Shear Stress that causes two adjacent parts of a body to slide past one another.

Sheeted dike complex A large group of nearly parallel dikes.

Sheet flow Runoff moving in unconfined thin sheets.

Sheeting A mechanical weathering process that is characterized by the splitting off of slablike sheets of rock.

Shelf break The point at which a rapid steepening of the gradient occurs, marking the outer edge of the continental shelf and the beginning of the continental slope.

Shield A large, relatively flat expanse of ancient igneous and metamorphic rocks within the craton.

Shield volcano A broad, gently sloping volcano built from fluid basaltic lavas.

Shock metamorphism Metamorphism that occurs when meteorites strike Earth's surface. Also called *impact metamorphism*.

Shoreline The line that marks the contact between land and sea. It migrates up and down as the tide rises and falls.

Silicate Any one of numerous minerals that have the silicon-oxygen tetrahedron as their basic structure.

Silicon–oxygen tetrahedron A structure composed of four oxygen atoms surrounding a silicon atom that constitutes the basic building block of silicate minerals.

Sill A tabular igneous body that was intruded parallel to the layering of preexisting rock.

Sink A depression produced in a region where soluble rock has been removed by groundwater. Also called a *sinkhole*.

Sinkhole A depression produced in a region where soluble rock has been removed by groundwater. Also called a *sink*.

Slab pull A mechanism that contributes to plate motion in which cool, dense oceanic crust sinks into the mantle and "pulls" the trailing lithosphere along.

Slate A very fine-grained metamorphic rock containing platy minerals and having excellent rock cleavage.

Slaty cleavage A type of foliation that is characteristic of slates, in which there is a parallel arrangement of fine-grained metamorphic minerals.

Slide A movement common to mass-wasting processes in which the material moving downslope remains fairly coherent and moves along a well-defined surface.

Slip face The steep, leeward surface of a sand dune that maintains a slope of about 34 degrees.

Slump The downward slipping of a mass of rock or unconsolidated material moving as a unit along a curved surface.

Snowline The lower limit of perennial snow.

Soil A combination of mineral and organic matter, water, and air; the portion of the regolith that supports plant growth.

Soil horizon A layer of soil that has identifiable characteristics produced by chemical weathering and other soil-forming processes.

Soil profile A vertical section through a soil, showing its succession of horizons and the underlying parent material.

Soil Taxonomy A soil classification system that consists of six hierarchical categories, based on observable soil characteristics. The system recognizes 12 soil orders.

Solar nebula The cloud of interstellar gas and/or dust from which the bodies of our solar system formed.

Solifluction The slow, downslope flow of water-saturated materials common to permafrost areas.

Solum The *O*, *A*, and *B* horizons in a soil profile. Living roots and other plant and animal life are largely confined to this zone.

Sonar An instrument that uses acoustic signals (sound energy) to measure water depths. Sonar is an acronym for *so*und *na*vigation and *r*anging.

Sorting The degree of similarity in particle size in sediment or sedimentary rock.

Source rock The sedimentary rocks in which petroleum and natural gas originate.

Specific gravity The ratio of a substance's weight to the weight of an equal volume of water.

Speleothem A collective term for the dripstone features found in caverns.

Spheroidal weathering Any weathering process that tends to produce a spherical shape from an initially blocky shape.

Spit An elongate ridge of sand that projects from the land into the mouth of an adjacent bay.

Spontaneous subduction A process that occurs at Mariana-type subduction zones in which old, dense lithosphere sinks into the mantle at a steep angle by its own weight, creating a deep trench.

Spreading center A boundary in which two plates move apart, resulting in upwelling of material from the mantle to create new seafloor. Also called *divergent plate boundary*.

Spring A flow of groundwater that emerges naturally at the ground surface.

Spring tide The highest tidal range. Occurs near the times of the new and full moons.

Stable platform That part of a carton that is mantled by relatively undeformed sedimentary rocks and underlain by a basement complex of igneous and metamorphic rocks.

Stalactite An icicle-like structure that hangs from the ceiling of a cavern.

Stalagmite A columnlike form that grows upward from the floor of a cavern.

Star dune An isolated hill of sand that exhibits a complex form and develops where wind directions are variable.

Steppe One of the two types of dry climate. A marginal and more humid variant of the desert that separates it from bordering humid climates.

Stock A pluton similar to but smaller than a batholith.

Storm surge The abnormal rise of the sea along a shore as a result of strong winds.

Strain An irreversible change in the shape and size of a rock body caused by stress.

Strata Parallel layers of sedimentary rock. Also called *beds*.

Stratified drift Sediments deposited by glacial-meltwater.

Stratosphere The layer of the atmosphere immediately above the troposphere, characterized by increasing temperatures with height due to the concentration of ozone.

Stratovolcano A volcano composed of both lava flows and pyroclastic material. Also called a *composite volcano*.

Streak The color of a mineral in powdered form.

Stream A general term to denote the flow of water within any natural channel. Thus, a small creek and a large river are both streams.

Stream terrace A flat, benchlike structure produced by a stream, which was left elevated as the stream cut downward.

Stream valley The channel, valley floor, and sloping valley walls of a stream.

Stress The force per unit area acting on any surface within a solid.

Striations (glacial) Scratches or grooves in a bedrock surface caused by the grinding action of a glacier and its load of sediment.

Strike The compass direction of the line of intersection created by a dipping bed or fault and a horizontal surface. A strike is always perpendicular to the direction of dip.

Strike-slip fault A fault along which movement occurs horizontally.

Stromatolite Distinctively layered mounds of calcium carbonate, which are fossil evidence for the existence of ancient microscopic bacteria.

Subduction The process by which oceanic lithosphere plunges into the mantle along a convergent zone.

Subduction erosion A process in subduction zones in which sediment and rock are scraped off the bottom of the overriding plate and transported into the mantle.

Subduction zone A long, narrow zone where one lithospheric plate descends beneath another. See also *convergent plate boundary.*

Subduction zone metamorphism High-pressure, low-temperature metamorphism that occurs where sediments are carried to great depths by a subducting plate.

Submarine canyon A seaward extension of a valley that was cut on the continental shelf during a time when sea level was lower, or a canyon carved into the outer continental shelf, slope, and rise by turbidity currents.

Submergent coast A coast whose form is largely a result of the partial drowning of a former land surface due to a rise of sea level or subsidence of the crust, or both.

Subsoil A term applied to the *B* horizon of a soil profile.

Sulfur dioxide A gas with the chemical formula SO_2, that is associated naturally with volcanic activity, and as a waste gas (air pollutant) with the burning of fossil fuels and various industrial processes.

Sunspot A dark area on the Sun that is associated with powerful magnetic storms that extend from the Sun's surface deep into the interior.

Supercontinent A large landmass that contains all, or nearly all, of the existing continents.

Supercontinent cycle The idea that the rifting and dispersal of one supercontinent is followed by a long period during which the fragments gradually reassemble into a new supercontinent.

Supernova An exploding star that increases its brightness many thousands of times.

Superplume Large mantle plumes that are thought to be responsible for creating basalt plateaus.

Superposed stream A stream that cuts through a ridge lying across its path. The stream established its course on uniform layers at a higher level without regard to underlying structures and subsequently downcut.

Superposition See *Principle of superposition.*

Surf A collective term for breakers; also the wave activity in the area between the shoreline and the outer limit of breakers.

Surface waves Seismic waves that travel along the outer layer of Earth.

Surge A period of rapid glacial advance. Surges are typically sporadic and short-lived.

Suspended load Fine sediment carried within the body of flowing water or air.

Suture A zone along which two crustal fragments are jointed together. For example, following a continental collision, the two continental blocks are sutured together.

Swells Wind-generated waves that have moved into an area of weaker winds or calm.

Syncline A linear downfold in sedimentary strata; the opposite of anticline.

System A group of interacting or interdependent parts that form a complex whole.

T

Tabular Describing a feature such as an igneous pluton that has two dimensions that are much longer than the third.

Talus slope An accumulation of rock debris at the base of a cliff. Also called *talus.*

Tarn A small lake in a cirque.

Tectonic plate See *Lithospheric plate.*

Tectonic structure A basic geologic feature, such as a fold, fault, or rock foliation, that results from forces associated with the interaction of tectonic plates. Also called a *geologic structure.*

Tectonics The study of the large-scale processes that collectively deform Earth's crust.

Temporary base level The level of a lake, resistant rock layer, or any other base level that stands above sea level. Also *local base level.*

Tenacity Describes a mineral's toughness or resistance to breaking or deforming.

Tensional stress The type of stress that tends to pull a body apart.

Tephra The volcanic rock ejected during an eruption. Pyroclastic materials include ash, bombs, and blocks. Also called *pyroclastic materials.*

Terminal moraine The end moraine that marks the farthest advance of a glacier.

Terrace A flat, benchlike structure produced by a stream, which was left elevated as the stream cut downward.

Terrane A crustal block bounded by faults, whose geologic history is distinct from the histories of adjoining crustal blocks.

Terrestrial planet One of the Earthlike planets: Mercury, Venus, Earth, and Mars. These planets have similar densities.

Terrigenous sediment Seafloor sediments derived from terrestrial weathering and erosion.

Texture The size, shape, and distribution of the particles that collectively constitute a rock.

Theory A well-tested and widely accepted view that explains certain observable facts.

Theory of plate tectonics A well-tested theory proposing that Earth's outer shell consists of individual plates that interact in various ways and thereby produce earthquakes, volcanoes, mountains, and the crust itself.

Thermal metamorphism. Changes in rock caused by the heat from a nearby magma body. Also called *contact metamorphism.*

Thermosphere The region of the atmosphere immediately above the mesosphere, which is characterized by increasing temperatures due to absorption of very short-wave solar energy by oxygen.

Thrust fault A low-angle reverse fault.

Tidal current The alternating horizontal movement of water associated with the rise and fall of the tide.

Tidal delta A deltalike feature created when a rapidly moving tidal current emerges from a narrow inlet and slows, depositing its load of sediment.

Tidal flat A marshy or muddy area that is alternately covered and uncovered by the rise and fall of the tide.

Tide The periodic change in the elevation of the ocean surface.

Till Unsorted sediment deposited directly by a glacier.

Tillite A rock formed when glacial till is lithified.

Tombolo A ridge of sand that connects an island to the mainland or to another island.

Tower karst Steep-sided hills formed in wet tropical and subtropical regions with thick beds of highly jointed limestone. The limestone is dissolved by groundwater, leaving residual towers.

Trace gases Gases present in Earth's atmosphere at concentrations much lower than that of carbon dioxide. Methane and nitrous oxide are important trace gases that absorb outgoing radiation and help warm the atmosphere.

Transform fault A major strike-slip fault that cuts through the lithosphere and accommodates motion between two plates.

Transform plate boundary A boundary in which two plates slide past one another without creating or destroying lithosphere. Also called a *transform fault* or *transform boundary.*

Transition zone The lowest portion of the upper mantle.

Transpiration The release of water vapor to the atmosphere by plants.

Transported soil Soil that forms on unconsolidated deposits.

Transverse dune Type of dune that forms a series of long ridges oriented at right angles to the prevailing wind; these dunes form where vegetation is sparse and sand is very plentiful.

Travertine A form of limestone ($CaCO_3$) that is deposited by hot springs or as a cave deposit.

Trellis pattern A system of streams in which nearly parallel tributaries occupy valleys cut in folded strata.

Trigger An event, such as an earthquake or heavy rainfall, that initiates a mass movement process.

Troposphere The lowermost layer of the atmosphere. It is generally characterized by a decrease in temperature with height.

Tsunami The Japanese word for a seismic sea wave.

Turbidite A turbidity current deposit characterized by graded bedding.

Turbidity current A downslope movement of dense, sediment-laden water created when sand and mud on the continental shelf and slope are dislodged and thrown into suspension.

Turbulent flow Erratic movement of water often characterized by swirling, whirlpool-like eddies. Most streamflow is of this type.

U

Ultimate base level Sea level; the lowest level to which stream erosion could lower the land.

Ultramafic Refers to a compositional group of igneous rocks containing mostly olivine and pyroxene.

Unconformity A surface that represents a break in the rock record, caused by erosion and nondeposition.

Uniformitarianism The concept that the processes that have shaped Earth in the geologic past are essentially the same as those operating today.

Unit cell The smallest group of atoms, ions, or molecules that form the building block of a crystal.

Unsaturated zone The area above the water table where openings in soil, sediment, and rock are not saturated but filled mainly with air. Also called the *vadose zone.*

V

Vadose zone The area above the water table where openings in soil, sediment, and rock are not saturated but filled mainly with air. Also called the *unsaturated zone.*

Valence electron The electrons involved in the bonding process; the electrons occupying the highest principal energy level of an atom.

Valley glacier A glacier confined to a mountain valley, which in most instances had previously been a stream valley. Also called an *alpine glacier*.

Valley train A relatively narrow body of stratified drift deposited on a valley floor by meltwater streams that issue from the terminus of an alpine glacier.

Vein deposit A mineral that fills a fracture or fault in a host rock. Such deposits have a sheetlike, or tabular, form.

Vent The surface opening of a conduit or pipe.

Ventifact A cobble or pebble polished and shaped by the sandblasting effect of wind.

Vesicles Spherical or elongated openings on the outer portion of a lava flow that were created by escaping gases.

Vesicular texture A term applied to aphanitic igneous rocks that contain many small cavities called vesicles.

Viscosity A measure of a fluid's resistance to flow.

Volatiles Gaseous components of magma dissolved in the melt. Volatiles will readily vaporize (form a gas) at surface pressures.

Volcanic Pertaining to the activities, structures, or rock types of a volcano.

Volcanic bomb A streamlined pyroclastic fragment ejected from a volcano while still semimolten.

Volcanic cone A cone-shaped structure built by successive eruptions of lava and/or pyroclastic materials.

Volcanic island A seamount that has grown large enough to rise above sea level.

Volcanic island arc A chain of volcanic islands generally located a few hundred kilometers from a trench where there is active subduction of one oceanic plate beneath another. Also called an *island arc*.

Volcanic neck An isolated, steep-sided, erosional remnant consisting of lava that once occupied the vent of a volcano. Also called a *volcanic plug*.

Volcanic plug An isolated, steep-sided, erosional remnant consisting of lava that once occupied the vent of a volcano. Also called *volcanic neck*.

Volcanic rock Igneous rock formed when magma solidifies at Earth's surface. Also *extrusive igneous rock*.

Volcano A mountain formed from lava and/or pyroclastics.

W

Water gap A pass through a ridge or mountain in which a stream flows.

Watershed The land area that contributes water to a stream. Also called a *drainage basin*.

Water table The upper level of the saturated zone of groundwater.

Wave height The vertical distance between the trough and crest of a wave.

Wave period The time interval between the passage of successive crests at a stationary point.

Wave refraction A change in direction of waves as they enter shallow water. The portion of the wave in shallow water is slowed, which causes the waves to bend and align with the underwater contours.

Wave-cut cliff A seaward-facing cliff along a steep shoreline formed by wave erosion at its base and mass wasting.

Wave-cut platform A bench or shelf along a shore at sea level, cut by wave erosion.

Wavelength The horizontal distance separating successive crests or troughs.

Weather The state of the atmosphere at any given time.

Weathering The disintegration and decomposition of rock at or near the surface of Earth.

Welded tuff A pyroclastic deposit composed of particles fused together by the combination of heat still contained in the deposit after it has come to rest and the weight of overlying material.

Well An opening bored into the zone of saturation.

Wetted perimeter The total distance in a linear cross-section of a stream that is in contact with water.

Wilson Cycle See *Supercontinent cycle*.

X

Xenolith An inclusion of unmelted country rock in an igneous pluton.

Y

Yardang A streamlined, wind-sculpted ridge that has the appearance of an inverted ship's hull that is oriented parallel to the prevailing wind.

Yazoo tributary A tributary that flows parallel to the main stream because a natural levee is present.

Z

Zone of accumulation The part of a glacier that is characterized by snow accumulation and ice formation. The outer limit of this zone is the snowline.

Zone of fracture The upper portion of a glacier consisting of brittle ice.

Zone of saturation The zone where all open spaces in sediment and rock are completely filled with water. Also called the *phreatic zone*.

Zone of soil moisture A zone in which water is held as a film on the surface of soil particles and may be used by plants or withdrawn by evaporation. The uppermost subdivision of the unsaturated zone.

Zone of wastage The part of a glacier beyond the snowline, where annually there is a net loss of ice.

INDEX

D

E